Trapezoid: A four-sided figure with one pair of parallel sides

Area: $A = \frac{1}{2}h(b_1 + b_2)$

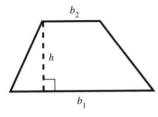

Parallelogram: A four-sided figure with opposite sides parallel

Area: $A = bh$

Rectangle: A four-sided figure with four right angles

Area: $A = LW$

Perimeter: $P = 2L + 2W$

Rhombus: A four-sided figure with four equal sides

Perimeter: $P = 4a$

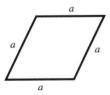

Square: A four-sided figure with four equal sides and four right angles

Area: $A = s^2$

Perimeter: $P = 4s$

Circle

Area: $A = \pi r^2$

Circumference: $C = 2\pi r$

Diameter: $d = 2r$

Value of pi: $\pi \approx 3.14$

Sphere

Volume: $V = \frac{4}{3}\pi r^3$

Surface Area: $S = 4\pi r^2$

Right Circular Cone

Volume: $V = \frac{1}{3}\pi r^2 h$

Lateral Surface Area: $S = \pi r\sqrt{r^2 + h^2}$

Right Circular Cylinder

Volume: $V = \pi r^2 h$

Lateral Surface Area: $S = 2\pi rh$

Rectangular Solid

Volume: $V = LWH$

Surface Area:
$A = 2LW + 2WH + 2LH$

Sixth edition

Algebra

for College Students

Mark Dugopolski
Southeastern Louisiana University

McGraw Hill

Connect
Learn
Succeed™

The McGraw-Hill Companies

McGraw Hill

Connect
Learn
Succeed™

ALGEBRA FOR COLLEGE STUDENTS, SIXTH EDITION

Published by McGraw-Hill, a business unit of The McGraw-Hill Companies, Inc., 1221 Avenue of the Americas, New York, NY 10020. Copyright © 2012 by The McGraw-Hill Companies, Inc. All rights reserved. Previous editions © 2009, 2006, and 2002. No part of this publication may be reproduced or distributed in any form or by any means, or stored in a database or retrieval system, without the prior written consent of The McGraw-Hill Companies, Inc., including, but not limited to, in any network or other electronic storage or transmission, or broadcast for distance learning.

Some ancillaries, including electronic and print components, may not be available to customers outside the United States.

This book is printed on acid-free paper.

1 2 3 4 5 6 7 8 9 0 RJE/RJE 1 0 9 8 7 6 5 4 3 2 1

ISBN 978–0–07–338434–4
MHID 0–07–338434–8

ISBN 978–0–07–735316–2 (Annotated Instructor's Edition)
MHID 0–07–735316–1

Vice President, Editor-in-Chief: *Marty Lange*
Vice President, EDP: *Kimberly Meriwether David*
Senior Director of Development: *Kristine Tibbetts*
Editorial Director: *Stewart K. Mattson*
Sponsoring Editor: *Mary Ellen Rahn*
Developmental Editor: *Adam Fischer*
Marketing Manager: *Peter A. Vanaria*
Lead Project Manager: *Peggy J. Selle*
Senior Buyer: *Sandy Ludovissy*
Senior Media Project Manager: *Jodi K. Banowetz*
Designer: *Tara McDermott*
Cover Designer: *Greg Nettles/Squarecrow Creative*
Cover Image: © *redbrickstock.com/Alamy*
Lead Photo Research Coordinator: *Carrie K. Burger*
Compositor: *Glyph International*
Typeface: *10.5/12 Times Roman*
Printer: *R. R. Donnelley*

All credits appearing on page or at the end of the book are considered to be an extension of the copyright page.

Photo Credits: Page 1: © Denkou Images/agefotostock RF; p. 61: © Digital Vision Vol. 185/Getty RF; p. 65: U.S. Army Corps of Engineers; p. 145: © Reuters NewMedia Inc./CORBIS; p. 236: © Vol. 128/Corbis RF; p. 292: © Getty RF; p. 307: © Associated Press/AP; p. 381: © Digital Vision Vol. 285/Getty RF; p. 457: © Daniel Novisedlak/Flickr/Getty RF; p. 759: © Stockdisc/Digital Vision RF; p. 730: © Vol. 168/Corbis RF; p. 456: © The McGraw-Hill Companies, Inc./Jill Braaten, photographer; p. 665: © Royalty Free/Corbis; p. 585: © C. Squared Studios/Getty RF; All other photos © PhotoDisc/Getty RF.

Library of Congress Cataloging-in-Publication Data

Dugopolski, Mark.
  Algebra for college students / Mark Dugopolski.—6th ed.
    p. cm.
  Includes index.
ISBN 978–0–07–338434–4—ISBN 0–07–338434–8 (hard copy: alk. paper) 1. Algebra. I. Title.
  QA152.3.D837 2012
  512.9—dc22
                                                                    2010024757

www.mhhe.com

About the Author

Mark Dugopolski was born and raised in Menominee, Michigan. He received a degree in mathematics education from Michigan State University and then taught high school mathematics in the Chicago area. While teaching high school, he received a master's degree in mathematics from Northern Illinois University. He then entered a doctoral program in mathematics at the University of Illinois in Champaign, where he earned his doctorate in topology in 1977. He was then appointed to the faculty at Southeastern Louisiana University, where he taught for 25 years. He is now professor emeritus of mathematics at SLU. He is a member of MAA and AMATYC. He has written many articles and numerous mathematics textbooks. He has a wife and two daughters. When he is not working, he enjoys gardening, hiking, bicycling, jogging, tennis, fishing, and motorcycling.

In loving memory of my parents,
Walter and Anne Dugopolski

connect™
|MATHEMATICS

McGraw-Hill Connect Mathematics McGraw-Hill conducted in-depth research to create a new and improved learning experience that meets the needs of today's students and instructors. The result is a reinvented learning experience rich in information, visually engaging, and easily accessible to both instructors and students. McGraw-Hill's Connect is a Web-based assignment and assessment platform that helps students connect to their coursework and prepares them to succeed in and beyond the course.

Connect Mathematics enables math instructors to create and share courses and assignments with colleagues and adjuncts with only a few clicks of the mouse. All exercises, learning objectives, videos, and activities are directly tied to text-specific material.

 You and your students want a fully integrated online homework and learning management system all in one place.

McGraw-Hill and Blackboard Inc. Partnership

- ▶ McGraw-Hill has partnered with Blackboard Inc. to offer the deepest integration of digital content and tools with Blackboard's teaching and learning platform.

- ▶ **Life simplified.** Now, all McGraw-Hill content (text, tools, & homework) can be accessed directly from within your Blackboard course. All with one sign-on.

- ▶ **Deep integration.** McGraw-Hill's content and content engines are seamlessly woven within your Blackboard course.

- ▶ **No more manual synching!** Connect assignments within Blackboard automatically (and instantly) feed grades directly to your Blackboard grade center. No more keeping track of two gradebooks!

Do More

2 *Your students want an assignment page that is easy to use and includes lots of extra resources for help.*

Efficient Assignment Navigation

- ▶ Students have access to immediate feedback and help while working through assignments.

- ▶ Students can view detailed step-by-step solutions for each exercise.

Connect. Learn. Succeed.

3 Your students want an interactive eBook rich with integrated functionality.

Integrated Media-Rich eBook

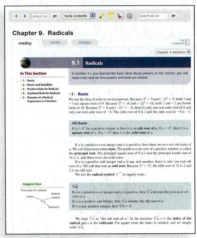

- ▶ A Web-optimized eBook is seamlessly integrated within ConnectPlus Mathematics for ease of use.
- ▶ Students can access videos, images, and other media in context within each chapter or subject area to enhance their learning experience.
- ▶ Students can highlight, take notes, or even access shared instructor highlights/notes to learn the course material.
- ▶ The integrated eBook provides students with a cost-saving alternative to traditional textbooks.

4 You want a more intuitive and efficient assignment creation process to accommodate your busy schedule.

Assignment Creation Process

- ▶ Instructors can select textbook-specific questions organized by chapter, section, and objective.
- ▶ Drag-and-drop functionality makes creating an assignment quick and easy.
- ▶ Instructors can preview their assignments for efficient editing.

5 You want a gradebook that is easy to use and provides you with flexible reports to see how your students are performing.

Flexible Instructor Gradebook

- ▶ Based on instructor feedback, Connect Mathematics' straightforward design creates an intuitive, visually pleasing grade management environment.
- ▶ View scored work immediately and track individual or group performance with various assignment and grade reports.

www.mcgrawhillconnect.com

McGraw Hill Connect Learn Succeed™

FROM THE AUTHOR

I would like to thank the many students and faculty that have used my books over the years. You have provided me with excellent feedback that has assisted me in writing a better, more student-focused book in each edition. Your comments are always taken seriously, and I have adjusted my focus on each revision to satisfy your needs.

Understandable Explanations

I originally undertook the task of writing my own book for the elementary and intermediate algebra course so I could explain mathematical concepts to students in language they would understand. Most books claim to do this, but my experience with a variety of texts had proven otherwise. What students and faculty will find in my book are **short, precise explanations** of terms and concepts that are written in **understandable language.** For example, when I introduce the Commutative Property of Addition, I make the concrete analogy that "the price of a hamburger plus a Coke is the same as the price of a Coke plus a hamburger," a mathematical fact in their daily lives that students can readily grasp. Math doesn't need to remain a mystery to students, and students reading my book will find other analogies like this one that connect abstractions to everyday experiences.

Detailed Examples Keyed to Exercises

My experience as a teacher has taught me two things about examples: they need to be detailed, and they need to help students do their homework. As a result, users of my book will find abundant examples with every step carefully laid out and explained where necessary so that students can follow along in class if the instructor is demonstrating an example on the board. Students will also be able to read them on their own later when they're ready to do the exercise sets. I have also included a **double cross-referencing** system between my examples and exercise sets so that, no matter which one students start with, they'll see the connection to the other. All examples in this edition refer to specific exercises by ending with a phrase such as "Now do Exercises 11–18" so that students will have the opportunity for immediate practice of that concept. If students work an exercise and find they are stumped on how to finish it, they'll see that for that group of exercises they're directed to a specific example to follow as a model. Either way, students will find my book's examples give them the guidance they need to succeed in the course.

Varied Exercises and Applications

A third goal of mine in writing this book was to give students **more variety** in the kinds of exercises they perform than I found in other books. Students won't find an intimidating page of endless drills in my book, but instead will see exercises in manageable groups with specific goals. They will also be able to augment their math proficiency using different formats (true/false, written response, multiple-choice) and different methods (discussion, collaboration, calculators). Not only is there an abundance of skill-building exercises, I have also researched a wide variety of **realistic applications** using **real data** so that those "dreaded word problems" will be seen as a useful and practical extension of what students have learned. Finally, every chapter ends with **critical thinking exercises** that go beyond numerical computation and call on students to employ their intuitive problem-solving skills to find the answers to mathematical puzzles in **fun and innovative** ways. With all of these resources to choose from, I am sure that instructors will be comfortable adapting my book to fit their course, and that students will appreciate having a text written for their level and to stimulate their interest.

Listening to Student and Instructor Concerns

McGraw-Hill has given me a wonderful resource for making my textbook more responsive to the immediate concerns of students and faculty. In addition to sending my manuscript out for review by instructors at many different colleges, several times a year McGraw-Hill holds symposia and focus groups with math instructors where the emphasis is *not* on selling products but instead on the **publisher listening** to the needs of faculty and their students. These encounters have provided me with a wealth of ideas on how to improve my chapter organization, make the page layout of my books more readable, and fine-tune exercises in every chapter. Consequently, students and faculty will feel comfortable using my book because it incorporates their specific suggestions and anticipates their needs. These events have particularly helped me in the shaping of the fourth edition.

Improvements in the Sixth Edition
OVERALL

- All Warm-Up exercise sets have been rewritten and now include a combination of fill-in-the-blank and true/false exercises. This was done to put a greater emphasis on vocabulary.
- Using a graphing calculator with this text is still optional. However, more Calculator Close-Ups and more graphing calculator required exercises have been included throughout the text for those instructors who prefer to emphasize graphing calculator use.
- Every chapter now includes a Mid-Chapter Quiz. This quiz can be used to assess student progress in the chapter.
- Numerous applications have been updated and rewritten.
- All Enriching Your Mathematical Word Power exercise sets have been expanded and rewritten as fill-in-the-blank exercises.
- All Making Connections exercise sets have been expanded so that they present a more comprehensive cumulative review.
- Teaching Tips are now included throughout the text, along with many new Helpful Hints.

CHAPTER 1:
- New material on set notation
- New explanation of irrational numbers
- One new example and two revised and updated applications
- Revised and updated Critical Thinking activity
- Exercise Sets: Revised and updated Section 1.2

CHAPTER 2:
- Exercise Sets: Revised and updated Section 2.1

CHAPTER 3:
- New definitions for Intercepts
- New material on the strategy for Graphing a Linear Inequality by the Test Point Method
- New example on compound inequalities

CHAPTER 4:
- Reorganization of the introduction to systems of equations
- New example on checking an ordered pair as a solution to a system of equations
- Five revised examples
- New summary of methods of solving systems of equations
- New material on Gauss-Jordan Elimination Method
- Exercise Sets: Revised and updated 4.1, 4.2, 4.3, 4.4, 4.5

CHAPTER 5:
- New explanation of reciprocal exponents
- Three revised and updated Caution Boxes
- Six revised and updated examples
- Simplified Section 5.5 by reorganizing material
- New material on factoring by grouping and the Pythagorean Theorem
- New strategy for factoring a four-term polynomial by grouping
- Rewritten explanation on factoring $ax^2 + bx + c$ with $a = 1$ and the difference or sum of two squares
- New strategy box for factoring $x^2 + bx + c$ by grouping and $ax^2 + bx + c$ by the ac method
- New strategy and explanation for factoring sum and difference of cubes
- Exercise Sets: Revised and updated Section 5.5, 5.6, 5.8

- End of Chapter: revised and updated review exercises and chapter test

CHAPTER 6:
- Exercise Sets: Revised and updated Section 6.1
- End of Chapter: revised and updated review exercises and chapter test

CHAPTER 7:
- New material on roots and variables
- New presentation of perfect squares, cubes and fourth powers
- Modified and improved Math at Work feature
- Exercise Sets: Revised Section 7.1
- End of Chapter: revised and updated the review exercises and chapter test

CHAPTER 8:
- Simplified Section 8.5 to focus exclusively on quadratic inequalities
- The sign-graph method of solving quadratic and rational inequalities has been removed and replaced with the more intuitive graphical method
- New definition of quadratic inequalities
- New strategies on solving a quadratic inequality graphically and with the Test-Point Method
- Four new examples on solving quadratic inequalities graphically and with the Test-Point Method
- Figures improved to help clarify graphing examples
- Exercise Sets: Revised Section 8.5
- End of Chapter: revised and updated summary, review exercises and chapter test

CHAPTER 9:
- Section 9.2 Transformations of Graphs has been rearranged in a more natural order
- New material on independent variables and their relationship to functions
- New material on Transformations of Graphs, Horizontal Translation, and Multiple Transformations

CHAPTER 10:
- Solving polynomial inequalities by the graphical method and test point method have been added to Section 10.4 after the graphs of polynomial functions are discussed

- Solving rational inequalities by the graphical method and test point method have been added to Section 10.5 after the graphs of rational functions are discussed
- Four revised and updated examples
- Exercise Sets: Revised Section 10.4, 10.5
- End of Chapter: revised and updated summary, review exercises and chapter test

CHAPTER 11:
- New simplified organization of Section 11.1, 11.2.
- New definition of domain

- New material on exponential functions
- New material on logarithmic functions
- Four updated applications

CHAPTER 12:
- Simplified organization of material on parabolas in Section 12.2

Manuscript Review Panels

Teachers and academics from across the country reviewed the various drafts of the manuscript to give feedback on content, design, pedagogy, and organization. This feedback was summarized by the book team and used to guide the direction of the text. I would like to thank the following professors for their participation in making this fourth edition.

Seth Daugherty, *Saint Louis CC–Forest Park*

Shing So, *University of Central Missouri*

Elsie Newman, *Owens Community College*

Patrick Ward, *Illinois Central College*

Sean Stewart, *Owens Community College*

Sharon Robertson, *University of Tennessee–Martin*

Randall Castleton, *University of Tennessee–Martin*

David Ray, *University of Tennessee–Martin*

Roland Trevino, *San Antonio College*

Irma Bakenhus, *San Antonio College*

Larry Green, *Lake Tahoe Community College*

Pinder Naidu, *Kennesaw State University*

Fereja Tahir, *Illinois Central College*

Brooke Lee, *San Antonio College*

Timothy McKenna, *University of Michigan–Dearborn*

Jean Peterson, *University of Wisconsin–Oshkosh*

Amy Young, *Navarro College*

Jenell Sargent, *Tennessee State University*

Mark Brenneman, *Mesa Community College*

Litsa St Amand, *Mesa Community College*

Jeff Igo, *University of Michigan–Dearborn*

Gerald Busald, *San Antonio College*

Bobbie Jo Hill, *Coastal Bend College*

Mary Kay Best, *Coastal Bend College*

Mary Frey, *Cincinnati State Tech and Community College*

Kimberly Bonacci, *Indiana University–Southeast*

Kenneth Thompson, *East Central Community College*

Joseph Sedlacek, *Kirkwood Community College*

Mildred Vernia, *Indiana University–Southeast*

Manuel Sanders, *University of South Carolina*

Randell Simpson, *Temple College*

Debra Pharo, *Northwestern Michigan College*

Carmen Buhler, *Minneapolis Community and Tech College*

Mary Peddycoart, *Kingwood College*

Tim McBride, *Spartanburg Technical College*

Glenn Robert Jablonski, *Triton College*

Derek Martinez, *Central New Mexico Community College*

Dennis Reissig, *Suffolk County Community College*

Charles Patterson, *Louisiana Tech University*

Rhoderick Fleming, *Wake Technical Community College*

Toni McCall, *Angelina College*

Suzanne Doviak, *Old Dominican University*

Judith Atkinson, *University of Alaska–Fairbanks*

Stephen Drake, *Northwestern Michigan College*

Michael Price, *University of Oregon*

Donald Munsey, *Louisiana Delta Community College*

Peggy Blanton, *Isothermal Community College*

Paul Diehl, *Indiana University Southeast*

Jinhua Tao, *University of Central Missouri*

Rajalakshmi Baradwaj, *University of Maryland–Baltimore County*

Jinfeng Wei, *Maryville University*

Dale Vanderwilt, *Dordt College*

Lori Wall, *University of England–Biddeford*

Hossein Behforooz, *Utica College*

Jan Butler, *CCC Online*

Paul Jones, *University of Cincinnati*

Joan Brown, *Eastern New Mexico University*

Kimberly Caldwell, *Volunteer State Community College*

Wendy Conway, *Oakland Community College–Highland Lakes*

Robert Diaz, *Fullerton College*

David French, *Tidewater Community College*

Teresa Houston, *East Mississippi Community College–Scooba*

Carla Monticelli, *Camden County College*

Madhu Motha, *Butler County Community College*

Chris Reisch, *Jamestown Community College*

Jill Rafael, *Sierra College*

Dan Rothe, *Alpena Community College*

Richard Rupp, *Del Mar College*

Kristina Sampson, *Lone Star College*

John Squires, *Chattanooga State Tech*

Jane Thompson, *Waubonsee Community College*

Richard Watkins, *Tidewater Community College*

Jackie Wing, *Angelina College*

I also want to express my sincere appreciation to my wife, Cheryl, for her invaluable patience and support.

Mark Dugopolski
Ponchatoula, Louisiana

Guided Tour

Features and Supplements

> "I was 'gripped' by the examples and introductions to the topics. These were interesting, current, and nicely written. Students will find these motivating to learn the material."
>
> *Timothy McKenna,*
> *University of Michigan–Dearborn*

Chapter Opener ❯

Each chapter opener features a real-world situation that can be modeled using mathematics. The application then refers students to a specific exercise in the chapter's exercise sets.

Chapter 5

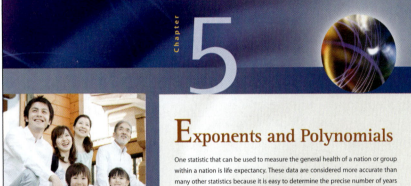

Exponents and Polynomials

One statistic that can be used to measure the general health of a nation or group within a nation is life expectancy. These data are considered more accurate than many other statistics because it is easy to determine the precise number of years in a person's lifetime.

According to the National Center for Health Statistics, an American born in 2006 has a life expectancy of 77.9 years. However, an American male born in 2006 has a life expectancy of only 75.0 years, whereas a female can expect 80.8 years. A male who makes it to 65 can expect to live 16.1 more years, whereas a female who makes it to 65 can expect 17.9 more years. In the next few years, thanks in part to advances in health care and science, longevity is expected to increase significantly worldwide. In fact, the World Health Organization predicts that by 2025 no country will have a life expectancy of less than 50 years.

In this chapter, we will see how functions involving exponents are used to model life expectancy.

- **5.1** Integral Exponents and Scientific Notation
- **5.2** The Power Rules
- **5.3** Polynomials and Polynomial Functions
- **5.4** Multiplying Binomials
- **5.5** Factoring Polynomials
- **5.6** Factoring $ax^2 + bx + c$
- **5.7** Factoring Strategy
- **5.8** Solving Equations by Factoring

In Exercises 87 and 88 of Section 5.2 you will see how exponents are used to determine the life expectancies of white men and women.

---

 88. *Life expectancy of white females.* Life expectancy improved more for females than for males during the 1940s and 1950s due to a dramatic decrease in maternal mortality rates. The function

$$L = 78.5(1.001)^a$$

can be used to model life expectancy L for U.S. white females with present age a.

a) To what age can a 20-year-old white female expect to live?   80.1 years

b) Bob, 30, and Ashley, 26, are an average white couple. How many years can Ashley expect to live as a widow?   7.9 years

c) Interpret the intersection of the life expectancy curves in the accompanying figure.   At 80 both males and females can expect about 5 more years.

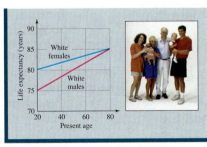

Figure for Exercises 87 and 88

Guided Tour

In This Section

The In This Section listing gives a preview of the topics to be covered in the section. These subsections have now been numbered for easier reference. In addition, these subsections are listed in the relevant places in the end-of-section exercises.

5.1 Integral Exponents and Scientific Notation

In This Section

⟨1⟩ Positive and Negative Exponents
⟨2⟩ The Product Rule for Exponents
⟨3⟩ Zero Exponent
⟨4⟩ Changing the Sign of an Exponent
⟨5⟩ The Quotient Rule for Exponents
⟨6⟩ Scientific Notation

In Chapter 1, we defined positive integral exponents and learned to evaluate expressions involving exponents. In this section we will extend the definition of exponents to include all integers and will learn some rules for working with integral exponents. In Chapter 7 we will see that any rational number can be used as an exponent.

⟨1⟩ Positive and Negative Exponents

We learned in Chapter 1 that a positive integral exponent indicates the number of times that the base is used as a factor. For example,

$$x^2 = x \cdot x \quad \text{and} \quad a^3 = a \cdot a \cdot a.$$

Examples

Examples refer directly to exercises, and those exercises in turn refer back to that example. This **double cross-referencing** helps students connect examples to exercises no matter which one they start with.

EXAMPLE 4

Negative powers of fractions

Simplify. Assume the variables are nonzero real numbers, and write the answers with positive exponents only.

a) $\left(\dfrac{3}{4}\right)^{-3}$ b) $\left(\dfrac{x^2}{5}\right)^{-2}$ c) $\left(-\dfrac{2y^3}{3}\right)^{-2}$

Solution

a) $\left(\dfrac{3}{4}\right)^{-3} = \left(\dfrac{4}{3}\right)^3$ The reciprocal of $\frac{3}{4}$ is $\frac{4}{3}$.

$= \dfrac{4^3}{3^3}$ Power of a quotient rule

$= \dfrac{64}{27}$ $4^3 = 64, 3^3 = 27$

b) There is more than one way to simplify these expressions. Taking the reciprocal of the fraction first we get

$$\left(\frac{x^2}{5}\right)^{-2} = \left(\frac{5}{x^2}\right)^2 = \frac{5^2}{(x^2)^2} = \frac{25}{x^4}.$$

Applying the power of a quotient rule first (as in Example 3) we get

$$\left(\frac{x^2}{5}\right)^{-2} = \frac{(x^2)^{-2}}{5^{-2}} = \frac{x^{-4}}{5^{-2}} = \frac{5^2}{x^4} = \frac{25}{x^4}.$$

c) $\left(-\dfrac{2y^3}{3}\right)^{-2} = \left(-\dfrac{3}{2y^3}\right)^2 = \dfrac{9}{4y^6}$

Now do Exercises 33–40

Simplify. See Example 4.

33. $\left(\dfrac{2}{5}\right)^{-2}$ 34. $\left(\dfrac{3}{4}\right)^{-2}$

35. $\left(-\dfrac{1}{2}\right)^{-2}$ 36. $\left(-\dfrac{2}{3}\right)^{-2}$

37. $\left(-\dfrac{2x}{3}\right)^{-3}$ 38. $\left(-\dfrac{ab}{c}\right)^{-1}$

39. $\left(\dfrac{2x^2}{3y}\right)^{-3}$ 40. $\left(\dfrac{ab^{-3}}{a^2b}\right)^{-2}$

"I really appreciate how the examples correlate with the homework sections. These specific examples are helpful to students that go onto college algebra and pre-calc math classes."
Sean Stewart, Owens Community College

"The worked out examples are clearly explained, no step is left out, and they progress in a fashion that eases the student from very basic to the somewhat complex."
Larry Green, Lake Tahoe Community College

Math at Work

The Math at Work feature appears in each chapter to reinforce the book's theme of real applications in the everyday world of work.

"Dugopolski uses language and context appropriate for the level of student for whom the text is written without sacrificing mathematical rigor or precision."

Irma Bakenhus,
San Antonio College

Math *at Work*   Kayak Design

Kayaks have been built by the Aleut and Inuit people for the past 4000 years. Today's builders have access to materials and techniques unavailable to the original kayak builders. Modern kayakers incorporate hydrodynamics and materials technology to create designs that are efficient and stable. Builders measure how well their designs work by calculating indicators such as prismatic coefficient, block coefficient, and the midship area coefficient, to name a few.

Even the fitting of a kayak to the paddler is done scientifically. For example, the formula

$$PL = 2 \cdot BL + BS \left(0.38 \cdot EE + 1.2 \sqrt{\left(\frac{BW}{2} - \frac{SW}{2}\right)^2 + (SL)^2} \right)$$

can be used to calculate the appropriate paddle length. BL is the length of the paddle's blade. BS is a boating style factor, which is 1.2 for touring, 1.0 for river running, and 0.95 for play boating. EE is the elbow to elbow distance with the paddler's arms straight out to the sides. BW is the boat width and SW is the shoulder width. SL is the spine length, which is the distance measured in a sitting position from the chair seat to the top of the paddler's shoulder. All lengths are in centimeters.

The degree of control a kayaker exerts over the kayak depends largely on the body contact with it. A kayaker wears the kayak. So the choice of a kayak should hinge first on the right body fit and comfort and second on the skill level or intended paddling style. So designing, building, and even fitting a kayak is a blend of art and science.

Strategy Boxes

The strategy boxes provide a handy reference for students to use when they review key concepts and techniques to prepare for tests and homework. They are now directly referenced in the end-of-section exercises where appropriate.

Strategy for Factoring Polynomials Completely

1. Factor out the GCF (with a negative coefficient if necessary).
2. When factoring a binomial, check to see whether it is a difference of two squares, a difference of two cubes, or a sum of two cubes. *A sum of two squares does not factor.*
3. When factoring a trinomial, check to see whether it is a perfect square trinomial.
4. If the polynomial has four terms, try factoring by grouping.
5. When factoring a trinomial that is not a perfect square, use the *ac* method or the trial-and-error method.
6. Check to see whether any of the factors can be factored again.

Margin Notes

Margin notes include **Helpful Hints,** which give advice on the topic they're adjacent to; **Calculator Close-Ups,** which provide advice on using calculators to verify students' work; and **Teaching Tips,** which are especially helpful in programs with new instructors who are looking for alternate ways to explain and reinforce material.

‹ Helpful Hint ›

Some students grow up believing that the only way to solve an equation is to "do the same thing to each side." Then along come quadratic equations and the zero factor property. For a quadratic equation, we write an equivalent compound equation that is not obtained by "doing the same thing to each side."

‹ Teaching Tip ›

Show students how to make up a problem like this example: If $x = 5$, then $(5 - 2)(5 + 7) = 36$. So one of the solutions to $(x - 2)(x + 7) = 36$ is 5. Now solve it to find both solutions.

‹ Calculator Close-Up ›

Your calculator can add signed numbers. Most calculators have a key for subtraction and a different key for the negative sign.

```
-5+13
              8
-5+.09
          -4.91
3/8+-5/6▶Frac
         -11/24
```

You should do the exercises in this section by hand and then check with a calculator.

Guided Tour
Features and Supplements

Exercises
Section exercises are preceded by true/false **Warm-Ups,** which can be used as quizzes or for class discussion.

Warm-Ups ▼

Fill in the blank.
1. To _____ means to write as a product.
2. A _____ number is an integer greater than 1 that has no factors besides itself and 1.
3. The _____ of two numbers is the largest number that is a factor of both.
4. All factoring can be checked by _____ the factors.

True or false?
5. There are only nine prime numbers.
6. The prime factorization of 32 is $2^3 \cdot 3$.
7. The integer 51 is a prime number.
8. The GCF for 12 and 16 is 4.
9. The GCF for $x^5y^3 - x^4y^7$ is x^4y^3.
10. We can factor out $2xy$ or $-2xy$ from $2x^2y - 6xy^2$.

Getting More Involved
concludes the exercise set with **Discussion, Writing, Exploration,** and **Cooperative Learning** activities for well-rounded practice in the skills for that section.

Applications
Solve each problem.

99. *Approach speed.* The formula $1211.1L = CA^2S$ is used to determine the approach speed for landing an aircraft, where L is the gross weight of the aircraft in pounds, C is the coefficient of lift, S is the surface area of the wings in square feet (ft^2), and A is approach speed in feet per second. Find A for the Piper Cheyenne, which has a gross weight of 8700 lb, a coefficient of lift of 2.81, and a wing surface area of 200 ft^2.

100. *Time to swing.* The period T (time in seconds for one complete cycle) of a simple pendulum is related to the length L (in feet) of the pendulum by the formula $8T^2 = \pi^2L$. If a

Getting More Involved
103. *Discussion*

Which of the following equations is not a quadratic equation? Explain your answer.
a) $\pi x^2 - \sqrt{5}x - 1 = 0$ b) $3x^2 - 1 = 0$
c) $4x + 5 = 0$ d) $0.009x^2 = 0$

104. *Exploration*

Solve $x^2 - 4x + k = 0$ for $k = 0, 4, 5,$ and 10.

a) When does the equation have only one solution?
b) For what values of k are the solutions real?
c) For what values of k are the solutions imaginary?

Calculator Exercises
Optional calculator exercises provide students with the opportunity to use scientific or graphing calculators to solve various problems.

33. $(y - 1)(y + 1)$
34. $(p + 2)(p - 2)$
35. $(3x - 8)(3x + 8)$
36. $(6x + 1)(6x - 1)$
37. $(r + s)(r - s)$
38. $(b - y)(b + y)$
39. $(8y - 3a)(8y + 3a)$
40. $(4u - 9v)(4u + 9v)$
41. $(5x^2 - 2)(5x^2 + 2)$
42. $(3y^2 + 1)(3y^2 - 1)$

74. $(0.1y + 0.5)^2$
75. $(a + b)^3$
76. $(2a - 3b)^3$
77. $(1.5x + 3.8)^2$
78. $(3.45a - 2.3)^2$
79. $(3.5t - 2.5)(3.5t + 2.5)$
80. $(4.5h + 5.7)(4.5h - 5.7)$

Video Exercises
A video icon indicates an exercise that has a video walking through how to solve it.

Mid-Chapter Quiz | Sections 6.1 through 6.4 | Chapter 6

Reduce to lowest terms.
1. $\dfrac{36}{84}$
2. $\dfrac{8x - 2}{8}$
3. $\dfrac{w^2 - 1}{2w + 2}$
4. $\dfrac{2a^2 - 10a + 12}{6 - 3a}$

Perform the indicated operation.
5. $\dfrac{6}{7} \cdot \dfrac{21}{10}$
6. $\dfrac{3xy^2}{5z} \cdot \dfrac{8x^2z^3}{8y^4}$

13. $\dfrac{5}{6} - \dfrac{5}{21}$
14. $\dfrac{4}{ab^3} + \dfrac{5}{a^2b}$
15. $\dfrac{3x}{x + 1} + \dfrac{x}{x^2 + 2x + 1}$
16. $\dfrac{y}{y + 5} - \dfrac{y}{y + 2}$
17. $\dfrac{1}{a} + \dfrac{1}{b} + \dfrac{1}{c}$

Mid-Chapter Quiz
Mid-Chapter Quizzes give students an earlier chance check their progress through the chapter allowing them to identify what past skills they need to practice as they move forward in their class.

"This text is very well written with good, detailed examples. It offers plenty of practice exercises in each section including several real world applications."

Randall Casleton, University of Tennessee–Martin

Wrap-Up

The extensive and varied review in the chapter Wrap-Up will help students prepare for tests. First comes the **Summary** with key terms and concepts illustrated by examples; then **Enriching Your Mathematical Word Power** enables students to test their recall of new terminology in a fill-in-the-blank format.

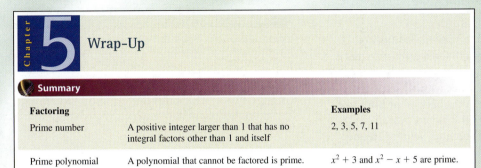

Chapter 5 Wrap-Up

Summary

Factoring		Examples
Prime number	A positive integer larger than 1 that has no integral factors other than 1 and itself	2, 3, 5, 7, 11
Prime polynomial	A polynomial that cannot be factored is prime.	$x^2 + 3$ and $x^2 - x + 5$ are prime.

Enriching Your Mathematical Word Power

Fill in the blank.

1. A _____ number is an integer greater than 1 that has no integral factors other than itself and 1.

2. An integer larger than 1 that is not prime is _____.

3. A polynomial that has no factors is a _____ polynomial.

4. Writing a polynomial as a product is _____.

5. Writing a polynomial as a product of primes is factoring _____.

8. The trinomial $a^2 + 2ab + b^2$ is a perfect _____ trinomial.

9. The polynomial $a^3 + b^3$ is a _____ of two cubes.

10. The polynomial $a^3 - b^3$ is a _____ of two cubes.

11. A _____ equation is an equation of the form $ax^2 + bx + c = 0$.

12. According to the _____ factor property, if $ab = 0$ then $a = 0$ or $b = 0$.

Review Exercises

5.1 Integral Exponents and Scientific Notation
Simplify each expression. Assume all variables represent nonzero real numbers. Write your answers with positive exponents.

1. $2 \cdot 2 \cdot 2^{-1}$

2. $5^{-1} \cdot 5$

3. $2^2 \cdot 3^2$

4. $3^2 \cdot 5^2$

33. $(m^2 n^3)^{-2} (m^{-3} n^2)^4$

34. $(w^{-3} xy)^{-1} (wx^{-3} y)^2$

35. $\left(\dfrac{2}{3}\right)^{-4}$

36. $\left(\dfrac{a^4}{3}\right)^{-2}$

37. $\left(\dfrac{1}{2} + \dfrac{1}{3}\right)^2$

38. $\left(\dfrac{1}{2} - \dfrac{1}{3}\right)^{-2}$

Next come **Review Exercises,** which are first linked back to the section of the chapter that they review, and then the exercises are mixed without section references in the **Miscellaneous** section.

Miscellaneous

Find all real and imaginary solutions to each equation.

63. $w^2 + 4 = 0$

64. $w^2 + 9 = 0$

65. $a^4 + 6a^2 + 8 = 0$

66. $b^4 + 13b^2 + 36 = 0$

67. $m^4 - 16 = 0$

68. $t^4 - 4 = 0$

69. $16b^4 - 1 = 0$

70. $b^4 - 81 = 0$

71. $x^3 + 1 = 0$

72. $x^3 - 1 = 0$

73. $x^3 + 8 = 0$

Photo for Exercise 81

Chapter Test

The test gives students additional practice to make sure they're ready for the real thing, with **all** answers provided at the back of the book and **all** solutions available in the Student's Solutions Manual.

Chapter 5 Test

Simplify each expression. Assume all variables represent nonzero real numbers. Exponents in your answers should be positive exponents.

1. 3^{-2}
2. $\dfrac{1}{6^{-2}}$
3. $\left(\dfrac{1}{2}\right)^{-3}$
4. $3x^4 \cdot 4x^3$

25. $12m^2 + 28m + 15$
26. $2x^{10} + 5x^5 - 12$
27. $2xa + 3a - 10x - 15$
28. $x^4 + 3x^2 - 4$
29. $a^4 - 1$

The **Making Connections**

feature following the Chapter Test is a cumulative review of all chapters up to and including the one just finished, helping to tie the course concepts together for students on a regular basis.

Making Connections | A Review of Chapters 1–5

Simplify each expression.

1. 4^2
2. $4(-2)$
3. 4^{-2}
4. $2^3 \cdot 4^{-1}$
5. $2^{-1} + 2^{-1}$
6. $2^{-1} \cdot 3^{-1}$
7. $2^{-2} \cdot 3^2$
8. $-3^4 \cdot 6^{-2}$

39. $\dfrac{6x^2 - 4x}{2x}$
40. $\dfrac{9x - 15}{3} - \dfrac{7x - 21}{7}$

Solve each problem.

41. The area of a rectangular field is 5525 square meters and the length is 85 meters. What is the width?
42. The perimeter of a rectangular canvas is 54 inches and its length is 15 inches. Find the width.

Critical Thinking

The Critical Thinking section that concludes every chapter encourages students to think creatively to solve unique and intriguing problems and puzzles.

Critical Thinking | For Individual or Group Work | Chapter 5

These exercises can be solved by a variety of techniques, which may or may not require algebra. So be creative and think critically. Explain all answers. Answers are in the Instructor's Edition of this text.

1. *Pile of pipes.* Ten pipes, each with radius a, are stacked as shown in the figure. What is the height of the pile?

Figure for Exercise 1

2. *Twin trains.* Two freight trains are approaching each other, each traveling at 40 miles per hour, on parallel tracks. If each train is $\frac{1}{2}$ mile long, then how long does it take for them to pass each other?

number of times that the digit above it appears in the second row.

0	1	2	3	4	5	6	7	8	9

Table for Exercise 6

7. *Presidential proof.* James Garfield, twentieth president of the United States, gave the following proof of the Pythagorean theorem. Start with a right triangle with legs a and b and hypotenuse c. Use the right triangle twice to make the trapezoid shown in the figure.

   a) Find the area of the trapezoid by using the formula $A = \frac{1}{2}h(b_1 + b_2)$.
   b) Find the area of each of the three triangles in the figure. Then find the sum of those areas.

"The critical thinking exercises at the end of the chapter are a good way to help students learn to work in groups and to write mathematically. Having to explain how and why you worked out a solution reinforces the thinking and writing skills necessary to be successful in today's world."

Mark Brenneman,
Mesa Community College

SUPPLEMENTS

Multimedia Supplements

MCGRAW-HILL HIGHER EDUCATION AND BLACKBOARD HAVE TEAMED UP.

Blackboard, the Web-based course-management system, has partnered with McGraw-Hill to better allow students and faculty to use online materials and activities to complement face-to-face teaching. Blackboard features exciting social learning and teaching tools that foster more logical, visually impactful and active learning opportunities for students. You'll transform your closed-door classrooms into communities where students remain connected to their educational experience 24 hours a day.

This partnership allows you and your students access to McGraw-Hill's Connect™ and Create™ right from within your Blackboard course—all with one single sign-on.

Not only do you get single sign-on with Connect™ and Create™, you also get deep integration of McGraw-Hill content and content engines right in Blackboard. Whether you're choosing a book for your course or building Connect™ assignments, all the tools you need are right where you want them—inside of Blackboard.

Gradebooks are now seamless. When a student completes an integrated Connect™ assignment, the grade for that assignment automatically (and instantly) feeds your Blackboard grade center.

McGraw-Hill and Blackboard can now offer you easy access to industry leading technology and content, whether your campus hosts it, or we do. Be sure to ask your local McGraw-Hill representative for details.

    www.mcgrawhillconnect.com

McGraw-Hill conducted in-depth research to create a new and improved learning experience that meets the needs of today's students and instructors. The result is a reinvented learning experience rich in information, visually engaging, and easily accessible to both instructors and students.

 McGraw-Hill's Connect is a Web-based assignment and assessment platform that helps students connect to their coursework and prepares them to succeed in and beyond the course. Connect Mathematics enables math instructors to create and share courses and assignments with colleagues and adjuncts with only a few clicks of the mouse. All exercises, learning objectives, videos, and activities are directly tied to text-specific material.

- Students have access to immediate feedback and help while working through assignments.

- A Web-optimized eBook is seamlessly integrated within ConnectPlus Mathematics.
- Instructors can select textbook-specific questions organized by chapter, section, and objective.
- Connect Mathematics' straightforward design creates and intuitive, visually pleasing grade management environment.

Instructors: To access Connect, request registration information from your McGraw-Hill sales representative.

Computerized Test Bank (CTB) Online (Instructors Only)

Available through Connect, this **computerized test bank,** utilizing Wimba Diploma® algorithm-based testing software, enables users to create customized exams quickly. This user-friendly program enables instructors to search for questions by topic, format, or difficulty level; to edit existing questions or to add new ones; and to scramble questions and answer keys for multiple versions of the same test. Hundreds of text-specific open-ended and multiple-choice questions are included in the question bank. Sample chapter tests in Microsoft Word® and PDF formats are also provided.

Online Instructor's Solutions Manual (Instructors Only)

Available on Connect, the Instructor's Solutions Manual provides comprehensive, **worked-out solutions** to all exercises in the text. The methods used to solve the problems in the manual are the same as those used to solve the examples in the textbook.

Video Lectures Available Online

In the videos, qualified teachers work through selected exercises from the textbook, following the solution methodology employed in the text. The video series is available online as an assignable element of Connect. The videos are closed-captioned for the hearing impaired, are subtitled in Spanish, and meet the Americans with Disabilities Act Standards for Accessible Design. Instructors may use them as resources in a learning center, for online courses, and/or to provide extra help for students who require extra practice.

ALEKS®   www.ALEKS.com

ALEKS (**A**ssessment and **LE**arning in **K**nowledge **S**paces) is a dynamic online learning system for mathematics education, available over the Web 24/7. ALEKS assesses students, accurately determines their knowledge, and then guides them to the material that they are most ready to learn. With a variety of reports, Textbook Integration Plus, quizzes, and homework assignment capabilities, ALEKS offers flexibility and ease of use for instructors.

- ALEKS uses artificial intelligence to determine exactly what each student knows and is ready to learn. ALEKS remediates student gaps and provides highly efficient learning and improved learning outcomes.
- ALEKS is a comprehensive curriculum that aligns with syllabi or specified textbooks. Used in conjunction with a McGraw-Hill text, students also receive links to text-specific videos, multimedia tutorials, and textbook pages.
- Textbook Integration Plus enables ALEKS to be automatically aligned with syllabi or specified McGraw-Hill textbooks with instructor-chosen dates, chapter goals, homework, and quizzes.
- ALEKS with AI-2 gives instructors increased control over the scope and sequence of student learning. Students using ALEKS demonstrate a steadily increasing mastery of the content of the course.
- ALEKS offers a dynamic classroom management system that enables instructors to monitor and direct student progress toward mastery of course objectives. See: www.aleks.com

Printed Supplements

Annotated Instructor's Edition (Instructors Only)

This ancillary contains answers to all exercises in the text. These answers are printed in a special color for ease of use by the instructor and are located on the appropriate pages throughout the text.

Student's Solutions Manual

The Student's Solutions Manual provides comprehensive, **worked-out solutions** to all of the odd-numbered section exercises and all exercises in the Mid-Chapter Quizzes, Chapter Tests, and Making Connections. The steps shown in the solutions match the style of solved examples in the textbook.

Contents

6

Chapter

Rational Expressions and Functions   381

7

Chapter

Radicals and Rational Exponents   457

Chapter 14

Counting and Probability   861

Appendix   A-1

Answers to Selected Exercises   A-3

Index   I-1

Applications Index

The Real Numbers

Everywhere you look people are running, riding, dancing, and exercising their way to fitness. In the past year more than $25 billion has been spent on sports equipment alone, and this amount is growing steadily.

Proponents of exercise claim that it can increase longevity, improve body image, decrease appetite, and generally enhance a person's health. While many sports activities can help you to stay fit, experts have found that aerobic, or dynamic, workouts provide the most fitness benefit. Some of the best aerobic exercises include cycling, running, and even jumping rope. Whatever athletic activity you choose, trainers recommend that you set realistic goals and work your way toward them consistently and slowly. To achieve maximum health benefits, experts suggest that you exercise three to five times a week for 15 to 60 minutes at a time.

There are many different ways to measure exercise. One is to measure the energy used, or the rate of oxygen consumption. Since heart rate rises as a function of increased oxygen, an easier measure of intensity of exercise is your heart rate during exercise. The desired heart rate, or target heart rate, for beneficial exercise varies for each individual depending on conditioning, age, and gender.

In Exercises 101 and 102 of Section 1.4 you will see how an algebraic expression can determine your target heart rate for beneficial exercise.

1.1   Sets

Every subject has its own terminology, and **algebra** is no different. In this section we will learn the basic terms and facts about sets.

〈1〉 Set Notation

A **set** is a collection of objects. At home you may have a set of dishes and a set of steak knives. In algebra, we generally discuss sets of numbers. For example, we refer to the numbers 1, 2, 3, 4, 5, and so on as the set of **counting numbers** or **natural numbers.** Of course, these are the numbers that we use for counting.

The objects or numbers in a set are called the **elements** or **members** of the set. To describe sets with a convenient notation, we use braces, { }, and name the sets with capital letters. For example,

$$A = \{1, 2, 3\}$$

means that set A is the set whose members are the natural numbers 1, 2, and 3. The letter N is used to represent the entire set of natural numbers.

A set that has a fixed number of elements such as $\{1, 2, 3\}$ is a **finite** set, whereas a set without a fixed number of elements such as the natural numbers is an **infinite** set. When listing the elements of a set, we use a series of three dots to indicate a continuing pattern. For example, the set of natural numbers is written as

$$N = \{1, 2, 3, \ldots\}.$$

The set of natural numbers *between* 4 and 40 can be written

$$\{5, 6, 7, 8, \ldots, 39\}.$$

Note that since the members of this set are *between* 4 and 40, it does not include 4 or 40.

Set-builder notation is another method of describing sets. In this notation, we use a variable to represent the numbers in the set. A **variable** is a letter that is used to stand for some numbers. The set is then built from the variable and a description of the numbers that the variable represents. For example, the set

$$B = \{1, 2, 3, \ldots, 49\}$$

is written in set-builder notation as

$$B = \{x \mid x \text{ is a natural number less than 50}\}.$$

$$\uparrow \uparrow \qquad\qquad\qquad \uparrow$$

The set of numbers   such that   condition for membership

This notation is read as "B is the set of numbers x such that x is a natural number less than 50." Notice that the number 50 is not a member of set B.

The symbol $\in$ is used to indicate that a specific number is a member of a set, and $\notin$ indicates that a specific number is not a member of a set. For example, the statement $1 \in B$ is read as "1 is a member of B," "1 belongs to B," "1 is in B," or "1 is an element of B." The statement $0 \notin B$ is read as "0 is not a member of B," "0 does not belong to B," "0 is not in B," or "0 is not an element of B."

Two sets are **equal** if they contain exactly the same members. Otherwise, they are said to be not equal. To indicate equal sets, we use the symbol $=$. For sets that are not equal we use the symbol $\neq$. So $\{3, 5, 6\} \neq \{3, 5, 7\}$, but

$$\{8, 9\} = \{8, 9\}, \{2, 4, 1\} = \{1, 4, 2\}, \text{ and } \{5, 5, 6, 7\} = \{5, 6, 7\}.$$

Note that the order of the elements in a set does not matter and it is not necessary to list an element of a set more than once.

EXAMPLE   1

Set notation

Let $A = \{1, 2, 3, 5\}$ and $B = \{x \mid x$ is an even natural number less than 10$\}$. Determine whether each statement is true or false.

a) $3 \in A$ b) $5 \in B$ c) $4 \notin A$ d) $A = N$

e) $A = \{x \mid x$ is a natural number less than 6$\}$ f) $B = \{2, 4, 6, 8\}$

Solution

a) True, because 3 is a member of set A.

b) False, because 5 is not an even natural number.

c) True, because 4 is not a member of set A.

d) False, because A does not contain all of the natural numbers.

e) False, because 4 is a natural number less than 6, and $4 \notin A$.

f) True, because the even counting numbers less than 10 are 2, 4, 6, and 8.

> Now do Exercises 1–10

⟨2⟩ Union of Sets

Any two sets A and B can be combined to form a new set called their union that consists of all elements of A together with all elements of B.

> **Union of Sets**
>
> If A and B are sets, the **union** of A and B, denoted $A \cup B$, is the set of all elements that are either in A, in B, or in both. In symbols,
>
> $$A \cup B = \{x \mid x \in A \text{ or } x \in B\}.$$

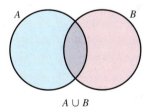

$A \cup B$

Figure 1.1

In mathematics the word "or" is always used in an inclusive manner (allowing the possibility of both alternatives). The diagram in Fig. 1.1 can be used to illustrate $A \cup B$. Any point that lies within circle A, circle B, or both is in $A \cup B$. Diagrams (like Fig. 1.1) that are used to illustrate sets are called **Venn diagrams.**

EXAMPLE   2

Union of sets

Let $A = \{0, 2, 3\}$, $B = \{2, 3, 7\}$, and $C = \{7, 8\}$. List the elements in each of these sets.

a) $A \cup B$ b) $A \cup C$

⟨ **Helpful Hint** ⟩

To remember what "union" means think of a labor union, which is a group formed by joining together many individuals.

Solution

a) $A \cup B$ is the set of numbers that are in A, in B, or in both A and B.

$$A \cup B = \{0, 2, 3, 7\}$$

b) $A \cup C = \{0, 2, 3, 7, 8\}$

> Now do Exercises 11–12

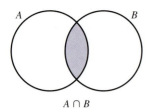

$A \cap B$

Figure 1.2

⟨ **Helpful Hint** ⟩

To remember the meaning of "inter-section," think of the intersection of two roads. At the intersection you are on both roads.

⟨**3**⟩ **Intersection of Sets**

Another way to form a new set from two known sets is by considering only those elements that the two sets have in common. The diagram shown in Fig. 1.2 illustrates the intersection of two sets A and B.

> **Intersection of Sets**
>
> If A and B are sets, the **intersection** of A and B, denoted $A \cap B$, is the set of all elements that are in both A and B. In symbols,
> $$A \cap B = \{x \mid x \in A \text{ and } x \in B\}.$$

It is possible for two sets to have no elements in common. A set with no members is called the **empty set** and is denoted by the symbol $\varnothing$. Note that $A \cup \varnothing = A$ and $A \cap \varnothing = \varnothing$ for any set A.

CAUTION It is not correct to use 0 or $\{0\}$ as the empty set. The number 0 is not a set and $\{0\}$ is a set with one member, the number 0. We use a special symbol $\varnothing$ for the empty set.

E X A M P L E 3

Intersection of sets

Let $A = \{0, 2, 3\}$, $B = \{2, 3, 7\}$, and $C = \{7, 8\}$. List the elements in each of these sets.

a) $A \cap B$ **b)** $B \cap C$ **c)** $A \cap C$

Solution

a) $A \cap B$ is the set of all numbers that are in both A and B. So $A \cap B = \{2, 3\}$.

b) $B \cap C = \{7\}$ **c)** $A \cap C = \varnothing$

> Now do Exercises 13–22

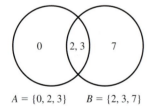

$A = \{0, 2, 3\}$ $B = \{2, 3, 7\}$

Figure 1.3

A Venn diagram can be used to illustrate the result of Example 3(a). Since 2 and 3 belong to both A and B, they are placed in the overlapping region of Fig. 1.3. Since 0 is in A but not in B, it is placed inside the circle for A but outside B. Likewise 7 is placed inside B but outside A.

E X A M P L E 4

Membership and equality

Let $A = \{1, 2, 3, 5\}$ and $B = \{2, 3, 7, 8\}$. Place one of the symbols $=$, $\neq$, $\in$, or $\notin$ in the blank to make each statement correct.

a) 5 _____ $A \cup B$ **b)** 5 _____ $A \cap B$

c) $A \cup B$ _____ $\{1, 2, 3, 5, 7, 8\}$ **d)** $A \cap B$ _____ $\{2\}$

Solution

a) $5 \in A \cup B$ because 5 is a member of A.

b) $5 \notin A \cap B$ because 5 must belong to *both* A and B to be a member of $A \cap B$.

c) $A \cup B = \{1, 2, 3, 5, 7, 8\}$ because the elements of A together with those of B are listed. Note that 2 and 3 are members of both sets but are listed only once.

d) $A \cap B \neq \{2\}$ because $A \cap B = \{2, 3\}$.

> Now do Exercises 23–32

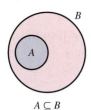

A ⊆ B

Figure 1.4

⟨**4**⟩ **Subsets**

If every member of set *A* is also a member of set *B*, then we write $A \subseteq B$ and say that *A* is a **subset** of *B*. See Fig. 1.4. For example,

$$\{2, 3\} \subseteq \{2, 3, 4\}$$

because $2 \in \{2, 3, 4\}$ and $3 \in \{2, 3, 4\}$. Note that the symbol for membership ($\in$) is used between a single element and a set, whereas the symbol for subset ($\subseteq$) is used between two sets. If *A* is not a subset of *B*, we write $A \not\subseteq B$.

CAUTION To claim that $A \not\subseteq B$, there *must* be an element of *A* that does *not* belong to *B*. For example,

$$\{1, 2\} \not\subseteq \{2, 3, 4\}$$

because 1 is a member of the first set but not of the second.

Is the empty set $\varnothing$ a subset of $\{2, 3, 4\}$? If we say that $\varnothing$ is *not* a subset of $\{2, 3, 4\}$, then there must be an element of $\varnothing$ that does not belong to $\{2, 3, 4\}$. But that cannot happen because $\varnothing$ is empty. So $\varnothing$ is a subset of $\{2, 3, 4\}$. In fact, by the same reasoning, *the empty set is a subset of every set.*

E X A M P L E **5**

Subsets

Determine whether each statement is true or false.

a) $\{1, 2, 3\}$ is a subset of the set of natural numbers.

b) The set of natural numbers is not a subset of $\{1, 2, 3\}$.

c) $\{1, 2, 3\} \not\subseteq \{2, 4, 6, 8\}$

d) $\{2, 6\} \subseteq \{1, 2, 3, 4, 5\}$

e) $\varnothing \subseteq \{2, 4, 6\}$

⟨ **Helpful Hint** ⟩

The symbols $\subseteq$ and $\subset$ are often used interchangeably. The symbol $\subseteq$ combines the subset symbol $\subset$ and the equal symbol $=$. We use it when sets are equal, $\{1, 2\} \subseteq \{1, 2\}$, and when they are not, $\{1\} \subseteq \{1, 2\}$. When sets are not equal, we could simply use $\subset$, as in $\{1\} \subset \{1, 2\}$.

Solution

a) True, because 1, 2, and 3 are natural numbers.

b) True, because 5, for example, is a natural number and $5 \notin \{1, 2, 3\}$.

c) True, because 1 is in the first set but not in the second.

d) False, because 6 is in the first set but not in the second.

e) True, because we cannot find anything in $\varnothing$ that fails to be in $\{2, 4, 6\}$.

Now do Exercises 33–44

⟨**5**⟩ **Combining Three or More Sets**

We know how to find the union and intersection of two sets. For three or more sets we use parentheses to indicate which pair of sets to combine first. In Example 6, notice that different results are obtained from different placements of the parentheses.

E X A M P L E **6**

Operations with three sets

Let $A = \{1, 2, 3, 4\}$, $B = \{2, 5, 6, 8\}$, and $C = \{4, 5, 7\}$. List the elements of each of these sets.

a) $(A \cup B) \cap C$ **b)** $A \cup (B \cap C)$

Solution

a) The parentheses indicate that the union of A and B is to be found first and then the result, $A \cup B$, is to be intersected with C.

$$A \cup B = \{1, 2, 3, 4, 5, 6, 8\}$$

Now examine $A \cup B$ and C to find the elements that belong to both sets:

$$A \cup B = \{1, 2, 3, 4, 5, 6, 8\}$$
$$C = \{4, 5, 7\}$$

The only numbers that are members of $A \cup B$ and C are 4 and 5. Thus,

$$(A \cup B) \cap C = \{4, 5\}.$$

b) In $A \cup (B \cap C)$, first find $B \cap C$:

$$B \cap C = \{5\}$$

Now $A \cup (B \cap C)$ consist of all members of A together with 5 from $B \cap C$:

$$A \cup (B \cap C) = \{1, 2, 3, 4, 5\}$$

> Now do Exercises 45–58

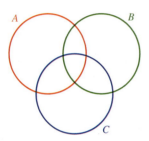

Figure 1.5

Every possibility for membership in three sets is shown in the Venn diagram in Fig. 1.5. Figure 1.6 shows the numbers from the three sets of Example 6 in the appropriate regions of this diagram. Since no number belongs to all three sets, there is no number in the center region of Fig. 1.6. Since 1 is in A, but is not in B or C, it is placed inside circle A but outside circles B and C. Since 2 is in A and B, but is not in C, it is placed in the intersection of A and B, but outside C. Check that the remaining numbers are in the appropriate regions. Now you can see from the diagram that the numbers that are in C and in $A \cup B$ are 4 and 5. So,

$$(A \cup B) \cap C = \{4, 5\}.$$

You can also see that $B \cap C = \{5\}$. The union of that set with A gives us

$$A \cup (B \cap C) = \{1, 2, 3, 4, 5\}.$$

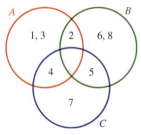

$A = \{1, 2, 3, 4\}$ $B = \{2, 5, 6, 8\}$
$C = \{4, 5, 7\}$

Figure 1.6

⟨6⟩ Applications

EXAMPLE 7

Using Venn diagrams

An instructor surveyed her class of 40 students and found that all of them either watched TV or surfed the Internet last evening. The number of students who watched TV but did not surf the Internet was 10. The number who surfed the Internet but did not watch TV was 16. Find the number who did both.

Solution

Draw a two-circle Venn diagram as shown in Fig. 1.7. Since 10 students watched TV, but did not surf, place 10 inside the TV circle, but outside the Internet circle. Since 16 surfed, but did not watch TV, place 16 inside the Internet circle, but outside the TV circle. Since the total in all three regions is 40, subtract 10 and 16 from 40 to get 14. So 14 is the number in the intersection of the two regions. So 14 students did both.

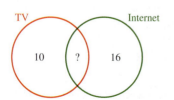

Figure 1.7

> Now do Exercises 83–84

The following box contains a summary of the symbols used in discussing sets in this section.

Set Symbols

$\in$	is a member of	$\notin$	is not a member of
$\subseteq$	is a subset of	$\not\subseteq$	is not a subset of
$=$	equal	$\neq$	not equal
$\cup$	union	$\cap$	intersection
$\varnothing$	empty set		

Warm-Ups ▼

Fill in the blank.

1. A _____ is a collection of objects.
2. A _____ set has a fixed number of elements.
3. A _____ diagram is used to illustrate sets.
4. A is a _____ of B if every element in A is in B.
5. The _____ set has no members.
6. The _____ of A and B consists of the elements that are in both A and B.
7. The _____ of A and B consists of the elements that are in either A or B.

True or false?

Let $A = \{1, 2, 3\}$, $B = \{3, 4\}$, and $C = \{3\}$.

8. A is the set of counting numbers.
9. B has an infinite number of elements.
10. $1 \in A \cap B$
11. $C \subseteq A$
12. $\varnothing \subseteq C$
13. $A \cap B = C$
14. $4 \in A \cup B$
15. $B \subseteq A$
16. $A \not\subseteq C$
17. $4 \notin A$

Exercises

1.1

‹ **Study Tips** ›

- Exercise sets are designed to increase gradually in difficulty. So start from the beginning and work lots of exercises.
- Find a group of students to work with outside of class. Explaining things to others improves your own understanding of the concepts.

‹1› **Set Notation**

Using the sets A, B, C, and N, determine whether each statement is true or false. Explain. See Example 1.

$A = \{1, 3, 5, 7, 9\}$ $B = \{2, 4, 6, 8\}$
$C = \{1, 2, 3, 4, 5\}$ $N = \{1, 2, 3, \ldots\}$

1. $3 \in A$ 2. $3 \in B$
3. $11 \notin A$ 4. $3 \notin C$
5. $C = N$ 6. $A = N$
7. $A \neq B$ 8. $C \neq N$
9. $99 \in N$ 10. $6 \in C$

‹2–3› **Union and Intersection of Sets**

Using the sets A, B, C, and N, list the elements in each set. If the set is empty, write $\varnothing$. See Examples 2 and 3.

$A = \{1, 3, 5, 7, 9\}$ $B = \{2, 4, 6, 8\}$
$C = \{1, 2, 3, 4, 5\}$ $N = \{1, 2, 3, \ldots\}$

11. $A \cup C$ 12. $A \cup B$

13. $A \cap C$ 14. $A \cap B$
15. $B \cup C$ 16. $B \cap C$

17. $A \cup \varnothing$ **18.** $B \cup \varnothing$
19. $A \cap \varnothing$ **20.** $B \cap \varnothing$
21. $A \cap N$ **22.** $A \cup N$

Use one of the symbols $\in$, $\notin$, $=$, $\neq$, $\cup$, or $\cap$ in each blank to make a true statement. See Example 4.
$A = \{1, 3, 5, 7, 9\}$ $B = \{2, 4, 6, 8\}$
$C = \{1, 2, 3, 4, 5\}$ $N = \{1, 2, 3, \ldots\}$

23. $A \cap B$ ____ $\varnothing$
24. $A \cap C$ ____ $\varnothing$
25. A ____ $B = \{1, 2, 3, 4, 5, 6, 7, 8, 9\}$
26. A ____ $B = \varnothing$
27. B ____ $C = \{2, 4\}$
28. B ____ $C = \{1, 2, 3, 4, 5, 6, 8\}$
29. 3 ____ $A \cap B$ **30.** 3 ____ $A \cap C$
31. 4 ____ $B \cap C$ **32.** 8 ____ $B \cup C$

⟨4⟩ Subsets

Determine whether each statement is true or false. Explain your answer. See Example 5.
$A = \{1, 3, 5, 7, 9\}$ $B = \{2, 4, 6, 8\}$
$C = \{1, 2, 3, 4, 5\}$ $N = \{1, 2, 3, \ldots\}$

33. $A \subseteq N$ **34.** $B \subseteq N$
35. $\{2, 3\} \subseteq C$ **36.** $C \subseteq A$
37. $B \not\subseteq C$ **38.** $C \not\subseteq A$
39. $\varnothing \subseteq B$ **40.** $\varnothing \subseteq C$
41. $A \subseteq \varnothing$ **42.** $B \subseteq \varnothing$
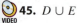 **43.** $A \cap B \subseteq C$ **44.** $B \cap C \subseteq \{2, 4, 6, 8\}$

⟨5⟩ Combining Three or More Sets

Using the sets D, E, and F, list the elements in each set. If the set is empty, write $\varnothing$. See Example 6.
$D = \{3, 5, 7\}$ $E = \{2, 4, 6, 8\}$ $F = \{1, 2, 3, 4, 5\}$

 45. $D \cup E$ **46.** $D \cap E$
 47. $D \cap F$ **48.** $D \cup F$
49. $E \cup F$ **50.** $E \cap F$

51. $(D \cup E) \cap F$ **52.** $(D \cup F) \cap E$

53. $D \cup (E \cap F)$ **54.** $D \cup (F \cap E)$

55. $(D \cap F) \cup (E \cap F)$ **56.** $(D \cap E) \cup (F \cap E)$

57. $(D \cup E) \cap (D \cup F)$ **58.** $(D \cup F) \cap (D \cup E)$

Miscellaneous

List the elements in each set.

59. $\{x \mid x$ is an even natural number less than $20\}$

60. $\{x \mid x$ is a natural number greater than $6\}$

61. $\{x \mid x$ is an odd natural number greater than $11\}$

62. $\{x \mid x$ is an odd natural number less than $14\}$

63. $\{x \mid x$ is an even natural number between 4 and $79\}$

64. $\{x \mid x$ is an odd natural number between 12 and $57\}$

Using the sets A, B, C, and D, list the elements in each set. If the set is empty, write $\varnothing$.
$A = \{1, 2, 3, 4, 5\}$ $B = \{4, 5, 6, 7, 8, 9\}$
$C = \{1, 3, 5, 7\}$ $D = \{2, 4, 6, 8\}$

65. $A \cup B$
66. $A \cup C$
67. $A \cap B$
68. $A \cap C$
69. $(A \cap B) \cup D$
70. $(A \cap C) \cup D$
71. $(A \cup B) \cup D$
72. $(B \cup C) \cap A$
73. $(B \cap C) \cap A$
74. $(B \cap D) \cap C$
75. $(B \cap D) \cup (C \cap D)$
76. $(A \cup B) \cap (C \cup D)$

Write each set using set-builder notation. Answers may vary.

77. $\{3, 4, 5, 6\}$

78. $\{1, 3, 5, 7\}$

79. $\{5, 7, 9, 11, \ldots\}$

80. $\{4, 5, 6, 7, \ldots\}$

81. $\{6, 8, 10, 12, \ldots, 82\}$

82. $\{9, 11, 13, 15, \ldots, 51\}$

⟨6⟩ Applications

Solve each problem. See Example 7.

83. In a class of 30 students all of them are either female or smokers, while only 5 are female and smokers. If there are 12 male smokers in the class, then how many female nonsmokers are in the class?

84. All of the 500 students at Tickfaw College take either math or English, while 300 of them take math and English. If only 50 take math but not English, then how many take English but not math?

Getting More Involved

85. *Discussion*

If A and B are finite sets, could $A \cup B$ be infinite? Explain.

86. *Cooperative learning*

Work with a small group to answer the following questions. If $A \subseteq B$ and $B \subseteq A$, then what can you conclude about A and B? If $(A \cup B) \subseteq (A \cap B)$, then what can you conclude about A and B?

87. *Discussion*

What is wrong with each statement? Explain.

a) $3 \subseteq \{1, 2, 3\}$
b) $\{3\} \in \{1, 2, 3\}$
c) $\varnothing = \{\varnothing\}$

88. *Exploration*

There are only two possible subsets of $\{1\}$, namely, $\varnothing$ and $\{1\}$.

a) List all possible subsets of $\{1, 2\}$. How many are there?
b) List all possible subsets of $\{1, 2, 3\}$. How many are there?
c) Guess how many subsets there are of $\{1, 2, 3, 4\}$. Verify your guess by listing all the possible subsets.
d) How many subsets are there for $\{1, 2, 3, \ldots, n\}$?

1.2   The Real Numbers

In This Section

⟨1⟩ **The Rational Numbers**
⟨2⟩ **Graphing on the Number Line**
⟨3⟩ **The Irrational Numbers**
⟨4⟩ **The Real Numbers**
⟨5⟩ **Intervals of Real Numbers**

The set of real numbers is the basic set of numbers used in algebra. There are many different types of real numbers. To understand better the set of real numbers, we will study some of the subsets of numbers that make up this set.

⟨1⟩ The Rational Numbers

We use the letter N to name the set of counting or natural numbers. The set of natural numbers together with the number 0 is the set of **whole numbers** (W). The whole numbers together with the negatives of the counting numbers form the set of **integers.** The letter Z (from *zahl*, the German word for number) is used for the integers:

$$N = \{1, 2, 3, \ldots\} \qquad \text{The natural numbers}$$
$$W = \{0, 1, 2, 3, \ldots\} \qquad \text{The whole numbers}$$
$$Z = \{\ldots, -3, -2, -1, 0, 1, 2, 3, \ldots\} \qquad \text{The integers}$$

Rational numbers are numbers that are written as ratios or as quotients of integers. We use the letter Q (for quotient) to name the set of rational numbers and write the set in set-builder notation as follows:

$$Q = \left\{ \frac{a}{b} \,\middle|\, a \text{ and } b \text{ are integers, with } b \neq 0 \right\} \qquad \text{The rational numbers}$$

⟨ **Helpful Hint** ⟩

A negative number can be used to represent a loss or a debt. The number -10 could represent a debt of \$10, a temperature of $10°$ below zero, or an altitude of 10 feet below sea level.

Examples of rational numbers are

$$7, \quad \frac{9}{4}, \quad -\frac{17}{10}, \quad 0, \quad \frac{0}{4}, \quad \frac{3}{1}, \quad -\frac{47}{3}, \quad \text{and} \quad \frac{-2}{-6}.$$

Note that the rational numbers are the numbers that can be expressed as a ratio (or quotient) of integers. The integer 7 is rational because we can write it as $\frac{7}{1}$.

Another way to describe rational numbers is by using their decimal form. To obtain the decimal form, we divide the denominator into the numerator. For some rational numbers the division terminates, and for others it continues indefinitely. These examples show some rational numbers and their equivalent decimal forms:

$$\frac{26}{100} = 0.26 \qquad \text{Terminating decimal}$$

$$\frac{4}{1} = 4.0 \qquad \text{Terminating decimal}$$

$$\frac{1}{4} = 0.25 \qquad \text{Terminating decimal}$$

$$\frac{2}{3} = 0.6666\ldots \qquad \text{The single digit 6 repeats.}$$

$$\frac{25}{99} = 0.252525\ldots \qquad \text{The pair of digits 25 repeats.}$$

$$\frac{4177}{990} = 4.2191919\ldots \qquad \text{The pair of digits 19 repeats.}$$

Rational numbers are defined as ratios of integers, but they can be described also by their decimal form. *The rational numbers are those decimal numbers whose digits either repeat or terminate.*

E X A M P L E   1

Subsets of the rational numbers

Determine whether each statement is true or false.

a) $0 \in W$ b) $N \subseteq Z$ c) $0.75 \in Z$ d) $Z \subseteq Q$

Solution

a) True, because 0 is a whole number.

b) True, because every natural number is also a member of the set of integers.

c) False, because the rational number 0.75 is not an integer.

d) True, because the rational numbers include the integers.

Now do Exercises 1–10

‹2› Graphing on the Number Line

To construct a number line, we draw a horizontal line and label any convenient point with the number 0. Now we choose any convenient length and use it to locate points to the right of 0 as points corresponding to the positive integers and points to the left of 0 as points corresponding to the negative integers. See Fig. 1.8. The numbers corresponding to the points on the line are called the **coordinates** of the points. The distance between two consecutive integers is called a **unit,** and it is the same for any two consecutive integers. The point with coordinate 0 is called the **origin.** The numbers on the number line increase in size from left to right. When we compare the size of any two numbers, the larger number lies to the right of the smaller one on the number line.

Figure 1.8

It is often convenient to illustrate sets of numbers on a number line. The set of integers, Z, is illustrated or **graphed** as in Fig. 1.9. The three dots to the right and left on the number line indicate that the integers go on indefinitely in both directions.

Figure 1.9

E X A M P L E   2

Graphing on the number line

List the elements of each set, and graph each set on a number line.

   a) $\{x \mid x$ is a whole number less than $4\}$

   b) $\{a \mid a$ is an integer between 3 and $9\}$

   c) $\{y \mid y$ is an integer greater than $-3\}$

Solution

   a) The whole numbers less than 4 are 0, 1, 2, and 3. Figure 1.10 shows the graph of this set.

   Figure 1.10

   b) The integers between 3 and 9 are 4, 5, 6, 7, and 8. The graph is shown in Fig. 1.11.

   Figure 1.11

   c) The integers greater than -3 are -2, -1, 0, 1, and so on. To indicate the continuing pattern, we use a series of dots on the graph in Fig. 1.12.

   Figure 1.12

> **Now do Exercises 11–18**

⟨3⟩ The Irrational Numbers

Some numbers can be expressed as ratios of integers and some can't. The set of **irrational numbers** (*I*) consists of those numbers that can't be expressed as a ratio of integers. The positive square root of 2 (in symbols $\sqrt{2}$) is an example of an irrational number. The square root of 2 is a number that you can multiply by itself to get 2. So we can write (using a raised dot to indicate multiplication)

$$\sqrt{2} \cdot \sqrt{2} = 2.$$

⟨ **Calculator Close-Up** ⟩

A calculator gives a 10-digit rational approximation for $\sqrt{2}$. Note that if the approximate value is squared, you do not get 2.

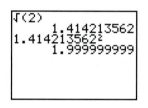

```
√(2)
         1.414213562
1.4142135622
         1.999999999
```

The screen shot that appears on this page and in succeeding pages may differ from the display on your calculator. You may have to consult your manual to get the desired results.

If we look for $\sqrt{2}$ on a calculator or in Appendix B, we find 1.414. But if we multiply 1.414 by itself, we get

$$(1.414)(1.414) = 1.999396.$$

So $\sqrt{2}$ is not equal to 1.414 (in symbols, $\sqrt{2} \neq 1.414$). The square root of 2 is approximately 1.414 (in symbols, $\sqrt{2} \approx 1.414$). There is no terminating or repeating decimal that will give exactly 2 when multiplied by itself. So $\sqrt{2}$ is an irrational number. It can be shown that other square roots such as $\sqrt{3}$, $\sqrt{5}$, and $\sqrt{7}$ are also irrational numbers. Of course, square roots such as $\sqrt{4}$ and $\sqrt{9}$ are rational numbers.

In decimal form the rational numbers either repeat or terminate. The irrational numbers neither repeat nor terminate. Examine each of these numbers to see that it has a continuing pattern that guarantees that its digits will neither repeat nor terminate:

$$0.606000600000600000006 \ldots$$
$$0.15115111511115 \ldots$$
$$3.12345678910111213 \ldots$$

So each of these numbers is an irrational number.

$$\pi = \frac{\text{Circumference}}{\text{Diameter}} \qquad \pi = \frac{C}{D}$$

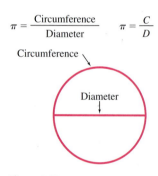

Figure 1.13

Since we generally work with rational numbers, the irrational numbers may seem to be unnecessary. However, irrational numbers occur in some very real situations. Over 2000 years ago people in the Orient and Egypt observed that the ratio of the circumference and diameter is the same for any circle. This constant value was proven to be an irrational number by Johann Heinrich Lambert in 1768. Like other irrational numbers, it does not have any convenient representation as a decimal number. This number has been given the name π (Greek letter pi). See Fig. 1.13. The value of π rounded to nine decimal places is 3.141592654. When using π in computations, we frequently use the rational number 3.14 as an approximate value for π.

E X A M P L E   **3**

Rational and irrational numbers

Determine whether each number is rational or irrational

a) $\sqrt{5}$ b) $\dfrac{1}{3}$ c) $5.121212 \ldots$

Solution

a) Square roots of numbers that are not perfect squares are irrational. So $\sqrt{5}$ is irrational.

b) Since $\dfrac{1}{3}$ is a ratio of integers, it is rational.

c) Since $5.121212 \ldots$ is a repeating decimal, it is rational.

> **Now do Exercises 19–28**

⟨4⟩ The Real Numbers

The set of irrational numbers I and the set of rational numbers Q have no numbers in common and together form the set of **real numbers** R. The set of real numbers can be visualized as the set of all points on the number line. Two real numbers are *equal* if they correspond to the same point on the number line. See Fig. 1.14.

Figure 1.14

Figure 1.15 illustrates the relationship between the set of real numbers and the various subsets that we have been discussing.

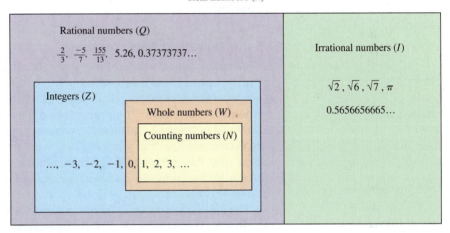

Figure 1.15

E X A M P L E   **4**

Classifying real numbers

Determine which elements of the set

$$\left\{ -\sqrt{7}, -\frac{1}{4}, 0, \sqrt{5}, \pi, 4.16, 12 \right\}$$

are members of each of these sets.

  a) Real numbers   **b)** Rational numbers   **c)** Integers

Solution

  a) All of the numbers are real numbers.

  b) The numbers $-\frac{1}{4}$, 0, 4.16, and 12 are rational numbers.

  c) The only integers in this set are 0 and 12.

> Now do Exercises 29–34

E X A M P L E   **5**

Subsets of the real numbers

Determine whether each of these statements is true or false.

  a) $\sqrt{7} \in Q$ **b)** $Z \subseteq W$ **c)** $I \cap Q = \varnothing$ **d)** $-3 \in N$

  e) $Z \cap I = \varnothing$ **f)** $Q \subseteq R$ **g)** $R \subseteq N$ **h)** $\pi \in R$

Solution

a) False, because Q represents the rationals and $\sqrt{7}$ is irrational.

b) False, because Z represents the integers and the negative integers are not in W (the set of whole numbers).

c) True, because there are no numbers in common to the rationals and the irrationals.

d) False, because there are no negative numbers in the set of natural numbers.

e) True, because Z is the integers and all integers are rational.

f) True, because the set of real numbers R contains both the rationals and irrationals.

g) False, because the set of real numbers R includes fractions, which are not in the set of natural numbers.

h) True, because π is an irrational number and R includes all rational and irrational numbers.

> Now do Exercises 35–48

⟨5⟩ Intervals of Real Numbers

Figure 1.16

Figure 1.17

An *interval* of time consists of the time between two times. For example, a professor has office hours for the time interval from 3 P.M. to 4 P.M. An **interval of real numbers** is the set of real numbers that lie between two real numbers, which are called the **endpoints** of the interval. **Interval notation** is used to represent intervals. For example, the interval notation (2, 3) is used to represent the real numbers that lie between 2 and 3 on the number line. The graph of (2, 3) is shown in Fig. 1.16. Parentheses are used to indicate that the endpoints do not belong to the interval, whereas brackets are used to indicate that the endpoints do belong to the interval. The interval [2, 3] consists of the real numbers between 2 and 3 including the endpoints. It is graphed in Fig. 1.17.

You may have graphed intervals in a previous course using a hollow circle to indicate that an endpoint is not in the interval and a solid circle to indicate that an endpoint is in the interval. With that method the intervals (2, 3) and [2, 3] would be drawn as in Fig. 1.18. The advantage of using parentheses and brackets on the graph is that they match the interval notation and the interval notation looks like an abbreviated version of the graph. So in this text we will use the parentheses and brackets.

Figure 1.18

Figure 1.19

Some time intervals do not have endpoints. For example, if a paper is turned in after 4 P.M., it is considered late. The infinity symbol is used to indicate that an interval does not end. For example, the interval (3, ∞) consists of the real numbers greater than 3 and extending infinitely far to the right on the number line. See Fig. 1.19. The interval (−∞, 3) consists of the real numbers less than 3, as shown in Fig. 1.20. The entire set of real numbers is written in interval notation as (−∞, ∞) and graphed as in Fig. 1.21. Note that the infinity symbol does not represent any particular real number and parentheses are always used next to −∞ and ∞.

Figure 1.20

Figure 1.21

EXAMPLE 6

Interval notation

Write each interval of real numbers in interval notation and graph it.

    a) The set of real numbers greater than or equal to 2

    b) The set of real numbers less than -3

    c) The set of real numbers between 1 and 5 inclusive

    d) The set of real numbers greater than or equal to 2 and less than 4

Solution

    a) The set of real numbers greater than or equal to 2 includes 2. It is written as $[2, \infty)$ and graphed in Fig. 1.22.

    b) The set of real numbers less than -3 does not include -3. It is written as $(-\infty, -3)$ and graphed in Fig. 1.23.

    c) The set of real numbers between 1 and 5 inclusive includes both 1 and 5. It is written as $[1, 5]$ and graphed in Fig. 1.24.

    d) The set of real numbers greater than or equal to 2 and less than 4 includes 2 but not 4. It is written as $[2, 4)$ and graphed in Fig. 1.25.

Figure 1.22   Figure 1.23

Figure 1.24   Figure 1.25

> Now do Exercises 71–80

The intersection of two intervals is the set of real numbers that belong to both intervals. The union of two intervals is the set of real numbers that belong to one, or the other, or both of the intervals.

EXAMPLE 7

Combining intervals

Write each union or intersection as a single interval.

    a) $(2, 4) \cup (3, 6)$

    b) $(2, 4) \cap (3, 6)$

    c) $(-1, 2) \cup [0, \infty)$

    d) $(-1, 2) \cap [0, \infty)$

Solution

    a) Graph $(2, 4)$ and $(3, 6)$ as in Fig. 1.26 on the next page. The union of the two intervals consists of the real numbers between 2 and 6, which is written as $(2, 6)$.

b) Examining the graphs in Fig. 1.26, we see that only the real numbers between 3 and 4 belong to both (2, 4) and (3, 6). So the intersection of (2, 4) and (3, 6) is (3, 4).

c) Graph (−1, 2) and [0, ∞) as in Fig. 1.27. The union of these intervals consists of the real numbers greater than −1, which is written as (−1, ∞).

d) Examining the graphs in Fig. 1.27, we see that the real numbers between 0 and 2 belong to both intervals. Note that 0 also belongs to both intervals but 2 does not. So the intersection is [0, 2).

Figure 1.26

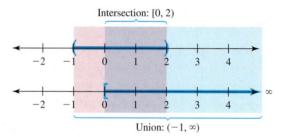

Figure 1.27

Now do Exercises 89–100

 Warm-Ups ▼

Fill in the blank.

1. The set of _____ is
{. . . , −3, −2, −1, 0, 1, 2, 3, . . . }.

2. The set of _____ numbers is {1, 2, 3, . . . }.

3. Every _____ number can be expressed as a ratio of integers.

4. _____ and _____ decimal numbers are rational numbers.

5. A decimal number that does not repeat and does not terminate is _____.

6. The union of the rationals and irrationals is the set of _____ numbers.

7. The ratio of the _____ and diameter of any circle is π.

8. The _____ of A and B consists of the elements that are in either A or B.

True or false?

9. The number π is rational.

10. Every rational number is a real number.

11. Zero is both rational and irrational.

12. Every real number is irrational.

13. Every irrational number corresponds to a point on the number line.

14. The number 0.444 . . . is rational.

15. The interval (4, 9) contains 8.

16. The interval (2, 6) contains 6.

17. The interval [3, 5] contains 3.

18. The interval (9, ∞) contains 88 trillion.

‹ **Study Tips** ›

• Note how the exercises are keyed to the examples and the examples are keyed to the exercises. If you get stuck on an exercise, study the corresponding example.
• The keys to success are desire and discipline. You must want success, and you must discipline yourself to do what it takes to get success.

‹1› The Rational Numbers

Determine whether each statement is true or false. Explain your answer. See Example 1.

1. $-2 \in N$ **2.** $-2 \in Q$

3. $0 \in Q$ **4.** $0 \in N$

5. $0.95 \in Q$ **6.** $0.333\ldots \in Q$

7. $Q \subseteq Z$ **8.** $Q \subseteq N$

9. $\dfrac{1}{2} \in Z$ **10.** $-99 \in Z$

‹2› Graphing on the Number Line

List the elements in each set and graph each set on a number line. See Example 2.

11. $\{x \mid x$ is a whole number smaller than $6\}$

12. $\{x \mid x$ is a natural number less than $7\}$

13. $\{a \mid a$ is an integer greater than $-5\}$

14. $\{z \mid z$ is an integer between 2 and $12\}$

15. $\{w \mid w$ is a natural number between 0 and $5\}$

16. $\{y \mid y$ is a whole number greater than $0\}$

17. $\{x \mid x$ is an integer between -3 and $5\}$

18. $\{y \mid y$ is an integer between -4 and $7\}$

‹3› The Irrational Numbers

Determine whether each number is rational or irrational. See Example 3.

19. $\sqrt{16}$

20. $\sqrt{7}$

21. $\dfrac{10}{8}$

22. $-\dfrac{1}{1000}$

23. $2.666\ldots$

24. $3.145145145\ldots$

25. $5.123456789101112\ldots$

26. $0.27476787107127147\ldots$

27. 3.4567

28. 2015

‹4› The Real Numbers

Determine which elements of the set

$$A = \left\{ -\sqrt{10},\, -3,\, -\frac{5}{2},\, -0.025,\, 0,\, \sqrt{2},\, 3\frac{1}{2},\, \frac{8}{2} \right\}$$

are members of these sets. See Example 4.

29. Real numbers

30. Natural numbers

31. Whole numbers

32. Integers

33. Rational numbers

34. Irrational numbers

Determine whether each statement is true or false. Explain.
See Example 5.

35. $Q \subseteq R$ **36.** $I \subseteq Q$

37. $I \cap Q = \{0\}$ **38.** $Z \subseteq Q$

39. $I \cup Q = R$ **40.** $Z \cap Q = \varnothing$

41. $0.2121121112\ldots \in Q$ **42.** $0.3333\ldots \in Q$

43. $3.252525\ldots \in I$ **44.** $3.1010010001\ldots \in I$

45. $0.999\ldots \in I$ **46.** $0.666\ldots \in Q$

47. $\pi \in I$ **48.** $\pi \in Q$

Place one of the symbols $\subseteq$, $\not\subseteq$, $\in$, or $\notin$ in each blank so that each statement is true.

49. N _____ W **50.** Z _____ Q

51. Z _____ N **52.** Q _____ W

53. Q _____ R **54.** I _____ R

55. $\varnothing$ _____ I **56.** $\varnothing$ _____ Q

57. N _____ R **58.** W _____ R

59. 5 _____ Z **60.** -6 _____ Z

61. 7 _____ Q **62.** 8 _____ Q

63. $\sqrt{2}$ _____ R **64.** $\sqrt{2}$ _____ I

65. 0 _____ I **66.** 0 _____ Q

67. $\{2, 3\}$ _____ Q **68.** $\{0, 1\}$ _____ N

69. $\{3, \sqrt{2}\}$ _____ R **70.** $\{3, \sqrt{2}\}$ _____ Q

⟨5⟩ Intervals of Real Numbers

Write each interval of real numbers in interval notation and graph it. See Example 6.

71. The set of real numbers greater than 1

72. The set of real numbers greater than -2

73. The set of real numbers less than -1

74. The set of real numbers less than 5

75. The set of real numbers between 3 and 4

76. The set of real numbers between -1 and 3

77. The set of real numbers between 0 and 2 inclusive

78. The set of real numbers between -1 and 1 inclusive

79. The set of real numbers greater than or equal to 1 and less than 3

80. The set of real numbers greater than 2 and less than or equal to 5.

Write the interval notation for the interval of real numbers shown in each graph.

81.
```
  +  +  [  ]  +  +  +
  3  4  5  6  7  8  9
```

82.
```
  +  [  +  +  )  +  +
  3  4  5  6  7  8  9
```

83.
```
  +  (  +  +  ]  +  +
 -4 -3 -2 -1  0  1  2
```

84.
```
  (  +  +  +  +  )
  3  4  5  6  7  8  9
```

85.
```
  +  +  [  +  +  +  +
 40 50 60 70 80 90 100
```

86.
```
  +  +  (  +  +  +  +
  3  4  5  6  7  8  9
```

87.
```
  +  +  +  +  )  +  +
 -9 -8 -7 -6 -5 -4 -3
```

88.
```
  +  +  +  +  ]  +  +
-12-11-10-9 -8 -7 -6
```

Write each union or intersection as a single interval. See Example 7.

89. $(1, 5) \cup (4, 9)$

90. $(-1, 2) \cup (0, 8)$

91. $(0, 3) \cap (2, 8)$

92. $(1, 8) \cap (2, 10)$

93. $(-2, 4) \cup (0, \infty)$

94. $(-\infty, 4) \cup (1, 5)$

95. $(-\infty, 2) \cap (0, 6)$

96. $(3, 6) \cap (0, \infty)$

97. $[2, 5) \cup (4, 9]$

98. $[-2, 2] \cup [2, 6)$

99. $[2, 6) \cap [2, 8)$

100. $[1, 5] \cap [2, 9]$

Determine whether each statement is true or false. Explain your answer.

101. $(1, 3) \subseteq [2, 4]$

102. $[1, 2] \subseteq (1, 2)$

103. $(5, 7) \subseteq R$

104. $6.3 \in (6, 7)$

105. $0 \in (0, 1)$

106. $(0, 2) \subseteq W$

107. $(0, 1) \cap [1, 2] = \varnothing$

108. $(-\infty, 0] \cup [0, \infty) = R$

Getting More Involved

109. *Writing*

What is the difference between a rational and an irrational number? Why is $\sqrt{9}$ rational and $\sqrt{3}$ irrational?

110. *Cooperative learning*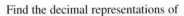

Work in a small group to make a list of the real numbers of the form $\sqrt{n}$, where n is a natural number between 1 and 100 inclusive. Decide on a method for determining which of these numbers are rational and find them. Compare your group's method and results with other groups' work.

111. *Exploration*

Find the decimal representations of

$$\frac{2}{9}, \quad \frac{2}{99}, \quad \frac{23}{99}, \quad \frac{23}{999}, \quad \frac{234}{999}, \quad \frac{23}{9999}, \text{ and } \frac{1234}{9999}.$$

a) What do these decimals have in common?

b) What is the relationship between each fraction and its decimal representation?

1.3 Operations on the Set of Real Numbers

Computations in algebra are performed with positive and negative numbers. In this section, we will extend the basic operations of arithmetic to the negative numbers.

⟨1⟩ Absolute Value

The real numbers are the coordinates of the points on the number line. However, we often refer to the points as numbers. For example, the numbers 5 and -5 are both five units away from 0 on the number line shown in Fig. 1.28. A number's *distance* from 0 on the number line is called the **absolute value** of the number. We write $|a|$ for "the absolute value of a." Therefore, $|5| = 5$ and $|-5| = 5$.

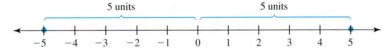

Figure 1.28

E X A M P L E　1

Absolute value

Find the value of $|4|$, $|-4|$, and $|0|$.

Solution

Because both 4 and -4 are four units from 0 on the number line, we have $|4| = 4$ and $|-4| = 4$. Because the distance from 0 to 0 on the number line is 0, we have $|0| = 0$.

Now do Exercises 1–6

‹ **Calculator Close-Up** ›

A graphing calculator uses abs for absolute value. Note that many calculators have a subtraction symbol for subtraction and a negative sign for indicating a negative number. You cannot use the subtraction symbol to indicate a negative number.

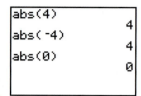

Note that $|a|$ represents distance, and distance is never negative. So $|a|$ is greater than or equal to zero for any number a.

Two numbers that are located on opposite sides of zero and have the same absolute value are called **opposites** of each other. The opposite of zero is zero. Every number has a unique opposite. The numbers 9 and -9 are opposites of one another. The minus sign, $-$, is used to signify "opposite" in addition to "negative." When the minus sign is used in front of a number, it is read as "negative." When it is used in front of parentheses or a variable, it is read as "opposite." For example,

$$-(9) = -9$$

is read as "the opposite of 9 is negative 9," and

$$-(-9) = 9$$

is read as "the opposite of negative 9 is 9." In general, $-a$ is read "the opposite of a." If a is positive, $-a$ is negative. If a is negative, $-a$ is positive. Opposites have the following property.

Opposite of an Opposite

For any number a,

$$-(-a) = a.$$

EXAMPLE   **2**

Opposite of an opposite
Evaluate.

  a) $-(-12)$ **b)** $-(-(-8))$

Solution

  a) The opposite of negative 12 is 12. So $-(-12) = 12$.
  b) The opposite of the opposite of -8 is -8. So $-(-(-8)) = -8$.

Now do Exercises 7–10

Remember that we have defined $|a|$ to be the distance between 0 and a on the number line. Using opposites, we can give a symbolic definition of absolute value.

Absolute Value

$$|a| = \begin{cases} a & \text{if } a \text{ is positive or zero} \\ -a & \text{if } a \text{ is negative} \end{cases}$$

Using this definition, we write

$$|7| = 7$$

because 7 is positive. To find the absolute value of -7, we use the second line of the definition and write

$$|-7| = -(-7) = 7.$$

‹2› Addition

A good way to understand positive and negative numbers is to think of the *positive numbers as assets* and the *negative numbers as debts.* For this illustration we can think of assets simply as cash. Think of debts as unpaid bills such as the electric bill and phone bill. If you have assets of $4 and $11 and no debts, then your net worth is $15. **Net worth** is the total of your debts and assets. If you have debts of $6 and $7 and no assets, then your net worth is −$13. In symbols,

$$(-6) \quad + \quad (-7) \quad = \quad -13.$$

$6 debt   Added to $7 debt $13 debt

We can think of this addition as adding the absolute values of −6 and −7 (that is, $6 + 7 = 13$) and then putting a negative sign on that result to get −13. These examples illustrate the next rule.

> **Sum of Two Numbers with Like Signs**
> To find the sum of two numbers with the same sign, add their absolute values. The sum has the same sign as the original numbers.

If you have a debt of $5 and have only $5 in cash, then your debts equal your assets (in absolute value), and your net worth is $0. In symbols,

$$-5 \quad + \quad 5 \quad = \quad 0.$$

Debt of $5   Asset of $5   Net worth

The number a and its opposite $-a$ have a sum of zero for any a. For this reason, a and $-a$ are called **additive inverses** of each other. Note that the words "negative," "opposite," and "additive inverse" are often used interchangeably.

> **Additive Inverse Property**
> For any real number a, there is a unique number $-a$ such that
> $$a + (-a) = -a + a = 0.$$

To understand the sum of a positive and a negative number, consider this situation. If you have a debt of $7 and a cash amount of $10, you may have $10 in hand, but your net worth is only $3. Your assets exceed your debts (in absolute value), and you have a positive net worth. In symbols,

$$-7 + 10 = 3.$$

Note that to get 3, we actually subtract 7 from 10. If you have a debt of $8 but have only $5 in cash, then your debts exceed your assets (in absolute value). You have a net worth of −$3. In symbols,

$$-8 + 5 = -3.$$

Note that to get the 3 in the answer, we subtract 5 from 8.

As you can see from these examples, the sum of a positive number and a negative number (with different absolute values) may be either positive or negative. These examples illustrate the rule for adding numbers with unlike signs and different absolute values.

The sum of two numbers with unlike signs and the same absolute value is zero because of the additive inverse property.

> ## Sum of Two Numbers with Unlike Signs (and Different Absolute Values)
>
> To find the sum of two numbers with unlike signs, subtract their absolute values. The sum is positive if the number with the larger absolute value is positive. The sum is negative if the number with the larger absolute value is negative.

E X A M P L E   3

Adding signed numbers
Find each sum.

a) $-6 + 13$ **b)** $-9 + (-7)$ **c)** $2 + (-2)$

d) $-35.4 + 2.51$ **e)** $-7 + 0.05$ **f)** $\dfrac{1}{5} + \left(-\dfrac{3}{4}\right)$

‹ **Calculator Close-Up** ›

A graphing calculator can add signed numbers in any form. If you use the fraction feature, the answer is given as a fraction.

```
-9+-7
              -16
-35.4+2.51
           -32.89
1/5+-3/4▶Frac
           -11/20
```

No one knows what calculators will be like in 10 or 20 years. So concentrate on understanding the mathematics and you will have no trouble with changing technology.

Solution

a) The absolute values of -6 and 13 are 6 and 13. Subtract 6 from 13 to get 7. Because the number with the larger absolute value is 13 and it is positive, the result is 7.

b) $-9 + (-7) = -16$

c) $2 + (-2) = 0$

d) Line up the decimal points and subtract 2.51 from 35.40 to get 32.89. Because 35.4 is larger than 2.51 and 35.4 has a negative sign, the answer is negative.

$$-35.4 + 2.51 = -32.89$$

e) Line up the decimal points and subtract 0.05 from 7.00 to get 6.95. Because 7.00 is larger than 0.05 and 7.00 has a negative sign, the answer is negative.

$$-7 + 0.05 = -6.95$$

f) $\dfrac{1}{5} + \left(-\dfrac{3}{4}\right) = \dfrac{4}{20} + \left(-\dfrac{15}{20}\right)$ The LCD for 5 and 4 is 20.

$\qquad\qquad = -\dfrac{11}{20}$ Add.

> **Now do Exercises 11–38**

‹3› Subtraction

Think of subtraction as removing debts or assets, and think of addition as receiving debts or assets. For example, if you have $10 in cash and $4 is taken from you, your resulting net worth is the same as if you have $10 and a water bill for $4 arrives in the mail. In symbols,

$$10 \quad - \quad 4 \quad = \quad 10 \quad + \quad (-4).$$

↑ Remove   ↑ Cash   ↑ Receive   ↑ Debt

Removing cash is equivalent to receiving a debt.

Suppose that you have $17 in cash but owe $7 in library fines. Your net worth is $10. If the debt of $7 is canceled or forgiven, your net worth will increase to $17, the same as if you received $7 in cash. In symbols,

$$10 \quad - \quad (-7) \quad = \quad 10 \quad + \quad 7.$$

↑ Remove   ↑ Debt   ↑ Receive   ↑ Cash

Removing a debt is equivalent to receiving cash.

Notice that each preceding subtraction problem is equivalent to an addition problem in which we add the opposite of what we were going to subtract. These examples illustrate the definition of subtraction.

Subtraction of Real Numbers

For any real numbers a and b,

$$a - b = a + (-b).$$

EXAMPLE 4

Subtracting signed numbers
Find each difference.

a) $-7 - 3$ **b)** $7 - (-3)$ **c)** $48 - 99$

d) $-3.6 - (-7)$ **e)** $0.02 - 7$ **f)** $\dfrac{1}{3} - \left(-\dfrac{1}{6}\right)$

‹ Calculator Close-Up ›

A graphing calculator can subtract signed numbers in any form. If your calculator has a subtraction symbol and a negative symbol, you will get an error message if you do not use them appropriately.

```
-3.6--7
            3.4
0.02-7
          -6.98
1/3--1/6►Frac
            1/2
```

Solution

a) To subtract 3 from -7, add the opposite of 3 and -7:

$$-7 - 3 = -7 + (-3) \quad \textcolor{blue}{a - b = a + (-b)}$$
$$= -10 \qquad \textcolor{blue}{\text{Add.}}$$

b) To subtract -3 from 7, add the opposite of -3 and 7. The opposite of -3 is 3:

$$7 - (-3) = 7 + (3) \quad \textcolor{blue}{a - b = a + (-b)}$$
$$= 10 \qquad \textcolor{blue}{\text{Add.}}$$

c) To subtract 99 from 48, add -99 and 48:

$$48 - 99 = 48 + (-99) \quad \textcolor{blue}{a - b = a + (-b)}$$
$$= -51 \qquad \textcolor{blue}{\text{Add.}}$$

d) $-3.6 - (-7) = -3.6 + 7 \quad \textcolor{blue}{a - b = a + (-b)}$
$$= 3.4 \qquad \textcolor{blue}{\text{Add.}}$$

e) $0.02 - 7 = 0.02 + (-7) \quad \textcolor{blue}{a - b = a + (-b)}$
$$= -6.98 \qquad \textcolor{blue}{\text{Add.}}$$

f) $\dfrac{1}{3} - \left(-\dfrac{1}{6}\right) = \dfrac{1}{3} + \left(\dfrac{1}{6}\right) \quad \textcolor{blue}{a - b = a + (-b)}$

$$= \dfrac{2}{6} + \dfrac{1}{6} \qquad \textcolor{blue}{\text{Get common denominators.}}$$

$$= \dfrac{3}{6} \qquad \textcolor{blue}{\text{Add.}}$$

$$= \dfrac{1}{2} \qquad \textcolor{blue}{\text{Reduce.}}$$

> **Now do Exercises 39–62**

⟨4⟩ Multiplication

The result of multiplying two numbers is called the **product** of the numbers. The numbers multiplied are called **factors**. In algebra, we use a raised dot to indicate multiplication or we place the symbols next to each other. One or both of the symbols may be enclosed in parentheses. For example, the product of a and b can be written as $a \cdot b$, ab, $a(b)$, $(a)b$, or $(a)(b)$.

Multiplication is just a short way to do repeated additions. Adding five 2's gives

$$2 + 2 + 2 + 2 + 2 = 10.$$

So we have the multiplication fact $5 \cdot 2 = 10$. Adding together five negative 2's gives

$$(-2) + (-2) + (-2) + (-2) + (-2) = -10.$$

So we must have $5(-2) = -10$. We can think of $5(-2) = -10$ as saying that taking on five debts of \$2 each is equivalent to a debt of \$10. Losing five debts of \$2 each is equivalent to gaining \$10, so we must have $-5(-2) = 10$.

The rules for multiplying signed numbers are easy to state and remember.

Product of Signed Numbers

To find the product of two nonzero real numbers, multiply their absolute values.
The product is *positive* if the numbers have the *same* sign.
The product is *negative* if the numbers have *unlike* signs.

For example, to multiply -4 and -5, we multiply their absolute values ($4 \cdot 5 = 20$). Since -4 and -5 have the same sign, $(-4)(-5) = 20$. To multiply -6 and 3, we multiply their absolute values ($6 \cdot 3 = 18$). Since -6 and 3 have unlike signs, $-6 \cdot 3 = -18$.

E X A M P L E   5

⟨ **Calculator Close-Up** ⟩

The products in Examples 5(b), 5(c), and 5(d) are shown here. The answer for $(-0.01)(-0.02)$ is given in scientific notation. The -4 after the E means that the decimal point belongs four places to the left. So the answer is -0.0002. See Section 5.1 for more information on scientific notation.

```
-4(10)
                -40
(-0.01)(0.02)
              -2E-4
4/9*-1/5▸Frac
             -4/45
```

Multiplying signed numbers

Find each product.

a) $(-3)(-6)$

b) $-4(10)$

c) $(-0.01)(0.02)$

d) $\dfrac{4}{9} \cdot \left(-\dfrac{1}{5}\right)$

Solution

a) First multiply the absolute values ($3 \cdot 6 = 18$). Because -3 and -6 have the same sign, we get $(-3)(-6) = 18$.

b) $-4(10) = -40$ Opposite signs, negative result

c) When multiplying decimals, we total the number of decimal places used in the numbers multiplied to get the number of decimal places in the answer. Thus, $(-0.01)(0.02) = -0.0002$.

d) $\dfrac{4}{9} \cdot \left(-\dfrac{1}{5}\right) = -\dfrac{4}{45}$ Opposite signs, negative result

Now do Exercises 63–70

⟨5⟩ Division

Just as every real number has an additive inverse or opposite, every nonzero real number a has a **multiplicative inverse** or **reciprocal** $\frac{1}{a}$. The reciprocal of 3 is $\frac{1}{3}$, and

$$3 \cdot \frac{1}{3} = 1.$$

Multiplicative Inverse Property

For any nonzero real number a, there is a unique number $\frac{1}{a}$ such that

$$a \cdot \frac{1}{a} = \frac{1}{a} \cdot a = 1.$$

EXAMPLE **6**

⟨ Helpful Hint ⟩

A doctor told a nurse to give a patient half the usual dose of a certain medicine. The nurse figured, "dividing in half means dividing by $\frac{1}{2}$, which means multiplying by 2." So the patient got four times the prescribed amount and died (true story). There is a big difference between dividing a quantity in half and dividing by one-half.

Finding multiplicative inverses

Find the multiplicative inverse (reciprocal) of each number.

a) -2 **b)** $\frac{3}{8}$ **c)** -0.2

Solution

a) The multiplicative inverse (reciprocal) of -2 is $-\frac{1}{2}$ because

$$-2\left(-\frac{1}{2}\right) = 1.$$

b) The reciprocal of $\frac{3}{8}$ is $\frac{8}{3}$ because

$$\frac{3}{8} \cdot \frac{8}{3} = 1.$$

c) First convert the decimal number -0.2 to a fraction:

$$-0.2 = -\frac{2}{10}$$

$$= -\frac{1}{5}$$

So the reciprocal of -0.2 is -5 and $-0.2(-5) = 1.$

Now do Exercises 71–76

Note that the reciprocal of any negative number is negative.

Earlier we defined subtraction for real numbers as addition of the additive inverse. We now define division for real numbers as multiplication by the multiplicative inverse (reciprocal).

Division of Real Numbers

For any real numbers a and b with $b \neq 0$,

$$a \div b = a \cdot \frac{1}{b}.$$

If $a \div b = c$, then a is called the **dividend,** b the **divisor,** and c the **quotient.** We also refer to $a \div b$ and $\frac{a}{b}$ as the quotient of a and b.

E X A M P L E   7

Dividing signed numbers

Find each quotient.

a) $-60 \div (-2)$

b) $-24 \div \dfrac{3}{8}$

c) $6 \div (-0.2)$

Solution

a) $-60 \div (-2) = -60 \cdot \left(-\dfrac{1}{2}\right)$ Multiply by $-\frac{1}{2}$, the reciprocal of -2.

$= 30$ Same sign, positive product

b) $-24 \div \dfrac{3}{8} = -24 \cdot \dfrac{8}{3}$ Multiply by $\frac{8}{3}$, the reciprocal of $\frac{3}{8}$.

$= -64$ Opposite signs, negative product

c) $6 \div (-0.2) = 6(-5)$ Multiply by -5, the reciprocal of -0.2.

$= -30$ Opposite signs, negative product

Now do Exercises 77–84

You can see from Examples 6 and 7 that a product or quotient is positive when the signs are the same and is negative when the signs are opposite:

$$\text{same signs} \leftrightarrow \text{positive result,}$$

$$\text{opposite signs} \leftrightarrow \text{negative result.}$$

‹ **Helpful Hint** ›

Some people remember that "two positives make a positive, a negative and a positive make a negative, and two negatives make a positive." Of course, that is true only for multiplication, division, and cute stories like this: If a good person comes to town, that's good. If a bad person comes to town, that's bad. If a good person leaves town, that's bad. If a bad person leaves town, that's good.

Even though all division can be done as multiplication by a reciprocal, we generally use reciprocals only when dividing fractions. Instead, we find quotients using our knowledge of multiplication and the fact that

$$a \div b = c \qquad \text{if and only if} \qquad c \cdot b = a.$$

For example, $-72 \div 9 = -8$ because $-8 \cdot 9 = -72$. Using long division or a calculator, you can get

$$-43.74 \div 1.8 = -24.3$$

and check that you have it correct by finding $-24.3 \cdot 1.8 = -43.74$.

We use the same rules for division when division is indicated by a fraction bar. For example,

$$\frac{-6}{3} = -2, \qquad \frac{6}{-3} = -2, \qquad \frac{-1}{3} = \frac{1}{-3} = -\frac{1}{3}, \qquad \text{and} \qquad \frac{-6}{-3} = 2.$$

Note that if one negative sign appears in a fraction, the fraction has the same value whether the negative sign is in the numerator, in the denominator, or in front of the fraction. If the numerator and denominator of a fraction are both negative, then the fraction has a positive value.

⟨6⟩ Division by Zero

Why do we omit division by zero from the definition of division? If we write $10 \div 0 = c$, we need to find c such that $c \cdot 0 = 10$. But there is no such number. If we write $0 \div 0 = c$, we need to find c such that $c \cdot 0 = 0$. But $c \cdot 0 = 0$ is true for any number c. Having $0 \div 0$ equal to any number would be confusing. Thus, $a \div b$ is defined only for $b \neq 0$. Quotients such as

$$5 \div 0, \qquad 0 \div 0, \qquad \frac{7}{0}, \qquad \text{and} \qquad \frac{0}{0}$$

are said to be *undefined*.

E X A M P L E   **8**

Division by zero

Find each quotient if possible.

a) $-3 \div 0$ **b)** $0 \div 2$

c) $\dfrac{-6}{0}$ **d)** $0 \div 0$

Solution

The quotients in (a), (c), and (d) are undefined because division by zero is undefined. For the quotient in (b) we have $0 \div 2 = 0$.

> Now do Exercises 85–100

Warm-Ups ▼

Fill in the blank.

1. The _____ inverse of a is $-a$.
2. The _____ value of a is the distance from a to 0 on the number line.
3. Two numbers on different sides of zero with the same absolute value are _____.
4. The absolute value of a is $-a$ if a is _____.
5. The ____ of a and $-a$ is 0.
6. The result of multiplying a and b is the _____ of a and b.
7. Two numbers that are multiplied are called _____.
8. The _____ of a is $\frac{1}{a}$.
9. If $a \div b = c$, then c is the _____.
10. Division by zero is _____.

True or false?

11. $-(-5) = 5$
12. $-6 + 6 = 0$
13. $|-12| = -12$
14. $a - b = a + (-b)$
15. If a is positive and b is negative, then ab is negative.
16. If a is positive and b is negative, then $a + b$ is always negative.
17. $-3 + 7 = 10$
18. $-8 - 4 = -12$
19. $-9 \div 0 = 0$
20. $0 \div 55 = 0$

1.3 Exercises

‹1› Absolute Value

Evaluate. See Examples 1 and 2.

1. $|-9|$ **2.** $|-12|$

3. $|7|$ **4.** $|0.5|$

5. $-|-4|$ **6.** $-|19|$

7. $-(-17)$ **8.** $-(-4)$

9. $-(-(-5))$ **10.** $-(-(-(-6)))$

‹2› Addition

Find each sum. See Example 3.

11. $(-5)+9$ **12.** $(-3)+10$

13. $(-4)+(-3)$ **14.** $(-15)+(-11)$

15. $-6+4$ **16.** $5+(-15)$

17. $7+(-17)$ **18.** $-8+13$

19. $(-11)+(-15)$ **20.** $-18+18$

21. $18+(-20)$ **22.** $7+(-19)$

23. $-14+9$ **24.** $-6+(-7)$

25. $-4+4$ **26.** $-7+9$

27. $-\dfrac{1}{10}+\dfrac{1}{5}$ **28.** $-\dfrac{1}{8}+\left(-\dfrac{1}{8}\right)$

29. $\dfrac{1}{2}+\left(-\dfrac{2}{3}\right)$ **30.** $\dfrac{3}{4}+\dfrac{1}{2}$

31. $-15+0.02$ **32.** $0.45+(-1.3)$

33. $-2.7+(-0.01)$ **34.** $0.8+(-1)$

35. $47.39+(-44.587)$ **36.** $0.65357+(-2.375)$

37. $0.2351+(-0.5)$ **38.** $-1.234+(-4.756)$

‹3› Subtraction

Find each difference. See Example 4.

39. $7-10$ **40.** $8-19$ **41.** $-4-7$

42. $-5-12$ **43.** $7-(-6)$ **44.** $3-(-9)$

45. $-1-5$ **46.** $-4-6$ **47.** $-12-(-3)$

48. $-15-(-6)$ **49.** $20-(-3)$ **50.** $50-(-70)$

51. $\dfrac{9}{10}-\left(-\dfrac{1}{10}\right)$ **52.** $\dfrac{1}{8}-\dfrac{1}{4}$ **53.** $1-\dfrac{3}{2}$

54. $-\dfrac{1}{2}-\left(-\dfrac{1}{3}\right)$ **55.** $2-0.03$ **56.** $-0.02-3$

57. $5.3-(-2)$ **58.** $-4.1-0.13$

59. $-2.44-48.29$ **60.** $-8.8-9.164$

61. $-3.89-(-5.16)$ **62.** $0-(-3.5)$

‹4› Multiplication

Find each product. See Example 5.

63. $(25)(-3)$ **64.** $(5)(-7)$

65. $\left(-\dfrac{1}{3}\right)\left(-\dfrac{1}{2}\right)$ **66.** $\left(-\dfrac{1}{2}\right)\left(-\dfrac{6}{7}\right)$

67. $(0.3)(-0.3)$ **68.** $(-0.1)(-0.5)$

69. $(-0.02)(-10)$ **70.** $(0.05)(-2.5)$

‹5› Division

Find the multiplicative inverse of each number. See Example 6.

71. 20 **72.** -5 **73.** $-\dfrac{6}{5}$

74. $-\dfrac{1}{8}$ **75.** -0.3 **76.** 0.125

Evaluate. See Example 7.

77. $-6\div3$ **78.** $84\div(-2)$

79. $30\div(-0.8)$ **80.** $(-9)(-6)$

81. $(-0.8)(0.1)$ **82.** $7\div(-0.5)$

83. $(-0.1) \div (-0.4)$ **84.** $(-18) \div (-0.9)$

⟨6⟩ Division by Zero

Evaluate. If a quotient is undefined, say so. See Example 8.

85. $0 \div 19$ **86.** $0 \div 99$

87. $-2 \div 0$ **88.** $33 \div 0$

89. $9 \div \left(-\dfrac{3}{4}\right)$ **90.** $-\dfrac{1}{3} \div \left(-\dfrac{5}{8}\right)$

91. $-\dfrac{2}{3}\left(-\dfrac{9}{10}\right)$ **92.** $\dfrac{1}{2}\left(-\dfrac{2}{5}\right)$

 93. $(0.25)(-365)$ **94.** $7.5 \div (-0.15)$

 95. $(-51) \div (-0.003)$ **96.** $(-2.8)(5.9)$

97. $0 \div 1.3422$ **98.** $0 \div 334.8$

99. $339.4 \div 0$ **100.** $0.667 \div 0$

Miscellaneous

Perform these computations.

101. $-62 + 13$ **102.** $-88 + 39$

103. $-32 - (-25)$ **104.** $-71 - (-19)$

105. $|-15|$ **106.** $-|-75|$

 107. $\dfrac{1}{2}(-684)$ **108.** $\dfrac{1}{3}(-123)$

109. $\dfrac{1}{2} - \left(-\dfrac{1}{4}\right)$ **110.** $\dfrac{1}{8} - \left(-\dfrac{1}{4}\right)$

 111. $-57 \div 19$ **112.** $0 \div (-36)$

 113. $|-17| + |-3|$ **114.** $64 - |-12|$

115. $0 \div (-0.15)$ **116.** $-20 \div \left(-\dfrac{8}{3}\right)$

117. $-63 + |8|$ **118.** $|-34| - 27$

119. $-\dfrac{1}{2} + \left(-\dfrac{1}{2}\right)$ **120.** $-\dfrac{2}{3} + \left(-\dfrac{2}{3}\right)$

121. $-\dfrac{1}{2} - 19$ **122.** $-\dfrac{1}{3} - 22$

123. $28 - 0.01$ **124.** $55 - 0.1$

125. $-29 - 0.3$ **126.** $-0.241 - 0.3$

127. $(-2)(0.35)$ **128.** $(-3)(0.19)$

Use an operation with signed numbers to solve each problem and identify the operation used.

129. *Net worth of a family.* The Jones family has a house that is worth $85,000, but they still owe $45,000 on the mortgage. They have $2300 in credit card debt, $1500 in other debts, $1200 in savings, and two cars worth $3500 each. What is the net worth of the Jones family?

130. *Net worth of a bank.* Just before the recession, First Federal Homestead had $15.6 million in mortgage loans, $23.3 million on deposit, and $8.5 million worth of real estate. After the recession started, the value of the real estate decreased to $4.8 million. What was the net worth of First Federal before the recession and after the recession started? (To a financial institution a loan is an asset and a deposit is a liability.)

131. *Warming up.* On January 11 the temperature at noon was 14°F in St. Louis and −6°F in Duluth. How much warmer was it in St. Louis?

Figure for Exercise 131

132. *Bitter cold.* On January 16 the temperature at midnight was −31°C in Calgary and −20°C in Toronto. How much warmer was it in Toronto?

133. *Below sea level.* The altitude of the floor of Death Valley is −282 feet (282 feet below sea level); the altitude of the shore of the Dead Sea is −1296 feet (*Rand McNally World Atlas*). How many feet above the shore of the Dead Sea is the floor of Death Valley?

134. *Highs and lows.* The altitude of the peak of Mt. Everest, the highest point on earth, is 29,028 feet. The world's greatest known ocean depth of $-36,201$ feet was recorded in the Marianas Trench (*Rand McNally World Atlas*). How many feet above the bottom of the Marianas Trench is a climber who has reached the top of Mt. Everest?

Getting More Involved

135. *Discussion*

Why is it necessary to learn addition of signed numbers before learning subtraction of signed numbers and to learn multiplication of signed numbers before division of signed numbers?

136. *Writing*

Explain why 0 is the only real number that does not have a multiplicative inverse.

Mt. Everest
29,028 ft

Marianas Trench
$-36,201$ ft

Figure for Exercise 134

Mid-Chapter **Quiz** | Sections 1.1 through 1.3 | **Chapter 1**

Let $A = \{1, 3, 5, 7, 9\}$, $B = \{2, 4, 6, 8\}$, $C = \{5, 6, 7, 8\}$, *and* $D = \{1, 2, 3, 4, 5, 6, 7, 8, 9\}$.
Determine whether each statement is true or false.

1. $A \cap B = \varnothing$

2. $A \cup C = D$

3. $1 \in B$

4. $6 \in B \cap C$

5. $3 \in B \cup C$

6. $A = C$

7. $B \subseteq D$

8. $D \subseteq A$

9. $B \not\subseteq C$

10. $(A \cap B) \cup C = C$

Graph each set of numbers on a number line.

11. The set of integers between 4 and 8

12. The real numbers between 4 and 8

13. The whole numbers less than or equal to 3

14. The real numbers less than or equal to 3

Write each union or intersection of intervals as a single interval.

15. $(2, 5) \cup (3, 8]$

16. $(1, 6) \cap [4, 9)$

Perform the indicated operations.

17. $\dfrac{1}{8} - \dfrac{1}{2}$

18. $-\dfrac{5}{6} + \dfrac{3}{4}$

19. $-\dfrac{3}{5}\left(-\dfrac{1}{6}\right)$

20. $-\dfrac{2}{3} \div \dfrac{8}{9}$

21. $-19 + 33$

22. $-6 - (-5)$

23. $-12(3)$

24. $-15 \div (-3)$

25. $-12 + (-5)$

26. $-60 - 40$

27. $-11(-6)$

28. $56 \div (-8)$

1.4 Evaluating Expressions and the Order of Operations

In algebra you will learn to work with variables. However, there is often nothing more important than finding a numerical answer to a question. This section is concerned with computation.

⟨1⟩ Arithmetic Expressions

The result of writing numbers in a meaningful combination with the ordinary operations of arithmetic is called an **arithmetic expression** or simply an **expression.** So $2 + 3$ is an expression. An expression that involves more than one operation is called a **sum, difference, product,** or **quotient** if the last operation to be performed is addition, subtraction, multiplication, or division, respectively. Parentheses are used as **grouping symbols** to indicate which operations are performed first. The expression

$$5 + (2 \cdot 3)$$

is a sum because the parentheses indicate that the product of 2 and 3 is to be found before the addition is performed. So we evaluate this expression as follows:

$$5 + (2 \cdot 3) = 5 + 6 = 11$$

If we write $(5 + 2)3$, the expression is a product and it has a different value.

$$(5 + 2)3 = 7 \cdot 3 = 21$$

Brackets [ ] are also used to indicate grouping. If an expression occurs within absolute value bars | |, it is evaluated before the absolute value is found. So absolute value bars also act as grouping symbols. We perform first the operations within the innermost grouping symbols.

E X A M P L E **1**

⟨ **Calculator Close-Up** ⟩

You can use parentheses to control the order in which your calculator performs the operations in an expression.

```
5((2*3)-8)
               -10
2((4*5)-abs(3-6)
)
               34
```

Grouping symbols

Evaluate each expression.

    a) $5[(2 \cdot 3) - 8]$ **b)** $2[(4 \cdot 5) - |\,3 - 6\,|]$

Solution

    a) $5[(2 \cdot 3) - 8] = 5[6 - 8]$ Innermost grouping first

                        $= 5[-2]$

                        $= -10$

    b) $2[(4 \cdot 5) - |\,3 - 6\,|] = 2[20 - |\,-3\,|]$ Innermost grouping first

                              $= 2[20 - 3]$

                              $= 2[17]$

                              $= 34$

                                          | Now do Exercises 1–6 |

⟨2⟩ Exponential Expressions

We use the notation of exponents to simplify the writing of repeated multiplication. The product $5 \cdot 5 \cdot 5 \cdot 5$ is written as 5^4. The number 4 in 5^4 is called the exponent, and it indicates the number of times that the factor 5 occurs in the product.

Exponential Expression

For any natural number n and real number a,
$$a^n = \underbrace{a \cdot a \cdot a \cdots a}_{n \text{ factors of } a}.$$

We call a the **base**, n the **exponent**, and a^n an **exponential expression.**

We read a^n as "the nth power of a" or "a to the nth power." The exponential expressions 3^5 and 10^6 are read as "3 to the fifth power" and "10 to the sixth power." We can also use the words "squared" and "cubed" for the second and third powers, respectively. For example, 5^2 and 2^3 are read as "5 squared" and "2 cubed," respectively.

E X A M P L E   2

Exponential expressions

Evaluate.

a) 2^3 b) $(-3)^4$ c) $\left(-\dfrac{1}{2}\right)^5$

Solution

a) $2^3 = 2 \cdot 2 \cdot 2$ The factor 2 is used three times.

$\qquad = 8$

b) $(-3)^4 = (-3)(-3)(-3)(-3)$ The factor -3 is used four times.

$\qquad\quad = 81$ Even number of negative signs, positive product

c) $\left(-\dfrac{1}{2}\right)^5 = \left(-\dfrac{1}{2}\right)\left(-\dfrac{1}{2}\right)\left(-\dfrac{1}{2}\right)\left(-\dfrac{1}{2}\right)\left(-\dfrac{1}{2}\right)$ The factor $-\frac{1}{2}$ is used five times.

$\qquad\quad = -\dfrac{1}{32}$ Odd number of negative signs, negative product

> **Now do Exercises 7–12**

⟨ Calculator Close-Up ⟩

Powers are indicated on a graphing calculator using a caret (^). Most calculators also have an x^2-key for squaring. Note that parentheses are necessary in $(-3)^4$. Without parentheses, your calculator should get $-3^4 = -81$. Try it.

```
2^3
                    8
(-3)^4
                   81
(-1/2)^5▶Frac
                -1/32
```

⟨3⟩ Square Roots

Because $3^2 = 9$ and $(-3)^2 = 9$, both 3 and -3 are square roots of 9. We use the **radical symbol** $\sqrt{}$ to indicate the nonnegative or principal square root of 9. We write $\sqrt{9} = 3$.

Square Roots

If $a^2 = b$, then a is a **square root** of b. If a is greater than or equal to 0, then a is the **principal square root** of b and we write $\sqrt{b} = a$.

The radical symbol is a grouping symbol. We perform all operations within the radical symbol before the square root is found.

E X A M P L E   3

Evaluating square roots
Evaluate.

   a) $\sqrt{64}$ **b)** $\sqrt{9 + 16}$ **c)** $\sqrt{3(17 - 5)}$

Solution

  a) Because $8^2 = 64$, we have $\sqrt{64} = 8$.

  b) Because the radical symbol is a grouping symbol, add 9 and 16 before finding the square root:

$$\sqrt{9 + 16} = \sqrt{25} \quad \text{Add first.}$$
$$= 5 \quad \text{Find the square root.}$$

     Note that $\sqrt{9} + \sqrt{16} = 3 + 4 = 7$. So $\sqrt{9 + 16} \neq \sqrt{9} + \sqrt{16}$.

  c) $\sqrt{3(17 - 5)} = \sqrt{3(12)} = \sqrt{36} = 6$

> **Now do Exercises 13–22**

‹ Calculator Close-Up ›

Because the radical symbol on most calculators cannot be extended, parentheses are used to group the expression that is inside the radical.

```
√(64)
                    8
√(9+16)
                    5
√(3(17-5))
                    6
```

⟨4⟩ Order of Operations

To simplify the writing of expressions, we often omit some grouping symbols. If we saw the expression

$$5 + 2 \cdot 3$$

written without parentheses, we would not know how to evaluate it unless we had a rule for which operations to perform first. Expressions in which some or all grouping symbols are omitted are evaluated consistently by using a rule called the **order of operations.**

> **Order of Operations**
>
> Evaluate inside any grouping symbols first. Where grouping symbols are missing use this order.
>
>   **1.** Evaluate each exponential expression (in order from left to right).
>
>   **2.** Perform multiplication and division (in order from left to right).
>
>   **3.** Perform addition and subtraction (in order from left to right).

"In order from left to right" means that we evaluate the operations in the order in which they are written. For example,

$$20 \cdot 3 \div 6 = 60 \div 6 = 10 \quad \text{and} \quad 10 - 3 + 6 = 7 + 6 = 13.$$

If an expression contains grouping symbols, we evaluate within the grouping symbols using the order of operations.

E X A M P L E   4

Order of operations
Evaluate each expression.

  a) $5 + 2 \cdot 3$ **b)** $9 \cdot 2^3$ **c)** $(6 - 4^2)^2$ **d)** $40 \div 8 \cdot 2 \div 5 \cdot 3$

Solution

  a) $5 + 2 \cdot 3 = 5 + 6$ Multiply first.

               $= 11$ Then add.

When parentheses are omitted, most (but not all) calculators follow the same order of operations that we use in this text. Try these computations on your calculator. To use a calculator effectively, you must practice with it.

```
5+2*3
              11
(6-4²)²
             100
40/8*2/5*3
               6
```

b) $9 \cdot 2^3 = 9 \cdot 8$ Evaluate the exponential expression first.

$ = 72$ Then multiply.

c) $(6 - 4^2)^2 = (6 - 16)^2$ Evaluate 4^2 within the parentheses first.

$ = (-10)^2$ Then subtract.

$ = 100$ $(-10)(-10) = 100$

d) Multiplication and division are done from left to right.

$$40 \div 8 \cdot 2 \div 5 \cdot 3 = 5 \cdot 2 \div 5 \cdot 3$$
$$= 10 \div 5 \cdot 3$$
$$= 2 \cdot 3$$
$$= 6$$

Now do Exercises 23–30

An expression that can cause confusion is -3^2. Is it 9 or -9? To eliminate the confusion we agree that *the exponent applies only to the 3 and the negative sign is handled last.* So,

$$-3^2 = -(3^2) = -(9) = -9.$$

Note that -3^2 is the opposite of 3^2. This rule also applies to other even powers. For example, $-1^2 = -1$, $-3^4 = -81$, and $-2^6 = -64$.

E X A M P L E   **5**

The order of negative signs

Evaluate each expression.

a) -2^4 **b)** -5^2 **c)** $(3 - 5)^2$ **d)** $-(5^2 - 4 \cdot 7)^2$

Solution

a) To evaluate -2^4, find 2^4 first and then take the opposite. So $-2^4 = -16$.

b) $-5^2 = -(5^2)$ The exponent applies to 5 only.

$ = -25$

c) Evaluate within the parentheses first; then square that result.

$$(3 - 5)^2 = (-2)^2$$ Evaluate within parentheses first.

$$ = 4$$ Square -2 to get 4.

d) $-(5^2 - 4 \cdot 7)^2 = -(25 - 28)^2$ Evaluate 5^2 within the parentheses first.

$ = -(-3)^2$ Then subtract.

$ = -9$ Square -3 to get 9, then take the opposite of 9 to get -9.

Now do Exercises 31–50

‹ **Helpful Hint** ›

"Everybody Loves My Dear Aunt Sally" is often used as a memory aid for the order of operations. Do **E**xponents and **L**ogarithms, **M**ultiplication and **D**ivision, and then **A**ddition and **S**ubtraction. Logarithms are discussed later in this text.

When an expression involves a fraction bar, the numerator and denominator are each treated as if they are in parentheses. Example 6 illustrates how the fraction bar groups the numerator and denominator.

E X A M P L E   **6**

Order of operations in fractions

Evaluate each quotient.

a) $\dfrac{10 - 8}{6 - 8}$ **b)** $\dfrac{-6^2 + 2 \cdot 7}{4 - 3 \cdot 2}$

‹ **Calculator Close-Up** ›

Some calculators use the built-up form for fractions $\left(\frac{1}{2}\right)$, but some do not $(1/2)$. If your calculator does not use the built-up form, then you must enclose numerators and denominators (that contain operations) in parentheses as shown here.

```
(10-8)/(6-8)
               -1
(-6²+2*7)/(4-3*2
)
                11
```

Solution

a) $\dfrac{10 - 8}{6 - 8} = \dfrac{2}{-2}$ ⟨Evaluate the numerator and denominator separately.⟩

$= -1$ ⟨Then divide.⟩

b) $\dfrac{-6^2 + 2 \cdot 7}{4 - 3 \cdot 2} = \dfrac{-36 + 14}{4 - 6}$

$= \dfrac{-22}{-2}$ ⟨Evaluate the numerator and denominator separately.⟩

$= 11$ ⟨Then divide.⟩

Now do Exercises 51–60

⟨5⟩ Algebraic Expressions

The result of combining numbers and variables with the ordinary operations of arithmetic (in some meaningful way) is called an **algebraic expression.** For example,

$$2x - 5y, \quad 5x^2, \quad (x - 3)(x + 2), \quad b^2 - 4ac, \quad 5, \quad \text{and} \quad \frac{x}{2}$$

are algebraic expressions, or simply **expressions.** An expression such as $2x - 5y$ has no definite value unless we assign values to x and y. For example, if $x = 3$ and $y = 4$, then the value of $2x - 5y$ is found by replacing x with 3 and y with 4 and evaluating:

$$2x - 5y = 2(3) - 5(4) = 6 - 20 = -14$$

Note the importance of the order of operations in evaluating an algebraic expression. To find the value of the difference $2x - 5y$ when $x = -2$ and $y = -3$, replace x and y by -2 and -3, respectively, and then evaluate:

$$2x - 5y = 2(-2) - 5(-3) = -4 - (-15) = -4 + 15 = 11$$

EXAMPLE 7

Value of an algebraic expression

Evaluate each expression for $a = 2$, $b = -3$, and $c = 4$.

a) $a - c^2$　　　**b)** $a - b^2$　　　**c)** $b^2 - 4ac$　　　**d)** $\dfrac{a - b}{c - b}$

‹ **Calculator Close-Up** ›

To evaluate $a - c^2$, first store the values for a and c using the STO key. Then enter the expression.

```
2→A
               2
4→C
               4
A-C²
             -14
```

Solution

a) Replace a by 2 and c by 4 in the expression $a - c^2$.

$$a - c^2 = 2 - 4^2 = 2 - 16 = -14$$

b) $a - b^2 = 2 - (-3)^2$ ⟨Let $a = 2$ and $b = -3$.⟩

$= 2 - 9$ ⟨Evaluate the exponential expression first.⟩

$= -7$ ⟨Then subtract.⟩

c) $b^2 - 4ac = (-3)^2 - 4(2)(4)$ ⟨Let $a = 2$, $b = -3$, and $c = 4$.⟩

$= 9 - 32$ ⟨Evaluate the exponential expression and product.⟩

$= -23$ ⟨Subtract last.⟩

d) $\dfrac{a - b}{c - b} = \dfrac{2 - (-3)}{4 - (-3)}$ ⟨Let $a = 2$, $b = -3$, and $c = 4$.⟩

$= \dfrac{5}{7}$ ⟨Evaluate the numerator and denominator.⟩

Now do Exercises 61–72

> **CAUTION** When you replace a variable by a negative number, be sure to use parentheses around the negative number. If we were to omit the parentheses in Example 7(c), we would get $-3^2 - 4(2)(4) = -41$ instead of -23.

Mathematical notation is readily available in scientific word processors. However, on Internet pages or in e-mail, multiplication is often written with a star (*), fractions are written with a slash (/), and exponents with a caret (^). For example, $\frac{x + y}{2x^3}$ is written as $(x + y)/(2*x\text{^}3)$. If the numerator or denominator contain more than one symbol, it is best to enclose them in parentheses to avoid confusion. An expression such as $1/2x$ is confusing. If your class evaluates it for $x = 4$, some students will probably assume that it is $1/(2x)$ and get $1/8$ and some will assume that it is $(1/2)x$ and get 2.

A symbol such as y_1 is treated like any other variable. We read y_1 as "y one" or "y sub one." The 1 is called a **subscript.** We can think of y_1 as the "first y" and y_2 as the "second y." We use the subscript notation in Example 8.

E X A M P L E 8

An algebraic expression with subscripts

Let $y_1 = -12$, $y_2 = -5$, $x_1 = -3$, and $x_2 = 4$. Find the value of $\frac{y_2 - y_1}{x_2 - x_1}$.

Solution

Substitute the appropriate values into the expression:

$$\frac{y_2 - y_1}{x_2 - x_1} = \frac{-5 - (-12)}{4 - (-3)} \quad \text{Let } y_1 = -12, y_2 = -5, \\ x_1 = -3, \text{ and } x_2 = 4.$$

$$= \frac{7}{7} = 1 \quad \text{Evaluate.}$$

> Now do Exercises 73–78

‹ **Helpful Hint** ›

Many of the expressions that we evaluate in this section are expressions that we will study later in this text. We use the expression in Example 8 to find the slope of a line in Chapter 3.

‹6› **Reading a Graph**

Algebraic expressions are essential in financial calculations. For example, the value of the algebraic expression $P(1 + r)^n$ is the amount after n years of an investment of P dollars earning interest rate r compounded annually. If $100 is invested at 5%, then the expression becomes $100(1 + 0.05)^n$. If this expression is evaluated for many values of n, then the results can be shown in a graph. A graph gives us a picture of the algebraic expression. Example 9 illustrates these ideas.

E X A M P L E 9

A financial expression

An investment of $100 is made at 5% compounded annually. The value in dollars of this investment after n years is given by the expression $100(1 + 0.05)^n$.

a) Find the value after 4 years for the $100 investment.

b) Use the accompanying graph to estimate the value of the expression after 30 years.

c) Use the accompanying graph to estimate how long it takes the investment to double.

Solution

a) Evaluating $100(1 + 0.05)^4$ yields approximately 121.55. So the value to the nearest cent is $121.55. Note that the result is very different if you fail to follow the order of operations. Multiplying $100 by $(1 + 0.05)$ first and then raising the result to the fourth power yields $121,550,625 for your $100 investment.

b) To find the value of the expression after 30 years, first draw a vertical line from 30 up to the graph as shown in Fig. 1.29. From the point of intersection, draw a horizontal line to the amount scale. So after 30 years, the investment of $100 is worth over $400.

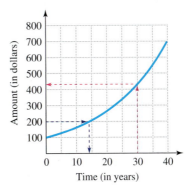

Figure 1.29

c) To find how long it takes for the investment to double, start at $200 on the amount scale and draw a horizontal line to the graph, as shown in the figure. From the point of intersection draw a vertical line down to the timescale. The time that it takes for the investment to double is about 15 years.

> Now do Exercises 101–108

Warm-Ups ▼

Fill in the blank.

1. The result of writing numbers in a meaningful combination with the ordinary operations of arithmetic is an _____ expression.

2. The result of writing numbers and variables in a meaningful combination with the ordinary operations of arithmetic is an _____ expression.

3. The expression $a + b$ is a _____ .

4. The expression $a - b$ is a _____ .

5. The expression $\dfrac{a}{b}$ is a _____ .

6. The expression a^n is an _____ expression.

7. In the expression 5^3, 5 is the _____ .

8. If $a^2 = b$, then a is a _____ root of b.

9. If $a^2 = b$ and $a \geq 0$, then a is the _____ square root of b.

10. In the expression x_3 the 3 is a _____ .

True or false?

11. $2^3 = 6$

12. $-2^2 = -4$

13. $(6 + 3) \cdot 2 = 81$

14. $6 + 3^2 = 15$

15. $\left| -3 - (-2) \right| = 5$

16. $-1 \cdot 2^2 = -4$

17. $6 + 3 \cdot 2 = 18$

18. $(6 + 3)^2 = 18$

19. $(-3)^3 = -3^3$

20. $\left| 7 - 8 \right| = \left| 7 \right| - \left| 8 \right|$

‹1› Arithmetic Expressions

Evaluate each expression. See Example 1.

1. $(-3 \cdot 4) - (2 \cdot 5)$

2. $|-3 - 2| - |2 - 6|$

3. $4[5 - |3 - (2 \cdot 5)|]$

4. $-2|(-3 \cdot 4) - 6|$

5. $(6 - 8)(|2 - 3| + 6)$

6. $-5(6 + [(5 - 7) - 4])$

‹2› Exponential Expressions

Evaluate each exponential expression. See Example 2.

7. 2^5 **8.** 3^4

9. $(-1)^4$ **10.** $(-1)^5$

11. $\left(-\dfrac{1}{3}\right)^2$ **12.** $\left(-\dfrac{1}{2}\right)^6$

‹3› Square Roots

Evaluate each radical. See Example 3.

13. $\sqrt{49}$ **14.** $\sqrt{100}$

15. $\sqrt{36 + 64}$ **16.** $\sqrt{25 - 9}$

17. $\sqrt{4(7 + 9)}$ **18.** $\sqrt{(11 + 2)(18 - 5)}$

19. $\sqrt{3 + 13} + 9$ **20.** $\sqrt{25 - 16} - 5$

21. $-2\sqrt{25 + 144}$ **22.** $-3\sqrt{169 - 144}$

‹4› Order of Operations

Evaluate each expression. See Examples 4 and 5.

23. $4 - 6 \cdot 2$ **24.** $8 - 3 \cdot 9$

25. $6/2 \cdot 3$ **26.** $-12/4 \cdot 3$

27. $5 - 6(3 - 5)$ **28.** $8 - 3(4 - 6)$

29. $\left(\dfrac{1}{3} - \dfrac{1}{2}\right)\left(\dfrac{1}{4} - \dfrac{1}{2}\right)$ **30.** $\left(\dfrac{1}{2} - \dfrac{1}{4}\right)\left(\dfrac{1}{2} - \dfrac{3}{4}\right)$

31. $-3^2 + (-8)^2 + 3$ **32.** $-6^2 + (-3)^3$

33. $-(2 - 7)^2$ **34.** $-(1 - 3 \cdot 2)^3$

35. $-5^2 \cdot 2^3$ **36.** $2^4 - 4^2$

37. $(-5)(-2)^3$ **38.** $(-1)(2 - 8)^3$

39. $-(3^2 - 4)^2$ **40.** $-(6 - 2^3)^4$

41. $8 + 2\sqrt{5^2 - 3^2}$ **42.** $6 + 3\sqrt{6^2 + 2 \cdot 2^5}$

43. $-60 \div 10 \cdot 3 \div 2 \cdot 5 \div 6$

44. $75 \div (-5)(-3) \div \dfrac{1}{2} \cdot 6$

45. $5.5 - 2.3^4$

46. $5.3^2 - 4 \cdot 6.1$

47. $(1.3 - 0.31)(2.9 - 4.88)$

48. $(6.7 - 9.88)^3$

49. $-388.8 \div (13.5)(9.6)$

50. $(-4.3)(5.5) \div (3.2)(-1.2)$

Evaluate. If an expression is undefined, say so. See Example 6.

51. $\dfrac{2 - 6}{9 - 7}$ **52.** $\dfrac{9 - 12}{4 - 5}$

53. $\dfrac{-3 - 5}{6 - (-2)}$ **54.** $\dfrac{-14 - (-2)}{-3 - 3}$

55. $\dfrac{4 + 2 \cdot 7}{3 \cdot 2 - 9}$ **56.** $\dfrac{-6 - 2(-3)}{8 - 3(-3)}$

57. $\dfrac{-3^2 - (-9)}{2 - 3^2}$ **58.** $\dfrac{-2^4 - 5}{3^2 - 2^4}$

59. $\dfrac{4 - 7}{-3 - (-3)}$ **60.** $\dfrac{4^2}{-4^2 - (-16)}$

‹5› Algebraic Expressions

Evaluate each expression for $a = -1$, $b = 3$, and $c = -4$. See Example 7.

61. $b^2 - 4ac$ **62.** $b^2 - 2b - 3$

63. $\dfrac{a - b}{a - c}$ **64.** $\dfrac{b - c}{b - a}$

65. $(a - b)(a + b)$ **66.** $(a - c)(a + c)$

67. $\sqrt{c^2 - 2c + 1}$ **68.** $\sqrt{a^2 - 4bc}$

69. $\dfrac{2}{a} + \dfrac{b}{c} - \dfrac{1}{c}$ **70.** $\dfrac{c}{a} + \dfrac{c}{b} - \dfrac{a}{b}$

71. $|a - b|$ **72.** $|b + c|$

Find the value of $\dfrac{y_2 - y_1}{x_2 - x_1}$ for each choice of y_1, y_2, x_1, and x_2. See Example 8.

73. $y_1 = 4$, $y_2 = -6$, $x_1 = 2$, $x_2 = -7$

74. $y_1 = -3$, $y_2 = -3$, $x_1 = 4$, $x_2 = -5$

75. $y_1 = -1$, $y_2 = 2$, $x_1 = -3$, $x_2 = 1$

76. $y_1 = -2$, $y_2 = 5$, $x_1 = 2$, $x_2 = 6$

77. $y_1 = 2.4$, $y_2 = 5.6$, $x_1 = 5.9$, $x_2 = 4.7$

78. $y_1 = -5.7$, $y_2 = 6.9$, $x_1 = 3.5$, $x_2 = 4.2$

Evaluate each expression without a calculator. Use a calculator to check.

79. $-2^2 + 5(3)^2$

80. $-3^2 + 3(6)^2$

81. $(-2 + 5)3^2$

82. $(-3 + 3)6^2$

83. $\sqrt{5^2 - 4(1)(6)}$

84. $\sqrt{6^2 - 4(2)(4)}$

85. $[13 + 2(-5)]^2$

86. $[6 + 2(-4)]^2$

87. $\dfrac{4 - (-1)}{-3 - 2}$

88. $\dfrac{2 - (-3)}{3 - 5}$

89. $3(-2)^2 - 5(-2) + 4$

90. $3(-1)^2 + 5(-1) - 6$

91. $-4\left(\dfrac{1}{2}\right)^2 + 3\left(\dfrac{1}{2}\right) - 2$

92. $8\left(\dfrac{1}{2}\right)^2 - 6\left(\dfrac{1}{2}\right) + 1$

93. $|6 - 3 \cdot 7| + |7 - 5|$

94. $|12 - 4| - |3 - 4 \cdot 5|$

95. $3 - 7[4 - (2 - 5)]$

96. $9 - 2[3 - (4 + 6)]$

97. $3 - 4(2 - |4 - 6|)$

98. $3 - (|-4| - |-5|)$

99. $[3 - (-1)]^2 + [-1 - 4]^2$

100. $[5 - (-3)]^2 + [4 - (-2)]^2$

‹6› Reading a Graph

Solve each problem. See Example 9.

101. *Female target heart rate.* The algebraic expression $0.65(220 - A)$ gives the target heart rate for beneficial exercise for women, where A is the age of the woman. How much larger is the target heart rate of a 25-year-old woman than that of a 65-year-old woman? Use the accompanying graph to estimate the age at which a woman's target heart rate is 115.

102. *Male target heart rate.* The algebraic expression $0.75(220 - A)$ gives the target heart rate for beneficial exercise for men, where A is the age of the man. Use the

algebraic expression to find the target heart rate for a 20-year-old and a 50-year-old man. Use the accompanying graph to estimate the age at which a man's target heart rate is 115.

Miscellaneous

Solve each problem.

103. *Perimeter of a pool.* The algebraic expression $2L + 2W$ gives the perimeter of a rectangle with length L and width W. Find the perimeter of a rectangular swimming pool that has length 34 feet and width 18 feet.

104. *Area of a lot.* The algebraic expression for the area of a trapezoid, $0.5h(b_1 + b_2)$, gives the area of the property shown in the figure. Find the area if $h = 150$ feet, $b_1 = 260$ feet, and $b_2 = 220$ feet.

Figure for Exercise 104

105. *Saving for retirement.* The expression $P(1 + r)^n$ gives the amount of an investment of P dollars invested for n years at interest rate r compounded annually. Long-term corporate bonds have had an average yield of 6.2% annually over the last 40 years (Fidelity Investments, www.fidelity.com).

   a) Use the accompanying graph to estimate the amount of a $10,000 investment in corporate bonds after 30 years.

   b) Use the given expression to calculate the value of a $10,000 investment after 30 years of growth at 6.2% compounded annually.

Figure for Exercises 101 and 102

Figure for Exercise 105

106. *Saving for college.* The average cost of a B.A. at a private college in 2030 will be $140,000 (U.S. Department of Education, www.ed.gov). The principal that must be invested at interest rate r compounded annually to have A dollars in n years is given by the algebraic expression

$$\frac{A}{(1 + r)^n}.$$

What amount must Melanie's generous grandfather invest in 2012 at 7% compounded annually so that Melanie will have $140,000 for her college education in 2030?

107. *Student loan.* A college student borrowed $4000 at 8% compounded annually in her freshman year and did not have to make payments until 4 years later. Use the accompanying graph to estimate the amount that she owes at the time the payments start. Use the expression

Figure for Exercise 107

$P(1 + r)^n$ to find the actual amount of the debt at the time the payments start.

108. *Nursing home costs.* The average cost of a 1-year stay in a nursing home in 2007 was $75,190 (www.medicare.gov). In n years from 2007 the average cost will be $75{,}190(1.05)^n$ dollars. Find the projected cost for a 1-year stay in 2020.

Getting More Involved

109. *Cooperative learning*

The sum of the integers from 1 through n is $\frac{n(n + 1)}{2}$. The sum of the squares of the integers from 1 through n is $\frac{n(n + 1)(2n + 1)}{6}$. The sum of the cubes of the integers from 1 through n is $\frac{n^2(n + 1)^2}{4}$. Use the appropriate expressions to find the following values.

a) The sum of the integers from 1 through 50
b) The sum of the squares of the integers from 1 through 40
c) The sum of the cubes of the integers from 1 through 30
d) The square of the sum of the integers from 1 through 20
e) The cube of the sum of the integers from 1 through 10

110. *Discussion*

Evaluate $5(5(5 \cdot 3 + 6) + 4) + 7$ and $3 \cdot 5^3 + 6 \cdot 5^2 + 4 \cdot 5 + 7$. Explain why these two expressions must have the same value.

1.5   Properties of the Real Numbers

In This Section

⟨1⟩ **Commutative Properties**
⟨2⟩ **Associative Properties**
⟨3⟩ **Distributive Property**
⟨4⟩ **Identity Properties**
⟨5⟩ **Inverse Properties**
⟨6⟩ **Multiplication Property of Zero**

You know that the price of a hamburger plus the price of a Coke is the same as the price of a Coke plus the price of a hamburger. But, do you know which property of the real numbers is at work in this situation? In arithmetic, we may be unaware when to use properties of the real numbers, but in algebra we need a better understanding of those properties. In this section, we will study the properties of the basic operations on the set of real numbers.

⟨1⟩   Commutative Properties

We get the same result whether we evaluate $3 + 7$ or $7 + 3$. With multiplication, we have $4 \cdot 5 = 5 \cdot 4$. These examples illustrate the commutative properties.

Commutative Property of Addition

For any real numbers a and b,

$$a + b = b + a.$$

Commutative Property of Multiplication

For any real numbers a and b,

$$ab = ba.$$

In writing the product of a number and a variable, it is customary to write the number first. We write $3x$ rather than $x3$. In writing the product of two variables, it is customary to write them in alphabetical order. We write cd rather than dc.

Addition and multiplication are commutative operations, but what about subtraction and division? Because $7 - 3 = 4$ and $3 - 7 = -4$, subtraction is not commutative. To see that division is not commutative, consider the amount each person gets when a \$1 million lottery prize is divided between two people and when a \$2 prize is divided among 1 million people.

⟨2⟩ Associative Properties

Consider the expression $2 + 3 + 7$. Using the order of operations, we add from left to right to get 12. If we first add 3 and 7 to get 10 and then add 2 and 10, we also get 12. So,

$$(2 + 3) + 7 = 2 + (3 + 7).$$

Now consider the expression $2 \cdot 3 \cdot 5$. Using the order of operations, we multiply from left to right to get 30. However, we can first multiply 3 and 5 to get 15 and then multiply by 2 to get 30. So,

$$(2 \cdot 3) \cdot 5 = 2 \cdot (3 \cdot 5).$$

These examples illustrate the associative properties.

Associative Property of Addition

For any real numbers a, b, and c,
$$(a + b) + c = a + (b + c).$$

Associative Property of Multiplication

For any real numbers a, b, and c,
$$(ab)c = a(bc).$$

Consider the expression

$$4 - 9 + 8 - 5 - 8 + 6 - 13.$$

According to the accepted order of operations, we could evaluate this expression by computing from left to right. However, if we use the definition of subtraction, we can rewrite this expression as

$$4 + (-9) + 8 + (-5) + (-8) + 6 + (-13).$$

The commutative and associative properties of addition enable us to add these numbers in any order we choose. A good way to add them is to add the positive numbers, add the negative numbers, and then combine the two totals:

$$4 + 8 + 6 + (-9) + (-5) + (-8) + (-13) = 18 + (-35)$$
$$= -17$$

For speed we usually do not rewrite the expression. We just sum the positive numbers and sum the negative numbers, and then combine their totals.

EXAMPLE 1

Using commutative and associative properties
Evaluate.

a) $4 - 7 + 10 - 5$ **b)** $6 - 5 - 9 + 7 - 2 + 5 - 8$

Solution

a) $4 - 7 + 10 - 5 = 14 + (-12) = 2$

         ↑   ↑

    Sum of the positive   Sum of the negative
    numbers   numbers

b) $6 - 5 - 9 + 7 - 2 + 5 - 8 = 18 + (-24)$ Add positive numbers;
                                  $= -6$ add negative numbers.

Now do Exercises 1–10

Not all operations are associative. Using subtraction, for example, we have

$$(8 - 4) - 1 \neq 8 - (4 - 1)$$

because $(8 - 4) - 1 = 3$ and $8 - (4 - 1) = 5$. For division we have

$$(8 \div 4) \div 2 \neq 8 \div (4 \div 2)$$

because $(8 \div 4) \div 2 = 1$ and $8 \div (4 \div 2) = 4$. So subtraction and division are not associative.

< Helpful Hint >

Imagine a parade in which 6 rows of horses are followed by 4 rows of horses with 3 horses in each row.

There are 10 rows of 3 horses or 30 horses, or there are 18 horses followed by 12 horses for a total of 30 horses.

⟨3⟩ Distributive Property

Using the order of operations, we evaluate the product $3(6 + 4)$ first by adding 6 and 4 and then multiplying by 3:

$$3(6 + 4) = 3 \cdot 10 = 30$$

Note that we also have

$$3 \cdot 6 + 3 \cdot 4 = 18 + 12 = 30.$$

Therefore,

$$3(6 + 4) = 3 \cdot 6 + 3 \cdot 4.$$

Note that multiplication by 3 from outside the parentheses is *distributed* over each term inside the parentheses. This example illustrates the distributive property.

Distributive Property

For any real numbers a, b, and c,

$$a(b + c) = ab + ac.$$

Because subtraction is defined in terms of addition, multiplication distributes over subtraction as well as over addition. For example,

$$3(x - 2) = 3(x + (-2))$$
$$= 3x + (-6)$$
$$= 3x - 6.$$

Because multiplication is commutative, we can write the multiplication before or after the parentheses. For example,

$$(y + 6)3 = 3(y + 6)$$
$$= 3y + 18.$$

The distributive property is used in two ways. If we start with the product $5(x + 4)$ and write

$$5(x + 4) = 5x + 20,$$

we are writing a product as a sum. We are removing the parentheses. If we start with the difference $6x - 18$ and write

$$6x - 18 = 6(x - 3),$$

we are using the distributive property to write a difference as a product.

EXAMPLE 2

Using the distributive property

Use the distributive property to rewrite each sum or difference as a product and each product as a sum or difference.

a) $9x - 9$ **b)** $b(2 - a)$

c) $3a + ac$ **d)** $-2(x - 3)$

Solution

a) $9x - 9 = 9(x - 1)$

b) $b(2 - a) = 2b - ab$ Note that $b \cdot 2 = 2b$ by the commutative property.

c) $3a + ac = a(3 + c)$

d) $-2(x - 3) = -2x - (-2)(3)$ Distributive property
$$= -2x - (-6)$$ Multiply.
$$= -2x + 6$$ $a - (-b) = a + b$

> **Now do Exercises 11–30**

⟨4⟩ Identity Properties

The numbers 0 and 1 have special properties. Addition of 0 to a number does not change the number, and multiplication of a number by 1 does not change the number. For this reason, 0 is called the **additive identity** and 1 is called the **multiplicative identity.**

Additive Identity Property

For any real number a,

$$a + 0 = 0 + a = a.$$

Multiplicative Identity Property

For any real number a,

$$a \cdot 1 = 1 \cdot a = a.$$

⟨5⟩ Inverse Properties

The ideas of *additive inverses* and *multiplicative inverses* were introduced in Section 1.3. Every real number a has a unique additive inverse or opposite, $-a$, such that $a + (-a) = 0$. Every nonzero real number a also has a unique multiplicative inverse (reciprocal), written $\frac{1}{a}$, such that $a\left(\frac{1}{a}\right) = 1$. For rational numbers the multiplicative inverse is easy to find. For example, the multiplicative inverse of $\frac{2}{5}$ is $\frac{5}{2}$ because

$$\frac{2}{5} \cdot \frac{5}{2} = \frac{10}{10} = 1.$$

Additive Inverse Property

For any real number a, there is a unique number $-a$ such that

$$a + (-a) = -a + a = 0.$$

Multiplicative Inverse Property

For any nonzero real number a, there is a unique number $\frac{1}{a}$ such that

$$a \cdot \frac{1}{a} = \frac{1}{a} \cdot a = 1.$$

EXAMPLE 3

Finding multiplicative inverses

Find the multiplicative inverse (or reciprocal) of each number.

a) $\dfrac{1}{8}$ b) -7 c) -0.26

Solution

a) Since $8 \cdot \frac{1}{8} = 1$, the multiplicative inverse of $\frac{1}{8}$ is 8.

b) Since $-7\left(-\frac{1}{7}\right) = 1$, the multiplicative inverse of -7 is $-\frac{1}{7}$.

c) Since $-0.26 = -\frac{26}{100} = -\frac{13}{50}$, the multiplicative inverse of -0.26 is $-\frac{50}{13}$.

> **Now do Exercises 31–42**

⟨6⟩ Multiplication Property of Zero

Zero has a property that no other number has. Multiplication involving zero always results in zero. It is the multiplication property of zero that prevents 0 from having a reciprocal.

Multiplication Property of Zero

For any real number a,

$$0 \cdot a = a \cdot 0 = 0.$$

EXAMPLE **4**

Recognizing properties

Identify the property that is illustrated in each case.

a) $5 \cdot 9 = 9 \cdot 5$

b) $3 \cdot \dfrac{1}{3} = 1$

c) $1 \cdot 865 = 865$

d) $3 + (5 + a) = (3 + 5) + a$

e) $4x + 6x = (4 + 6)x$

f) $7 + (x + 3) = 7 + (3 + x)$

g) $4567 \cdot 0 = 0$

h) $239 + 0 = 239$

i) $-8 + 8 = 0$

j) $-4(x - 5) = -4x + 20$

Solution

a) Commutative property of multiplication

b) Multiplicative inverse property

c) Multiplicative identity property

d) Associative property of addition

e) Distributive property

f) Commutative property of addition

g) Multiplication property of zero

h) Additive identity property

i) Additive inverse property

j) Distributive property

Now do Exercises 47–66

Warm-Ups ▼

Fill in the blank.

1. The _____ property says that
   $a + b = b + a$ for any real numbers a and b.

2. The _____ property says that $(a + b) + c = a + (b + c)$ for any real numbers a, b, and c.

3. The _____ property says that $a(b + c) = ab + ac$ for any real numbers a, b, and c.

4. The number 0 is the _____ identity.

5. The number 1 is the _____ identity.

6. The sum of a and $-a$ is the _____ identity.

7. The product of a and $\dfrac{1}{a}$ is the _____ identity.

8. The _____ of any number and 0 is 0.

True or false?

9. $77 + 35 = 35 + 77$

10. $17 \cdot 12 = 12 \cdot 17$

11. $(16 + 13) + 12 = 16 + (13 + 12)$

12. $10 \div 2 = 2 \div 10$

13. $8 \div (4 \div 2) = (8 \div 4) \div 2$

14. $15 - 3 = 3 - 15$

15. $4(6 + 3) = 24 + 12$

16. $3(10 \div 2) = 30 \div 6$

17. $0.02 \cdot 50 = 1$

18. $99 + 0 = 0$

‹1–2› Commutative and Associative Properties

Evaluate. See Example 1.

1. $9 - 4 + 6 - 10$

2. $-3 + 4 - 12 + 9$

3. $6 - 10 + 5 - 8 - 7$

4. $5 - 11 + 6 - 9 + 12 - 2$

5. $-4 - 11 + 6 - 8 + 13 - 20$

6. $-8 + 12 - 9 - 15 + 6 - 22 + 3$

7. $-3.2 + 1.4 - 2.8 + 4.5 - 1.6$

8. $4.4 - 5.1 + 3.6 - 2.3 + 8.1$

9. $3.27 - 11.41 + 5.7 - 12.36 - 5$

10. $4.89 - 2.1 + 7.58 - 9.06 - 5.34$

‹3› Distributive Property

Use the distributive property to write each product as a sum or difference. See Example 2.

11. $4(x - 6)$ **12.** $5(a - 1)$

13. $a(3 + t)$ **14.** $b(y + w)$

15. $-2(w - 5)$ **16.** $-4(m - 7)$

17. $-1(-2x - y)$ **18.** $-1(-4y - w)$

19. $\frac{1}{2}(4x + 8)$ **20.** $\frac{1}{3}(3x + 6)$

Use the distributive property to write each sum or difference as a product. See Example 2.

21. $2m + 10$ **22.** $3y + 9$

23. $5x - 5$ **24.** $3y + 3$

25. $3y - 15$ **26.** $5x + 10$

27. $3a + 9$ **28.** $7b - 49$

29. $bw + w$ **30.** $3ax + a$

‹4–5› Identity and Inverse Properties

Find the multiplicative inverse (reciprocal) of each number. See Example 3.

31. $\frac{1}{2}$ **32.** $\frac{1}{3}$ **33.** 1

34. -1 **35.** 6 **36.** 8

37. 0.25 **38.** 0.75 **39.** -0.7

40. -0.9 **41.** -1.8 **42.** -2.6

Use a calculator to evaluate each expression. Round answers to four decimal places.

43. $\dfrac{1}{2.3} + \dfrac{1}{5.4}$ **44.** $\dfrac{1}{13.5} - \dfrac{1}{4.6}$

45. $\dfrac{\frac{1}{4.3}}{\frac{1}{5.6} + \frac{1}{7.2}}$ **46.** $\dfrac{\frac{1}{4.5} - \frac{1}{5.6}}{\frac{1}{3.2} + \frac{1}{2.7}}$

‹6› Multiplication Property of Zero

Name the property that is illustrated in each case. See Example 4.

47. $3 + x = x + 3$

48. $x \cdot 5 = 5x$

49. $5(x - 7) = 5x - 35$

50. $a(3b) = (a \cdot 3)b$

51. $3(xy) = (3x)y$

52. $3(x - 1) = 3x - 3$

53. $4(0.25) = 1$

54. $0.3 + 9 = 9 + 0.3$

55. $y^3 x = xy^3$

56. $0 \cdot 52 = 0$

57. $1 \cdot x = x$

58. $(0.1)(10) = 1$

59. $2x + 3x = (2 + 3)x$

60. $8 + 0 = 8$

61. $7 + (-7) = 0$

62. $1 \cdot y = y$

63. $(36 + 79)0 = 0$

64. $5x + 5 = 5(x + 1)$

65. $xy + x = x(y + 1)$

66. $ab + 3ac = a(b + 3c)$

Miscellaneous

Complete each statement using the property named.

67. $5 + w =$ _____ , commutative property of addition

68. $2x + 2 =$ _____ , distributive property

69. $5(xy) =$ _____ , associative property of multiplication

70. $x + \dfrac{1}{2} =$ _____ , commutative property of addition

71. $\dfrac{1}{2}x - \dfrac{1}{2} =$ _____ , distributive property

72. $3(x - 7) =$ _____ , distributive property

73. $6x + 9 =$ _____ , distributive property

74. $(x + 7) + 3 =$ _____ , associative property of addition

75. $8(0.125) =$ ___ , multiplicative inverse property

76. $-1(a - 3) =$ _____ , distributive property

77. $0 = 5(\_\_\_)$, multiplication property of zero

78. $8 \cdot (\_\_\_) = 8$, multiplicative identity property

79. $0.25 (\_\_\_) = 1$, multiplicative inverse property

80. $45(1) = \_\_\_$, multiplicative identity property

Getting More Involved

81. *Discussion*

Does the order in which your groceries are placed on the checkout counter make any difference in your total bill? Which properties are at work here?

82. *Discussion*

Suppose that you just bought 10 grocery items and paid a total bill that included 6% sales tax. Would there be any difference in your total bill if you purchased the items one at a time? Which property is at work here?

Math *at* Work   Blood Pressure

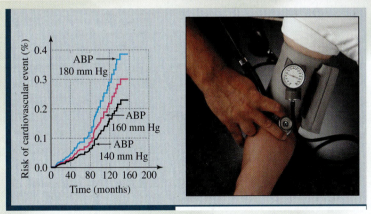

Blood pressure, the force of blood against the walls of arteries, is recorded as the systolic pressure (as the heart beats) over the diastolic pressure (as the heart relaxes between beats). The measurement is written like a fraction, with the systolic number over the diastolic number. For example, a blood pressure of 120/80 mm Hg (millimeters of mercury) is expressed verbally as "120 over 80." Normal blood pressure is 120/80 or less. High blood pressure or arterial hypertension is defined as a systolic blood pressure over 140 and a diastolic pressure over 90. Many studies have shown that above 140/90 the risk for the cardiovascular system is significant.

The most reliable method for measuring blood pressure is to place a probe directly into the artery. This technique is sometimes carried out during a cardiological examination or when a patient is in intensive care and permanent monitoring of blood pressure is required. The most useful method of measuring blood pressure is to use a sphygmomanometer with a cuff. The principle of measurement consists in recording the arterial counter pressure by squeezing the artery on which the pressure is measured. The most reliable readings of blood pressure are done with a device that is fitted to the patient and measures blood pressure over a 24-hour period, ambulatory blood pressure (ABP). Usually measurements are taken every 15 minutes during the day and every 30 minutes during the night. One study of ABP resulted in the accompanying graph, which shows the risk of a cardiovascular event over time for several values of ABP.

1.6   Using the Properties

The properties of the real numbers can be helpful when we are doing computations. In this section, we will see how the properties can be applied in arithmetic and algebra.

⟨1⟩   Using the Properties in Computation

Consider the product of 36 and 200. Using the associative property of multiplication, we can write

$$(36)(200) = (36)(2 \cdot 100) = (36 \cdot 2)(100).$$

To find this product mentally, first multiply 36 by 2 to get 72, then multiply 72 by 100 to get 7200.

E X A M P L E   **1**

Using properties in computation

Evaluate each expression mentally by using an appropriate property.

    a) $536 + 25 + 75$

    b) $5 \cdot 426 \cdot \dfrac{1}{5}$

    c) $7 \cdot 45 + 3 \cdot 45$

Solution

    a) To perform this addition mentally, the associative property of addition can be applied as follows:

$$536 + (25 + 75) = 536 + 100 = 636$$

    b) Use the commutative and associative properties of multiplication to rearrange mentally this product.

$$5 \cdot 426 \cdot \frac{1}{5} = 426 \cdot 5 \cdot \frac{1}{5} \quad \text{Commutative property of multiplication}$$

$$= 426\left(5 \cdot \frac{1}{5}\right) \quad \text{Associative property of multiplication}$$

$$= 426 \cdot 1 \quad \text{Multiplicative inverse property}$$

$$= 426$$

    c) Use the distributive property to rewrite the expression; then evaluate it.

$$7 \cdot 45 + 3 \cdot 45 = (7 + 3)45 = 10 \cdot 45 = 450$$

Now do Exercises 1–24

⟨2⟩ Combining Like Terms

The properties of the real numbers are used also with algebraic expressions. Simple algebraic expressions such as

$$-2, \quad 4x, \quad -5x^2y, \quad b, \quad \text{and} \quad -abc$$

are called terms. A **term** is a single number or the product of a number and one or more variables raised to powers. The number preceding the variables in a term is called the **coefficient.** In the term $4x$ the coefficient of x is 4. In the term $-5x^2y$ the coefficient of x^2y is -5. In the term b the coefficient of b is 1, and in the term $-abc$ the coefficient of abc is -1. If two terms contain the same variables with the same powers, they are called **like terms.** For example, $3x^2$ and $-5x^2$ are like terms, whereas $3x^2$ and $-2x^3$ are not like terms.

We can combine any two like terms involved in a sum by using the distributive property. For example,

$$2x + 5x = (2 + 5)x \quad \text{Distributive property}$$
$$= 7x. \quad\quad\quad \text{Add 2 and 5.}$$

Because the distributive property is valid for any real numbers, we have $2x + 5x = 7x$ for any real number x.

We can also use the distributive property to combine any two like terms involved in a difference. For example,

$$-3xy - (-2xy) = [-3 - (-2)]xy \quad \text{Distributive property}$$
$$= -1xy \quad\quad\quad\quad\quad \text{Subtract.}$$
$$= -xy. \quad\quad\quad\quad\quad \text{Multiplying by } -1 \text{ is the same as taking the opposite.}$$

Of course, we do not want to write out these steps every time we combine like terms. We can combine like terms as easily as we can add or subtract their coefficients.

EXAMPLE   2

Combining like terms

Perform the indicated operation.

a) $b + 3b$

b) $5x^2 - 7x^2$

c) $5xy - (-13xy)$

d) $-2a + (-9a)$

Solution

a) $b + 3b = 1b + 3b = 4b$

b) $5x^2 - 7x^2 = -2x^2$

c) $5xy - (-13xy) = 18xy$

d) $-2a + (-9a) = -11a$

> **Now do Exercises 25–38**

CAUTION   The distributive property enables us to combine only *like* terms. Expressions such as

$$3xw + 5, \quad 7xy + 9t, \quad 5b + 6a, \quad \text{and} \quad 6x^2 + 7x$$

do not contain like terms, so their terms cannot be combined.

⟨3⟩ **Multiplying and Dividing Terms**

We can use the associative property of multiplication to simplify the product of two terms. For example,

$$4(7x) = (4 \cdot 7)x \quad \text{Associative property of multiplication}$$
$$= (28)x$$
$$= 28x. \quad \text{Remove unnecessary parentheses.}$$

> **CAUTION** Multiplication does not distribute over multiplication. For example, $2(3 \cdot 4) \neq 6 \cdot 8$ because $2(3 \cdot 4) = 2(12) = 24$.

⟨ **Helpful Hint** ⟩

Did you know that the line separating the numerator and denominator in a fraction is called the *vinculum*?

In the next example we use the fact that dividing by 3 is equivalent to multiplying by $\frac{1}{3}$, the reciprocal of 3:

$$3\left(\frac{x}{3}\right) = 3\left(x \cdot \frac{1}{3}\right) \quad \text{Definition of division}$$

$$= 3\left(\frac{1}{3} \cdot x\right) \quad \text{Commutative property of multiplication}$$

$$= \left(3 \cdot \frac{1}{3}\right)x \quad \text{Associative property of multiplication}$$

$$= 1 \cdot x \qquad 3 \cdot \frac{1}{3} = 1 \text{ (Multiplicative inverse property)}$$

$$= x \qquad \text{Multiplicative identity property}$$

To find the product $(3x)(5x)$, we use both the commutative and associative properties of multiplication:

$$(3x)(5x) = (3x \cdot 5)x \qquad \text{Associative property of multiplication}$$
$$= (3 \cdot 5x)x \qquad \text{Commutative property of multiplication}$$
$$= (3 \cdot 5)(x \cdot x) \qquad \text{Associative property of multiplication}$$
$$= (15)\left(x^2\right) \qquad \text{Simplify.}$$
$$= 15x^2 \qquad \text{Remove unnecessary parentheses.}$$

All of the steps in finding the product $(3x)(5x)$ are shown here to illustrate that every step is justified by a property. However, you should write $(3x)(5x) = 15x^2$ without doing any intermediate steps.

EXAMPLE 3

Multiplying terms
Find each product.

a) $(-5)(6x)$ b) $(-3a)(-8a)$ c) $(-4y)(-6)$ d) $(-5a)\left(\dfrac{b}{5}\right)$

Solution

a) $-30x$ b) $24a^2$ c) $24y$ d) $-ab$

Now do Exercises 39–54

In Example 4 we use the properties to find quotients. Try to identify the property that is used at each step.

EXAMPLE 4

Dividing terms

Find each quotient.

a) $\dfrac{5x}{5}$ **b)** $\dfrac{4x + 8}{2}$

Solution

a) First use the definition of division to change the division by 5 to multiplication by $\frac{1}{5}$.

$$\frac{5x}{5} = 5x \cdot \frac{1}{5} = \left(\frac{1}{5} \cdot 5\right)x = 1 \cdot x = x$$

b) First use the definition of division to change division by 2 to multiplication by $\frac{1}{2}$.

$$\frac{4x + 8}{2} = (4x + 8) \cdot \frac{1}{2} = \frac{1}{2} \cdot (4x + 8) = 2x + 4$$

Since both $4x$ and 8 are divided by 2, we could have written

$$\frac{4x + 8}{2} = \frac{4x}{2} + \frac{8}{2} = 2x + 4.$$

Now do Exercises 55–62

CAUTION It is not correct to divide a number into just one term of a sum. For example, $\frac{2 + 7}{2} \neq 1 + 7$ because $\frac{2 + 7}{2} = \frac{9}{2}$ and $1 + 7 = 8$.

⟨4⟩ **Removing Parentheses**

Multiplying a number by -1 merely changes the sign of the number. For example,

$$(-1)(6) = -6 \qquad \text{and} \qquad (-1)(-15) = 15.$$

Thus, -1 times a number is the same as the *opposite* of the number. Using variables, we have

$$(-1)x = -x \qquad \text{or} \qquad -1(a + 2) = -(a + 2).$$

When a negative sign appears in front of a sum, we can think of it as multiplication by -1 and use the distributive property. For example,

$$\begin{aligned}
-(a + 2) &= -1(a + 2) \\
&= (-1)a + (-1)2 \quad \text{Distributive property} \\
&= -a + (-2) \\
&= -a - 2.
\end{aligned}$$

If a negative sign occurs in front of a difference, we can rewrite the expression as a sum. For example,

$$\begin{aligned}
-(x - 5) &= -1(x - 5) \qquad\quad -a = -1 \cdot a \\
&= (-1)x - (-1)5 \quad \text{Distributive property} \\
&= -x + 5. \qquad\qquad \text{Simplify.}
\end{aligned}$$

Because subtraction is defined as adding the opposite, a minus sign in front of parentheses has the same effect as a negative sign:

$$7 - (x - 5) = 7 + (-(x - 5)) \quad \text{Subtraction means add the opposite.}$$
$$= 7 + (-x + 5) \quad \text{Change the sign of each term.}$$
$$= -x + 12 \quad \text{Add like terms.}$$

A minus or negative sign in front of parentheses affects each term in the parentheses, changing the sign of each term.

E X A M P L E   5

Removing parentheses

Simplify each expression.

    a) $6 - (x + 8)$ **b)** $4x - 6 - (7x - 4)$ **c)** $3x - (-x + 7)$

Solution

  a) $6 - (x + 8) = 6 - x - 8$ Change the sign of each term in parentheses.
$$= 6 - 8 - x \quad \text{Rearrange the terms.}$$
$$= -2 - x \quad \text{Combine like terms.}$$

  b) $4x - 6 - (7x - 4) = 4x - 6 - 7x + 4$ Remove parentheses.
$$= 4x - 7x - 6 + 4 \quad \text{Rearrange the terms.}$$
$$= -3x - 2 \quad \text{Combine like terms.}$$

  c) $3x - (-x + 7) = 3x + x - 7$ Remove parentheses.
$$= 4x - 7 \quad \text{Combine like terms.}$$

> **Now do Exercises 63–74**

The commutative and associative properties of addition enable us to rearrange the terms so that we can combine like terms. However, it is not necessary actually to write down the rearrangement. We can identify like terms and combine them without rearranging.

E X A M P L E   6

More parentheses and like terms

Simplify each expression.

    a) $(-5x + 7) + (2x - 9)$ **b)** $-4x + 7x + 3(2 - 5x)$
    c) $-3x(4x - 9) - (x - 5)$ **d)** $x - 0.03(x + 300)$

Solution

  a) $(-5x + 7) + (2x - 9) = -3x - 2$ Combine like terms.

  b) $-4x + 7x + 3(2 - 5x) = -4x + 7x + 6 - 15x$ Distributive property
$$= -12x + 6 \quad \text{Combine like terms.}$$

  c) $-3x(4x - 9) - (x - 5) = -12x^2 + 27x - x + 5$ Remove parentheses.
$$= -12x^2 + 26x + 5 \quad \text{Combine like terms.}$$

  d) $x - 0.03(x + 300) = 1x - 0.03x - 9$ Distributive property: $(-0.03)(300) = -9$
$$= 0.97x - 9 \quad \text{Combine like terms: } 1.00 - 0.03 = 0.97$$

> **Now do Exercises 75–92**

To **simplify** an expression means to write an equivalent expression that looks simpler, but *simplify* is not a precisely defined term. An expression that uses fewer symbols is usually considered simpler, but we should not be too picky with this idea. So $5x$ is clearly simpler than $2x + 3x$, but we would not say that $\frac{x}{2}$ is simpler than $\frac{1}{2}x$. Since $2ax + 2ay$ and $2a(x + y)$ both have seven symbols, either is an acceptable answer if the directions just read "simplify." If you are asked to write $2a(x + y)$ as a sum or to remove the parentheses rather than to simplify it, then it is clear that the answer should be $2ax + 2ay$.

Warm-Ups ▼

Fill in the blank.

1. A _____ is a single number or the product of a number and one or more variables raised to powers.

2. Terms that contain the same variables with the same powers are _____ terms.

3. The _____ property is used to combine like terms.

4. To _____ an expression means to write an equivalent expression that looks simpler.

5. The number preceding the variable(s) in a term is the _____ .

True or false?

A statement involving variables is true only if it is true for all values of the variable.

6. $5(x + 7) = 5x + 35$
7. $-1(a - 3) = -(a - 3)$
8. $(2x)(5x) = 10x$
9. $a + a = a^2$
10. $1 + 7x = 8x$
11. $(3x - 4) - (x - 1) = 2x - 3$
12. $-4x + 8 = -4(x + 8)$
13. $5y + 4y = 9y$
14. $-2t(5t - 3) = -10t^2 + 6t$
15. $b \cdot b = 2b$

Exercises
1.6

‹ **Study Tips** ›

- The review exercises at the end of this chapter are keyed to the sections in this chapter. If you have trouble with the review exercises, go back and study the corresponding section.
- Work the sample test at the end of this chapter to see if you are ready for your instructor's chapter test. Your instructor might not ask the same questions, but you will get a good idea of your test readiness.

‹1› Using the Properties in Computation

Perform each computation. Make use of appropriate rules to simplify each problem. See Example 1.

1. $45(200)$

2. $25(300)$

3. $\frac{4}{3}(0.75)$

4. $5(0.2)$

5. $(427 + 68) + 32$

6. $(194 + 78) + 22$

7. $47 \cdot 4 + 47 \cdot 6$

8. $53 \cdot 3 + 53 \cdot 7$

9. $19 \cdot 5 \cdot 2 \cdot \frac{1}{5}$

10. $17 \cdot 4 \cdot 2 \cdot \frac{1}{4}$

11. $(120)(400)$

12. $150 \cdot 300$

13. $13 \cdot 377(-5 + 5)$

14. $(456 \cdot 8)\frac{1}{8}$

15. $(348 + 5) + 45$

16. $(135 + 38) + 12$

17. $\frac{2}{3}(1.5)$

18. $(1.25)(0.8)$

19. $17 \cdot 101 - 17 \cdot 1$

20. $33 \cdot 2 - 12 \cdot 33$

21. $354 + 7 + 8 + 3 + 2$

22. $564 + 35 + 65 + 72 + 28$

23. $(567 + 874)(-2 \cdot 4 + 8)$

24. $(567^2 + 48)[3(-5) + 15]$

⟨2⟩ Combining Like Terms

Combine like terms. See Example 2.

25. $-4n + 6n$

26. $-3a + 15a$

27. $3w - (-4w)$

28. $3b - (-7b)$

29. $4mw^2 - 15mw^2$

30. $2b^2x - 16b^2x$

31. $-5x - (-2x)$

32. $-19m - (-3m)$

33. $4ay + 5ya$

34. $3ab + 7ba$

35. $9mn - mn$

36. $3cm - cm$

37. $-kz^6 - kz^6$

38. $s^4t - 5s^4t$

⟨3⟩ Multiplying and Dividing Terms

Find each product or quotient. See Examples 3 and 4.

39. $4(7t)$

40. $-3(4r)$

41. $(-2x)(-5x)$

42. $(-3h)(-7h)$

43. $(-h)(-h)$

44. $x(-x)$

45. $7w(-4)$

46. $-5t(-1)$

47. $-x(1 - x)$

48. $-p(p - 1)$

49. $(5k)(5k)$

50. $(-4y)(-4y)$

51. $3\left(\frac{y}{3}\right)$

52. $5z\left(\frac{z}{5}\right)$

53. $9\left(\frac{2y}{9}\right)$

54. $8\left(\frac{y}{8}\right)$

55. $\frac{6x^3}{2}$

56. $\frac{-8x^2}{4}$

57. $\frac{3x^2y + 15x}{3}$

58. $\frac{6xy^2 - 8w}{2}$

59. $\frac{2x - 4}{-2}$

60. $\frac{-6x - 9}{-3}$

61. $\frac{-xt + 10}{-2}$

62. $\frac{-2xt^2 + 8}{-4}$

⟨4⟩ Removing Parentheses

Simplify each expression. See Example 5.

63. $a - (4a - 1)$

64. $5x - (2x - 7)$

65. $6 - (x - 4)$

66. $9 - (w - 5)$

67. $4m + 6 - (m + 5)$

68. $5 - 6t - (3t + 4)$

69. $-5b - (-at + 7b)$

70. $-4x^2 - (-7x^2 + 2y)$

71. $t^2 - 5w - (-2w - t^2)$

72. $n^2 - 6m - (-n^2 - 2m)$

73. $x^2 - (x^2 - y^2 - z)$

74. $5w - (6w - 3xy - yz)$

Simplify each expression. See Example 6.

75. $(2x + 3) + (7x + 5)$

76. $(3x + 5) + (4x + 12)$

77. $(-3x + 4) - (6x - 6)$

78. $(-2x - 3) - (x - 7)$

79. $3(5x + 2) - 2(2x + 4) + x$

80. $-2(x - 3) - 4(2x + 1) - x$

81. $3x^2 - 2(x^2 - 5) - 5(2x^2 + 1)$

82. $2(-2x^2 + 1) - 4(x^2 - 3) + x^2$

83. $5(x^2 - 6x + 4) + 4(x^2 + 3x - 1)$

84. $-7(x^2 - x - 1) + 3(2x^2 - 4x + 2)$

85. $8 - 7(k^3 + 3) - 4$

86. $6 + 5(k^3 - 2) - k^3 + 5$

87. $x - 0.04(x + 50)$

88. $x - 0.03(x + 500)$

89. $0.1(x + 5) - 0.04(x + 50)$

90. $0.06x + 0.14(x + 200)$

91. $3k + 5 - 2(3k - 4) - k + 3$

92. $5w - 2 + 4(w - 3) - 6(w - 1)$

Miscellaneous

Simplify.

93. $3(1 - xy) - 2(xy - 5) - (35 - xy)$

94. $2(x^2 - 3) - (6x^2 - 2) + 2(-7x^2 - 4)$

95. $w \cdot 3w + 5w(-6w) - w(2w)$

96. $3w^3 + 5w^3 - 4w^3 + 12w^3 - 2w^2$

97. $3a^2w^2 - 5w^2 \cdot a^2 - 2aw \cdot 2aw$

98. $-3(aw^2 + 5a^2w) - 2(-a^2w - a^2w)$

99. $\dfrac{1}{6} - \dfrac{1}{3}\left(-6x^2y - \dfrac{1}{2}\right)$

100. $-\dfrac{1}{2}bc - \dfrac{1}{2}bc(3 - a)$

101. $-\dfrac{1}{2}m\left(-\dfrac{1}{2}m\right) - \dfrac{1}{2}m - \dfrac{1}{2}m$

102. $\dfrac{4wyt}{4} + \dfrac{-8wyt}{2} - \dfrac{-2wy}{2}$

103. $\dfrac{-8t^3 - 6t^2 + 2}{-2}$

104. $\dfrac{7x^3 - 5x^3 - 4}{-2}$

105. $\dfrac{-6xyz - 3xy + 9z}{-3}$

106. $\dfrac{20a^2b^4 - 10a^2b^4 + 5}{-5}$

Write an algebraic expression for each problem.

107. Triangle. The lengths of the sides of a triangular flower bed are s feet, $s + 2$ feet, and $s + 4$ feet. What is its perimeter?

108. Parallelogram. The lengths of the sides of a lot in the shape of a parallelogram are w feet and $w + 50$ feet. What is its perimeter? Is it possible to find the area from this information?

Figure for Exercise 108

109. Parthenon. To obtain a pleasing rectangular shape, the ancient Greeks constructed buildings with a length that was about $\frac{1}{6}$ longer than the width. If the width of the Parthenon is x meters and its length is $x + \frac{1}{6}x$ meters, then what is the perimeter? What is the area?

110. Square. If the length of each side of a square sign is x inches, then what are the perimeter and area of the square?

Figure for Exercise 110

Chapter 1

Wrap-Up

Summary

Sets		Examples
Set-builder notation	Notation for describing a set using variables.	$C = \{x \mid x$ is a natural number smaller than 4$\}$ $D = \{3, 4\}$
Membership	The symbol $\in$ means "is an element of."	$1 \in C, 4 \notin C$
Union	$A \cup B = \{x \mid x \in A$ or $x \in B\}$	$C \cup D = \{1, 2, 3, 4\}$
Intersection	$A \cap B = \{x \mid x \in A$ and $x \in B\}$	$C \cap D = \{3\}$
Subset	A is a subset of B if every element of A is also an element of B. The symbol $\subseteq$ means "is a subset of." $\varnothing \subseteq A$ for any set A.	$\{1, 2\} \subseteq C$ $\varnothing \subseteq C, \varnothing \subseteq D$

Real Numbers		Examples
Rational numbers	$Q = \left\{ \dfrac{a}{b} \mid a$ and b are integers with $b \neq 0 \right\}$	$\dfrac{3}{2}, 5, -6, 0, 0.25252525\ldots$
Irrational numbers	$I = \{x \mid x$ is a real number that is not rational$\}$	$\sqrt{2}, \sqrt{3}, \pi, 0.1515515551\ldots$
Real numbers	$R = \{x \mid x$ is the coordinate of a point on the number line$\}$. $R = Q \cup I$	$\dfrac{3}{2}, 5, -6, 0, 0.25252525\ldots$ $\sqrt{2}, \sqrt{3}, \pi, 0.1515515551\ldots$
Intervals of real numbers	An interval of real numbers is the set of real numbers that lie between two real numbers, which are called the endpoints of the interval. We may use $-\infty$ or ∞ as endpoints.	The real numbers between 3 and 4: $(3, 4)$ The real numbers greater than or equal to 6: $[6, \infty)$

Operations with Real Numbers		Examples
Absolute value	$\lvert a \rvert = \begin{cases} a & \text{if } a \text{ is positive or zero} \\ -a & \text{if } a \text{ is negative} \end{cases}$	$\lvert 6 \rvert = 6, \lvert 0 \rvert = 0$ $\lvert -6 \rvert = 6$
Addition and subtraction	To find the sum of two numbers with the same sign, add their absolute values. The sum has the same sign as the original numbers.	$-2 + (-7) = -9$

To find the sum of two numbers with unlike signs, subtract their absolute values. The sum is positive if the number with the larger absolute value is positive. The sum is negative if the number with the larger absolute value is negative.

$-6 + 9 = 3$
$-9 + 6 = -3$

	Subtraction: $a - b = a + (-b)$ (Change the sign and add.)	$4 - 7 = 4 + (-7) = -3$ $5 - (-3) = 5 + 3 = 8$
Multiplication and division	To find the product or quotient of two numbers, multiply or divide their absolute values: Same signs $\leftrightarrow$ positive result Opposite signs $\leftrightarrow$ negative result	$(-4)(-2) = 8, (-4)(2) = -8$ $-8 \div (-2) = 4, -8 \div 2 = -4$
Exponential expressions	In the expression a^n, a is the base and n is the exponent.	$2^3 = 2 \cdot 2 \cdot 2 = 8$
Square roots	If $a^2 = b$, then a is a square root of b. If a is greater than or equal to 0 and $a^2 = b$, then $\sqrt{b} = a$.	Both 3 and -3 are square roots of 9. Because $3 \geq 0$, $\sqrt{9} = 3$.
Order of operations	In an expression without parentheses or absolute value: 1. Evaluate exponential expressions. 2. Perform multiplication and division. 3. Perform addition and subtraction. With parentheses or absolute value: First evaluate within each set of parentheses or absolute value, using the preceding order.	$7 + 2^3 = 7 + 8 = 15$ $3 + 4 \cdot 6 = 3 + 24 = 27$ $5 + 4 \cdot 3^2 = 5 + 4 \cdot 9 = 5 + 36 = 41$ $(2 + 4)(5 - 9) = -24$ $3 + 4\,|2 - 3| = 7$

Properties of the Real Numbers		**Examples**
	For any real numbers a, b, and c:	
Commutative property of addition multiplication	$a + b = b + a$ $ab = ba$	$3 + 7 = 7 + 3$ $4 \cdot 3 = 3 \cdot 4$
Associative property of addition multiplication	$(a + b) + c = a + (b + c)$ $(ab)c = a(bc)$	$(1 + 3) + 5 = 1 + (3 + 5)$ $(3 \cdot 5)7 = 3(5 \cdot 7)$
Distributive property	$a(b + c) = ab + ac$	$3(4 + x) = 12 + 3x$ $5x - 10 = 5(x - 2)$
Additive identity property	$a + 0 = 0 + a = a$	$6 + 0 = 0 + 6 = 6$
Multiplicative identity property	$1 \cdot a = a \cdot 1 = a$	$1 \cdot 6 = 6 \cdot 1 = 6$
Additive inverse property	$a + (-a) = -a + a = 0$	$8 + (-8) = -8 + 8 = 0$

| Multiplicative inverse property | $a \cdot \dfrac{1}{a} = \dfrac{1}{a} \cdot a = 1$ for $a \neq 0$ | $8 \cdot \dfrac{1}{8} = 1,\ -2\left(-\dfrac{1}{2}\right) = 1$ |
| Multiplication property of zero | $0 \cdot a = a \cdot 0 = 0$ | $9 \cdot 0 = 0$
 $(0)(-4) = 0$ |

Algebraic Concepts		**Examples**
Algebraic expressions	Any meaningful combination of numbers, variables, and operations	$x^2 + y^2,\ -5abc$
Term	An expression containing a number or the product of a number and one or more variables raised to powers	$3x^2,\ -7x^2y,\ 8$
Like terms	Terms with identical variable parts	$4bc - 8bc = -4bc$

Enriching Your Mathematical Word Power

Fill in the blank.

1. A _____ is an expression containing a number or the product of a number and one or more variables raised to powers.

2. Terms that have the same variables with the same exponents are _____ terms.

3. A letter that is used to represent some numbers is a _____.

4. If a is a real number, then $-a$ is the _____ inverse of a.

5. The _____ of operations is the order in which operations are to be performed in the absence of grouping symbols.

6. The _____ value of a number is its distance from 0 on the number line.

7. The numbers $\{\ldots, -3, -2, -1, 0, 1, 2, 3, \ldots\}$ are the _____.

8. The numbers $\{1, 2, 3, 4, \ldots\}$ are the _____ or counting numbers.

9. The numbers $\{0, 1, 2, 3, 4, \ldots\}$ are the _____ numbers.

10. The real numbers that can be expressed as a ratio of two integers are the _____ numbers.

11. The real numbers that cannot be expressed as a ratio of two integers are the _____ numbers.

12. The number 0 is the _____ identity.

13. The number 1 is the _____ identity.

14. In the division $a \div b = c$, a is the _____, b is the _____, and c is the _____.

Review Exercises

1.1 Sets

Let $A = \{1, 2, 3\}$, $B = \{3, 4, 5\}$, $C = \{1, 2, 3, 4, 5\}$, $D = \{3\}$, and $E = \{4, 5\}$. Determine whether each statement is true or false.

1. $A \cap B = D$

2. $A \cap B = E$

3. $A \cup B = E$

4. $A \cup B = C$

5. $B \cup C = C$

6. $A \cap C = B$

7. $A \cap \varnothing = A$

8. $A \cup \varnothing = \varnothing$

9. $(A \cap B) \cup E = B$

10. $(C \cap B) \cap A = D$

11. $B \subseteq C$

12. $A \subseteq E$

13. $A = B$

14. $B = C$

15. $3 \in D$

16. $5 \notin A$

17. $0 \in E$

18. $D \subseteq \varnothing$

19. $\varnothing \subseteq E$

20. $1 \in A$

1.2 The Real Numbers

Which elements of the set

$$\left\{-\sqrt{2}, -1, 0, 1, 1.732, \sqrt{3}, \pi, \frac{22}{7}, 31\right\}$$

are members of these sets?

21. Whole numbers

22. Natural numbers

23. Integers

24. Rational numbers

25. Irrational numbers

26. Real numbers

Write each interval of real numbers in interval notation and graph it.

27. The set of real numbers greater than 0

28. The set of real numbers less than 4

29. The set of real numbers between 5 and 6

30. The set of real numbers between 5 and 6 inclusive

31. The set of real numbers greater than or equal to -1 and less than 2

32. The set of real numbers greater than 3 and less than or equal to 6

Write each union or intersection as a single interval.

33. $(0, 2) \cup (1, 5)$ **34.** $(0, 2) \cap (1, 5)$

35. $(2, 4) \cap (3, \infty)$ **36.** $(-\infty, 3) \cup (1, 6)$

37. $[2, 6) \cup (4, 8)$ **38.** $[-2, 1] \cap [0, 5)$

1.3 Operations on the Set of Real Numbers

Evaluate.

39. $-4 + 9$ **40.** $-3 + (-5)$

41. $25 - 37$ **42.** $-6 - 10$

43. $(-4)(6)$ **44.** $(-7)(-6)$

45. $(-8) \div (-4)$ **46.** $40 \div (-8)$

47. $-\dfrac{1}{4} + \dfrac{1}{12}$ **48.** $\dfrac{1}{3} - \left(-\dfrac{1}{12}\right)$

49. $\dfrac{-20}{-2}$ **50.** $\dfrac{30}{-6}$

51. $-0.04 + 10$ **52.** $-0.05 + (-3)$

53. $-6 - (-2)$ **54.** $-0.2 - (-0.04)$

55. $-0.5 + 0.5$ **56.** $-0.04 \div 0.2$

57. $3.2 \div (-0.8)$ **58.** $(0.2)(-0.9)$

59. $0 \div (-0.3545)$ **60.** $(-6)(-0.5)$

1.4 Evaluating Expressions and the Order of Operations

Evaluate each expression. If the expression is undefined, say so.

61. $4 + 7(5)$ **62.** $(4 + 7)5$

63. $20/2 \cdot 5$ **64.** $-18/2 \cdot 3$

65. $(4 + 7)^2$ **66.** $4 + 7^2$

67. $6 - (7 - 8)$ **68.** $(6 - 8) - (5 - 9)$

69. $5 - 6 - 8 - 10$ **70.** $3 - 5(6 - 2 \cdot 5)$

71. $4^2 - 9 + 3^2$ **72.** $5^2 - (6 + 5)^2$

73. $5 + 3 \cdot |6 - 4 \cdot 3|$ **74.** $|3 - 4 \cdot 2| - |5 - 8|$

75. $\sqrt{3^2 + 4^2}$ **76.** $\sqrt{13^2 - 5^2}$

77. $\dfrac{-4 - 5}{7 - (-2)}$ **78.** $\dfrac{5 - 9}{2 - 4}$

79. $\dfrac{-12 + 2(6)}{4 - (-3)}$ **80.** $\dfrac{-6 - 2(-3)}{7 - 9}$

81. $\dfrac{-1 - (-6)}{-4 - (-4)}$

82. $\dfrac{-10 - 5(-2)}{8 - 2(4)}$

83. $1 - (0.8)(0.3)$

84. $5 - (0.2)(0.1)$

85. $(-3)^2 - (4)(-1)(-2)$

86. $3^2 - 4(1)(-3)$

Let $a = -2$, $b = 3$, and $c = -1$. Find the value of each algebraic expression.

87. $\sqrt{b^2 - 4ac}$

88. $\sqrt{a^2 + 4b}$

89. $(c - b)(c + b)$

90. $(a + b)(a - b)$

91. $a^2 + 2ab + b^2$

92. $a^2 - 2ab + b^2$

93. $a^3 - b^3$

94. $a^3 + b^3$

95. $\dfrac{b + c}{a + b}$

96. $\dfrac{b - c}{2b - a}$

97. $|a - b|$

98. $|b - a|$

99. $(a + b)c$

100. $ac + bc$

1.5 Properties of the Real Numbers
Name the property that justifies each equation.

101. $a + x = x + a$

102. $0 \cdot 5 = 0$

103. $3(x - 1) = 3x - 3$

104. $10 + (-10) = 0$

105. $5(2x) = (5 \cdot 2)x$

106. $w + y = y + w$

107. $1 \cdot y = y$

108. $4 \cdot \dfrac{1}{4} = 1$

109. $5(0.2) = 1$

110. $3 \cdot 1 = 3$

111. $12 \cdot 0 = 0$

112. $x + 1 = 1 + x$

113. $18 + 0 = 18$

114. $2w + 2m = 2(w + m)$

115. $-5 + 5 = 0$

116. $2 + (3 + 4) = (2 + 3) + 4$

Use the distributive property to write each expression as a sum or difference.

117. $3(w + 1)$

118. $2(m + 14)$

119. $-1(x + 5)$

120. $-1(a + b)$

121. $-3(-2x - 5)$

122. $-2a(5 - 4b)$

Use the distributive property to write each expression as a product.

123. $3x - 6a$

124. $5x - 15y$

125. $7x + 7$

126. $6w - 3$

127. $p - pt$

128. $wx - x$

129. $ab + a$

130. $3xy + y$

1.6 Using the Properties
Simplify each expression.

131. $3a + 7 + 4a - 5$

132. $2m + 6 + m - 2$

133. $5(t - 4) - 3(2t - 6)$

134. $2(x - 3) + 2(3 - x)$

135. $-(a - 2) + 2 - a$

136. $-(w - y) + 3(y - w)$

137. $5 - 3(x - 2) + 7(x + 4) - 6$

138. $7 - 2(x - 7) + 7 - x$

139. $0.2(x + 0.1) - (x + 0.5)$

140. $0.1(x - 0.2) - (x + 0.1)$

141. $0.05(x + 3) - 0.1(x + 20)$

142. $0.02(x + 100) - 0.2(x + 50)$

143. $\dfrac{1}{2}(x + 4) - \dfrac{1}{4}(x - 8)$

144. $\dfrac{1}{2}(2x - 1) + \dfrac{1}{4}(x + 1)$

145. $\dfrac{-9x^2 - 6x + 3}{3}$

146. $\dfrac{4x - 2}{-2} + \dfrac{4x + 2}{2}$

Miscellaneous
Evaluate these expressions for $w = 24$, $x = -6$, $y = 6$, and $z = 4$. Name the property or properties used.

147. $32z(x + y)$

148. $(wz)\dfrac{1}{w}$

149. $768z + 768y$

150. $28z + 28y$

151. $(12z + x) + y$

152. $(42 + x) + y$

153. $752x + 752y$

154. $37y + 37x$

155. $(47y)\dfrac{z}{w}$

156. $3w + 3y$

157. $(xw)\dfrac{1}{y}$

158. $(xz)\dfrac{1}{x}$

159. $5(x + y)(z + w)$

160. $(4x + 7y)(w + xz)$

Solve each problem.

161. *Perimeter and area.* The width of a rectangle is x feet and its length is $x + 3$ feet. Write algebraic expressions for the perimeter and area.

162. *Carpeting costs.* Write an algebraic expression for the cost of carpeting a rectangular room that is x yards by $x + 2$ yards if carpeting costs \$20 per square yard.

163. *Inflationary spiral.* If car prices increase 5% annually, then in n years a car that currently costs P dollars will cost $P(1.05)^n$ dollars.

  a) Use this algebraic expression to predict the price of a new 2015 Hummer H2, if the price of the 2009 model was \$63,090 (www.edmunds.com).

  b) Use the accompanying graph to predict the first year in which the price of this car will be over \$100,000.

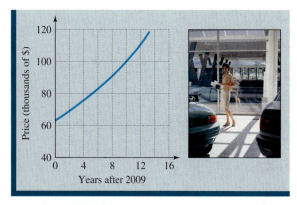

Figure for Exercise 163

164. *Lots of water.* The volume of water in a round swimming pool with radius r feet and depth h feet is $7.5\pi r^2 h$ gallons. Find the volume of water in a pool that has diameter 24 feet and depth 3 feet.

Figure for Exercise 164

Chapter 1 Test

Let A = {2, 4, 6, 8, 10}, B = {3, 4, 5, 6, 7}, and C = {6, 7, 8, 9, 10}. List the elements in each of these sets.

1. $A \cup B$

2. $B \cap C$

3. $A \cap (B \cup C)$

Which elements of $\left\{-4, -\sqrt{3}, -\dfrac{1}{2}, 0, 1.65, \sqrt{5}, \pi, 8\right\}$ *are members of these sets?*

4. Whole numbers

5. Integers

6. Rational numbers

7. Irrational numbers

Graph each of these sets.

8. The integers between −3 and 5

9. The interval (−3, 5]

Write each union or intersection as a single interval.

10. $(-\infty, 2) \cup (1, 4)$

11. $(2, 8) \cap [4, 9)$

Evaluate each expression.

12. $6 + 3(-5)$

13. $\sqrt{(-2)^2 - 4(3)(-5)}$

14. $-5 + 6 - 12$

15. $0.02 - 2$

16. $\dfrac{-3 - (-7)}{3 - 5}$

17. $\dfrac{-6 - 2}{4 - 2}$

18. $\left(\dfrac{2}{3} - 1\right)\left(\dfrac{1}{3} - \dfrac{1}{2}\right)$

19. $-\dfrac{4}{7} - \dfrac{1}{2}\left(24 - \dfrac{8}{7}\right)$

20. $|3 - 5(2)|$

21. $5 - 2|6 - 10|$

22. $(452 + 695)[2(-4) + 8]$

23. $478(8) + 478(2)$

24. $-8 \cdot 3 - 4(6 - 9 \cdot 2^3)$

Evaluate each expression for a = −3, b = −4, and c = 2.

25. $b^2 - 4ac$

26. $\dfrac{a^2 - b^2}{b - a}$

27. $\dfrac{ab - 6c}{b^2 - c^2}$

Identify the property that justifies each equation.

28. $2(5 + 7) = 10 + 14$

29. $57 \cdot 4 = 4 \cdot 57$

30. $2 + (6 + x) = (2 + 6) + x$

31. $-6 + 6 = 0$

32. $1 \cdot (-6) = (-6) \cdot 1$

Simplify each expression.

33. $3(m - 5) - 4(-2m - 3)$

34. $x + 3 - 0.05(x + 2)$

35. $\dfrac{1}{2}(x - 4) + \dfrac{1}{4}(x + 3)$

36. $-3(x^2 - 2y) - 2(3y - 4x^2)$

37. $\dfrac{-6x^2 - 4x + 2}{-2}$

Use the distributive property to rewrite each expression as a product.

38. $5x - 40$

39. $7t - 7$

Solve each problem.

40. The rectangular table for table tennis is x feet long and $x - 4$ feet wide. Write algebraic expressions for the perimeter and the area of the table. Find the actual perimeter and area using $x = 9$.

 41. If the population of the earth grows at 3% annually, then in n years the present population P will grow to $P(1.03)^n$. Assuming an annual growth rate of 3% and a present population of 6 billion people, what will the population be in 25 years?

Critical **Thinking** │ For Individual or Group Work │ Chapter 1

These exercises can be solved by a variety of techniques, which may or may not require algebra. So be creative and think critically. Explain all answers. Answers are in the Instructor's Edition of this text.

1. *Four squares.* Arrange the digits from 1 through 9 in a three by three table so that each three-digit number reading across from left to right as well as the three-digit number on the diagonal from the top left to the bottom right is a perfect square. Use all 9 digits.

Table for Exercise 1

2. *Two ones.* Write two different expressions using the digits 1, 2, 3, 4, 5, 6, 7, 8, and 9 once and only once so that the value of each expression is one. You may use any other mathematical symbols.

3. *Increasing numbers.* An *increasing number* is a positive integer in which each digit is less than the digit to its right. For example, 259 is increasing. How many increasing numbers are there between 9 and 1000?

4. *Postage reform.* A new postage system is being discussed. Postage for a letter would depend on its weight and would be a whole number of cents. Proponents claim that any appropriate postage amount could be achieved using only 5 cent and 11 cent stamps. The Postal Service would have to print only two types of stamps and these stamps would work even as rates go up. What is the largest whole number amount of postage that could not be formed using only these two types of stamps? Explain why every amount thereafter could be formed using only these two types of stamps.

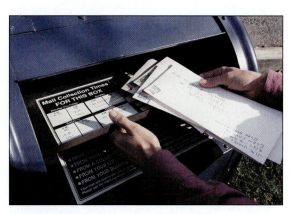

Photo for Exercise 4

5. *Half and half.* Each letter in the following addition problem represents a unique digit. Determine values of the letters that would make the addition problem correct. Find two solutions.

$$\begin{array}{r} \text{HALF} \\ + \text{HALF} \\ \hline \text{WHOLE} \end{array}$$

6. *Checkered flag.* A checkered flag used for racing is a square flag containing 64 alternating white and black squares. How many squares on the checkered flag contain an equal number of white and black squares?

7. *Largest expression.* For each integer n determine which of the expressions $\frac{2}{n}$, $n - 2$, and $2 - n$ has the largest value.

8. *Making change.* A man cashed a check for $63. The bank teller gave him six bills, but no one-dollar bills and no change. What did she give him?

9. *Big square.* Find the exact area of a square that is 111,111,111 feet on each side.

10. *Spelling bee.* If you spell out the counting numbers starting at one, then what is the first counting number for which you will use the letter "a"?

Chapter

2

Linear Equations and Inequalities in One Variable

On April 13, 1992, the headline of the *Chicago Tribune* read, "Flood Cripples Loop Businesses." Workers driving pilings around a bridge had ruptured an abandoned freight tunnel under the Chicago River. Water was gushing into the 40 miles of open tunnels below the 12 square blocks of Chicago's downtown area called the Loop. The rapidly rising water entered basements, saturated foundations, and quickly forced the shutdown of most utilities. Some subway lines were closed, and eventually thousands of workers were evacuated. While divers were used to survey the problem, the Army Corps of Engineers was called in. Their solution was to seal off the portion of the tunnel that was ruptured, using a steel-reinforced concrete plug. Once the plug was in place, the engineers worked on reversing the flow of the water. For over a month, millions of gallons of water were drained off to a water reclamation plant, and the Loop slowly returned to normal.

In this chapter, we will study algebraic equations and formulas.

In Exercises 85 and 86 of Section 2.2 you will see how the engineers used very simple algebraic formulas to calculate the amount of force the water would have on the plug.

2.1   Linear Equations in One Variable

The applications of algebra often lead to equations. The skills that you learned in Chapter 1, such as combining like terms and performing operations with algebraic expressions, will now be used to solve equations.

⟨1⟩ Equations

An **equation** is a sentence that expresses the equality of two algebraic expressions. Consider the equation

$$2x + 1 = 7.$$

Because $2(3) + 1 = 7$ is true, we say that 3 **satisfies** the equation. No other number in place of x will make the statement $2x + 1 = 7$ true. However, an equation might be satisfied by more than one number. For example, both 3 and -3 satisfy $x^2 = 9$. Any number that satisfies an equation is called a **solution** or **root** to the equation.

> **Solution Set**
>
> The set of all solutions to an equation is called the **solution set** to the equation.

The solution set to $2x + 1 = 7$ is $\{3\}$ and the solution set to $x^2 = 9$ is $\{-3, 3\}$. Note that enclosing all of the solutions to an equation in braces is not absolutely necessary. It is simply a formal way of saying "This is my final answer."

To determine whether a number is in the solution set to an equation, we replace the variable by the number and see whether the equation is correct.

EXAMPLE   1

Satisfying an equation

Determine whether each equation is satisfied by the number following the equation.

a) $3x + 7 = -8$, -5 **b)** $2(x - 1) = 2x + 3$,   4

Solution

a) Replace x by -5 and evaluate each side of the equation.

$$3x + 7 = -8$$
$$3(-5) + 7 = -8$$
$$-15 + 7 = -8$$
$$-8 = -8 \quad \text{Correct}$$

Because both sides of the equation have the same value, -5 satisfies the equation.

b) Replace x by 4 and evaluate each side of the equation.

$$2(x - 1) = 2x + 3$$
$$2(4 - 1) = 2(4) + 3 \quad \text{Replace } x \text{ by 4.}$$
$$2(3) = 8 + 3$$
$$6 = 11 \quad \text{Incorrect}$$

The two sides of the equation have different values when $x = 4$. So 4 does *not* satisfy the equation.

> **Now do Exercises 1–6**

‹ **Helpful Hint** ›

Think of an equation like a balance scale. To keep the scale in balance, what you add to one side you must add to the other side.

‹2› Solving Equations

To **solve** an equation means to find its solution set. It is easy to determine whether a given number is in the solution set of an equation as in Example 1, but that example does not provide a method for *solving* equations. The most basic method for solving equations involves the **properties of equality.**

Properties of Equality

Addition Property of Equality
Adding the same number to both sides of an equation does not change the solution set to the equation. In symbols, if $a = b$, then $a + c = b + c$.

Multiplication Property of Equality
Multiplying both sides of an equation by the same nonzero number does not change the solution set to the equation. In symbols, if $a = b$ and $c \neq 0$, then $ca = cb$.

Because subtraction is defined in terms of addition, the addition property of equality also enables us to subtract the same number from both sides. For example, subtracting 3 from both sides is equivalent to adding -3 to both sides. Because division is defined in terms of multiplication, the multiplication property of equality also enables us to divide both sides by the same nonzero number. For example, dividing both sides by 2 is equivalent to multiplying both sides by $\frac{1}{2}$.

Equations that have the same solution set are called **equivalent equations.** In Example 2, we use the properties of equality to solve an equation by writing an equivalent equation with x isolated on one side of the equation.

E X A M P L E   2

Using the properties of equality
Solve the equation $6 - 3x = 8 - 2x$.

Solution
We want to obtain an equivalent equation with only a single x on the left-hand side and a number on the other side.

$$6 - 3x = 8 - 2x$$
$$6 - 3x - 6 = 8 - 2x - 6 \qquad \text{Subtract 6 from each side.}$$
$$-3x = 2 - 2x \qquad \text{Simplify.}$$
$$-3x + 2x = 2 - 2x + 2x \qquad \text{Add } 2x \text{ to each side.}$$
$$-x = 2 \qquad \text{Combine like terms.}$$
$$-1 \cdot (-x) = -1 \cdot 2 \qquad \text{Multiply each side by } -1.$$
$$x = -2$$

Replacing x by -2 in the original equation gives us

$$6 - 3(-2) = 8 - 2(-2),$$

which is correct. So the solution set to the original equation is $\{-2\}$.

Now do Exercises 7–24

‹ **Helpful Hint** ›

Solving equations is like playing football. In football, you run a play and then pick up the injured players, regroup, and get ready for the next play. In equations, you apply a property of equality and then remove parentheses, do arithmetic, simplify, and get ready to apply another property of equality.

The addition property of equality enables us to add $2x$ to each side of the equation in Example 2 because $2x$ represents a real number.

CAUTION If you add an expression to each side that does not always represent a real number, then the equations might not be equivalent. For example,

$$x = 0 \qquad \text{and} \qquad x + \frac{1}{x} = 0 + \frac{1}{x}$$

are *not* equivalent because 0 satisfies the first equation but not the second one. (The expression $\frac{1}{x}$ is not defined if x is 0.)

To solve some equations, we must simplify the equation before using the properties of equality.

EXAMPLE 3

Simplifying the equation first
Solve the equation $2(x - 4) + 5x = 34$.

Solution
Before using the properties of equality, we simplify the expression on the left-hand side of the equation:

$$2(x - 4) + 5x = 34$$

$$2x - 8 + 5x = 34 \qquad \text{Distributive property}$$

$$7x - 8 = 34 \qquad \text{Combine like terms.}$$

$$7x - 8 + 8 = 34 + 8 \qquad \text{Add 8 to each side.}$$

$$7x = 42 \qquad \text{Simplify.}$$

$$\frac{7x}{7} = \frac{42}{7} \qquad \text{Divide each side by 7 to get a single } x \text{ on the left side.}$$

$$x = 6$$

To check, we replace x by 6 in the original equation and simplify:

$$2(6 - 4) + 5 \cdot 6 = 34$$

$$2(2) + 30 = 34$$

$$34 = 34$$

The solution set to the equation is {6}.

Now do Exercises 25–30

When an equation involves fractions, we can simplify it by multiplying each side by a number that is evenly divisible by all of the denominators. The smallest such number is called the **least common denominator (LCD).** Multiplying each side of the equation by the LCD will eliminate all of the fractions.

EXAMPLE   **4**

An equation with fractions

Find the solution set for the equation

$$\frac{x}{2} - \frac{1}{3} = \frac{x}{3} + \frac{5}{6}.$$

Solution

To solve this equation, we multiply each side by 6, the LCD for 2, 3, and 6:

$$6\left(\frac{x}{2} - \frac{1}{3}\right) = 6\left(\frac{x}{3} + \frac{5}{6}\right)$$ Multiply each side by 6.

$$6 \cdot \frac{x}{2} - 6 \cdot \frac{1}{3} = 6 \cdot \frac{x}{3} + 6 \cdot \frac{5}{6}$$ Distributive property

$$3x - 2 = 2x + 5$$ Simplify.

$$3x - 2 - 2x = 2x + 5 - 2x$$ Subtract $2x$ from each side.

$$x - 2 = 5$$ Combine like terms.

$$x - 2 + 2 = 5 + 2$$ Add 2 to each side.

$$x = 7$$ Combine like terms.

Check 7 in the original equation. The solution set is $\{7\}$.

> **Now do Exercises 31–44**

Equations that involve decimal numbers can be solved like equations involving fractions. If we multiply a decimal number by 10, 100, or 1000, the decimal point is moved one, two, or three places to the right, respectively. If the decimal points are all moved far enough to the right, the decimal numbers will be replaced by whole numbers. Example 5 shows how to use the multiplication property of equality to eliminate decimal numbers in an equation.

EXAMPLE   **5**

An equation with decimals

Solve the equation $x - 0.1x = 0.75x + 4.5$.

‹ Calculator Close-Up ›

To check 30 in Example 5 you can calculate the value of each side of the equation as shown here. Another way to check is to display the whole equation and then press ENTER. (Look in the TEST menu for the "=" symbol.) The calculator returns a 1 if the equation is correct or a 0 if the equation is incorrect.

```
30-.1*30
              27
.75*30+4.5
              27
30-.1*30=.75*30+
4.5
               1
```

Solution

Because the number with the most decimal places in this equation is 0.75 (75 hundredths), multiplying each side by 100 will eliminate all decimals.

$$100(x - 0.1x) = 100(0.75x + 4.5)$$ Multiply each side by 100.

$$100x - 10x = 75x + 450$$ Distributive property

$$90x = 75x + 450$$ Combine like terms.

$$90x - 75x = 75x + 450 - 75x$$ Subtract $75x$ from each side.

$$15x = 450$$ Combine like terms.

$$\frac{15x}{15} = \frac{450}{15}$$ Divide each side by 15.

$$x = 30$$

Check that 30 satisfies the original equation. The solution set is $\{30\}$.

> **Now do Exercises 45–50**

The equation in Examples 4 and 5 could be solved with fewer steps if the fractions and decimals are not eliminated in the first step. You should try this. Of course, you will have to perform operations such as

$$\frac{x}{2} - \frac{x}{3} = \frac{1}{2}x - \frac{1}{3}x = \frac{1}{6}x \quad \text{and} \quad x - 0.1x = 0.9x.$$

⟨3⟩ Types of Equations

We often think of an equation such as $3x + 4x = 7x$ as an "addition fact" because the equation is satisfied by all real numbers. However, some equations that we think of as facts are not satisfied by all real numbers. For example, $\frac{x}{x} = 1$ is satisfied by every real number except 0 because $\frac{0}{0}$ is undefined. The equation $x + 1 = x + 1$ is satisfied by all real numbers because both sides are identical. All of these equations are called *identities*.

The equation $2x + 1 = 7$ is true only on the condition that we choose $x = 3$. For this reason, it is called a *conditional equation*. The equations in Examples 2 through 5 are conditional equations.

Some equations are false no matter what value is used to replace the variable. For example, no number satisfies $x = x + 1$. The solution set to this *inconsistent* equation is the empty set, $\varnothing$.

Identity, Conditional Equation, Inconsistent Equation

An **identity** is an equation that is satisfied by every number for which both sides are defined.

A **conditional equation** is an equation that is satisfied by at least one number but is not an identity.

An **inconsistent equation** is an equation whose solution set is the empty set.

It is easy to classify $2x = 2x$ as an identity and $x = x + 2$ as an inconsistent equation, but some equations must be simplified before they can be classified.

E X A M P L E   6

An inconsistent equation and an identity
Solve each equation.

   a) $8 - 3(x - 5) + 7 = 3 - (x - 5) - 2(x - 11)$

   b) $5 - 3(x - 6) = 4(x - 9) - 7x$

‹ Helpful Hint ›

Removing parentheses with the distributive property and combining like terms was discussed in Chapter 1. If you are having trouble with these equations, your problem might be in Chapter 1.

Solution

   a) First simplify each side. Note that you cannot subtract 3 from 8. Because of the order of operations you must first multiply -3 and $x - 5$.

$$8 - 3(x - 5) + 7 = 3 - (x - 5) - 2(x - 11)$$
$$8 - 3x + 15 + 7 = 3 - x + 5 - 2x + 22 \quad \text{Distributive property}$$
$$30 - 3x = 30 - 3x \quad \text{Combine like terms.}$$

This last equation is satisfied by any value of x because the two sides are identical. Because the last equation is equivalent to the original equation, the original equation is satisfied by any value of x and is an identity. The solution set is R, the set of all real numbers.

b) First simplify each side of the equation.

$$5 - 3(x - 6) = 4(x - 9) - 7x$$

$$5 - 3x + 18 = 4x - 36 - 7x \qquad \text{Distributive property}$$

$$23 - 3x = -36 - 3x \qquad \text{Combine like terms.}$$

$$23 - 3x + 3x = -36 - 3x + 3x \qquad \text{Add } 3x \text{ to each side.}$$

$$23 = -36 \qquad \text{Combine like terms.}$$

The equation $23 = -36$ is false for any choice of x. Because these equations are all equivalent, the original equation is also false for any choice of x. The solution set to this inconsistent equation is the empty set, $\varnothing$.

> **Now do Exercises 51–66**

⟨4⟩ Strategy for Solving Linear Equations

The most basic equations of algebra are linear equations. In Chapter 3 we will see a connection between linear equations and straight lines.

Linear Equation in One Variable

A **linear equation in one variable** x is an equation of the form $ax = b$, where a and b are real numbers, with $a \neq 0$.

The equations in Examples 2 through 5 are called linear equations in one variable, or simply linear equations, because they could all be rewritten in the form $ax = b$. At first glance the equations in Example 6 appear to be linear equations. However, they cannot be written in the form $ax = b$, with $a \neq 0$, so they are not linear equations. A linear equation has exactly one solution. The strategy that we use for solving linear equations is summarized in the following box.

Strategy for Solving a Linear Equation

1. If fractions are present, multiply each side by the LCD to eliminate them. If decimals are present, multiply each side by a power of 10 to eliminate them.
2. Use the distributive property to remove parentheses.
3. Combine any like terms.
4. Use the addition property of equality to get all variables on one side and numbers on the other side.
5. Use the multiplication property of equality to get a single variable on one side.
6. Check by replacing the variable in the original equation with your solution.

Note that not all equations require all of the steps.

<table>
<tr><td>

E X A M P L E   **7**

</td><td>

Using the equation-solving strategy

Solve the equation $\frac{y}{2} - \frac{y-4}{5} = \frac{23}{10}$.

</td></tr>
</table>

‹ Calculator Close-Up ›

You can use the fraction feature of a graphing calculator to check that 5 satisfies the equation. If you make a mistake entering an expression, you can recall the expression by pressing the ENTRY key and modify the expression.

```
5/2-(5-4)/5▶Frac
              23/10
```

Solution

We first multiply each side of the equation by 10, the LCD for 2, 5, and 10. However, we do not have to write down that step. We can simply use the distributive property to multiply each term of the equation by 10.

$$\frac{y}{2} - \frac{y-4}{5} = \frac{23}{10}$$

$$\overset{5}{\cancel{10}}\left(\frac{y}{2}\right) - \overset{2}{\cancel{10}}\left(\frac{y-4}{5}\right) = \cancel{10}\left(\frac{23}{\cancel{10}}\right)$$ Multiply each side by 10.

$$5y - 2(y-4) = 23$$ Divide each denominator into 10 to eliminate fractions.

$$5y - 2y + 8 = 23$$ Be careful to change all signs: $-2(y-4) = -2y + 8$

$$3y + 8 = 23$$ Combine like terms.

$$3y + 8 - 8 = 23 - 8$$ Subtract 8 from each side.

$$3y = 15$$ Simplify.

$$\frac{3y}{3} = \frac{15}{3}$$ Divide each side by 3.

$$y = 5$$

Check that 5 satisfies the original equation. The solution set is {5}.

> Now do Exercises 67–78

‹5› Techniques

Writing down every step when solving an equation is not always necessary. Solving an equation is often part of a larger problem, and anything that we can do to make the process more efficient will make solving the entire problem faster and easier. For example, we can combine some steps.

Combining Steps	*Writing Every Step*

$$\begin{array}{ll} 4x - 5 = 23 & \\ 4x = 28 & \text{Add 5 to each side.} \\ x = 7 & \text{Divide each side by 4.} \end{array}$$

$$\begin{array}{c} 4x - 5 = 23 \\ 4x - 5 + 5 = 23 + 5 \\ 4x = 28 \\ \frac{4x}{4} = \frac{28}{4} \\ x = 7 \end{array}$$

The same steps are used in each of the solutions. However, when 5 is added to each side in the solution on the left, only the result is written. When each side is divided by 4, only the result is written.

To solve $-x = -5$ we must multiply each side by -1, but it is not necessary to actually show that step. We can simply write the answer:

$$-x = -5$$
$$x = 5 \qquad \text{Multiply each side by } -1.$$

Sometimes it is simpler to isolate x on the right-hand side of the equation:

$$3x + 1 = 4x - 5$$
$$6 = x \qquad \text{Subtract } 3x \text{ from each side and add 5 to each side.}$$

You can rewrite $6 = x$ as $x = 6$ or leave it as is. Either way, 6 is the solution.

For some equations with fractions it is more efficient to multiply by a multiplicative inverse instead of multiplying by the LCD:

$$-\frac{2}{3}x = \frac{1}{2}$$

$$-\frac{3}{2}\left(-\frac{2}{3}x\right) = -\frac{3}{2}\left(\frac{1}{2}\right) \qquad \text{Multiply each side by } -\frac{3}{2}, \text{ the reciprocal of } -\frac{2}{3}.$$

$$x = -\frac{3}{4}$$

The techniques shown here should not be attempted until you have become proficient at solving equations by writing out every step. The more efficient techniques shown here are not a requirement of algebra, but they can be a labor-saving tool that will be useful when we solve more complicated problems.

E X A M P L E   8

Efficient solutions

Solve each equation.

  a) $3x + 4 = 0$

  b) $2 - (x + 5) = -2(3x - 1) + 6x$

Solution

  a) Combine steps to solve the equation efficiently.

$$3x + 4 = 0$$
$$3x = -4 \qquad \text{Subtract 4 from each side.}$$
$$x = -\frac{4}{3} \qquad \text{Divide each side by 3.}$$

Check $-\frac{4}{3}$ in the original equation:

$$3\left(-\frac{4}{3}\right) + 4 = 0 \qquad \text{Correct}$$

The solution set is $\left\{-\frac{4}{3}\right\}$.

b) $2 - (x + 5) = -2(3x - 1) + 6x$

$\qquad -x - 3 = 2$ Simplify each side.

$\qquad\qquad -x = 5$ Add 3 to each side.

$\qquad\qquad\quad x = -5$ Multiply each side by -1.

Check that -5 satisfies the original equation. The solution set is $\{-5\}$.

> Now do Exercises 79–90

⟨6⟩ Applications

In Example 9 we show how a linear equation can occur in an application.

EXAMPLE 9

Completing high school

The percentage of persons 25 years and over who had completed 4 years of high school was only 25% in 1940 (Census Bureau, www.census.gov). See Fig. 2.1. The expression $0.96n + 25$ gives the percentage of persons 25 and over who have completed 4 years of high school in the year $1940 + n$, where n is the number of years since 1940.

    a) What was the percentage in 1990?

    b) When will the percentage reach 95%?

Figure 2.1

Solution

    a) Since 1990 is 50 years after 1940, $n = 50$ and

$$0.96(50) + 25 = 73.$$

So in 1990 approximately 73% of persons 25 and over had completed 4 years of high school.

    b) To find when the percentage will reach 95%, solve this equation:

$$0.96n + 25 = 95$$

$$0.96n = 70$$

$$n = \frac{70}{0.96} \approx 73$$

So 73 years after 1940 or in 2013 the percentage will reach 95%.

> Now do Exercises 97–98

Warm-Ups ▼

Fill in the blank.

1. An _____ is a sentence that expresses the equality of two algebraic expressions.

2. A number _____ an equation if the equation is true when the variable is replaced by the number.

3. Equations that have the same solution set are _____.

4. A _____ equation in one variable has the form $ax = b$, with $a \neq 0$.

5. To solve an equation involving fractions, multiply each side by the _____.

6. An _____ is an equation that is satisfied by all real numbers for which both sides are defined.

7. A _____ equation has at least one solution but is not an identity.

8. An _____ equation has no solution.

True or false?

9. The equations $-2x + 3 = 8$ and $-2x = 11$ are equivalent.

10. The equations $x - (x - 3) = 5x$ and $3 = 5x$ are equivalent.

11. To solve $\dfrac{3}{4}x = 12$, multiply each side by $\dfrac{3}{4}$.

12. The equations $-x = -6$ and $x = 6$ are equivalent.

13. To solve $\dfrac{1}{3}x + 2 = \dfrac{1}{2}x$, multiply each side by 6.

14. The solution set to $3x + 5 = 7$ is $\left\{\dfrac{2}{3}\right\}$.

15. The equation $2(3x + 4) = 6x + 12$ is inconsistent.

16. The equation $4(x + 3) = x + 3$ is conditional.

17. The equation $x - 0.2x = 0.8x$ is an identity.

18. The equation $3x = 7$ is a linear equation.

Exercises
2.1

‹ **Study Tips** ›

- Don't stay up all night cramming for a test. Prepare for a test well in advance, and get a good night's sleep before a test.
- Do your homework on a regular basis so that there is no need to cram.

‹1› **Equations**

Determine whether each equation is satisfied by the given number. See Example 1.

1. $3x + 7 = -5$, -4

2. $-3x - 5 = 13$, -6

3. $\dfrac{1}{2}x - 4 = \dfrac{1}{3}x - 2$, 12

4. $\dfrac{y - 7}{2} - \dfrac{1}{3} = \dfrac{y - 7}{3}$, 9

5. $0.2(x - 50) = 20 - 0.05x$, 200

6. $0.1x - 30 = 16 - 0.06x$, 80

‹2› **Solving Equations**

Solve each linear equation. Show your work and check your answer. See Examples 2 and 3.

7. $x + 3 = 24$

8. $x - 5 = 12$

9. $5x = 20$

10. $3x = 51$

11. $2x - 3 = 25$

12. $3x + 5 = 26$

13. $-72 - x = 15$

14. $51 - x = -9$

15. $-3x - 19 = 5 - 2x$

16. $-5x + 4 = -9 - 4x$

17. $2x - 3 = 0$

18. $5x + 7 = 0$

19. $-2x + 5 = 7$

20. $-3x - 4 = 11$

21. $-12x - 15 = 21$

22. $-13x + 7 = -19$

23. $26 = 4x + 16$

24. $14 = -5x - 21$

25. $-3(x - 16) = 12 - x$

26. $-2(x + 17) = 13 - x$

27. $2(x + 9) - x = 36$

28. $3(x - 13) - x = 9$

29. $2 + 3(x - 1) = x - 1$

30. $x + 9 = 1 - 4(x - 2)$

Solve each equation. See Example 4.

31. $-\dfrac{3}{7}x = 4$

32. $\dfrac{5}{6}x = -2$

33. $-\dfrac{5}{7}x - 1 = 3$

34. $4 - \dfrac{3}{5}x = -6$

35. $\dfrac{x}{3} + \dfrac{1}{2} = \dfrac{7}{6}$

36. $\dfrac{1}{4} + \dfrac{1}{5} = \dfrac{x}{2}$

37. $\dfrac{2}{3}x + 5 = -\dfrac{1}{3}x + 17$

38. $\dfrac{1}{4}x - 6 = -\dfrac{3}{4}x + 14$

39. $\dfrac{1}{2}x + \dfrac{1}{4} = \dfrac{1}{4}(x - 6)$

40. $\dfrac{1}{3}(x - 2) = \dfrac{2}{3}x - \dfrac{13}{3}$

41. $8 - \dfrac{x - 2}{2} = \dfrac{x}{4}$

42. $\dfrac{x}{3} - \dfrac{x - 5}{5} = 3$

43. $\dfrac{y - 3}{3} - \dfrac{y - 2}{2} = -1$

44. $\dfrac{x - 2}{2} - \dfrac{x - 3}{4} = \dfrac{7}{4}$

Solve each equation. See Example 5.

45. $x - 0.2x = 72$

46. $x - 0.1x = 63$

47. $0.03(x + 200) + 0.05x = 86$

48. $0.02(x - 100) + 0.06x = 62$

49. $0.1x + 0.05(x - 300) = 105$

50. $0.2x - 0.05(x - 100) = 35$

⟨3⟩ Types of Equations

Solve each equation. Identify each as a conditional equation, an inconsistent equation, or an identity. See Example 6.

51. $2(x + 1) = 2(x + 3)$

52. $2x + 3x = 6x$

53. $x + x = 2x$

54. $4x - 3x = x$

55. $x + x = 2$

56. $4x - 3x = 5$

57. $\dfrac{4x}{4} = x$

58. $5x \div 5 = x$

59. $x \cdot x = x^2$

60. $\dfrac{2x}{2x} = 1$

61. $2(x + 3) - 7 = 5(5 - x) + 7(x + 1)$

62. $2(x + 4) - 8 = 2x + 1$

63. $2\left(\dfrac{1}{2}x + \dfrac{3}{2}\right) - \dfrac{7}{2} = \dfrac{3}{2}(x + 1) - \left(\dfrac{1}{2}x + 2\right)$

64. $2\left(\dfrac{1}{4}x + 1\right) - 2 = \dfrac{1}{2}x$

65. $2(0.5x + 1.5) - 3.5 = 3(0.5x + 0.5)$

66. $2(0.25x + 1) - 2 = 0.75x - 1.75$

⟨4⟩ Strategy for Solving Linear Equations

Solve each equation. See Example 7. See the Strategy for Solving a Linear Equation box on page 71.

67. $4 - 6(2x - 3) + 1 = 3 + 2(5 - x)$

68. $3x - 5(6 - 2x) = 4(x - 8) + 3$

69. $5x - 2(3x + 6) = 4 - (2 + x) + 7$

70. $-1 + 5(2x - 3) = 16x - 2(3x + 8)$

71. $\dfrac{2x - 5}{4} - \dfrac{3x - 1}{6} = -\dfrac{13}{12}$

72. $\dfrac{x - 1}{2} - \dfrac{3x - 4}{6} = \dfrac{1}{3}$

73. $\dfrac{1}{2}\left(y - \dfrac{1}{6}\right) + \dfrac{2}{3} = \dfrac{5}{6} + \dfrac{1}{3}\left(\dfrac{1}{2} - 3y\right)$

74. $\dfrac{3}{4} - \dfrac{1}{3}\left(\dfrac{1}{2}y - 2\right) = 3\left(y - \dfrac{1}{4}\right)$

75. $\dfrac{40x - 5}{2} + \dfrac{5}{2} = \dfrac{33 - 2x}{3} - 11$

76. $\dfrac{a-3}{4} - \dfrac{2a-5}{2} = \dfrac{a+1}{3} - \dfrac{1}{6}$

77. $1.3 - 0.2(6 - 3x) = 0.1(0.2x + 3)$

78. $0.01(500 - 30x) = 5.4x + 200$

⟨5⟩ Techniques

Solve each equation. Practice combining some steps. Look for more efficient ways to solve each equation. See Example 8.

79. $3x - 9 = 0$　　　　　**80.** $5x + 1 = 0$

81. $7 - z = -9$　　　　**82.** $-3 - z = 3$

83. $\dfrac{2}{3}x = \dfrac{1}{2}$　　　　　**84.** $\dfrac{3}{2}x = -\dfrac{9}{5}$

85. $-\dfrac{3}{5}y = 9$　　　　**86.** $-\dfrac{2}{7}w = 4$

87. $3y + 5 = 4y - 1$　　**88.** $2y - 7 = 3y + 1$

89. $5x + 10(x + 2) = 110$

90. $1 - 3(x - 2) = 4(x - 1) - 3$

Miscellaneous

Use a calculator to assist you in solving each equation. Round approximate answers to two decimal places.

91. $45.8x - 32.7 = 102.1$

92. $-3.24 + 4.3x = -23.7$

93. $a - 0.06a = 50,000$

94. $b - 0.05b = 800$

95. $-3.48c + 6.981 = 4.329c - 6.851$

96. $4.5 - 0.62(d - 8.36) = 5.688$

⟨6⟩ Applications

Solve each problem. See Example 9.

97. *Public school enrollment.* The expression

$$0.45x + 39.05$$

can be used to approximate in millions the total enrollment in public elementary and secondary schools in the year 1985 + *x* (National Center for Education Statistics, www.nces.ed.gov).

a) What was the public school enrollment in 1992?

b) In which year will enrollment reach 50 million students?

c) Judging from the accompanying graph, is enrollment increasing or decreasing?

Figure for Exercise 97

98. *Teacher's average salary.* The expression

$$553.7x + 27,966$$

can be used to approximate the average annual salary in dollars of public school teachers in the year 1985 + *x* (National Center for Education Statistics, www.nces.ed.gov).

a) What was the average teacher's salary in 1993?

b) In which year will the average salary reach $45,000?

Getting More Involved

99. *Writing*

Explain how to eliminate the decimals in an equation that involves numbers with decimal points. Would you use the same technique when using a calculator?

100. *Discussion*

Explain why the multiplication property of equality does not allow us to multiply each side of an equation by zero.

2.2   Formulas and Functions

In this section, we will combine our knowledge of evaluating expressions from Chapter 1 with the equation-solving skills of Section 2.1 in studying formulas and functions.

⟨1⟩ Solving for a Variable

A **formula** or **literal equation** is an equation involving two or more variables. For example, the formula

$$A = LW$$

expresses the known relationship among the length L, width W, and area A of a rectangle. The formula

$$C = \frac{5}{9}(F - 32)$$

expresses the relationship between the Fahrenheit and Celsius measurements of temperature. The Celsius temperature is determined by the Fahrenheit temperature. For example, if the Fahrenheit temperature F is 95°, we can use the formula to find the Celsius temperature C as follows:

$$C = \frac{5}{9}(95 - 32) = \frac{5}{9}(63) = 35$$

A temperature of 95°F is equivalent to 35°C.

   The formula $C = \frac{5}{9}(F - 32)$ expresses *C in terms of F* or *is solved for C*. It is useful for finding C when F is known. If we want to find F when C is known, it is better to have the formula solved for F, as shown in Example 1. When a formula is **solved** for one of its variables, that variable must occur by itself on one side and must not occur on the other side.

E X A M P L E   1

Helpful Hint ⟩

There is more than one way to solve for F in Example 1. We could have first used the distributive property to remove the parentheses or multiplied by 9 to eliminate fractions. Try solving this formula for F by using these approaches.

Solving for a variable

Solve the formula $C = \frac{5}{9}(F - 32)$ for F.

Solution

To solve the formula for F, we isolate F on one side of the equation. We can eliminate both the 9 and the 5 from the right-hand side of the equation by multiplying by $\frac{9}{5}$, the reciprocal of $\frac{5}{9}$:

$$C = \frac{5}{9}(F - 32)$$

$$\frac{9}{5}C = \frac{9}{5} \cdot \frac{5}{9}(F - 32) \quad \text{Multiply each side by } \tfrac{9}{5}.$$

$$\frac{9}{5}C = F - 32 \qquad\qquad \tfrac{9}{5} \cdot \tfrac{5}{9} = 1$$

$$\frac{9}{5}C + 32 = F \qquad\qquad \text{Add 32 to each side.}$$

So the formula $F = \frac{9}{5}C + 32$ expresses F in terms of C. With this formula, we can use the value of C to determine the corresponding value of F.

> **Now do Exercises 1–12**

Note that both $F = \frac{9}{5}C + 32$ and $C = \frac{5}{9}(F - 32)$ express the relationship between C and F. The formula $F = \frac{9}{5}C + 32$ gives F in terms of C and $C = \frac{5}{9}(F - 32)$ gives C in terms of F. If we substitute 35 for C in $F = \frac{9}{5}C + 32$, we get

$$F = \frac{9}{5}(35) + 32 = 63 + 32 = 95.$$

⟨2⟩ The Language of Functions

The formula $C = \frac{5}{9}(F - 32)$ is a *rule* for determining the Celsius temperature from the Fahrenheit temperature. (The rule is to subtract 32 from F and then multiply by $\frac{5}{9}$.) We say that this formula expresses C *as a function of F* and that the formula is a function. The formula $F = \frac{9}{5}C + 32$ expresses F as a function of C. Using the formula $A = LW$, we can determine the area of a rectangle from its length L and width W, and we say that A *is a function of L and W.*

Function

A **function** is a rule for determining uniquely the value of one variable a from the value(s) of one or more other variables. We say that a **is a function of** the other variable(s).

If y is uniquely determined by x, then there is only one y-value for any given x-value. The plus-or-minus symbol $\pm$ is sometimes used in formulas, as in $y = \pm x$. In this case, there are two y-values for each nonzero x. Since y is not uniquely determined by x, y is *not* a function of x.

The function concept is one of the central ideas in algebra. In Chapter 3 you will see that a formula is not the only way to express a function.

E X A M P L E   2

Expressing one variable as a function of another

Suppose that $3a - 2b = 6$. Write a formula that expresses a as a function of b and one that expresses b as a function of a.

Solution

Solve the equation for a:

$$3a - 2b = 6$$

$$3a = 2b + 6 \qquad \text{Add } 2b \text{ to each side.}$$

$$\frac{3a}{3} = \frac{2b + 6}{3} \qquad \text{Divide each side by 3.}$$

$$a = \frac{2b}{3} + \frac{6}{3} \qquad \text{Distributive property}$$

$$a = \frac{2}{3}b + 2 \qquad \text{Simplify.}$$

So either $a = \frac{2}{3}b + 2$ or $a = \frac{2b + 6}{3}$ expresses a as a function of b. Now solve the equation for b:

$$3a - 2b = 6$$

$$-2b = -3a + 6 \qquad \text{Subtract } 3a \text{ from each side.}$$

$$\frac{-2b}{-2} = \frac{-3a + 6}{-2} \qquad \text{Divide each side by } -2.$$

$$b = \frac{-3a}{-2} + \frac{6}{-2} \qquad \text{Distributive property}$$

$$b = \frac{3}{2}a - 3 \qquad \text{Simplify.}$$

The formula $b = \frac{3}{2}a - 3$ expresses b as a function of a.

Now do Exercises 13–20

The amount A of an investment is a function of the principal P, the simple interest rate r, and the time in years t. This function is expressed by the formula $A = P + Prt$, in which P occurs twice. To express P as a function of A, r, and t we use the distributive property, as shown in Example 3.

E X A M P L E   **3**   **Solving for a variable that occurs twice**
Suppose that $A = P + Prt$. Write a formula that expresses P as a function of A, r, and t.

Solution

We can use the distributive property to write the sum $P + Prt$ as a product of P and $1 + rt$:

$$A = P + Prt$$

$$A = P \cdot 1 + P \cdot rt \qquad \text{Express } P \text{ as } P \cdot 1.$$

$$A = P(1 + rt) \qquad \text{Distributive property}$$

$$\frac{A}{1 + rt} = \frac{P(1 + rt)}{1 + rt} \qquad \text{Divide each side by } 1 + rt.$$

$$\frac{A}{1 + rt} = P$$

The formula $P = \frac{A}{1 + rt}$ expresses P as a function of A, r, and t. Note that parentheses are not needed around the expression $1 + rt$ in the denominator because the fraction bar acts as a grouping symbol.

Now do Exercises 21–24

CAUTION   If you write $A = P + Prt$ as $P = A - Prt$, then you have not solved the formula for P. When a formula is solved for a specified variable, that variable must be isolated on one side, and it must not occur on the other side.

When the variable for which we are solving occurs on opposite sides of the equation, we must move all terms involving that variable to the same side and then use the distributive property to write the expression as a product.

EXAMPLE 4

Specified variable occurring on both sides
Suppose $3a + 7 = -5ab + b$. Solve for a.

Solution

Get all terms involving a onto one side and all other terms onto the other side:

$$3a + 7 = -5ab + b$$

$3a + 5ab + 7 = b$	Add $5ab$ to each side.
$3a + 5ab = b - 7$	Subtract 7 from each side.
$a(3 + 5b) = b - 7$	Use the distributive property to write the left-hand side as a product.
$\dfrac{a(3 + 5b)}{3 + 5b} = \dfrac{b - 7}{3 + 5b}$	Divide each side by $3 + 5b$.
$a = \dfrac{b - 7}{3 + 5b}$	

‹ **Helpful Hint** ›

If you do the steps in Example 4 in a different way, you might end up with

$$a = \frac{7 - b}{-3 - 5b}.$$

This answer is correct because a is isolated. However, we usually prefer to see fewer negative signs. So we multiply this numerator and denominator by -1 and get the answer in Example 4.

Now do Exercises 25–28

When solving an equation in one variable that contains many decimal numbers, we usually use a calculator for the arithmetic. However, if you use a calculator at every step and round off the result of every computation, the final answer can differ greatly from the correct answer. Example 5 shows how to avoid this problem. The numbers are treated as if they were variables and no arithmetic is performed until all of the numbers are on one side of the equation. This technique is similar to solving an equation for a specified variable.

EXAMPLE 5

Doing computations last
Solve $3.24x - 6.78 = 6.31(x + 23.45)$.

Solution

Use the distributive property on the right-hand side, but simply write $(6.31)(23.45)$ rather than the result obtained on a calculator.

$3.24x - 6.78 = 6.31(x + 23.45)$	
$3.24x - 6.78 = 6.31x + (6.31)(23.45)$	Distributive property
$3.24x - 6.31x = (6.31)(23.45) + 6.78$	Get all x-terms on the left.
$(3.24 - 6.31)x = (6.31)(23.45) + 6.78$	Distributive property
$x = \dfrac{(6.31)(23.45) + 6.78}{3.24 - 6.31}$	Divide each side by $(3.24 - 6.31)$.
≈ -50.407	Round to three decimal places.

‹ **Calculator Close-Up** ›

A graphing calculator enables you to enter the entire expression in Example 5 and to evaluate it in one step. The ANS key holds the last value calculated. If we use ANS for x in the original equation, the calculator returns a 1, indicating that the equation is satisfied.

Check -50.407 in the original equation. When you check an approximate answer, you will get approximately the same value for each side of the equation. The solution set is $\{-50.407\}$.

Now do Exercises 29–34

‹ 3 › **Finding the Value of a Variable**

If we know the values of all of the variables in a formula except one, we can usually determine the unknown value.

E X A M P L E   **6**

Finding the value of a variable

Use the formula $-2x + 3y = 9$ to find y given that $x = -3$.

Solution

To find y, we first write y as a function of x.

$$-2x + 3y = 9 \qquad \text{Original equation}$$
$$3y = 2x + 9 \qquad \text{Add } 2x \text{ to each side.}$$
$$y = \frac{2}{3}x + 3 \qquad \text{Divide each side by 3.}$$

Now replace x by -3:

$$y = \frac{2}{3}(-3) + 3$$
$$y = 1$$

Now do Exercises 35–54

‹ **Helpful Hint** ›

It doesn't matter what form to use when solving for y here. If you use

$$y = \frac{2x + 9}{3}$$

and let $x = -3$, you get $y = 1$.

Many of the formulas used in the examples and exercises can be found inside the front and back covers of this book. Example 7 involves the simple interest formula from the back cover.

E X A M P L E   **7**

Finding the interest rate

The simple interest on a loan is \$50, the principal is \$500, and the time is 2 years. What is the annual simple interest rate?

Solution

The formula $I = Prt$ expresses the interest I as a function of the principal P, the annual simple interest rate r, and the time t. To find the rate, first express r as a function of I, P, and t. Then insert values for I, P, and t:

$$Prt = I$$
$$\frac{Prt}{Pt} = \frac{I}{Pt} \qquad \text{Divide each side by } Pt.$$
$$r = \frac{I}{Pt} \qquad \text{This formula expresses } r \text{ as a function of } I, P, \text{ and } t.$$
$$r = \frac{50}{500(2)} \qquad \text{Substitute values for } I, P, \text{ and } t.$$
$$r = 0.05$$
$$= 5\% \qquad \text{A rate is usually written as a percent.}$$

Now do Exercises 55–60

In Example 7 we solved the formula for r and then inserted the values of the other variables. If we had to find the interest rate for many different loans, this method would be the most direct. But we could also have inserted the values of I, P, and t into the original formula and then solved for r. Examples 8 and 9 illustrate this second approach.

⟨4⟩ Geometric Formulas

Some geometric formulas that will be useful in problems that involve geometric shapes are provided inside the front cover of the book. In geometry it is common to use variables with subscripts. A subscript is a slightly lowered number following the variable. For example, the areas of two triangles might be referred to as A_1 and A_2. (We read A_1 as "A sub one" or simply "A one.") You will see subscripts in Example 8.

E X A M P L E 8

Figure 2.2

Area of a trapezoid

The wildlife sanctuary shown in Fig. 2.2 has a trapezoidal shape with an area of 30 square kilometers. If one base, b_1, of the trapezoid is 10 kilometers and its height is 5 kilometers, find the length of the other base, b_2.

Solution

In any geometric problem, it is helpful to have a diagram, as in Fig. 2.2. The area of a trapezoid is a function of its height, lower base, and upper base. The formula $A = \frac{1}{2}h(b_1 + b_2)$ can be found inside the front cover of this book. Substitute the given values into the formula and then solve for b_2:

$$A = \frac{1}{2}h(b_1 + b_2) \qquad \text{The area is a function of } h, b_1, \text{ and } b_2.$$

$$30 = \frac{1}{2} \cdot 5(10 + b_2) \qquad \text{Substitute given values into the formula for the area of a trapezoid.}$$

$$60 = 5(10 + b_2) \qquad \text{Multiply each side by 2.}$$

$$12 = 10 + b_2 \qquad \text{Divide each side by 5.}$$

$$2 = b_2 \qquad \text{Subtract 10 from each side.}$$

The length of the base b_2 is 2 kilometers.

> **Now do Exercises 61–66**

E X A M P L E 9

Figure 2.3

Volume of a rectangular solid

Millie has just completed pouring 14 cubic yards of concrete to construct a rectangular driveway. If the concrete is 4 inches thick and the driveway is 18 feet wide, then how long is her driveway?

Solution

First draw a diagram as in Fig. 2.3. From a geometric point of view, the driveway is a rectangular solid. The volume of a rectangular solid is a function of its length L, width W, and height H. The formula $V = LWH$ can be found inside the front cover of this book. Before we insert the values of the variables into the formula, we must convert all of them to the same unit of measurement. We will convert feet and inches to yards:

$$4 \text{ inches} = 4 \text{ in.} \cdot \frac{1 \text{ yd}}{36 \text{ in.}} = \frac{1}{9} \text{ yard}$$

$$18 \text{ feet} = 18 \text{ ft} \cdot \frac{1 \text{ yd}}{3 \text{ ft}} = 6 \text{ yards}$$

Now replace W, H, and V by the appropriate values:

$$V = LWH$$ The volume is a function of the length, width, and height.

$$14 = L \cdot 6 \cdot \frac{1}{9}$$

$$\frac{9}{6} \cdot 14 = L$$ Multiply each side by $\frac{9}{6}$.

$$21 = L$$

The length of the driveway is 21 yards, or 63 feet.

> **Now do Exercises 67–90**

Warm-Ups ▼

Fill in the blank.

1. A _____ is an equation involving two or more variables.

2. _____ for a variable means to rewrite a formula by isolating the specified variable.

3. If the value of A can be determined from the value of x, then A is a _____ of x.

4. A formula is also called a _____ equation.

5. The _____ of a rectangular solid is the product of the length, width, and height.

6. The formula $C = \frac{5}{9}(F - 32)$ expresses C in _____ of F.

True or false?

7. If $A = P + Prt$ is solved for P, then $P = A - Prt$.

8. To solve $A = P + Prt$ for P, first apply the distributive property.

9. If $I = Prt$ is solved for t, then $t = \frac{I}{Pr}$.

10. If $a = \frac{bh}{2}$, $b = 5$, and $h = 6$, then $a = 15$.

11. The perimeter of a rectangle is a function of its length and width.

12. The area of a circle is a function of its radius.

13. The area of a rectangle is the distance around the outside edge.

14. The equation $x - y = 5$ express y in terms of x.

15. If $x = -3$ and $y = -2x - 4$, then $y = 2$.

16. The area of a trapezoid is given by $A = \frac{1}{2}(b_1 + b_2)$.

2.2 Exercises

‹ Study Tips ›

- When you get a test back, don't simply file it in your notebook. Rework all of the problems that you missed.
- Being a full-time student is a full-time job. A successful student spends 2 to 4 hours studying outside of class for every hour spent in the classroom.

‹ 1 › Solving for a Variable

Solve each formula for the specified variable. See Example 1.

1. $I = Prt$ for t

2. $d = rt$ for r

3. $F = \frac{9}{5}C + 32$ for C

4. $A = \frac{1}{2}bh$ for h

5. $A = LW$ for W

6. $C = 2\pi r$ for r

7. $A = \frac{1}{2}(b_1 + b_2)$ for b_1

8. $A = \frac{1}{2}(b_1 + b_2)$ for b_2

9. $P = 2L + 2W$ for L

10. $P = 2L + 2W$ for W

11. $V = \pi r^2 h$ for h

12. $V = \frac{1}{3}\pi r^2 h$ for h

⟨2⟩ The Language of Functions

For each formula, express y as a function of x. See Example 2.

13. $2x + 3y = 9$ **14.** $4y + 5x = 8$

15. $x - y = 4$ **16.** $y - x = 6$

17. $\frac{1}{2}x - \frac{1}{3}y = 2$ **18.** $\frac{1}{3}x - \frac{1}{4}y = 1$

19. $y - 2 = \frac{1}{2}(x - 3)$ **20.** $y - 3 = \frac{1}{3}(x - 4)$

Solve for the specified variable. See Examples 3 and 4.

21. $A = P + Prt$ for t **22.** $A = P + Prt$ for r

23. $ab + a = 1$ for a **24.** $y - wy = m$ for y

25. $xy + 5 = y - 7$ for y **26.** $xy + 5 = x + 7$ for x

27. $xy^2 + xz^2 = xw^2 - 6$ for x

28. $xz^2 + xw^2 = xy^2 + 5$ for x

 Solve each equation. Use a calculator only on the last step. Round answers to three decimal places and use your calculator to check your answer. See Example 5.

29. $3.35x - 54.6 = 44.3 - 4.58x$

30. $-4.487x - 33.41 = 55.83 - 22.49x$

31. $4.59x - 66.7 = 3.2(x - 5.67)$

32. $457(36x - 99) = 34(28x - 239)$

33. $\frac{x}{19} - \frac{3}{23} = \frac{4}{31} - \frac{3x}{7}$

34. $\frac{1}{8} - \frac{5}{7}\left(x - \frac{5}{22}\right) = \frac{4x}{9} + \frac{1}{12}$

⟨3⟩ Finding the Value of a Variable

Find y given that x = 3. See Example 6.

35. $2x - 3y = 5$ **36.** $-3x - 4y = 4$

37. $-4x + 2y = 1$ **38.** $x - y = 7$

39. $y = -2x + 5$ **40.** $y = -3x - 6$

41. $-x + 2y = 5$ **42.** $-x - 3y = 6$

43. $y - 1.046 = 2.63(x - 5.09)$

44. $y - 2.895 = -1.07(x - 2.89)$

Find x in each formula given that y = 2, z = −3, and w = 4. See Example 6.

45. $wxy = 5$ **46.** $wxz = 4$

47. $x + xz = 7$ **48.** $xw - x = 3$

49. $w(x - z) = y(x - 4)$

50. $z(x - y) = y(x + 5)$

51. $w = \frac{1}{2}xz$ **52.** $y = \frac{1}{2}wx$

53. $\frac{1}{w} + \frac{1}{x} = \frac{1}{y}$ **54.** $\frac{1}{w} + \frac{1}{y} = \frac{1}{x}$

Solve each problem. See Example 7.

55. *Simple interest rate.* If the simple interest on $1000 for 2 years is $300, then what is the rate?

56. *Simple interest rate.* If the simple interest on $20,000 for 5 years is $2,000, then what is the rate?

57. *Payday loan.* The Payday Loan Company lends you $500. After 2 weeks you pay back $520. What is the simple interest rate? Note that the time is a fraction of a year.

58. *Check holding.* You can write a check for $219 to USA Check and receive $200 in cash. After 2 weeks USA Check cashes your $219 check. What is the simple interest rate on this loan?

59. *Finding the time.* If the simple interest on $2000 at 18% is $180, then what is the time?

60. Finding the time. If the simple interest on $10,000 at 6% is $3000, then what is the time?

⟨4⟩ Geometric Formulas

Find the geometric formula that expresses each given function. See Example 8.

61. The area of a circle is a function of its radius.

62. The circumference of a circle is a function of its diameter.

63. The radius of a circle is a function of its circumference.

64. The diameter of a circle is a function of its circumference.

65. The width of a rectangle is a function of its length and perimeter.

66. The length of a rectangle is a function of its width and area.

Solve each problem. See Examples 8 and 9.

67. Rectangular floor. The area of a rectangular floor is 23 square yards. The width is 4 yards. Find the length.

68. Rectangular garden. The area of a rectangular garden is 55 square meters. The length is 7 meters. Find the width.

69. Ice sculpture. The volume of a rectangular block of ice is 36 cubic feet. The bottom is 2 feet by 2.5 feet. Find the height of the block.

70. Cardboard box. A shipping box has a volume of 2.5 cubic meters. The box measures 1 meter high by 1.25 meters wide. How long is the box?

71. Fish tank. The volume of a rectangular aquarium is 900 gallons. The bottom is 4 feet by 6 feet. Find the height of the tank. (*Hint:* There are 7.5 gallons per cubic foot.)

Figure for Exercise 71

72. Reflecting pool. A rectangular reflecting pool with a horizontal bottom holds 60,000 gallons of water. If the pool is 40 feet by 100 feet, how deep is the water?

73. Area of a triangle. The area of a triangle is 30 square feet. If the base is 4 feet, then what is the height?

74. Larger triangle. The area of a triangle is 40 square meters. If the height is 10 meters, then what is the length of the base?

75. Second base. The area of a trapezoid is 300 square inches. If the height is 20 inches and the lower base is 16 inches, then what is the length of the upper base?

Figure for Exercise 75

76. Height of a trapezoid. The area of a trapezoid is 200 square centimeters. The bases are 16 centimeters and 24 centimeters. Find the height.

77. Fencing. If it takes 600 feet of fence to enclose a rectangular lot that is 132 feet wide, then how deep is the lot?

78. Football. The perimeter of a football field in the NFL, excluding the end zones, is $306\frac{2}{3}$ yards. How many feet wide is the field?

Figure for Exercise 78

79. Radius of a circle. If the circumference of a circle is 3π meters, then what is the radius?

80. Diameter of a circle. If the circumference of a circle is 12π inches, then what is the diameter?

81. Radius of the earth. If the circumference of the earth is 25,000 miles, then what is the radius?

82. *Altitude of a satellite.* If a satellite travels 26,000 miles in each circular orbit of the earth, then how high above the earth is the satellite orbiting? See Exercise 81 and the figure.

Figure for Exercise 82

83. *Height of a can.* If the volume of a can is 30 cubic inches and the diameter of the top is 3 inches, then what is the height of the can?

Figure for Exercise 83

84. *Height of a cylinder.* If the volume of a cylinder is 6.3 cubic meters and the diameter of the lid is 1.2 meters, then what is the height of the cylinder?

85. *Great Chicago flood.* The great Chicago flood of April 1992 occurred when an old freight tunnel connecting buildings in the Loop ruptured. As shown in the figure, engineers plugged the tunnel with concrete on each side of the hole. They used the formula $F = WDA$ to find the force F of the water on the plug. In this formula the weight of water W is 62 pounds per cubic foot (lb/ft^3), the average depth D of the tunnel below the surface of the river is 32 ft, and the cross-sectional area A of the tunnel is 48 ft^2. Find the force on the plug.

Figure for Exercise 85

86. *Will it hold?* To plug the tunnel described in Exercise 85, engineers drilled a 5-foot-diameter shaft down to the tunnel. The concrete plug was made so that it extended up into the shaft. For the plug to remain in place, the shear strength of the concrete in the shaft would have to be greater than the force of the water. The amount of force F that it would take for the water to shear the concrete in the shaft is given by $F = s\pi r^2$, where s is the shear strength of concrete and r is the radius of the shaft in inches. If the shear strength of concrete is 38 lb/in.^{2}, then what force of water would shear the concrete in the shaft? Use the result from Exercise 85 to determine whether the concrete would be strong enough to hold back the water.

87. *Distance between streets.* Harold Johnson lives on a four-sided, 50,000-square-foot lot that is bounded on two sides by parallel streets. The city has assessed him $1000 for curb repair, $2 for each foot of property bordering on these two streets. How far apart are the streets?

88. *Assessed for repairs.* Harold's sister, Maude, lives next door on a triangular lot of 25,000 square feet that also extends from street to street but has frontage on only one street. What will her assessment be? (See Exercise 87.)

Figure for Exercises 87–89

89. *Juniper's lot.* Harold's other sister, Juniper, lives on the other side of him on a lot of 60,000 square feet in the shape of a parallelogram. What will her assessment be? (See Exercise 87.)

90. *Mother's driveway.* Harold's mother, who lives across the street, is pouring a concrete driveway, 12 feet wide and 4 inches thick, from the street straight to her house. This is too much work for Harold to do in one day, so his mother has agreed to buy 4 cubic yards of concrete each Saturday for three consecutive Saturdays. How far is it from the street to her house?

Miscellaneous

Solve each problem.

91. *Exercise times.* For Isabel's exercise program she jogs for 1 minute on August 1, 2 minutes on August 2, 3 minutes on August 3, and so on. What is the total number of minutes that she jogs during August? The formula $S = \frac{n(n + 1)}{2}$ gives the sum of the integers from 1 through n.

92. *Modern art.* Nicholas is painting black squares of various sizes on one white wall of his living room. The first square has 1-in. sides, the second has 2-in. sides, the third has 3-in. sides, and so on. If he does 40 squares, then how much area (in square feet) will they cover? The formula $S = \frac{n(n + 1)(2n + 1)}{6}$ gives the sum of the squares of the integers from 1 through n. Will all of these squares fit on an 8-foot by 15-foot wall?

93. *Estimating armaments.* During World War II the Allies captured some German tanks on which the smallest serial number was S and the biggest was B. Assuming the entire production of tanks was numbered 1 through N, the Allies used the function $N = B + S - 1$ to estimate the number of tanks in the German army.

   a) Find N if $B = 2003$ and $S = 455$.
   b) If this formula was used to estimate $N = 1452$ and the largest serial number was 1033, what was the smallest serial number?

94. *Cigarette usage.* The percentage of Americans 18 to 25 who use cigarettes has been decreasing at an approximately constant rate since 1974 (National Institute on Drug Abuse, www.nida.nih.gov). The function

$$P = 47.9 - 0.94n$$

can be used to estimate the percentage of smokers in this age group n years after 1974.

   a) Use the formula to find the percentage of smokers in this age group in 2006.
   b) Use the accompanying graph to estimate the year in which smoking will be eliminated from this age group.
   c) Use the formula to find the year in which smoking will be eliminated from this age group.

Figure for Exercise 94

Getting More Involved

95. *Exploration*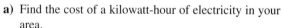

   Electric companies often point out the low cost of electricity in performing common household tasks.

   a) Find the cost of a kilowatt-hour of electricity in your area.
   b) Write a formula for finding the cost of electricity for a household appliance to perform a certain task and explain what each variable represents.
   c) Use your formula to find the cost in your area for baking a $1\frac{1}{2}$-pound loaf of bread for 5 hours in a 750-watt Welbilt breadmaker.

2.3   Applications

In This Section

⟨1⟩ **Writing Algebraic Expressions**

⟨2⟩ **Solving Problems**

⟨3⟩ **Geometric Problems**

⟨4⟩ **Investment Problems**

⟨5⟩ **Mixture Problems**

⟨6⟩ **Uniform Motion Problems**

⟨7⟩ **Commission Problems**

We often use algebra to solve problems by translating them into algebraic equations. Sometimes we can use formulas such as those inside the front cover. More often we have to set up a new equation that represents or **models** the problem. We begin with translating verbal expressions into algebraic expressions.

⟨1⟩ Writing Algebraic Expressions

Consider the three consecutive integers 5, 6, and 7. Note that each integer is 1 larger than the previous integer. To represent three *unknown* consecutive integers, we let

$$x = \text{the first integer},$$
$$x + 1 = \text{the second integer},$$
and $$x + 2 = \text{the third integer}.$$

Consider the three consecutive odd integers 7, 9, and 11. Note that each odd integer is 2 larger than the previous odd integer. To represent three *unknown* consecutive odd integers, we let

$$x = \text{the first odd integer},$$
$$x + 2 = \text{the second odd integer},$$
and $$x + 4 = \text{the third odd integer}.$$

Note that consecutive even integers as well as consecutive odd integers differ by 2. So the same expressions are used in either case.

How would we represent two numbers that have a sum of 8? If one of the numbers is 2, the other is certainly 6, or $8 - 2$. So if x is one of the numbers, then $8 - x$ is the other number. The expressions x and $8 - x$ have a sum of 8 for any value of x.

E X A M P L E   1

Writing algebraic expressions

Write algebraic expressions to represent each verbal expression.

  a) Two numbers that differ by 12

  b) Two consecutive even integers

  c) Two investments that total $5000

  d) The length of a rectangle if the width is x meters and the perimeter is 10 meters

Solution

  a) The expressions x and $x + 12$ differ by 12. Note that we could also use x and $x - 12$ for two numbers that differ by 12.

  b) The expressions x and $x + 2$ represent two consecutive even integers if x is even.

  c) If x represents the amount of one investment, then $5000 - x$ represents the amount of the other investment.

  d) Because the perimeter is 10 meters and $P = 2L + 2W = 2(L + W)$, the sum of the length and width is 5 meters. Because the width is x meters, the length is $5 - x$ meters.

  | Now do Exercises 1–12 |

Many verbal phrases occur repeatedly in applications. This list of some frequently occurring verbal phrases and their translations into algebraic expressions will help you to translate words into algebra.

Summary: Verbal Phrases and Algebraic Expressions

	Verbal Phrase	Algebraic Expression
Addition:	The sum of a number and 8	$x + 8$
	Five is added to a number	$x + 5$
	Two more than a number	$x + 2$
	A number increased by 3	$x + 3$
Subtraction:	Four is subtracted from a number	$x - 4$
	Three less than a number	$x - 3$
	The difference between 7 and a number	$7 - x$
	Some number decreased by 2	$x - 2$
	A number less 5	$x - 5$
Multiplication:	The product of 5 and a number	$5x$
	Seven times a number	$7x$
	Twice a number	$2x$
	One-half of a number	$\frac{1}{2}x \left(\text{or } \frac{x}{2}\right)$
Division:	The ratio of a number to 6	$\frac{x}{6}$
	The quotient of 5 and a number	$\frac{5}{x}$
	Three divided by some number	$\frac{3}{x}$

More than one operation can be combined in a single expression. For example, 7 less than twice a number is written as $2x - 7$.

⟨2⟩ Solving Problems

We will now see how algebraic expressions can be used to form an equation. If the equation correctly models a problem, then we may be able to solve the equation to get the solution to the problem. Some problems in this section could be solved without using algebra. However, the purpose of this section is to gain experience in setting up equations and using algebra to solve problems. We will show a complete solution to each problem so that you can gain the experience needed to solve more complex problems. We begin with a simple number problem.

EXAMPLE 2

A number problem

The sum of three consecutive integers is 228. Find the integers.

Solution

We first represent the unknown quantities with variables. The unknown quantities are the three consecutive integers. Let

$$x = \text{the first integer,}$$
$$x + 1 = \text{the second integer,}$$

and
$$x + 2 = \text{the third integer.}$$

Now do Exercises 13–24

‹ Helpful Hint ›

Making a guess can be a good way to become familiar with the problem. For example, let's guess that the answers to Example 2 are 50, 51, and 52. Since $50 + 51 + 52 = 153$, these are not the correct numbers. But now we realize that we should use x, $x + 1$, and $x + 2$ and that the equation should be

$$x + x + 1 + x + 2 = 228.$$

Since the sum of these three expressions for the consecutive integers is 228, we can write the following equation and solve it:

$$x + (x + 1) + (x + 2) = 228 \quad \text{The sum of the integers is 228.}$$
$$3x + 3 = 228$$
$$3x = 225$$
$$x = 75$$
$$x + 1 = 76 \quad \text{Identify the other unknown quantities.}$$
$$x + 2 = 77$$

To verify that these values are the correct integers, we compute

$$75 + 76 + 77 = 228.$$

The three consecutive integers that have a sum of 228 are 75, 76, and 77.

The steps to follow in providing a complete solution to a verbal problem can be stated as follows.

Strategy for Solving Word Problems

1. Read the problem until you understand the problem. Making a guess and checking it will help you to understand the problem.
2. If possible, draw a diagram to illustrate the problem.
3. Choose a variable and write down what it represents.
4. Represent any other unknowns in terms of that variable.
5. Write an equation that models the situation.
6. Solve the equation.
7. Be sure that your solution answers the question posed in the original problem.
8. Check your answer by using it to solve the original problem (not the equation).

We will now see how this strategy can be applied to various types of problems.

‹3› Geometric Problems

Any problem that involves a geometric figure may be referred to as a **geometric problem.** For geometric problems, the equation is often a geometric formula.

EXAMPLE 3

Finding the length and width of a rectangle

The length of a rectangular piece of property is 1 foot more than twice the width. If the perimeter is 302 feet, find the length and width.

Solution

First draw a diagram as in Fig. 2.4. Because the length is 1 foot more than twice the width, we let

$$x = \text{the width in feet}$$

and

$$2x + 1 = \text{the length in feet.}$$

x

$2x + 1$

Figure 2.4

< **Helpful Hint** >

To become familiar with the problem, let's guess that the width is 20 feet. The length would be 41 feet (1 foot more than twice the width). The perimeter of a 20-foot by 41-foot rectangle is 2(20) + 2(41) or 122 feet, which is not correct, but now we understand the problem.

The perimeter of a rectangle is modeled by the equation $2L + 2W = P$:

$$2L + 2W = P$$

$2(2x + 1) + 2(x) = 302$ Replace L by $2x + 1$ and W by x.

$4x + 2 + 2x = 302$ Remove the parentheses.

$$6x = 300$$

$$x = 50$$

$2x + 1 = 101$ Because $2(50) + 1 = 101$

Because $P = 2(101) + 2(50) = 302$ and 101 is 1 more than twice 50, we can be sure that the answer is correct. So the length is 101 feet, and the width is 50 feet.

> Now do Exercises 25–36

‹4› Investment Problems

Investment problems involve sums of money invested at various interest rates. In this chapter we consider simple interest only.

E X A M P L E   **4**

Investing at two rates

Greg Smith invested some money in a certificate of deposit (CD) with an annual yield of 9%. He invested twice as much money in a mutual fund with an annual yield of 12%. His interest from the two investments at the end of the year was $396. How much money was invested at each rate?

< **Helpful Hint** >

To become familiar with the problem, let's guess that he invested $400 in a CD at 9% and $800 (twice as much) in a mutual fund at 12%. His total interest is

$$0.09(400) + 0.12(800)$$

or $132, which is not correct, but now we understand the problem.

Solution

Recall the formula $I = Prt$. In this problem the time t is 1 year, so $I = Pr$. If we let x represent the amount invested at the 9% rate, then $2x$ is the amount invested at 12%. The interest on these investments is the principal times the rate, or $0.09x$ and $0.12(2x)$. It is often helpful to make a table for the unknown quantities.

	Principal	Rate	Interest
Certificate of deposit	x dollars	9%	$0.09x$ dollars
Mutual fund	$2x$ dollars	12%	$0.12(2x)$ dollars

The fact that the total interest from the investments was $396 is expressed in this equation:

$$0.09x + 0.12(2x) = 396$$

$0.09x + 0.24x = 396$ We could multiply each side by 100 to eliminate the decimals.

$$0.33x = 396$$

$$x = \frac{396}{0.33}$$

$$x = 1200$$

$$2x = 2400$$

To check this answer, we find that $0.09(\$1200) = \$108$ and $0.12(\$2400) = \$288$. Now $\$108 + \$288 = \$396$. So Greg invested $1200 at 9% and $2400 at 12%.

> Now do Exercises 37–44

‹5› Mixture Problems

Mixture problems involve solutions containing various percentages of a particular ingredient.

EXAMPLE 5

Mixing milk

How many gallons of milk containing 5% butterfat must be mixed with 90 gallons of 1% milk to obtain 2% milk?

Solution

If x represents the number of gallons of 5% milk, then $0.05x$ represents the amount of fat in that milk. If we mix x gallons of 5% milk with 90 gallons of 1% milk, we will have $x + 90$ gallons of 2% milk. See Fig. 2.5. We can make a table to classify all of the unknown amounts.

	Amount of Milk	% Fat	Amount of Fat
5% milk	x gal	5	$0.05x$ gal
1% milk	90 gal	1	$0.01(90)$ gal
2% milk	$x + 90$ gal	2	$0.02(x + 90)$ gal

In mixture problems, we always write an equation that accounts for one of the ingredients in the process. In this case, we write an equation to express the fact that the total amount of fat from the first two types of milk is the same as the amount of fat in the mixture.

$$0.05x + 0.01(90) = 0.02(x + 90)$$
$$0.05x + 0.9 = 0.02x + 1.8 \quad \text{Remove parentheses.}$$
$$0.03x = 0.9 \quad \text{Note that we chose to work with the decimals}$$
$$x = 30 \quad \text{rather than eliminate them.}$$

In 30 gallons of 5% milk there are 1.5 gallons of fat because $0.05(30) = 1.5$. In 90 gallons of 1% milk there is 0.9 gallon of fat and in 120 gallons of 2% milk there are 2.4 gallons of fat. Since $1.5 + 0.9 = 2.4$, we can be sure that the correct answer is to use 30 gallons of 5% milk.

Now do Exercises 45–48

‹ **Helpful Hint** ›

To become familiar with the problem, let's guess that 100 gallons of 5% milk should be mixed with 90 gallons of 1% milk. The total amount of fat would be $0.05(100) + 0.01(90)$ or 5.9 gallons of fat. But 2% of 190 is 3.8 gallons of fat. Since the amounts of fat should be equal, our guess is incorrect.

Figure 2.5

EXAMPLE 6

Blending fruit juice

A food chemist wants to mix some Tropical Sensation, which contains 10% juice, with some Berry Good, which contains 20% juice, to obtain 10 gallons of a new drink that will contain 14% juice. How many gallons of each should be used?

Solution

Let x represent the number of gallons of Tropical Sensation. Then $10 - x$ represents the number of gallons of Berry Good. Classify all of the information in a table as follows:

	Amount of Drink	% Juice	Amount of Juice
Tropical Sensation	x gal	10%	$0.1x$ gal
Berry Good	$10 - x$ gal	20%	$0.2(10 - x)$ gal
Mixture	10 gal	14%	$0.14(10)$ gal

We can write an equation using the amounts in the last column. The amount of juice in the Tropical Sensation plus the amount of juice in the Berry Good is equal to the amount of juice in the mixture:

$$0.1x + 0.2(10 - x) = 0.14(10)$$

$$0.1x + 2 - 0.2x = 1.4 \qquad \text{Remove parentheses.}$$

$$-0.1x = -0.6 \qquad \text{Combine like terms.}$$

$$x = \frac{-0.6}{-0.1} = 6 \qquad \text{Divide each side by } -0.1.$$

If $x = 6$, then $10 - x = 4$. So the chemist should mix 6 gallons of Tropical Sensation and 4 gallons of Berry Good to obtain a mix with 14% juice.

> Now do Exercises 49–52

⟨6⟩ Uniform Motion Problems

Problems that involve motion at a constant rate are called **uniform motion problems.** However, motion at a constant rate is rather difficult in real-life driving. We usually give an average speed when describing a driving situation, but we often omit the word "average." For uniform motion problems we will need the formula $D = RT$ (distance equals rate times time).

EXAMPLE 7

Uniform motion

Jennifer drove for 3 hours and 30 minutes in a dust storm. When the skies cleared, she increased her speed by 35 miles per hour and drove for 4 more hours. If she traveled a total of 365 miles, then how fast did she travel during the dust storm?

Solution

If x was Jennifer's speed during the dust storm, then her speed under clear skies was $x + 35$. Make a table for the given information. Note that the time of 3 hours and 30 minutes must be expressed in hours as 3.5 hours. The entries for distance come from the product of the rate and the time ($D = RT$).

	Rate	Time	Distance
Dust storm	x mph	3.5 hr	$3.5x$ mi
Clear skies	$x + 35$ mph	4 hr	$4(x + 35)$ mi

This equation expresses the fact that the total distance was 365 miles:

$$3.5x + 4(x + 35) = 365$$

$$3.5x + 4x + 140 = 365$$

$$7.5x = 225$$

$$x = 30$$

So Jennifer drove 30 miles per hour during the dust storm. To check, calculate the total distance using 30 miles per hour for 3.5 hours and 65 miles per hour for 4 hours. Since $30(3.5) + 65(4) = 365$, we can be sure that the answer is correct.

> Now do Exercises 53–60

⟨7⟩ Commission Problems

When property is sold, the percentage of the selling price that the selling agent receives is the **commission.**

E X A M P L E   8

Selling price of a house

Sonia is selling her house through a real estate agent whose commission is 6% of the selling price. What should be the selling price so that Sonia can get $84,600?

⟨ **Helpful Hint** ⟩

To become familiar with the problem, let's guess that the selling price is $100,000. The commission is 6% of the selling price: 0.06(100,000) or $6,000, so Sonia receives $94,000, which is incorrect.

Solution

Let x be the selling price. The commission is 6% of x (not 6% of $84,600). Sonia receives the selling price less the sales commission.

$$\text{Selling price} - \text{commission} = \text{Sonia's share}$$

$$x - 0.06x = 84{,}600$$

$$0.94x = 84{,}600$$

$$x = \frac{84{,}600}{0.94}$$

$$= \$90{,}000$$

The commission is 0.06($90,000), or $5,400. Sonia's share is $90,000 − $5,400, or $84,600. The house should sell for $90,000.

> **Now do Exercises 61–64**

Warm-Ups ▼

Fill in the blank.

1. We represent two unknown _____ integers with x and $x + 1$.
2. Two unknown consecutive _____ or _____ integers are represented by x and $x + 2$.
3. The _____ of a rectangle is given by $P = 2L + 2W$.
4. The words "sum" and "more than" indicate the operation of _____.
5. _____ motion is motion at a constant rate.
6. The words "difference" and "less than" indicate the operation of _____.
7. The word "product" indicates the operation of _____.

True or false?

8. The first step in solving a word problem is to write an equation.
9. We must always indicate what the variable represents.
10. Two consecutive odd integers are represented as x and $x + 1$.
11. Two numbers with a sum of 6 are represented as x and $x + 6$.
12. Two numbers that differ by 5 are represented as x and $x - 5$.
13. If the commission is 8% of the selling price x, then the commission is $0.08x$.
14. A 4% acid solution mixed with a 5% acid solution yields a 9% acid solution.

Exercises

‹ Study Tips ›

- Get to know your classmates whether you are an online student or in a classroom.
- Talk about what you are learning. Verbalizing ideas helps you get them straight in your mind.

‹1› Writing Algebraic Expressions

Find an algebraic expression for each verbal expression. See Example 1. See the Summary of Verbal Phrases and Algebraic Expressions box on page 90.

1. Two consecutive even integers
2. Two consecutive odd integers
3. Two numbers with a sum of 10
4. Two numbers with a sum of -6
5. Two numbers with a difference of 2
6. Two numbers with a difference of 3
7. Eighty-five percent of the selling price
8. The product of a number and 3
9. The distance traveled in 3 hours at x miles per hour
10. The time it takes to travel 100 miles at $x + 5$ miles per hour
11. The perimeter of a rectangle if the width is x feet and the length is 5 feet longer than the width
12. The width of a rectangle if the length is x meters and the perimeter is 20 meters

‹2› Solving Problems

Show a complete solution for each number problem. See Example 2.

13. The sum of three consecutive integers is 84. Find the integers.
14. Find three consecutive integers whose sum is 171.
15. Find three consecutive even integers whose sum is 252.
16. Find three consecutive even integers whose sum is 84.
17. Two consecutive odd integers have a sum of 128. What are the integers?
18. Four consecutive odd integers have a sum of 56. What are the integers?
19. The sum of a number and 5 is -8. What is the number?
20. The sum of a number and -12 is 6. What is the number?
21. Twice a number increased by 6 is 52. What is the number?
22. Twice a number decreased by 3 is 31. What is the number?
23. One-sixth of a number minus one-seventh of the same number is 1. What is the number?
24. One-fifth of a number plus one-sixth of the same number is 33. What is the number?

‹3› Geometric Problems

Solve each geometric problem. See Example 3. See the Strategy for Solving Word Problems box on page 91.

25. *Rectangular closet.* If the perimeter of a rectangular closet is 16 feet and the length is 2 feet longer than the width, then what are the length and width?
26. *Dimensions of a frame.* A frame maker made a large picture frame using 10 feet of frame molding. If the length of the finished frame was 2 feet more than the width, then what were the dimensions of the frame?
27. *Rectangular glass.* If the perimeter of a rectangular piece of glass is 26 in. and the length is 4 in. longer than twice the width, then what are the length and width?
28. *Rectangular courtyard.* If the perimeter of a rectangular courtyard is 34 yd and the length is 2 yd longer than twice the width, then what are the length and width?
29. *Rectangular sign.* If the perimeter of a rectangular sign is 44 cm and the width is 2 cm shorter than half the length, then what are the length and width?
30. *Rectangular closet.* If the perimeter of a rectangular closet is 40 ft and the width is 1 ft less than half the length, then what are the length and width?
31. *Rabbit region.* Fabian plans to fence a rectangular area for rabbits alongside his house. So he will use 14 feet of fencing to fence only three sides of the rectangle. If the side that runs parallel to the house is 2 feet longer than either of the other two sides, then what are the dimensions of the rectangular area?
32. *Door trim.* A carpenter used 30 feet of trim to finish three sides of the opening for a garage door. If the side parallel to the ground is 9 feet longer than either of the other two sides, then what are the dimensions of the doorway?
33. *Hog heaven.* Farmer Hodges has 50 feet of fencing to make a rectangular hog pen beside a very large barn. He needs to fence only three sides because the barn will form the fourth side. Studies have shown that under those conditions the side

parallel to the barn should be 5 feet longer than twice the width. If Farmer Hodges uses all of the fencing, what should the dimensions be?

Figure for Exercise 33

34. Doorway dimensions. A carpenter made a doorway that is 1 foot taller than twice the width. If she used three pieces of door edge molding with a total length of 17 feet, then what are the approximate dimensions of the doorway?

35. Perimeter of a lot. Having finished fencing the perimeter of a triangular piece of land, Lance observed that the second side was just 10 feet short of being twice as long as the first side, and the third side was exactly 50 feet longer than the first side. If he used 684 feet of fencing, what are the lengths of the three sides?

36. Isosceles triangle. A flag in the shape of an isosceles triangle has a base that is 3.5 inches shorter than either of the equal sides. If the perimeter of the triangle is 49 inches, what is the length of the equal sides?

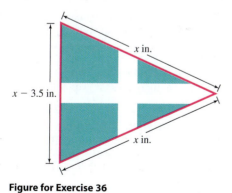

Figure for Exercise 36

‹4› Investment Problems

Solve each investment problem. See Example 4. See the Strategy for Solving Word Problems box on page 91.

37. Bob's bucks. Bob invested some money at 5% simple interest and some money at 9% simple interest. The amount invested at the higher rate was twice the amount invested at the lower rate. If the total interest on the investments for

1 year was $920, then how much did he invest at each rate?

38. Danny's dough. Danny invested some money at 3% simple interest and some money at 7% simple interest. The amount invested at the higher rate was $3000 more than the amount invested at the lower rate. If the total interest on the investments for 1 year was $810, then how much did he invest at each rate?

39. Investing money. Mr. and Mrs. Jackson invested some money at 6% simple interest and some money at 10% simple interest. In the second investment they put $1000 more than they put in the first. If the income from both investments for 1 year was $340, then how much did they invest at each rate?

40. Sibling rivalry. Samantha lent her brother some money at 9% simple interest and her sister one-half as much money at 16% simple interest. If she received a total of 34 cents in interest, then how much did she lend to each one?

41. Investing inheritance. Norman invested one-half of his inheritance in a CD that had a 10% annual yield. He lent one-quarter of his inheritance to his brother-in-law at 12% simple interest. His income from these two investments was $6400 for 1 year. How much was the inheritance?

42. Insurance settlement. Gary invested one-third of his insurance settlement in a CD that yielded 12%. He also invested one-third in Tara's computer business. Tara paid Gary 15% on this investment. If Gary's total income from these investments was $10,800 for 1 year, then what was the amount of his insurance settlement?

43. Claudette's cash. Claudette invested one-half of her inheritance in a CD paying 5%, one-third in a mutual fund paying 6%, and spent the rest on a new car. If the total income on the investments after 1 year was $9000, then what was the amount of her inheritance?

44. Wanda's windfall. Wanda invested one-half of her lottery winnings in a corporate bond that returned 8% after 1 year, one-fourth in a mutual fund that returned 3% after 1 year, and put the rest in a CD that returned 4% after 1 year. If the total income on the investments after 1 year was $5750, then what was the amount of her lottery winnings?

‹5› Mixture Problems

Solve each mixture problem. See Examples 5 and 6. See the Strategy for Solving Word Problems box on page 91.

45. Acid solutions. How many gallons of 5% acid solution should be mixed with 20 gallons of a 10% acid solution to obtain an 8% acid solution?

46. Alcohol solutions. How many liters of a 10% alcohol solution should be mixed with 12 liters of a 20% alcohol solution to obtain a 14% alcohol solution?

47. *Aaron's apricots.* Aaron mixes 12 pounds of dried apricots that sell for $5 per pound with some dried cherries that sell for $8 per pound. If he wants the mix to be worth $7 per pound, then how many pounds of cherries should he use?

48. *Cathy's cranberries.* Cathy mixes 5 pounds of dried cranberries that sell for $4 per pound with some dried peaches that sell for $12 per pound. If she wants the mix to be worth $10 per pound, then how many pounds of peaches should she use?

49. *Six-gallon solution.* Armond has two solutions available in the lab, one with 5% alcohol and another with 13% alcohol. How much of each should he mix together to obtain 6 gallons of a solution that contains 8% alcohol?

50. *Sharon's solution.* Sharon has two solutions available in the lab, one with 6% alcohol and another with 14% alcohol. How much of each should she mix together to obtain 10 gallons of a solution that contains 12% alcohol?

51. *Increasing acidity.* A gallon of Del Monte White Vinegar is labeled 5% acidity. How many fluid ounces of pure acid must be added to get 6% acidity?

52. *Chlorine bleach.* A gallon of Clorox bleach is labeled "5.25% sodium hypochlorite by weight." If a gallon of bleach weighs 8.3 pounds, then how many ounces of sodium hypochlorite must be added so that the bleach will be 6% sodium hypochlorite?

⟨6⟩ **Uniform Motion Problems**

Show a complete solution to each uniform motion problem. See Example 7.

53. *Driving in a fog.* Carlo drove for 3 hours in fog and then increased his speed by 30 miles per hour (mph) and drove 6 more hours. If his total trip was 540 miles, then what was his speed in the fog?

54. *Walk, don't run.* Louise walked for 2 hours and then ran for $1\frac{1}{2}$ hours. If she runs twice as fast as she walks and the total trip was 20 miles, then how fast does she run?

55. *Commuting to work.* A commuter bus takes 2 hours to get downtown; an express bus, averaging 25 mph faster, takes 45 minutes to cover the same route. What is the average speed for the commuter bus?

x + 25 mph   *x* mph

Figure for Exercise 55

56. *Passengers versus freight.* A freight train takes $1\frac{1}{4}$ hours to get to the city; a passenger train averaging 40 mph faster takes only 45 minutes to cover the same distance. What is the average speed of the passenger train?

57. *Terri's trip.* Terri drove for 3 hours before lunch. After lunch she drove 4 more hours and averaged 15 mph more than before lunch. If her total distance was 410 miles, then what was her average speed before lunch?

58. *Jerry's journey.* Jerry drove for 4 hours before lunch. After lunch he drove 5 more hours and averaged 10 mph less than he did before lunch. If his total distance was 688 miles, then what was his average speed after lunch?

59. *Candy's commute.* Candy drives to class in 30 minutes. By averaging 10 mph more, her roommate Fran drives to the same class in 20 minutes. What is Candy's average speed?

60. *Violet's vacation.* Violet drives to the beach in 45 minutes. By averaging 10 mph less, her roommate Veronica drives the same route to the beach in 54 minutes. What is Veronica's average speed?

⟨7⟩ **Commission Problems**

Show a complete solution to each problem. See Example 8.

61. *Listing a house.* Karl wants to get $80,000 for his house. The real estate agent charges 8% of the selling price for selling the house. What should the selling price be?

62. *Hot tamales.* Martha sells hot tamales at a sidewalk stand. Her total receipts including the 5% sales tax were $915.60. What amount of sales tax did she collect?

63. *Mustang Sally.* Sally bought a used Mustang. The selling price plus the 7% state sales tax was $9041.50. What was the selling price?

64. *Choosing a selling price.* Roy is selling his car through a broker. Roy wants to get $3000 for himself, but the broker gets a commission of 10% of the selling price. What should the selling price be?

Miscellaneous

Show a complete solution to each problem.

65. *Tennis.* The distance from the baseline to the service line on a tennis court is 3 feet less than the distance from the service line to the net. If the distance from the baseline to the net is 39 feet, then what is the distance from the service line to the net?

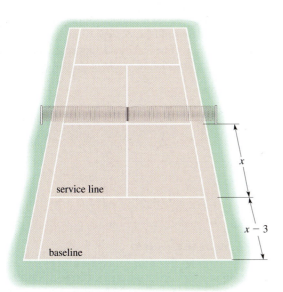

Figure for Exercise 65

66. *Mixed doubles.* The doubles court in tennis is one-third wider than the singles court. If the doubles court is 36 feet wide, then what is the width of the singles court?

67. *Rectangular room.* If the perimeter of a rectangular room is 38 m and the length is 1 m longer than twice the width, then what are the length and width?

68. *Rectangular lawn.* If the perimeter of a rectangular lawn is 116 m and the width is 6 m less than the length, then what are the length and width?

69. *Cruising America.* Suzie and Scott drive together for American Freight. One day Suzie averaged 54 mph and Scott averaged 58 mph, but Scott drove for 3 more hours than Suzie. If together they drove 734 miles, then for how many hours did Scott drive?

70. *Coast to coast.* Sam and Dave are driving together across the United States. On the first day Sam averaged 60 mph and Dave averaged 57 mph, but Sam drove for 3 hours less than Dave. If they drove a total of 873 miles that day, then for how many hours did Sam drive?

71. *Millie's mix.* Millie blends 3 pounds of Brazil nuts that sell for $8 per pound with 2 pounds of cashews that sell for $6 per pound. What should be the price per pound of the mixed nuts?

72. *Bill's blend.* Bill blends $\frac{1}{2}$ pound of peanuts that sell for $2.50 per pound with $\frac{3}{4}$ pound of walnuts that sell for $5.50 per pound. What should be the price per pound of the mixed nuts?

73. *Rectangular picture.* If the perimeter of a rectangular picture is 48 cm and the width is 3 cm shorter than half the length, then what are the length and width?

74. *Length and width.* If the perimeter of a rectangle is 278 meters and the length is 1 meter longer than twice the width, then what are the length and width?

75. *First Super Bowl.* In the first Super Bowl game in the Los Angeles Coliseum in 1967, the Green Bay Packers outscored the Kansas City Chiefs by 25 points. If 45 points were scored in that game, then what was the final score?

76. *Toy sales.* In 2009 Toys "R" Us and Wal-Mart together held 42% of the toy market share (www.fortune.com). If the market share for Toys "R" Us was 10 percentage points lower than the market share for Wal-Mart, then what was the market share for each company?

77. *Blending coffee.* Mark blends $\frac{3}{4}$ of a pound of premium Brazilian coffee with $1\frac{1}{2}$ pounds of standard Colombian coffee. If the Brazilian coffee sells for $10 per pound and the Colombian coffee sells for $8 per pound, then what should be the price per pound be for the blended coffee?

Figure for Exercise 77

78. *'Tis the seasoning.* Cheryl's Famous Pumpkin Pie Seasoning consists of a blend of cinnamon, nutmeg, and cloves. When Cheryl mixes up a batch, she uses 200 ounces of cinnamon, 100 ounces of nutmeg, and 100 ounces of cloves. If cinnamon sells for $1.80 per ounce, nutmeg sells for $1.60 per ounce, and cloves sell for $1.40 per ounce, what should be the price per ounce of the mixture?

79. *Health food mix.* Dried bananas sell for $0.80 per quarter-pound, and dried apricots sell for $1.00 per quarter-pound. How many pounds of apricots should be mixed with 10 pounds of bananas to get a mixture that sells for $0.95 per quarter-pound?

80. *Mixed nuts.* Cashews sell for $1.20 per quarter-pound, and Brazil nuts sell for $1.50 per quarter-pound. How many pounds of cashews should be mixed with 20 pounds of Brazil nuts to get a mix that sells for $1.30 per quarter-pound?

81. *Antifreeze mixture.* A mechanic finds that a car with a 20-quart radiator has a mixture containing 30% antifreeze in it. How much of this mixture would he have to drain out and replace with pure antifreeze to get a 50% antifreeze mixture?

82. *Increasing the percentage.* A mechanic has found that a car with a 16-quart radiator has a 40% antifreeze mixture in the radiator. She has on hand a 70% antifreeze solution. How much of the 40% solution would she have to replace with the 70% solution to get the solution in the radiator up to 50%?

83. *GE profit.* General Electric posted a third quarter profit of 44 cents per share. This profit was 15.8% greater than the third quarter profit of the previous year. What was the profit per share in the third quarter of the previous year?

84. *GE market value.* The 2009 market value for General Electric was $129 billion, which was 71.7% less than the market value of Exxon Mobil. What was the market value of Exxon Mobil in 2009? Round to the nearest billion.

85. *Dividing the estate.* Uncle Albert's estate is to be divided among his three nephews. The will specifies that Daniel receive one-half of the amount that Brian receives and that Raymond receive $1000 less than one-third of the amount that Brian receives. If the estate amounts to $25,400, then how much does each inherit?

86. *Mary's assets.* Mary Hall's will specifies that her lawyer is to liquidate her assets and divide the proceeds among her three sisters. Lena's share is to be one-half of Lisa's, and Lisa's share is to be one-half of Lauren's. If the lawyer has agreed to a fee that is equal to 10% of the largest share and the proceeds amount to $164,428, then how much does each person get?

87. *Missing integers.* If the larger of two consecutive integers is subtracted from twice the smaller integer, then the result is 21. Find the integers.

88. *Really odd integers.* If the smaller of two consecutive odd integers is subtracted from twice the larger one, then the result is 13. Find the integers.

89. *Highway miles.* Berenice and Jarrett drive a rig for Continental Freightways. In 1 day Berenice averaged 50 mph and Jarrett averaged 56 mph, but Berenice drove for 2 more hours than Jarrett. If together they covered 683 miles, then for how many hours did Berenice drive?

90. *Spring break.* Fernell and Dabney shared the driving to Florida for spring break. Fernell averaged 50 mph, and Dabney averaged 64 mph. If Fernell drove for 3 hours longer than Dabney but covered 18 miles less than Dabney, then for how many hours did Fernell drive?

91. *Stacy's square.* Stacy has 70 meters of fencing and plans to make a square pen. In one side she is going to leave an opening that is one-half the length of the side. If she uses all 70 meters of fencing, how large can the square be?

92. *Shawn's shed.* Shawn is building a tool shed with a square foundation and has enough siding to cover 32 linear feet of walls. If he leaves a 4-foot space for a door, then what size foundation would use up all of his siding?

Figure for Exercise 92

93. *Splitting investments.* Joan had $3000 to invest. She invested part of it in an investment paying 8% and the remainder in an investment paying 10%. If the total income on these investments was $290, then how much did she invest at each rate?

94. *Financial independence.* Dorothy had $8000 to invest. She invested part of it in an investment paying 6% and the rest in an investment paying 9%. If the total income from these investments was $690, then how much did she invest at each rate?

95. *Alcohol solutions.* Amy has two solutions available in the laboratory, one with 5% alcohol and the other with 10% alcohol. How much of each should she mix together to obtain 5 gallons of an 8% solution?

96. *Alcohol and water.* Joy has a solution containing 12% alcohol. How much of this solution and how much water must she use to get 6 liters of a solution containing 10% alcohol?

97. *Making E85.* How much ethanol should be added to 90 gallons of gasoline to get a mixture that is 85% ethanol?

98. *Making E85.* How much gasoline should be added to 765 gallons of ethanol to get a mixture that is 85% ethanol?

99. *Chance meeting.* In 6 years Todd will be twice as old as Darla was when they met 6 years ago. If their ages total 78 years, then how old are they now?

100. *Centennial Plumbing Company.* The three Hoffman brothers advertise that together they have a century of plumbing experience. Bart has twice the experience of Al, and in 3 years Carl will have twice the experience that Al had a year ago. How many years of experience does each of them have?

Math *at Work*   Nutritional Needs of Burn Patients

Providing adequate calories and nutrients is a difficult task when treating burn victims. Yet proper nutrition is essential to the healing process. The Harris-Benedict equation developed in 1919 addresses this problem. This formula is designed to calculate the basic caloric needs of adults. The basal energy expenditure in calories (*MB*) is a function of weight, height, and age. For men the function is

$$MB = 66.5 + 13.75w + 5.003h - 6.775a,$$

and for women the function is

$$WB = 655.1 + 9.563w + 1.850h - 4.676a,$$

where *w* is weight in kilograms, *h* is height in centimeters, and *a* is age in years. The value of *MB* or *WB* gives the basic number of calories per day necessary to sustain a healthy individual. To determine the caloric need of a patient, *MB* or *WB* is calculated and then multiplied by the activity plus stress factor *A* + *S*, where *A* is 1.2 for a patient confined to bed and 1.25 for an active patient. The value of *S* depends on the severity of the burns and ranges from 0.1 for mild infection to 1 for burns over 40% of the body.

The Harris-Benedict formula is just one of the tools used for determining nutritional needs of burn victims. Doctors also use the Galveston formula for children or the Curreri formula, which applies to adults and children. Recent studies have shown that these formulas can overestimate the caloric needs of patients by as much as 150%. Because no formula can accurately determine caloric needs, doctors use these formulas along with a close monitoring of the patient to ensure proper nutrition and a speedy recovery for a burn victim.

Mid-Chapter **Quiz** | Sections 2.1 through 2.3 | **Chapter 2**

Solve each equation.

1. $2x + 9 = -11$

2. $\frac{3}{4}m + 5 = \frac{1}{2}m$

3. $-9x + 3 = 5 - 11x$

4. $6 - 3(x + 2) = 4(x - 7)$

5. $0.8x + 120 = x - 70$

6. $0.09x + 3.4 = 0.4x + 65.4$

Identify each equation as a conditional equation, an inconsistent equation, or an identity.

7. $7x - 12x = -5x$

8. $7x - 12x = -5$

9. $7x - 12x = 6x$

10. $7x - 12x = -5x + 4$

Solve each equation for x.

11. $ax + b = c$

12. $m(x - a) = 3(x - b)$

Miscellaneous.

13. Express the area of a square *A* as a function of the length of its side *s*.

14. What was the original price of a car that sold for $13,904 after a 12% discount?

15. If the perimeter of a rectangle is 54 yards and the length is 3 yards longer than the width, then what is the width?

16. If $x = 8$ and $3x - 4y = 12$, then what is *y*?

17. Find three consecutive integers whose sum is 129.

18. Bob has a $50,000 inheritance that he wants to split between two investments. He anticipates that one will yield 15% and the other less risky investment will yield 4%. How should he invest the money so that his total return after one year is $3100?

2.4   Inequalities

In This Section

⟨1⟩ Inequality Symbols
⟨2⟩ Interval Notation and Graphs
⟨3⟩ Solving Linear Inequalities
⟨4⟩ Applications

An equation is a statement that indicates that two algebraic expressions are equal. An **inequality** is a statement that indicates that two algebraic expressions are not equal in a specific way, one expression being greater than or less than the other.

⟨1⟩ Inequality Symbols

The inequality symbols that we will be using are listed along with their meanings in the box.

Inequality Symbols

Symbol	Meaning
$<$	Is less than
$\leq$	Is less than or equal to
$>$	Is greater than
$\geq$	Is greater than or equal to

It is clear that 5 is less than 10, but how do we compare -5 and -10? If we think of negative numbers as debts, we would say that -10 is the larger debt. However, in algebra the size of a number is determined only by its position on the number line. For two numbers a and b we say that a *is less than* b if and only if a is to the *left* of b on the number line. To compare -5 and -10, we locate each point on the number line in Fig. 2.6. Because -10 is to the left of -5 on the number line, we say that -10 is less than -5. In symbols,

$$-10 < -5.$$

Figure 2.6

We say that a is greater than b if and only if a is to the *right* of b on the number line. Thus, we can also write

$$-5 > -10.$$

The statement $a \leq b$ is true if a is less than b or if a is equal to b. The statement $a \geq b$ is true if a is greater than b or if a equals b. For example, the statement $3 \leq 5$ is true, and so is the statement $5 \leq 5$.

Note that when two different numbers are written with an inequality symbol, the inequality symbol always points to the smaller number. For example, $4 < 9$. Note also that an inequality symbol can be read in either direction. We can read $4 < 9$ as "4 is less than 9" or "9 is greater than 4." If an inequality involves a variable, it is usually clearer to read the variable first. For example, read $-4 < x$ as "x is greater than -4."

EXAMPLE 1

Inequalities

Determine whether each statement is true or false.

a) $-5 < 3$ b) $-9 > -6$

c) $-3 \le 2$ d) $4 \ge 4$

Solution

a) The statement $-5 < 3$ is true because -5 is to the left of 3 on the number line. In fact, any negative number is less than any positive number.

b) The statement $-9 > -6$ is false because -9 lies to the left of -6.

c) The statement $-3 \le 2$ is true because -3 is less than 2.

d) The statement $4 \ge 4$ is true because $4 = 4$ is true.

Now do Exercises 1–14

‹ **Calculator Close-Up** ›

We can use a calculator to check whether an inequality is satisfied in the same manner that we check equations. The calculator returns a 1 if the inequality is correct or a 0 if it is not correct.

‹2› Interval Notation and Graphs

If an inequality involves a variable, then which real numbers can be used in place of the variable to obtain a correct statement? The set of all such numbers is the **solution set** to the inequality. For example, $x < 3$ is correct if x is replaced by any number that lies to the left of 3 on the number line:

$$1.5 < 3, \qquad 0 < 3, \qquad \text{and} \qquad -2 < 3$$

Using interval notation and the infinity symbol from Section 1.2, the solution set to $x < 3$ is the interval of real numbers $(-\infty, 3)$. The graph of the solution set or the graph of $x < 3$ is shown in Fig. 2.7.

Figure 2.7

An inequality such as $x \ge 1$ is satisfied by 1 and any real number that lies to the right of 1 on the number line. So the solution set to $x \ge 1$ is the interval $[1, \infty)$. Its graph is shown in Fig. 2.8.

Figure 2.8

The solution set and graph for each of the four basic inequalities are given in the box on the next page. Note that a bracket indicates that a number is included in the solution set and a parenthesis indicates that a number is not included.

Basic Interval Notation (k is any real number)

Inequality	Solution Set with Interval Notation	Graph
$x > k$	(k, ∞)	
$x \geq k$	$[k, \infty)$	
$x < k$	$(-\infty, k)$	
$x \leq k$	$(-\infty, k]$	

EXAMPLE   2

Interval notation and graphs

Write the solution set to each inequality in interval notation and graph it.

a) $x > -5$ **b)** $x \leq 2$

Solution

a) Every real number to the right of -5 satisfies $x > -5$. So the solution set is the interval $(-5, \infty)$. The graph is shown in Fig. 2.9.

Figure 2.9

b) The inequality $x \leq 2$ is satisfied by 2 and every real number to the left of 2. So the solution set is $(-\infty, 2]$. The graph is in Fig. 2.10.

Figure 2.10

Now do Exercises 15–22

⟨3⟩ Solving Linear Inequalities

In Section 2.1 we defined a linear equation as an equation of the form $ax = b$. If we replace the equality symbol in a linear equation with an inequality symbol, we have a linear inequality.

Linear Inequality

A **linear inequality** in one variable x is any inequality of the form $ax < b$, where a and b are real numbers, with $a \neq 0$. In place of $<$ we may also use $\leq$, $>$, or $\geq$.

Inequalities that can be *rewritten* in the form of a linear inequality are also called linear inequalities.

Before we solve linear inequalities, let's examine the results of performing various operations on each side of an inequality. If we start with the inequality $2 < 6$ and add 2 to each side, we get the true statement $4 < 8$. Examine the results in the table shown here.

Perform these operations on each side of $2 < 6$:

	Add 2	**Subtract 2**	**Multiply by 2**	**Divide by 2**
Resulting inequality	$4 < 8$	$0 < 4$	$4 < 12$	$1 < 3$

All of the resulting inequalities are correct. However, if we perform operations on each side of $2 < 6$ using -2, the situation is not as simple. For example, $-2 \cdot 2 = -4$ and $-2 \cdot 6 = -12$, but -4 is greater than -12. To get a correct inequality when each side is multiplied or divided by -2, we must reverse the inequality symbol, as shown in this table.

Perform these operations on each side of $2 < 6$:

	Add −2	**Subtract −2**	**Multiply by −2**	**Divide by −2**
Resulting inequality	$0 < 4$	$4 < 8$	$-4 > -12$	$-1 > -3$

Inequality reverses

Multiplying or dividing by a negative number changes the inequality because it changes the relative position of the results on the number line as shown in Fig. 2.11.

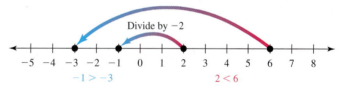

Figure 2.11

These examples illustrate the properties that we use for solving inequalities.

Properties of Inequality

Addition Property of Inequality
If the same number is added to both sides of an inequality, then the solution set to the inequality is unchanged.

Multiplication Property of Inequality
If both sides of an inequality are multiplied by the same *positive number*, then the solution set to the inequality is unchanged.

If both sides of an inequality are multiplied by the same *negative number* and *the inequality symbol is reversed*, then the solution set to the inequality is unchanged.

Because subtraction is defined in terms of addition, the addition property of inequality also allows us to subtract the same number from both sides. Because division is defined in terms of multiplication, the multiplication property of inequality also enables us to divide both sides by the same nonzero number *as long as we reverse the inequality symbol when dividing by a negative number.*

Equivalent inequalities are inequalities with the same solution set. We find the solution to a linear inequality by using the properties to convert it into an equivalent inequality with an obvious solution set, just as we do when solving equations.

E X A M P L E   3

Solving inequalities

Solve each inequality. State the solution set in interval notation and graph it.

    a) $2x - 7 < -1$ **b)** $5 - 3x < 11$ **c)** $6 - x \geq 4$

Solution

    a) We proceed exactly as we do when solving equations:

$$2x - 7 < -1 \quad \text{Original inequality}$$
$$2x < 6 \quad \text{Add 7 to each side.}$$
$$x < 3 \quad \text{Divide each side by 2.}$$

The solution set is the interval $(-\infty, 3)$. The graph is shown in Fig. 2.12.

    b) We divide by a negative number to solve this inequality.

$$5 - 3x < 11 \quad \text{Original inequality}$$
$$-3x < 6 \quad \text{Subtract 5 from each side.}$$
$$x > -2 \quad \text{Divide each side by } -3 \text{ and reverse the inequality symbol.}$$

The solution set is the interval $(-2, \infty)$. The graph is shown in Fig. 2.13.

    c) Note how the inequality symbol is reversed when each side is multiplied by -1:

$$6 - x \geq 4$$
$$-x \geq -2 \quad \text{Subtract 6 from each side.}$$
$$x \leq 2 \quad \text{Multiply each side by } -1 \text{ and reverse the inequality symbol.}$$

The solution set is the interval $(-\infty, 2]$. The graph is shown in Fig. 2.14.

> **Now do Exercises 23–46**

Figure 2.12

Figure 2.13

Figure 2.14

    Since the solution set to an inequality can contain infinitely many numbers, checking the solution is not as simple as checking the solution to an equation. However, you can check. For example, to check Example 3(b), first consider $x = -2$. This "boundary value" should satisfy the equation $5 - 3x = 11$, which it does. Now pick a number in $(-2, \infty)$ and one not in $(-2, \infty)$; say $x = -1$ and $x = -3$. Since $5 - 3(-1) < 11$ is correct and $5 - 3(-3) < 11$ is incorrect, we can be quite sure that the solution set is $(-2, \infty)$. The following Calculator Close-Up shows how to do a similar check with a graphing calculator.

⟨ **Calculator Close-Up** ⟩

To check the solution to Example 3(b), press the Y = key and let $y_1 = 5 - 3x$.

Press TBLSET to set the starting point for x and the distance between the x-values.

Now press TABLE and scroll through values of x until y_1 gets smaller than 11.

This table supports the conclusion that if $x > -2$, then $5 - 3x < 11$.

EXAMPLE 4

Inequalities involving fractions

Solve each inequality. State and graph the solution set.

a) $\dfrac{8 + 3x}{-5} \geq -4$

b) $\dfrac{1}{2}x - \dfrac{2}{3} \leq x + \dfrac{4}{3}$

Solution

a) First multiply each side by -5 to eliminate the fraction:

$$\dfrac{8 + 3x}{-5} \geq -4 \qquad \text{Original inequality}$$

$$-5\left(\dfrac{8 + 3x}{-5}\right) \leq -5(-4) \qquad \text{Multiply each side by } -5 \text{ and reverse the inequality symbol.}$$

$$8 + 3x \leq 20 \qquad \text{Simplify.}$$

$$3x \leq 12 \qquad \text{Subtract 8 from each side.}$$

$$x \leq 4 \qquad \text{Divide each side by 3.}$$

The solution set is $(-\infty, 4]$, and its graph is shown in Fig. 2.15.

Figure 2.15

⟨ **Helpful Hint** ⟩

Notice that we use the same strategy for solving inequalities as we do for solving equations. But we must remember to reverse the inequality symbol when we multiply or divide by a negative number. For inequalities it is usually best to isolate the variable on the left-hand side.

b) First multiply each side by 6, the LCD:

$$\dfrac{1}{2}x - \dfrac{2}{3} \leq x + \dfrac{4}{3} \qquad \text{Original inequality}$$

$$6\left(\dfrac{1}{2}x - \dfrac{2}{3}\right) \leq 6\left(x + \dfrac{4}{3}\right) \qquad \text{Multiplying by positive 6 does not reverse the inequality.}$$

$$3x - 4 \leq 6x + 8 \qquad \text{Distributive property}$$

$$3x \leq 6x + 12 \qquad \text{Add 4 to each side.}$$

$$-3x \leq 12 \qquad \text{Subtract } 6x \text{ from each side.}$$

$$x \geq -4 \qquad \text{Divide each side by } -3 \text{ and reverse the inequality.}$$

The solution set is the interval $[-4, \infty)$. Its graph is shown in Fig. 2.16.

Figure 2.16

Now do Exercises 47–56

In Example 5 we see an inequality that is satisfied by all real numbers and one that has no solution.

EXAMPLE 5

All or nothing

Solve each inequality and graph the solution set.

a) $6 - 4x < -4x + 7$ **b)** $2(4x - 5) \geq 4(2x - 1)$

Solution

a) Adding $4x$ to each side will greatly simplify the inequality:

$$6 - 4x < -4x + 7 \qquad \text{Original inequality}$$
$$6 < 7 \qquad \text{Add } 4x \text{ to each side.}$$

Since $6 < 7$ is correct no matter what real number is used in place of x, the solution set is the set of all real numbers $(-\infty, \infty)$. Its graph is shown in Fig. 2.17.

b) Start by simplifying each side of the inequality.

$$2(4x - 5) \geq 4(2x - 1) \qquad \text{Original inequality}$$
$$8x - 10 \geq 8x - 4 \qquad \text{Distributive property}$$
$$-10 \geq -4 \qquad \text{Subtract } 8x \text{ from each side.}$$

Since $-10 \geq -4$ is false no matter what real number is used in place of x, the solution set is the empty set $\varnothing$ and there is no graph to draw.

Now do Exercises 57–68

Figure 2.17

⟨4⟩ Applications

There are a variety of ways to express inequalities verbally. Some of the most common are illustrated in this table.

Verbal Sentence	Inequality
x is greater than 6; x is more than 6	$x > 6$
y is smaller than 0; y is less than 0	$y < 0$
w is at least 9; w is not less than 9	$w \geq 9$
m is at most 7; m is not greater than 7	$m \leq 7$

EXAMPLE **6**

Writing inequalities

Identify the variable and write an inequality that describes the situation.

    a) Chris paid more than $200 for a suit.

    b) A candidate for president must be at least 35 years old.

    c) The capacity of an elevator is at most 1500 pounds.

    d) The company must hire no fewer than 10 programmers.

Solution

    a) If c is the cost of the suit in dollars, then $c > 200$.

    b) If a is the age of the candidate in years, then $a \geq 35$.

    c) If x is the capacity of the elevator in pounds, then $x \leq 1500$.

    d) If n represents the number of programmers and n is not less than 10, then $n \geq 10$.

> **Now do Exercises 69–80**

    In Example 6(d) we knew that n was not less than 10. So there were exactly two other possibilities: n was greater than 10 or equal to 10. The fact that there are only three possible ways to position two real numbers on a number line is called the **trichotomy property.**

> **Trichotomy Property**
>
> For any two real numbers a and b, exactly one of these is true:
>
> $$a < b, \qquad a = b, \qquad \text{or} \qquad a > b$$

    We follow the same steps to solve problems involving inequalities as we do to solve problems involving equations.

EXAMPLE **7**

Price range

Lois plans to spend less than $500 on an electric dryer, including the 9% sales tax and a $64 setup charge. In what range is the selling price of the dryer that she can afford?

Solution

If we let x represent the selling price in dollars for the dryer, then the amount of sales tax is $0.09x$. Because her total cost must be less than $500, we can write the following inequality:

$$x + 0.09x + 64 < 500$$

$$1.09x < 436 \qquad \text{Subtract 64 from each side.}$$

$$x < \frac{436}{1.09} \qquad \text{Divide each side by 1.09.}$$

$$x < 400$$

The selling price of the dryer must be less than $400.

> **Now do Exercises 81–82**

Note that if we had written the equation $x + 0.09x + 64 = 500$ for the last example, we would have gotten $x = 400$. We could then have concluded that the selling price must be less than $400. This would certainly solve the problem, but it would not illustrate the use of inequalities. The original problem describes an inequality, and we should solve it as an inequality.

E X A M P L E   8

Paying off the mortgage

Tessie owns a piece of land on which she owes $12,760 to a bank. She wants to sell the land for enough money to at least pay off the mortgage. The real estate agent gets 6% of the selling price, and her city has a $400 real estate transfer tax paid by the seller. What should the range of the selling price be for Tessie to get at least enough money to pay off her mortgage?

Solution

If x is the selling price in dollars, then the commission is $0.06x$. We can write an inequality expressing the fact that the selling price minus the real estate commission minus the $400 tax must be at least $12,760:

$$x - 0.06x - 400 \geq 12,760$$
$$0.94x - 400 \geq 12,760 \quad \text{1 − 0.06 = 0.94}$$
$$0.94x \geq 13,160 \quad \text{Add 400 to each side.}$$
$$x \geq \frac{13,160}{0.94} \quad \text{Divide each side by 0.94.}$$
$$x \geq 14,000$$

The selling price must be at least $14,000 for Tessie to pay off the mortgage.

> **Now do Exercises 83–86**

E X A M P L E   9

Final average

The final average in History 101 is one-third of the midterm exam score plus two-thirds of the final exam score. To get an A, the final average must be greater than or equal to 90. If a student scored 62 on the midterm, then for what range of final exam scores would the student get an A?

Solution

If x is the final exam score, then one-third of 62 plus two-thirds of x must be greater than or equal to 90:

$$\frac{1}{3}(62) + \frac{2}{3}x \geq 90$$
$$62 + 2x \geq 270 \quad \text{Multiply each side by 3.}$$
$$2x \geq 208 \quad \text{Subtract 62 from each side.}$$
$$x \geq 104 \quad \text{Divide each side by 2.}$$

So the final exam score must be greater than or equal to 104. Whether the student can actually get an A depends on how many points are possible on the final exam.

> **Now do Exercises 87–90**

Warm-Ups ▼

Fill in the blank.

1. If a is _____ than b, then a lies to the left of b on the number line.

2. If $a \leq 5$, then a is _____ 5 or a is _____ to 5.

3. The inequality $ax < b$ is a _____ inequality.

4. If you divided each side of an inequality by a negative number, then the inequality symbol is _____.

5. If Bob's hourly wage x is at least \$16, then x __ 16.

6. The interval $(-\infty, 9)$ is the _____ to the inequality $x < 9$.

7. The solution set to the inequality $x \geq 2$ is the interval _____.

8. _____ inequalities have the same solution set.

9. If x is not greater than 3, then by the _____ property x must be less than or equal to 3.

True or false?

10. $0 < 0$

11. $-3000 > -3$

12. $89 \leq 89$

13. The inequalities $x > 5$ and $2x > 10$ are equivalent.

14. The inequalities $-3x > 12$ and $x < -4$ are equivalent.

15. The solution set to $-x > 6$ is $(-\infty, -6)$.

16. The solution set to $5x \geq 15$ is $[3, \infty)$.

17. The solution set to $x - 6 > 3$ includes 9.

18. If x is not larger than 8, then $x \leq 8$.

19. If w is at least 14, then $w \leq 14$.

Exercises

2.4

‹ **Study Tips** ›

• Don't simply work exercises to get answers. Keep reminding yourself of what you are actually doing.
• Look for the big picture. Where have we come from? Where are we going next? When will the picture be complete?

‹ 1 › Inequality Symbols

Determine whether each inequality is true or false. See Example 1.

1. $-3 < -9$

2. $-8 > -7$

3. $0 \leq 8$

4. $-6 \geq -8$

5. $(-3)20 > (-3)40$

6. $(-1)(-3) < (-1)(5)$

7. $9 - (-3) \leq 12$

8. $(-4)(-5) + 2 \geq 21$

Determine whether each inequality is satisfied by the given number.

9. $2x - 4 < 8, -3$

10. $5 - 3x > -1, 6$

11. $2x - 3 \leq 3x - 9, 5$

12. $6 - 3x \geq 10 - 2x, -4$

13. $5 - x < 4 - 2x, -1$

14. $3x - 7 \geq 3x - 10, 9$

‹ 2 › Interval Notation and Graphs

Write the solution set in interval notation and graph it. See Example 2.

15. $x \leq -1$

16. $x \geq -7$

17. $x > 20$

18. $x < 30$

19. $3 \le x$

20. $-2 > x$

21. $x < 2.3$

22. $x \le 4.5$

⟨3⟩ Solving Linear Inequalities

Fill in the blank with an inequality symbol so that the two statements are equivalent.

23. $x + 5 > 12$ **24.** $2x - 3 \le -4$ **25.** $-x < 6$
    x ___ 7 $2x$ ___ -1 x ___ -6

26. $-5 \ge -x$ **27.** $-2x \ge 8$ **28.** $-5x > -10$
    5 ___ x x ___ -4 x ___ 2

29. $4 < x$ **30.** $-3 \ge x$ **31.** $-9 \le -x$
    x ___ 4 x ___ -3 x ___ 9

32. $6 > -x$
    x ___ -6

Solve each of these inequalities. Express the solution set in interval notation and graph it. See Examples 3 and 4.

33. $x + 3 < 5$

34. $x - 9 > -6$

35. $7x > -14$

36. $4x \le -8$

37. $-3x \le 12$

38. $-2x > -6$

39. $-x > 2$

40. $-x < -3$

41. $2x - 3 > 7$

42. $3x - 2 < 6$

43. $4 - x \le 3$

44. $-2 - x \ge -1$

45. $18 \ge 3 - 5x$

46. $19 \le 5 - 4x$

47. $\dfrac{x - 3}{-5} < -2$

48. $\dfrac{2x - 3}{4} > 6$

49. $2 \ge \dfrac{5 - 3x}{4}$

50. $-1 \le \dfrac{7 - 5x}{-2}$

51. $3 - \dfrac{1}{4}x \ge 2$

52. $5 - \dfrac{1}{3}x > 2$

53. $\dfrac{1}{4}x - \dfrac{1}{2} < \dfrac{1}{2}x - \dfrac{2}{3}$

54. $\dfrac{1}{3}x - \dfrac{1}{6} < \dfrac{1}{6}x - \dfrac{1}{2}$

55. $\dfrac{y - 3}{2} > \dfrac{1}{2} - \dfrac{y - 5}{4}$

56. $\dfrac{y - 1}{3} - \dfrac{y + 1}{5} > 1$

Solve each inequality and graph the solution set. See Example 5.

57. $x - 3 > x$

58. $5 - x < 1 - x$

59. $x \geq x$

60. $x - 5 \leq x + 5$

61. $3(x + 2) \leq 9 + 3x$

62. $2x + 3 > 2(x - 4)$

63. $-2(5x - 1) \leq -5(5 + 2x)$

64. $-4(2x - 5) \leq 2(6 - 4x)$

65. $3x - (4 - 2x) < 5 - (2 - 5x)$

66. $6 - (5 - 3x) > 7x - (3 + 4x)$

67. $\frac{1}{2}x + \frac{1}{4}x < \frac{1}{8}(6x - 4)$

68. $\frac{3}{8}x - \frac{1}{4}x < \frac{1}{6}\left(\frac{3}{4}x - 6\right)$

‹4› Applications

Identify the variable and write an inequality that describes each situation. See Example 6.

69. Tony is taller than 6 feet.

70. Glenda is under 60 years old.

 71. Wilma makes less than $80,000 per year.

72. Bubba weighs over 80 pounds.

73. The maximum speed for the Concorde is 1450 miles per hour (mph).

74. The minimum speed on the freeway is 45 mph.

75. Julie can afford at most $400 per month.

76. Fred must have at least a 3.2 grade point average.

77. Burt is no taller than 5 feet.

78. Ernie cannot run faster than 10 mph.

79. Tina makes no more than $8.20 per hour.

80. Rita will not take less than $12,000 for the car.

Solve each problem by using an inequality. See Examples 7–9.

81. *Car shopping.* Jennifer is shopping for a new car. In addition to the price of the car, there is an 8% sales tax and a $172 title and license fee. If Jennifer decides that she will spend less than $10,000 total, then what is the price range for the car?

82. *Sewing machines.* Charles wants to buy a sewing machine in a city with a 10% sales tax. He has at most $700 to spend. In what price range should he look?

83. *Truck shopping.* Linda and Bob are shopping for a new truck in a city with a 9% sales tax. There is also an $80 title and license fee to pay. They want to get a good truck and plan to spend at least $10,000. What is the price range for the truck?

84. *DVD rental.* For $19.95 per month you can rent an unlimited number of DVD movies through an Internet rental service. You can rent the same DVDs at a local store for $3.98 each. How many movies would you have to rent per month for the Internet service to be the better deal?

85. *Declining birthrate.* The graph shows the number of births per 1000 women per year since 1980 in the United States (www.census.gov).

a) Has the number of births per 1000 women been increasing or decreasing since 1980?

b) The formula $B = -0.52n + 71.1$ can be used to approximate the number of births per 1000 women, where n is the number of years since 1980. What is the first year in which the number of births will be less than 55?

Figure for Exercise 85

86. *Bachelor's degrees.* The number of bachelor's degrees in thousands awarded in the United States can be approximated using the formula

$$B = 16.45n + 980.2,$$

where n is the number of years since 1985 (National Center for Education Statistics, www.nces.ed.gov). What is the first year in which the number of bachelor's degrees will exceed 1.5 million?

87. *Weighted average.* Professor Jorgenson gives only a midterm exam and a final exam. The semester average is computed by taking $\frac{1}{3}$ of the midterm exam score plus $\frac{2}{3}$ of the final exam score. The grade is determined from the semester average by using the grading scale given in the table. If Stanley scored only 56 on the midterm, then for what range of scores on the final exam would he get a C or better in the course?

88. *C or better.* Professor Brown counts her midterm as $\frac{2}{3}$ of the grade and her final as $\frac{1}{3}$ of the grade. Wilbert scored only 56 on the midterm. If Professor Brown also uses the grading scale given in the table, then what range of scores on the final exam would give Wilbert a C or better in the course?

Grading	Scale
90–100	A
80–89	B
70–79	C
60–69	D

Table for Exercises 87 and 88

89. *Designer jeans.* A pair of ordinary jeans at A-Mart costs $50 less than a pair of designer jeans at Enrico's. In fact, you can buy four pairs of A-Mart jeans for less than one pair of Enrico's jeans. What is the price range for a pair of A-Mart jeans?

90. *United Express.* Al and Rita both drive parcel delivery trucks for United Express. Al averages 20 mph less than Rita. In fact, Al is so slow that in 5 hours he covered fewer miles than Rita did in 3 hours. What are the possible values for Al's rate of speed?

Getting More Involved

91. *Discussion*

If 3 is added to every number in $(4, \infty)$, the resulting set is $(7, \infty)$. In each of the following cases, write the resulting set of numbers in interval notation. Explain your results.

a) The number -6 is subtracted from every number in $[2, \infty)$.
b) Every number in $(-\infty, -3)$ is multiplied by 2.
c) Every number in $(8, \infty)$ is divided by 4.
d) Every number in $(6, \infty)$ is multiplied by -2.
e) Every number in $(-\infty, -10)$ is divided by -5.

92. *Writing*

Explain why saying that x is *at least* 9 is equivalent to saying that x is *greater than or equal to* 9. Explain why saying that x is *at most* 5 is equivalent to saying that x is *less than or equal to* 5.

2.5   Compound Inequalities

In this section, we will use our knowledge of inequalities from Section 2.4 to solve compound inequalities. We will use also the ideas of union and intersection of intervals from Section 1.2. You may wish to review that section at this time.

In This Section

⟨1⟩ Compound Inequalities
⟨2⟩ Graphing the Solution Set
⟨3⟩ Applications

⟨1⟩ Compound Inequalities

The inequalities that we studied in Section 2.4 are referred to as **simple inequalities.** If we join two simple inequalities with the connective "and" or the connective "or," we get a **compound inequality.** A compound inequality using the connective "and" is true if and only if *both* simple inequalities are true. If at least one of the simple inequalities is false, then the compound inequality is false.

EXAMPLE 1

Compound inequalities using the connective "and"

Determine whether each compound inequality is true.

a) $3 > 2$ and $3 < 5$ **b)** $6 > 2$ and $6 < 5$

Solution

a) The compound inequality is true because $3 > 2$ is true and $3 < 5$ is true.

b) The compound inequality is false because $6 < 5$ is false.

Now do Exercises 1–3

A compound inequality using the connective "or" is true if one or the other or both of the simple inequalities are true. It is false only if both simple inequalities are false.

EXAMPLE 2

Compound inequalities using the connective "or"

Determine whether each compound inequality is true.

a) $2 < 3$ or $2 > 7$ **b)** $4 < 3$ or $4 \geq 7$

Solution

a) The compound inequality is true because $2 < 3$ is true.

b) The compound inequality is false because both $4 < 3$ and $4 \geq 7$ are false.

Now do Exercises 4–6

⟨ **Helpful Hint** ⟩

There is a big difference between "and" and "or." To get money from an automatic teller you must have a bank card *and* know a secret number (PIN). There would be a lot of problems if you could get money by having a bank card *or* knowing a PIN.

If a compound inequality involves a variable, then we are interested in the solution set to the inequality. The solution set to an "and" inequality consists of all numbers that satisfy both simple inequalities, whereas the solution set to an "or" inequality consists of all numbers that satisfy at least one of the simple inequalities.

EXAMPLE 3

Solutions of compound inequalities

Determine whether 5 satisfies each compound inequality.

a) $x < 6$ and $x < 9$ **b)** $2x - 9 \leq 5$ or $-4x \geq -12$

Solution

a) Because $5 < 6$ and $5 < 9$ are both true, 5 satisfies the compound inequality.

b) Because $2 \cdot 5 - 9 \leq 5$ is true, it does not matter that $-4 \cdot 5 \geq -12$ is false. So 5 satisfies the compound inequality.

Now do Exercises 7–14

⟨2⟩ Graphing the Solution Set

The solution set to a compound inequality using the connective "and" is the intersection of the two solution sets, because it consists of all real numbers that satisfy both simple inequalities.

E X A M P L E   **4**

Graphing compound inequalities

Graph the solution set to the compound inequality $x > 2$ and $x < 5$.

Solution

First sketch the graph of $x > 2$ and then the graph of $x < 5$, as shown in Fig. 2.18. The intersection of these two solution sets is the portion of the number line that is shaded on both graphs, just the part between 2 and 5, not including the endpoints. In symbols, $(2, \infty) \cap (-\infty, 5) = (2, 5)$. So the solution set is the interval $(2, 5)$ and its graph is shown in Fig. 2.19.

Figure 2.18

Figure 2.19

Now do Exercises 15–18

The solution set to a compound inequality using the connective "or" is the union of the two solution sets, because it consists of all real numbers that satisfy one or the other or both simple inequalities.

E X A M P L E   **5**

Graphing compound inequalities

Graph the solution set to the compound inequality $x > 4$ or $x < -1$.

Solution

First graph the solution sets to the simple inequalities as shown in Fig. 2.20. The union of these two intervals is shown in Fig. 2.21. Since the union does not simplify to a single interval, the solution set is written using the symbol for union as $(-\infty, -1) \cup (4, \infty)$.

Figure 2.20

Figure 2.21

Now do Exercises 19–20

CAUTION When graphing the intersection of two simple inequalities, do not draw too much. For the intersection, graph only numbers that satisfy *both* inequalities. Omit numbers that satisfy one but not the other inequality. Graphing a union is usually easier because we can simply draw both solution sets on the same number line.

It is not always necessary to graph the solution set to each simple inequality before graphing the solution set to the compound inequality. We can save time and work if we learn to think of the two preliminary graphs but draw only the final one.

EXAMPLE 6

Overlapping intervals

Sketch the graph and write the solution set in interval notation to each compound inequality.

a) $x < 3$ and $x < 5$ **b)** $x > 4$ or $x > 0$

Solution

a) Figure 2.22 shows $x < 3$ and $x < 5$ on the same number line. The intersection of these two intervals consists of the numbers that are less than 3. Numbers between 3 and 5 are not shaded twice and do not satisfy both inequalities. In symbols, $(-\infty, 3) \cap (-\infty, 5) = (-\infty, 3)$. So $x < 3$ and $x < 5$ is equivalent to $x < 3$. The solution set is $(-\infty, 3)$ and its graph is shown in Fig. 2.23.

Figure 2.22

Figure 2.23

b) Figure 2.24 shows the graph of $x > 4$ and the graph of $x > 0$ on the same number line. The union of these two intervals consists of everything that is shaded in Fig. 2.24. In symbols, $(4, \infty) \cup (0, \infty) = (0, \infty)$. So $x > 4$ or $x > 0$ is equivalent to $x > 0$. The solution set to the compound inequality is $(0, \infty)$ and its graph is shown in Fig. 2.25.

Figure 2.24

Figure 2.25

Now do Exercises 21–22

Example 7 shows a compound inequality that has no solution and one that is satisfied by every real number.

EXAMPLE 7

All or nothing

Sketch the graph and write the solution set in interval notation to each compound inequality.

a) $x < 2$ and $x > 6$ **b)** $x < 3$ or $x > 1$

Solution

a) A number satisfies $x < 2$ and $x > 6$ if it is both less than 2 *and* greater than 6. There are no such numbers. The solution set is the empty set, $\varnothing$. In symbols, $(-\infty, 2) \cap (6, \infty) = \varnothing$.

b) To graph $x < 3$ or $x > 1$, we shade both regions on the same number line as shown in Fig. 2.26. Since the two regions cover the entire line, the solution set is the set of all real numbers $(-\infty, \infty)$. In symbols, $(-\infty, 3) \cup (1, \infty) = (-\infty, \infty)$.

Figure 2.26

> Now do Exercises 23–28

If we start with a more complicated compound inequality, we first simplify each part of the compound inequality and then find the union or intersection.

EXAMPLE 8

Intersection

Solve $x + 2 > 3$ and $x - 6 < 7$. Graph the solution set.

Solution

First simplify each simple inequality:

$$x + 2 - 2 > 3 - 2 \qquad \text{and} \qquad x - 6 + 6 < 7 + 6$$
$$x > 1 \qquad \qquad \text{and} \qquad \qquad x < 13$$

The intersection of these two solution sets is the set of numbers between (but not including) 1 and 13. Its graph is shown in Fig. 2.27. The solution set is written in interval notation as $(1, 13)$.

Figure 2.27

> Now do Exercises 29–32

‹ Calculator Close-Up ›

To check Example 8, press Y= and let $y_1 = x + 2$ and $y_2 = x - 6$. Now scroll through a table of values for y_1 and y_2. From the table you can see that y_1 is greater than 3 and y_2 is less than 7 precisely when x is between 1 and 13.

X	Y1	Y2
1	3	-5
3	5	-3
5	7	-1
7	9	1
9	11	3
11	13	5
13	15	7

Y1◻X+2

E X A M P L E   9

‹ **Calculator Close-Up** ›

To check Example 9, press Y = and let $y_1 = 5 - 7x$ and $y_2 = 3x - 2$. Now scroll through a table of values for y_1 and y_2. From the table you can see that either $y_1 \geq 12$ or $y_2 < 7$ is true for $x < 3$. Note also that for $x \geq 3$ both $y_1 \geq 12$ and $y_2 < 7$ are incorrect. The table supports the conclusion of Example 9.

X	Y1	Y2
1	-2	1
2	-9	4
3	-16	7
4	-23	10
5	-30	13
6	-37	16
7	-44	19

Y1 ■5-7X

Union

Graph the solution set to the inequality

$$5 - 7x \geq 12 \qquad \text{or} \qquad 3x - 2 < 7.$$

Solution

First solve each of the simple inequalities:

$$5 - 7x - 5 \geq 12 - 5 \qquad \text{or} \qquad 3x - 2 + 2 < 7 + 2$$
$$-7x \geq 7 \qquad \text{or} \qquad 3x < 9$$
$$x \leq -1 \qquad \text{or} \qquad x < 3$$

The union of the two solution intervals is $(-\infty, 3)$. The graph is shown in Fig. 2.28.

Figure 2.28

> **Now do Exercises 33–40**

An inequality may be read from left to right or from right to left. Consider the inequality $1 < x$. If we read it in the usual way, we say, "1 is less than x." The meaning is clearer if we read the variable first. Reading from right to left, we say, "x is greater than 1."

Another notation is commonly used for the compound inequality

$$x > 1 \qquad \text{and} \qquad x < 13.$$

This compound inequality can also be written as

$$1 < x < 13.$$

Reading from left to right, we read $1 < x < 13$ as "1 is less than x is less than 13." The meaning of this inequality is clearer if we read the variable first and read the first inequality symbol from right to left. Reading the variable first, $1 < x < 13$ is read as "x is greater than 1 and less than 13." So x is between 1 and 13, and reading x first makes it clear.

CAUTION We write $a < x < b$ only if $a < b$, and we write $a > x > b$ only if $a > b$. Similar rules hold for $\leq$ and $\geq$. So $4 < x < 9$ and $-6 \geq x \geq -8$ are correct uses of this notation, but $5 < x < 2$ is not correct. Also, the inequalities should *not* point in opposite directions as in $5 < x > 7$.

E X A M P L E   10

Another notation

Solve the inequality and graph the solution set:

$$-2 \leq 2x - 3 < 7$$

Solution

This inequality could be written as the compound inequality

$$2x - 3 \geq -2 \qquad \text{and} \qquad 2x - 3 < 7.$$

‹ **Calculator Close-Up** ›

Do not use a table on your calculator as a method for solving an inequality. Use a table to check your algebraic solution and you will get a better understanding of inequalities.

However, there is no need to rewrite the inequality because we can solve it in its original form.

$$-2 + 3 \leq 2x - 3 + 3 < 7 + 3 \quad \text{Add 3 to each part.}$$

$$1 \leq 2x < 10$$

$$\frac{1}{2} \leq \frac{2x}{2} < \frac{10}{2} \qquad\qquad \text{Divide each part by 2.}$$

$$\frac{1}{2} \leq x < 5$$

The solution set is $\left[\frac{1}{2}, 5\right)$, and its graph is shown in Fig. 2.29.

Figure 2.29

> Now do Exercises 41–44

E X A M P L E   **11**

Solving a compound inequality

Solve the inequality $-1 < 3 - 2x < 9$ and graph the solution set.

Solution

$$-1 - 3 < 3 - 2x - 3 < 9 - 3 \quad \text{Subtract 3 from each part of the inequality.}$$

$$-4 < -2x < 6$$

$$2 > x > -3 \qquad\qquad \text{Divide each part by } -2 \text{ and reverse both inequality symbols.}$$

$$-3 < x < 2 \qquad\qquad \text{Rewrite the inequality with the smallest number on the left.}$$

The solution set is $(-3, 2)$, and its graph is shown in Fig. 2.30.

Figure 2.30

> Now do Exercises 45–52

‹ **Calculator Close-Up** ›

Let $y_1 = 3 - 2x$ and make a table. Scroll through the table to see that y_1 is between -1 and 9 when x is between -3 and 2. The table supports the conclusion of Example 11.

X	Y1
-3	9
-2	7
-1	5
0	3
1	1
2	-1
3	-3

Y1◻3-2X

‹3› Applications

When final exams are approaching, students are often interested in finding the final exam score that would give them a certain grade for a course.

E X A M P L E   **12**

Final exam scores

Fiana made a score of 76 on her midterm exam. For her to get a B in the course, the average of her midterm exam and final exam must be between 80 and 89 inclusive. What possible scores on the final exam would give Fiana a B in the course? The final exam has a maximum of 100 points possible.

‹ **Helpful Hint** ›

When you use two inequality symbols as in Example 12, they must both point in the same direction. In fact, we usually have them both point to the left so that the numbers increase in size from left to right.

Solution

Let x represent her final exam score. Between 80 and 89 inclusive means that an average between 80 and 89 as well as an average of exactly 80 or 89 will get a B. So the average of the two scores must be greater than or equal to 80 and less than or equal to 89.

$$80 \leq \frac{x + 76}{2} \leq 89$$

$$160 \leq x + 76 \leq 178 \quad \text{Multiply by 2.}$$

$$160 - 76 \leq x \leq 178 - 76 \quad \text{Subtract 76.}$$

$$84 \leq x \leq 102$$

Note that the maximum possible score on the final is 100. If Fiana scores between 84 and 100 inclusive, she will get a B in the course.

Now do Exercises 77–88

Warm-Ups ▼

Fill in the blank.

1. A _____ inequality consists of two simple inequalities connected by "and" or "or."
2. The compound inequality $x > 5$ ____ $x > 8$ is true only if both simple inequalities are true.
3. The compound inequality $x > 5$ ___ $x > 8$ is true if either simple inequality is true.
4. If $a < b < c$, then $a < b$ ____ $b < c$.
5. The solution set to $x > 3$ and $x < 9$ is the _____ of the intervals $(3, \infty)$ and $(-\infty, 9)$.
6. The solution set to $x > 3$ or $x < 9$ is the _____ of the intervals $(3, \infty)$ and $(-\infty, 9)$.

True or false?

7. $3 < 5$ and $3 \leq 10$
8. $3 < 5$ or $3 < 10$
9. $3 > 5$ and $3 < 10$
10. $3 \geq 5$ or $3 \leq 10$
11. $4 < 8$ and $4 > 2$
12. $4 < 8$ or $4 > 2$
13. $-3 < 0 < -2$
14. $(3, \infty) \cup [8, \infty) = [8, \infty)$
15. $(3, \infty) \cap [8, \infty) = [8, \infty)$
16. $(-2, \infty) \cap (-\infty, 9) = (-2, 9)$

Exercises 2.5

‹ **Study Tips** ›

- What's on the final exam? If your instructor thinks a problem is important enough for a test or quiz, it is probably important enough for the final exam. You should be thinking of the final exam all semester.
- Write all of the test and quiz questions on note cards, one to a card. To prepare for the final, shuffle the cards and try to answer the questions in a random order.

‹1› Compound Inequalities

Determine whether each compound inequality is true. See Examples 1 and 2.

1. $-6 < 5$ and $-6 > -3$
2. $4 \leq 4$ and $-4 \leq 0$
3. $1 < 5$ and $1 > -3$
4. $3 < 5$ or $0 < -3$
5. $6 < 5$ or $-4 > -3$
6. $4 \leq -4$ or $0 \leq 0$

Determine whether -4 satisfies each compound inequality. See Example 3.

7. $x < 5$ and $x > -3$

8. $x > -5$ and $x < 0$

9. $x < 5$ or $x > -3$

10. $x < -9$ or $x > 0$

11. $x - 3 \geq -7$ or $x + 1 > 1$

12. $2x \leq -8$ and $5x \leq 0$

13. $2x - 1 < -7$ or $-2x > 18$

14. $-3x > 0$ and $3x - 4 < 11$

⟨2⟩ Graphing the Solution Set

Graph the solution set to each compound inequality. See Examples 4–7.

15. $x > -1$ and $x < 4$

16. $x \leq 5$ and $x \geq 4$

17. $x \leq 3$ and $x \leq 0$

18. $x > 2$ and $x > 0$

19. $x \geq 2$ or $x \geq 5$

20. $x < -1$ or $x < 3$

21. $x \leq 6$ or $x > -2$

22. $x > -2$ and $x \leq 4$

23. $x \leq 6$ and $x > 9$

24. $x < 7$ or $x > 0$

25. $x \leq 6$ or $x > 9$

26. $x \geq 4$ and $x \leq -4$

27. $x \geq 6$ and $x \leq 1$

28. $x > 3$ or $x < -3$

Solve each compound inequality. Write the solution set using interval notation and graph it. See Examples 8 and 9.

29. $x - 3 > 7$ or $3 - x > 2$

30. $x - 5 > 6$ or $2 - x > 4$

31. $3 < x$ and $1 + x > 10$

32. $-0.3x < 9$ and $0.2x > 2$

33. $\frac{1}{2}x > 5$ or $-\frac{1}{3}x < 2$

34. $5 < x$ or $3 - \frac{1}{2}x < 7$

35. $2x - 3 \leq 5$ and $x - 1 > 0$

36. $\frac{3}{4}x < 9$ and $-\frac{1}{3}x \leq -15$

37. $\frac{1}{2}x - \frac{1}{3} \geq -\frac{1}{6}$ or $\frac{2}{7}x \leq \frac{1}{10}$

38. $\frac{1}{4}x - \frac{1}{3} > -\frac{1}{5}$ and $\frac{1}{2}x < 2$

39. $0.5x < 2$ and $-0.6x < -3$

40. $0.3x < 0.6$ or $0.05x > -4$

Solve each compound inequality. Write the solution set in interval notation and graph it. See Examples 10 and 11.

41. $-3 < x + 1 < 3$

42. $-4 \leq x - 4 \leq 1$

43. $5 < 2x - 3 < 11$

44. $-2 < 3x + 1 < 10$

45. $-1 < 5 - 3x \le 14$

46. $-1 \le 3 - 2x < 11$

47. $-3 < \dfrac{3m + 1}{2} \le 5$

48. $0 \le \dfrac{3 - 2x}{2} < 5$

 49. $-2 < \dfrac{1 - 3x}{-2} < 7$

50. $-3 < \dfrac{2x - 1}{3} < 7$

51. $3 \le 3 - 5(x - 3) \le 8$

52. $2 \le 4 - \dfrac{1}{2}(x - 8) \le 10$

Write each union or intersection of intervals as a single interval if possible.

53. $(2, \infty) \cup (4, \infty)$ **54.** $(-3, \infty) \cup (-6, \infty)$

55. $(-\infty, 5) \cap (-\infty, 9)$ **56.** $(-\infty, -2) \cap (-\infty, 1)$

57. $(-\infty, 4] \cap [2, \infty)$ **58.** $(-\infty, 8) \cap [3, \infty)$

59. $(-\infty, 5) \cup [-3, \infty)$ **60.** $(-\infty, -2] \cup (2, \infty)$

61. $(3, \infty) \cap (-\infty, 3]$ **62.** $[-4, \infty) \cap (-\infty, -6]$

 63. $(3, 5) \cap [4, 8)$ **64.** $[-2, 4] \cap (0, 9]$

65. $[1, 4) \cup (2, 6]$ **66.** $[1, 3) \cup (0, 5)$

Write either a simple or a compound inequality that has the given graph as its solution set.

67.

68.

69.

70.

71.

72.

73.

74.

75.

76.

⟨3⟩ **Applications**

Solve each problem by using a compound inequality. See Example 12.

77. *Aiming for a C.* Professor Johnson gives only a midterm exam and a final exam. The semester average is computed by taking $\frac{1}{3}$ of the midterm exam score plus $\frac{2}{3}$ of the final exam score. To get a C, Beth must have a semester average between 70 and 79 inclusive. If Beth scored only 64 on the midterm, then for what range of scores on the final exam would Beth get a C?

78. *Two tests only.* Professor Davis counts his midterm as $\frac{2}{3}$ of the grade, and his final as $\frac{1}{3}$ of the grade. Jason scored only 64 on the midterm. What range of scores on the final exam would put Jason's average between 70 and 79 inclusive? Assume that the maximum possible score on the final is 100.

79. *Car costs.* The function $C = 0.0004x + 20$ gives the cost in cents per mile for operating a company car and the function $V = 20{,}000 - 0.2x$ gives the value of the car in dollars, where x is the number of miles on the odometer. A car is replaced if the operating cost is greater than 40 cents per mile *and* the value is less than \$12,000. For what values of x is a car replaced? Use interval notation.

80. *Changing plans.* If the company in Exercise 79 replaces any car for which the operating cost is greater than 40 cents per mile *or* the value is less than $12,000, then for what values of x is a car replaced? Use interval notation.

81. *Supply and demand.* The function $S = 20 + 0.1x$ gives the amount of oil in millions of barrels per day that will be supplied to a small country and the function $D = 30 - 0.5x$ gives the demand for oil in millions of barrels per day, where x is the price of oil in dollars per barrel. The president worries if the supply is less than 22 million barrels per day or if demand is less than 15 million barrels per day. For what values of x does the president worry? Use interval notation.

82. *Predicting recession.* The country of Exercise 81 will be in recession if the supply of oil is greater than 23 million barrels per day *and* the demand is less than 14 million barrels per day. For what values of x will the country be in recession? Use interval notation.

83. *Keep on truckin'.* Abdul is shopping for a new truck in a city with an 8% sales tax. There is also an $84 title and license fee to pay. He wants to get a good truck and plans to spend at least $12,000 but no more than $15,000. What is the price range for the truck?

84. *Selling-price range.* Renee wants to sell her car through a broker who charges a commission of 10% of the selling price. The book value of the car is $14,900, but Renee still owes $13,104 on it. Although the car is in only fair condition and will not sell for more than the book value, Renee must get enough to at least pay off the loan. What is the range of the selling price?

85. *Hazardous to her health.* Trying to break her smoking habit, Jane calculates that she smokes only three full cigarettes a day, one after each meal. The rest of the time she smokes on the run and smokes only half of the cigarette. She estimates that she smokes the equivalent of 5 to 12 cigarettes per day. How many times a day does she light up on the run?

86. *Possible width.* The length of a rectangle is 20 meters longer than the width. The perimeter must be between 80 and 100 meters. What are the possible values for the width of the rectangle?

87. *Higher education.* The functions

$$B = 16.45n + 1062.45$$
and
$$M = 7.79n + 326.82$$

can be used to approximate the number of bachelor's and master's degrees in thousands, respectively, awarded per

year, n years after 1990 (National Center for Educational Statistics, www.nces.ed.gov).

a) How many bachelor's degrees were awarded in 2000?

b) In what year will the number of bachelor's degrees that are awarded reach 1.4 million?

c) What is the first year in which both B is greater than 1.4 million and M is greater than 0.55 million?

d) What is the first year in which either B is greater than 1.4 million or M is greater than 0.55 million?

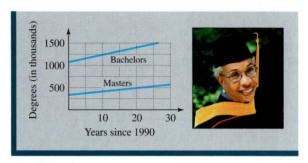

Figure for Exercise 87

88. *Senior citizens.* The number of senior citizens (65 and over) in the United States in millions n years after 1990 can be estimated by using the function

$$s = 0.38n + 31.2$$

(U.S. Bureau of the Census, www.census.gov). The percentage of senior citizens living below the poverty level n years after 1990 can be estimated by using the function

$$p = -0.25n + 12.2.$$

a) How many senior citizens were there in 2000?

b) In what year will the percentage of seniors living below the poverty level reach 7%?

c) What is the first year in which we can expect both the number of seniors to be greater than 40 million and fewer than 7% living below the poverty level?

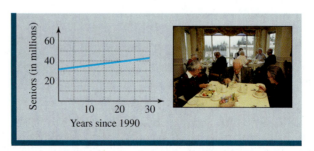

Figure for Exercise 88

Getting More Involved

89. *Discussion*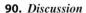

If $-x$ is between a and b, then what can you say about x?

90. *Discussion*

For which of the inequalities is the notation used correctly?

a) $-2 \leq x < 3$ **b)** $-4 \geq x < 7$ **c)** $-1 \leq x > 0$
d) $6 < x \leq -8$
e) $5 \geq x \geq -9$

91. *Discussion*

In each case, write the resulting set of numbers in interval notation. Explain your answers.

a) Every number in $(3, 8)$ is multiplied by 4.
b) Every number in $[-2, 4)$ is multiplied by -5.

c) Three is added to every number in $(-3, 6)$.
d) Every number in $[3, 9]$ is divided by -3.

92. *Discussion*

Write the solution set using interval notation for each of these inequalities in terms of s and t. State any restrictions on s and t. For what values of s and t is the solution set empty?

a) $x > s$ and $x < t$

b) $x > s$ and $x > t$

2.6   Absolute Value Equations and Inequalities

In This Section

⟨1⟩ **Absolute Value Equations**
⟨2⟩ **Absolute Value Inequalities**
⟨3⟩ **All or Nothing**
⟨4⟩ **Applications**

In Chapter 1, we learned that absolute value measures the distance of a number from 0 on the number line. In this section we will learn to solve equations and inequalities involving absolute value.

⟨1⟩ Absolute Value Equations

The equation $|x| = 5$ means that x is a number whose distance from 0 on the number line is 5 units. Since both -5 and 5 are 5 units away from 0 as shown in Fig. 2.31, the solution set is $\{-5, 5\}$. So $|x| = 5$ is equivalent to $x = 5$ or $x = -5$.

⟨ Helpful Hint ⟩

Some students grow up believing that the only way to solve an equation is to "do the same thing to each side." Then along come absolute value equations. For an absolute value equation, we write an equivalent compound equation that is not obtained by "doing the same thing to each side."

Figure 2.31

The equation $|x| = 0$ is equivalent to the equation $x = 0$ because 0 is the only number whose distance from 0 is zero. The solution set to $|x| = 0$ is $\{0\}$.

The equation $|x| = -7$ is inconsistent because absolute value measures distance, and distance is never negative. So the solution set is empty. These ideas are summarized as follows.

Summary of Basic Absolute Value Equations

Absolute Value Equation	Equivalent Equation	Solution Set
$\lvert x \rvert = k \; (k > 0)$	$x = k$ or $x = -k$	$\{k, -k\}$
$\lvert x \rvert = 0$	$x = 0$	$\{0\}$
$\lvert x \rvert = k \; (k < 0)$		$\varnothing$

We can use these ideas to solve more complicated absolute value equations.

EXAMPLE 1

Absolute value equal to a positive number

Solve each equation.

a) $\lvert x - 7 \rvert = 2$
b) $\lvert 3x - 5 \rvert = 7$

Solution

a) First rewrite $\lvert x - 7 \rvert = 2$ without absolute value:

$$x - 7 = 2 \quad \text{or} \quad x - 7 = -2 \quad \text{Equivalent equation}$$
$$x = 9 \quad \text{or} \quad x = 5$$

The solution set is $\{5, 9\}$. The distance from 5 to 7 or from 9 to 7 is 2 units.

b) First rewrite $\lvert 3x - 5 \rvert = 7$ without absolute value:

$$3x - 5 = 7 \quad \text{or} \quad 3x - 5 = -7 \quad \text{Equivalent equation}$$
$$3x = 12 \quad \text{or} \quad 3x = -2$$
$$x = 4 \quad \text{or} \quad x = -\frac{2}{3}$$

The solution set is $\left\{-\frac{2}{3}, 4\right\}$.

Now do Exercises 1–8

< **Calculator Close-Up** >

Use Y= to set $y_1 = \text{abs}(x - 7)$. Make a table to see that y_1 has value 2 when $x = 5$ or $x = 9$. The table supports the conclusion of Example 1(a).

X	Y1
5	2
6	1
7	0
8	1
9	2
10	3
11	4

Y1■abs(X-7)

EXAMPLE 2

Absolute value equal to zero

Solve $\lvert 2(x - 6) + 7 \rvert = 0$.

Solution

Since 0 is the only number whose absolute value is 0, the expression within the absolute value bars must be 0.

$$2(x - 6) + 7 = 0 \quad \text{Equivalent equation}$$
$$2x - 12 + 7 = 0$$
$$2x - 5 = 0$$
$$2x = 5$$
$$x = \frac{5}{2}$$

The solution set is $\left\{\frac{5}{2}\right\}$.

Now do Exercises 9–10

< **Helpful Hint** >

Examples 1, 2, and 3 show the three basic types of absolute value equations—absolute value equal to a positive number, zero, or a negative number. These equations have 2, 1, and no solutions, respectively.

EXAMPLE   3

Absolute value equal to a negative number
Solve each equation.

a) $|x - 9| = -6$ b) $-5|3x - 7| + 4 = 14$

Solution

a) The equation indicates that $|x - 9| = -6$. However, the absolute value of any quantity is greater than or equal to zero. So there is no solution to the equation.

b) First subtract 4 from each side to isolate the absolute value expression:

$$-5|3x - 7| + 4 = 14 \quad \text{Original equation}$$
$$-5|3x - 7| = 10 \quad \text{Subtract 4 from each side.}$$
$$|3x - 7| = -2 \quad \text{Divide each side by } -5.$$

There is no solution because no quantity has a negative absolute value.

> **Now do Exercises 11–28**

The equation in Example 4 has an absolute value on both sides.

EXAMPLE   4

Absolute value on both sides
Solve $|2x - 1| = |x + 3|$.

Solution

Two quantities have the same absolute value only if they are equal or opposites. So we can write an equivalent compound equation:

$$2x - 1 = x + 3 \quad \text{or} \quad 2x - 1 = -(x + 3)$$
$$x - 1 = 3 \quad \text{or} \quad 2x - 1 = -x - 3$$
$$x = 4 \quad \text{or} \quad 3x = -2$$
$$x = 4 \quad \text{or} \quad x = -\frac{2}{3}$$

Check 4 and $-\frac{2}{3}$ in the original equation. The solution set is $\left\{-\frac{2}{3}, 4\right\}$.

> **Now do Exercises 29–34**

‹2› Absolute Value Inequalities

Since absolute value measures distance from 0 on the number line, $|x| > 5$ indicates that x is more than five units from 0. Any number on the number line to the right of 5 or to the left of -5 is more than five units from 0. So $|x| > 5$ is equivalent to

$$x > 5 \quad \text{or} \quad x < -5.$$

The solution set to this inequality is the union of the solution sets to the two simple inequalities. The solution set is $(-\infty, -5) \cup (5, \infty)$. The graph of $|x| > 5$ is shown in Fig. 2.32.

Figure 2.32

The inequality $|x| \leq 3$ indicates that x is less than or equal to three units from 0. Any number between -3 and 3 inclusive satisfies that condition. So $|x| \leq 3$ is equivalent to

$$-3 \leq x \leq 3.$$

The graph of $|x| \leq 3$ is shown in Fig. 2.33. These examples illustrate the basic types of absolute value inequalities.

Figure 2.33

Summary of Basic Absolute Value Inequalities ($k > 0$)

Absolute Value Inequality	Equivalent Inequality	Solution Set	Graph of Solution Set
$|x| > k$	$x > k$ or $x < -k$	$(-\infty, -k) \cup (k, \infty)$	
$|x| \geq k$	$x \geq k$ or $x \leq -k$	$(-\infty, -k] \cup [k, \infty)$	
$|x| < k$	$-k < x < k$	$(-k, k)$	
$|x| \leq k$	$-k \leq x \leq k$	$[-k, k]$	

We can solve more complicated inequalities in the same manner as simple ones.

E X A M P L E   5

⟨ **Calculator Close-Up** ⟩

Use Y= to set $y_1 = \text{abs}(x - 9)$. Make a table to see that $y_1 < 2$ when x is between 7 and 11.

Absolute value inequality

Solve $|x - 9| < 2$ and graph the solution set.

Solution

Because $|x| < k$ is equivalent to $-k < x < k$, we can rewrite $|x - 9| < 2$ as follows:

$$-2 < x - 9 < 2$$
$$-2 + 9 < x - 9 + 9 < 2 + 9 \quad \text{Add 9 to each part of the inequality.}$$
$$7 < x < 11$$

The graph of the solution set $(7, 11)$ is shown in Fig. 2.34. Note that the graph consists of all real numbers that are within two units of 9.

Figure 2.34

> **Now do Exercises 35–36**

EXAMPLE   **6**

Absolute value inequality

Solve $|3x + 5| > 2$ and graph the solution set.

Solution

$$3x + 5 > 2 \qquad \text{or} \qquad 3x + 5 < -2 \qquad \text{Equivalent compound inequality}$$
$$3x > -3 \qquad \text{or} \qquad 3x < -7$$
$$x > -1 \qquad \text{or} \qquad x < -\frac{7}{3}$$

The solution set is $\left(-\infty, -\frac{7}{3}\right) \cup (-1, \infty)$, and its graph is shown in Fig. 2.35.

Figure 2.35

Now do Exercises 37–38

EXAMPLE   **7**

Absolute value inequality

Solve $|5 - 3x| \le 6$ and graph the solution set.

‹ Calculator Close-Up ›

Use Y= to set $y_1 = \text{abs}(5 - 3x)$. The table supports the conclusion that $y_1 \le 6$ when x is between $-\frac{1}{3}$ and $\frac{11}{3}$ even though $-\frac{1}{3}$ and $\frac{11}{3}$ do not appear in the table. For more accuracy, make a table in which the change in x is $\frac{1}{3}$.

X	Y1
-1	8
0	5
1	2
2	1
3	4
4	7
5	10

Y1◻abs(5-3X)

Solution

$$-6 \le 5 - 3x \le 6 \qquad \text{Equivalent inequality}$$
$$-11 \le -3x \le 1 \qquad \text{Subtract 5 from each part.}$$
$$\frac{11}{3} \ge x \ge -\frac{1}{3} \qquad \text{Divide by } -3 \text{ and reverse each inequality symbol.}$$
$$-\frac{1}{3} \le x \le \frac{11}{3} \qquad \text{Write } -\frac{1}{3} \text{ on the left because it is smaller than } \frac{11}{3}.$$

The solution set is $\left[-\frac{1}{3}, \frac{11}{3}\right]$ and its graph is shown in Fig. 2.36.

Figure 2.36

Now do Exercises 39–42

‹3› All or Nothing

The solution to an absolute value inequality can be all real numbers or no real numbers. To solve such inequalities you must remember that the absolute value of any real number is greater than or equal to zero.

EXAMPLE 8

All real numbers

Solve $3 + |7 - 2x| \geq 3$.

Solution

Subtract 3 from each side to isolate the absolute value expression.

$$|7 - 2x| \geq 0$$

Because the absolute value of any real number is greater than or equal to 0, the solution set is R, the set of all real numbers.

> Now do Exercises 67–72

EXAMPLE 9

No real numbers

Solve $|5x - 12| < -2$.

Solution

We write an equivalent inequality only when the value of k is positive. With -2 on the right-hand side, we do not write an equivalent inequality. Since the absolute value of any quantity is greater than or equal to 0, no value for x can make this absolute value less than -2. The solution set is $\varnothing$, the empty set.

> Now do Exercises 73–76

⟨4⟩ Applications

A simple example will show how absolute value inequalities can be used in applications.

EXAMPLE 10

Controlling water temperature

The water temperature in a certain manufacturing process must be kept at 143°F. The computer is programmed to shut down the process if the water temperature is more than 7° away from what it is supposed to be. For what temperature readings is the process shut down?

Solution

If we let x represent the water temperature, then $x - 143$ represents the difference between the actual temperature and the desired temperature. The quantity $x - 143$ could be positive or negative. The process is shut down if the absolute value of $x - 143$ is greater than 7.

$$|x - 143| > 7$$

$$x - 143 > 7 \qquad \text{or} \qquad x - 143 < -7$$

$$x > 150 \qquad \text{or} \qquad x < 136$$

The process is shut down for temperatures greater than 150°F or less than 136°F.

> Now do Exercises 85–92

Warm-Ups ▼

Fill in the blank.

1. The _____ of x is the distance from x to 0 on the number line.
2. The equation $|x| = 4$ has ____ solutions.
3. The equation $|x| = -4$ has ____ solutions.
4. The equation $|x| = 0$ has ____ solution.
5. ____ real numbers satisfy $|x| \geq 0$.
6. ____ real numbers satisfy $|x| < 0$.
7. The solution set to $|x| < 3$ is _____.
8. The inequality $|x| > 3$ is _____ to $x > 3$ or $x < -3$.

True or false?

9. If $|x| = 2$, then $x = 2$ or $x = -2$.
10. If $|x - 1| = 7$, then $x - 1 = 7$ or $x + 1 = 7$.
11. If $|x| > 5$, then $x > 5$ or $x < -5$.
12. If $|x| = -4$, then $x = -4$.
13. If $|2x - 8| = 0$, then $x = 4$.
14. If $-3 < x < 3$, then $|x| < 3$.
15. If $5 < x < 9$, then $x > 5$ and $x < 9$.
16. If x is any real number, then $|x| \geq 0$.
17. If $|x| + 1 = 5$, then $x + 1 = 5$ or $x + 1 = -5$.
18. If $|3x - 99| \leq 0$, then $x = 33$.

Exercises

2.6

‹ Study Tips ›

- When studying for an exam, start by working the exercises in the Chapter Review. They are grouped by section so that you can go back and review any topics that you have trouble with.
- Never leave an exam early. Most papers turned in early contain careless errors that could be found and corrected. Every point counts.

‹1› Absolute Value Equations

Solve each absolute value equation. See Examples 1–3. See the Summary of Basic Absolute Value Equations box on page 126.

1. $|a| = 5$
2. $|x| = 2$
3. $|x - 3| = 1$
4. $|x - 5| = 2$
5. $|3 - x| = 6$
6. $|7 - x| = 6$
7. $|3x - 4| = 12$
8. $\left| 3 - \frac{3}{4}x \right| = \frac{1}{4}$
9. $\left| \frac{2}{3}x - 8 \right| = 0$
10. $|5 - 0.1x| = 0$
11. $|5x + 2| = -3$
12. $|7(x - 6)| = -3$

13. $|6 - 0.2x| = 10$
14. $|2(a + 3)| = 15$
15. $|2(x - 4) + 3| = 5$
16. $|3(x - 2) + 7| = 6$
17. $|7.3x - 5.26| = 4.215$
18. $|5.74 - 2.17x| = 10.28$

Solve each absolute value equation. See Examples 3 and 4.

19. $3 + |x| = 5$
20. $|x| - 10 = -3$
21. $3|a| - 6 = 21$
22. $-2|b| + 3 = -9$
23. $3|w + 1| - 2 = 7$
24. $2|y - 3| - 11 = -1$

25. $2 - |x + 3| = -6$

 26. $4 - 3|x - 2| = -8$

27. $5 - \dfrac{|3 - 2x|}{3} = 4$

28. $3 - \dfrac{1}{2}\left|\dfrac{1}{2}x - 4\right| = 2$

29. $|x - 5| = |2x + 1|$

30. $|w - 6| = |3 - 2w|$

31. $\left|\dfrac{5}{2} - x\right| = \left|2 - \dfrac{x}{2}\right|$

32. $\left|x - \dfrac{1}{4}\right| = \left|\dfrac{1}{2}x - \dfrac{3}{4}\right|$

33. $|x - 3| = |3 - x|$

34. $|a - 6| = |6 - a|$

⟨2⟩ Absolute Value Inequalities

Write an absolute value inequality whose solution set is shown by the graph. See Examples 5–7. See the Summary of Basic Absolute Value Inequalities box on page 128.

35.

36.
```
 -6 -5 -4 -3 -2 -1  0  1  2  3  4  5  6
```

37.
```
 -6 -5 -4 -3 -2 -1  0  1  2  3  4  5  6
```

38.
```
 -8 -7 -6 -5 -4 -3 -2 -1  0  1  2  3  4  5  6  7  8
```

39.
```
 -6 -5 -4 -3 -2 -1  0  1  2  3  4  5  6
```

40.
```
 -6 -5 -4 -3 -2 -1  0  1  2  3  4  5  6
```

41.
```
 -6 -5 -4 -3 -2 -1  0  1  2  3  4  5  6
```

42.

Determine whether each absolute value inequality is equivalent to the inequality following it. See Examples 5–7.

43. $|x| < 3,\ x < 3$

44. $|x| > 3,\ x > 3$

45. $|x - 3| > 1,\ x - 3 > 1$ or $x - 3 < -1$

46. $|x - 3| \leq 1,\ -1 \leq x - 3 \leq 1$

47. $|x - 3| \geq 1,\ x - 3 \geq 1$ or $x - 3 \leq 1$

48. $|x - 3| > 0,\ x - 3 > 0$

Solve each absolute value inequality and graph the solution set. See Examples 5–7.

49. $|x| > 6$

50. $|w| \geq 3$

51. $|t| \leq 2$

52. $|b| < 4$

53. $|2a| < 6$

54. $|3x| < 21$

55. $|x - 2| \geq 3$

56. $|x - 5| \geq 1$

57. $3|a| - 3 > 21$

58. $-2|b| + 5 < -9$

59. $3|w + 1| - 5 \leq 7$

60. $2|y - 3| - 7 \leq -1$

61. $\dfrac{1}{5}|2x - 4| < 1$

62. $\dfrac{1}{3}|2x - 1| < 1$

63. $-2|5 - x| \geq -14$

64. $-3|6 - x| \geq -3$

65. $2|3 - 2x| - 6 \geq 18$

66. $2|5 - 2x| - 15 \geq 5$

⟨3⟩ All or Nothing

Solve each absolute value inequality and graph the solution set. See Examples 8 and 9.

67. $|x| > 0$

68. $|x - 2| > 0$

69. $|x| \leq 0$

70. $|x| < 0$
71. $|x - 5| \geq 0$

72. $|3x - 7| \geq -3$

73. $-2|3x - 7| > 6$
74. $-3|7x - 42| > 18$
75. $|2x + 3| + 6 > 0$

76. $|5 - x| + 5 > 5$

Solve each inequality. Write the solution set using interval notation.

77. $1 < |x + 2|$
78. $5 \geq |x - 4|$
79. $5 > |x| + 1$
80. $4 \leq |x| - 6$
81. $3 - 5|x| > -2$
82. $1 - 2|x| < -7$
83. $|5.67x - 3.124| < 1.68$
84. $|4.67 - 3.2x| \geq 1.43$

⟨4⟩ Applications

Solve each problem by using an absolute value equation or inequality. See Example 10.

85. *Famous battles.* In the Hundred Years' War, Henry V defeated a French army in the battle of Agincourt and Joan of Arc defeated an English army in the battle of Orleans (*The*

Doubleday Almanac). Suppose you know only that these two famous battles were 14 years apart and that the battle of Agincourt occurred in 1415. Use an absolute value equation to find the possibilities for the year in which the battle of Orleans occurred.

86. *World records.* In July 1985 Steve Cram of Great Britain set a world record of 3 minutes 29.67 seconds for the 1500-meter race and a world record of 3 minutes 46.31 seconds for the 1-mile race (*The Doubleday Almanac*). Suppose you know only that these two events occurred 11 days apart and that the 1500-meter record was set on July 16. Use an absolute value equation to find the possible dates for the 1-mile record run.

87. *Weight difference.* Research at a major university has shown that identical twins generally differ by less than 6 pounds in body weight. If Kim weighs 127 pounds, then in what range is the weight of her identical twin sister Kathy?

88. *Intelligence quotient.* Jude's IQ score is more than 15 points away from Sherry's. If Sherry scored 110, then in what range is Jude's score?

89. *Approval rating.* According to a Fox News survey, the presidential approval rating is 39% plus or minus 5 percentage points.

  a) In what range is the percentage of people who approve of the president?

  b) Let *x* represent the actual percentage of people who approve of the president. Write an absolute value inequality for *x*.

90. *Time of death.* According to the coroner the time of death was 3 A.M. plus or minus 2 hours.

  a) In what range is the actual time of death?

  b) Let *x* represent the actual time of death. Write an absolute value inequality for *x*.

91. *Unidentified flying objects.* The function

$$S = -16t^2 + v_0 t + s_0$$

gives the height in feet above the earth at time *t* seconds for an object projected into the air with an initial velocity of v_0 feet per second (ft/sec) from an initial height of s_0 feet. Two balls are tossed into the air simultaneously, one from the ground at 50 ft/sec and one from a height of 10 feet at 40 ft/sec. See the accompanying graph.

  a) Use the graph to estimate the time at which the balls are at the same height.

  b) Find the time from part (a) algebraically.

c) For what values of t will their heights above the ground differ by less than 5 feet (while they are both in the air)?

Figure for Exercise 91

92. *Playing catch.* A circus clown at the top of a 60-foot platform is playing catch with another clown on the ground. The clown on the platform drops a ball at the same time as the one on the ground tosses a ball upward at 80 ft/sec. For what length of time is the distance between the balls less than or equal to 10 feet? (*Hint:* Use the formula given in Exercise 91. The initial velocity of a ball that is dropped is 0 ft/sec.) See the accompanying figure.

Figure for Exercise 92

Getting More Involved

93. *Discussion*

For which real numbers m and n is each equation satisfied?

a) $|m - n| = |n - m|$

b) $|mn| = |m| \cdot |n|$

c) $\left| \dfrac{m}{n} \right| = \dfrac{|m|}{|n|}$

94. *Exploration*

a) Evaluate $|m + n|$ and $|m| + |n|$ for

i) $m = 3$ and $n = 5$

ii) $m = -3$ and $n = 5$

iii) $m = 3$ and $n = -5$

iv) $m = -3$ and $n = -5$

b) What can you conclude about the relationship between $|m + n|$ and $|m| + |n|$?

2 Wrap-Up

Summary

Equations		Examples
Solution set	The set of all numbers that satisfy an equation (or inequality)	$x + 2 = 6$ has solution set $\{4\}$.
Equivalent equations	Equations with the same solution set	$2x + 1 = 5$ $2x = 4$
Properties of equality	We may perform the same operation $(+, -, \cdot, \div)$ with the same real number on each side of an equation without changing the solution set (excluding multiplication and division by 0).	$x = 4$ $x + 1 = 5$ $x - 1 = 3$ $2x = 8$ $\dfrac{x}{2} = 2$
Identity	An equation that is satisfied by every number for which both sides are defined	$x + x = 2x$
Conditional equation	An equation whose solution set contains at least one real number but is not an identity	$5x - 10 = 0$
Inconsistent equation	An equation whose solution set is $\varnothing$	$x = x + 1$
Linear equation in one variable	An equation of the form $ax = b$ with $a \neq 0$ or an equation that can be rewritten in this form	$3x + 8 = 0$ $5x - 1 = 2x - 9$
Strategy for solving a linear equation	1. If fractions are present, multiply each side by the LCD to eliminate the fractions. 2. Use the distributive property to remove parentheses. 3. Combine any like terms. 4. Use the addition property of equality to get all variables on one side and numbers on the other side. 5. Use the multiplication property of equality to get a single variable on one side. 6. Check by replacing the variable in the original equation with your solution.	
Strategy for solving word problems	1. Read the problem until you understand the problem. 2. If possible, draw a diagram to illustrate the problem. 3. Choose a variable and write down what it represents. 4. Represent any other unknowns in terms of that variable.	

5. Write an equation that models the situation.
6. Solve the equation.
7. Be sure that your solution answers the question posed in the original problem.
8. Check your answer by using it to solve the original problem (not the equation).

Inequalities   Examples

Linear inequality in one variable	A linear inequality has the form $ax < b$ with $a \neq 0$ or can be rewritten in this form. In place of $<$ we can use $\leq$, $>$ or $\geq$.	$2x + 9 < 0$ $x - 2 \geq 7$ $-3x - 1 \geq 2x + 5$
Properties of inequality	We may perform the same operation $(+, -, \cdot, \div)$ on each side of an inequality just as we do in solving equations, with one exception: When multiplying or dividing by a negative number, the inequality symbol is reversed.	$-3x > 6$ $x < -2$
Trichotomy property	For any two real numbers a and b, exactly one of the following statements is true: $a < b$, $a = b$, or $a > b$.	If w is not greater than 7, then $w \leq 7$.
Compound inequality	Two simple inequalities connected with the word "and" or "or" *And* corresponds to *intersection.* *Or* corresponds to *union.*	$x > 1$ and $x < 5$ $x > 3$ or $x < 1$

Absolute Value

	Absolute Value Equation	**Equivalent Equation**	**Solution Set**
Basic absolute value equations	$\|x\| = k \quad (k > 0)$ $\|x\| = 0$ $\|x\| = k \quad (k < 0)$	$x = k$ or $x = -k$ $x = 0$	$\{k, -k\}$ $\{0\}$ $\varnothing$

	Absolute Value Inequality	**Equivalent Inequality**	**Solution Set**	**Graph of Solution Set**
Basic absolute value inequalities $(k > 0)$	$\|x\| > k$	$x > k$ or $x < -k$	$(-\infty, -k) \cup (k, \infty)$	
	$\|x\| \geq k$	$x \geq k$ or $x \leq -k$	$(-\infty, -k] \cup [k, \infty)$	
	$\|x\| < k$	$-k < x < k$	$(-k, k)$	
	$\|x\| \leq k$	$-k \leq x \leq k$	$[-k, k]$	

Enriching Your Mathematical Word Power

Fill in the blank.

1. A(n) _____ is a sentence that expresses the equality of two algebraic expressions.

2. A _____ equation has the form $ax = b$ with $a \neq 0$.

3. A(n) _____ is satisfied by all real numbers for which both sides are defined.

4. A _____ equation has at least one solution but is not an identity.

5. A(n) _____ equation has no solutions.

6. Equations that have the same solution are _____ equations.

7. An equation involving two or more variables is a _____ equation or _____.

8. A _____ is a rule by which the value of one variable is determined from the value(s) of another variable(s).

9. Motion at a constant rate is _____ motion.

10. The smallest common multiple of some denominators is the _____ common denominator.

11. A statement that uses $<$, $\leq$, $>$, or $\geq$ is an _____.

12. Inequalities that have the same solution set are _____.

13. A _____ inequality consists of two simple inequalities joined with "and" or "or."

Review Exercises

2.1 Linear Equations in One Variable

Solve each equation.

1. $2x - 7 = 9$

2. $5x - 7 = 38$

3. $11 = 5 - 4x$

4. $-8 = 7 - 3x$

5. $x - 6 - (x - 6) = 0$

6. $x - 6 - 2(x - 3) = 0$

7. $2(x - 3) - 5 = 5 - (3 - 2x)$

8. $2(x - 4) + 5 = -(3 - 2x)$

9. $\dfrac{3}{17}x = 0$

10. $-\dfrac{3}{8}x = \dfrac{1}{2}$

11. $\dfrac{1}{4}x - \dfrac{1}{5} = \dfrac{1}{5}x + \dfrac{4}{5}$

12. $\dfrac{1}{2}x - 1 = \dfrac{1}{3}x$

13. $\dfrac{t}{2} - \dfrac{t - 2}{3} = \dfrac{3}{2}$

14. $\dfrac{y + 1}{4} - \dfrac{y - 1}{6} = y + 5$

15. $1 - 0.4(x - 4) + 0.6(x - 7) = -0.6$

16. $0.04x - 0.06(x - 8) = 0.1x$

2.2 Formulas and Functions

Solve each equation for x.

17. $ax + b = 0$

18. $mx + c = d$

19. $ax + 2 = cx$

20. $mx = 3 - x$

21. $mwx = P$

22. $xyz = 2$

23. $\dfrac{x}{2} + \dfrac{a}{6} = \dfrac{x}{3}$

24. $\dfrac{x}{4} - \dfrac{x}{3} = \dfrac{a}{2}$

Write y in terms of x.

25. $3x - 2y = -6$

26. $4x - 3y + 9 = 0$

27. $y - 2 = -\dfrac{1}{3}(x - 6)$

28. $y + 6 = \dfrac{1}{2}(x - 4)$

29. $\dfrac{1}{2}x - \dfrac{1}{4}y = 5$

30. $-\dfrac{x}{3} + \dfrac{y}{2} = \dfrac{5}{8}$

2.3 Applications

Solve each problem.

31. *Legal pad.* If the perimeter of a legal-size note pad is 45 inches and the pad is 5.5 inches longer than it is wide, then what are its length and width?

32. *Area of a trapezoid.* The height of a trapezoid is 5 feet, and the upper base is 2 feet shorter than the lower base. If the area of the trapezoid is 45 square feet, then how long is the lower base?

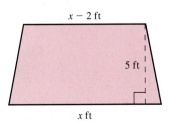

$x - 2$ ft

5 ft

x ft

Figure for Exercise 32

33. *Saving for retirement.* Roy makes $8000 more per year than his wife. Roy saves 10% of his income for retirement, and his wife saves 8%. Together they save $5660 per year. How much does each make?

34. *Charitable contributions.* Duane makes $1000 less per year than his wife. Duane gives 5% of his income to charity, and his wife gives 10% of her income to charity. Together they contribute $2500 to charity. How much does each make?

35. *Dealer discounts.* Sarah is buying a car for $7600. The dealer gave her a 20% discount off the list price. What was the list price?

36. *Gold sale.* At 25% off, a jeweler is selling a gold chain for $465. What was the list price?

37. *Nickels and dimes.* Rebecca has 15 coins consisting of nickels and dimes. The total value of the coins is $0.95. How many of each does she have?

38. *Nickels, dimes, and quarters.* Camille has 19 coins consisting of nickels, dimes, and quarters. The value of the coins is $1.60. If she has six times as many nickels as quarters, then how many of each does she have?

39. *Tour de desert.* On a recent bicycle trip across the desert Barbara rode for 5 hours. Her bicycle then developed mechanical difficulties, and she walked the bicycle for 3 hours to the nearest town. Altogether, she covered 85 miles. If she rides 9 miles per hour (mph) faster than she walks, then how far did she walk?

40. *Motor city.* Delmas flew to Detroit in 90 minutes and drove his new car back home in 6 hours. If he drove 150 mph slower than he flew, then how fast did he fly?

Speed to Detroit = x mph

Speed from Detroit = $x - 150$ mph

Figure for Exercise 40

2.4 Inequalities

Solve each inequality. State the solution set using interval notation and graph it.

41. $3 - 4x < 15$

42. $5 - 6x > 35$

43. $-3 - x > 2$

44. $5 - x < -3$

45. $2(x - 3) > -6$

46. $4(5 - x) < 20$

47. $-\dfrac{3}{4}x \geq 6$

48. $-\dfrac{2}{3}x \leq 4$

49. $3(x + 2) > 5(x - 1)$

50. $4 - 2(x - 3) < 0$

51. $\dfrac{1}{2}x + 7 \leq \dfrac{3}{4}x - 5$

52. $\dfrac{5}{6}x - 3 \geq \dfrac{2}{3}x + 7$

2.5 Compound Inequalities

Solve each compound inequality. State the solution set using interval notation and graph it.

53. $x + 2 > 3$ or $x - 6 < -10$

54. $x - 2 > 5$ or $x - 2 < -1$

55. $x > 0$ and $x - 6 < 3$

56. $x \le 0$ and $x + 6 > 3$

57. $6 - x < 3$ or $-x < 0$

58. $-x > 0$ or $x + 2 < 7$

59. $2x < 8$ and $2(x - 3) < 6$

60. $\frac{1}{3}x > 2$ and $\frac{1}{4}x > 2$

61. $x - 6 > 2$ and $6 - x > 0$

62. $-\frac{1}{2}x < 6$ or $\frac{2}{3}x < 4$

63. $0.5x > 10$ or $0.1x < 3$

64. $0.02x > 4$ and $0.2x < 3$

65. $-2 \le \frac{2x - 3}{10} \le 1$

66. $-3 < \frac{4 - 3x}{5} < 2$

Write each union or intersection of intervals as a single interval.

67. $[1, 4) \cup (2, \infty)$

68. $(2, 5) \cup (-1, \infty)$

69. $(3, 6) \cap [2, 8]$

70. $[-1, 3] \cap [0, 8]$

71. $(-\infty, 5) \cup [5, \infty)$

72. $(-\infty, 1) \cup (0, \infty)$

73. $(-3, -1] \cap [-2, 5]$

74. $[-2, 4] \cap (4, 7]$

2.6 Absolute Value Equations and Inequalities
Solve each absolute value equation and graph the solution set.

75. $|x| + 2 = 16$

76. $\left|\frac{x}{2}\right| - 5 = -1$

77. $|4x - 12| = 0$

78. $|2x - 8| = 0$

79. $|x| = -5$

80. $\left|\frac{x}{2} - 5\right| = -1$

81. $|2x - 1| - 3 = 0$

82. $|5 - x| - 2 = 0$

Solve each absolute value inequality and graph the solution set.

83. $|2x| \ge 8$

84. $|5x - 1| \le 14$

85. $\left|1 - \frac{x}{5}\right| > \frac{9}{5}$

86. $\left|1 - \frac{1}{6}x\right| < \frac{1}{2}$

87. $|x - 3| < -3$ **88.** $|x - 7| \le -4$

89. $|x + 4| \ge -1$

90. $|6x - 1| \ge 0$

91. $1 - \frac{3}{2}|x - 2| < -\frac{1}{2}$

92. $1 > \frac{1}{2}|6 - x| - \frac{3}{4}$

Miscellaneous

Solve each problem by using equations or inequalities.

93. *Rockbuster video.* Stephen plans to open a video rental store in Edmonton. Industry statistics show that 45% of the rental price goes for overhead. If the maximum that anyone will pay to rent a video is $5 and Stephen wants a profit of at least $1.65 per video, then in what range should the rental price be?

94. *Working girl.* Regina makes $6.80 per hour working in the snack bar. To keep her grant, she may not earn more than $51 per week. What is the range of the number of hours per week that she may work?

95. *Skeletal remains.* Forensic scientists use the function $h = 60.089 + 2.238F$ to predict the height h (in centimeters) for a male whose femur measures F centimeters. (See the accompanying figure.) In what range is the length of the femur for males between 150 centimeters and 180 centimeters in height? Round to the nearest tenth of a centimeter.

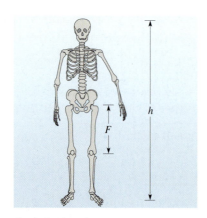

Figure for Exercise 95

96. *Female femurs.* Forensic scientists use the function $h = 61.412 + 2.317F$ to predict the height h in centimeters for a female whose femur measures F centimeters.

a) Use the accompanying graph to estimate the femur length for a female with height of 160 centimeters.

b) In what range is the length of the femur for females who are over 170 centimeters tall?

Figure for Exercise 96

97. *Car trouble.* Dane's car was found abandoned at mile marker 86 on the interstate. If Dane was picked up by the police on the interstate exactly 5 miles away, then at what mile marker was he picked up?

98. *Comparing scores.* Scott scored 72 points on the midterm, and Katie's score was more than 16 points away from Scott's. What was Katie's score?

99. *Year-end bonus.* A law firm has agreed to distribute 20% of its profits to its employees as a year-end bonus. To the firm's accountant, the bonus is an expense that must be used to determine the profit. That is, bonus = 20% × (profit before bonus − bonus). Given that the profit before the bonus is $300,000, find the amount of the bonus using the accountant's method. How does this answer compare to 20% of $300,000, which is what the employees want?

100. *Higher rate.* Suppose that the employees in Exercise 99 got the bonus that they wanted. To the accountant, what percent of the profits was given in bonuses?

101. *Dairy cattle.* Thirty percent of the dairy cattle in Washington County are Holsteins, whereas 60% of the dairy cattle in neighboring Cade County are Holsteins. In the combined two-county area, 50% of the 3600 dairy cattle are Holsteins. How many dairy cattle are in each county?

102. *Profitable business.* United Home Improvement (UHI) makes 20% profit on its good grade of vinyl siding, 30% profit on its better grade, and 60% profit on its best grade. So far this year, UHI has $40,000 in sales of good siding and $50,000 in sales of better siding. The company goal is to have at least an overall profit of 50% of total sales. What would the sales figures for the best grade of siding have to be to reach this goal?

For each graph in Exercises 103–120, write an equation or inequality that has the solution set shown by the graph. Use absolute value when possible.

103.

104.

105.

$$-6\ -5\ -4\ -3\ -2\ -1\quad 0\quad 1\quad 2\quad 3\quad 4\quad 5\quad 6$$

106.

$$-2\ -1\quad 0\quad 1\quad 2\quad 3\quad 4\quad 5\quad 6\quad 7\quad 8\quad 9\quad 10$$

107.

$$-6\ -5\ -4\ -3\ -2\ -1\quad 0\quad 1\quad 2\quad 3\quad 4\quad 5\quad 6$$

108.

$$-2\ -1\quad 0\quad 1\quad 2\quad 3\quad 4\quad 5\quad 6\quad 7\quad 8\quad 9\quad 10$$

109.

$$-6\ -5\ -4\ -3\ -2\ -1\quad 0\quad 1\quad 2\quad 3\quad 4\quad 5\quad 6$$

110.

$$-6\ -5\ -4\ -3\ -2\ -1\quad 0\quad 1\quad 2\quad 3\quad 4\quad 5\quad 6$$

111.

$$-6\ -5\ -4\ -3\ -2\ -1\quad 0\quad 1\quad 2\quad 3\quad 4\quad 5\quad 6$$

112.

$$-6\ -5\ -4\ -3\ -2\ -1\quad 0\quad 1\quad 2\quad 3\quad 4\quad 5\quad 6$$

113.

$$-1\quad 0\quad 1\quad 2\quad 3\quad 4\quad 5\quad 6\quad 7\quad 8\quad 9\quad 10\quad 11$$

114.

$$-6\ -5\ -4\ -3\ -2\ -1\quad 0\quad 1\quad 2\quad 3\quad 4\quad 5\quad 6$$

115.

$$-6\ -5\ -4\ -3\ -2\ -1\quad 0\quad 1\quad 2\quad 3\quad 4\quad 5\quad 6$$

116.

$$-5\ -4\ -3\ -2\ -1\quad 0\quad 1\quad 2\quad 3\quad 4\quad 5\quad 6\quad 7$$

117.

$$0\quad 1\quad 2\quad 3\quad 4\quad 5\quad 6\quad 7\quad 8\quad 9\quad 10\quad 11\quad 12$$

118.

$$-6\ -5\ -4\ -3\ -2\ -1\quad 0\quad 1\quad 2\quad 3\quad 4\quad 5\quad 6$$

119.

$$-6\ -5\ -4\ -3\ -2\ -1\quad 0\quad 1\quad 2\quad 3\quad 4\quad 5\quad 6$$

120.

$$-6\ -5\ -4\ -3\ -2\ -1\quad 0\quad 1\quad 2\quad 3\quad 4\quad 5\quad 6$$

Chapter 2 Test

Solve each equation.

1. $-10x - 5 + 4x = -4x + 3$

2. $\dfrac{y}{2} - \dfrac{y-3}{3} = \dfrac{y+6}{6}$

3. $|w| + 3 = 9$

4. $|3 - 2(5 - x)| = 3$

Express y as a function of x.

5. $2x - 5y = 20$ **6.** $y = 3xy + 5$

Solve each inequality. State the solution set using interval notation and graph the solution set.

7. $|m - 6| \le 2$

8. $2|x - 3| - 5 > 15$

9. $2 - 3(w - 1) < -2w$

10. $2 < \dfrac{5 - 2x}{3} < 7$

11. $3x - 2 < 7$ and $-3x \le 15$

12. $\dfrac{2}{3}y < 4$ or $y - 3 < 12$

Solve each equation or inequality.

13. $|2x - 7| = -3$

14. $x - 4 > 1$ or $x < 12$

15. $3x < 0$ and $x - 5 > 2$

16. $|2x - 5| \le 0$

17. $|x - 3| < 0$

18. $x + 3x = 4x$

19. $2(x + 7) = 2x + 9$

20. $|x - 6| > -6$

21. $x - 0.04(x - 10) = 96.4$

Write a complete solution to each problem.

22. The perimeter of a rectangle is 84 meters. If the width is 16 meters less than the length, then what is the width of the rectangle?

23. If the area of a triangle is 21 square inches and the base is 3 inches, then what is the height?

24. Joan bought a gold chain marked 30% off. If she paid $210, then what was the original price?

25. How many liters of an 11% alcohol solution should be mixed with 60 liters of a 5% alcohol solution to obtain a mixture that is 7% alcohol?

26. Al and Brenda do the same job, but their annual salaries differ by more than $3000. Assume Al makes $28,000 per year and write an absolute value inequality to describe this situation. What are the possibilities for Brenda's salary?

*Making*Connections | A Review of Chapters 1–2

Simplify each expression.

1. $5x + 6x$

2. $5x \cdot 6x$

3. $\dfrac{6x + 2}{2}$

4. $5 - 4(2 - x)$

5. $(30 - 1)(30 + 1)$

6. $(30 + 1)^2$

7. $(30 - 1)^2$

8. $(2 + 3)^2$

9. $2^2 + 3^2$

10. $(8 - 3)(3 - 8)$

11. $(-1)(3 - 8)$

12. -2^2

13. $3x + 8 - 5(x - 1)$

14. $(-6)^2 - 4(-3)2$

15. $3^2 \cdot 2^3$

16. $4(-6) - (-5)(3)$

17. $-3x \cdot x \cdot x$

18. $(-1)(-1)(-1)(-1)(-1)(-1)$

Solve each equation.

19. $5x + 6x = 8x$

20. $5x + 6x = 11x$

21. $5x + 6x = 0$

22. $5x + 6 = 11x$

23. $3x + 1 = 0$

24. $5 - 4(2 - x) = 1$

25. $3x + 6 = 3(x + 2)$

26. $x - 0.01x = 990$

27. $|5x + 6| = 11$

28. $\dfrac{1}{2}x + \dfrac{1}{6}x = 3 + \dfrac{2}{3}x$

Evaluate each expression if $x = -2$ and $y = 5$.

29. $5x + 6x$ **30.** $11x$

31. $(y - x)(y + x)$ **32.** $y^2 - x^2$

33. $(x - y)^2$ **34.** $x^2 - 2xy + y^2$

35. $(3x + y)^2$ **36.** $9x^2 + 6xy + y^2$

Write each union or intersection of intervals as a single interval.

37. $(1, 3) \cup (2, 8)$
38. $(-1, 5) \cup [5, 7)$
39. $(-20, 40) \cap (30, 99)$
40. $[-23, 88] \cap (55, 91)$
41. $(-6, \infty) \cap (7, \infty)$
42. $[3, \infty) \cap (5, 9)$
43. $(-\infty, 6) \cup (-\infty, 10)$
44. $(-\infty, 3) \cup (0, \infty)$

Solve the problem.

45. *Cost analysis.* Diller Electronics can rent a copy machine for 5 years from American Business Supply for $75 per month plus 6 cents per copy. The same copier can be purchased for $8000, but then it costs only 2 cents per copy for supplies and maintenance. The purchased copier has no value after 5 years.

a) Use the accompanying graph to estimate the number of copies for 5 years for which the cost of renting would equal the cost of buying.
b) Write a formula for the 5-year cost under each plan.
c) Algebraically find the number of copies for which the 5-year costs would be equal.
d) If Diller makes 120,000 copies in 5 years, which plan is cheaper and by how much?
e) For what range of copies do the two plans differ by less than $500?

Figure for Exercise 45

*Critical***Thinking** | **For Individual or Group Work** | **Chapter 2**

These exercises can be solved by a variety of techniques, which may or may not require algebra. So be creative and think critically. Explain all answers. Answers are in the Instructor's Edition of this text.

1. ***Lost billion.*** If the pattern of integers in the accompanying table were continued, then in what column would you find one billion?

A	B	C
1	8	27
64	125	216

Table for Exercise 1

2. ***Stooge party.*** Larry, Curly, and Moe are sitting around a table along with an elephant. On the table is a bowl of peanuts. Larry gives one peanut to the elephant, eats one-third of what is left, and passes the bowl to Curly. Curly gives one peanut to the elephant, eats one-third of what is left, and passes the bowl to Moe. Moe gives one peanut to the elephant, eats one-third of what is left, and then passes the bowl to the elephant. The elephant then divides the remaining peanuts equally among Larry, Curly, and Moe, with each of them getting at least one peanut. What is the minimum number of peanuts that could have been in the bowl originally?

3. ***Divisibility by units.*** The number 36 is divisible by 6, which is its units digit. How many numbers are there between 0 and 100 that are divisible by their units digit?

4. ***Two-digit number.*** A positive two-digit number is twice as large as the product of its digits. Find the number.

5. ***Highest possible score.*** Nine students take a test, none get the same score, and each score is a positive integer. If their mean score is 14, then what is the highest possible score on the test?

6. ***Slobs and globs.*** The state of Slobovia has a simple tax system. If a Slob makes x globs per week, then the Slobovian government takes x% of the Slob's globs for income tax. What is the maximum amount of take-home pay in globs for a Slob?

Photo for Exercise 6

7. ***Rain gauge.*** Vira uses an old bottle as a rain gauge. After hurricane Zoe, the water in the bottle was 1.5 in. deep. If the inside diameter of the opening to the bottle is 0.75 in. and the inside diameter at the bottom of the bottle is 2 in., then what amount of rain had fallen?

0.75 in.

1.5 in.

2 in.

Figure for Exercise 7

8. ***Painting a cube.*** A large cube is made up of smaller cubes that are all identical in size. If some of the sides of the large cube are painted completely and the large cube is taken apart, then 24 of the smaller cubes have no paint on them. How many smaller cubes have paint on them and which sides of the large cube were painted?

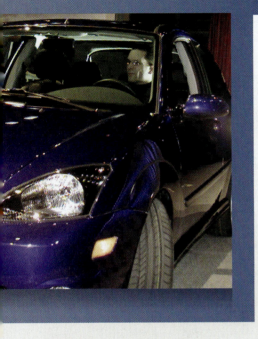

Linear Equations and Inequalities in Two Variables

The first self-propelled automobile to carry passengers was built in 1801 by the British inventor Richard Trevithick. By 1911 about 600,000 automobiles were operated in the United States alone. Some were powered by steam and some by electricity, but most were powered by gasoline. In 1913, to meet the ever growing demand, Henry Ford increased production by introducing a moving assembly line to carry automobile parts. Today the United States is a nation of cars. Over 11 million automobiles are produced here annually, and total car registrations number over 114 million.

Prices for new cars rise every year. Today the most basic Ford Focus sells for $13,000 to $15,000, whereas Henry Ford's early model T sold for $850. Unfortunately, the moment you buy your new car its value begins to decrease. Much of the behavior of automobile prices can be modeled with linear functions.

In Exercises 77 and 78 of Section 3.1 you will use linear functions to find increasing new car prices and depreciating used car prices.

3.1   Graphing Lines in the Coordinate Plane

In Chapter 2, we used the number line to illustrate the solution sets to equations and inequalities in one variable. In this chapter, we will use a new coordinate system made from a pair of number lines to illustrate the solution sets to equations and inequalities in two variables.

⟨1⟩   Graphing Ordered Pairs

A single number is used to describe the location of a point on the number line, but a single number cannot be used to locate a point in a plane. Points on the earth are located using longitude and latitude. Highway maps typically use a letter and a number for locating cities. In mathematics, we position two number lines at a right angle to each other, as shown in Fig. 3.1. The horizontal number line is the ***x*-axis** and the vertical number line is the ***y*-axis.** Every point in the plane corresponds to a pair of numbers—its location with respect to the *x*-axis and its location with respect to the *y*-axis. This system is called the **Cartesian coordinate system.** It is named after the French mathematician Renè Descartes (1596–1650). It is also called the **rectangular coordinate system.**

The intersection of the axes is the **origin.** The axes divide the **coordinate plane** or ***xy*-plane** into four regions called **quadrants.** The quadrants are numbered as shown in Fig. 3.1, and they do not include any points on the axes. Locating a point in the *xy*-plane that corresponds to a pair of real numbers is referred to as **plotting** or **graphing** the point.

⟨ **Helpful Hint** ⟩

In this chapter, you will be doing a lot of graphing. Using graph paper will help you understand the concepts and recognize errors. For your convenience, a page of graph paper can be found on page 219 of this text. Make as many copies of it as you wish.

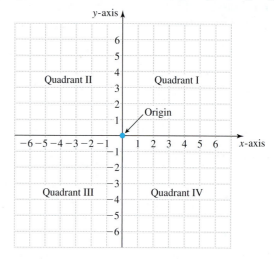

Figure 3.1

E X A M P L E   **1**

Plotting points

Graph the points corresponding to the pairs (2, 4), (4, 2), (−2, −3), (−1, 3), (0, −4), and (4, −2).

Solution

To plot (2, 4), start at the origin, move two units to the right, and then up four units, as shown in Fig. 3.2. To plot (4, 2), start at the origin, move four units to the right, and then two units up, as shown in Fig. 3.2. To plot (−2, −3), start at the origin, move two units to the left, and then down three units. All six points are shown in Fig. 3.2.

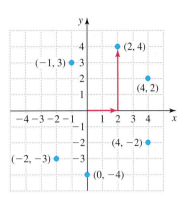

Figure 3.2

Now do Exercises 1–14

A pair of numbers, such as (2, 4), is called an **ordered pair** because the order of the numbers is important. The pairs (4, 2) and (2, 4) correspond to different points in Fig. 3.2. The first number in an ordered pair is the ***x*-coordinate** and the second number is the ***y*-coordinate.**

Note that we use the same notation for ordered pairs that we use for intervals of real numbers, but the meaning should always be clear from the context. The ordered pair (2, 4) represents a single pair of real numbers and a single point in the *xy*-plane, whereas the interval (2, 4) represents all real numbers between 2 and 4. Since ordered pairs correspond to points, we often refer to them as points.

⟨2⟩ Graphing a Linear Equation in Two Variables

In Chapter 2, we defined a linear equation in one variable as an equation of the form $ax = b$, where $a \neq 0$. Every linear equation in one variable has a single real number in its solution set. The graph of the solution set is a single point on the number line. A linear equation in two variables is defined similarly.

Linear Equation in Two Variables

A **linear equation in two variables** is an equation of the form

$$Ax + By = C,$$

where A and B are not both zero.

Consider the linear equation $-2x + y = 3$. It is simpler to find ordered pairs that satisfy the equation if it is solved for y as $y = 2x + 3$. Now if x is replaced by 4, we get $y = 2 \cdot 4 + 3 = 11$. So the ordered pair (4, 11) satisfies this equation. Replacing x with 5 yields $y = 2 \cdot 5 + 3 = 13$. So (5, 13) is also in the solution set. Since there are infinitely many real numbers that could be used for x, there are infinitely many ordered pairs in the solution set. The solution set is written in set notation as $\{(x, y) \mid y = 2x + 3\}$.

The solution set to a linear equation in two variables consists of infinitely many ordered pairs. To get a better understanding of the solution set to a linear equation we look at its graph. It can be proved that *the graph of the solution set is a straight line in the coordinate plane.* We will not prove this statement. Proving it requires a geometric definition of a straight line and is beyond the scope of this text. However, it is easy to graph the straight line by simply plotting a selection of points from the solution set and drawing a straight line through the points as shown in Example 2.

EXAMPLE   2

Graphing a linear equation

Graph the solution set to $y = 2x + 3$.

Solution

We arbitrarily choose some values for x and find corresponding y-values:

Choose x,	then	$y = 2(x) + 3$.
If $x = -4$,	then	$y = 2(-4) + 3 = -5$.
If $x = -3$,	then	$y = 2(-3) + 3 = -3$.
If $x = -2$,	then	$y = 2(-2) + 3 = -1$.
If $x = -1$,	then	$y = 2(-1) + 3 = 1$.
If $x = 0$,	then	$y = 2(0) + 3 = 3$.
If $x = 1$,	then	$y = 2(1) + 3 = 5$.

We can display the corresponding x- and y-values in this table:

x	-4	-3	-2	-1	0	1
$y = 2x + 3$	-5	-3	-1	1	3	5

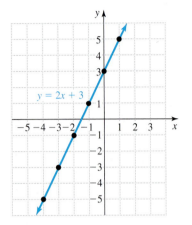

Figure 3.3

Now plot the points $(-4, -5)$, $(-3, -3)$, $(-2, -1)$, $(-1, 1)$, $(0, 3)$, and $(1, 5)$, as shown in Fig. 3.3, and draw a line through them. There are infinitely many ordered pairs that satisfy $y = 2x + 3$, but they all lie on this line. The arrows on the ends of the line indicate that it extends without bound in both directions. The line in Fig. 3.3 is the graph of the solution set to $y = 2x + 3$ or simply the graph of $y = 2x + 3$.

> **Now do Exercises 15–18**

Since the value of y in $y = 2x + 3$ is determined by the value of x, y is a function of x. Because the graph of $y = 2x + 3$ is a line, the equation is a **linear equation** and y is a **linear function** of x. Since the second coordinate in an ordered pair is usually determined by or dependent on the first coordinate, the variable corresponding to the second coordinate is the **dependent variable** and the variable corresponding to the first coordinate is the **independent variable.**

When we draw any graph, we are attempting to put on paper an image that exists in our minds. The line for $y = 2x + 3$ that we have in mind has no thickness, is perfectly straight, and extends infinitely. All attempts to draw it on paper fall short. The best we can do is to use a sharp pencil to keep it as thin as possible, a ruler to make it as straight as possible, and arrows to indicate that it does not end.

EXAMPLE   3

Graphing a linear equation

Graph $y + 2x = 1$. Plot at least four points.

Solution

First express y as a function of x. Subtracting $2x$ from each side of $y + 2x = 1$ yields $y = -2x + 1$. Now arbitrarily select some values for x (the independent variable) and determine the corresponding y-values:

If $x = -1$,	then	$y = -2(-1) + 1 = 3$.
If $x = 0$,	then	$y = -2(0) + 1 = 1$.
If $x = 1$,	then	$y = -2(1) + 1 = -1$.
If $x = 2$,	then	$y = -2(2) + 1 = -3$.

‹ **Calculator Close-Up** ›

To graph $y + 2x = 1$ with a graphing calculator, first press Y= and enter $y_1 = -2x + 1$.

Next press WINDOW to set the viewing window as follows:

Xmin = −10, Xmax = 10, Xscl = 1, Ymin = −10, Ymax = 10, Yscl = 1

These settings are referred to as the standard window.

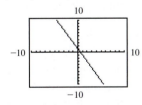

Press GRAPH to draw the graph. Even though the calculator does not draw a very good straight line, it supports our conclusion that Fig. 3.4 is the graph of $y + 2x = 1$.

We can display the corresponding x- and y-coordinates in a table:

x	−1	0	1	2
$y = -2x + 1$	3	1	−1	−3

Plot the points $(-1, 3)$, $(0, 1)$, $(1, -1)$, and $(2, -3)$. The line through these points shown in Fig. 3.4 is the graph of the linear equation or linear function.

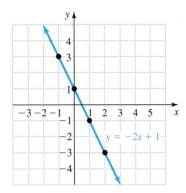

Figure 3.4

Now do Exercises 19–20

If the coefficient of a variable in a linear equation is 0, then that variable is usually omitted from the equation. For example, the equation $y = 0 \cdot x + 2$ is written as $y = 2$. Because x is multiplied by 0, any value of x can be used as long as y is 2. Because the y-coordinates are all the same, the graph is a horizontal line.

E X A M P L E   **4**

Graphing a horizontal line
Graph $y = 2$. Plot at least four points.

Solution

This table gives four points that satisfy $y = 2$, or $y = 0 \cdot x + 2$. Note that it is easy to determine y in this case because y is always 2.

x	−2	−1	0	1
$y = 0 \cdot x + 2$	2	2	2	2

The horizontal line through these points is shown in Fig. 3.5.

Now do Exercises 21–22

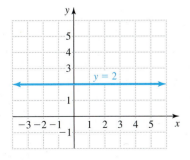

Figure 3.5

If the coefficient of y is 0 in a linear equation, then the graph is a vertical line.

Graphing a vertical line

Graph $x = 4$. Plot at least four points.

Solution

We can think of the equation $x = 4$ as $x = 4 + 0 \cdot y$. Now arbitrarily select some y-values and find the corresponding x-values:

$$\text{If } y = -2, \qquad \text{then} \qquad x = 4 + 0(-2) = 4.$$
$$\text{If } y = -1, \qquad \text{then} \qquad x = 4 + 0(-1) = 4.$$
$$\text{If } y = 0, \qquad \text{then} \qquad x = 4 + 0(0) = 4.$$

Because y is multiplied by 0, the equation is satisfied by every ordered pair with an x-coordinate of 4:

x	4	4	4	4	4	4
y	-2	-1	0	1	2	3

Graphing these points produces the vertical line shown in Fig. 3.6.

‹ Calculator Close-Up ›

You cannot graph $x = 4$ using the $y=$ feature. However, you can "trick" your calculator. Try $y = 9999(x - 4)$ in the standard window. Why does this work?

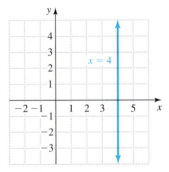

Figure 3.6

Now do Exercises 25–40

‹3› Using Intercepts for Graphing

For many lines, the easiest points to locate are the points where the line crosses the axes.

Intercepts

The **x-intercept** is the point at which a line crosses the x-axis.
The **y-intercept** is the point at which a line crosses the y-axis.

The second coordinate of the x-intercept is 0, and the first coordinate of the y-intercept is 0. If a line has two distinct intercepts, they can be used as two points that determine the location of the line.

Using intercepts to graph

Use the intercepts to graph the line $3x - 4y = 6$.

Solution

Let $x = 0$ in $3x - 4y = 6$ to find the y-intercept:

$$3(0) - 4y = 6$$
$$-4y = 6$$
$$y = -\frac{3}{2}$$

Let $y = 0$ in $3x - 4y = 6$ to find the x-intercept:

$$3x - 4(0) = 6$$
$$3x = 6$$
$$x = 2$$

The y-intercept is $\left(0, -\frac{3}{2}\right)$, and the x-intercept is $(2, 0)$. The line through the intercepts is shown in Fig. 3.7. To check, find another point that satisfies the equation. The point $(-2, -3)$ satisfies the equation and is on the line in Fig. 3.7.

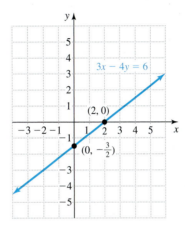

Figure 3.7

Now do Exercises 41–52

CAUTION Even though two points determine the location of a line, finding at least three points will help you to avoid errors.

⟨4⟩ Applications

When we use the variables x and y, the independent variable is x and the dependent variable is y. When other variables are used, we usually have one variable (the dependent variable) written as a function of another (the independent variable). For example, if $W = 3n - 7$, then W is the dependent variable and n is the independent variable. A graph of this function would have n on the horizontal axis and W on the vertical axis.

E X A M P L E   7

Graphing a linear function in an application

The cost per week C (in dollars) of producing n pairs of shoes for the Reebop Shoe Company is given by the linear function $C = 2n + 8000$. Graph the function for n between 0 and 800 inclusive ($0 \leq n \leq 800$).

Solution

Make a table of values for n and C as follows:

n	0	200	400	600	800
$C = 2n + 8000$	8000	8400	8800	9200	9600

Graph the line as shown in Fig. 3.8. Notice how the scale is changed to accommodate the large numbers. The wave in the C-axis is used to indicate that the scale does not start at zero. Starting the C-axis at \$8000 tends to exaggerate the difference between \$8000 and \$9600.

Now do Exercises 77–82

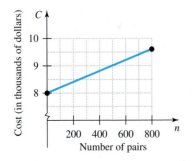

Figure 3.8

EXAMPLE   8

Writing a linear equation

A store manager is ordering shirts at $20 each and jackets at $30 each. The total cost of the order must be $1200. Write an equation for the total cost and graph it. If she orders 15 shirts, then how many jackets can she order?

Solution

Let s be the number of shirts in the order and j be the number of jackets in the order. Since the total cost is $1200, we can write

$$20s + 30j = 1200.$$

If $s = 0$, then $30j = 1200$ or $j = 40$. If $j = 0$, then $20s = 1200$ or $s = 60$. Graph the line through $(0, 40)$ and $(60, 0)$ as in Fig. 3.9. Note that we arbitrarily put s on the horizontal axis and j on the vertical axis.

If $s = 15$, find j as follows:

$$20(15) + 30j = 1200$$
$$300 + 30j = 1200$$
$$30j = 900$$
$$j = 30$$

If she orders 15 shirts, then she must order 30 jackets.

In this example, the numbers of shirts and jackets in the order must be whole numbers and every possible order corresponds to a point on the line in Fig. 3.9. However, points on the line whose coordinates are not whole numbers do not correspond to possible orders. Even though the line does not show the possible orders exactly, we draw it because it is more convenient than finding every possible order and then plotting just those points.

Figure 3.9

The graph shows the line $20s + 30j = 1200$.

> **Now do Exercises 83–86**

Warm-Ups ▼

Fill in the blank.

1. The point at the intersection of the x-axis and y-axis is the _____ .

2. Every point in the xy-plane corresponds to an _____ of real numbers.

3. The point at which a line crosses the x-axis is the _____ .

4. The point at which a line crosses the y-axis is the _____ .

5. The graph of an equation of the form $y = k$ is a _____ line.

6. The graph of an equation of the form $x = k$ is a _____ line.

7. The _____ variable usually goes on the vertical axis.

8. A _____ equation in two variables has the form $Ax + By = C$.

True or false?

9. The point $(2, 5)$ satisfies $3y - 2x = -4$.

10. The point $(8, -1)$ satisfies $y = \frac{3}{4}x - 7$.

11. The point $(0, 0)$ is in quadrant I.

12. The point $(-3, -4)$ is in quadrant II.

13. The point $(0, 1)$ is on the y-axis.

14. The independent variable in $A = \pi r^2$ is r.

15. The line $x = 7$ has no y-intercept.

16. If $C = 3n + 4$, then $C = 10$ when $n = 2$.

17. If $x - y = -1$, then $y = -4$ when $x = 3$.

18. The y-intercept for $y = -3x - 9$ is $(0, -9)$.

‹1› Graphing Ordered Pairs

Plot the following points in a rectangular coordinate system. For each point, name the quadrant in which it lies or the axis on which it lies. See Example 1.

1. $(2, 5)$

2. $(-5, 1)$

 3. $\left(-3, -\dfrac{1}{2}\right)$

4. $(-2, -6)$

5. $(0, 4)$

6. $(0, 2)$

7. $(\pi, -1)$

8. $\left(\dfrac{4}{3}, 0\right)$

9. $(-4, 3)$

10. $(0, -3)$

11. $\left(\dfrac{3}{2}, 0\right)$

12. $(3, 2)$

13. $\left(0, -\dfrac{7}{3}\right)$

14. $\left(4, -\dfrac{10}{3}\right)$

17. $y = -2x + 3$

18. $y = 2x - 3$

19. $y = x$

20. $y = -x$

21. $y = 3$

22. $y = -2$

23. $y = 1 - x$

24. $y = 2 - x$

‹2› Graphing a Linear Equation in Two Variables

Graph each linear equation. Plot four points for each line. See Examples 2–5.

15. $y = x + 1$

16. $y = x - 1$

25. $x = 2$

26. $x = -3$

27. $y = \dfrac{1}{2}x - 1$

28. $y = \dfrac{1}{3}x - 2$

39. $y = 2x - 20$

40. $y = 40 - 2x$

29. $y + 3 = 0$

30. $y + 4 = 0$

⟨**3**⟩ **Using Intercepts for Graphing**

Find the x- and y-intercepts for each line and use them to graph the line. See Example 6.

41. $4x - 3y = 12$

42. $2x + 5y = 20$

31. $x - 4 = 0$

32. $x + 5 = 0$

43. $x - y + 5 = 0$

44. $x + y + 7 = 0$

33. $y = \dfrac{1}{2}x$

34. $y = -\dfrac{2}{3}x$

45. $2x + 3y = 5$

46. $3x - 4y = 7$

35. $3x + y = 5$

36. $x + 2y = 4$

37. $6x + 3y = 0$

38. $2x + 4y = 0$

47. $2x - 3y = 60$

48. $2x + 3y = 30$

49. $y = 2x - 4$

50. $y = -3x + 6$

51. $y = -\dfrac{1}{2}x - 20$

52. $y = \dfrac{1}{3}x + 10$

Complete the given ordered pairs so that each ordered pair satisfies the given equation.

73. $(2, \ \ ), (\ \ , -3), \quad y = -3x + 6$

74. $(-1, \ \ ), (\ \ , 4), \quad y = \dfrac{1}{2}x + 2$

75. $(-4, \ \ ), (\ \ , 6), \quad \dfrac{1}{2}x - \dfrac{1}{3}y = 9$

76. $(3, \ \ ), (\ \ , -1), \quad 2x - 3y = 5$

⟨**4**⟩ **Applications**

Solve each problem. See Examples 7 and 8.

77. *Ford F150 inflation.* The rising base price P (in dollars) for a new Ford F150 can be modeled by the function $P = 624n + 15{,}950$, where n is the number of years since 2000 (www.edmunds.com).

   a) What will be the base price for a new Ford F150 in 2015?

   b) By what amount is the price increasing annually?

   c) Graph the equation for $0 \leq n \leq 20$.

Graph each equation on a graphing calculator using a window that shows both intercepts. Then use the appropriate feature of your calculator to find the intercepts.

53. $y = -3x + 1$

54. $y = 2 - 3x$

55. $y = 400x - 2$

56. $y = 800x + 8$

57. $x + 2y = 600$

58. $3x - 2y = 1500$

59. $y = 4.26x + 23.54$

60. $y = 30.6 - 3.6x$

Find all intercepts for each line. Some of these lines have only one intercept.

61. $x + y = 50$

62. $x + 2y = 100$

63. $3x - 5y = 15$

64. $9x + 8y = 72$

65. $y = 5x$

66. $y = -4x$

67. $6x + 3 = 0$

68. $40x + 5 = 0$

69. $12 + 18y = 0$

70. $2 - 10y = 0$

71. $2 - 4y = 8x$

72. $9x + 3 = 12y$

78. *Toyota Camry depreciation.* The 2009 average retail price P (in dollars) for an n-year-old Toyota Camry can be modeled by the function $P = 24{,}215 - 1832n$, where $0 \leq n \leq 4$ (www.edmunds.com).

   a) What was the average retail price of a 4-year-old Camry in 2009?

   b) By what amount does this model depreciate annually?

   c) Graph the equation for $0 \leq n \leq 4$.

79. *Rental cost.* For a one-day car rental the X-press Car Company charges C dollars, where C is determined by the function $C = 0.26m + 42$ and m is the number of miles driven.

  a) What is the charge for a car driven 400 miles?
  b) Sketch a graph of the equation for m ranging from 0 to 1000.

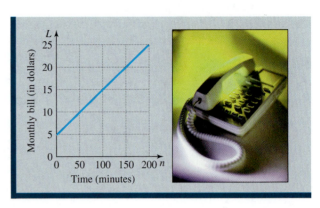

Figure for Exercise 82

80. *Measuring risk.* The Friendly Bob Loan Company gives each applicant a rating, t, from 0 to 10 according to the applicant's ability to repay, a higher rating indicating higher risk. The interest rate, r, is then determined by the function $r = 0.02t + 0.15$.

  a) If your rating were 8, then what would be your interest rate?
  b) Sketch the graph of the equation for t ranging from 0 to 10.

the total cost and graph it. If he orders 30 note pads, then how many binders must he order?

84. *Tacos and burritos.* Jessenda is ordering tacos at $0.75 each and burritos at $2 each for a large group. She must spend $300. Write an equation for the total cost and graph it. If she orders 200 tacos, then how many burritos must she order?

81. *Little Chicago pizza.* The function $C = 0.50t + 8.95$ gives the customer's cost in dollars for a pan pizza, where t is the number of toppings.

  a) Find the cost of a five-topping pizza.
  b) Find t if $C = 14.45$ and interpret your result.

82. *Long distance charges.* The function $L = 0.10n + 4.95$ gives the monthly bill in dollars for AT&T's one-rate plan, where n is the number of minutes of long distance used during the month.

  a) Find n if the long distance charge is $23.45.
  b) Find L for 120 minutes.
  c) Estimate the L-intercept from the accompanying graph.

  d) Use the formula to find the L-intercept.
  e) Use the formula to find the n-intercept.

83. *Note pads and binders.* An office manager is placing an order for note pads at $1 each and binders at $2 each. The total cost of the order must be $100. Write an equation for

85. *Cost, revenue, and profit.* Hillary sells roses at a busy Los Angeles intersection. The functions

$$C = 0.55x + 50,$$
$$R = 1.50x,$$

and

$$P = 0.95x - 50$$

give her weekly cost, revenue, and profit in terms of x, where x is the number of roses that she sells in one week.

  a) Find C, R, and P if $x = 850$. Interpret your results.
  b) Find x if $P = 995$ and interpret your result.

c) Find $R - C$ if $x = 1100$ and interpret your result.

d) If the ball takes the same time going up as it does coming down, then what is its velocity as it hits the ground?

86. *Velocity of a pop up.* A pop up off the bat of Mark McGwire goes straight into the air at 88 feet per second (ft/sec). The function $v = -32t + 88$ gives the velocity of the ball in feet per second, t seconds after the ball is hit.

a) Find the velocity for $t = 2$ and $t = 3$ seconds. What does a negative velocity mean?

b) For what value of t is $v = 0$? Where is the ball at this time?

c) What are the two intercepts on the accompanying graph? Interpret this answer.

Figure for Exercise 86

3.2   Slope of a Line

In This Section

⟨1⟩ **Slope**

⟨2⟩ **The Coordinate Formula for Slope**

⟨3⟩ **Parallel Lines**

⟨4⟩ **Perpendicular Lines**

⟨5⟩ **Applications of Slope**

In Section 3.1, we saw some equations whose graphs were straight lines. In this section, we look at graphs of straight lines in more detail and study the concept of slope of a line.

⟨1⟩ Slope

If a highway has a 6% grade, then in 100 feet (measured horizontally) the road rises 6 feet (measured vertically). See Fig. 3.10. The ratio of 6 to 100 is 6%. If a roof rises 9 feet in a horizontal distance (or run) of 12 feet, then the roof has a 9–12 pitch. A roof with a 9–12 pitch is steeper than a roof with a 6–12 pitch. The grade of a road and the pitch of a roof are measurements of steepness. In each case the measurement is a ratio of rise (vertical change) to run (horizontal change).

We measure the steepness of a line in the same way that we measure steepness of a road or a roof. The slope of a line is the ratio of the change in y-coordinate, or the **rise,** to the change in x-coordinate, or the **run,** between two points on the line.

Figure 3.10

> ## Slope
>
> $$\text{Slope} = \frac{\text{change in } y\text{-coordinate}}{\text{change in } x\text{-coordinate}} = \frac{\text{rise}}{\text{run}}$$

Consider the line in Fig. 3.11(a). In going from (0, 1) to (1, 3), there is a change of $+1$ in the x-coordinate and a change of $+2$ in the y-coordinate, or a run of 1 and a rise of 2. So the slope is $\frac{2}{1}$ or 2. If we move from (1, 3) to (0, 1) as in Fig. 3.11(b), the rise is -2 and the run is -1. So the slope is $\frac{-2}{-1}$ or 2. If we start at either point and move to the other point, we get the same slope.

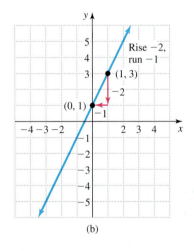

(a)   (b)

Figure 3.11

E X A M P L E   1

Finding the slope from a graph

Find the slope of each line by going from point A to point B.

a)

b)

c)

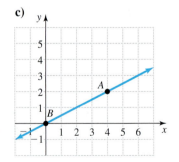

Solution

a) A is located at (0, 3) and B at (2, 0). In going from A to B, the change in y is -3 and the change in x is 2. So,

$$\text{slope} = \frac{-3}{2} = -\frac{3}{2}.$$

b) In going from $A(2, 1)$ to $B(6, 3)$, we must rise 2 and run 4. So,

$$\text{slope} = \frac{2}{4} = \frac{1}{2}.$$

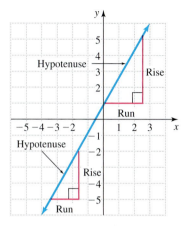

Figure 3.12

c) In going from $A(4, 2)$ to $B(0, 0)$, we find that the rise is -2 and the run is -4. So,

$$\text{slope} = \frac{-2}{-4} = \frac{1}{2}.$$

> Now do Exercises 1–12

Note that in Example 1(c) we found the slope of the line of Example 1(b) by using two different points. The slope is the ratio of the lengths of the two legs of a right triangle whose hypotenuse is on the line. See Fig. 3.12. As long as one leg is vertical and the other leg is horizontal, all such triangles for a given line have the same shape: They are similar triangles. Because ratios of corresponding sides in similar triangles are equal, the slope has the same value no matter which two points of the line are used to find it.

⟨2⟩ The Coordinate Formula for Slope

In Example 1 we obtained the slope by counting the amount of rise and run between two points on a graph. If we know the coordinates of the points, we can obtain the rise and run without looking at the graph. If (x_1, y_1) and (x_2, y_2) are two points on a line, then the rise is $y_2 - y_1$ and the run is $x_2 - x_1$, as shown in Fig. 3.13. So we have the following coordinate formula for slope.

Slope Using Coordinates

The slope m of the line containing the points (x_1, y_1) and (x_2, y_2) is given by

$$m = \frac{y_2 - y_1}{x_2 - x_1}, \quad \text{provided that } x_2 - x_1 \neq 0.$$

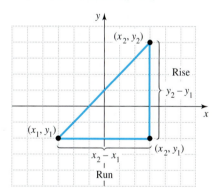

Figure 3.13

E X A M P L E 2

Finding slope from coordinates

Find the slope of each line.

a) The line through $(2, 5)$ and $(6, 3)$

b) The line through $(-2, 3)$ and $(-5, -1)$

c) The line through $(-6, 4)$ and the origin

Solution

a) Let $(x_1, y_1) = (2, 5)$ and $(x_2, y_2) = (6, 3)$ in the slope formula:

$$m = \frac{y_2 - y_1}{x_2 - x_1} = \frac{3 - 5}{6 - 2} = \frac{-2}{4} = -\frac{1}{2}$$

It does not matter which point is called (x_1, y_1) and which is called (x_2, y_2). If $(x_1, y_1) = (6, 3)$ and $(x_2, y_2) = (2, 5)$, we get

$$m = \frac{y_2 - y_1}{x_2 - x_1} = \frac{5 - 3}{2 - 6} = \frac{2}{-4} = -\frac{1}{2}.$$

Note that in either case, the coordinates in $(2, 5)$ and $(6, 3)$ are aligned vertically in the expressions $\frac{3-5}{6-2}$ or $\frac{5-3}{2-6}$.

b) Let $(x_1, y_1) = (-5, -1)$ and $(x_2, y_2) = (-2, 3)$:

$$m = \frac{y_2 - y_1}{x_2 - x_1} = \frac{3 - (-1)}{-2 - (-5)} = \frac{4}{3}$$

c) Let $(x_1, y_1) = (0, 0)$ and $(x_2, y_2) = (-6, 4)$:

$$m = \frac{y_2 - y_1}{x_2 - x_1} = \frac{4 - 0}{-6 - 0} = \frac{4}{-6} = -\frac{2}{3}$$

> **Now do Exercises 13–26**

CAUTION Do not reverse the order of subtraction from numerator to denominator when finding the slope. If you divide $y_2 - y_1$ by $x_1 - x_2$, you will get the wrong sign for the slope.

E X A M P L E 3

Slope for horizontal and vertical lines

Find the slope of each line.

a)

b)

Zero slope

Small slope

Solution

a) Using $(-3, 2)$ and $(4, 2)$ to find the slope of the horizontal line, we get

$$m = \frac{2 - 2}{-3 - 4}$$
$$= \frac{0}{-7} = 0.$$

b) Using $(1, -4)$ and $(1, 2)$ to find the slope of the vertical line, we get $x_2 - x_1 = 0$. Because the definition of slope using coordinates says that $x_2 - x_1$ must be nonzero, the slope is undefined for this line.

> **Now do Exercises 27–32**

Larger slope

Undefined slope

In Example 3, the horizontal line has slope 0 and slope is undefined for the vertical line. These results hold in general. Since the *y*-coordinates are equal for any two points on a horizontal line, the rise is 0 between any two points and the slope is 0. So all horizontal lines have slope 0. Since the *x*-coordinates are equal for any two points on a vertical line, the run is 0 between any two points and the slope is undefined. So lines such as $y = 2$, $y = -9$, and $y = 200$ have slope 0. Slope is undefined for lines such as $x = 1$, $x = -99$, and $x = 7$.

Horizontal and Vertical Lines

The slope of any horizontal line is 0.
Slope is undefined for any vertical line.

CAUTION Do not say that a vertical line has no slope because "no slope" could be confused with 0 slope, the slope of a horizontal line.

As you move the tip of your pencil from left to right along a line with positive slope, the *y*-coordinates are increasing. As you move the tip of your pencil from left to right along a line with negative slope, the *y*-coordinates are decreasing. See Fig. 3.14.

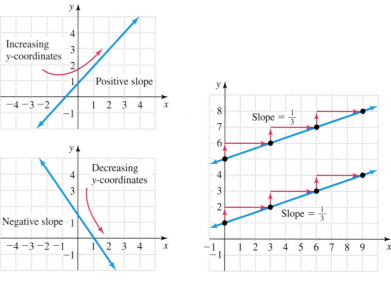

Figure 3.14

Figure 3.15

⟨3⟩ Parallel Lines

Two lines in a coordinate plane that do not intersect are **parallel.** Consider the two lines with slope $\frac{1}{3}$ shown in Fig. 3.15. At the *y*-axis, these lines are 4 units apart, measured vertically. A slope of $\frac{1}{3}$ means that you can forever rise 1 and run 3 to get to another point on the line. So the lines will always be 4 units apart vertically and they will never intersect. This example illustrates the following fact.

Parallel Lines

Two lines with slopes m_1 and m_2 are parallel if and only if $m_1 = m_2$.

For lines that have slope, the slopes can be used to determine whether the lines are parallel. The only lines that do not have slope are vertical lines. Of course, any two vertical lines are parallel.

E X A M P L E   **4**

Parallel lines
Line l goes through the origin and is parallel to the line through $(-2, 3)$ and $(4, -5)$. Find the slope of line l.

Solution
The line through $(-2, 3)$ and $(4, -5)$ has slope

$$m = \frac{-5 - 3}{4 - (-2)} = \frac{-8}{6} = -\frac{4}{3}.$$

Because line l is parallel to a line with slope $-\frac{4}{3}$, the slope of line l is $-\frac{4}{3}$ also.

Now do Exercises 33–34

⟨4⟩ Perpendicular Lines
Figure 3.16 shows line l_1 with positive slope m_1. The rise m_1 and the run 1 are the sides of a right triangle. If l_1 and the triangle are rotated 90° clockwise, then l_1 will coincide with line l_2 and the slope of l_2 can be determined from the triangle in its new position. Starting at the point of intersection, the run for l_2 is m_1 and the rise is -1 (moving downward). So if m_2 is the slope of l_2, then $m_2 = -\frac{1}{m_1}$. *The slope of l_2 is the opposite of the reciprocal of the slope of l_1.* This result can be stated also as $m_1 m_2 = -1$ or as follows.

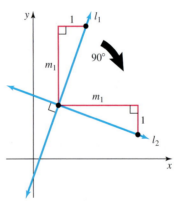

Figure 3.16

Perpendicular Lines
Two lines with slopes m_1 and m_2 are perpendicular if and only if

$$m_1 = -\frac{1}{m_2}.$$

Of course, any vertical line and any horizontal line are perpendicular, but their slopes do not satisfy this equation because slope is undefined for vertical lines.

E X A M P L E   **5**

Perpendicular lines

Line l contains the point $(1, 6)$ and is perpendicular to the line through $(-4, 1)$ and $(3, -2)$. Find the slope of line l.

Solution

The line through $(-4, 1)$ and $(3, -2)$ has slope

$$m = \frac{1 - (-2)}{-4 - 3} = \frac{3}{-7} = -\frac{3}{7}.$$

Because line l is perpendicular to a line with slope $-\frac{3}{7}$, the slope of line l is $\frac{7}{3}$.

> **Now do Exercises 35–44**

‹5› Applications of Slope

When a geometric figure is located in a coordinate system, we can use slope to determine whether it has any parallel or perpendicular sides.

E X A M P L E   **6**

Using slope with geometric figures

Determine whether $(-3, 2)$, $(-2, -1)$, $(4, 1)$, and $(3, 4)$ are the vertices of a rectangle.

Solution

Figure 3.17 shows the quadrilateral determined by these points. If a parallelogram has at least one right angle, then it is a rectangle. Calculate the slope of each side.

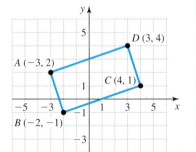

Figure 3.17

$$m_{AB} = \frac{2 - (-1)}{-3 - (-2)} \qquad m_{BC} = \frac{-1 - 1}{-2 - 4}$$

$$= \frac{3}{-1} = -3 \qquad = \frac{-2}{-6} = \frac{1}{3}$$

$$m_{CD} = \frac{1 - 4}{4 - 3} \qquad m_{AD} = \frac{2 - 4}{-3 - 3}$$

$$= \frac{-3}{1} = -3 \qquad = \frac{-2}{-6} = \frac{1}{3}$$

Because the opposite sides have the same slope, they are parallel, and the figure is a parallelogram. Because $\frac{1}{3}$ is the opposite of the reciprocal of -3, the intersecting sides are perpendicular. Therefore, the figure is a rectangle.

> **Now do Exercises 55–60**

The slope of a line is a rate. The slope tells us how much the dependent variable changes for a change of 1 in the independent variable. For example, if the horizontal axis is hours and the vertical axis is miles, then the slope is miles per hour (mph). If the horizontal axis is days and the vertical axis is dollars, then the slope is dollars per day.

E X A M P L E   **7**

Interpreting slope

Worldwide carbon dioxide (CO_2) emissions have increased from 14 billion tons in 1970 to 26 billion tons in 2000 (World Resources Institute, www.wri.org).

a) Find and interpret the slope of the line in Fig. 3.18.

b) Use the slope to predict the amount of worldwide CO_2 emissions in 2010.

Figure 3.18

Solution

a) Find the slope of the line through (1970, 14) and (2000, 26):

$$m = \frac{26 - 14}{2000 - 1970} = \frac{12}{30} = 0.4$$

So CO_2 emissions are increasing 0.4 billion tons per year.

b) If the CO_2 emissions keep increasing 0.4 billion tons per year, then in 10 years the level will go up 10(0.4) or 4 billion tons. So in 2010 CO_2 emissions will be 30 billion tons.

> Now do Exercises 61–62

Warm-Ups ▼

Fill in the blank.

1. _____ is a measure of the steepness of a line.
2. The ___ is the change in y-coordinates.
3. The ___ is the change in x-coordinates.
4. A _____ line has zero slope.
5. Slope is undefined for _____ lines.
6. Slope is the ratio of the ___ to the ___.
7. If m_1 and m_2 are the slopes of _____ lines, then $m_1 m_2 = -1$.
8. If m_1 and m_2 are the slopes of _____ lines, then $m_1 = m_2$.

True or false?

9. The line through $(-2, 1)$ and $(5, 1)$ has zero slope.
10. The line through $(0, 1)$ and $(1, 0)$ has undefined slope.
11. The line through $(4, 4)$ and $(5, 5)$ has slope 1.
12. Any two parallel lines have equal slopes.
13. Lines with slopes $\frac{2}{5}$ and $-\frac{5}{2}$ are perpendicular.
14. If a line contains points in quadrants I and III, then its slope is positive.
15. Slope cannot be negative.
16. The slope of the line through $(0, -2)$ and $(5, 0)$ is $\frac{2}{5}$.

‹ 1 › **Slope**

Determine the slope of each line. See Example 1.

1.

2.

3.

4.

5.

6.

7.

8.

9.

10.

11.

12.
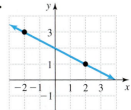

‹ 2 › **The Coordinate Formula for Slope**

Find the slope of the line that contains each of the following pairs of points. See Examples 2 and 3.

13. $(2, 6), (5, 1)$

14. $(3, 4), (6, 10)$

15. $(-3, -1), (4, 3)$

16. $(-2, -3), (1, 3)$

17. $(-2, 2), (-1, 7)$

18. $(-3, 5), (1, -6)$

19. $(3, -5), (0, 0)$

20. $(0, 0), (-2, -1)$

21. $(0, 3), (5, 0)$

22. $(3, 0), (0, 10)$

23. $\left(\frac{3}{4}, -1\right), \left(-\frac{1}{2}, -\frac{1}{2}\right)$

24. $\left(\frac{1}{2}, 2\right), \left(\frac{1}{4}, \frac{1}{2}\right)$

25. $(6, 212), (7, 209)$

26. $(1988, 306), (1990, 315)$

27. $(4, 7), (-12, 7)$

28. $(5, -3), (9, -3)$

29. $(2, 6), (2, -6)$

30. $(-3, 2), (-3, 0)$

31. $(24.3, 11.9), (3.57, 8.4)$

32. $(-2.7, 19.3), (5.46, -3.28)$

⟨3–4⟩ Parallel and Perpendicular Lines

In each case find the slope of line l and graph both lines that are mentioned. See Examples 4 and 5.

33. Line *l* contains $(2, 5)$ and is parallel to the line through $(2, 3)$ and $(4, 9)$.

34. Line *l* contains $(0, 0)$ and is parallel to the line through $(-1, 4)$ and $(2, -3)$.

35. Line *l* contains the point $(3, 4)$ and is perpendicular to the line through $(-5, 1)$ and $(3, -2)$.

36. Line *l* goes through $(-3, -5)$ and is perpendicular to the line through $(-2, 6)$ and $(5, 3)$.

37. Line *l* goes through $(2, 5)$ and is parallel to the line through $(-3, -2)$ and $(4, 1)$.

38. Line *l* goes through the origin and is parallel to the line through $(-3, -5)$ and $(4, -1)$.

39. Line *l* contains $(1, 4)$ and is perpendicular to the line through $(0, 0)$ and $(2, 4)$.

40. Line *l* contains $(-1, 1)$ and is perpendicular to the line through $(0, 0)$ and $(-3, 5)$.

41. Line *l* is perpendicular to a line with slope $\frac{4}{5}$. Both lines contain the origin.

42. Line *l* is perpendicular to a line with slope -5. Both lines contain the origin.

43. Line *l* passes through $(0, 4)$ and is parallel to a line through the origin with slope 2.

44. Line *l* passes through $(2, 0)$ and is parallel to a line through the origin with slope 1.

Determine whether the lines l_1 and l_2 are parallel, perpendicular, or neither. See Examples 4 and 5.

45. l_1 goes through $(1, 2)$ and $(4, 8)$, l_2 goes through $(0, 3)$ and $(2, 2)$.

46. l_1 goes through $(-2, 5)$ and $(3, 7)$, l_2 goes through $(8, 4)$ and $(13, 6)$.

47. l_1 goes through $(0, 4)$ and $(-1, 6)$, l_2 goes through $(7, 7)$ and $(8, 9)$.

48. l_1 goes through $(3, 6)$ and $(4, 9)$, l_2 goes through $(5, 3)$ and $(6, 0)$.

49. l_1 goes through $(0, 2)$ and $(7, 9)$, l_2 goes through $(0, -3)$ and $(1, -2)$.

50. l_1 goes through $(4, 3)$ and $(2, 6)$, l_2 goes through $(0, 0)$ and $(3, 2)$.

51. l_1 goes through $(0, 0)$ and $(-2, 5)$, l_2 goes through $(0, 1)$ and $(2, 6)$.

52. l_1 goes through $(0, 3)$ and $(4, 17)$, l_2 goes through $(-1, -4)$ and $(1, 3)$.

53. l_1 goes through $(-2, -3)$ and $(4, 1)$, l_2 goes through $(-1, 7)$ and $(1, 4)$.

54. l_1 goes through $(0, 5)$ and $(1, 4)$, l_2 goes through $(-3, -5)$ and $(-5, -7)$.

⟨**5**⟩ **Applications of Slope**

Solve each geometric figure problem. See Example 6.

55. If the opposite sides of a quadrilateral are parallel, then it is a parallelogram. Use slope to determine whether the points $(-6, 1)$, $(-2, -1)$, $(0, 3)$, and $(4, 1)$ are the vertices of a parallelogram.

56. Use slope to determine whether the points $(-7, 0)$, $(-1, 6)$, $(-1, -2)$, and $(6, 5)$ are the vertices of a parallelogram. See Exercise 55.

57. A trapezoid is a quadrilateral with one pair of parallel sides. Use slope to determine whether the points $(-3, 2)$, $(-1, -1)$, $(3, 6)$, and $(6, 4)$ are the vertices of a trapezoid.

58. A parallelogram with at least one right angle is a rectangle. Determine whether the points $(-4, 4)$, $(-1, -2)$, $(0, 6)$, and $(3, 0)$ are the vertices of a rectangle.

59. If a triangle has one right angle, then it is a right triangle. Use slope to determine whether the points $(-3, 3)$, $(-1, 6)$, and $(0, 0)$ are the vertices of a right triangle.

60. Use slope to determine whether the points $(0, -1)$, $(2, 5)$, and $(5, 4)$ are the vertices of a right triangle. See Exercise 59.

Solve each problem. See Example 7.

61. ***Pricing the Taurus.*** The list price for a new Ford Taurus four-door sedan was \$20,120 in 2000 and \$27,170 in 2010 (www.edmunds.com).

a) Find the slope of the line shown in the accompanying figure.

Figure for Exercise 61

b) Use the accompanying figure to predict the price in 2015.

c) Use the slope to predict the price in 2015.

62. *Depreciating Malibu.* In 2009 the average retail price of a 1-year-old Chevrolet Malibu was $19,335, whereas the average retail price of a 4-year-old Malibu was $12,525 (www.edmunds.com).

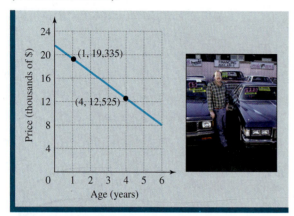

Figure for Exercise 62

a) Use the accompanying graph to estimate the average retail price of a 3-year-old car in 2009.

b) Find the slope of the line in the figure.

c) Use the slope to predict the price of a 3-year-old car in 2009.

Miscellaneous

63. The points $(3, \quad)$ and $(\quad, -7)$ are on the line that passes through $(2, 1)$ and has slope 4. Find the missing coordinates of the points.

64. If a line passes through $(5, 2)$ and has slope $\frac{2}{3}$, then what is the value of y on this line when $x = 8$, $x = 11$, and $x = 12$?

65. Find k so that the line through $(2, k)$ and $(-3, -5)$ has slope $\frac{1}{2}$.

66. Find k so that the line through $(k, 3)$ and $(-2, 0)$ has slope 3.

67. What is the slope of a line that is perpendicular to a line with slope 0.247?

68. What is the slope of a line that is perpendicular to the line through $(3.27, -1.46)$ and $(-5.48, 3.61)$?

Getting More Involved

69. *Writing*

What is the difference between zero slope and undefined slope?

70. *Writing*

Is it possible for a line to be in only one quadrant? Two quadrants? Write a rule for determining whether a line has positive, negative, zero, or undefined slope from knowing in which quadrants the line is found.

71. *Exploration*

A rhombus is a quadrilateral with four equal sides. Draw a rhombus with vertices $(-3, -1)$, $(0, 3)$, $(2, -1)$, and $(5, 3)$. Find the slopes of the diagonals of the rhombus. What can you conclude about the diagonals of this rhombus?

72. *Exploration*

Draw a square with vertices $(-5, 3)$, $(-3, -3)$, $(1, 5)$, and $(3, -1)$. Find the slopes of the diagonals of the square. What can you conclude about the diagonals of this square?

 Graphing Calculator Exercises

73. Graph $y = 1x$, $y = 2x$, $y = 3x$, and $y = 4x$ together in the standard viewing window. These equations are all of the form $y = mx$. What effect does increasing m have on the graph of the equation? What are the slopes of these four lines?

74. Graph $y = -1x$, $y = -2x$, $y = -3x$, and $y = -4x$ together in the standard viewing window. These equations are all of the form $y = mx$. What effect does decreasing m have on the graph of the equation? What are the slopes of these four lines?

3.3   Three Forms for the Equation of a Line

In This Section

⟨1⟩ **Slope-Intercept Form**

⟨2⟩ **Using Slope-Intercept Form for Graphing**

⟨3⟩ **Standard Form**

⟨4⟩ **Point-Slope Form**

⟨5⟩ **Applications**

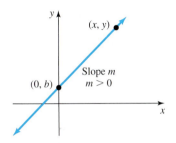

Figure 3.19

⟨ **Calculator Close-Up** ⟩

With slope-intercept form and a graphing calculator, it is easy to see how the slope affects the steepness of a line. The graphs of $y_1 = \frac{1}{2}x$, $y_2 = x$, $y_3 = 2x$, and $y_4 = 3x$ are all shown on the accompanying screen.

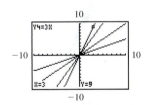

In Section 3.1, you learned how to graph a straight line corresponding to a linear equation. The line contains all of the points that satisfy the equation. In this section, we start with a line or a description of a line and write an equation corresponding to the line.

⟨1⟩ Slope-Intercept Form

Suppose that (x, y) is an arbitrary point on a line that has slope m and y-intercept $(0, b)$. If m is positive, the line would look like the one shown in Fig. 3.19. If we use the two points (x, y) and $(0, b)$ in the coordinate formula for slope, we must get m:

$$\frac{y - b}{x - 0} = m \qquad \text{Coordinate formula for slope}$$

$$\frac{y - b}{x} = m \qquad \text{Simplify.}$$

$$y - b = mx \qquad \text{Multiply each side by } x.$$

$$y = mx + b \qquad \text{Add } b \text{ to each side.}$$

Since (x, y) was arbitrary, $y = mx + b$ is the equation for this line. Since the slope and y-intercept are apparent from this equation, it is called the *slope-intercept form*.

Slope-Intercept Form

The equation of a line in **slope-intercept form** is

$$y = mx + b,$$

where m is the slope and $(0, b)$ is the y-intercept.

So if you know the slope and y-intercept for a line, you can use slope-intercept form to write its equation, as shown in Example 1.

E X A M P L E   1

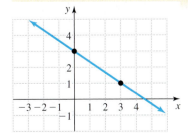

Figure 3.20

Writing an equation given the slope and *y*-intercept

Find the equation of each line in slope-intercept form:

a) The line that goes through $(0, -6)$ and has slope -2

b) The line shown in Fig. 3.20

Solution

a) Since the y-intercept is $(0, -6)$, the equation is $y = -2x - 6$.

b) From Fig. 3.20 we see that the y-intercept is $(0, 3)$. If we start at the y-intercept and move down 2 units and 3 units to the right, we get to another point on the line. So the slope is $-\frac{2}{3}$ and the equation in slope-intercept form is $y = -\frac{2}{3}x + 3$.

> Now do Exercises 1–12

To find the slope and y-intercept from an equation, simply rewrite the equation in slope-intercept form, that is, solve it for y. Of course, if there is no y-term in the equation, it is a vertical line and slope is undefined.

E X A M P L E   2

Changing to slope-intercept form
Find the slope and y-intercept of the line $3x - 2y = 5$.

Solution
Solve for y to get slope-intercept form:

$$3x - 2y = 5 \qquad \text{Original equation}$$
$$-2y = -3x + 5 \qquad \text{Subtract } 3x \text{ from each side.}$$
$$y = \frac{3}{2}x - \frac{5}{2} \qquad \text{Divide each side by } -2.$$

The slope is $\frac{3}{2}$, and the y-intercept is $\left(0, -\frac{5}{2}\right)$.

> Now do Exercises 13–24

‹ **Helpful Hint** ›

Note that every term in a linear equation in two variables is either a constant or a multiple of a variable. That is why equations in one variable of the form $ax = b$ were called linear equations in Chapter 2.

‹2› Using Slope-Intercept Form for Graphing

In the slope-intercept form, a point on the line (the y-intercept) and the slope are readily available. To graph a line, we can start at the y-intercept and count off the rise and run to get a second point on the line.

E X A M P L E   3

Graphing a line using the slope and y-intercept
Graph the line $y = \frac{2}{3}x + 1$ by using its slope and y-intercept.

Solution
A slope of $\frac{2}{3}$ means that the line rises 2 units in a run of 3 units. So starting at the y-intercept $(0, 1)$, rise 2 and run 3 to locate a second point on the line as shown in Fig. 3.21. The second point is $(3, 3)$. You can draw the line through these points or rise 2 and run 3 from $(3, 3)$ to locate a third point $(6, 5)$. You need only two points to determine the location of the line, but locating more points will improve the accuracy of your graph.

> Now do Exercises 25–40

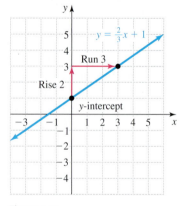

Figure 3.21

Graphing the line $y = mx + b$ by counting off the slope as in Example 3 reinforces the idea of slope. This method works best when b is an integer and m is a simple rational number. If the equation is more complicated, you can always graph it as we did in Section 3.1, by simply finding some points that satisfy the equation. As we saw in Section 3.1, a couple of good points to locate are the x- and y-intercepts.

‹3› Standard Form

In Section 3.1 we defined a linear equation in two variables as an equation of the form $Ax + By = C$, where A and B are not both zero. The form $Ax + By = C$ is called the *standard form* of the equation of a line. This form of a linear equation is common in applications. For example, if x students paid \$5 each and y adults paid \$7 each to attend a play for which the ticket sales totaled \$1900, then $5x + 7y = 1900$.

> **Standard Form**
>
> The equation of a line in **standard form** is
>
> $$Ax + By = C,$$
>
> where A, B, and C are real numbers with A and B not both zero.

Standard form is not unique. Multiplying each side of an equation in standard form by the same nonzero number will produce another equivalent equation in standard form. For example, $2x - 3y = 7$, $-6x + 9y = -21$, and $x - \frac{3}{2}y = \frac{7}{2}$ are all standard form for the same line. To simplify this situation we prefer only integers for A, B, and C with A being positive and as small as possible. So $2x - 3y = 7$ would be the preferred standard form.

EXAMPLE 4

‹ Helpful Hint ›

Solve $Ax + By = C$ for y, to get

$$y = \frac{-A}{B}x + \frac{C}{B}.$$

So the slope of $Ax + By = C$ is $-\frac{A}{B}$. This fact can be used in checking standard form. The slope of $2x - 4y = 3$ in Example 4 is $\frac{-2}{-4}$ or $\frac{1}{2}$, which is the slope of the original equation.

Changing to standard form

Write the equation $y = \frac{1}{2}x - \frac{3}{4}$ in standard form using only integers and a positive coefficient for x.

Solution

Use the properties of equality to get the equation in the form $Ax + By = C$:

$$y = \frac{1}{2}x - \frac{3}{4} \qquad \text{Original equation}$$

$$-\frac{1}{2}x + y = -\frac{3}{4} \qquad \text{Subtract } \tfrac{1}{2}x \text{ from each side.}$$

$$4\left(-\frac{1}{2}x + y\right) = 4\left(-\frac{3}{4}\right) \qquad \begin{array}{l}\text{Multiply each side by 4 to}\\ \text{get integral coefficients.}\end{array}$$

$$-2x + 4y = -3 \qquad \text{Distributive property}$$

$$2x - 4y = 3 \qquad \begin{array}{l}\text{Multiply by } -1 \text{ to make the}\\ \text{coefficient of } x \text{ positive.}\end{array}$$

> **Now do Exercises 41–48**

Since slope is undefined for a vertical line, we can't write the equation of a vertical line in slope-intercept form. However, the equations of vertical lines are included in standard form. For example, the vertical line $x = 4$ is just a simplified version $1 \cdot x + 0 \cdot y = 4$, which is standard form. *Every line has an equation in standard form*, but slope-intercept form is only for lines that have a defined slope.

‹4› Point-Slope Form

Suppose that (x, y) is an arbitrary point on a line through (x_1, y_1) with slope m. If m is positive, the line would look like the one shown in Fig. 3.22. If we use the two points (x, y) and (x_1, y_1) in the coordinate formula for slope, we must get m:

$$\frac{y - y_1}{x - x_1} = m \qquad \text{Coordinate formula for slope}$$

$$y - y_1 = m(x - x_1) \qquad \text{Multiply each side by } x - x_1.$$

Since (x, y) was arbitrary, $y - y_1 = m(x - x_1)$ is the equation for this line. So if you know a specific point (x_1, y_1) on a line with slope m, you can write its equation in this form, the *point-slope form.*

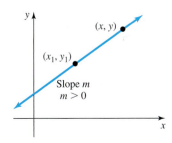

Figure 3.22

> **Point-Slope Form**
>
> The equation of the line through (x_1, y_1) with slope m in **point-slope form** is
> $$y - y_1 = m(x - x_1).$$

To write the equation of a line in slope-intercept form you must know the slope and the y-intercept. The point-slope form is more general. It enables you to write the equation of a line if you know the slope and *any* point on the line.

E X A M P L E   5

Writing an equation for a line given a point and the slope

Find an equation for the line through $(-2, 5)$ with slope -3 and solve it for y.

Solution

Use $x_1 = -2$, $y_1 = 5$, and $m = -3$ in the point-slope form:
$$y - 5 = -3(x - (-2))$$

Now solve the equation for y:
$$y - 5 = -3(x + 2)$$
$$y - 5 = -3x - 6$$
$$y = -3x - 1$$

> Now do Exercises 49–56

⟨ **Calculator Close-Up** ⟩

Graph $y = -3x - 1$ and check that the line goes through $(-2, 5)$. On a TI-83 press TRACE and then enter the x-coordinate -2. The calculator will show $x = -2$ and $y = 5$.

Two points determine a line. If you know two points on a line, you can graph the line and you can write an equation for the line. To get the equation, you must find the slope from the two points and then use the slope along with one of the points in the point-slope form, as shown in Example 6.

E X A M P L E   6

Writing an equation for a line given two points on the line

Find the equation of the line through the given pair of points and solve it for y.

a) $(3, 5)$ and $(4, 7)$ b) $(3, -2)$ and $(-1, 1)$

Solution

a) First find the slope:
$$m = \frac{y_2 - y_1}{x_2 - x_1} = \frac{7 - 5}{4 - 3} = 2$$

Now use the slope and one point, say $(3, 5)$, in the point-slope form:

$$y - y_1 = m(x - x_1) \qquad \text{Point-slope form}$$
$$y - 5 = 2(x - 3) \qquad \text{Substitute } m = 2, (x_1, y_1) = (3, 5).$$
$$y - 5 = 2x - 6 \qquad \text{Distributive property}$$
$$y = 2x - 1 \qquad \text{Solve for } y.$$

Because $(3, 5)$ and $(4, 7)$ both satisfy $y = 2x - 1$, we can be sure that we have the correct equation.

b) First find the slope of the line through $(3, -2)$ and $(-1, 1)$:

$$m = \frac{1 - (-2)}{-1 - 3}$$

$$= \frac{3}{-4} = -\frac{3}{4}$$

Now use this slope and one of the points, say $(3, -2)$, to write the equation in point-slope form:

$$y - (-2) = -\frac{3}{4}(x - 3) \qquad \text{Point-slope form}$$

$$y + 2 = -\frac{3}{4}x + \frac{9}{4} \qquad \text{Distributive property}$$

$$y = -\frac{3}{4}x + \frac{1}{4} \qquad \text{Solve for } y.$$

Note that we would get the same equation if we had used slope $-\frac{3}{4}$ and the other point $(-1, 1)$. Try it.

> **Now do Exercises 57–66**

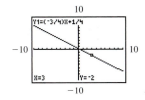

We know that if a line has slope m, then the slope of any line perpendicular to it is $-\frac{1}{m}$, provided $m \neq 0$. We also know that nonvertical parallel lines have equal slopes. These facts are used in Example 7.

EXAMPLE 7

Writing equations of perpendicular and parallel lines

In each case find an equation for line l and then solve it for y.

    a) Line l goes through $(2, 0)$ and is perpendicular to the line through $(5, -1)$ and $(-1, 3)$.

    b) Line l goes through $(-1, 6)$ and is parallel to the line through $(2, 4)$ and $(7, -11)$.

Solution

    a) First find the slope of the line through $(5, -1)$ and $(-1, 3)$:

$$m = \frac{3 - (-1)}{-1 - 5} = \frac{4}{-6} = -\frac{2}{3}$$

Because line l is perpendicular to this line, line l has slope $\frac{3}{2}$. Now use $(2, 0)$ and the slope $\frac{3}{2}$ in the point-slope formula to get the equation of line l:

$$y - 0 = \frac{3}{2}(x - 2)$$

$$y = \frac{3}{2}x - 3 \qquad \text{Distributive property}$$

    b) First find the slope of the line through $(2, 4)$ and $(7, -11)$:

$$m = \frac{-11 - 4}{7 - 2} = \frac{-15}{5} = -3$$

Since parallel lines have equal slopes, use the slope -3 and the point $(-1, 6)$:

$$y - 6 = -3(x - (-1)) \qquad \text{Point-slope form}$$

$$y - 6 = -3(x + 1) \qquad \text{Simplify.}$$

$$y - 6 = -3x - 3 \qquad \text{Distributive property}$$

$$y = -3x + 3 \qquad \text{Solve for } y.$$

> **Now do Exercises 67–70**

EXAMPLE   **8**

Finding an equation of a line

Write an equation in standard form with integral coefficients for the line l through (2, 5) that is perpendicular to the line $2x + 3y = 1$.

Solution

First solve the equation $2x + 3y = 1$ for y to find its slope:

$$2x + 3y = 1$$
$$3y = -2x + 1$$
$$y = -\frac{2}{3}x + \frac{1}{3} \quad \text{The slope is } -\frac{2}{3}.$$

The slope of line l is the opposite of the reciprocal of $-\frac{2}{3}$. So line l has slope $\frac{3}{2}$ and goes through (2, 5). Now use the point-slope form to write the equation:

$$y - 5 = \frac{3}{2}(x - 2) \quad \text{Point-slope form}$$
$$y - 5 = \frac{3}{2}x - 3 \quad \text{Distributive property}$$
$$y = \frac{3}{2}x + 2$$
$$-\frac{3}{2}x + y = 2$$
$$3x - 2y = -4 \quad \text{Multiply each side by } -2.$$

So $3x - 2y = -4$ is an equation in standard form of the line through (2, 5) that is perpendicular to $2x + 3y = 1$.

> **Now do Exercises 71–74**

‹ **Calculator Close-Up** ›

Graph $y_1 = \left(-\frac{2}{3}\right)x + \frac{1}{3}$ and $y_2 = \left(\frac{3}{2}\right)x + 2$ to check that y_2 is perpendicular to y_1 and that y_2 goes through (2, 5). The lines will look perpendicular only if the same unit length is used on both axes.

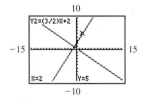

Some calculators have a feature that adjusts the window to get the same unit length on both axes.

The three forms for the equation of a line are summarized as follows.

Form	General Equation	Notes
Slope-intercept	$y = mx + b$	Slope is m and $(0, b)$ is y-intercept. Good for graphing. Does not include vertical lines.
Standard	$Ax + By = C$	Includes all lines. Intercepts are easy to find.
Point-slope	$y - y_1 = m(x - x_1)$	Used to write an equation when given a point and a slope or two points.

‹5› Applications

The linear equation $y = mx + b$ with $m \neq 0$ expresses y as a linear function of x. In Example 9, we will use the point-slope formula to find the well-known formula that expresses Fahrenheit temperature as a linear function of Celsius temperature.

EXAMPLE   **9**

Finding a linear function given two points

Fahrenheit temperature F is a linear function of Celsius temperature C. Water freezes at 0°C or 32°F and boils at 100°C or 212°F. Find the linear function.

Solution

We want an equation of the line that contains the points (0, 32) and (100, 212) as shown in Fig. 3.23. Use C as the independent variable (x) and F as the dependent variable (y). The

slope of the line is

$$m = \frac{F_2 - F_1}{C_2 - C_1} = \frac{212 - 32}{100 - 0} = \frac{180}{100} = \frac{9}{5}.$$

Figure 3.23

Using a slope of $\frac{9}{5}$ and the point $(100, 212)$ in the point-slope formula, we get

$$F - 212 = \frac{9}{5}(C - 100).$$

We can solve this equation for F to get the familiar linear function that is used to determine Fahrenheit temperature from Celsius temperature:

$$F = \frac{9}{5}C + 32$$

Because we knew the intercept $(0, 32)$, we could have used it and the slope $\frac{9}{5}$ in slope-intercept form to write $F = \frac{9}{5}C + 32$.

Now do Exercises 97–102

Warm-Ups ▼

Fill in the blank.

1. The _____ form is $y = mx + b$.
2. The _____ is $Ax + By = C$.
3. The _____ form is $y - y_1 = m(x - x_1)$.
4. The _____ of the line $y = 3x - 8$ is 3.
5. The _____ of the line $y = 4x + 1$ is $(0, 1)$.
6. Every line has an equation in _____ form.
7. _____ form can be used for all lines except vertical ones.

True or false?

8. There is exactly one line through a given point with a given slope.
9. The line $y - 2 = 4(x - 5)$ goes through $(2, 5)$ and has slope 4.

10. The equation of the line through $(0, 5)$ with slope m is $y = mx + 5$.
11. The x-coordinate of the y-intercept of a nonvertical line is 0.
12. The y-coordinate of the x-intercept of $y = 2x - 1$ is 0.

13. The line $2x + 3y = 7$ has slope $\frac{2}{3}$.
14. The lines $x - y = 1$ and $x + y = 1$ are perpendicular.

15. The lines $2x + 3y = 4$ and $2x + 3y = 5$ are parallel.

‹ 1 › Slope-Intercept Form

Write an equation in slope-intercept form (if possible) for each of the lines shown. See Example 1.

1.

2.

3.

4.

5.

6.

7.

8.

9.

10.

11.

12.

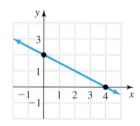

Write each equation in slope-intercept form, and identify the slope and y-intercept. See Example 2.

13. $2x + 5y = 1$

14. $3x - 3y = 2$

15. $3x - y - 2 = 0$

16. $5 - x - 2y = 0$

17. $y + 3 = 5$

18. $y - 9 = 0$

19. $y - 2 = 3(x - 1)$

20. $y + 4 = -2(x - 5)$

21. $y - \dfrac{1}{2} = \dfrac{1}{3}\left(x + \dfrac{1}{4}\right)$

22. $y - \dfrac{1}{3} = -\dfrac{1}{2}\left(x - \dfrac{1}{4}\right)$

23. $y - 6000 = 0.01(x + 5700)$

24. $y - 5000 = 0.05(x - 1990)$

⟨2⟩ Using Slope-Intercept Form for Graphing

Graph each line. Use the slope and y-intercept. See Example 3.

25. $y = \dfrac{1}{2}x$

26. $y = -\dfrac{2}{3}x$

35. $y = 3x + 1$
   $y = -\dfrac{1}{3}x + 1$

36. $y = 2x + 3$
   $y = -\dfrac{1}{2}x + 3$

27. $y = 2x - 3$

28. $y = -x + 1$

37. $y = \dfrac{2}{3}x - 1$
   $y = \dfrac{2}{3}x + 1$

38. $y = -2x + 4$
   $y = -2x + 2$

29. $y = -\dfrac{2}{3}x + 2$

30. $y = 3x - 4$

39. $y = \dfrac{3}{4}x + 3$
   $y = -\dfrac{4}{3}x + 1$

40. $y = \dfrac{2}{3}x + 1$
   $y = -\dfrac{3}{2}x + 3$

31. $3y + x = 0$

32. $4y - x = 0$

Graph each pair of lines in the same coordinate system using the slope and y-intercept.

33. $y = x + 3$
   $y = x + 2$

34. $y = -x + 2$
   $y = -x - 2$

⟨3⟩ Standard Form

Write each equation in standard form using only integers and a positive coefficient for x. See Example 4.

41. $y = \dfrac{1}{3}x - 2$

42. $y = \frac{1}{2}x + 7$

43. $y - 5 = \frac{1}{2}(x + 3)$

44. $y - 1 = \frac{1}{4}(x - 6)$

45. $y + \frac{1}{2} = \frac{1}{3}(x - 4)$

46. $y + \frac{1}{3} = \frac{1}{4}(x - 3)$

47. $0.05x + 0.06y - 8.9 = 0$

48. $0.03x - 0.07y = 2$

⟨4⟩ Point-Slope Form

Find an equation of the line that goes through the given point and has the given slope. Give the answer in slope-intercept form. See Example 5.

49. $(2, -3)$ with slope 2

50. $(-3, -1)$ with slope 6

51. $(-2, 3)$ with slope $-\frac{1}{2}$

52. $(3, 5)$ with slope $\frac{2}{3}$

53. $(5, 12)$ with slope 0

54. $(-9, -4)$ with slope 0

55. $(3, 60)$ with slope 20

56. $(-5, 150)$ with slope -30

Find an equation of the line through each given pair of points. Give the answer in slope-intercept form. See Example 6.

57. $(-1, -11)$ and $(4, 4)$

58. $(-2, 12)$ and $(1, -3)$

59. $(2, 2)$ and $(-1, 1)$

60. $(2, 3)$ and $(-5, 6)$

61. $(8, 0)$ and $(-6, 7)$

62. $(6, 0)$ and $(9, 1)$

63. $(2, 13)$ and $(4, 26)$

64. $(3, 120)$ and $(-2, -80)$

65. $(-5, 6)$ and $(14, 6)$

66. $(3, -9)$ and $(-4, -9)$

Find an equation of line l and solve it for y. See Example 7.

67. Line l goes through $(-1, -12)$ and is perpendicular to the line through $(-3, 1)$ and $(5, -1)$.

68. Line l goes through $(0, 0)$ and is perpendicular to the line through $(0, 6)$ and $(-5, 0)$.

69. Line l goes through $(0, 0)$ and is parallel to the line through $(9, -3)$ and $(-3, 6)$.

70. Line l goes through $(2, -4)$ and is parallel to the line through $(6, 2)$ and $(-2, 6)$.

Find the equation of line l. Write the answer in standard form with integral coefficient with a positive coefficient for x. See Example 8.

71. Line l goes through $(3, 2)$ and is perpendicular to $3x - 12y = 1$.

72. Line l goes through $(-2, 5)$ and is perpendicular to $6x + 3y = 7$.

73. Line l goes through $(4, -2)$ and is parallel to $4x + 2y = 5$.

74. Line l goes through $(-6, 2)$ and is parallel to $3x - 9y = 7$.

Miscellaneous

Find the equation of line l in each case and then write it in standard form with integral coefficients.

75. Line l has slope $\frac{1}{2}$ and goes through $(0, 5)$.

76. Line l has slope 5 and goes through $\left(0, \frac{1}{2}\right)$.

77. Line l has x-intercept $(2, 0)$ and y-intercept $(0, 4)$.

78. Line l has y-intercept $(0, 5)$ and x-intercept $(4, 0)$.

79. Line l goes through $(-3, -1)$ and is parallel to $y = 2x + 6$.

80. Line l goes through $(1, -3)$ and is parallel to $y = -3x - 5$.

81. Line l is parallel to $2x + 4y = 1$ and goes through $(-3, 5)$.

82. Line l is parallel to $3x - 5y = -7$ and goes through $(-8, -2)$.

83. Line l goes through $(1, 1)$ and is perpendicular to $y = \frac{1}{2}x - 3$.

84. Line l goes through $(-1, -2)$ and is perpendicular to $y = -3x + 7$.

85. Line l goes through $(-4, -3)$ and is perpendicular to $x + 3y = 4$.

86. Line l is perpendicular to $2y + 5 - 3x = 0$ and goes through $(2, 7)$.

87. Line l goes through $(2, 5)$ and is parallel to the x-axis.

88. Line l goes through $(-1, 6)$ and is parallel to the y-axis.

Determine whether each pair of lines is parallel, perpendicular, or neither.

89. $y = 3x - 8, x + 3y = 7$

90. $y = \frac{1}{2}x - 4, \frac{1}{2}x + \frac{1}{4}y = 1$

91. $2x - 4y = 9, \frac{1}{3}x = \frac{2}{3}y - 8$

92. $\frac{1}{4}x - \frac{1}{6}y = \frac{1}{3}, \frac{1}{3}y = \frac{1}{2}x - 2$

93. $2y = x + 6, y - 2x = 4$

94. $y - 3x = 5, 3x + y = 7$

95. $x - 6 = 9, y - 4 = 12$

96. $9 - x = 3, \frac{1}{2}x = 8$

‹5› Applications

Solve each problem. See Example 9.

97. *Heating water.* The temperature of a cup of water is a linear function of the time that it is in the microwave. The temperature at 0 seconds is 60°F and the temperature at 120 seconds is 200°F.

    a) Express the linear function in the form $t = ms + b$ where t is the Fahrenheit temperature and s is the time in seconds. [*Hint:* Write the equation of the line through $(0, 60)$ and $(120, 200)$.]

    b) Use the linear function to determine the temperature at 30 seconds.

    c) Graph the linear function.

98. *Making circuit boards.* The weekly cost of making circuit boards is a linear function of the number of boards made. The cost is \$1500 for 1000 boards and \$2000 for 2000 boards.

    a) Express the linear function in the form $C = mn + b$, where C is the cost in dollars and n is the number of boards.

    b) What is the cost if only one circuit board is made in a week?

    c) Graph the linear function.

99. *Carbon dioxide emission.* Worldwide emission of carbon dioxide (CO_2) increased linearly from 14 billion tons in 1970 to 26 billion tons in 2000 (World Resources Institute, www.wri.org).

    a) Express the emission as a linear function of the year in the form $y = mx + b$, where y is in billions of tons and x is the year. [*Hint:* Write the equation of the line through $(1970, 14)$ and $(2000, 26)$.]

    b) Use the function from part (a) to predict the worldwide emission of CO_2 in 2015.

100. *World energy use.* Worldwide energy use in all forms increased linearly from the equivalent of 3.5 billion tons of oil in 1970 to the equivalent of 6.5 billion tons of oil in 2000 (World Resources Institute, www.wri.org).

    a) Express the energy use as a function of the year in the form $y = mx + b$ where x is the year and y is the energy use in billions of tons of oil.

b) Use the function from part (a) to predict the worldwide energy use in 2015.

101. ***Depth and flow.*** When the depth of the water in the Tangipahoa River at Robert, Louisiana, is 9.14 feet, the flow is 1230 cubic feet per second (ft³/sec). When the depth is 7.84 feet, the flow is 826 ft³/sec. (U.S. Geological Survey, www.usgs.gov.) Let w represent the flow in cubic feet per second and d represent the depth in feet.

   a) Write the equation of the line through (9.14, 1230) and (7.84, 826) and express w in terms of d. Round to two decimal places.

   b) What is the flow when the depth is 8.25 ft?

   c) Is the flow increasing or decreasing as the depth increases?

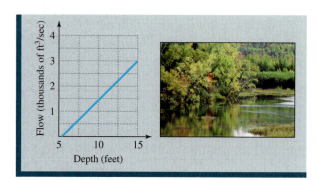

Figure for Exercise 101

102. ***Buying stock.*** A mutual fund manager spent $329,175 on x shares of Exxon Mobil Stock at $66.50 per share and y shares of Walmart stock at $49.50 per share.

   a) Write a linear equation that models this situation.

   b) If 990 shares of Exxon Mobil were purchased, then how many shares of Walmart were purchased?

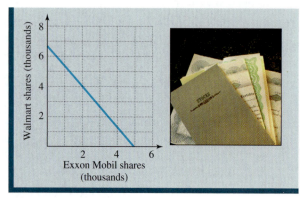

Figure for Exercise 102

c) Find and interpret the intercepts of the graph of the linear equation.

d) As the number of shares of Exxon Mobil increases, does the number of shares of Walmart increase or decrease?

Getting More Involved

103. ***Exploration***

   The **intercept form** for the equation of a line is

   $$\frac{x}{a} + \frac{y}{b} = 1$$

   where neither a nor b is zero.

   a) Find the x- and y-intercepts for $\frac{x}{4} + \frac{y}{6} = 1$.

   b) Find the x- and y-intercepts for $\frac{x}{a} + \frac{y}{b} = 1$.

   c) Write the equation of the line through (0, 3) and (−5, 0) in intercept form.

   d) Which lines cannot be written in intercept form?

Graphing Calculator Exercises

104. Graph the equation $y = 0.5x - 1$ using the standard viewing window. Adjust the range of y-values so that the line goes from the lower left corner of your viewing window to the upper right corner.

105. Graph $y = x - 3000$, using a viewing window that shows both the x-intercept and the y-intercept.

106. Graph $y = 2x - 400$ and $y = -0.5x + 1$ on the same screen, using the viewing window $-500 \leq x \leq 500$ and $-1000 \leq y \leq 1000$. Should these lines be perpendicular? Explain.

107. The lines $y = 2x - 3$ and $y = 1.9x + 2$ are not parallel. Find a viewing window in which the lines intersect. Estimate the point of intersection.

Saturday, July 4, 2043

Math *at Work* | Day of the Week Calculator

Did you know that July 4, 2043, will be a Saturday? Here is how you can find the day of the week for any date. First select a date; say, July 4, 2043, or 7/4/2043. So month = 7, day = 4, and year = 2043. Next find a and use it to find y and m, where

$$a = \frac{14 - \text{month}}{12}, \quad y = \text{year} - a, \quad \text{and} \quad m = \text{month} + 12a - 2.$$

All divisions, unless noted otherwise, are integer divisions, in which we keep the quotient and discard the remainder. The quotient for $\frac{14-7}{12}$ is 0 and the remainder is 7. So $a = 0$, $y = 2043$, and $m = 7 + 12(0) - 2 = 5$. Next, plug the values of y and m into the following formula to calculate d:

$$d = \left(\text{day} + y + \frac{y}{4} - \frac{y}{100} + \frac{y}{400} + \frac{31m}{12} \right) \text{mod } 7$$

For the divisions within the parentheses we keep the quotient. But "mod 7" means that we divide by 7 but keep the remainder as the value of d. For example, 30 mod 7 is 2 because the remainder of 30 divided by 7 is 2. For 7/4/2043 we have

$$\left(4 + 2043 + \frac{2043}{4} - \frac{2043}{100} + \frac{2043}{400} + \frac{31(5)}{12} \right) = 4 + 2043 + 510 - 20 + 5 + 12 = 2554.$$

Now 2554 mod 7 is 6 (the remainder of division by 7). So $d = 6$. The value of d corresponds to a day of the week with 0 = Sunday, 1 = Monday, 2 = Tuesday, 3 = Wednesday, 4 = Thursday, 5 = Friday, and 6 = Saturday. So July 4, 2043, is a Saturday. Now try the current date to check this out.

Mid-Chapter **Quiz** | Sections 3.1 through 3.3 | Chapter 3

Use the given equation to find the missing coordinates in the given table.

1. $2x - 3y = 12$

x	−3		3	
y		4		0

2. $y = \frac{3}{4}x + 3$

x	−4		8	
y		12		−6

5. $x = 4$

6. $y = 2$

Graph each equation in the rectangular coordinate system.

3. $y = -x$

4. $y = 5 - \frac{1}{2}x$

7. $2x - 30y = 600$

8. $y = \frac{5}{3}x - 4$

Find the slope and y-intercept for each line.

9. $y = -5x + 2$ **10.** $y = 6$

11. $3x - 8y = 16$

Write each equation in standard form using only integers.

12. $y = -0.02x + 5$

13. $\frac{1}{2}y - \frac{1}{3}x - 9 = 0$

Find the equation in slope-intercept form for each line.

14. The line through $(6, 3)$ with slope $\frac{1}{3}$

15. The line through $(0, 6)$ that is parallel to $y = -5x + 12$

16. The line through $(0, -4)$ that is perpendicular to $3x - 5y = 9$

17. The line through $(4, 5)$ that is parallel to the x-axis

18. The line through $(-2, 1)$ and $(3, 5)$

Miscellaneous.

19. Draw the graph of a line through the origin with slope $\frac{3}{4}$.

20. Is the line through $(-2, 1)$ and $(3, 5)$ parallel or perpendicular to the line through $(-4, 0)$ and $(0, -5)$?

3.4   Linear Inequalities and Their Graphs

In This Section

⟨1⟩ **Graphing Linear Inequalities**

⟨2⟩ **The Test Point Method**

⟨3⟩ **Graphing Compound Inequalities**

⟨4⟩ **Absolute Value Inequalities**

⟨5⟩ **Inequalities with No Solution**

⟨6⟩ **Applications**

In Sections 3.1–3.3, you studied linear equations. We now turn our attention to linear inequalities.

⟨1⟩ Graphing Linear Inequalities

A linear inequality is a linear equation with the equal sign replaced by an inequality symbol.

> **Linear Inequality**
>
> If A, B, and C are real numbers with A and B not both zero, then
>
> $$Ax + By \leq C$$
>
> is called a **linear inequality.** In place of $\leq$, we can also use $\geq$, $<$, or $>$.

The graph of $y = mx + b$ is a nonvertical line. If the y-coordinate of a point on this line is increased, then the new point will be above the line and will satisfy $y > mx + b$. Likewise, if the y-coordinate of a point on the line is decreased, then the new point will be below the line and will satisfy $y < mx + b$. All points above the line $y = mx + b$ satisfy the inequality $y > mx + b$. Since there are infinitely many points in the solution set to $y > mx + b$, the solution set is best illustrated with a graph as shown in Fig. 3.24. The solution set or graph for $y > mx + b$ is the shaded region above the line $y = mx + b$. The boundary line is dashed to indicate that the line is not part of the solution set. If the inequality symbol is $\leq$ or $\geq$, a solid boundary line is used to indicate that the line is part of the solution set.

The only lines that do not have an equation of the form $y = mx + b$ are the vertical lines. The equation for a vertical line is of the form $x = k$, where k is a real number. So the graph of $x > k$ is the region to the right of the line $x = k$ and the graph of $x < k$ is the region to the left.

Figure 3.24

Strategy for Graphing a Linear Inequality

1. Solve the inequality for y, and then graph $y = mx + b$.

> $y > mx + b$ is satisfied above the line.
> $y = mx + b$ is satisfied on the line itself.
> $y < mx + b$ is satisfied below the line.

2. If the inequality involves x and not y, then graph the vertical line $x = k$.

> $x > k$ is satisfied to the right of the line.
> $x = k$ is satisfied on the line itself.
> $x < k$ is satisfied to the left of the line.

E X A M P L E   1

Graphing linear inequalities

Graph each inequality.

a) $y < \dfrac{1}{2}x - 1$ **b)** $y \geq -2x + 1$ **c)** $3x - 2y < 6$

⟨ **Helpful Hint** ⟩

Why do we keep drawing graphs? When we solve $2x + 1 = 7$, we don't bother to draw a graph showing 3 because the solution set is so simple. However, the solution set to a linear inequality is a very large set of ordered pairs. Graphing gives us a way to visualize the solution set.

Solution

a) The set of points satisfying this inequality is the region below the line $y = \frac{1}{2}x - 1$. To show this region, we first graph the boundary line $y = \frac{1}{2}x - 1$. The slope of the line is $\frac{1}{2}$, and the y-intercept is $(0, -1)$. Start at $(0, -1)$ on the y-axis, and then rise 1 and run 2 to get a second point of the line. We draw the line dashed because points on the line do not satisfy this inequality. The solution set to the inequality is the shaded region shown in Fig. 3.25.

b) Because the inequality symbol is $\geq$, every point on or above the line $y = -2x + 1$ satisfies $y \geq -2x + 1$. To graph the line use y-intercept $(0, 1)$ and slope -2. Start at $(0, 1)$ and find a second point on the line using a rise of -2 and a run of 1. Draw a solid line through $(0, 1)$ and $(1, -1)$ to show that it is included in the solution set to the inequality. Shade above the line as in Fig. 3.26.

Figure 3.25

Figure 3.26

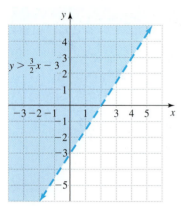

Figure 3.27

c) First solve for y:

$$3x - 2y < 6$$
$$-2y < -3x + 6$$
$$y > \frac{3}{2}x - 3 \quad \text{Divide by } -2 \text{ and reverse the inequality.}$$

To graph this inequality, first graph the boundary $y = \frac{3}{2}x - 3$ using its y-intercept $(0, -3)$ and slope $\frac{3}{2}$ or graph the line using its intercepts $(0, -3)$ and $(2, 0)$. Use a dashed line for the boundary, and shade the region above the line as in Fig. 3.27.

Now do Exercises 1–12

CAUTION In Example 1(c) we solved the inequality for y before graphing the line. We did that because $<$ corresponds to the region below the line and $>$ corresponds to the region above the line only when the inequality is solved for y.

E X A M P L E 2

Inequalities with horizontal and vertical boundaries
Graph the inequalities.

a) $y \leq 5$ **b)** $x > 4$

Solution

a) The line $y = 5$ is the horizontal line with y-intercept $(0, 5)$. Draw a solid horizontal line and shade below it as in Fig. 3.28.

b) The points that satisfy $x > 4$ lie to the right of the vertical line $x = 4$. The solution set is shown in Fig. 3.29.

Now do Exercises 13–16

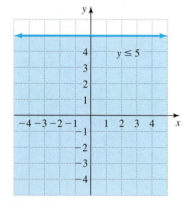

Figure 3.28

⟨2⟩ The Test-Point Method

We have seen that every point above the line $y = mx + b$ satisfies $y > mx + b$ and every point below the line satisfies $y < mx + b$. Likewise, every point on one side of the line $Ax + By = C$ satisfies $Ax + By > C$ and every point on the other side satisfies $Ax + By < C$. However, in this case we can't tell which side is which from the inequality symbol. We simply test a point not on the line to see if it satisfies the inequality. Here is the strategy.

Strategy for Graphing a Linear Inequality by the Test Point Method

1. Graph $Ax + By = C$. (Using the x- and y-intercepts is usually best.)
2. Test any point *not* on the line to see if it satisfies the inequality.
3. If the test point satisfies the inequality, shade the region containing the test point. If not, shade the other region.

Figure 3.29

The test point method works also on inequalities that are solved for y, but it is usually used for inequalities that are not solved for y.

EXAMPLE **3**

Using the test point method

Graph the inequality $3x - 4y > 7$.

Solution

First graph the equation $3x - 4y = 7$ using the x-intercept and the y-intercept. If $x = 0$, then $y = -\frac{7}{4}$. If $y = 0$, then $x = \frac{7}{3}$. Use the x-intercept $\left(\frac{7}{3}, 0\right)$ and the y-intercept $\left(0, -\frac{7}{4}\right)$ to graph the line as shown in Fig. 3.30(a). Select a point on one side of the line, say $(0, 1)$, to test in the inequality. Because

$$3(0) - 4(1) > 7$$

is false, the region on the other side of the line satisfies the inequality. The graph of $3x - 4y > 7$ is shown in Fig. 3.30(b).

(a)   (b)

Figure 3.30

Now do Exercises 17–26

⟨3⟩ Graphing Compound Inequalities

We can write compound inequalities with two variables just as we do for one variable using the connectives *and* or *or*. Remember that a compound statement involving *and* is true only if both parts are true. A compound statement involving *or* is true if one, or the other, or both parts are true.

EXAMPLE **4**

Satisfying compound inequalities

Determine whether the ordered pair $(-1, 3)$ satisfies each compound or absolute value inequality.

a) $y > 3$ and $x < 1$

b) $x + y > 4$ or $x - y < -2$

c) $|2x - 3y| > 5$

Solution

a) Let $x = -1$ and $y = 3$ in each inequality:

$$y > 3 \qquad\qquad\qquad x < 1$$
$$3 > 3 \quad \text{Incorrect.} \qquad -1 < 1 \quad \text{Correct.}$$

Since only one inequality is satisfied, $(-1, 3)$ does not satisfy $y > 3$ and $x < 1$.

b) Let $x = -1$ and $y = 3$ in each inequality:

$$x + y > 4 \qquad\qquad\qquad x - y < -2$$
$$-1 + 3 > 4 \quad \text{Incorrect.} \qquad -1 - 3 < -2 \quad \text{Correct.}$$

Since one inequality is satisfied, $(-1, 3)$ satisfies the compound inequality $x + y > 4$ or $x - y < -2$.

c) Let $x = -1$ and $y = 3$ in $|2x - 3y| > 5$:

$$|2(-1) - 3(3)| > 5$$
$$|-11| > 5 \quad \text{Correct.}$$

So $(-1, 3)$ satisfies $|2x - 3y| > 5$.

Now do Exercises 27–36

Example 5 illustrates two methods for solving a compound inequality with *and*.

E X A M P L E **5**

Graphing a compound inequality with *and*

Graph the compound inequality $y > x - 3$ and $y < -\frac{1}{2}x + 2$.

Solution

The Intersection Method

Start by graphing the lines $y = x - 3$ and $y = -\frac{1}{2}x + 2$. Points that satisfy $y > x - 3$ lie above the line $y = x - 3$ and points that satisfy $y < -\frac{1}{2}x + 2$ lie below the line $y = -\frac{1}{2}x + 2$, as shown in Fig. 3.31(a). Since the connectivity is *and,* only points that are shaded with both colors (the intersection of the two regions) satisfy the compound inequality. The solution set to the compound inequality is shown in Fig. 3.31(b). Dashed lines are used because the inequalities are $>$ and $<$.

(a)

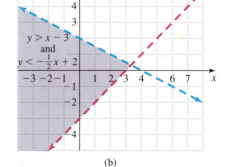

(b)

Figure 3.31

The Test Point Method

Again graph the lines, but this time select a point in each of the four regions determined by the lines as shown in Fig. 3.32(a). Test each of the four points $(3, 3)$, $(0, 0)$, $(4, -5)$, and $(5, 0)$ to see if it satisfies the compound inequality:

$$y > x - 3 \qquad \text{and} \qquad y < -\tfrac{1}{2}x + 2$$

$$3 > 3 - 3 \qquad \text{and} \qquad 3 < -\tfrac{1}{2} \cdot 3 + 2 \qquad \text{\color{blue}{Second inequality is incorrect.}}$$

$$0 > 0 - 3 \qquad \text{and} \qquad 0 < -\tfrac{1}{2} \cdot 0 + 2 \qquad \text{\color{blue}{Both inequalities are correct.}}$$

$$-5 > 4 - 3 \qquad \text{and} \qquad -5 < -\tfrac{1}{2} \cdot 4 + 2 \qquad \text{\color{blue}{First inequality is incorrect.}}$$

$$0 > 5 - 3 \qquad \text{and} \qquad 0 < -\tfrac{1}{2} \cdot 5 + 2 \qquad \text{\color{blue}{Both inequalities are incorrect.}}$$

The only point that satisfies both inequalities is $(0, 0)$. So the solution set to the compound inequality consists of all points in the region containing $(0, 0)$ as shown in Fig. 3.32(b).

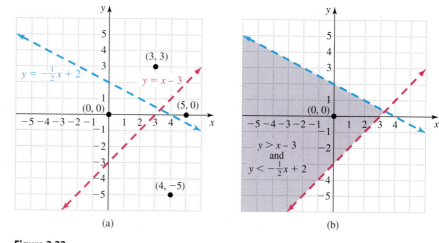

(a)   (b)

Figure 3.32

Now do Exercises 37–38

Example 6 involves a compound inequality using *or*. Remember that a compound sentence with *or* is true if one, the other, or both parts of it are true. The solution set to a compound inequality with *or* is the union of the two solution sets.

EXAMPLE 6

Graphing a compound inequality with *or*

Graph the compound inequality $2x - 3y \leq -6$ or $x + 2y \geq 4$.

Solution

The Union Method

Graph the line $2x - 3y = -6$ through its intercepts $(0, 2)$ and $(-3, 0)$. Since $(0, 0)$ does not satisfy this inequality, shade the region above this line as shown in Fig. 3.33(a). Graph the line $x + 2y = 4$ through $(0, 2)$ and $(4, 0)$. Since $(0, 0)$ does not satisfy this inequality, shade the region above the line as shown in Fig. 3.33(a). The union of these two solution sets consists of everything that is shaded as shown in Fig. 3.33(b). The boundary lines are solid because of the inequality symbols $\leq$ and $\geq$.

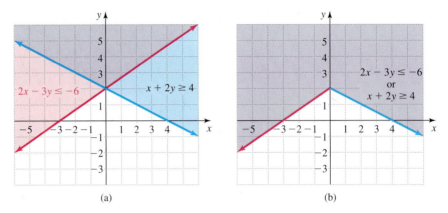

(a)   (b)

Figure 3.33

The Test Point Method

Graph the lines, and select a point in each of the four regions determined by the lines as shown in Fig. 3.34(a). Test each of the four points $(0, 0)$, $(-3, 2)$, $(0, 5)$, and $(3, 2)$ to see if it satisfies the compound inequality:

$2x - 3y \leq -6$	or	$x + 2y \geq 4$	
$2(0) - 3(0) \leq -6$	or	$0 + 2(0) \geq 4$	False
$2(-3) - 3(2) \leq -6$	or	$-3 + 2(2) \geq 4$	True
$2(0) - 3(5) \leq -6$	or	$0 + 2(5) \geq 4$	True
$2(3) - 3(2) \leq -6$	or	$3 + 2(2) \geq 4$	True

The solution set to the compound inequality consists of the three regions containing the test points that satisfy the compound inequality as shown in Fig. 3.34(b).

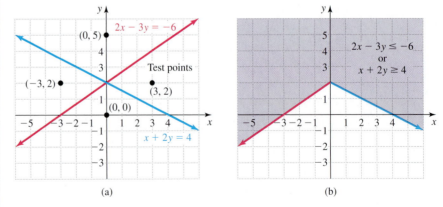

(a)   (b)

Figure 3.34

Now do Exercises 39–58

⟨4⟩ **Absolute Value Inequalities**

In Section 2.6 we learned that the absolute value inequality $|x| > 2$ is equivalent to the compound inequality $x < -2$ or $x > 2$. The absolute value inequality $|x| < 2$ is equivalent to the compound inequality $x > -2$ and $x < 2$. We can also write $|x| < 2$ as $-2 < x < 2$. We use these ideas with inequalities in two variables in Example 7.

E X A M P L E 7

Graphing absolute value inequalities

Graph each absolute value inequality.

    a) $|y - 2x| \leq 3$ **b)** $|x - y| > 1$

Solution

a) The inequality $|y - 2x| \leq 3$ is equivalent to $-3 \leq y - 2x \leq 3$, which is equivalent to the compound inequality

$$y - 2x \leq 3 \qquad \text{and} \qquad y - 2x \geq -3.$$

First graph the lines $y - 2x = 3$ and $y - 2x = -3$ as shown in Fig. 3.35(a). These lines divide the plane into three regions. Test a point from each region in the original inequality, say $(-5, 0)$, $(0, 1)$, and $(5, 0)$:

$$|0 - 2(-5)| \leq 3 \qquad |1 - 2 \cdot 0| \leq 3 \qquad |0 - 2 \cdot 5| \leq 3$$
$$10 \leq 3 \qquad\qquad 1 \leq 3 \qquad\qquad 10 \leq 3$$

Only $(0, 1)$ satisfies the original inequality. So the region satisfying the absolute value inequality is the shaded region containing $(0, 1)$ as shown in Fig. 3.35(b). The boundary lines are solid because of the $\leq$ symbol.

⟨ Helpful Hint ⟩

Remember that the absolute value of a quantity is its distance from 0 (Section 2.6). If $|w| < 3$, then w is less than 3 units from 0:

$$-3 < w < 3$$

If $|w| > 1$, then w is more than 1 unit away from 0:

$$w > 1 \quad \text{or} \quad w < -1$$

In Example 7 we are using an expression in place of w.

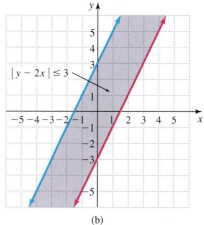

(a)   (b)

Figure 3.35

b) The inequality $|x - y| > 1$ is equivalent to

$$x - y > 1 \qquad \text{or} \qquad x - y < -1.$$

First graph the lines $x - y = 1$ and $x - y = -1$ as shown in Fig. 3.36(a). Test a point from each region in the original inequality, say $(-4, 0)$, $(0, 0)$, and $(4, 0)$:

$$|-4 - 0| > 1 \qquad |0 - 0| > 1 \qquad |4 - 0| > 1$$
$$4 > 1 \qquad\qquad 0 > 1 \qquad\qquad 4 > 1$$

Because $(-4, 0)$ and $(4, 0)$ satisfy the inequality, we shade those regions as shown in Fig. 3.36(b). The boundary lines are dashed because of the $>$ symbol.

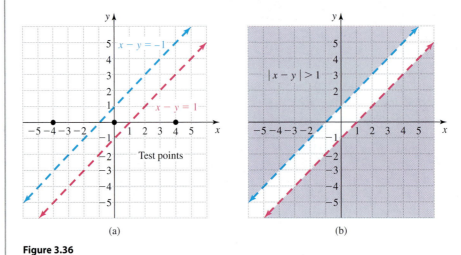

(a)   (b)

Figure 3.36

Now do Exercises 59–74

⟨5⟩ Inequalities with No Solution

The solution set to a compound inequality using *or* is the union of the individual solution sets. So the solution set to an *or* inequality is not empty unless all of the individual inequalities are inconsistent. However, the solution set to an *and* inequality can be empty even when the solution sets to the individual inequalities are not empty.

EXAMPLE 8

Compound inequalities with no solution

Solve each inequality.

a) $y > x + 1$ and $y < x - 2$ **b)** $x \geq 1$ and $x \leq 0$ **c)** $|x - y| \leq -3$

Solution

a) The solution set to $y > x + 1$ is the region above the line $y = x + 1$, and the solution set to $y < x - 2$ is the region below the line $y = x - 2$, as shown in Fig. 3.37(a). A point that satisfies the compound inequality would be in the intersection of these regions. Because the lines are parallel these regions do not intersect. So the solution set to the compound inequality is the empty set $\varnothing$.

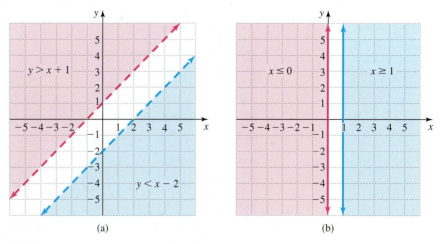

Figure 3.37

b) The solution set to $x \geq 1$ is the region on or to the right of the line $x = 1$, and the solution set to $x \leq 0$ is the region on or to the left of the line $x = 0$ as shown in Fig. 3.37(b). Because these lines are parallel these regions do not intersect and no points satisfy $x \geq 1$ and $x \leq 0$. The solution set is the empty set $\varnothing$.

c) Since the absolute value of any real number is nonnegative, there are no ordered pairs that satisfy $|x - y| \leq -3$. The solution set is the empty set, $\varnothing$.

> **Now do Exercises 75–90**

⟨6⟩ Applications

In real situations, x and y often represent quantities or amounts, which cannot be negative. In this case our graphs are restricted to the first quadrant, where x and y are both nonnegative.

E X A M P L E   9

Inequalities in business

The manager of a furniture store can spend a maximum of $3000 on advertising per week. It costs $50 to run a 30-second ad on an AM radio station and $75 to run the ad on an FM station. Graph the region that shows the possible numbers of AM and FM ads that can be purchased, and identify some possibilities.

Solution

If x represents the number of AM ads and y represents the number of FM ads, then x and y must satisfy the inequality $50x + 75y \leq 3000$. Because the number of ads cannot be negative, we also have $x \geq 0$ and $y \geq 0$. So we graph only points in the first quadrant that satisfy $50x + 75y \leq 3000$. The line $50x + 75y = 3000$ goes through $(0, 40)$ and $(60, 0)$. The inequality is satisfied below this line. The region showing the possible numbers of AM ads and FM ads is shown in Fig. 3.38. We shade the entire region in Fig. 3.38, but only points in the shaded region in which both coordinates are whole numbers actually satisfy the given condition. For example, 40 AM ads and 10 FM ads could be purchased. Other possibilities are 30 AM ads and 20 FM ads, or 10 AM ads and 10 FM ads.

Figure 3.38

> **Now do Exercises 91–98**

Warm-Ups ▼

Fill in the blank.

1. An inequality of the form $Ax + By \leq C$ is a _____ inequality.

2. The graph of $x + y > 3$ has a _____ boundary line.

3. The graph of $x - y \geq 5$ has a _____ boundary line.

4. The inequality $y > 2x - 3$ is satisfied _____ the line $y = 2x - 3$.

5. The inequality $y < -3x + 6$ is satisfied _____ the line $y = -3x + 6$.

6. The inequality $x < 5$ is satisfied to the _____ of the line $x = 5$.

7. The inequality $x > 9$ is satisfied to the _____ of the line $x = 9$.

True or false?

8. The point $(2, -3)$ satisfies $y > -3x + 2$.

9. The graph of $3x - y > 2$ is the region above $3x - y = 2$.

10. The graph of $3x + y < 5$ is the region below $y = -3x + 5$.

11. The graph of $x \leq 3$ includes the line $x = 3$.

12. The graph of $x > 5$ and $y < 6$ is the intersection of two regions.

13. The graph of $x < 8$ or $y > 5$ is the union of two regions.

14. The point $(2, -5)$ satisfies $y > -3x + 5$ and $y < 2x - 3$.

15. The point $(-3, 2)$ satisfies $y \leq 3x - 6$ or $y \leq x + 5$.

16. The inequality $|x - y| > 2$ is equivalent to $x - y > 2$ or $x - y < -2$.

17. The inequality $|x + y| \leq 1$ is equivalent to $-1 < x + y < 1$.

‹ **Study Tips** ›

• Be careful not to spend too much time on a single problem when taking a test. If a problem seems to be taking too much time, you might be on the wrong track. Be sure to finish the test.
• Before you take a test on this chapter, work the test given in this book at the end of this chapter. This will give you a good idea of your test readiness. ‹

1 › Graphing Linear Inequalities

Graph each linear inequality. See Examples 1 and 2. See the Strategy for Graphing a Linear Inequality box on page 183.

1. $y < x + 2$

2. $y < x - 1$

3. $y \leq -2x + 1$

4. $y \geq -3x + 4$

5. $x + y > 3$

6. $x + y \leq -1$

7. $2x + 3y < 9$

8. $-3x + 2y > 6$

9. $3x - 4y \leq 8$

10. $4x - 5y > 10$

11. $x - y > 0$

12. $2x - y < 0$

13. $x \geq 1$

14. $x < 0$

15. $y < 3$

16. $y > -1$

⟨2⟩ The Test Point Method

Graph each linear inequality by using a test point. See Example 3.

17. $2x - 3y < 5$

18. $5x - 4y > 3$

19. $x + y + 3 \geq 0$

20. $x - y - 6 \leq 0$

21. $y - 2x \leq 0$

22. $2y - x > 0$

23. $3x - 2y > 0$

24. $6x - 2y \leq 0$

25. $\frac{1}{2}x + \frac{1}{3}y < 1$

26. $2 - \frac{2}{5}y > \frac{1}{2}x$

⟨3⟩ Graphing Compound Inequalities

Determine which of the ordered pairs $(1, 3)$, $(-2, 5)$, $(-6, -4)$, *and* $(7, -8)$ *satisfy each compound or absolute value inequality. See Example 4.*

27. $y > 4$ or $x < 1$

28. $y < 2$ or $x < 0$

29. $y > 4$ and $x < 1$

30. $y < 2$ and $x < 0$

31. $y > 5x$ and $y < -x$

32. $y > 5x$ and $y > -x$

33. $y > -x + 1$ or $y > 4x$

34. $y > -x + 1$ or $y < 4x$

35. $|x + y| < 3$

36. $|x - y| > 2$

Graph each compound inequality. See Examples 5 and 6.

37. $y > x$ and
$\quad\ y > -2x + 3$

38. $y < x$ and
$\quad\ y < -3x + 2$

39. $y < x + 3$ or
$\quad\ y > -x + 2$

40. $y \geq x - 5$ or
$\quad\ y \leq -2x + 1$

41. $x - 4y < 0$ and
$\quad\ 3x + 2y \geq 6$

42. $x \geq -2y$ and
$\quad\ x - 3y < 6$

43. $x + y \leq 5$ and
$\quad x - y \leq 3$

44. $2x - y < 3$ and
$\quad 3x - y > 0$

53. $y > x - 1$ and
$\quad y < x + 3$

54. $y > x - 1$ and
$\quad y < 2x + 5$

55. $0 \leq y \leq x$ and $x \leq 1$

56. $x \leq y \leq 1$ and $x \geq 0$

45. $x - 2y \leq 4$ or
$\quad 2x - 3y \leq 6$

46. $4x - 3y \leq 3$ or
$\quad 2x + y \geq 2$

57. $1 \leq x \leq 3$ and
$\quad 2 \leq y \leq 5$

58. $-1 < x < 1$ and
$\quad -1 < y < 1$

47. $y > 2$ and $x < 3$

48. $x \leq 5$ and $y \geq -1$

⟨4⟩ Absolute Value Inequalities

Graph the absolute value inequalities. See Example 7.

49. $y \geq x$ and $x \leq 2$

50. $y < x$ and $y > 0$

59. $|x + y| < 2$

60. $|2x + y| < 1$

51. $2x < y + 3$ or
$\quad y > 2 - x$

52. $3 - x < y + 2$ or
$\quad x > y + 5$

61. $|2x + y| \geq 1$

62. $|x + 2y| \geq 6$

63. $|y - x| > 2$

64. $|2y - x| > 6$

65. $|x - 2y| \leq 4$

66. $|x - 3y| \leq 6$

67. $|x| > 2$

68. $|x| \leq 3$

69. $|y| < 1$

70. $|y| \geq 2$

71. $|x| < 2$ and $|y| < 3$

72. $|x| \geq 3$ or $|y| \geq 1$

73. $|x - 3| < 1$ and
$|y - 2| < 1$

74. $|x - 2| \geq 3$ or
$|y - 5| \geq 2$

⟨5⟩ Inequalities with No Solution

Determine whether or not the solution set to each compound or absolute value inequality is the empty set. See Example 8.

75. $y > x$ and $x < 1$

76. $y > x$ and $x > 1$

77. $y < 2x - 5$ and $y > 2x + 5$

78. $y \geq 3x$ and $y \leq 3x - 1$

79. $y < 2x - 5$ or $y > 2x + 5$

80. $y \geq 3x$ or $y \leq 3x - 1$

81. $y < 2x$ and $y > 3x$

82. $y < 2x$ or $y > 3x$

83. $y < x$ and $x < y$

84. $y > 3$ and $y < 1$

85. $|y + 2x| < 0$

86. $|x - 2y| < 0$

87. $|3x + 2y| \leq -4$

88. $|x - 2y| < -9$

89. $|x + y| > -4$

90. $|2x + 3y| < 4$

⟨6⟩ Applications

Solve each problem. See Example 9.

91. *Budget planning.* The Highway Patrol can spend a maximum of $120,000 on new vehicles this year. They can get a fully equipped compact car for $15,000 or a fully equipped full-size car for $20,000. Graph the region that shows the number of cars of each type that could be purchased.

92. *Allocating resources.* A furniture maker has a shop that can employ 12 workers for 40 hours per week at its maximum capacity. The shop makes tables and chairs. It takes 16 hours of labor to make a table and 8 hours of labor to make a chair. Graph the region that shows the possibilities for the number of tables and chairs that could be made in one week.

93. *More restrictions.* In Exercise 91, add the condition that the number of full-size cars must be greater than or equal to the number of compact cars. Graph the region showing the possibilities for the number of cars of each type that could be purchased.

94. *Chairs per table.* In Exercise 92, add the condition that the number of chairs must be at least four times the number of tables and at most six times the number of tables. Graph the region showing the possibilities for the number of tables and chairs that could be made in one week.

95. *Building fitness.* To achieve cardiovascular fitness, you should exercise so that your target heart rate is between 70% and 85% of its maximum rate. Your target heart rate h depends on your age a. For building fitness, you should have $h \leq 187 - 0.85a$ and $h \geq 154 - 0.70a$ (NordicTrack brochure). Graph this compound inequality for $20 \leq a \leq 75$ to see the heart rate target zone for building fitness.

96. *Waist-to-hip ratio.* A study by Dr. Aaron R. Folsom concluded that waist-to-hip ratios are a better predictor of 5-year survival than more traditional height-to-weight ratios. Dr. Folsom concluded that for good health the waist size of a woman aged 50 to 69 should be less than or equal to 80% of her hip size, $w \leq 0.80h$. Make a graph showing possible waist and hip sizes for good health for women in this age group for which hip size is no more than 50 inches.

97. *Advertising dollars.* A restaurant manager can spend at most $9000 on advertising per month and has two choices for advertising. The manager can purchase an ad in the *Daily Chronicle* (a 7-day-per-week newspaper) for $300 per day or a 30-second ad on WBTU television for $1000 each time the ad is aired. Graph the region that shows the possible number of days that an ad can be run in the newspaper and the possible number of times that an ad can be aired on television.

98. *Shipping restrictions.* The accompanying graph shows all of the possibilities for the number of refrigerators and the number of TVs that will fit into an 18-wheeler.

   a) Write an inequality to describe this region.
   b) Will the truck hold 71 refrigerators and 118 TVs?
   c) Will the truck hold 51 refrigerators and 176 TVs?

Figure for Exercise 98

Getting More Involved

99. *Writing*

Explain the difference between a compound inequality using the word *and* and a compound inequality using the word *or*.

100. *Discussion*

Explain how to write an absolute value inequality as a compound inequality.

3.5   Functions and Relations

In Section 2.2 we defined a function as a rule by which the value of one variable can be determined from the value(s) of one or more other variables. And we have been using functions in that chapter and this chapter. However, we have not yet seen an example of a rule that fails to be a function. In this section, we see many such examples as we study functions in more detail. Functions in mathematics are like automobiles in society. You cannot get along without them and you can use them without knowing all of their inner workings, but the more you know the better.

⟨1⟩ The Concept of a Function

We stated in Section 2.2 that if the value of y is determined by the value of x, then y **is a function of** x. But what exactly does "determined" mean? Here "determined" means "uniquely determined." There can be *only one* value of y for any particular value of x. There can be no ambiguity. We know that $y = 2x + 5$ is a function, because y is determined uniquely from this formula for any given value of x. However, the inequality $y > 2x + 5$ is *not* a function because there are infinitely many y-values that satisfy the inequality for any given x-value.

The x-value is thought of as *input* and the y-value as *output*. If y is a function of x, then there is only one output for any input. For example, after a shopper places an order on the Internet, the shopper is asked to *input* a ZIP code so that the shipping cost (*output*) can be determined. The shopper expects that the shipping cost is a function of ZIP code for that order. Note that many different ZIP codes can correspond to the same output. However, if any ZIP code caused the computer to output more than one shipping cost, then shipping cost is not a function of ZIP code and the shopper is confused. See Fig. 3.39.

Figure 3.39

E X A M P L E   **1**

Deciding if y is a function of x

In each case, determine whether y is a function of x.

a) Consider all possible circles. Let y represent the area of a circle and x represent its radius.

b) Consider all possible first-class letters mailed today in the United States. Let y represent the weight of a letter and x represent the amount of postage on the letter.

c) Consider all students at Pasadena City College. Let y represent the weight of a student to the nearest pound and x represent the height of the same student to the nearest inch.

d) Consider all possible rectangles. Let y represent the area of a rectangle and x represent the width.

e) Consider all cars sold at Bill Hood Ford this year where the sales tax rate is 9%. Let y represent the amount of sales tax and x represent the selling price of the car.

Solution

a) Can the area of a circle be determined from its radius? The well-known formula $A = \pi r^2$ (or in this case $y = \pi x^2$) indicates exactly how to determine the area if the radius is known. So there is only one area for any given radius and y is a function of x.

b) Can the weight of a letter be determined if the amount of postage on the letter is known? There are certainly letters that have the same amount of postage and different weights. Since the weight cannot be determined conclusively from the postage, the weight is *not* a function of the postage and y is *not* a function of x.

c) Can the weight of a student be determined from the height of the student? Imagine that we have a list containing the weights and heights for all students. There will certainly be two 5 ft 9 in. students with different weights. So weight cannot be determined from the height and y is *not* a function of x.

d) Can the area of a rectangle be determined from the width? Among all possible rectangles there are infinitely many rectangles with width 1 ft and different areas. So the area is not determined by the width and y is *not* a function of x.

e) Can the amount of sales tax be determined from the price of the car? The formula $y = 0.09x$ is used to determine the amount of tax. For example, the tax on every \$20,000 car is \$1800. So y is a function of x.

> Now do Exercises 1–8

⟨2⟩ Functions Expressed by Formulas

In Section 2.2 we defined a function as a rule. We will rephrase that definition here, concentrating on functions of one variable.

> **Function (as a Rule)**
>
> A function is a rule by which any allowable value of one variable (the **independent variable**) determines a *unique* value of a second variable (the **dependent variable**).

There are many ways to express a rule. A rule can be given verbally, with a formula, a table, or a graph. Of course, in mathematics we prefer the preciseness of a formula or equation. Since a formula such as $A = \pi r^2$ is a rule for obtaining the unique value of the dependent variable A from the value of the independent variable r, we say that the formula is a function. In Example 2, we convert a verbal rule for a function into a formula.

EXAMPLE 2

Writing a formula for a function

A carpet layer charges $25 plus $4 per square yard for installing carpet. Write the total charge C as a function of the number n of square yards of carpet installed.

Solution

At $4 per square yard, n square yards installed cost $4n$ dollars. If we include the $25 charge, then the total cost is $4n + 25$ dollars. Thus the equation

$$C = 4n + 25$$

expresses C as a function of n. Since $C = 4n + 25$ has the form $y = mx + b$, C is a linear function of n.

> **Now do Exercises 9–12**

Any formula that has the form $y = mx + b$ with $m \neq 0$ is a **linear function.** If $m = 0$, then $y = b$ and the y-values do not change. So $y = b$ is a **constant function.**

EXAMPLE 3

A function in geometry

Express the area of a circle as a function of its diameter.

Solution

The area of a circle is given by $A = \pi r^2$. Because the radius of a circle is one-half of the diameter, we have $r = \frac{d}{2}$. Now replace r by $\frac{d}{2}$ in the formula $A = \pi r^2$:

$$A = \pi \left(\frac{d}{2}\right)^2$$

$$= \frac{\pi d^2}{4}$$

So $A = \frac{\pi}{4}d^2$ expresses the area of a circle as a function of its diameter.

> **Now do Exercises 13–18**

⟨3⟩ Functions Expressed by Tables

Tables are often used to provide a rule for pairing the value of one variable with the value of another. For a table to define a function, each value of the independent variable must correspond to only one value of the dependent variable.

EXAMPLE 4

Functions defined by tables

Determine whether each table expresses y as a function of x.

a)

Weight (lb) x	Cost ($) y
0 to 10	4.60
11 to 30	12.75
31 to 79	32.90
80 to 99	55.82

b)

Weight (lb) x	Cost ($) y
0 to 15	4.60
10 to 30	12.75
31 to 79	32.90
80 to 99	55.82

c)

x	y
1	1
−1	1
2	2
−2	2
3	3

Solution

a) For each allowable weight, this table gives a unique cost. So the cost is a function of the weight and y is a function of x.

b) Using this table a weight of say 12 pounds would correspond to a cost of $4.60 and also to $12.75. Either the table has an error or perhaps there is some other factor that is being used to determine cost. In any case the weight does not determine a unique cost and y is *not* a function of x.

c) In this table every allowable value for x corresponds to a unique y-value so y is a function of x. Note that different values of x corresponding to the same y-value are permitted in a function.

> **Now do Exercises 19–26**

‹ Helpful Hint ›

In a function, every value for the independent variable determines conclusively a corresponding value for the dependent variable. If there is more than one possible value for the dependent variable, then the set of ordered pairs is not a function.

‹4› Functions Expressed by Ordered Pairs

A computer at your grocery store determines the price of each item by searching a long list of ordered pairs in which the first coordinate is the universal product code and the second coordinate is the price of the item with that code. For each product code there is a unique price. This process certainly satisfies the rule definition of a function. Since the set of ordered pairs is the essential part of this rule, we say that the set of ordered pairs is a function.

Function (as a Set of Ordered Pairs)

A function is a set of ordered pairs of real numbers such that no two ordered pairs have the same first coordinates and different second coordinates.

Note the importance of the phrase "no two ordered pairs have the same first coordinates and different second coordinates." Imagine the problems at the grocery store if the computer gave two different prices for the same universal product code. Note also that the product code is an identification number and it cannot be used in calculations. So the computer can use a function defined by a formula to determine the amount of tax, but it cannot use a formula to determine the price from the product code.

Any set of ordered pairs is called a **relation.** A **function** is a special relation.

EXAMPLE 5

Relations given as lists of ordered pairs
Determine whether each relation is a function.

a) $\{(1, 2), (1, 5), (3, 7)\}$ **b)** $\{(4, 5), (3, 5), (2, 6), (1, 7)\}$

Solution

a) This relation is not a function because $(1, 2)$ and $(1, 5)$ have the same first coordinate but different second coordinates.

b) This relation is a function. Note that the same second coordinate with different first coordinates is permitted in a function.

> **Now do Exercises 27–34**

The solution set to an equation involving x and y is a set of ordered pairs of the form (x, y). Since the equation corresponds to a set of ordered pairs, we say that the equation is a relation. The variables are related simply by the fact that they are in the same equation. If there are two ordered pairs with the same first coordinates and different second coordinates, then the equation is not a function. Note that when we ask whether an equation involving x and y is a function, we are asking whether y is a function of x or if y can be determined from x.

EXAMPLE 6

Relations given as equations

Determine whether each relation is a function. (Determine whether y is a function of x.)

a) $x = y^2$ **b)** $y = 2x$ **c)** $x = |y|$

⟨ **Helpful Hint** ⟩

To determine whether an equation expresses y as a function of x, always select a number for x (the independent variable) and then see if there is more than one corresponding value for y (the dependent variable). If there is more than one corresponding y-value, then y is not a function of x.

Solution

a) Is it possible to find two ordered pairs with the same first coordinate and different second coordinates that satisfy $x = y^2$? Since $(1, 1)$ and $(1, -1)$ both satisfy $x = y^2$, this relation is not a function.

b) The equation $y = 2x$ indicates that the y-coordinate is always twice the x-coordinate. Ordered pairs such as $(0, 0)$, $(2, 4)$, and $(3, 6)$ satisfy $y = 2x$. It is not possible to find two ordered pairs with the same first coordinate and different second coordinates. So $y = 2x$ is a function.

c) The equation $x = |y|$ is satisfied by ordered pairs such as $(2, 2)$ and $(2, -2)$ because $2 = |2|$ and $2 = |-2|$ are both correct. So this relation is not a function.

> **Now do Exercises 35–62**

⟨5⟩ The Vertical-Line Test

Since every graph illustrates a set of ordered pairs, every graph is a relation. To determine whether a graph is a function, we must see whether there are two (or more) ordered pairs on the graph that have the same first coordinate and different second coordinates. Two points with the same first coordinate lie on a vertical line that crosses the graph.

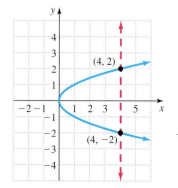

Figure 3.40

The Vertical-Line Test

A graph is the graph of a function if and only if there is no vertical line that crosses the graph more than once.

If there is a vertical line that crosses a graph twice (or more) as in Fig. 3.40, then we have two points with the same x-coordinate and different y-coordinates, and the graph is not the graph of a function. If you mentally consider every possible vertical line and none of them crosses the graph more than once, then you can conclude that the graph is the graph of a function.

EXAMPLE **7**

Using the vertical-line test
Which of these graphs are graphs of functions?

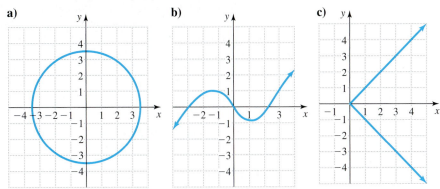

a)   b)   c)

Solution
Neither (a) nor (c) is the graph of a function, since we can draw vertical lines that cross these graphs twice. The graph (b) is the graph of a function, since no vertical line crosses it more than once.

| Now do Exercises 63–68 |

The vertical-line test illustrates the visual difference between a set of ordered pairs that is a function and one that is not. Because graphs are not precise, the vertical-line test might be inconclusive.

⟨6⟩ Domain and Range

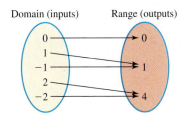

Domain (inputs)   Range (outputs)

Figure 3.41

A relation (or function) is a set of ordered pairs. The set of all first coordinates of the ordered pairs is the **domain** of the relation (or function). The set of all second coordinates of the ordered pairs is the **range** of the relation (or function). A function is a rule that pairs each member of the domain (the inputs) with a unique member of the range (the outputs). See Fig. 3.41. If a function is given as a table or a list of ordered pairs, then the domain and range are determined by simply reading them from the table or list. More often, a relation or function is given by an equation, with no domain stated. In this case, *the domain consists of all real numbers that, when substituted for the independent variable, produce real numbers for the dependent variable.*

EXAMPLE **8**

Identifying the domain and range
Determine the domain and range of each relation.

a) $\{(2, 5), (2, 7), (4, 3)\}$ b) $y = 2x$ c) $y = \sqrt{x - 1}$

Solution
a) The domain is the set of first coordinates, $\{2, 4\}$. The range is the set of second coordinates, $\{3, 5, 7\}$.

b) Since any real number can be used in place of x in $y = 2x$, the domain is $(-\infty, \infty)$. Since any real number can be used in place of y in $y = 2x$, the range is also $(-\infty, \infty)$.

c) Since the square root of a negative number is not a real number, we must have $x - 1 \geq 0$ or $x \geq 1$. So the domain is the interval $[1, \infty)$. Since the square root of a nonnegative real number is a nonnegative real number, we must have $y \geq 0$. So the range is the interval $[0, \infty)$.

> Now do Exercises 69–80

⟨7⟩ Function Notation

If y is a function of x, we can use the notation $f(x)$ to represent y. The expression $f(x)$ is read as "f of x." The notation $f(x)$ is called **function notation.** So if x is the independent variable, then either y or $f(x)$ is the dependent variable. For example, the function $y = 2x + 3$ can be written as

$$f(x) = 2x + 3.$$

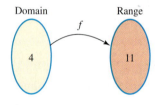

Figure 3.42

We use y and $f(x)$ interchangeably. We can think of f as the name of the function. We may use letters other than f. For example $g(x) = 2x + 3$ is the same function as $f(x) = 2x + 3$. The ordered pairs for each function are identical. Note that $f(x)$ does not mean f times x. The expression $f(x)$ represents the second coordinate when the first coordinate is x.

If $f(x) = 2x + 3$, then $f(4) = 2(4) + 3 = 11$. So the second coordinate is 11 if the first coordinate is 4. The ordered pair $(4, 11)$ is an ordered pair in the function f. Figure 3.42 illustrates this situation.

E X A M P L E   9

Using function notation

Let $f(x) = 3x - 2$ and $g(x) = x^2 - x$. Evaluate each expression.

a) $f(-5)$ **b)** $g(-5)$ **c)** $f(0) + g(3)$

Solution

a) Replace x by -5 in the equation defining the function f:
$$f(x) = 3x - 2$$
$$f(-5) = 3(-5) - 2$$
$$= -17$$
So $f(-5) = -17$.

b) Replace x by -5 in the equation defining the function g:
$$g(x) = x^2 - x$$
$$g(-5) = (-5)^2 - (-5) = 30$$
So $g(-5) = 30$.

c) Since $f(0) = 3(0) - 2 = -2$ and $g(3) = 3^2 - 3 = 6$, we have $f(0) + g(3) = -2 + 6 = 4$.

> Now do Exercises 81–96

E X A M P L E   **10**

An application of function notation

To determine the cost of an in-home repair, a computer technician uses the linear function $C(n) = 40n + 30$, where n is the time in hours and $C(n)$ is the cost in dollars. Find $C(2)$ and $C(4)$.

Solution

Replace n with 2 to get

$$C(2) = 40(2) + 30 = 110.$$

Replace n with 4 to get

$$C(4) = 40(4) + 30 = 190.$$

So for 2 hours the cost is \$110 and for 4 hours the cost is \$190.

> Now do Exercises 97–104

In this section, we studied functions of one variable. However, a variable can be a function of another variable or a function of many other variables. For example, your grade on the next test is not a function of the number of hours that you study for it. Your grade is a function of many variables: study time, sleep time, work time, your mother's IQ, and so on. Even though study time alone does not determine your grade, it is the variable that has the most influence on your grade.

‹ Calculator Close-Up ›

A graphing calculator has function notation built in. To find $C(2)$ and $C(4)$ with a graphing calculator, enter $y_1 = 40x + 30$ as shown here:

To find $C(2)$ and $C(4)$, enter $y_1(2)$ and $y_1(4)$ as shown here:

Warm-Ups ▼

Fill in the blank.

1. Any set of ordered pairs is a _____.

2. A _____ is a set of ordered pairs in which no two have the same first coordinates and different second coordinates.

3. If m is a _____ of w, then m is uniquely determined by w.

4. The _____ of a relation is the set of all first coordinates of the ordered pairs.

5. The _____ of a relation is the set of all second coordinates of the ordered pairs.

6. In function notation $f(x)$ is used for the _____ variable.

True or false?

7. The set $\{(1, 2), (3, 0), (9, 0)\}$ is a function.

8. The set $\{(2, 1), (0, 3), (0, 9)\}$ is a function.

9. The diameter of a circle is a function of the radius.

10. The equation $y = x^2$ is a function.

11. Every relation is a function.

12. The domain of $\{(2, 1), (0, 3), (0, 9)\}$ is $\{0, 2\}$.

13. The range of $\{(2, 1), (0, 3), (0, 9)\}$ is $\{0, 1, 3, 9\}$.

14. The domain of $f(x) = \sqrt{x}$ is $[0, \infty)$.

15. If $h(x) = x^2 - 3$, then $h(-2) = 1$.

Exercises

‹1› The Concept of a Function

In each situation determine whether y is a function of x. Explain your answer. See Example 1.

1. Consider all gas stations in your area. Let x represent the price per gallon of regular unleaded gasoline and y represent the number of gallons that you can get for $10.

2. Consider all items at Sears. Let x represent the universal product code for an item and y represent the price of that item.

3. Consider all students taking algebra at your school. Let x represent the number of hours (to the nearest hour) a student spent studying for the first test and y represent the student's score on the test.

4. Consider all students taking algebra at your school. Let x represent a student's height to the nearest inch and y represent the student's IQ.

5. Consider the air temperature at noon today in every town in the United States. Let x represent the Celsius temperature for a town and y represent the Fahrenheit temperature.

6. Consider all first-class letters mailed within the United States today. Let x represent the weight of a letter and y represent the amount of postage on the letter.

7. Consider all items for sale at the nearest Walmart. Let x represent the cost of an item and y represent the universal product code for the item.

8. Consider all packages shipped by UPS. Let x represent the weight of a package and y represent the cost of shipping that package.

‹2› Functions Expressed by Formulas

Write a formula that describes the function. See Examples 2 and 3.

9. A small pizza costs $5.00 plus 50 cents for each topping. Express the total cost C as a function of the number of toppings t.

10. A developer prices condominiums in Florida at $20,000 plus $40 per square foot of living area. Express the cost C as a function of the number of square feet of living area s.

11. The sales tax rate on groceries in Mayberry is 9%. Express the total cost T (including tax) as a function of the total price of the groceries S.

12. With a GM MasterCard, 5% of the amount charged is credited toward a rebate on the purchase of a new car. Express the rebate R as a function of the amount charged A.

13. Express the circumference of a circle as a function of its radius.

14. Express the circumference of a circle as a function of its diameter.

15. Express the perimeter P of a square as a function of the length s of a side.

16. Express the perimeter P of a rectangle with width 10 ft as a function of its length L.

17. Express the area A of a triangle with a base of 10 m as a function of its height h.

18. Express the area A of a trapezoid with bases 12 cm and 10 cm as a function of its height h.

‹3› Functions Expressed by Tables

Determine whether each table expresses the second variable as a function of the first variable. See Example 4.

19.

x	y
1	1
4	2
9	3
16	4
25	5
36	6
49	8

20.

x	y
2	4
3	9
4	16
5	25
8	36
9	49
10	100

21.

t	v
2	2
−2	2
3	3
−3	3
4	4
−4	4
5	5

22.

s	W
5	17
6	17
−1	17
−2	17
−3	17
7	17
8	17

23.

a	P
2	2
2	-2
3	3
3	-3
4	4
4	-4
5	5

24.

n	r
17	5
17	6
17	-1
17	-2
17	-3
17	-4
17	-5

25.

b	q
1970	0.14
1972	0.18
1974	0.18
1976	0.22
1978	0.25
1980	0.28

26.

c	h
345	0.3
350	0.4
355	0.5
360	0.6
365	0.7
370	0.8
380	0.9

⟨4⟩ Functions Expressed by Ordered Pairs

Determine whether each relation is a function. See Example 5.

27. $\{(2, 4), (3, 4), (4, 5)\}$

28. $\{(2, -5), (2, 5), (3, 10)\}$

29. $\{(-2, 4), (-2, 6), (3, 6)\}$

30. $\{(3, 6), (6, 3)\}$

31. $\{(\pi, -1), (\pi, 1)\}$

32. $\{(-0.3, -0.3), (-0.2, 0), (-0.3, 1)\}$

33. $\left\{\left(\frac{1}{2}, \frac{1}{2}\right)\right\}$

34. $\left\{\left(\frac{1}{3}, 7\right)\left(-\frac{1}{3}, 7\right)\left(\frac{1}{6}, 7\right)\right\}$

Find two ordered pairs that satisfy each equation and have the same x-coordinate but different y-coordinates. Answers may vary. See Example 6.

35. $x = 2y^2$ **36.** $x^2 = y^2$

37. $x = |2y|$ **38.** $|x| = |y|$

39. $x^2 + y^2 = 1$ **40.** $x^2 + y^2 = 4$

41. $x = y^4$ **42.** $x^4 = y^4$

43. $x - 2 = |y|$ **44.** $x + 5 = |y|$

Determine whether each relation is a function. See Example 6.

45. $y = x^2$ **46.** $y = x^2 + 3$

47. $x = |y| + 1$ **48.** $|x| = |y + 1|$

49. $y = x$ **50.** $x = y + 4$

51. $x = y^4 + 1$ **52.** $x^4 = y^2$

53. $y = \sqrt{x}$ **54.** $x = \sqrt{y}$

55. $|x| = |2y|$ **56.** $|4x| = |2y|$

57. $x^2 + y^2 = 9$ **58.** $x^2 + y^4 = 1$

59. $x = 2\sqrt{y}$ **60.** $y = \sqrt{x - 5}$

61. $x + 5 = |y|$ **62.** $x - 2 = |y|$

⟨5⟩ The Vertical-Line Test

Use the vertical-line test to determine which of the graphs are graphs of functions. See Example 7.

63.

64.

65.

66.

67.

68.

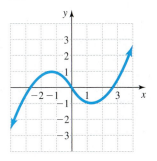

⟨6⟩ Domain and Range

Determine the domain and range of each relation.
See Example 8.

69. $\{(4, 1), (7, 1)\}$

70. $\{(0, 2), (3, 5)\}$

71. $\{(2, 3), (2, 5), (2, 7)\}$

72. $\{(3, 1), (5, 1), (4, 1)\}$

73. $y = x + 1$

74. $y = 3x + 1$

75. $y = 5 - x$

76. $y = -2x + 1$

77. $y = \sqrt{x - 2}$

78. $y = \sqrt{x + 4}$

79. $y = \sqrt{2x}$

80. $y = \sqrt{2x - 4}$

⟨7⟩ Function Notation

Let $f(x) = 3x - 2$, $g(x) = -x^2 + 3x - 2$, and $h(x) = |x + 2|$.
Evaluate each expression. See Example 9.

81. $f(0)$

82. $f(1)$

83. $f(4)$

84. $f(100)$

85. $g(-2)$

86. $g(-3)$

87. $h(-3)$

88. $h(-19)$

89. $h(-4.236)$

90. $h(-1.99)$

91. $f(2) + g(3)$

92. $f(1) - g(0)$

93. $\dfrac{g(2)}{h(-3)}$

94. $\dfrac{h(-10)}{f(2)}$

95. $f(-1) \cdot h(-4)$

96. $h(0) \cdot g(0)$

Solve each problem. See Example 10.

97. Height. If a ball is dropped from the top of a 256-ft building, then the formula

$$h(t) = 256 - 16t^2$$

expresses its height $h(t)$ in feet as a function of the time t in seconds.

a) Find $h(2)$, the height of the ball 2 seconds after it is dropped.

b) Find $h(4)$.

98. Velocity. If a ball is dropped from a height of 256 ft, then the formula

$$v(t) = -32t$$

expresses its velocity $v(t)$ in feet per second as a function of time t in seconds.

a) Find $v(0)$, the velocity of the ball at time $t = 0$.

b) Find $v(4)$.

99. Area of a square. Find a formula that expresses the area of a square A as a function of the length of its side s.

100. Perimeter of a square. Find a formula that expresses the perimeter of a square P as a function of the length of its side s.

101. Cost of fabric. If a certain fabric is priced at $3.98 per yard, express the cost $C(x)$ as a function of the number of yards x. Find $C(3)$.

102. Earned income. If Mildred earns $14.50 per hour, express her total pay $P(h)$ as a function of the number of hours worked h. Find $P(40)$.

103. Cost of pizza. A pizza parlor charges $14.95 for a pizza plus $0.50 for each topping. Express the total cost of a pizza $C(n)$ in dollars as a function of the number of toppings n. Find $C(6)$.

104. Cost of gravel. A gravel dealer charges $50 plus $30 per cubic yard for delivering a truckload of gravel. Express the total cost $C(n)$ in dollars as a function of the number of cubic yards delivered n. Find $C(12)$.

Getting More Involved

105. Writing

Consider $y = x + 2$ and $y > x + 2$. Explain why one of these relations is a function and the other is not.

106. Writing

Consider the graphs of $y = 2$ and $x = 3$ in the rectangular coordinate system. Explain why one of these relations is a function and the other is not.

Wrap-Up

Summary

Rectangular Coordinate System		Examples
x-intercept	The point where a nonhorizontal line intersects the x-axis	For the line $2x + y = 6$, the x-intercept is $(3, 0)$ and the y-intercept is $(0, 6)$.
y-intercept	The point where a nonvertical line intersects the y-axis	

Slope		Examples
Slope of a line	$$\text{Slope} = \frac{\text{change in } y\text{-coordinate}}{\text{change in } x\text{-coordinate}}$$ $$= \frac{\text{rise}}{\text{run}}$$	

| Slope using coordinates | Slope of line through (x_1, y_1) and (x_2, y_2) is $m = \dfrac{y_2 - y_1}{x_2 - x_1}$, provided that $x_2 - x_1 \neq 0$. | If $(x_1, y_1) = (4, -2)$ and $(x_2, y_2) = (3, -6)$, then $$m = \frac{-6 - (-2)}{3 - 4} = 4.$$ |

Types of slope	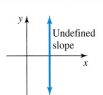

Perpendicular lines	The slope of one line is the opposite of the reciprocal of the slope of the other line.	The lines $y = -\frac{1}{3}x + 5$ and $y = 3x - 9$ are perpendicular.
Parallel lines	Nonvertical parallel lines have equal slopes.	The lines $y = 2x - 3$ and $y = 2x + 7$ are parallel.

Forms of Linear Equations		Examples
Point-slope form	$y - y_1 = m(x - x_1)$ (x_1, y_1) is a point on the line, and m is the slope.	Line through $(5, -3)$ with slope 2: $y + 3 = 2(x - 5)$

Slope-intercept form	$y = mx + b$	Line through $(0, -3)$ with slope 2:
	m is the slope, $(0, b)$ is the y-intercept.	$y = 2x - 3$
Standard form	$Ax + By = C$	$3x - 2y = 12$
	A and B are not both 0.	
Vertical line	$x = k$, where k is any real number.	$x = 5$
	Slope is undefined for vertical lines.	
Horizontal line	$y = k$, where k is any real number.	$y = -2$
	Slope is zero for horizontal lines.	

Graphing Linear Equations

Examples

Point-plotting	Arbitrarily select some points that satisfy the equation, and draw a line through them.	For $y = 2x + 1$, draw a line through $(0, 1)$, $(1, 3)$, and $(2, 5)$.
Intercepts	Find the x- and y-intercepts (provided that they are not the origin), and draw a line through them.	For $x + y = 4$, the intercepts are $(0, 4)$ and $(4, 0)$.
y-intercept and slope	Start at the y-intercept and use the slope to locate a second point; then draw a line through the two points.	For $y = 3x - 2$, start at $(0, -2)$ and rise 3 and run 1 to get to $(1, 1)$. Draw a line through the two points.

Linear Inequalities

Examples

Linear inequality	$Ax + By \leq C$, where A and B are not both zero. The symbols $<$, $>$, and $\geq$ are also used.	$2x - 3y \leq 7$ $x - y > 6$
Graphing linear inequalities	Solve for y, and then graph the line $y = mx + b$. $\quad y > mx + b$ is the region above the line. $\quad y < mx + b$ is the region below the line.	Graph of $y = x + 2$ is a line. $y > x + 2$ is above $y = x + 2$. $y < x + 2$ is below $y = x + 2$.
	For inequalities without y, graph $x = k$. $\quad x > k$ is the region to the right of $x = k$. $\quad x < k$ is the region to the left of $x = k$.	The graph of $x > 5$ is to the right of the vertical line $x = 5$, and the graph of $x < 5$ is to the left of $x = 5$.
Test points	A linear inequality may also be graphed by graphing the corresponding line and then testing a point to determine which region satisfies the inequality.	

Compound Inequalities

Examples

In one variable (from Section 2.5)	Two simple inequalities in one variable connected with the word *and* or *or*	$x > 1$ and $x < 5$
	The solution set for an *and* inequality is the intersection of the solution sets.	$x > 3$ or $x < 1$
	The solution set for an *or* inequality is the union of the solution sets.	

In two variables	Two simple inequalities in two variables connected with the word *and* or *or*	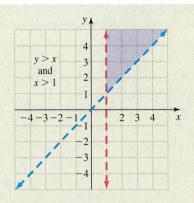
	The solution set for an *and* inequality is the intersection of the solution sets.	
	The solution set for an *or* inequality is the union of the solution sets.	
	Note that the graph of $x > 1$ (an inequality containing only one variable) in the rectangular coordinate system is the region to the right of the vertical line $x = 1$.	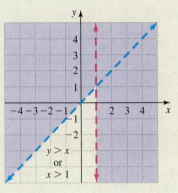

Relations and Functions

Examples

Relation	Any set of ordered pairs of real numbers	$\{(1, 2), (1, 3)\}$
Function	A relation in which no two ordered pairs have the same first coordinate and different second coordinates.	$\{(1, 2), (3, 5), (4, 5)\}$
	If y is a function of x, then y is uniquely determined by x. A function may be defined by a table, a listing of ordered pairs, or an equation.	
Domain	The set of first coordinates of the ordered pairs	Function: $y = x^2$, Domain: $(-\infty, \infty)$
Range	The set of second coordinates of the ordered pairs	Function: $y = x^2$, Range: $[0, \infty)$
Function notation	If y is a function of x, the expression $f(x)$ is used in place of y.	$y = 2x + 3$ $f(x) = 2x + 3$
Vertical-line test	If a graph can be crossed more than once by a vertical line, then it is not the graph of a function.	
Linear function	A function of the form $f(x) = mx + b$ with $m \neq 0$	$f(x) = 3x - 7$ $f(x) = -2x + 5$
Constant function	A function of the form $f(x) = b$, where b is a real number	$f(x) = 2$

Enriching Your Mathematical Word Power

Fill in the blank.

1. The _____ of an equation is an illustration in the coordinate plane that shows all ordered pairs that satisfy the equation.

2. The _____ is the point at the intersection of the x-axis and y-axis.

3. The first number in an ordered pair is the _____.

4. A point at which a graph intersects the y-axis is the

   _____ .

5. The _____ variable corresponds to the first coordinate of an ordered pair.

6. The _____ variable corresponds to the second coordinate of an ordered pair.

7. The _____ of a line is the change in y-coordinates divided by the change in x-coordinates.

8. $Ax + By = C$ is the _____ form for the equation of a line.

9. The _____ -intercept form for the equation of a line is $y = mx + b$.

10. The _____ -slope form for the equation of a line is $y - y_1 = m(x - x_1)$.

11. An inequality of the form $Ax + By > 0$ is a _____ inequality in two variables.

12. A _____ is a set of ordered pairs in which no two have the same first coordinates and different second coordinates.

13. Any set of ordered pairs is a _____.

14. The _____ is the set of first coordinates of a relation.

15. The _____ is the set of second coordinates of a relation.

16. The notation in which $f(x)$ is used for the dependent variable is _____ notation.

Review Exercises

3.1 Graphing Lines in the Coordinate Plane

For each point, name the quadrant in which it lies or the axis on which it lies.

1. $(-3, -2)$

2. $(0, \pi)$

3. $(\pi, 0)$

4. $(-5, 4)$

5. $(0, -1)$

6. $\left(\dfrac{\pi}{2}, 1\right)$

7. $\left(\sqrt{2}, -3\right)$

8. $(6, -3)$

Complete the given ordered pairs so that each ordered pair satisfies the given equation.

9. $(0, \ ), (\ , 0), (4, \ ), (\ , -3), y = -3x + 2$

10. $(0, \ ), (\ , 0), (-6, \ ), (\ , 5), 2x + 3y = 5$

13. $(4, 1), (-3, -2)$ 14. $(6, 0), (0, -3)$

Solve each problem.

15. What is the slope of any line that is parallel to the line through $(-3, -4)$ and $(5, -1)$?

16. What is the slope of the line through $(4, 6)$ that is parallel to the line through $(-2, 1)$ and $(7, 1)$?

17. What is the slope of any line that is perpendicular to the line through $(-3, 5)$ and $(4, -6)$?

18. What is the slope of the line through $(1, 2)$ that is perpendicular to the line through $(5, 4)$ and $(5, -2)$?

In each case find the slope of line l and graph both lines that are mentioned.

19. Line l contains the origin and is perpendicular to the line through $(2, 2)$ and $(3, 3)$.

3.2 Slope of a Line

Find the slope of the line through each pair of points.

11. $(-5, 6), (-2, 9)$ 12. $(-2, 7), (3, -4)$

20. Line l contains the origin and is perpendicular to the line through $(-1, 4)$ and $(-3, 5)$.

21. Line l passes through $(0, 3)$ and is parallel to a line through $(1, 0)$ with slope 2.

22. Line l passes through $(2, 0)$ and is parallel to a line through the origin with slope -1.

23. Line l is perpendicular to a line with slope $\frac{2}{3}$, and both lines pass through $(0, 0)$.

24. Line l is perpendicular to a line with slope -4, and both lines pass through $(0, 0)$.

3.3 Three Forms for the Equation of a Line

Find the slope and y-intercept for each line.

25. $y = -3x + 4$

26. $2y - 3x + 1 = 0$

27. $y - 3 = \frac{2}{3}(x - 1)$

28. $y - 3 = 5$

Write each equation in standard form with integral coefficients.

29. $y = \frac{2}{3}x - 4$

30. $y = -0.05x + 0.26$

31. $y - 1 = \frac{1}{2}(x + 3)$

32. $\frac{1}{2}x - \frac{1}{3}y = \frac{1}{4}$

Write the equation of the line containing the given point and having the given slope. Rewrite each equation in standard form with integral coefficients.

33. $(1, -3)$, $m = \frac{1}{2}$

34. $(0, 2)$, $m = 3$

35. $(-2, 6)$, $m = -\frac{3}{4}$

36. $\left(2, \frac{1}{2}\right)$, $m = \frac{1}{4}$

37. $(3, 5)$, $m = 0$

38. $(0, 0)$, $m = -1$

Graph each equation.

39. $y = 2x - 3$ **40.** $y = \frac{2}{3}x + 1$

41. $3x - 2y = -6$ **42.** $4x + 5y = 10$

43. $y - 3 = 10$

44. $2x = 8$

53. $x - y \leq 5$

54. $2x + y > 1$

45. $5x - 3y = 7$

46. $3x + 4y = -1$

55. $3x > 2$

56. $x + 2 \leq 0$

47. $5x + 4y = 100$

48. $2x - y = 120$

57. $4y \leq 0$

58. $4y - 4 > 0$

49. $x - 80y = 400$

50. $75x + y = 300$

59. $4x - 2y \geq 6$

60. $-5x - 3y > 6$

3.4 Linear Inequalities and Their Graphs

Graph each linear inequality.

51. $y > 3x - 2$

52. $y \leq 2x + 3$

61. $5x - 2y < 9$

62. $3x + 4y \leq -1$

Graph each compound or absolute value inequality.

63. $y > 3$ and
$y - x < 5$

64. $x + y \leq 1$ or
$y \leq 4$

65. $3x + 2y \geq 8$ or
$3x - 2y \leq 6$

66. $x + 8y > 8$ and
$x - 2y < 10$

67. $|x + 2y| < 10$

68. $|x - 3y| \geq 9$

69. $|x| \leq 5$

70. $|y| > 6$

71. $|y - x| > 2$

72. $|x - y| \leq 1$

3.5 Functions and Relations

Determine whether each relation is a function.

73. $\{(5, 7), (5, 10), (5, 3)\}$

74. $\{(1, 3), (4, 7), (1, 6)\}$

75. $\{(1, 1), (2, 1), (3, 3)\}$

76. $\{(2, 4), (4, 6), (6, 8)\}$

77. $y = x^2$

78. $x^2 = 1 + y^2$

79. $x = y^4$

80. $y = \sqrt{x - 1}$

Determine the domain and range of each relation.

81. $\{(3, 5), (4, 9), (5, 1)\}$

82. $\{(2, 6), (6, 7), (8, 9)\}$

83. $y = x + 1$

84. $y = 2x - 3$

85. $y = \sqrt{x + 5}$

86. $y = \sqrt{x - 1}$

Let $f(x) = 2x - 5$ and $g(x) = x^2 + x - 6$. Evaluate each expression.

87. $f(0)$

88. $f(-3)$

89. $g(0)$

90. $g(-2)$

91. $g\left(\dfrac{1}{2}\right)$

92. $g\left(-\dfrac{1}{2}\right)$

Miscellaneous

Write an equation in standard form with integral coefficients for each line described.

93. The line that crosses the *x*-axis at (2, 0) and the *y*-axis at (0, −6)

94. The line with an *x*-intercept of (4, 0) and slope $-\dfrac{1}{2}$

95. The line through (−1, 4) with slope $-\dfrac{1}{2}$

96. The line through (2, −3) with slope 0

97. The line through (2, −6) and (2, 5)

98. The line through (−3, 6) and (4, 2)

99. The line through (0, 0) perpendicular to $x = 5$

100. The line through $(2, -3)$ perpendicular to $y = -3x + 5$

101. The line through $(-1, 4)$ parallel to $y = 2x + 1$

102. The line through $(2, 1)$ perpendicular to $y = 10$

Write an equation in standard form with integral coefficients for each line.

103.

104.

105.

106.

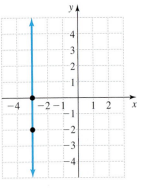

Use slope to solve each geometric problem.

107. Show that the points $(-5, -5)$, $(-3, -1)$, $(6, 2)$, and $(4, -2)$ are the vertices of a parallelogram.

108. Show that the points $(-5, -5)$, $(4, -2)$, and $(3, 1)$ are the vertices of a right triangle.

109. Show that the points $(-2, 2)$, $(0, 0)$, $(2, 6)$, and $(4, 4)$ are the vertices of a rectangle.

110. Determine whether the points $(2, 1)$, $(4, 7)$, and $(-3, -14)$ lie on a straight line.

Solve each problem.

111. *Maximum heart rate.* The maximum heart rate during exercise for a 20-year-old is 200 beats per minute, and the maximum heart rate for a 70-year-old is 150 (NordicTrack brochure) as shown in the accompanying figure.

a) Write the maximum heart rate h as a linear function of age a.

b) What is the maximum heart rate for a 40-year-old?

c) Does your maximum heart rate increase or decrease as you get older?

Figure for Exercise 111

112. *Resting heart rate.* A subject is given 3 milligrams (mg) of an experimental drug, and a resting heart rate of 82 is recorded. Another subject is given 5 mg of the same drug, and a resting heart rate of 89 is recorded. If we assume the heart rate, h, is a linear function of the dosage, d, find the linear equation expressing h in terms of d. If a subject is given 10 mg of the drug, what would be the expected heart rate?

113. *Rental costs.* The charge, C, in dollars, for renting an air hammer from the Tools Is Us Rental Company is determined from the formula $C = 26 + 17d$, where d is the number of days in the rental period. Graph this function for d from 1 to 30. If the air hammer is worth \$1080, then in how many days would the rental charge equal the value of the air hammer?

114. *Waist-to-hip ratio.* Dr. Aaron R. Folsom, from the University of Minnesota School of Public Health, has concluded that for a man aged 50 to 69 to be in good health, his waist size w should be less than or equal to 95% of his hip size h as shown in the figure.

   a) Write an inequality that describes the region shown in the figure.

   b) Is a man in this group with a 36-inch waist and 37-inch hips in good health?

   c) If a man in this group has a waist of 38 inches, then what is his minimum hip size for good health?

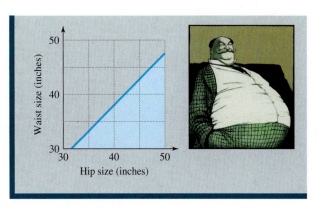

Figure for Exercise 114

Chapter 3 Test

Complete each ordered pair so that it satisfies the given equation.

1. $(0, \ ), (\ , 0), (\ , -8), 2x + y = 5$

Solve each problem.

2. Find the slope of the line through $(-3, 7)$ and $(2, 1)$.

3. Determine the slope and y-intercept for the line $8x - 5y = -10$.

4. Show that $(-1, -2), (0, 0), (6, 2),$ and $(5, 0)$ are the vertices of a parallelogram.

5. Suppose the value, V, in dollars, of a boat is a linear function of its age, a, in years. If a boat was valued at \$22,000 brand new and it is worth \$16,000 when it is 3 years old, find the linear equation that expresses V in terms of a.

For each line described, write its equation in standard form with integral coefficients.

6. The line with y-intercept $(0, 3)$ and slope $-\frac{1}{2}$

7. The line through $(-3, 5)$ with slope -4

8. The line through $(2, 3)$ that is perpendicular to $3x - 5y = 7$

9. The line shown in the graph:

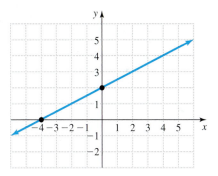

Figure for Exercise 9

Sketch the graph of each equation in the rectangular coordinate system.

10. $y = 4$

11. $x = 3$

12. $3x + 4y = 12$

16. $|2x + y| \geq 3$

13. $y = \dfrac{2}{3}x - 2$

Solve each problem.

17. Determine whether $\{(0, 5), (9, 5), (4, 5)\}$ is a function.

18. Let $f(x) = -2x + 5$. Find $f(-3)$.

Sketch the graph of each inequality.

19. Find the domain and range of the function $y = \sqrt{x - 7}$.

14. $y > -\dfrac{1}{2}x + 3$

20. A mail-order firm charges its customers a shipping and handling fee of $3.00 plus $0.50 per pound for each order shipped. Express the shipping and handling fee S as a function of the weight of the order n.

21. If a ball is tossed into the air from a height of 6 feet with a velocity of 32 feet per second, then its altitude at time t (in seconds) can be described by the function

$$A(t) = -16t^2 + 32t + 6.$$

Find the altitude of the ball at 2 seconds.

15. $x > 2$ and $x + y > 0$

Graph Paper

Use these grids for graphing. Make as many copies of this page as you need. If you have access to a computer, you can download this page from www.mhhe.com/dugopolski and print it.

*Making*Connections | **A Review of Chapters 1–3**

Evaluate each expression.

1. $2^3 \cdot 4^2$

2. $2^7 - 2^6$

3. $3^2 - 4(5)(-2)$

4. $3 - 2|5 - 7 \cdot 3|$

5. $\dfrac{2 - (-3)}{5 - 6}$

6. $\dfrac{-3 - 7}{-1 - (-3)}$

Simplify each expression.

7. $3t \cdot 4t$

8. $3t + 4t$

9. $\dfrac{4x + 8}{4}$

10. $\dfrac{-8y}{-4} - \dfrac{10y}{-2}$

11. $3(x - 4) - 4(5 - x)$ **12.** $-2(3x^2 - x) + 3(2x - 5x^2)$

Solve each equation.

13. $15(b - 27) = 0$ **14.** $0.05a - 0.04(a - 50) = 4$

15. $|3v - 7| = 0$ **16.** $|3u - 7| = 3$

17. $|3x - 7| = -77$ **18.** $|3x - 7| + 1 = 8$

Graph the solution set to each inequality or compound inequality in one variable on the number line.

19. $2x - 1 > 7$ **20.** $5 - 3x \le -1$

21. $x - 5 \le 4$ and $3x - 1 < 8$ **22.** $2x \le -6$ or $5 - 2x < -7$

23. $|x - 3| < 2$ **24.** $|1 - 2x| \ge 7$

Graph the solution set to each linear inequality or compound inequality in a rectangular coordinate system.

25. $y < 2x - 1$ **26.** $3x - y \le 2$

27. $y > x$ and $y < 5 - 3x$ **28.** $y \le 2$ or $x \ge -3$

In each case, express a as a function of b.

29. $ab = 10$

30. $a - b = 10$

31. $3a + 5b = 10$

32. $ab + 3 = 5a$

Solve each problem.

33. The perimeter of a rectangular field is 660 yards. If the width is 20 yards less than the length, then what is the length?

34. The area of a rectangular tabletop is 1392 square inches. If the length is 48 inches, then what is the width?

35. A motor home is advertised at 30% off the manufacturer's suggested retail price (msrp). If the advertised price is $131,600, then what is the msrp?

36. A retailer has 10,000 beads in stock of which 12% are Swarovski crystal beads. How many Swarovski crystal beads must she add to her stock so she can say that 20% of her stock is Swarovski crystal beads?

37. *Social Security.* A person retiring after 2005 who earned a lifetime average annual salary of $25,000 receives a benefit based on age (Social Security Administration, www.ssa.gov). For ages 62 through 64 the benefit in dollars is determined by $b = 7000 + 500(a - 62)$, for ages 65 through 67 the benefit is determined by $b = 10,000 + 667(a - 67)$, and for ages 68 through 70 the benefit is determined by $b = 10,000 + 800(a - 67)$.

  a) Write each benefit formula in slope-intercept form.
  b) What will be the annual Social Security benefit for a person who retires at age 64?

c) If a person retires and gets an $11,600 benefit, then what is the age of that person at retirement?

d) Find the slope of each line segment in the accompanying figure and interpret your results.

Figure for Exercise 37

Critical **Thinking** | For Individual or Group Work | Chapter 3

These exercises can be solved by a variety of techniques, which may or may not require algebra. So be creative and think critically. Explain all answers. Answers are in the Instructor's Edition of this text.

1. **Trailblazing.** How many different paths are there from point *A* to point *B* in the accompanying hexagonal figure? You can only move in a downward direction along the line segments.

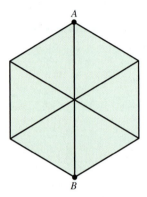

A

B

Figure for Exercise 1

2. **Sticky cubes.** Three wooden cubes have edges of 2 cm, 6 cm, and 8 cm. What is the minimum possible surface area after the three cubes are glued together?

3. **Adding letters.** Each letter in the following addition problem represents a unique digit.

$$\begin{array}{r} \text{USSR} \\ + \text{ USA} \\ \hline \text{PEACE} \end{array}$$

Determine values of the letters that would make the addition problem correct.

4. **Polygonal diagonals.** A convex quadrilateral has 2 diagonals. A convex pentagon has 5 diagonals. If a convex polygon has 324 diagonals, then how many sides does this polygon have?

5. **Rational pairs.** Find three different pairs of positive rational numbers such that the product of each pair is equal to its sum.

6. **Summing ages.** The sum of Anne's, Ben's, Curt's, and Deb's ages is 125. If you add 4 to Anne's age, subtract 4 from Ben's age, multiply Curt's age by 4, or divide Deb's age by 4, you get the same number. What are their ages?

7. **Changing dimensions.** The Smiths are debating whether to change the size of the living room on the plans for their new house. If they add 3 feet to the width and 2 feet to the length, the area will be 240 ft². If they add 2 feet to the width and 3 feet to the length, the area will be 238 ft². What is the original size of the living room?

8. **Vegi tales.** A farmer laid out a large rectangular garden. He then divided it into four rectangular sections by running a string in the north-south direction and another string in the east-west direction. He planted corn in the 216-m² northwest rectangle, beans in the 144-m² northeast rectangle, squash in the 192-m² southeast rectangle, and okra in the southwest rectangle. What was the total area of the garden?

Photo for Exercise 8

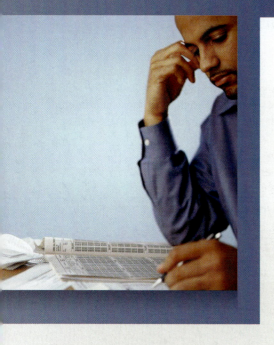

Systems of Linear Equations

In his letter to M. Leroy in 1789 Benjamin Franklin said, "In this world nothing is certain but death and taxes." Since that time, taxes have become not only inevitable, but also intricate and complex.

Each year the U.S. Congress revises parts of the Federal Income Tax Code. To help clarify these revisions, the Internal Revenue Service issues frequent revenue rulings. In addition, there are seven tax courts that further interpret changes and revisions, sometimes in entirely different ways. Is it any wonder that tax preparation has become complicated and many individuals do not prepare their own taxes? Both corporate and individual tax preparation is a growing business, and there are over 500,000 tax counselors helping more than 60 million taxpayers to file their returns correctly.

Everyone knows that doing taxes involves a lot of arithmetic, but not everyone knows that computing taxes can also involve algebra. In fact, to find state and federal taxes for certain corporations, you must solve a system of equations.

You will see an example of using algebra to find amounts of income taxes in Exercises 103 and 104 of Section 4.1.

4.1 Solving Systems by Graphing and Substitution

In This Section

⟨1⟩ Systems of Equations

⟨2⟩ Solving a System by Graphing

⟨3⟩ Solving by Substitution

⟨4⟩ Applications

In Chapter 3, we studied linear equations in two variables, but we usually considered only one equation at a time. In this chapter we will see problems that involve more than one equation.

⟨1⟩ Systems of Equations

Any collection of two or more equations is a **system** of equations. We will first study systems of two equations in two variables. A **solution** to such a system is an ordered pair that satisfies both equations. The **solution set** consists of all ordered pairs that satisfy both equations of the system.

EXAMPLE 1

Checking whether an ordered pair satisfies a system

Determine whether the given ordered pair is in the solution set to the given system.

    a) $(2, 5)$ **b)** $(-3, 6)$

      $x + y = 7$ $y = -2x$

      $3x - 5y = -19$ $y - 3x = 14$

Solution

a) To check $(2, 5)$, let $x = 2$ and $y = 5$ in each equation:

$$x + y = 7 \qquad\qquad 3x - 5y = -19$$
$$2 + 5 = 7 \qquad\qquad 3(2) - 5(5) = -19$$

     Correct   Correct

Since both equations are correct, $(2, 5)$ is in the solution set to the given system.

b) To check $(-3, 6)$, let $x = -3$ and $y = 6$ in each equation:

$$y = -2x \qquad\qquad y - 3x = 14$$
$$6 = -2(-3) \qquad\quad 6 - 3(-3) = 14$$

     Correct   Incorrect

Since $(-3, 6)$ does not satisfy the second equation, it is not in the solution set to the given system.

                                    Now do Exercises 1–8

⟨2⟩ Solving a System by Graphing

Because the graph of each linear equation is a line, points that satisfy both equations lie on both lines. For some systems these points can be found by graphing.

EXAMPLE 2

A system with only one solution

Solve the system by graphing:

$$y = x + 2$$
$$x + y = 4$$

‹ **Calculator Close-Up** ›

To check Example 2, graph

$$y_1 = x + 2$$

and

$$y_2 = -x + 4.$$

From the CALC menu, choose intersect to have the calculator locate the point of intersection of the two lines. After choosing intersect, you must indicate which two lines you want to intersect and then guess the point of intersection.

Solution

First write the equations in slope-intercept form:

$$y = x + 2$$
$$y = -x + 4$$

Use the y-intercept and the slope to graph each line. The graph of the system is shown in Fig. 4.1. From the graph it appears that these lines intersect at $(1, 3)$. To be certain, we can check that $(1, 3)$ satisfies both equations. Let $x = 1$ and $y = 3$ in $y = x + 2$ to get

$$3 = 1 + 2.$$

Let $x = 1$ and $y = 3$ in $x + y = 4$ to get

$$1 + 3 = 4.$$

Because $(1, 3)$ satisfies both equations, the solution set to the system is $\{(1, 3)\}$.

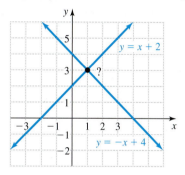

Figure 4.1

Now do Exercises 9–16

EXAMPLE **3**

A system with infinitely many solutions

Solve the system by graphing:

$$2(y + 2) = x$$
$$x - 2y = 4$$

Solution

Write each equation in slope-intercept form:

$$
\begin{aligned}
2(y + 2) &= x & x - 2y &= 4 \\
2y + 4 &= x & -2y &= -x + 4 \\
y &= \tfrac{1}{2}x - 2 & y &= \tfrac{1}{2}x - 2
\end{aligned}
$$

Because the equations have the same slope-intercept form, the original equations are equivalent. Their graphs are the same straight line as shown in Fig. 4.2. Every point on the line satisfies both equations of the system. There are infinitely many points in the solution set. The solution set is $\{(x, y) \mid x - 2y = 4\}$.

Now do Exercises 17–18

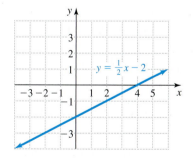

Figure 4.2

EXAMPLE **4**

A system with no solution

Solve the system by graphing:

$$2x - 3y = 6$$
$$3y - 2x = 3$$

Solution

First write each equation in slope-intercept form:

$$2x - 3y = 6 \qquad\qquad 3y - 2x = 3$$
$$-3y = -2x + 6 \qquad\qquad 3y = 2x + 3$$
$$y = \frac{2}{3}x - 2 \qquad\qquad y = \frac{2}{3}x + 1$$

The graph of the system is shown in Fig. 4.3. Because these lines both have slope $\frac{2}{3}$ the lines are parallel and there is no ordered pair that satisfies both equations. The solution set to the system is the empty set $\varnothing$. Of course, it is not really necessary to graph the equations to make this conclusion. Once you see that the slopes are the same and the y-intercepts are different, there is no solution to the system.

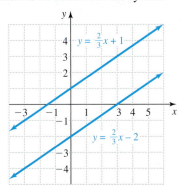

Figure 4.3

> Now do Exercises 19–22

A system of equations that has at least one solution is **consistent** (Examples 2 and 3). A system with no solutions is **inconsistent** (Example 4). There are two types of consistent systems. A consistent system with exactly one solution is **independent** (Example 2), and a consistent system with infinitely many solutions is **dependent** (Example 3). These ideas are summarized in Fig. 4.4.

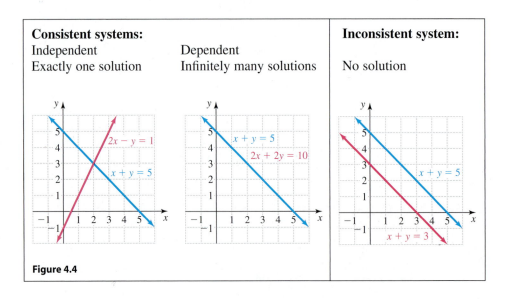

Consistent systems:		Inconsistent system:
Independent Exactly one solution	Dependent Infinitely many solutions	No solution

Figure 4.4

‹3› **Solving by Substitution**

Solving a system by graphing is certainly limited by the accuracy of the graph. If the lines intersect at a point whose coordinates are not integers, then it is difficult to determine those coordinates from the graph. The method of solving a system by **substitution** does not depend on a graph and is totally accurate. For substitution, we replace a variable in one equation with an equivalent expression obtained from the other equation. Our intention in this substitution step is to eliminate a variable and to give us an equation involving only one variable.

E X A M P L E   5

An independent system solved by substitution

Solve the system by substitution.

$$y = 2x - 3$$
$$y = x + 5$$

Solution

We can eliminate y by replacing y in the second equation with $2x - 3$ from the first equation:

$$y = x + 5 \quad \text{Second equation}$$
$$2x - 3 = x + 5 \quad \text{Replace } y \text{ with } 2x - 3.$$
$$x - 3 = 5 \quad \text{Subtract } x \text{ from each side.}$$
$$x = 8 \quad \text{Add 3 to each side.}$$

Since $y = x + 5$ and $x = 8$, we have $y = 8 + 5 = 13$. Check $x = 8$ and $y = 13$ in both equations:

$$13 = 2 \cdot 8 - 3 \quad \text{Correct}$$
$$13 = 8 + 5 \quad \text{Correct}$$

The solution set to the system is $\{(8, 13)\}$, and the equations are independent.

> **Now do Exercises 37–48**

‹ **Calculator Close-Up** ›

Check Example 5 by graphing

$$y_1 = 2x - 3$$

and

$$y_2 = x + 5.$$

In Example 6, we must isolate a variable in one equation before we can substitute it into the other.

E X A M P L E   6

An independent system solved by substitution

Solve the system by substitution:

$$2x + 3y = 8$$
$$y + 2x = 6$$

Solution

Either equation could be solved for either variable. However, it is simplest to solve $y + 2x = 6$ for y to get $y = -2x + 6$. Now replace y in the first equation by $-2x + 6$:

$$2x + 3y = 8$$
$$2x + 3(-2x + 6) = 8 \quad \text{Substitute } -2x + 6 \text{ for } y.$$
$$2x - 6x + 18 = 8$$
$$-4x = -10$$
$$x = \frac{5}{2}$$

‹ **Calculator Close-Up** ›

To check Example 6, graph

$$y_1 = (8 - 2x)/3$$

and

$$y_2 = -2x + 6.$$

From the CALC menu, choose intersect to have the calculator locate the point of intersection of the two lines.

To find y, we let $x = \frac{5}{2}$ in the equation $y = -2x + 6$:

$$y = -2\left(\frac{5}{2}\right) + 6 = -5 + 6 = 1$$

The next step is to check $x = \frac{5}{2}$ and $y = 1$ in each equation. If $x = \frac{5}{2}$ and $y = 1$ in $2x + 3y = 8$, we get

$$2\left(\frac{5}{2}\right) + 3(1) = 8.$$

If $x = \frac{5}{2}$ and $y = 1$ in $y + 2x = 6$, we get

$$1 + 2\left(\frac{5}{2}\right) = 6.$$

Because both of these equations are true, the solution set to the system is $\left\{\left(\frac{5}{2}, 1\right)\right\}$. The equations of this system are independent.

> **Now do Exercises 49–54**

E X A M P L E   7

A dependent system solved by substitution

Solve by substitution:

$$2x + 3y = 5 + x + 4y$$
$$y = x - 5$$

Solution

Substitute $y = x - 5$ into the first equation:

$$2x + 3(x - 5) = 5 + x + 4(x - 5)$$
$$2x + 3x - 15 = 5 + x + 4x - 20$$
$$5x - 15 = 5x - 15$$

Because the last equation is an identity, any ordered pair that satisfies $y = x - 5$ will also satisfy $2x + 3y = 5 + x + 4y$. The equations of this system are dependent. The solution set to the system is the set of all points that satisfy $y = x - 5$. We write the solution set in set notation as

$$\{(x, y) \mid y = x - 5\}.$$

We can verify this result by writing $2x + 3y = 5 + x + 4y$ in slope-intercept form:

$$2x + 3y = 5 + x + 4y$$
$$3y = -x + 5 + 4y$$
$$-y = -x + 5$$
$$y = x - 5$$

Because this slope-intercept form is identical to the slope-intercept form of the other equation, they are two different equations for the same straight line.

> **Now do Exercises 55–56**

‹ **Helpful Hint** ›

The purpose of Example 7 is to show what happens when a dependent system is solved by substitution. If we had first written the first equation in slope-intercept form, we would have known that the equations are dependent and would not have done substitution.

If a system is dependent, then an identity will result after the substitution. If the system is inconsistent, then a false equation will result after the substitution.

EXAMPLE **8**

An inconsistent system solved by substitution
Solve by substitution:

$$x - 2y = 3$$
$$2x - 4y = 7$$

Solution

Since the coefficient of x is 1 in $x - 2y = 3$, the simplest way to proceed is to solve this equation for x to get $x = 2y + 3$. Now substitute $2y + 3$ for x in the second equation:

$$2x - 4y = 7$$
$$2(2y + 3) - 4y = 7$$
$$4y + 6 - 4y = 7$$
$$6 = 7$$

> **‹ Helpful Hint ›**
>
> The purpose of Example 8 is to show what happens when you try to solve an inconsistent system by substitution. If we had first written the equations in slope-intercept form, we would have known that the lines are parallel and the solution set is the empty set.

Because $6 = 7$ is incorrect no matter what values are chosen for x and y, there is no solution to this system of equations. The equations are inconsistent. To check, we write each equation in slope-intercept form:

$$x - 2y = 3 \qquad\qquad 2x - 4y = 7$$
$$-2y = -x + 3 \qquad\qquad -4y = -2x + 7$$
$$y = \frac{1}{2}x - \frac{3}{2} \qquad\qquad y = \frac{1}{2}x - \frac{7}{4}$$

The graphs of these equations are parallel lines with different y-intercepts. The solution set to the system is the empty set, $\varnothing$.

> Now do Exercises 57–64

The strategy for solving an independent system by substitution follows.

Strategy for the Substitution Method

1. Solve one of the equations for one variable in terms of the other. Choose the equation that is easiest to solve for x or y.
2. Substitute into the other equation to get an equation in one variable.
3. Solve for the remaining variable (if possible).
4. Insert the value just found into one of the original equations to find the value of the other variable.
5. Check the two values in both equations.

‹4› Applications

Many of the problems that we solved in previous chapters involved more than one unknown quantity. To solve them, we wrote expressions for all of the unknowns in terms of one variable. Now we can solve problems involving two unknowns by using two variables and writing a system of equations.

E X A M P L E   9

Perimeter of a rectangle

The length of a rectangular swimming pool is twice the width. If the perimeter is 120 feet, then what are the length and width?

Solution

Draw a diagram as shown in Fig. 4.5. If L represents the length and W represents the width, then we can write the following system.

$$L = 2W$$
$$2L + 2W = 120$$

Since $L = 2W$, we can replace L in $2L + 2W = 120$ with $2W$:

$$2(2W) + 2W = 120$$
$$4W + 2W = 120$$
$$6W = 120$$
$$W = 20$$

So the width is 20 feet and the length is 2(20) or 40 feet.

> Now do Exercises 81–92

Figure 4.5

E X A M P L E   10

Tale of two investments

Last year Belinda made $2160 income on a $20,000 investment. This year she wants to split the $20,000 between two funds. She expects that one of the funds will have a 10% return and the other more risky fund will have a 12% return. How much of the $20,000 should she invest in each fund so that her total income is again $2160?

Solution

Let x be the amount invested at 10% and y be the amount invested at 12%. We can summarize all of the given information in a table:

	Amount	Rate	Interest
First investment	x	10%	$0.10x$
Second investment	y	12%	$0.12y$

We can write one equation about the amounts invested and another about the interest from the investments:

$$x + y = 20,000 \quad \text{Total amount invested}$$
$$0.10x + 0.12y = 2160 \quad \text{Total interest}$$

Solve the first equation for x to get $x = 20,000 - y$. Substitute $20,000 - y$ for x in the second equation:

$$0.10x + 0.12y = 2160$$
$$0.10(20,000 - y) + 0.12y = 2160 \quad \text{Replace } x \text{ by } 20,000 - y.$$
$$2000 - 0.10y + 0.12y = 2160 \quad \text{Solve for } y.$$
$$0.02y = 160$$
$$y = 8000$$
$$x = 12,000 \quad \text{Because } x = 20,000 - y$$

‹ **Helpful Hint** ›

In Chapter 2 we would have done Example 10 with one variable by letting x represent the amount invested at 10% and $20,000 - x$ represent the amount invested at 12%.

‹ **Calculator Close-Up** ›

To check Example 10, graph
$$y_1 = 20,000 - x$$
and
$$y_2 = (2160 - 0.1x)/0.12.$$
The viewing window needs to be large enough to contain the point of intersection. Use the intersection feature to find the point of intersection.

To check this answer, find 10% of $12,000 and 12% of $8000:

$$0.10(12,000) = 1200$$

$$0.12(8000) = 960$$

Because $1200 + $960 = $2160 and $8000 + $12,000 = $20,000, we can be certain that Belinda should invest $12,000 at 10% and $8000 at 12%.

Now do Exercises 93–108

Warm-Ups ▼

Fill in the blank.

1. The solution to an _____ linear system is the point of intersection of the two lines.
2. If the two lines are _____, then there is no solution to the linear system.
3. The disadvantage of solving a system by _____ is inaccuracy.
4. In the _____ method we eliminate a variable by substituting one equation into the other.
5. If substitution in a linear system results in a _____ equation, then the system has exactly one solution.
6. If substitution results in an identity, then the system is _____.
7. If substitution results in an _____ equation, then the system has no solution.

True or false?

8. The point $(1, 2)$ satisfies $2x + y = 4$.
9. The point $(1, 2)$ satisfies $2x + y = 4$ and $3x - y = 6$.
10. The point $(2, 3)$ satisfies $4x - y = 5$ and $4x - y = -5$.
11. If two distinct lines in a plane are not parallel, then they intersect at exactly one point.
12. Substituting $y = 2x$ into $x + 3y = 11$ yields $x + 6x = 11$.
13. No ordered pair satisfies $y = 3x - 5$ and $y = 3x + 1$.
14. The equations $y = 3x - 6$ and $y = 2x - 6$ are dependent.

Exercises

‹ Study Tips ›

- It is a good idea to work with others, but don't be misled. Working a problem with help is not the same as working a problem on your own.
- Math is personal. Make sure that you can do it.

‹ 1 › Systems of Equations

Determine whether the given ordered pair is in the solution set to the given system. See Example 1.

1. $(4, 5)$
$$x + y = 9$$
$$5x - 4y = 0$$

2. $(-6, 8)$
$$2x + y = -4$$
$$x - y = -14$$

3. $(3, -2)$
$$y = \frac{1}{3}x - 3$$
$$x = \frac{1}{2}y + 2$$

4. $(10, 20)$
$$0.2x + 0.5y = 11$$
$$y = 2x$$

5. $(-3, -7)$
$$x + y = -10$$
$$2x + 2y = -20$$

6. $(14, -16)$
$$2x + 4y = -36$$
$$-3x - 6y = 54$$

7. $(4, 12)$
$x + y = 15$
$2x + 2y = 32$

8. $(1, -9)$
$x - y = 10$
$y = x - 11$

⟨2⟩ Solving a System by Graphing

Solve each system by graphing. See Examples 2–4.

9. $y = 2x$
$y = -x + 3$

10. $y = x - 3$
$y = -x + 1$

11. $y = 2x - 1$
$2y = x - 2$

12. $y = 2x + 1$
$x + y = -2$

 13. $y = x - 3$
$x - 2y = 4$

14. $y = -3x$
$x + y = 2$

15. $y = 2x + 4$
$3x + y = -1$

16. $3x - 2y = 6$
$3x + 2y = 6$

17. $y = -\dfrac{1}{2}x + 4$
$x + 2y = 8$

18. $2x - 3y = 6$
$y = \dfrac{2}{3}x - 2$

19. $2y - 2x = 2$
$2y - 2x = 6$

20. $3y - 3x = 9$
$x - y = 1$

21. $y = -\dfrac{1}{4}x$
$x + 4y = 8$

22. $y = -\dfrac{2}{3}x$
$2x + 3y = 5$

The graphs of the following systems are given in (a) through (d). Match each system with the correct graph.

23. $5x + 4y = 7$
$x - 3y = 9$

24. $3x - 5y = -9$
$5x - 6y = -8$

25. $4x - 5y = -2$
$3y - x = -3$

26. $4x + 5y = -2$
$4y - x = 11$

a)

b)

c)

d)

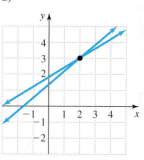

Graph each pair of equations on a graphing calculator using a window that shows the point of intersection. Use the intersect feature to find the solution.

27. $y = 4x - 22.35$
$y = -6x + 50.15$

28. $y = 10x - 31.32$
$y = 5x - 13.27$

29. $y = -3x + 51$
$y = 2x - 9$

30. $y = 5x + 98$
$y = -7x - 70$

31. $3.5x + y = 66$
$7.5x - y = 506$

32. $y - 12.5x = 1266$
$4.5x + y = -230$

33. $3x - 2y = -158$
$5x + 4y = 1526$

34. $x - 7y = -270$
$2x + 15y = -1120$

35. $y = 3.1x + 452$
$y = 3.2x + 443.6$

36. $y = 1.99x + 0.2$
$y = 1.98x + 0.7$

⟨3⟩ Solving by Substitution

Solve each system by substitution. Determine whether the equations are independent, dependent, or inconsistent. See Examples 5–8. See the Strategy for the Substitution Method box on page 229.

37. $y = 4x - 1$
$y = x + 8$

38. $y = -3x + 19$
$y = 2x - 1$

39. $y = -2x$
$y = 4x + 12$

40. $y = -4x - 7$
$y = 3x$

41. $y = \dfrac{1}{3}x + 2$
$y = -\dfrac{1}{2}x + 7$

42. $y = -\dfrac{2}{5}x - 2$
$y = \dfrac{3}{2}x + 17$

43. $y = x - 5$
$2x - 5y = 1$

44. $y = x + 4$
$3y - 5x = 6$

45. $x = 2y - 7$
$3x + 2y = -5$

46. $x = y + 3$
$3x - 2y = 4$

47. $y = 2x - 30$
$\frac{1}{5}x - \frac{1}{2}y = -1$

48. $3x - 5y = 4$
$y = \frac{3}{4}x - 2$

49. $2x + y = 9$
$2x - 5y = 15$

50. $3y - x = 0$
$x - 4y = -2$

51. $x - y = 0$
$2x + 3y = 35$

52. $2y = x + 6$
$-3x + 2y = -2$

53. $x + y = 40$
$0.2x + 0.8y = 23$

54. $x - y = 10$
$0.1x + 0.5y = 13$

55. $y = 2x - 5$
$y + 1 = 2(x - 2)$

56. $2x - y = 3$
$2y = 4x - 6$

57. $x - y = 5$
$2x = 2y + 14$

58. $2x - y = 4$
$2x - y = 3$

59. $y = \frac{5}{7}x$
$x = -\frac{2}{3}y$

60. $7y = 9x$
$-3x = 4y$

61. $3(y - 1) = 2(x - 3)$
$3y - 2x = -3$

62. $y = 3(x - 4)$
$3x - y = 12$

63. $y = 3x$
$y = 3x + 1$

64. $y = 3x - 4$
$y = 3x + 4$

Solve each system by the substitution method.

65. $y = \frac{5}{2}x$
$x + 3y = 3$

66. $6x - 3y = 3$
$10x = y + 7$

67. $x + y = 4$
$x - y = 5$

68. $3x - 6y = 5$
$2y = 4x - 6$

69. $2x - 4y = 0$
$6x + 8y = 5$

70. $-3x + 10y = 4$
$6x - 5y = 1$

71. $3x + y = 2$
$-x - 3y = 6$

72. $x + 3y = 2$
$-x + y = 1$

73. $-9x + 6y = 3$
$18x + 30y = 1$

74. $x + 6y = -2$
$5x - 20y = 5$

75. $y = -2x$
$3y - x = 1$

76. $y = 2x$
$15x - 10y = -2$

77. $x = -6y + 1$
$2y = -5x$

78. $x = -3y + 2$
$7y = 3x$

79. $x - y = 0.1$
$2x - 3y = -0.5$

80. $y - 2x = -7.5$
$3x - 5y = 3.2$

⟨4⟩ **Applications**

Write a system of two equations in two unknowns for each problem. Solve each system by substitution. See Examples 9 and 10.

81. **Rectangular patio.** The length of a rectangular patio is 12 feet greater than the width. If the perimeter is 84 feet, then what are the length and width?

82. **Rectangular notepad.** The length of a rectangular notepad is 2 cm longer than twice the width. If the perimeter is 34 cm, then what are the length and width?

83. **Rectangular table.** The width of a rectangular table is 1 ft less than half of the length. If the perimeter is 28 ft, then what are the length and width?

84. **Rectangular painting.** The width of a rectangular painting is two-thirds of its length. If the perimeter is 60 in., then what are the length and width?

85. ***Sum and difference.*** The sum of two numbers is 10 and their difference is 3. Find the numbers.

86. ***Sum and difference.*** The sum of two numbers is 51 and their difference is 26. Find the numbers.

87. ***Sum and difference.*** The sum of two numbers is 1 and their difference is 20. Find the numbers.

88. ***Sum and difference.*** The sum of two numbers is 5 and their difference is 30. Find the numbers.

89. *Flying to Vegas.* Two hundred people were on a charter flight to Las Vegas. Some paid $200 for their tickets and some paid $250. If the total revenue for the flight was $44,000, then how many tickets of each type were sold?

90. *Annual concert.* A total of 150 tickets were sold for the annual concert to students and nonstudents. Student tickets were $5 and nonstudent tickets were $8. If the total revenue for the concert was $930, then how many tickets of each type were sold?

91. *Annual play.* There were twice as many tickets sold to nonstudents than to students for the annual play. Student tickets were $6 and nonstudent tickets were $11. If the total revenue for the play was $1540, then how many tickets of each type were sold?

92. *Soccer game.* There were 1000 more students at the soccer game than nonstudents. Student tickets were $8.50 and nonstudent tickets were $13.25. If the total revenue for the game was $75,925, then how many tickets of each type were sold?

93. *Mixing investments.* Helen wants to invest a total of $40,000 in two investments. She expects that one will return 5% and the other 8%. If she wants a return of $2300 after one year, then how much should she put into each investment?

94. *Investing her bonus.* Donna wants to invest her $33,000 bonus in two investments. She expects that one will return 4% and the other 2.25%. How much should she put in each investment so that her total interest is $970 after one year?

95. *Balancing risk.* Doug has $100,000 to invest. He is looking at a risky gold fund that is expected to return 30% and a stock index fund that is expected to return 8% in the next year. If he wants a 15% return on his money, then how should he invest his money?

96. *Different growth rates.* In 2000 the total population of East Springfield and West Springfield was 30,000. In 2010 the population of East Springfield was 14% greater and the population of West Springfield was 8% greater, and the total population was 10% greater. What was the population of each city in 2000?

97. *Mixing acid.* A chemist wants to mix a 5% acid solution with a 25% acid solution to obtain 50 liters of a 20% acid solution. How many liters of each solution should be used?

98. *Mixing fertilizer.* A farmer wants to mix a liquid fertilizer that contains 2% nitrogen with one that contains 10% nitrogen to obtain 40 gallons of a fertilizer that contains 8% nitrogen. How many gallons of each fertilizer should be used?

99. *Finding numbers.* The sum of two numbers is 2, and their difference is 26. Find the numbers.

100. *Finding more numbers.* The sum of two numbers is -16, and their difference is 8. Find the numbers.

101. *Toasters and vacations.* During one week a land developer gave away Florida vacation coupons or toasters to 100 potential customers who listened to a sales presentation. It costs the developer $6 for a toaster and $24 for a Florida vacation coupon. If his bill for prizes that week was $708, then how many of each prize did he give away?

102. *Ticket sales.* Tickets for a concert were sold to adults for $3 and to students for $2. If the total receipts were $824 and twice as many adult tickets as student tickets were sold, then how many of each were sold?

103. *Corporate taxes.* According to Bruce Harrell, CPA, the amount of federal income tax for a class C corporation is deductible on the Louisiana state tax return, and the amount of state income tax for a class C corporation is deductible on the federal tax return. So for a state tax rate of 5% and a federal tax rate of 30%, we have

state tax $= 0.05$(taxable income $-$ federal tax)

and

federal tax $= 0.30$(taxable income $-$ state tax).

Find the amounts of state and federal income taxes for a class C corporation that has a taxable income of $100,000.

104. *More taxes.* Use the information given in Exercise 103 to find the amounts of state and federal income taxes for a class C corporation that has a taxable income of $300,000. Use a state tax rate of 6% and a federal tax rate of 40%.

105. *Cost accounting.* The problems presented in this exercise and Exercise 106 are encountered in cost accounting. A company has agreed to distribute 20% of its net income N to its employees as a bonus: $B = 0.20N$. If the company has an income of $120,000 before the bonus, the bonus B is deducted from the $120,000 as an expense to determine net income; $N = 120,000 - B$. Solve the system of two equations in N and B to find the amount of the bonus.

106. *Bonus and taxes.* A company has an income of $100,000 before paying taxes and a bonus. The bonus B is to be 20% of the income after deducting income taxes T but before deducting the bonus. So,

$$B = 0.20(100,000 - T).$$

Because the bonus is a deductible expense, the amount of income tax T at a 40% rate is 40% of the income after deducting the bonus. So,

$$T = 0.40(100,000 - B).$$

a) Use the accompanying graph to estimate the values of T and B that satisfy both equations.

b) Solve the system algebraically to find the bonus and the amount of tax.

Figure for Exercise 106

107. *Textbook case.* The accompanying graph shows the cost of producing textbooks and the revenue from the sale of those textbooks.

Figure for Exercise 107

a) What is the cost of producing 10,000 textbooks?
b) What is the revenue when 10,000 textbooks are sold?

c) For what number of textbooks is the cost equal to the revenue?

d) The cost of producing zero textbooks is called the *fixed cost*. Find the fixed cost.

108. *Free market.* The function $S = 5000 + 200x$ and $D = 9500 - 100x$ express the supply S and the demand D, respectively, for a popular compact disc brand as a function of its price x (in dollars).

a) Graph the functions on the same coordinate system.
b) What happens to the supply as the price increases?
c) What happens to the demand as the price increases?
d) The price at which supply and demand are equal is called the *equilibrium price*. What is the equilibrium price?

Getting More Involved

109. *Discussion*

Which of the following equations is not equivalent to $2x - 3y = 6$?

a) $3y - 2x = 6$ **b)** $y = \dfrac{2}{3}x - 2$

c) $x = \dfrac{3}{2}y + 3$ **d)** $2(x - 5) = 3y - 4$

110. *Discussion*

Which of the following equations is inconsistent with the equation $3x + 4y = 8$?

a) $y = \dfrac{3}{4}x + 2$

b) $6x + 8y = 16$

c) $y = -\dfrac{3}{4}x + 8$

d) $3x - 4y = 8$

Math *at Work* | Circuit Breakers

Electricity is the flow of electrons through a circuit. It is measured in volts, amps, and watts. Volts measure the force that causes the electricity or electrons to flow. Amps measure the amount of electric current. Watts measure the amount of work done by a certain amount of current at a certain force or voltage. The basic relationship is watts = amps · volts or $W = A \cdot V$.

A circuit breaker is used as a safety device in a circuit. If the amperage exceeds a certain level, the breaker trips and prevents damage to the system. For example, suppose that 8 strings of Christmas lights each containing 25 bulbs that are 7 watts each are all plugged into one 120-volt circuit containing a 15-amp breaker. Will the breaker trip? The total wattage is $8 \cdot 25 \cdot 7$ or 1400 watts. Use $A = W/V$ to get $A = 1400/120 \approx 11.7$. So the lights will not blow a 15-amp fuse. See the accompanying figure.

While houses use standard single-phase electricity, electric power companies may supply power for large users to transformers through three-phase lines. The power in a three-phase system is measured in volt-amps. The formula used here is volt-amps = $\sqrt{3} \cdot A \cdot V$. For example, suppose a large shopping mall has a 1,000,000 volt-amp transformer and the power company provides 25,000 volts to the mall's transformer. Will this power trip a 20-amp breaker? Because $A = \text{volt-amps}/(\sqrt{3} \cdot V)$, we have $A = 1,000,000/(\sqrt{3} \cdot 25,000) \approx 23.1$ amps. So the 20-amp breaker will blow.

4.2   The Addition Method

In Section 4.1, you used substitution to eliminate a variable in a system of equations. In this section, we see another method for eliminating a variable in a system of equations.

In This Section

⟨1⟩ **The Addition Method**

⟨2⟩ **Equations Involving Fractions or Decimals**

⟨3⟩ **Applications**

⟨1⟩ The Addition Method

In the **addition method** we eliminate a variable by adding the equations.

E X A M P L E   1

An independent system solved by addition

Solve the system by the addition method:

$$3x - 5y = -9$$
$$4x + 5y = 23$$

Solution

The addition property of equality allows us to add the same number to each side of an equation. We can also use the addition property of equality to add the two left sides and add the two right sides:

$$3x - 5y = -9$$
$$\underline{4x + 5y = 23}$$
$$7x \qquad = 14 \qquad \text{Add.}$$
$$x = 2$$

‹ Calculator Close-Up ›

To check Example 1, graph

$$y_1 = (-9 - 3x)/-5$$

and

$$y_2 = (23 - 4x)/5.$$

Use the intersect feature to find the point of intersection of the two lines.

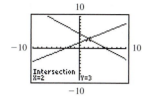

The y-term was eliminated when we added the equations because the coefficients of the y-terms were opposites. Now use $x = 2$ in one of the original equations to find y. It does not matter which original equation we use. In this example we will use both equations to see that we get the same y in either case.

$$3x - 5y = -9 \qquad\qquad\qquad\qquad 4x + 5y = 23$$
$$3(2) - 5y = -9 \quad \text{Replace } x \text{ by 2.} \qquad 4(2) + 5y = 23$$
$$6 - 5y = -9 \quad \text{Solve for } y. \qquad 8 + 5y = 23$$
$$-5y = -15 \qquad\qquad\qquad\qquad 5y = 15$$
$$y = 3 \qquad\qquad\qquad\qquad\qquad y = 3$$

Because $3(2) - 5(3) = -9$ and $4(2) + 5(3) = 23$ are both true, $(2, 3)$ satisfies both equations. The solution set is $\{(2, 3)\}$.

> **Now do Exercises 1–8**

Actually the addition method can be used to eliminate any variable whose coefficients are opposites. If neither variable has coefficients that are opposites, then we use the multiplication property of equality to change the coefficients of the variables, as shown in Examples 2 and 3.

E X A M P L E   2

‹ Calculator Close-Up ›

Check Example 2 by graphing

$$y_1 = (-13 - 2x)/(-3)$$

and

$$y_2 = (-46 - 5x)/(-12).$$

Using multiplication and addition

Solve the system by the addition method:

$$2x - 3y = -13$$
$$5x - 12y = -46$$

Solution

First examine the system to find the simplest way to eliminate a variable. Note that 5 is not a multiple of 2, but 12 is a multiple of 3. So if we multiply both sides of the first equation by -4, the coefficients of y will be 12 and -12, and y will be eliminated by addition:

$$(-4)(2x - 3y) = (-4)(-13) \quad \text{Multiply each side by } -4.$$
$$5x - 12y = -46$$

$$\begin{array}{r} -8x + 12y = 52 \\ 5x - 12y = -46 \quad \text{Add.} \\ \hline -3x = 6 \\ x = -2 \end{array}$$

Replace x by -2 in one of the original equations to find y:

$$2x - 3y = -13$$
$$2(-2) - 3y = -13$$
$$-4 - 3y = -13$$
$$-3y = -9$$
$$y = 3$$

Because $2(-2) - 3(3) = -13$ and $5(-2) - 12(3) = -46$ are both true, the solution set is $\{(-2, 3)\}$.

> **Now do Exercises 9–12**

EXAMPLE 3

Multiplying both equations before adding

Solve each system by the addition method:

a) $-2x + 3y = 6$ b) $-2x + 3y = 0$
   $3x - 5y = -11$ $3x - 5y = 0$

Solution

a) Examine the coefficients. Since 3 is not a multiple of 2 and 5 is not a multiple of 3, we can't eliminate a variable by multiplying only one equation. However, multiplying the first equation by 3 and the second by 2 will give us $-6x$ and $6x$:

$$3(-2x + 3y) = 3(6) \qquad \text{Multiply each side by 3.}$$

$$2(3x - 5y) = 2(-11) \qquad \text{Multiply each side by 2.}$$

$$\begin{aligned} -6x + 9y &= 18 \\ \underline{6x - 10y} &= \underline{-22} \qquad \text{Add.} \\ -y &= -4 \\ y &= 4 \end{aligned}$$

Note that we could have eliminated y by multiplying by 5 and 3. Now insert $y = 4$ into one of the original equations to find x:

$$\begin{aligned} -2x + 3(4) &= 6 \quad \text{Let } y = 4 \text{ in } -2x + 3y = 6. \\ -2x + 12 &= 6 \\ -2x &= -6 \\ x &= 3 \end{aligned}$$

Check that (3, 4) satisfies both equations. The solution set is $\{(3, 4)\}$.

b) Multiplying the first equation by 3, the second by 2, and then adding will eliminate x as it did in part (a):

$$\begin{array}{lll} -2x + 3y = 0 & \text{Multiply by 3} & -6x + 9y = 0 \\ 3x - 5y = 0 & \text{Multiply by 2} & \underline{6x - 10y = 0} \\ & & -y = 0 \\ & & y = 0 \end{array}$$

If $y = 0$ in $-2x + 3y = 0$, we get $-2x = 0$ or $x = 0$. So the solution set is $\{(0, 0)\}$. Note that the graphs of these two equations intersect at the origin.

> **Now do Exercises 13–18**

The strategy for solving an independent system by addition follows.

Strategy for the Addition Method

1. Write both equations in the same form (usually $Ax + By = C$).

2. If necessary, multiply one or both equations by the appropriate integer to obtain opposite coefficients on one of the variables.

3. Add the equations to get an equation in one variable.

4. Solve the equation in one variable.

5. Substitute the value obtained for one variable into one of the original equations to obtain the value of the other variable.

6. Check the two values in both of the original equations.

We can identify dependent and inconsistent systems in the same way that we did for the substitution method. If the result of the addition is an identity, the system is dependent and there are infinitely many solutions. If the result of the addition is a false equation, the system is inconsistent and there are no solutions. When you use addition, make sure that the equations are in the same form with the variables and equal signs aligned.

E X A M P L E 4

Solving dependent and inconsistent systems by addition

Solve each system by addition:

a) $2x - 3y = 9$
    $6y = 4x - 18$

b) $-4y = 5x + 7$
    $4y + 5x = 12$

Solution

a) First rewrite $6y = 4x - 18$ as $-4x + 6y = -18$ so that it is in the same form as the first equation:

$$2x - 3y = 9$$
$$-4x + 6y = -18$$

Now examine the coefficients. Multiply the first equation by 2 to get $4x - 6y = 18$ and add:

$$
\begin{array}{ll}
4x - 6y = 18 & \text{First equation multiplied by 2} \\
\underline{-4x + 6y = -18} & \text{Second equation rewritten} \\
0 = 0 &
\end{array}
$$

Because the result of the addition is an identity, the equations are dependent and there are infinitely many solutions. The solution set is $\{(x, y) \mid 2x - 3y = 9\}$.

b) Rewrite the first equation $-4y = 5x + 7$ as $-4y - 5x = 7$ to get the same form as the second. Now add:

$$
\begin{array}{l}
-4y - 5x = 7 \\
\underline{4y + 5x = 12} \\
0 = 19
\end{array}
$$

Because the result of the addition is a false equation, the system is inconsistent. There are no solutions to the system. The solution set is the empty set, $\varnothing$.

> **Now do Exercises 25–30**

‹ **Calculator Close-Up** ›

To check Example 4(b), graph

$$y_1 = (5x + 7)/-4$$

and

$$y_2 = (-5x + 12)/4.$$

Since the lines appear to be parallel, the graph supports the conclusion that the system is inconsistent.

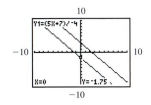

‹ 2 › Equations Involving Fractions or Decimals

When a system of equations involves fractions or decimals, we can use the multiplication property of equality to eliminate the fractions or decimals.

E X A M P L E 5

A system with fractions

Solve the system:

$$\frac{1}{2}x - \frac{2}{3}y = 7$$

$$\frac{2}{3}x - \frac{3}{4}y = 11$$

‹ Calculator Close-Up ›

To check Example 5, graph

$$y_1 = (7 - (1/2)x)/(-2/3)$$

and

$$y_2 = (11 - (2/3)x)/(-3/4).$$

The lines appear to intersect at $(30, 12)$.

Solution

Since 2 and 3 both divide evenly into 6, multiplying the first equation by 6 will eliminate its fractions. Since 3 and 4 both divide evenly into 12, multiplying the second equation by 12 will eliminate its fractions:

$$6\left(\frac{1}{2}x - \frac{2}{3}y\right) = 6(7) \qquad \rightarrow \qquad 3x - 4y = 42$$

$$12\left(\frac{2}{3}x - \frac{3}{4}y\right) = 12(11) \qquad \rightarrow \qquad 8x - 9y = 132$$

Now examine the coefficients in the two new equations. Both equations will have to be multiplied to eliminate x or y. To eliminate x, multiply the first by -8 and the second by 3:

$$-8(3x - 4y) = -8(42) \qquad \rightarrow \qquad -24x + 32y = -336$$

$$3(8x - 9y) = 3(132) \qquad \rightarrow \qquad \underline{24x - 27y = 396}$$

$$5y = 60$$

$$y = 12$$

Substitute $y = 12$ into the first of the original equations:

$$\frac{1}{2}x - \frac{2}{3}(12) = 7$$

$$\frac{1}{2}x - 8 = 7$$

$$\frac{1}{2}x = 15$$

$$x = 30$$

Check $(30, 12)$ in the original system. The solution set is $\{(30, 12)\}$.

> **Now do Exercises 31–38**

Note that there are many ways to proceed in Example 5. We multiplied first to eliminate the fractions and second to eliminate a variable. That is usually the simplest approach. However, if you multiply the first equation by -48 and the second by 36, you would only have to multiply once.

E X A M P L E   6

A system with decimals

Solve the system:

$$0.05x + 0.7y = 40$$

$$x + 0.4y = 120$$

Solution

Multiplying by 10 or 100 moves the decimal point one or two places to the right, respectively. So multiplying the first equation by 100 and the second by 10 will eliminate all of the decimals:

$$100(0.05x + 0.7y) = 100(40) \qquad \rightarrow \qquad 5x + 70y = 4000$$

$$10(x + 0.4y) = 10(120) \qquad \rightarrow \qquad 10x + 4y = 1200$$

Examine the coefficients. Since 10 is a multiple of 5, we can eliminate x by multiplying the first equation by -2:

$$-2(5x + 70y) = -2(4000) \qquad \rightarrow \qquad -10x - 140y = -8000$$

$$10x + 4y = 1200 \qquad \rightarrow \qquad \underline{10x + 4y = 1200}$$

$$-136y = -6800$$

$$y = 50$$

‹ Calculator Close-Up ›

Check Example 6 by graphing

$$y_1 = (40 - 0.05x)/(0.7)$$

and

$$y_2 = (120 - x)/(0.4).$$

Use $y = 50$ in $x + 0.4y = 120$ to find x:

$$x + 0.4(50) = 120$$
$$x + 20 = 120$$
$$x = 100$$

Check $(100, 50)$ in the original system. The solution set is $\{(100, 50)\}$.

> **Now do Exercises 39–46**

We have seen three methods for solving a system of two linear equations in two variables. For some systems the method you choose can make a difference. The following summary should help you decide which method to use.

Summary of the Methods

Graphing	It is impossible to identify the solution from a graph unless it is very simple. Graphing helps us understand the difference between independent, dependent, and inconsistent systems. Graphing works well with a graphing calculator.
Substitution	Substitution is used when one of the equations is solved for one of the variables or when it is easy to isolate one of the variables in an equation.
Addition	Addition is used when both equations are in the same form and it is easy to eliminate a variable by multiplying and adding. The addition method is the basis for the Gauss-Jordan method in Section 4.4 where a graphing calculator can do all of the work.

‹3› Applications

Any system of two linear equations in two variables can be solved by either the addition method or substitution. In applications we use whichever method appears to be the simpler for the problem at hand.

E X A M P L E   7

‹ Helpful Hint ›

You can see from Example 7 that the standard form $Ax + By = C$ occurs naturally in accounting. This form will occur whenever we have the price of each item and a quantity of two items and want to express the total cost.

Fajitas and burritos

At the Cactus Cafe the total price for four fajita dinners and three burrito dinners is $48, and the total price for three fajita dinners and two burrito dinners is $34. What is the price of each type of dinner?

Solution

Let x represent the price (in dollars) of a fajita dinner, and let y represent the price (in dollars) of a burrito dinner. We can write two equations to describe the given information:

$$4x + 3y = 48$$
$$3x + 2y = 34$$

Because 12 is the least common multiple of 4 and 3 (the coefficients of x), we multiply the first equation by -3 and the second by 4:

$$-3(4x + 3y) = -3(48) \quad \text{Multiply each side by } -3.$$
$$4(3x + 2y) = 4(34) \quad \text{Multiply each side by 4.}$$

$$\begin{array}{r} -12x - 9y = -144 \\ \underline{12x + 8y = 136} \quad \text{Add.} \\ -y = -8 \\ y = 8 \end{array}$$

To find x, use $y = 8$ in the first equation $4x + 3y = 48$:

$$4x + 3(8) = 48$$
$$4x + 24 = 48$$
$$4x = 24$$
$$x = 6$$

So the fajita dinners are $6 each, and the burrito dinners are $8 each. Check this solution in the original problem.

> **Now do Exercises 65–70**

E X A M P L E   8

Mixing cooking oil

Canola oil is 7% saturated fat, and corn oil is 14% saturated fat. Crisco sells a blend, Crisco Canola and Corn Oil, which is 11% saturated fat. How many gallons of each type of oil must be mixed to get 280 gallons of this blend?

Solution

Let x represent the number of gallons of canola oil, and let y represent the number of gallons of corn oil. Make a table to summarize all facts:

	Amount (gallons)	% fat	Amount of Fat (gallons)
Canola oil	x	7	$0.07x$
Corn oil	y	14	$0.14y$
Canola and corn oil	280	11	0.11(280) or 30.8

Since the total amount of oil is 280 gallons, we have $x + y = 280$. Since the total amount of fat is 30.8 gallons, we have $0.07x + 0.14y = 30.80$. Since we can easily solve $x + y = 280$ for y, we choose substitution to solve the system. Substitute $y = 280 - x$ into the second equation:

$$0.07x + 0.14(280 - x) = 30.80 \quad \text{Substitution}$$
$$0.07x + 39.2 - 0.14x = 30.80 \quad \text{Distributive property}$$
$$-0.07x = -8.4$$
$$x = \frac{-8.4}{-0.07} = 120$$

If $x = 120$ and $y = 280 - x$, then $y = 280 - 120 = 160$. Check that

$$0.07(120) + 0.14(160) = 30.8.$$

So it takes 120 gallons of canola oil and 160 gallons of corn oil to make 280 gallons of Crisco Canola and Corn Oil.

> **Now do Exercises 71–78**

Warm-Ups ▼

Fill in the blank.

1. In the _____ method we eliminate a variable by adding the equations.

2. If addition in a linear system results in a _____ equation, then the system has exactly one solution.

3. If addition results in an identity, then the system is
_____ .

4. If addition results in an _____ equation, then the
system has no solution.

5. If addition results in a _____ equation, then the
two lines intersect at exactly one point.

6. If addition results in an _____ then the two lines
have the same graph.

7. If addition results in an _____ equation, then the
two lines are parallel.

True or false?

8. To solve $3x - y = 9$ and $3x + y = 6$ by addition we
simply add the equations.

9. To solve $2x + 7y = 5$ and $3x + 2y = 8$ by addition we
multiply the first equation by 3, the second by 2, and
then add.

10. Both $(0, -10)$ and $(5, 0)$ satisfy $4x - 2y = 20$ and
$4x + 2y = 20$.

11. The system $4x - 5y = 9$ and $-4x + 5y = -9$ has no
solution.

12. Either addition or substitution could be used to solve
$2x - y = 5$ and $3x + 2y = 9$.

13. To eliminate fractions, multiply both sides of the
equation by the least common denominator.

14. Either variable can be eliminated by the addition
method.

Exercises 4.2

‹ 1 › **The Addition Method**

*Solve each system by addition. See Examples 1–3. See the
Strategy for the Addition Method box on page 238.*

1. $x - y = 1$
$\quad x + y = 7$

2. $x + y = 7$
$\quad x - y = 9$

3. $\quad 3x - 4y = 11$
$\quad -3x + 2y = -7$

4. $7x - 5y = -1$
$\quad -3x + 5y = 9$

5. $\quad x - y = 12$
$\quad 2x + y = 3$

6. $\quad x - 2y = -1$
$\quad -x + 5y = 4$

7. $3x - y = 5$
$\quad 5x + y = -2$

8. $-x + 2y = 4$
$\quad x - 5y = 1$

9. $2x - \quad y = -5$
$\quad 3x + 2y = 3$

10. $3x + 5y = -11$
$\quad x - 2y = 11$

11. $-3x + 5y = 1$
$\quad 9x - 3y = 5$

12. $7x - 4y = -3$
$\quad x + 2y = 3$

13. $2x - 5y = 13$
$\quad 3x + 4y = -15$

14. $3x + 4y = -5$
$\quad 5x + 6y = -7$

15. $2x = 3y + 11$
$\quad 7x - 4y = 6$

16. $2x = 2 - y$
$\quad 3x + y = -1$

17. $\quad x + \quad y = 48$
$\quad 12x + 14y = 628$

18. $\quad x + \quad y = 13$
$\quad 22x + 36y = 356$

Use a calculator to check whether the given ordered pair satisfies both equations of the given system.

19. $(-45, 16)$
$3x + 2y = -103$
$5x - 8y = -353$

20. $(502, 388)$
$-3x + 5y = 434$
$6x - 7y = 296$

21. $(42, 99)$
$\dfrac{2}{3}x + \dfrac{5}{11}y = 73$
$-\dfrac{1}{3}x + \dfrac{2}{9}y = 9$

22. $(16.5, 25.6)$
$\dfrac{1}{3}x + \dfrac{5}{2}y = 69.5$
$\dfrac{4}{5}x - \dfrac{3}{4}y = 6$

23. $(34.56, 59.66)$
$0.02x + 0.03y = 2.481$
$0.8x + 0.9y = 81.342$

24. $(40,000, 120,000)$
$0.08x + 0.12y = 17,600$
$x + y = 160,000$

Solve each system by the addition method. Determine whether the equations are independent, dependent, or inconsistent. See Example 4.

25. $3x - 4y = 9$
$-3x + 4y = 12$

26. $x - y = 3$
$-6x + 6y = 17$

27. $5x - y = 1$
$10x - 2y = 2$

28. $4x + 3y = 2$
$-12x - 9y = -6$

29. $2x - y = 5$
$2x + y = 5$

30. $-3x + 2y = 8$
$3x + 2y = 8$

‹2› Equations Involving Fractions or Decimals

Solve each system by the addition method. See Examples 5 and 6.

31. $\dfrac{1}{4}x + \dfrac{1}{3}y = 5$
$x - y = 6$

32. $\dfrac{3x}{2} - \dfrac{2y}{3} = 10$
$\dfrac{1}{2}x + \dfrac{1}{2}y = -1$

33. $\dfrac{x}{4} - \dfrac{y}{3} = -4$
$\dfrac{x}{8} + \dfrac{y}{6} = 0$

34. $\dfrac{x}{3} - \dfrac{y}{2} = -\dfrac{5}{6}$
$\dfrac{x}{5} - \dfrac{y}{3} = -\dfrac{3}{5}$

35. $\dfrac{1}{8}x + \dfrac{1}{4}y = 5$
$\dfrac{1}{16}x + \dfrac{1}{2}y = 7$

36. $\dfrac{3}{7}x + \dfrac{5}{9}y = 27$
$\dfrac{1}{9}x + \dfrac{2}{7}y = 7$

37. $\dfrac{1}{3}x + \dfrac{1}{2}y = \dfrac{1}{3}$
$\dfrac{5}{6}x - \dfrac{3}{4}y = \dfrac{1}{6}$

38. $\dfrac{2}{3}x + \dfrac{5}{6}y = \dfrac{1}{4}$
$\dfrac{1}{5}x - \dfrac{1}{10}y = -\dfrac{1}{10}$

39. $0.05x + 0.10y = 1.30$
$x + y = 19$

40. $0.1x + 0.06y = 9$
$0.09x + 0.5y = 52.7$

41. $x + y = 1200$
$0.12x + 0.09y = 120$

42. $x - y = 100$
$0.20x + 0.06y = 150$

43. $1.5x - 2y = -0.25$
$3x + 1.5y = 6.375$

44. $3x - 2.5y = 7.125$
$2.5x - 3y = 7.3125$

45. $0.24x + 0.6y = 0.58$
$0.8x - 0.12y = 0.52$

46. $0.18x + 0.27y = 0.09$
$0.06x - 0.54y = -0.04$

Miscellaneous

Solve each system by substitution or addition, whichever is easier.

47. $y = x + 1$
$2x - 5y = -20$

48. $y = 3x - 4$
$x + y = 32$

49. $x - y = 19$
$2x + y = -13$

50. $x + y = 3$
$7x - y = 29$

51. $2y = x + 2$
$x = y - 1$

52. $2y - x = 3$
$x = 3y - 5$

53. $2y - 3x = -1$
$5y + 3x = 29$

54. $y - 5 = 2x$
$y - 9 = -2x$

55. $6x + 3y = 4$
$y = \dfrac{2}{3}x$

56. $3x - 2y = 2$
$x = \dfrac{2}{9}y$

57. $y = 3x + 1$
$x = \dfrac{1}{3}y + 5$

58. $y = -\dfrac{2}{3}x - 3$
$x = -\dfrac{3}{2}y + 9$

59. $x - y = 0$
$x + y = 2x$

60. $5x - 4y = 9$
$8y - 10x = -18$

For each system find the value of a so that the solution set to the system is $\{(2, 3)\}$.

61. $x + y = 5$
$x - y = a$

62. $2x - y = 1$
$ax + y = 13$

For each system find the values of a and b so that the solution set to the system is $\{(5, 12)\}$.

63. $y = ax + 2$
$y = bx + 17$

64. $y = 3x + a$
$y = -2x + b$

‹3› Applications

Write a system of two equations in two unknowns for each problem. Solve each system by the method of your choice. See Examples 7 and 8.

65. *Coffee and doughnuts.* On Monday, Archie paid $3.40 for three doughnuts and two coffees. On Tuesday he paid $3.60 for two doughnuts and three coffees. On Wednesday he was tired of paying the tab and went out for coffee by himself. What was his bill for one doughnut and one coffee?

Figure for Exercise 65

66. *Books and magazines.* At Gwen's garage sale, all books were one price, and all magazines were another price. Harriet bought four books and three magazines for $1.45, and June bought two books and five magazines for $1.25. What was the price of a book and what was the price of a magazine?

67. *Boys and girls.* One-half of the boys and one-third of the girls of Freemont High attended the homecoming game, whereas one-third of the boys and one-half of the girls attended the homecoming dance. If there were 570 students at the game and 580 at the dance, then how many students are there at Freemont High?

68. *Girls and boys.* There are 385 surfers in Surf City. Two-thirds of the boys are surfers and one-twelfth of the girls are surfers. If there are two girls for every boy, then how many boys and how many girls are there in Surf City?

69. *Nickels and dimes.* Winborne has 35 coins consisting of dimes and nickels. If the value of his coins is $3.30, then how many of each type does he have?

70. *Pennies and nickels.* Wendy has 52 coins consisting of nickels and pennies. If the value of the coins is $1.20, then how many of each type does she have?

71. *Blending fudge.* The Chocolate Factory in Vancouver blends its double-dark-chocolate fudge, which is 35% fat, with its peanut butter fudge, which is 25% fat, to obtain double-dark-peanut fudge, which is 29% fat.

  a) Use the accompanying graph to estimate the number of pounds of each type that must be mixed to obtain 50 pounds of double-dark-peanut fudge.
  b) Write a system of equations and solve it algebraically to find the exact amount of each type that should be used to obtain 50 pounds of double-dark-peanut fudge.

Figure for Exercise 71

72. *Low-fat yogurt.* Ziggy's Famous Yogurt blends regular yogurt that is 3% fat with its no-fat yogurt to obtain low-fat yogurt that is 1% fat. How many pounds of regular yogurt and how many pounds of no-fat yogurt should be mixed to obtain 60 pounds of low-fat yogurt?

73. *Keystone state.* Judy averaged 42 miles per hour (mph) driving from Allentown to Harrisburg and 51 mph driving from Harrisburg to Pittsburgh. See the accompanying figure. If she drove a total of 288 miles in 6 hours, then how long did it take her to drive from Harrisburg to Pittsburgh?

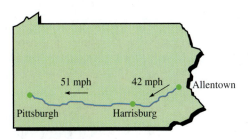

Figure for Exercise 73

74. *Empire state.* Spike averaged 45 mph driving from Rochester to Syracuse and 49 mph driving from Syracuse to

Albany. If he drove a total of 237 miles in 5 hours, then how far is it from Syracuse to Albany?

75. *Probability of rain.* The probability of rain tomorrow is four times the probability that it does not rain tomorrow. The probability that it rains plus the probability that it does not rain is 1. What is the probability that it rains tomorrow?

76. *Super Bowl contender.* The probability that San Francisco plays in the next Super Bowl is nine times the probability that they do not play in the next Super Bowl. The probability that San Francisco plays in the next Super Bowl plus the probability that they do not play is 1. What is the probability that San Francisco plays in the next Super Bowl?

77. *Rectangular lot.* The width of a rectangular lot is 75% of its length. If the perimeter is 700 meters, then what are the length and width?

78. *Fence painting.* Darren and Douglas must paint the 792-foot fence that encircles their family home. Because Darren is older, he has agreed to paint 20% more than Douglas. How much of the fence will each boy paint?

Getting More Involved

79. *Discussion*

Explain how you decide whether it is easier to solve a system by substitution or addition.

80. *Exploration*

a) Write a linear equation in two variables that is satisfied by $(-3, 5)$.
b) Write another linear equation in two variables that is satisfied by $(-3, 5)$.
c) Are your equations independent or dependent?
d) Explain how to select the second equation so that it will be independent of the first.

81. *Exploration*

a) Make up a system of two linear equations in two variables such that both $(-1, 2)$ and $(4, 5)$ are in the solution set.
b) Are your equations independent or dependent?
c) Is it possible to find an independent system that is satisfied by both ordered pairs? Explain.

4.3   Systems of Linear Equations in Three Variables

The techniques that you learned in Section 4.2 can be extended to systems of equations in more than two variables. In this section, we use elimination of variables to solve systems of equations in three variables.

In This Section

⟨1⟩ **Solving a System by Elimination**

⟨2⟩ **Dependent and Inconsistent Systems**

⟨3⟩ **Applications**

⟨1⟩ Solving a System by Elimination

The equation $5x - 4y = 7$ is called a linear equation in two variables because its graph is a straight line. The equation $2x + 3y - 4z = 12$ is similar in form, and so it is a linear equation in three variables. An equation in three variables is graphed in a three-dimensional coordinate system. The graph of a linear equation in three variables is a plane, not a line. We will not graph equations in three variables in this text, but we can solve systems without graphing. In general, we make the following definition.

> **Linear Equation in Three Variables**
>
> If A, B, C, and D are real numbers, with A, B, and C not all zero, then
>
> $$Ax + By + Cz = D$$
>
> is called a **linear equation in three variables**.

A solution to an equation in three variables is an **ordered triple** such as $(-2, 1, 5)$, where the first coordinate is the value of x, the second coordinate is the value of y, and the third coordinate is the value of z. There are infinitely many solutions to a linear equation in three variables.

The solution to a system of equations in three variables is the set of all ordered triples that satisfy all of the equations of the system. The techniques for solving a system of linear equations in three variables are similar to those used on systems of linear equations in two variables. We eliminate variables by either substitution or addition.

E X A M P L E   1

A linear system with a single solution

Solve the system:

$$(1) \qquad x + y - z = -1$$
$$(2) \qquad 2x - 2y + 3z = 8$$
$$(3) \qquad 2x - y + 2z = 9$$

Solution

We can eliminate y from Eqs. (1) and (2) by multiplying Eq. (1) by 2 and adding it to Eq. (2):

$$2x + 2y - 2z = -2 \quad \text{Eq. (1) multiplied by 2}$$
$$\underline{2x - 2y + 3z = 8} \qquad \text{Eq. (2)}$$
$$(4) \qquad 4x + z = 6$$

Now we must eliminate the same variable, y, from another pair of equations. Eliminate y from Eqs. (1) and (3) by simply adding them:

$$x + y - z = -1 \quad \text{Eq. (1)}$$
$$\underline{2x - y + 2z = 9} \qquad \text{Eq. (3)}$$
$$(5) \qquad 3x + z = 8$$

Equations (4) and (5) give us a system with two variables. We now solve this system. Eliminate z by multiplying Eq. (4) by -1 and adding the equations:

$$-4x - z = -6 \quad \text{Eq. (4) multiplied by } -1$$
$$\underline{3x + z = 8} \qquad \text{Eq. (5)}$$
$$-x = 2$$
$$x = -2$$

Now that we have x, we can replace x by -2 in Eq. (5) to find z:

$$3x + z = 8 \quad \text{Eq. (5)}$$
$$3(-2) + z = 8$$
$$-6 + z = 8$$
$$z = 14$$

Now replace x by -2 and z by 14 in Eq. (1) to find y:

$$x + y - z = -1 \quad \text{Eq. (1)}$$
$$-2 + y - 14 = -1 \quad x = -2, z = 14$$
$$y - 16 = -1$$
$$y = 15$$

Check that $(-2, 15, 14)$ satisfies all three of the original equations. The solution set is $\{(-2, 15, 14)\}$.

‹ Calculator Close-Up ›

You can use a calculator to check that $(-2, 15, 14)$ satisfies all three equations of the original system.

```
-2+15-14
                -1
2*-2-2*15+3*14
                 8
2*-2-15+2*14
                 9
```

Now do Exercises 1–4

Note that we could have eliminated any one of the three variables in Example 1 to get a system of two equations in two variables. We chose to eliminate y first because it was the easiest to eliminate. The strategy that we follow for solving a system of three linear equations in three variables is stated as follows:

Strategy for Solving a System in Three Variables

1. Use substitution or addition to eliminate any one of the variables from a pair of equations of the system. Look for the easiest variable to eliminate.
2. Eliminate the same variable from another pair of equations of the system.
3. Solve the resulting system of two equations in two unknowns.
4. After you have found the values of two of the variables, substitute into one of the original equations to find the value of the third variable.
5. Check the three values in all of the original equations.

In Example 2, we use a combination of addition and substitution.

EXAMPLE 2

Using addition and substitution

Solve the system:

$$(1) \quad x + y = 4$$
$$(2) \quad 2x - 3z = 14$$
$$(3) \quad 2y + z = 2$$

Solution

From Eq. (1) we get $y = 4 - x$. If we substitute $y = 4 - x$ into Eq. (3), then Eqs. (2) and (3) will be equations involving x and z only.

$$(3) \qquad 2y + z = 2$$
$$2(4 - x) + z = 2 \quad \text{Replace } y \text{ by } 4 - x.$$
$$8 - 2x + z = 2 \quad \text{Simplify.}$$
$$(4) \qquad -2x + z = -6$$

Now solve the system consisting of Eqs. (2) and (4) by addition:

$$\begin{array}{ll} 2x - 3z = 14 & \text{Eq. (2)} \\ \underline{-2x + z = -6} & \text{Eq. (4)} \\ -2z = 8 & \\ z = -4 & \end{array}$$

Use Eq. (3) to find y:

$$2y + z = 2 \quad \text{Eq. (3)}$$
$$2y + (-4) = 2 \quad \text{Let } z = -4.$$
$$2y = 6$$
$$y = 3$$

Use Eq. (1) to find x:

$$x + y = 4 \quad \text{Eq. (1)}$$
$$x + 3 = 4 \quad \text{Let } y = 3.$$
$$x = 1$$

Check that $(1, 3, -4)$ satisfies all three of the original equations. The solution set is $\{(1, 3, -4)\}$.

‹ **Helpful Hint** ›

In Example 2 we chose to eliminate y first. Try solving Example 2 by first eliminating z. Write $z = 2 - 2y$ and then substitute $2 - 2y$ for z in Eqs. (1) and (2).

‹ **Calculator Close-Up** ›

Check that $(1, 3, -4)$ satisfies all three equations in Example 2.

```
1+3
            4
2(1)-3(-4)
           14
2(3)+(-4)
            2
```

Now do Exercises 5–20

CAUTION In solving a system in three variables it is essential to keep your work organized and neat. Writing short notes that explain your steps (as was done in the examples) will allow you to go back and check your work.

‹2› Dependent and Inconsistent Systems

The graph of any equation in three variables can be drawn on a three-dimensional coordinate system. The graph of a linear equation in three variables is a plane. To solve a system of three linear equations in three variables by graphing, we would have to draw the three planes and then identify the points that lie on all three of them. This method would be difficult even when the points have simple coordinates. So we will not attempt to solve these systems by graphing.

In Section 4.1 we classified systems of two linear equations in two variables as *consistent* if the system had at least one solution and *inconsistent* if the system had no solutions. A consistent system with exactly one solution is *independent*, and a consistent system with infinitely many solutions is *dependent*. We use the same terminology with systems of three linear equations in three variables. The only difference here is that there are more possibilities for the graphs of the dependent and inconsistent systems. Even though we don't solve systems in three variables by graphing, the graphs in Fig. 4.6 will help you to better understand these systems.

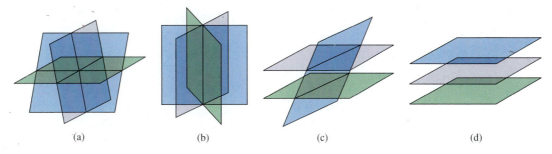

(a)   (b)   (c)   (d)

Figure 4.6

In most of the problems that we will solve, the planes intersect at a single point, as in Fig. 4.6(a). The solution set contains exactly one ordered triple, and the system is **independent.**

If the intersection of the three planes is a line or a plane, then the solution set is infinite and the system is **dependent.** There are three possibilities. All three planes could intersect along a line as shown in Fig. 4.6(b). All three planes could be the same, in which case all points on that plane satisfy the system. We could also have two equations for the same plane with the third plane intersecting it along a line.

If there are no points in common to all three planes, then the system is **inconsistent.** The system will be inconsistent if at least two of the planes are parallel as shown in Fig. 4.6(c) and (d). There is one other configuration for an inconsistent system that is not shown here. See if you can find it.

We will not solve systems corresponding to all of the possible configurations described for the planes. Examples 3 and 4 illustrate two of these cases.

E X A M P L E   **3**

A system with infinitely many solutions

Solve the system:

$$
\begin{aligned}
(1) \qquad & 2x - 3y - \ z = 4 \\
(2) \qquad & -6x + 9y + 3z = -12 \\
(3) \qquad & 4x - 6y - 2z = 8
\end{aligned}
$$

If you recognize that multiplying Eq. (1) by -3 will produce Eq. (2), and multiplying Eq. (1) by 2 will produce Eq. (3), then you can conclude that all three equations are equivalent and there is no need to add the equations.

Solution

We will first eliminate x from Eqs. (1) and (2). Multiply Eq. (1) by 3 and add the resulting equation to Eq. (2):

$$
\begin{array}{ll}
6x - 9y - 3z = 12 & \text{Eq. (1) multiplied by 3} \\
\underline{-6x + 9y + 3z = -12} & \text{Eq. (2)} \\
0 = 0 &
\end{array}
$$

The last statement is an identity. The identity occurred because Eq. (2) is a multiple of Eq. (1). In fact, Eq. (3) is also a multiple of Eq. (1). These equations are dependent. They are all equations for the same plane. The solution set is the set of all points on that plane,

$$\{(x, y, z) \mid 2x - 3y - z = 4\}.$$

> Now do Exercises 21–22

E X A M P L E   4

A system with no solutions

Solve the system:

$$
\begin{array}{rl}
(1) & x + y - z = 5 \\
(2) & 3x - 2y + z = 8 \\
(3) & 2x + 2y - 2z = 7
\end{array}
$$

Solution

We can eliminate the variable z from Eqs. (1) and (2) by adding them:

$$
\begin{array}{ll}
x + y - z = 5 & \text{Eq. (1)} \\
\underline{3x - 2y + z = 8} & \text{Eq. (2)} \\
4x - y = 13 &
\end{array}
$$

To eliminate z from Eqs. (1) and (3), multiply Eq. (1) by -2 and add the resulting equation to Eq. (3):

$$
\begin{array}{ll}
-2x - 2y + 2z = -10 & \text{Eq. (1) multiplied by } -2 \\
\underline{2x + 2y - 2z = 7} & \text{Eq. (3)} \\
0 = -3 &
\end{array}
$$

Because the last equation is false, the system is inconsistent. The solution set is the empty set.

> Now do Exercises 23–34

‹3› Applications

Problems involving three unknown quantities can often be solved by using a system of three equations in three variables.

E X A M P L E   5

Finding three unknown rents

Theresa took in a total of $1240 last week from the rental of three condominiums. She had to pay 10% of the rent from the one-bedroom condo for repairs, 20% of the rent from the two-bedroom condo for repairs, and 30% of the rent from the three-bedroom condo for repairs. If the three-bedroom condo rents for twice as much as the one-bedroom condo and her total repair bill was $276, then what is the rent for each condo?

‹ **Helpful Hint** ›

A problem involving two unknowns can often be solved with one variable as in Chapter 2. Likewise, you can often solve a problem with three unknowns using only two variables. Solve Example 5 by letting a, b, and $2a$ be the rent for a one-bedroom, two-bedroom, and a three-bedroom condo.

‹ **Calculator Close-Up** ›

Check that $(280, 400, 560)$ satisfies all three equations in Example 5.

```
280+400+560
              1240
.1*280+.2*400+.3
*560
              276
2*280
              560
```

Solution

Let x, y, and z represent the rent on the one-bedroom, two-bedroom, and three-bedroom condos, respectively. We can write one equation for the total rent, another equation for the total repairs, and a third equation expressing the fact that the rent for the three-bedroom condo is twice that for the one-bedroom condo:

$$x + y + z = 1240$$
$$0.1x + 0.2y + 0.3z = 276$$
$$z = 2x$$

Substitute $z = 2x$ into both of the other equations to eliminate z:

$$x + y + 2x = 1240$$
$$0.1x + 0.2y + 0.3(2x) = 276$$

$$3x + \quad y = 1240$$
$$0.7x + 0.2y = 276$$

$$-2(3x + y) = -2(1240) \qquad \text{Multiply each side by } -2.$$
$$10(0.7x + 0.2y) = 10(276) \qquad \text{Multiply each side by } 10.$$

$$-6x - 2y = -2480$$
$$\underline{7x + 2y = 2760} \qquad \text{Add.}$$
$$x \quad\quad = 280$$

$$z = 2(280) = 560 \qquad \text{Because } z = 2x$$
$$280 + y + 560 = 1240 \qquad \text{Because } x + y + z = 1240$$
$$y = 400$$

Check that $(280, 400, 560)$ satisfies all three of the original equations. The condos rent for $280, $400, and $560 per week.

> **Now do Exercises 51–64**

Warm-Ups ▼

Fill in the blank.

1. An equation of the form $Ax + By + Cz = D$, where A, B, and C are not all zero, is a _____ equation in three variables.
2. The _____ triple (a, b, c) corresponds to a point in a three-dimensional coordinate system.
3. A _____ to a linear system in three variables is an ordered triple that satisfies all of the equations.
4. To solve a linear system in three variables use _____ or _____ to eliminate variables.
5. The graph of a linear equation in three variables is a _____ in a three-dimensional coordinate system.
6. For an _____ system of three linear equations in three variables the planes intersect at a single point.

True or false?

7. The point $(1, -2, 3)$ satisfies $x + y - z = 4$.
8. The point $(4, 1, 1)$ is the only solution to $x + y - z = 4$.
9. The point $(1, -1, 2)$ satisfies $x + y + z = 2$ and $2x + y - z = -1$.
10. Two distinct planes are either parallel or intersect at a single point.
11. The equations $3x + 2y - 6z = 4$ and $-6x - 4y + 12z = -8$ are dependent.
12. The graph of $y = 2x - 3z + 4$ is a line.
13. The value of x nickels, y dimes, and z quarters is $0.05x + 0.10y + 0.25x$ cents.
14. If $x = -2$, $z = 3$, and $x + y + z = 6$, then $y = 7$.

Exercises

‹ Study Tips ›

- Finding out what happened in class and attending class are not the same. Attend every class and be attentive.
- Don't just take notes and let your mind wander. Use class time as a learning time.

‹ 1 › Solving a System by Elimination

Solve each system of equations. See Examples 1 and 2. See the Strategy for Solving a System in Three Variables box on page 248.

1. $x + y + z = 9$
$y + z = 7$
$z = 4$

2. $x + y - z = 4$
$y = 6$
$y + z = 13$

3. $x + y + z = 10$
$x - y = -1$
$x + y = 5$

4. $x + y - z = 6$
$y + z = 11$
$y - z = 3$

5. $x + y + z = 6$
$x - y + z = 2$
$x - y - z = -4$

6. $x + y + z = 0$
$x + y - z = 2$
$x - y + z = 0$

7. $x + y + z = 2$
$x + 2y - z = 6$
$2x + y - z = 5$

8. $2x - y + 3z = 14$
$x + y - 2z = -5$
$3x + y - z = 2$

9. $x - 2y + 4z = 3$
$x + 3y - 2z = 6$
$x - 4y + 3z = -5$

10. $2x + 3y + z = 13$
$-3x + 2y + z = -4$
$4x - 4y + z = 5$

11. $2x - y + z = 10$
$3x - 2y - 2z = 7$
$x - 3y - 2z = 10$

12. $x - 3y + 2z = -11$
$2x - 4y + 3z = -15$
$3x - 5y - 4z = 5$

13. $2x - 3y + z = -9$
$-2x + y - 3z = 7$
$x - y + 2z = -5$

14. $3x - 4y + z = 19$
$2x + 4y + z = 0$
$x - 2y + 5z = 17$

15. $2x - 5y + 2z = 16$
$3x + 2y - 3z = -19$
$4x - 3y + 4z = 18$

16. $-2x + 3y - 4z = 3$
$3x - 5y + 2z = 4$
$-4x + 2y - 3z = 0$

17. $x + y = 4$
$y - z = -2$
$x + y + z = 9$

18. $x + y - z = 0$
$x - y = -2$
$y + z = 10$

19. $x + y = 7$
$y - z = -1$
$x + 3z = 18$

20. $2x - y = -8$
$y + 3z = 22$
$x - z = -8$

‹ 2 › Dependent and Inconsistent Systems

Solve each system. See Examples 3 and 4.

21. $x + y - z = 2$
$-x - y + z = -2$
$2x + 2y - 2z = 4$

22. $x + y + z = 1$
$2x + 2y + 2z = 2$
$4x + 4y + 4z = 4$

23. $x + y - z = 2$
$x + y + z = 8$
$x + y - z = 6$

24. $x + y + z = 6$
$2z + 2y + 2z = 9$
$3x + 3y + 3z = 12$

25. $x + y + z = 9$
$x + y = 5$
$z = 1$

26. $x - y + z = 2$
$y - z = 3$
$x = 4$

27. $x - y + 2z = 3$
$2x + y - z = 5$
$3x - 3y + 6z = 4$

28. $4x - 2y - 2z = 5$
$2x - y - z = 7$
$-4x + 2y + 2z = 6$

29. $2x - 4y + 6z = 12$
$6x - 12y + 18z = 36$
$-x + 2y - 3z = -6$

30. $3x - y + z = 5$
$9x - 3y + 3z = 15$
$-12x + 4y - 4z = -20$

31. $x - y = 3$
$y + z = 8$
$2x + 2z = 7$

32. $2x - y = 6$
$2y + z = -4$
$8x + 2z = 3$

33. $0.10x + 0.08y - 0.04z = 3$
$5x + 4y - 2z = 150$
$0.3x + 0.24y - 0.12z = 9$

34. $0.06x - 0.04y + z = 6$
$3x - 2y + 50z = 300$
$0.03x - 0.02y + 0.5z = 3$

Use a calculator to solve each system.

35. $3x + 2y - 0.4z = 0.1$
$3.7x - 0.2y + 0.05z = 0.41$
$-2x + 3.8y - 2.1z = -3.26$

36. $3x - 0.4y + 9z = 1.668$
$0.3x + 5y - 8z = -0.972$
$5x - 4y - 8z = 1.8$

Use a calculator to check whether the given ordered triple satisfies all three equations of the given system.

37. (45, 32, 12)
$3x - 2y + z = 83$
$x + 5y - z = 193$
$5x + y - 6z = 185$

38. (−16, 45, 19)
$7x + 6y - 3z = 101$
$3x + 4y - 9z = -39$
$-x + 5y + 8z = 393$

39. (244, 386, 122)
$0.1x + 0.3y - 0.12z = 125.56$
$0.9x + 0.4y - 0.25z = 343.5$
$0.5x + 0.2y + 0.15z = 181.0$

40. (66, 72, 84)
$$\frac{1}{2}x + \frac{1}{3}y - \frac{5}{7}z = -3$$
$$-\frac{1}{3}x - \frac{1}{4}y + \frac{5}{12}z = -5$$
$$\frac{5}{6}x - \frac{5}{4}y + \frac{5}{14}z = 5$$

Miscellaneous

Solve each system. State whether the system is independent, dependent, or inconsistent.

41. $x - 2y = -12$
$2x + 3y = 4$

42. $x - 2y = 3$
$-2x + 4y = 6$

43. $x + y = 4$
$2x + 2y = 8$

44. $-x + y = 12$
$5x + 4y = -6$

45. $x + 2y - 3z = 6$
$-2x - 4y + 6z = 10$

46. $x - y - z = 4$
$2x + y + 3z = 6$
$2x - 2y - 2z = 10$

47. $x - y - z = 0$
$x + y + 3z = 8$
$x + 3y + z = 10$

48. $x + y - z = 0$
$3x - y - z = -10$
$2x + y + z = 35$

49. $5x - 5y + 5z = 5$
$x - y + z = 1$
$3x - 3y + 3z = 3$

50. $4x - 7y + 3z = 5$
$8x - 14y + 6z = 10$

‹3› Applications

Solve each problem by using a system of three equations in three unknowns. See Example 5.

51. ***Three cars.*** The town of Springfield purchased a Chevrolet, a Ford, and a Toyota for a total of $66,000. The Ford was $2,000 more than the Chevrolet and the Toyota was $2,000 more than the Ford. What was the price of each car?

52. ***Buying texts.*** Melissa purchased an English text, a math text, and a chemistry text for a total of $276. The English text was $20 more than the math text, and the chemistry text was twice the price of the math text. What was the price of each text?

53. ***Three-day drive.*** In three days, Carter drove 2196 miles in 36 hours behind the wheel. The first day he averaged 64 mph, the second day 62 mph, and the third day 58 mph. If he drove 4 more hours on the third day than on the first day, then how many hours did he drive each day?

54. ***Three-day trip.*** In three days, Katy traveled 146 miles down the Mississippi River in her kayak with 30 hours of paddling. The first day she averaged 6 mph, the second day 5 mph, and the third day 4 mph. If her distance on the third day was equal to her distance on the first day, then for how many hours did she paddle each day?

55. *Diversification.* Ann invested a total of $12,000 in stocks, bonds, and a mutual fund. She received a 10% return on her stock investment, an 8% return on her bond investment, and a 12% return on her mutual fund. Her total return was $1230. If the total investment in stocks and bonds equaled her mutual fund investment, then how much did she invest in each?

56. *Paranoia.* Fearful of a bank failure, Norman split his life savings of $60,000 among three banks. He received 5%, 6%, and 7% on the three deposits. In the account earning 7% interest, he deposited twice as much as in the account earning 5% interest. If his total earnings were $3760, then how much did he deposit in each account?

57. *Weighing in.* Anna, Bob, and Chris will not disclose their weights but agree to be weighed in pairs. Anna and Bob together weigh 226 pounds. Bob and Chris together weigh 210 pounds. Anna and Chris together weigh 200 pounds. How much does each student weigh?

Anna & Bob   Bob & Chris   Anna & Chris

Figure for Exercise 57

58. *Big tipper.* On Monday Headley paid $1.70 for two cups of coffee and one doughnut, including the tip. On Tuesday he paid $1.65 for two doughnuts and a cup of coffee, including the tip. On Wednesday he paid $1.30 for one coffee and one doughnut, including the tip. If he always tips the same amount, then what is the amount of each item?

59. *Three coins.* Nelson paid $1.75 for his lunch with 13 coins, consisting of nickels, dimes, and quarters. If the number of dimes was twice the number

of nickels, then how many of each type of coin did he use?

60. *Pocket change.* Harry has $2.25 in nickels, dimes, and quarters. If he had twice as many nickels, half as many dimes, and the same number of quarters, he would have $2.50. If he has 27 coins altogether, then how many of each does he have?

61. *Working overtime.* To make ends meet, Ms. Farnsby works three jobs. Her total income last year was $48,000. Her income from teaching was just $6000 more than her income from house painting. Royalties from her textbook sales were one-seventh of the total money she received from teaching and house painting. How much did she make from each source last year?

62. *Lunch-box special.* Salvador's Fruit Mart sells variety packs. The small pack contains three bananas, two apples, and one orange for $1.80. The medium pack contains four bananas, three apples, and three oranges for $3.05. The family size contains six bananas, five apples, and four oranges for $4.65. What price should Salvador charge for his lunch-box special that consists of one banana, one apple, and one orange?

63. *Three generations.* Edwin, his father, and his grandfather have an average age of 53. One-half of his grandfather's age, plus one-third of his father's age, plus one-fourth of Edwin's age is 65. If 4 years ago, Edwin's grandfather was four times as old as Edwin, then how old are they all now?

64. *Three-digit number.* The sum of the digits of a three-digit number is 11. If the digits are reversed, the new number is 46 more than five times the old number. If the hundreds digit plus twice the tens digit is equal to the units digit, then what is the number?

Getting More Involved

65. *Exploration*

Draw diagrams showing the possible ways to position three planes in three-dimensional space.

66. *Discussion*

Make up a system of three linear equations in three variables for which the solution set is {(0, 0, 0)}. A system with this solution set is called a *homogeneous* system. Why do you think it is given that name?

Mid-Chapter **Quiz** │ Sections 4.1 through 4.3 │ **Chapter 4**

Determine whether $(1, -2)$ is in the solution set to each system.

1. $x - y = 3$
$2x + y = 0$

2. $x + y = -1$
$3x - y = 8$

3. $5x + 12y = -19$
$5x + 12y = 6$

Solve by graphing.

4. $y = 2x - 4$
$x + y = 5$

5. $x - y = 8$
$x + y = 0$

Solve by substitution.

6. $y = 3x - 5$
$2x + 5y = 9$

7. $x + y = 6$
$3x - 5y = 26$

8. $5x - y = 8$
$35x - 6 = 7y$

Solve by addition.

9. $5x - y = 17$
$2x + y = 18$

10. $4x + 2y = 2$
$3x - 5y = 8$

11. $4x = 16 - 8y$
$2x + 4y = 8$

Determine whether each system is independent, dependent, or inconsistent.

12. $y = \dfrac{1}{2}x - 7$
$y = \dfrac{1}{2}x + 5$

13. $y = 5x - 12$
$y = 3x + 7$

14. $y = \dfrac{3}{4}x + 1$
$4y = 3x + 4$

Solve by the method of your choice.

15. $\dfrac{2}{3}x - \dfrac{3}{4}y = 2$
$\dfrac{5}{6}x - \dfrac{1}{8}y = 9$

16. $0.02x + 0.1y = 5$
$0.4x - 0.05y = 18$

Solve each system in three variables.

17. $x - y + z = 4$
$2x + y + z = 10$
$3x - y + 3z = 16$

18. $x + y - z = 2$
$x - y + 3z = 6$
$2x + 2y - 2z = 8$

19. $3x - 3y + 9z = 12$
$x - y + 3z = 4$
$-6x + 6y - 18z = -24$

4.4   Solving Linear Systems Using Matrices

In This Section

⟨1⟩ **Matrices**

⟨2⟩ **The Augmented Matrix**

⟨3⟩ **The Gauss-Jordan Elimination Method**

⟨4⟩ **Dependent and Inconsistent Systems**

You solved linear systems in two variables by substitution and addition in Sections 4.1 and 4.2. Those methods are done differently on each system. In this section, you will learn the Gauss-Jordan elimination method, which is related to the addition method. The Gauss-Jordan elimination method is performed in the same way on every system. We first need to introduce some new terminology.

⟨1⟩ **Matrices**

A **matrix** is a rectangular array of numbers enclosed in brackets. The **rows** of a matrix run horizontally, and the **columns** of a matrix run vertically. A matrix with m rows and n columns has **size** $m \times n$ (read "m by n"). Each number in a matrix is called an **element** or **entry** of the matrix.

E X A M P L E   **1**

Size of a matrix

Determine the size of each matrix.

a) $\begin{bmatrix} -1 & 2 \\ 5 & \sqrt{2} \\ 0 & 3 \end{bmatrix}$ b) $\begin{bmatrix} 2 & 3 \\ -1 & 5 \end{bmatrix}$

c) $\begin{bmatrix} 1 & 2 & 3 \\ 4 & 5 & 6 \\ -1 & 0 & 2 \end{bmatrix}$ d) $\begin{bmatrix} 1 & 3 & 6 \end{bmatrix}$

Solution

Because matrix (a) has 3 rows and 2 columns, its size is 3×2. Matrix (b) is a 2×2 matrix, matrix (c) is a 3×3 matrix, and matrix (d) is a 1×3 matrix.

> Now do Exercises 1–8

⟨2⟩ The Augmented Matrix

The solution to a system of linear equations such as

$$x - 2y = -5$$
$$3x + y = 6$$

depends on the coefficients of x and y and the constants on the right-hand side of the equation. The matrix of coefficients for this system is the 2×2 matrix

$$\begin{bmatrix} 1 & -2 \\ 3 & 1 \end{bmatrix}.$$

If we insert the constants from the right-hand side of the system into the matrix of coefficients, we get the 2×3 matrix

$$\left[\begin{array}{cc|c} 1 & -2 & -5 \\ 3 & 1 & 6 \end{array}\right].$$

We use a vertical line between the coefficients and the constants to represent the equal signs. This matrix is the **augmented matrix** of the system. Two systems of linear equations are **equivalent** if they have the same solution set. Two augmented matrices are **equivalent** if the systems they represent are equivalent.

E X A M P L E   **2**

Writing the augmented matrix

Write the augmented matrix for each system of equations.

a) $3x - 5y = 7$
   $x + y = 4$

b) $x + y - z = 5$
   $2x + z = 3$
   $2x - y + 4z = 0$

c) $x + y = 1$
   $y + z = 6$
   $z = -5$

Solution

a) $\begin{bmatrix} 3 & -5 & | & 7 \\ 1 & 1 & | & 4 \end{bmatrix}$ **b)** $\begin{bmatrix} 1 & 1 & -1 & | & 5 \\ 2 & 0 & 1 & | & 3 \\ 2 & -1 & 4 & | & 0 \end{bmatrix}$ **c)** $\begin{bmatrix} 1 & 1 & 0 & | & 1 \\ 0 & 1 & 1 & | & 6 \\ 0 & 0 & 1 & | & -5 \end{bmatrix}$

Now do Exercises 9–12

EXAMPLE **3**

Writing the system

Write the system of equations represented by each augmented matrix.

a) $\begin{bmatrix} 1 & 4 & | & -2 \\ 1 & -1 & | & 3 \end{bmatrix}$ **b)** $\begin{bmatrix} 1 & 0 & | & 5 \\ 0 & 1 & | & 1 \end{bmatrix}$ **c)** $\begin{bmatrix} 2 & 3 & 4 & | & 6 \\ -1 & 0 & 5 & | & -2 \\ 1 & -2 & 3 & | & 1 \end{bmatrix}$

Solution

a) Use the first two numbers in each row as the coefficients of x and y and the last number as the constant to get the following system:

$$x + 4y = -2$$
$$x - y = 3$$

b) Use the first two numbers in each row as the coefficients of x and y and the last number as the constant to get the following system:

$$x = 5$$
$$y = 1$$

c) Use the first three numbers in each row as the coefficients of x, y, and z and the last number as the constant to get the following system:

$$2x + 3y + 4z = 6$$
$$-x \qquad + 5z = -2$$
$$x - 2y + 3z = 1$$

Now do Exercises 13–16

⟨3⟩ **The Gauss-Jordan Elimination Method**

When we solve a single equation, we write simpler and simpler equivalent equations to get an equation whose solution is obvious. In the **Gauss-Jordan elimination method** we write simpler and simpler equivalent augmented matrices until we get an augmented matrix [like the one in Example 3(b)] in which the solution to the corresponding system is obvious.

Because each row of an augmented matrix represents an equation, we can perform the operations on the rows of the augmented matrix. These **row operations,** which follow, correspond to the usual operations with equations used in the addition method.

Row Operations

The following row operations on an augmented matrix give an equivalent augmented matrix:

1. Interchange two rows of the matrix.

2. Multiply every element in a row by a nonzero real number.

3. Add to a row a multiple of another row.

If a matrix has three rows, we use R_1, R_2, and R_3 to name them. The notation $3R_1$ is the row that results when every entry in R_1 is multiplied by 3. $R_1 + R_2$ is the row that results from adding the corresponding entries in R_1 and R_2. To indicate which row operations have been performed we use a double arrow ($\leftrightarrow$) for "interchange" and a single arrow ($\rightarrow$) for "replace." For example, $R_1 \leftrightarrow R_2$ indicates that R_1 and R_2 have been interchanged. The notation $\frac{1}{2}R_1 \rightarrow R_1$ indicates that $\frac{1}{2}R_1$ has replaced R_1.

EXAMPLE 4

Row operations

Perform the indicated row operation on the given augmented matrix.

a) $R_1 \leftrightarrow R_3$

$$\begin{bmatrix} 0 & 0 & 3 & | & 1 \\ 0 & 1 & 7 & | & 2 \\ 1 & 0 & 0 & | & 4 \end{bmatrix}$$

b) $\frac{1}{3}R_2 \rightarrow R_2$

$$\begin{bmatrix} 1 & 0 & | & 5 \\ 0 & 3 & | & -6 \end{bmatrix}$$

c) $-2R_1 + R_2 \rightarrow R_2$

$$\begin{bmatrix} 1 & 1 & | & 4 \\ 2 & 3 & | & 6 \end{bmatrix}$$

Solution

a) For $R_1 \leftrightarrow R_3$, R_1 and R_3 are interchanged. So the resulting matrix is

$$\begin{bmatrix} 1 & 0 & 0 & | & 4 \\ 0 & 1 & 7 & | & 2 \\ 0 & 0 & 3 & | & 1 \end{bmatrix}.$$

b) For $\frac{1}{3}R_2 \rightarrow R_2$ every entry in R_2 is multiplied by $\frac{1}{3}$, which yields $[0 \quad 1 \quad -2]$. Replacing the original R_2 with those results yields

$$\begin{bmatrix} 1 & 0 & | & 5 \\ 0 & 1 & | & -2 \end{bmatrix}.$$

c) For $-2R_1 + R_2 \rightarrow R_2$ multiply each entry of R_1 by -2 to get $[-2 \quad -2 \quad -8]$. Now add these results to the corresponding entries in R_2 to get $[0 \quad 1 \quad -2]$. This result then replaces the original entries in R_2, yielding

$$\begin{bmatrix} 1 & 1 & | & 4 \\ 0 & 1 & | & -2 \end{bmatrix}.$$

> **Now do Exercises 17–34**

In the Gauss-Jordan elimination method our goal is to use row operations to obtain an augmented matrix that has ones on the **diagonal** in its matrix of coefficients and zeros elsewhere:

$$\begin{bmatrix} 1 & 0 & | & a \\ 0 & 1 & | & b \end{bmatrix}$$

The system corresponding to this augmented matrix is $x = a$ and $y = b$. So the solution set to the system is $\{(a, b)\}$.

EXAMPLE 5

Gauss–Jordan elimination with two equations in two variables

Use the Gauss-Jordan elimination method to solve the system:

$$x - 3y = 11$$
$$2x + y = 1$$

Solution

Start with the augmented matrix:

$$\begin{bmatrix} 1 & -3 & | & 11 \\ 2 & 1 & | & 1 \end{bmatrix}$$

‹ Calculator Close-Up ›

Use MATRX EDIT to enter the matrix A into the calculator.

Under the MATRX MATH menu select rref (row-reduced echelon form). Choose A from the MATRX NAMES menu, and the calculator does all of the calculations in Example 5.

```
[A]
   [[1 -3 11]
    [2 1  1 ]]
rref([A])
   [[1 0  2 ]
    [0 1 -3]]
```

To get a 0 in the first position of the second row (R_2), multiply the first row (R_1) by -2 and add the result to R_2. In symbols, $-2R_1 + R_2 \rightarrow R_2$. Because $-2R_1 = [-2, 6, -22]$ and $R_2 = [2, 1, 1]$, we add corresponding entries to get $-2R_1 + R_2 = [0, 7, -21]$. Note that with this operation the coefficient of x in the second equation is 0 and we get the following matrix:

$$\begin{bmatrix} 1 & -3 & | & 11 \\ 0 & 7 & | & -21 \end{bmatrix} \quad -2R_1 + R_2 \rightarrow R_2$$

Multiply each element of row 2 by $\frac{1}{7}$ (in symbols, $\frac{1}{7}R_2 \rightarrow R_2$):

$$\begin{bmatrix} 1 & -3 & | & 11 \\ 0 & 1 & | & -3 \end{bmatrix} \quad \frac{1}{7}R_2 \rightarrow R_2$$

Multiply row 2 by 3 and add the result to row 1. Because $3R_2 = [0, 3, -9]$ and $R_1 = [1, -3, 11]$, $3R_2 + R_1 = [1, 0, 2]$. Note that the coefficient of y in the first equation is now 0. We get the following matrix:

$$\begin{bmatrix} 1 & 0 & | & 2 \\ 0 & 1 & | & -3 \end{bmatrix} \quad 3R_2 + R_1 \rightarrow R_1$$

This augmented matrix represents the system $x = 2$ and $y = -3$. So the solution set to the system is $\{(2, -3)\}$. Check in the original system.

> Now do Exercises 35–48

In Example 6 we use the row operations on the augmented matrix of a system of three linear equations in three variables.

EXAMPLE 6

Gauss-Jordan elimination with three equations in three variables

Use the Gauss-Jordan elimination method to solve the following system:

$$2x - y + z = -3$$
$$x + y - z = 6$$
$$3x - y - z = 4$$

‹ Helpful Hint ›

It is not necessary to perform the row operations in exactly the same order as is shown in Example 6. As long as you use the legitimate row operations and get to the final form, you will get the solution to the system. Of course, you must double check your arithmetic at every step if you want to be successful at Gauss-Jordan elimination.

Solution

Start with the augmented matrix and interchange the first and second rows to get a 1 in the upper left position in the matrix:

$$\begin{bmatrix} 2 & -1 & 1 & | & -3 \\ 1 & 1 & -1 & | & 6 \\ 3 & -1 & -1 & | & 4 \end{bmatrix} \quad \text{The augmented matrix}$$

$$\begin{bmatrix} 1 & 1 & -1 & | & 6 \\ 2 & -1 & 1 & | & -3 \\ 3 & -1 & -1 & | & 4 \end{bmatrix} \quad R_1 \leftrightarrow R_2$$

Now multiply the first row by -2 and add the result to the second row. Multiply the first row by -3 and add the result to the third row. These two steps eliminate the variable x from the second and third rows:

$$\begin{bmatrix} 1 & 1 & -1 & | & 6 \\ 0 & -3 & 3 & | & -15 \\ 0 & -4 & 2 & | & -14 \end{bmatrix} \quad \begin{matrix} -2R_1 + R_2 \rightarrow R_2 \\ -3R_1 + R_3 \rightarrow R_3 \end{matrix}$$

Multiply the second row by $-\frac{1}{3}$ to get 1 in the second position on the diagonal:

$$\begin{bmatrix} 1 & 1 & -1 & | & 6 \\ 0 & 1 & -1 & | & 5 \\ 0 & -4 & 2 & | & -14 \end{bmatrix} \quad -\frac{1}{3}R_2 \rightarrow R_2$$

Use the second row to eliminate the variable y from the first and third rows:

$$\begin{bmatrix} 1 & 0 & 0 & | & 1 \\ 0 & 1 & -1 & | & 5 \\ 0 & 0 & -2 & | & 6 \end{bmatrix} \quad \begin{matrix} -1R_2 + R_1 \rightarrow R_1 \\ 4R_2 + R_3 \rightarrow R_3 \end{matrix}$$

Multiply the third row by $-\frac{1}{2}$ to get a 1 in the third position on the diagonal:

$$\begin{bmatrix} 1 & 0 & 0 & | & 1 \\ 0 & 1 & -1 & | & 5 \\ 0 & 0 & 1 & | & -3 \end{bmatrix} \quad -\frac{1}{2}R_3 \rightarrow R_3$$

Use the third row to eliminate the variable z from the second row:

$$\begin{bmatrix} 1 & 0 & 0 & | & 1 \\ 0 & 1 & 0 & | & 2 \\ 0 & 0 & 1 & | & -3 \end{bmatrix} \quad R_3 + R_2 \rightarrow R_2$$

This last augmented matrix represents the system $x = 1$, $y = 2$, and $z = -3$. So the solution set to the system is $\{(1, 2, -3)\}$.

> **Now do Exercises 49–64**

‹ **Calculator Close-Up** ›

Use MATRX EDIT to enter the 3×4 matrix A from Example 6 into the calculator. Under the MATRX MATH menu select rref (row-reduced echelon form). Choose A from the MATRX NAMES menu and the calculator does all of the calculations in Example 6.

‹4› **Dependent and Inconsistent Systems**

It is easy to recognize dependent and inconsistent systems using the Gauss-Jordan elimination method.

EXAMPLE   7

Gauss–Jordan elimination with infinitely many solutions

Solve the system:

$$3x + y = 1$$
$$6x + 2y = 2$$

Solution

Start with the augmented matrix:

$$\begin{bmatrix} 3 & 1 & | & 1 \\ 6 & 2 & | & 2 \end{bmatrix}$$

Multiply row 1 by -2 and add the result to row 2. We get the following matrix:

$$\begin{bmatrix} 3 & 1 & | & 1 \\ 0 & 0 & | & 0 \end{bmatrix} \quad -2R_1 + R_2 \rightarrow R_2$$

In the second row of the augmented matrix we have the equation $0 = 0$. So the equations are dependent. Every ordered pair that satisfies the first equation satisfies both equations. The solution set is $\{(x, y) \mid 3x + y = 1\}$.

> **Now do Exercises 65–68**

E X A M P L E   **8**

Gauss-Jordan elimination with no solution

Solve the system:

$$\begin{aligned} x - y &= 1 \\ -3x + 3y &= 4 \end{aligned}$$

Solution

Start with the augmented matrix:

$$\left[\begin{array}{rr|r} 1 & -1 & 1 \\ -3 & 3 & 4 \end{array}\right]$$

Multiply row 1 by 3 and add the result to row 2. We get the following matrix:

$$\left[\begin{array}{rr|r} 1 & -1 & 1 \\ 0 & 0 & 7 \end{array}\right] \quad 3R_1 + R_2 \to R_2$$

The second row of the augmented matrix corresponds to the equation $0 = 7$. So the equations are inconsistent, and there is no solution to the system.

> **Now do Exercises 69–74**

< **Helpful Hint** ›

The point of Example 8 is to recognize an inconsistent system with Gauss-Jordan elimination. We could also observe that -3 times the first equation yields

$$-3x + 3y = -3,$$

which is inconsistent with

$$-3x + 3y = 4.$$

The Gauss-Jordan elimination method may be applied to a system of n linear equations in n unknowns, where $n \geq 2$. However, it is a rather tedious method to perform when n is greater than 2, especially when fractions are involved. Computers are programmed to work with matrices, and the Gauss-Jordan elimination method is a popular method for computers.

Warm-Ups ▼

Fill in the blank.

1. A _____ is a rectangular array of numbers.

2. A _____ of a matrix runs horizontally.

3. A _____ of a matrix runs vertically.

4. The _____ of a matrix is the number of rows and columns.

5. An _____ of a matrix is a number that occupies a position in the matrix.

6. An _____ matrix is a matrix made up of the coefficients and the constants in a linear system.

7. The _____ method is used to solve a linear system using a matrix.

True or false?

8. The augmented matrix for $x + 3y = 5$ and $-x - 3y = 2$ is $\left[\begin{array}{rr|r} 1 & 3 & 5 \\ -1 & -3 & 2 \end{array}\right]$.

9. The system corresponding to the augmented matrix $\left[\begin{array}{rr|r} 2 & 4 & 6 \\ 0 & 0 & 3 \end{array}\right]$ is dependent.

10. The system corresponding to the augmented matrix $\left[\begin{array}{rr|r} 1 & 0 & 8 \\ 0 & 1 & 5 \end{array}\right]$ is $x = 8$ and $y = 5$.

11. The system corresponding to the augmented matrix $\left[\begin{array}{rr|r} 2 & 4 & 6 \\ 0 & 0 & 0 \end{array}\right]$ is inconsistent.

12. The notation $2R_1 + R_3 \to R_3$ means to replace R_3 with $2R_1 + R_3$.

13. The notation $R_1 \leftrightarrow R_3$ means to replace R_3 with R_1.

‹1› Matrices

Determine the size of each matrix. See Example 1.

1. $\begin{bmatrix} 5 & 0 \\ -2 & 3 \end{bmatrix}$

2. $\begin{bmatrix} 1 & 3 & 6 \\ -7 & 0 & 2 \end{bmatrix}$

3. $\begin{bmatrix} a & c \\ 0 & d \\ 3 & w \end{bmatrix}$

4. $\begin{bmatrix} 0 & a & b \\ 5 & 7 & -8 \\ a & b & 2 \end{bmatrix}$

5. $[2 \quad -3 \quad 5]$

6. $[0 \quad 0 \quad 0 \quad 6]$

7. $\begin{bmatrix} 1 \\ 0 \end{bmatrix}$

8. $\begin{bmatrix} 22 \\ 33 \\ -45 \end{bmatrix}$

‹2› The Augmented Matrix

Write the augmented matrix for each system of equations. See Example 2.

9. $\begin{aligned} 2x - 3y &= 9 \\ -3x + y &= -1 \end{aligned}$

10. $\begin{aligned} x - y &= 4 \\ 2x + y &= 3 \end{aligned}$

11. $\begin{aligned} x - y + z &= 1 \\ x + y - 2z &= 3 \\ y - 3z &= 4 \end{aligned}$

12. $\begin{aligned} x + y &= 2 \\ y - 3z &= 5 \\ -3x + 2z &= 8 \end{aligned}$

Write the system of equations represented by each augmented matrix. See Example 3.

13. $\begin{bmatrix} 5 & 1 & | & -1 \\ 2 & -3 & | & 0 \end{bmatrix}$

14. $\begin{bmatrix} 1 & 0 & | & 4 \\ 0 & 1 & | & -3 \end{bmatrix}$

15. $\begin{bmatrix} 1 & 0 & 0 & | & 6 \\ -1 & 0 & 1 & | & -3 \\ 1 & 1 & 0 & | & 1 \end{bmatrix}$

16. $\begin{bmatrix} 1 & 0 & 4 & | & 3 \\ 0 & 2 & 1 & | & -1 \\ 1 & 1 & 1 & | & 1 \end{bmatrix}$

‹3› The Gauss-Jordan Elimination Method

Perform the indicated row operation on the given augmented matrix. See Example 4.

17. $R_1 \leftrightarrow R_2$

$\begin{bmatrix} 0 & 2 & 4 & | & 1 \\ 1 & 1 & 7 & | & 2 \\ 0 & 0 & 5 & | & 10 \end{bmatrix}$

18. $R_1 \leftrightarrow R_2$

$\begin{bmatrix} -4 & 2 & | & 4 \\ 1 & 3 & | & 6 \end{bmatrix}$

19. $\frac{1}{2}R_1 \rightarrow R_1$

$\begin{bmatrix} 2 & 6 & | & 8 \\ 3 & 5 & | & 9 \end{bmatrix}$

20. $\frac{1}{4}R_2 \rightarrow R_2$

$\begin{bmatrix} 1 & 5 & | & 6 \\ 0 & 4 & | & 12 \end{bmatrix}$

21. $-2R_2 + R_1 \rightarrow R_1$

$\begin{bmatrix} 1 & 2 & | & 4 \\ 0 & 1 & | & 3 \end{bmatrix}$

22. $-3R_3 + R_2 \rightarrow R_2$

$\begin{bmatrix} 1 & 0 & 4 & | & 1 \\ 0 & 1 & 3 & | & 9 \\ 0 & 0 & 1 & | & 5 \end{bmatrix}$

Fill in the blanks in the augmented matrices using the indicated row operations. See Example 4.

23. $\begin{bmatrix} 0 & 2 & | & 4 \\ 1 & 0 & | & 6 \end{bmatrix}$

$\begin{bmatrix} 1 & 0 & | & 6 \\ & & | & \end{bmatrix}$ $R_1 \leftrightarrow R_2$

24. $\begin{bmatrix} 0 & 3 & | & 6 \\ 1 & 2 & | & 5 \end{bmatrix}$

$\begin{bmatrix} & & | & \\ 0 & 3 & | & 6 \end{bmatrix}$ $R_1 \leftrightarrow R_2$

25. $\begin{bmatrix} 4 & 12 & | & 16 \\ 2 & -4 & | & 3 \end{bmatrix}$

$\begin{bmatrix} & & | & \\ 2 & -4 & | & 3 \end{bmatrix}$ $\frac{1}{4}R_1 \rightarrow R_1$

26. $\begin{bmatrix} 1 & 0 & | & -9 \\ 0 & -3 & | & 6 \end{bmatrix}$

$\begin{bmatrix} 1 & 0 & | & -9 \\ & & | & \end{bmatrix}$ $-\frac{1}{3}R_2 \rightarrow R_2$

27. $\begin{bmatrix} 1 & 0 & | & -3 \\ -1 & 2 & | & 4 \end{bmatrix}$

$\begin{bmatrix} 1 & 0 & | & -3 \\ & & | & \end{bmatrix}$ $R_1 + R_2 \rightarrow R_2$

28. $\begin{bmatrix} 1 & 2 & | & 7 \\ 0 & -2 & | & 6 \end{bmatrix}$

$\begin{bmatrix} & & | & \\ 0 & -2 & | & 6 \end{bmatrix}$ $R_2 + R_1 \rightarrow R_1$

29. $\begin{bmatrix} 1 & 2 & | & 3 \\ -2 & 3 & | & 5 \end{bmatrix}$

$\begin{bmatrix} 1 & 2 & | & 3 \\ & & | & \end{bmatrix}$ $2R_1 + R_2 \rightarrow R_2$

30. $\begin{bmatrix} 1 & 3 & | & 7 \\ 0 & 1 & | & 4 \end{bmatrix}$

$\begin{bmatrix} & & | & \\ 0 & 1 & | & 4 \end{bmatrix}$ $-3R_2 + R_1 \rightarrow R_1$

Determine the row operation that was used to convert each given augmented matrix into the equivalent augmented matrix that follows it. See Example 4.

31. $\begin{bmatrix} 3 & 2 & | & 12 \\ 1 & -1 & | & -1 \end{bmatrix}, \begin{bmatrix} 1 & -1 & | & -1 \\ 3 & 2 & | & 12 \end{bmatrix}$

32. $\begin{bmatrix} 1 & -1 & | & -1 \\ 3 & 2 & | & 12 \end{bmatrix}, \begin{bmatrix} 1 & -1 & | & -1 \\ 0 & 5 & | & 15 \end{bmatrix}$

33. $\begin{bmatrix} 1 & -1 & | & -1 \\ 0 & 5 & | & 15 \end{bmatrix}, \begin{bmatrix} 1 & -1 & | & -1 \\ 0 & 1 & | & 3 \end{bmatrix}$

34. $\begin{bmatrix} 1 & -1 & | & -1 \\ 0 & 1 & | & 3 \end{bmatrix}, \begin{bmatrix} 1 & 0 & | & 2 \\ 0 & 1 & | & 3 \end{bmatrix}$

Solve each system using the Gauss-Jordan elimination method. See Examples 5 and 6.

35. $x - y = -3$
$\quad\quad y = 4$

36. $x + y = -3$
$\quad\quad y = 6$

37. $x + y = -6$
$\quad\quad 3y = 6$

38. $x - y = -7$
$\quad\quad 4y = 12$

39. $x - y = 7$
$\quad -x - y = -3$

40. $x + y = 6$
$\quad -x + y = 8$

41. $x + y = 3$
$\quad -3x + y = -1$

42. $x - y = -1$
$\quad 2x - y = 2$

43. $2x - y = 3$
$\quad x + y = 9$

44. $3x - 4y = -1$
$\quad x - y = 0$

45. $3x - y = 4$
$\quad 2x + y = 1$

46. $2x - y = -3$
$\quad 3x + y = -2$

47. $6x - 7y = 0$
$\quad 2x + y = 20$

48. $2x + y = 11$
$\quad 2x - y = 1$

49. $x + y - z = 4$
$\quad\quad y + z = 6$
$\quad\quad\quad z = 2$

50. $x - y + z = 5$
$\quad\quad y + z = 8$
$\quad\quad\quad z = 3$

51. $x + y + z = 6$
$\quad x - y + z = 2$
$\quad\quad 2y - z = 1$

52. $x - y - z = 0$
$\quad -x - y + z = -4$
$\quad -x + y - z = -2$

53. $2x + y + z = 4$
$\quad x + y - z = 1$
$\quad x - y + 2z = 2$

54. $3x - y = 1$
$\quad x + y + z = 4$
$\quad x + 2z = 3$

55. $2x - y + z = 0$
$\quad x + y - 3z = 3$
$\quad x - y + z = -1$

56. $x - y - z = 0$
$\quad -x - y + 2z = -1$
$\quad -x + y - 2z = -3$

57. $-x + 3y + z = 0$
$\quad x - y - 4z = -3$
$\quad x + y + 2z = 3$

58. $-x + z = -2$
$\quad 2x - y = 5$
$\quad y + 3z = 9$

Solve each system by using a graphing calculator to find the row-reduced echelon form (rref) of the augmented matrix.

59. $3x + 2y = -108$
$\quad 5x - 8y = -316$

60. $-3x + 5y = -266$
$\quad 6x - 7y = 646$

61. $3x - 2y + z = -9$
$\quad x + 5y - z = 165$
$\quad 5x + y - 6z = 47$

62. $7x + 6y - 3z = 12$
$\quad 3x + 4y - 9z = -19.2$
$\quad -x + 5y + 8z = 39.6$

63. $2a + 3b - 2c + 4d = 30$
$\quad 9a + 4b - 5c - 3d = -32$
$\quad 5a + 2b + 5c + d = 42$
$\quad 3a - 4b + c - 7d = -56$

64. $2a + 3b - 2c + 4d = 36$
$\quad 9a + 4b - 5c - 3d = -20$
$\quad 5a + 2b + 5c + d = 56$
$\quad 3a - 4b + c - 7d = -60$

⟨4⟩ Dependent and Inconsistent Systems

*Solve each system using the Gauss-Jordan elimination method.
See Examples 7 and 8.*

65. $x - 5y = 11$
$-2x + 10y = -22$

66. $-3x + 12y = 3$
$x - 4y = -1$

67. $2x - 3y = 4$
$-2x + 3y = 5$

68. $x - 3y = 8$
$2x - 6y = 1$

69. $x + 2y = 1$
$3x + 6y = 3$

70. $2x - 3y = 1$
$-6x + 9y = -3$

71. $x - y + z = 1$
$2x - 2y + 2z = 2$
$-3x + 3y - 3z = -3$

72. $4x - 2y + 2z = 2$
$2x - y + z = 1$
$-2x + y - z = -1$

73. $x + y - z = 2$
$2x - y + z = 1$
$3x + 3y - 3z = 8$

74. $x + y + z = 5$
$x - y - z = 8$
$-x + y + z = 2$

Miscellaneous

*Solve each system by using the Gauss-Jordan elimination
method. State whether the system is independent, dependent,
or inconsistent.*

75. $x + 2y = 2$
$2x + 3y = 1$

76. $-x + 3y = 4$
$2x - 6y = 8$

77. $-x + 3y = 4$
$2x - 6y = -8$

78. $-2x + y = 15$
$x - y = -10$

79. $2x + y - 5z = 6$
$-4x - 2y + 10z = 8$

80. $x + y + z = 4$
$x - y + 3z = 6$
$2x + 2y + 2z = 10$

81. $x - y - z = 0$
$-x + y + 3z = -2$
$-x + 2y + z = -2$

82. $x + y - z = 0$
$5x - y - z = 0$
$2x - y + z = 30$

83. $10x - 10y - 10z = 10$
$x - y - z = 1$
$5x - 5y - 5z = 5$

84. $3x - y + z = 5$
$18x - 6y + 6z = 30$

Applications

*Solve each problem using a system of linear equations and the
Gauss-Jordan elimination method.*

85. *Two numbers.* The sum of two numbers is 12 and their
difference is 2. Find the numbers.

86. *Two more numbers.* The sum of two numbers is 11 and
their difference is 6. Find the numbers.

87. *Paper size.* The length of a rectangular piece of paper is
2.5 inches greater than the width. The perimeter is 39 inches.
Find the length and width.

88. *Photo size.* The length of a rectangular photo is 2 inches
greater than the width. The perimeter is 20 inches. Find
the length and width.

89. *Buy and sell.* Cory buys and sells baseball cards on eBay.
He always buys at the same price and then sells the cards
for $2 more than he buys them. One month he broke even
after buying 56 cards and selling 49. Find his buying price
and selling price.

90. *Jay Leno's garage.* Jay Leno's collection of cars and
motorcycles totals 187. When he checks the air in the
tires, he has 588 tires to check. How many cars and how
many motorcycles does he own? Assume that the cars
all have four tires and the motorcycles have two.

91. *Parking lot boredom.* A late-night parking lot attendant
counted 50 vehicles on the lot consisting of four-wheel
cars, three-wheel cars, and two-wheel motorcycles. She
then counted 192 tires touching the ground and observed
that the number of four-wheel cars was nine times the total
of the other vehicles on the lot. How many of each type of
vehicle were on the lot?

92. *Happy meals.* The total price of a hamburger, an order of
fries, and a Coke at a fast food restaurant is $3.00. The
price of a hamburger minus the price of an order of fries is
$0.20, and the price of an order of fries minus the price of
a Coke is also $0.20. Find the price of each item.

93. *Hot dogs.* A street vendor on Constitution Avenue sells
hot dogs, hamburgers, and Cokes. The number sold for

each item and the total receipts for the first three days of a week are shown in the accompanying table. Let x be the price of a hot dog, y be the price of a hamburger, and z be the price of a coke. Find x, y, and z by using a graphing calculator to find the row-reduced echelon form of an augmented matrix.

	Dogs	Burgers	Cokes	Receipts
Mon	120	88	230	$1380
Tues	98	122	196	$1394
Wed	155	132	225	$1730

Table for Exercise 93

 94. *Gas sales*. A station sells regular, plus, and premium Shell gasoline. The number of gallons sold for each type and the total receipts for the 4th of July weekend are shown in the accompanying table. Let x be the price per gallon of regular, y be the price per gallon of plus, and z be the price per gallon of supreme. Find x, y, and z by using a graphing calculator to find the row-reduced echelon form of an augmented matrix.

	Regular	Plus	Supreme	Receipts
Fri	700	500	100	$3578.70
Sat	600	400	300	$3608.70
Sun	900	600	400	$5268.10

Table for Exercise 94

Getting More Involved

95. *Cooperative learning*

Write a step-by-step procedure for solving any system of two linear equations in two variables by the Gauss-Jordan elimination method. Have a classmate evaluate your procedure by using it to solve a system.

96. *Cooperative learning*

Repeat Exercise 95 for a system of three linear equations in three variables.

4.5   Determinants and Cramer's Rule

In This Section

⟨1⟩ **Determinants**

⟨2⟩ **Cramer's Rule (2 × 2)**

⟨3⟩ **Minors**

⟨4⟩ **Evaluating a 3 × 3 Determinant**

⟨5⟩ **Cramer's Rule (3 × 3)**

The Gauss-Jordan elimination method of Section 4.4 can be performed the same way on every system. Another method that is applied the same way for every system is Cramer's rule, which we study in this section. Before you learn Cramer's rule, we need to introduce a new number associated with a matrix, called a *determinant*.

⟨1⟩   Determinants

The determinant of a square matrix is a real number corresponding to the matrix. For a 2 × 2 matrix the determinant is defined as follows.

> **Determinant of a 2 × 2 Matrix**
>
> The **determinant** of the matrix $\begin{bmatrix} a & b \\ c & d \end{bmatrix}$ is defined to be the real number $ad - bc$. We write
>
> $$\begin{vmatrix} a & b \\ c & d \end{vmatrix} = ad - bc.$$

Note that the symbol for the determinant is a pair of vertical lines similar to the absolute value symbol, while a matrix is enclosed in brackets.

E X A M P L E 1

Using the definition of determinant

Find the determinant of each matrix.

a) $\begin{bmatrix} 1 & 3 \\ -2 & 5 \end{bmatrix}$

b) $\begin{bmatrix} 2 & 4 \\ 6 & 12 \end{bmatrix}$

Solution

a) $\begin{vmatrix} 1 & 3 \\ -2 & 5 \end{vmatrix} = 1 \cdot 5 - 3(-2)$

$= 5 + 6$

$= 11$

b) $\begin{vmatrix} 2 & 4 \\ 6 & 12 \end{vmatrix} = 2 \cdot 12 - 4 \cdot 6$

$= 24 - 24$

$= 0$

Now do Exercises 1–8

‹ Calculator Close-Up ›

With a graphing calculator you can define matrix A using MATRX EDIT.

Then use the determinant function (det) found in MATRX MATH and the A from MATRX NAMES to find its determinant.

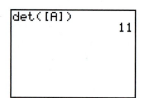

‹ Helpful Hint ›

Notice that Cramer's rule gives us a precise formula for finding the solution to an independent system. The addition and substitution methods are more like guidelines under which we choose the best way to proceed.

‹2› Cramer's Rule (2 × 2)

To understand Cramer's rule, we first solve a general system of two linear equations in two variables. Consider the system

(1) $a_1x + b_1y = c_1$

(2) $a_2x + b_2y = c_2$

where a_1, b_1, c_1, a_2, b_2, and c_2 represent real numbers. To eliminate y, we multiply Eq. (1) by b_2 and Eq. (2) by $-b_1$:

$a_1b_2x + b_1b_2y = c_1b_2$ Eq. (1) multiplied by b_2

$-a_2b_1x - b_1b_2y = -c_2b_1$ Eq. (2) multiplied by $-b_1$

$a_1b_2x - a_2b_1x \qquad = c_1b_2 - c_2b_1$ Add.

$(a_1b_2 - a_2b_1)x = c_1b_2 - c_2b_1$

$x = \dfrac{c_1b_2 - c_2b_1}{a_1b_2 - a_2b_1}$ Provided that $a_1b_2 - a_2b_1 \neq 0$

Using similar steps to eliminate x from the system, we get

$y = \dfrac{a_1c_2 - a_2c_1}{a_1b_2 - a_2b_1},$

provided that $a_1b_2 - a_2b_1 \neq 0$. These formulas for x and y can be written by using determinants. In the determinant form they are known as **Cramer's rule.**

Cramer's Rule

The solution to the system

$a_1x + b_1y = c_1$

$a_2x + b_2y = c_2$

is given by $x = \dfrac{D_x}{D}$ and $y = \dfrac{D_y}{D}$, where

$D = \begin{vmatrix} a_1 & b_1 \\ a_2 & b_2 \end{vmatrix}, \qquad D_x = \begin{vmatrix} c_1 & b_1 \\ c_2 & b_2 \end{vmatrix}, \qquad \text{and} \qquad D_y = \begin{vmatrix} a_1 & c_1 \\ a_2 & c_2 \end{vmatrix},$

provided that $D \neq 0$.

Note that D is the determinant made up of the original coefficients of x and y. D is used in the denominator for both x and y. D_x is obtained by replacing the first (or x) column of D by the constants c_1 and c_2. D_y is found by replacing the second (or y) column of D by the constants c_1 and c_2.

E X A M P L E   **2**

Solving an independent system with Cramer's rule
Use Cramer's rule to solve the system:

$$3x - 2y = 4$$
$$2x + \ y = -3$$

Solution
First find the determinants D, D_x, and D_y:

$$D = \begin{vmatrix} 3 & -2 \\ 2 & 1 \end{vmatrix} = 3 - (-4) = 7$$

$$D_x = \begin{vmatrix} 4 & -2 \\ -3 & 1 \end{vmatrix} = 4 - 6 = -2, \qquad D_y = \begin{vmatrix} 3 & 4 \\ 2 & -3 \end{vmatrix} = -9 - 8 = -17$$

By Cramer's rule, we have

$$x = \frac{D_x}{D} = -\frac{2}{7} \qquad \text{and} \qquad y = \frac{D_y}{D} = -\frac{17}{7}.$$

Check in the original equations. The solution set is $\left\{ \left(-\frac{2}{7}, -\frac{17}{7} \right) \right\}$.

> **Now do Exercises 9–22**

‹ **Calculator Close-Up** ›

Use MATRX EDIT to define D, D_x, and D_y as A, B, and C. Now use Cramer's rule on the home screen to find x and y.

```
det([B])/det([A]
)▶Frac
                -2/7
det([C])/det([A]
)▶Frac
               -17/7
```

CAUTION   Cramer's rule works *only* when the determinant D is *not* equal to zero. Cramer's rule solves only those systems that have a single point in their solution set. If $D = 0$, we use elimination to determine whether the solution set is empty or contains all points of a line.

‹3› Minors
To each element of a 3×3 matrix there corresponds a 2×2 matrix that is obtained by deleting the row and column of that element. The determinant of the 2×2 matrix is called the **minor** of that element.

E X A M P L E   **3**

Finding minors
Find the minors for the elements 2, 3, and -6 of the 3×3 matrix

$$\begin{bmatrix} 2 & -1 & -8 \\ 0 & -2 & 3 \\ 4 & -6 & 7 \end{bmatrix}.$$

Solution
To find the minor for 2, delete the first row and first column of the matrix:

$$\begin{bmatrix} 2 & -1 & -8 \\ 0 & -2 & 3 \\ 4 & -6 & 7 \end{bmatrix}$$

Now find the determinant of $\begin{bmatrix} -2 & 3 \\ -6 & 7 \end{bmatrix}$:

$$\begin{vmatrix} -2 & 3 \\ -6 & 7 \end{vmatrix} = (-2)(7) - (-6)(3) = 4$$

The minor for 2 is 4. To find the minor for 3, delete the second row and third column of the matrix:

$$\begin{bmatrix} 2 & -1 & -8 \\ 0 & -2 & 3 \\ 4 & -6 & 7 \end{bmatrix}$$

Now find the determinant of $\begin{bmatrix} 2 & -1 \\ 4 & -6 \end{bmatrix}$:

$$\begin{vmatrix} 2 & -1 \\ 4 & -6 \end{vmatrix} = (2)(-6) - (4)(-1) = -8$$

The minor for 3 is -8. To find the minor for -6, delete the third row and the second column of the matrix:

$$\begin{bmatrix} 2 & -1 & -8 \\ 0 & -2 & 3 \\ 4 & -6 & 7 \end{bmatrix}$$

Now find the determinant of $\begin{bmatrix} 2 & -8 \\ 0 & 3 \end{bmatrix}$:

$$\begin{vmatrix} 2 & -8 \\ 0 & 3 \end{vmatrix} = (2)(3) - (0)(-8) = 6$$

The minor for -6 is 6.

> **Now do Exercises 31–38**

‹4› Evaluating a 3 × 3 Determinant

The determinant of a 3 × 3 matrix is defined in terms of the determinants of minors.

Determinant of a 3 × 3 Matrix

The determinant of a 3 × 3 matrix is defined as follows:

$$\begin{vmatrix} a_1 & b_1 & c_1 \\ a_2 & b_2 & c_2 \\ a_3 & b_3 & c_3 \end{vmatrix} = a_1 \cdot \begin{vmatrix} b_2 & c_2 \\ b_3 & c_3 \end{vmatrix} - a_2 \cdot \begin{vmatrix} b_1 & c_1 \\ b_3 & c_3 \end{vmatrix} + a_3 \cdot \begin{vmatrix} b_1 & c_1 \\ b_2 & c_2 \end{vmatrix}$$

Note that the determinants following a_1, a_2, and a_3 are the minors for a_1, a_2, and a_3, respectively. Writing the determinant of a 3 × 3 matrix in terms of minors is called **expansion by minors.** In the definition we expanded by minors about the first column. Later we will see how to expand by minors using any row or column and get the same value for the determinant.

EXAMPLE 4

Determinant of a 3 × 3 matrix

Find the determinant of the matrix by expansion by minors about the first column.

$$\begin{bmatrix} 1 & 3 & -5 \\ -2 & 4 & 6 \\ 0 & -7 & 9 \end{bmatrix}$$

Solution

$$\begin{vmatrix} 1 & 3 & -5 \\ -2 & 4 & 6 \\ 0 & -7 & 9 \end{vmatrix} = 1 \cdot \begin{vmatrix} 4 & 6 \\ -7 & 9 \end{vmatrix} - (-2) \cdot \begin{vmatrix} 3 & -5 \\ -7 & 9 \end{vmatrix} + 0 \cdot \begin{vmatrix} 3 & -5 \\ 4 & 6 \end{vmatrix}$$

$$= 1 \cdot [36 - (-42)] + 2 \cdot (27 - 35) + 0 \cdot [18 - (-20)]$$

$$= 1 \cdot 78 + 2 \cdot (-8) + 0$$

$$= 78 - 16$$

$$= 62$$

> **Now do Exercises 39–46**

In Example 5 we evaluate a determinant using expansion by minors about the second row. In expanding about any row or column, the signs of the coefficients of the minors alternate according to the **sign array** that follows:

$$\begin{bmatrix} + & - & + \\ - & + & - \\ + & - & + \end{bmatrix}$$

The sign array is easily remembered by observing that there is a "+" sign in the upper left position and then alternating signs for all of the remaining positions.

E X A M P L E   5

Determinant of a 3 × 3 matrix

Evaluate the determinant of the matrix by expanding by minors about the second row.

$$\begin{bmatrix} 1 & 3 & -5 \\ -2 & 4 & 6 \\ 0 & -7 & 9 \end{bmatrix}$$

Solution

For expansion using the second row we prefix the signs "$- + -$" from the second row of the sign array to the corresponding numbers in the second row of the matrix, -2, 4, and 6. Note that the signs from the sign array are used in addition to any signs that occur on the numbers in the second row.

From the sign array, second row

$$\begin{vmatrix} 1 & 3 & -5 \\ -2 & 4 & 6 \\ 0 & -7 & 9 \end{vmatrix} = -(-2) \cdot \begin{vmatrix} 3 & -5 \\ -7 & 9 \end{vmatrix} + 4 \cdot \begin{vmatrix} 1 & -5 \\ 0 & 9 \end{vmatrix} - 6 \cdot \begin{vmatrix} 1 & 3 \\ 0 & -7 \end{vmatrix}$$

$$= 2(27 - 35) + 4(9 - 0) - 6(-7 - 0)$$

$$= 2(-8) + 4(9) - 6(-7)$$

$$= -16 + 36 + 42$$

$$= 62$$

Note that 62 is the same value that was obtained for this determinant in Example 4.

> **Now do Exercises 47–50**

‹ **Calculator Close-Up** ›

A calculator is very useful for finding the determinant of a 3 × 3 matrix. Define *A* using MATRX EDIT.

```
MATRIX[A] 3 ×3
[ 1    3    -5 ]
[ -2   4    6  ]
[ 0    -7   9  ]
```

Now use the determinant function from MATRX MATH and the *A* from MATRX NAMES to find the determinant.

```
det([A])
            62
```

It can be shown that expanding by minors using any row or column prefixed by the corresponding signs from the sign array yields the same value for the determinant. Because we can use any row or column to evaluate a determinant of a 3 × 3 matrix, we can choose a row or column that makes the work easier. We can shorten the work considerably by picking a row or column with zeros in it.

E X A M P L E **6**

Choosing the simplest row or column

Find the determinant of the matrix

$$\begin{bmatrix} 3 & -5 & 0 \\ 4 & -6 & 0 \\ 7 & 9 & 2 \end{bmatrix}.$$

Solution

We choose to expand by minors about the third column of the matrix because the third column contains two zeros. Prefix the third-column entries $0, 0, 2$ by the signs "$+ - +$" from the third column of the sign array:

$$\begin{vmatrix} 3 & -5 & 0 \\ 4 & -6 & 0 \\ 7 & 9 & 2 \end{vmatrix} = 0 \cdot \begin{vmatrix} 4 & -6 \\ 7 & 9 \end{vmatrix} - 0 \cdot \begin{vmatrix} 3 & -5 \\ 7 & 9 \end{vmatrix} + 2 \cdot \begin{vmatrix} 3 & -5 \\ 4 & -6 \end{vmatrix}$$

$$= 0 - 0 + 2[-18 - (-20)]$$

$$= 4$$

Now do Exercises 51–54

⟨5⟩ **Cramer's Rule (3 × 3)**

An independent system of three linear equations in three variables can be solved by using determinants and Cramer's rule.

Cramer's Rule for Three Equations in Three Unknowns

The solution to the system

$$a_1 x + b_1 y + c_1 z = d_1$$
$$a_2 x + b_2 y + c_2 z = d_2$$
$$a_3 x + b_3 y + c_3 z = d_3$$

is given by $x = \frac{D_x}{D}$, $y = \frac{D_y}{D}$, and $z = \frac{D_z}{D}$, where

$$D = \begin{vmatrix} a_1 & b_1 & c_1 \\ a_2 & b_2 & c_2 \\ a_3 & b_3 & c_3 \end{vmatrix}, \qquad D_x = \begin{vmatrix} d_1 & b_1 & c_1 \\ d_2 & b_2 & c_2 \\ d_3 & b_3 & c_3 \end{vmatrix},$$

$$D_y = \begin{vmatrix} a_1 & d_1 & c_1 \\ a_2 & d_2 & c_2 \\ a_3 & d_3 & c_3 \end{vmatrix}, \qquad D_z = \begin{vmatrix} a_1 & b_1 & d_1 \\ a_2 & b_2 & d_2 \\ a_3 & b_3 & d_3 \end{vmatrix},$$

provided that $D \neq 0$.

Note that D_x, D_y, and D_z are obtained from D by replacing the x-, y-, or z-column with the constants d_1, d_2, and d_3.

EXAMPLE **7**

‹ **Calculator Close-Up** ›

When you see the amount of arithmetic required to solve the system in Example 7 by Cramer's rule, you can understand why computers and calculators have been programmed to perform this method. Some calculators can find determinants for matrices as large as 10×10. Try to solve Example 7 with a graphing calculator that has determinants.

Solving an independent system with Cramer's rule

Use Cramer's rule to solve the system:

$$x + y + z = 4$$
$$x - y \quad\quad = -3$$
$$x + 2y - z = 0$$

Solution

We first calculate $D, D_x, D_y,$ and D_z. To calculate D, expand by minors about the third column because the third column has a zero in it:

$$D = \begin{vmatrix} 1 & 1 & 1 \\ 1 & -1 & 0 \\ 1 & 2 & -1 \end{vmatrix} = 1 \cdot \begin{vmatrix} 1 & -1 \\ 1 & 2 \end{vmatrix} - 0 \cdot \begin{vmatrix} 1 & 1 \\ 1 & 2 \end{vmatrix} + (-1) \cdot \begin{vmatrix} 1 & 1 \\ 1 & -1 \end{vmatrix}$$

$$= 1 \cdot [2 - (-1)] - 0 + (-1)[-1 - 1]$$
$$= 3 - 0 + 2$$
$$= 5$$

For D_x, expand by minors about the first column:

$$D_x = \begin{vmatrix} 4 & 1 & 1 \\ -3 & -1 & 0 \\ 0 & 2 & -1 \end{vmatrix} = 4 \cdot \begin{vmatrix} -1 & 0 \\ 2 & -1 \end{vmatrix} - (-3) \cdot \begin{vmatrix} 1 & 1 \\ 2 & -1 \end{vmatrix} + 0 \cdot \begin{vmatrix} 1 & 1 \\ -1 & 0 \end{vmatrix}$$

$$= 4 \cdot (1 - 0) + 3 \cdot (-1 - 2) + 0$$
$$= 4 - 9 + 0 = -5$$

For D_y, expand by minors about the third row:

$$D_y = \begin{vmatrix} 1 & 4 & 1 \\ 1 & -3 & 0 \\ 1 & 0 & -1 \end{vmatrix} = 1 \cdot \begin{vmatrix} 4 & 1 \\ -3 & 0 \end{vmatrix} - 0 \cdot \begin{vmatrix} 1 & 1 \\ 1 & 0 \end{vmatrix} + (-1) \cdot \begin{vmatrix} 1 & 4 \\ 1 & -3 \end{vmatrix}$$

$$= 1 \cdot 3 - 0 + (-1)(-7) = 10$$

To get D_z, expand by minors about the third row:

$$D_z = \begin{vmatrix} 1 & 1 & 4 \\ 1 & -1 & -3 \\ 1 & 2 & 0 \end{vmatrix} = 1 \cdot \begin{vmatrix} 1 & 4 \\ -1 & -3 \end{vmatrix} - 2 \cdot \begin{vmatrix} 1 & 4 \\ 1 & -3 \end{vmatrix} + 0 \cdot \begin{vmatrix} 1 & 1 \\ 1 & -1 \end{vmatrix}$$

$$= 1 \cdot 1 - 2(-7) + 0 = 15$$

Now, by Cramer's rule,

$$x = \frac{D_x}{D} = \frac{-5}{5} = -1, \quad y = \frac{D_y}{D} = \frac{10}{5} = 2, \quad \text{and} \quad z = \frac{D_z}{D} = \frac{15}{5} = 3.$$

Check $(-1, 2, 3)$ in the original equations. The solution set is $\{(-1, 2, 3)\}$.

> **Now do Exercises 55–64**

If $D = 0$, Cramer's rule does not apply. Cramer's rule provides the solution only to a system of three equations with three variables that has a single point in the solution set. If $D = 0$, then the solution set either is empty or consists of infinitely many points, and we can use the methods discussed in Sections 4.3 or 4.4 to find the solution.

Warm-Ups ▼

Fill in the blank.

1. A _____ is a real number associated with a square matrix.
2. _____ rule is used to solve a system of linear equations using determinants.
3. Cramer's rule works only for _____ systems.
4. A _____ for an element of a 3×3 matrix is the determinant of a 2×2 matrix.
5. The _____ array indicates what signs to use in the expansion by minors.

True or false?

6. $\begin{vmatrix} -1 & 2 \\ 3 & -5 \end{vmatrix} = -1$

7. $\begin{vmatrix} 2 & 4 \\ -4 & 8 \end{vmatrix} = 0$

8. Cramer's rule solves any linear system of two equations in two variables.
9. The determinant of a 3×3 matrix is found by using minors.
10. Expansion by minors can be done about any row or column.
11. If $D = 0$, then there is no solution to the system.

4.5

Exercises

‹ Study Tips ›

- Get in a habit of checking your work. Don't look in the back of the book for the answer until after you have checked your work.
- You will not always have an answer section for your problems.

‹1› Determinants

Find the value of each determinant. See Example 1.

1. $\begin{vmatrix} 2 & 5 \\ 3 & 7 \end{vmatrix}$

2. $\begin{vmatrix} -1 & 0 \\ 1 & 1 \end{vmatrix}$

3. $\begin{vmatrix} 0 & 3 \\ 1 & 5 \end{vmatrix}$

4. $\begin{vmatrix} 2 & 4 \\ 6 & 12 \end{vmatrix}$

5. $\begin{vmatrix} -3 & -2 \\ -4 & 2 \end{vmatrix}$

6. $\begin{vmatrix} -2 & 2 \\ -3 & -5 \end{vmatrix}$

7. $\begin{vmatrix} 0.05 & 0.06 \\ 10 & 20 \end{vmatrix}$

8. $\begin{vmatrix} 0.02 & -0.5 \\ 30 & 50 \end{vmatrix}$

‹2› Cramer's Rule (2 × 2)

Solve each system using Cramer's rule. See Example 2.

9. $x - y = -4$
   $ 2y = 12$

10. $x + y = -2$
    $ 3y = 3$

11. $x + y = 0$
    $2x = -16$

12. $x - y = 0$
    $3x = -3$

13. $2x - y = 5$
    $3x + 2y = -3$

14. $3x + y = -1$
    $x + 2y = 8$

15. $3x - 5y = -2$
    $2x + 3y = 5$

16. $x - y = 1$
    $3x - 2y = 0$

17. $4x - 3y = 5$
    $2x + 5y = 7$

18. $2x - y = 2$
    $3x - 2y = 1$

19. $0.5x + 0.2y = 8$
    $0.4x - 0.6y = -5$

20. $0.6x + 0.5y = 18$
    $0.5x - 0.25y = 7$

21. $\frac{1}{2}x + \frac{1}{4}y = 5$ **22.** $\frac{1}{2}x + \frac{2}{3}y = 4$

$\frac{1}{3}x - \frac{1}{2}y = -1$ $\frac{3}{4}x + \frac{1}{3}y = -2$

Evaluate the determinant of each 3×3 matrix using expansion by minors about the row or column of your choice. See Examples 5 and 6.

47. $\begin{bmatrix} 3 & 1 & 5 \\ 2 & 0 & 6 \\ 4 & 0 & 1 \end{bmatrix}$ **48.** $\begin{bmatrix} 2 & 1 & 2 \\ 1 & 2 & 5 \\ 3 & 0 & 0 \end{bmatrix}$

Use the determinant feature on your graphing calculator to find each determinant.

23. $\begin{vmatrix} 2.3 & -1.6 \\ 4.8 & 5.1 \end{vmatrix}$ **24.** $\begin{vmatrix} 3.9 & 4.7 \\ -8.1 & 1.3 \end{vmatrix}$

25. $\begin{vmatrix} 1/3 & 1/4 \\ 1/8 & 1/6 \end{vmatrix}$ **26.** $\begin{vmatrix} 5/6 & 3/7 \\ 3/4 & -1/2 \end{vmatrix}$

49. $\begin{bmatrix} -2 & 1 & 3 \\ 0 & 1 & -1 \\ 2 & -4 & -3 \end{bmatrix}$ **50.** $\begin{bmatrix} -2 & 0 & 1 \\ -3 & 2 & -5 \\ 4 & -2 & 6 \end{bmatrix}$

51. $\begin{bmatrix} -2 & -3 & 0 \\ 4 & -1 & 0 \\ 0 & 3 & 5 \end{bmatrix}$ **52.** $\begin{bmatrix} -2 & 6 & 3 \\ 0 & 4 & 0 \\ -1 & -4 & 5 \end{bmatrix}$

Use Cramer's rule and a graphing calculator to solve each system. Round approximate answers to two decimal places.

27. $3.2x - 5.7y = 6.24$ **28.** $3.2x - 5.7y = -26.36$
$4.6x + 7.1y = 33.44$ $4.6x + 7.1y = 78.34$

29. $\sqrt{2}x - \sqrt{3}y = -11$ **30.** $\sqrt{2}x - \sqrt{3}y = -7$
$\sqrt{8}x + \sqrt{12}y = 38$ $\sqrt{8}x + \sqrt{12}y = -14$

53. $\begin{bmatrix} 2 & 1 & 1 \\ 0 & 0 & 5 \\ 5 & 0 & 4 \end{bmatrix}$ **54.** $\begin{bmatrix} 2 & 3 & 0 \\ 6 & 4 & 1 \\ 1 & 2 & 0 \end{bmatrix}$

⟨**5**⟩ **Cramer's Rule (3 × 3)**

Use Cramer's rule to solve each system. See Example 7.

55. $x + y + z = 6$ **56.** $x + y + z = 2$
$x - y + z = 2$ $x - y - 2z = -3$
$2x + y + z = 7$ $2x - y + z = 7$

⟨**3**⟩ **Minors**

Find the indicated minors using the following matrix. See Example 3.

$\begin{bmatrix} 3 & -2 & 5 \\ 4 & -3 & 7 \\ 0 & 1 & -6 \end{bmatrix}$

57. $x - 3y + 2z = 0$ **58.** $3x + 2y + 2z = 0$
$x + y + z = 2$ $x - y + z = 1$
$x - y + z = 0$ $x + y - z = 3$

31. Minor for 3   **32.** Minor for -2
33. Minor for 5   **34.** Minor for -3
35. Minor for 7   **36.** Minor for 0
37. Minor for 1   **38.** Minor for -6

59. $x + y = -1$ **60.** $x - y = 8$
$2y - z = 3$ $x - 2z = 0$
$x + y + z = 0$ $x + y - z = 1$

⟨**4**⟩ **Evaluating a 3 × 3 Determinant**

Find the determinant of each 3×3 matrix by using expansion by minors about the first column. See Example 4.

39. $\begin{bmatrix} 1 & 1 & 2 \\ 2 & 3 & 1 \\ 3 & 1 & 5 \end{bmatrix}$ **40.** $\begin{bmatrix} 2 & 1 & 3 \\ 1 & 1 & 2 \\ 3 & 4 & 6 \end{bmatrix}$

61. $x + y - z = 0$ **62.** $x + y + z = 1$
$2x + 2y + z = 6$ $5x - y = 0$
$x - 3y = 0$ $3x + y + 2z = 0$

41. $\begin{bmatrix} 2 & 1 & 0 \\ 1 & 0 & 1 \\ 3 & 1 & 2 \end{bmatrix}$ **42.** $\begin{bmatrix} 1 & 0 & 2 \\ 2 & 1 & 3 \\ 4 & 3 & 0 \end{bmatrix}$

43. $\begin{bmatrix} -2 & 1 & 2 \\ -3 & 3 & 1 \\ -5 & 4 & 0 \end{bmatrix}$ **44.** $\begin{bmatrix} -2 & 1 & 3 \\ -1 & 4 & 2 \\ 2 & 1 & 1 \end{bmatrix}$

63. $x + y + z = 0$ **64.** $x + z = 0$
$2y + 2z = 0$ $x - 3y = 1$
$3x - y = -1$ $4y - 3z = 3$

Use Cramer's rule and a graphing calculator to solve each system.

45. $\begin{bmatrix} 1 & 1 & 5 \\ 0 & 3 & 2 \\ 0 & 2 & 3 \end{bmatrix}$ **46.** $\begin{bmatrix} 1 & 0 & 6 \\ 0 & 1 & 4 \\ 0 & 0 & 9 \end{bmatrix}$

65. $1.3x - 1.4y + 1.5z = 1.7$
$2.4x + 3.1y - 5.6z = -0.92$
$3.7x - 1.5y + 4.8z = 8.51$

66. $1.3x - 1.4y + 1.5z = 3.4$
$2.4x + 3.1y - 5.6z = -1.84$
$3.7x - 1.5y + 4.8z = 17.02$

Applications

Solve each problem by using two equations in two variables and Cramer's rule.

67. *Peas and beets.* One serving of canned peas contains 3 grams of protein and 11 grams of carbohydrates. One serving of canned beets contains 1 gram of protein and 8 grams of carbohydrates. A dietitian wants to determine the number of servings of each that would provide 38 grams of protein and 187 grams of carbohydrates.

  a) Use the accompanying graph to estimate the number of servings of each.

  b) Use Cramer's rule to find the number of servings of each.

Figure for Exercise 67

68. *Protein and carbohydrates.* One serving of Cornies breakfast cereal contains 2 grams of protein and 25 grams of carbohydrates. One serving of Oaties breakfast cereal contains 4 grams of protein and 20 grams of carbohydrates. How many servings of each would provide exactly 24 grams of protein and 210 grams of carbohydrates?

69. *Milk and a magazine.* Althia bought a gallon of milk and a magazine for a total of $4.65, excluding tax. Including the tax, the bill was $4.95. If there is a 5% sales tax on milk and an 8% sales tax on magazines, then what was the price of each item?

70. *Washing machines and refrigerators.* A truck carrying 3600 cubic feet of cargo consisting of washing machines and refrigerators was hijacked. The washing machines are worth $300 each and are shipped in 36-cubic-foot cartons. The refrigerators are worth $900 each and are shipped in 45-cubic-foot cartons. If the total value of the cargo was $51,000, then how many of each were there on the truck?

71. *Singles and doubles.* Windy's Hamburger Palace sells singles and doubles. Toward the end of the evening, Windy himself noticed that he had on hand only 32 patties

and 34 slices of tomatoes. If a single takes 1 patty and 2 slices, and a double takes 2 patties and 1 slice, then how many more singles and doubles must Windy sell to use up all of his patties and tomato slices?

72. *Valuable wrenches.* Carmen has a total of 28 wrenches, all of which are either box wrenches or open-end wrenches. For insurance purposes she values the box wrenches at $3.00 each and the open-end wrenches at $2.50 each. If the value of her wrench collection is $78, then how many of each type does she have?

73. *Gary and Harry.* Gary is 5 years older than Harry. Twenty-nine years ago, Gary was twice as old as Harry. How old are they now?

74. *Acute angles.* One acute angle of a right triangle is 3° more than twice the other acute angle. What are the sizes of the acute angles?

Figure for Exercise 74

75. *Equal perimeters.* A rope of length 80 feet is to be cut into two pieces. One piece will be used to form a square, and the other will be used to form an equilateral triangle. If the figures are to have equal perimeters, then what should be the length of a side of each?

Figure for Exercise 75

76. *Coffee and doughnuts.* For a cup of coffee and a doughnut, Thurrel spent $2.25, including a tip. Later he spent $4.00 for two coffees and three doughnuts, including a tip. If he always tips $1.00, then what is the price of a cup of coffee?

77. *Chlorine mixture.* A 10% chlorine solution is to be mixed with a 25% chlorine solution to obtain 30 gallons of 20% solution. How many gallons of each must be used?

78. *Safe drivers.* Emily and Camille started from the same city and drove in opposite directions on the freeway. After

3 hours they were 354 miles apart. If they had gone in the same direction, Emily would have been 18 miles ahead of Camille. How fast did each woman drive?

Write a system of three equations in three variables for each word problem. Use Cramer's rule to solve each system.

79. *Weighing dogs.* Cassandra wants to determine the weights of her two dogs, Mimi and Mitzi. However, neither dog will sit on the scale by herself. Cassandra, Mimi, and Mitzi altogether weigh 175 pounds. Cassandra and Mimi together weigh 143 pounds. Cassandra and Mitzi together weigh 139 pounds. How much does each weigh individually?

175   143   139

Cassandra   Cassandra   Cassandra
Mimi   Mimi   Mitzi
Mitzi

Figure for Exercise 79

80. *Nickels, dimes, and quarters.* Bernard has 41 coins consisting of nickels, dimes, and quarters, and they are worth a total of $4.00. If the number of dimes plus the number of quarters is one more than the number of nickels, then how many of each does he have?

81. *Finding three angles.* If the two acute angles of a right triangle differ by 12°, then what are the measures of the three angles of this triangle?

82. *Two acute and one obtuse.* The obtuse angle of a triangle is twice as large as the sum of the two acute angles. If the smallest angle is only one-eighth as large as the sum of the other two, then what is the measure of each angle?

83. *Apples.* Cream of the Crop sells red delicious, golden delicious, and gala apples. The number of pounds of each type sold and the total receipts for the apples for the first three days of a week are shown in the accompanying table. Let x be the price per pound of red delicious, y be the price per pound of golden delicious, and z be the price per pound of gala apples. Find x, y, and z by using a graphing calculator and Cramer's rule.

	Red	Golden	Gala	Receipts
Mon	20	30	10	$ 75.40
Tues	30	10	40	$106.20
Wed	50	40	20	$139.90

Table for Exercise 83

84. *Tolls.* Vehicles crossing the Washington Bridge are classified as small (2 axles), medium (3 or 4 axles), and large (5 or more axles). A different toll is charged for each class of vehicle. The number of vehicles of each type crossing the bridge for the first three hours of a day and the total receipts for the hour are shown in the accompanying table. Let x be the toll for a small vehicle, y be the toll for a medium vehicle, and z be the toll for a large vehicle. Find x, y, and z by using a graphing calculator and Cramer's rule.

Hour	Small	Medium	Large	Receipts
1	345	632	685	$29,365
2	309	877	265	$22,252
3	298	661	455	$23,674

Table for Exercise 84

Getting More Involved

85. *Writing*

Explain what to do when you are trying to use Cramer's rule and $D = 0$.

86. *Exploration*

For what values of a does the system
$$ax - y = 3$$
$$x + 2y = 1$$
have a single solution?

87. *Exploration*

Can Cramer's rule be used to solve this system? Explain.
$$2x^2 - y = 3$$
$$3x^2 + 2y^2 = 22$$

88. *Writing*

For what values of a, b, c, and d is the determinant of the matrix
$$\begin{bmatrix} a & b & 0 \\ c & d & 0 \\ b & a & 0 \end{bmatrix}$$
equal to zero? Explain your answer.

4.6   Linear Programming

In This Section

⟨1⟩ **Graphing the Constraints**

⟨2⟩ **Maximizing or Minimizing**

In this section we graph the solution set to a system of several linear inequalities in two variables as in Section 3.4. We then use the solution set to the inequalities to determine the maximum or minimum value of another variable. The method that we use is called **linear programming.**

⟨1⟩ Graphing the Constraints

In linear programming we have two variables that must satisfy several linear inequalities. These inequalities are called the **constraints** because they restrict the variables to only certain values. A graph in the coordinate plane is used to indicate the points that satisfy all of the constraints.

E X A M P L E   1

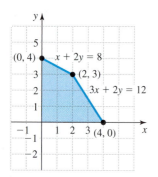

Figure 4.7

Graphing the constraints

Graph the solution set to the system of inequalities and identify each vertex of the region:

$$x \geq 0, \quad y \geq 0$$
$$3x + 2y \leq 12$$
$$x + 2y \leq 8$$

Solution

The points on or to the right of the y-axis satisfy $x \geq 0$. The points on or above the x-axis satisfy $y \geq 0$. The points on or below the line $3x + 2y = 12$ satisfy $3x + 2y \leq 12$. The points on or below the line $x + 2y = 8$ satisfy $x + 2y \leq 8$. Graph each straight line and shade the region that satisfies all four inequalities as shown in Fig. 4.7. Three of the vertices are easily identified as $(0, 0)$, $(0, 4)$, and $(4, 0)$. The fourth vertex is found by solving the system $3x + 2y = 12$ and $x + 2y = 8$. The fourth vertex is $(2, 3)$.

> **Now do Exercises 1–10**

In linear programming the constraints usually come from physical limitations in some problem. In Example 2, we write the constraints and then graph the points in the coordinate plane that satisfy all of the constraints.

E X A M P L E   2

Writing the constraints

Jules is in the business of constructing dog houses. A small dog house requires 8 square feet (ft^2) of plywood and 6 ft^2 of insulation. A large dog house requires 16 ft^2 of plywood and 3 ft^2 of insulation. Jules has available only 48 ft^2 of plywood and 18 ft^2 of insulation. Write the constraints on the number of small and large dog houses that he can build with the available supplies and graph the solution set to the system of constraints.

Solution

Let x represent the number of small dog houses and y represent the number of large dog houses. We have two natural constraints $x \geq 0$ and $y \geq 0$ since he cannot build a negative

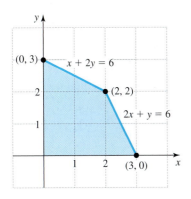

Figure 4.8

number of dog houses. Since the total plywood available for use is at most 48 ft^2, $8x + 16y \leq 48$. Since the total insulation available is at most 18 ft^2, $6x + 3y \leq 18$. Simplify the inequalities to get the following constraints:

$$x \geq 0, \quad y \geq 0$$
$$x + 2y \leq 6$$
$$2x + y \leq 6$$

The graph of the solution set to the system of inequalities is shown in Fig. 4.8.

> Now do Exercises 11–12

‹2› Maximizing or Minimizing

If a small dog house sells for $15 and a large sells for $20, then the total revenue in dollars from the sale of x small and y large dog houses is given by $R = 15x + 20y$. Since R is determined by or *is a function of x and y*, we use the function notation that was introduced in Section 3.5 and write $R(x, y)$ in place of R. The equation $R(x, y) = 15x + 20y$ is called a *linear function* of x and y. Any ordered pair within the region shown in Fig. 4.8 is a possibility for the number of dog houses of each type that could be built and so it is the *domain* of the function R. (We will study functions in general in Chapter 9.)

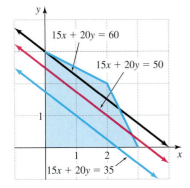

Figure 4.9

Linear Function of Two Variables

An equation of the form $f(x, y) = Ax + By + C$, where A, B, and C are fixed real numbers, is called a **linear function of two variables** (x and y).

Naturally, we are interested in the maximum revenue subject to the constraints on x and y. To investigate some possible revenues, replace R in $R = 15x + 20y$ with, say 35, 50, and 60. The graphs of the parallel lines $15x + 20y = 35$, $15x + 20y = 50$, and $15x + 20y = 60$ are shown in Fig. 4.9. The revenue at any point on the line $15x + 20y = 35$ is $35. We get a larger revenue on a higher revenue line (and lower revenue on a lower line). The maximum revenue possible will be on the highest revenue line that still intersects the region. Because the sides of the region are straight-line segments, the intersection of the highest (or lowest) revenue line with the region must include a vertex of the region. This is the fundamental principle behind linear programming.

The Principle of Linear Programming

The maximum or minimum value of a linear function subject to linear constraints occurs at a vertex of the region determined by the constraints.

EXAMPLE 3

Maximizing a linear function with linear constraints

A small dog house requires 8 ft^2 of plywood and 6 ft^2 of insulation. A large dog house requires 16 ft^2 of plywood and 3 ft^2 of insulation. Only 48 ft^2 of plywood and 18 ft^2 of insulation are available. If a small dog house sells for $15 and a large dog house sells for $20, then how many dog houses of each type should be built to maximize the revenue and to satisfy the constraints?

Solution

Let x be the number of small dog houses and y be the number of large dog houses. We wrote and graphed the constraints for this problem in Example 2, so we will not repeat that here. The graph in Fig. 4.8 has four vertices: $(0, 0)$, $(0, 3)$, $(3, 0)$, and $(2, 2)$. The revenue function is $R(x, y) = 15x + 20y$. Since the maximum value of this function must occur at a vertex, we evaluate the function at each vertex:

$$R(0, 0) = 15(0) + 20(0) = \$0$$
$$R(0, 3) = 15(0) + 20(3) = \$60$$
$$R(3, 0) = 15(3) + 20(0) = \$45$$
$$R(2, 2) = 15(2) + 20(2) = \$70$$

From this list we can see that the maximum revenue is $70 when two small and two large dog houses are built. We also see that the minimum revenue is $0 when no dog houses of either type are built.

Now do Exercises 13–32

Use the following strategy for solving linear programming problems.

Strategy for Linear Programming

Use the following steps to find the maximum or minimum value of a linear function subject to linear constraints.

1. Graph the region that satisfies all of the constraints.

2. Determine the coordinates of each vertex of the region.

3. Evaluate the function at each vertex of the region.

4. Identify which vertex gives the maximum or minimum value of the function.

In Example 4, we solve another linear programming problem.

EXAMPLE   4

Minimizing a linear function with linear constraints

One serving of food A contains 2 grams of protein and 6 grams of carbohydrates. One serving of food B contains 4 grams of protein and 3 grams of carbohydrates. A dietitian wants a meal that contains at least 12 grams of protein and at least 18 grams of carbohydrates. If the cost of food A is 9 cents per serving and the cost of food B is 20 cents per serving, then how many servings of each food would minimize the cost and satisfy the constraints?

Solution

Let x equal the number of servings of food A and y equal the number of servings of food B. If the meal is to contain at least 12 grams of protein, then $2x + 4y \geq 12$. If the meal is to contain at least 18 grams of carbohydrates, then $6x + 3y \geq 18$. Simplify each inequality and use the two natural constraints to get the following system:

$$x \geq 0, \quad y \geq 0$$
$$x + 2y \geq 6$$
$$2x + y \geq 6$$

The graph of the constraints is shown in Fig. 4.10. The vertices are $(0, 6)$, $(6, 0)$, and $(2, 2)$. The cost in cents for x servings of A and y servings of B is $C(x, y) = 9x + 20y$. Evaluate

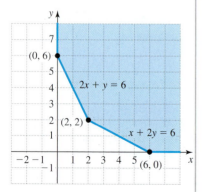

Figure 4.10

the cost at each vertex:

$$C(0, 6) = 9(0) + 20(6) = 120 \text{ cents}$$
$$C(6, 0) = 9(6) + 20(0) = 54 \text{ cents}$$
$$C(2, 2) = 9(2) + 20(2) = 58 \text{ cents}$$

The minimum cost of 54 cents is attained by using six servings of food A and no servings of food B.

Now do Exercises 33–38

Warm-Ups ▼

Fill in the blank.

1. A _____ is an inequality that restricts the values of the variables.

2. _____ is a process used to maximize or minimize a linear function subject to linear constraints.

3. A _____ function of two variables has the form $f(x, y) = Ax + By + C$.

4. The maximum or minimum of a linear function subject to linear constraints occurs at a _____ of the region determined by the constraints.

True or false?

5. The graph of $x \geq 0$ consists of the points on or above the x-axis.

6. The graph of $y \geq 0$ consists of the points on or to the right of the y-axis.

7. The graph of $x + y \leq 6$ consists of points on or below the line $x + y = 6$.

8. The graph of $2x + 3y = 30$ has x-intercept $(15, 0)$ and y-intercept $(0, 10)$.

9. The value of $R(x, y) = 3x + 5y$ at $(2, 4)$ is 26.

10. If $C(x, y) = 12x + 10y$, then $C(0, 5) = 62$.

Exercises

4.6

‹ Study Tips ›

• Working problems 1 hour per day every day of the week is better than working problems for 7 hours on one day of the week. Spread out your study time. Avoid long study sessions.

• No two students learn in exactly the same way or at the same speed. Figure out what works for you.

‹1› Graphing the Constraints

Graph the solution set to each system of inequalities and identify each vertex of the region. See Example 1.

1. $x \geq 0, y \geq 0$
$x + y \leq 5$

2. $x \geq 0, y \geq 0$
$y \leq 5, y \geq x$

VIDEO
3. $x \geq 0, y \geq 0$
$2x + y \leq 4$
$x + y \leq 3$

4. $x \geq 0, y \geq 0$
$x + y \leq 4$
$x + 2y \leq 6$

5. $x \geq 0, y \geq 0$
$2x + y \geq 3$
$x + y \geq 2$

6. $x \geq 0, y \geq 0$
$3x + 2y \geq 12$
$2x + y \geq 7$

7. $x \geq 0, y \geq 0$
$x + 3y \leq 15$
$2x + y \leq 10$

8. $x \geq 0, y \geq 0$
$2x + 3y \leq 15$
$x + y \leq 7$

9. $x \geq 0, y \geq 0$
$x + y \geq 4$
$3x + y \geq 6$

10. $x \geq 0, y \geq 0$
$x + 3y \geq 6$
$2x + y \geq 7$

For each problem, write the constraints and graph the solution set to the system of constraints. See Example 2.

11. *Making guitars.* A company makes an acoustic and an electric guitar. Each acoustic guitar requires $100 in materials and 20 hours of labor. Each electric guitar requires $200 in materials and 15 hours of labor. The company has at most $3000 for materials and 300 hours of labor available. Let x represent the possible number of acoustic guitars and y represent the possible number of electric guitars that can be made.

12. *Making boats.* A company makes kayaks and canoes. Each kayak requires $80 in materials and 60 hours of labor. Each canoe requires $120 in materials and 40 hours of labor. The company has at most $12,000 available for materials and at most 4800 hours of labor. Let x represent the possible number of kayaks and y represent the possible number of canoes that can be built.

‹2› Maximizing or Minimizing

Let $P(x, y) = 6x + 8y$, $R(x, y) = 11x + 20y$, and $C(x, y) = 5x + 12y$. Evaluate each expression. See Example 3.

13. $P(1, 5)$

14. $P(3, 8)$

15. $R(8, 0)$

16. $R(5, 10)$

17. $C(4, 9)$

18. $C(0, 6)$

Determine the maximum value of the given linear function on the given region. See Example 3.

19. $P(x, y) = 2x + 3y$

20. $W(x, y) = 6x + 7y$

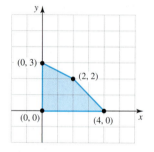

21. $R(x, y) = 9x + 8y$

22. $F(x, y) = 3x + 10y$

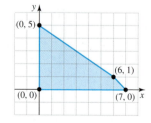

Determine the minimum value of the given function on the given region.

23. $C(x, y) = 11x + 10y$

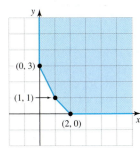

24. $H(x, y) = 4x + 7y$

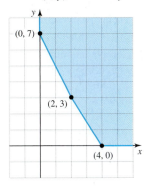

25. $A(x, y) = 9x + 3y$

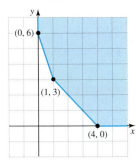

26. $R(x, y) = 5x + 4y$

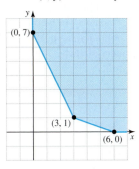

Solve each problem. See Examples 2–4. See the Strategy for Linear Programming box on page 278.

 27. *Phase I advertising.* The publicity director for Mercy Hospital is planning to bolster the hospital's image by running a TV ad and a radio ad. Due to budgetary and other constraints, the number of times that she can run the TV ad, x, and the number of times that she can run the radio ad, y, must be in the region shown in the figure. The function

$$A = 9000x + 4000y$$

gives the total number of people reached by the ads.

  a) Find the total number of people reached by the ads at each vertex of the region.

  b) What mix of TV and radio ads maximizes the number of people reached?

28. *Phase II advertising.* Suppose the radio station in Exercise 27 starts playing country music and the function for the total number of people changes to

$$A = 9000x + 2000y.$$

  a) Find A at each vertex of the region using this function.

Figure for Exercises 27 and 28

  b) What mix of TV and radio ads maximizes the number of people reached?

29. At Burger Heaven a double contains 2 meat patties and 6 pickles, whereas a triple contains 3 meat patties and 3 pickles. Near closing time one day, only 24 meat patties and 48 pickles are available. If a double burger sells for $1.20 and a triple burger sells for $1.50, then how many of each should be made to maximize the total revenue?

30. Sam and Doris manufacture rocking chairs and porch swings in the Ozarks. Each rocker requires 3 hours of work from Sam and 2 hours from Doris. Each swing requires 2 hours of work from Sam and 2 hours from Doris. Sam cannot work more than 48 hours per week, and Doris cannot work more than 40 hours per week. If a rocker sells for $160 and a swing sells for $100, then how many of each should be made per week to maximize the revenue?

31. If a double burger sells for $1.00 and a triple burger sells for $2.00, then how many of each should be made to maximize the total revenue subject to the constraints of Exercise 29?

32. If a rocker sells for $120 and a swing sells for $100, then how many of each should be made to maximize the total revenue subject to the constraints of Exercise 30?

33. One cup of Doggie Dinner contains 20 grams of protein and 40 grams of carbohydrates. One cup of Puppy Power contains 30 grams of protein and 20 grams of carbohydrates. Susan wants her dog to get at least 200 grams of protein and 180 grams of carbohydrates per day. If Doggie Dinner costs 16 cents per cup and Puppy Power costs 20 cents per cup, then how many cups of each would satisfy the constraints and minimize the total cost?

34. Mammoth Muffler employs supervisors and helpers. According to the union contract, a supervisor does 2 brake jobs and 3 mufflers per day, whereas a helper does 6 brake jobs and 3 mufflers per day. The home office requires enough staff for at least 24 brake jobs and for at least 18 mufflers per day. If a supervisor makes $90 per day and a helper makes $100 per day, then how many of each should be employed to satisfy the constraints and to minimize the daily labor cost?

35. Suppose in Exercise 33 Doggie Dinner costs 4 cents per cup and Puppy Power costs 10 cents per cup. How many cups of each would satisfy the constraints and minimize the total cost?

36. Suppose in Exercise 34 the supervisor makes $110 per day and the helper makes $100 per day. How many of each should be employed to satisfy the constraints and to minimize the daily labor cost?

37. Anita has at most $24,000 to invest in her brother-in-law's laundromat and her nephew's car wash. Her brother-in-law has high blood pressure and heart disease but he will pay 18%, whereas her nephew is healthier but will pay only 12%. So the amount she will invest in the car wash will be at least twice the amount that she will invest in the laundromat but not more than three times as much. How much should she invest in each to maximize her total income from the two investments?

38. Herbert assembles computers in his shop. The parts for each economy model are shipped to him in a carton with a volume of 2 cubic feet (ft^3) and the parts for each deluxe model are shipped to him in a carton with a volume of 3 ft^3. After assembly, each economy model is shipped out in a carton with a volume of 4 ft^3, and each deluxe model is shipped out in a carton with a volume of 4 ft^3. The truck that delivers the parts has a maximum capacity of 180 ft^3, and the truck that takes out the completed computers has a maximum capacity of 280 ft^3. He can receive only one shipment of parts and send out one shipment of computers per week. If his profit on an economy model is $60 and his profit on a deluxe model is $100, then how many of each should he order per week to maximize his profit?

4 Wrap-Up

Summary

Systems of Linear Equations		Examples
Methods for solving systems in two variables	Graphing: Sketch the graphs to see the solution.	The graphs of $y = x - 1$ and $x + y = 3$ intersect at $(2, 1)$.
	Substitution: Solve one equation for one variable in terms of the other, and then substitute into the other equation.	Substitution: $x + (x - 1) = 3$
	Addition: Multiply each equation as necessary to eliminate a variable upon addition of the equations.	$\begin{aligned} -x + y &= -1 \\ \underline{x + y} &= \underline{3} \\ 2y &= 2 \end{aligned}$
Types of linear systems in two variables	Independent: One point in solution set. The lines intersect at one point.	$y = x - 5$ $y = 2x + 3$
	Dependent: Infinite solution set. The lines are the same.	$2x + 3y = 4$ $4x + 6y = 8$
	Inconsistent: Empty solution set. The lines are parallel.	$2x + y = 1$ $2x + y = 5$
Linear equation in three variables	$Ax + By + Cz = D$ In a three-dimensional coordinate system the graph is a plane.	$2x - y + 3z = 5$
Linear systems in three variables	Use substitution or addition to eliminate variables in the system. The solution set may be a single point, the empty set, or an infinite set of points.	$\begin{aligned} x + y - z &= 3 \\ 2x - 3y + z &= 2 \\ x - y - 4z &= 14 \end{aligned}$

Matrices and Determinants		Examples
Matrix	A rectangular array of real numbers. An $m \times n$ matrix has m rows and n columns.	$\begin{bmatrix} 1 & -3 \\ 2 & 5 \end{bmatrix}, \begin{bmatrix} 1 & 0 & 1 \\ 2 & 1 & 4 \end{bmatrix}$
Augmented matrix	The matrix of coefficients and constants from a system of linear equations	$x - 3y = -7$ $2x + 5y = 19$ Augmented matrix: $\begin{bmatrix} 1 & -3 & \mid & -7 \\ 2 & 5 & \mid & 19 \end{bmatrix}$

| Gauss-Jordan elimination method | Use the row operations to get ones on the diagonal and zeros elsewhere for the coefficients in the augmented matrix. | $\begin{bmatrix} 1 & 0 & \vert & 2 \\ 0 & 1 & \vert & 3 \end{bmatrix}$ $x = 2$ and $y = 3$ |

| Determinant | A real number corresponding to a square matrix | |

| Determinant of a 2 × 2 matrix | $\begin{vmatrix} a_1 & b_1 \\ a_2 & b_2 \end{vmatrix} = a_1 b_2 - a_2 b_1$ | $\begin{vmatrix} 1 & -3 \\ 2 & 5 \end{vmatrix} = 5 - (-6) = 11$ |

| Determinant of a 3 × 3 matrix | Expand by minors about any row or column, using signs from the sign array. $\begin{vmatrix} a_1 & b_1 & c_1 \\ a_2 & b_2 & c_2 \\ a_3 & b_3 & c_3 \end{vmatrix} = a_1 \cdot \begin{vmatrix} b_2 & c_2 \\ b_3 & c_3 \end{vmatrix} - a_2 \cdot \begin{vmatrix} b_1 & c_1 \\ b_3 & c_3 \end{vmatrix}$ $+ a_3 \cdot \begin{vmatrix} b_1 & c_1 \\ b_2 & c_2 \end{vmatrix}$ | Sign array: $\begin{bmatrix} + & - & + \\ - & + & - \\ + & - & + \end{bmatrix}$ |

Cramer's Rules

| Two linear equations in two variables | The solution to the system $$a_1 x + b_1 y = c_1$$ $$a_2 x + b_2 y = c_2$$ is given by $x = \dfrac{D_x}{D}$ and $y = \dfrac{D_y}{D}$, where $D = \begin{vmatrix} a_1 & b_1 \\ a_2 & b_2 \end{vmatrix}$, $\qquad D_x = \begin{vmatrix} c_1 & b_1 \\ c_2 & b_2 \end{vmatrix}$, $\qquad$ and $\qquad D_y = \begin{vmatrix} a_1 & c_1 \\ a_2 & c_2 \end{vmatrix}$ provided that $D \neq 0$. |

| Three linear equations in three variables | The solution to the system $$a_1 x + b_1 y + c_1 z = d_1$$ $$a_2 x + b_2 y + c_2 z = d_2$$ $$a_3 x + b_3 y + c_3 z = d_3$$ is given by $x = \dfrac{D_x}{D}$, $y = \dfrac{D_y}{D}$, and $z = \dfrac{D_z}{D}$, where $D = \begin{vmatrix} a_1 & b_1 & c_1 \\ a_2 & b_2 & c_2 \\ a_3 & b_3 & c_3 \end{vmatrix}$, $\qquad D_x = \begin{vmatrix} d_1 & b_1 & c_1 \\ d_2 & b_2 & c_2 \\ d_3 & b_3 & c_3 \end{vmatrix}$, $D_y = \begin{vmatrix} a_1 & d_1 & c_1 \\ a_2 & d_2 & c_2 \\ a_3 & d_3 & c_3 \end{vmatrix}$, and $D_z = \begin{vmatrix} a_1 & b_1 & d_1 \\ a_2 & b_2 & d_2 \\ a_3 & b_3 & d_3 \end{vmatrix}$, provided that $D \neq 0$. |

Linear Programming

Use the following steps to find the maximum or minimum value of a linear function subject to linear constraints.
1. Graph the region that satisfies all of the constraints.
2. Determine the coordinates of each vertex of the region.
3. Evaluate the function at each vertex of the region.
4. Identify which vertex gives the maximum or minimum value of the function.

Enriching Your Mathematical Word Power

Fill in the blank.

1. A collection of two or more equations is a _____ of equations.

2. A(n) _____ system has exactly one solution.

3. A(n) _____ system has no solutions.

4. A(n) _____ system has infinitely many solutions.

5. A(n) _____ system has at least one solution.

6. With the _____ method a variable is eliminated by adding the equations.

7. $Ax + By + Cz = D$ is a _____ equation in three variables.

8. A(n) _____ is a rectangular array of numbers.

9. A(n) _____ matrix contains the coefficients and constants of a linear system.

10. The _____ of a matrix is the number of rows and columns in a matrix.

11. A(n) _____ is a number derived from a square matrix.

12. A(n) _____ array is a matrix of $+$ and $-$ signs used in computing a determinant.

13. The inequalities in a linear programming problem are _____.

14. A method for minimizing or maximizing a linear function of two variables subject to linear constraints is _____ programming.

Review Exercises

4.1 Solving Systems by Graphing and Substitution

Solve by graphing. Indicate whether each system is independent, dependent, or inconsistent.

1. $y = 2x - 1$
$x + y = 2$

2. $y = 3x - 4$
$y = -2x + 1$

3. $x + 2y = 4$
$y = -\dfrac{1}{2}x + 2$

4. $2x - 3y = 12$
$3y - 2x = -12$

5. $y = -x$
$y = -x + 3$

6. $3x - y = 4$
$3x - y = 0$

Solve by substitution. Indicate whether each system is independent, dependent, or inconsistent.

7. $y = 3x + 11$
$2x + 3y = 0$

8. $x - y = 3$
$3x - 2y = 3$

9. $x = y + 5$
$2x - 2y = 12$

10. $3y = x + 5$
$3x - 9y = -10$

11. $2x - y = 3$
$6x - 9 = 3y$

12. $y = \dfrac{1}{2}x - 9$
$3x - 6y = 54$

25

13. $y = \dfrac{1}{2}x - 3$

$y = \dfrac{1}{3}x + 2$

14. $x = \dfrac{1}{8}y - 1$

$y = \dfrac{1}{4}x + 39$

15. $x + 2y = 1$

$8x + 6y = 4$

16. $x - 5y = 4$

$4x + 8y = -5$

4.2 The Addition Method

Solve by addition. Indicate whether each system is independent, dependent, or inconsistent.

17. $5x - 3y = -20$

$3x + 2y = 7$

18. $-3x + y = 3$

$2x - 3y = 5$

19. $2(y - 5) + 4 = 3(x - 6)$

$3x - 2y = 12$

20. $x + 3(y - 1) = 11$

$2(x - y) + 8y = 28$

21. $3x - 4(y - 5) = x + 2$

$2y - x = 7$

22. $4(1 - x) + y = 3$

$3(1 - y) - 4x = -4y$

23. $\dfrac{1}{4}x + \dfrac{3}{8}y = \dfrac{3}{8}$

$\dfrac{5}{2}x - 6y = 7$

24. $\dfrac{1}{3}x - \dfrac{1}{6}y = \dfrac{1}{3}$

$\dfrac{1}{6}x + \dfrac{1}{4}y = 0$

25. $0.4x + 0.06y = 11.6$

$0.8x - 0.05y = 13$

26. $0.08x + 0.7y = 37.4$

$0.06x - 0.05y = -0.7$

4.3 Systems of Linear Equations in Three Variables

Solve each system by elimination of variables.

27. $x - y + z = 4$

$-x + 2y - z = 0$

$-x + y - 3z = -16$

28. $2x - y + z = 5$

$x + y - 2z = -4$

$3x - y + 3z = 10$

29. $2x - y - z = 3$

$3x + y + 2z = 4$

$4x + 2y - z = -4$

30. $2x + 3y - 2z = -11$

$3x - 2y + 3z = 7$

$x - 4y + 4z = 14$

31. $x + y - z = 4$

$y + z = 6$

$x + 2y = 8$

32. $x - 3y + z = 5$

$2x - 4y - z = 7$

$2x - 6y + 2z = 6$

33. $x - 2y + z = 8$

$-x + 2y - z = -8$

$2x - 4y + 2z = 16$

34. $x - y + z = 1$

$2x - 2y + 2z = 2$

$-3x + 3y - 3z = -3$

4.4 Solving Linear Systems Using Matrices
Solve each system by the Gauss-Jordan elimination method.

35. $x + y = 7$
$-x + 2y = 5$

36. $-x + y = 1$
$2x - 3y = -7$

37. $2x + y = 0$
$x - 3y = 14$

38. $2x - y = 8$
$3x + 2y = -2$

39. $x + y - z = 0$
$x - y + 2z = 4$
$2x + y - z = 1$

40. $2x - y + 2z = 9$
$x + 3y = 5$
$3x + z = 9$

4.5 Determinants and Cramer's Rule
Evaluate each determinant.

41. $\begin{vmatrix} 1 & 3 \\ 0 & 2 \end{vmatrix}$

42. $\begin{vmatrix} -1 & 2 \\ -3 & 5 \end{vmatrix}$

43. $\begin{vmatrix} 0.01 & 0.02 \\ 50 & 80 \end{vmatrix}$

44. $\begin{vmatrix} \dfrac{1}{2} & \dfrac{1}{3} \\ \dfrac{1}{4} & \dfrac{1}{5} \end{vmatrix}$

Solve each system. Use Cramer's rule.

45. $2x - y = 0$
$3x + y = -5$

46. $3x - 2y = 14$
$2x + 3y = -8$

47. $y = 2x - 3$
$3x - 2y = 4$

48. $y = 2x - 5$
$y = 3x - 3y$

Evaluate each determinant.

49. $\begin{vmatrix} 2 & 3 & 1 \\ -1 & 2 & 4 \\ 6 & 1 & 1 \end{vmatrix}$

50. $\begin{vmatrix} 1 & -1 & 0 \\ -2 & 0 & 0 \\ 3 & 1 & 5 \end{vmatrix}$

51. $\begin{vmatrix} 2 & 3 & -2 \\ 2 & 0 & 4 \\ -1 & 0 & 3 \end{vmatrix}$

52. $\begin{vmatrix} 3 & -1 & 4 \\ 2 & -1 & 1 \\ -2 & 0 & 1 \end{vmatrix}$

Solve each system. Use Cramer's rule.

53. $x + y = 3$
$x + y + z = 0$
$x - y - z = 2$

54. $2x - y + z = 0$
$4x + 6y - 2z = 0$
$x - 2y - z = -9$

4.6 Linear Programming
Graph each system of inequalities and identify each vertex of the region.

55. $x \geq 0, y \geq 0$
$x + 2y \leq 6$
$x + y \leq 5$

56. $x \geq 0, y \geq 0$
$3x + 2y \geq 12$
$x + 2y \geq 8$

Solve each problem by linear programming.

57. Find the maximum value of the function $R(x, y) = 6x + 9y$ subject to the following constraints:

$$x \geq 0, y \geq 0$$
$$2x + y \leq 6$$
$$x + 2y \leq 6$$

58. Find the minimum value of the function $C(x, y) = 9x + 10y$ subject to the following constraints:

$$x \geq 0, y \geq 0$$
$$x + y \geq 4$$
$$3x + y \geq 6$$

Miscellaneous
Use a system of equations in two or three variables to solve each problem. Solve by the method of your choice.

59. *Perimeter of a rectangle.* The length of a rectangular swimming pool is 15 feet longer than the width. If the perimeter is 82 feet, then what are the length and width?

60. *Household income.* Alkena and Hsu together earn $84,326 per year. If Alkena earns $12,468 more per year than Hsu, then how much does each of them earn per year?

61. *Two-digit number.* The sum of the digits in a two-digit number is 15. When the digits are reversed, the new number is 9 more than the original number. What is the original number?

62. *Two-digit number.* The sum of the digits in a two-digit number is 8. When the digits are reversed, the new number is 18 less than the original number. What is the original number?

63. *Traveling by boat.* Alonzo can travel from his camp downstream to the mouth of the river in 30 minutes. If it takes him 45 minutes to come back, then how long would it take him to go that same distance in the lake with no current?

Time with current = 30 min
Time against current = 45 min

Figure for Exercise 63

64. *Driving and dating.* In 4 years Gasper will be old enough to drive. His parents said that he must have a driver's license for 2 years before he can date. Three years ago, Gasper's age was only one-half of the age necessary to date. How old must Gasper be to drive, and how old is he now?

65. *Three solutions.* A chemist has three solutions of acid that must be mixed to obtain 20 liters of a solution that is 38% acid. Solution A is 30% acid, solution B is 20% acid, and solution C is 60% acid. Because of another chemical in these solutions, the chemist must keep the ratio of solution

C to solution A at 2 to 1. How many liters of each should she mix together?

66. *Mixing investments.* Darlene invested a total of $20,000. The part that she invested in Dell Computer stock returned 70%, and the part that she invested in U.S. Treasury bonds returned 5%. Her total return on these two investments was $9580.

a) Use the graph to estimate the amount that she put into each investment.

b) Solve a system of equations to find the exact amount that she put into each investment.

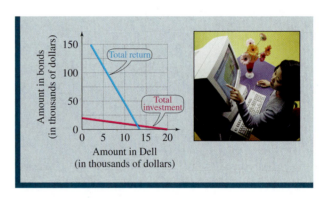

Amount in bonds (in thousands of dollars)

Total return

Total investment

Amount in Dell (in thousands of dollars)

Figure for Exercise 66

67. *Beets and beans.* One serving of canned beets contains 1 gram of protein and 6 grams of carbohydrates. One serving of canned red beans contains 6 grams of protein and 20 grams of carbohydrates. How many servings of each would it take to get exactly 21 grams of protein and 78 grams of carbohydrates?

Chapter 4 Test

Solve the system by graphing.

1. $x + y = 4$
$y = 2x + 1$

Solve each system by substitution.

2. $y = 2x - 8$
$4x + 3y = 1$

3. $y = x - 5$
$3x - 4(y - 2) = 28 - x$

Solve each system by the addition method.

4. $3x + 2y = 3$
$4x - 3y = -13$

5. $3x - y = 5$
$-6x + 2y = 1$

Determine whether each system is independent, dependent, or inconsistent.

6. $y = 3x - 5$
$y = 3x + 2$

7. $2x + 2y = 8$
$x + y = 4$

8. $y = 2x - 3$
  $y = 5x - 14$

Solve the following system by elimination of variables.

9. $x + y - z = 2$
  $2x - y + 3z = -5$
  $x - 3y + z = 4$

Solve by the Gauss-Jordan elimination method.

10. $3x - y = 1$ **11.** $x - y - z = 1$
  $x + 2y = 12$ $-x - y + 2z = -2$
                             $-x - 3y + z = -5$

Evaluate each determinant.

12. $\begin{vmatrix} 2 & 3 \\ 4 & -3 \end{vmatrix}$

13. $\begin{vmatrix} 1 & -2 & -1 \\ 2 & 3 & 1 \\ 1 & 1 & 0 \end{vmatrix}$

Solve each system by using Cramer's rule.

14. $2x - y = -4$ **15.** $x + y = 0$
  $3x + y = -1$ $x - y + 2z = 6$
                                $2x + y - z = 1$

For each problem, write a system of equations in two or three variables. Use the method of your choice to solve each system.

16. One night the manager of the Sea Breeze Motel rented 5 singles and 12 doubles for a total of $1583. The next night he rented 9 singles and 10 doubles for a total of $1701. What is the rental charge for each type of room?

17. Jill, Karen, and Betsy studied a total of 93 hours last week. Jill's and Karen's study time totaled only one-half as much as Betsy's. If Jill studied 3 hours more than Karen, then how many hours did each one of the girls spend studying?

Solve the following problem by linear programming.

18. Find the maximum value of the function

$$P(x, y) = 8x + 10y$$

subject to the following constraints:

$$x \geq 0, y \geq 0$$
$$2x + 3y \leq 12$$
$$x + y \leq 5$$

Making **Connections** | A Review of Chapters 1–4

Simplify each expression.

1. -3^4

2. $\dfrac{1}{3}(3) + 6$

3. $(-5)^2 - 4(-2)(6)$

4. $6 - (0.2)(0.3)$

5. $5(t - 3) - 6(t - 2)$

6. $0.1(x - 1) - (x - 1)$

7. $\dfrac{-9x^2 - 6x + 3}{-3}$

8. $\dfrac{4y - 6}{2} - \dfrac{3y - 9}{3}$

Solve each equation for y.

9. $3x - 5y = 7$

10. $Cx - Dy = W$

11. $Cy = Wy - K$

12. $A = \dfrac{1}{2}b(w - y)$

Solve each equation.

13. $\dfrac{5}{17}x = \dfrac{10}{51}$

14. $x + \dfrac{3}{8} = \dfrac{9}{20}$

15. $0.35x = 0.4x + 1.6$

16. $\dfrac{0.05x - 6.9}{15} = 0.2(x - 20)$

17. $5 - 6(4x + 12) = 1 - 3(8x + 1)$

18. $5 - 6(4x + 12) = -64 - 3(8x + 1)$

19. $|x| = 1000$

20. $|3x - 8| = 7$

21. $|9 - 2x| = 0$

22. $|7x - 12| + 8 = 6$

Solve each inequality. State the solution set using interval notation.

23. $3x - 12 \geq 6$

24. $20 - 9x < 2$

25. $2x > 6$ and $x - 9 < 12$

26. $2x > 6$ or $x - 9 < 12$

27. $-6 < 3x < 12$

28. $-3 \leq \dfrac{5 - 2x}{9} < 7$

29. $|8 - x| > 7$

30. $|12x - 1| \leq 11$

31. $\left| \dfrac{5}{13}x - \dfrac{3}{4} \right| \geq -\dfrac{3}{2}$

32. $|17x - 88| + 215 < 109$

Solve each system.

33. $y = x - 5$
    $2x + 3y = 5$

34. $0.05x + 0.06y = 67$
    $x + y = 1200$

35. $3x - 15y = -51$
    $x + 17 = 5y$

36. $0.07a + 0.3b = 6.70$
    $7a + 30b = 67$

Find the equation of each line.

37. The line through $(0, 55)$ and $(-99, 0)$

38. The line through $(2, -3)$ and $(-4, 8)$

39. The line through $(-4, 6)$ that is parallel to $y = 5x$

40. The line through $(4, 7)$ that is perpendicular to $y = -2x + 1$

41. The line through $(3, 5)$ that is parallel to the x-axis

42. The line through $(-7, 0)$ that is perpendicular to the x-axis

Solve.

43. *Comparing copiers.* A self-employed consultant has prepared the accompanying graph to compare the total cost of purchasing and using two different copy machines.

a) Which machine has the larger purchase price?

b) What is the per copy cost for operating each machine, not including the purchase price?

c) Find the slope of each line and interpret your findings.

d) Find the equation of each line.

e) Find the number of copies for which the total cost is the same for both machines.

Figure for Exercise 43

*Critical*Thinking | **For Individual or Group Work** | **Chapter 4**

These exercises can be solved by a variety of techniques, which may or may not require algebra. So be creative and think critically. Explain all answers. Answers are in the Instructor's Edition of this text.

1. *Counting columns.* Consider the accompanying table of counting numbers. If the pattern were continued, then in which column would 1001 appear and why?

	2	3	4	5
9	8	7	6	
	10	11	12	13
17	16	15	14	
	18	19	20	21
25	24	23	22	
	26	27	28	29
33	32	31	30	

Table for Exercise 1

2. *Five plus four is ten.* Add 5 more straight line segments to the 4 vertical line segments shown here to make 10.

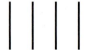

3. *Summing reciprocals.* The sum of the positive divisors of 960 is 3048. Use this fact to find the sum of the reciprocals of the positive divisors.

4. *Strange coincidence.* In Arial font, some capital letters are formed using straight line segments only. For example, FIVE is formed using 10 line segments but FOUR is not formed using only straight line segments. There is one number whose value is the same as the number of line segments used to write it with capital letters in Arial font. What is that number? Do not count a hyphen as a line segment.

5. *Megadigits.* How many digits are in the number $2^{2002} \cdot 5^{1995}$?

6. *Passing freights.* Two freight trains are traveling in the same direction on parallel tracks, one at 60 mph and the

Photo for Exercise 6

other at 40 mph. If each train is $\frac{1}{2}$ mile long, then how long does it take for the faster train to pass the slower train?

7. *Multiplying flowers.* Each letter in the following multiplication problem represents a unique digit. Determine values of the letters that would make the multiplication problem correct.

$$
\begin{array}{r}
1PANSY \\
\times \qquad 3 \\
\hline
PANSY1
\end{array}
$$

8. *Pass and fail.* The mean score on the last exam for Professor Habibi's algebra class of 25 students was 68. A score of 60 or above is required to pass the exam. The students who passed the exam had a mean of 75, and the students who did not pass had a mean of 50. How many students passed the exam?

Exponents and Polynomials

One statistic that can be used to measure the general health of a nation or group within a nation is life expectancy. These data are considered more accurate than many other statistics because it is easy to determine the precise number of years in a person's lifetime.

According to the National Center for Health Statistics, an American born in 2006 has a life expectancy of 77.9 years. However, an American male born in 2006 has a life expectancy of only 75.0 years, whereas a female can expect 80.8 years. A male who makes it to 65 can expect to live 16.1 more years, whereas a female who makes it to 65 can expect 17.9 more years. In the next few years, thanks in part to advances in health care and science, longevity is expected to increase significantly worldwide. In fact, the World Health Organization predicts that by 2025 no country will have a life expectancy of less than 50 years.

In this chapter, we will see how functions involving exponents are used to model life expectancy.

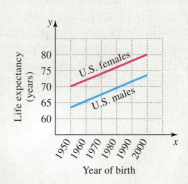

In Exercises 87 and 88 of Section 5.2 you will see how exponents are used to determine the life expectancies of white men and women.

5.1   **Integral Exponents and Scientific Notation**

In Chapter 1, we defined positive integral exponents and learned to evaluate expressions involving exponents. In this section we will extend the definition of exponents to include all integers and will learn some rules for working with integral exponents. In Chapter 7 we will see that any rational number can be used as an exponent.

⟨1⟩ **Positive and Negative Exponents**

We learned in Chapter 1 that a positive integral exponent indicates the number of times that the base is used as a factor. For example,

$$x^2 = x \cdot x \qquad \text{and} \qquad a^3 = a \cdot a \cdot a.$$

If n is a positive integer, a^n indicates that a is used as a factor n times. We define a^{-n} as the reciprocal of a^n. For example,

$$x^{-2} = \frac{1}{x^2} \qquad \text{and} \qquad a^{-3} = \frac{1}{a^3}.$$

Negative Integral Exponents

If a is a nonzero real number and n is a positive integer, then

$$a^{-n} = \frac{1}{a^n}. \qquad \text{If } n \text{ is positive, } -n \text{ is negative.}$$

Note that $a^{-1} = \frac{1}{a^1} = \frac{1}{a}$. The exponent -1 simply indicates reciprocal. So $\left(\frac{2}{3}\right)^{-1} = \frac{3}{2}$. Since $a^n \cdot a^{-n} = 1$, we can write $\frac{1}{a^{-n}} = a^n$. (The reciprocal of the reciprocal of a^n is a^n.)

A negative exponent indicates a power and a reciprocal. The result is the same regardless of which is performed first. For example, $2^{-3} = \frac{1}{2^3} = \left(\frac{1}{2}\right)^3 = \frac{1}{8}$. Note that

$$\left(\frac{2}{3}\right)^{-2} = \left(\frac{3}{2}\right)^2 = \frac{9}{4} \qquad \text{and} \qquad \left(\frac{2}{3}\right)^{-2} = \left(\frac{4}{9}\right)^{-1} = \frac{9}{4}.$$

Remember that if the negative sign in a negative exponent is deleted, then you must find a reciprocal. Four situations where this idea occurs are summarized in the following box. Don't think of this as four more rules to be memorized. Remember the idea.

Rules for Negative Exponents

The following rules hold if a and b are nonzero real numbers and n is a positive integer:

1. $a^{-1} = \dfrac{1}{a}$

2. $\dfrac{1}{a^{-n}} = a^n$

3. $a^{-n} = \left(\dfrac{1}{a}\right)^n$

4. $\left(\dfrac{a}{b}\right)^{-n} = \left(\dfrac{b}{a}\right)^n$

E X A M P L E   1

Negative exponents

Evaluate each expression.

a) 3^{-1} **b)** $\dfrac{1}{5^{-3}}$ **c)** 3^{-2}

d) $(-3)^{-2}$ **e)** -3^{-2} **f)** $\left(\dfrac{3}{4}\right)^{-3}$

Solution

a) $3^{-1} = \dfrac{1}{3}$ First rule for negative exponents

b) $\dfrac{1}{5^{-3}} = 5^3 = 125$ Second rule for negative exponents

c) Using the definition of negative exponents we have

$$3^{-2} = \frac{1}{3^2} = \frac{1}{9}.$$

Using the third rule for negative exponents we have

$$3^{-2} = \left(\frac{1}{3}\right)^2 = \frac{1}{3} \cdot \frac{1}{3} = \frac{1}{9}.$$

d) $(-3)^{-2} = \dfrac{1}{(-3)^2} = \dfrac{1}{9}$ Since $(-3)^2 = (-3)(-3) = 9$

e) $-3^{-2} = -\dfrac{1}{3^2}$ The exponent applies to 3 only.

$$= -\frac{1}{9}$$

f) $\left(\dfrac{3}{4}\right)^{-3} = \left(\dfrac{4}{3}\right)^3$ Fourth rule for negative exponents

$$= \frac{4}{3} \cdot \frac{4}{3} \cdot \frac{4}{3}$$ Definition of positive exponents

$$= \frac{64}{27}$$

Note that the same result is obtained if you cube first and then find the reciprocal: $\left(\dfrac{3}{4}\right)^3 = \dfrac{27}{64}$ and the reciprocal of $\dfrac{27}{64}$ is $\dfrac{64}{27}$.

Now do Exercises 1–12

‹ **Calculator Close-Up** ›

You can evaluate expressions with negative exponents using a graphing calculator. Use the fraction feature to get fractional answers.

```
3^-2▶Frac
                1/9
(-3)^-2▶Frac
                1/9
-3^-2▶Frac
               -1/9
```

‹ **Helpful Hint** ›

A negative exponent does not cause an expression to have a negative value. The negative exponent "causes" the reciprocal:

$$2^{-3} = \frac{1}{2^3} = \frac{1}{8}$$

$$(-3)^{-4} = \frac{1}{(-3)^4} = \frac{1}{81}$$

$$(-4)^{-3} = \frac{1}{(-4)^3} = -\frac{1}{64}$$

> **CAUTION** In Chapter 1, we agreed to evaluate -3^2 by squaring 3 first and then taking the opposite. So $-3^2 = -9$, whereas $(-3)^2 = 9$. The same agreement also holds for negative exponents. That is why the answer to Example 1(e) is negative.

⟨2⟩ The Product Rule for Exponents

To find the product of the exponential expressions 2^3 and 2^5 we could simply count the number of times 2 appears in the product:

$$2^3 \cdot 2^5 = \overbrace{(2 \cdot 2 \cdot 2)}^{3 \text{ factors}}\underbrace{\overbrace{(2 \cdot 2 \cdot 2 \cdot 2 \cdot 2)}^{5 \text{ factors}}}_{8 \text{ factors}} = 2^8$$

Instead of counting to find that 2 occurs eight times, it is easier to add 3 and 5 to get 8.

Now consider the product of 2^{-3} and 2^5:

$$2^{-3} \cdot 2^5 = \left(\frac{1}{2}\right)^3 \cdot 2^5 = \frac{1}{2} \cdot \frac{1}{2} \cdot \frac{1}{2} \cdot 2 \cdot 2 \cdot 2 \cdot 2 \cdot 2 = 2^2$$

The exponent in 2^2 is the sum of the exponents -3 and 5.

These examples illustrate the **product rule for exponents.**

> **Product Rule for Exponents**
>
> If $a \neq 0$ and m and n are integers, then
>
> $$a^m \cdot a^n = a^{m+n}.$$

The product rule for exponents applies only when the bases are identical.

EXAMPLE 2

Using the product rule

Simplify each expression. Write answers with positive exponents and assume all variables represent nonzero real numbers.

a) $3^4 \cdot 3^6$ b) $4x^{-3} \cdot 5x$ c) $-2y^{-3}\left(-5y^{-4}\right)$

⟨ **Helpful Hint** ⟩

The definitions of the different types of exponents are a really clever mathematical invention. The fact that we have rules for performing arithmetic with those exponents makes the notation of exponents even more amazing.

Solution

a) $3^4 \cdot 3^6 = 3^{4+6} = 3^{10}$ Product rule: $4 + 6 = 10$

b) $4x^{-3} \cdot 5x = 4 \cdot 5 \cdot x^{-3} \cdot x^1$

$\quad\quad\quad = 20x^{-2}$ Product rule: $-3 + 1 = -2$

$\quad\quad\quad = \dfrac{20}{x^2}$ Definition of negative exponent

c) $-2y^{-3}\left(-5y^{-4}\right) = (-2)(-5)y^{-3}y^{-4}$

$\quad\quad\quad = 10y^{-7}$ Product rule: $-3 + (-4) = -7$

$\quad\quad\quad = \dfrac{10}{y^7}$ Definition of negative exponent

Now do Exercises 13–18

CAUTION   The product rule cannot be applied to $2^3 \cdot 3^2$ because the bases are not identical. Even when the bases are identical, we do not multiply the bases. For example, $2^5 \cdot 2^4 \neq 4^9$. Using the rule correctly, we get $2^5 \cdot 2^4 = 2^9$.

⟨3⟩ Zero Exponent

We have used positive and negative integral exponents, but we have not yet seen the integer 0 used as an exponent. Note that the product rule was stated to hold for *any* integers m and n. If we use the product rule on $2^3 \cdot 2^{-3}$, we get

$$2^3 \cdot 2^{-3} = 2^0.$$

However, $2^3 \cdot 2^{-3} = 2^3 \cdot \frac{1}{2^3} = 8 \cdot \frac{1}{8} = 1$. So for consistency we define 2^0 and the zero power of any nonzero number to be 1.

> **Zero Exponent**
>
> If a is any nonzero real number, then $a^0 = 1$.

EXAMPLE 3

⟨ **Helpful Hint** ⟩

Defining a^0 to be 1 gives a consistent pattern to exponents:

$$3^{-2} = \frac{1}{9}$$

$$3^{-1} = \frac{1}{3}$$

$$3^0 = 1$$

$$3^1 = 3$$

$$3^2 = 9$$

If the exponent is increased by 1 (with base 3), the value of the expression is multiplied by 3.

Using zero as an exponent

Simplify each expression. Write answers with positive exponents and assume all variables represent nonzero real numbers.

a) -3^0 b) $\left(\frac{1}{4} - \frac{3}{2}\right)^0$

c) $-2a^5 b^{-6} \cdot 3a^{-5} b^2$

Solution

a) To evaluate -3^0, we find 3^0 and then take the opposite. So $-3^0 = -1$.

b) $\left(\frac{1}{4} - \frac{3}{2}\right)^0 = 1$ Definition of zero exponent

c) $-2a^5 b^{-6} \cdot 3a^{-5} b^2 = -6a^5 \cdot a^{-5} \cdot b^{-6} \cdot b^2$

$= -6a^0 b^{-4}$ Product rule: $5 + (-5) = 0$, $-6 + 2 = -4$

$= -\frac{6}{b^4}$ Definitions of negative and zero exponents

> Now do Exercises 19–26

⟨4⟩ Changing the Sign of an Exponent

Because a^{-n} and a^n are reciprocals of each other, we know that

$$a^{-n} = \frac{1}{a^n} \qquad \text{and} \qquad \frac{1}{a^{-n}} = a^n.$$

So a negative exponent in the numerator or denominator can be changed to positive by relocating the exponential expression. In Example 4, we use these facts to remove negative exponents from exponential expressions.

E X A M P L E   **4**

Simplifying expressions with negative exponents

Write each expression without negative exponents and simplify. All variables represent nonzero real numbers.

a) $\dfrac{5a^{-3}}{a^2 \cdot 2^{-2}}$

b) $\dfrac{-2x^{-3}}{y^{-2}z^3}$

Solution

a) $\dfrac{5a^{-3}}{a^2 \cdot 2^{-2}} = 5 \cdot a^{-3} \cdot \dfrac{1}{a^2} \cdot \dfrac{1}{2^{-2}}$ Rewrite division as multiplication.

$= 5 \cdot \dfrac{1}{a^3} \cdot \dfrac{1}{a^2} \cdot 2^2$ Change the signs of the negative exponents.

$= \dfrac{20}{a^5}$ Product rule: $a^3 \cdot a^2 = a^5$

Note that in $5a^{-3}$ the negative exponent applies only to a.

b) $\dfrac{-2x^{-3}}{y^{-2}z^3} = -2 \cdot x^{-3} \cdot \dfrac{1}{y^{-2}} \cdot \dfrac{1}{z^3}$ Rewrite as multiplication.

$= -2 \cdot \dfrac{1}{x^3} \cdot y^2 \cdot \dfrac{1}{z^3}$ Definition of negative exponent

$= \dfrac{-2y^2}{x^3z^3}$ Simplify.

Now do Exercises 27–34

In Example 4, we showed more steps than are necessary. For instance, in part (b) we could simply write

$$\frac{-2x^{-3}}{y^{-2}z^3} = \frac{-2y^2}{x^3z^3}.$$

Exponential expressions (that are factors) can be moved from numerator to denominator (or vice versa) as long as we change the sign of the exponent.

CAUTION   If an exponential expression is *not* a factor, you *cannot* move it from numerator to denominator (or vice versa). For example,

$$\frac{2^{-1} + 3^{-1}}{4^{-1}} \neq \frac{4}{2 + 3}.$$

Because $2^{-1} = \frac{1}{2}$, $3^{-1} = \frac{1}{3}$, and $4^{-1} = \frac{1}{4}$, we get

$$\frac{2^{-1} + 3^{-1}}{4^{-1}} = \frac{\frac{1}{2} + \frac{1}{3}}{\frac{1}{4}} = \frac{\frac{5}{6}}{\frac{1}{4}} = \frac{5}{6} \cdot \frac{4}{1} = \frac{10}{3} \quad \text{and} \quad \frac{4}{2 + 3} = \frac{4}{5}.$$

‹5› The Quotient Rule for Exponents

By the product rule for exponents we have

$$a^{m-n} \cdot a^n = a^{m-n+n} = a^m.$$

‹ **Calculator Close-Up** ›

A graphing calculator cannot prove that the quotient rule is correct, but it can provide numerical support for the quotient rule.

```
2^15/2^5
              1024
2^10
              1024
```

Dividing each side of this equation by a^n yields $\frac{a^m}{a^n} = a^{m-n}$, which is the **quotient rule for exponents:**

Quotient Rule for Exponents

If m and n are any integers and $a \neq 0$, then

$$\frac{a^m}{a^n} = a^{m-n}.$$

If you want to "see" the quotient rule at work, consider dividing 2^5 by 2^3:

$$\frac{2^5}{2^3} = \frac{2 \cdot 2 \cdot 2 \cdot 2 \cdot 2}{2 \cdot 2 \cdot 2} = 2^2$$

There are five 2's in the numerator and three in the denominator. After dividing, two 2's remain. The exponent in 2^2 can be obtained by subtracting 3 from 5.

CAUTION   Do not divide the bases when using the quotient rule. We cannot apply the quotient rule to $\frac{6^5}{2^4}$ even though 6 is divisible by 2.

E X A M P L E   **5**

Using the quotient rule

Simplify each expression. Write answers with positive exponents only. All variables represent nonzero real numbers.

a) $\dfrac{2^9}{2^4}$ b) $\dfrac{m^5}{m^{-3}}$ c) $\dfrac{y^{-4}}{y^{-2}}$

Solution

a) $\dfrac{2^9}{2^4} = 2^{9-4} \qquad = 2^5$ Quotient rule: $9 - 4 = 5$

b) $\dfrac{m^5}{m^{-3}} = m^{5-(-3)} \quad = m^8$ Quotient rule: $5 - (-3) = 8$

c) $\dfrac{y^{-4}}{y^{-2}} = y^{-4-(-2)} \quad = y^{-2}$ Quotient rule: $-4 - (-2) = -2$

$\qquad\qquad\qquad\qquad = \dfrac{1}{y^2}$ Rewrite with a positive exponent.

> Now do Exercises 35–42

E X A M P L E   **6**

Using the product and quotient rules

Use the rules of exponents to simplify each expression. Write answers with positive exponents only. All variables represent nonzero real numbers.

a) $\dfrac{2x^{-7}}{x^{-7}}$ b) $\dfrac{w(10w^{-4})}{6w^{-2}}$ c) $\dfrac{x^{-1}x^{-3}y^5}{x^{-2}y^2}$

Solution

a) $\dfrac{2x^{-7}}{x^{-7}} = 2x^0$ Quotient rule: $-7 - (-7) = 0$

   $\quad\quad\quad = 2$ Definition of zero exponent

b) $\dfrac{w(10w^{-4})}{6w^{-2}} = \dfrac{10w^{-3}}{6w^{-2}}$ Product rule: $w^1 \cdot w^{-4} = w^{-3}$

   $\quad\quad\quad = \dfrac{10w^{-1}}{6}$ Quotient rule: $-3 - (-2) = -1$

   $\quad\quad\quad = \dfrac{5}{3w}$ Reduce and rewrite w^{-1} with a positive exponent.

c) $\dfrac{x^{-1}x^{-3}y^5}{x^{-2}y^2} = \dfrac{x^{-4}y^5}{x^{-2}y^2}$ Product rule: $-1 + (-3) = -4$

   $\quad\quad\quad = x^{-2}y^3$ Quotient rule: $-4 - (-2) = -2, 5 - 2 = 3$

   $\quad\quad\quad = \dfrac{y^3}{x^2}$ Rewrite x^{-2} with a positive exponent.

> Now do Exercises 43–46

Note that in Examples 5 and 6 we could first eliminate all negative exponents as we did in Example 4. However, that approach is not necessary and would be more work than simply applying the product and quotient rules. Remember that the bases must be identical for the product or the quotient rule.

‹6› Scientific Notation

Many of the numbers that are encountered in science are either very large or very small. For example, the distance from the earth to the sun is 93,000,000 miles, and a hydrogen atom weighs 0.0000000000000000000000017 gram. Scientific notation provides a convenient way of writing very large and very small numbers. In scientific notation, the distance from the earth to the sun is 9.3×10^7 miles and a hydrogen atom weighs 1.7×10^{-24} gram. In scientific notation the times symbol, $\times$, is used to indicate multiplication.

> **Scientific Notation**
>
> A number is in **scientific notation** if it is written in the form
>
> $$a \times 10^n$$
>
> where $1 \le a < 10$ and n is a positive or negative integer.

Converting a number from scientific notation to standard notation is simply a matter of multiplication.

E X A M P L E   **7**

Scientific notation to standard notation

Write each number using standard notation.

a) 7.62×10^5 **b)** 6.35×10^{-4}

Solution

a) Multiplying a number by 10^5 moves the decimal point five places to the right:

$$7.62 \times 10^5 = 762000. = 762,000$$

b) Multiplying a number by 10^{-4} or 0.0001 moves the decimal point four places to the left:

$$6.35 \times 10^{-4} = 0.000635 = 0.000635$$

Now do Exercises 71–78

‹ **Calculator Close-Up** ›

In normal mode, display a number in scientific notation and press ENTER to convert to standard notation. Use the EE key to get the exponent for the built-in scientific notation.

```
7.62E5
             762000
7.62*10^5
             762000
```

The procedure for converting a number from scientific notation to standard notation is summarized as follows.

Strategy for Converting to Standard Notation

1. Determine the number of places to move the decimal point by examining the exponent on the 10.

2. Move to the right for a positive exponent and to the left for a negative exponent.

A number in scientific notation is written as a product of a number between 1 and 10, and a power of 10. Numbers in scientific notation are written with only one digit to the left of the decimal point. A number larger than 10 is written with a positive power of 10, and a positive number smaller than 1 is written with a negative power of 10. Note that 1000 (a power of 10) could be written as 1×10^3 or simply 10^3. Numbers between 1 and 10 are usually not written in scientific notation. To convert to scientific notation, we reverse the strategy for converting from scientific notation.

Strategy for Converting to Scientific Notation

1. Count the number of places (n) that the decimal point must be moved so that it will follow the first nonzero digit of the number.

2. If the original number was larger than 10, use 10^n.

3. If the original number was smaller than 1, use 10^{-n}.

E X A M P L E   **8**

Standard notation to scientific notation

Convert each number to scientific notation.

a) 934,000,000   **b)** 0.0000025

Solution

a) In 934,000,000 the decimal point must be moved eight places to the left to get it to
follow 9, the first nonzero digit.

$$934{,}000{,}000 = 9.34 \times 10^8 \quad \text{Use 8 because } 934{,}000{,}000 > 10.$$

b) The decimal point in 0.0000025 must be moved six places to the right to get the
2 to the left of the decimal point.

$$0.0000025 = 2.5 \times 10^{-6} \quad \text{Use } -6 \text{ because } 0.0000025 < 1.$$

Now do Exercises 79–86

We can perform computations with numbers in scientific notation by using the
rules of exponents on the powers of 10.

E X A M P L E   9

Using scientific notation in computations

Evaluate each expression without using a calculator. Express each answer in scientific
notation.

a) $(2 \times 10^7)(6.3 \times 10^{-11})$ **b)** $\dfrac{7 \times 10^{13}}{2 \times 10^6}$ **c)** $\dfrac{(10{,}000)(0.000025)}{0.000005}$

Solution

a) $(2 \times 10^7)(6.3 \times 10^{-11}) = 2 \cdot 6.3 \cdot 10^7 \cdot 10^{-11}$

$\qquad\qquad\qquad\qquad = 12.6 \times 10^{-4} \quad$ Commutative and associative properties

$\qquad\qquad\qquad\qquad = 1.26 \times 10^1 \times 10^{-4} \quad$ Write 12.6 in scientific notation.

$\qquad\qquad\qquad\qquad = 1.26 \times 10^{-3}$

b) $\dfrac{7 \times 10^{13}}{2 \times 10^6} = \dfrac{7}{2} \cdot \dfrac{10^{13}}{10^6} = 3.5 \times 10^7$

c) $\dfrac{(10{,}000)(0.000025)}{0.000005} = \dfrac{\left(1 \times 10^4\right)\left(2.5 \times 10^{-5}\right)}{5 \times 10^{-6}}$

$\qquad\qquad\qquad\qquad\quad = \dfrac{2.5}{5} \cdot \dfrac{10^4 \cdot 10^{-5}}{10^{-6}} \quad$ Commutative and associative properties

$\qquad\qquad\qquad\qquad\quad = 0.5 \times 10^5$

$\qquad\qquad\qquad\qquad\quad = 5 \times 10^{-1} \times 10^5 \quad$ Write 0.5 in scientific notation.

$\qquad\qquad\qquad\qquad\quad = 5 \times 10^4$

Now do Exercises 87–94

E X A M P L E   10

Counting hydrogen atoms

If the weight of hydrogen is 1.7×10^{-24} gram per atom, then how many hydrogen atoms
are there in one kilogram of hydrogen?

<Helpful Hint>

You can divide 1×10^3 by 1.7×10^{-24} without a calculator by dividing 1 by 1.7 to get 0.59 and 10^3 by 10^{-24} to get 10^{27}. Then convert

$$0.59 \times 10^{27}$$

to

$$5.9 \times 10^{-1} \times 10^{27}$$

or

$$5.9 \times 10^{26}.$$

Solution

There are 1000 or 1×10^3 grams in one kilogram. So to find the number of hydrogen atoms in one kilogram of hydrogen, we divide 1×10^3 by 1.7×10^{-24}:

$$\frac{1 \times 10^3 \text{ g/kg}}{1.7 \times 10^{-24} \text{ g/atom}} \approx 5.9 \times 10^{26} \text{ atom per kilogram (atom/kg)}$$

To divide by grams per atom, we invert and multiply: $\frac{g}{kg} \cdot \frac{atom}{g} = \frac{atom}{kg}$. Keeping track of the units as we did here helps us to be sure that we performed the correct operation. So there are approximately 5.9×10^{26} hydrogen atoms in one kilogram of hydrogen.

> Now do Exercises 101–106

Warm-Ups ▼

Fill in the blank.

1. An _____ expression has the form a^n.
2. According to the _____ rule, $a^m a^n = a^{m+n}$.
3. According to the _____ rule, $a^m/a^n = a^{m-n}$.
4. In _____ notation numbers are written as a product of a number between 1 and 10 and a power of 10.
5. Any nonzero number to the power ____ is 1.

True or false?

6. $3^5 \cdot 3^4 = 3^9$
7. If $x \neq 0$, then $2x^{-4} = \frac{1}{2x^4}$.

8. $10^{-3} = 0.0001$
9. If $x \neq 0$, then $\frac{x^5}{x^{-2}} = x^3$.
10. $\frac{2^5}{2^{-2}} = 2^7$
11. $2^3 \cdot 5^2 = 10^5$
12. $-2^{-2} = -\frac{1}{4}$
13. $46.7 \times 10^5 = 4.67 \times 10^6$
14. $0.512 \times 10^{-3} = 5.12 \times 10^{-4}$
15. $\frac{8 \times 10^{30}}{2 \times 10^{-5}} = 4 \times 10^{25}$

Exercises 5.1

<Study Tips>

- Students who have difficulty with a subject often schedule a class that meets one day per week so that they do not have to see it too often. It is better to be in a class that meets more often for shorter time periods.
- Students who explain things to others often learn from it. If you must work on math alone, try explaining things to yourself.

<1> Positive and Negative Exponents

Each of Exercises 1–12 contains six similar expressions. For each exercise evaluate the six expressions and note their similarities and differences. See Example 1.

1. $2^2, -2^2, (-2)^2, 2^{-2}, -2^{-2}, (-2)^{-2}$

2. $3^2, -3^2, (-3)^2, 3^{-2}, -3^{-2}, (-3)^{-2}$

3. $2^3, -2^3, (-2)^3, 2^{-3}, -2^{-3}, (-2)^{-3}$

4. $4^3, -4^3, (-4)^3, 4^{-3}, -4^{-3}, (-4)^{-3}$

5. $\dfrac{1}{5^2}, \dfrac{1}{-5^2}, \dfrac{1}{(-5)^2}, \dfrac{1}{5^{-2}}, \dfrac{1}{-5^{-2}}, \dfrac{1}{(-5)^{-2}}$

6. $\dfrac{1}{4^2}, \dfrac{1}{-4^2}, \dfrac{1}{(-4)^2}, \dfrac{1}{4^{-2}}, \dfrac{1}{-4^{-2}}, \dfrac{1}{(-4)^{-2}}$

7. $7^1, -7^1, (-7)^1, 7^{-1}, -7^{-1}, (-7)^{-1}$

8. $\left(\dfrac{1}{6}\right)^1, \left(-\dfrac{1}{6}\right)^1, -\left(\dfrac{1}{6}\right)^1, \left(\dfrac{1}{6}\right)^{-1}, \left(-\dfrac{1}{6}\right)^{-1}, -\left(\dfrac{1}{6}\right)^{-1}$

9. $\left(\dfrac{1}{2}\right)^2, \left(-\dfrac{1}{2}\right)^2, -\left(\dfrac{1}{2}\right)^2, \left(\dfrac{1}{2}\right)^{-2}, \left(-\dfrac{1}{2}\right)^{-2}, -\left(\dfrac{1}{2}\right)^{-2}$

10. $\left(\dfrac{1}{3}\right)^2, \left(-\dfrac{1}{3}\right)^2, -\left(\dfrac{1}{3}\right)^2, \left(\dfrac{1}{3}\right)^{-2}, \left(-\dfrac{1}{3}\right)^{-2}, -\left(\dfrac{1}{3}\right)^{-2}$

11. $\left(\dfrac{2}{3}\right)^3, \left(-\dfrac{2}{3}\right)^3, -\left(\dfrac{2}{3}\right)^3, \left(\dfrac{2}{3}\right)^{-3}, \left(-\dfrac{2}{3}\right)^{-3}, -\left(\dfrac{2}{3}\right)^{-3}$

12. $\left(\dfrac{3}{5}\right)^3, \left(-\dfrac{3}{5}\right)^3, -\left(\dfrac{3}{5}\right)^3, \left(\dfrac{3}{5}\right)^{-3}, \left(-\dfrac{3}{5}\right)^{-3}, -\left(\dfrac{3}{5}\right)^{-3}$

⟨2⟩ The Product Rule for Exponents

For Exercises 13–62 assume that the variables represent nonzero real numbers and use only positive exponents in your answers.

Simplify. See Example 2.

13. $2^5 \cdot 2^{12}$

14. $3^{15} \cdot 3^{-3}$

15. $-2x^{-7} \cdot 3x$

16. $-5a \cdot 6a^{-12}$

17. $-7b^{-7}(-3b^{-3})$

18. $-\dfrac{1}{2}w^{-4} \cdot (-6w^{-2})$

⟨3⟩ Zero Exponent

Simplify each expression. See Example 3.

19. $3^0, -3^0, (-3)^0, -(-3)^0$

20. $2a^0, -2a^0, (-2a)^0, -2(-a)^0$

21. $(2 + 3)^0, 2^0 + 3^0, (2^0 + 3)^0$

22. $(4 - 9)^0, 4^0 - 9^0, 4^0 - 9$

23. $3st^0, 3(st)^0, (3st)^0$

24. $-4xy^0, -4x^0y, -(4xy)^0$

25. $2w^{-3}(w^7 \cdot w^{-4})$

26. $5y^2z(y^{-3}z^{-1})$

⟨4⟩ Changing the Sign of an Exponent

Simplify. See Example 4.

27. $\dfrac{2}{4^{-2}}$

28. $\dfrac{5}{10^{-3}}$

29. $\dfrac{3^{-1}}{10^{-2}}$

30. $\dfrac{2y^{-2}}{3^{-1}}$

31. $\dfrac{2x^{-3}(4x)}{5y^{-2}}$

32. $\dfrac{5^{-2}xy^{-3}}{3x^{-2}}$

33. $\dfrac{4^{-2}x^3x^{-6}}{3x^{-3}x^2}$

34. $\dfrac{3y^{-4}y^{-6}}{2^{-3}y^2y^{-7}}$

⟨5⟩ The Quotient Rule for Exponents

Simplify each expression. See Examples 5 and 6.

35. $\dfrac{x^5}{x^3}$

36. $\dfrac{a^8}{a^3}$

37. $\dfrac{3^6}{3^{-2}}$

38. $\dfrac{6^2}{6^{-5}}$

39. $\dfrac{4a^{-5}}{12a^{-2}}$

40. $\dfrac{-3a^{-3}}{-21a^{-4}}$

41. $\dfrac{-6w^{-5}}{2w^3}$

42. $\dfrac{10x^{-6}}{-2x^2}$

43. $\dfrac{3^3 w^{-2} w^5}{3^{-5} w^{-3}}$ **44.** $\dfrac{2^{-3} w^5}{2^5 w^3 w^{-7}}$

45. $\dfrac{3x^{-6} \cdot x^2 y^{-1}}{6x^{-5} y^{-2}}$ **46.** $\dfrac{2r^{-3} t^{-1}}{10 r^5 t^2 \cdot t^{-3}}$

Miscellaneous

Use the rules of exponents to simplify each expression.

47. $3^{-1}\left(\dfrac{1}{3}\right)^{-3}$ **48.** $2^{-2}\left(\dfrac{1}{4}\right)^{-3}$

49. $-2^4 + \left(\dfrac{1}{2}\right)^{-1}$ **50.** $-3^4 - (-3)^4$

51. $-(-2)^{-3} \cdot 2^{-1}$ **52.** $-(-3)^{-1} \cdot 9^{-1}$

53. $7 \cdot 2^{-3} - 2 \cdot 4^{-1}$ **54.** $5 \cdot 3^{-2} - 2 \cdot 5^0 \cdot 3^{-1}$

55. $\left(1 + 2^{-1}\right)^{-2}$ **56.** $\left(2^{-1} + 2^{-1}\right)^{-2}$

57. $2x^2 \cdot 5x^{-5}$ **58.** $2x^2 \cdot 5y^{-5}$

59. $\dfrac{-3a^5\left(-2a^{-1}\right)}{6a^3}$

60. $\dfrac{6a\left(-ab^{-2}\right)}{-2a^2 b^{-3}}$

61. $\dfrac{\left(-3x^3 y^2\right)\left(-2xy^{-3}\right)}{-9x^2 y^{-5}}$

62. $\dfrac{\left(-2x^{-5} y\right)\left(-3xy^6\right)}{-6x^{-6} y^2}$

For each equation, find the integer that can be used as the exponent to make the equation correct.

63. $8 = 2^?$ **64.** $27 = 3^?$

65. $\dfrac{1}{4} = 2^?$ **66.** $\dfrac{1}{125} = 5^?$

67. $16 = \left(\dfrac{1}{2}\right)^?$ **68.** $81 = \left(\dfrac{1}{3}\right)^?$

69. $10^? = 0.001$ **70.** $10^? = 10,000$

⟨6⟩ Scientific Notation

Write each number in standard notation. See Example 7. See the Strategy for Converting to Standard Notation box on page 301.

71. 4.86×10^8 **72.** 3.80×10^2

73. 2.37×10^{-6} **74.** 1.62×10^{-3}

75. 4×10^6 **76.** 496×10^3

77. 5×10^{-6} **78.** 48×10^{-3}

Write each number in scientific notation. See Example 8. See the Strategy for Converting to Scientific Notation box on page 301.

79. $320,000$ **80.** $43,298,000$

81. 0.00000071 **82.** 0.00000894

83. 0.0000703 **84.** $8,200,100$

85. 205×10^5 **86.** 0.403×10^{-9}

Evaluate each expression using scientific notation without a calculator. See Example 9.

87. $(4000)(5000)(0.0003)$

88. $(50,000)(0.00002)(100)$

89. $\dfrac{(5,000,000)(0.0003)}{2000}$

90. $\dfrac{(6000)(0.00004)}{(30,000)(0.002)}$

91. $\dfrac{6 \times 10^{40}}{2 \times 10^{18}}$

92. $\dfrac{4.6 \times 10^{12}}{2.3 \times 10^5}$

93. $\dfrac{\left(-4 \times 10^5\right)\left(6 \times 10^{-9}\right)}{2 \times 10^{-16}}$

94. $\dfrac{\left(4.8 \times 10^{-3}\right)\left(5 \times 10^{-8}\right)}{\left(1.2 \times 10^{-6}\right)\left(2 \times 10^{12}\right)}$

Evaluate each expression using a calculator. Write answers in scientific notation. Round the decimal part to three decimal places.

95. $(4.3 \times 10^9)(3.67 \times 10^{-5})$

96. $(2.34 \times 10^6)(8.7 \times 10^5)$

97. $(4.37 \times 10^{-6}) + (8.75 \times 10^{-5})$

98. $(6.72 \times 10^5) + (8.98 \times 10^6)$

99. $\dfrac{(5.6 \times 10^{14})^2(3.2 \times 10^{-6})}{(6.4 \times 10^{-3})^3}$

100. $\dfrac{(3.51 \times 10^{-6})^3(4000)^5}{2\pi}$

Solve each problem. Round to three decimal places. See Example 10.

101. *Distance to the sun.* The distance from the earth to the sun is 93 million miles. Express this distance in feet using scientific notation (1 mile = 5280 feet).

Figure for Exercise 101

102. *Traveling time.* The speed of light is 9.83569×10^8 feet per second. How long does it take light to get from the sun to the earth? (See Exercise 101.)

103. *Space travel.* How long does it take a spacecraft traveling 1.2×10^5 kilometers per second to travel 4.6×10^{12} kilometers?

104. *Diameter of a dot.* If the circumference of a very small circle is 2.35×10^{-8} meter, then what is the diameter of the circle?

105. *Solid waste per person.* In 1960 the 1.80863×10^8 people in the United States generated 8.71×10^7 tons of municipal solid waste (Environmental Protection Agency, www.epa.gov). How many pounds per person per day were generated in 1960? Round to the nearest tenth.

106. *An increasing problem.* According to the EPA, in 2009 the 3.0631×10^8 people in the United States generated 5.2×10^{11} pounds of solid municipal waste.

a) How many pounds per person per day were generated in 2009? Round to the nearest tenth.

b) Use the graph to predict the number of pounds per person per day that will be generated in the year 2020.

Figure for Exercises 105 and 106

Getting More Involved

107. *Exploration*

a) Using pairs of integers, find values for m and n for which $2^m \cdot 3^n = 6^{m+n}$.

b) For which values of m and n is it true that $2^m \cdot 3^n \neq 6^{m+n}$?

108. *Cooperative learning*

Work in a group to find the units digit of 3^{99} and explain how you found it.

109. *Discussion*

What is the difference between $-a^n$ and $(-a)^n$, where n is an integer? For which values of a and n do they have the same value, and for which values of a and n do they have different values?

110. *Exploration*

If $a + b = a$, then what can you conclude about b? Use scientific notation on your calculator to find $5 \times 10^{20} + 3 \times 10^6$. Explain why your calculator displays the answer that it gets.

You have probably experienced the reflection time of sound waves in the form of an echo. For example, if you shout in a large auditorium, the sound takes a noticeable amount of time to reach a distant wall and travel back to your ear. We know that sound travels about 1000 feet per second. So if you could measure the amount of time that it takes for the sound to return to your ear, you could use the simple formula $D = RT$ to determine how far the sound had traveled. This is the same principle used in laser speed guns, one of the newest instruments used by police to catch speeders.

A laser speed gun measures the amount of time for light to reach a car and reflect back to the gun. Light from a laser speed gun travels at 9.8×10^8 feet per second. A laser speed gun shoots a very short burst of infrared laser light and then waits for it to reflect off the vehicle. The gun counts the number of nanoseconds it takes for the round trip, and by dividing by 2 it can use $D = RT$ to calculate the distance to the car. But that does not give the speed of the car. The gun must send a second burst of light and calculate the distance again. Using $R = D/T$, the gun divides the change in distance by the amount of time between light bursts to get the speed of the car. Actually, the gun takes about 1000 samples per second, each time dividing the change in distance by the change in time to determine the speed with a very high degree of accuracy.

The advantage of a laser speed gun is that the width of the laser beam is very small. Even at a range of about 1000 feet the beam is only 3 feet wide. So the laser gun can target a specific vehicle and it cannot be detected by radar detectors. The disadvantage is that the officer has to aim a laser speed gun. A radar speed gun does not need to be aimed.

5.2   The Power Rules

In Section 5.1, you learned some of the basic rules for working with exponents. All of the rules of exponents are designed to make it easier to work with exponential expressions. In this section, we will extend our list of rules to include three new ones.

⟨1⟩ Raising an Exponential Expression to a Power

An expression such as $(x^3)^2$ consists of the exponential expression x^3 raised to the power 2. We can use known rules to simplify this expression.

$$(x^3)^2 = x^3 \cdot x^3 \quad \text{Exponent 2 indicates two factors of } x^3.$$
$$= x^6 \qquad \text{Product rule: } 3 + 3 = 6$$

Note that the exponent 6 is the *product* of the exponents 2 and 3. This example illustrates the **power of a power rule.**

> **Power of a Power Rule**
>
> If *m* and *n* are any integers and $a \neq 0$, then
>
> $$(a^m)^n = a^{mn}.$$

‹ Calculator Close-Up ›

A graphing calculator cannot prove that the power of a power rule is correct, but it can provide numerical support for it.

```
(2^3)^5
            3.2768E4
2^15
            3.2768E4
```

E X A M P L E   1

Using the power of a power rule

Use the rules of exponents to simplify each expression. Write the answer with positive exponents only. Assume all variables represent nonzero real numbers.

a) $(2^3)^5$

b) $(x^2)^{-6}$

c) $3(y^{-3})^{-2}y^{-5}$

d) $\dfrac{(x^2)^{-1}}{(x^{-3})^3}$

Solution

a) $(2^3)^5 = 2^{15}$ Power of a power rule: $3 \cdot 5 = 15$

b) $(x^2)^{-6} = x^{-12}$ Power of a power rule: $2(-6) = -12$

$\quad = \dfrac{1}{x^{12}}$ Definition of a negative exponent

c) $3(y^{-3})^{-2}y^{-5} = 3y^6 y^{-5}$ Power of a power rule: $(-3)(-2) = 6$

$\quad = 3y$ Product rule for exponents: $6 + (-5) = 1$

d) $\dfrac{(x^2)^{-1}}{(x^{-3})^3} = \dfrac{x^{-2}}{x^{-9}}$ Power of a power rule: $2(-1) = -2$, $-3(3) = -9$

$\quad = x^7$ Quotient rule for exponents: $-2 - (-9) = 7$

> **Now do Exercises 1–12**

‹ Calculator Close-Up ›

You can use a graphing calculator to illustrate the power of a product rule.

```
(2*3)^5
            7776
2^5*3^5
            7776
```

‹2› Raising a Product to a Power

Consider how we would simplify a product raised to a positive power and a product raised to a negative power using known rules.

$$\overset{\text{3 factors of } 2x}{\overbrace{(2x)^3 = 2x \cdot 2x \cdot 2x}} = 2^3 \cdot x^3 = 8x^3$$

$$(ay)^{-3} = \frac{1}{(ay)^3} = \frac{1}{(ay)(ay)(ay)} = \frac{1}{a^3 y^3} = a^{-3} y^{-3}$$

In each of these cases the original exponent is applied to each factor of the product. These examples illustrate the **power of a product rule.**

> **Power of a Product Rule**
>
> If *a* and *b* are nonzero real numbers and *n* is any integer, then
>
> $$(ab)^n = a^n \cdot b^n.$$

EXAMPLE   2

Using the power of a product rule

Simplify. Assume the variables represent nonzero real numbers. Write the answers with positive exponents only.

   a) $(-3x)^4$ **b)** $(-2x^2)^3$ **c)** $(3x^{-2}y^3)^{-2}$

Solution

  a) $(-3x)^4 = (-3)^4 x^4$ Power of a product rule

          $= 81x^4$

  b) $(-2x^2)^3 = (-2)^3 (x^2)^3$ Power of a product rule

          $= -8x^6$ Power of a power rule: $2 \cdot 3 = 6$

  c) $(3x^{-2}y^3)^{-2} = (3)^{-2}(x^{-2})^{-2}(y^3)^{-2}$ Power of a product rule

          $= \dfrac{1}{9}x^4 y^{-6}$ Power of a power rule: $-2(-2) = 4$, $3(-2) = -6$

          $= \dfrac{x^4}{9y^6}$ Rewrite y^{-6} with a positive exponent.

> Now do Exercises 13–24

‹ **Calculator Close-Up** ›

You can use a graphing calculator to illustrate the power of a quotient rule.

‹3› **Raising a Quotient to a Power**

Now consider an example of applying known rules to a power of a quotient:

$$\left(\frac{x}{5}\right)^3 = \frac{x}{5} \cdot \frac{x}{5} \cdot \frac{x}{5} = \frac{x^3}{5^3}$$

We get a similar result with a negative power:

$$\left(\frac{x}{5}\right)^{-3} = \left(\frac{5}{x}\right)^3 = \frac{5}{x} \cdot \frac{5}{x} \cdot \frac{5}{x} = \frac{5^3}{x^3} = \frac{x^{-3}}{5^{-3}}$$

In each of these cases the original exponent applies to both the numerator and denominator. These examples illustrate **the power of a quotient rule.**

Power of a Quotient Rule

If a and b are nonzero real numbers and n is any integer, then

$$\left(\frac{a}{b}\right)^n = \frac{a^n}{b^n}.$$

EXAMPLE   3

Using the power of a quotient rule

Use the rules of exponents to simplify each expression. Write your answers with positive exponents only. Assume the variables are nonzero real numbers.

  a) $\left(\dfrac{x}{2}\right)^3$ **b)** $\left(-\dfrac{2x^3}{3y^2}\right)^3$ **c)** $\left(\dfrac{x^{-2}}{2^3}\right)^{-1}$ **d)** $\left(-\dfrac{3}{4x^3}\right)^{-2}$

‹ **Helpful Hint** ›

The exponent rules in this section apply to expressions that involve only multiplication and division. This is not too surprising since exponents, multiplication, and division are closely related. Recall that $a^3 = a \cdot a \cdot a$ and $a \div b = a \cdot b^{-1}$.

Solution

a) $\left(\dfrac{x}{2}\right)^3 = \dfrac{x^3}{2^3}$ Power of a quotient rule

$\phantom{\left(\dfrac{x}{2}\right)^3} = \dfrac{x^3}{8}$

b) $\left(-\dfrac{2x^3}{3y^2}\right)^3 = \dfrac{(-2)^3 x^9}{3^3 y^6}$ Because $(x^3)^3 = x^9$ and $(y^2)^3 = y^6$

$\phantom{\left(-\dfrac{2x^3}{3y^2}\right)^3} = \dfrac{-8x^9}{27y^6}$ $(-2)^3 = -8$, $3^3 = 27$

$\phantom{\left(-\dfrac{2x^3}{3y^2}\right)^3} = -\dfrac{8x^9}{27y^6}$ Rewrite with a negative sign in front.

c) $\left(\dfrac{x^{-2}}{2^3}\right)^{-1} = \dfrac{x^2}{2^{-3}} = 8x^2$ $\dfrac{1}{2^{-3}} = 2^3 = 8$

d) $\left(-\dfrac{3}{4x^3}\right)^{-2} = \dfrac{(-3)^{-2}}{4^{-2}x^{-6}} = \dfrac{4^2 x^6}{(-3)^2} = \dfrac{16x^6}{9}$

> Now do Exercises 25–32

A fraction to a negative power can be simplified by using the power of a quotient rule, as in Example 3. Another method is to find the reciprocal of the fraction first and then use the power of a quotient rule, as shown in Example 4.

E X A M P L E   4

Negative powers of fractions

Simplify. Assume the variables are nonzero real numbers, and write the answers with positive exponents only.

a) $\left(\dfrac{3}{4}\right)^{-3}$ **b)** $\left(\dfrac{x^2}{5}\right)^{-2}$ **c)** $\left(-\dfrac{2y^3}{3}\right)^{-2}$

Solution

a) $\left(\dfrac{3}{4}\right)^{-3} = \left(\dfrac{4}{3}\right)^3$ The reciprocal of $\dfrac{3}{4}$ is $\dfrac{4}{3}$.

$\phantom{\left(\dfrac{3}{4}\right)^{-3}} = \dfrac{4^3}{3^3}$ Power of a quotient rule

$\phantom{\left(\dfrac{3}{4}\right)^{-3}} = \dfrac{64}{27}$ $4^3 = 64$, $3^3 = 27$

b) There is more than one way to simplify these expressions. Taking the reciprocal of the fraction first we get

$$\left(\dfrac{x^2}{5}\right)^{-2} = \left(\dfrac{5}{x^2}\right)^2 = \dfrac{5^2}{(x^2)^2} = \dfrac{25}{x^4}.$$

Applying the power of a quotient rule first (as in Example 3) we get

$$\left(\dfrac{x^2}{5}\right)^{-2} = \dfrac{(x^2)^{-2}}{5^{-2}} = \dfrac{x^{-4}}{5^{-2}} = \dfrac{5^2}{x^4} = \dfrac{25}{x^4}.$$

c) $\left(-\dfrac{2y^3}{3}\right)^{-2} = \left(-\dfrac{3}{2y^3}\right)^2 = \dfrac{9}{4y^6}$

> Now do Exercises 33–40

‹4› **Variable Exponents**

So far, we have used the rules of exponents only on expressions with integral exponents. However, we can use the rules to simplify expressions having variable exponents that represent integers.

E X A M P L E   5

Expressions with variables as exponents
Simplify. Assume the variables represent integers.

     a) $3^{4y} \cdot 3^{5y}$ **b)** $\left(5^{2x}\right)^{3x}$ **c)** $\left(\dfrac{2^n}{3^m}\right)^{5n}$

Solution

  a) $3^{4y} \cdot 3^{5y} = 3^{9y}$ Product rule: $4y + 5y = 9y$

  b) $\left(5^{2x}\right)^{3x} = 5^{6x^2}$ Power of a power rule: $2x \cdot 3x = 6x^2$

  c) $\left(\dfrac{2^n}{3^m}\right)^{5n} = \dfrac{\left(2^n\right)^{5n}}{\left(3^m\right)^{5n}}$ Power of a quotient rule

                $= \dfrac{2^{5n^2}}{3^{5mn}}$ Power of a power rule: $n(5n) = 5n^2$, $m(5n) = 5mn$

> **Now do Exercises 41–48**

‹5› **Summary of the Rules**

The definitions and rules that were introduced in the last two sections are summarized in the following box.

Rules for Integral Exponents

For these rules, m and n are integers and a and b are nonzero real numbers.

1. $a^{-n} = \dfrac{1}{a^n}$ Definition of negative exponent

2. $a^{-n} = \left(\dfrac{1}{a}\right)^n$, $a^{-1} = \dfrac{1}{a}$, and $\dfrac{1}{a^{-n}} = a^n$ Negative exponent rules

3. $a^0 = 1$ Definition of zero exponent

4. $a^m a^n = a^{m+n}$ Product rule for exponents

5. $\dfrac{a^m}{a^n} = a^{m-n}$ Quotient rule for exponents

6. $\left(a^m\right)^n = a^{mn}$ Power of a power rule

7. $(ab)^n = a^n b^n$ Power of a product rule

8. $\left(\dfrac{a}{b}\right)^n = \dfrac{a^n}{b^n}$ Power of a quotient rule

‹ **Helpful Hint** ›

In this section we use the amount formula for interest compounded annually only. But you probably have money in a bank where interest is compounded daily. In this case r represents the daily rate (APR/365) and n is the number of days that the money is on deposit.

‹6› Applications

Both positive and negative exponents occur in formulas used in investment situations. The amount of money invested is the **principal,** and the value of the principal after a certain time period is the **amount.** Interest rates are annual percentage rates (APRs).

Amount Formula

The amount A of an investment of P dollars with interest rate r compounded annually for n years is given by the formula

$$A = P(1 + r)^n.$$

E X A M P L E   6

Finding the amount

According to Fidelity Investments of Boston, U.S. common stocks have returned an average of 10% annually since 1926. If your great-grandfather had invested $100 in the stock market in 1926 and obtained the average increase each year, then how much would the investment be worth in the year 2016 after 90 years of growth?

Solution

Use $n = 90$, $P = \$100$, and $r = 0.10$ in the amount formula:

$$A = P(1 + r)^n$$
$$A = 100(1 + 0.10)^{90}$$
$$= 100(1.1)^{90}$$
$$\approx 531,302.26$$

So $100 invested in 1926 will amount to $531,302.26 in 2016.

Now do Exercises 83–84

‹ **Calculator Close-Up** ›

With a graphing calculator you can enter $100(1 + 0.10)^{90}$ almost as it appears in print.

```
100(1+.10)^90
       531302.2612
```

When we are interested in the principal that must be invested today to grow to a certain amount, the principal is called the **present value** of the investment. We can find a formula for present value by solving the amount formula for P:

$$A = P(1 + r)^n$$
$$P = \frac{A}{(1 + r)^n} \qquad \text{Divide each side by } (1 + r)^n.$$
$$P = A(1 + r)^{-n} \qquad \text{Definition of a negative exponent}$$

Present Value Formula

The present value P that will amount to A dollars after n years with interest compounded annually at annual interest rate r is given by

$$P = A(1 + r)^{-n}.$$

E X A M P L E   7

Finding the present value

If your great-grandfather wanted you to have $1,000,000 in 2016, then how much could he have invested in the stock market in 1926 to achieve this goal? Assume he could get the average annual return of 10% (from Example 6) for 90 years.

Solution

Use $r = 0.10$, $n = 90$, and $A = 1,000,000$ in the present value formula:

$$P = A(1 + r)^{-n}$$
$$= 1,000,000(1 + 0.10)^{-90}$$
$$= 1,000,000(1.1)^{-90}$$
$$\approx 188.22$$

An investment of $188.22 in 1926 would grow to $1,000,000 in 90 years at a rate of 10% compounded annually.

> Now do Exercises 85–88

Warm-Ups ▼

Fill in the blank.

1. According to the power of a _____ rule $(a^m)^n = a^{mn}$.
2. According to the power of a _____ rule $(ab)^m = a^m b^m$.
3. According to the power of a _____ rule $(a/b)^m = a^m/b^m$.
4. The _____ is the amount of money invested initially.
5. If interest is compounded _____, where i is the annual interest rate and P is the principal, then the amount after n years is $P(1 + i)^n$.
6. The _____ value of A dollars in n years at an annual interest rate i compounded annually is $A(1 + i)^{-n}$.

True or false?

7. $(2^2)^3 = 2^5$
8. $(2^{-3})^{-1} = 8$
9. If $x \neq 0$, then $(x^{-3})^3 = x^{-9}$.
10. $2^3 \cdot 2^3 = (2^3)^2$
11. $(2x)^3 = 6x^3$
12. $(-3y^3)^2 = 9y^9$
13. $\left(\dfrac{2}{3}\right)^{-1} = \dfrac{3}{2}$
14. $\left(\dfrac{2}{3}\right)^3 = \dfrac{8}{27}$
15. $\left(\dfrac{x^2}{2}\right)^3 = \dfrac{x^6}{8}$
16. If $x \neq 0$, then $\left(\dfrac{2}{x}\right)^{-2} = \dfrac{x^2}{4}$.

Exercises

5.2

‹ Study Tips ›

- Keep reviewing. When you are done with your current assignment, go back and work a few problems from the past. You will be amazed at how much your knowledge will improve with a regular review.
- Play offensive math not defensive math. A student who takes an active approach and knows the usual questions and answers is playing offensive math. Don't wait for a question to hit you on the head.

‹1› Raising an Exponential Expression to a Power

For Exercises 1–68 assume that the variables represent nonzero real numbers and use only positive exponents in your answers.

Use the rules of exponents to simplify. See Example 1.

1. $(2^2)^3$
2. $(3^2)^2$
3. $(y^2)^5$
4. $(x^6)^2$
5. $(x^2)^{-4}$
6. $(x^{-2})^7$

7. $\left(m^{-3}\right)^{-6}$

8. $\left(a^{-3}\right)^{-3}$

9. $\left(x^{-2}\right)^3\left(x^{-3}\right)^{-2}$

10. $\left(m^{-3}\right)^{-1}\left(m^2\right)^{-4}$ **11.** $\dfrac{\left(x^3\right)^{-4}}{\left(x^2\right)^{-5}}$

12. $\dfrac{\left(a^2\right)^{-3}}{\left(a^{-2}\right)^4}$

43. $\left(2^{-3w}\right)^{-2w}$

44. $6^{8x}\cdot\left(6^{2x}\right)^{-3}$

45. $\dfrac{7^{2m+6}}{7^{m+3}}$

46. $\dfrac{4^{-3p}}{4^{-4p}}$

47. $8^{2a-1}\cdot\left(8^{a+4}\right)^3$

48. $\left(5^{4-3y}\right)^3\left(5^{y-2}\right)^2$

⟨2⟩ Raising a Product to a Power

Simplify. See Example 2.

13. $(-9y)^2$

14. $(-2a)^3$

 15. $\left(-5w^3\right)^2$

16. $\left(-2w^{-5}\right)^3$

17. $\left(x^3y^{-2}\right)^3$

18. $\left(a^2b^{-3}\right)^2$

19. $\left(3ab^{-1}\right)^{-2}$

20. $\left(2x^{-1}y^2\right)^{-3}$

21. $\dfrac{2xy^{-2}}{\left(3x^2y\right)^{-1}}$

22. $\dfrac{3ab^{-1}}{\left(5ab^2\right)^{-1}}$

23. $\dfrac{(2ab)^{-2}}{2ab^2}$

24. $\dfrac{(3xy)^{-3}}{3xy^3}$

⟨3⟩ Raising a Quotient to a Power

Simplify. See Example 3.

25. $\left(\dfrac{w}{2}\right)^3$

26. $\left(\dfrac{m}{5}\right)^2$

27. $\left(-\dfrac{3a}{4}\right)^3$

28. $\left(-\dfrac{2}{3b}\right)^4$

29. $\left(\dfrac{2x^{-1}}{y}\right)^{-2}$

30. $\left(\dfrac{2a^2b}{3}\right)^{-3}$

 31. $\left(\dfrac{-3x^3}{y}\right)^{-2}$

32. $\left(\dfrac{-2y^2}{x}\right)^{-3}$

Simplify. See Example 4.

33. $\left(\dfrac{2}{5}\right)^{-2}$

34. $\left(\dfrac{3}{4}\right)^{-2}$

35. $\left(-\dfrac{1}{2}\right)^{-2}$

36. $\left(-\dfrac{2}{3}\right)^{-2}$

37. $\left(-\dfrac{2x}{3}\right)^{-3}$

38. $\left(-\dfrac{ab}{c}\right)^{-1}$

39. $\left(\dfrac{2x^2}{3y}\right)^{-3}$

40. $\left(\dfrac{ab^{-3}}{a^2b}\right)^{-2}$

⟨4⟩ Variable Exponents

Simplify each expression. Assume that the variables represent integers. See Example 5.

41. $5^{2t}\cdot5^{4t}$

42. $3^{2n-3}\cdot3^{4-2n}$

⟨5⟩ Summary of the Rules

Use the rules of exponents to simplify each expression. If possible, write down only the answer.

49. $3x^4\cdot2x^5$

50. $\left(3x^4\right)^2$

51. $\left(-2x^2\right)^3$

52. $3x^2\cdot2x^{-4}$

53. $\dfrac{3x^{-2}y^{-1}}{z^{-1}}$

54. $\dfrac{2^{-1}x^2}{y^{-2}}$

55. $\left(\dfrac{-2}{3}\right)^{-1}$

56. $\left(\dfrac{-1}{5}\right)^{-1}$

57. $\left(\dfrac{2x^3}{3}\right)^2$

58. $\left(\dfrac{-2y^4}{x}\right)^3$

59. $\left(-2x^{-2}\right)^{-1}$

60. $\left(-3x^{-2}\right)^3$

Use the rules of exponents to simplify each expression.

61. $\left(\dfrac{2x^2y}{xy^2}\right)^{-3}$

62. $\left(\dfrac{2x^3y^2}{3xy^3}\right)^{-1}$

63. $\dfrac{\left(5a^{-1}b^2\right)^3}{\left(5ab^{-2}\right)^4}$

64. $\dfrac{\left(2m^2n^{-3}\right)^4}{mn^5}$

65. $\dfrac{\left(2x^{-2}y\right)^{-3}}{\left(2xy^{-1}\right)^2}\left(2x^2y^{-7}\right)$

66. $\dfrac{\left(3x^{-1}y^3\right)^{-2}}{\left(3xy^{-1}\right)^3}\left(9x^{-9}y^5\right)$

67. $\left(\dfrac{6a^{-2}b^3}{2c^4}\right)^{-2}\left(3a^{-1}b^2\right)^3$

68. $\left(7xz^2\right)^{-4}\left(\dfrac{7xy^{-1}}{z}\right)^3$

Miscellaneous

Write each expression as 2 raised to a power. Assume that the variables represent integers.

69. $32\cdot64$

70. 8^{20}

71. $81\cdot6^{-4}$

72. $10^{-6}\cdot20^6$

73. 4^{3n}

74. $6^{n-5}\cdot3^{5-n}$

Use a calculator to evaluate each expression. Round approximate answers to three decimal places.

75. $\dfrac{1}{5^{-2}}$

76. $\dfrac{(2.5)^{-3}}{(2.5)^{-5}}$

77. $2^{-1}+2^{-2}$

78. $\left(\dfrac{2}{3}\right)^{-1}+2^{-1}$

79. $(0.036)^{-2}+(4.29)^3$

80. $3(4.71)^2-5(0.471)^{-3}$

81. $\dfrac{(5.73)^{-1}+(4.29)^{-1}}{(3.762)^{-1}}$

82. $\left[5.29+(0.374)^{-1}\right]^3$

⟨6⟩ Applications

Solve each problem. See Examples 6 and 7.

83. ***Deeper in debt.*** Melissa borrowed $40,000 at 12% compounded annually and made no payments for 3 years. How much did she owe the bank at the end of the 3 years? (Use the compound interest formula.)

84. ***Comparing stocks and bonds.*** Historically, the average annual return on stocks is 10%, whereas the average annual return on bonds is 7%.

    a) If you had invested $10,000 in bonds in 2000 and achieved the average annual return, then what would you expect your investment to be worth in 2015?

    b) How much more would your $10,000 investment be worth in 2015 if you had invested in stocks in 2000?

Figure for Exercise 84

85. ***Saving for college.*** Mr. Watkins wants to have $10,000 in a savings account when his little Wanda is ready for college. How much must he deposit today in an account paying 7% compounded annually to have $10,000 in 18 years?

86. ***Saving for retirement.*** Wilma wants to have $2,000,000 when she retires in 45 years. Assuming that she can average 4.5% return annually in Treasury Bills, then how much must she invest now in Treasury Bills to achieve her goal?

87. ***Life expectancy of white males.*** Strange as it may seem, your life expectancy increases as you get older. The function

$$L = 72.2(1.002)^a$$

can be used to model life expectancy L for U.S. white males with present age a (National Center for Health Statistics, www.cdc.gov/nchswww).

    a) To what age can a 20-year-old white male expect to live?

    b) To what age can a 60-year-old white male expect to live? (See also Chapter Review Exercises 149 and 150.)

88. ***Life expectancy of white females.*** Life expectancy improved more for females than for males during the 1940s and 1950s due to a dramatic decrease in maternal mortality rates. The function

$$L = 78.5(1.001)^a$$

can be used to model life expectancy L for U.S. white females with present age a.

    a) To what age can a 20-year-old white female expect to live?

    b) Bob, 30, and Ashley, 26, are an average white couple. How many years can Ashley expect to live as a widow?

    c) Interpret the intersection of the life expectancy curves in the accompanying figure.

Figure for Exercises 87 and 88

Getting More Involved

89. ***Discussion***

For which values of a and b is it true that $(ab)^{-1} = a^{-1}b^{-1}$? Find a pair of nonzero values for a and b for which $(a + b)^{-1} \neq a^{-1} + b^{-1}$.

90. ***Writing***

Explain how to evaluate $\left(-\frac{2}{3}\right)^{-3}$ in three different ways.

91. ***Discussion***

Which of the following expressions has a value different from the others? Explain.

    a) -1^{-1} **b)** -3^0 **c)** $-2^{-1} - 2^{-1}$

    d) $(-1)^{-2}$ **e)** $(-1)^{-3}$

92. *True or False?* Explain your answer.

a) The square of a product is the product of the squares.
b) The square of a sum is the sum of the squares.

Graphing Calculator Exercises

93. At 12% compounded annually, the value of an investment of $10,000 after x years is given by

$$y = 10{,}000(1.12)^x.$$

a) Graph $y = 10{,}000(1.12)^x$ and the function $y = 20{,}000$ on a graphing calculator. Use a viewing window that shows the intersection of the two graphs.

b) Use the intersect feature of your calculator to find the point of intersection.

c) The x-coordinate of the point of intersection is the number of years that it will take for the $10,000 investment to double. What is that number of years?

94. The function $y = 72.2(1.002)^x$ gives the life expectancy y of a U.S. white male with present age x. (See Exercise 87.)

a) Graph $y = 72.2(1.002)^x$ and $y = 86$ on a graphing calculator. Use a viewing window that shows the intersection of the two graphs.

b) Use the intersect feature of your calculator to find the point of intersection.

c) What does the x-coordinate of the point of intersection tell you?

5.3   Polynomials and Polynomial Functions

In This Section

⟨1⟩ Polynomials
⟨2⟩ Evaluating Polynomials and Polynomial Functions
⟨3⟩ Addition and Subtraction of Polynomials
⟨4⟩ Multiplication of Polynomials

A polynomial is a particular type of algebraic expression that serves as a fundamental building block in algebra. We used polynomials in Chapters 1 and 2, but we did not identify them as polynomials. In this section, you will learn to recognize polynomials and to add, subtract, and multiply them.

⟨1⟩   Polynomials

In Chapter 1 we defined a **term** as a single number or the product of a number and one or more variables raised to powers. What those powers are is not specified.

Polynomial

A **polynomial** is a single term or a finite sum of terms in which the powers of the variables are positive integers.

For example, the expressions $3x^3$, $-15x^2$, $7x$, and -2 could be used as terms of a polynomial. The number preceding the variable in each term is the **coefficient** of that term. The coefficient of the x^3-term is 3, the coefficient of the x^2-term is -15, and the coefficient of the x-term is 7. In algebra, a number is often referred to as a **constant,** and so the term -2 is called a **constant term.** So the expression

$$3x^3 + \left(-15x^2\right) + 7x + (-2)$$

is a polynomial in one variable with four terms. For simplicity we will write this polynomial as $3x^3 - 15x^2 + 7x - 2$.

EXAMPLE   **1**

Identifying polynomials

Determine whether each algebraic expression is a polynomial.

a) -3 **b)** $3x + 2^{-1}$ **c)** $3x^{-2} + 4x^2$

d) $\dfrac{1}{x} + \dfrac{1}{x^2}$ **e)** $x^{49} - 8x^2 + 11x - 2$

Solution

a) The number -3 is a polynomial of one term, a constant term.

b) Since $3x + 2^{-1}$ can be written as $3x + \frac{1}{2}$, it is a polynomial of two terms.

c) The expression $3x^{-2} + 4x^2$ is not a polynomial because x has a negative exponent.

d) If this expression is rewritten as $x^{-1} + x^{-2}$, then it fails to be a polynomial because of the negative exponents. So a polynomial does not have variables in denominators, and

$$\frac{1}{x} + \frac{1}{x^2}$$

is not a polynomial.

e) The expression $x^{49} - 8x^2 + 11x - 2$ is a polynomial.

> Now do Exercises 1–8

For simplicity we usually write polynomials in one variable with the exponents in decreasing order from left to right. Thus, we would write

$$3x^3 - 15x^2 + 7x - 2 \qquad \text{rather than} \qquad -15x^2 - 2 + 7x + 3x^3.$$

When a polynomial is written in decreasing order, the coefficient of the first term is called the **leading coefficient.**

Certain polynomials have special names depending on the number of terms. A **monomial** is a polynomial that has one term, a **binomial** is a polynomial that has two terms, and a **trinomial** is a polynomial that has three terms.

The **degree** of a polynomial in one variable is the highest power of the variable in the polynomial. The number 0 is considered to be a monomial without degree because $0 = 0x^n$, where n could be any number.

EXAMPLE   **2**

Identifying coefficients and degree

State the degree of each polynomial and the coefficient of x^2. Determine whether the polynomial is monomial, binomial, or trinomial.

a) $\dfrac{x^2}{3} - 5x^3 + 7$ **b)** $x^{48} - x^2$ **c)** 6

Solution

a) The degree of this trinomial is 3, and the coefficient of x^2 is $\frac{1}{3}$.

b) The degree of this binomial is 48, and the coefficient of x^2 is -1.

c) Because $6 = 6x^0$, the number 6 is a monomial with degree 0. Because x^2 does not appear in this polynomial, the coefficient of x^2 is 0.

> **Now do Exercises 9–16**

Although we are mainly concerned here with polynomials in one variable, we will also encounter polynomials in more than one variable, such as

$$4x^2 - 5xy + 6y^2, \qquad x^2 + y^2 + z^2, \qquad \text{and} \qquad ab^2 - c^2.$$

In a term containing more than one variable, the coefficient of a variable consists of all other numbers and variables in the term. For example, the coefficient of x in $-5xy$ is $-5y$, and the coefficient of y is $-5x$. The degree of a term with more than one variable is the sum of the powers of the variables. So $-5xy$ has degree 2. The degree of a polynomial in more than one variable is equal to the highest degree of any of its terms. So $ab^2 - c^2$ has degree 3.

⟨2⟩ Evaluating Polynomials and Polynomial Functions

We learned how to evaluate algebraic expressions in Chapter 1. Since a polynomial is an algebraic expression, it can be evaluated just like any other algebraic expression. If one variable is expressed in terms of another using a polynomial, then we have a **polynomial function.** In Example 3, we use function notation from Section 3.5.

E X A M P L E   3

Evaluating a polynomial and a polynomial function

a) Find the value of the polynomial $x^3 - 3x + 5$ when $x = 2$.

b) Find $P(2)$ if $P(x) = x^3 - 3x + 5$.

‹ Calculator Close-Up ›

To check Example 3 use the Y= key to enter

$$y_1 = x^3 - 3x + 5.$$

Then press VARS, Y-VARS, Function, and Y1 to get y_1 on the home screen. Follow with a 2 in parentheses.

Solution

a) Let $x = 2$ in $x^3 - 3x + 5$ to get

$$(2)^3 - 3(2) + 5 = 8 - 6 + 5 = 7.$$

So if $x = 2$, then the value of the polynomial is 7.

b) This is simply a repeat of part (a) using function notation. Replace x with 2 in $P(x) = x^3 - 3x + 5$:

$$P(2) = (2)^3 - 3(2) + 5 = 7$$

Note that $P(2)$ is the value of the polynomial when $x = 2$. The equation $P(2) = 7$ contains the value of the polynomial and the number that was used for x.

> **Now do Exercises 17–22**

⟨3⟩ Addition and Subtraction of Polynomials

In Section 1.6, we learned that like terms are terms that have the same variables with the same exponents. For example, $3x^2$ and $-5x^2$ are like terms. The distributive property enables us to add or subtract like terms. For example,

$$-5x^2 + 3x^2 = (-5 + 3)x^2 = -2x^2$$

and

$$-5x^2 - 3x^2 = (-5 - 3)x^2 = -8x^2.$$

To add two polynomials, we simply add the like terms.

E X A M P L E   4

Adding polynomials

Find the sums.

a) $(x^2 - 5x - 7) + (7x^2 - 4x + 10)$ **b)** $(3x^3 - 5x^2 - 7) + (4x^2 - 2x + 3)$

Solution

a) $(x^2 - 5x - 7) + (7x^2 - 4x + 10) = 8x^2 - 9x + 3$ Combine like terms.

b) For illustration we will write this addition vertically:

$$\begin{array}{r} 3x^3 - 5x^2 \quad\quad - 7 \\ \underline{4x^2 - 2x + 3} \\ 3x^3 - \ x^2 - 2x - 4 \end{array}$$

Line up like terms.

Add.

> **Now do Exercises 23–24**

< **Helpful Hint** >

When we perform operations with polynomials and write the results as equations, those equations are identities. For example,

$(2x + 1) + (3x + 7) = 5x + 8$

is an identity.

When we add or subtract polynomials, we add or subtract the like terms. Because $a - b = a + (-b)$, we often perform subtraction by changing the signs and adding. We usually perform addition and subtraction horizontally, but vertical subtraction is used in dividing polynomials in Section 6.5.

E X A M P L E   5

Subtracting polynomials

Find the differences.

a) $(x^2 - 7x - 2) - (5x^2 + 6x - 4)$ **b)** $(6y^3z - 5yz + 7) - (4y^2z - 3yz - 9)$

Solution

a) We find the first difference horizontally:

$$(x^2 - 7x - 2) - (5x^2 + 6x - 4) = x^2 - 7x - 2 - 5x^2 - 6x + 4 \quad \text{Change signs.}$$
$$= -4x^2 - 13x + 2 \quad\quad\quad\quad\quad \text{Combine like terms.}$$

b) For illustration we write $(6y^3z - 5yz + 7) - (4y^2z - 3yz - 9)$ vertically:

$$\begin{array}{r} 6y^3z \quad\quad\quad - 5yz + 7 \\ \underline{-4y^2z + 3yz + 9} \\ 6y^3z - 4y^2z - 2yz + 16 \end{array}$$

Change signs.

Add.

> **Now do Exercises 25–38**

< **Helpful Hint** >

For subtraction, write the original problem and then rewrite it as addition with the signs changed. Many students have trouble when they write the original problem and then overwrite the signs. Vertical subtraction is essential for performing long division of polynomials in Section 6.5.

It is certainly not necessary to write out all of the steps shown in Examples 4 and 5, but we must use the following rule.

> **Addition and Subtraction of Polynomials**
>
> To add two polynomials, add the like terms.
> To subtract two polynomials, subtract the like terms.

⟨4⟩ **Multiplication of Polynomials**

We learned how to multiply monomials when we learned the product rule in Section 5.1. For example,

$$-2x^3 \cdot 4x^2 = -8x^5.$$

To multiply a monomial and a polynomial of two or more terms, we apply the distributive property. For example,

$$3x(x^3 - 5) = 3x^4 - 15x.$$

E X A M P L E   6

Multiplying by a monomial

Find the products.

a) $2ab^2 \cdot 3a^2b$ **b)** $(-1)(5 - x)$

c) $(x^3 - 5x + 2)(-3x)$

Solution

a) $2ab^2 \cdot 3a^2b = 6a^3b^3$

b) $(-1)(5 - x) = -5 + x = x - 5$

c) Each term of $x^3 - 5x + 2$ is multiplied by $-3x$:

$$(x^3 - 5x + 2)(-3x) = -3x^4 + 15x^2 - 6x$$

> **Now do Exercises 39–46**

Note what happened to the binomial in Example 6(b) when we multiplied it by -1. If we multiply any difference by -1, we get the same type of result:

$$-1(a - b) = -a + b = b - a.$$

Because multiplying by -1 is the same as taking the opposite, we can write this equation as

$$-(a - b) = b - a.$$

This equation says that $a - b$ and $b - a$ are opposites or additive inverses of each other.

CAUTION The opposite of $a + b$ is $-a - b$, not $a - b$.

To multiply a binomial and a trinomial, we can use the distributive property or set it up like multiplication of whole numbers.

E X A M P L E   7

Multiplying a binomial and a trinomial

Find the product $(x + 2)(x^2 + 3x - 5)$.

Solution

We can find this product by applying the distributive property twice. First we multiply the binomial and each term of the trinomial:

$$
\begin{aligned}
(x + 2)(x^2 + 3x - 5) &= (x + 2)x^2 + (x + 2)3x + (x + 2)(-5) && \text{Distributive property} \\
&= x^3 + 2x^2 + 3x^2 + 6x - 5x - 10 && \text{Distributive property} \\
&= x^3 + 5x^2 + x - 10 && \text{Combine like terms.}
\end{aligned}
$$

⟨ **Helpful Hint** ⟩

Many students find vertical multiplication easier than applying the distributive property twice horizontally. However, you should learn both methods because horizontal multiplication will help you with factoring by grouping in Section 5.6.

We could have found this product vertically:

$$x^2 + 3x - 5$$
$$\underline{x + 2}$$
$$2x^2 + 6x - 10 \quad 2(x^2 + 3x - 5) = 2x^2 + 6x - 10$$
$$\underline{x^3 + 3x^2 - 5x \quad\quad x(x^2 + 3x - 5) = x^3 + 3x^2 - 5x}$$
$$x^3 + 5x^2 + \ x - 10 \quad \text{Add.}$$

> **Now do Exercises 47–50**

Multiplication of Polynomials

To multiply polynomials, multiply each term of the first polynomial by each term of the second polynomial and then combine like terms.

In Example 8, we multiply binomials.

E X A M P L E 8

Multiplying binomials

Find the products.

a) $(x + y)(z + 4)$ **b)** $(x - 3)(2x + 5)$

Solution

a) $(x + y)(z + 4) = (x + y)z + (x + y)4$ Distributive property

$\quad\quad\quad\quad\quad\quad\quad = xz + yz + 4x + 4y$ Distributive property

Notice that this product does not have any like terms to combine.

b) Multiply:

$$x - 3$$
$$\underline{2x + 5}$$
$$5x - 15 \quad\quad 5(x - 3) = 5x - 15$$
$$\underline{2x^2 - 6x \quad\quad\quad 2x(x - 3) = 2x^2 - 6x}$$
$$2x^2 - \ x - 15$$

> **Now do Exercises 51–58**

Warm-Ups ▼

Fill in the blank.

1. A _____ of a polynomial is a single number or the product of a number and one or more variables raised to whole number powers.

2. The number preceding the variable in each term is the _____ of that term.

3. The _____ term is just a number.

4. A _____ is a single term or a finite sum of terms.

5. The _____ of a polynomial in one variable is the highest power of the variable in the polynomial.

6. To multiply a binomial and trinomial we use the _____ property.

True or false?

7. The expression $3x^{-2} - 5x + 2$ is a trinomial.

8. In $3x^2 - 5x + 3$ the coefficient of x is 5.

9. The degree of $x^2 + 3x - 5x^3 + 4$ is 2.

10. If $C(x) = x^2 - 3$, then $C(5) = 22$.

11. If $P(t) = 30t + 10$, then $P(0) = 40$.

12. If x is any real number, then $(x^3 - 3x) + (x^2 + 5x) = x^3 + x^2 + 2x$.

13. If x is any real number, then $(x^2 - 5) - (x^2 - x) = x - 5$.

14. If x is any real number, then $-2x(3x - 4x^2) = -6x^2 + 8x^3$.

15. For any value of y, the opposite of $y + 5$ is $y - 5$.

‹ **1** › **Polynomials**

Determine whether each algebraic expression is a polynomial. See Example 1.

1. $3x$

2. -9

3. $x^{-1} + 4$

4. $3x^{-3} + 4x - 1$

5. $x^2 - 3x + \sqrt{5}$

6. $\dfrac{x^3}{3} - \dfrac{3x^2}{5} + 0.2x$

7. $\dfrac{1}{x} + x - 3$

8. $x^5 - \dfrac{9}{x^2}$

State the degree of each polynomial and the coefficient of x^3. Determine whether each polynomial is a monomial, binomial, or trinomial. See Example 2.

9. $x^4 - 8x^3$

10. $15 - x^3$

11. -8

12. 17

13. $\dfrac{x^7}{15}$

14. $5x^4$

15. $x^3 + 3x^4 - 5x^6$

16. $\dfrac{x^3}{2} + \dfrac{5x}{2} - 7$

‹ **2** › **Evaluating Polynomials and Polynomial Functions**

For each given polynomial, find the indicated value of the polynomial. See Example 3.

17. $P(x) = x^4 - 1$, $P(3)$

18. $P(x) = x^2 - x - 2$, $P(-1)$

19. $M(x) = -3x^2 + 4x - 9$, $M(-2)$

20. $C(w) = 3w^2 - w$, $C(0)$

21. $R(x) = x^5 - x^4 + x^3 - x^2 + x - 1$, $R(1)$

22. $T(a) = a^7 + a^6$, $T(-1)$

‹ **3** › **Addition and Subtraction of Polynomials**

Perform the indicated operations. See Examples 4 and 5.

23. $(2a - 3) + (a + 5)$

24. $(2w - 6) + (w + 5)$

25. $(7xy + 30) - (2xy + 5)$

26. $(5ab + 7) - (3ab + 6)$

27. $(x^2 - 3x) + (-x^2 + 5x - 9)$

28. $(2y^2 - 3y - 8) + (y^2 + 4y - 1)$

29. $(2x^3 - 4x - 3) - (x^2 - 2x + 5)$

30. $(2x - 5) - (x^2 - 3x + 2)$

Perform the indicated operations vertically. See Examples 4 and 5.

31. Add.
$$\begin{array}{r} x^3 + 3x^2 - 5x - 2 \\ -x^3 + 8x^2 + 3x - 7 \\ \hline \end{array}$$

32. Add.
$$\begin{array}{r} x^2 - 3x + 7 \\ -2x^2 - 5x + 2 \\ \hline \end{array}$$

33. Subtract.
$$\begin{array}{r} 5x + 2 \\ 4x - 3 \\ \hline \end{array}$$

34. Subtract.
$$\begin{array}{r} 4x + 3 \\ 2x - 6 \\ \hline \end{array}$$

35. Subtract.
$$\begin{array}{r} -x^2 + 3x - 5 \\ 5x^2 - 2x - 7 \\ \hline \end{array}$$

36. Subtract.
$$\begin{array}{r} -3x^2 + 5x - 2 \\ x^2 - 5x - 6 \\ \hline \end{array}$$

37. Add.
$$\begin{array}{r} x - y \\ x + y \\ \hline \end{array}$$

38. Add.
$$\begin{array}{r} -w + 4 \\ 2w - 3 \\ \hline \end{array}$$

‹ **4** › **Multiplication of Polynomials**

Find each product. See Examples 6–8.

39. $-3x^2 \cdot 5x^4$

40. $(-ab^5)(-2a^2b)$

41. $x^2(x - 2)$

42. $-2x(x^3 - x)$

43. $-1(3x - 2)$

44. $-1(-x^2 + 3x - 9)$

45. $5x^2y^3(3x^2y - 4x)$

46. $3y^4z(8y^2z^2 - 3yz + 2y)$

47. $(x - 2)(x + 2)$

48. $(x - 1)(x + 1)$

49. $(x^2 + x + 2)(2x - 3)$

50. $(x^2 - 3x + 2)(x - 4)$

Find each product vertically. See Examples 6–8.

51. Multiply.

$$\begin{array}{r} 2x - 3 \\ -5x \\ \hline \end{array}$$

52. Multiply.

$$\begin{array}{r} 3a^3 - 5a^2 + 7 \\ -2a \\ \hline \end{array}$$

53. Multiply.

$$\begin{array}{r} x + 5 \\ x + 5 \\ \hline \end{array}$$

54. Multiply.

$$\begin{array}{r} a + b \\ a - b \\ \hline \end{array}$$

55. Multiply.

$$\begin{array}{r} x + 6 \\ 2x - 3 \\ \hline \end{array}$$

56. Multiply.

$$\begin{array}{r} 3x^2 + 2 \\ 2x^2 - 5 \\ \hline \end{array}$$

57. Multiply.

$$\begin{array}{r} x^2 + xy + y^2 \\ x - y \\ \hline \end{array}$$

58. Multiply.

$$\begin{array}{r} a^2 - ab + b^2 \\ a + b \\ \hline \end{array}$$

Miscellaneous

Perform the indicated operations.

59. $(x - 7) + (2x - 3) + (5 - x)$

60. $(5x - 3) + (x^3 + 3x - 2) + (-2x - 3)$

 61. $(a^2 - 5a + 3) + (3a^2 - 6a - 7)$

62. $(w^2 - 3w + 2) + (2w - 3 + w^2)$

 63. $(w^2 - 7w - 2) - (w - 3w^2 + 5)$

64. $(a^3 - 3a) - (1 - a - 2a^2)$

65. $(x - 2)(x^2 + 2x + 4)$

66. $(a - 3)(a^2 + 3a + 9)$

67. $(x - w)(z + 2w)$

68. $(w^2 - a)(t^2 + 3)$

69. $(x^2 - x + 2)(x^2 + x - 2)$

70. $(a^2 + a + b)(a^2 - a - b)$

 Perform the following operations using a calculator.

71. $(2.31x - 5.4)(6.25x + 1.8)$

72. $(x - 0.28)(x^2 - 34.6x + 21.2)$

73. $(3.759x^2 - 4.71x + 2.85) + (11.61x^2 + 6.59x - 3.716)$

74. $(43.19x^3 - 3.7x^2 - 5.42x + 3.1)$
$- (62.7x^3 - 7.36x - 12.3)$

Perform the indicated operations.

75. $\left(\dfrac{1}{2}x + 2\right) + \left(\dfrac{1}{4}x - \dfrac{1}{2}\right)$

76. $\left(\dfrac{1}{3}x + 1\right) + \left(\dfrac{1}{3}x - \dfrac{3}{2}\right)$

77. $\left(\dfrac{1}{2}x^2 + \dfrac{1}{3}x - \dfrac{1}{5}\right) - \left(x^2 - \dfrac{2}{3}x - \dfrac{1}{5}\right)$

78. $\left(\dfrac{2}{3}x^2 - \dfrac{1}{3}x + \dfrac{1}{6}\right) - \left(-\dfrac{1}{3}x^2 + x + 1\right)$

79. $[x^2 - 3 - (x^2 + 5x - 4)] - [x - 3(x^2 - 5x)]$

80. $[x^3 - 4x(x^2 - 3x + 2) - 5x] + [x^2 - 5(4 - x^2) + 3]$

81. $[5x - 4(x - 3)][3x - 7(x + 2)]$

82. $[x^2 - (5x - 2)][x^2 + (5x - 2)]$

83. $[x^2 - (m + 2)][x^2 + (m + 2)]$

84. $[3x^2 - (x - 2)][3x^2 + (x + 2)]$

Perform the indicated operations. A variable used in an exponent represents an integer; a variable used as a base represents a nonzero real number.

85. $(a^{2m} + 3a^m - 3) + (-5a^{2m} - 7a^m + 8)$

86. $(b^{3z} - 6) - (4b^{3z} - b^{2z} - 7)$

87. $(x^n - 1)(x^n + 3)$

88. $(2y^t - 3)(4y^t + 7)$

89. $z^{3w} - z^{2w}(z^{1-w} - 4z^w)$

90. $(w^p - 1)(w^{2p} + w^p + 1)$

91. $(x^{2r} + y)(x^{4r} - x^{2r}y + y^2)$

92. $(2x^a - z)(2x^a + z)$

Applications

Solve each problem.

93. *Cost of gravel.* The cost in dollars of x cubic yards of gravel is given by the function

$$C(x) = 20x + 15.$$

Find $C(3)$, the cost of 3 cubic yards of gravel.

94. *Annual bonus.* Sheila's annual bonus in dollars for selling n life insurance policies is given by the function

$$B(n) = 0.1n^2 + 3n + 50.$$

Find $B(20)$, her bonus for selling 20 policies.

95. *Marginal cost.* A company uses the function $C(n) = 50n - 0.01n^4$ to find the daily cost in dollars of manufacturing n aluminum windows. The *marginal cost* of the nth window is the additional cost incurred for manufacturing that window. For example, the marginal cost of the third window is $C(3) - C(2)$. Find the marginal cost for manufacturing the third window. What is the marginal cost for manufacturing the tenth window?

96. *Marginal profit.* A company uses the function $P(n) = 4n + 0.9n^3$ to estimate its daily profit in dollars for producing n automatic garage door openers. The *marginal profit* of the nth opener is the amount of additional profit made for that opener. For example, the marginal profit for the fourth opener is $P(4) - P(3)$. Find the marginal profit for the fourth opener. What is the marginal profit for the tenth opener? Use the bar graph to explain why the marginal profit increases as production goes up.

Figure for Exercise 96

 97. *Male and female life expectancy.* Since 1950 the life expectancies of U.S. males and females born in year y can be modeled by the functions

$$M(y) = 0.16252y - 251.91$$

and

$$F(y) = 0.18268y - 284.98,$$

respectively (National Center for Health Statistics, www.cdc.gov/nchswww).

a) How much greater was the life expectancy of a female born in 1950 than a male born in 1950?

b) Are the lines in the accompanying figure parallel?

c) In what year will female life expectancy be 8 years greater than male life expectancy?

 98. *More life expectancy.* Use the functions from Exercise 97 for these questions.

a) A male born in 1975 does not want his future wife to outlive him. What should be the year of birth for his wife so that they both can be expected to die in the same year?

b) Find $\dfrac{M(y) + F(y)}{2}$ to get a formula for the life expectancy of a person born in year y.

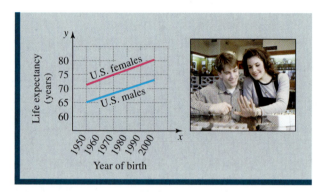

Figure for Exercises 97 and 98

Getting More Involved

99. *Discussion*

Is it possible for a binomial to have degree 4? If so, give an example.

100. *Discussion*

Give an example of two fourth-degree trinomials whose sum is a third-degree binomial.

101. *Cooperative learning*

Work in a group to find the product $(a + b)(c + d)$. How many terms does it have? Find the product $(a + b)(c + d)(e + f)$. How many terms does it have? How many terms are there in a product of four binomials in which there are no like terms to combine? How many terms are there in a product of n binomials in which there are no like terms?

5.4 Multiplying Binomials

In Section 5.3, you learned to multiply polynomials. In this section, you will learn rules to make multiplication of binomials simpler.

In This Section

⟨1⟩ **The FOIL Method**

⟨2⟩ **The Square of a Binomial**

⟨3⟩ **Product of a Sum and a Difference**

⟨4⟩ **Higher Powers of Binomials**

⟨5⟩ **Polynomial Functions**

⟨1⟩ **The FOIL Method**

Consider how we find the product of two binomials $x + 3$ and $x + 5$ using the distributive property twice:

$$
\begin{aligned}
(x + 3)(x + 5) &= (x + 3)x + (x + 3)5 &&\text{Distributive property} \\
&= x^2 + 3x + 5x + 15 &&\text{Distributive property} \\
&= x^2 + 8x + 15 &&\text{Combine like terms.}
\end{aligned}
$$

⟨ **Helpful Hint** ⟩

The product of two binomials always has four terms before combining like terms. The product of two trinomials always has nine terms before combining like terms. How many terms are there in the product of a binomial and trinomial?

There are four terms in the product. The term x^2 is the product of the first term of each binomial. The term $5x$ is the product of the two outer terms, 5 and x. The term $3x$ is the product of the two inner terms, 3 and x. The term 15 is the product of the last two terms in each binomial, 3 and 5. It may be helpful to connect the terms multiplied by lines.

$$
(x + 3)(x + 5)
$$

F = First terms
O = Outer terms
I = Inner terms
L = Last terms

So instead of writing out all of the steps in using the distributive property, we can get the result by finding the products of the first, outer, inner, and last terms. This method is called the **FOIL method.**

For example, let's apply FOIL to the product $(x - 3)(x + 4)$:

$$
(x - 3)(x + 4) = x^2 + 4x - 3x - 12 = x^2 + x - 12
$$

If the outer and inner products are like terms, you can save a step by writing down only their sum. Note that FOIL is simply a way to "speed up" the distributive property.

E X A M P L E   1

Multiplying binomials

Use FOIL to find the products of the binomials.

a) $(2x - 3)(3x + 4)$ **b)** $(2x^3 + 5)(2x^3 - 5)$

c) $(m + w)(2m - w)$ **d)** $(a + b)(a - 3)$

Solution

a) $(2x - 3)(3x + 4) \overset{\text{F \quad O \quad I \quad L}}{=} 6x^2 + 8x - 9x - 12$

$\qquad\qquad\qquad\quad = 6x^2 - x - 12$ Combine like terms.

b) $(2x^3 + 5)(2x^3 - 5) \overset{\text{F \quad O \quad I \quad L}}{=} 4x^6 - 10x^3 + 10x^3 - 25$

$\qquad\qquad\qquad\qquad = 4x^6 - 25$ Combine like terms.

c) $(m + w)(2m - w) \overset{\text{F \quad O \quad I \quad L}}{=} 2m^2 - mw + 2mw - w^2$

$\qquad\qquad\qquad\quad = 2m^2 + mw - w^2$ Combine like terms.

d) $(a + b)(a - 3) = a^2 - 3a + ab - 3b$ There are no like terms.

> Now do Exercises 1–18

‹ **Helpful Hint** ›

To visualize the square of a sum, draw a square with sides of length $a + b$ as shown.

The area of the large square is $(a + b)^2$. It comes from four terms as stated in the rule for the square of a sum.

‹2› The Square of a Binomial

To find $(a + b)^2$, the square of a sum, we can use FOIL on $(a + b)(a + b)$:

$$(a + b)(a + b) = a^2 + ab + ab + b^2$$
$$= a^2 + 2ab + b^2$$

You can use the result $a^2 + 2ab + b^2$ that we obtained from FOIL to quickly find the square of *any* sum. *To square a sum, we square the first term* (a^2), *add twice the product of the two terms* $(2ab)$, *and then add the square of the last term* (b^2).

Rule for the Square of a Sum

$$(a + b)^2 = a^2 + 2ab + b^2$$

In general, the square of a sum $(a + b)^2$ is not equal to the sum of the squares $a^2 + b^2$. The square of a sum has the middle term $2ab$.

E X A M P L E   **2**

Squaring a binomial

Square each sum, using the new rule.

a) $(x + 5)^2$ **b)** $(2w + 3)^2$ **c)** $(2y^4 + 3)^2$

Solution

a) $(x + 5)^2 = x^2 + 2(x)(5) + 5^2 = x^2 + 10x + 25$

$\qquad\qquad\quad\;\; \uparrow \qquad \uparrow \qquad\;\; \uparrow$

$\qquad\qquad\;\; \text{Square} \;\; \text{Twice} \;\; \text{Square}$
$\qquad\qquad\qquad \text{of} \qquad \text{the} \qquad \text{of}$
$\qquad\qquad\;\; \text{first} \;\; \text{product} \;\; \text{last}$

b) $(2w + 3)^2 = (2w)^2 + 2(2w)(3) + 3^2$ Use the rule $(a + b)^2 = a^2 + 2ab + b^2$.

$\qquad\qquad\qquad = 4w^2 + 12w + 9$

c) $(2y^4 + 3)^2 = (2y^4)^2 + 2(2y^4)(3) + 3^2$ Use the rule $(a + b)^2 = a^2 + 2ab + b^2$.

$\qquad\qquad\qquad = 4y^8 + 12y^4 + 9$

> Now do Exercises 19–20

CAUTION Squaring $x + 5$ correctly, as in Example 2(a), gives us the identity

$$(x + 5)^2 = x^2 + 10x + 25,$$

which is satisfied by any x. If you forget the middle term and write $(x + 5)^2 = x^2 + 25$, then you have an equation that is satisfied only if $x = 0$.

To find $(a - b)^2$, the square of a difference, we can use FOIL:

$$(a - b)(a - b) = a^2 - ab - ab + b^2$$
$$= a^2 - 2ab + b^2$$

⟨ **Helpful Hint** ⟩

Many students keep using FOIL to find the square of a sum or a difference. However, you will be greatly rewarded if you learn the new rules for squaring a sum or a difference.

As in squaring a sum, it is simply better to remember the result of using FOIL. *To square a difference, square the first term, subtract twice the product of the two terms, and add the square of the last term.*

> **Rule for the Square of a Difference**
> $$(a - b)^2 = a^2 - 2ab + b^2$$

EXAMPLE 3

Squaring a binomial
Square each difference, using the new rule.

   a) $(x - 6)^2$ **b)** $(3w - 5y)^2$ **c)** $(-4 - st)^2$ **d)** $(3 - 5a^3)^2$

Solution

a) $(x - 6)^2 = x^2 - 2(x)(6) + 6^2$ For the middle term, subtract twice
$$= x^2 - 12x + 36$$ the product: $2(x)(6)$.

b) $(3w - 5y)^2 = (3w)^2 - 2(3w)(5y) + (5y)^2$ Use the rule $(a - b)^2 = a^2 - 2ab + b^2$.
$$= 9w^2 - 30wy + 25y^2$$

c) $(-4 - st)^2 = (-4)^2 - 2(-4)(st) + (st)^2$
$$= 16 + 8st + s^2t^2$$

d) $(3 - 5a^3)^2 = 3^2 - 2(3)(5a^3) + (5a^3)^2$
$$= 9 - 30a^3 + 25a^6$$

> Now do Exercises 21–30

⟨ **Helpful Hint** ⟩

A red rectangle with sides b and $a - b$ is added in two different ways to a blue rectangle whose sides are $a - b$ and a, as shown in the figure. The area of the region with the dashed boundary is $(a - b)(a + b)$. The area of the region with the solid boundary is $a^2 - b^2$. Since these areas are equal, $(a - b)(a + b) = a^2 - b^2$.

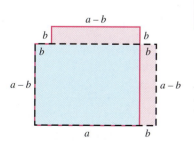

⟨**3**⟩ **Product of a Sum and a Difference**

If we multiply the sum $a + b$ and the difference $a - b$ by using FOIL, we get

$$(a + b)(a - b) = a^2 - ab + ab - b^2$$
$$= a^2 - b^2.$$

The inner and outer products add up to zero, canceling each other out. So *the product of a sum and a difference is the difference of two squares,* as shown in the following rule.

> **Rule for the Product of a Sum and a Difference**
> $$(a + b)(a - b) = a^2 - b^2$$

E X A M P L E   4

Finding the product of a sum and a difference
Find the products.

  a) $(x + 3)(x - 3)$
  b) $(a^3 + 8)(a^3 - 8)$
  c) $(3x^2 - y^3)(3x^2 + y^3)$

Solution

  a) $(x + 3)(x - 3) = x^2 - 9$ Use the rule $(a + b)(a - b) = a^2 - b^2$.
  b) $(a^3 + 8)(a^3 - 8) = a^6 - 64$
  c) $(3x^2 - y^3)(3x^2 + y^3) = 9x^4 - y^6$ Note that $(a - b)(a + b) = a^2 - b^2$.

> **Now do Exercises 31–40**

The square of a sum, the square of a difference, and the product of a sum and a difference are referred to as **special products.** Although the special products can be found by using the distributive property or FOIL, they occur so frequently in algebra that it is essential to learn the new rules. In Example 5 we use the special product rules to multiply two trinomials and to square a trinomial.

E X A M P L E   5

Using special product rules to multiply trinomials
Find the products.

  a) $[(x + y) + 3][(x + y) - 3]$
  b) $[(m - n) + 5]^2$

Solution

  a) Use the rule $(a + b)(a - b) = a^2 - b^2$ with $a = x + y$ and $b = 3$:
  $$[(x + y) + 3][(x + y) - 3] = (x + y)^2 - 3^2$$
  $$= x^2 + 2xy + y^2 - 9$$
  b) Use the rule $(a + b)^2 = a^2 + 2ab + b^2$ with $a = m - n$ and $b = 5$:
  $$[(m - n) + 5]^2 = (m - n)^2 + 2(m - n)5 + 5^2$$
  $$= m^2 - 2mn + n^2 + 10m - 10n + 25$$

> **Now do Exercises 41–48**

‹4› Higher Powers of Binomials

To find a power of a binomial that is higher than 2, we can use the rule for squaring a binomial along with the method of multiplying binomials using the distributive property. Finding the second or higher power of a binomial is called **expanding the binomial** because the result has more terms than the original.

EXAMPLE 6

Higher powers of a binomial
Expand each binomial.

a) $(x + 2)^3$ **b)** $(a - 4)^4$

Solution

a)
$$
\begin{aligned}
(x + 2)^3 &= (x + 2)^2(x + 2) \\
&= (x^2 + 4x + 4)(x + 2) && \text{Square } x + 2. \\
&= (x^2 + 4x + 4)x + (x^2 + 4x + 4)2 && \text{Distributive property} \\
&= x^3 + 4x^2 + 4x + 2x^2 + 8x + 8 && \text{Distributive property} \\
&= x^3 + 6x^2 + 12x + 8 && \text{Combine like terms.}
\end{aligned}
$$

b)
$$
\begin{aligned}
(a - 4)^4 &= (a - 4)^2(a - 4)^2 \\
&= (a^2 - 8a + 16)(a^2 - 8a + 16) \\
&= (a^2 - 8a + 16)a^2 + (a^2 - 8a + 16)(-8a) + (a^2 - 8a + 16)16 \\
&= a^4 - 8a^3 + 16a^2 - 8a^3 + 64a^2 - 128a + 16a^2 - 128a + 256 \\
&= a^4 - 16a^3 + 96a^2 - 256a + 256
\end{aligned}
$$

> **Now do Exercises 49–60**

CAUTION   In general, the expansion of the fourth power of a binomial has five terms just like the expansion of $(a - 4)^4$ in Example 6(b). The expansion of $(a + b)^4$ is *not* $a^4 + b^4$.

⟨5⟩ Polynomial Functions
In Example 7 we find a formula for a polynomial function by multiplying binomials.

EXAMPLE 7

Writing a polynomial function
The width of a rectangular box is x inches. Its length is 2 inches greater than the width, and its height is 4 inches greater than the width. Write a polynomial function $V(x)$ that gives the volume in cubic inches.

Solution

The width of the box is x inches, the length is $x + 2$ inches, and the height is $x + 4$ inches as shown in Fig. 5.1. The function $V(x)$ is the product of the length, width, and height:

$$
\begin{aligned}
V(x) &= x(x + 2)(x + 4) \\
&= x(x^2 + 6x + 8) \\
&= x^3 + 6x^2 + 8x
\end{aligned}
$$

So $V(x) = x^3 + 6x^2 + 8x$ gives the volume in cubic inches.

> **Now do Exercises 99–104**

Figure 5.1

Warm-Ups ▼

Fill in the blank.

1. We can use the _____ property to multiply two binomials.
2. _____ stands for First, Outer, Inner, Last.
3. The _____ method gives the product of two binomials quickly.
4. The square of a sum, the square of a difference, and the product of a sum and a difference are referred to as _____ products.
5. The product of a sum and a difference is equal to the _____ of two squares.
6. Finding the second or higher power of a binomial is called _____ the binomial.

True or false?

7. If x is any real number, then $(x + 2)(x + 5) = x^2 + 7x + 10$.
8. If x is any real number, then $(2x - 3)(3x + 5) = 6x^2 + x - 15$.
9. If x is any real number, then $(x + 7)^2 = x^2 + 14x + 49$.

10. $(2 + 3)^2 = 2^2 + 3^2$
11. $(8 - 3)^2 = 64 - 9$
12. $(60 - 1)(60 + 1) = 3600 - 1$
13. If a and b are any real number, then $(a + b)^2 = a^2 + b^2$.

14. If x is any real number, then $(x - 1)^2 = x^2 - 2x + 1$.

5.4 Exercises

‹ Study Tips ›

- Relax and don't worry about grades. If you are doing everything that you can and should be doing, then there is no reason to worry.
- Be active in class. Don't be embarrassed to ask questions or answer questions. You can often learn more from giving a wrong answer than a right one.

‹1› The FOIL Method

Find each product. When possible, write down only the answer. See Example 1.

1. $(x + 3)(x + 5)$
2. $(x + 7)(x + 3)$
3. $(x - 2)(x + 4)$
4. $(x - 3)(x + 5)$
5. $(1 + 2x)(3 + x)$
6. $(3 + 2y)(y + 2)$
7. $(-2a - 3)(-a + 5)$
8. $(-3x - 5)(-x + 6)$
9. $(2x^2 - 7)(2x^2 + 7)$
10. $(3y^3 + 8)(3y^3 - 8)$
11. $(2x^3 - 1)(x^3 + 4)$
12. $(3t^2 - 4)(2t^2 + 3)$
13. $(6z + w)(w - z)$

14. $(4y + w)(w - 2y)$
15. $(3k - 2t)(4t + 3k)$ **VIDEO**
16. $(7a - 2x)(x + a)$
17. $(x - 3)(y + w)$
18. $(z - 1)(y + 2)$

‹2› The Square of a Binomial

Find the square of each sum or difference. When possible, write down only the answer. See Examples 2 and 3.

19. $(m + 3)^2$
20. $(a + 2)^2$
21. $(4 - a)^2$
22. $(3 - b)^2$
23. $(2w + 1)^2$
24. $(3m + 4)^2$
25. $(3t - 5u)^2$ **VIDEO**

26. $(3w - 2x)^2$

27. $(-x - 1)^2$

28. $(-d - 5)^2$

29. $(a - 3y^3)^2$

30. $(3m - 5n^3)^2$

⟨3⟩ Product of a Sum and a Difference

Find each product. See Example 4.

31. $(w - 9)(w + 9)$

32. $(m - 4)(m + 4)$

33. $(w^3 + y)(w^3 - y)$

34. $(a^3 - x)(a^3 + x)$

35. $(7 - 2x)(7 + 2x)$

36. $(3 + 5x)(3 - 5x)$

37. $(3x^2 - 2)(3x^2 + 2)$

38. $(4y^2 + 1)(4y^2 - 1)$

39. $(5a^3 - 2b)(5a^3 + 2b)$

40. $(6w^4 + 5y^3)(6w^4 - 5y^3)$

Use the special product rules to find each product.
See Example 5.

41. $[(m + t) + 5][(m + t) - 5]$

42. $[(2x + 3) - y][(2x + 3) + y]$

43. $[y - (r + 5)][y + (r + 5)]$

44. $[x + (3 - k)][x - (3 - k)]$

45. $[(2y - t) + 3]^2$

46. $[(u - 3v) - 4]^2$

47. $[3h + (k - 1)]^2$

48. $[2p - (3q + 6)]^2$

⟨4⟩ Higher Powers of Binomials

Expand each binomial. See Example 6.

49. $(x + 1)^3$

50. $(a + 3)^3$

51. $(w - 2)^3$

52. $(m - 4)^3$

53. $(2x + 1)^3$

54. $(3x + 2)^3$

55. $(3x - 1)^3$

56. $(5x - 2)^3$

57. $(x + 1)^4$

58. $(x + 2)^4$

59. $(h - 3)^4$

60. $(b - 5)^4$

Miscellaneous

Perform the operations and simplify.

61. $(x - 6)(x + 9)$

62. $(2x^2 - 3)(3x^2 + 4)$

63. $(5 - x)(5 + x)$

64. $(4 - ab)(4 + ab)$

65. $(3x - 4a)(2x + 5a)$

66. $(x^5 + 2)(x^5 - 2)$

67. $(2t - 3)(t + w)$

68. $(5x - 9)(ax + b)$

69. $(3x^2 + 2y^3)^2$

70. $(5a^4 - 2b)^2$

71. $(2 + 2y)(3y - 5)$

72. $(3b - 3)(3 + 2b)$

73. $(2m - 7)^2$

74. $(5a + 4)^2$

75. $(3 + 7x)^2$

76. $(1 - pq)^2$

77. $4y\left(3y + \dfrac{1}{2}\right)^2$

78. $25y\left(2y - \dfrac{1}{5}\right)^2$

79. $(a + h)^2 - a^2$

80. $\dfrac{(x + h)^2 - x^2}{h}$

81. $(x + 2)(x + 2)^2$

82. $(a + 1)^2(a + 1)^2$

83. $(y + 3)^3$

84. $(2x - 3y)^3$

85. $4x - 3(x - 5)^2$

86. $2(x + 3)(x - 2) - (2x - 1)^2$

Use a calculator to help you perform the following operations.

87. $(3.2x - 4.5)(5.1x + 3.9)$

88. $(5.3x - 9.2)^2$

89. $(3.6y + 4.4)^2$

90. $(3.3a - 7.9b)(3.3a + 7.9b)$

Find the products. Assume all variables are nonzero and variables used in exponents represent integers.

91. $(x^m + 2)(x^{2m} + 3)$

92. $(a^n - b)(a^n + b)$

93. $a^{n+1}(a^{2n} + a^n - 3)$

94. $x^{3b}(x^{-3b} + 3x^{-b} + 5)$

95. $(a^m + a^n)^2$

96. $(x^w - x^t)^2$

97. $(5y^m + 8z^k)(3y^{2m} + 4z^{3-k})$

98. $(4x^{a-1} + 3y^{b+5})(x^{2a-3} - 2y^{4-b})$

⟨5⟩ Polynomial Functions

Solve each problem. See Example 7.

99. *Area of a room.* The length of a rectangular room is $x + 3$ meters, and its width is $x + 1$ meters. Find a polynomial function $A(x)$ that gives the area in square meters.

100. *House plans.* Barbie and Ken planned to build a square house that was x feet on each side. Then they revised the plan so that one side was lengthened by 20 feet and the other side was shortened by 6 feet, as shown in the accompanying figure. Find a polynomial function $A(x)$ that gives the area of the revised house in square feet.

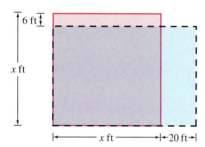

Figure for Exercise 100

101. *Available habitat.* The available habitat for a wild animal excludes an area of uniform width on the edge of an 8-kilometer by 10-kilometer rectangular forest preserve as shown in the figure.

a) Find a trinomial function $A(x)$ that gives the area of the available habitat in square kilometers (km^2).

b) The value of x depends on the animal. What is the available habitat for a bobcat for which $x = 0.4$ kilometer?

Figure for Exercise 101

102. *Cubic coating.* Frozen specimens are stored in a cubic metal box that is x inches on each side. The box is surrounded by a 2-inch-thick layer of Styrofoam insulation.
a) Find a polynomial function $V(x)$ that gives the total volume in cubic inches for the box and insulation.

b) Find the total volume if x is 10 inches.

103. *Overflow pan.* A metalworker makes an overflow pan by cutting equal squares with sides of length x feet from the corners of a 4-foot by 6-foot piece of aluminum, as shown in the figure. The sides are then folded up and the corners sealed.
a) Find a polynomial function $V(x)$ that gives the volume of the pan in cubic feet (ft^3).

b) Find the volume of the pan (to the nearest tenth of a cubic foot) if the height is 4 inches.

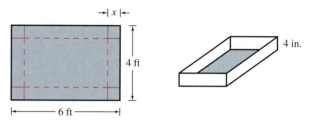

Figure for Exercise 103

104. *Square pan.* Suppose that the pan in Exercise 103 is formed from a square piece of aluminum that is 6 feet on each side.
a) Find a polynomial function $V(x)$ that gives the volume in cubic feet.

b) The cost is \$0.50 per square foot of aluminum used in the finished pan. Find a polynomial function $C(x)$ that gives the cost in dollars.

Getting More Involved

105. *Exploration*
a) Find $(a + b)^3$ by multiplying $(a + b)^2$ by $a + b$.

b) Next find $(a + b)^4$ and $(a + b)^5$.

c) How many terms are in each of these powers of $a + b$ after combining like terms?

d) Make a general statement about the number of terms in $(a + b)^n$.

106. *Cooperative learning*

Make a four-column table with columns for a, b, $(a + b)^2$, and $a^2 + b^2$. Work with a group to fill in the table with five pairs of numbers for a and b for which $(a + b)^2 \neq a^2 + b^2$. For what values of a and b does $(a + b)^2 = a^2 + b^2$?

107. *Discussion*

The area of the large square shown in the figure is $(a + b)^2$. Find the area of each of the four smaller regions in the figure, and then find the sum of those areas. What conclusion can you draw from these areas about $(a + b)^2$?

Figure for Exercise 107

Mid-Chapter **Quiz** | Sections 5.1 through 5.4 | **Chapter 5**

Simplify. All variables represent nonzero real numbers. Use only positive exponents in your answers.

1. -2^4

2. $(-2)^{-4}$

3. $-2^0 \cdot 5^{-2} \cdot 5^1$

4. $x^3 \cdot x^2 \cdot x$

5. $\dfrac{a^9b^2}{a^2b^{-3}}$

6. $(2a^{-5}b^3)^5$

7. $\left(\dfrac{ab^{-6}}{a^7b^4}\right)^3$

8. $(2xy^{-3})^2(3x^3y^{-4})^{-3}$

9. $(3.5 \times 10^3)(4 \times 10^{-4})^2$

10. $\dfrac{4.5 \times 10^{-4}}{(3 \times 10^{-5})^2}$

Perform the indicated operations.

11. $(5x^2 - 3x) + (-8x^2 - 2x + 6)$

12. $(5x^2 - 3x) - (-8x^2 - 2x + 6)$

13. $(-5x^3y)(2xy^3)$

14. $-6x^2(2x - 3)$

15. $(w + 2)(w^2 - 2w + 4)$

16. $2 - 3x(4x - 7)$

17. $(2 - 3x)(4x - 7)$

18. $(4a - 3b)^2$

19. $(9a + 5)(9a - 5)$

20. $(a + 5)^3$

Miscellaneous

21. Find the value of the polynomial $3x^3 - 5x^2 + 6x - 9$ when $x = -2$.

22. Find $P(-1)$ if $P(x) = -8x^4 + 9x^3 - 7x^2 + 5$.

5.5　Factoring Polynomials

In This Section

⟨1⟩ **Factoring Out the Greatest Common Factor (GCF)**

⟨2⟩ **Factoring by Grouping**

⟨3⟩ **Factoring the Difference of Two Squares**

⟨4⟩ **Factoring Perfect Square Trinomials**

⟨5⟩ **Factoring a Polynomial Completely**

In Sections 5.3 and 5.4, we multiplied polynomials. In this section and in Sections 5.6 and 5.7 we will be factoring polynomials. We begin with some special cases and do more general factoring in Section 5.6. Factoring will then be used to solve equations and problems in Section 5.8.

⟨1⟩ Factoring Out the Greatest Common Factor (GCF)

A natural number larger than 1 that has no factors other than itself and 1 is called a **prime number.** The numbers

$$2, 3, 5, 7, 11, 13, 17, 19, 23$$

are the first nine prime numbers. There are infinitely many prime numbers.

To factor a natural number **completely** means to write it as a product of prime numbers. In factoring 12 we might write $12 = 4 \cdot 3$. However, 12 is not factored completely as $4 \cdot 3$ because 4 is not a prime. To factor 12 completely, we write $12 = 2 \cdot 2 \cdot 3$ (or $2^2 \cdot 3$).

We use the distributive property to multiply a monomial and a binomial:

$$6x(2x - 1) = 12x^2 - 6x$$

If we start with $12x^2 - 6x$, we can use the distributive property to get

$$12x^2 - 6x = 6x(2x - 1).$$

We have **factored out** $6x$, which is a common factor of $12x^2$ and $-6x$. We could have factored out just 3 to get

$$12x^2 - 6x = 3(4x^2 - 2x),$$

but this would not be factoring out the *greatest* common factor. The **greatest common factor** (GCF) is a monomial that includes every number or variable that is a factor of all of the terms of the polynomial.

We can use the following strategy for finding the greatest common factor of a group of terms.

Strategy for Finding the Greatest Common Factor (GCF)

1. Factor each term completely.

2. Write a product using each factor that is common to all of the terms.

3. On each of these factors, use an exponent equal to the smallest exponent that appears on that factor in any of the terms.

E X A M P L E   1

The greatest common factor
Find the greatest common factor (GCF) for each group of terms.

a) $8x^2y,\ 20xy^3$ **b)** $30a^2,\ 45a^3b^2,\ 75a^4b$

Solution

a) First factor each term completely:

$$8x^2y = 2^3x^2y$$
$$20xy^3 = 2^2 \cdot 5xy^3$$

The factors common to both terms are 2, x, and y. In the GCF we use the smallest exponent that appears on each factor in either of the terms. So the GCF is 2^2xy or $4xy$.

b) First factor each term completely:

$$30a^2 = 2 \cdot 3 \cdot 5a^2$$
$$45a^3b^2 = 3^2 \cdot 5a^3b^2$$
$$75a^4b = 3 \cdot 5^2a^4b$$

The GCF is $3 \cdot 5a^2$ or $15a^2$.

> Now do Exercises 1–6

To factor out the GCF from a polynomial, find the GCF for the terms and then use the distributive property to factor it out.

E X A M P L E   2

Factoring out the greatest common factor
Factor each polynomial by factoring out the GCF.

a) $5x^4 - 10x^3 + 15x^2$ **b)** $8xy^2 + 20x^2y$

c) $60x^5 + 24x^3 + 36x^2$

Solution

a) First factor each term completely:

$$5x^4 = 5x^4 \qquad 10x^3 = 2 \cdot 5x^3 \qquad 15x^2 = 3 \cdot 5x^2$$

The GCF of the three terms is $5x^2$. Now factor $5x^2$ out of each term:

$$5x^4 - 10x^3 + 15x^2 = 5x^2(x^2 - 2x + 3)$$

b) The GCF for $8xy^2$ and $20x^2y$ is $4xy$:

$$8xy^2 + 20x^2y = 4xy(2y + 5x)$$

c) First factor each coefficient in $60x^5 + 24x^3 + 36x^2$:

$$60 = 2^2 \cdot 3 \cdot 5 \qquad 24 = 2^3 \cdot 3 \qquad 36 = 2^2 \cdot 3^2$$

The GCF of the three terms is $2^2 \cdot 3x^2$ or $12x^2$:

$$60x^5 + 24x^3 + 36x^2 = 12x^2(5x^3 + 2x + 3)$$

> **Now do Exercises 7–14**

Once you determine the greatest common factor for a group of terms, you have a choice. You can factor out the GCF or its opposite. Sometimes it is necessary to factor out the opposite of the GCF (see Example 5). You can factor out the opposite of the GCF by simply changing all of the signs, as shown in Example 3.

E X A M P L E   3

Factoring out the opposite of the GCF

Factor each polynomial twice. First factor out the GCF, and then factor out the opposite of the GCF.

a) $5x - 5y$ **b)** $-x^2 - 3$ **c)** $-x^3 + 3x^2 - 5x$

Solution

a) $5x - 5y = 5(x - y)$ Factor out 5.

$\qquad\qquad = -5(-x + y)$ Factor out -5.

b) $-x^2 - 3 = 1(-x^2 - 3)$ The GCF is 1.

$\qquad\qquad = -1(x^2 + 3)$ Factor out -1.

c) $-x^3 + 3x^2 - 5x = x(-x^2 + 3x - 5)$ Factor out x.

$\qquad\qquad\qquad = -x(x^2 - 3x + 5)$ Factor out $-x$.

> **Now do Exercises 15–22**

Sometimes the common factor is not a monomial. In Example 4, we factor out a binomial.

E X A M P L E   4

Factoring out a binomial

Factor.

a) $(x + 3)w + (x + 3)a$ **b)** $x(x - 9) - 4(x - 9)$

Solution

a) We treat $x + 3$ like a common monomial when factoring:

$$(x + 3)w + (x + 3)a = (x + 3)(w + a)$$

b) Factor out the common binomial $x - 9$:

$$x(x - 9) - 4(x - 9) = (x - 4)(x - 9)$$

Now do Exercises 23–30

⟨2⟩ Factoring by Grouping

The product of two binomials may have four terms. For example,

$$(x + a)(x + 2) = (x + a)x + (x + a)2$$
$$= x^2 + ax + 2x + 2a.$$

To factor $x^2 + ax + 2x + 2a$, we simply reverse the steps we used to find the product. Factor out the common factor x from the first two terms and the common factor 2 from the last two terms:

$$x^2 + ax + 2x + 2a = x(x + a) + 2(x + a) \quad \text{Factor out } x \text{ and } 2.$$
$$= (x + 2)(x + a) \quad \text{Factor out } x + a.$$

This method of factoring is called **factoring by grouping.** To factor by grouping it is sometimes necessary to factor out the opposite of the greatest common factor, as was shown in Example 3.

Strategy for Factoring a Four-Term Polynomial by Grouping

1. Factor out the GCF from the first group of two terms.

2. Factor out the GCF from the last group of two terms.

3. Factor out the common binomial.

EXAMPLE 5

Factoring by grouping

Factor each four-term polynomial by grouping.

a) $2x + 2y + ax + ay$ **b)** $wa - wb + a - b$ **c)** $4am - 4an - bm + bn$

Solution

a) The first group of two terms has 2 as a common factor, and the second group of two terms has a as a common factor:

$$2x + 2y + ax + ay = 2(x + y) + a(x + y)$$
$$= (2 + a)(x + y)$$

b) Factor w out of the first two terms and 1 out of the last two terms:

$$wa - wb + a - b = w(a - b) + 1(a - b)$$
$$= (w + 1)(a - b)$$

c) Factor out $4a$ from the first two terms and $-b$ (the opposite of the greatest common factor) from the last two terms:

$$4am - 4an - bm + bn = 4a(m - n) - b(m - n)$$
$$= (4a - b)(m - n)$$

Now do Exercises 31–40

⟨3⟩ Factoring the Difference of Two Squares

A first-degree polynomial in one variable, such as $3x - 5$, is called a linear polynomial. (The equation $3x - 5 = 0$ is a linear equation.)

Linear Polynomial

If a and b are real numbers with $a \neq 0$, then $ax + b$ is called a **linear polynomial.**

A second-degree polynomial such as $x^2 + 5x - 6$ is called a quadratic polynomial.

Quadratic Polynomial

If a, b, and c are real numbers with $a \neq 0$, then $ax^2 + bx + c$ is called a **quadratic polynomial.**

One of the main goals of this chapter is to write a quadratic polynomial (when possible) as a product of linear factors.

Consider the quadratic polynomial $x^2 - 25$. We recognize that $x^2 - 25$ is a difference of two squares, $x^2 - 5^2$. We recall that the product of a sum and a difference is a difference of two squares: $(a + b)(a - b) = a^2 - b^2$. If we reverse this special product rule, we get a rule for factoring the difference of two squares.

Factoring the Difference of Two Squares

$$a^2 - b^2 = (a + b)(a - b)$$

The difference of two squares factors as the product of a sum and a difference. To factor $x^2 - 25$, we replace a by x and b by 5 to get

$$x^2 - 25 = (x + 5)(x - 5).$$

This equation expresses a quadratic polynomial as a product of two linear factors.

EXAMPLE **6**

Factoring the difference of two squares

Factor each polynomial.

   a) $y^2 - 36$ **b)** $9x^2 - 1$ **c)** $4x^2 - y^2$

⟨ **Helpful Hint** ⟩

Using the power of a power rule, we can see that any even power is a perfect square:

$$x^{2n} = (x^n)^2$$

Solution

Each of these binomials is a difference of two squares. Each binomial factors into a product of a sum and a difference.

   a) $y^2 - 36 = (y + 6)(y - 6)$ We could also write $(y - 6)(y + 6)$ because the factors can be written in any order.

   b) $9x^2 - 1 = (3x + 1)(3x - 1)$

   c) $4x^2 - y^2 = (2x + y)(2x - y)$

Now do Exercises 41–48

CAUTION   We can factor $a^2 - b^2$, but what about $a^2 + b^2$? The polynomial $a^2 + b^2$ (a sum of two squares) cannot be factored. You will see why when we study the general question of whether a polynomial can be factored in Section 5.7.

⟨4⟩ Factoring Perfect Square Trinomials

The trinomial that results from squaring a binomial is called a **perfect square trinomial.** We can reverse the rules from Section 5.4 for the square of a sum or a difference to get rules for factoring.

Factoring Perfect Square Trinomials
$$a^2 + 2ab + b^2 = (a + b)^2$$ $$a^2 - 2ab + b^2 = (a - b)^2$$

Consider the polynomial $x^2 + 6x + 9$. If we recognize that

$$x^2 + 6x + 9 = x^2 + 2 \cdot x \cdot 3 + 3^2,$$

then we can see that it is a perfect square trinomial. It fits the rule if $a = x$ and $b = 3$:

$$x^2 + 6x + 9 = (x + 3)^2$$

Perfect square trinomials can be identified by using the following strategy.

Strategy for Identifying Perfect Square Trinomials

A trinomial is a perfect square trinomial if

1. the first and last terms are of the form a^2 and b^2,

2. the middle term is 2 or -2 times the product of a and b.

We use this strategy in Example 7.

EXAMPLE   7

Factoring perfect square trinomials

Factor each polynomial.

a) $x^2 - 8x + 16$ **b)** $a^2 + 14a + 49$ **c)** $4x^2 + 12x + 9$

Solution

a) Because the first term is x^2, the last is 4^2, and $-2(x)(4)$ is equal to the middle term $-8x$, the trinomial $x^2 - 8x + 16$ is a perfect square trinomial:

$$x^2 - 8x + 16 = (x - 4)^2$$

b) Because $49 = 7^2$ and $14a = 2(a)(7)$, we have a perfect square trinomial:

$$a^2 + 14a + 49 = (a + 7)^2$$

c) Because $4x^2 = (2x)^2$, $9 = 3^2$, and the middle term $12x$ is equal to $2(2x)(3)$, the trinomial $4x^2 + 12x + 9$ is a perfect square trinomial:

$$4x^2 + 12x + 9 = (2x + 3)^2$$

Now do Exercises 49–54

⟨5⟩ Factoring a Polynomial Completely

Polynomials that cannot be factored are called **prime polynomials.** Because binomials such as $x + 5$, $a - 6$, and $3x + 1$ cannot be factored, they are prime polynomials. A polynomial is **factored completely** when it is written as a product of prime polynomials. To factor completely, always factor out the GCF (or its opposite) first. Then continue to factor until all of the factors are prime.

EXAMPLE 8

Factoring completely

Factor each polynomial completely.

a) $5x^2 - 20$ **b)** $3a^3 - 30a^2 + 75a$

c) $-2b^3 - 16b^2 - 32b$

Solution

a) $5x^2 - 20 = 5(x^2 - 4)$ Greatest common factor

              $= 5(x - 2)(x + 2)$ Difference of two squares

b) $3a^3 - 30a^2 + 75a = 3a(a^2 - 10a + 25)$ Greatest common factor

                       $= 3a(a - 5)^2$ Perfect square trinomial

c) $-2b^3 - 16b^2 - 32b = -2b(b^2 + 8b + 16)$ Factor out $-2b$ to make the next step easier.

                      $= -2b(b + 4)^2$ Perfect square trinomial

> **Now do Exercises 55–84**

Warm-Ups ▼

Fill in the blank.

1. A _____ number is a natural number greater than 1 that has no factors other than itself and 1.

2. A natural number is factored _____ when it is expressed as a product of prime numbers.

3. The _____ for a polynomial is a monomial that includes every number or variable that is a factor of all terms of the polynomial.

4. A _____ polynomial is a polynomial of the form $ax + b$ where $a \neq 0$.

5. A _____ polynomial is a polynomial of the form $ax^2 + bx + c$ where $a \neq 0$.

6. A _____ polynomial is a polynomial that cannot be factored.

True or false?

7. Both $3xy$ and $-3xy$ are factors of $3x^2y - 6xy^2$.

8. The GCF for $8a^3 - 15b^2$ is 1.

9. For any x, $2x - 4 = -2(2 - x)$.

10. The polynomial $x^2 + 6x + 9$ is a perfect square trinomial.

11. The polynomial $x^2 + 16$ is a perfect square trinomial.

12. The binomial $x + 1$ is a factor of $x^2 - 1$.

13. For any x, $x^3 - 9 = (x - 3)(x^2 + 3)$.

14. For any x, $x^2 - 81 = (x - 9)^2$.

‹1› Factoring Out the Greatest Common Factor (GCF)

Find the greatest common factor for each group of terms. See Example 1. See the Strategy for Finding the GCF box on page 334.

1. $48, 36x$
2. $42a, 28a^2$
3. $9wx, 21wy, 15xy$
4. $70x^2, 84x, 42x^3$
5. $24x^2y, 42xy^2, 66xy^3$
6. $60a^2b^5, 140a^9b^2, 40a^3b^6$

Factor out the greatest common factor in each expression. See Example 2.

7. $x^3 - 5x$
8. $10x^2 - 20y^3$
9. $48wx + 36wy$
10. $42wz + 28wa$
11. $2x^3 - 4x^2 + 6x$
12. $6x^3 - 12x^2 + 18x$
13. $36a^3b^6 - 24a^4b^2 + 60a^5b^3$
14. $44x^8y^6z - 110x^6y^9z^2$

Factor out the greatest common factor, and then factor out the opposite of the greatest common factor. See Example 3.

15. $2x - 2y$
16. $-3x + 6$
17. $6x^2 - 3x$
18. $10x^2 + 5x$
19. $-w^3 + 3w^2$
20. $-2w^4 + 6w^3$
21. $-a^3 + a^2 - 7a$
22. $-2a^4 - 4a^3 + 6a^2$

Factor each expression by factoring out a binomial or a power of a binomial. See Example 4.

23. $(x - 6)a + (x - 6)b$
24. $(y - 4)3 + (y - 4)b$
25. $(y - 4)x - (y - 4)3$
26. $(a + 1)b - (a + 1)c$
27. $(y - 1)^2y + (y - 1)^2z$
28. $(w - 2)^2 \cdot w + (w - 2)^2 \cdot 3$
29. $a(a - b)^2 - b(a - b)^2$
30. $x(x + y)^2 + y(x + y)^2$

‹2› Factoring by Grouping

Factor each polynomial by grouping. See Example 5. See the Strategy for Factoring a Four-Term Polynomial by Grouping on page 336.

31. $ax + ay + 3x + 3y$
32. $2a + 2b + wa + wb$
33. $xy - 3y + x - 3$
34. $2wt - 2wa + t - a$
35. $4a - 4b - ca + cb$
36. $pr - 2r - ap + 2a$
37. $xy - y - 6x + 6$
38. $-3a + 3 + ax - x$
39. $x^2 + 8x - 2x - 16$
40. $a^2 - 6a + 3a - 18$

‹3› Factoring the Difference of Two Squares

Factor each polynomial. See Example 6.

41. $x^2 - 100$
42. $81 - y^2$
43. $4y^2 - 49$
44. $16b^2 - 1$
45. $9x^2 - 25a^2$
46. $121a^2 - b^2$
47. $144w^2z^2 - h^2$
48. $x^2y^2 - 9c^2$

‹4› Factoring Perfect Square Trinomials

Factor each polynomial. See Example 7. See the Strategy for Identifying Perfect Square Trinomials box on page 338.

49. $x^2 - 20x + 100$
50. $y^2 + 10y + 25$
51. $4m^2 - 4m + 1$
52. $9t^2 + 30t + 25$

53. $w^2 - 2wt + t^2$
54. $4r^2 + 20rt + 25t^2$

⟨5⟩ **Factoring a Polynomial Completely**

Factor each polynomial completely. See Example 8.

55. $2x^2 - 8$
56. $3x^3 - 27x$
57. $x^3 + 10x^2 + 25x$
58. $5a^4m - 45a^2m$
59. $4x^2 + 4x + 1$
60. $ax^2 - 8ax + 16a$
61. $(x + 3)x + (x + 3)7$
62. $(x - 2)x - (x - 2)5$
63. $6y^2 + 3y$
64. $4y^2 - y$
65. $4x^2 - 20x + 25$
66. $a^3x^3 - 6a^2x^2 + 9ax$
67. $(2x - 3)x - (2x - 3)2$
68. $(2x + 1)x + (2x + 1)3$
69. $9a^3 - aw^2$
70. $2bn^2 - 4b^2n + 2b^3$
71. $-5a^2 + 30a - 45$
72. $-2x^2 + 50$
73. $-3y^3 - 18y^2 - 27y$
74. $-2m^2n - 8mn - 8n$
75. $-7a^2b^2 + 7$
76. $-17a^2 - 17a$
77. $7x - 7h - hx + h^2$
78. $6a - 6y - ax + xy$
79. $a^2x + 3a^2 - 4x - 12$
80. $x^2y - 2x^2 - y + 2$
81. $ab^2 + 3b^2 - 9a - 27$
82. $w^2x + 3w^2 - 75 - 25x$
83. $4a^2b - 4b + 8a^2 - 8$
84. $2x^2y^2 + 6xy^2 - 24x - 8x^2$

Miscellaneous

Replace k in each trinomial by a number that makes the trinomial a perfect square trinomial.

85. $x^2 + 6x + k$ **86.** $y^2 - 8y + k$
87. $4a^2 - ka + 25$ **88.** $9u^2 + kuv + 49v^2$
89. $km^2 - 24m + 9$ **90.** $kz^2 + 40z + 16$

Applications

Solve each problem.

91. *Volume of a bird cage.* A company makes rectangular shaped bird cages with height b inches and square bottoms. The volume of these cages is given by the function
$$V = b^3 - 6b^2 + 9b.$$

a) Find an expression for the length of each side of the square bottom by factoring the expression on the right side of the function.

b) Use the function to find the volume of a cage with a height of 18 inches.

c) Use the accompanying graph to estimate the height of a cage for which the volume is 20,000 cubic inches.

Figure for Exercise 91

92. *Pyramid power.* A powerful crystal pyramid has a square base and a volume of $3y^3 + 12y^2 + 12y$ cubic centimeters. If its height is y centimeters, then what polynomial represents the length of a side of the square base? $\left(\text{The volume of a pyramid with a square base of area } a^2 \text{ and height } h \text{ is given by } V = \frac{ha^2}{3}.\right)$

Figure for Exercise 92

Getting More Involved

93. *Cooperative learning*

List the perfect square trinomials corresponding to $(x + 1)^2, (x + 2)^2, (x + 3)^2, \ldots, (x + 12)^2$. Use your list to quiz a classmate. Read a perfect square trinomial at random from your list, and ask your classmate to write its factored form. Repeat until both of you have mastered these 12 perfect square trinomials.

5.6   Factoring $ax^2 + bx + c$

In Section 5.5, you learned to factor certain special polynomials. In this section, you will learn to factor general quadratic polynomials. We first factor $ax^2 + bx + c$ with $a = 1$, and then we consider the case $a \neq 1$.

In This Section

⟨1⟩ Factoring $ax^2 + bx + c$ with $a = 1$

⟨2⟩ Factoring $ax^2 + bx + c$ with $a \neq 1$

⟨3⟩ Trial and Error

⟨4⟩ Factoring by Substitution

⟨1⟩ Factoring $ax^2 + bx + c$ with $a = 1$

Suppose $x^2 + bx + c$ factors as $(x + m)(x + n)$. Then the product of $x + m$ and $x + n$ must equal $x^2 + bx + c$:

$$(x + m)(x + n) = (x + m)x + (x + m)n \quad \text{Distributive property}$$
$$= x^2 + mx + nx + mn \quad \text{Distributive property}$$
$$= x^2 + (m + n)x + mn \quad \text{Combine like terms.}$$
$$= x^2 + bx + c$$

So $c = mn$ and $b = m + n$. To factor $x^2 + bx + c$ we simply find two integers (m and n) whose product is c and whose sum is b and then reverse the steps that we just used in finding the product $(x + m)(x + n)$. We summarize these ideas with the following strategy.

Strategy for Factoring $x^2 + bx + c$ by Grouping

To factor $x^2 + bx + c$:

1. Find two integers that have a product of c and a sum equal to b.

2. Replace bx by the sum of two terms whose coefficients are the two numbers found in step 1.

3. Factor the resulting four-term polynomial by grouping.

E X A M P L E   **1**

Factoring $ax^2 + bx + c$ with $a = 1$ by grouping
Factor each trinomial by grouping.

a) $x^2 + 9x + 18$

b) $x^2 - 2x - 24$

Solution

a) We need to find two integers with a product of 18 and a sum of 9. For a product of 18 we could use 1 and 18, 2 and 9, or 3 and 6. Only 3 and 6 have a sum of 9. So we replace $9x$ with $3x + 6x$ and factor by grouping:

$$x^2 + 9x + 18 = x^2 + 3x + 6x + 18 \quad \text{Replace } 9x \text{ by } 3x + 6x.$$
$$= (x + 3)x + (x + 3)6 \quad \text{Factor out common factors.}$$
$$= (x + 3)(x + 6) \quad \text{Check by using FOIL.}$$

b) We need to find two integers with a product of -24 and a sum of -2. For a product of 24 we have 1 and 24, 2 and 12, 3 and 8, or 4 and 6. To get a product of -24 and a sum of -2, we must use 4 and -6:

$$x^2 - 2x - 24 = x^2 - 6x + 4x - 24 \quad \text{\color{blue} Replace } -2x \text{ with } -6x + 4x.$$
$$= (x - 6)x + (x - 6)4 \quad \text{\color{blue} Factor out common factors.}$$
$$= (x - 6)(x + 4) \quad \text{\color{blue} Check by using FOIL.}$$

> **Now do Exercises 1–6**

We factored the trinomials in Example 1 by grouping to show that we could reverse the steps in the multiplication of binomials. However, it is not necessary to write down all of the details shown in Example 1. In Example 2 we simply write the factors.

EXAMPLE   **2**   **A simpler way to factor $ax^2 + bx + c$ with $a = 1$**

Factor each quadratic polynomial.

a) $x^2 + 4x + 3$ **b)** $x^2 + 3x - 10$ **c)** $a^2 - 5a + 6$

Solution

a) Two integers with a product of 3 and a sum of 4 are 1 and 3:

$$x^2 + 4x + 3 = (x + 1)(x + 3)$$

Check by using FOIL.

b) Two integers with a product of -10 and a sum of 3 are 5 and -2:

$$x^2 + 3x - 10 = (x + 5)(x - 2)$$

Check by using FOIL.

c) Two integers with a product of 6 and a sum of -5 are -3 and -2:

$$a^2 - 5a + 6 = (a - 3)(a - 2)$$

Check by using FOIL.

> **Now do Exercises 7–12**

⟨2⟩ Factoring $ax^2 + bx + c$ with $a \neq 1$

If the leading coefficient of $ax^2 + bx + c$ is not 1, there are two ways to proceed: the *ac* method or the trial-and-error method. The *ac* method is a slight variation of the grouping method shown in Example 1, and there is a definite procedure to follow. The trial-and-error method is not as definite. We write down possible factors. If they check, you are done, and if not, you try again. We will present the *ac* method first.

Suppose $ax^2 + bx + c$ factors as $(sx + m)(tx + n)$. Then the product of $sx + m$ and $tx + n$ must equal $ax^2 + bx + c$:

$$(sx + m)(tx + n) = stx^2 + (sn + mt)x + mn$$
$$= ax^2 + bx + c$$

So to factor $ax^2 + bx + c$ we must find two integers (sn and mt) whose sum is b and whose product is ac ($sn \cdot mt = stmn = ac$). Then we proceed to factor by grouping as we did in Example 1. This method is known as the **ac method.** It is a generalization of the grouping method used in Example 1. We summarize it with the following strategy.

Strategy for Factoring $ax^2 + bx + c$ by the ac Method

To factor $ax^2 + bx + c$:

1. Find two integers that have a product of ac and a sum equal to b.

2. Replace bx by the sum of two terms whose coefficients are the two numbers found in step 1.

3. Factor the resulting four-term polynomial by grouping.

E X A M P L E   **3**

Factoring $ax^2 + bx + c$ with $a \neq 1$

Factor each trinomial.

a) $2x^2 + 9x + 4$ **b)** $2x^2 + 5x - 12$

c) $6w^2 - w - 15$

Solution

a) Because $ac = 2 \cdot 4 = 8$, we need two numbers with a product of 8 and a sum of 9. The numbers are 1 and 8. Replace $9x$ by $x + 8x$ and factor by grouping:

$$2x^2 + 9x + 4 = 2x^2 + x + 8x + 4$$
$$= (2x + 1)x + (2x + 1)4$$
$$= (2x + 1)(x + 4) \quad \text{Check by FOIL.}$$

Note that if you start with $2x^2 + 8x + x + 4$, and factor by grouping, you get the same result.

b) Because $ac = 2(-12) = -24$, we need two numbers with a product of -24 and a sum of 5. The pairs of numbers with a product of 24 are 1 and 24, 2 and 12, 3 and 8, and 4 and 6. To get a product of -24, one of the numbers must be negative and the other positive. To get a sum of positive 5, we need -3 and 8:

$$2x^2 + 5x - 12 = 2x^2 - 3x + 8x - 12$$
$$= (2x - 3)x + (2x - 3)4$$
$$= (2x - 3)(x + 4) \quad \text{Check by FOIL.}$$

c) To factor $6w^2 - w - 15$ we first find that $ac = 6(-15) = -90$. Now we need two numbers that have a product of -90 and a sum of -1. The numbers are -10 and 9. Replace $-w$ with $-10w + 9w$ and factor by grouping:

$$6w^2 - w - 15 = 6w^2 - 10w + 9w - 15$$
$$= 2w(3w - 5) + 3(3w - 5)$$
$$= (2w + 3)(3w - 5) \quad \text{Check by FOIL.}$$

> Now do Exercises 13–24

❬3❭ Trial and Error

After we have gained some experience at factoring by grouping, we can often find the factors without going through the steps of grouping. Consider the polynomial

$$2x^2 - 7x + 6.$$

‹ **Helpful Hint** ›

Trial and error can save time, but it can also give you a wrong answer quickly. Be sure to double check your answer using FOIL.

The factors of $2x^2$ can only be $2x$ and x. The factors of 6 could be 2 and 3 or 1 and 6. We can list all of the possibilities that give the correct first and last terms without putting in the signs:

$$(2x \quad 2)(x \quad 3) \qquad (2x \quad 6)(x \quad 1)$$
$$(2x \quad 3)(x \quad 2) \qquad (2x \quad 1)(x \quad 6)$$

Before actually trying these out, we make an important observation. If $(2x \quad 2)$ or $(2x \quad 6)$ were one of the factors, then there would be a common factor 2 in the original trinomial, but there is not. *If the original trinomial has no common factor, there can be no common factor in either of its linear factors.* Since 6 is positive and the middle term is $-7x$, both of the missing signs must be negative. So the only possibilities are $(2x - 1)(x - 6)$ and $(2x - 3)(x - 2)$. The middle term of the first product is $-13x$, and the middle term of the second product is $-7x$. So we have found the factors:

$$2x^2 - 7x + 6 = (2x - 3)(x - 2)$$

Even though there may be many possibilities in some factoring problems, often we find the correct factors without writing down every possibility. We can use a bit of guesswork in factoring trinomials. *Try* whichever possibility you think might work. *Check* it by multiplying. If it is not right, then *try again*. That is why this method is called **trial and error.**

EXAMPLE 4

Trial and error

Factor each quadratic trinomial using trial and error.

a) $2x^2 + 5x - 3$ **b)** $3x^2 - 11x + 6$ **c)** $6m^2 + 17m + 10$

Solution

a) Because $2x^2$ factors only as $2x \cdot x$ and 3 factors only as $1 \cdot 3$, there are only two possible ways to factor this trinomial to get the correct first and last terms:

$$(2x \quad 1)(x \quad 3) \qquad \text{and} \qquad (2x \quad 3)(x \quad 1)$$

Because the last term of the trinomial is negative, one of the missing signs must be $+$, and the other must be $-$. Now we try the various possibilities until we get the correct middle term:

$$(2x + 1)(x - 3) = 2x^2 - 5x - 3$$
$$(2x + 3)(x - 1) = 2x^2 + x - 3$$
$$(2x - 1)(x + 3) = 2x^2 + 5x - 3$$

Since the last product has the correct middle term, the trinomial is factored as

$$2x^2 + 5x - 3 = (2x - 1)(x + 3).$$

b) There are four possible ways to factor $3x^2 - 11x + 6$:

$$(3x \quad 1)(x \quad 6) \qquad (3x \quad 2)(x \quad 3)$$
$$(3x \quad 6)(x \quad 1) \qquad (3x \quad 3)(x \quad 2)$$

Because the last term is positive and the middle term is negative, both signs must be negative. Observe that we can eliminate $(3x \quad 6)$ and $(3x \quad 3)$ as possible factors since 3 is not a common factor in the original trinomial. Now try possible factors until we get the correct middle term:

$$(3x - 1)(x - 6) = 3x^2 - 19x + 6$$
$$(3x - 2)(x - 3) = 3x^2 - 11x + 6$$

The trinomial is factored correctly as

$$3x^2 - 11x + 6 = (3x - 2)(x - 3).$$

c) Because all of the signs in $6m^2 + 17m + 10$ are positive, all of the signs in the factors are positive. There are eight possible products that will start with $6m^2$ and end with 10:

$(2m + 2)(3m + 5)$	$(6m + 2)(m + 5)$
$(2m + 5)(3m + 2)$	$(6m + 5)(m + 2)$
$(2m + 1)(3m + 10)$	$(6m + 1)(m + 10)$
$(2m + 10)(3m + 1)$	$(6m + 10)(m + 1)$

We can eliminate $(2m + 2)$, $(6m + 2)$, $(2m + 10)$, and $(6m + 10)$ as factors since 2 is not a common factor in the original trinomial. Only $(6m + 5)(m + 2)$ has a middle term of $17m$. So,

$$6m^2 + 17m + 10 = (6m + 5)(m + 2).$$

> Now do Exercises 25–36

⟨4⟩ Factoring by Substitution

So far, the polynomials that we have factored, without common factors, have all been of degree 2 or 3. Some polynomials of higher degree can be factored by substituting a single variable for a variable with a higher power. After factoring, we replace the single variable by the higher-power variable. This method is called **substitution.**

E X A M P L E   5

Factoring by substitution
Factor each polynomial.

a) $x^4 - 9$

b) $y^8 - 14y^4 + 49$

Solution

a) We recognize $x^4 - 9$ as a difference of two squares in which $x^4 = (x^2)^2$ and $9 = 3^2$. If we let $w = x^2$, then $w^2 = x^4$. So we can replace x^4 by w^2 and factor:

$$x^4 - 9 = w^2 - 9 \qquad \text{Replace } x^4 \text{ by } w^2.$$
$$= (w + 3)(w - 3) \qquad \text{Difference of two squares}$$
$$= (x^2 + 3)(x^2 - 3) \qquad \text{Replace } w \text{ by } x^2.$$

b) We recognize $y^8 - 14y^4 + 49$ as a perfect square trinomial in which $y^8 = (y^4)^2$ and $49 = 7^2$. We let $w = y^4$ and $w^2 = y^8$:

$$y^8 - 14y^4 + 49 = w^2 - 14w + 49 \qquad \text{Replace } y^4 \text{ by } w \text{ and } y^8 \text{ by } w^2.$$
$$= (w - 7)^2 \qquad \text{Perfect square trinomial}$$
$$= (y^4 - 7)^2 \qquad \text{Replace } w \text{ by } y^4.$$

> Now do Exercises 37–46

CAUTION Polynomials that we factor by substitution must contain just the right exponents. We can factor $y^8 - 14y^4 + 49$ because y^8 is a perfect square: $y^8 = (y^4)^2$. Note that even powers such as x^4, y^{14}, and w^{20} are perfect squares, because $x^4 = (x^2)^2$, $y^{14} = (y^7)^2$, and $w^{20} = (w^{10})^2$.

In Example 6, we use substitution to factor polynomials that have variables as exponents.

E X A M P L E 6

Polynomials with variable exponents

Factor completely. The variables used in the exponents represent positive integers.

a) $x^{2m} - y^2$

b) $z^{2n+1} - 6z^{n+1} + 9z$

Solution

a) Notice that $x^{2m} = (x^m)^2$. So if we let $w = x^m$, then $w^2 = x^{2m}$:

$$x^{2m} - y^2 = w^2 - y^2 \qquad \text{Substitution}$$
$$= (w + y)(w - y) \qquad \text{Difference of two squares}$$
$$= (x^m + y)(x^m - y) \qquad \text{Replace } w \text{ by } x^m.$$

b) First factor out the common factor z:

$$z^{2n+1} - 6z^{n+1} + 9z = z(z^{2n} - 6z^n + 9)$$
$$= z(a^2 - 6a + 9) \qquad \text{Let } a = z^n.$$
$$= z(a - 3)^2 \qquad \text{Perfect square trinomial}$$
$$= z(z^n - 3)^2 \qquad \text{Replace } a \text{ by } z^n.$$

> **Now do Exercises 47–54**

It is not absolutely necessary to use substitution in factoring polynomials with higher degrees or variable exponents, as we did in Examples 5 and 6. In Example 7, we use trial and error to factor similar polynomials. Remember to always look for a common factor first.

E X A M P L E 7

Higher-degree and variable exponent trinomials

Factor each polynomial completely. Variables used as exponents represent positive integers.

a) $x^8 - 2x^4 - 15$

b) $-18y^7 + 21y^4 + 15y$

c) $2u^{2m} - 5u^m - 3$

Solution

a) To factor by trial and error, notice that $x^8 = x^4 \cdot x^4$. Now 15 is $3 \cdot 5$ or $1 \cdot 15$. Using 1 and 15 will not give the required -2 for the coefficient of the middle term. So choose 3 and -5 to get the -2 in the middle term:

$$x^8 - 2x^4 - 15 = (x^4 - 5)(x^4 + 3)$$

b) $-18y^7 + 21y^4 + 15y = -3y(6y^6 - 7y^3 - 5)$ Factor out the common factor $-3y$ first.

$$= -3y(2y^3 + 1)(3y^3 - 5)$$ Factor the trinomial by trial and error.

c) Notice that $2u^{2m} = 2u^m \cdot u^m$ and $3 = 3 \cdot 1$. Using trial and error, we get

$$2u^{2m} - 5u^m - 3 = (2u^m + 1)(u^m - 3).$$

Now do Exercises 55–70

Warm-Ups ▼

Fill in the blank.

1. To factor $x^2 + bx + c$ find two integers whose _____ is b and whose _____ is c.

2. To factor $ax^2 + bx + c$ find two integers whose _____ is b and whose _____ is ac.

3. When factoring a polynomial, first factor out the _____ .

4. Some higher-degree polynomials can be factored by _____ .

True or false?

Answer true if the factorization is correct and false otherwise.

5. $x^2 + 9x + 18 = (x + 3)(x + 6)$

6. $y^2 + 2y - 35 = (y + 5)(y - 7)$

7. $x^2 + 4 = (x + 2)(x + 2)$

8. $x^2 - 5x - 6 = (x - 3)(x - 2)$

9. $x^2 - 4x - 12 = (x - 6)(x + 2)$

10. $x^2 + 15x + 36 = (x + 4)(x + 9)$

11. $3x^2 + 4x - 15 = (3x + 5)(x - 3)$

12. $4x^2 + 4x - 3 = (4x - 1)(x + 3)$

13. $4x^2 - 4x - 3 = (2x + 1)(2x - 3)$

5.6 Exercises

‹ **Study Tips** ›

• Effective time management will allow adequate time for school, work, social life, and free time. However, at times you will have to sacrifice to do well.
• Everyone has different attention spans. Start by studying 10 to 15 minutes at a time and then build up to longer periods. Be realistic. When you can no longer concentrate, take a break.

‹1› **Factoring $ax^2 + bx + c$ with $a = 1$**

Factor each polynomial. See Examples 1 and 2. See the Strategy for Factoring $x^2 + bx + c$ by Grouping on page 342.

1. $x^2 + 4x + 3$

2. $y^2 + 5y + 6$

3. $a^2 + 15a + 50$

4. $t^2 + 11t + 24$

5. $y^2 - 5y - 14$

6. $x^2 - 3x - 18$

7. $x^2 - 6x + 8$

8. $y^2 - 13y + 30$

9. $a^2 - 12a + 27$

10. $x^2 - x - 30$

 11. $a^2 + 7a - 30$

12. $w^2 + 29w - 30$

⟨2⟩ Factoring $ax^2 + bx + c$ with $a \neq 1$

Factor each polynomial. See Example 3. See the Strategy for Factoring $ax^2 + bx + c$ by the ac Method box on page 344.

13. $6w^2 + 5w + 1$

14. $4x^2 + 11x + 6$

15. $2x^2 - 5x - 3$

16. $2a^2 + 3a - 2$

17. $4x^2 + 16x + 15$

18. $6y^2 + 17y + 12$

19. $6x^2 - 5x + 1$

20. $6m^2 - m - 12$

21. $12y^2 + y - 1$

22. $12x^2 + 5x - 2$

23. $6a^2 + a - 5$

24. $30b^2 - b - 3$

⟨3⟩ Trial and Error

Factor each polynomial. See Example 4.

25. $2x^2 + 15x - 8$

26. $3a^2 + 20a + 12$

27. $3b^2 - 16b - 35$

28. $2y^2 - 17y + 21$

29. $6w^2 + w - 12$

30. $15x^2 - x - 6$

31. $4x^2 - 5x + 1$

32. $4x^2 + 7x + 3$

33. $5m^2 + 13m - 6$

34. $5t^2 - 9t - 2$

 35. $6y^2 - 7y - 20$

36. $7u^2 + 11u - 6$

⟨4⟩ Factoring by Substitution

Factor each polynomial completely. See Example 5.

37. $x^{10} - 9$

38. $y^8 - 4$

39. $z^{12} - 6z^6 + 9$

40. $a^6 + 10a^3 + 25$

41. $2x^7 + 8x^4 + 8x$

42. $x^{13} - 6x^7 + 9x$

43. $4x^5 + 4x^3 + x$

44. $18x^6 + 24x^3 + 8$

45. $ax^4 + 8ax^2 + 12a$

46. $a^2b^4 + 6a^2b^2 + 5a^2$

Factor each polynomial completely. The variables used as exponents represent positive integers. See Example 6.

47. $a^{2n} - 1$

48. $b^{4n} - 9$

49. $a^{2r} + 6a^r + 9$

50. $u^{6n} - 4u^{3n} + 4$

51. $3a^{2w} - 30a^w + 75$

52. $4b^{4x} + 16b^{2x} + 16$

53. $k^{2w+1} - 10k^{w+1} + 25k$

54. $4a^{2t+1} + 4a^{t+1} + a$

Factor each polynomial completely. See Example 7. The variables used in exponents represent positive integers.

 55. $x^6 - 2x^3 - 35$

56. $x^4 + 7x^2 - 30$

57. $a^{20} - 20a^{10} + 100$

58. $b^{16} + 22b^8 + 121$

59. $-12a^5 - 10a^3 - 2a$

60. $-4b^7 + 4b^4 + 3b$

61. $x^{2a} + 2x^a - 15$

62. $y^{2b} + y^b - 20$

63. $x^{2a} - y^{2b}$

64. $w^{4m} - a^2$

65. $x^8 - x^4 - 6$

66. $m^{10} - 5m^5 - 6$

67. $x^{a+2} - x^a$

68. $y^{2a+1} - y$

69. $x^{2a} + 6x^a + 9$

70. $x^{2a} - 2x^a y^b + y^{2b}$

Miscellaneous

Factor each polynomial completely.

71. $2x^2 + 20x + 50$

72. $3a^2 + 6a + 3$

73. $a^3 - 36a$

74. $x^3 + 5x^2 - 6x$

75. $10a^2 + 55a - 30$

76. $6a^2 + 22a - 84$

77. $2x^2 - 128y^2$

78. $a^3 - 6a^2 + 9a$

79. $-9x^2 + 33x + 12$

80. $2xy^2 - 27xy + 70x$

81. $m^5 + 20m^4 + 100m^3$

82. $4a^2 - 16a + 16$

83. $6x^2 + 23x + 20$

84. $2y^2 - 13y + 6$

85. $9y^3 - 24y^2 + 16y$

86. $25m^3 - 10m^2 + m$

87. $r^2 - 6rs + 8s^2$

88. $7z^2 + 15zy + 2y^2$

89. $m^3 + 2m + 3m^2$

90. $7w^2 - 18w + w^3$

91. $6m^3 - m^2n - 2mn^2$

92. $3a^3 + 3a^2b - 18ab^2$

93. $9m^2 - 25n^2$

94. $m^2n^2 - 2mn^3 + n^4$

95. $5a^2 + 20a - 60$

96. $-3y^2 + 9y + 30$

97. $-2w^2 + 18w + 20$

98. $x^2z + 2xyz + y^2z$

99. $w^2x^2 - 100x^2$

100. $9x^2 + 30x + 25$

101. $81x^2 - 9$

102. $12w^2 - 38w - 72$

103. $8x^2 - 2x - 15$

104. $4w^2 + 12w + 9$

117. $xr - yr - xw + yw$

118. $y^2 - ay - by + ab$

119. $xt - t^2 - ax + at$

120. $2nw - 5w - 2n + 5$

121. $2qh - h + 8q - 4$

122. $2m^2 - 3am - 3at + 2mt$

123. $6t - 3ty + awy - 2aw$

124. $n^2b + 3b + 15 + 5n^2$

125. $x^3 + 7x - 7a - ax^2$

126. $t^2a - 3a - 3 + t^2$

127. $m^4 - 5m^2 + m^2p - 5p$

128. $x^2a^2 + a^2 + x^2 + 1$

Getting More Involved

105. *Discussion*

Which of the following are not perfect square trinomials? Explain.

a) $4a^6 - 6a^3b^4 + 9b^8$ b) $1000x^2 + 200ax + a^2$
c) $900y^4 - 60y^2 + 1$ d) $36 - 36z^7 + 9z^{14}$

106. *Discussion*

Which of the following is not a difference of two squares? Explain.

a) $16a^8y^4 - 25c^{12}$ b) $a^9 - b^4$
c) $t^{90} - 1$ d) $x^2 - 196$

107. *Writing*

Factor each polynomial and explain how you decided which method to use.

a) $x^2 + 10x + 25$
b) $x^2 - 10x + 25$
c) $x^2 + 26x + 25$
d) $x^2 - 25$
e) $x^2 + 25$

108. *Discussion*

On an exam, a student factored $2x^2 - 6x + 4$ as $(2x - 4)(x - 1)$. Even though the student carefully checked that
$$(2x - 4)(x - 1) = 2x^2 - 6x + 4,$$
the student lost some points. What went wrong?

Extra Factoring Exercises

Factor each polynomial by grouping.

109. $ax + 3x + 4a + 12$

110. $wb + 3w + 12 + 4b$

111. $ax - 2a + 4x - 8$

112. $ma - 3a + 8m - 24$

113. $bm - 4b - 5m + 20$

114. $ax - 4x - 7a + 28$

115. $nx - ax - ac + nc$

116. $bn - yb - yp + np$

Factor each polynomial using the trial-and-error method.

129. $y^2 + 3y + 2$

130. $a^2 + 7a + 10$

131. $x^2 + 10x + 21$

132. $p^2 + 9p + 20$

133. $a^2 + 15a + 54$

134. $b^2 + 14b + 40$

135. $y^2 + 3y - 10$

136. $a^2 + a - 30$

137. $w^2 - 2w - 15$

138. $m^2 + 8m - 9$

139. $b^2 + 6b - 16$

140. $z^2 - 2z - 35$

141. $a^2 - 8a - 33$

142. $b^2 + 2b - 48$

143. $a^2 - 9a + 18$

144. $b^2 - 4b + 3$

145. $x^2 - 11x + 24$

146. $x^2 - 8x + 12$

147. $y^2 - 23y + 130$

148. $a^2 - 20a + 96$

149. $2w^2 + 7w + 3$

150. $2b^2 + 5b - 3$

151. $2x^2 - 9x - 5$

152. $2x^2 - 9x + 4$

153. $3x^2 + 25x + 8$

154. $3x^2 - 25x + 8$

155. $3x^2 + 26x - 9$

156. $3x^2 - 17x - 6$

157. $5y^2 + 16y + 3$

158. $5y^2 - 14y - 3$

159. $5y^2 - 21y + 4$

160. $5y^2 + 14y - 3$

161. $7a^2 + 6a - 1$

162. $7a^2 - 6a - 1$

163. $7a^2 - 8a + 1$

164. $7a^2 + 8a + 1$

165. $2w^2 + 23w + 11$

166. $2w^2 - 23w + 11$

167. $2w^2 + 21w - 11$

168. $2w^2 - 21w - 11$

5.7 Factoring Strategy

In previous sections, we established the general idea of factoring and many special cases. In this section, we will see that a polynomial can have as many factors as its degree, and we will factor higher-degree polynomials completely. We will also see a general strategy for factoring polynomials.

⟨1⟩ Prime Polynomials

A polynomial that cannot be factored is a prime polynomial. Binomials with no common factors, such as $2x + 1$ and $a - 3$, are prime polynomials. To determine whether a polynomial such as $x^2 + 1$ is a prime polynomial, we must try all possibilities for factoring it. If $x^2 + 1$ could be factored as a product of two binomials, the only possibilities that would give a first term of x^2 and a last term of 1 are $(x + 1)(x + 1)$ and $(x - 1)(x - 1)$. However,

$$(x + 1)(x + 1) = x^2 + 2x + 1 \qquad \text{and} \qquad (x - 1)(x - 1) = x^2 - 2x + 1.$$

Both products have an x-term. Of course, $(x + 1)(x - 1)$ has no x-term, but

$$(x + 1)(x - 1) = x^2 - 1.$$

Because none of these possibilities results in $x^2 + 1$, it is a prime polynomial. Note that $x^2 + 1$ is a sum of two squares. There are many other sums of squares that are prime. For example,

$$x^2 + 4, \quad a^2 + 9, \quad b^2 + 16, \quad \text{and} \quad 4y^2 + 25$$

are prime. However, not every sum of two squares is prime. For example, $4x^2 + 16$ is a sum of two squares that is not prime because $4x^2 + 16 = 4(x^2 + 4)$.

> **Difference or Sum of Two Squares**
>
> The difference of two squares $a^2 - b^2$ is *not* prime: $[a^2 - b^2 = (a - b)(a + b)]$. The sum of two squares $a^2 + b^2$ is prime, but not every sum of two squares is prime.

E X A M P L E   **1**

Prime polynomials
Determine whether the polynomial $x^2 + 3x + 4$ is a prime polynomial.

Solution

To factor $x^2 + 3x + 4$, we must find two integers with a product of 4 and a sum of 3. The only pairs of positive integers with a product of 4 are 1 and 4, and 2 and 2. Because the product is positive 4, both numbers must be negative or both positive. Under these conditions it is impossible to get a sum of positive 3. The polynomial is prime.

Now do Exercises 1–12

⟨2⟩ Factoring a Difference or Sum of Two Cubes

We know that $a^2 - b^2$ can be factored, but $a^2 + b^2$ is prime. Since we have not yet tried to factor $a^3 - b^3$ or $a^3 + b^3$, you might think they are prime. However, we can factor them if we learn two new rules.

A monomial is a *perfect cube* or simply a *cube* if it is the cube of another monomial whose coefficient is an integer. For example, 8, $27x^3$, and $64a^3b^6$ are perfect cubes because $8 = 2^3$, $27x^3 = (3x)^3$, and $64a^3b^6 = (4ab^2)^3$. Any *difference or sum of two cubes can be factored*. To see how, examine the following product that results in a difference of two cubes:

$$(a - b)(a^2 + ab + b^2) = a(a^2 + ab + b^2) - b(a^2 + ab + b^2)$$
$$= a^3 + a^2b + ab^2 - a^2b - ab^2 - b^3$$
$$= a^3 - b^3$$

To factor $a^3 + b^3$, a sum of two cubes, examine the following product:

$$(a + b)(a^2 - ab + b^2) = a(a^2 - ab + b^2) + b(a^2 - ab + b^2)$$
$$= a^3 - a^2b + ab^2 + a^2b - ab^2 + b^3$$
$$= a^3 + b^3$$

By finding these products, we have verified the following formulas for factoring $a^3 - b^3$ and $a^3 + b^3$.

Factoring a Difference or a Sum of Two Cubes
$$a^3 - b^3 = (a - b)(a^2 + ab + b^2)$$
$$a^3 + b^3 = (a + b)(a^2 - ab + b^2)$$

EXAMPLE 2

Factoring a difference or a sum of two cubes
Factor each polynomial.

a) $x^3 - 8$ b) $y^3 + 1$ c) $8z^3 - 27$

Solution

a) Because $8 = 2^3$, we can use the formula for factoring the difference of two cubes. In the formula $a^3 - b^3 = (a - b)(a^2 + ab + b^2)$, let $a = x$ and $b = 2$:
$$x^3 - 8 = (x - 2)(x^2 + 2x + 4)$$

b) $y^3 + 1 = y^3 + 1^3$ Recognize a sum of two cubes.
$$= (y + 1)(y^2 - y + 1)$$ Let $a = y$ and $b = 1$ in the formula for the sum of two cubes.

c) $8z^3 - 27 = (2z)^3 - 3^3$ Recognize a difference of two cubes.
$$= (2z - 3)(4z^2 + 6z + 9)$$ Let $a = 2z$ and $b = 3$ in the formula for a difference of two cubes.

Now do Exercises 13–24

⟨3⟩ Factoring Polynomials Completely

So far, a typical polynomial has been a product of two factors, with possibly a common factor removed first. However, it is possible that the factors can still be factored again. A polynomial in a single variable may have as many factors as its degree. We have factored a polynomial completely when all of the factors are prime polynomials.

EXAMPLE 3

Factoring higher-degree polynomials completely

Factor $x^4 + x^2 - 2$ completely.

Solution

Two numbers with a product of -2 and a sum of 1 are 2 and -1:

$$x^4 + x^2 - 2 = (x^2 + 2)(x^2 - 1)$$
$$= (x^2 + 2)(x - 1)(x + 1) \quad \text{Difference of two squares}$$

Since $x^2 + 2$, $x - 1$, and $x + 1$ are prime, the polynomial is factored completely.

> Now do Exercises 25–28

In Example 4, we factor a sixth-degree polynomial.

EXAMPLE 4

Factoring completely

Factor $3x^6 - 3$ completely.

Solution

To factor $3x^6 - 3$, we must first factor out the common factor 3 and then recognize that x^6 is a perfect square: $x^6 = (x^3)^2$:

$$3x^6 - 3 = 3(x^6 - 1) \qquad\qquad \text{Factor out the common factor.}$$
$$= 3((x^3)^2 - 1) \qquad\qquad \text{Write } x^6 \text{ as a perfect square.}$$
$$= 3(x^3 - 1)(x^3 + 1) \qquad\quad \text{Difference of two squares}$$
$$= 3(x - 1)(x^2 + x + 1)(x + 1)(x^2 - x + 1) \quad \text{Difference of two cubes and sum of two cubes}$$

Since $x^2 + x + 1$ and $x^2 - x + 1$ are prime, the polynomial is factored completely.

> Now do Exercises 29–34

In Example 4, we recognized $x^6 - 1$ as a difference of two squares. However, $x^6 - 1$ is also a difference of two cubes, and we can factor it using the rule for the difference of two cubes:

$$x^6 - 1 = (x^2)^3 - 1 = (x^2 - 1)(x^4 + x^2 + 1)$$

Now we can factor $x^2 - 1$, but it is difficult to see how to factor $x^4 + x^2 + 1$. (It is not prime.) Although x^6 can be thought of as a perfect square or a perfect cube, in this case thinking of it as a perfect square is better.

In Example 5, we use substitution to simplify the polynomial before factoring. This fourth-degree polynomial has four factors.

E X A M P L E 5

Using substitution to simplify

Factor $(w^2 - 1)^2 - 11(w^2 - 1) + 24$ completely.

Solution

Let $a = w^2 - 1$ to simplify the polynomial:

$$
\begin{aligned}
(w^2 - 1)^2 - 11(w^2 - 1) + 24 &= a^2 - 11a + 24 && \text{Replace } w^2 - 1 \text{ by } a. \\
&= (a - 8)(a - 3) && \text{Factor.} \\
&= (w^2 - 1 - 8)(w^2 - 1 - 3) && \text{Replace } a \text{ by } w^2 - 1. \\
&= (w^2 - 9)(w^2 - 4) \\
&= (w + 3)(w - 3)(w + 2)(w - 2) && \text{Factor completely.}
\end{aligned}
$$

> Now do Exercises 35–44

Example 6 shows two four-term polynomials that can be factored by grouping. In the first part, the terms must be rearranged before the polynomial can be factored by grouping. In the second part the polynomial is grouped in a new manner.

E X A M P L E 6

More factoring by grouping

Factor completely.

a) $x^2 - 3w - 3x + xw$ **b)** $x^2 - 6x + 9 - y^2$

Solution

a) Since the first two terms do not have a common factor, we rearrange the terms as follows:

$$
\begin{aligned}
x^2 - 3w - 3x + xw &= x^2 - 3x + xw - 3w \\
&= x(x - 3) + w(x - 3) && \text{Factor out } x \text{ and } w. \\
&= (x + w)(x - 3) && \text{Factor out } x - 3.
\end{aligned}
$$

b) We cannot factor this polynomial by grouping pairs of terms. However, $x^2 - 6x + 9$ is a perfect square, $(x - 3)^2$. So we can group the first three terms and then factor the difference of two squares:

$$
\begin{aligned}
x^2 - 6x + 9 - y^2 &= (x - 3)^2 - y^2 && \text{Perfect square trinomial} \\
&= (x - 3 + y)(x - 3 - y) && \text{Difference of two squares}
\end{aligned}
$$

> Now do Exercises 45–54

‹4› Strategy for Factoring Polynomials

A strategy for factoring polynomials is given in the following box.

Strategy for Factoring Polynomials

1. If there are any common factors, factor them out first.

2. When factoring a binomial, look for the special cases: difference of two squares, difference of two cubes, and sum of two cubes. Remember that a sum of two squares $a^2 + b^2$ is prime.

3. When factoring a trinomial, check to see whether it is a perfect square trinomial.

4. When factoring a trinomial that is not a perfect square, use grouping or trial and error.

5. When factoring a polynomial of high degree, use substitution to get a polynomial of degree 2 or 3, or use trial and error.

6. If the polynomial has four terms, try factoring by grouping.

E X A M P L E   **7**

Using the factoring strategy

Factor each polynomial completely.

a) $3w^3 - 3w^2 - 18w$ **b)** $10x^2 + 160$

c) $16a^2b - 80ab + 100b$ **d)** $aw + mw + az + mz$

e) $a^4b + 125ab$ **f)** $12x^2y - 26xy - 30y$

‹ **Helpful Hint** ›

When factoring integers, we write $4 = 2 \cdot 2$. However, when factoring polynomials, we usually do not factor any of the integers that appear. So we say that $4b(2a - 5)^2$ is factored completely.

Solution

a) The greatest common factor (GCF) for the three terms is $3w$:

$$3w^3 - 3w^2 - 18w = 3w(w^2 - w - 6) \quad \text{Factor out } 3w.$$
$$= 3w(w - 3)(w + 2) \quad \text{Factor completely.}$$

b) The GCF in $10x^2 + 160$ is 10:

$$10x^2 + 160 = 10(x^2 + 16)$$

Because $x^2 + 16$ is prime, the polynomial is factored completely.

c) The GCF in $16a^2b - 80ab + 100b$ is $4b$:

$$16a^2b - 80ab + 100b = 4b(4a^2 - 20a + 25)$$
$$= 4b(2a - 5)^2$$

d) The polynomial has four terms, and we can factor it by grouping:

$$aw + mw + az + mz = w(a + m) + z(a + m)$$
$$= (w + z)(a + m)$$

e) The GCF in $a^4b + 125ab$ is ab:

$$a^4b + 125ab = ab(a^3 + 125) \qquad \text{Factor out } ab.$$
$$= ab(a + 5)(a^2 - 5a + 25) \quad \text{Factor the sum of two cubes.}$$

f) The GCF in $12x^2y - 26xy - 30y$ is $2y$:

$$12x^2y - 26xy - 30y = 2y(6x^2 - 13x - 15) \quad \text{Factor out } 2y.$$
$$= 2y(x - 3)(6x + 5) \quad \text{Trial and error}$$

Now do Exercises 55–108

Warm-Ups ▼

Fill in the blank.

1. A _____ has the form $a^2 - b^2$.

2. A _____ has the form $a^3 - b^3$.

3. A _____ has the form $a^3 + b^3$.

4. Some four-term polynomials can be factored by

   _____.

5. A _____ trinomial has the form $a^2 + 2ab + b^2$.

6. A _____ has the form $a^2 + b^2$.

True or false?

7. For any x, $x^2 - 9 = (x - 3)^2$.

8. The polynomial $4x^2 + 12x + 9$ is a perfect square trinomial.

9. The sum of two squares $m^2 + 16$ is prime.

10. The binomial $x^4 - 16$ is factored completely as $(x^2 - 4)(x^2 + 4)$.

11. For any y, $y^3 - 27 = (y + 3)(y^2 + 3y - 9)$.

12. The binomial $y^6 - 1$ is a difference of two squares.

13. The trinomial $2x^2 + 2x - 12$ is factored completely as $(2x - 4)(x + 3)$.

14. The trinomial $x^2 - 4x - 4$ is prime.

15. The binomial $a^6 - 1$ is a difference of two cubes.

5.7 Exercises

‹ Study Tips ›

- Set short-term goals and reward yourself for accomplishing them. When you have solved 10 problems, take a short break and listen to your favorite music.
- Study in a clean comfortable well-lit place, but don't get too comfortable. Study at a desk, not in bed.

‹1› Prime Polynomials

Determine whether each polynomial is a prime polynomial. See Example 1.

1. $y^2 + 100$

2. $3x^2 + 27$

3. $-9w^2 - 9$

4. $25y^2 + 36$

5. $x^2 - 2x - 3$

6. $x^2 - 2x + 3$

7. $x^2 + 2x + 3$

8. $x^2 + 4x + 3$

9. $x^2 - 4x - 3$

10. $x^2 + 4x - 3$

VIDEO 11. $6x^2 + 3x - 4$

12. $4x^2 - 5x - 3$

‹2› Factoring a Difference or Sum of Two Cubes

Factor. See Example 2.

13. $a^3 - 1$

14. $w^3 + 1$

15. $w^3 + 27$

16. $x^3 - 64$

17. $8x^3 - 1$

18. $27x^3 + 1$

19. $64x^3 + 125$

20. $27a^3 + 1000$

21. $8a^3 - 27b^3$

22. $27w^3 - 125y^3$

23. $x^6 - 8$

24. $y^6 - 27$

⟨3⟩ Factoring Polynomials Completely

Factor each polynomial completely. See Examples 3 and 4.

25. $a^4 - 10a^2 + 25$

26. $9y^4 + 12y^2 + 4$

27. $x^4 - 6x^2 + 8$

28. $x^6 + 2x^3 - 3$

 29. $3m^4 - 24m$

30. $6w^3z + 6z$

31. $2y^6 - 128$

32. $6 - 6y^6$

33. $32a^4 - 18$

34. $2a^4 - 32$

Factor each polynomial completely. See Example 5.

 35. $(3x - 5)^2 - 1$

36. $(2x + 1)^2 - 4$

37. $x^4 - (x - 6)^2$

38. $y^4 - (2y + 1)^2$

39. $(m + 2)^2 + 2(m + 2) - 3$

40. $(2w - 3)^2 - 2(2w - 3) - 15$

41. $3(y - 1)^2 + 11(y - 1) - 20$

42. $2(w + 2)^2 + 5(w + 2) - 3$

43. $(y^2 - 3)^2 - 4(y^2 - 3) - 12$

44. $(m^2 - 8)^2 - 4(m^2 - 8) - 32$

Factor completely. See Example 6.

45. $x^2 - 2b - 2x + bx$

46. $y^2 - c - y + cy$

47. $x^2 - ay - xy + ax$

48. $ax - by + bx - ay$

49. $x^2 + 2x + 1 - a^2$

50. $x^2 + 10x + 25 - b^2$

51. $x^2 - 4x + 4 - w^2$

52. $x^2 - 8x + 16 - c^2$

53. $x^2 - z^2 + 4x + 4$

54. $x^2 + 36 - m^2 - 12x$

⟨4⟩ Strategy for Factoring Polynomials

Factor each polynomial completely. See Example 7. See the Strategy for Factoring Polynomials box on pages 354–355.

55. $9x^2 - 24x + 16$

56. $-3x^2 + 18x + 48$

57. $12x^2 - 13x + 3$

58. $2x^2 - 3x - 6$

59. $3a^4 + 81a$

60. $-a^3 + 25a$

61. $32 + 2x^2$

62. $x^3 + 4x^2 + 4x$

63. $6x^2 - 5x + 12$

64. $x^4 + 2x^3 - x - 2$

65. $(x + y)^2 - 1$

66. $x^3 + 9x$

67. $a^3b - ab^3$

68. $2m^3 - 250n^3$

69. $x^4 - 16$

70. $a^4 - 81$

 71. $x^4 + 2x^3 - 8x - 16$

72. $(x + 5)^2 - 4$

73. $m^2n + 2mn^2 + n^3$

74. $a^2b - 2ab^2 + b^3$

75. $2m + wn + 2n + wm$

76. $aw - 5b + bw - 5a$

77. $4w^2 + 4w - 3$

78. $4w^2 + 8w - 63$

79. $t^4 + 4t^2 - 21$

80. $m^4 + 5m^2 + 4$

81. $-a^3 - 7a^2 + 30a$

82. $2y^4 + 3y^3 - 20y^2$

83. $a^4 - w^4$

84. $m^4 - 16n^4$

85. $(y + 5)^2 - 2(y + 5) - 3$

86. $(2t - 1)^2 + 7(2t - 1) + 10$

87. $-2w^4 + 1250$

88. $5a^5 - 5a$

89. $4a^2 + 16$

90. $9w^2 + 81$

91. $8a^3 + 8a$

92. $awx + ax$

93. $(w + 5)^2 - 9$

94. $(a - 6)^2 - 1$

95. $4aw^2 - 12aw + 9a$

96. $9an^3 + 15an^2 - 14an$

97. $x^2 - 6xy + 9y^2$

98. $x^3 + 12x^2y + 36xy^2$

99. $3x^4 - 75x^2$

100. $3x^2 + 9x + 12$

101. $m^3n - n$

102. $m^4 + 16m^2$

 103. $12x^2 + 2x - 30$

104. $90x^2 + 3x - 60$

105. $2a^3 - 32$

106. $12x^2 - 28x + 15$

107. $x^6 - y^6$

108. $a^6 - 64$

Factor completely. Assume variables used as exponents represent positive integers.

109. $a^{3m} - 1$

110. $x^{6a} + 8$

111. $a^{3w} - b^{6n}$

112. $x^{2n} - 9$

113. $x^{3m} - 8$

114. $y^{3n} + 1$

115. $a^{3m} - b^3$

116. $r^{3m} + 8t^3$

117. $t^{4n} - 16$

118. $a^{3n+2} + a^2$

119. $a^{2n+1} - 2a^{n+1} - 15a$

120. $x^{3m} + x^{2m} - 6x^m$

121. $a^{2n} - 3a^n + a^nb - 3b$

122. $x^mz + 5z + x^{m+1} + 5x$

Getting More Involved

123. *Cooperative learning*

Write down 10 trinomials of the form $ax^2 + bx + c$ "at random" using integers for a, b, and c. What percent of your 10 trinomials are prime? Would you say that prime trinomials are the exception or the rule? Compare your results with those of your classmates.

124. *Writing*

The polynomial

$$x^3 + 5x^2 + 7x + 3$$

is a product of three factors of the form $x \pm n$, where n is a natural number smaller than 4. Factor this polynomial completely and explain your procedure.

Extra Factoring Exercises

Factor each polynomial completely.

125. $a^2 + 8a + 16$ **126.** $b^2 + 6b + 9$

127. $36 - y^2$ **128.** $9 - z^2$

129. $a^2 - 16$ **130.** $b^2 - y^2$

131. $w^2 + 18w + 81$ **132.** $z^2 - 49$

133. $a^2 - 2a$ **134.** $ab - ay + 3b - 3y$

135. $4w^2 + 36w + 81$ **136.** $z^3 - 1$

137. $x^2 - 1$ **138.** $ab - ay + 2bz - 2zy$

139. $w^3 + 27$ **140.** $w^3 - 8$

141. $aw - 4w - ab + 4b$ **142.** $a^2 - y^4$

143. $zw - 3w - 5z + 15$ **144.** $am - m + 2a - 2$

145. $4b^2 - 4ab + a^2$ **146.** $z^2 - 14z + 49$

147. $a^3 - 27$

148. $w^2 - yw - y + w$

149. $3z^2 - 30z + 75$

150. $3a^3 - 24a^2 + 48a$

151. $2b^3 - 16$

152. $-5z^2 - 50z - 125$

153. $-2a^3 + 36a^2 - 162a$

154. $xb - ab + ax - a^2$

155. $z^4 - 16$

156. $a^3 - 3a^2 + 9a - 27$

157. $a^3 + ab^2$

158. $x^4 - 1$

159. $w^2 - 2w + 3aw - 6a$

160. $x^4 - y^4$

161. $a^2 + 10a + 25$

162. $3a^2 - 22a + 7$

163. $25b^2 + 30b + 9$

164. $5x^2 - 26x + 5$

165. $144 - y^2$

166. $y^2 - 16y + 48$

167. $9a^2 - z^2$

168. $7a^2 - 10a + 3$

169. $3w^2 - 38w - 13$

170. $a^2z^2 - 16$

171. $m^2 + 4m - 21$

172. $t^2 + 4t - 45$

173. $b^4 - y^2$

174. $9w^2 + 30w + 25$

175. $z^6 - 49$

176. $11x^2 - 12x + 1$

177. $a^5 + 4a^3$

178. $ab^2 - ay + 2b^2 - 2y$

179. $75w^2 + 120w + 48$

180. $a^3 - 64$

181. $a^2x^2 - b^2$

182. $-2x^2 - 4x + 16$

183. $bx - xy + bz - zy$

184. $a^3 + 27$

185. $z^3 - 125$

186. $6x^2 + 11x - 35$

187. $27x^2 - 6x - 1$

188. $aw - 3w + 3b - ab$

189. $b^2n^2 - y^4$

190. $z^2 - 9z + 18$

191. $-6a^2 + 33a - 15$

192. $ax - 8x + 2a - 16$

193. $4a^2b^2 - 4abw + w^2$

194. $4q^2 - 28q + 49$

195. $t^3 - 27$

196. $w^4 - wy + w^3 - y$

197. $3z^6 - 30z^3 + 75$

198. $3a^2b^2 - 24ab^2 + 48b^2$

199. $5b^3 - 40$

200. $-2z^2 - 16z - 32$

201. $-2a^4 + 20a^3 - 50a^2$

202. $xb - 2ab + 3ax - 6a^2$

203. $a^4 - 16$

204. $a^3 - 7a^2 + 9a - 63$

205. $a^2b^2 + b^4$

206. $t^4 - 1$

207. $x^4 + 2x^2 - 15$

208. $a^6 - 8$

209. $w^4 - 2w^2 - 6a + 3aw^2$

210. $x^4 - 6561$

5.8   Solving Equations by Factoring

In This Section

⟨1⟩ **The Zero Factor Property**

⟨2⟩ **Applications**

The techniques of factoring can be used to solve equations involving polynomials that cannot be solved by the other methods that you have learned. After you learn to solve equations by factoring, we will use this technique to solve some new applied problems in this section and in Chapters 6 and 8.

‹ **Helpful Hint** ›

Note that the zero factor property is our second example of getting an equivalent equation without "doing the same thing to each side." What was the first?

‹1› The Zero Factor Property

The equation $ab = 0$ indicates that the product of two unknown numbers is 0. But the product of two real numbers is zero only when one or the other of the numbers is 0. So even though we do not know exactly the values of a and b from $ab = 0$, we do know that $a = 0$ or $b = 0$. This idea is called the **zero factor property.**

> **Zero Factor Property**
>
> The equation $ab = 0$ is equivalent to the compound equation
> $$a = 0 \qquad \text{or} \qquad b = 0.$$

CAUTION   Note that the *zero* factor property does not work with any number other than zero. For example, if $ab = 6$, we can't make a definite conclusion about a or b. There are infinitely many possibilities: $1 \cdot 6 = 6, 2 \cdot 3 = 6, (0.2)(30) = 6$, and so on.

Example 1 shows how to use the zero factor property to solve an equation in one variable.

E X A M P L E   1

Using the zero factor property

Solve $x^2 + x - 12 = 0$.

Solution

We factor the left-hand side of the equation to get a product of two factors that are equal to 0. Then we write an equivalent equation using the zero factor property.

$$x^2 + x - 12 = 0$$
$$(x + 4)(x - 3) = 0 \quad \text{Factor the left-hand side.}$$
$$x + 4 = 0 \quad \text{or} \quad x - 3 = 0 \quad \text{Zero factor property}$$
$$x = -4 \quad \text{or} \quad x = 3 \quad \text{Solve each part of the compound equation.}$$

Check that both -4 and 3 satisfy $x^2 + x - 12 = 0$. If $x = -4$, we get

$$(-4)^2 + (-4) - 12 = 16 - 4 - 12 = 0.$$

If $x = 3$, we get

$$(3)^2 + 3 - 12 = 9 + 3 - 12 = 0.$$

So the solution set is $\{-4, 3\}$.

> Now do Exercises 1–12

‹ **Calculator Close-Up** ›

To check Example 1 simply calculate the value of the left side of the equation.

```
(-4)²+(-4)-12
            0
(3)²+(3)-12
            0
```

The zero factor property is used only in solving polynomial equations that have zero on one side and a polynomial that can be factored on the other side. The polynomials that we factored most often were the quadratic polynomials. The equations that we will solve most often using the zero factor property will be quadratic equations.

> ### Quadratic Equation
>
> If a, b, and c are real numbers, with $a \neq 0$, then the equation
>
> $$ax^2 + bx + c = 0$$
>
> is called a **quadratic equation.**

In Chapter 8 we will study quadratic equations further and solve quadratic equations that cannot be solved by factoring. Keep the following strategy in mind when solving equations by factoring.

> ## Strategy for Solving Equations by Factoring
>
> **1.** Write the equation with 0 on one side.
> **2.** Factor the other side completely.
> **3.** Use the zero factor property to get simpler equations. (Set each factor equal to 0.)
> **4.** Solve the simpler equations.
> **5.** Check the answers in the original equation.
> **6.** State the solution set to the original equation.

EXAMPLE   2

Solving a quadratic equation by factoring

Solve each equation.

   a) $10x^2 = 5x$ **b)** $3x - 6x^2 = -9$ **c)** $(x - 4)(x + 1) = 14$

Solution

a) Use the steps in the strategy for solving equations by factoring:

$$10x^2 = 5x \qquad \text{Original equation}$$
$$10x^2 - 5x = 0 \qquad \text{Rewrite with zero on the right-hand side.}$$
$$5x(2x - 1) = 0 \qquad \text{Factor the left-hand side.}$$
$$5x = 0 \quad \text{or} \quad 2x - 1 = 0 \qquad \text{Zero factor property}$$
$$x = 0 \quad \text{or} \quad x = \frac{1}{2} \qquad \text{Solve for } x.$$

The solution set is $\left\{0, \frac{1}{2}\right\}$. Check each solution in the original equation.

b) First rewrite the equation with 0 on the right-hand side and the left-hand side in order of descending exponents:

$$3x - 6x^2 = -9 \qquad \text{Original equation}$$
$$-6x^2 + 3x + 9 = 0 \qquad \text{Add 9 to each side.}$$
$$2x^2 - x - 3 = 0 \qquad \text{Divide each side by } -3.$$
$$(2x - 3)(x + 1) = 0 \qquad \text{Factor.}$$
$$2x - 3 = 0 \quad \text{or} \quad x + 1 = 0 \qquad \text{Zero factor property}$$
$$x = \frac{3}{2} \quad \text{or} \quad x = -1 \qquad \text{Solve for } x.$$

The solution set is $\left\{-1, \frac{3}{2}\right\}$. Check each solution in the original equation.

‹ **Calculator Close-Up** ›

To check Example 2(b) use the equals sign from the test menu. The calculator returns a 1 if the equation you enter is correct and a 0 if it is incorrect.

```
3(-1)-6(-1)²=-9
                1
3(3/2)-6(3/2)²=-
9
                1
```

c) The zero factor property applies only to equations that have a product equal to zero. So we must rewrite the equation:

$$(x - 4)(x + 1) = 14 \quad \text{Original equation}$$
$$x^2 - 3x - 4 = 14 \quad \text{Multiply the binomials.}$$
$$x^2 - 3x - 18 = 0 \quad \text{Subtract 14 from each side.}$$
$$(x - 6)(x + 3) = 0 \quad \text{Factor.}$$
$$x - 6 = 0 \quad \text{or} \quad x + 3 = 0 \quad \text{Zero factor property}$$
$$x = 6 \quad \text{or} \quad x = -3$$

Checking -3 and 6 in the original equation yields

$$(-3 - 4)(-3 + 1) = 14 \quad \text{and} \quad (6 - 4)(6 + 1) = 14,$$

which are both correct. So the solution set is $\{-3, 6\}$.

Now do Exercises 13–30

CAUTION In Example 2a, if we divide each side of $10x^2 = 5x$ by $5x$, we get $2x = 1$, or $x = \frac{1}{2}$. We do not get $x = 0$. By dividing by $5x$ we have lost one of the factors and one of the solutions.

In Example 3, there are more than two factors, but we can still write an equivalent equation by setting each factor equal to 0.

EXAMPLE 3

Solving a cubic equation by factoring
Solve $2x^3 - 3x^2 - 8x + 12 = 0$.

Solution

First notice that the first two terms have the common factor x^2 and the last two terms have the common factor -4.

$$x^2(2x - 3) - 4(2x - 3) = 0 \quad \text{Factor by grouping.}$$
$$(x^2 - 4)(2x - 3) = 0 \quad \text{Factor out } 2x - 3.$$
$$(x - 2)(x + 2)(2x - 3) = 0 \quad \text{Factor completely.}$$
$$x - 2 = 0 \quad \text{or} \quad x + 2 = 0 \quad \text{or} \quad 2x - 3 = 0 \quad \text{Set each factor equal to 0.}$$
$$x = 2 \quad \text{or} \quad x = -2 \quad \text{or} \quad x = \frac{3}{2}$$

The solution set is $\left\{-2, \frac{3}{2}, 2\right\}$. Check each solution in the original equation.

Now do Exercises 31–38

‹ Calculator Close-Up ›

To check Example 3, use Y= to enter

$$y_1 = 2x^3 - 3x^2 - 8x + 12.$$

Then use the variables feature (VARS) to find $y_1(-2)$, $y_1(3/2)$, and $y_1(2)$.

```
Y₁(-2)
              0
Y₁(3/2)
              0
Y₁(2)
              0
```

The equation in Example 4 involves absolute value.

EXAMPLE 4

To check Example 4 use the Y= key to enter $y_1 = abs(x^2 - 2x - 16)$. Then make a table and scroll through it to see that the value of y_1 is 8 when x is $-4, -2, 4,$ or 6.

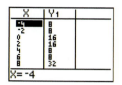

Solving an absolute value equation by factoring

Solve $|x^2 - 2x - 16| = 8$.

Solution

First write an equivalent compound equation without absolute value:

$$x^2 - 2x - 16 = 8 \quad \text{or} \quad x^2 - 2x - 16 = -8$$
$$x^2 - 2x - 24 = 0 \quad \text{or} \quad x^2 - 2x - 8 = 0$$
$$(x - 6)(x + 4) = 0 \quad \text{or} \quad (x - 4)(x + 2) = 0$$
$$x - 6 = 0 \quad \text{or} \quad x + 4 = 0 \quad \text{or} \quad x - 4 = 0 \quad \text{or} \quad x + 2 = 0$$
$$x = 6 \quad \text{or} \quad x = -4 \quad \text{or} \quad x = 4 \quad \text{or} \quad x = -2$$

The solution set is $\{-4, -2, 4, 6\}$. Check each solution in the original equation.

> **Now do Exercises 39–46**

In Example 5 we use factoring to solve equations involving more than one letter.

EXAMPLE 5

Solving for a specified variable

Solve each equation for y.

a) $y^2 - ay - by + ab = 0$ **b)** $y^2 + 6ay + 9a^2 = 0$

Solution

a) Factor the left side by grouping:

$$y^2 - ay - by + ab = 0$$
$$y(y - a) - b(y - a) = 0$$
$$(y - a)(y - b) = 0$$
$$y - a = 0 \quad \text{or} \quad y - b = 0$$
$$y = a \quad \text{or} \quad y = b$$

The solution set is $\{a, b\}$.

b) $y^2 + 6ay + 9a^2 = 0$
$$(y + 3a)^2 = 0 \quad \text{Perfect square trinomial}$$
$$y + 3a = 0$$
$$y = -3a$$
The solution set is $\{-3a\}$.

> **Now do Exercises 65–74**

‹2› Applications

Many applied problems can be solved by using equations such as those we have been solving.

EXAMPLE 6

Area of a room

Ronald's living room is 2 feet longer than it is wide, and its area is 168 square feet. What are the dimensions of the room?

Solution

Let x be the width and $x + 2$ be the length. See Fig. 5.2. Because the area of a rectangle is the length times the width, we can write the equation

$$x(x + 2) = 168.$$

Figure 5.2

We solve the equation by factoring:

$$x^2 + 2x - 168 = 0$$

$$(x - 12)(x + 14) = 0$$

$$x - 12 = 0 \quad \text{or} \quad x + 14 = 0$$

$$x = 12 \quad \text{or} \quad x = -14$$

Because the width of a room is a positive number, we disregard the solution $x = -14$. We use $x = 12$ and get a width of 12 feet and a length of 14 feet. Check this answer by multiplying 12 and 14 to get 168.

> Now do Exercises 75–80

Applications involving quadratic equations often require a theorem called the **Pythagorean theorem.** This theorem states that *in any right triangle the sum of the squares of the lengths of the legs is equal to the length of the hypotenuse squared.*

The Pythagorean Theorem

The triangle shown is a right triangle if and only if

$$a^2 + b^2 = c^2.$$

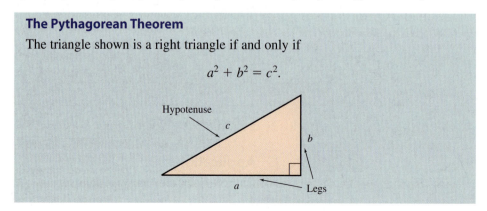

To prove the Pythagorean theorem draw two squares with sides of length $a + b$ and partition them as shown in Fig. 5.3. All eight triangles shown in the figure are identical and have equal areas. Now erase all of the triangles. The remaining squares have areas a^2, b^2, and c^2. Since we started with two squares with equal areas, we must have $a^2 + b^2 = c^2$.

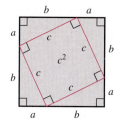

Figure 5.3

We use the Pythagorean theorem in Example 7.

E X A M P L E   **7**

Using the Pythagorean theorem

Shirley used 14 meters of fencing to enclose a rectangular region. To be sure that the region was a rectangle, she measured the diagonals and found that they were 5 meters each. (If the opposite sides of a quadrilateral are equal and the diagonals are equal, then the quadrilateral is a rectangle.) What are the length and width of the rectangle?

Solution

The perimeter of a rectangle is twice the length plus twice the width, $P = 2L + 2W$. Because the perimeter is 14 meters, the sum of one length and one width is 7 meters. If we let x represent the width, then $7 - x$ is the length. We use the Pythagorean theorem to get a relationship among the length, width, and diagonal. See Fig. 5.4.

$$x^2 + (7 - x)^2 = 5^2 \quad \text{Pythagorean theorem}$$
$$x^2 + 49 - 14x + x^2 = 25 \quad \text{Simplify.}$$
$$2x^2 - 14x + 24 = 0 \quad \text{Simplify.}$$
$$x^2 - 7x + 12 = 0 \quad \text{Divide each side by 2.}$$
$$(x - 3)(x - 4) = 0 \quad \text{Factor the left-hand side.}$$
$$x - 3 = 0 \quad \text{or} \quad x - 4 = 0 \quad \text{Zero factor property}$$
$$x = 3 \quad \text{or} \quad x = 4$$
$$7 - x = 4 \quad \text{or} \quad 7 - x = 3$$

Solving the equation gives two possible rectangles: a 3 by 4 rectangle or a 4 by 3 rectangle. However, those are identical rectangles. The rectangle is 3 meters by 4 meters.

Figure 5.4

Now do Exercises 81–82

Example 8 involves a formula from physics for the height of a projectile where the only force acting on the object is gravity. If an object is projected upward at v_0 feet/sec from h_0 feet above the ground, then its height in feet at time t in seconds is given by $h(t) = -16t^2 + v_0 t + h_0$.

EXAMPLE 8

Height of a projectile

A construction worker accidentally fires a nail gun upward from a height of 144 feet. The nail is propelled upward at 128 feet/sec, as shown in Fig. 5.5. The height of the nail in feet at time t in seconds is given by the function $h(t) = -16t^2 + 128t + 144$. How long does it take for the nail to fall to the ground?

Solution

On the ground the height is 0 feet. So we want to solve the quadratic equation $-16t^2 + 128t + 144 = 0$:

$$-16t^2 + 128t + 144 = 0$$
$$-16(t^2 - 8t - 9) = 0 \quad \text{Factor out the GCF.}$$
$$-16(t - 9)(t + 1) = 0 \quad \text{Factor the trinomial.}$$
$$t - 9 = 0 \quad \text{or} \quad t + 1 = 0 \quad \text{Zero factor property}$$
$$t = 9 \quad \text{or} \quad t = -1$$

Since $t = -1$ does not make sense, the nail takes 9 seconds to fall to the ground.

Now do Exercises 83–88

144 ft

Figure 5.5

Warm-Ups ▼

Fill in the blank.

1. According to the _____, $ab = 0$ is equivalent to $a = 0$ or $b = 0$.

2. A _____ equation has the form $ax^2 + bx + c = 0$ where $a \neq 0$.

3. The _____ of a right triangle is the side opposite the right angle.

4. According to the _____ theorem, the sum of the squares of the legs of a right triangle is equal to the square of the hypotenuse.

5. The zero factor property is used when solving a quadratic equation by _____.

True or false?

6. The equation $x(x + 5) = 12$ is equivalent to $x = 3$ or $x + 5 = 4$.

7. The equation $(x + 4)(x - 5) = 0$ is equivalent to $x + 4 = 0$ or $x - 5 = 0$.

8. The equation $|x^2 - 4| = 5$ is equivalent to $x^2 - 4 = 5$ or $x^2 - 4 = -5$.

9. Two numbers that have a sum of 8 can be represented by x and $8 - x$.

10. The only solutions to $x(x - 1)(x - 2) = 0$ are 1 and 2.

11. The solutions to $3(x + 2)(x - 5) = 0$ are 3, −2, and 5.

12. If the perimeter of a rectangle is 38 feet, then the sum of the length and width is 19 feet.

‹ 1 › The Zero Factor Property

Solve each equation. See Examples 1–3. See the Strategy for Solving Equations by Factoring box on page 361.

1. $(x - 5)(x + 4) = 0$

2. $(a - 6)(a + 5) = 0$

3. $(2x - 5)(3x + 4) = 0$

4. $(3k + 8)(4k - 3) = 0$

5. $4(x - 2)(x + 5) = 0$

6. $8(x - 9)(x + 9) = 0$

7. $x(x - 5)(x + 5) = 0$

8. $x(x - 4)(x + 7) = 0$

9. $w^2 + 5w - 14 = 0$

10. $t^2 - 6t - 27 = 0$

11. $m^2 - 7m = 0$

12. $h^2 - 5h = 0$

13. $a^2 - a = 20$

14. $p^2 - p = 42$

15. $x^2 - 6x + 9 = 0$

16. $x^2 + 8x + 16 = 0$

17. $2a^2 + 7a = 15$

18. $6p^2 + p = 1$

19. $(x - 3)(x + 2) = 14$

20. $(x - 6)(x + 1) = 18$

21. $(x - 8)(x - 2) = -5$

22. $(x - 7)(x + 2) = -8$

23. $(x - 6)(x - 2) = -4$

24. $(x + 7)(x - 5) = -36$

25. $10a^2 + 38a - 8 = 0$

26. $-48b^2 + 28b + 6 = 0$

27. $3x^2 - 3x - 36 = 0$

28. $-2x^2 - 16x - 24 = 0$

29. $z^2 + \dfrac{3}{2}z = 10$

30. $m^2 + \dfrac{11}{3}m = -2$

31. $x^3 - 4x = 0$

32. $16x - x^3 = 0$

33. $-4x^3 + x = 3x^2$

34. $2x - 11x^2 = 6x^3$

35. $w^3 + 4w^2 - 25w - 100 = 0$

36. $a^3 + 2a^2 - 16a - 32 = 0$

37. $n^3 - 2n^2 - n + 2 = 0$

38. $w^3 - w^2 - 25w + 25 = 0$

Solve each equation. See Example 4.

39. $|x^2 - 5| = 4$

40. $|x^2 - 17| = 8$

41. $|x^2 + 2x - 36| = 12$

42. $|x^2 + 2x - 19| = 16$

43. $|x^2 + 4x + 2| = 2$

44. $|x^2 + 8x + 8| = 8$

45. $|x^2 + 6x + 1| = 8$

46. $|x^2 - x - 21| = 9$

Solve each equation.

47. $2x^2 - x = 6$

48. $3x^2 + 14x = 5$

49. $|x^2 + 5x| = 6$

50. $|x^2 + 6x - 4| = 12$

51. $x^2 + 5x = 6$

52. $x + 5x = 6$

53. $(x + 2)(x + 1) = 12$

54. $(x + 2)(x + 3) = 20$

55. $y^3 + 9y^2 + 20y = 0$

56. $m^3 - 2m^2 - 3m = 0$

57. $5a^3 = 45a$

58. $5x^3 = 125x$

59. $(2x - 1)(x^2 - 9) = 0$

60. $(3x - 5)(25x^2 - 4) = 0$

61. $(2x - 1)(3x + 1)(4x - 1) = 0$

62. $(x - 1)(x + 3)(x - 9) = 0$

63. $4x^2 - 12x + 9 = 0$

64. $16x^2 + 8x + 1 = 0$

Solve each equation for y. Assume a and b are nonzero real numbers. See Example 5.

65. $y^2 + ay + by + ab = 0$

66. $aby^2 - by - a^2y + a = 0$

67. $9y^2 + 6ay + a^2 = 0$

68. $4y^2 + 4by + b^2 = 0$

69. $y^2 + by = 0$

70. $by^2 = ay$

71. $y^2 - b^2 = 0$

72. $a^2y^2 - b^2 = 0$

73. $ay^2 + 3y - ay = 3$

74. $a^2y^2 + 2aby + b^2 = 0$

⟨2⟩ Applications

Solve each problem. See Examples 6–8.

75. *Color print.* The length of a new "super size" color print is 2 inches more than the width. If the area is 24 square inches, what are the length and width?

76. *Tennis court dimensions.* In singles competition, each player plays on a rectangular area of 117 square yards. Given that the length of that area is 4 yards greater than its width, find the length and width.

77. *Missing numbers.* The sum of two numbers is 13 and their product is 36. Find the numbers.

78. *More missing numbers.* The sum of two numbers is 6.5, and their product is 9. Find the numbers.

79. *Bodyboarding.* The Seamas Channel pro bodyboard shown in the figure has a length that is 21 inches greater than its width. Any rider weighing up to 200 pounds can use it because its surface area is 946 square inches. Assume that it is rectangular in shape and find the length and width.

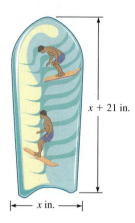

$x + 21$ in.

x in.

Figure for Exercise 79

80. *New dimensions in gardening.* Mary Gold has a rectangular flower bed that measures 4 feet by 6 feet. If she wants to increase the length and width by the same amount to have a flower bed of 48 square feet, then what will be the new dimensions?

x ft

4 ft

6 ft

x ft

Figure for Exercise 80

81. *Yolanda's closet.* The length of Yolanda's closet is 2 feet longer than twice its width. If the diagonal measures 13 feet, then what are the length and width?

13 ft

$2x + 2$ ft

x ft

Figure for Exercise 81

82. Ski jump. The base of a ski ramp forms a right triangle. One leg of the triangle is 2 meters longer than the other. If the hypotenuse is 10 meters, then what are the lengths of the legs?

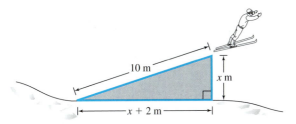

Figure for Exercise 82

83. Shooting arrows. An archer shoots an arrow straight upward at 64 feet per second. The height of the arrow $h(t)$ (in feet) at time t seconds is given by the function

$$h(t) = -16t^2 + 64t.$$

a) Use the accompanying graph to estimate the amount of time that the arrow is in the air.

b) Algebraically find the amount of time that the arrow is in the air.

c) Use the accompanying graph to estimate the maximum height reached by the arrow.

d) At what time does the arrow reach its maximum height?

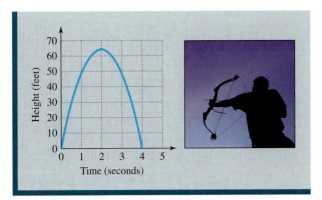

Figure for Exercise 83

84. Time until impact. If an object is dropped from a height of s_0 feet, then its altitude after t seconds is given by the formula $S = -16t^2 + s_0$. If a pack of emergency supplies is dropped from an airplane at a height of 1600 feet, then how long does it take for it to reach the ground?

85. Firing an M-16. If an M-16 is fired straight upward, then the height $h(t)$ of the bullet in feet at time t in seconds is given by

$$h(t) = -16t^2 + 325t.$$

a) What is the height of the bullet 5 seconds after it is fired?

b) How long does it take for the bullet to return to the earth?

86. Firing a howitzer. If an 8-in. (diameter) howitzer is fired straight into the air, then the height $h(t)$ of the projectile in feet at time t in seconds is given by

$$h(t) = -16t^2 + 1332t.$$

a) What is the height of the projectile 10 seconds after it is fired?

b) How long does it take for the projectile to return to earth?

87. Tossing a ball. A boy tosses a ball upward at 32 feet per second from a window that is 48 feet above the ground. The height of the ball above the ground (in feet) at time t (in seconds) is given by

$$h(t) = -16t^2 + 32t + 48.$$

Find the time at which the ball strikes the ground.

88. Firing a slingshot. A girl uses a slingshot to propel a stone upward at 64 feet per second from a window that is 80 feet above the ground. The height of the stone above the ground (in feet) at time t (in seconds) is given by

$$h(t) = -16t^2 + 64t + 80.$$

Find the time at which the stone strikes the ground.

89. Trimming a gate. A total of 34 feet of 1×4 lumber is used around the perimeter of the gate shown in the figure. If the diagonal brace is 13 feet long, then what are the length and width of the gate?

Figure for Exercise 89

90. *Maria's kids.* The sum of the squares of the ages of Maria's two kids is 289. If the boy is seven years older than the girl, then what are their ages?

91. *Leaning ladder.* A 15-foot ladder is leaning against a wall. If the distance from the top of the ladder to the ground is 3 feet more than the distance from the bottom of the ladder to the wall, then what is the distance from the top of the ladder to the ground?

92. *Laying tile.* Lorinda is planning to redo the floor in her bedroom, which has an area of 192 square feet. If the width of the rectangular room is 4 feet less than the length, then what are its dimensions?

93. *Finding numbers.* If the square of a number decreased by the number is 12, then what is the number?

94. *Consecutive integers.* The sum of the squares of two consecutive integers is 25. Find the integers.

95. *Perimeter of a rectangle.* The perimeter of a rectangle is 28 inches, and the diagonal measures 10 inches. What are the length and width of the rectangle?

96. *Pete's garden.* Each row in Pete's garden is 3 feet wide. If the rows run north and south, he can have two more rows than if they run east and west. If the area of Pete's garden is 135 square feet, then what are the length and width?

97. *House plans.* In the plans for their dream house the Baileys have a master bedroom that is 240 square feet in area. If they increase the width by 3 feet, they must decrease the length by 4 feet to keep the original area. What are the original dimensions of the bedroom?

98. *Arranging the rows.* Mr. Converse has 112 students in his algebra class with an equal number in each row. If he arranges the desks so that he has one fewer rows, he will have two more students in each row. How many rows did he have originally?

99. *Broken Bamboo I.* A bamboo stalk 10 chi high is broken by the wind. The top touches the ground 3 chi from its base as shown in the accompanying figure. At what height did the stalk break? This problem appeared in a book by Chinese mathematician Yang Hui in 1261.

|←————— 3 chi —————→|

Figure for Exercises 99 and 100

100. *Broken Bamboo II.* A section of bamboo that is 5 chi in length is broken from a stalk of bamboo of unknown height. If the broken section touches the ground 3 chi from the base as in Exercise 99, then what was the original height of the bamboo stalk?

Getting More Involved

101. *Writing*

If you divide each side of $x^2 = x$ by x, you get $x = 1$. If you subtract x from each side of $x^2 = x$, you get $x^2 - x = 0$, which has two solutions. Which method is correct? Explain.

102. *Cooperative learning*

Work with a group to examine the following solution to $x^2 - 2x = -1$:

$$x(x - 2) = -1$$

$x = -1$	or	$x - 2 = -1$
$x = -1$	or	$x = 1$

Is this method correct? Explain.

103. *Cooperative learning*

Work with a group to examine the following steps in the solution to $5x^2 - 5 = 0$.

$$5(x^2 - 1) = 0$$
$$5(x - 1)(x + 1) = 0$$

$x - 1 = 0$	or	$x + 1 = 0$
$x = 1$	or	$x = -1$

What happened to the 5? Explain.

Wrap-Up

Summary

Definitions		Examples
Definition of negative integral exponents	If a is a nonzero real number and n is a positive integer, then $$a^{-n} = \frac{1}{a^n}.$$	$2^{-3} = \frac{1}{2^3} = \frac{1}{8}$
Definition of zero exponent	If a is any nonzero real number, then $a^0 = 1$. The expression 0^0 is undefined.	$3^0 = 1$

Rules of Exponents

If a and b are nonzero real numbers and m and n are integers, then the following rules hold.

		Examples
Negative exponent rules	$a^{-n} = \left(\frac{1}{a}\right)^n, \quad a^{-1} = \frac{1}{a}, \quad$ and $\quad \frac{1}{a^{-n}} = a^n$ Find the power and reciprocal in either order.	$5^{-1} = \frac{1}{5}, \frac{1}{5^{-3}} = 5^3$ $\left(\frac{2}{3}\right)^{-2} = \left(\frac{3}{2}\right)^2$
Product rule	$a^m \cdot a^n = a^{m+n}$	$3^5 \cdot 3^7 = 3^{12}, 2^{-3} \cdot 2^{10} = 2^7$
Quotient rule	$\frac{a^m}{a^n} = a^{m-n}$	$\frac{x^8}{x^5} = x^3, \frac{5^4}{5^{-7}} = 5^{11}$
Power of a power rule	$\left(a^m\right)^n = a^{mn}$	$\left(5^2\right)^3 = 5^6$
Power of a product rule	$(ab)^n = a^n b^n$	$(2x)^3 = 8x^3$ $\left(2x^3\right)^4 = 16x^{12}$
Power of a quotient rule	$\left(\frac{a}{b}\right)^n = \frac{a^n}{b^n}$	$\left(\frac{x}{3}\right)^2 = \frac{x^2}{9}$

Scientific Notation

		Examples
Converting from scientific notation	1. Determine the number of places to move the decimal point by examining the exponent on the 10. 2. Move to the right for a positive exponent and to the left for a negative exponent.	$4 \times 10^3 = 4000$ $3 \times 10^{-4} = 0.0003$

Converting to scientific notation	1. Count the number of places (n) that the decimal point must be moved so that it will follow the first nonzero digit of the number. 2. If the original number was larger than 10, use 10^n. 3. If the original number was smaller than 1, use 10^{-n}.	$67{,}000 = 6.7 \times 10^4$ $0.009 = 9 \times 10^{-3}$

Polynomials		**Examples**
Term of a polynomial	The product of a number (coefficient) and one or more variables raised to whole number powers	$3x^4,\ -2xy^2,\ 5$
Polynomial	A single term or a finite sum of terms	$x^5 - 3x^2 + 7$
Adding or subtracting polynomials	Add or subtract the like terms.	$(x + 3) + (x - 7) = 2x - 4$ $(x^2 - 2x) - (3x^2 - x) = -2x^2 - x$
Multiplying two polynomials	Multiply each term of the first polynomial by each term of the second polynomial, and then combine like terms.	$(x^2 + 2x + 3)(x + 1)$ $= (x^2 + 2x + 3)x + (x^2 + 2x + 3)1$ $= x^3 + 2x^2 + 3x + x^2 + 2x + 3$ $= x^3 + 3x^2 + 5x + 3$

Shortcuts for Multiplying Two Binomials		**Examples**
FOIL	The product of two binomials can be found quickly by multiplying their **F**irst, **O**uter, **I**nner, and **L**ast terms.	$(x + 2)(x + 3) = x^2 + 5x + 6$
Square of a sum	$(a + b)^2 = a^2 + 2ab + b^2$	$(x + 5)^2 = x^2 + 10x + 25$
Square of a difference	$(a - b)^2 = a^2 - 2ab + b^2$	$(m - 3)^2 = m^2 - 6m + 9$
Product of a sum and a difference	$(a + b)(a - b) = a^2 - b^2$	$(x + 3)(x - 3) = x^2 - 9$

Factoring		**Examples**
Factoring a polynomial	Write a polynomial as a product of two or more polynomials. A polynomial is factored completely if it is a product of prime polynomials.	$3x^2 - 3 = 3(x^2 - 1)$ $\qquad\quad = 3(x + 1)(x - 1)$
Common factors	Factor out the greatest common factor (GCF).	$2x^3 - 6x = 2x(x^2 - 3)$
Difference of two squares	$a^2 - b^2 = (a + b)(a - b)$ (The sum of two squares $a^2 + b^2$ is prime.)	$m^2 - 25 = (m + 5)(m - 5)$ $m^2 + 25$ is prime.
Perfect square trinomials	$a^2 + 2ab + b^2 = (a + b)^2$ $a^2 - 2ab + b^2 = (a - b)^2$	$x^2 + 10x + 25 = (x + 5)^2$ $x^2 - 6x + 9 = (x - 3)^2$

Difference of two cubes	$a^3 - b^3 = (a - b)(a^2 + ab + b^2)$	$x^3 - 8 = (x - 2)(x^2 + 2x + 4)$
Sum of two cubes	$a^3 + b^3 = (a + b)(a^2 - ab + b^2)$	$x^3 + 27 = (x + 3)(x^2 - 3x + 9)$

Grouping	Factor out common factors from groups of terms.	$3x + 3w + bx + bw$ $= 3(x + w) + b(x + w)$ $= (3 + b)(x + w)$

Factoring $ax^2 + bx + c$	By the *ac* method: 1. Find two numbers that have a product equal to *ac* and a sum equal to *b*. 2. Replace *bx* by the sum of two terms using the two new numbers as coefficients. 3. Factor the resulting four-term polynomial by grouping. By trial and error: Try possibilities by considering factors of the first term and factors of the last term. Check them by FOIL.	$2x^2 + 7x + 3$ $ac = 6, b = 7, 1 \cdot 6 = 6, 1 + 6 = 7$ $2x^2 + 7x + 3$ $= 2x^2 + x + 6x + 3$ $= (2x + 1)x + (2x + 1)3$ $= (2x + 1)(x + 3)$ $12x^2 + 19x - 18$ $= (3x - 2)(4x + 9)$

Substitution	Use substitution on higher-degree polynomials to reduce the degree to 2 or 3.	$x^4 - 3x^2 - 18$ Let $a = x^2$. $a^2 - 3a - 18$

Solving Equations by Factoring   **Examples**

Strategy	1. Write the equation with 0 on one side. 2. Factor the other side. 3. Set each factor equal to 0. 4. Solve the simpler equations. 5. Check the answers in the original equation.	$x^2 - 3x - 18 = 0$ $(x - 6)(x + 3) = 0$ $x - 6 = 0$ or $x + 3 = 0$ $x = 6$ or $x = -3$ $6^2 - 3(6) - 18 = 0$ $(-3)^2 - 3(-3) - 18 = 0$

Enriching Your Mathematical Word Power

Fill in the blank.

1. A _____ is a single term or a finite sum of terms.

2. The _____ of a polynomial is the highest degree of any of its terms.

3. The coefficient of the first term of a polynomial when it is written in order of decreasing exponents is the _____ coefficient.

4. A polynomial with one term is a _____.

5. A polynomial with two terms is a _____.

6. A polynomial with three terms is a _____.

7. The _____ method is a procedure for multiplying two binomials quickly.

8. A _____ number is an integer greater than 1 that has no integral factors other than itself and 1.

9. A polynomial that has no factors is a _____ polynomial.

10. Writing a polynomial as a product is _____.

11. Writing a polynomial as a product of primes is factoring _____.

12. The largest integer that is a factor of two or more integers is the _____ common factor.

13. A notation for expressing large or small numbers using powers of 10 is _____ notation.

14. The polynomial $a^3 + b^3$ is a ____ of two cubes.

15. The polynomial $a^3 - b^3$ is a _____ of two cubes.

16. A _____ equation is an equation of the form $ax^2 + bx + c = 0$.

17. According to the _____ factor property, if $ab = 0$ then $a = 0$ or $b = 0$.

Review Exercises

5.1 Integral Exponents and Scientific Notation

Simplify each expression. Assume all variables represent nonzero real numbers. Write your answers with positive exponents.

1. $2 \cdot 2 \cdot 2^{-1}$

2. $5^{-1} \cdot 5$

3. $2^2 \cdot 3^2$

4. $3^2 \cdot 5^2$

5. $(-3)^{-3}$

6. $(-2)^{-2}$

7. $-(-1)^{-3}$

8. $3^4 \cdot 3^7$

9. $2x^3 \cdot 4x^{-6}$

10. $-3a^{-3} \cdot 4a^{-4}$

11. $\dfrac{y^{-5}}{y^{-3}}$

12. $\dfrac{w^3}{w^{-3}}$

13. $\dfrac{a^5 \cdot a^{-2}}{a^{-4}}$

14. $\dfrac{2m^3 \cdot m^6}{2m^{-2}}$

15. $\dfrac{6x^{-2}}{3x^2}$

16. $\dfrac{-5y^2x^{-3}}{5y^{-2}x^7}$

Write each number in standard notation.

17. 8.36×10^6

18. 3.4×10^7

19. 5.7×10^{-4}

20. 4×10^{-3}

Write each number in scientific notation.

21. 8,070,000

22. 90,000

23. 0.000709

24. 0.0000005

Perform each computation without a calculator. Write the answer in scientific notation.

25. $(5(2 \times 10^4))^3$

26. $(6(2 \times 10^{-3}))^2$

27. $\dfrac{(2 \times 10^{-9})(3 \times 10^7)}{5 \cdot (6 \times 10^{-4})}$

28. $\dfrac{(3 \times 10^{12})(5 \times 10^4)}{30 \times 10^{-9}}$

29. $\dfrac{(4,000,000,000)(0.0000006)}{(0.000012)(2,000,000)}$

30. $\dfrac{(1.2 \times 10^{32})(2 \times 10^{-5})}{4 \times 10^{-7}}$

5.2 The Power Rules

Simplify each expression. Assume all variables represent nonzero real numbers. Write your answers with positive exponents.

31. $(a^{-3})^{-2} \cdot a^{-7}$

32. $(-3x^{-2}y)^{-4}$

33. $(m^2n^3)^{-2}(m^{-3}n^2)^4$

34. $(w^{-3}xy)^{-1}(wx^{-3}y)^2$

35. $\left(\dfrac{2}{3}\right)^{-4}$

36. $\left(\dfrac{a^4}{3}\right)^{-2}$

37. $\left(\dfrac{1}{2} + \dfrac{1}{3}\right)^2$

38. $\left(\dfrac{1}{2} - \dfrac{1}{3}\right)^{-2}$

39. $\left(-\dfrac{3a}{4b^{-1}}\right)^{-1}$

40. $\left(-\dfrac{4x^5}{5y^{-3}}\right)^{-1}$

41. $\dfrac{(a^{-3}b)^4}{(ab^2)^{-5}}$

42. $\dfrac{(2x^3)^3}{(3x^2)^{-2}}$

Simplify each expression. Assume that the variables represent integers.

43. $5^{2w} \cdot 5^{4w} \cdot 5^{-1}$

44. $3^y(3^{2y})^3$

45. $\left(\dfrac{7^{3a}}{7^8}\right)^5$

46. $\left(\dfrac{2^{6-k}}{2^{2-3k}}\right)^3$

5.3 Polynomials and Polynomial Functions

Perform the indicated operations.

47. $(2w - 3) + (6w + 5)$

48. $(3a - 2xy) + (5xy - 7a)$

49. $(x^2 - 3x - 4) - (x^2 + 3x - 7)$

50. $(7 - 2x - x^2) - (x^2 - 5x + 6)$

51. $(x^2 - 2x + 4)(x - 2)$

52. $(x + 5)(x^2 - 2x + 10)$

53. $xy + 7z - 5(xy - 3z)$

54. $7 - 4(x - 3)$

55. $m^2(5m^3 - m + 2)$

56. $(a + 2)^3$

5.4 Multiplying Binomials

Perform the following operations mentally. Write down only the answers.

57. $(x - 3)(x + 7)$

58. $(k - 5)(k + 4)$

59. $(z - 5y)(z + 5y)$

60. $(m - 3)(m + 3)$

61. $(m + 8)^2$

62. $(b + 2a)^2$

63. $(w - 6x)(w - 4x)$

64. $(2w - 3)(w + 6)$

65. $(k - 3)^2$

66. $(n - 5)^2$

67. $(m^2 - 5)(m^2 + 5)$

68. $(3k^2 - 5t)(2k^2 + 6t)$

5.5 Factoring Polynomials

Complete the factoring by filling in the parentheses.

69. $3x - 6 = 3($ $)$

70. $7x^2 - x = x($ $)$

71. $4a - 20 = -4($ $)$

72. $w^2 - w = -w($ $)$

73. $3w - w^2 = -w($ $)$

74. $3x - 6 = ($ $)(2 - x)$

Factor each polynomial.

75. $y^2 - 81$

76. $r^2t^2 - 9v^2$

77. $4x^2 + 28x + 49$

78. $y^2 - 20y + 100$

79. $t^2 - 18t + 81$

80. $4w^2 + 4ws + s^2$

5.6 Factoring $ax^2 + bx + c$

Factor each polynomial.

81. $x^2 + 14x + 40$

82. $a^2 + 10a + 24$

83. $x^2 - 7x - 30$

84. $y^2 + 4y - 32$

85. $w^2 - 3w - 28$

86. $6t^2 - 5t + 1$

87. $2m^2 + 5m - 7$

88. $12x^2 - 17x + 6$

89. $m^7 - 3m^4 - 10m$

90. $6w^5 - 7w^3 - 5w$

5.7 Factoring Strategy

Factor each polynomial completely.

91. $5x^3 + 40$

92. $w^3 - 6w^2 + 9w$

93. $9x^2 + 9x + 2$

94. $ax^3 + a$

95. $x^3 + x^2 - x - 1$

96. $16x^2 - 4x - 2$

97. $-x^2y + 16y$

98. $-5m^2 + 5$

99. $-a^3b^2 + 2a^2b^2 - ab^2$

100. $-2w^2 - 16w - 32$

101. $x^3 - x^2 + 9x - 9$

102. $w^4 + 2w^2 - 3$

103. $x^4 - x^2 - 12$

104. $8x^3 - 1$

105. $a^6 - a^3$

106. $a^2 - ab + 2a - 2b$

107. $t^3 - 125$

108. $8y^3 + 1$

109. $16 - 54x^3$

110. $27x^2y - 64x^2y^4$

111. $5x^3y^2 - y^5$

112. $2m^4 - 2mn^3$

113. $-8m^2 - 24m - 18$

114. $-3x^2 - 9x + 30$

115. $(2x - 3)^2 - 16$

116. $(m - 6)^2 - (m - 6) - 12$

117. $x^6 + 7x^3 - 8$

118. $32a^5 - 2a$

119. $(a^2 - 9)^2 - 5(a^2 - 9) + 6$

120. $x^3 - 9x + x^2 - 9$

Factor each polynomial completely. Variables used as exponents represent positive integers.

121. $x^{2k} - 49$

122. $x^{6k} - 1$

123. $m^{2a} - 2m^a - 3$

124. $2y^{2n} - 7y^n + 6$

125. $9z^{2k} - 12z^k + 4$

126. $25z^{6m} + 20z^{3m} + 4$

127. $y^{2a} - by^a + cy^a - bc$

128. $x^3y^b - xy^b + 2x^3 - 2x$

5.8 Solving Equations by Factoring

Solve each equation.

129. $x^3 - 5x^2 = 0$

130. $2m^2 + 10m + 12 = 0$

131. $(a - 2)(a - 3) = 6$

132. $(w - 2)(w + 3) = 50$

133. $2m^2 - 9m - 5 = 0$

134. $m^3 + 4m^2 - 9m - 36 = 0$

135. $w^3 + 5w^2 - w - 5 = 0$

136. $12x^2 + 5x - 3 = 0$

137. $|x^2 - 5| = 4$

138. $|x^2 - 3x - 7| = 3$

Solve each equation for x.

139. $x^2 - ax + bx - ab = 0$

140. $x^2 - 2ax - bx + 2ab = 0$

141. $x^2 - 2abx + a^2b^2 = 0$

142. $x^2 + 4ax + 4a^2 = 0$

Miscellaneous

Solve each problem.

143. *Roadrunner and the coyote.* The roadrunner has just taken a position atop a giant saguaro cactus. While positioning a 10-foot Acme ladder against the cactus, Wile E. Coyote notices a warning label on the ladder. For safety, Acme recommends that the distance from the ground to the top of the ladder, measured vertically along the cactus, must be 2 feet longer than the distance between the bottom of the ladder and the cactus. How far from the cactus should he place the bottom of this ladder?

144. *Three consecutive integers.* Find three consecutive integers such that the sum of their squares is 50.

145. *Playground dimensions.* It took 32 meters of fencing to enclose the rectangular playground at Kiddie Kare. If the area of the playground is 63 square meters, then what are its dimensions?

146. *Landscape design.* Rico is planting red tulips in a rectangular flower bed that is 2 feet longer than it is wide. He plans to surround the tulips with a border of daffodils that is 2 feet wide. If the total area is 224 square feet and he plants 36 daffodils per square foot, then how many daffodils does he need?

147. *Panoramic screen.* Engineers are designing a new 25-inch diagonal measure television. The new rectangular screen will have a length that is 17 inches larger than its width. What are the dimensions of the screen?

148. *Less panoramic.* The engineers are also experimenting with a 25-inch diagonal measure television that has a width that is 5 inches less than the length. What are the dimensions of this rectangular screen?

149. *Life expectancy of black males.* The age at which people die is precisely measured and provides an indication of the health of the population as a whole. The formula

$$L = 64.3(1.0033)^a$$

can be used to model life expectancy L for U.S. black males with present age a (National Center for Health Statistics, www.cdc.gov/nchswww).

a) To what age can a 20-year-old black male expect to live?

b) How many more years is a 20-year-old white male expected to live than a 20-year-old black male? (See Section 5.2 Exercise 87.)

150. *Life expectancy of black females.* The formula

$$L = 72.9(1.002)^a$$

can be used to model life expectancy for U.S. black females with present age a. How long can a 20-year-old black female expect to live?

151. *Golden years.* A person earning $80,000 per year should expect to receive 21% of her retirement income from Social Security and the rest from personal savings. To calculate the amount of regular savings, we use the formula

$$S = R \cdot \frac{(1 + i)^n - 1}{i},$$

where S is the amount at the end of n years of n investments of R dollars each year earning interest rate i compounded annually.

a) Use the graph below to estimate the interest rate needed to get an investment of $1 per year for 20 years to amount to $100.

b) Use the formula to determine the annual savings for 20 years that would amount to $500,000 at 7% compounded annually.

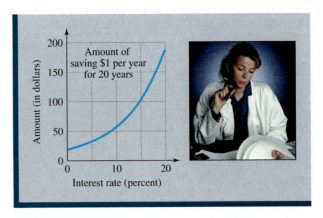

Figure for Exercise 151

152. *Costly education.* The cost of attending Tulane University for one year is approximately $35,414 (www.tulane.edu). Use the formula in Exercise 151 to find the annual savings for 18 years that would amount to $35,414 with an annual return of 8%.

Chapter 5 Test

Simplify each expression. Assume all variables represent nonzero real numbers. Exponents in your answers should be positive exponents.

1. 3^{-2}

2. $\dfrac{1}{6^{-2}}$

3. $\left(\dfrac{1}{2}\right)^{-3}$

4. $3x^4 \cdot 4x^3$

5. $\dfrac{8y^9}{2y^{-3}}$

6. $\left(4a^2b\right)^3$

7. $\left(\dfrac{x^2}{3}\right)^{-3}$

8. $\dfrac{\left(2^{-1}a^2b\right)^{-3}}{4a^{-9}}$

Convert to standard notation.

9. 3.24×10^9

10. 8.673×10^{-4}

Perform each computation by converting each number to scientific notation. Give the answer in scientific notation.

11. $\dfrac{(80,000)(0.0006)}{2,000,000}$

12. $\dfrac{(0.00006)^2(500)}{(30,000)^2(0.01)}$

Perform the indicated operations.

13. $\left(3x^3 - x^2 + 6\right) + \left(4x^2 - 2x - 3\right)$

14. $\left(x^2 - 6x - 7\right) - \left(3x^2 + 2x - 4\right)$

15. $\left(x^2 - 3x + 7\right)(x - 2)$

16. $(x - 2)^3$

Find the products.

17. $(x - 7)(2x + 3)$

18. $(x - 6)^2$

19. $(2x + 5)^2$

20. $\left(3y^2 - 5\right)\left(3y^2 + 5\right)$

Factor completely.

21. $a^2 - 2a - 24$

22. $4x^2 + 28x + 49$

23. $3m^3 - 24$

24. $2x^2y - 32y$

25. $12m^2 + 28m + 15$

26. $2x^{10} + 5x^5 - 12$

27. $2xa + 3a - 10x - 15$

28. $x^4 + 3x^2 - 4$

29. $a^4 - 1$

Solve each equation.

30. $2m^2 + 7m - 15 = 0$

31. $a^2 + 10a + 25 = 0$

32. $x^3 - 4x = 0$

33. $\left| x^2 + x - 9 \right| = 3$

Solve for w.

34. $w^2 - aw + 2bw - 2ab = 0$

Write a complete solution for each problem.

35. A portable television is advertised as having a 10-inch diagonal measure screen. If the width of the screen is 2 inches more than the height, then what are the dimensions of the screen?

36. The infant mortality rate for the United States, the number of deaths per 100,000 live births, has decreased dramatically since 1950. The formula

$$d = \left(1.8 \times 10^{28}\right)(1.032)^{-y}$$

gives the infant mortality rate d as a function of the year y (National Center for Health Statistics, www.cdc.gov/nchswww). Find the infant mortality rates in 1950, 1990, and 2000.

37. If a boy uses a slingshot to propel a stone straight upward, then the height $h(t)$ of the stone in feet at time t in seconds is given by

$$h(t) = -16t^2 + 80t.$$

a) What is the height of the stone at 2 seconds and at 3 seconds?

b) For how long is the stone in the air?

38. *Room dimensions.* The perimeter of the den in the Bailey's house is 88 feet. If the area is 480 square feet, then what are the dimensions of this rectangular room?

*Making*Connections | A Review of Chapters 1–5

Simplify each expression.

1. 4^2

2. $4(-2)$

3. 4^{-2}

4. $2^3 \cdot 4^{-1}$

5. $2^{-1} + 2^{-1}$

6. $2^{-1} \cdot 3^{-1}$

7. $2^{-2} \cdot 3^2$

8. $-3^4 \cdot 6^{-2}$

9. $(-2)^3 \cdot 6^{-1}$

10. $8^{-3} \cdot 8^3$

11. $\left(\dfrac{2^{-2}}{2} + \dfrac{1}{2}\right)^2$

12. $\left(\dfrac{2^2 + 1}{2^{-2} + 1}\right)^{-3}$

13. $\left(\dfrac{3 - 6}{8 - 4}\right)^{-2}$

14. $\left(\dfrac{6 - 9}{14 - 20}\right)^{-3}$

15. $\left(\dfrac{1}{2^{-3} + 1}\right)^{-1}$

16. $(2^{-1} + 1)^2$

17. $3^{-1} - 2^{-2}$

18. $3^2 - 4(5)(-2)$

19. $2^7 - 2^6$

20. $0.08(32) + 0.08(68)$

21. $3 - 2\,|\,5 - 7 \cdot 3\,|$

22. $5^{-1} + 6^{-1}$

Solve each equation.

23. $0.05a - 0.04(a - 50) = 4$

24. $15b - 27 = 0$

25. $2c^2 + 15c - 27 = 0$

26. $2t^2 + 15t = 0$

27. $|\,15u - 27\,| = 3$

28. $|\,15v - 27\,| = 0$

29. $|\,15x - 27\,| = -78$

30. $|\,x^2 + x - 4\,| = 2$

31. $(2x - 1)(x + 5) = 0$

32. $|\,3x - 1\,| + 6 = 9$

33. $(1.5 \times 10^{-4})w - 5 \times 10^5 = 7 \times 10^6$

34. $(3 \times 10^7)(y - 5 \times 10^3) = 6 \times 10^{12}$

Perform the indicated operations.

35. $-6(3x - 4)$

36. $5x - 6(3x - 4)$

37. $(5x - 6)(3x - 4)$

38. $(5x - 6)(3x^2 - 4x + 1)$

39. $\dfrac{6x^2 - 4x}{2x}$

40. $\dfrac{9x - 15}{3} - \dfrac{7x - 21}{7}$

Solve each problem.

41. The area of a rectangular field is 5525 square meters and the length is 85 meters. What is the width?

42. The perimeter of a rectangular canvas is 54 inches and its length is 15 inches. Find the width.

43. The area of a rectangular balcony is 84 square feet. If the length is 5 feet more than the width, what is the length?

44. A glass shop charges six cents per square inch for a rectangular mirror. If the width is 4 inches less than the length and the charge is $5.76, then what is the width?

45. The diagonal measure of a new television screen is 1 inch greater than the length and 8 inches greater than the width. Find the length and width.

46. *Negative income tax.* In a negative income tax proposal, the function

$$D = 0.75E + 5000$$

is used to determine the disposable income D (the amount available for spending) for an earned income E (the amount earned). If $E > D$, then the difference is paid in federal taxes. If $D > E$, then the difference is paid to the wage earner by Uncle Sam.

a) Find the amount of tax paid by a person who earns $100,000.

b) Find the amount received from Uncle Sam by a person who earns $10,000.

c) The accompanying graph shows the lines $D = 0.75E + 5000$ and $D = E$. Find the intersection of these lines.

d) How much tax does a person pay whose earned income is at the intersection found in part (c)?

Figure for Exercise 46

These exercises can be solved by a variety of techniques, which may or may not require algebra. So be creative and think critically. Explain all answers. Answers are in the Instructor's Edition of this text.

1. *Pile of pipes.* Ten pipes, each with radius *a*, are stacked as shown in the figure. What is the height of the pile?

Figure for Exercise 1

2. *Twin trains.* Two freight trains are approaching each other, each traveling at 40 miles per hour, on parallel tracks. If each train is $\frac{1}{2}$ mile long, then how long does it take for them to pass each other?

3. *Peculiar number.* A given two-digit number is seven times the sum of its digits. After the digits are reversed, the new number is also an integral multiple of the sum of its digits. What is the multiple?

4. *Billion dollar sales.* A new company forecasts its sales at $1 on the first day of business, $2 on the second day of business, $3 on the third day of business, and so on. On which day will the total sales first reach a nine-digit number?

5. *Standing in line.* Hector is in line to buy tickets to a playoff game. There are six more people ahead of him in line than behind him. One-third of the people in line are behind him. How many people are ahead of him?

6. *Delightful digits.* Use the digits 0 through 9 to fill in the second row of the table. You may use a digit more than once, but each digit in the second row must indicate the number of times that the digit above it appears in the second row.

0	1	2	3	4	5	6	7	8	9

Table for Exercise 6

7. *Presidential proof.* James Garfield, twentieth president of the United States, gave the following proof of the Pythagorean theorem. Start with a right triangle with legs *a* and *b* and hypotenuse *c*. Use the right triangle twice to make the trapezoid shown in the figure.

a) Find the area of the trapezoid by using the formula $A = \frac{1}{2}h(b_1 + b_2)$.

b) Find the area of each of the three triangles in the figure. Then find the sum of those areas.

c) Set the answer to part (a) equal to the answer to part (b) and simplify. What do you get?

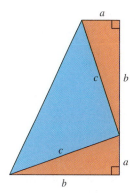

Figure for Exercise 7

8. *Prime time.* Prove or disprove. The expression $n^2 - n + 41$ produces a prime number for every positive integer *n*.

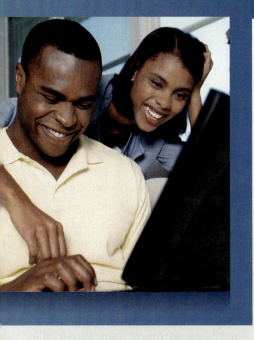

Rational Expressions and Functions

Information is everywhere—in the newspapers and magazines we read, the televisions we watch, and the computers we use. And now people are talking about the information superhighway, which delivers vast amounts of information directly to consumers' homes. In the future, the combination of telephone, television, and computer will give us on-the-spot health care recommendations, video conferences, home shopping, and perhaps even electronic voting and driver's license renewal, to name just a few.

Some experts are concerned that the consumer will give up privacy for this technology. Others worry about regulation, access, and content of the enormous international computer network.

Whatever the future of this technology, few people understand how all their electronic devices work. However, this vast array of electronics rests on physical principles, which are described by mathematical formulas.

In Exercises 45 and 46 of Section 6.7 we will see that the formula governing resistance for receivers connected in parallel involves rational expressions, which are the subject of this chapter.

6.1   Properties of Rational Expressions and Functions

A rational number is a quotient of two integers, and a rational expression is a quotient of two polynomials. Rational expressions are as fundamental to algebra as rational numbers are to arithmetic. In this section, we will see how the properties of rational numbers extend to properties of rational expressions.

⟨1⟩ Rational Expressions

A **rational expression** is the quotient or ratio of two polynomials with the denominator not equal to zero. For example,

$$\frac{2}{3}, \qquad 3a + 5, \qquad \frac{x - 3}{2x^2 - 2}, \qquad \frac{y + 2}{5y}, \qquad \text{and} \qquad \frac{x - 2}{x + 1}$$

are rational expressions. The rational number $\frac{2}{3}$ is a rational expression because 2 and 3 are monomials and $\frac{2}{3}$ is a ratio of two monomials. If the denominator of a rational expression is 1, it is usually omitted, as in the expression $3a + 5$.

The **domain** of any expression involving a variable is the set of all real numbers that can be used in place of the variable. Because division by zero is undefined, a rational expression is undefined for any real number that causes the denominator to be zero. So numbers that cause the denominator to have a value of zero cannot be used for the variable.

E X A M P L E   **1**

Domain

Find the domain of each rational expression.

a) $\dfrac{x - 2}{x + 9}$ **b)** $\dfrac{y + 2}{5y}$ **c)** $\dfrac{x - 3}{2x^2 - 2}$

⟨ **Helpful Hint** ⟩

If the domain consists of all real numbers except -5, some people write $R - \{-5\}$ for the domain. Even though there are several ways to indicate the domain, you should keep practicing interval notation because it is used in algebra, trigonometry, and calculus.

Solution

a) The denominator is zero if $x + 9 = 0$ or $x = -9$. So the domain is the set of all real numbers except -9. This set is written in set notation as $\{x \mid x \neq -9\}$ and in interval notation as

$$(-\infty, -9) \cup (-9, \infty).$$

b) The denominator is zero if $5y = 0$ or $y = 0$. So the domain is the set of all real numbers except 0. This set is written in set notation as $\{y \mid y \neq 0\}$ and in interval notation as

$$(-\infty, 0) \cup (0, \infty).$$

c) The denominator is zero if $2x^2 - 2 = 0$. Solve this equation.

$$2x^2 - 2 = 0$$
$$2(x^2 - 1) = 0 \quad \text{Factor out 2.}$$
$$2(x + 1)(x - 1) = 0 \quad \text{Factor completely.}$$
$$x + 1 = 0 \qquad \text{or} \qquad x - 1 = 0 \quad \text{Zero factor property}$$
$$x = -1 \qquad \text{or} \qquad x = 1$$

The domain is the set of all real numbers except -1 and 1. This set is written as $\{x \mid x \neq -1 \text{ and } x \neq 1\}$, or in interval notation as

$$(-\infty, -1) \cup (-1, 1) \cup (1, \infty).$$

> **Now do Exercises 1–12**

CAUTION The numbers that you find when you set the denominator equal to zero and solve for x are *not* in the domain of the rational expression. The solutions to that equation are excluded from the domain.

⟨ **Helpful Hint** ⟩

Most students learn to convert $\frac{2}{3}$ into $\frac{4}{6}$ by dividing 3 into 6 to get 2 and then multiplying 2 by 2 to get 4. In algebra it is better to do this conversion by multiplying the numerator and denominator of $\frac{2}{3}$ by 2 as shown here.

⟨2⟩ Reducing to Lowest Terms

Each rational number can be written in infinitely many equivalent forms. For example,

$$\frac{2}{3} = \frac{4}{6} = \frac{6}{9} = \frac{8}{12} = \frac{10}{15} = \cdots.$$

Each equivalent form of $\frac{2}{3}$ is obtained from $\frac{2}{3}$ by multiplying both numerator and denominator by the same nonzero number. For example,

$$\frac{2}{3} = \frac{2}{3} \cdot 1 = \frac{2}{3} \cdot \frac{2}{2} = \frac{4}{6} \quad \text{and} \quad \frac{2}{3} = \frac{2}{3} \cdot \frac{3}{3} = \frac{6}{9}.$$

Note that we are actually multiplying $\frac{2}{3}$ by equivalent forms of 1, the multiplicative identity.

If we start with $\frac{4}{6}$ and convert it into $\frac{2}{3}$, we are simplifying by *reducing* $\frac{4}{6}$ to its *lowest terms*. We can reduce as follows:

$$\frac{4}{6} = \frac{2 \cdot 2}{2 \cdot 3} = \frac{2}{2} \cdot \frac{2}{3} = 1 \cdot \frac{2}{3} = \frac{2}{3}$$

A rational number is expressed in its lowest terms when the numerator and denominator have no common factors other than 1. In reducing $\frac{4}{6}$, we divide the numerator and denominator by the common factor 2, or "divide out" the common factor 2. We can multiply or divide both numerator and denominator of a rational number by the same nonzero number without changing the value of the rational number. This fact is called the **basic principle of rational numbers.**

> **Basic Principle of Rational Numbers**
>
> If $\frac{a}{b}$ is a rational number and c is a nonzero real number, then
>
> $$\frac{ac}{bc} = \frac{a}{b}.$$

CAUTION Although it is true that

$$\frac{5}{6} = \frac{2+3}{2+4},$$

we cannot divide out the 2's in this expression because the 2's are not factors. We can divide out only common *factors* when reducing fractions.

Just as a rational number has infinitely many equivalent forms, a rational expression also has infinitely many equivalent forms. To reduce rational expressions to lowest terms, we follow exactly the same procedure as we do for rational numbers: *Factor the numerator and denominator completely, and then divide out all common factors.*

E X A M P L E 2

Reducing

Reduce each rational expression to its lowest terms.

a) $\dfrac{18}{42}$ b) $\dfrac{-2a^7b}{a^2b^3}$

⟨ **Helpful Hint** ⟩

A negative sign in a fraction can be placed in three locations:

$$\frac{-1}{2} = \frac{1}{-2} = -\frac{1}{2}$$

The same goes for rational expressions:

$$\frac{-3x^2}{5y} = \frac{3x^2}{-5y} = -\frac{3x^2}{5y}$$

Solution

a) Factor 18 as $2 \cdot 3^2$ and 42 as $2 \cdot 3 \cdot 7$:

$$\frac{18}{42} = \frac{2 \cdot 3^2}{2 \cdot 3 \cdot 7} \quad \text{Factor.}$$

$$= \frac{3}{7} \quad \text{Divide out the common factors.}$$

b) Because this expression is already factored, we use the quotient rule for exponents to reduce:

$$\frac{-2a^7b}{a^2b^3} = \frac{-2a^5}{b^2}$$

Now do Exercises 13–24

In Example 3 we use the techniques for factoring polynomials that we learned in Chapter 5.

E X A M P L E 3

Reducing

Reduce each rational expression to its lowest terms.

a) $\dfrac{2x^2 - 18}{x^2 + x - 6}$ b) $\dfrac{w - 2}{2 - w}$ c) $\dfrac{2a^3 - 16}{16 - 4a^2}$

Solution

a) $\dfrac{2x^2 - 18}{x^2 + x - 6} = \dfrac{2(x^2 - 9)}{(x - 2)(x + 3)}$ Factor.

$$= \frac{2(x - 3)(x + 3)}{(x - 2)(x + 3)} \quad \text{Factor completely.}$$

$$= \frac{2x - 6}{x - 2} \quad \text{Divide out the common factors.}$$

b) Factor out -1 from the numerator to get a common factor:

$$\frac{w - 2}{2 - w} = \frac{-1(2 - w)}{(2 - w)} = -1$$

c) $\dfrac{2a^3 - 16}{16 - 4a^2} = \dfrac{2(a^3 - 8)}{-4(a^2 - 4)}$ Factoring out -4 will give the common factor $a - 2$.

$= \dfrac{2(a - 2)(a^2 + 2a + 4)}{-2 \cdot 2(a - 2)(a + 2)}$ Difference of two cubes, difference of two squares

$= -\dfrac{a^2 + 2a + 4}{2(a + 2)}$ Divide out common factors.

> **Now do Exercises 25–42**

The rational expressions in Example 3(a) are equivalent because they have the same value for any replacement of the variables, provided that the replacement is in the domain of both expressions. In other words, the equation

$$\frac{2x^2 - 18}{x^2 + x - 6} = \frac{2x - 6}{x - 2}$$

is an identity. It is true for any value of x except 2 and -3.

Note that in Example 3(c) there are several ways to write the answer:

$$-\frac{a^2 + 2a + 4}{2(a + 2)} = -\frac{a^2 + 2a + 4}{2a + 4} = \frac{-a^2 - 2a - 4}{2(a + 2)} = \frac{-a^2 - 2a - 4}{2a + 4}$$

If the form of the denominator is not specified, then in this text we will write the denominator factored and the numerator not factored. The reason for this practice is that common denominators for addition or subtraction are determined from factored form. However, the answer is certainly correct if $2a + 4$ is written rather than $2(a + 2)$. Note also that the negative sign is usually placed in the numerator or in front of the rational expression. In this case, it was placed in front to make the expression look a bit simpler.

The main points to remember for reducing rational expressions are summarized as follows.

⟨ **Helpful Hint** ⟩

Since $-1(a - b) = b - a$, placement of a negative sign in a rational expression changes the appearance of the expression:

$$-\frac{3 - x}{x - 2} = \frac{-(3 - x)}{x - 2}$$

$$= \frac{x - 3}{x - 2}$$

$$\frac{3 - x}{x - 2} = \frac{3 - x}{-(x - 2)}$$

$$= \frac{3 - x}{2 - x}$$

Strategy for Reducing Rational Expressions

1. Factor the numerator and denominator completely and look for common factors.
2. Divide out the common factors.
3. In a ratio of two monomials with exponents, the quotient rule for exponents is used to divide out the common factors.
4. You may have to use a negative sign with the greatest common factor to get any identical factors.

CAUTION   Step 1 in this strategy is absolutely essential. You can cancel out only common factors. An expression such as $\dfrac{x}{x + 4}$ does not reduce to $\dfrac{1}{4}$, $\dfrac{1}{5}$, or anything else. If you claim that $\dfrac{x}{x + 4}$ reduces to $\dfrac{1}{4}$, then $\dfrac{x}{x + 4} = \dfrac{1}{4}$ must be an identity. But if $x = 0$ in this equation, you get $0 = \dfrac{1}{4}$, which is wrong.

⟨3⟩ Building Up the Denominator

In Section 6.3 we will see that only rational expressions with identical denominators can be added or subtracted. Fractions without identical denominators can be converted to equivalent fractions with a common denominator by reversing the procedure for reducing fractions to lowest terms. This procedure is called **building up the denominator.**

Consider converting the fraction $\frac{1}{3}$ into an equivalent fraction with a denominator of 51. Any fraction that is equivalent to $\frac{1}{3}$ can be obtained by multiplying the numerator and denominator of $\frac{1}{3}$ by the same nonzero number. Because $51 = 3 \cdot 17$, we multiply the numerator and denominator of $\frac{1}{3}$ by 17 to get an equivalent fraction with a denominator of 51:

$$\frac{1}{3} = \frac{1}{3} \cdot 1 = \frac{1}{3} \cdot \frac{17}{17} = \frac{17}{51}$$

E X A M P L E   4

Building up the denominator

Convert each rational expression into an equivalent rational expression that has the indicated denominator.

a) $\frac{2}{7}, \frac{?}{42}$ **b)** $\frac{5}{3a^2b}, \frac{?}{9a^3b^4}$

‹ Helpful Hint ›

Notice that reducing and building up are exactly the opposite of each other. In reducing you remove a factor that is common to the numerator and denominator, and in building up you put a common factor into the numerator and denominator.

Solution

a) Factor 42 as $42 = 2 \cdot 3 \cdot 7$, and then multiply the numerator and denominator of $\frac{2}{7}$ by the missing factors, 2 and 3:

$$\frac{2}{7} = \frac{2 \cdot 2 \cdot 3}{7 \cdot 2 \cdot 3} = \frac{12}{42}$$

b) Because $9a^3b^4 = 3ab^3 \cdot 3a^2b$, we multiply the numerator and denominator by $3ab^3$:

$$\frac{5}{3a^2b} = \frac{5 \cdot 3ab^3}{3a^2b \cdot 3ab^3}$$

$$= \frac{15ab^3}{9a^3b^4}$$

> **Now do Exercises 43–46**

When building up a denominator to match a more complicated denominator, we factor both denominators completely to see which factors are missing from the simpler denominator. Then we multiply the numerator and denominator of the simpler expression by the missing factors.

E X A M P L E   5

Building up the denominator

Convert each rational expression into an equivalent rational expression that has the indicated denominator.

a) $\frac{5}{2a - 2b}, \frac{?}{6b - 6a}$ **b)** $\frac{x + 2}{x + 3}, \frac{?}{x^2 + 7x + 12}$

Solution

a) Factor both $2a - 2b$ and $6b - 6a$ to see which factor is missing in $2a - 2b$. Note that we factor out -6 from $6b - 6a$ to get the factor $a - b$:

$$2a - 2b = 2(a - b)$$

$$6b - 6a = -6(a - b) = -3 \cdot 2(a - b)$$

‹ **Helpful Hint** ›

Multiplying the numerator and denominator of a rational expression by -1 changes the appearance of the expression:

$$\frac{6-x}{x-7} = \frac{-1(6-x)}{-1(x-7)}$$

$$= \frac{x-6}{7-x}$$

$$\frac{y-5}{-4-y} = \frac{-1(y-5)}{-1(-4-y)}$$

$$= \frac{5-y}{4+y}$$

Now multiply the numerator and denominator by the missing factor, -3:

$$\frac{5}{2a-2b} = \frac{5(-3)}{(2a-2b)(-3)} = \frac{-15}{6b-6a}$$

b) Because $x^2 + 7x + 12 = (x+3)(x+4)$, multiply the numerator and denominator by $x+4$:

$$\frac{x+2}{x+3} = \frac{(x+2)(x+4)}{(x+3)(x+4)} = \frac{x^2+6x+8}{x^2+7x+12}$$

> Now do Exercises 47–62

‹4› Rational Functions

A rational expression can be used to determine the value of a variable. For example, if

$$y = \frac{3x-1}{x^2-4},$$

then we say that y is a **rational function** of x. We can also use function notation, as shown in Example 6. The domain of a rational function is the same as the domain of the rational expression used to define the function.

E X A M P L E   **6**

Evaluating a rational function

Find the domain of the rational function $R(x) = \frac{3x-1}{x^2-4}$. Then find $R(3)$, $R(-1)$, and $R(2)$.

Solution

To find the domain, solve $x^2 - 4 = 0$ factoring:

$$x^2 - 4 = 0$$
$$(x-2)(x+2) = 0$$
$$x - 2 = 0 \quad \text{or} \quad x + 2 = 0$$
$$x = 2 \quad \text{or} \quad x = -2$$

Note that -2 and 2 cause the denominator to be 0. So the domain is the set of all real numbers except -2 and 2, which is written in interval notation as $(-\infty, -2) \cup (-2, 2) \cup (2, \infty)$. To find $R(3)$, replace x by 3 in the formula:

$$R(3) = \frac{3 \cdot 3 - 1}{3^2 - 4} = \frac{8}{5}$$

To find $R(-1)$, replace x by -1 in the formula:

$$R(-1) = \frac{3(-1) - 1}{(-1)^2 - 4}$$

$$= \frac{-4}{-3} = \frac{4}{3}$$

We can't find $R(2)$ because 2 is not in the domain of the rational function. We say that $R(2)$ is undefined.

> Now do Exercises 63–68

‹ **Calculator Close-Up** ›

To check, use Y= to enter

$$y_1 = (3x - 1)/(x^2 - 4).$$

Then use the variables feature (VARS) to find $y_1(3)$ and $y_1(-1)$.

```
Y₁(3)▶Frac
              8/5
Y₁(-1)▶Frac
              4/3
```

⟨5⟩ **Applications**

A rational expression can occur in finding an average cost. The average cost of making a product is the total cost divided by the number of products made.

EXAMPLE **7**

Average cost function

A car maker spent $700 million to develop a new SUV, which will sell for $40,000. If the cost of manufacturing the SUV is $30,000 each, then what rational function gives the average cost of developing and manufacturing x vehicles? Compare the average cost per vehicle for manufacturing levels of 10,000 vehicles and 100,000 vehicles. See Fig. 6.1.

Figure 6.1

Solution

The polynomial $30{,}000x + 700{,}000{,}000$ gives the cost in dollars of developing and manufacturing x vehicles. The average cost per vehicle is given by the rational function

$$AC(x) = \frac{30{,}000x + 700{,}000{,}000}{x}.$$

If $x = 10{,}000$, then

$$AC(10{,}000) = \frac{30{,}000(10{,}000) + 700{,}000{,}000}{10{,}000} = 100{,}000.$$

If $x = 100{,}000$, then

$$AC(100{,}000) = \frac{30{,}000(100{,}000) + 700{,}000{,}000}{100{,}000} = 37{,}000.$$

The average cost per vehicle when 10,000 vehicles are made is $100,000, whereas the average cost per vehicle when 100,000 vehicles are made is $37,000.

> **Now do Exercises 87–94**

Warm-Ups ▼

Fill in the blank.

1. A _____ is a ratio of two polynomials with the denominator not equal to zero.

2. The _____ of a rational expression is all real numbers except those that cause the denominator to be zero.

3. The _____ of rational numbers says that $\frac{ab}{ac} = \frac{b}{c}$ provided a and c are not zero.

4. To _____ a rational expression, factor the numerator and denominator and divide out the common factors.

5. _____ the denominator of a rational expression by multiplying the numerator and denominator by the same expression.

True or false?

6. The number $\frac{1}{2}$ is a rational expression.

7. The domain of $\frac{3}{x-2}$ is $\{2\}$.

8. The domain of $\frac{2x+5}{(x-9)(2x+1)}$ is $\left\{x \mid x \neq 9 \text{ and } x \neq -\frac{1}{2}\right\}$.

9. The domain of $\frac{x-1}{x+2}$ is $(-\infty, -2) \cup (-2, 1) \cup (1, \infty)$.

10. The expression $\frac{5x+2}{15}$ reduces to $\frac{x+2}{3}$.

11. The expression $\frac{2}{3-x}$ is equivalent to $\frac{-2}{x-3}$.

12. The equation $\frac{4x^3}{6x} = \frac{2x^2}{3}$ is an identity.

13. The expression $\frac{x^2-y^2}{x-y}$ reduces to $x+y$.

Exercises 6.1

‹ Study Tips ›

- Eliminate the obvious distractions when you study. Disconnect the telephone and put away newspapers, magazines, and unfinished projects.
- The sight of a textbook from another class might be a distraction if you have a lot of work to do in that class.

‹1› Rational Expressions

Find the domain of each rational expression. See Example 1.

1. $\frac{3x}{x-1}$

2. $\frac{x}{x+5}$

3. $\frac{2z-5}{7z}$

4. $\frac{z-12}{4z}$

 5. $\frac{5y-1}{y^2-4}$

6. $\frac{2y-1}{y^2-9}$

7. $\frac{x-1}{x^2+4}$

8. $\frac{y+5}{y^2+9}$

Which real numbers cannot be used in place of the variable in each rational expression?

9. $\frac{2a-3}{a^2+5a+6}$

10. $\frac{3b+1}{b^2-3b-4}$

11. $\frac{x+1}{x^3+x^2-6x}$

12. $\frac{x^2-3x-4}{2x^5-2x}$

‹2› Reducing to Lowest Terms

Reduce each rational expression to its lowest terms. See Examples 2 and 3. See the Strategy for Reducing Rational Expressions box on page 385.

13. $\frac{6}{57}$

14. $\frac{14}{91}$

15. $\dfrac{42}{210}$

16. $\dfrac{242}{154}$

17. $\dfrac{2x + 2}{4}$

18. $\dfrac{3a + 3}{3}$

19. $\dfrac{3x - 6y}{10y - 5x}$

20. $\dfrac{5b - 10a}{2a - b}$

21. $\dfrac{ab^2}{a^3 b}$

22. $\dfrac{36y^3 z^8}{54y^2 z^9}$

23. $\dfrac{-2w^2 x^3 y}{6wx^5 y^2}$

24. $\dfrac{6a^3 b^{12} c^5}{-8ab^4 c^9}$

25. $\dfrac{a^3 b^2}{a^3 + a^4}$

26. $\dfrac{b^8 - ab^5}{ab^5}$

27. $\dfrac{a - b}{2b - 2a}$

28. $\dfrac{2m - 2n}{4n - 4m}$

29. $\dfrac{3x + 6}{3x}$

30. $\dfrac{7x - 14}{7x}$

31. $\dfrac{a^3 - b^3}{a - b}$

32. $\dfrac{27x^3 + y^3}{6x + 2y}$

33. $\dfrac{4x^2 - 4}{4x^2 + 4}$

34. $\dfrac{2a^2 - 2b^2}{2a^2 + 2b^2}$

35. $\dfrac{12x^2 - 26x - 10}{4x^2 - 25}$

36. $\dfrac{9x^2 - 15x - 6}{81x^2 - 9}$

37. $\dfrac{x^3 + 7x^2 - 4x}{x^3 - 16x}$

38. $\dfrac{2x^4 - 32}{4x - 8}$

39. $\dfrac{2ab + 2by + 3a + 3y}{2b^2 - 7b - 15}$

40. $\dfrac{3m^2 + 3mn + m + n}{12m^2 - 5m - 3}$

41. $\dfrac{4x^2 - 10x - 6}{2x^2 + 11x + 5}$

42. $\dfrac{6x^2 + x - 1}{8x^2 - 2x - 3}$

⟨3⟩ Building Up the Denominator

Convert each rational expression into an equivalent rational expression that has the indicated denominator. See Examples 4 and 5.

43. $\dfrac{1}{5}, \dfrac{?}{50}$

44. $\dfrac{2}{3}, \dfrac{?}{9}$

45. $\dfrac{1}{xy}, \dfrac{?}{3x^2 y^3}$

46. $\dfrac{3}{ab^2}, \dfrac{?}{a^3 b^5}$

47. $\dfrac{5}{x - 1}, \dfrac{?}{x^2 - 2x + 1}$

48. $\dfrac{7}{2x + 1}, \dfrac{?}{4x^2 + 4x + 1}$

49. $\dfrac{3}{2x - 5}, \dfrac{?}{4x^2 - 25}$

50. $\dfrac{x}{x - 3}, \dfrac{?}{x^2 - 9}$

51. $\dfrac{1}{2x + 2}, \dfrac{?}{-6x - 6}$

52. $\dfrac{-2}{-3x + 4}, \dfrac{?}{15x - 20}$

53. $5, \dfrac{?}{a}$

54. $3, \dfrac{?}{a + 1}$

55. $\dfrac{x + 2}{x + 3}, \dfrac{?}{x^2 + 2x - 3}$

56. $\dfrac{x}{x - 5}, \dfrac{?}{x^2 - x - 20}$

57. $\dfrac{7}{x - 1}, \dfrac{?}{1 - x}$

58. $\dfrac{1}{a - b}, \dfrac{?}{2b - 2a}$

59. $\dfrac{3}{x + 2}, \dfrac{?}{x^3 + 8}$

60. $\dfrac{x}{x - 2}, \dfrac{?}{x^3 - 8}$

61. $\dfrac{x + 2}{3x - 1}, \dfrac{?}{6x^2 + 13x - 5}$

62. $\dfrac{a}{2a + 1}, \dfrac{?}{4a^2 - 16a - 9}$

⟨4⟩ Rational Functions

Find the domain of each rational function. Then find the indicated values of the function if possible. See Example 6.

63. $R(x) = \dfrac{3x - 5}{x + 4}$, $R(3)$, $R(-4)$

64. $T(x) = \dfrac{5 - x}{x - 5}$, $T(5)$, $T(-9)$

65. $H(y) = \dfrac{y^2 - 5}{3y - 4}$, $H(-2)$, $H\left(\dfrac{4}{3}\right)$

66. $G(a) = \dfrac{3 - 5a}{2a + 7}$, $G(5)$, $G\left(-\dfrac{7}{2}\right)$

67. $W(b) = \dfrac{4b^3 - 1}{b^2 - b - 6}$, $W(-2)$, $W(3)$, $W(1)$

68. $N(x) = \dfrac{x + 3}{x^3 - 2x^2 - x + 2}$, $N(-1)$, $N(-3)$, $N(2)$

Miscellaneous

In place of each question mark in Exercises 69–86, put an expression that will make the rational expressions equivalent.

69. $\dfrac{1}{3} = \dfrac{?}{21}$

70. $4 = \dfrac{?}{3}$

71. $5 = \dfrac{10}{?}$

72. $\dfrac{3}{4} = \dfrac{12}{?}$

73. $\dfrac{3}{a} = \dfrac{?}{a^2}$

74. $\dfrac{5}{y} = \dfrac{10}{?}$

75. $\dfrac{2}{a - b} = \dfrac{?}{b - a}$

76. $\dfrac{3}{x - 4} = \dfrac{?}{4 - x}$

77. $\dfrac{2}{x - 1} = \dfrac{?}{x^2 - 1}$

78. $\dfrac{5}{x + 3} = \dfrac{?}{x^2 - 9}$

79. $\dfrac{2}{w - 3} = \dfrac{-2}{?}$

80. $\dfrac{-2}{5 - x} = \dfrac{2}{?}$

81. $\dfrac{2x + 4}{6} = \dfrac{?}{3}$

82. $\dfrac{2x - 3}{4x - 6} = \dfrac{1}{?}$

83. $\dfrac{3a + 3}{3a} = \dfrac{?}{a}$

84. $\dfrac{x - 3}{x^2 - 9} = \dfrac{1}{?}$

85. $\dfrac{1}{x - 1} = \dfrac{?}{x^3 - 1}$

86. $\dfrac{x^2 + 2x + 4}{x + 2} = \dfrac{?}{x^2 - 4}$

⟨5⟩ Applications

Solve each problem. See Example 7.

87. *Driving speed.* Jeremy drives 500 miles in x hours. Find a rational function $S(x)$ that gives his average speed in miles per hour.

88. *Travel time.* Marsha traveled 400 miles with an average speed of x miles per hour. Find a rational function $T(x)$ that gives her travel time in hours.

89. *Average cost.* Bobby spent $150 on x pieces of clothing for her child.

a) Find a rational function $C(x)$ that gives the average cost in dollars for an item of clothing.

b) Find $C(5)$, $C(10)$, and $C(30)$.

90. *Flying high.* A flying club has x members who plan to share equally the cost of a $300,000 airplane.

a) Find a rational function $C(x)$ that gives the cost in dollars per member.

b) Find $C(10)$, $C(30)$, and $C(500)$.

91. *Wedding bells.* Wheeler Printing Co. charges $45 plus $0.50 per invitation to print wedding invitations.

a) Write a rational function that gives the average cost in dollars per invitation for printing n invitations.

b) How much less does it cost per invitation to print 300 invitations rather than 200 invitations?

c) As the number of invitations increases, does the average cost per invitation increase or decrease?

d) As the number of invitations increases, does the total cost of the invitations increase or decrease?

Figure for Exercise 91

92. *Rose Bowl bound.* A travel agent offers a Rose Bowl package including hotel, tickets, and transportation. It costs the travel agent $50,000 plus $300 per person to charter the airplane. Find a rational function that gives the average cost in dollars per person for the charter flight. How much lower is the average cost per person when 200 people go compared to 100 people?

93. *Solid waste recovery.* The amount of municipal solid waste generated in the United States in the year 1960 + *n* is given by the polynomial

$$3.43n + 87.24,$$

whereas the amount recycled is given by the polynomial

$$0.053n^2 - 0.64n + 6.71,$$

where the amounts are in millions of tons (U.S. Environmental Protection Agency, www.epa.gov).

a) Write a rational function $p(n)$ that gives the fraction of solid waste that is recycled in the year 1960 + *n*.
b) Find $p(0)$, $p(30)$, and $p(50)$.

94. *Higher education.* The total number of degrees awarded in U.S. higher education in the year 1990 + *n* is given in thousands by the polynomial $41.7n + 1429$, whereas the number of bachelor's degrees awarded is given in thousands by the polynomial $25.2n + 1069$ (National Center for Education Statistics, www.nces.ed.gov).

a) Write a rational function $p(n)$ that gives the percentage of bachelor's degrees among the total number of degrees conferred for the year 1990 + *n*.

b) What percentage of the degrees awarded in 2015 will be bachelor's degrees?

Getting More Involved

95. *Exploration*

Use a calculator to find $R(2)$, $R(30)$, $R(500)$, $R(9000)$, and $R(80,000)$ for the rational function

$$R(x) = \frac{x - 3}{2x + 1}.$$

Round answers to four decimal places. What can you conclude about the value of $R(x)$ as x gets larger and larger without bound?

96. *Exploration*

Use a calculator to find $H(1000)$, $H(100,000)$, $H(1,000,000)$, and $H(10,000,000)$ for the rational function

$$H(x) = \frac{7x - 50}{3x + 91}.$$

Round answers to four decimal places. What can you conclude about the value of $H(x)$ as x gets larger and larger without bound?

6.2   Multiplication and Division

In Chapter 5, you learned to add, subtract, multiply, and divide polynomials. In this chapter, you will learn to perform the same operations with rational expressions. We begin in this section with multiplication and division.

In This Section

⟨1⟩ **Multiplying Rational Expressions**

⟨2⟩ **Dividing Rational Expressions**

⟨1⟩ Multiplying Rational Expressions

We multiply two rational numbers by multiplying their numerators and multiplying their denominators. For example,

$$\frac{6}{7} \cdot \frac{14}{15} = \frac{84}{105} = \frac{21 \cdot 4}{21 \cdot 5} = \frac{4}{5}.$$

Instead of reducing the rational number after multiplying, it is often easier to reduce before multiplying. We first factor all terms, then divide out the common factors, and then multiply:

$$\frac{6}{7} \cdot \frac{14}{15} = \frac{2 \cdot 3 \cdot 2 \cdot 7}{7 \cdot 3 \cdot 5} = \frac{4}{5}.$$

When we multiply rational numbers, we use the following definition.

Multiplication of Rational Numbers

If $\frac{a}{b}$ and $\frac{c}{d}$ are rational numbers, then $\frac{a}{b} \cdot \frac{c}{d} = \frac{ac}{bd}$.

We multiply rational expressions in the same way that we multiply rational numbers: Factor all polynomials, divide out the common factors, and then multiply the remaining factors.

E X A M P L E   1

Multiplying rational expressions

Find each product of rational expressions.

a) $\dfrac{3a^8b^3}{6b} \cdot \dfrac{10a}{a^2b^6}$

b) $\dfrac{x^2 + 7x + 12}{x^2 + 3x} \cdot \dfrac{x^2}{x^2 - 16}$

Solution

a) First factor the coefficients in each numerator and denominator:

$$\frac{3a^8b^3}{6b} \cdot \frac{10a}{a^2b^6} = \frac{3a^8b^3}{2 \cdot 3b} \cdot \frac{2 \cdot 5a}{a^2b^6} \qquad \text{Factor.}$$

$$= \frac{5a^9b^3}{a^2b^7} \qquad \text{Divide out the common factors.}$$

$$= \frac{5a^7}{b^4} \qquad \text{Quotient rule for exponents}$$

b) $\dfrac{x^2 + 7x + 12}{x^2 + 3x} \cdot \dfrac{x^2}{x^2 - 16} = \dfrac{(x+3)(x+4)}{x(x+3)} \cdot \dfrac{x \cdot x}{(x-4)(x+4)} = \dfrac{x}{x-4}$

> **Now do Exercises 1–8**

CAUTION   Do not attempt to divide out the x in $\frac{x}{x-4}$. This expression cannot be reduced because x is not a factor of *both* terms in the denominator. Compare this expression to the following:

$$\frac{3x}{x - xy} = \frac{x \cdot 3}{x(1 - y)} = \frac{3}{1 - y}$$

In Example 2(a) we will multiply a rational expression and a polynomial. For Example 2(b) we will use the rule for factoring the difference of two cubes.

E X A M P L E   2

Multiplying rational expressions

Find each product.

a) $(a^2 - 1) \cdot \dfrac{6}{2a^2 + 4a + 2}$

b) $\dfrac{a^3 - b^3}{b - a} \cdot \dfrac{6}{2a^2 + 2ab + 2b^2}$

Solution

a) First factor the polynomials completely:

$$(a^2 - 1) \cdot \frac{6}{2a^2 + 4a + 2} = \frac{(a + 1)(a - 1)}{1} \cdot \frac{2 \cdot 3}{2(a + 1)(a + 1)}$$

$$= \frac{3(a - 1)}{a + 1} \qquad \text{Divide out the common factors.}$$

$$= \frac{3a - 3}{a + 1} \qquad \text{Multiply.}$$

b) First factor the polynomials completely:

$$\frac{a^3 - b^3}{b - a} \cdot \frac{6}{2a^2 + 2ab + 2b^2} = \frac{(a - b)(a^2 + ab + b^2)}{b - a} \cdot \frac{2 \cdot 3}{2(a^2 + ab + b^2)}$$

$$= \frac{(a - b)3}{b - a}$$

$$= \frac{-1(b - a)3}{b - a} \qquad \begin{array}{l}\text{Factor out } -1 \text{ to get} \\ \text{a common } b - a.\end{array}$$

$$= -3$$

Now do Exercises 9–14

Since $a - b = -1(b - a)$, we have $\frac{a - b}{b - a} = -1$. So instead of factoring out -1 as in Example 2(b) we can simply divide $a - b$ by $b - a$ to get -1, as shown in Example 3.

E X A M P L E   3

Dividing $a - b$ by $b - a$

Find the product:

$$\frac{m - 4}{3} \cdot \frac{6}{4 - m}$$

Solution

Instead of factoring out -1 from $m - 4$, we divide $m - 4$ by $4 - m$ to get -1:

$$\frac{m - 4}{3} \cdot \frac{6}{4 - m} = \frac{\overset{-1}{m - 4}}{3} \cdot \frac{\overset{2}{6}}{4 - m} \qquad \text{Note that } (m - 4) \div (4 - m) = -1.$$

$$= -2$$

Now do Exercises 15–20

⟨ **Helpful Hint** ⟩

Note how all of the operations with rational expressions are performed according to the rules for fractions. So keep thinking of how you perform operations with fractions and you will improve your skills with both fractions and rational expressions.

⟨2⟩ Dividing Rational Expressions

We divide rational numbers by multiplying by the reciprocal or multiplicative inverse of the divisor. For example,

$$\frac{3}{4} \div \frac{15}{2} = \frac{3}{4} \cdot \frac{2}{15} = \frac{3}{2 \cdot 2} \cdot \frac{2 \cdot 1}{3 \cdot 5} = \frac{1}{10}.$$

When we divide rational numbers, we use the following definition.

> **Division of Rational Numbers**
>
> If $\frac{a}{b}$ and $\frac{c}{d}$ are rational numbers with $\frac{c}{d} \neq 0$, then
>
> $$\frac{a}{b} \div \frac{c}{d} = \frac{a}{b} \cdot \frac{d}{c}.$$

We use the same method to divide rational expressions: We invert the divisor and multiply.

EXAMPLE   4

Dividing rational expressions

Find each quotient.

a) $\dfrac{10}{3x} \div \dfrac{6}{5x}$ **b)** $\dfrac{5a^2b^8}{c^3} \div \left(4ab^3c\right)$

Solution

a) The reciprocal of the divisor $\frac{6}{5x}$ is $\frac{5x}{6}$.

$$\frac{10}{3x} \div \frac{6}{5x} = \frac{10}{3x} \cdot \frac{5x}{6} \qquad \text{Invert and multiply.}$$

$$= \frac{2 \cdot 5}{3x} \cdot \frac{5x}{2 \cdot 3} = \frac{25}{9}$$

b) The reciprocal of $4ab^3c$ is $\frac{1}{4ab^3c}$.

$$\frac{5a^2b^8}{c^3} \div \left(4ab^3c\right) = \frac{5a^2b^8}{c^3} \cdot \frac{1}{4ab^3c} = \frac{5ab^5}{4c^4} \qquad \text{Quotient rule for exponents}$$

> **Now do Exercises 21–26**

In Example 5, we factor the polynomials in the rational expressions.

EXAMPLE   5

Dividing rational expressions

Find the quotient.

a) $\dfrac{a^2 - 4}{a^2 + a - 2} \div \dfrac{2a - 4}{3a - 3}$ **b)** $\dfrac{25 - x^2}{x^2 + x} \div \dfrac{x - 5}{x^2 - 1}$

Solution

a) $\dfrac{a^2 - 4}{a^2 + a - 2} \div \dfrac{2a - 4}{3a - 3} = \dfrac{a^2 - 4}{a^2 + a - 2} \cdot \dfrac{3a - 3}{2a - 4}$ Invert and multiply.

$$= \frac{(a + 2)(a - 2)}{(a + 2)(a - 1)} \cdot \frac{3(a - 1)}{2(a - 2)} \qquad \text{Factor.}$$

$$= \frac{3}{2} \qquad\qquad\qquad\qquad\qquad \text{Divide out the common factors.}$$

b) $\dfrac{25 - x^2}{x^2 + x} \div \dfrac{x - 5}{x^2 - 1} = \dfrac{25 - x^2}{x^2 + x} \cdot \dfrac{x^2 - 1}{x - 5}$ Invert and multiply.

$$= \dfrac{\overset{-1}{\cancel{(5 - x)}}(5 + x)}{x(x + 1)} \cdot \dfrac{(x + 1)(x - 1)}{\cancel{x - 5}}$$ $(5 - x) \div (x - 5) = -1.$

Divide out the common factors.

$$= \dfrac{-1(5 + x)(x - 1)}{x}$$

$$= \dfrac{-x^2 - 4x + 5}{x}$$ Multiply the factors in the numerator.

Now do Exercises 27–32

CAUTION   When dividing rational expressions, you can factor the polynomials at any time, but do not reduce until after you have inverted the divisor.

In Example 6, division is indicated by a fraction bar.

EXAMPLE 6

Dividing rational expressions

Perform the operations indicated.

a) $\dfrac{\dfrac{a + b}{3}}{\dfrac{1}{2}}$ **b)** $\dfrac{\dfrac{x^2 - 4}{2}}{\dfrac{x - 2}{3}}$ **c)** $\dfrac{\dfrac{m^2 + 1}{5}}{3}$

Solution

a) $\dfrac{\dfrac{a + b}{3}}{\dfrac{1}{2}} = \dfrac{a + b}{3} \div \dfrac{1}{2}$

$$= \dfrac{a + b}{3} \cdot \dfrac{2}{1}$$ Invert the divisor.

$$= \dfrac{2a + 2b}{3}$$ Multiply.

b) $\dfrac{\dfrac{x^2 - 4}{2}}{\dfrac{x - 2}{3}} = \dfrac{x^2 - 4}{2} \cdot \dfrac{3}{x - 2}$ Invert and multiply.

$$= \dfrac{(x - 2)(x + 2)}{2} \cdot \dfrac{3}{x - 2}$$ Factor.

$$= \dfrac{3x + 6}{2}$$ Reduce.

c) $\dfrac{\dfrac{m^2 + 1}{5}}{3} = \dfrac{m^2 + 1}{5} \cdot \dfrac{1}{3} = \dfrac{m^2 + 1}{15}$ Multiply by $\frac{1}{3}$, the reciprocal of 3.

Now do Exercises 33–40

Warm-Ups ▼

Fill in the blank.

1. To _____ rational numbers, multiply the numerators and multiply the denominators.

2. The expressions $a - b$ and $b - a$ are _____.

3. To divide rational numbers, invert the _____ and multiply.

True or false?

4. To multiply fractions, they must have identical denominators.

5. $\dfrac{2}{7} \cdot \dfrac{3}{7} = \dfrac{6}{7}$

6. If $b \neq 0$, then $a \div b = a \cdot \dfrac{1}{b}$.

7. One-half of one-third is one-sixth.

8. For any nonzero x, $\dfrac{1}{2x} \cdot 8x^2 = 4x$.

9. One-third divided by one-half is two-thirds.

10. If $z - w \neq 0$, then $\dfrac{z - w}{w - z} = 1$.

11. $\dfrac{x}{3} \div 2 = \dfrac{x}{6}$

12. For any nonzero a and b, $\dfrac{a}{b} \div \dfrac{b}{a} = 1$.

Exercises 6.2

‹ Study Tips ›

- Studying in an environment similar to the one in which you will be tested can increase your chances of recalling information.
- If possible, do some studying in the classroom where you will be taking the test.

‹1› Multiplying Rational Expressions

Perform the indicated operations. See Examples 1–3.

1. $\dfrac{12}{42} \cdot \dfrac{35}{22}$

2. $\dfrac{3}{8} \cdot \dfrac{20}{21}$

3. $\dfrac{2x}{3} \cdot \dfrac{6}{5x}$

4. $\dfrac{3}{4a} \cdot \dfrac{2a}{9}$

5. $\dfrac{3a}{10b} \cdot \dfrac{5b^2}{6}$

6. $\dfrac{3x}{7y} \cdot \dfrac{14y^2}{9x}$

7. $\dfrac{3x - 3}{6} \cdot \dfrac{x}{x^2 - x}$

8. $\dfrac{-2x - 4}{2} \cdot \dfrac{6}{3x + 6}$

9. $\dfrac{10x + 5}{5x^2 + 5} \cdot \dfrac{2x^2 + x - 1}{4x^2 - 1}$

10. $\dfrac{x^3 + x}{5} \cdot \dfrac{5x - 5}{x^3 - x}$

VIDEO 11. $\dfrac{ax + aw + bx + bw}{x^2 - w^2} \cdot \dfrac{x - w}{a^2 - b^2}$

12. $\dfrac{3a - 3y}{3a - 3y - ab + by} \cdot \dfrac{b^2 - 9}{6b + 18}$

13. $\dfrac{a^2 - 2a + 4}{a^3 + 8} \cdot \dfrac{(a + 2)^3}{2a + 4}$

14. $\dfrac{w^3 - 1}{(w - 1)^2} \cdot \dfrac{w^2 - 1}{w^2 + w + 1}$

15. $\dfrac{x - 9}{12y} \cdot \dfrac{8y}{9 - x}$

16. $\dfrac{19x^2}{12y - 1} \cdot \dfrac{1 - 12y}{3x}$

17. $(a^2 - 4) \cdot \dfrac{7}{2 - a}$

18. $\dfrac{10x - 4x^2}{4x^2 - 20x + 25} \cdot \left(2x^3 - 5x^2\right)$

19. $\dfrac{(3x + 1)^3}{2x - 1} \cdot \dfrac{4x^2 - 4x + 1}{9x^2 + 6x + 1}$

20. $\dfrac{a^2 - 2ab + b^2}{a - b} \cdot \dfrac{(a + b)^3}{a^2 + 2ab + b^2}$

⟨2⟩ Dividing Rational Expressions

Perform the indicated operations. See Examples 4 and 5.

21. $\dfrac{15}{17} \div \dfrac{10}{17}$

22. $\dfrac{3}{4} \div \dfrac{1}{8}$

23. $\dfrac{36x}{5y} \div \dfrac{20x}{35y}$

24. $\dfrac{18a^3b^4}{c^9} \div \dfrac{12ab^6}{7c^2}$

25. $\dfrac{24a^5b^2}{5c^3} \div \left(4a^5bc^5\right)$

26. $\dfrac{60x^9y^2}{z} \div \left(48x^4y^3\right)$

27. $(w + 1) \div \dfrac{w^2 - 1}{w}$

28. $(a - 3) \div \dfrac{9 - a^2}{4}$

29. $\dfrac{x - y}{5} \div \dfrac{x^2 - 2xy + y^2}{10}$ [VIDEO]

30. $\dfrac{x^2 + 6x + 9}{18} \div \dfrac{(x + 3)^2}{36}$

31. $\dfrac{4x - 2}{x^2 - 5x} \div \dfrac{2x^2 + 9x - 5}{x^2 - 25}$

32. $\dfrac{2x^2 - 5x - 12}{6 + 4x} \div \dfrac{x^2 - 16}{2}$

Perform the indicated operations. See Example 6.

33. $\dfrac{\dfrac{x - y}{3}}{\dfrac{1}{6}}$

34. $\dfrac{\dfrac{2a - b}{10}}{\dfrac{1}{5}}$

35. $\dfrac{\dfrac{x^2 - 25}{3}}{\dfrac{x - 5}{6}}$ [VIDEO]

36. $\dfrac{\dfrac{3x^2 + 3}{5}}{\dfrac{3x + 3}{5}}$

37. $\dfrac{\dfrac{a - b}{2}}{3}$

38. $\dfrac{\dfrac{10}{a + b}}{5}$

39. $\dfrac{\dfrac{a^2 - b^2}{a + b}}{3}$

40. $\dfrac{\dfrac{x^2 + 5x + 6}{x + 2}}{x + 3}$

Miscellaneous

Perform the indicated operations. When possible, write down only the answer.

41. $\dfrac{5x}{2} \div 3$

42. $\dfrac{x}{a} \div 2$

43. $\dfrac{3}{4} \div \dfrac{1}{4}$

44. $\dfrac{1}{4} \div \dfrac{1}{2}$

45. One-half of $\dfrac{1}{6}$

46. One-half of $\dfrac{b}{a}$

47. One-half of $\dfrac{4x}{3}$

48. One-third of $\dfrac{6x}{y}$

49. $(a - b) \div (b - a)$

50. $\left(x^2 - y^2\right) \div \left(y^2 - x^2\right)$

51. $(a - b) \div (-1)$

52. $\left(x^2 + y^2\right) \div (-1)$

53. $\dfrac{x - y}{3} \cdot \dfrac{6}{y - x}$

54. $\dfrac{5x - 5y}{x} \cdot \dfrac{1}{x - y}$

55. $\dfrac{2a + 2b}{a} \cdot \dfrac{1}{2}$

56. $\dfrac{x - y}{y - x} \cdot \dfrac{1}{2}$

57. $-1\left(\dfrac{9 - x}{2}\right)$

58. $\dfrac{-1}{x - 1} \cdot \dfrac{1 - x}{2}$

59. $\dfrac{4}{y - 7} \div \dfrac{2}{7 - y}$

60. $\dfrac{1}{3 - m} \div \dfrac{1}{2m - 6}$

61. $\dfrac{a + b}{\dfrac{1}{2}}$

62. $\dfrac{x + 3}{\dfrac{1}{3}}$

63. $\dfrac{\dfrac{3x}{5}}{y}$

64. $\dfrac{\dfrac{b^2 - 4a}{2}}{a}$

65. $\dfrac{\dfrac{3a}{5b}}{2}$

66. $\dfrac{\dfrac{6x}{a}}{x}$

Perform the indicated operations.

67. $\dfrac{3x^2 + 13x - 10}{x} \cdot \dfrac{x^3}{9x^2 - 4} \cdot \dfrac{7x - 35}{x^2 - 25}$

68. $\dfrac{x^2 + 5x + 6}{x} \cdot \dfrac{x^2}{3x + 6} \cdot \dfrac{9}{x^2 - 4}$

69. $\dfrac{\left(a^2b^3c\right)^2}{\left(-2ab^2c\right)^3} \cdot \dfrac{\left(a^3b^2c\right)^3}{(abc)^4}$

70. $\dfrac{\left(-wy^2\right)^3}{3w^2y} \cdot \dfrac{(2wy)^2}{4wy^3}$

71. $\dfrac{(2mn)^3}{6mn^2} \div \dfrac{2m^2n^3}{\left(m^2n\right)^4}$

72. $\dfrac{(rt)^3}{rt^4} \div \dfrac{\left(rt^2\right)^3}{r^2t^3}$

73. $\dfrac{2x^2 + 7x - 15}{4x^2 - 100} \cdot \dfrac{2x^2 - 9x - 5}{4x^2 - 1}$ [VIDEO]

74. $\dfrac{x^3 + 1}{x^2 - 1} \cdot \dfrac{3x - 3}{x^3 - x^2 + x}$

75. $\dfrac{2h^2 - 5h - 3}{5h^2 - 4h - 1} \div \dfrac{2h^2 + 7h + 3}{h^2 + 2h - 3}$

76. $\dfrac{9w^2 - 64}{3w^2 - 5w - 8} \cdot \dfrac{5w^2 + 3w - 2}{25w^2 - 4}$

77. $\dfrac{9a - 3}{1 - 9a^2} \cdot \dfrac{9a^2 + 6a + 1}{6}$

78. $\dfrac{5 - 10k}{k^2 - 2k} \div \dfrac{2k^2 + 7k - 4}{k^2 + 2k - 8}$

79. $\dfrac{k^2 + 2km + m^2}{k^2 - 2km + m^2} \cdot \dfrac{m^2 + 3m - mk - 3k}{m^2 + mk + 3m + 3k}$

80. $\dfrac{a^2 + 2ab + b^2}{ac + bc - ad - bd} \div \dfrac{ac + ad - bc - bd}{c^2 - d^2}$

Perform the indicated operations. Variables in exponents represent integers.

81. $\dfrac{x^a}{y^2} \cdot \dfrac{y^{b+2}}{x^{2a}}$ **82.** $\dfrac{x^{3a+1}}{y^{2b-3}} \cdot \dfrac{y^{3b+4}}{x^{2a-1}}$

83. $\dfrac{x^{2a} + x^a - 6}{x^{2a} + 6x^a + 9} \div \dfrac{x^{2a} - 4}{x^{2a} + 2x^a - 3}$

84. $\dfrac{w^{2b} + 2w^b - 8}{w^{2b} + 3w^b - 4} \div \dfrac{w^{2b} - w^b - 2}{w^{2b} - 1}$

85. $\dfrac{m^k v^k + 3v^k - 2m^k - 6}{m^{2k} - 9} \cdot \dfrac{m^{2k} - 2m^k - 3}{v^k m^k - 2m^k + 2v^k - 4}$

86. $\dfrac{m^{3k} - 1}{m^{3k} + 1} \cdot \dfrac{m^{2k+1} - m^{k+1} + m}{m^{3k} + m^{2k} + m^k}$

Applications

Solve each problem.

87. *School enrollment.* In 2009, $\frac{1}{50}$ of the children enrolled in U.S. schools were enrolled in private secondary schools (National Center for Education Statistics, www.nces.ed.gov). Use the accompanying figure to determine the percentage

2009 distribution of students
in U.S. schools

Secondary schools $\frac{7}{25}$

Elementary schools $\frac{18}{25}$

Figure for Exercise 87

of secondary school children who were in private schools in 2009.

88. *The golden state.* In 2000, $\frac{3}{25}$ of the U.S. population was living in California (U.S. Census Bureau). Use the figure to determine the percentage of the population of the western region living in California in 2000.

Distribution of U.S. Population in 2000

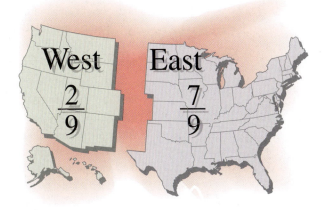

West $\dfrac{2}{9}$ East $\dfrac{7}{9}$

Figure for Exercise 88

89. *Distance traveled.* Bonita drove 100 miles in x hours. Assuming she continued to drive at the same speed, write a rational expression for the distance that she traveled in the next $\frac{3}{4}$ of an hour.

90. *Increasing speed.* Before lunch Avonda drove 200 miles at x miles per hour (mph). After lunch she drove 250 miles in the same amount of time. Write a rational expression for her speed after lunch.

Getting More Involved

91. *Discussion*

Which of the following expressions is not equivalent to $x \div \frac{y}{z}$? Explain.

a) $x \cdot \dfrac{z}{y}$ **b)** $\dfrac{x}{y} \cdot z$ **c)** $zx \cdot \dfrac{1}{y}$ **d)** $\dfrac{x}{\frac{y}{z}}$ **e)** $\dfrac{xy}{z}$

92. *Discussion*

Which of the following equations is not an identity? Explain.

a) $\dfrac{x^2 - 1}{2} \cdot \dfrac{2}{x - 1} = x + 1$ **b)** $\dfrac{x - 1}{x^2 - 1} = x + 1$

c) $x^2 - 1 = (x - 1)(x + 1)$

d) $\dfrac{1}{x^2 - 1} \div \dfrac{1}{x + 1} = \dfrac{1}{x - 1}$

| **Math** *at Work* | **Pediatric Dosing Rules** |

A drug is generally tested on adults, and an appropriate adult dose (*AD*) of the drug is determined. When a drug is given to a child, a doctor determines an appropriate child's dose (*CD*) using pediatric dosing rules. However, no single rule works for all children. Determining a child's dose also involves common sense and experience.

Clark's rule is based on the ratio of the child's body weight to the mean weight of an adult, 150 pounds. By Clark's rule, $CD = \dfrac{\text{child's weight in pounds}}{150 \text{ lb}} \cdot AD$. A dose determined by body weight alone might be too little to be effective in a small child.

Young's rule is based on the assumption that age approximates body weight for patients over 2 years old. Of course, there is a great variability of body weight of children of any given age. By Young's rule, $CD = \dfrac{\text{age in years}}{\text{age in years} + 12} \cdot AD$. See the accompanying figure.

The area rule is often used for drugs required in radioactive imaging. It is based on the idea that (body mass)$^{2/3}$ is approximately the body surface area. For radioactive imaging the adult's body mass (*MA*) often determines *AD*. By the area rule, $CD = \dfrac{(MC)^{2/3}}{(MA)^{2/3}} \cdot AD$, where *MC* is the child's body mass.

Webster's rule uses age to approximate the ratio in the area rule and agrees well with the area rule until age 11 or 12. By Webster's rule $CD = \dfrac{\text{age} + 1}{\text{age} + 7} \cdot AD$.

Fried's rule is generally used for patients less than one year old. By Fried's rule $CD = \dfrac{\text{age in months}}{150} \cdot AD$.

6.3 Addition and Subtraction

In This Section

We can multiply or divide any rational expressions, but we add or subtract only rational expressions with identical denominators. So when the denominators are not the same, we must find equivalent forms of the expressions that have identical denominators. In this section, we will review the idea of the least common denominator and will learn to use it for addition and subtraction of rational expressions.

⟨1⟩ Adding and Subtracting with Identical Denominators

It is easy to add or subtract fractions with identical denominators. For example,

$$\frac{1}{7} + \frac{3}{7} = \frac{4}{7} \qquad \text{and} \qquad \frac{3}{5} - \frac{2}{5} = \frac{1}{5}.$$

In general, we have the following definition.

> **Addition and Subtraction of Rational Numbers**
>
> If $b \neq 0$, then
>
> $$\frac{a}{b} + \frac{c}{b} = \frac{a+c}{b} \qquad \text{and} \qquad \frac{a}{b} - \frac{c}{b} = \frac{a-c}{b}.$$

Rational expressions with identical denominators are added or subtracted in the same manner as fractions.

E X A M P L E   1

Identical denominators

Perform the indicated operations.

a) $\dfrac{3}{2x} + \dfrac{5}{2x}$ **b)** $\dfrac{5x-3}{x-1} + \dfrac{5-7x}{x-1}$ **c)** $\dfrac{x^2+4x+7}{x^2-1} - \dfrac{x^2-2x+1}{x^2-1}$

‹ Helpful Hint ›

You can remind yourself of the difference between addition and multiplication of fractions with a simple example: If you and your spouse each own 1/7 of Microsoft, then together you own 2/7 of Microsoft. If you own 1/7 of Microsoft, and give 1/7 of your stock to your child, then your child owns 1/49 of Microsoft.

Solution

a) $\dfrac{3}{2x} + \dfrac{5}{2x} = \dfrac{8}{2x}$ Add the numerators.

       $= \dfrac{4}{x}$ Reduce.

b) $\dfrac{5x-3}{x-1} + \dfrac{5-7x}{x-1} = \dfrac{5x-3+5-7x}{x-1}$ Add the numerators.

       $= \dfrac{-2x+2}{x-1}$ Combine like terms.

       $= \dfrac{-2(x-1)}{x-1}$ Factor.

       $= -2$ Reduce to lowest terms.

c) The polynomials in the numerators are treated as if they were in parentheses:

$$\frac{x^2+4x+7}{x^2-1} - \frac{x^2-2x+1}{x^2-1} = \frac{x^2+4x+7-(x^2-2x+1)}{x^2-1}$$

$$= \frac{x^2+4x+7-x^2+2x-1}{x^2-1}$$

$$= \frac{6x+6}{x^2-1}$$

$$= \frac{6(x+1)}{(x+1)(x-1)} = \frac{6}{x-1}$$

> **Now do Exercises 1–12**

‹2› Least Common Denominator

To add fractions with denominators that are not identical, we use the basic principle of rational numbers to build up the denominators to the **least common denominator (LCD)**. For example,

$$\frac{1}{4} + \frac{1}{6} = \frac{1 \cdot 3}{4 \cdot 3} + \frac{1 \cdot 2}{6 \cdot 2} = \frac{3}{12} + \frac{2}{12} = \frac{5}{12}.$$

The LCD 12 is the **least common multiple (LCM)** of the numbers 4 and 6.

Finding the LCM for a pair of large numbers such as 24 and 126 will help you to understand the procedure for finding the LCM for any polynomials. First factor the numbers completely:

$$24 = 2^3 \cdot 3$$
$$126 = 2 \cdot 3^2 \cdot 7$$

‹ **Helpful Hint** ›

The product of 24 and 126 is 3024, and 3024 is a common multiple but not the least common multiple of 24 and 126. If you divide 3024 by 6, the greatest common factor of 24 and 126, you get 504.

Any number that is a multiple of both 24 and 126 must have all of the factors of 24 and all of the factors of 126 in its factored form. So in the LCM we use the factors 2, 3, and 7, and for each factor we use the highest power that appears on that factor. The highest power of 2 is 3, the highest power of 3 is 2, and the highest power of 7 is 1. So the LCM is $2^3 \cdot 3^2 \cdot 7$. If we write this product without exponents, we can see clearly that it is a multiple of both 24 and 126:

$$\overbrace{2 \cdot 2 \cdot 2 \cdot 3 \cdot 3 \cdot 7}^{126} = 504 \quad \begin{array}{l} 504 = 126 \cdot 4 \\ 504 = 24 \cdot 21 \end{array}$$

$$\underbrace{}_{24}$$

The strategy for finding the LCM for a group of polynomials can be stated as follows.

Strategy for Finding the LCM for Polynomials

1. Factor each polynomial completely. Use exponents to express repeated factors.
2. Write the product of all of the different factors that appear in the polynomials.
3. For each factor, use the highest power of that factor in any of the polynomials.

EXAMPLE   **2**

Finding the LCM

Find the least common multiple for each group of polynomials.

a) $4x^2y, 6y$ **b)** a^2bc, ab^3c^2, a^3bc **c)** $x^2 + 5x + 6, x^2 + 6x + 9$

Solution

a) Factor $4x^2y$ and $6y$ as follows:

$$4x^2y = 2^2 \cdot x^2y, \qquad 6y = 2 \cdot 3y$$

To get the LCM, we use 2, 3, x, and y the maximum number of times that each appears in either of the expressions. The LCM is $2^2 \cdot 3 \cdot x^2y$, or $12x^2y$.

b) The expressions a^2bc, ab^3c^2, and a^3bc are already factored. To get the LCM, we use a, b, and c the maximum number of times that each appears in any of the expressions. The LCM is $a^3b^3c^2$.

c) Factor $x^2 + 5x + 6$ and $x^2 + 6x + 9$ completely:

$$x^2 + 5x + 6 = (x + 2)(x + 3), \qquad x^2 + 6x + 9 = (x + 3)^2$$

The LCM is $(x + 2)(x + 3)^2$.

Now do Exercises 13–30

⟨3⟩ Adding and Subtracting with Different Denominators

To add or subtract rational expressions with different denominators, we must build up each rational expression to equivalent forms with identical denominators, as we did in Section 6.1. Of course, it is most efficient to use the LCD as in Examples 3 through 5.

EXAMPLE   3

Different denominators

Perform the indicated operations.

a) $\dfrac{3}{a^2b} + \dfrac{5}{ab^3}$

b) $\dfrac{x+1}{6} - \dfrac{2x-3}{4}$

Solution

a) The LCD for a^2b and ab^3 is a^2b^3. To build up each denominator to a^2b^3, multiply the numerator and denominator of the first expression by b^2, and multiply the numerator and denominator of the second expression by a:

$$\frac{3}{a^2b} + \frac{5}{ab^3} = \frac{3(b^2)}{a^2b(b^2)} + \frac{5(a)}{ab^3(a)} \qquad \text{Build up each denominator to the LCD.}$$

$$= \frac{3b^2}{a^2b^3} + \frac{5a}{a^2b^3}$$

$$= \frac{3b^2 + 5a}{a^2b^3} \qquad \text{Add the numerators.}$$

b)
$$\frac{x+1}{6} - \frac{2x-3}{4} = \frac{(x+1)(2)}{6(2)} - \frac{(2x-3)(3)}{4(3)} \qquad \begin{array}{l}\text{Build up each denominator}\\ \text{to the LCD 12.}\end{array}$$

$$= \frac{2x+2}{12} - \frac{6x-9}{12} \qquad \text{Distributive property}$$

$$= \frac{2x+2-(6x-9)}{12} \qquad \begin{array}{l}\text{Subtract the numerators.}\\ \text{Note that } 6x-9 \text{ is put in}\\ \text{parentheses.}\end{array}$$

$$= \frac{2x+2-6x+9}{12} \qquad \text{Remove the parentheses.}$$

$$= \frac{-4x+11}{12} \qquad \text{Combine like terms.}$$

> **Now do Exercises 31–46**

CAUTION Before you add or subtract rational expressions, they must be written with identical denominators. For multiplication and division it is not necessary to have identical denominators.

In Example 4 we must first factor polynomials to find the LCD.

EXAMPLE   4

Different denominators

Perform the indicated operations.

a) $\dfrac{1}{x^2-1} + \dfrac{2}{x^2+x}$

b) $\dfrac{5}{a-2} - \dfrac{3}{2-a}$

‹ **Helpful Hint** ›

It is not actually necessary to identify the LCD. Once the denominators are factored, simply look at each denominator and ask, "What factor does the other denominator have that is missing from this one?" Then use the missing factor to build up the denominator and you will obtain the LCD.

Solution

a) Because $x^2 - 1 = (x + 1)(x - 1)$ and $x^2 + x = x(x + 1)$, the LCD is $x(x - 1)(x + 1)$. The first denominator is missing the factor x, and the second denominator is missing the factor $x - 1$.

$$\underbrace{\frac{1}{x^2 - 1}}_{} + \frac{2}{x^2 + x} = \underbrace{\frac{1}{(x - 1)(x + 1)}}_{\text{Missing } x} + \underbrace{\frac{2}{x(x + 1)}}_{\text{Missing } x - 1}$$ The LCD is $x(x - 1)(x + 1)$.

$$= \frac{1(x)}{(x - 1)(x + 1)(x)} + \frac{2(x - 1)}{x(x + 1)(x - 1)}$$ Build up the denominators to the LCD.

$$= \frac{x}{x(x - 1)(x + 1)} + \frac{2x - 2}{x(x - 1)(x + 1)}$$

$$= \frac{3x - 2}{x(x - 1)(x + 1)}$$ Add the numerators.

For this type of answer we usually leave the denominator in factored form. That way, if we need to work with the expression further, we do not have to factor the denominator again.

b) Because $-1(2 - a) = a - 2$, we can convert the denominator $2 - a$ to $a - 2$.

$$\frac{5}{a - 2} - \frac{3}{2 - a} = \frac{5}{a - 2} - \frac{3(-1)}{(2 - a)(-1)}$$

$$= \frac{5}{a - 2} - \frac{-3}{a - 2}$$ The LCD is $a - 2$.

$$= \frac{5 - (-3)}{a - 2}$$ Subtract the numerators.

$$= \frac{8}{a - 2}$$ Simplify.

Note that we get an equivalent answer if we multiply the numerator and denominator by -1:

$$\frac{8}{a - 2} = \frac{8(-1)}{(a - 2)(-1)} = \frac{-8}{2 - a}$$

This is the answer that we would have gotten if we had used $2 - a$ as the common denominator in the beginning.

Now do Exercises 47–48

If the rational expressions in a sum or difference are not in lowest terms, then they should be reduced before finding the least common denominator.

E X A M P L E   **5**

Reducing before finding the LCD

Perform the indicated operations.

a) $\dfrac{2xy}{4x} + \dfrac{x^2}{xy}$

b) $\dfrac{8x - 8}{4x^2 - 4} - \dfrac{9x}{3x^2 - 3x - 6}$

Solution

a) Notice that the rational expressions can be reduced:

$$\frac{2xy}{4x} + \frac{x^2}{xy} = \frac{y}{2} + \frac{x}{y}$$ Reduce each rational expression.

$$= \frac{y \cdot y}{2 \cdot y} + \frac{x \cdot 2}{y \cdot 2}$$ Build up each denominator to $2y$.

$$= \frac{y^2 + 2x}{2y}$$ Add the rational expressions.

b) Notice that $3x^2 - 3x - 6 = 3(x^2 - x - 2) = 3(x - 2)(x + 1)$ and $4x^2 - 4 = 4(x^2 - 1) = 4(x - 1)(x + 1)$:

$$\frac{8x - 8}{4x^2 - 4} - \frac{9x}{3x^2 - 3x - 6}$$

$$= \frac{8(x - 1)}{4(x - 1)(x + 1)} - \frac{9x}{3(x - 2)(x + 1)}$$ Factor.

$$= \frac{2}{x + 1} - \frac{3x}{(x - 2)(x + 1)}$$ Reduce.

$$= \frac{2(x - 2)}{(x + 1)(x - 2)} - \frac{3x}{(x - 2)(x + 1)}$$ Build up to get the LCD.

$$= \frac{2x - 4 - 3x}{(x + 1)(x - 2)}$$ Subtract the expressions.

$$= \frac{-x - 4}{(x + 1)(x - 2)}$$ Simplify. Leave denominator factored.

> Now do Exercises 49–62

⟨4⟩ Shortcuts

Consider the following addition:

$$\frac{a}{b} + \frac{c}{d} = \frac{a(d)}{b(d)} + \frac{c(b)}{d(b)} = \frac{ad + bc}{bd}$$ The LCD is bd.

We can use this result as a rule for adding simple fractions in which the LCD is the product of the denominators. A similar rule works for subtraction.

Adding or Subtracting Simple Fractions

If $b \neq 0$ and $d \neq 0$, then

$$\frac{a}{b} + \frac{c}{d} = \frac{ad + bc}{bd}$$ and $$\frac{a}{b} - \frac{c}{d} = \frac{ad - bc}{bd}.$$

EXAMPLE 6

Adding and subtracting simple fractions

Use the rules for adding and subtracting simple fractions to find the sums and differences.

a) $\dfrac{1}{2} + \dfrac{1}{3}$ **b)** $\dfrac{1}{a} - \dfrac{1}{x}$

c) $\dfrac{a}{5} + \dfrac{a}{3}$ **d)** $x - \dfrac{2}{3}$

Solution

a) For the numerator, compute $ad + bc = 1 \cdot 3 + 2 \cdot 1 = 5$. Use $2 \cdot 3$ or 6 for the denominator:

$$\frac{1}{2} + \frac{1}{3} = \frac{5}{6}$$

b) $\dfrac{1}{a} - \dfrac{1}{x} = \dfrac{1 \cdot x - 1 \cdot a}{ax} = \dfrac{x - a}{ax}$ c) $\dfrac{a}{5} + \dfrac{a}{3} = \dfrac{3a + 5a}{15} = \dfrac{8a}{15}$

d) $x - \dfrac{2}{3} = \dfrac{x}{1} - \dfrac{2}{3} = \dfrac{3x - 2}{3}$

> Now do Exercises 63–78

CAUTION The rules for adding or subtracting simple fractions can be applied to any rational expressions, but they work best when the LCD is the product of the two denominators. Always make sure that the answer is in its lowest terms. If the product of the two denominators is too large, these rules are not helpful because then reducing can be difficult.

⟨5⟩ Applications

Rational expressions occur often in expressing rates. For example, if you can process one application in 2 hours, then you are working at the rate of $\frac{1}{2}$ of an application per hour. If you can complete one task in x hours, then you are working at the rate of $\frac{1}{x}$ task per hour.

EXAMPLE 7

Work rates

Susan takes an average of x hours to process a mortgage application, whereas Betty's average is 1 hour longer. Write a rational function $A(x)$ that gives the number of applications that they can process together in 40 hours. Find $A(4)$.

⟨ **Helpful Hint** ⟩

Notice that a work rate is the same as a slope from Chapter 3. The only difference is that the work rates here can contain a variable.

Solution

The number of applications processed by Susan is the product of her rate and her time:

$$\frac{1}{x} \frac{\text{application}}{\text{hr}} \cdot 40 \text{ hr} = \frac{40}{x} \text{ applications}$$

The number of applications processed by Betty is the product of her rate and her time:

$$\frac{1}{x + 1} \frac{\text{application}}{\text{hr}} \cdot 40 \text{ hr} = \frac{40}{x + 1} \text{ applications}$$

Find the sum of the rational expressions:

$$\frac{40}{x} + \frac{40}{x + 1} = \frac{40x + 40 + 40x}{x(x + 1)} = \frac{80x + 40}{x(x + 1)}$$

So the function $A(x) = \frac{80x + 40}{x(x + 1)}$ gives the number of applications that they can process in 40 hours. Substituting 4 for x yields $A(4) = 18$. Together they can process 18 applications in 40 hours.

> Now do Exercises 99–104

Warm-Ups ▼

Fill in the blank.

1. The _____ of two numbers is the smallest number that is a multiple of both numbers.

2. The _____ for two fractions is the least common multiple of their denominators.

3. To add fractions, they must have equal _____.

4. We do not need to have equal denominators for _____ and _____.

True or false?

5. The LCM of 6 and 10 is 60.

6. The LCM of $6a^2b$ and $8ab^3$ is $24ab$.

7. The LCM of $x^2 - 1$ and $x - 1$ is $x^2 - 1$.

8. $\dfrac{3}{8} + \dfrac{1}{8} = \dfrac{1}{2}$

9. $\dfrac{1}{2} + \dfrac{2}{3} = \dfrac{3}{5}$

10. $\dfrac{2}{3} + \dfrac{3}{4} = \dfrac{17}{12}$

11. If $x \neq 0$, then $5 + \dfrac{1}{x} = \dfrac{6}{x}$.

12. If $a \neq 0$, then $\dfrac{7}{a} + 3 = \dfrac{7 + 3a}{a}$.

Exercises 6.3

⟨ **Study Tips** ⟩

- Don't try to get everything done before you start studying. Since the average attention span for a task is only 20 minutes, it is better to study and take breaks from studying to do other duties.
- Your mood for studying should match the mood in which you are tested. Being too relaxed in studying will not match the increased anxiety that you feel during a test.

⟨1⟩ Adding and Subtracting with Identical Denominators

Perform the indicated operations. Reduce answers to their lowest terms. See Example 1.

1. $\dfrac{3x}{2} + \dfrac{5x}{2}$

2. $\dfrac{5x^2}{3} + \dfrac{4x^2}{3}$

3. $\dfrac{7x}{2} - \dfrac{9x}{2}$

4. $\dfrac{3a}{4} - \dfrac{a}{4}$

5. $\dfrac{x - 3}{2x} - \dfrac{3x - 5}{2x}$

6. $\dfrac{9 - 4y}{3y} - \dfrac{6 - y}{3y}$

7. $\dfrac{3x - 4}{2x - 4} + \dfrac{2x - 6}{2x - 4}$

8. $\dfrac{a^3}{a + b} + \dfrac{b^3}{a + b}$

 9. $\dfrac{x^2 + 4x - 6}{x^2 - 9} - \dfrac{x^2 + 2x - 12}{x^2 - 9}$

10. $\dfrac{x^2 + 3x - 3}{x - 4} - \dfrac{x^2 + 4x - 7}{x - 4}$

11. $\dfrac{2x^2 - 8x - 4}{2x^2 + 7x + 3} + \dfrac{4x^2 + x - 1}{2x^2 + 7x + 3}$

12. $\dfrac{5x^2 - 2x - 5}{2x^2 + 3x - 2} - \dfrac{x^2 - 9x - 3}{2x^2 + 3x - 2}$

⟨2⟩ Least Common Denominator

Find the least common multiple for each group of polynomials. See Example 2. See the Strategy for Finding the LCM for Polynomials box on page 402.

13. 24, 20

14. 15, 18

15. 12, 18, 22

16. 8, 20, 28

17. $10x^3y, 15x$

18. $12a^3b^2, 18ab^5$

19. a^3b, ab^4c, ab^5c^2

20. x^2yz, xy^2z^3, xy^6

21. $x, x + 2, x - 2$

22. $y, y - 5, y + 2$

23. $4a + 8, 6a + 12$

24. $4a - 6, 2a^2 - 3a$

25. $x^2 - 1, x^2 + 2x + 1$

26. $y^2 - 2y - 15, y^2 + 6y + 9$

27. $x^2 - 4x, x^2 - 16, x^2 + 6x + 8$

28. $z^2 - 25, 5z - 25, 5z + 25$

29. $6x^2 + 17x + 12, 9x^2 - 16$

30. $16x^2 - 8x - 3, 4x^2 - 7x + 3$

⟨3⟩ Adding and Subtracting with Different Denominators

Perform the indicated operations. Reduce answers to lowest terms. See Examples 3–5.

31. $\dfrac{1}{28} + \dfrac{3}{35}$

32. $\dfrac{7}{24} - \dfrac{4}{15}$

33. $\dfrac{7}{48} - \dfrac{5}{36}$

34. $\dfrac{7}{52} + \dfrac{3}{40}$

35. $\dfrac{3}{wz^2} + \dfrac{5}{w^2z}$

36. $\dfrac{2}{a^2b} - \dfrac{3}{ab^2}$

37. $\dfrac{2x - 3}{8} - \dfrac{x - 2}{6}$

38. $\dfrac{a - 5}{10} + \dfrac{3 - 2a}{15}$

39. $\dfrac{xa^3}{2a^4} + \dfrac{21x^2}{35ax}$

40. $\dfrac{5x^2}{30xy} - \dfrac{30x}{80y}$

41. $\dfrac{9}{4y} - x$

42. $\dfrac{b^2}{4a} - c$

43. $\dfrac{5}{a + 2} - \dfrac{7}{a}$

44. $\dfrac{2}{x + 1} - \dfrac{3}{x}$

45. $\dfrac{1}{a - b} + \dfrac{2}{a + b}$

46. $\dfrac{5}{x + 2} + \dfrac{3}{x - 2}$

47. $\dfrac{1}{a - b} + \dfrac{1}{b - a}$

48. $\dfrac{3}{x - 5} + \dfrac{7}{5 - x}$

49. $\dfrac{1 - 2x}{x^2 + 3x + 2} + \dfrac{5}{x + 2}$

50. $\dfrac{30 - 4x}{x^2 - 9} + \dfrac{7}{x + 3}$

51. $\dfrac{2x^2}{2x^3 - 18x} + \dfrac{15}{5x - 15}$

52. $\dfrac{5x}{5x^2 - 125} + \dfrac{5x - 5}{x^2 - 6x + 5}$

53. $\dfrac{6x - 11}{x^2 + x - 12} - \dfrac{5}{x + 4}$

54. $\dfrac{2}{x - 2} - \dfrac{10}{x^2 + x - 6}$

55. $\dfrac{10}{4x - 8} - \dfrac{15}{10 - 5x}$

56. $\dfrac{8}{6x - 18} - \dfrac{14}{6 - 2x}$

57. $\dfrac{5}{x^2 + x - 2} - \dfrac{6}{x^2 + 2x - 3}$

58. $\dfrac{2}{x^2 - 4} - \dfrac{5}{x^2 - 3x - 10}$

59. $\dfrac{x}{2x^2 + x - 1} + \dfrac{3}{3x^2 + 2x - 1}$

60. $\dfrac{x + 1}{3x^2 - 2x - 1} + \dfrac{x - 1}{3x^2 + 4x + 1}$

61. $\dfrac{1}{x} + \dfrac{2}{x - 1} - \dfrac{3}{x + 2}$

62. $\dfrac{2}{a} - \dfrac{3}{a + 1} + \dfrac{5}{a - 1}$

⟨4⟩ Shortcuts

Perform the following operations. Write down only the answer. See Example 6.

63. $\dfrac{1}{3} + \dfrac{1}{4}$

64. $\dfrac{3}{5} + \dfrac{1}{4}$

65. $\dfrac{1}{8} - \dfrac{3}{5}$

66. $\dfrac{a}{2} + \dfrac{5}{3}$

67. $\dfrac{x}{3} + \dfrac{x}{2}$

68. $\dfrac{y}{4} - \dfrac{y}{3}$

69. $\dfrac{a}{b} - \dfrac{2}{3}$

70. $\dfrac{3}{x} + \dfrac{1}{9}$

71. $a + \dfrac{2}{3}$

72. $\dfrac{m}{3} + y$

73. $\dfrac{3}{a} + 1$

74. $\dfrac{1}{x} + 1$

75. $\dfrac{3 + x}{x} - 1$

76. $\dfrac{a + 2}{a} + 3$

77. $\dfrac{2}{3} + \dfrac{1}{4x}$

78. $\dfrac{1}{5} + \dfrac{1}{5x}$

Miscellaneous

Perform the indicated operations.

79. $\dfrac{w^2 - 3w + 6}{w - 5} + \dfrac{9 - w^2}{w - 5}$

80. $\dfrac{2z^2 - 3z + 6}{z^2 - 1} - \dfrac{z^2 - 5z + 9}{z^2 - 1}$

81. $\dfrac{1}{3x - 6} - \dfrac{6}{5x - 10}$

82. $\dfrac{3}{6x^2 - 4x} - \dfrac{x - 2}{9x - 6}$

83. $\dfrac{x - 1}{2x^2 + 3x + 1} - \dfrac{x + 1}{2x^2 - x - 1}$

84. $\dfrac{2x + 1}{6x^2 - 5x + 1} + \dfrac{2x - 1}{6x^2 + x - 1}$

85. $\dfrac{(a^2 b^3)^4}{(ab^4)^3} \cdot \dfrac{(ab)^3}{(a^4 b)^2}$

86. $\dfrac{(ab)^2}{(a + b)^2} \cdot \dfrac{(a + b)^3}{(ab)^3}$

87. $\dfrac{x^2 + 4}{25x^2 - 20x + 4} + \dfrac{10x + 4}{25x^2 - 4}$

88. $\dfrac{8a}{2a^2 + 4a + 2} - \dfrac{3a - 3}{a^2 - 1}$

89. $\dfrac{4x^2 + 9}{4x^2 - 9} \cdot \dfrac{4x^2 + 12x + 9}{2x^2 + 3x}$

90. $\dfrac{3a^2 - 2a - 16}{2a^2 + 3a - 2} \cdot \dfrac{6a + 16}{9a^2 - 64}$

91. $\dfrac{w^2 - 3}{3w^3 + 81} - \dfrac{2}{6w + 18} - \dfrac{w - 4}{w^2 - 3w + 9}$

92. $\dfrac{a - 3}{a^3 + 8} - \dfrac{2}{a + 2} - \dfrac{a - 3}{a^2 - 2a + 4}$

93. $\dfrac{a^2 - 6a + 9}{a^3 - 8} \div \dfrac{a^2 - a - 6}{a^2 - 4}$

94. $\dfrac{1}{z^2 + 4} \div \dfrac{z^3 - 8}{z^4 - 16}$

95. $\dfrac{w^2 + 3}{w^3 - 8} - \dfrac{2w}{w^2 - 4}$

96. $\dfrac{x + 5}{x^3 + 27} - \dfrac{x - 1}{x^2 - 9}$

97. $\dfrac{1}{x^3 - 1} - \dfrac{1}{x^2 - 1} + \dfrac{1}{x - 1}$

98. $\dfrac{x - 4}{x^3 - 1} + \dfrac{x - 2}{x^2 - 1}$

⟨5⟩ Applications

Solve each problem. See Example 7.

99. *Processing claims.* Joe takes x hours on the average to process a claim, whereas Ellen's average is 1 hour longer.

  a) Write a rational function $C(x)$ that gives the number of claims that they process while working together for an 8-hour shift.

  b) Find $C(2)$.

100. *Attaching shingles.* Bill attaches one bundle of shingles in an average of x minutes using a hammer, whereas Julio attaches one bundle in 6 minutes less time using a pneumatic stapler.

  a) Write a rational function $B(x)$ that gives the number of bundles that they attach while working together for 10 hours.

  b) Find $B(30)$.

Photo for Exercise 100

101. *Telemarketing.* George sells one magazine subscription every 20 minutes, whereas Theresa sells one every x minutes.

a) Write a rational function $M(x)$ that gives the number of magazine subscriptions that they sell when working together for 1 hour.

b) Find $M(5)$.

102. *Housepainting.* Harry can paint his house by himself in 6 days. His wife Judy can paint the house by herself in x days.

a) Write a rational expression $F(x)$ that gives the portion of the house that they paint when working together for 2 days.

b) Find $F(12)$

103. *Driving time.* Joan drove for 100 miles at x miles per hour. Then she increased her speed by 5 miles per hour and drove 200 additional miles.

a) Write a rational function $T(x)$ that gives her total travel time in hours.

b) Find $T(65)$ to the nearest minute.

104. *Running time.* Willard jogged for 3 miles at x miles per hour. Then he doubled his speed and jogged an additional mile.

a) Write a rational function $R(x)$ that gives his total running time in hours.

b) Find $R(3.5)$.

Getting More Involved

105. *Discussion*

Explain why fractions must have common denominators for addition but not for multiplication.

106. *Discussion*

Find each "infinite sum" and explain your answer.

a) $\dfrac{3}{10} + \dfrac{3}{10^2} + \dfrac{3}{10^3} + \dfrac{3}{10^4} + \cdots$

b) $\dfrac{9}{10} + \dfrac{9}{10^2} + \dfrac{9}{10^3} + \dfrac{9}{10^4} + \cdots$

6.4   Complex Fractions

In This Section

⟨ 1 ⟩ **Simplifying Complex Fractions**

⟨ 2 ⟩ **Simplifying Expressions with Negative Exponents**

⟨ 3 ⟩ **Applications**

In this section, we will use the techniques of Section 6.3 to simplify complex fractions. As their name suggests, complex fractions are rather messy-looking expressions.

⟨ 1 ⟩ Simplifying Complex Fractions

A **complex fraction** is a fraction that has rational expressions in the numerator, the denominator, or both. For example,

$$\frac{\dfrac{1}{2} + \dfrac{1}{3}}{\dfrac{1}{4} + \dfrac{1}{5}}, \qquad \frac{3 - \dfrac{2}{x}}{\dfrac{1}{x^2} - \dfrac{1}{4}}, \qquad \text{and} \qquad \frac{\dfrac{x+2}{x^2-9}}{\dfrac{x}{x^2-6x+9} + \dfrac{4}{x-3}}$$

are complex fractions. There are two methods for simplifying complex fractions. Method A is simply the order of operations. We compute the values of the numerator and denominator, and then divide the results. For Method B, we use the fact that the value of a fraction is not changed when its numerator and denominator are multiplied by the same nonzero number. So we multiply the numerator and the denominator of the complex fraction by the LCD for all of the denominators of the simple fractions. This method is not as natural as Method A, but it has the advantage of greatly simplifying the complex fraction in one step.

EXAMPLE 1

A complex fraction without variables

Simplify $\dfrac{\frac{1}{2} + \frac{1}{3}}{\frac{1}{4} + \frac{1}{5}}$.

⟨ **Calculator Close-Up** ⟩

You can use a calculator to find the value of a complex fraction.

```
(1/2+1/3)/(1/4+1
/5)▶Frac
               50/27
```

Solution

Method A   For this method we perform the computations of the numerator and denominator separately and then divide:

$$\frac{\frac{1}{2} + \frac{1}{3}}{\frac{1}{4} + \frac{1}{5}} = \frac{\frac{5}{6}}{\frac{9}{20}} = \frac{5}{6} \div \frac{9}{20} = \frac{5}{6} \cdot \frac{20}{9} = \frac{5 \cdot 2 \cdot 10}{2 \cdot 3 \cdot 9} = \frac{50}{27}$$

Method B   For this method we find the LCD for all of the fractions in the complex fraction. Then we multiply the numerator and denominator of the complex fraction by the LCD. The LCD for the denominators 2, 3, 4, and 5 is 60. So we multiply the numerator and denominator of the complex fraction by 60:

$$\frac{\frac{1}{2} + \frac{1}{3}}{\frac{1}{4} + \frac{1}{5}} = \frac{\left(\frac{1}{2} + \frac{1}{3}\right)60}{\left(\frac{1}{4} + \frac{1}{5}\right)60} = \frac{30 + 20}{15 + 12} = \frac{50}{27} \qquad \begin{array}{l} \frac{1}{2} \cdot 60 = 30, \frac{1}{3} \cdot 60 = 20 \\[4pt] \frac{1}{4} \cdot 60 = 15, \frac{1}{5} \cdot 60 = 12 \end{array}$$

> **Now do Exercises 1–6**

In most cases, Method B of Example 1 is the faster method for simplifying complex fractions, and we will continue to use it.

EXAMPLE 2

A complex fraction with variables

Simplify $\dfrac{3 - \frac{2}{x}}{\frac{1}{x^2} - \frac{1}{4}}$.

⟨ **Helpful Hint** ⟩

When students see addition or subtraction in a complex fraction, they often convert all of the fractions to the same denominator. This is not wrong, but it is not necessary. Simply multiplying every fraction by the LCD eliminates the denominators of the original fractions.

Solution

The LCD of x, x^2, and 4 is $4x^2$. Multiply the numerator and denominator by $4x^2$:

$$\frac{3 - \frac{2}{x}}{\frac{1}{x^2} - \frac{1}{4}} = \frac{\left(3 - \frac{2}{x}\right)(4x^2)}{\left(\frac{1}{x^2} - \frac{1}{4}\right)(4x^2)} \qquad \text{Multiply numerator and denominator by } 4x^2.$$

$$= \frac{3(4x^2) - \dfrac{2}{x}(4x^2)}{\dfrac{1}{x^2}(4x^2) - \dfrac{1}{4}(4x^2)} \qquad \text{Distributive property}$$

$$= \frac{12x^2 - 8x}{4 - x^2} \qquad \text{Simplify.}$$

$$= \frac{12x^2 - 8x}{(2 - x)(2 + x)} \qquad \begin{array}{l}\text{Answer with denominator}\\ \text{in factored form.}\end{array}$$

> **Now do Exercises 7–20**

E X A M P L E 3

More complicated denominators

Simplify $\dfrac{\dfrac{x + 2}{x^2 - 9}}{\dfrac{x}{x^2 - 6x + 9} + \dfrac{4}{x - 3}}$.

Solution

Because $x^2 - 9 = (x - 3)(x + 3)$ and $x^2 - 6x + 9 = (x - 3)^2$, the LCD is $(x - 3)^2(x + 3)$. Multiply the numerator and denominator by the LCD:

$$\frac{\dfrac{x + 2}{x^2 - 9}}{\dfrac{x}{x^2 - 6x + 9} + \dfrac{4}{x - 3}} = \frac{\dfrac{x + 2}{(x - 3)(x + 3)}(x - 3)^2(x + 3)}{\dfrac{x}{(x - 3)^2}(x - 3)^2(x + 3) + \dfrac{4}{x - 3}(x - 3)^2(x + 3)}$$

$$= \frac{(x + 2)(x - 3)}{x(x + 3) + 4(x - 3)(x + 3)} \qquad \text{Simplify.}$$

$$= \frac{(x + 2)(x - 3)}{(x + 3)[x + 4(x - 3)]} \qquad \text{Factor out } x + 3.$$

$$= \frac{(x + 2)(x - 3)}{(x + 3)(5x - 12)}$$

> **Now do Exercises 21–38**

⟨2⟩ **Simplifying Expressions with Negative Exponents**

Consider the expression

$$\frac{3a^{-1} - 2^{-1}}{1 - b^{-1}}.$$

Using the definition of negative exponents, we can rewrite this expression as a complex fraction:

$$\frac{3a^{-1} - 2^{-1}}{1 - b^{-1}} = \frac{\dfrac{3}{a} - \dfrac{1}{2}}{1 - \dfrac{1}{b}}$$

The LCD for the complex fraction is $2ab$. Note that $2ab$ could be obtained from a^{-1}, 2^{-1}, and b^{-1} in the original expression. To simplify the complex fraction we can multiply the numerator and denominator of either of the two preceding expressions by $2ab$. To gain more experience with negative exponents, we will work with the first expression in Example 4.

E X A M P L E   4

A complex fraction with negative exponents

Simplify the complex fraction $\frac{3a^{-1} - 2^{-1}}{1 - b^{-1}}$.

Solution

Multiply the numerator and denominator by $2ab$, the LCD of the fractions. Remember that $a^{-1} \cdot a = a^0 = 1$.

$$\frac{3a^{-1} - 2^{-1}}{1 - b^{-1}} = \frac{\left(3a^{-1} - 2^{-1}\right)2ab}{\left(1 - b^{-1}\right)2ab}$$

$$= \frac{3a^{-1}(2ab) - 2^{-1}(2ab)}{1(2ab) - b^{-1}(2ab)} \qquad \text{Distributive property}$$

$$= \frac{6b - ab}{2ab - 2a}$$

$$= \frac{6b - ab}{2a(b - 1)}$$

> **Now do Exercises 39–42**

E X A M P L E   5

A complex fraction with negative exponents

Simplify the complex fraction $\frac{a^{-1} + b^{-2}}{ab^{-2} + ba^{-3}}$.

Solution

If we rewrote a^{-1}, b^{-2}, b^{-2}, and a^{-3}, then the denominators would be a, b^2, b^2, and a^3. So the LCD is a^3b^2. If we multiply the numerator and denominator by a^3b^2, the negative exponents will be eliminated:

$$\frac{a^{-1} + b^{-2}}{ab^{-2} + ba^{-3}} = \frac{\left(a^{-1} + b^{-2}\right)a^3b^2}{\left(ab^{-2} + ba^{-3}\right)a^3b^2}$$

$$= \frac{a^{-1} \cdot a^3b^2 + b^{-2} \cdot a^3b^2}{ab^{-2} \cdot a^3b^2 + ba^{-3} \cdot a^3b^2} \qquad \text{Distributive property}$$

$$= \frac{a^2b^2 + a^3}{a^4 + b^3} \qquad \begin{array}{l} b^{-2}b^2 = b^0 = 1 \\ a^{-3}a^3 = a^0 = 1 \end{array}$$

Note that the positive exponents of a^3b^2 are just large enough to eliminate all of the negative exponents when we multiply.

> **Now do Exercises 43–44**

‹ **Helpful Hint** ›

In Examples 4, 5, and 6 we are simplifying the expressions without first removing the negative exponents, to gain experience in working with negative exponents. Of course, each expression with a negative exponent could be rewritten with a positive exponent and then the complex fraction could be simplified, as in Examples 2 and 3.

Example 6 is not exactly a complex fraction, but we can use the same technique as in Example 5.

EXAMPLE 6

More negative exponents

Eliminate negative exponents and simplify $p + p^{-1}q^{-2}$.

Solution

If we multiply the numerator and denominator by pq^2, we will eliminate the negative exponents:

$$p + p^{-1}q^{-2} = \frac{\left(p + p^{-1}q^{-2}\right)}{1} \cdot \frac{pq^2}{pq^2}$$

$$= \frac{p^2q^2 + 1}{pq^2} \quad \begin{array}{l} p \cdot pq^2 = p^2q^2 \\ p^{-1}q^{-2} \cdot pq^2 = 1 \end{array}$$

> Now do Exercises 45–58

⟨3⟩ Applications

Complex fractions are called *complex* for a reason. A situation that is modeled by a complex fraction will necessarily be a *complex* situation. So keep that in mind as you study Example 7.

EXAMPLE 7

An application of complex fractions

Eastside Elementary has the same number of students as Westside Elementary. One-half of the students at Eastside ride buses to school, and two-thirds of the students at Westside ride buses to school. One-sixth of the students at Eastside are female, and one-third of the students at Westside are female. If all of the female students ride the buses, then what percentage of the students who ride the buses are female?

Solution

To find the required percentage, we must divide the number of females who ride the buses by the total number of students who ride the buses. Let

$$x = \text{the number of students at Eastside.}$$

Because the number of students at Westside is also x, we have

$$\frac{1}{2}x + \frac{2}{3}x = \text{the total number of students who ride the buses}$$

and

$$\frac{1}{6}x + \frac{1}{3}x = \text{the total number of female students.}$$

Because all of the female students ride the buses, we can express the percentage of riders who are female by the following rational expression:

$$\frac{\dfrac{1}{6}x + \dfrac{1}{3}x}{\dfrac{1}{2}x + \dfrac{2}{3}x}$$

Multiply the numerator and denominator by 6, the LCD for 2, 3, and 6:

$$\frac{\left(\frac{1}{6}x + \frac{1}{3}x\right)6}{\left(\frac{1}{2}x + \frac{2}{3}x\right)6} = \frac{x + 2x}{3x + 4x} = \frac{3x}{7x} = \frac{3}{7} \approx 0.43 = 43\%$$

So 43% of the students who ride the buses are female.

> **Now do Exercises 63–66**

 Warm-Ups ▼

Fill in the blank.

1. A _____ is a fraction that contains fractions in the numerator, denominator, or both.

2. To simplify a complex fraction, we can multiply the numerator and denominator by the _____ of all of the denominators.

True or false?

3. The LCM of 2, x, 6, and x^2 is $6x^3$.

4. The LCM of $a - b$, $2b - 2a$, and 6 is $6a - 6b$.

5. $\dfrac{1 + \frac{1}{2}}{2 + \frac{1}{2}} = \dfrac{3}{2} \div \dfrac{5}{2}$

6. $\dfrac{\frac{1}{12} + \frac{5}{12}}{\frac{11}{12} - \frac{7}{12}} = \dfrac{1 + 5}{11 - 7}$

7. $\dfrac{\frac{1}{2} + \frac{1}{3}}{\frac{1}{6} + \frac{1}{3}} = \dfrac{3 + 2}{1 + 2}$

8. $\dfrac{5^{-1} + 2^{-3}}{3^{-2} + 4^{-1}} = \dfrac{\frac{1}{5} + \frac{1}{2^3}}{\frac{1}{3^2} + \frac{1}{4}}$

9. $\dfrac{(3^{-1} + 2^{-2})3^2 2^2}{(3^{-2} + 2^{-1})3^2 2^2} = \dfrac{3 \cdot 2^2 + 3^2}{2^2 + 3^2 \cdot 2}$

Exercises   6.4

‹1› Simplifying Complex Fractions

Simplify each complex fraction. Use either method. See Example 1.

1. $\dfrac{\frac{1}{2} + \frac{1}{4}}{\frac{1}{2} + \frac{1}{8}}$

2. $\dfrac{\frac{1}{3} + \frac{1}{4}}{\frac{1}{5} + \frac{1}{6}}$

 5. $\dfrac{\frac{2}{3} + \frac{5}{6} - \frac{1}{2}}{\frac{1}{8} - \frac{1}{3} + \frac{1}{12}}$

3. $\dfrac{\frac{1}{2} - \frac{1}{3}}{\frac{1}{4} - \frac{1}{5}}$

4. $\dfrac{\frac{1}{2} - \frac{1}{4}}{\frac{1}{6} - \frac{1}{8}}$

6. $\dfrac{\frac{2}{5} - \frac{x}{9} - \frac{1}{3}}{\frac{1}{3} + \frac{x}{5} + \frac{2}{15}}$

Simplify the complex fractions. Use Method B. See Example 2.

7. $\dfrac{\dfrac{x}{2} - \dfrac{1}{3}}{\dfrac{x}{2} + \dfrac{3}{4}}$

8. $\dfrac{\dfrac{b}{3} + \dfrac{b}{5}}{\dfrac{b}{5} - 1}$

27. $\dfrac{\dfrac{w+2}{w-1} - \dfrac{w-3}{w}}{\dfrac{w+4}{w} + \dfrac{w-2}{w-1}}$

28. $\dfrac{\dfrac{x-1}{x+2} - \dfrac{x-2}{x+3}}{\dfrac{x-3}{x+3} + \dfrac{x+1}{x+2}}$

9. $\dfrac{a + \dfrac{3}{b}}{\dfrac{b}{a} + \dfrac{1}{b}}$

10. $\dfrac{m - \dfrac{2}{n}}{\dfrac{1}{m} - \dfrac{3}{n}}$

29. $\dfrac{\dfrac{1}{a-b} - \dfrac{3}{a+b}}{\dfrac{2}{b-a} + \dfrac{4}{b+a}}$

30. $\dfrac{\dfrac{3}{2+x} - \dfrac{4}{2-x}}{\dfrac{1}{x+2} - \dfrac{3}{x-2}}$

11. $\dfrac{\dfrac{a+b}{b}}{\dfrac{a-b}{ab}}$

12. $\dfrac{\dfrac{m-n}{m^2}}{\dfrac{m-3}{mn^3}}$

13. $\dfrac{\dfrac{x-3y}{xy}}{\dfrac{1}{x} + \dfrac{1}{y}}$

14. $\dfrac{\dfrac{2}{w} + \dfrac{3}{t}}{\dfrac{w-t}{4wt}}$

31. $\dfrac{\dfrac{4}{y} - \dfrac{y+4}{y-3}}{\dfrac{2}{y-3} + \dfrac{y+1}{y}}$

32. $\dfrac{\dfrac{x+4}{x+1} + \dfrac{4}{x}}{\dfrac{x+1}{x} - \dfrac{1}{x+1}}$

15. $\dfrac{3 - \dfrac{m-2}{6}}{\dfrac{4}{9} + \dfrac{2}{m}}$

16. $\dfrac{6 - \dfrac{2-z}{z}}{\dfrac{1}{3z} - \dfrac{1}{6}}$

33. $\dfrac{3 - \dfrac{4}{a-1}}{5 - \dfrac{3}{1-a}}$

34. $\dfrac{\dfrac{x}{3} - \dfrac{x-1}{9-x}}{\dfrac{x}{6} - \dfrac{2-x}{x-9}}$

17. $\dfrac{\dfrac{a^2 - b^2}{a^2 b^3}}{\dfrac{a+b}{a^3 b}}$

18. $\dfrac{\dfrac{4x^2 - 1}{x^2 y}}{\dfrac{4x - 2}{xy^2}}$

19. $\dfrac{\dfrac{1}{x^2 y^2} + \dfrac{1}{xy^3}}{\dfrac{1}{x^3 y} - \dfrac{1}{xy}}$

20. $\dfrac{\dfrac{1}{2a^3 b} - \dfrac{1}{ab^4}}{\dfrac{1}{6a^2 b^2} + \dfrac{1}{3a^4 b}}$

35. $\dfrac{\dfrac{2}{m-3} + \dfrac{4}{m}}{\dfrac{3}{m-2} + \dfrac{1}{m}}$

36. $\dfrac{\dfrac{1}{y+2} - \dfrac{4}{3y}}{\dfrac{3}{y} - \dfrac{2}{y+3}}$

Simplify each complex fraction. See Examples 1–3.

21. $\dfrac{x + \dfrac{4}{x+4}}{x - \dfrac{4x+4}{x+4}}$

22. $\dfrac{x - \dfrac{x+6}{x+2}}{x - \dfrac{4x+15}{x+2}}$

37. $\dfrac{\dfrac{3}{x^2 - 1} - \dfrac{x-2}{x^3 - 1}}{\dfrac{3}{x^2 + x + 1} + \dfrac{x-3}{x^3 - 1}}$

38. $\dfrac{\dfrac{2}{a^3 + 8} - \dfrac{3}{a^2 - 2a + 4}}{\dfrac{4}{a^2 - 4} + \dfrac{a-3}{a^3 + 8}}$

23. $\dfrac{1 - \dfrac{1}{y-1}}{3 + \dfrac{1}{y+1}}$

24. $\dfrac{2 - \dfrac{3}{a-2}}{4 - \dfrac{1}{a+2}}$

⟨2⟩ Simplifying Expressions with Negative Exponents

Simplify. See Examples 4–6.

39. $\dfrac{w^{-1} + y^{-1}}{z^{-1} + y^{-1}}$

40. $\dfrac{a^{-1} - b^{-1}}{a^{-1} + b^{-1}}$

25. $\dfrac{\dfrac{2}{3-x} - 4}{\dfrac{1}{x-3} - 1}$

26. $\dfrac{\dfrac{x}{x-5} - 2}{\dfrac{2x}{5-x} - 1}$

41. $\dfrac{1 - x^{-1}}{1 - x^{-2}}$

42. $\dfrac{4 - a^{-2}}{2 - a^{-1}}$

43. $\dfrac{a^{-2} + b^{-2}}{a^{-1}b}$ **44.** $\dfrac{m^{-3} + n^{-3}}{mn^{-2}}$

45. $1 - a^{-1}$ **46.** $m^{-1} - a^{-1}$

47. $\dfrac{x^{-1} + x^{-2}}{x + x^{-2}}$ **48.** $\dfrac{x - x^{-2}}{1 - x^{-2}}$

49. $\dfrac{2m^{-1} - 3m^{-2}}{m^{-2}}$ **50.** $\dfrac{4x^{-3} - 6x^{-5}}{2x^{-5}}$

51. $\dfrac{a^{-1} - b^{-1}}{a - b}$ **52.** $\dfrac{a^2 - b^2}{a^{-2} - b^{-2}}$

53. $\dfrac{x^3 - y^3}{x^{-3} - y^{-3}}$ **54.** $\dfrac{(a - b)^2}{a^{-2} - b^{-2}}$

55. $\dfrac{1 - 8x^{-3}}{x^{-1} + 2x^{-2} + 4x^{-3}}$ **56.** $\dfrac{a + 27a^{-2}}{1 - 3a^{-1} + 9a^{-2}}$

57. $\left(x^{-1} + y^{-1}\right)^{-1}$ **58.** $\left(a^{-1} - b^{-1}\right)^{-2}$

 Use a calculator to evaluate each complex fraction. Round answers to four decimal places. If your calculator does fractions, then also find the fractional answer.

59. $\dfrac{\dfrac{5}{3} - \dfrac{4}{5}}{\dfrac{1}{3} - \dfrac{5}{6}}$ **60.** $\dfrac{\dfrac{1}{12} + \dfrac{1}{2} - \dfrac{3}{4}}{\dfrac{3}{5} + \dfrac{5}{6}}$

61. $\dfrac{4^{-1} - 9^{-1}}{2^{-1} + 3^{-1}}$ **62.** $\dfrac{2^{-1} + 3^{-1} - 6^{-1}}{3^{-1} - 5^{-1} + 4^{-1}}$

⟨3⟩ **Applications**

Solve each problem. See Example 7.

63. *Racial balance.* Clarksville has three elementary schools. Northside has one-half as many students as Central, and Southside has two-thirds as many students as Central. One-third of the students at Northside are African-American, three-fourths of the students at Central are African-American, and one-sixth of the students at Southside are African-American. What percent of the city's elementary students are African-American?

64. *Explosive situation.* All of the employees at Acme Explosives are in either development, manufacturing, or sales. One-fifth of the employees in development are women, one-third of the employees in manufacturing are women, and one-half of the employees in sales are women. Use the accompanying figure to determine the percentage of workers at Acme who are women. What percent of the women at Acme are in sales?

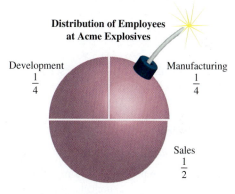

Distribution of Employees at Acme Explosives

Development $\dfrac{1}{4}$

Manufacturing $\dfrac{1}{4}$

Sales $\dfrac{1}{2}$

Figure for Exercise 64

65. *Average speed.* Mary drove from Clarksville to Leesville at 45 miles per hour (mph). At Leesville she discovered that she had forgotten her purse. She immediately returned to Clarksville at 55 mph. What was her average speed for the entire trip? (The answer is *not* 50 mph.)

 66. *Average price.* On her way to New York, Jenny spent the same amount for gasoline each time she stopped for gas. She paid 239.9 cents per gallon the first time, 249.9 cents per gallon the second time, and 259.9 cents per gallon the third time. What was the average price per gallon to the nearest tenth of a cent for the gasoline that she bought?

Photo for Exercise 66

67. *Harmonic mean.* The **harmonic mean** of two numbers x_1 and x_2 is defined as

$$\frac{2}{\dfrac{1}{x_1} + \dfrac{1}{x_2}}.$$

Find the harmonic mean for the two speeds used in Exercise 65.

68. *Harmonic mean.* The **harmonic mean** of three numbers x_1, x_2, and x_3 is defined as

$$\frac{3}{\dfrac{1}{x_1} + \dfrac{1}{x_2} + \dfrac{1}{x_3}}.$$

Find the harmonic mean to the nearest tenth of a cent for the three gas prices in Exercise 66.

Getting More Involved

69. *Cooperative learning*

Write a step-by-step strategy for simplifying complex fractions with negative exponents. Have a classmate use your strategy to simplify some complex fractions from Exercises 39–58.

70. *Discussion*

a) Find the exact value of each expression.

i)
$$\frac{1}{1 + \dfrac{1}{1 + \dfrac{1}{1 + \dfrac{1}{2}}}}$$

ii)
$$\frac{1}{1 + \dfrac{1}{1 + \dfrac{1}{1 + \dfrac{1}{3}}}}$$

b) Explain why in each case the exact value must be less than 1.

71. *Cooperative learning*

Work with a group to simplify the complex fraction. For what values of x is the complex fraction undefined?

$$\frac{1}{1 + \dfrac{1}{1 + \dfrac{1}{1 + \dfrac{1}{x}}}}$$

Mid-Chapter Quiz | Sections 6.1 through 6.4 | Chapter 6

Reduce to lowest terms.

1. $\dfrac{36}{84}$

2. $\dfrac{8x - 2}{8}$

3. $\dfrac{w^2 - 1}{2w + 2}$

4. $\dfrac{2a^2 - 10a + 12}{6 - 3a}$

Perform the indicated operation.

5. $\dfrac{6}{7} \cdot \dfrac{21}{10}$

6. $\dfrac{3xy^2}{5z} \cdot \dfrac{8x^2z^3}{8y^4}$

7. $\dfrac{a^2 - 9}{2a + 4} \cdot \dfrac{5a + 10}{2a - 6}$

8. $\dfrac{b^2}{3} \div \dfrac{b^6}{21}$

9. $\dfrac{3x - 9}{8} \div \dfrac{x^2 - 6x + 9}{12}$

10. $\dfrac{s^2}{3} + \dfrac{s^2}{21}$

11. $\dfrac{m^2 - 8m + 7}{2m} \div (m - 7)$ ——

12. $\dfrac{5}{6} - \dfrac{5}{21}$

13. $\dfrac{4}{ab^3} + \dfrac{5}{a^2b}$

14. $\dfrac{3x}{x + 1} + \dfrac{x}{x^2 + 2x + 1}$

15. $\dfrac{y}{y + 5} - \dfrac{y}{y + 2}$

16. $\dfrac{1}{a} + \dfrac{1}{b} + \dfrac{1}{c}$

Simplify the complex fractions.

17. $\dfrac{\dfrac{1}{2} + \dfrac{3}{5}}{\dfrac{1}{10} - \dfrac{1}{15}}$

18. $\dfrac{3 - \dfrac{1}{x + 2}}{1 + \dfrac{1}{x^2 - 4}}$

Miscellaneous.

19. What number(s) can't be used in place of x in $\dfrac{3x - 6}{2x + 1}$?

20. Find the value of $\dfrac{3x - 6}{2x + 1}$ when $x = -2$.

21. Find $R(-1)$ if $R(x) = \dfrac{6x^2 + 3}{5x - 1}$.

22. Write the domain of the function $R(x) = \dfrac{x^2 - 1}{x - 9}$ in interval notation.

6.5 Division of Polynomials

We began our study of polynomials in Section 5.3 by learning how to add, subtract, and multiply polynomials. In this section, we will study division of polynomials.

⟨1⟩ Dividing a Polynomial by a Monomial

You learned how to divide monomials in Section 5.1. For example,

$$6x^3 \div (3x) = \frac{6x^3}{3x} = 2x^2.$$

We check by multiplying. Because $2x^2 \cdot 3x = 6x^3$, this answer is correct. Recall that $a \div b = c$ if and only if $c \cdot b = a$. We call a the **dividend,** b the **divisor,** and c the **quotient.** We may also refer to $a \div b$ and $\frac{a}{b}$ as quotients.

We can use the distributive property to find that

$$3x(2x^2 + 5x - 4) = 6x^3 + 15x^2 - 12x.$$

So if we divide $6x^3 + 15x^2 - 12x$ by the monomial $3x$, we must get $2x^2 + 5x - 4$. We can perform this division by dividing $3x$ into *each term* of $6x^3 + 15x^2 - 12x$:

$$\frac{6x^3 + 15x^2 - 12x}{3x} = \frac{6x^3}{3x} + \frac{15x^2}{3x} - \frac{12x}{3x}$$
$$= 2x^2 + 5x - 4$$

In this case the divisor is $3x$, the dividend is $6x^3 + 15x^2 - 12x$, and the quotient is $2x^2 + 5x - 4$.

E X A M P L E 1

Dividing a polynomial by a monomial
Find the quotient.

a) $-12x^5 \div (2x^3)$ **b)** $(-20x^6 + 8x^4 - 4x^2) \div (4x^2)$

⟨ **Helpful Hint** ⟩

Recall that the order of operations gives multiplication and division an equal ranking and says to do them in order from left to right. So without parentheses,

$$-12x^5 \div 2x^3$$

actually means

$$\frac{-12x^5}{2} \cdot x^3.$$

Solution

a) When dividing x^5 by x^3, we subtract the exponents:

$$-12x^5 \div (2x^3) = \frac{-12x^5}{2x^3} = -6x^2$$

The quotient is $-6x^2$. Check:

$$-6x^2 \cdot 2x^3 = -12x^5$$

b) Divide each term of $-20x^6 + 8x^4 - 4x^2$ by $4x^2$:

$$\frac{-20x^6 + 8x^4 - 4x^2}{4x^2} = \frac{-20x^6}{4x^2} + \frac{8x^4}{4x^2} - \frac{4x^2}{4x^2}$$
$$= -5x^4 + 2x^2 - 1$$

The quotient is $-5x^4 + 2x^2 - 1$. Check:

$$4x^2(-5x^4 + 2x^2 - 1) = -20x^6 + 8x^4 - 4x^2$$

Now do Exercises 1–12

In Example 1 we found the quotient of a polynomial and a monomial and the remainder was zero. If the remainder is zero, then the dividend is equal to the divisor times the quotient. If the remainder is not zero, then the degree of the remainder must be less than the degree of the divisor and

$$\textbf{dividend} = (\textbf{divisor})(\textbf{quotient}) + (\textbf{remainder}).$$

If we divide each side of this equation by the divisor, we get

$$\frac{\textbf{dividend}}{\textbf{divisor}} = \textbf{quotient} + \frac{\textbf{remainder}}{\textbf{divisor}}.$$

The remainder in Example 1 was zero, because the degree of the monomial denominator was less than or equal to the degree of every term in the numerator. If the degree of the monomial denominator is larger than the degree of at least one term in the numerator, then there will be a remainder, as shown in Example 2.

E X A M P L E   2

Dividing a polynomial by a monomial (remainder ≠ 0)
Find the quotient and remainder.

a) $\dfrac{6x - 1}{2x}$

b) $\dfrac{x^3 - 4x^2 + 5x - 3}{2x^2}$

Solution

a) Divide each term of $6x - 1$ by $2x$:

$$\frac{6x - 1}{2x} = \frac{6x}{2x} + \frac{-1}{2x} = 3 + \frac{-1}{2x}$$

Now $3 + \frac{-1}{2x}$ has the form quotient $+ \frac{\text{remainder}}{\text{divisor}}$. So the quotient is 3 and the remainder is -1. Check that (divisor)(quotient) + remainder = dividend:

$$(2x)(3) + (-1) = 6x - 1$$

b) The first two terms in the numerator have a degree that is greater than or equal to the degree of the denominator. So divide the first two terms of the numerator by the monomial denominator:

$$\frac{x^3 - 4x^2 + 5x - 3}{2x^2} = \frac{x^3}{2x^2} - \frac{4x^2}{2x^2} + \frac{5x - 3}{2x^2}$$

$$= \frac{1}{2}x - 2 + \frac{5x - 3}{2x^2}$$

The last expression has the form quotient $+ \frac{\text{remainder}}{\text{divisor}}$. So the quotient is $\frac{1}{2}x - 2$ and the remainder is $5x - 3$. Check that

$$(2x^2)\left(\frac{1}{2}x - 2\right) + 5x - 3 = x^3 - 4x^2 + 5x - 3.$$

> **Now do Exercises 13–20**

⟨2⟩ Dividing a Polynomial by a Binomial

To divide a polynomial by a binomial we use a method that is very similar to the long division algorithm for whole numbers.

E X A M P L E   **3**

Dividing a polynomial by binomial

Find the quotient and remainder when $x^2 + 3x - 9$ is divided by $x - 2$.

Solution

The setup is like long division of whole numbers. Divide x^2 by x to get x, and then multiply and subtract:

$$
\begin{array}{r}
x \phantom{{}+3x-9} \\
x - 2 \overline{)x^2 + 3x - 9} \\
\underline{x^2 - 2x} \\
5x
\end{array}
$$

$x^2 \div x = x$

Multiply: $x(x - 2) = x^2 - 2x$.

Subtract: $3x - (-2x) = 5x$.

Now bring down -9. We get the second term of the quotient by dividing the first term of $x - 2$ into the first term of $5x - 9$:

$$
\begin{array}{r}
x + 5 \\
x - 2 \overline{)x^2 + 3x - 9} \\
\underline{x^2 - 2x} \\
5x - 9 \\
\underline{5x - 10} \\
1
\end{array}
$$

$5x \div x = 5$

Multiply: $5(x - 2) = 5x - 10$.

Subtract: $-9 - (-10) = 1$.

So the quotient is $x + 5$ and the remainder is 1. Check by multiplying the quotient and divisor and adding on the remainder:

$$(x + 5)(x - 2) + 1 = x^2 + 3x - 10 + 1 = x^2 + 3x - 9$$

> Now do Exercises 21–22

When dividing polynomials, we must write the terms of the divisor and the dividend in descending order of the exponents. If any terms are missing, as in Example 4, we insert terms with a coefficient of 0 as placeholders. When dividing polynomials, we stop the process when the degree of the remainder is smaller than the degree of the divisor.

E X A M P L E   **4**

Dividing polynomials

Find the quotient and remainder for $(3x^4 - 2 - 5x) \div (x^2 - 3x)$.

Solution

Rearrange $3x^4 - 2 - 5x$ as $3x^4 - 5x - 2$ and insert the terms $0x^3$ and $0x^2$:

‹ Helpful Hint ›

Students usually have the most difficulty with the subtraction part of long division. So pay particular attention to that step and double check your work.

Place $3x^2$ in the quotient because $3x^4 \div x^2 = 3x^2$.

Place $9x$ in the quotient because $9x^3 \div x^2 = 9x$.

Place 27 in the quotient because $27x^2 \div x^2 = 27$.

$$
\begin{array}{r}
3x^2 + 9x + 27 \\
x^2 - 3x \overline{)3x^4 + 0x^3 + 0x^2 - 5x - 2} \\
\underline{3x^4 - 9x^3} \\
9x^3 + 0x^2 \\
\underline{9x^3 - 27x^2} \\
27x^2 - 5x \\
\underline{27x^2 - 81x} \\
76x - 2
\end{array}
$$

$3x^2(x^2 - 3x) = 3x^4 - 9x^3$

$0x^3 - (-9x^3) = 9x^3$

$9x(x^2 - 3x) = 9x^3 - 27x^2$

$0x^2 - (-27x^2) = 27x^2$

$27(x^2 - 3x) = 27x^2 - 81x$

$-5x - (-81x) = 76x$

The quotient is $3x^2 + 9x + 27$, and the remainder is $76x - 2$. Note that the degree of the remainder is 1, and the degree of the divisor is 2. To check, verify that

$$(x^2 - 3x)(3x^2 + 9x + 27) + 76x - 2 = 3x^4 - 5x - 2.$$

> **Now do Exercises 23–42**

EXAMPLE 5

Rewriting a ratio of two polynomials

Write each rational expression in the form quotient $+ \frac{\text{remainder}}{\text{divisor}}$.

a) $\dfrac{3x}{x - 6}$
                                            b) $\dfrac{4x^3 - x - 9}{2x - 3}$

Solution

a) Divide $3x$ by $x - 6$:

$$
\begin{array}{r}
3 \\
x - 6\overline{)3x} \\
\underline{3x - 18} \\
18
\end{array}
$$

      $3x \div x = 3$

      $3(x - 6) = 3x - 18$

      $0 - (-18) = 18$

The quotient is 3 and the remainder is 18. Since $3(x - 6) + 18 = 3x - 18 + 18 = 3x$, we can be sure that the division is correct. So,

$$\frac{3x}{x - 6} = 3 + \frac{18}{x - 6}.$$

b) Divide $4x^3 - x - 9$ by $2x - 3$. Insert $0 \cdot x^2$ for the missing term.

$$
\begin{array}{r}
2x^2 + 3x + 4 \\
2x - 3\overline{)4x^3 + 0x^2 - x - 9} \\
\underline{4x^3 - 6x^2} \\
6x^2 - x \\
\underline{6x^2 - 9x} \\
8x - 9 \\
\underline{8x - 12} \\
3
\end{array}
$$

$4x^3 \div (2x) = 2x^2$

$2x^2(2x - 3) = 4x^3 - 6x^2$

$0x^2 - (-6x^2) = 6x^2$

$3x(2x - 3) = 6x^2 - 9x$

$-x - (-9x) = 8x$

$4(2x - 3) = 8x - 12$

$-9 - (-12) = 3$

Since the quotient is $2x^2 + 3x + 4$ and the remainder is 3, we have

$$\frac{4x^3 - x - 9}{2x - 3} = 2x^2 + 3x + 4 + \frac{3}{2x - 3}.$$

To check the answer, we must verify that

$$(2x - 3)(2x^2 + 3x + 4) + 3 = 4x^3 - x - 9.$$

> **Now do Exercises 43–58**

⟨3⟩ Synthetic Division

Synthetic division is an abbreviated version of ordinary division. To divide $x^3 - 5x^2 + 4x - 3$ by $x - 2$ with synthetic division we use only the coefficients of $x^3 - 5x^2 + 4x - 3$, which are 1, -5, 4, and -3, and we use only 2 from the divisor $x - 2$. Compare the two types of division side by side before we go through the details of how to perform synthetic division. Synthetic division certainly looks simpler.

The bottom row of synthetic division gives the coefficients of the quotient and the remainder:

Ordinary Long Division　　　　　　**Synthetic Division**

$$x - 2 \overline{)\begin{array}{c} x^2 - 3x - 2 \\ x^3 - 5x^2 + 4x - 3 \end{array}}$$

$$
\begin{array}{r}
x^3 - 2x^2 \\ \hline
-3x^2 + 4x \\
-3x^2 + 6x \\ \hline
-2x - 3 \\
-2x + 4 \\ \hline
-7
\end{array}
$$

$$
\begin{array}{r|rrrr}
2 & 1 & -5 & 4 & -3 \\
  &   & 2 & -6 & -4 \\ \hline
  & 1 & -3 & -2 & -7
\end{array}
$$

Quotient　Remainder

To divide $x^3 - 5x^2 + 4x - 3$ by $x - 2$ we start with the following arrangement:

$$
\begin{array}{r|rrr}
2 & 1 & -5 \ \ 4 \ \ -3
\end{array}
\qquad (1 \cdot x^3 - 5x^2 + 4x - 3) \div (x - 2)
$$

Next we bring the first coefficient, 1, straight down:

$$
\begin{array}{r|rrr}
2 & 1 & -5 & 4 & -3 \\
  & \downarrow & \text{Bring down} \\ \hline
  & 1
\end{array}
$$

We then multiply the 1 by the 2 from the divisor, place the answer under the -5, and then add that column. Using 2 for $x - 2$ enables us to add the column rather than subtract as in ordinary division:

$$
\begin{array}{r|rrrr}
2 & 1 & -5 & 4 & -3 \\
  &   & 2 & & \text{Add} \\ \hline
\text{Multiply} & 1 & -3
\end{array}
$$

We then repeat the multiply-and-add step for each of the remaining columns:

$$
\begin{array}{r|rrrr}
2 & 1 & -5 & 4 & -3 & \quad -6 = 2(-3) \\
  &   & 2 & -6 & -4 & \quad -4 = 2(-2) \\ \hline
\text{Multiply} & 1 & -3 & -2 & -7 & \ \leftarrow \text{Remainder}
\end{array}
$$

Quotient

From the bottom row we can read the quotient and remainder. Since the degree of the quotient is one less than the degree of the dividend, the quotient is $1x^2 - 3x - 2$. The remainder is -7.

The strategy for getting the quotient $Q(x)$ and remainder R by synthetic division can be stated as follows.

Strategy for Synthetic Division

1. List the coefficients of the polynomial (the dividend).
2. Be sure to include zeros for any missing terms in the dividend.
3. For dividing by $x - c$, place c to the left.
4. Bring the first coefficient down.
5. Multiply by c and add for each column.
6. Read $Q(x)$ and R from the bottom row.

CAUTION Synthetic division is used only for dividing a polynomial by the binomial $x - c$, where c is a constant. If the binomial is $x - 7$, then $c = 7$. For the binomial $x + 7$, we have $x + 7 = x - (-7)$ and $c = -7$.

E X A M P L E   **6**

Using synthetic division

Find the quotient and remainder when $2x^4 - 5x^2 + 6x - 9$ is divided by $x + 2$.

Solution

Since $x + 2 = x - (-2)$, we use -2 for the divisor. Because x^3 is missing in the dividend, use a zero for the coefficient of x^3:

$$
\begin{array}{r|rrrrr}
-2 & 2 & 0 & -5 & 6 & -9 \\
   &   & -4 & 8 & -6 & 0 \\
\hline
   & 2 & -4 & 3 & 0 & -9
\end{array}
$$

$\left.\begin{array}{}\\ \end{array}\right\}$ Add $\leftarrow 2x^4 + 0 \cdot x^3 - 5x^2 + 6x - 9$

Multiply $\leftarrow$ Quotient and remainder

Because the degree of the dividend is 4, the degree of the quotient is 3. The quotient is $2x^3 - 4x^2 + 3x$, and the remainder is -9. We can also express the results of this division in the form quotient $+ \frac{\text{remainder}}{\text{divisor}}$:

$$\frac{2x^4 - 5x^2 + 6x - 9}{x + 2} = 2x^3 - 4x^2 + 3x + \frac{-9}{x + 2}$$

Now do Exercises 59–72

⟨4⟩ Division and Factoring

To **factor** a polynomial means to write it as a product of two or more simpler polynomials. If we divide two polynomials and get 0 remainder, then we can write

$$\text{dividend} = (\text{divisor})(\text{quotient})$$

and we have factored the dividend. *The dividend factors as the divisor times the quotient if and only if the remainder is* 0. We can use division to help us discover factors of polynomials. To use this idea, however, we must know a factor or a possible factor to use as the divisor.

E X A M P L E   **7**

Using synthetic division to determine factors

Is $x - 1$ a factor of $6x^3 - 5x^2 - 4x + 3$?

Solution

We can use synthetic division to divide $6x^3 - 5x^2 - 4x + 3$ by $x - 1$:

$$
\begin{array}{r|rrrr}
1 & 6 & -5 & -4 & 3 \\
  & \downarrow & 6 & 1 & -3 \\
\hline
  & 6 & 1 & -3 & 0
\end{array}
$$

Because the remainder is 0, $x - 1$ is a factor, and

$$6x^3 - 5x^2 - 4x + 3 = (x - 1)(6x^2 + x - 3).$$

Now do Exercises 73–82

⟨5⟩ The Remainder Theorem

If a polynomial $P(x)$ is divided by $x - c$, we get a quotient and a remainder that satisfy

$$P(x) = (x - c)(\text{quotient}) + \text{remainder}$$

Now replace x by c:

$$P(c) = (c - c)(\text{quotient}) + \text{remainder}$$
$$= 0(\text{quotient}) + \text{remainder}$$
$$= \text{remainder}$$

This computation proves the remainder theorem.

> **The Remainder Theorem**
>
> If the polynomial $P(x)$ is divided by $x - c$, then the remainder is equal to $P(c)$.

The remainder theorem gives us a new way to evaluate a polynomial. Note that we could use long division or synthetic division to find the remainder, but it is easier to use synthetic division.

EXAMPLE 8

Using synthetic division to evaluate a polynomial

Use synthetic division to find $P(2)$ when $P(x) = 4x^3 - 5x^2 + 6x - 7$.

Solution

Find the remainder when $P(x)$ is divided by $x - 2$:

$$
\begin{array}{r|rrrr}
2 & 4 & -5 & 6 & -7 \\
  & \downarrow & 8 & 6 & 24 \\
\hline
  & 4 & 3 & 12 & 17
\end{array}
$$

Since 17 is the remainder, $P(2) = 17$. Check by replacing x with 2 in $P(x)$:

$$P(2) = 4 \cdot 2^3 - 5 \cdot 2^2 + 6 \cdot 2 - 7$$
$$= 32 - 20 + 12 - 7$$
$$= 17$$

> Now do Exercises 83–88

 ## Warm-Ups ▼

Fill in the blank.

1. If $a \div b = c$, then a is the _____, b is the _____, and c is the _____.

2. For long division polynomials should be written with exponents in _____ order.

3. If the term x^n is missing from the _____, we can insert the term $0 \cdot x^n$ for the missing term.

4. The long division process for polynomials stops when the degree of the remainder is _____ the degree of the divisor.

5. _____ division is used only for dividing a polynomial by a binomial of the form $x - c$.

6. The remainder in polynomial division is zero if and only if the _____ is a factor of the _____.

True or false?

7. The quotient times the divisor plus the remainder is equal to the dividend.

8. For any x, $(x + 2)(x + 3) + 1 = x^2 + 5x + 7$.

9. The quotient for $(x^2 + 5x + 7) \div (x + 2)$ is $x + 3$.

10. We use 2 in the synthetic division when dividing $x^2 + 5x + 7$ by $x + 2$.

11. We can use synthetic division to divide $x^3 - 4x + 1$ by $x^2 - 5$.

12. If the remainder is zero, then the quotient is a factor of the dividend.

13. If $x^4 - 6$ is divided by $x - 2$, then the degree of the quotient is less than 4 and the degree of the remainder is 0.

Exercises

‹ Study Tips ›

• As you study a chapter make a list of topics and questions that you would put on the test if you were to write it.
• Write about what you read in the text. Sum things up in your own words.

‹1› Dividing a Polynomial by a Monomial

Find the quotient. See Example 1.

1. $36x^7 \div (3x^3)$

2. $-30x^3 \div (-5x)$

3. $16x^2 \div (-8x^2)$

4. $-22a^3 \div (11a^2)$

5. $(6b - 9) \div 3$

6. $(8x^2 - 6x) \div 2$

7. $(3x^2 + 3x) \div (3x)$

8. $(5x^3 - 10x^2 - 5x) \div (5x)$

9. $(10x^4 - 8x^3 + 6x^2) \div (-2x^2)$

10. $(-9x^3 + 6x^2 - 12x) \div (-3x)$

11. $(7x^3 - 4x^2) \div (2x)$

12. $(6x^3 - 5x^2) \div (4x^2)$

Find the quotient and remainder. See Example 2.

13. $(8x - 4) \div 2$

14. $(8x^2 - 4x) \div (4x)$

15. $(8x - 3) \div (4x)$

16. $(9x - 5) \div (3x)$

17. $(2x^3 + x^2 + 4x - 3) \div (3x^2)$

18. $(5x^3 - 6x^2 + 4x - 1) \div (5x^2)$

19. $(-10x^4 - 5x^3 - 6x - 7) \div (-5x^2)$

20. $(-12x^5 - 8x^4 - 6x^2 + x - 4) \div (3x^3)$

‹2› Dividing a Polynomial by a Binomial

Find the quotient and remainder as in Examples 3 and 4. Check by using the formula

$$dividend = (divisor)(quotient) + remainder.$$

21. $(x^2 + 8x + 13) \div (x + 3)$

22. $(x^2 + 5x + 7) \div (x + 3)$

23. $(x^2 - 2x) \div (x + 2)$

24. $(3x) \div (x - 1)$

25. $(x^3 + 8) \div (x + 2)$

26. $(y^3 - 1) \div (y - 1)$

27. $(a^3 + 4a - 5) \div (a - 2)$

28. $(w^3 + w^2 - 3) \div (w - 2)$

29. $(x^3 - x^2 + x - 3) \div (x + 1)$

30. $(a^3 - a^2 + a - 4) \div (a + 2)$

31. $(x^4 - x + x^3 - 1) \div (x - 2)$

32. $(3x^4 + 6 - x^2 + 3x) \div (x + 2)$

33. $(5x^2 - 3x^4 + x - 2) \div (x^2 - 2)$

34. $(x^4 - 2 + x^3) \div (x^2 + 3)$

35. $(6x^2 + x - 16) \div (2x - 3)$

36. $(12x^2 - x - 9) \div (3x + 2)$

37. $(10b^2 - 17b - 22) \div (5b + 4)$

38. $(20a^2 + 2a - 7) \div (4a - 2)$

39. $(2x^3 + 3x^2 - 3x - 2) \div (2x + 1)$

40. $(6x^3 - 7x^2 + 5x + 6) \div (3x - 2)$

41. $(x^3 - 4x^2 - 3x - 10) \div (x^2 + x + 2)$

42. $(2x^3 - 3x^2 + 7x - 3) \div (x^2 - x + 3)$

Write each expression in the form

$$quotient + \frac{remainder}{divisor}.$$

See Example 5.

43. $\dfrac{2x}{x - 5}$

44. $\dfrac{x}{x - 1}$

45. $\dfrac{2x^2 - x}{2x + 1}$

46. $\dfrac{8x^2 - 3}{4x - 6}$

47. $\dfrac{x^3}{x+2}$

48. $\dfrac{x^3 - 1}{x - 2}$

49. $\dfrac{x^3 + 2x}{x^2}$

50. $\dfrac{2x^2 + 3}{2x}$

51. $\dfrac{2x^2 - 11x - 4}{2x + 1}$

52. $\dfrac{5x^2 - 13x + 13}{5x - 3}$

53. $\dfrac{3x^3 - 4x^2 + 7}{x - 1}$

54. $\dfrac{-2x^3 + x^2 - 3}{x + 2}$

55. $\dfrac{6x^3 - 4x + 5}{x - 2}$

56. $\dfrac{-x^3 - x + 2}{x + 3}$

57. $\dfrac{x^3 + x}{x + 1}$

58. $\dfrac{x^3 - x}{x - 3}$

⟨**3**⟩ **Synthetic Division**

Use synthetic division to find the quotient and remainder when the first polynomial is divided by the second. See Example 6. See the Strategy for Synthetic Division box on page 423.

59. $x^2 + x + 7, x - 1$

60. $x^2 + 2x + 5, x - 2$

61. $2x^2 - 4x + 5, x + 1$

62. $3x^2 - 7x + 4, x + 2$

63. $x^3 - 5x^2 + 6x - 3, x - 2$

64. $x^3 + 6x^2 - 3x - 5, x - 3$

65. $3x^4 - 15x^2 + 7x - 9, x - 3$

66. $-2x^4 + 3x^2 - 5, x - 2$

67. $x^5 - 1, x - 1$

68. $x^6 - 1, x + 1$

69. $x^3 - 5x + 6, x + 2$

70. $x^3 - 3x - 7, x - 4$

71. $x^3 - 3x^2 + 5, x - 5$

72. $3x^3 + 20x^2 + 2, x + 7$

⟨**4**⟩ **Division and Factoring**

For each pair of polynomials, use division to determine whether the first polynomial is a factor of the second. Use synthetic division when possible. If the first polynomial is a factor, then factor the second polynomial. See Example 7.

73. $x + 4, x^3 + x^2 - 11x + 8$

74. $x - 1, x^3 + 3x^2 - 5x$

75. $x - 4, x^2 - 6x + 8$

76. $x + 8, x^2 + 3x - 40$

77. $w - 3, w^3 - 27$

78. $w + 5, w^3 + 125$

79. $2x - 3, 2x^3 - 3x^2 - 4x + 7$

80. $3x - 5, 3x^3 + x^2 - 7x + 6$

81. $y - 2, y^3 - 4y^2 + 6y - 4$

82. $z + 1, 2z^3 + 5z + 7$

⟨**5**⟩ **The Remainder Theorem**

Use synthetic division and the remainder theorem to find $P(c)$ for the given polynomial and given value of c. See Example 8.

83. $P(x) = x^2 - 5x - 9, c = 3$

84. $P(x) = x^3 + 3x^2 - 7x + 2, c = 2$

85. $P(y) = 4y^3 - 6y + 7, c = -1$

86. $P(y) = 2y^3 - y^2 - 1, c = -4$

87. $P(w) = -w^3 + 5w^2 + 3w, c = 4$

88. $P(w) = -2w^3 - w^2 - 15, c = -3$

Applications

Solve each problem.

Figure for Exercise 92

89. *Average cost.* The total cost in dollars for manufacturing x professional racing bicycles in one week is given by the polynomial function

$$C(x) = 0.03x^2 + 300x.$$

The average cost per bicycle is given by

$$AC(x) = \frac{C(x)}{x}.$$

a) Find a formula for $AC(x)$.

b) Is $AC(x)$ a constant function?

c) Why does the average cost look constant in the accompanying figure?

where a^2 is the area of the square base, b^2 is the area of the square top, and H is the distance from the base to the top. Find the volume of a truncated pyramid that has a base of 900 square meters, a top of 400 square meters, and a height H of 10 meters.

Figure for Exercise 89

Figure for Exercise 93

90. *Average profit.* The weekly profit in dollars for manufacturing x bicycles is given by the polynomial $P(x) = 100x + 2x^2$. The average profit per bicycle is given by $AP(x) = \frac{P(x)}{x}$. Find $AP(x)$. Find the average profit per bicycle when 12 bicycles are manufactured.

94. *Egyptian pyramid formula.* Simplify the expression in Exercise 93.

91. *Area of a poster.* The area of a rectangular poster advertising a Pearl Jam concert is $x^2 - 1$ square feet. If the length is $x + 1$ feet, then what is the width?

92. *Volume of a box.* The volume of a shipping crate is $h^3 + 5h^2 + 6h$. If the height is h and the length is $h + 2$, then what is the width?

93. *Volume of a pyramid.* Ancient Egyptian pyramid builders knew that the volume of the truncated pyramid shown in the figure is given by

$$V = \frac{H(a^3 - b^3)}{3(a - b)},$$

Getting More Involved

95. *Discussion*

On a test a student divided $3x^3 - 5x^2 - 3x + 7$ by $x - 3$ and got a quotient of $3x^2 + 4x$ and remainder $9x + 7$. Verify that the divisor times the quotient plus the remainder is equal to the dividend. Why was the student's answer incorrect?

96. *Exploration*

Use synthetic division to find the quotient when $x^5 - 1$ is divided by $x - 1$ and the quotient when $x^6 - 1$ is divided by $x - 1$. Observe the pattern in the first two quotients, and then write the quotient for $x^9 - 1$ divided by $x - 1$ without dividing.

6.6   Solving Equations Involving Rational Expressions

Many problems in algebra are modeled by equations involving rational expressions. In this section, you will learn how to solve equations that have rational expressions, and in Section 6.7, we will solve problems using these equations.

⟨1⟩ Multiplying by the LCD

To solve equations having rational expressions, we multiply each side of the equation by the LCD of the rational expressions.

EXAMPLE 1

An equation with rational expressions

Solve $\frac{1}{x} + \frac{1}{4} = \frac{1}{6}$.

Solution

First note that $\frac{1}{x}$ is undefined if $x = 0$. So 0 cannot be a solution. Since the LCD for the denominators 4, 6, and x is $12x$, multiply each side by $12x$:

$$12x\left(\frac{1}{x} + \frac{1}{4}\right) = 12x\left(\frac{1}{6}\right) \quad \text{Multiply each side by } 12x.$$

$$12x \cdot \frac{1}{x} + \overset{3}{12}x \cdot \frac{1}{4} = \overset{2}{12}x \cdot \frac{1}{6} \quad \text{Distributive property}$$

$$12 + 3x = 2x \qquad \text{Divide out the common factors.}$$

$$12 + x = 0$$

$$x = -12$$

Check -12 in the original equation. The solution set is $\{-12\}$.

> **Now do Exercises 1–6**

CAUTION To solve an equation with rational expressions, we do *not* convert the rational expressions to ones with a common denominator. Instead, we multiply each side by the LCD to *eliminate* the denominators.

EXAMPLE 2

An equation with two solutions

Solve $\frac{10}{x} + \frac{14}{x+2} = 4$.

Solution

First note that 0 and -2 cannot be solutions because replacing x with 0 or -2 would result in an undefined expression. Since the LCD for x and $x + 2$ is $x(x + 2)$, multiply each side by $x(x + 2)$:

$$x(x + 2)\left(\frac{10}{x} + \frac{14}{x + 2}\right) = x(x + 2)4 \qquad \text{Multiply each side by } x(x + 2).$$

$$x(x + 2)\frac{10}{x} + x(x + 2)\frac{14}{x + 2} = x(x + 2)4 \qquad \text{Distributive property}$$

$$(x + 2)10 + x(14) = (x^2 + 2x)4 \qquad \text{Simplify.}$$

$$10x + 20 + 14x = 4x^2 + 8x$$

$$0 = 4x^2 - 16x - 20 \qquad \text{Combine like terms.}$$

$$0 = x^2 - 4x - 5 \qquad \text{Divide each side by 4.}$$

$$0 = (x - 5)(x + 1)$$

$$x - 5 = 0 \qquad \text{or} \qquad x + 1 = 0$$

$$x = 5 \qquad \text{or} \qquad x = -1$$

Checking 5 and -1 in the original equation yields

$$\frac{10}{5} + \frac{14}{5 + 2} = 4 \qquad \text{and} \qquad \frac{10}{-1} + \frac{14}{-1 + 2} = 4,$$

which are both correct. So the solution set is $\{-1, 5\}$.

> Now do Exercises 7–10

If a number appears as a solution to an equation, but does not satisfy the original equation, then it is called an **extraneous** solution. A solution to an equation is also called a **root** to the equation. So an extraneous solution is also called an **extraneous root.** The equation in Example 3 has an extraneous root.

E X A M P L E   3

An equation with an extraneous root

Solve $\frac{3}{x} + \frac{6}{x - 2} = \frac{12}{x^2 - 2x}$.

Solution

First note that 0 and 2 cannot be solutions because replacing x with 0 or 2 would result in a zero denominator. Because $x^2 - 2x = x(x - 2)$, the LCD for x, $x - 2$, and $x^2 - 2x$ is $x(x - 2)$:

$$x(x - 2)\frac{3}{x} + x(x - 2)\frac{6}{x - 2} = x(x - 2)\frac{12}{x(x - 2)} \qquad \text{Multiply each side by } x(x - 2).$$

$$3(x - 2) + 6x = 12$$

$$3x - 6 + 6x = 12$$

$$9x - 6 = 12$$

$$9x = 18$$

$$x = 2$$

Since 2 cannot be a solution to this equation, 2 is an extraneous root and the solution set is the empty set $\varnothing$.

> Now do Exercises 11–12

E X A M P L E   4

An equation with an extraneous root

Solve $x + 4 - \dfrac{x}{x-4} = \dfrac{-4}{x-4}$.

Solution

First note that 4 can't be a solution because replacing x with 4 would result in a zero denominator. Because the LCD is $x - 4$, multiply each side by $x - 4$:

$$(x-4)(x+4) - (x-4)\frac{x}{x-4} = (x-4)\frac{-4}{x-4}$$

$$x^2 - 16 - x = -4$$

$$x^2 - x - 12 = 0$$

$$(x-4)(x+3) = 0$$

$$x - 4 = 0 \quad \text{or} \quad x + 3 = 0$$

$$x = 4 \quad \text{or} \quad x = -3$$

Since 4 can't be a solution, 4 is an extraneous root. Check that -3 satisfies the original equation:

$$-3 + 4 - \frac{-3}{-3-4} = \frac{-4}{-3-4}$$

$$1 - \frac{3}{7} = \frac{4}{7} \quad \text{Correct.}$$

The solution set is $\{-3\}$.

> **Now do Exercises 13–18**

⟨2⟩ Proportions

An equation that expresses the equality of two rational expressions is called a **proportion.** The equation

$$\frac{a}{b} = \frac{c}{d}$$

is a proportion. The terms in the position of b and c are called the **means.** The terms in the position of a and d are called the **extremes.** If we multiply this proportion by the LCD, bd, we get

$$bd \cdot \frac{a}{b} = bd \cdot \frac{c}{d}$$

or

$$ad = bc.$$

⟨ **Helpful Hint** ⟩

The extremes-means property is often referred to as *cross-multiplying.* Whatever you call it, remember that it is nothing new. You can accomplish the same thing by multiplying each side of the equation by the LCD.

The equation $ad = bc$ says that *the product of the extremes is equal to the product of the means.* When solving a proportion, we can omit multiplication by the LCD and just remember the result, $ad = bc$, as the **extremes-means property.**

Extremes-Means Property

If $\dfrac{a}{b} = \dfrac{c}{d}$, then $ad = bc$.

The extremes-means property makes it easier to solve proportions.

E X A M P L E   5

A proportion with one solution

Solve $\frac{20}{x} = \frac{30}{x + 20}$.

Solution

Rather than multiplying by the LCD, we use the extremes-means property to eliminate the denominators:

$$\frac{20}{x} = \frac{30}{x + 20}$$

$$20(x + 20) = 30x \quad \text{Extremes-means property}$$

$$20x + 400 = 30x$$

$$400 = 10x$$

$$40 = x$$

Check 40 in the original equation. The solution set is {40}.

> **Now do Exercises 19–26**

E X A M P L E   6

A proportion with two solutions

Solve $\frac{2}{x} = \frac{x + 3}{5}$.

Solution

Use the extremes-means property to write an equivalent equation:

$$x(x + 3) = 2 \cdot 5 \quad \text{Extremes-means property}$$

$$x^2 + 3x = 10$$

$$x^2 + 3x - 10 = 0$$

$$(x + 5)(x - 2) = 0 \quad \text{Factor.}$$

$$x + 5 = 0 \quad \text{or} \quad x - 2 = 0 \quad \text{Zero factor property}$$

$$x = -5 \quad \text{or} \quad x = 2$$

Both -5 and 2 satisfy the original equation. The solution set is $\{-5, 2\}$.

> **Now do Exercises 27–36**

CAUTION The extremes-means property works best on simple proportions. It is easier to solve $\frac{5x}{x^4 - 16} = \frac{1}{x^2 - 4}$ by multiplying each side by the LCD. Also, be careful not to apply it to equations that are not proportions, such as $\frac{3}{x} = \frac{2}{x + 1} + 5$.

‹3› Applications

E X A M P L E   7

Ratios and proportions

The ratio of men to women in a poker tournament is 13 to 2. If there are 8 women, then how many men are in the tournament?

Solution

If x is the number of men, then we can write and solve the following proportion:

$$\frac{x}{8} = \frac{13}{2} \qquad \text{The ratio of men to women is 13 to 2.}$$

$$2x = 104 \qquad \text{Extremes-means property}$$

$$x = 52 \qquad \text{Divide each side by 2.}$$

The number of men in the tournament is 52.

> **Now do Exercises 79–82**

E X A M P L E   **8**

Ratios and proportions

The ratio of men to women at a football game was 4 to 3. If there were 12,000 more men than women in attendance, then how many men and how many women were in attendance?

Solution

Let x represent the number of men in attendance and $x - 12{,}000$ represent the number of women in attendance. Because the ratio of men to women was 4 to 3, we can write the following proportion:

$$\frac{4}{3} = \frac{x}{x - 12{,}000}$$

$$4x - 48{,}000 = 3x$$

$$x = 48{,}000$$

So there were 48,000 men and 36,000 women at the game.

> **Now do Exercises 83–90**

 ## Warm-Ups ▼

Fill in the blank.

1. The first step in solving an equation involving rational expressions is to multiply both sides by the _____.

2. An equation involving rational expressions can have _____ roots.

3. A _____ is an equation expressing equality of two rational expressions.

4. In $a/b = c/d$ the _____ are b and c.

5. In $a/b = c/d$ the _____ are a and d.

6. The _____ property says that if $a/b = c/d$, then $ad = bc$.

True or false?

7. To solve $\dfrac{1}{x} + \dfrac{1}{2x} = \dfrac{1}{3}$ multiply each side by 6.

8. To solve $\dfrac{1}{x-2} + 3 = \dfrac{1}{x+2}$ multiply each side by $x^2 - 4$.

9. The solution set to $\dfrac{5}{x-1} + 8 = \dfrac{7}{x-3}$ is $\{1, 3\}$.

10. The solution set to $\dfrac{3}{x} = \dfrac{2}{5}$ is $\left\{\dfrac{15}{2}\right\}$.

11. The solution set to $(2x - 3)(3x + 4) = 0$ is $\left\{\dfrac{3}{2}, \dfrac{4}{3}\right\}$.

12. The equations $x^2 = x$ and $x = 1$ are equivalent.

Exercises

6.6

‹ **Study Tips** ›

- Put important facts on note cards. Work on memorizing the note cards when you have a few spare minutes.
- Post some note cards on your refrigerator door. Make this course a part of your life.

‹ 1 › Multiplying by the LCD

Find the solution set to each equation. See Examples 1–4.

1. $\dfrac{1}{x} + \dfrac{1}{6} = \dfrac{1}{8}$

2. $\dfrac{3}{x} + \dfrac{1}{5} = \dfrac{1}{2}$

3. $\dfrac{2}{3x} + \dfrac{1}{15x} = \dfrac{1}{2}$

4. $\dfrac{5}{6x} - \dfrac{1}{8x} = \dfrac{17}{24}$

5. $\dfrac{3}{x-2} + \dfrac{5}{x} = \dfrac{10}{x}$

6. $\dfrac{5}{x-1} + \dfrac{1}{2x} = \dfrac{1}{x}$

7. $\dfrac{x}{x-2} + \dfrac{3}{x} = 2$

8. $\dfrac{x}{x-5} + \dfrac{5}{x} = \dfrac{11}{6}$

9. $\dfrac{100}{x} = \dfrac{150}{x+5} - 1$

10. $\dfrac{30}{x} = \dfrac{50}{x+10} + \dfrac{1}{2}$

11. $\dfrac{3x-5}{x-1} = 2 - \dfrac{2x}{x-1}$

12. $\dfrac{x-3}{x+2} = 3 - \dfrac{1-2x}{x+2}$

13. $x + 1 + \dfrac{2x-5}{x-5} = \dfrac{x}{x-5}$

14. $\dfrac{x-3}{2} - \dfrac{1}{x-3} = \dfrac{8-3x}{x-3}$

15. $5 + \dfrac{9}{x-2} = 2 + \dfrac{x+7}{x-2}$

16. $3 + \dfrac{x+1}{x-3} = 2 - \dfrac{5-3x}{x-3}$

17. $\dfrac{2}{x+2} + \dfrac{x}{x-3} + \dfrac{1}{x^2-x-6} = 0$

18. $\dfrac{x-4}{x^2+2x-15} = 2 - \dfrac{2}{x-3}$

‹ 2 › Proportions

Find the solution set to each equation. See Examples 5 and 6.

19. $\dfrac{2}{x} = \dfrac{3}{4}$

20. $\dfrac{5}{x} = \dfrac{7}{9}$

21. $\dfrac{a}{3} = \dfrac{-1}{4}$

22. $\dfrac{b}{5} = \dfrac{-3}{7}$

23. $-\dfrac{5}{7} = \dfrac{2}{x}$

24. $-\dfrac{3}{8} = \dfrac{5}{x}$

25. $\dfrac{10}{x} = \dfrac{20}{x+20}$

26. $\dfrac{x}{5} = \dfrac{x+2}{3}$

27. $\dfrac{2}{x+1} = \dfrac{x-1}{4}$

28. $\dfrac{3}{x-2} = \dfrac{x+2}{7}$

29. $\dfrac{x}{6} = \dfrac{5}{x-1}$

30. $\dfrac{x+5}{2} = \dfrac{3}{x}$

31. $\dfrac{x+7}{x+4} = \dfrac{x+1}{x-2}$

32. $\dfrac{x+1}{x-5} = \dfrac{x+2}{x-4}$

33. $\dfrac{x-2}{x-3} = \dfrac{x+5}{x+2}$

34. $\dfrac{a-5}{a+6} = \dfrac{a-7}{a+8}$

35. $\dfrac{3w}{3w-5} = \dfrac{w}{w+2}$

36. $\dfrac{x}{x+5} = \dfrac{x}{x-2}$

Miscellaneous

Solve each equation.

37. $\dfrac{a}{9} = \dfrac{4}{a}$

38. $\dfrac{y}{3} = \dfrac{27}{y}$

39. $\dfrac{x}{9} = \dfrac{-20}{9x} + 1$

40. $\dfrac{y}{3} = \dfrac{4}{3} - \dfrac{1}{y}$

41. $\dfrac{1}{2x-4} + \dfrac{1}{x-2} = \dfrac{1}{4}$

42. $\dfrac{7}{3x-9} - \dfrac{1}{x-3} = \dfrac{4}{9}$

43. $\dfrac{x-2}{4} = \dfrac{x-2}{x}$

44. $\dfrac{y+5}{2} = \dfrac{y+5}{y}$

45. $\dfrac{5}{2x+4} - \dfrac{1}{x-1} = \dfrac{3}{x+2}$

46. $\dfrac{5}{2w+6} - \dfrac{1}{w-1} = \dfrac{1}{w+3}$

47. $\dfrac{5}{x-3} = \dfrac{x}{x-3}$

48. $\dfrac{6}{a+2} = \dfrac{a}{a+2}$

49. $\dfrac{w}{6} = \dfrac{3}{2w}$

50. $\dfrac{2m}{5} = \dfrac{10}{m}$

51. $\dfrac{5}{4x-2} - \dfrac{1}{1-2x} = \dfrac{7}{3x+6}$

52. $\dfrac{5}{x+1} - \dfrac{1}{1-x} = \dfrac{1}{x^2-1}$

53. $\dfrac{5}{x} = \dfrac{2}{5}$

54. $\dfrac{-3}{2x} = \dfrac{1}{-5}$

55. $\dfrac{x}{x-2} - \dfrac{x+2}{x^2-2x} = \dfrac{1}{x}$

56. $\dfrac{x-2}{x-6} - \dfrac{4}{x} = \dfrac{24}{x^2-6x}$

57. $\dfrac{5}{x^2-9} + \dfrac{2}{x+3} = \dfrac{1}{x-3}$

58. $\dfrac{1}{x-2} - \dfrac{2}{x+3} = \dfrac{11}{x^2+x-6}$

59. $\dfrac{9}{x^3-1} - \dfrac{1}{x-1} = \dfrac{2}{x^2+x+1}$

60. $\dfrac{x+4}{x^3+8} + \dfrac{x+2}{x^2-2x+4} = \dfrac{11}{2x+4}$

Either solve the given equation or perform the indicated operation(s), whichever is appropriate.

61. $\dfrac{4}{x} = \dfrac{3}{4}$

62. $\dfrac{5}{h} = \dfrac{h}{5}$

63. $\dfrac{4}{x} + \dfrac{3}{4}$

64. $\dfrac{5}{h} - \dfrac{h}{5}$

65. $\dfrac{2}{x} - \dfrac{3}{4} = \dfrac{1}{2}$

66. $\dfrac{1}{2x} - \dfrac{5}{3x} = \dfrac{1}{4}$

67. $\dfrac{2}{x} - \dfrac{3}{4} - \dfrac{1}{2}$

68. $\dfrac{1}{2x} - \dfrac{5}{3x} + \dfrac{1}{4}$

Solve each equation. Identify each equation as a conditional equation, an inconsistent equation, or an identity. State the solution sets to the identities using interval notation.

69. $\dfrac{1}{x} + \dfrac{1}{2x} = \dfrac{3}{2x}$

70. $\dfrac{1}{x} + \dfrac{1}{x^2} = \dfrac{x+1}{x^2}$

71. $\dfrac{1}{x} + \dfrac{1}{2x} = \dfrac{5}{4x}$

72. $\dfrac{1}{x} + \dfrac{1}{x^2} = \dfrac{x+2}{x^2}$

73. $\dfrac{1}{x} + \dfrac{1}{2x} = \dfrac{x+2}{2x}$

74. $\dfrac{1}{x} + \dfrac{1}{x^2} = \dfrac{6}{x^3}$

75. $\dfrac{1}{x} + \dfrac{1}{x-1} = \dfrac{2x-1}{x^2-x}$

76. $\dfrac{2}{x+1} + \dfrac{3}{x-1} = \dfrac{5x+1}{x^2-1}$

77. $\dfrac{1}{x} + \dfrac{1}{x-1} = \dfrac{2x}{x^2-x}$

78. $\dfrac{2}{x+1} + \dfrac{3}{x-1} = \dfrac{5x+2}{x^2-1}$

⟨3⟩ Applications

Solve each problem. See Examples 7 and 8.

79. *Aspect ratio.* The aspect ratio for a television screen is the ratio of the width to the height of the screen. For a high definition TV it is 16 to 9. What is the height of a screen that is 48 inches wide?

80. *Hybrids.* The ratio of the number of cars with gasoline engines to hybrids sold at a car dealer is 9 to 2. If the dealer sells 10 hybrids one month, then how many gasoline cars did the dealer sell?

81. *Maritime losses.* The amount paid to an insured party by the American Insurance Company is computed by using the proportion

$$\frac{\text{value shipped}}{\text{amount of loss}} = \frac{\text{amount of declared premium}}{\text{amount insured party gets paid}}.$$

If the value shipped was \$300,000, the amount of loss was \$250,000, and the amount of declared premium was \$200,000, then what amount is paid to the insured party?

82. *Maritime losses.* Suppose the value shipped was \$400,000, the amount of loss was \$300,000, and the amount that the insured party got paid was \$150,000. Use the proportion of Exercise 81 to find the amount of declared premium.

83. *Capture-recapture method.* To estimate the size of the grizzly bear population in a national park, rangers tagged and released 12 bears. Later it was observed that in 23 sightings of grizzly bears, only two had been tagged. Assuming the proportion of tagged bears in the later sightings is the same as the proportion of tagged bears in the population, estimate the number of bears in the population.

84. *Please rewind.* In a sample of 24 returned videotapes, it was found that only 3 were rewound as requested. If

872 videos are returned in a day, then how many of them would you expect to find that are not rewound?

85. *Pleasing painting.* The ancient Greeks often used the ratio of length to width for a rectangle as 7 to 6 to give the rectangle a pleasing shape. If the length of a pleasantly shaped Greek painting is 22 centimeters (cm) longer than its width, then what are its length and width?

86. *Pickups and cars.* The ratio of pickups to cars sold at a dealership is 2 to 3. If the dealership sold 142 more cars than pickups in 2006, then how many of each did it sell?

87. *Cleaning up the river.* Pollution in the Tickfaw River has been blamed primarily on pesticide runoff from area farms. The formula

$$C = \frac{4,000,000p}{100 - p}$$

Figure for Exercise 87

has been used to model the cost in dollars for removing $p\%$ of the pollution in the river. If the state gets a $1 million federal grant for cleaning up the river, then what percentage of the pollution can be removed? Use the bar graph to estimate the percentage that can be cleaned up with a $100 million grant.

88. *Campaigning for governor.* A campaign manager for a gubernatorial candidate estimates that the cost in dollars for an advertising campaign that will get his candidate $p\%$ of the votes is given by

$$C = \frac{1,000,000 + 2,000,000p}{100 - p}.$$

If the candidate can spend only $2 million for advertising, then what percentage of the votes can she expect to receive? Use the bar graph to estimate the percentage of votes expected if $4 million is spent.

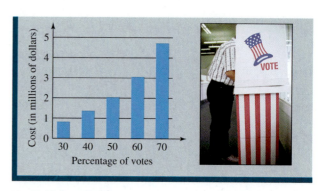

Figure for Exercise 88

89. *Wealth-building portfolio.* Misty decided to invest her annual bonus in a wealth-building portfolio as shown in the accompanying figure (www.fidelity.com).

a) If the amount that she invested in stocks was $20,000 greater than her investment in bonds, then how much did she invest in bonds?

b) What was the amount of her annual bonus?

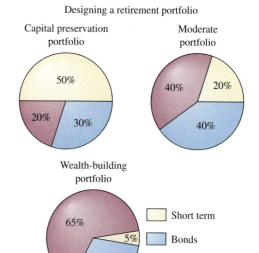

Figure for Exercise 89

90. *Estimating weapons.* When intelligence agents obtain enemy weapons marked with serial numbers, they use the formula $N = (1 + 1/C)B - 1$ to estimate the total number of such weapons N that the enemy has produced. B is the biggest serial number obtained, and C is the number of weapons obtained. It is assumed the weapons are numbered 1 through N.

a) Find N if agents obtain five nerve gas containers numbered 45, 143, 258, 301, and 465.

b) Find C if agents estimate that the enemy has 255 tanks from a group of captured tanks on which the biggest serial number is 224.

and using the LCD to find the sum

$$\frac{3}{x-2}+\frac{7}{x+2}.$$

Getting More Involved

91. *Writing*

In this chapter the LCD is used to add rational expressions and to solve equations. Explain the difference between using the LCD to solve the equation

$$\frac{3}{x-2}+\frac{7}{x+2}=2$$

92. *Discussion*

For each equation, find the values for x that *cannot* be solutions to the equation. Do not solve the equations.

a) $\dfrac{1}{x}+\dfrac{1}{x-1}=\dfrac{1}{2}$ b) $\dfrac{x}{x-1}=\dfrac{1}{2}$

c) $\dfrac{1}{x^2+1}=\dfrac{1}{x+1}$

6.7   Applications

In This Section

⟨1⟩ **Formulas**

⟨2⟩ **Uniform Motion Problems**

⟨3⟩ **Work Problems**

⟨4⟩ **Miscellaneous Problems**

In this section we will use the techniques of Section 6.6 to rewrite formulas involving rational expressions and to solve some problems.

⟨1⟩ Formulas

In Section 2.2 we solved formulas for a specified variable, but we did not encounter any formulas involving rational expressions with variables in the denominator. Example 1 involves such formulas.

E X A M P L E   1

Solving a formula for a specified variable

Solve each formula for y.

a) $\dfrac{1}{2y}+\dfrac{1}{x}=\dfrac{a}{3}$ b) $\dfrac{a-b}{2y}=\dfrac{6}{a+b}$

Solution

a) The least common denominator for $2y$, x, and 3 is $6xy$. Multiplying each side of the equation by the LCD eliminates all of the rational expressions:

$$\frac{1}{2y}+\frac{1}{x}=\frac{a}{3} \qquad \text{Original equation}$$

$$6xy\left(\frac{1}{2y}+\frac{1}{x}\right)=6xy\cdot\frac{a}{3} \qquad \text{Multiply each side by the LCD.}$$

$$3x+6y=2axy \qquad \text{Simplify.}$$

$$3x=2axy-6y \qquad \text{Get all terms involving } y \text{ onto one side.}$$

$$3x=y(2ax-6) \qquad \text{Factor out } y.$$

$$\frac{3x}{2ax-6}=y \qquad \text{Divide each side by } 2ax-6.$$

The formula solved for y is $y=\dfrac{3x}{2ax-6}$.

b) Since this formula is actually a proportion, we can apply the extremes-means property:

$$\frac{a-b}{2y} = \frac{6}{a+b} \qquad \text{Original formula}$$

$$6 \cdot 2y = (a-b)(a+b) \qquad \text{Extremes-means property}$$

$$12y = a^2 - b^2 \qquad \text{Simplify.}$$

$$y = \frac{a^2 - b^2}{12}$$

Now do Exercises 1–20

In Example 2 we find the value of one variable when given the values of the remaining variables.

E X A M P L E   2

Evaluating a formula

Find x if $x_1 = 2$, $y_1 = -3$, $y = -1$, $m = \frac{1}{2}$, and

$$\frac{y - y_1}{x - x_1} = m.$$

Solution

Substitute all of the values into the formula and solve for x:

$$\frac{-1 - (-3)}{x - 2} = \frac{1}{2} \qquad \text{Substitute.}$$

$$\frac{2}{x - 2} = \frac{1}{2}$$

$$x - 2 = 4 \qquad \text{Extremes-means property}$$

$$x = 6 \qquad \text{Check in the original formula.}$$

Now do Exercises 21–28

⟨2⟩ Uniform Motion Problems

The uniform motion problems here are similar to those of Chapter 2, but in this chapter the equations have rational expressions.

E X A M P L E   3

Uniform motion

Michele drove her empty rig 300 miles to Salina to pick up a load of cattle. When her rig was fully loaded, her average speed was 10 miles per hour less than when the rig was empty. If the return trip took her 1 hour longer, then what was her average speed with the rig empty? (See Fig. 6.2.)

Solution

Let x be Michele's average speed empty and let $x - 10$ be her average speed full. Because the time can be determined from the distance and the rate, $T = \frac{D}{R}$, we can make the following table.

300 mi
Speed = x mph

300 mi
Speed = $x - 10$ mph

Figure 6.2

	Rate	Time	Distance
Empty	$x \, \frac{\text{mi}}{\text{hr}}$	$\frac{300}{x}$ hr	300 mi
Full	$x - 10 \, \frac{\text{mi}}{\text{hr}}$	$\frac{300}{x - 10}$ hr	300 mi

We now write an equation expressing the fact that her time empty was 1 hour less than her time full:

$$\frac{300}{x} = \frac{300}{x - 10} - 1$$

$$x(x - 10)\frac{300}{x} = x(x - 10)\frac{300}{x - 10} - x(x - 10)1 \qquad \text{Multiply each side by } x(x - 10).$$

$$300x - 3000 = 300x - x^2 + 10x \qquad \text{Reduce.}$$

$$-3000 = -x^2 + 10x$$

$$x^2 - 10x - 3000 = 0 \qquad \text{Get 0 on one side.}$$

$$(x + 50)(x - 60) = 0 \qquad \text{Factor.}$$

$$x + 50 = 0 \qquad \text{or} \qquad x - 60 = 0 \qquad \text{Zero factor property}$$

$$x = -50 \qquad \text{or} \qquad x = 60$$

The equation is satisfied if $x = -50$, but because -50 is negative, it cannot be the speed of the truck. Michele's average speed empty was 60 miles per hour (mph). Checking this answer, we find that if she traveled 300 miles at 60 mph, it would take her 5 hours. If she traveled 300 miles at 50 mph with the loaded rig, it would take her 6 hours. Because Michele's time with the empty rig was 1 hour less than her time with the loaded rig, 60 mph is the correct answer.

> **Now do Exercises 29–36**

⟨3⟩ Work Problems

Problems involving different rates for completing a task are referred to as **work problems.** We did not solve work problems earlier because they usually require equations with rational expressions. Work problems are similar to uniform motion problems in which $RT = D$. The product of a person's time and rate is the amount of work completed. For example, if your puppy gains 1 pound every 3 days, then he is growing at the rate of $\frac{1}{3}$ pound per day. If he grows at the rate of $\frac{1}{3}$ pound per day for a period of 30 days, then he gains 10 pounds.

E X A M P L E   4

Working together

Bill can do the inventory by himself in 8 hours. Hilda can do the same job by herself in 6 hours. Assuming they do not interfere with each other, how long would it take them to do the inventory working together?

Solution

Let x represent the number of hours that it takes them to do the inventory together. Since Bill can do the entire inventory in 8 hours, his rate is $\frac{1}{8}$ of the inventory per hour. The amount of work done by Bill is the product of his rate and his time. If he works for x hours at a rate of $\frac{1}{8}$

of the inventory/hour, then the amount of work he does is $\frac{1}{8} \cdot x$ or $\frac{x}{8}$ of the inventory. Likewise, Hilda works for x hours at a rate of $\frac{1}{6}$ of the inventory/hour and does $\frac{x}{6}$ of the inventory. We can classify all of this information in a table just like the one in Example 3.

	Rate	Time	Amount of Work
Bill	$\frac{1}{8}\frac{\text{inventory}}{\text{hr}}$	x hr	$\frac{x}{8}$ inventory
Hilda	$\frac{1}{6}\frac{\text{inventory}}{\text{hr}}$	x hr	$\frac{x}{6}$ inventory

Since one inventory is completed in x hours, the portions of the inventory (in the work column) must have a sum of one:

$$\frac{x}{8} + \frac{x}{6} = 1 \qquad \text{One inventory completed}$$
$$24 \cdot \frac{x}{8} + 24 \cdot \frac{x}{6} = 24 \cdot 1 \qquad \text{Multiply each side by the LCD 24.}$$
$$3x + 4x = 24$$
$$7x = 24$$
$$x = \frac{24}{7}$$

So the time together is exactly $\frac{24}{7}$ hours or approximately 3 hours 26 minutes.

> **Now do Exercises 37–38**

The methods in Examples 3 and 4 are the same. One table uses $RT = D$ and the other uses $RT = W$. Making a table will help you understand work problems. Note that Example 4 could be solved using only the rates. Since Bill's rate is $\frac{1}{8}$ inventory/hr and Hilda's rate is $\frac{1}{6}$ inventory/hr, together their rate is $\left(\frac{1}{8} + \frac{1}{6}\right)$ inventory/hr. Since their rate together is also $\frac{1}{x}$ inventory/hr, we have $\frac{1}{8} + \frac{1}{6} = \frac{1}{x}$. Solving this equation gives the same result.

In Example 5, we know the time it takes to complete a task, but we do not know one of the individual times.

EXAMPLE 5 Working together

Linda can mow a certain lawn with her riding lawn mower in 4 hours. When Linda uses the riding mower and Rebecca operates the push mower, it takes them 3 hours to mow the lawn. How long would it take Rebecca to mow the lawn by herself using the push mower?

‹ **Helpful Hint** ›

The secret to work problems is remembering that the individual amounts of work or the individual rates can be added when people work together. If your painting rate is $\frac{1}{10}$ of the house per day and your helper's rate is $\frac{1}{5}$ of the house per day, then your rate together will be $\frac{3}{10}$ of the house per day.

Solution

If x is the number of hours it takes for Rebecca to complete the lawn alone, then her rate is $\frac{1}{x}$ of the lawn per hour. Because Linda can mow the entire lawn in 4 hours, her rate is $\frac{1}{4}$ of the lawn per hour. In the 3 hours that they work together, Rebecca completes $\frac{3}{x}$ of the lawn while Linda completes $\frac{3}{4}$ of the lawn. We can classify all of the necessary information in a table.

	Rate	Time	Amount of Work
Linda	$\frac{1}{4}\frac{\text{lawn}}{\text{hr}}$	3 hr	$\frac{3}{4}$ lawn
Rebecca	$\frac{1}{x}\frac{\text{lawn}}{\text{hr}}$	3 hr	$\frac{3}{x}$ lawn

Because the lawn is finished in 3 hours, the two portions of the lawn (in the work column) mowed by each girl have a sum of 1:

$$\frac{3}{4} + \frac{3}{x} = 1$$

$$4x \cdot \frac{3}{4} + 4x \cdot \frac{3}{x} = 4x \cdot 1 \quad \text{Multiply each side by } 4x.$$

$$3x + 12 = 4x$$

$$12 = x$$

If $x = 12$, then in the 3 hours that they work together, Rebecca does $\frac{3}{12}$ or $\frac{1}{4}$ of the job while Linda does $\frac{3}{4}$ of the job. So it would take Rebecca 12 hours to mow the lawn by herself using the push mower.

> **Now do Exercises 39–42**

⟨4⟩ Miscellaneous Problems

EXAMPLE 6

Hamburger and steak

Patrick bought 50 pounds of meat consisting of hamburger and steak. Steak costs twice as much per pound as hamburger. If he bought $30 worth of hamburger and $90 worth of steak, then how many pounds of each did he buy?

Solution

Let x be the number of pounds of hamburger and $50 - x$ be the number of pounds of steak. Because Patrick got x pounds of hamburger for $30, he paid $\frac{30}{x}$ dollars per pound for the hamburger. We can classify all of the given information in a table.

	Price per Pound	Amount	Total Price
Hamburger	$\frac{30}{x}$ $\frac{\text{dollars}}{\text{lb}}$	x lb	30 dollars
Steak	$\frac{90}{50-x}$ $\frac{\text{dollars}}{\text{lb}}$	$50 - x$ lb	90 dollars

Because the price per pound of steak is twice that of hamburger, we can write the following equation:

$$2\left(\frac{30}{x}\right) = \frac{90}{50 - x}$$

$$\frac{60}{x} = \frac{90}{50 - x}$$

$$90x = 3000 - 60x \quad \text{The extremes-means property}$$

$$150x = 3000$$

$$x = 20$$

$$50 - x = 30$$

Patrick purchased 20 pounds of hamburger and 30 pounds of steak. Check this answer.

> **Now do Exercises 43–54**

Warm-Ups ▼

Fill in the blank.

1. A _____ is an equation involving two or more variables.

2. If a formula is _____ for a specified variable, then that variable is isolated on one side and does not occur on the other side.

3. _____ motion is motion at a constant rate.

4. Problems that involve different rates for completing a task are referred to as _____ problems.

True or false?

5. The formula $w = \dfrac{1-t}{t}$ solved for t is $t = \dfrac{1-t}{w}$.

6. To solve $\dfrac{1}{p} + \dfrac{1}{q} = \dfrac{1}{s}$ multiply each side by pqs.

7. If 50 pounds of steak cost x dollars, then the price is $\dfrac{50}{x}$ dollars per pound

8. If Sue drives x miles in 3 hours, then her rate is $\dfrac{x}{3}$ miles per hour.

9. If Bob mows his entire lawn in x hours, then his rate is $\dfrac{1}{x}$ of the lawn per hour.

10. If x hours is one hour less than y hours, then $x - 1 = y$.

6.7   Exercises

‹ Study Tips ›

- Pay particular attention to the examples that your instructor works in class or presents to you online.
- The examples and homework assignments should give you a good idea of what your instructor expects from you.

‹1› Formulas

Solve each equation for y. See Example 1.

1. $\dfrac{y-5}{x+3} = -\dfrac{4}{3}$

2. $\dfrac{y+1}{x-9} = \dfrac{3}{4}$

3. $M = \dfrac{1}{y}$

4. $L = \dfrac{ay}{w}$

5. $\dfrac{1}{y} = \dfrac{a}{w} + \dfrac{w}{a}$

6. $\dfrac{1}{n} = \dfrac{a}{y} - \dfrac{w}{a}$

7. $h = \dfrac{b}{y} + 3$

8. $z = \dfrac{y}{m} + a$

Solve each formula for the indicated variable. See Example 1.

9. $M = \dfrac{F}{f}$ for f

10. $P = \dfrac{A}{1+rt}$ for A

11. $\dfrac{1}{a} + \dfrac{1}{b} = \dfrac{1}{2}$ for a

12. $\dfrac{2}{x} = \dfrac{3}{y} - w$ for y

13. $\dfrac{1}{2x} + \dfrac{1}{2} - \dfrac{2}{y} = 0$ for x

14. $\dfrac{1}{x} - \dfrac{2}{3y} + z = 0$ for y

15. $\dfrac{1}{R} = \dfrac{1}{R_1} + \dfrac{1}{R_2}$ for R_1 **16.** $\dfrac{1}{R} = \dfrac{1}{R_1} + \dfrac{1}{R_2}$ for R_2

17. $\dfrac{P_1V_1}{T_1} = \dfrac{P_2V_2}{T_2}$ for T_1 **18.** $\dfrac{P_1V_1}{T_1} = \dfrac{P_2V_2}{T_2}$ for P_2

19. $V = \dfrac{4}{3}\pi r^2 h$ for h **20.** $h = \dfrac{S - 2\pi r^2}{2\pi r}$ for S

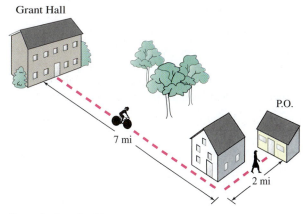

Figure for Exercise 29

Find the value of the indicated variable. Round approximate answers to three decimal places. See Example 2.

21. Find f if $M = 10$, $F = 5$, and $M = \dfrac{F}{f}$.

22. Find r if $A = 550$, $P = 500$, $t = 2$, and $P = \dfrac{A}{1 + rt}$.

23. Find h if $r = 3$, $V = 12\pi$, and $r^2 = \dfrac{V}{\pi h}$.

24. Find k if $F = 32$, $r = 4$, $m_1 = 6$, $m_2 = 8$, and $m_1 = \dfrac{Fr^2}{km_2}$.

25. Find r if $F = 10$, $m = 8$, $v = 6$, and $F = \dfrac{mv^2}{r}$.

26. Find p if $f = 2.3$, $q = 1.7$, and $\dfrac{1}{p} + \dfrac{1}{q} = \dfrac{1}{f}$.

27. Find R_2 if $R = 1.29$, $R_1 = 0.045$, and $\dfrac{1}{R} = \dfrac{1}{R_1} + \dfrac{1}{R_2}$.

28. Find S if $h = 3.6$, $r = 2.45$, and $h = \dfrac{S - 2\pi r^2}{2\pi r}$.

⟨2⟩ Uniform Motion Problems

Solve each problem. See Example 3.

29. *Walking and riding.* Karen can ride her bike from home to school in the same amount of time as she can walk from home to the post office. She rides 10 miles per hour (mph) faster than she walks. The distance from her home to school is 7 miles, and the distance from her home to the post office is 2 miles. How fast does Karen walk?

30. *Fast driving.* Beverly can drive 600 miles in the same time as it takes Susan to drive 500 miles. If Beverly drives 10 mph faster than Susan, then how fast does Beverly drive?

31. *Driving speed.* John can drive 240 miles in the same time as it takes George to drive 220 miles. If John drives 5 mph faster than George, then how fast does John drive?

32. *Commuting speed.* Bill and Bob both drive 60 miles to work. By averaging 10 miles per hour faster than Bob, Bill gets to work 12 minutes earlier than Bob. How fast does each one drive?

33. *Faster driving.* Patrick drives 40 miles to work, and Guy drives 60 miles to work. Guy claims that he drives at the same speed as Patrick, but it takes him only 12 minutes longer to get to work. If this is true, then how long does it take each of them to get to work? What are their speeds? Do you think that Guy's claim is correct?

34. *Route drivers.* David and Keith are route drivers for a fast-photo company. David's route is 80 miles, and Keith's is 100 miles. Keith averages 10 mph more than David and finishes his route 10 minutes before David. What is David's speed?

35. *Physically fit.* Every morning, Yong Yi runs 5 miles, then walks 1 mile. He runs 6 mph faster than he walks. If his total time yesterday was 45 minutes, then how fast did he run?

36. *Row, row, row your boat.* Norma can row her boat 12 miles in the same time as it takes Marietta to cover 36 miles in her motorboat. If Marietta's boat travels 15 mph faster than Norma's boat, then how fast is Norma rowing her boat?

⟨3⟩ **Work Problems**

Solve each problem. See Examples 4 and 5.

37. *Pumping out the pool.* A large pump can drain an 80,000-gallon pool in 3 hours. With a smaller pump also operating, the job takes only 2 hours. How long would it take the smaller pump to drain the pool by itself?

38. *Trimming hedges.* Lourdes can trim the hedges around her property in 8 hours by using an electric hedge trimmer. Rafael can do the same job in 15 hours by using a manual trimmer. How long would it take them to trim the hedges working together?

39. *Filling the tub.* It takes 10 minutes to fill Alisha's bathtub and 12 minutes to drain the water out. How long would it take to fill it with the drain accidentally left open?

Figure for Exercise 39

40. *Eating machine.* Charles can empty the cookie jar in $1\frac{1}{2}$ hours. It takes his mother 2 hours to bake enough cookies to fill it. If the cookie jar is full when Charles comes home from school, and his mother continues baking and restocking the cookie jar, then how long will it take him to empty the cookie jar?

41. *Filing the invoices.* It takes Gina 90 minutes to file the monthly invoices. If Hilda files twice as fast as Gina does, how long will it take them working together?

42. *Painting alone.* Julie can paint a fence by herself in 12 hours. With Betsy's help, it takes only 5 hours. How long would it take Betsy by herself?

⟨4⟩ **Miscellaneous Problems**

Solve each problem. See Example 6.

43. *Buying fruit.* Molly bought $5.28 worth of oranges and $8.80 worth of apples. She bought 2 more pounds of oranges than apples. If apples cost twice as much per pound as oranges, then how many pounds of each did she buy?

44. *Raising rabbits.* Luke raises rabbits and raccoons to sell for meat. The price of raccoon meat is three times the price of rabbit meat. One day Luke sold 160 pounds of meat, $72 worth of each type. What is the price per pound of each type of meat?

45. *Total resistance.* If two receivers with resistances R_1 and R_2 are connected in parallel, then the formula

$$\frac{1}{R} = \frac{1}{R_1} + \frac{1}{R_2}$$

relates the total resistance for the circuit R with R_1 and R_2. Given that R_1 is 3 ohms and R is 2 ohms, find R_2.

Figure for Exercise 45

46. *More resistance.* Use the formula from Exercise 45 to find R_1 and R_2 given that the total resistance is 1.2 ohms and R_1 is 1 ohm larger than R_2.

47. *Thin lens.* The thin lens equation

$$\frac{1}{S_o} + \frac{1}{S_i} = \frac{1}{F}$$

relates the object distance S_o, the image distance S_i, and the focal length F for a thin lens. If the object distance is 500 mm and the focal length is 100 mm, then what is the image distance?

48. *Another thin lens.* Use the thin lens equation from Exercise 47 to find S_o and S_i if S_o is twice as large as S_i and F is 50 mm.

49. *Office party.* A group of coworkers are planning to share the $1000 cost of an office party. If they can get three more people to join them in sharing the cost, then the cost per person will go down by $75. How many workers are in the original group?

50. *Sailing party.* A group of sailors is planning to share equally the cost of a $40,000 sailboat. When two sailors dropped out of the group, the cost per sailor increased by $1000. How many sailors were in the original group?

51. *Las Vegas vacation.* Brenda of Horizon Travel has arranged for a group of gamblers to share the $24,000 cost of a charter flight to Las Vegas. If Brenda can get 40 more people to share the cost, then the cost per person will decrease by $100. See the figure.

Figure for Exercise 51

a) How many people were in the original group?

b) Write the cost per person as a function of the number of people sharing the cost.

52. *White-water rafting.* Adventures, Inc., has a $1500 group rate for an overnight rafting trip on the Colorado River. For the last trip five people failed to show, causing the price per person to increase by $25. How many were originally scheduled for the trip?

53. *Doggie bag.* Muffy can eat a 25-pound bag of dog food in 28 days, whereas Missy eats a 25-pound bag in 23 days. How many days would it take them together to finish a 50-pound bag of dog food?

54. *Rodent food.* A pest control specialist has found that 6 rats can eat an entire box of sugar-coated breakfast cereal in 13.6 minutes, and it takes a dozen mice 34.7 minutes to devour the same size box of cereal. How long would it take all 18 rodents, in a cooperative manner, to finish off a box of cereal?

Wrap-Up

Summary

Rational Expressions		Examples
Rational expression	The ratio of two polynomials with the denominator not equal to zero	$\dfrac{x^2 - 1}{2x - 3}$
Domain of a rational expression	The set of all possible numbers that can be used as replacements for the variable	$D = \left\{ x \mid x \neq \dfrac{3}{2} \right\}$

Operations with Rational Expressions		Examples
Basic principle of rational numbers	If $\dfrac{a}{b}$ is a rational number and c is a nonzero real number, then $$\frac{a}{b} = \frac{ac}{bc}.$$	Used for reducing: $$\frac{14}{16} = \frac{2 \cdot 7}{2 \cdot 8} = \frac{7}{8}$$ Used for building: $$\frac{2}{x} = \frac{2 \cdot 3}{x \cdot 3} = \frac{6}{3x}$$
Multiplication of rational numbers	If $\dfrac{a}{b}$ and $\dfrac{c}{d}$ are rational numbers, then $$\frac{a}{b} \cdot \frac{c}{d} = \frac{ac}{bd}.$$	$$\frac{3}{x} \cdot \frac{6}{x^2} = \frac{18}{x^3}$$
Division of rational numbers	If $\dfrac{a}{b}$ and $\dfrac{c}{d}$ are rational numbers with $\dfrac{c}{d} \neq 0$, then $$\frac{a}{b} \div \frac{c}{d} = \frac{a}{b} \cdot \frac{d}{c}. \quad \text{(Invert and multiply.)}$$	$$\frac{a}{x} \div \frac{5}{4x} = \frac{a}{x} \cdot \frac{4x}{5} = \frac{4a}{5}$$
Least common multiple	The LCM is the product of all of the different factors that appear in the polynomials. The exponent on each factor is the highest power that occurs on that factor in any of the polynomials.	$4a^3b, \, 6ab^2$ $\text{LCM} = 12a^3b^2$
Least common denominator	The LCD for a group of denominators is the LCM of the denominators.	$\dfrac{1}{4a^3b} + \dfrac{1}{6ab^2}$ $\text{LCD} = 12a^3b^2$

Addition and subtraction	If $b \neq 0$, then $$\frac{a}{b} + \frac{c}{b} = \frac{a+c}{b} \quad \text{and} \quad \frac{a}{b} - \frac{c}{b} = \frac{a-c}{b}.$$ If the denominators are not identical, we must build up each fraction to an equivalent fraction with the LCD as denominator.	$$\frac{2x}{x-3} + \frac{7x}{x-3} = \frac{9x}{x-3}$$ $$\frac{1}{2} + \frac{1}{x} = \frac{x}{2x} + \frac{2}{2x} = \frac{x+2}{2x}$$
Rules for adding and subtracting simple fractions	If $b \neq 0$ and $d \neq 0$, then $$\frac{a}{b} + \frac{c}{d} = \frac{ad+bc}{bd}$$ and $$\frac{a}{b} - \frac{c}{d} = \frac{ad-bc}{bd}.$$	$$\frac{1}{2} + \frac{1}{3} = \frac{5}{6}$$ $$\frac{2}{5} - \frac{3}{7} = \frac{-1}{35}$$
Simplifying complex fractions	Multiply the numerator and denominator by the LCD.	$$\frac{\left(\frac{1}{2} + \frac{1}{x}\right)6x}{\left(\frac{1}{x} - \frac{1}{3}\right)6x} = \frac{3x+6}{6-2x} = \frac{3x+6}{2(3-x)}$$

Division of Polynomials

Ordinary or long division	dividend = (quotient)(divisor) + remainder $$\frac{\text{dividend}}{\text{divisor}} = \text{quotient} + \frac{\text{remainder}}{\text{divisor}}$$	$$\begin{array}{r} x-7 \\ x+2{\overline{\smash{\big)}\,x^2-5x-14}} \\ \underline{x^2+2x} \\ -7x-14 \\ \underline{-7x-14} \\ 0 \end{array}$$
Synthetic division	A condensed version of long division, used only for dividing by a polynomial of the form $x - c$ If the remainder is 0, then the dividend factors as dividend = (quotient)(divisor).	$$\begin{array}{r} -2 \,\big|\, \begin{array}{rrr} 1 & -5 & -14 \\ & -2 & 14 \end{array} \\ \hline \begin{array}{rrr} 1 & -7 & 0 \end{array} \end{array}$$ $$x^2 - 5x - 14 = (x-7)(x+2)$$
Remainder theorem	If the polynomial $P(x)$ is divided by $x - c$, then the remainder is equal to $P(c)$.	$$P(x) = x^2 - 2x + 7$$ $$\begin{array}{r} 3 \,\big|\, \begin{array}{rrr} 1 & -2 & 7 \\ & 3 & 3 \end{array} \\ \hline \begin{array}{rrr} 1 & 1 & 10 \end{array} \end{array}$$ $$P(3) = 10$$

Equations with Rational Expressions

		Examples
Solving equations with rational expressions	Multiply each side by the LCD to eliminate all denominators.	$\dfrac{1}{x} - \dfrac{1}{3} = \dfrac{1}{2x} - \dfrac{1}{6}$

$$6x\left(\frac{1}{x} - \frac{1}{3}\right) = 6x\left(\frac{1}{2x} - \frac{1}{6}\right)$$

$$6 - 2x = 3 - x$$

| Solving proportions by the extremes-means property | If $\dfrac{a}{b} = \dfrac{c}{d}$, then $ad = bc$. | $\dfrac{2}{x-3} = \dfrac{5}{6}$ |

$$12 = 5x - 15$$

Enriching Your Mathematical Word Power

Fill in the blank.

1. A _____ expression is a ratio of two polynomials with the denominator not equal to zero.

2. The _____ of a rational expression is the set of all real numbers that can be used in place of the variable.

3. A rational expression is in _____ terms when the numerator and denominator have no common factors.

4. When common factors are divided out of the numerator and denominator of a rational expression, the rational expression is _____.

5. Two fractions that represent the same number are _____ fractions.

6. A _____ fraction has rational expressions in its numerator or denominator or both.

7. The opposite of reducing a fraction is _____ a fraction.

8. The smallest number that is a common multiple of a group of denominators is the _____ common denominator.

9. A number that appears to be a solution to an equation but does not satisfy the equation is an _____ root.

10. The expression a/b is the _____ of a to b.

11. _____ division is a quick method for dividing by $x - c$.

12. A _____ is a statement expressing the equality of two rational expressions.

13. The numbers a and d in $a/b = c/d$ are the _____.

14. The numbers b and c in $a/b = c/d$ are the _____.

15. If $a/b = c/d$, then $ad = bc$ is the _____ property.

Review Exercises

6.1 Properties of Rational Expressions and Functions
State the domain of each rational expression.

1. $\dfrac{5-x}{3x-3}$

2. $\dfrac{x-4}{x^2-25}$

3. $\dfrac{x}{x^2-x-2}$

4. $\dfrac{1}{x^3-x^2}$

Find the indicated values of the given rational function if possible.

5. $R(x) = \dfrac{5-x}{3x-3}$, $R(5)$, $R(1)$

6. $T(x) = \dfrac{x-4}{x^2-25}$, $T(4)$, $T(-5)$

7. $H(x) = \dfrac{x}{x^2-x-2}$, $H(-2)$, $H(-1)$

8. $G(x) = \dfrac{1}{x^3-x^2}$, $G(0)$, $G(-1)$

Reduce each rational expression to its lowest terms.

9. $\dfrac{a^3bc^3}{a^5b^2c}$

10. $\dfrac{x^4-1}{3x^2-3}$

11. $\dfrac{68x^3}{51xy}$

12. $\dfrac{5x^2-15x+10}{5x-10}$

6.2 Multiplication and Division

Perform the indicated operations.

13. $\dfrac{a^3b^2}{b^3a} \cdot \dfrac{ab - b^2}{ab - a^2}$

14. $\dfrac{x^3 - 1}{3x} \cdot \dfrac{6x^2}{x - 1}$

15. $\dfrac{w - 4}{3w} \div \dfrac{2w - 8}{9w}$

16. $\dfrac{x^3 - xy^2}{y} \div \dfrac{x^3 + 2x^2y + xy^2}{3y}$

6.3 Addition and Subtraction

Find the least common multiple for each group of polynomials.

17. $6x, 3x - 6, x^2 - 2x$

18. $x^3 - 8, x^2 - 4, 2x + 4$

19. $6ab^3, 4a^5b^2$

20. $4x^2 - 9, 4x^2 + 12x + 9$

Perform the indicated operations.

21. $\dfrac{3}{2x - 6} + \dfrac{1}{x^2 - 9}$

22. $\dfrac{3}{x - 3} - \dfrac{5}{x + 4}$

23. $\dfrac{w}{ab^2} - \dfrac{5}{a^2b}$

24. $\dfrac{x}{x - 1} + \dfrac{3x}{x^2 - 1}$

25. $\dfrac{-x - 17}{x^2 + 6x + 5} + \dfrac{4}{x + 1}$

26. $\dfrac{10x + 11}{2x^2 + 5x + 2} - \dfrac{4}{2x + 1}$

6.4 Complex Fractions

Simplify the complex fractions.

27. $\dfrac{\dfrac{3}{2x} - \dfrac{4}{5x}}{\dfrac{1}{3} - \dfrac{2}{x}}$

28. $\dfrac{\dfrac{5}{x - 2} - \dfrac{4}{4 - x^2}}{\dfrac{3}{x + 2} - \dfrac{1}{2 - x}}$

29. $\dfrac{\dfrac{1}{y - 2} - 3}{\dfrac{5}{y - 2} + 4}$

30. $\dfrac{\dfrac{a}{b^2} - \dfrac{b}{a^3}}{\dfrac{a}{b} + \dfrac{b}{a^2}}$

31. $\dfrac{a^{-2} - b^{-3}}{a^{-1}b^{-2}}$

32. $p^{-1} + pq^{-2}$

6.5 Division of Polynomials

Find the quotient and remainder.

33. $(x^3 + x^2 - 11x + 10) \div (x - 2)$

34. $(2x^3 + 5x^2 + 9) \div (x + 3)$

35. $(m^4 - 1) \div (m + 1)$

36. $(x^4 - 1) \div (x - 1)$

37. $(a^9 - 8) \div (a^3 - 2)$

38. $(a^2 - b^2) \div (a - b)$

39. $(3m^3 + 6m^2 - 18m) \div (3m)$

40. $(w - 3) \div (3 - w)$

Rewrite each expression in the form

$$quotient + \dfrac{remainder}{divisor}.$$

Use synthetic division.

41. $\dfrac{x^2 - 5}{x - 1}$

42. $\dfrac{x^2 + 3x + 2}{x + 3}$

43. $\dfrac{3x}{x - 2}$

44. $\dfrac{4x}{x - 5}$

Use division to determine whether the first polynomial is a factor of the second. Use synthetic division when possible.

45. $x + 2, \quad x^3 - 2x^2 + 3x + 22$

46. $x - 2, \quad x^3 + x - 10$

47. $x - 5, \quad x^3 - x - 120$

48. $x + 3, \quad x^3 + 2x + 15$

49. $x - 1, \quad x^3 + x^2 - 3$

50. $x - 1, \quad x^3 + 1$

51. $x^2 + 2, \quad x^4 + x^3 + 5x^2 + 2x + 6$

52. $x^2 + 1, \quad x^4 - 1$

6.6 Solving Equations Involving Rational Expressions

Solve each equation.

53. $\dfrac{-3}{8} = \dfrac{2}{x}$

54. $\dfrac{2}{x} + \dfrac{5}{2x} = 1$

55. $5 + \dfrac{x + 1}{x - 1} = 3 + \dfrac{5x - 3}{x - 1}$

56. $2 + \dfrac{7}{x - 5} = 3 + \dfrac{x + 2}{x - 5}$

57. $\dfrac{15}{a^2 - 25} + \dfrac{1}{a - 5} = \dfrac{6}{a + 5}$

58. $2 + \dfrac{3}{x - 5} = \dfrac{x - 1}{x - 5}$

6.7 Applications

Solve each formula for the indicated variable.

59. $\dfrac{y - b}{m} = x$ for y

60. $\dfrac{2A}{h} = b_1 + b_2$ for A

61. $\dfrac{1}{x} + \dfrac{1}{2} = w$ for x

62. $\dfrac{1}{x} + \dfrac{1}{a} = 2$ for x

63. $F = \dfrac{mv^2}{r}$ for m

64. $P = \dfrac{A}{1 + rt}$ for r

65. $A = \dfrac{2}{3}\pi rh$ for r

66. $\dfrac{a}{w^2} = \dfrac{2}{b}$ for b

67. $\dfrac{y + 3}{x - 7} = 2$ for y

68. $\dfrac{y - 5}{x + 4} = \dfrac{-1}{2}$ for y

69. $\dfrac{1}{p} + \dfrac{1}{q} = \dfrac{1}{f}$ for q

70. $\dfrac{2}{a} + \dfrac{1}{3b} = \dfrac{1}{2}$ for a

Solve each problem.

71. *AIDS by gender.* The ratio of new reported male AIDS cases to female AIDS cases in 2008 was 15 to 7 (Centers for Disease Control, www.cdc.gov). See the accompanying figure. If there were 16,000 more male AIDS cases than female AIDS cases, then how many reported male AIDS cases were there in 2008?

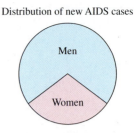

Distribution of new AIDS cases

Men

Women

Figure for Exercise 71

72. *Aggressive portfolio.* In an aggressive portfolio the ratio of money invested in stocks to money invested in bonds should be 5 to 1. If Halle has an aggressive portfolio with $20,972 more invested in stocks than bonds, then how much does she have in her portfolio?

73. *Just passing through.* Nikita drove 310 miles on his way to Louisville in the same amount of time that he drove 360 miles after passing through Louisville. If his average speed after passing Louisville was 10 miles per hour (mph) more than his average speed on his way to Louisville, then for how many hours did he drive?

310 mi
Average speed = x mph

Louisville

360 mi
Average speed = $x + 10$ mph

Figure for Exercise 73

74. *Pushing a barge.* A tug can push a barge 144 miles down the Mississippi River in the same time that it takes to push the barge 84 miles in the Gulf of Mexico. If the tug's speed is 5 mph greater going down the river, then what is its speed in the Gulf of Mexico?

Photo for Exercise 74

75. *Quilting bee.* Debbie can make a hand-sewn quilt in 2000 hours, and Rosalina can make an identical quilt in 1000 hours. If Cheryl works just as fast as Rosalina, then how long will it take all three of them working together to make one quilt?

76. *Blood out of a turnip.* A small pump can pump all of the blood out of an average turnip in 30 minutes. A larger pump can pump all of the blood from the same turnip in 20 minutes. If both pumps are hooked to the turnip, then how long would it take to get all of the blood out?

77. *Picking apples.* Trine picks a bushel of apples in x minutes on the average, whereas Thud's average is 6 minutes less per bushel.

    a) Write a rational function $B(x)$ that gives the number of bushels that they pick in two hours while working together.

    b) Find $B(12)$.

78. *Hiking time.* Lines hiked for 3 miles uphill averaging x miles per hour. For the next 4 miles he was going downhill and his average speed increased by 1 mile per hour.

    a) Write a rational function $T(x)$ that gives the total time in hours for this hike.

    b) Find $T(3.5)$ to the nearest minute.

Miscellaneous

Either perform the indicated operation or solve the equation, whichever is appropriate.

79. $\dfrac{5x}{3x^2 y} + \dfrac{7a^2}{6a^2 x}$

80. $\dfrac{4}{2x - 4} - \dfrac{15}{5x}$

81. $\dfrac{5}{a - 5} - \dfrac{3}{-a - 5}$

82. $\dfrac{2}{x - 2} - \dfrac{3}{x} = \dfrac{-1}{5x}$

83. $\dfrac{1}{x - 2} - \dfrac{1}{x + 2} = \dfrac{1}{15}$

84. $\dfrac{2}{x - 3} \cdot \dfrac{6x - 18}{30}$

85. $\dfrac{-3}{x + 2} \cdot \dfrac{5x + 10}{10}$

86. $\dfrac{x}{10} = \dfrac{10}{x}$

87. $\dfrac{x}{-3} = \dfrac{-27}{x}$

88. $\dfrac{x^2 - 4}{x} \div \dfrac{x^3 - 8}{x}$

89. $\dfrac{wx + wm + 3x + 3m}{w^2 - 9} \div \dfrac{x^2 - m^2}{w - 3}$

90. $\dfrac{-5}{7} = \dfrac{3}{x}$

91. $\dfrac{5}{a^2 - 25} + \dfrac{3}{a^2 - 4a - 5}$

92. $\dfrac{3}{w^2 - 1} + \dfrac{2}{2w + 2}$

93. $\dfrac{-7}{2a^2 - 18} - \dfrac{4}{a^2 + 5a + 6}$

94. $\dfrac{-5}{3a^2 - 12} - \dfrac{1}{a^2 - 3a + 2}$

95. $\dfrac{7}{a^2 - 1} + \dfrac{2}{1 - a} = \dfrac{1}{a + 1}$

96. $2 + \dfrac{4}{x - 1} = \dfrac{3x + 1}{x - 1}$

97. $\dfrac{2x}{x - 3} + \dfrac{3}{x - 2} = \dfrac{6}{(x - 2)(x - 3)}$

98. $\dfrac{a - 3}{a + 3} \div \dfrac{9 - a^2}{3}$

99. $\dfrac{x - 2}{6} \div \dfrac{2 - x}{2}$

100. $\dfrac{x}{x+4} - \dfrac{2}{x+1} = \dfrac{-2}{(x+1)(x+4)}$

101. $\dfrac{x-3}{x^2+3x+2} \cdot \dfrac{x^2-4}{3x-9}$

102. $\dfrac{x^2-1}{x^2+2x+1} \cdot \dfrac{x^3+1}{2x-2}$

103. $\dfrac{a+4}{a^3-8} - \dfrac{3}{2-a}$

104. $\dfrac{x+2}{5} = \dfrac{3}{x}$

105. $\dfrac{x^3-9x}{1-x^2} \div \dfrac{x^3+6x^2+9x}{x-1}$

106. $\dfrac{x+3}{2x+3} = \dfrac{x-3}{x-1}$

107. $\dfrac{a^2+3a+3w+aw}{a^2+6a+8} \cdot \dfrac{a^2-aw-2w+2a}{a^2+3a-3w-aw}$

108. $\dfrac{3}{4-2y} + \dfrac{6}{y^2-4} + \dfrac{3}{2+y}$

109. $\dfrac{5}{x} - \dfrac{4}{x+2} = \dfrac{1}{5} + \dfrac{1}{5x}$

110. $\dfrac{1}{x} + \dfrac{1}{x-5} = \dfrac{2x+1}{x^2-25} + \dfrac{9}{x^2+5x}$

Replace each question mark by an expression that makes the equation an identity.

111. $\dfrac{6}{x} = \dfrac{?}{3x}$

112. $\dfrac{?}{a} = \dfrac{8}{4a}$

113. $\dfrac{3}{a-b} = \dfrac{?}{b-a}$

114. $\dfrac{-2}{a-x} = \dfrac{2}{?}$

115. $4 = \dfrac{?}{x}$

116. $5a = \dfrac{?}{b}$

117. $5x \div \dfrac{1}{2} = ?$

118. $3a \div \dfrac{1}{a} = ?$

119. $4a \div ? = 12a$

120. $14x \div ? = 28x^2$

121. $\dfrac{a-3}{a^2-9} = \dfrac{1}{?}$

122. $\dfrac{?}{x^2-4} = \dfrac{1}{x-2}$

123. $\dfrac{1}{2} - \dfrac{1}{5} = ?$

124. $\dfrac{1}{4} - \dfrac{1}{5} = ?$

125. $\dfrac{a}{3} + \dfrac{a}{2} = ?$

126. $\dfrac{x}{5} + \dfrac{x}{3} = ?$

Chapter 6 Test

State the domain of each rational expression.

1. $\dfrac{5}{4-3x}$

2. $\dfrac{2x-1}{x^2-9}$

3. $\dfrac{17}{x^2+9}$

Find the indicated values of the given rational function if possible.

4. $R(x) = \dfrac{2x-1}{x^2-9}$, $R\left(\dfrac{1}{2}\right)$, $R(3)$

Reduce to lowest terms.

5. $\dfrac{12a^9b^8}{(2a^2b^3)^3}$

6. $\dfrac{y^2-x^2}{2x^2-4xy+2y^2}$

Perform the indicated operations. Write answers in lowest terms.

7. $\dfrac{5y}{12y} - \dfrac{4x}{9x}$

8. $\dfrac{3}{y} + 7y$

9. $\dfrac{4}{a-9} - \dfrac{1}{9-a}$

10. $\dfrac{1}{6ab^2} + \dfrac{1}{8a^2b}$

11. $\dfrac{3a^3b}{20ab} \cdot \dfrac{2a^2b}{9ab^3}$

12. $\dfrac{a-b}{7} \div \dfrac{b^2 - a^2}{21}$

13. $\dfrac{x-3}{x-1} \div \left(x^2 - 2x - 3\right)$

14. $\dfrac{2}{x^2 - 4} - \dfrac{6}{x^2 - 3x - 10}$

15. $\dfrac{m^3 - 1}{(m-1)^2} \cdot \dfrac{m^2 - 1}{3m^2 + 3m + 3}$

Find the solution set to each equation.

16. $\dfrac{3}{x} = \dfrac{7}{4}$

17. $\dfrac{x}{x-2} - \dfrac{5}{x} = \dfrac{3}{4}$

18. $\dfrac{3m}{2} = \dfrac{6}{m}$

Solve each formula for the indicated variable.

19. $W = \dfrac{a^2}{t}$ for t

20. $\dfrac{1}{a} + \dfrac{1}{b} = \dfrac{1}{2}$ for b

Simplify.

21. $\dfrac{\dfrac{1}{x} + \dfrac{1}{3x}}{\dfrac{3}{4x} - \dfrac{1}{2}}$

22. $\dfrac{m^{-2} - w^{-2}}{m^{-2}w^{-1} + m^{-1}w^{-2}}$

23. $\dfrac{\dfrac{a^2b^3}{4a}}{\dfrac{ab^3}{6a^2}}$

Find the quotient and remainder.

24. $\left(6x^2 + 7x - 6\right) \div (2x + 1)$

25. $(x - 3) \div (3 - x)$

Rewrite each expression in the form

$$\text{quotient} + \frac{\text{remainder}}{\text{divisor}}.$$

Use synthetic division.

26. $\dfrac{5x}{x+3}$

27. $\dfrac{x^2 + 3x - 6}{x - 2}$

Solve each problem.

28. When Jane's wading pool was new, it could be filled in 6 minutes with water from the hose. Now that the pool has several leaks, it takes only 8 minutes for all of the water to leak out of a full pool. How long does it take to fill the leaky pool?

29. Milton and Bonnie are hiking the Appalachian Trail together. Milton averages 4 miles per hour (mph), and Bonnie averages 3 mph. If they start out together in the morning, but Milton gets to camp 2 hours and 30 minutes ahead of Bonnie, then how many miles did they each hike that day?

30. A group of sailors plans to share equally the cost and use of a $72,000 boat. If they can get three more sailors to join their group, then the cost per person will be reduced by $2000. How many sailors are in the original group?

31. *Biking time.* Katherine biked on the Katy Trail for 20 miles before lunch averaging x miles per hour. After lunch she biked for 30 miles and averaged 3 miles per hour less.

a) Write a rational function $T(x)$ that gives the total time in hours for this bike ride.

b) Find $T(12)$.

*Making***Connections** | **A Review of Chapters 1–6**

Find the solution set to each equation.

1. $\dfrac{3}{x} = \dfrac{4}{5}$

2. $\dfrac{2}{x} = \dfrac{x}{8}$

3. $\dfrac{x}{3} = \dfrac{4}{5}$

4. $\dfrac{3}{x} = \dfrac{x+3}{6}$

5. $\dfrac{1}{x} = 4$

6. $\dfrac{2}{3}x = 4$

7. $2x + 3 = 4$

8. $2x + 3 = 4x$

9. $\dfrac{2a}{3} = \dfrac{6}{a}$

10. $\dfrac{12}{x} - \dfrac{14}{x+1} = \dfrac{1}{2}$

11. $|6x - 3| = 1$

12. $\dfrac{x}{2x+9} = \dfrac{3}{x}$

13. $4(6x - 3)(2x + 9) = 0$

14. $\dfrac{x-1}{x+2} - \dfrac{1}{5(x+2)} = 1$

Solve each equation for y. Assume A, B, and C are constants for which all expressions are defined.

15. $Ax + By = C$

16. $\dfrac{y-3}{x+5} = -\dfrac{1}{3}$

17. $Ay = By + C$

18. $\dfrac{A}{y} = \dfrac{y}{A}$

19. $\dfrac{A}{y} - \dfrac{1}{2} = \dfrac{B}{y}$

20. $\dfrac{A}{y} - \dfrac{1}{2} = \dfrac{B}{C}$

21. $3x - 4y = 6$

22. $y^2 - 2y - Ay + 2A = 0$

23. $A = \dfrac{1}{2}B(C + y)$

24. $y^2 + Cy = BC + By$

Simplify each expression.

25. $3x^5 \cdot 4x^8$

26. $3x^2(x^3 + 5x^6)$

27. $(5x^6)^2$

28. $(3a^3b^2)^3$

29. $\dfrac{12a^9b^4}{-3a^3b^{-2}}$

30. $\left(\dfrac{x^{-2}}{2}\right)^5$

31. $\left(\dfrac{2x^{-4}}{3y^5}\right)^{-3}$

32. $\left(-2a^{-1}b^3c\right)^{-2}$

33. $\dfrac{a^{-1} + b^3}{a^{-2} + b^{-1}}$

34. $\dfrac{(a+b)^{-1}}{(a+b)^{-2}}$

Factor each expression completely.

35. $4p^4 + 12p^3 + 32p^2$

36. $15m^3 - 24m^2 + 9m$

37. $-12a^2 + 84a - 147$

38. $-2b^2 + 288$

39. $aq + qw + 3w + 3a$

40. $2am + 4bm - 3an - 6bn$

41. $-7x^3 - 7$

42. $2a^3 - 54$

Solve each problem.

43. The sum of the squares of two consecutive positive even integers is 164. What are the integers?

44. The difference of the squares of two consecutive positive odd integers is 56. What are the integers?

45. A cannonball is fired straight upward. Its height above the ground in feet at time t seconds is given by $h(t) = -16t^2 + 320t$.

    a) What is the height of the cannonball 2 seconds after it is fired?

    b) For how long is the cannonball in the air?

46. *Basic energy requirement.* Clinical dietitians must design diets that meet patients' energy requirements and are suitable for the condition of their health (*Snapshots of Applications in Mathematics*). The basic energy requirement B (in calories) for a male is a function of three variables,

$$B = 655 + 9.56W + 1.85H - 4.68A,$$

where W is the patient's weight in kilograms, H is the height in centimeters, and A is the age in years.

a) Find the basic energy requirement for former Chicago Bulls' center Luc Longley when he was 30 years old, had a height of 7 ft 2 in., and a weight of 292 pounds (www.nba.com). (1 in. $\approx$ 2.54 cm, 1 kg $\approx$ 2.2 lb.)

b) The accompanying graph shows the basic energy requirement for a 7 ft 2 in. male at age 30 as a function of his weight. As the weight increases, does the basic energy requirement increase or decrease?

c) What is the equation for the line in the accompanying figure?

d) Write the basic energy requirement for Luc Longley as a function of his age for $20 \leq A \leq 70$. Assume his size stays fixed.

Figure for Exercise 46

Critical **Thinking** | **For Individual or Group Work** | **Chapter 6**

These exercises can be solved by a variety of techniques, which may or may not require algebra. So be creative and think critically. Explain all answers. Answers are in the Instructor's Edition of this text.

1. ***Stacking balls.*** Identical balls, each with radius a, can be stacked in the shape of a tetrahedron as shown in the accompanying figure.

   a) Find the height of the stack if five balls are used (four on the first level and one on the second level).

   b) Find the height if 14 balls are used (nine on the first level, four on the second level, and one on the top level).

   Photo for Exercise 1

2. ***Shaq walk.*** Assume that the earth is a perfect sphere with a radius of 4000 miles, and Shaquille O'Neal who is 7 ft 1 in. tall walks around it at the equator. Would his head travel farther than his feet? If so, how much farther? If not, why not?

3. ***Passing trains.*** Two freight trains are traveling in opposite directions on parallel tracks, one at 60 mph and the other at 40 mph. If the length of the faster train is $\frac{1}{2}$ mile and the length of the slower train is $\frac{1}{4}$ mile, then how long does it take for them to pass each other?

4. ***Seventy-five.*** How many four-digit numbers contain the digit pattern 75 once and only once?

5. ***Squares and cubes.*** For some pairs of positive integers, the difference between their squares is a perfect cube. For example, $3^2 - 1^2 = 2^3$, $6^2 - 3^2 = 3^3$, and $10^2 - 6^2 = 4^3$.

   a) Find two pairs of integers whose squares differ by 5^3.

   b) Find two pairs of integers whose squares differ by 6^3.

6. ***Differing means.*** Consider the following sequence of numbers:

   $$-1, 2, -3, 4, -5, 6, -7, 8, \ldots$$

   What is the difference of the mean of the first 400 terms of the sequence and the mean of the first 500 terms of the sequence?

7. ***Shady squares.*** Divide a square into four equal squares and shade one of them. Now divide one of the unshaded squares into four equal squares and shade one of them. Keep repeating this process forever. What percent of the original square is shaded?

8. ***Inscribed square.*** A square with sides of length s is inscribed in an equilateral triangle with sides of length t, as shown in the accompanying figure. Find the exact ratio of the area of the equilateral triangle to the area of the square.

   Figure for Exercise 8

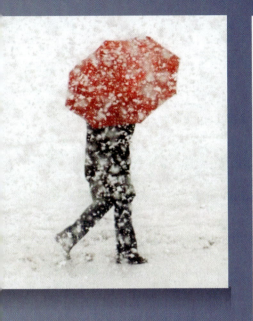

Radicals and Rational Exponents

Just how cold is it in Fargo, North Dakota, in winter? According to local meteorologists, the mercury hit a low of −33°F on January 18, 1994. But air temperature alone is not always a reliable indicator of how cold you feel. On the same date, the average wind velocity was 13.8 miles per hour. This dramatically affected how cold people felt when they stepped outside. High winds along with cold temperatures make exposed skin feel colder because the wind significantly speeds up the loss of body heat. Meteorologists use the terms "wind chill factor," "wind chill index," and "wind chill temperature" to take into account both air temperature and wind velocity.

Through experimentation in Antarctica, Paul A. Siple developed a formula in the 1940s that measures the wind chill from the velocity of the wind and the air temperature. His complex formula involving the square root of the velocity of the wind is still used today to calculate wind chill temperatures. Siple's formula is unlike most scientific formulas in that it is not based on theory. Siple experimented with various formulas involving wind velocity and temperature until he found a formula that seemed to predict how cold the air felt.

Wind chill temperature (°F) for 25°F air temperature

Wind velocity (mph)

Siple's formula is stated and used in Exercises 105 and 106 of Section 7.1.

7.1   Radicals

In This Section

⟨1⟩ **Roots**

⟨2⟩ **Roots and Variables**

⟨3⟩ **Product Rule for Radicals**

⟨4⟩ **Quotient Rule for Radicals**

⟨5⟩ **Domain of a Radical Function**

In Section 5.1, you learned the basic facts about powers. In this section, you will study roots and see how powers and roots are related.

⟨1⟩ Roots

We use the idea of roots to reverse powers. Because $3^2 = 9$ and $(-3)^2 = 9$, both 3 and -3 are square roots of 9. Because $2^4 = 16$ and $(-2)^4 = 16$, both 2 and -2 are fourth roots of 16. Because $2^3 = 8$ and $(-2)^3 = -8$, there is only one real cube root of 8 and only one real cube root of -8. The cube root of 8 is 2 and the cube root of -8 is -2.

> **nth Roots**
>
> If $a = b^n$ for a positive integer n, then b is an **nth root of a.** If $a = b^2$, then b is a **square root** of a. If $a = b^3$, then b is the **cube root** of a.

If n is a positive even integer and a is positive, then there are two real nth roots of a. We call these roots **even roots.** The positive even root of a positive number is called the **principal root.** The principal square root of 9 is 3 and the principal fourth root of 16 is 2, and these roots are even roots.

If n is a positive odd integer and a is any real number, there is only one real nth root of a. We call that root an **odd root.** Because $2^5 = 32$, the fifth root of 32 is 2 and 2 is an odd root.

We use the **radical symbol** $\sqrt{}$ to signify roots.

⟨ Helpful Hint ⟩

The parts of a radical:

Index — Radical symbol — $\sqrt[n]{a}$ — Radicand

> $\sqrt[n]{a}$
>
> If n is a positive *even* integer and a is positive, then $\sqrt[n]{a}$ denotes the *principal nth root of a.*
>
> If n is a positive *odd* integer, then $\sqrt[n]{a}$ denotes the *nth root of a.*
>
> If n is any positive integer, then $\sqrt[n]{0} = 0$.

We read $\sqrt[n]{a}$ as "the nth root of a." In the notation $\sqrt[n]{a}$, n is the **index of the radical** and a is the **radicand.** For square roots the index is omitted, and we simply write $\sqrt{a}$.

EXAMPLE   1

Evaluating radical expressions

Find the following roots:

a) $\sqrt{25}$

b) $\sqrt[3]{-27}$

c) $\sqrt[6]{64}$

d) $-\sqrt{4}$

Solution

a) Because $5^2 = 25$, $\sqrt{25} = 5$.

b) Because $(-3)^3 = -27$, $\sqrt[3]{-27} = -3$.

c) Because $2^6 = 64$, $\sqrt[6]{64} = 2$.

d) Because $\sqrt{4} = 2$, $-\sqrt{4} = -(\sqrt{4}) = -2$.

> Now do Exercises 1–16

CAUTION In radical notation, $\sqrt{4}$ represents the *principal square root of* 4, so $\sqrt{4} = 2$. Note that -2 is also a square root of 4, but $\sqrt{4} \neq -2$.

Note that even roots of negative numbers are omitted from the definition of nth roots because even powers of real numbers are never negative. So no real number can be an even root of a negative number. Expressions such as

$$\sqrt{-9}, \quad \sqrt[4]{-81}, \quad \text{and} \quad \sqrt[6]{-64}$$

are not real numbers. Square roots of negative numbers will be discussed in Section 7.6 when we discuss the imaginary numbers.

‹2› Roots and Variables

A whole number is a perfect square if it is the square of another whole number. So 9 is a perfect square because $3^2 = 9$. Likewise, an exponential expression is a perfect square if it is the square of another exponential expression. So x^{10} is a perfect square because $(x^5)^2 = x^{10}$. The exponent in a perfect square must be divisible by 2. An exponential expression is a perfect cube if it is the cube of another exponential expression. So x^{21} is a perfect cube because $(x^7)^3 = x^{21}$. The exponent in a perfect cube must be divisible by 3. The exponent in a perfect fourth power is divisible by 4, and so on.

Perfect squares	$x^2, x^4, x^6, x^8, x^{10}, x^{12}, \ldots$	Exponent divisible by 2
Perfect cubes	$x^3, x^6, x^9, x^{12}, x^{15}, x^{18}, \ldots$	Exponent divisible by 3
Perfect fourth powers	$x^4, x^8, x^{12}, x^{16}, x^{20}, x^{24}, \ldots$	Exponent divisible by 4

To find the square root of a perfect square, divide the exponent by 2. If x is *non-negative*, we have

$$\sqrt{x^2} = x, \quad \sqrt{x^4} = x^2, \quad \sqrt{x^6} = x^3, \quad \text{and so on.}$$

We specified that x was nonnegative because $\sqrt{x^2} = x$ and $\sqrt{x^6} = x^3$ are not correct if x is negative. If x is negative, x and x^3 are negative, but the radical symbol with an even root must be a positive number. Using absolute value symbols we can say that $\sqrt{x^2} = |x|$ and $\sqrt{x^6} = |x^3|$ for any real numbers.

To find the cube root of a perfect cube, divide the exponent by 3. If x is *any real number*, we have

$$\sqrt[3]{x^3} = x, \quad \sqrt[3]{x^6} = x^2, \quad \sqrt[3]{x^9} = x^3, \quad \text{and so on.}$$

Note that both sides of each of these equations have the same sign whether x is positive or negative. For cube roots and other odd roots, we will not need absolute value symbols to make statements that are true for any real numbers. *We need absolute value symbols only when the result of an even root has an odd exponent.* For example, $\sqrt[6]{m^{30}} = |m^5|$ for any real number m.

EXAMPLE 2

Roots of exponential expressions

Find each root. Assume that the variables can represent any real numbers. Use absolute value symbols when necessary.

a) $\sqrt{a^2}$ b) $\sqrt{x^{22}}$ c) $\sqrt[4]{w^{40}}$ d) $\sqrt[3]{t^{18}}$ e) $\sqrt[5]{s^{30}}$

Solution

a) For a square root, divide the exponent by 2. But if a is negative, $\sqrt{a^2} = a$ is not correct, because the square root symbol represents the nonnegative square root. So if a is any real number, $\sqrt{a^2} = |a|$.

b) Divide the exponent by 2. But if x is negative, $\sqrt{x^{22}} = x^{11}$ is not correct because x^{11} is negative and $\sqrt{x^{22}}$ is positive. So if x is any real number, $\sqrt{x^{22}} = |x^{11}|$.

c) For a fourth root, divide the exponent by 4. So $\sqrt[4]{w^{40}} = w^{10}$. We don't need absolute value symbols because both sides of this equation have the same sign whether w is positive or negative.

d) For a cube root, divide the exponent by 3. So $\sqrt[3]{t^{18}} = t^6$. We don't need absolute value symbols because both sides of this equation have the same sign whether t is positive or negative.

e) For a fifth root, divide the exponent by 5. So $\sqrt[5]{s^{30}} = s^6$. We don't need absolute value symbols because both sides of this equation have the same sign whether s is positive or negative.

> Now do Exercises 28–32

‹ **Calculator Close-Up** ›

You can illustrate the product rule for radicals with a calculator.

```
√(2)*√(3)
        2.449489743
√(6)
        2.449489743
```

‹3› Product Rule for Radicals

Consider the expression $\sqrt{2} \cdot \sqrt{3}$. If we square this product, we get

$$(\sqrt{2} \cdot \sqrt{3})^2 = (\sqrt{2})^2(\sqrt{3})^2 \quad \text{Power of a product rule}$$
$$= 2 \cdot 3 \quad (\sqrt{2})^2 = 2 \text{ and } (\sqrt{3})^2 = 3$$
$$= 6.$$

The number $\sqrt{6}$ is the unique positive number whose square is 6. Because we squared $\sqrt{2} \cdot \sqrt{3}$ and obtained 6, we must have $\sqrt{6} = \sqrt{2} \cdot \sqrt{3}$. This example illustrates the product rule for radicals.

Product Rule for Radicals

The *n*th root of a product is equal to the product of the *n*th roots. In symbols,

$$\sqrt[n]{ab} = \sqrt[n]{a} \cdot \sqrt[n]{b},$$

provided all of these roots are real numbers.

EXAMPLE **3**

Using the product rule for radicals to simplify

Simplify each radical. Assume that all variables represent nonnegative real numbers.

a) $\sqrt{4y}$ b) $\sqrt{3y^8}$ c) $\sqrt[3]{125w^2}$

Solution

a) $\sqrt{4y} = \sqrt{4} \cdot \sqrt{y}$ Product rule for radicals

$= 2\sqrt{y}$ Simplify.

b) $\sqrt{3y^8} = \sqrt{3} \cdot \sqrt{y^8}$ Product rule for radicals

$= \sqrt{3} \cdot y^4$ Simplify.

$= y^4\sqrt{3}$ A radical is usually written last in a product.

c) $\sqrt[3]{125w^2} = \sqrt[3]{125} \cdot \sqrt[3]{w^2} = 5\sqrt[3]{w^2}$

> Now do Exercises 33–44

In Example 4, we simplify by factoring the radicand before applying the product rule.

EXAMPLE **4**

Using the product rule to simplify

Simplify each radical.

a) $\sqrt{12}$ b) $\sqrt[3]{54}$ c) $\sqrt[4]{80}$ d) $\sqrt[5]{64}$

Solution

a) Since $12 = 4 \cdot 3$ and 4 is a perfect square, we can factor and then apply the product rule:

$$\sqrt{12} = \sqrt{4 \cdot 3} = \sqrt{4} \cdot \sqrt{3} = 2\sqrt{3}$$

b) Since $54 = 27 \cdot 2$ and 27 is a perfect cube, we can factor and then apply the product rule:

$$\sqrt[3]{54} = \sqrt[3]{27 \cdot 2} = \sqrt[3]{27} \cdot \sqrt[3]{2} = 3\sqrt[3]{2}$$

c) Since $80 = 16 \cdot 5$ and 16 is a perfect fourth power, we can factor and then apply the product rule:

$$\sqrt[4]{80} = \sqrt[4]{16 \cdot 5} = \sqrt[4]{16} \cdot \sqrt[4]{5} = 2\sqrt[4]{5}$$

d) $\sqrt[5]{64} = \sqrt[5]{32 \cdot 2} = \sqrt[5]{32} \cdot \sqrt[5]{2} = 2\sqrt[5]{2}$

> Now do Exercises 45–58

In general, we simplify radical expressions of index n by using the product rule to remove any perfect nth powers from the radicand. In Example 5, we use the product rule to simplify more radicals involving variables. Remember x^n is a perfect square if n is divisible by 2, a perfect cube if n is divisible by 3, and so on.

EXAMPLE 5

Using the product rule to simplify

Simplify each radical. Assume that all variables represent nonnegative real numbers.

a) $\sqrt{20x^3}$ b) $\sqrt[3]{40a^8}$ c) $\sqrt[4]{48a^4b^{11}}$ d) $\sqrt[5]{w^7}$

Solution

a) Factor $20x^3$ so that all possible perfect squares are inside one radical:

$$\sqrt{20x^3} = \sqrt{4x^2 \cdot 5x} \quad \text{Factor out perfect squares.}$$
$$= \sqrt{4x^2} \cdot \sqrt{5x} \quad \text{Product rule}$$
$$= 2x\sqrt{5x} \quad \text{Simplify.}$$

b) Factor $40a^8$ so that all possible perfect cubes are inside one radical:

$$\sqrt[3]{40a^8} = \sqrt[3]{8a^6 \cdot 5a^2} \quad \text{Factor out perfect cubes.}$$
$$= \sqrt[3]{8a^6} \cdot \sqrt[3]{5a^2} \quad \text{Product rule}$$
$$= 2a^2\sqrt[3]{5a^2} \quad \text{Simplify.}$$

c) Factor $48a^4b^{11}$ so that all possible perfect fourth powers are inside one radical:

$$\sqrt[4]{48a^4b^{11}} = \sqrt[4]{16a^4b^8 \cdot 3b^3} \quad \text{Factor out perfect fourth powers.}$$
$$= \sqrt[4]{16a^4b^8} \cdot \sqrt[4]{3b^3} \quad \text{Product rule}$$
$$= 2ab^2\sqrt[4]{3b^3} \quad \text{Simplify.}$$

d) $\sqrt[5]{w^7} = \sqrt[5]{w^5 \cdot w^2} = \sqrt[5]{w^5} \cdot \sqrt[5]{w^2} = w\sqrt[5]{w^2}$

> **Now do Exercises 59–72**

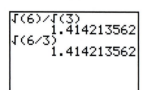

‹4› Quotient Rule for Radicals

Because $\sqrt{2} \cdot \sqrt{3} = \sqrt{6}$, we have $\sqrt{6} \div \sqrt{3} = \sqrt{2}$, or

$$\sqrt{2} = \sqrt{\frac{6}{3}} = \frac{\sqrt{6}}{\sqrt{3}}.$$

This example illustrates the quotient rule for radicals.

> **Quotient Rule for Radicals**
>
> The nth root of a quotient is equal to the quotient of the nth roots. In symbols,
>
> $$\sqrt[n]{\frac{a}{b}} = \frac{\sqrt[n]{a}}{\sqrt[n]{b}},$$
>
> provided that all of these roots are real numbers and $b \neq 0$.

EXAMPLE **6**

Using the quotient rule for radicals

Simplify each radical. Assume that all variables represent positive real numbers.

a) $\sqrt{\dfrac{25}{9}}$ **b)** $\dfrac{\sqrt{15}}{\sqrt{3}}$ **c)** $\sqrt[3]{\dfrac{b}{125}}$ **d)** $\sqrt[3]{\dfrac{x^{21}}{y^6}}$

Solution

a) $\sqrt{\dfrac{25}{9}} = \dfrac{\sqrt{25}}{\sqrt{9}}$ Quotient rule for radicals

$\phantom{\sqrt{\dfrac{25}{9}}} = \dfrac{5}{3}$ Simplify.

b) $\dfrac{\sqrt{15}}{\sqrt{3}} = \sqrt{\dfrac{15}{3}}$ Quotient rule for radicals

$\phantom{\dfrac{\sqrt{15}}{\sqrt{3}}} = \sqrt{5}$ Simplify.

c) $\sqrt[3]{\dfrac{b}{125}} = \dfrac{\sqrt[3]{b}}{\sqrt[3]{125}} = \dfrac{\sqrt[3]{b}}{5}$

d) $\sqrt[3]{\dfrac{x^{21}}{y^6}} = \dfrac{\sqrt[3]{x^{21}}}{\sqrt[3]{y^6}} = \dfrac{x^7}{y^2}$

> **Now do Exercises 73–84**

In Example 7, we use the product and quotient rule to simplify radical expressions.

EXAMPLE **7**

Using the product and quotient rules for radicals

Simplify each radical. Assume that all variables represent positive real numbers.

a) $\sqrt{\dfrac{50}{49}}$ **b)** $\sqrt[3]{\dfrac{x^5}{8}}$ **c)** $\sqrt[4]{\dfrac{a^5}{b^8}}$

Solution

a) $\sqrt{\dfrac{50}{49}} = \dfrac{\sqrt{25} \cdot \sqrt{2}}{\sqrt{49}}$ Product and quotient rules for radicals

$\phantom{\sqrt{\dfrac{50}{49}}} = \dfrac{5\sqrt{2}}{7}$ Simplify.

b) $\sqrt[3]{\dfrac{x^5}{8}} = \dfrac{\sqrt[3]{x^3} \cdot \sqrt[3]{x^2}}{\sqrt[3]{8}} = \dfrac{x\sqrt[3]{x^2}}{2}$

c) $\sqrt[4]{\dfrac{a^5}{b^8}} = \dfrac{\sqrt[4]{a^4} \cdot \sqrt[4]{a}}{\sqrt[4]{b^8}} = \dfrac{a\sqrt[4]{a}}{b^2}$

> **Now do Exercises 85–96**

⟨5⟩ Domain of a Radical Function

A function defined using a radical is called a **radical function.** So $f(x) = \sqrt{x}$ and $g(x) = \sqrt[3]{x}$ are radical functions. The domain of any function is the set of all real numbers that can be used in place of x. Since $\sqrt{x}$ is a real number only if $x \geq 0$, the domain of $f(x) = \sqrt{x}$ is the set of nonnegative real numbers, $[0, \infty)$. Since every real number has a real cube root, the domain of $g(x) = \sqrt[3]{x}$ is the set of all real numbers, $(-\infty, \infty)$. The radicand in an odd root can be any real number, but in an even root the radicand must be nonnegative. So the domain of a radical function depends on the radicand and whether the root is even or odd.

E X A M P L E   8

Finding the domain of a radical function

Find the domain of each function.

a) $f(x) = \sqrt{x - 5}$ b) $f(x) = \sqrt[3]{x + 7}$ c) $f(x) = \sqrt[4]{2x + 6}$

Solution

a) Since the radicand in a square root must be nonnegative, $x - 5$ must be nonnegative:

$$x - 5 \geq 0$$
$$x \geq 5$$

So only values of x that are 5 or larger can be used for x. The domain is $[5, \infty)$.

b) Since any real number has a cube root, any real number can be used in place of x in $\sqrt[3]{x + 7}$. So the domain is $(-\infty, \infty)$.

c) Since the radicand in a fourth root must be nonnegative, $2x + 6$ must be nonnegative:

$$2x + 6 \geq 0$$
$$2x \geq -6$$
$$x \geq -3$$

So the domain of $f(x) = \sqrt[4]{2x + 6}$ is $[-3, \infty)$.

> **Now do Exercises 97–104**

Warm-Ups ▼

Fill in the blank.

1. If $b^n = a$, then b is an _____ of a.

2. If n is even and $a > 0$, then $\sqrt[n]{a}$ is the _____ nth root of a.

3. According to the _____ rule for radicals $\sqrt[n]{a} \cdot \sqrt[n]{b} = \sqrt[n]{ab}$ provided all of the roots are real.

4. According to the _____ rule for radicals $\sqrt[n]{a}/\sqrt[n]{b} = \sqrt[n]{a/b}$ provided all of the roots are real.

True or false?

5. $\sqrt{2} \cdot \sqrt{2} = 2$

6. $\sqrt[3]{2} \cdot \sqrt[3]{2} = 2$

7. $\sqrt[3]{-27} = -3$

8. $\sqrt[4]{16} = 2$

9. $\sqrt{9} = \pm 3$

10. $\sqrt{2} \cdot \sqrt{7} = \sqrt{14}$

11. $\dfrac{\sqrt{6}}{\sqrt{2}} = \sqrt{3}$

12. $\dfrac{\sqrt{10}}{2} = \sqrt{5}$

‹1› Roots

Find each root. See Example 1.

1. $\sqrt{36}$ 2. $\sqrt{49}$

3. $\sqrt{100}$ 4. $\sqrt{81}$

5. $-\sqrt{9}$ 6. $-\sqrt{25}$

7. $\sqrt[3]{8}$ 8. $\sqrt[3]{27}$

9. $\sqrt[3]{-8}$ 10. $\sqrt[3]{-1}$

11. $\sqrt[5]{32}$ 12. $\sqrt[4]{81}$

13. $\sqrt[3]{1000}$ 14. $\sqrt[4]{16}$

15. $\sqrt[4]{-16}$ 16. $\sqrt{-1}$

‹2› Roots and Variables

Find each root. See Example 2. Variables represent any real numbers. Use absolute value when necessary.

17. $\sqrt{m^2}$ 18. $\sqrt{m^6}$

19. $\sqrt{x^{16}}$ 20. $\sqrt{y^{36}}$

21. $\sqrt[5]{y^{15}}$ 22. $\sqrt[4]{m^8}$

23. $\sqrt[3]{y^{15}}$ 24. $\sqrt{m^8}$

25. $\sqrt[3]{m^3}$ 26. $\sqrt[4]{x^4}$

27. $\sqrt[4]{w^{12}}$ 28. $\sqrt[5]{a^{30}}$

29. $\sqrt{b^{18}}$ 30. $\sqrt[6]{m^{42}}$

31. $\sqrt[4]{y^{24}}$ 32. $\sqrt{t^{44}}$

‹3› Product Rule for Radicals

Use the product rule for radicals to simplify each expression. See Example 3. All variables represent nonnegative real numbers.

33. $\sqrt{9y}$ 34. $\sqrt{16n}$

35. $\sqrt{4a^2}$ 36. $\sqrt{36n^2}$

37. $\sqrt{x^4y^2}$ 38. $\sqrt{w^6t^2}$

39. $\sqrt{5m^{12}}$ 40. $\sqrt{7z^{16}}$

41. $\sqrt[3]{8y}$ 42. $\sqrt[3]{27z^2}$

43. $\sqrt[3]{3a^6}$ 44. $\sqrt[3]{5b^9}$

Use the product rule to simplify. See Example 4.

45. $\sqrt{20}$ 46. $\sqrt{18}$

47. $\sqrt{50}$ 48. $\sqrt{45}$

49. $\sqrt{72}$ 50. $\sqrt{98}$

51. $\sqrt[3]{40}$ 52. $\sqrt[3]{24}$

53. $\sqrt[3]{81}$ 54. $\sqrt[3]{250}$

55. $\sqrt[4]{48}$ 56. $\sqrt[4]{32}$

57. $\sqrt[5]{96}$ 58. $\sqrt[5]{2430}$

Use the product rule to simplify. See Example 5. All variables represent nonnegative real numbers.

59. $\sqrt{a^3}$ 60. $\sqrt{b^5}$

61. $\sqrt{18a^6}$ 62. $\sqrt{12x^8}$

63. $\sqrt{20x^5y}$ 64. $\sqrt{8w^3y^3}$

65. $\sqrt[3]{24m^4}$ 66. $\sqrt[3]{54ab^5}$

67. $\sqrt[4]{32a^5}$ 68. $\sqrt[4]{162b^4}$

69. $\sqrt[5]{64x^6}$ 70. $\sqrt[5]{96a^8}$

71. $\sqrt{48x^3y^8z^7}$ 72. $\sqrt[3]{48x^3y^8z^7}$

‹4› Quotient Rule for Radicals

Simplify each radical. See Example 6. All variables represent positive real numbers.

73. $\sqrt{\dfrac{t}{4}}$ 74. $\sqrt{\dfrac{w}{36}}$

75. $\sqrt{\dfrac{625}{16}}$ 76. $\sqrt{\dfrac{9}{144}}$

77. $\dfrac{\sqrt{30}}{\sqrt{3}}$ 78. $\dfrac{\sqrt{50}}{\sqrt{2}}$

79. $\sqrt[3]{\dfrac{t}{8}}$ 80. $\sqrt[3]{\dfrac{a}{27}}$

81. $\sqrt[3]{\dfrac{-8x^6}{y^3}}$ 82. $\sqrt[3]{\dfrac{-27y^{36}}{1000}}$

83. $\sqrt{\dfrac{4a^6}{9}}$

84. $\sqrt{\dfrac{9a^2}{49b^4}}$

Use the product and quotient rules to simplify. See Example 7. All variables represent positive real numbers.

85. $\sqrt{\dfrac{12}{25}}$

86. $\sqrt{\dfrac{8}{81}}$

87. $\sqrt{\dfrac{27}{16}}$

88. $\sqrt{\dfrac{98}{9}}$

89. $\sqrt[3]{\dfrac{a^4}{125}}$

90. $\sqrt[3]{\dfrac{b^7}{1000}}$

91. $\sqrt[3]{\dfrac{81}{8b^3}}$

92. $\sqrt[3]{\dfrac{a^3b^4}{125}}$

93. $\sqrt[4]{\dfrac{x^7}{y^8}}$

94. $\sqrt[4]{\dfrac{x^5y^4}{z^{12}}}$

95. $\sqrt[4]{\dfrac{a^5}{16b^{12}}}$

96. $\sqrt[4]{\dfrac{a^7b}{81c^{16}}}$

‹5› Domain of a Radical Function

Find the domain of each radical function. See Example 8.

97. $f(x) = \sqrt{x - 2}$

98. $f(x) = \sqrt{2 - x}$

99. $f(x) = \sqrt[3]{3x - 7}$

100. $f(x) = \sqrt[3]{5 - 4x}$

101. $f(x) = \sqrt[4]{9 - 3x}$

102. $f(x) = \sqrt[4]{4x - 8}$

103. $f(x) = \sqrt{2x + 1}$

104. $f(x) = \sqrt{4x - 1}$

Applications

Solve each problem.

105. *Wind chill.* The wind chill temperature W (how cold the air feels) is a function of the air temperature t and the wind velocity v. Through experimentation in Antarctica, Paul Siple developed a formula for W:

$$W = 91.4 - \frac{(10.5 + 6.7\sqrt{v} - 0.45v)(457 - 5t)}{110},$$

where W and t are in degrees Fahrenheit and v is in miles per hour (mph).

a) Find W to the nearest whole degree when $t = 25°F$ and $v = 20$ mph.

b) Use the accompanying graph to estimate W when $t = 25°F$ and $v = 30$ mph.

Figure for Exercise 105

106. *Comparing wind chills.* Use the formula from Exercise 105 to determine who will feel colder: a person in Minneapolis at 10°F with a 15-mph wind or a person in Chicago at 20°F with a 25-mph wind.

 107. *Diving time.* The time t (in seconds) that it takes for a cliff diver to reach the water is a function of the height h (in feet) from which he dives:

$$t = \sqrt{\frac{h}{16}}$$

Figure for Exercise 107

a) Use the properties of radicals to simplify this formula.

b) Find the exact time (according to the formula) that it takes for a diver to hit the water when diving from a height of 40 feet.

c) Use the graph to estimate the height if a diver takes 2.5 seconds to reach the water.

108. *Sky diving.* The formula in Exercise 107 accounts for the effect of gravity only on a falling object. According to that formula, how long would it take a sky diver to reach the earth when jumping from 17,000 feet? (A sky diver can actually get about twice as much falling time by spreading out and using the air to slow the fall.)

 109. *Maximum sailing speed.* To find the maximum possible speed in knots (nautical miles per hour) for a sailboat, sailors use the function $M = 1.3\sqrt{w}$, where w is the length of the waterline in feet. If the waterline for the sloop *Golden Eye* is 20 feet, then what is the maximum speed of the *Golden Eye*?

 110. *America's Cup.* Since 1988 basic yacht dimensions for the America's Cup competition have satisfied the inequality

$$L + 1.25\sqrt{S} - 9.8\sqrt[3]{D} \le 16.296,$$

where L is the boat's length in meters (m), S is the sail area in square meters, and D is the displacement in cubic meters (www.sailing.com). A team of naval architects is planning to build a boat with a displacement of 21.44 cubic meters (m³), a sail area of 320.13 square meters (m²), and a length of 21.22 m. Does this boat satisfy the inequality? If the length and displacement of this boat cannot be changed, then how many square meters of sail area must be removed so that the boat satisfies the inequality?

 111. *Landing speed.* The proper landing speed for an airplane V (in feet per second) is a function of the gross weight of the aircraft L (in pounds), the coefficient of lift C, and the wing surface area S (in square feet), given by

$$V = \sqrt{\frac{841L}{CS}}.$$

a) Find V (to the nearest tenth) for the Piper Cheyenne, for which $L = 8700$ lb, $C = 2.81$, and $S = 200$ ft².

b) Find V in miles per hour (to the nearest tenth).

112. *Landing speed and weight.* Because the gross weight of the Piper Cheyenne depends on how much fuel and cargo are on board, the proper landing speed (from Exercise 111) is not always the same. The function $V = \sqrt{1.496L}$ gives the landing speed as a function of the gross weight only.

a) Find the landing speed if the gross weight is 7000 lb.

b) What gross weight corresponds to a landing speed of 115 ft/sec?

Getting More Involved

113. *Cooperative learning*

Work in a group to determine whether each equation is an identity. Explain your answers.

a) $\sqrt{x^2} = |x|$ **b)** $\sqrt[3]{x^3} = |x|$

c) $\sqrt{x^4} = x^2$ **d)** $\sqrt[4]{x^4} = |x|$

For which values of n is $\sqrt[n]{x^n} = x$ an identity?

114. *Cooperative learning*

Work in a group to determine whether each inequality is correct.

a) $\sqrt{0.9} > 0.9$

b) $\sqrt{1.01} > 1.01$

c) $\sqrt[3]{0.99} > 0.99$

d) $\sqrt[3]{1.001} > 1.001$

For which values of x and n is $\sqrt[n]{x} > x$?

115. *Discussion*

If your test scores are 80 and 100, then the arithmetic mean of your scores is 90. The geometric mean of the scores is a number h such that

$$\frac{80}{h} = \frac{h}{100}.$$

Are you better off with the arithmetic mean or the geometric mean?

Math *at Work*   Deficit and Debt

Have you ever heard politicians talk about budget surpluses and lowering the deficit, while the national debt keeps increasing? The national debt has increased every year since 1967 and stood at $11.3 trillion in 2009. Confusing? Not if you know the definitions of these words. If the federal government spends more than it collects in taxes in a particular year, then it has a *deficit.* The amount that is overspent must be borrowed, and that adds to the *national debt,* which is the total amount that the federal government owes. Interest alone on the national debt was $676 billion in 2009 and is the second largest expense in the federal budget.

To get an idea of the size of the national debt, divide the $11.3 trillion debt in 2009 by the U.S. population of 306 million to get about $37,000 per person. The national debt went from $2.4 trillion in 1987 to $11.3 trillion in 2009. We can calculate the average annual percentage increase in the debt for these 22 years using the formula $i = \sqrt[n]{A/P} - 1$, which yields $i = \sqrt[22]{11.3/2.4} - 1 \approx 7.3\%$. With the U.S. population increasing an average of 1% per year and the debt increasing 7.3% per year, in 25 years the debt will be $11.3(1 + 0.073)^{25}$ or about $65.8 trillion while the population will increase to $306(1 + 0.01)^{25}$ or about 392 million. See the accompanying figure. So in 25 years the debt will be about $168,000 per person. Since only one person in three is a wage earner, the debt will be about one-half of a million dollars per wage earner!

7.2   Rational Exponents

You have learned how to use exponents to express powers of numbers and radicals to express roots. In this section, you will see that roots can be expressed with exponents also. The advantage of using exponents to express roots is that the rules of exponents can be applied to the expressions.

In This Section

⟨1⟩ **Rational Exponents**

⟨2⟩ **Using the Rules of Exponents**

⟨3⟩ **Simplifying Expressions Involving Variables**

⟨1⟩ Rational Exponents

Cubing and cube root are inverse operations. For example, if we start with 2 and apply both operations, we get back 2: $\sqrt[3]{2^3} = 2$. If we were to use an exponent for cube root, then we must have $(2^3)^? = 2$. The only exponent that is consistent with the power of a power rule is $\frac{1}{3}$ because $(2^3)^{1/3} = 2^1 = 2$. So we make the following definition.

⟨ **Calculator Close-Up** ⟩

You can find the fifth root of 2 using radical notation or exponent notation. Note that the fractional exponent 1/5 must be in parentheses.

```
5 ˣ √ ( 2 )
               1.148698355
2^(1/5)
               1.148698355
2^.2
               1.148698355
```

Definition of $a^{1/n}$

If n is any positive integer, then

$$a^{1/n} = \sqrt[n]{a},$$

provided that $\sqrt[n]{a}$ is a real number.

Later in this section we will see that using exponent $1/n$ for the nth root is compatible with the rules for integral exponents that we already know.

EXAMPLE 1

Radicals or exponents

Write each radical expression using exponent notation and each exponential expression using radical notation.

   a) $\sqrt[3]{35}$ **b)** $\sqrt[4]{xy}$ **c)** $5^{1/2}$ **d)** $a^{1/5}$

Solution

   a) $\sqrt[3]{35} = 35^{1/3}$ **b)** $\sqrt[4]{xy} = (xy)^{1/4}$

   c) $5^{1/2} = \sqrt{5}$ **d)** $a^{1/5} = \sqrt[5]{a}$

> Now do Exercises 1–8

In Example 2, we evaluate some exponential expressions.

EXAMPLE 2

Finding roots

Evaluate each expression.

   a) $4^{1/2}$ **b)** $(-8)^{1/3}$ **c)** $81^{1/4}$

   d) $(-9)^{1/2}$ **e)** $-9^{1/2}$

Solution

   a) $4^{1/2} = \sqrt{4} = 2$

   b) $(-8)^{1/3} = \sqrt[3]{-8} = -2$

   c) $81^{1/4} = \sqrt[4]{81} = 3$

   d) Because $(-9)^{1/2}$ or $\sqrt{-9}$ is an even root of a negative number, it is not a real number.

   e) Because the exponent in $-a^n$ is applied only to the base a (Section 1.4), we have $-9^{1/2} = -\sqrt{9} = -3$.

> Now do Exercises 9–16

We now extend the definition of exponent $1/n$ to include any rational number as an exponent. The numerator of the rational number indicates the power, and the denominator indicates the root. For example, the expression

$$8^{2/3} \quad \begin{array}{l} \leftarrow \text{Power} \\ \leftarrow \text{Root} \end{array}$$

represents the square of the cube root of 8. So we have

$$8^{2/3} = \left(8^{1/3}\right)^2 = (2)^2 = 4.$$

‹ Helpful Hint ›

Note that in $a^{m/n}$ we do not require m/n to be reduced. As long as the nth root of a is real, then the value of $a^{m/n}$ is the same whether or not m/n is in lowest terms.

Definition of $a^{m/n}$

If m and n are positive integers and $a^{1/n}$ is a real number, then

$$a^{m/n} = \left(a^{1/n}\right)^m.$$

Using radical notation, $a^{m/n} = \left(\sqrt[n]{a}\right)^m$.

By definition $a^{m/n}$ is the mth power of the nth root of a. However, $a^{m/n}$ is also equal to the nth root of the mth power of a. For example,

$$8^{2/3} = (8^2)^{1/3} = 64^{1/3} = 4.$$

> **Evaluating $a^{m/n}$ in Either Order**
>
> If m and n are positive integers and $a^{1/n}$ is a real number, then
>
> $$a^{m/n} = (a^{1/n})^m = (a^m)^{1/n}.$$
>
> Using radical notation, $a^{m/n} = (\sqrt[n]{a})^m = \sqrt[n]{a^m}$.

A negative rational exponent indicates a reciprocal:

> **Definition of $a^{-m/n}$**
>
> If m and n are positive integers, $a \neq 0$, and $a^{1/n}$ is a real number, then
>
> $$a^{-m/n} = \frac{1}{a^{m/n}}.$$
>
> Using radical notation, $a^{-m/n} = \dfrac{1}{(\sqrt[n]{a})^m}$.

EXAMPLE 3

Radicals to exponents

Write each radical expression using exponent notation.

a) $\sqrt[3]{x^2}$

b) $\dfrac{1}{\sqrt[4]{m^3}}$

Solution

a) $\sqrt[3]{x^2} = x^{2/3}$

b) $\dfrac{1}{\sqrt[4]{m^3}} = \dfrac{1}{m^{3/4}} = m^{-3/4}$

> Now do Exercises 17–20

EXAMPLE 4

Exponents to radicals

Write each exponential expression using radicals.

a) $5^{2/3}$

b) $a^{-2/5}$

Solution

a) $5^{2/3} = \sqrt[3]{5^2}$

b) $a^{-2/5} = \dfrac{1}{\sqrt[5]{a^2}}$

> Now do Exercises 21–24

To evaluate an expression with a negative rational exponent, remember that the denominator indicates root, the numerator indicates power, and the negative sign indicates reciprocal:

The root, power, and reciprocal can be evaluated in any order. However, it is usually simplest to use the following strategy.

Strategy for Evaluating $a^{-m/n}$

1. Find the nth root of a.

2. Raise your result to the mth power.

3. Find the reciprocal.

For example, to evaluate $8^{-2/3}$, we find the cube root of 8 (which is 2), square 2 to get 4, then find the reciprocal of 4 to get $\frac{1}{4}$. In print $8^{-2/3}$ could be written for evaluation as $\left((8^{1/3})^2\right)^{-1}$ or $\frac{1}{(8^{1/3})^2}$.

EXAMPLE 5

Rational exponents
Evaluate each expression.

    a) $27^{2/3}$ **b)** $4^{-3/2}$ **c)** $81^{-3/4}$ **d)** $(-8)^{-5/3}$

Solution

a) Because the exponent is 2/3, we find the cube root of 27 and then square it:

$$27^{2/3} = (27^{1/3})^2 = 3^2 = 9$$

b) Because the exponent is $-3/2$, we find the square root of 4, cube it, and find the reciprocal:

$$4^{-3/2} = \frac{1}{(4^{1/2})^3} = \frac{1}{2^3} = \frac{1}{8}$$

c) Because the exponent is $-3/4$, we find the fourth root of 81, cube it, and find the reciprocal:

$$81^{-3/4} = \frac{1}{(81^{1/4})^3} = \frac{1}{3^3} = \frac{1}{27} \quad \text{\color{blue}Definition of negative exponent}$$

d) $(-8)^{-5/3} = \dfrac{1}{((-8)^{1/3})^5} = \dfrac{1}{(-2)^5} = \dfrac{1}{-32} = -\dfrac{1}{32}$

> **Now do Exercises 25–36**

‹ **Calculator Close-Up** ›

A negative fractional exponent indicates a reciprocal, a root, and a power. To find $4^{-3/2}$ you can find the reciprocal first, the square root first, or the third power first as shown here.

```
(1/4)^(3/2)
              .125
(√(4))^-3
              .125
(4³)^(-1/2)
              .125
```

CAUTION An expression with a negative base and a negative exponent can have a positive or a negative value. For example,

$$(-8)^{-5/3} = -\frac{1}{32} \quad \text{and} \quad (-8)^{-2/3} = \frac{1}{4}.$$

⟨2⟩ Using the Rules of Exponents

All of the rules for integral exponents that you learned in Sections 5.1 and 5.2 hold for rational exponents as well. We restate those rules in the following box. Note that some expressions with rational exponents [such as $(-3)^{3/4}$] are not real numbers and the rules do not apply to such expressions.

Rules for Rational Exponents

The following rules hold for any nonzero real numbers a and b and rational numbers r and s for which the expressions represent real numbers.

1. $a^r a^s = a^{r+s}$ Product rule
2. $\dfrac{a^r}{a^s} = a^{r-s}$ Quotient rule
3. $(a^r)^s = a^{rs}$ Power of a power rule
4. $(ab)^r = a^r b^r$ Power of a product rule
5. $\left(\dfrac{a}{b}\right)^r = \dfrac{a^r}{b^r}$ Power of a quotient rule

---

We can use the product rule to add rational exponents. For example,

$$16^{1/4} \cdot 16^{1/4} = 16^{2/4}.$$

The fourth root of 16 is 2, and 2 squared is 4. So $16^{2/4} = 4$. Because we also have $16^{1/2} = 4$, we see that a rational exponent can be reduced to its lowest terms. If an exponent can be reduced, it is usually simpler to reduce the exponent before we evaluate the expression. We can simplify $16^{1/4} \cdot 16^{1/4}$ as follows:

$$16^{1/4} \cdot 16^{1/4} = 16^{2/4} = 16^{1/2} = 4$$

E X A M P L E   6

Using the product and quotient rules with rational exponents

Simplify each expression.

a) $27^{1/6} \cdot 27^{1/2}$

b) $\dfrac{5^{3/4}}{5^{1/4}}$

Solution

a) $27^{1/6} \cdot 27^{1/2} = 27^{1/6+1/2}$ Product rule for exponents

$\qquad\qquad\qquad = 27^{2/3}$

$\qquad\qquad\qquad = 9$

b) $\dfrac{5^{3/4}}{5^{1/4}} = 5^{3/4-1/4} = 5^{2/4} = 5^{1/2} = \sqrt{5}$ We used the quotient rule to subtract the exponents.

Now do Exercises 37–44

EXAMPLE **7**

Using the power rules with rational exponents
Simplify each expression.

a) $3^{1/2} \cdot 12^{1/2}$ **b)** $(3^{10})^{1/2}$ **c)** $\left(\dfrac{2^6}{3^9}\right)^{-1/3}$

Solution

a) Because the bases 3 and 12 are different, we cannot use the product rule to add the exponents. Instead, we use the power of a product rule to place the 1/2 power outside the parentheses:

$$3^{1/2} \cdot 12^{1/2} = (3 \cdot 12)^{1/2} = 36^{1/2} = 6$$

b) Use the power of a power rule to multiply the exponents:

$$(3^{10})^{1/2} = 3^5$$

c) $\left(\dfrac{2^6}{3^9}\right)^{-1/3} = \dfrac{(2^6)^{-1/3}}{(3^9)^{-1/3}}$ Power of a quotient rule

$\qquad\qquad = \dfrac{2^{-2}}{3^{-3}}$ Power of a power rule

$\qquad\qquad = \dfrac{3^3}{2^2}$ Definition of a negative exponent

$\qquad\qquad = \dfrac{27}{4}$

> Now do Exercises 45–54

‹ Helpful Hint ›

We usually think of squaring and taking a square root as inverse operations, which they are as long as we stick to positive numbers. We can square 3 to get 9, and then find the square root of 9 to get 3—what we started with. We don't get back to where we began if we start with −3.

‹3› Simplifying Expressions Involving Variables

When simplifying expressions involving rational exponents and variables, we must be careful to write equivalent expressions. For example, in the equation

$$(x^2)^{1/2} = x$$

it looks as if we are correctly applying the power of a power rule. However, this statement is false if x is negative because the 1/2 power on the left-hand side indicates the positive square root of x^2. For example, if $x = -3$, we get

$$[(-3)^2]^{1/2} = 9^{1/2} = 3,$$

which is not equal to −3. To write a simpler equivalent expression for $(x^2)^{1/2}$, we use absolute value as follows.

Square Root of x^2

For any real number x,

$$(x^2)^{1/2} = |x| \quad \text{and} \quad \sqrt{x^2} = |x|.$$

Note that both $(x^2)^{1/2} = |x|$ and $\sqrt{x^2} = |x|$ are identities. They are true whether x is positive, negative, or zero.

It is also necessary to use absolute value when writing identities for other even roots of expressions involving variables.

EXAMPLE 8

Using absolute value symbols with roots

Simplify each expression. Assume the variables represent any real numbers and use absolute value symbols as necessary.

a) $\left(x^8 y^4\right)^{1/4}$

b) $\left(\dfrac{x^9}{8}\right)^{1/3}$

Solution

a) Apply the power of a product rule to get the equation $\left(x^8 y^4\right)^{1/4} = x^2 y$. The left-hand side is nonnegative for any choices of x and y, but the right-hand side is negative when y is negative. So for any real values of x and y we have

$$\left(x^8 y^4\right)^{1/4} = x^2 |y|.$$

Note that the absolute value symbols could also be placed around the entire expression: $\left(x^8 y^4\right)^{1/4} = |x^2 y|$.

b) Using the power of a quotient rule, we get

$$\left(\frac{x^9}{8}\right)^{1/3} = \frac{x^3}{2}.$$

This equation is valid for every real number x, so no absolute value signs are used.

> **Now do Exercises 55–64**

Because there are no real even roots of negative numbers, the expressions

$$a^{1/2}, \quad x^{-3/4}, \quad \text{and} \quad y^{1/6}$$

are not real numbers if the variables have negative values. To simplify matters, we sometimes assume the variables represent only positive numbers when we are working with expressions involving variables with rational exponents. That way we do not have to be concerned with undefined expressions and absolute value.

EXAMPLE 9

Expressions involving variables with rational exponents

Use the rules of exponents to simplify the following. Write your answers with positive exponents. Assume all variables represent *positive* real numbers.

a) $x^{2/3} x^{4/3}$

b) $\dfrac{a^{1/2}}{a^{1/4}}$

c) $\left(x^{1/2} y^{-3}\right)^{1/2}$

d) $\left(\dfrac{x^2}{y^{1/3}}\right)^{-1/2}$

Solution

a) $x^{2/3} x^{4/3} = x^{6/3}$ Use the product rule to add the exponents.

$\qquad\qquad = x^2$ Reduce the exponent.

b) $\dfrac{a^{1/2}}{a^{1/4}} = a^{1/2 - 1/4}$ Use the quotient rule to subtract the exponents.

$\qquad\quad = a^{1/4}$ Simplify.

c) $\left(x^{1/2} y^{-3}\right)^{1/2} = \left(x^{1/2}\right)^{1/2}\left(y^{-3}\right)^{1/2}$ Power of a product rule

$\qquad\qquad\quad = x^{1/4} y^{-3/2}$ Power of a power rule

$\qquad\qquad\quad = \dfrac{x^{1/4}}{y^{3/2}}$ Definition of a negative exponent

d) Because this expression is a negative power of a quotient, we can first find the reciprocal of the quotient and then apply the power of a power rule:

$$\left(\frac{x^2}{y^{1/3}}\right)^{-1/2} = \left(\frac{y^{1/3}}{x^2}\right)^{1/2} = \frac{y^{1/6}}{x} \qquad \frac{1}{3} \cdot \frac{1}{2} = \frac{1}{6}$$

> **Now do Exercises 65–76**

Warm-Ups ▼

Fill in the blank.

1. The notation $a^{1/n}$ represents the _____ of a.

2. The notation $a^{m/n}$ represents the _____ of the nth root.

3. The expression $a^{-m/n}$ is the _____ of $a^{m/n}$.

4. The expression $a^{-m/n}$ is a real number except when n is ____ and a is _____, or when $a = 0$.

True or false?

5. $9^{1/3} = \sqrt[3]{9}$

6. $8^{5/3} = \sqrt[5]{8^3}$

7. $(-16)^{1/2} = -16^{1/2}$

8. $9^{-3/2} = \dfrac{1}{27}$

9. $6^{-1/2} = \dfrac{\sqrt{6}}{6}$

10. $2^{1/2} \cdot 2^{1/2} = 4^{1/2}$

11. $6^{1/6} \cdot 6^{1/6} = 6^{1/3}$

12. $(2^8)^{3/4} = 2^6$

Exercises 7.2

‹ **Study Tips** ›

• Avoid cramming. When you have limited time to study for a test, start with class notes and homework assignments. Work one or two problems of each type.

• Don't get discouraged if you cannot work the hardest problems. Instructors often ask some relatively easy questions to see if you understand the basics.

‹ 1 › **Rational Exponents**

Write each radical expression using exponent notation. See Example 1.

1. $\sqrt[4]{7}$

2. $\sqrt[3]{cbs}$

3. $\sqrt{5x}$

4. $\sqrt{3y}$

Write each exponential expression using radical notation. See Example 1.

5. $9^{1/5}$

6. $3^{1/2}$

7. $a^{1/2}$

8. $(-b)^{1/5}$

Evaluate each expression. See Example 2.

9. $25^{1/2}$

10. $16^{1/2}$

11. $(-125)^{1/3}$

12. $(-32)^{1/5}$

13. $16^{1/4}$

14. $8^{1/3}$

15. $(-4)^{1/2}$

16. $(-16)^{1/4}$

Write each radical expression using exponent notation. See Example 3.

17. $\sqrt[3]{w^7}$

18. $\sqrt{a^5}$

19. $\dfrac{1}{\sqrt[3]{2^{10}}}$

20. $\sqrt[3]{\dfrac{1}{a^2}}$

Write each exponential expression using radical notation. See Example 4.

21. $w^{-3/4}$

22. $6^{-5/3}$

23. $(ab)^{3/2}$

24. $(3m)^{-1/5}$

Evaluate each expression. See Example 5. See the Strategy for Evaluating $a^{-m/n}$ box on page 471.

25. $125^{2/3}$

26. $1000^{2/3}$

27. $25^{3/2}$

28. $16^{3/2}$

29. $27^{-4/3}$

30. $16^{-3/4}$

31. $16^{-3/2}$

32. $25^{-3/2}$

33. $(-27)^{-1/3}$

34. $(-8)^{-4/3}$

35. $(-16)^{-1/4}$

36. $(-100)^{-3/2}$

⟨2⟩ Using the Rules of Exponents

Use the rules of exponents to simplify each expression. See Examples 6 and 7.

37. $3^{1/3}3^{1/4}$

38. $2^{1/2}2^{1/3}$

39. $3^{1/3}3^{-1/3}$

40. $5^{1/4}5^{-1/4}$

41. $\dfrac{8^{1/3}}{8^{2/3}}$

42. $\dfrac{27^{-2/3}}{27^{-1/3}}$

43. $4^{3/4} \div 4^{1/4}$

44. $9^{1/4} \div 9^{3/4}$

45. $18^{1/2}2^{1/2}$

46. $8^{1/2}2^{1/2}$

47. $(2^6)^{1/3}$

48. $(3^{10})^{1/5}$

49. $(3^8)^{1/2}$

50. $(3^{-6})^{1/3}$

51. $(2^{-4})^{1/2}$

52. $(5^4)^{1/2}$

53. $\left(\dfrac{3^4}{2^6}\right)^{1/2}$

54. $\left(\dfrac{5^4}{3^6}\right)^{1/2}$

⟨3⟩ Simplifying Expressions Involving Variables

Simplify each expression. Assume the variables represent any real numbers, and use absolute value as necessary. See Example 8.

55. $(x^4)^{1/4}$

56. $(y^6)^{1/6}$

57. $(a^8)^{1/2}$

58. $(b^{10})^{1/2}$

59. $(y^3)^{1/3}$

60. $(w^9)^{1/3}$

61. $(9x^6y^2)^{1/2}$

62. $(16a^8b^4)^{1/4}$

63. $\left(\dfrac{81x^{12}}{y^{20}}\right)^{1/4}$

64. $\left(\dfrac{144a^8}{9y^{18}}\right)^{1/2}$

Simplify. Assume all variables represent positive numbers. Write answers with positive exponents only. See Example 9.

65. $x^{1/2}x^{1/4}$

66. $y^{1/3}y^{1/3}$

67. $(x^{1/2}y)(x^{-3/4}y^{1/2})$

68. $(a^{1/2}b^{-1/3})(ab)$

69. $\dfrac{w^{1/3}}{w^3}$

70. $\dfrac{a^{1/2}}{a^2}$

71. $(144x^{16})^{1/2}$

72. $(125a^8)^{1/3}$

73. $\left(\dfrac{a^{-1/2}}{b^{-1/4}}\right)^{-4}$

74. $\left(\dfrac{2a^{1/2}}{b^{1/3}}\right)^6$

75. $\left(\dfrac{2w^{1/3}}{w^{-3/4}}\right)^3$

76. $\left(\dfrac{a^{-1/2}}{3a^{2/3}}\right)^{-3}$

Miscellaneous

Simplify each expression. Write your answers with positive exponents. Assume that all variables represent positive real numbers.

77. $(9^2)^{1/2}$

78. $(4^{16})^{1/2}$

79. $-16^{-3/4}$

80. $-25^{-3/2}$

81. $125^{-4/3}$

82. $27^{-2/3}$

83. $2^{1/2}2^{-1/4}$

84. $9^{-1}9^{1/2}$

85. $3^{0.26}3^{0.74}$

86. $2^{1.5}2^{0.5}$

87. $3^{1/4}27^{1/4}$

88. $3^{2/3}9^{2/3}$

89. $\left(-\dfrac{8}{27}\right)^{2/3}$

90. $\left(-\dfrac{8}{27}\right)^{-1/3}$

91. $\left(-\dfrac{1}{16}\right)^{-3/4}$

92. $\left(-\dfrac{5}{9}\right)^{-7/2}$

93. $\left(\dfrac{9}{16}\right)^{-1/2}$

94. $\left(\dfrac{16}{81}\right)^{-1/4}$

95. $-\left(\dfrac{25}{36}\right)^{-3/2}$

96. $\left(-\dfrac{27}{8}\right)^{-4/3}$

97. $\left(9x^9\right)^{1/2}$

98. $\left(-27x^9\right)^{1/3}$

99. $\left(3a^{-2/3}\right)^{-3}$

100. $\left(5x^{-1/2}\right)^{-2}$

101. $\left(a^{1/2}b\right)^{1/2}\left(ab^{1/2}\right)$

102. $\left(m^{1/4}n^{1/2}\right)^2\left(m^2n^3\right)^{1/2}$

103. $\left(km^{1/2}\right)^3\left(k^3m^5\right)^{1/2}$

104. $\left(tv^{1/3}\right)^2\left(t^2v^{-3}\right)^{-1/2}$

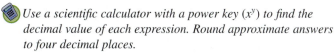 *Use a scientific calculator with a power key (x^y) to find the decimal value of each expression. Round approximate answers to four decimal places.*

105. $2^{1/3}$

106. $5^{1/2}$

107. $-2^{1/2}$

108. $(-3)^{1/3}$

109. $1024^{1/10}$

110. $7776^{0.2}$

111. $\left(\dfrac{64}{15{,}625}\right)^{-1/6}$

112. $\left(\dfrac{32}{243}\right)^{-3/5}$

Simplify each expression. Assume a and b are positive real numbers and m and n are rational numbers.

113. $a^{m/2} \cdot a^{m/4}$

114. $b^{n/2} \cdot b^{-n/3}$

115. $\dfrac{a^{-m/5}}{a^{-m/3}}$

116. $\dfrac{b^{-n/4}}{b^{-n/3}}$

117. $\left(a^{-1/m}b^{-1/n}\right)^{-mn}$

118. $\left(a^{-m/2}b^{-n/3}\right)^{-6}$

119. $\left(\dfrac{a^{-3m}b^{-6n}}{a^{9m}}\right)^{-1/3}$

120. $\left(\dfrac{a^{-3/m}b^{6/n}}{a^{-6/m}b^{9/n}}\right)^{-1/3}$

Applications

Solve each problem. Round answers to two decimal places when necessary.

 121. *Diagonal of a box.* The length of the diagonal of a box D is a function of its length L, width W, and height H:

$$D = \left(L^2 + W^2 + H^2\right)^{1/2}$$

a) Find D for the box shown in the accompanying figure.

b) Find D if $L = W = H = 1$ inch.

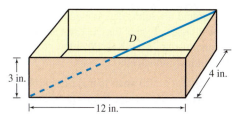

Figure for Exercise 121

122. *Radius of a sphere.* The radius of a sphere is a function of its volume, given by the formula

$$r = \left(\dfrac{0.75V}{\pi}\right)^{1/3},$$

where V is its volume. Find the radius of a spherical tank that has a volume of $\dfrac{32\pi}{3}$ cubic meters.

Figure for Exercise 122

 123. *Maximum sail area.* According to the new International America's Cup Class Rules, the maximum sail area in square meters for a yacht in the America's Cup race is given by the function

$$S = \left(13.0368 + 7.84D^{1/3} - 0.8L\right)^2,$$

where D is the displacement in cubic meters (m^3), and L is the length in meters (m) (www.sailing.com). Find the maximum sail area for a boat that has a displacement of 18.42 m^3 and a length of 21.45 m. See the figure on the next page.

Figure for Exercise 123

124. *Orbits of the planet.* According to Kepler's third law of planetary motion, the average radius R of the orbit of a planet around the sun is determined by $R = T^{2/3}$, where T is the number of years for one orbit and R is measured in astronomical units or AUs (Windows to the Universe, www.windows.umich.edu).

a) It takes Mars 1.881 years to make one orbit of the sun. What is the average radius (in AUs) of the orbit of Mars?

b) The average radius of the orbit of Saturn is 9.5 AU. Use the accompanying graph to estimate the number of years it takes Saturn to make one orbit of the sun.

Figure for Exercise 124

125. *Top stock fund.* The average annual return r is a function of the initial investment P, the number of years n, and the amount S that it is worth after n years:

$$r = \left(\frac{S}{P}\right)^{1/n} - 1$$

An investment of $10,000 in the T. Rowe Price Latin America Fund in 2004 was worth $20,733 in 2009

(www.money.com). Find the 5-year average annual return.

126. *Top bond fund.* An investment of $10,000 in the Templeton Global Bond Fund in 2004 was worth $14,789 in 2009 (www.money.com). Use the formula from Exercise 125 to find the 5-year average annual return.

127. *Overdue loan payment.* In 1777 a wealthy Pennsylvania merchant, Jacob DeHaven, lent $450,000 to the Continental Congress to rescue the troops at Valley Forge. The loan was not repaid. In 1990 DeHaven's descendants filed suit for $141.6 billion (*New York Times*, May 27, 1990). What average annual rate of return were they using to calculate the value of the debt after 213 years? (See Exercise 125.)

128. *California growin'.* The population of California grew from 19.9 million in 1970 to 32.5 million in 2000 (U.S. Census Bureau, www.census.gov). Find the average annual rate of growth for that time period. (Use the formula from Exercise 125 with P being the initial population and S being the population n years later.)

Figure for Exercise 128

Getting More Involved

129. *Discussion*

Determine whether each equation is an identity. Explain.

a) $\left(w^2 x^2\right)^{1/2} = |w| \cdot |x|$

b) $\left(w^2 x^2\right)^{1/2} = |wx|$

c) $\left(w^2 x^2\right)^{1/2} = w|x|$

7.3 Adding, Subtracting, and Multiplying Radicals

In this section, we will use the ideas of Section 7.1 in performing arithmetic operations with radical expressions.

In This Section

⟨1⟩ Adding and Subtracting Radicals

⟨2⟩ Multiplying Radicals

⟨3⟩ Conjugates

⟨4⟩ Multiplying Radicals with Different Indices

⟨1⟩ Adding and Subtracting Radicals

To find the sum of $\sqrt{2}$ and $\sqrt{3}$, we can use a calculator to get $\sqrt{2} \approx 1.414$ and $\sqrt{3} \approx 1.732$. (The symbol $\approx$ means "is approximately equal to.") We can then add the decimal numbers and get

$$\sqrt{2} + \sqrt{3} \approx 1.414 + 1.732 = 3.146.$$

We cannot write an exact decimal form for $\sqrt{2} + \sqrt{3}$; the number 3.146 is an approximation of $\sqrt{2} + \sqrt{3}$. To represent the exact value of $\sqrt{2} + \sqrt{3}$, we just use the form $\sqrt{2} + \sqrt{3}$. This form cannot be simplified any further. However, a sum of like radicals can be simplified. **Like radicals** are radicals that have the same index and the same radicand.

To simplify the sum $3\sqrt{2} + 5\sqrt{2}$, we can use the fact that $3x + 5x = 8x$ is true for any value of x. Substituting $\sqrt{2}$ for x gives us $3\sqrt{2} + 5\sqrt{2} = 8\sqrt{2}$. So like radicals can be combined just as like terms are combined.

EXAMPLE 1

Adding and subtracting like radicals

Simplify the following expressions. Assume the variables represent positive numbers.

a) $3\sqrt{5} + 4\sqrt{5}$ **b)** $\sqrt[4]{w} - 6\sqrt[4]{w}$

c) $\sqrt{3} + \sqrt{5} - 4\sqrt{3} + 6\sqrt{5}$ **d)** $3\sqrt[3]{6x} + 2\sqrt[3]{x} + \sqrt[3]{6x} + \sqrt[3]{x}$

Solution

a) $3\sqrt{5} + 4\sqrt{5} = 7\sqrt{5}$ **b)** $\sqrt[4]{w} - 6\sqrt[4]{w} = -5\sqrt[4]{w}$

c) $\sqrt{3} + \sqrt{5} - 4\sqrt{3} + 6\sqrt{5} = -3\sqrt{3} + 7\sqrt{5}$ Only like radicals are combined.

d) $3\sqrt[3]{6x} + 2\sqrt[3]{x} + \sqrt[3]{6x} + \sqrt[3]{x} = 4\sqrt[3]{6x} + 3\sqrt[3]{x}$

> Now do Exercises 1–12

Remember that *only radicals with the same index and same radicand can be combined by addition or subtraction.* If the radicals are not in simplified form, then they must be simplified before you can determine whether they can be combined.

EXAMPLE 2

Simplifying radicals before combining

Perform the indicated operations. Assume the variables represent positive numbers.

a) $\sqrt{8} + \sqrt{18}$ **b)** $\sqrt{2x^3} - \sqrt{4x^2} + 5\sqrt{18x^3}$

c) $\sqrt[3]{16x^4y^3} - \sqrt[3]{54x^4y^3}$

Solution

a) $\sqrt{8} + \sqrt{18} = \sqrt{4} \cdot \sqrt{2} + \sqrt{9} \cdot \sqrt{2}$

$\qquad\qquad = 2\sqrt{2} + 3\sqrt{2}$ Simplify each radical.

$\qquad\qquad = 5\sqrt{2}$ Add like radicals.

$\qquad$ Note that $\sqrt{8} + \sqrt{18} \neq \sqrt{26}$.

b) $\sqrt{2x^3} - \sqrt{4x^2} + 5\sqrt{18x^3} = \sqrt{x^2} \cdot \sqrt{2x} - 2x + 5 \cdot \sqrt{9x^2} \cdot \sqrt{2x}$

$\qquad\qquad\qquad = x\sqrt{2x} - 2x + 15x\sqrt{2x}$ Simplify each radical.

$\qquad\qquad\qquad = 16x\sqrt{2x} - 2x$ Add like radicals only.

c) $\sqrt[3]{16x^4y^3} - \sqrt[3]{54x^4y^3} = \sqrt[3]{8x^3y^3} \cdot \sqrt[3]{2x} - \sqrt[3]{27x^3y^3} \cdot \sqrt[3]{2x}$

$\qquad\qquad\qquad = 2xy\sqrt[3]{2x} - 3xy\sqrt[3]{2x}$ Simplify each radical.

$\qquad\qquad\qquad = -xy\sqrt[3]{2x}$

> Now do Exercises 13–28

‹2› **Multiplying Radicals**

The product rule for radicals, $\sqrt[n]{a} \cdot \sqrt[n]{b} = \sqrt[n]{ab}$, allows multiplication of radicals with the same index, such as

$$\sqrt{5} \cdot \sqrt{3} = \sqrt{15}, \qquad \sqrt[3]{2} \cdot \sqrt[3]{5} = \sqrt[3]{10}, \qquad \text{and} \qquad \sqrt[5]{x^2} \cdot \sqrt[5]{x} = \sqrt[5]{x^3}.$$

CAUTION   The product rule does not allow multiplication of radicals that have different indices. We cannot use the product rule to multiply $\sqrt{2}$ and $\sqrt[3]{5}$.

E X A M P L E   **3**

Multiplying radicals with the same index

Multiply and simplify the following expressions. Assume the variables represent positive numbers.

a) $5\sqrt{6} \cdot 4\sqrt{3}$ $\qquad\qquad\qquad$ **b)** $\sqrt{3a^2} \cdot \sqrt{6a}$

c) $\sqrt[3]{4} \cdot \sqrt[3]{4}$ $\qquad\qquad\qquad$ **d)** $\sqrt[4]{\dfrac{x^3}{2}} \cdot \sqrt[4]{\dfrac{x^2}{8}}$

Solution

a) $5\sqrt{6} \cdot 4\sqrt{3} = 5 \cdot 4 \cdot \sqrt{6} \cdot \sqrt{3}$

$\qquad\qquad = 20\sqrt{18}$ Product rule for radicals

$\qquad\qquad = 20 \cdot 3\sqrt{2}$ $\sqrt{18} = \sqrt{9} \cdot \sqrt{2} = 3\sqrt{2}$

$\qquad\qquad = 60\sqrt{2}$

b) $\sqrt{3a^2} \cdot \sqrt{6a} = \sqrt{18a^3}$ Product rule for radicals

$\qquad\qquad = \sqrt{9a^2} \cdot \sqrt{2a}$

$\qquad\qquad = 3a\sqrt{2a}$ Simplify.

c) $\sqrt[3]{4} \cdot \sqrt[3]{4} = \sqrt[3]{16}$

$\qquad = \sqrt[3]{8} \cdot \sqrt[3]{2}$ Simplify.

$\qquad = 2\sqrt[3]{2}$

d) $\sqrt[4]{\dfrac{x^3}{2}} \cdot \sqrt[4]{\dfrac{x^2}{8}} = \sqrt[4]{\dfrac{x^5}{16}}$ Product rule for radicals

$\qquad = \dfrac{\sqrt[4]{x^4} \cdot \sqrt[4]{x}}{\sqrt[4]{16}}$ Product and quotient rules for radicals

$\qquad = \dfrac{x\sqrt[4]{x}}{2}$ Simplify.

> Now do Exercises 29–42

We find a product such as $3\sqrt{2}(4\sqrt{2} - \sqrt{3})$ by using the distributive property as we do when multiplying a monomial and a binomial. A product such as $(2\sqrt{3} + \sqrt{5})(3\sqrt{3} - 2\sqrt{5})$ can be found by using FOIL as we do for the product of two binomials.

EXAMPLE 4

Multiplying radicals

Multiply and simplify.

a) $3\sqrt{2}(4\sqrt{2} - \sqrt{3})$ 　　　　　　**b)** $\sqrt[3]{a}(\sqrt[3]{a} - \sqrt[3]{a^2})$

c) $(2\sqrt{3} + \sqrt{5})(3\sqrt{3} - 2\sqrt{5})$ 　　**d)** $(3 + \sqrt{x-9})^2$

Solution

a) $3\sqrt{2}(4\sqrt{2} - \sqrt{3}) = 3\sqrt{2} \cdot 4\sqrt{2} - 3\sqrt{2} \cdot \sqrt{3}$ Distributive property

$\qquad\qquad\qquad\qquad = 12 \cdot 2 - 3\sqrt{6}$ 　Because $\sqrt{2} \cdot \sqrt{2} = 2$ and $\sqrt{2} \cdot \sqrt{3} = \sqrt{6}$

$\qquad\qquad\qquad\qquad = 24 - 3\sqrt{6}$

b) $\sqrt[3]{a}(\sqrt[3]{a} - \sqrt[3]{a^2}) = \sqrt[3]{a^2} - \sqrt[3]{a^3}$ 　Distributive property

$\qquad\qquad\qquad\qquad = \sqrt[3]{a^2} - a$

c) $(2\sqrt{3} + \sqrt{5})(3\sqrt{3} - 2\sqrt{5})$

$\qquad\qquad\qquad \overset{\text{F}}{\overbrace{\quad\quad}} \quad \overset{\text{O}}{\overbrace{\quad\quad}} \quad \overset{\text{I}}{\overbrace{\quad\quad}} \quad \overset{\text{L}}{\overbrace{\quad\quad}}$

$= 2\sqrt{3} \cdot 3\sqrt{3} - 2\sqrt{3} \cdot 2\sqrt{5} + \sqrt{5} \cdot 3\sqrt{3} - \sqrt{5} \cdot 2\sqrt{5}$

$= 18 - 4\sqrt{15} + 3\sqrt{15} - 10$

$= 8 - \sqrt{15}$ 　Combine like radicals.

d) To square a sum, we use $(a + b)^2 = a^2 + 2ab + b^2$:

$(3 + \sqrt{x-9})^2 = 3^2 + 2 \cdot 3\sqrt{x-9} + (\sqrt{x-9})^2$

$\qquad\qquad\qquad = 9 + 6\sqrt{x-9} + x - 9$

$\qquad\qquad\qquad = x + 6\sqrt{x-9}$

> Now do Exercises 43–56

CAUTION We can't simplify $\sqrt{x - 9}$ in Example 4(d), because in general $\sqrt{a - b}$ $\neq \sqrt{a} - \sqrt{b}$. For example, $\sqrt{25 - 16} = \sqrt{9} = 3$ and $\sqrt{25} - \sqrt{16} = 1$. Find an example where $\sqrt{a + b} \neq \sqrt{a} + \sqrt{b}$.

⟨3⟩ Conjugates

Recall the special product rule $(a + b)(a - b) = a^2 - b^2$. The product of the sum $4 + \sqrt{3}$ and the difference $4 - \sqrt{3}$ can be found by using this rule:

$$(4 + \sqrt{3})(4 - \sqrt{3}) = 4^2 - (\sqrt{3})^2 = 16 - 3 = 13$$

The product of the irrational number $4 + \sqrt{3}$ and the irrational number $4 - \sqrt{3}$ is the rational number 13. For this reason the expressions $4 + \sqrt{3}$ and $4 - \sqrt{3}$ are called **conjugates** of one another. We will use conjugates in Section 7.4 to rationalize some denominators.

E X A M P L E 5

Multiplying conjugates

Find the products. Assume the variables represent positive real numbers.

a) $(2 + 3\sqrt{5})(2 - 3\sqrt{5})$

b) $(\sqrt{3} - \sqrt{2})(\sqrt{3} + \sqrt{2})$

c) $(\sqrt{2x} - \sqrt{y})(\sqrt{2x} + \sqrt{y})$

Solution

a) $(2 + 3\sqrt{5})(2 - 3\sqrt{5}) = 2^2 - (3\sqrt{5})^2$ $(a + b)(a - b) = a^2 - b^2$

$\phantom{(2 + 3\sqrt{5})(2 - 3\sqrt{5})} = 4 - 45$ $(3\sqrt{5})^2 = 9 \cdot 5 = 45$

$\phantom{(2 + 3\sqrt{5})(2 - 3\sqrt{5})} = -41$

b) $(\sqrt{3} - \sqrt{2})(\sqrt{3} + \sqrt{2}) = 3 - 2$

$\phantom{(\sqrt{3} - \sqrt{2})(\sqrt{3} + \sqrt{2})} = 1$

c) $(\sqrt{2x} - \sqrt{y})(\sqrt{2x} + \sqrt{y}) = 2x - y$

Now do Exercises 57–66

⟨4⟩ Multiplying Radicals with Different Indices

The product rule for radicals applies only to radicals with the same index. To multiply radicals with different indices we convert the radicals into exponential expressions with rational exponents. If the exponential expressions have the same base, apply the product rule for exponents $(a^m \cdot a^n = a^{m+n})$ to get a single exponential expression, and then convert back to a radical [Example 6(a)]. If the bases of the exponential expressions are different, get a common denominator for the rational exponents, convert back to radicals, and then apply the product rule for radicals $\left(\sqrt[n]{a} \cdot \sqrt[n]{b} = \sqrt[n]{ab}\right)$ to get a single radical expression [Example 6(b)].

E X A M P L E 6

⟨ **Calculator Close-Up** ⟩

Check that

$\sqrt[3]{2} \cdot \sqrt[4]{2} = \sqrt[12]{128}$.

Multiplying radicals with different indices

Write each product as a single radical expression.

a) $\sqrt[3]{2} \cdot \sqrt[4]{2}$ **b)** $\sqrt[3]{2} \cdot \sqrt{3}$

Solution

a) $\sqrt[3]{2} \cdot \sqrt[4]{2} = 2^{1/3} \cdot 2^{1/4}$ Write in exponential notation.

$= 2^{7/12}$ Product rule for exponents: $\frac{1}{3} + \frac{1}{4} = \frac{7}{12}$

$= \sqrt[12]{2^7}$ Write in radical notation.

$= \sqrt[12]{128}$

b) $\sqrt[3]{2} \cdot \sqrt{3} = 2^{1/3} \cdot 3^{1/2}$ Write in exponential notation.

$= 2^{2/6} \cdot 3^{3/6}$ Write the exponents with the LCD of 6.

$= \sqrt[6]{2^2} \cdot \sqrt[6]{3^3}$ Write in radical notation.

$= \sqrt[6]{2^2 \cdot 3^3}$ Product rule for radicals

$= \sqrt[6]{108}$ $2^2 \cdot 3^3 = 4 \cdot 27 = 108$

Now do Exercises 67–74

CAUTION Because the bases in $2^{1/3} \cdot 2^{1/4}$ are identical, we can add the exponents [Example 6(a)]. Because the bases in $2^{2/6} \cdot 3^{3/6}$ are not the same, we cannot add the exponents [Example 6(b)]. Instead, we write each factor as a sixth root and use the product rule for radicals.

 Warm-Ups ▼

Fill in the blank.

1. _____ radicals have the same index and the same radicand.

2. The _____ property is used to combine like radicals.

3. The product rule for radicals is used to multiply radicals with the same _____.

4. The _____ of $2 - \sqrt{3}$ is $2 + \sqrt{3}$.

True or false?

5. $\sqrt{3} + \sqrt{3} = \sqrt{6}$

6. $\sqrt{8} + \sqrt{2} = 3\sqrt{2}$

7. $2\sqrt{3} \cdot 3\sqrt{3} = 6\sqrt{3}$

8. $2\sqrt{5} \cdot 3\sqrt{2} = 6\sqrt{10}$

9. $\sqrt[3]{2} \cdot \sqrt[3]{2} = 2$

10. $\sqrt{2}(\sqrt{3} - \sqrt{2}) = \sqrt{6} - 2$

11. $(\sqrt{2} + \sqrt{3})^2 = 2 + 3$

12. $(\sqrt{3} - \sqrt{2})(\sqrt{3} + \sqrt{2}) = 1$

Exercises

‹1› Adding and Subtracting Radicals

All variables in the following exercises represent positive numbers. Simplify the sums and differences. Give exact answers. See Example 1.

1. $\sqrt{3} - 2\sqrt{3}$

2. $\sqrt{5} - 3\sqrt{5}$

3. $5\sqrt{7x} + 4\sqrt{7x}$

4. $3\sqrt{6a} + 7\sqrt{6a}$

5. $2\sqrt[3]{2} + 3\sqrt[3]{2}$

6. $\sqrt[3]{4} + 4\sqrt[3]{4}$

7. $\sqrt{3} - \sqrt{5} + 3\sqrt{3} - \sqrt{5}$

8. $\sqrt{2} - 5\sqrt{3} - 7\sqrt{2} + 9\sqrt{3}$

9. $\sqrt[3]{2} + \sqrt[3]{x} - \sqrt[3]{2} + 4\sqrt[3]{x}$

10. $\sqrt[3]{5y} - 4\sqrt[3]{5y} + \sqrt[3]{x} + \sqrt[3]{x}$

11. $\sqrt[3]{x} - \sqrt{2x} + \sqrt[3]{x}$

12. $\sqrt[3]{ab} + \sqrt{a} + 5\sqrt{a} + \sqrt[3]{ab}$

Simplify each expression. Give exact answers. See Example 2.

13. $\sqrt{8} + \sqrt{28}$

14. $\sqrt{12} + \sqrt{24}$

15. $\sqrt{8} + \sqrt{18}$

16. $\sqrt{12} + \sqrt{27}$

17. $2\sqrt{45} - 3\sqrt{20}$

18. $3\sqrt{50} - 2\sqrt{32}$

19. $\sqrt{2} - \sqrt{8}$

20. $\sqrt{20} - \sqrt{125}$

21. $\sqrt{45x^3} - \sqrt{18x^2} + \sqrt{50x^2} - \sqrt{20x^3}$

22. $\sqrt{12x^5} - \sqrt{18x} - \sqrt{300x^5} + \sqrt{98x}$

23. $2\sqrt[3]{24} + \sqrt[3]{81}$

24. $5\sqrt[3]{24} + 2\sqrt[3]{375}$

25. $\sqrt[4]{48} - 2\sqrt[4]{243}$

26. $\sqrt[5]{64} + 7\sqrt[5]{2}$

27. $\sqrt[3]{54t^4y^3} - \sqrt[3]{16t^4y^3}$

28. $\sqrt[3]{2000w^2z^5} - \sqrt[3]{16w^2z^5}$

‹2› Multiplying Radicals

Simplify the products. Give exact answers. See Examples 3 and 4.

29. $\sqrt{3} \cdot \sqrt{5}$

30. $\sqrt{5} \cdot \sqrt{7}$

31. $2\sqrt{5} \cdot 3\sqrt{10}$

32. $(3\sqrt{2})(-4\sqrt{10})$

33. $2\sqrt{7a} \cdot 3\sqrt{2a}$

34. $2\sqrt{5c} \cdot 5\sqrt{5}$

35. $\sqrt[4]{9} \cdot \sqrt[4]{27}$

36. $\sqrt[3]{5} \cdot \sqrt[3]{100}$

37. $(2\sqrt{3})^2$

38. $(-4\sqrt{2})^2$

39. $\sqrt{5x^3} \cdot \sqrt{8x^4}$

40. $\sqrt{3b^3} \cdot \sqrt{6b^5}$

41. $\sqrt[4]{\dfrac{x^5}{3}} \cdot \sqrt[4]{\dfrac{x^2}{27}}$

42. $\sqrt[3]{\dfrac{a^4}{2}} \cdot \sqrt[3]{\dfrac{a^3}{4}}$

43. $2\sqrt{3}(\sqrt{6} + 3\sqrt{3})$

44. $2\sqrt{5}(\sqrt{3} + 3\sqrt{5})$

45. $\sqrt{5}(\sqrt{10} - 2)$

46. $\sqrt{6}(\sqrt{15} - 1)$

47. $\sqrt[3]{3t}(\sqrt[3]{9t} - \sqrt[3]{t^2})$

48. $\sqrt[3]{2}(\sqrt[3]{12x} - \sqrt[3]{2x})$

49. $(\sqrt{3} + 2)(\sqrt{3} - 5)$

50. $(\sqrt{5} + 2)(\sqrt{5} - 6)$

51. $(\sqrt{11} - 3)(\sqrt{11} + 3)$

52. $(\sqrt{2} + 5)(\sqrt{2} + 5)$

53. $(2\sqrt{5} - 7)(2\sqrt{5} + 4)$

54. $(2\sqrt{6} - 3)(2\sqrt{6} + 4)$

55. $(2\sqrt{3} - \sqrt{6})(\sqrt{3} + 2\sqrt{6})$

56. $(3\sqrt{3} - \sqrt{2})(\sqrt{2} + \sqrt{3})$

‹3› Conjugates

Find the product of each pair of conjugates. See Example 5.

57. $(\sqrt{3} - 2)(\sqrt{3} + 2)$

58. $(7 - \sqrt{3})(7 + \sqrt{3})$

59. $(\sqrt{5} + \sqrt{2})(\sqrt{5} - \sqrt{2})$

60. $(\sqrt{6} + \sqrt{5})(\sqrt{6} - \sqrt{5})$

 61. $(2\sqrt{5} + 1)(2\sqrt{5} - 1)$

62. $(3\sqrt{2} - 4)(3\sqrt{2} + 4)$

63. $(3\sqrt{2} + \sqrt{5})(3\sqrt{2} - \sqrt{5})$

64. $(2\sqrt{3} - \sqrt{7})(2\sqrt{3} + \sqrt{7})$

65. $(5 - 3\sqrt{x})(5 + 3\sqrt{x})$

66. $(4\sqrt{y} + 3\sqrt{z})(4\sqrt{y} - 3\sqrt{z})$

‹4› Multiplying Radicals with Different Indices

Write each product as a single radical expression. See Example 6.

67. $\sqrt[3]{3} \cdot \sqrt{3}$ **68.** $\sqrt{3} \cdot \sqrt[4]{3}$

69. $\sqrt[3]{5} \cdot \sqrt[4]{5}$ **70.** $\sqrt[3]{2} \cdot \sqrt[5]{2}$

71. $\sqrt[3]{2} \cdot \sqrt{5}$ **72.** $\sqrt{6} \cdot \sqrt[3]{2}$

73. $\sqrt[3]{2} \cdot \sqrt[4]{3}$ **74.** $\sqrt[3]{3} \cdot \sqrt[4]{2}$

Miscellaneous

Simplify each expression.

75. $\sqrt{300} + \sqrt{3}$ **76.** $\sqrt{50} + \sqrt{2}$

77. $2\sqrt{5} \cdot 5\sqrt{6}$ **78.** $3\sqrt{6} \cdot 5\sqrt{10}$

79. $(3 + 2\sqrt{7})(\sqrt{7} - 2)$

80. $(2 + \sqrt{7})(\sqrt{7} - 2)$ **81.** $4\sqrt{w} \cdot 4\sqrt{w}$

82. $3\sqrt{m} \cdot 5\sqrt{m}$ **83.** $\sqrt{3x^3} \cdot \sqrt{6x^2}$

84. $\sqrt{2t^5} \cdot \sqrt{10t^4}$

85. $(2\sqrt{5} + \sqrt{2})(3\sqrt{5} - \sqrt{2})$

86. $(3\sqrt{2} - \sqrt{3})(2\sqrt{2} + 3\sqrt{3})$

87. $\dfrac{\sqrt{2}}{3} + \dfrac{\sqrt{2}}{5}$

88. $\dfrac{\sqrt{2}}{4} + \dfrac{\sqrt{3}}{5}$

89. $(5 + 2\sqrt{2})(5 - 2\sqrt{2})$

90. $(3 - 2\sqrt{7})(3 + 2\sqrt{7})$

91. $(3 + \sqrt{x})^2$

92. $(1 - \sqrt{x})^2$

93. $(5\sqrt{x} - 3)^2$

94. $(3\sqrt{a} + 2)^2$

95. $(1 + \sqrt{x + 2})^2$

96. $(\sqrt{x - 1} + 1)^2$

97. $\sqrt{4w} - \sqrt{9w}$

98. $10\sqrt{m} - \sqrt{16m}$

99. $2\sqrt{a^3} + 3\sqrt{a^3} - 2a\sqrt{4a}$

100. $5\sqrt{w^2y} - 7\sqrt{w^2y} + 6\sqrt{w^2y}$

101. $\sqrt{x^5} + 2x\sqrt{x^3}$

102. $\sqrt{8x^3} + \sqrt{50x^3} - x\sqrt{2x}$

103. $\sqrt[3]{-16x^4} + 5x\sqrt[3]{54x}$

104. $\sqrt[3]{3x^5y^7} - \sqrt[3]{24x^5y^7}$

105. $\sqrt[3]{2x} \cdot \sqrt{2x}$ **106.** $\sqrt[3]{2m} \cdot \sqrt[4]{2n}$

Applications

Solve each problem.

107. *Area of a rectangle.* Find the exact area of a rectangle that has a length of $\sqrt{6}$ feet and a width of $\sqrt{3}$ feet.

108. *Volume of a cube.* Find the exact volume of a cube with sides of length $\sqrt{3}$ meters.

 109. *Area of a trapezoid.* Find the exact area of a trapezoid with a height of $\sqrt{6}$ feet and bases of $\sqrt{3}$ feet and $\sqrt{12}$ feet.

Figure for Exercise 108   **Figure for Exercise 109**

110. *Area of a triangle.* Find the exact area of a triangle with a base of $\sqrt{30}$ meters and a height of $\sqrt{6}$ meters.

Figure for Exercise 110

Getting More Involved

111. *Discussion*

Is $\sqrt{a} + \sqrt{b} = \sqrt{a + b}$ for all values of a and b?

112. *Discussion*

Which of the following equations are identities? Explain your answers.

a) $\sqrt{9x} = 3\sqrt{x}$ **b)** $\sqrt{9 + x} = 3 + \sqrt{x}$

c) $\sqrt{x - 4} = \sqrt{x} - 2$ **d)** $\sqrt{\dfrac{x}{4}} = \dfrac{\sqrt{x}}{2}$

113. *Exploration*

Because 3 is the square of $\sqrt{3}$, a binomial such as $y^2 - 3$ is a difference of two squares.

a) Factor $y^2 - 3$ and $2a^2 - 7$ using radicals.

b) Use factoring with radicals to solve the equation $x^2 - 8 = 0$.

c) Assuming a is a positive real number, solve the equation $x^2 - a = 0$.

Mid-Chapter **Quiz** | Sections 7.1 through 7.3 | **Chapter 7**

Simplify each expression.

1. $\sqrt{64}$

2. $\sqrt[3]{-27}$

3. $\sqrt{120}$

4. $\sqrt[3]{56}$

5. $\sqrt{12x^7}$

6. $\sqrt[3]{24a^3b^{13}}$

7. $\sqrt{\dfrac{w}{16}}$

8. $\sqrt{\dfrac{8x^3}{9}}$

9. $81^{1/2}$

10. $100^{3/2}$

11. $-16^{3/2}$

12. $\left(\dfrac{5^{1/3}}{5^{-2/3}}\right)^{-3}$

Perform the indicated operations.

13. $2\sqrt{3} - 5\sqrt{6} - 4\sqrt{3} + \sqrt{6}$

14. $9\sqrt{20} - 3\sqrt{45}$

15. $3\sqrt{10} \cdot 2\sqrt{14}$

16. $(8 + \sqrt{10})(8 - \sqrt{10})$

17. $\sqrt[3]{8x^5} + \sqrt[3]{27x^5}$

Miscellaneous.

18. Find the domain of the function $f(x) = \sqrt{6 - 3x}$.

19. Find the solution set to $\sqrt{x^2} = x$.

20. Find the solution set to $(x^4)^{1/4} = |x|$.

21. Write the product $\sqrt{2} \cdot \sqrt[3]{2}$ as a single radical expression.

7.4   **Quotients, Powers, and Rationalizing Denominators**

In this section, we will continue studying operations with radicals. We will first learn how to rationalize denominators, and then we will find quotients and powers with radicals.

In This Section

⟨1⟩ **Rationalizing the Denominator**

⟨2⟩ **Simplifying Radicals**

⟨3⟩ **Dividing Radicals**

⟨4⟩ **Rationalizing Denominators Using Conjugates**

⟨5⟩ **Powers of Radical Expressions**

⟨1⟩ **Rationalizing the Denominator**

Square roots such as $\sqrt{2}$, $\sqrt{3}$, and $\sqrt{5}$ are irrational numbers. If roots of this type appear in the denominator of a fraction, it is customary to rewrite the fraction with a rational number in the denominator, or **rationalize** it. We rationalize a denominator by multiplying both the numerator and denominator by another radical that makes the denominator rational.

You can find products of radicals in two ways. By definition, $\sqrt{2}$ is the positive number that you multiply by itself to get 2. So,

$$\sqrt{2} \cdot \sqrt{2} = 2.$$

By the product rule, $\sqrt{2} \cdot \sqrt{2} = \sqrt{4} = 2$. Note that $\sqrt[3]{2} \cdot \sqrt[3]{2} = \sqrt[3]{4}$ by the product rule, but $\sqrt[3]{4} \neq 2$. By definition of a cube root,

$$\sqrt[3]{2} \cdot \sqrt[3]{2} \cdot \sqrt[3]{2} = 2.$$

E X A M P L E   1

Rationalizing the denominator

Rewrite each expression with a rational denominator.

a) $\dfrac{\sqrt{3}}{\sqrt{5}}$ b) $\dfrac{3}{\sqrt[3]{2}}$

‹ **Helpful Hint** ›

If you are going to compute the value of a radical expression with a calculator, it does not matter if the denominator is rational. However, rationalizing the denominator provides another opportunity to practice building up the denominator of a fraction and multiplying radicals.

Solution

a) Because $\sqrt{5} \cdot \sqrt{5} = 5$, multiplying both the numerator and denominator by $\sqrt{5}$ will rationalize the denominator:

$$\frac{\sqrt{3}}{\sqrt{5}} = \frac{\sqrt{3}}{\sqrt{5}} \cdot \frac{\sqrt{5}}{\sqrt{5}} = \frac{\sqrt{15}}{5} \quad \text{By the product rule, } \sqrt{3} \cdot \sqrt{5} = \sqrt{15}.$$

b) We must build up the denominator to be the cube root of a perfect cube. So we multiply by $\sqrt[3]{4}$ to get $\sqrt[3]{4} \cdot \sqrt[3]{2} = \sqrt[3]{8}$:

$$\frac{3}{\sqrt[3]{2}} = \frac{3}{\sqrt[3]{2}} \cdot \frac{\sqrt[3]{4}}{\sqrt[3]{4}} = \frac{3\sqrt[3]{4}}{\sqrt[3]{8}} = \frac{3\sqrt[3]{4}}{2}$$

> **Now do Exercises 1–8**

CAUTION To rationalize a denominator with a single square root, you simply multiply by that square root. If the denominator has a cube root, you build the denominator to a cube root of a perfect cube, as in Example 1(b). For a fourth root you build to a fourth root of a perfect fourth power, and so on.

‹2› Simplifying Radicals

When simplifying a radical expression, we have three specific conditions to satisfy. First, we use the product rule to factor out perfect nth powers from the radicand in nth roots. That is, we factor out perfect squares in square roots, perfect cubes in cube roots, and so on. For example,

$$\sqrt{72} = \sqrt{36} \cdot \sqrt{2} = 6\sqrt{2} \qquad \text{and} \qquad \sqrt[3]{24} = \sqrt[3]{8} \cdot \sqrt[3]{3} = 2\sqrt[3]{3}.$$

Second, we use the quotient rule to remove all fractions from inside a radical. For example,

$$\sqrt{\frac{2}{3}} = \frac{\sqrt{2}}{\sqrt{3}}.$$

Third, we remove radicals from denominators by rationalizing the denominator:

$$\sqrt{\frac{2}{3}} = \frac{\sqrt{2} \cdot \sqrt{3}}{\sqrt{3} \cdot \sqrt{3}} = \frac{\sqrt{6}}{3}.$$

A radical expression that satisfies the following three conditions is in *simplified radical form.*

Simplified Radical Form for Radicals of Index *n*

A radical expression of index *n* is in **simplified radical form** if it has

1. *no* perfect *n*th powers as factors of the radicand,
2. *no* fractions inside the radical, and
3. *no* radicals in the denominator.

E X A M P L E   2

Writing radical expressions in simplified radical form

Simplify.

a) $\dfrac{\sqrt{10}}{\sqrt{6}}$

b) $\sqrt[3]{\dfrac{5}{9}}$

Solution

a) To rationalize the denominator, multiply the numerator and denominator by $\sqrt{6}$:

$$\frac{\sqrt{10}}{\sqrt{6}} = \frac{\sqrt{10}}{\sqrt{6}} \cdot \frac{\sqrt{6}}{\sqrt{6}} \qquad \text{Rationalize the denominator.}$$

$$= \frac{\sqrt{60}}{6}$$

$$= \frac{\sqrt{4}\sqrt{15}}{6} \qquad \text{Remove the perfect square from } \sqrt{60}.$$

$$= \frac{2\sqrt{15}}{6}$$

$$= \frac{\sqrt{15}}{3} \qquad \text{Reduce } \frac{2}{6} \text{ to } \frac{1}{3}. \text{ Note that } \sqrt{15} \div 3 \neq \sqrt{5}.$$

b) To rationalize the denominator, build up the denominator to a cube root of a perfect cube. Because $\sqrt[3]{9} \cdot \sqrt[3]{3} = \sqrt[3]{27} = 3$, we multiply by $\sqrt[3]{3}$:

$$\sqrt[3]{\frac{5}{9}} = \frac{\sqrt[3]{5}}{\sqrt[3]{9}} \qquad \text{Quotient rule for radicals}$$

$$= \frac{\sqrt[3]{5}}{\sqrt[3]{9}} \cdot \frac{\sqrt[3]{3}}{\sqrt[3]{3}} \qquad \text{Rationalize the denominator.}$$

$$= \frac{\sqrt[3]{15}}{\sqrt[3]{27}}$$

$$= \frac{\sqrt[3]{15}}{3}$$

> **Now do Exercises 9–18**

E X A M P L E   3

Rationalizing the denominator with variables

Simplify each expression. Assume all variables represent positive real numbers.

a) $\sqrt{\dfrac{a}{b}}$ **b)** $\sqrt{\dfrac{x^3}{y^5}}$ **c)** $\sqrt[3]{\dfrac{x}{y}}$

Solution

a) $\sqrt{\dfrac{a}{b}} = \dfrac{\sqrt{a}}{\sqrt{b}}$ Quotient rule for radicals

$$= \frac{\sqrt{a} \cdot \sqrt{b}}{\sqrt{b} \cdot \sqrt{b}} \qquad \text{Rationalize the denominator.}$$

$$= \frac{\sqrt{ab}}{b}$$

b) $\sqrt{\dfrac{x^3}{y^5}} = \dfrac{\sqrt{x^3}}{\sqrt{y^5}}$ Quotient rule for radicals

$$= \frac{\sqrt{x^2} \cdot \sqrt{x}}{\sqrt{y^4} \cdot \sqrt{y}} \qquad \text{Product rule for radicals}$$

$$= \frac{x\sqrt{x}}{y^2\sqrt{y}} \qquad \text{Simplify.}$$

$$= \frac{x\sqrt{x} \cdot \sqrt{y}}{y^2\sqrt{y} \cdot \sqrt{y}} \qquad \text{Rationalize the denominator.}$$

$$= \frac{x\sqrt{xy}}{y^2 \cdot y} = \frac{x\sqrt{xy}}{y^3}$$

c) Multiply by $\sqrt[3]{y^2}$ to rationalize the denominator:

$$\sqrt[3]{\frac{x}{y}} = \frac{\sqrt[3]{x}}{\sqrt[3]{y}} = \frac{\sqrt[3]{x}}{\sqrt[3]{y}} \cdot \frac{\sqrt[3]{y^2}}{\sqrt[3]{y^2}} = \frac{\sqrt[3]{xy^2}}{\sqrt[3]{y^3}} = \frac{\sqrt[3]{xy^2}}{y}$$

> **Now do Exercises 19–28**

⟨3⟩ **Dividing Radicals**

In Section 7.3 you learned how to add, subtract, and multiply radical expressions. To divide two radical expressions, simply write the quotient as a ratio and then simplify. In general, we have

$$\sqrt[n]{a} \div \sqrt[n]{b} = \frac{\sqrt[n]{a}}{\sqrt[n]{b}} = \sqrt[n]{\frac{a}{b}},$$

provided that all expressions represent real numbers. Note that the quotient rule is applied only to radicals that have the same index.

E X A M P L E   **4**

Dividing radicals with the same index

Divide and simplify. Assume the variables represent positive numbers.

    a) $\sqrt{10} \div \sqrt{5}$ **b)** $(3\sqrt{2}) \div (2\sqrt{3})$ **c)** $\sqrt[3]{10x^2} \div \sqrt[3]{5x}$

Solution

  a) $\sqrt{10} \div \sqrt{5} = \dfrac{\sqrt{10}}{\sqrt{5}}$ $a \div b = \frac{a}{b}$, provided that $b \neq 0$.

             $= \sqrt{\dfrac{10}{5}}$ Quotient rule for radicals

             $= \sqrt{2}$ Reduce.

  b) $(3\sqrt{2}) \div (2\sqrt{3}) = \dfrac{3\sqrt{2}}{2\sqrt{3}}$

                 $= \dfrac{3\sqrt{2}}{2\sqrt{3}} \cdot \dfrac{\sqrt{3}}{\sqrt{3}}$ Rationalize the denominator.

                 $= \dfrac{3\sqrt{6}}{2 \cdot 3}$

                 $= \dfrac{\sqrt{6}}{2}$ Note that $\sqrt{6} \div 2 \neq \sqrt{3}$.

  c) $\sqrt[3]{10x^2} \div \sqrt[3]{5x} = \dfrac{\sqrt[3]{10x^2}}{\sqrt[3]{5x}}$

                 $= \sqrt[3]{\dfrac{10x^2}{5x}}$ Quotient rule for radicals

                 $= \sqrt[3]{2x}$ Reduce.

> **Now do Exercises 29–36**

Note that in Example 4(a) we applied the quotient rule to get $\sqrt{10} \div \sqrt{5} = \sqrt{2}$. In Example 4(b) we did not use the quotient rule because 2 is not evenly divisible by 3. Instead, we rationalized the denominator to get the result in simplified form.

When working with radicals, it is usually best to write them in simplified radical form before doing any operations with the radicals.

E X A M P L E   **5**

Simplifying before dividing

Divide and simplify. Assume the variables represent positive numbers.

a) $\sqrt{12} \div \sqrt{72x}$ **b)** $\sqrt[4]{16a} \div \sqrt[4]{a^5}$

Solution

a) $\sqrt{12} \div \sqrt{72x} = \dfrac{\sqrt{4} \cdot \sqrt{3}}{\sqrt{36} \cdot \sqrt{2x}}$ Factor out perfect squares.

$= \dfrac{2\sqrt{3}}{6\sqrt{2x}}$ Simplify.

$= \dfrac{\sqrt{3} \cdot \sqrt{2x}}{3\sqrt{2x} \cdot \sqrt{2x}}$ Reduce $\frac{2}{6}$ to $\frac{1}{3}$ and rationalize.

$= \dfrac{\sqrt{6x}}{6x}$ Multiply the radicals.

b) $\sqrt[4]{16a} \div \sqrt[4]{a^5} = \dfrac{\sqrt[4]{16} \cdot \sqrt[4]{a}}{\sqrt[4]{a^4} \cdot \sqrt[4]{a}}$ Factor out perfect fourth powers.

$= \dfrac{\sqrt[4]{16}}{\sqrt[4]{a^4}}$ Reduce.

$= \dfrac{2}{a}$ Simplify the radicals.

> **Now do Exercises 37–44**

In Chapter 8 it will be necessary to simplify expressions of the type found in Example 6.

E X A M P L E   **6**

Simplifying radical expressions

Simplify.

a) $\dfrac{4 - \sqrt{12}}{4}$ **b)** $\dfrac{-6 + \sqrt{20}}{-2}$

‹ **Helpful Hint** ›

The expressions in Example 6 are the types of expressions that you must simplify when learning the quadratic formula in Chapter 8.

Solution

a) First write $\sqrt{12}$ in simplified form. Then simplify the expression.

$\dfrac{4 - \sqrt{12}}{4} = \dfrac{4 - 2\sqrt{3}}{4}$ Simplify $\sqrt{12}$.

$= \dfrac{\cancel{2}(2 - \sqrt{3})}{\cancel{2} \cdot 2}$ Factor.

$= \dfrac{2 - \sqrt{3}}{2}$ Divide out the common factor.

b) $\dfrac{-6 + \sqrt{20}}{-2} = \dfrac{-6 + 2\sqrt{5}}{-2}$

$= \dfrac{-\cancel{2}(3 - \sqrt{5})}{-\cancel{2}}$

$= 3 - \sqrt{5}$

> **Now do Exercises 45–48**

CAUTION   To simplify the expressions in Example 6, you must simplify the radical, factor the numerator, and then divide out the common factors. You cannot simply "cancel" the 4's in $\frac{4 - \sqrt{12}}{4}$ or the 2's in $\frac{2 - \sqrt{3}}{2}$ because they are not common factors.

⟨4⟩ Rationalizing Denominators Using Conjugates

A simplified expression involving radicals does not have radicals in the denominator. If an expression such as $4 - \sqrt{3}$ appears in a denominator, we can multiply both the numerator and denominator by its conjugate $4 + \sqrt{3}$ to get a rational number in the denominator.

E X A M P L E   7

Rationalizing the denominator using conjugates

Write in simplified form.

a) $\dfrac{2 + \sqrt{3}}{4 - \sqrt{3}}$ **b)** $\dfrac{\sqrt{5}}{\sqrt{6} + \sqrt{2}}$

Solution

a) $\dfrac{2 + \sqrt{3}}{4 - \sqrt{3}} = \dfrac{(2 + \sqrt{3})(4 + \sqrt{3})}{(4 - \sqrt{3})(4 + \sqrt{3})}$ Multiply the numerator and denominator by $4 + \sqrt{3}$.

$\qquad = \dfrac{8 + 6\sqrt{3} + 3}{13}$ $(4 - \sqrt{3})(4 + \sqrt{3}) = 16 - 3 = 13$

$\qquad = \dfrac{11 + 6\sqrt{3}}{13}$ Simplify.

b) $\dfrac{\sqrt{5}}{\sqrt{6} + \sqrt{2}} = \dfrac{\sqrt{5}(\sqrt{6} - \sqrt{2})}{(\sqrt{6} + \sqrt{2})(\sqrt{6} - \sqrt{2})}$ Multiply the numerator and denominator by $\sqrt{6} - \sqrt{2}$.

$\qquad = \dfrac{\sqrt{30} - \sqrt{10}}{4}$ $(\sqrt{6} + \sqrt{2})(\sqrt{6} - \sqrt{2}) = 6 - 2 = 4$

> **Now do Exercises 49–58**

⟨5⟩ Powers of Radical Expressions

In Example 8, we use the power of a product rule $[(ab)^n = a^n b^n]$ and the power of a power rule $[(a^m)^n = a^{mn}]$ with radical expressions. We also use the fact that a root and a power can be found in either order. That is, $(\sqrt[n]{a})^m = \sqrt[n]{a^m}$.

E X A M P L E   8

Finding powers of rational expressions

Simplify. Assume the variables represent positive numbers.

a) $(5\sqrt{2})^3$ **b)** $(2\sqrt{x^3})^4$ **c)** $(3w\sqrt[3]{2w})^3$ **d)** $(2t\sqrt[4]{3t})^3$

Solution

a) $(5\sqrt{2})^3 = 5^3(\sqrt{2})^3$ Power of a product rule

$\qquad = 125\sqrt{8}$ $(\sqrt{2})^3 = \sqrt{2^3} = \sqrt{8}$

$\qquad = 125 \cdot 2\sqrt{2}$ $\sqrt{8} = \sqrt{4}\sqrt{2} = 2\sqrt{2}$

$\qquad = 250\sqrt{2}$

b) $(2\sqrt{x^3})^4 = 2^4(\sqrt{x^3})^4$ Power of a product rule

$= 2^4\sqrt{(x^3)^4}$ $(\sqrt[n]{a})^m = \sqrt[n]{a^m}$

$= 16\sqrt{x^{12}}$ $(a^m)^n = a^{mn}$

$= 16x^6$

c) $(3w\sqrt[3]{2w})^3 = 3^3w^3(\sqrt[3]{2w})^3$

$= 27w^3(2w)$

$= 54w^4$

d) $(2t\sqrt[4]{3t})^3 = 2^3t^3(\sqrt[4]{3t})^3 = 8t^3\sqrt[4]{27t^3}$

> Now do Exercises 59–70

Warm-Ups ▼

Fill in the blank.

1. The numbers $\sqrt{2}$, $\sqrt{3}$, and $\sqrt{5}$ are _____ numbers.

2. Writing a fraction with an irrational denominator as an equivalent fraction with a rational denominator is _____ the denominator.

3. A simplified square root expression has no perfect _____ as factors of the radicand.

4. A simplified radical expression has no fractions inside the _____.

5. A simplified radical expression has no radicals in the _____.

True or false?

6. $\dfrac{\sqrt{6}}{\sqrt{2}} = \sqrt{3}$

7. $\dfrac{2}{\sqrt{2}} = \sqrt{2}$

8. $\dfrac{4 - \sqrt{10}}{2} = 2 - \sqrt{10}$

9. $\dfrac{4 - \sqrt{10}}{2} = 2 - \sqrt{5}$

10. $\dfrac{1}{\sqrt{3}} = \dfrac{\sqrt{3}}{3}$

11. $(2\sqrt{4})^2 = 16$

12. $(3\sqrt{5})^3 = 27\sqrt{125}$

Exercises

7.4

‹ Study Tips ›

• Personal issues can have a tremendous effect on your progress in any course. If you need help, get it.
• Most schools have counseling centers that can help you to overcome personal issues that are affecting your studies.

‹ 1 › Rationalizing the Denominator

All variables in the following exercises represent positive numbers. Rewrite each expression with a rational denominator. See Example 1.

1. $\dfrac{2}{\sqrt{5}}$

2. $\dfrac{5}{\sqrt{3}}$

 3. $\dfrac{\sqrt{3}}{\sqrt{7}}$

4. $\dfrac{\sqrt{6}}{\sqrt{5}}$

5. $\dfrac{1}{\sqrt[3]{4}}$

6. $\dfrac{7}{\sqrt[3]{3}}$

7. $\dfrac{\sqrt[3]{6}}{\sqrt[3]{5}}$

8. $\dfrac{\sqrt[4]{2}}{\sqrt[4]{27}}$

‹2› Simplifying Radicals

Write each radical expression in simplified radical form.
See Example 2.

9. $\dfrac{\sqrt{5}}{\sqrt{12}}$ **10.** $\dfrac{\sqrt{7}}{\sqrt{18}}$

11. $\dfrac{\sqrt{3}}{\sqrt{12}}$ **12.** $\dfrac{\sqrt{2}}{\sqrt{18}}$

13. $\sqrt{\dfrac{1}{2}}$ **14.** $\sqrt{\dfrac{3}{8}}$

15. $\sqrt[3]{\dfrac{2}{3}}$ **16.** $\sqrt[3]{\dfrac{3}{5}}$

 17. $\sqrt[3]{\dfrac{7}{4}}$ **18.** $\sqrt[4]{\dfrac{1}{5}}$

Simplify. See Example 3.

19. $\sqrt{\dfrac{x}{y}}$ **20.** $\sqrt{\dfrac{x^2}{a}}$

21. $\sqrt{\dfrac{a^3}{b^7}}$ **22.** $\sqrt{\dfrac{w^5}{y^3}}$

23. $\sqrt{\dfrac{a}{3b}}$ **24.** $\sqrt{\dfrac{5x}{2y}}$

25. $\sqrt[3]{\dfrac{a}{b}}$ **26.** $\sqrt[3]{\dfrac{4a}{b}}$

27. $\sqrt[3]{\dfrac{5}{2b^2}}$ **28.** $\sqrt[3]{\dfrac{3}{4a^2}}$

‹3› Dividing Radicals

Divide and simplify. See Examples 4 and 5.

29. $\sqrt{15} \div \sqrt{5}$ **30.** $\sqrt{14} \div \sqrt{7}$

31. $\sqrt{3} \div \sqrt{5}$ **32.** $\sqrt{5} \div \sqrt{7}$

33. $(3\sqrt{3}) \div (5\sqrt{6})$ **34.** $(2\sqrt{2}) \div (4\sqrt{10})$

35. $(2\sqrt{3}) \div (3\sqrt{6})$ **36.** $(5\sqrt{12}) \div (4\sqrt{6})$

37. $\sqrt{24a^2} \div \sqrt{72a}$ **38.** $\sqrt{32x^3} \div \sqrt{48x^2}$

39. $\sqrt[3]{20} \div \sqrt[3]{2}$ **40.** $\sqrt[3]{8x^7} \div \sqrt[3]{2x}$

41. $\sqrt[4]{48} \div \sqrt[4]{3}$ **42.** $\sqrt[4]{4a^{10}} \div \sqrt[4]{2a^2}$

43. $\sqrt[4]{16w} \div \sqrt[4]{w^5}$ **44.** $\sqrt[4]{81b^5} \div \sqrt[4]{b}$

Simplify. See Example 6.

45. $\dfrac{6 + \sqrt{45}}{3}$ **46.** $\dfrac{10 + \sqrt{50}}{5}$

47. $\dfrac{-2 + \sqrt{12}}{-2}$ **48.** $\dfrac{-6 + \sqrt{72}}{-6}$

‹4› Rationalizing Denominators Using Conjugates

Simplify each expression by rationalizing the denominator.
See Example 7.

49. $\dfrac{4}{2 + \sqrt{8}}$

50. $\dfrac{6}{3 - \sqrt{18}}$

51. $\dfrac{3}{\sqrt{11} - \sqrt{5}}$

52. $\dfrac{6}{\sqrt{5} - \sqrt{14}}$

53. $\dfrac{1 + \sqrt{2}}{\sqrt{3} - 1}$

54. $\dfrac{2 - \sqrt{3}}{\sqrt{2} + \sqrt{6}}$

55. $\dfrac{\sqrt{2}}{\sqrt{6} + \sqrt{3}}$

56. $\dfrac{5}{\sqrt{7} - \sqrt{5}}$

57. $\dfrac{2\sqrt{3}}{3\sqrt{2} - \sqrt{5}}$

58. $\dfrac{3\sqrt{5}}{5\sqrt{2} + \sqrt{6}}$

‹5› Powers of Radical Expressions

Simplify. See Example 8.

59. $(2\sqrt{2})^5$ **60.** $(3\sqrt{3})^4$

61. $(\sqrt{x})^5$ **62.** $(2\sqrt{y})^3$

63. $(-3\sqrt{x^3})^3$ **64.** $(-2\sqrt{x^3})^4$

65. $(2x\sqrt[3]{x^2})^3$ **66.** $(2y\sqrt[3]{4y})^3$

67. $(-2\sqrt[3]{5})^2$ **68.** $(-3\sqrt[3]{4})^2$

69. $(\sqrt[3]{x^2})^6$ **70.** $(2\sqrt[4]{y^3})^3$

Miscellaneous

Simplify.

71. $\dfrac{\sqrt{3}}{\sqrt{2}} + \dfrac{2}{\sqrt{2}}$

72. $\dfrac{2}{\sqrt{7}} + \dfrac{5}{\sqrt{7}}$

73. $\dfrac{\sqrt{3}}{\sqrt{2}} + \dfrac{3\sqrt{6}}{2}$

74. $\dfrac{\sqrt{3}}{2\sqrt{2}} + \dfrac{\sqrt{5}}{3\sqrt{2}}$

75. $\dfrac{\sqrt{6}}{2} \cdot \dfrac{1}{\sqrt{3}}$

76. $\dfrac{\sqrt{6}}{\sqrt{7}} \cdot \dfrac{\sqrt{14}}{\sqrt{3}}$

77. $\dfrac{8 - \sqrt{32}}{20}$

78. $\dfrac{4 - \sqrt{28}}{6}$

79. $\dfrac{5 + \sqrt{75}}{10}$

80. $\dfrac{3 + \sqrt{18}}{6}$

81. $\sqrt{a}(\sqrt{a} - 3)$

82. $3\sqrt{m}(2\sqrt{m} - 6)$

83. $4\sqrt{a}(a + \sqrt{a})$

84. $\sqrt{3ab}(\sqrt{3a} + \sqrt{3})$

85. $(2\sqrt{3m})^2$

86. $(-3\sqrt{4y})^2$

87. $\left(-2\sqrt{xy^2z}\right)^2$

88. $(5a\sqrt{ab})^2$

89. $\sqrt[3]{m}\left(\sqrt[3]{m^2} - \sqrt[3]{m^5}\right)$

90. $\sqrt[4]{w}\left(\sqrt[4]{w^3} - \sqrt[4]{w^7}\right)$

91. $\sqrt[3]{8x^4} + \sqrt[3]{27x^4}$

92. $\sqrt[3]{16a^4} + a\sqrt[3]{2a}$

93. $\left(2m\sqrt[4]{2m^2}\right)^3$

94. $\left(-2t\sqrt[6]{2t^2}\right)^5$

95. $\dfrac{x - 9}{\sqrt{x} - 3}$

96. $\dfrac{x - y}{\sqrt{x} - \sqrt{y}}$

97. $\dfrac{3\sqrt{k}}{\sqrt{k} + \sqrt{7}}$

98. $\dfrac{\sqrt{hk}}{\sqrt{h} + 3\sqrt{k}}$

 99. $\dfrac{5}{\sqrt{2} - 1} + \dfrac{3}{\sqrt{2} + 1}$

100. $\dfrac{\sqrt{3}}{\sqrt{6} - 1} - \dfrac{\sqrt{3}}{\sqrt{6} + 1}$

101. $\dfrac{1}{\sqrt{2}} + \dfrac{1}{\sqrt{3}}$

102. $\dfrac{4}{2\sqrt{3}} + \dfrac{1}{\sqrt{5}}$

103. $\dfrac{3}{\sqrt{2} - 1} + \dfrac{4}{\sqrt{2} + 1}$

104. $\dfrac{3}{\sqrt{5} - \sqrt{3}} - \dfrac{2}{\sqrt{5} + \sqrt{3}}$

105. $\dfrac{\sqrt{x}}{\sqrt{x} + 2} + \dfrac{3\sqrt{x}}{\sqrt{x} - 2}$

106. $\dfrac{\sqrt{5}}{3 - \sqrt{y}} - \dfrac{\sqrt{5y}}{3 + \sqrt{y}}$

107. $\dfrac{1}{\sqrt{x}} + \dfrac{1}{1 - \sqrt{x}}$

108. $\dfrac{\sqrt{x}}{\sqrt{x} - 3} + \dfrac{5}{\sqrt{x}}$

Getting More Involved

109. *Exploration*

A polynomial is prime if it cannot be factored by using integers, but many prime polynomials can be factored if we use radicals.

a) Find the product $(x - \sqrt[3]{2})(x^2 + \sqrt[3]{2}x + \sqrt[3]{4})$.

b) Factor $x^3 + 5$ using radicals.

c) Find the product
$$(\sqrt[3]{5} - \sqrt[3]{2})(\sqrt[3]{25} + \sqrt[3]{10} + \sqrt[3]{4}).$$

d) Use radicals to factor $a + b$ as a sum of two cubes and $a - b$ as a difference of two cubes.

110. *Discussion*

Which one of the following expressions is not equivalent to the others?

a) $(\sqrt[3]{x})^4$ **b)** $\sqrt[4]{x^3}$ **c)** $\sqrt[3]{x^4}$

d) $x^{4/3}$ **e)** $(x^{1/3})^4$

7.5　Solving Equations with Radicals and Exponents

One of our goals in algebra is to keep increasing our knowledge of solving equations because the solutions to equations can give us the answers to various applied questions. In this section, we will apply our knowledge of radicals and exponents to solving some new types of equations.

⟨1⟩ The Odd-Root Property

Because $(-2)^3 = -8$ and $2^3 = 8$, the equation $x^3 = 8$ is equivalent to $x = 2$. The equation $x^3 = -8$ is equivalent to $x = -2$. Because there is only one real odd root of each real number, there is a simple rule for writing an equivalent equation in this situation.

Odd-Root Property

If n is an odd positive integer,

$$x^n = k \qquad \text{is equivalent to} \qquad x = \sqrt[n]{k}$$

for any real number k.

Note that $x^n = k$ is equivalent to $x = \sqrt[n]{k}$ means that these two equations have the same *real* solutions. So $x^3 = 1$ and $x = \sqrt[3]{1}$ each have only one real solution.

E X A M P L E   1

Using the odd-root property

Solve each equation.

a) $x^3 = 27$　　　　　**b)** $x^5 + 32 = 0$　　　　　**c)** $(x - 2)^3 = 24$

Solution

a) $x^3 = 27$

$\quad x = \sqrt[3]{27}$　　Odd-root property

$\quad x = 3$

Check 3 in the original equation. The solution set is $\{3\}$.

b) $x^5 + 32 = 0$

$\quad\quad x^5 = -32$　　Isolate the variable.

$\quad\quad x = \sqrt[5]{-32}$　　Odd-root property

$\quad\quad x = -2$

Check -2 in the original equation. The solution set is $\{-2\}$.

c) $(x - 2)^3 = 24$

$\quad x - 2 = \sqrt[3]{24}$　　　　Odd-root property

$\quad x = 2 + 2\sqrt[3]{3}$　　$\sqrt[3]{24} = \sqrt[3]{8} \cdot \sqrt[3]{3} = 2\sqrt[3]{3}$

Check. The solution set is $\{2 + 2\sqrt[3]{3}\}$.

> **Now do Exercises 1–8**

⟨2⟩ **The Even-Root Property**

In solving the equation $x^2 = 4$, you might be tempted to write $x = 2$ as an equivalent equation. But $x = 2$ is not equivalent to $x^2 = 4$ because $2^2 = 4$ and $(-2)^2 = 4$. So the solution set to $x^2 = 4$ is $\{-2, 2\}$. The equation $x^2 = 4$ is equivalent to the compound sentence $x = 2$ or $x = -2$, which we can abbreviate as $x = \pm 2$. The equation $x = \pm 2$ is read "x equals positive or negative 2."

Equations involving other even powers are handled like the squares. Because $2^4 = 16$ and $(-2)^4 = 16$, the equation $x^4 = 16$ is equivalent to $x = \pm 2$. So $x^4 = 16$ has two real solutions. Note that $x^4 = -16$ has no real solutions. The equation $x^6 = 5$ is equivalent to $x = \pm \sqrt[6]{5}$. We can now state a general rule.

Even-Root Property

Suppose n is a positive even integer.
  If $k > 0$, then $x^n = k$ is equivalent to $x = \pm \sqrt[n]{k}$.
  If $k = 0$, then $x^n = k$ is equivalent to $x = 0$.
  If $k < 0$, then $x^n = k$ has no real solution.

Note that $x^n = k$ for $k > 0$ is equivalent to $x = \pm \sqrt[n]{k}$ means that these two equations have the same *real* solutions.

E X A M P L E **2**

Using the even-root property
Solve each equation.

  a) $x^2 = 10$ **b)** $w^8 = 0$ **c)** $x^4 = -4$

⟨ **Helpful Hint** ⟩

We do not say, "take the square root of each side." We are not doing the same thing to each side of $x^2 = 9$ when we write $x = \pm 3$. This is the third time that we have seen a rule for obtaining an equivalent equation without "doing the same thing to each side." (What were the other two?) Because there is only one odd root of every real number, you can actually take an odd root of each side.

Solution

  a) $x^2 = 10$
      $x = \pm\sqrt{10}$ Even-root property
      The solution set is $\{-\sqrt{10}, \sqrt{10}\}$, or $\{\pm\sqrt{10}\}$.
  b) $w^8 = 0$
      $w = 0$ Even-root property
      The solution set is $\{0\}$.
  c) By the even-root property, $x^4 = -4$ has no real solution. (The fourth power of any real number is nonnegative.)

Now do Exercises 9–14

Whether an equation has a solution depends on the domain of the variable. For example, $2x = 5$ has no solution in the set of integers and $x^2 = -9$ has no solution in the set of real numbers. We can say that the solution set to both of these equations is the empty set, $\varnothing$, as long as the domain of the variable is clear. In Section 7.6 we introduce a new set of numbers, the *imaginary numbers,* in which $x^2 = -9$ will have two solutions. So in this section it is best to say that $x^2 = -9$ has no real solution, because in Section 7.6 its solution set will *not* be $\varnothing$. An equation such as $x = x + 1$ never has a solution, and so saying that its solution set is $\varnothing$ is clear.

In Example 3, the even-root property is used to solve some equations that are a bit more complicated than those of Example 2.

EXAMPLE 3

Using the even-root property

Solve each equation.

a) $(x - 3)^2 = 4$ **b)** $2(x - 5)^2 - 7 = 0$ **c)** $x^4 - 1 = 80$

Solution

a) $(x - 3)^2 = 4$

$$x - 3 = 2 \quad \text{or} \quad x - 3 = -2 \quad \text{Even-root property}$$

$$x = 5 \quad \text{or} \quad x = 1 \quad \text{Add 3 to each side.}$$

The solution set is $\{1, 5\}$.

b) $2(x - 5)^2 - 7 = 0$

$$2(x - 5)^2 = 7 \quad \text{Add 7 to each side.}$$

$$(x - 5)^2 = \frac{7}{2} \quad \text{Divide each side by 2.}$$

$$x - 5 = \sqrt{\frac{7}{2}} \quad \text{or} \quad x - 5 = -\sqrt{\frac{7}{2}} \quad \text{Even-root property}$$

$$x = 5 + \frac{\sqrt{14}}{2} \quad \text{or} \quad x = 5 - \frac{\sqrt{14}}{2} \qquad \sqrt{\frac{7}{2}} = \frac{\sqrt{7} \cdot \sqrt{2}}{\sqrt{2} \cdot \sqrt{2}} = \frac{\sqrt{14}}{2}$$

$$x = \frac{10 + \sqrt{14}}{2} \quad \text{or} \quad x = \frac{10 - \sqrt{14}}{2}$$

The solution set is $\left\{ \frac{10 + \sqrt{14}}{2}, \frac{10 - \sqrt{14}}{2} \right\}$.

c) $x^4 - 1 = 80$

$$x^4 = 81$$

$$x = \pm\sqrt[4]{81} = \pm 3$$

The solution set is $\{-3, 3\}$.

> Now do Exercises 15–24

     In Chapter 5 we solved quadratic equations by factoring. The quadratic equations that we encounter in this chapter can be solved by using the even-root property as in parts (a) and (b) of Example 3. In Chapter 8 you will learn general methods for solving any quadratic equation.

‹3› Equations Involving Radicals

     If we start with the equation $x = 3$ and square both sides, we get $x^2 = 9$, which has the solution set $\{-3, 3\}$. But the solution set to $x = 3$ is $\{3\}$. Squaring both sides produced an equation with more solutions than the original. We call the extra solutions **extraneous solutions.** The same problem can occur when we raise each side to any even power. Note that we don't always get extraneous solutions. We *might* get one or more of them.

     Raising each side to an odd power does not cause extraneous solutions. For example, if we cube each side of $x = 3$, we get $x^3 = 27$. The solution set to both equations is $\{3\}$. Likewise, $x = -3$ and $x^3 = -27$ both have solution set $\{-3\}$.

> **Raising each side of an equation to a power**
>
> If n is odd, then $a = b$ and $a^n = b^n$ are equivalent equations.
>
> If n is even, then $a = b$ and $a^n = b^n$ may not be equivalent. However, the solution set to $a^n = b^n$ contains all of the solutions to $a = b$.

It has always been important to check solutions any time you solve an equation. When raising each side to a power, it is even more important. We use these ideas most often with equations involving radicals as shown in Example 4.

EXAMPLE 4

Raising each side to a power to eliminate radicals

Solve each equation.

a) $\sqrt{2x - 3} - 5 = 0$ **b)** $\sqrt[3]{3x + 5} = \sqrt[3]{x - 1}$ **c)** $\sqrt{3x + 18} = x$

‹ Calculator Close-Up ›

If 14 satisfies the equation
$$\sqrt{2x - 3} - 5 = 0,$$
then (14, 0) is an x-intercept for the graph of
$$y = \sqrt{2x - 3} - 5.$$
So the calculator graph shown here provides visual support for the conclusion that 14 is the only solution to the equation.

Solution

a) Eliminate the square root by raising each side to the power 2:

$$\sqrt{2x - 3} - 5 = 0 \qquad \text{Original equation}$$
$$\sqrt{2x - 3} = 5 \qquad \text{Isolate the radical.}$$
$$(\sqrt{2x - 3})^2 = 5^2 \qquad \text{Square both sides.}$$
$$2x - 3 = 25$$
$$2x = 28$$
$$x = 14$$

Check by evaluating $x = 14$ in the original equation:

$$\sqrt{2(14) - 3} - 5 = 0$$
$$\sqrt{28 - 3} - 5 = 0$$
$$\sqrt{25} - 5 = 0$$
$$0 = 0$$

The solution set is $\{14\}$.

b)
$$\sqrt[3]{3x + 5} = \sqrt[3]{x - 1} \qquad \text{Original equation}$$
$$(\sqrt[3]{3x + 5})^3 = (\sqrt[3]{x - 1})^3 \qquad \text{Cube each side.}$$
$$3x + 5 = x - 1$$
$$2x = -6$$
$$x = -3$$

Check $x = -3$ in the original equation:

$$\sqrt[3]{3(-3) + 5} = \sqrt[3]{-3 - 1}$$
$$\sqrt[3]{-4} = \sqrt[3]{-4}$$

Note that $\sqrt[3]{-4}$ is a real number. The solution set is $\{-3\}$. In this example, we checked for arithmetic mistakes. There was no possibility of extraneous solutions here because we raised each side to an odd power.

c)

$$\sqrt{3x + 18} = x \qquad \text{Original equation}$$
$$(\sqrt{3x + 18})^2 = x^2 \qquad \text{Square both sides.}$$
$$3x + 18 = x^2 \qquad \text{Simplify.}$$
$$-x^2 + 3x + 18 = 0 \qquad \text{Subtract } x^2 \text{ from each side to get zero on one side.}$$
$$x^2 - 3x - 18 = 0 \qquad \text{Multiply each side by } -1 \text{ for easier factoring.}$$
$$(x - 6)(x + 3) = 0 \qquad \text{Factor.}$$
$$x - 6 = 0 \quad \text{or} \quad x + 3 = 0 \quad \text{Zero factor property}$$
$$x = 6 \quad \text{or} \quad x = -3$$

Because we squared both sides, we must check for extraneous solutions. If $x = -3$ in the original equation $\sqrt{3x + 18} = x$, we get

$$\sqrt{3(-3) + 18} = -3$$
$$\sqrt{9} = -3$$
$$3 = -3,$$

which is not correct. If $x = 6$ in the original equation, we get

$$\sqrt{3(6) + 18} = 6,$$

which is correct. The solution set is $\{6\}$.

> **Now do Exercises 25–44**

In Example 5, the radicals are not eliminated after squaring both sides of the equation. In this case, we must square both sides a second time. Note that we square the side with two terms the same way we square a binomial.

E X A M P L E   5

Squaring both sides twice

Solve $\sqrt{5x - 1} - \sqrt{x + 2} = 1$.

Solution

It is easier to square both sides if the two radicals are not on the same side.

$$\sqrt{5x - 1} - \sqrt{x + 2} = 1 \qquad \text{Original equation}$$
$$\sqrt{5x - 1} = 1 + \sqrt{x + 2} \qquad \text{Add } \sqrt{x + 2} \text{ to each side.}$$
$$(\sqrt{5x - 1})^2 = (1 + \sqrt{x + 2})^2 \qquad \text{Square both sides.}$$
$$5x - 1 = 1 + 2\sqrt{x + 2} + x + 2 \qquad \text{Square the right side like a binomial.}$$
$$5x - 1 = 3 + x + 2\sqrt{x + 2} \qquad \text{Combine like terms on the right side.}$$
$$4x - 4 = 2\sqrt{x + 2} \qquad \text{Isolate the square root.}$$
$$2x - 2 = \sqrt{x + 2} \qquad \text{Divide each side by 2.}$$
$$(2x - 2)^2 = (\sqrt{x + 2})^2 \qquad \text{Square both sides.}$$
$$4x^2 - 8x + 4 = x + 2 \qquad \text{Square the binomial on the left side.}$$
$$4x^2 - 9x + 2 = 0$$
$$(4x - 1)(x - 2) = 0$$
$$4x - 1 = 0 \quad \text{or} \quad x - 2 = 0$$
$$x = \frac{1}{4} \quad \text{or} \quad x = 2$$

Check to see whether $\sqrt{5x-1} - \sqrt{x+2} = 1$ for $x = \frac{1}{4}$ and for $x = 2$:

$$\sqrt{5 \cdot \frac{1}{4} - 1} - \sqrt{\frac{1}{4} + 2} = \sqrt{\frac{1}{4}} - \sqrt{\frac{9}{4}} = \frac{1}{2} - \frac{3}{2} = -1$$

$$\sqrt{5 \cdot 2 - 1} - \sqrt{2 + 2} = \sqrt{9} - \sqrt{4} = 3 - 2 = 1$$

So the original equation is not satisfied for $x = \frac{1}{4}$ but is satisfied for $x = 2$. Since 2 is the only solution to the equation, the solution set is $\{2\}$.

> Now do Exercises 45–60

⟨4⟩ Equations Involving Rational Exponents

Equations involving rational exponents can be solved by combining the methods that you just learned for eliminating radicals and integral exponents. For equations involving rational exponents, always eliminate the root first and the power second.

E X A M P L E 6

Eliminating the root, then the power

Solve each equation.

a) $x^{2/3} = 4$

b) $(w - 1)^{-2/5} = 4$

‹ Helpful Hint ›

Note how we eliminate the root first by raising each side to an integer power, and then apply the even-root property to get two solutions in Example 6(a). A common mistake is to raise each side to the 3/2 power and get $x = 4^{3/2} = 8$. If you do not use the even-root property, you can easily miss the solution -8.

Solution

a) Because the exponent 2/3 indicates a cube root, raise each side to the power 3:

$$x^{2/3} = 4 \qquad \text{Original equation}$$
$$(x^{2/3})^3 = 4^3 \qquad \text{Cube each side.}$$
$$x^2 = 64 \qquad \text{Multiply the exponents: } \frac{2}{3} \cdot 3 = 2.$$
$$x = 8 \qquad \text{or} \qquad x = -8 \quad \text{Even-root property}$$

All of the equations are equivalent. Check 8 and -8 in the original equation. The solution set is $\{-8, 8\}$.

b)
$$(w - 1)^{-2/5} = 4 \qquad \text{Original equation}$$
$$[(w - 1)^{-2/5}]^{-5} = 4^{-5} \qquad \text{Raise each side to the power } -5 \text{ to eliminate the negative exponent.}$$
$$(w - 1)^2 = \frac{1}{1024} \qquad \text{Multiply the exponents: } -\frac{2}{5}(-5) = 2.$$
$$w - 1 = \pm\sqrt{\frac{1}{1024}} \qquad \text{Even-root property}$$
$$w - 1 = \frac{1}{32} \qquad \text{or} \qquad w - 1 = -\frac{1}{32}$$
$$w = \frac{33}{32} \qquad \text{or} \qquad w = \frac{31}{32}$$

‹ Calculator Close-Up ›

Check that 31/32 and 33/32 satisfy the original equation.

```
(31/32-1)^(-2/5)
                4
(33/32-1)^(-2/5)
                4
```

Check the values in the original equation. The solution set is $\left\{\frac{31}{32}, \frac{33}{32}\right\}$.

> Now do Exercises 61–72

An equation with a rational exponent might not have a real solution because all even powers of real numbers are nonnegative.

E X A M P L E 7

An equation with no solution

Solve $(2t - 3)^{-2/3} = -1$.

Solution

Raise each side to the power -3 to eliminate the root and the negative sign in the exponent:

$$(2t - 3)^{-2/3} = -1 \qquad \text{Original equation}$$

$$\left[(2t - 3)^{-2/3}\right]^{-3} = (-1)^{-3} \qquad \text{Raise each side to the } -3 \text{ power.}$$

$$(2t - 3)^2 = -1 \qquad \text{Multiply the exponents: } -\frac{2}{3}(-3) = 2$$

By the even-root property this equation has no real solution. The square of every real number is nonnegative.

Now do Exercises 73–74

The three most important rules for solving equations with exponents and radicals are restated here.

Strategy for Solving Equations with Exponents and Radicals

1. In raising each side of an equation to an even power, we can create an equation that gives extraneous solutions. We must check all possible solutions in the original equation.

2. When applying the even-root property, remember that there is a positive and a negative even root for any positive real number.

3. For equations with rational exponents, raise each side to a positive or negative integral power first and then apply the even- or odd-root property. (Positive fraction—raise to a positive power; negative fraction—raise to a negative power.)

⟨5⟩ Applications

The square of the hypotenuse of any right triangle is equal to the sum of the squares of the legs (the Pythagorean theorem). In Example 8 we use this fact and the even-root property to find a distance on a baseball diamond.

E X A M P L E 8

Diagonal of a baseball diamond

A baseball diamond is actually a square, 90 feet on each side. What is the distance from third base to first base?

Solution

First make a sketch as in Fig. 7.1. The distance x from third base to first base is the length of the diagonal of the square shown in Fig. 7.1. The Pythagorean theorem can be applied to

Figure 7.1

the right triangle formed from the diagonal and two sides of the square. The sum of the squares of the sides is equal to the diagonal squared:

$$x^2 = 90^2 + 90^2$$
$$x^2 = 8100 + 8100$$
$$x^2 = 16{,}200$$
$$x = \pm\sqrt{16{,}200} = \pm90\sqrt{2}$$

The length of the diagonal of a square must be positive, so we disregard the negative solution. Checking the answer in the original equation verifies that the *exact* length of the diagonal is $90\sqrt{2}$ feet.

Now do Exercises 95–110

Warm-Ups ▼

Fill in the blank.

1. If n is an ____ positive integer, then $x^n = k$ is equivalent to $x = \sqrt[n]{k}$ for any real number k.

2. If n is an ____ positive integer, then $x^n = k$ is equivalent to $x = \pm\sqrt[n]{k}$ for $k > 0$.

3. An _____ solution is a solution that appears when solving an equation, but does not satisfy the original equation.

4. Raising each side of an equation to an ____ power can produce an extraneous solution.

True or false?

5. The equations $x^2 = 4$ and $x = 2$ are equivalent.

6. The equation $x^2 = -25$ has no real solution.

7. The equation $x^2 = 0$ has no solution.

8. The equation $x^3 = 8$ is equivalent to $x = 2$.

9. Squaring both sides of $\sqrt{x} = -7$ yields an equation with an extraneous solution.

10. The equations $x^2 - 6 = 0$ and $x = \pm\sqrt{6}$ are equivalent.

Exercises
7.5

‹ Study Tips ›

• Try changing subjects or tasks every hour when you study. The brain does not easily assimilate the same material hour after hour.
• You will learn more from working on a subject 1 hour per day than 7 hours on Saturday.

‹1› The Odd-Root Property

Solve each equation. See Example 1.

1. $x^3 = -1000$

2. $y^3 = 125$

3. $32m^5 - 1 = 0$

4. $243a^5 + 1 = 0$

5. $(y - 3)^3 = -8$

6. $(x - 1)^3 = -1$

7. $\dfrac{1}{2}x^3 + 4 = 0$

8. $3(x - 9)^7 = 0$

‹2› The Even-Root Property

Find all real solutions to each equation. See Examples 2 and 3.

9. $x^2 = 25$

10. $x^2 = 36$

11. $x^2 - 20 = 0$ **12.** $a^2 - 40 = 0$

13. $x^2 + 9 = 0$ **14.** $w^2 + 49 = 0$

15. $(x - 3)^2 = 16$ **16.** $(a - 2)^2 = 25$

17. $(x + 1)^2 - 8 = 0$

18. $(w + 3)^2 - 12 = 0$

19. $\dfrac{1}{2}x^2 = 5$ **20.** $\dfrac{1}{3}x^2 = 6$

21. $(y - 3)^4 = 0$ **22.** $(2x - 3)^6 = 0$

23. $2x^6 = 128$ **24.** $3y^4 = 48$

⟨3⟩ Equations Involving Radicals

Solve each equation and check for extraneous solutions.
See Example 4.

25. $\sqrt{x - 3} - 3 = 4$ **26.** $\sqrt{a - 1} - 5 = 1$

27. $2\sqrt{w + 4} = 5$ **28.** $3\sqrt{w + 1} = 6$

29. $\sqrt[3]{2x + 3} = \sqrt[3]{x + 12}$ **30.** $\sqrt[3]{a + 3} = \sqrt[3]{2a - 7}$

31. $\sqrt{2t - 4} = \sqrt{t - 1}$ **32.** $\sqrt{w - 3} = \sqrt{4w - 15}$

33. $\sqrt{4x^2 + x - 3} = 2x$ **34.** $\sqrt{x^2 - 5x + 2} = x$

35. $\sqrt{x^2 + 2x - 6} = 3$ **36.** $\sqrt{x^2 - x - 4} = 4$

37. $\sqrt{2x^2 - 1} = x$ **38.** $\sqrt{2x^2 - 3x - 10} = x$

39. $\sqrt{2x^2 + 5x + 6} = x$ **40.** $\sqrt{2x^2 + 6x + 9} = x$

41. $\sqrt{x - 1} = x - 1$ **42.** $\sqrt{2x - 1} = 2x - 1$

43. $x + \sqrt{x - 9} = 9$ **44.** $\sqrt{3x - 1} + 3x = 1$

Solve each equation and check for extraneous solutions. See Example 5.

45. $\sqrt{x} + \sqrt{x - 3} = 3$

46. $\sqrt{x} + \sqrt{x + 3} = 3$

47. $\sqrt{x + 2} + \sqrt{x - 1} = 3$

48. $\sqrt{x} + \sqrt{x - 5} = 5$

49. $\sqrt{x + 3} - \sqrt{x - 2} = 1$

50. $\sqrt{2x + 1} - \sqrt{x} = 1$

51. $\sqrt{3x + 1} - \sqrt{2x - 1} = 1$

52. $\sqrt{4x + 1} - \sqrt{3x - 2} = 1$

53. $\sqrt{2x + 2} - \sqrt{x - 3} = 2$

54. $\sqrt{3x} - \sqrt{x - 2} = 4$

55. $\sqrt{4 - x} - \sqrt{x + 6} = 2$

56. $\sqrt{6 - x} - \sqrt{x - 2} = 2$

57. $\sqrt{x - 5} - \sqrt{x} = 3$

58. $\sqrt{2x} - \sqrt{2x - 12} = 6$

59. $\sqrt{3x + 1} + \sqrt{2x + 4} = 3$

60. $\sqrt{2x + 5} + \sqrt{x + 2} = 1$

⟨4⟩ Equations Involving Rational Exponents

Solve each equation. See Examples 6 and 7.

61. $x^{2/3} = 3$ **62.** $a^{2/3} = 2$

63. $y^{-2/3} = 9$ **64.** $w^{-2/3} = 4$

65. $w^{1/3} = 8$ **66.** $a^{1/3} = 27$

67. $t^{-1/2} = 9$ **68.** $w^{-1/4} = \dfrac{1}{2}$

69. $(3a - 1)^{-2/5} = 1$ **70.** $(r - 1)^{-2/3} = 1$

71. $(t - 1)^{-2/3} = 2$ **72.** $(w + 3)^{-1/3} = \dfrac{1}{3}$

73. $(x - 3)^{2/3} = -4$

74. $(x + 2)^{3/2} = -1$

Miscellaneous

Solve each equation. See the Strategy for Solving Equations with Exponents and Radicals box on page 502.

75. $2x^2 + 3 = 7$

76. $3x^2 - 5 = 16$

77. $\sqrt[3]{2w + 3} = \sqrt[3]{w - 2}$

78. $\sqrt[3]{2 - w} = \sqrt[3]{2w - 28}$

79. $(w + 1)^{2/3} = -3$

80. $(x - 2)^{4/3} = -2$

81. $(a + 1)^{1/3} = -2$

82. $(a - 1)^{1/3} = -3$

83. $(4y - 5)^7 = 0$

84. $(5x)^9 = 0$

85. $\sqrt{5x^2 + 4x + 1} - x = 0$

86. $3 + \sqrt{x^2 - 8x} = 0$

87. $\sqrt{4x^2} = x + 2$

88. $\sqrt{9x^2} = x + 6$

89. $(t + 2)^4 = 32$

90. $(w + 1)^4 = 48$

91. $\sqrt{x^2 - 3x} = x$

92. $\sqrt[4]{4x^4 - 48} = -x$

93. $x^{-3} = 8$

94. $x^{-2} = 4$

⟨5⟩ Applications

Solve each problem by writing an equation and solving it. Find the exact answer, and simplify it using the rules for radicals. See Example 8.

95. *Side of a square.* Find the length of the side of a square whose diagonal is 8 feet.

96. *Diagonal of a patio.* Find the length of the diagonal of a square patio with an area of 40 square meters.

97. *Side of a sign.* Find the length of the side of a square sign whose area is 50 square feet.

98. *Side of a cube.* Find the length of the side of a cubic box whose volume is 80 cubic feet.

99. *Diagonal of a rectangle.* If the sides of a rectangle are 30 feet and 40 feet in length, find the length of the diagonal of the rectangle.

100. *Diagonal of a sign.* What is the length of the diagonal of a rectangular billboard whose sides are 5 meters and 12 meters?

101. *A 30-60-90 triangle.* In a 30°-60°-90° triangle, the side opposite the 30° angle is half the length of the hypotenuse. See the accompanying figure.
  a) Find the length of the hypotenuse if the side opposite the 30° angle is 1.
  b) Find the length of the side opposite 60° if the side opposite 30° is 1.
  c) Find the length of the side opposite 60° if the length of the hypotenuse is 1.

102. *An isosceles right triangle.* An isosceles right triangle has two 45° angles. The sides opposite those angles are equal in length. See the accompanying figure.
  a) Find the length of the hypotenuse if the length of each of the equal sides is 1.
  b) Find the length of each of the equal sides if the length of the hypotenuse is 1.

Figure for Exercises 101 and 102

103. *Sailboat stability.* To be considered safe for ocean sailing, the capsize screening value C should be less than 2 (www.sailing.com). For a boat with a beam (or width) b in feet and displacement d in pounds, C is determined by the function

$$C = 4d^{-1/3}b.$$

Figure for Exercise 103

a) Find the capsize screening value for the Tartan 4100, which has a displacement of 23,245 pounds and a beam of 13.5 feet.

b) Write d as a function of b and C.

c) The accompanying graph shows C as a function of d for the Tartan 4100 ($b = 13.5$). For what displacement is the Tartan 4100 safe for ocean sailing?

104. *Sailboat speed.* The sail area-displacement ratio S provides a measure of the sail power available to drive a boat. For a boat with a displacement of d pounds and a sail area of A square feet, S is determined by the function
$$S = 16Ad^{-2/3}.$$

a) Find S to the nearest tenth for the Tartan 4100, which has a sail area of 810 square feet and a displacement of 23,245 pounds.

b) Write d as a function of A and S.

105. *Diagonal of a side.* Find the length of the diagonal of a side of a cubic packing crate whose volume is 2 cubic meters.

106. *Volume of a cube.* Find the volume of a cube on which the diagonal of a side measures 2 feet.

107. *Length of a road.* An architect designs a public park in the shape of a trapezoid. Find the length of the diagonal road marked a in the figure.

108. *Length of a boundary.* Find the length of the border of the park marked b in the trapezoid shown in the figure.

Figure for Exercises 107 and 108

109. *Average annual return.* In Exercise 125 of Section 7.2, the function
$$r = \left(\frac{S}{P}\right)^{1/n} - 1$$
was used to find the average annual return for an investment.

a) Write S as a function of r, P, and n.

b) Write P as a function of r, S, and n.

110. *Surface area of a cube.* The function $A = 6V^{2/3}$ gives the surface area of a cube in terms of its volume V. What

is the volume of a cube with surface area 12 square feet?

111. *Kepler's third law.* According to Kepler's third law of planetary motion, the ratio $\frac{T^2}{R^3}$ has the same value for every planet in our solar system. R is the average radius of the orbit of the planet measured in astronomical units (AU), and T is the number of years it takes for one complete orbit of the sun. Jupiter orbits the sun in 11.86 years with an average radius of 5.2 AU, whereas Saturn orbits the sun in 29.46 years. Find the average radius of the orbit of Saturn. (One AU is the distance from the earth to the sun.)

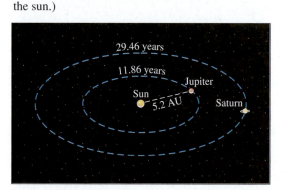

Figure for Exercise 111

112. *Orbit of Venus.* If the average radius of the orbit of Venus is 0.723 AU, then how many years does it take for Venus to complete one orbit of the sun? Use the information in Exercise 111.

Use a calculator to find approximate solutions to the following equations. Round your answers to three decimal places.

113. $x^2 = 3.24$

114. $(x + 4)^3 = 7.51$

115. $\sqrt{x - 2} = 1.73$

116. $\sqrt[3]{x - 5} = 3.7$

117. $x^{2/3} = 8.86$

118. $(x - 1)^{-3/4} = 7.065$

Getting More Involved

119. *Cooperative learning*

Work in a small group to write a formula that gives the side of a cube in terms of the volume of the cube, and explain the formula to the other groups.

120. *Cooperative learning*

Work in a small group to write a formula that gives the side of a square in terms of the diagonal of the square, and explain the formula to the other groups.

7.6 Complex Numbers

In Chapter 1, we discussed the real numbers and the various subsets of the real numbers. In this section, we define a set of numbers that has the real numbers as a subset. This new set of numbers is the set of *complex numbers*. Although it is hard to imagine numbers beyond the real numbers, the complex numbers are used to model some very real phenomena in physics and electrical engineering. These applications are beyond the scope of this text, but if you want a better understanding of them, search the Internet for "applications of complex numbers."

⟨1⟩ Definition

The equation $2x = 1$ has no solution in the set of integers, but in the set of rational numbers, $2x = 1$ has a solution. The situation is similar for the equation $x^2 = -4$. It has no solution in the set of real numbers because the square of every real number is nonnegative. However, in the set of complex numbers $x^2 = -4$ has two solutions. The complex numbers were developed so that equations such as $x^2 = -4$ would have solutions.

The complex numbers are based on the symbol $\sqrt{-1}$. In the real number system this symbol has no meaning. In the set of complex numbers this symbol is given meaning. We call it i. We make the definition that

$$i = \sqrt{-1} \quad \text{and} \quad i^2 = -1.$$

Complex Numbers

The set of **complex numbers** is the set of all numbers of the form

$$a + bi,$$

where a and b are real numbers, $i = \sqrt{-1}$, and $i^2 = -1$.

In the complex number $a + bi$, a is called the **real part** and b is called the **imaginary part.** If $b \neq 0$, the number $a + bi$ is called an **imaginary number.** If $b = 0$, then the complex number $a + 0i$ is the real number a.

In dealing with complex numbers, we treat $a + bi$ as if it were a binomial, with i being a variable. Thus, we would write $2 + (-3)i$ as $2 - 3i$. We agree that $2 + i3$, $3i + 2$, and $i3 + 2$ are just different ways of writing $2 + 3i$ (the standard form). Some examples of complex numbers are

$$-3 - 5i, \quad \frac{2}{3} - \frac{3}{4}i, \quad 1 + i\sqrt{2}, \quad 9 + 0i, \quad \text{and} \quad 0 + 7i.$$

For simplicity we write only $7i$ for $0 + 7i$. The complex number $9 + 0i$ is the real number 9, and $0 + 0i$ is the real number 0. Any complex number with $b = 0$ is a real number. For any real number a,

$$a + 0i = a.$$

The set of real numbers is a subset of the set of complex numbers. See Fig. 7.2.

Complex numbers	
Real numbers	Imaginary numbers
$3, \pi, \frac{5}{2}, 0, -9, \sqrt{2}$	$i, 2 + 3i, \sqrt{-5}, -3 - 8i$

Figure 7.2

‹2› Addition, Subtraction, and Multiplication

Addition and subtraction of complex numbers are performed as if the complex numbers were algebraic expressions with i being a variable.

EXAMPLE 1

Addition and subtraction of complex numbers

Find the sums and differences.

a) $(2 + 3i) + (6 + i)$
c) $(3 + 5i) - (1 + 2i)$

b) $(-2 + 3i) + (-2 - 5i)$
d) $(-2 - 3i) - (1 - i)$

Solution

a) $(2 + 3i) + (6 + i) = 8 + 4i$

b) $(-2 + 3i) + (-2 - 5i) = -4 - 2i$

c) $(3 + 5i) - (1 + 2i) = 2 + 3i$

d) $(-2 - 3i) - (1 - i) = -3 - 2i$

Now do Exercises 1–8

Informally, we add and subtract complex numbers as in Example 1. Formally, we use the following symbolic definition. We include this definition for completeness, but you don't need to memorize it. Just add or subtract as in Example 1.

Addition and Subtraction of Complex Numbers

The sum and difference of $a + bi$ and $c + di$ are defined as follows:

$$(a + bi) + (c + di) = (a + c) + (b + d)i$$
$$(a + bi) - (c + di) = (a - c) + (b - d)i$$

Complex numbers are multiplied as if they were algebraic expressions. Whenever i^2 appears, we replace it by -1.

EXAMPLE 2

Products of complex numbers

Find each product.

a) $2i(1 + i)$ b) $(2 + 3i)(4 + 5i)$ c) $(3 + i)(3 - i)$

Solution

a) $2i(1 + i) = 2i + 2i^2$ Distributive property

   $= 2i + 2(-1)$ $i^2 = -1$

   $= -2 + 2i$

‹ **Calculator Close-Up** ›

Many graphing calculators can perform operations with complex numbers.

```
2i(1+i)
              -2+2i
(2+3i)(4+5i)
              -7+22i
(3+i)(3-i)
                 10
```

b) Use the FOIL method to find the product:

$$(2 + 3i)(4 + 5i) = 8 + 10i + 12i + 15i^2$$
$$= 8 + 22i + 15(-1) \quad \text{Replace } i^2 \text{ by } -1.$$
$$= 8 + 22i - 15$$
$$= -7 + 22i$$

c) This product is the product of a sum and a difference.

$$(3 + i)(3 - i) = 9 - 3i + 3i - i^2$$
$$= 9 - (-1) \quad i^2 = -1$$
$$= 10$$

> Now do Exercises 9–26

For completeness we give the following symbolic definition of multiplication of complex numbers. However, it is simpler to find products as we did in Examples 2 and 3 than to use this definition.

> **Multiplication of Complex Numbers**
>
> The complex numbers $a + bi$ and $c + di$ are multiplied as follows:
>
> $$(a + bi)(c + di) = (ac - bd) + (ad + bc)i$$

We can find powers of i using the fact that $i^2 = -1$. For example,

$$i^3 = i^2 \cdot i = -1 \cdot i = -i.$$

The value of i^4 is found from the value of i^3:

$$i^4 = i^3 \cdot i = -i \cdot i = -i^2 = 1$$

Using $i^2 = -1$, $i^3 = -i$, and $i^4 = 1$, you can actually find any power of i by factoring out all of the fourth powers. For example,

$$i^{13} = (i^4)^3 \cdot i = (1)^3 \cdot i = i \quad \text{and} \quad i^{18} = (i^4)^4 \cdot i^2 = (1)^4 \cdot i^2 = -1.$$

E X A M P L E   **3**

Powers of imaginary numbers

Write each expression in the form $a + bi$.

a) $(2i)^2$ **b)** $(-2i)^4$ **c)** i^6

d) i^{22} **e)** i^{19}

Solution

a) $(2i)^2 = 2^2 \cdot i^2 = 4(-1) = -4$

b) $(-2i)^4 = (-2)^4 \cdot i^4 = 16 \cdot 1 = 16$

c) $i^6 = i^2 \cdot i^4 = -1 \cdot 1 = -1$

d) $i^{22} = (i^4)^5 \cdot i^2 = (1)^5 \cdot i^2 = -1$

e) $i^{19} = (i^4)^4 \cdot i^3 = (1)^4 \cdot i^3 = -i$

> Now do Exercises 27–38

⟨3⟩ Division of Complex Numbers

To divide a complex number by a real number, divide each term by the real number, just as we would divide a binomial by a number. For example,

$$\frac{4 + 6i}{2} = \frac{2(2 + 3i)}{2}$$

$$= 2 + 3i.$$

⟨ **Helpful Hint** ⟩

Here is that word "conjugate" again. It is generally used to refer to two things that go together in some way.

To understand division by a complex number, we first look at imaginary numbers that have a real product. The product of the two imaginary numbers in Example 2(c) is a real number:

$$(3 + i)(3 - i) = 10$$

We say that $3 + i$ and $3 - i$ are complex conjugates of each other.

Complex Conjugates

The complex numbers $a + bi$ and $a - bi$ are called **complex conjugates** of one another. Their product is the real number $a^2 + b^2$.

E X A M P L E **4**

Products of conjugates

Find the product of the given complex number and its conjugate.

a) $2 + 3i$ **b)** $5 - 4i$

Solution

a) The conjugate of $2 + 3i$ is $2 - 3i$.

$$(2 + 3i)(2 - 3i) = 4 - 9i^2$$

$$= 4 + 9$$

$$= 13$$

b) The conjugate of $5 - 4i$ is $5 + 4i$.

$$(5 - 4i)(5 + 4i) = 25 + 16$$

$$= 41$$

Now do Exercises 39–46

We use complex conjugates to divide complex numbers. The process is the same as rationalizing the denominator. We multiply the numerator and denominator of the quotient by the complex conjugate of the denominator. If we were to use $\sqrt{-1}$ instead of i, then Example 5 here would look just like Example 7 in Section 7.4.

E X A M P L E **5**

Dividing complex numbers

Find each quotient. Write the answer in the form $a + bi$.

a) $\dfrac{5}{3 - 4i}$ **b)** $\dfrac{3 - i}{2 + i}$

c) $\dfrac{3 + 2i}{i}$

Solution

a) Multiply the numerator and denominator by $3 + 4i$, the conjugate of $3 - 4i$:

$$\frac{5}{3 - 4i} = \frac{5(3 + 4i)}{(3 - 4i)(3 + 4i)}$$

$$= \frac{15 + 20i}{9 - 16i^2}$$

$$= \frac{15 + 20i}{25} \qquad 9 - 16i^2 = 9 - 16(-1) = 25$$

$$= \frac{15}{25} + \frac{20}{25}i$$

$$= \frac{3}{5} + \frac{4}{5}i$$

b) Multiply the numerator and denominator by $2 - i$, the conjugate of $2 + i$:

$$\frac{3 - i}{2 + i} = \frac{(3 - i)(2 - i)}{(2 + i)(2 - i)}$$

$$= \frac{6 - 5i + i^2}{4 - i^2}$$

$$= \frac{6 - 5i - 1}{4 - (-1)}$$

$$= \frac{5 - 5i}{5}$$

$$= 1 - i$$

c) Multiply the numerator and denominator by $-i$, the conjugate of i:

$$\frac{3 + 2i}{i} = \frac{(3 + 2i)(-i)}{i(-i)}$$

$$= \frac{-3i - 2i^2}{-i^2}$$

$$= \frac{-3i + 2}{1}$$

$$= 2 - 3i$$

> **Now do Exercises 47–58**

The symbolic definition of division of complex numbers follows.

Division of Complex Numbers

We divide the complex number $a + bi$ by the complex number $c + di$ as follows:

$$\frac{a + bi}{c + di} = \frac{(a + bi)(c - di)}{(c + di)(c - di)}$$

⟨4⟩ Square Roots of Negative Numbers

In the complex number system, negative numbers have two square roots. Because $i^2 = -1$ and $(-i)^2 = -1$, both i and $-i$ are square roots of -1. Because $(2i)^2 = -4$ and $(-2i)^2 = -4$, both $2i$ and $-2i$ are square roots of -4. We use the radical symbol only for the square root that has the positive coefficient, as in $\sqrt{-4} = 2i$.

> **Square Root of a Negative Number**
>
> For any positive real number b,
>
> $$\sqrt{-b} = i\sqrt{b}.$$

For example, $\sqrt{-9} = i\sqrt{9} = 3i$ and $\sqrt{-7} = i\sqrt{7}$. Note that the expression $\sqrt{7}i$ could easily be mistaken for the expression $\sqrt{7i}$, where i is under the radical. For this reason, when the coefficient of i is a radical, we write i preceding the radical.

Note that the product rule $(\sqrt{a} \cdot \sqrt{b} = \sqrt{ab})$ does not apply to negative numbers. For example $\sqrt{-2} \cdot \sqrt{-3} \neq \sqrt{6}$:

$$\sqrt{-2} \cdot \sqrt{-3} = i\sqrt{2} \cdot i\sqrt{3} = i^2\sqrt{6} = -\sqrt{6}$$

Square roots of negative numbers must be written in terms of i before operations are performed.

EXAMPLE **6**

Square roots of negative numbers
Write each expression in the form $a + bi$, where a and b are real numbers.

a) $3 + \sqrt{-9}$

b) $\sqrt{-12} + \sqrt{-27}$

c) $\dfrac{-1 - \sqrt{-18}}{3}$

d) $\sqrt{-4} \cdot \sqrt{-9}$

Solution

a) $3 + \sqrt{-9} = 3 + i\sqrt{9}$
$\qquad\qquad\quad = 3 + 3i$

b) $\sqrt{-12} + \sqrt{-27} = i\sqrt{12} + i\sqrt{27}$
$\qquad\qquad\qquad\quad = 2i\sqrt{3} + 3i\sqrt{3}$ $\begin{aligned}\sqrt{12} &= \sqrt{4}\,\sqrt{3} = 2\sqrt{3} \\ \sqrt{27} &= \sqrt{9}\,\sqrt{3} = 3\sqrt{3}\end{aligned}$
$\qquad\qquad\qquad\quad = 5i\sqrt{3}$

c) $\dfrac{-1 - \sqrt{-18}}{3} = \dfrac{-1 - i\sqrt{18}}{3}$

$\qquad\qquad\quad = \dfrac{-1 - 3i\sqrt{2}}{3}$

$\qquad\qquad\quad = -\dfrac{1}{3} - i\sqrt{2}$

d) $\sqrt{-4} \cdot \sqrt{-9} = i\sqrt{4} \cdot i\sqrt{9} = 2i \cdot 3i = 6i^2 = -6$

> **Now do Exercises 59–78**

⟨5⟩ Imaginary Solutions to Equations

In the complex number system the even-root property can be restated so that $x^2 = k$ is equivalent to $x = \pm\sqrt{k}$ for any $k \neq 0$. So an equation such as $x^2 = -9$ that has no real solutions has two imaginary solutions in the complex numbers.

EXAMPLE   **7**

Imaginary solutions to equations

Find the imaginary solutions to each equation.

a) $x^2 = -9$

b) $3x^2 + 2 = 0$

Solution

a) First apply the even-root property:

$$x^2 = -9$$
$$x = \pm\sqrt{-9} \quad \text{Even-root property}$$
$$= \pm i\sqrt{9}$$
$$= \pm 3i$$

Check these solutions in the original equation:

$$(3i)^2 = 9i^2 = 9(-1) = -9$$
$$(-3i)^2 = 9i^2 = -9$$

The solution set is $\{\pm 3i\}$.

b) First solve the equation for x^2:

$$3x^2 + 2 = 0$$
$$x^2 = -\frac{2}{3}$$
$$x = \pm\sqrt{-\frac{2}{3}} = \pm i\sqrt{\frac{2}{3}} = \pm i\frac{\sqrt{6}}{3}$$

Check these solutions in the original equation. The solution set is $\left\{\pm i\dfrac{\sqrt{6}}{3}\right\}$.

> **Now do Exercises 79–86**

The basic facts about complex numbers are listed in the following box.

Complex Numbers

1. Definition of i: $i = \sqrt{-1}$, and $i^2 = -1$.

2. A complex number has the form $a + bi$, where a and b are real numbers.

3. The complex number $a + 0i$ is the real number a.

4. If b is a positive real number, then $\sqrt{-b} = i\sqrt{b}$.

5. The numbers $a + bi$ and $a - bi$ are called complex conjugates of each other. Their product is the real number $a^2 + b^2$.

6. Add, subtract, and multiply complex numbers as if they were algebraic expressions with i being the variable, and replace i^2 by -1.

7. Divide complex numbers by multiplying the numerator and denominator by the conjugate of the denominator.

8. In the complex number system, $x^2 = k$ for any real number k is equivalent to $x = \pm\sqrt{k}$.

Warm-Ups ▼

Fill in the blank.

1. A(n) _____ number is a number of the form $a + bi$ where a and b are real numbers.

2. A(n) _____ number is a complex number in which $b \neq 0$.

3. The union of the real numbers and the imaginary numbers is the set of _____ numbers.

4. The _____ of $a + bi$ is $a - bi$.

5. The _____ of $a + bi$ and $a - bi$ is $a^2 + b^2$.

6. If $b = 0$, then $a + bi$ is a ___ number.

7. The set of real numbers is a _____ of the set of complex numbers.

True or false?

8. $\sqrt{-1} = i$

9. $2 - \sqrt{-6} = 2 - 6i$

10. $\sqrt{-9} = \pm 3$

11. $2 - 3i - (4 - 2i) = -2 - i$

12. $i^4 = 1$

13. $i^3 = i$

14. $i^{48} = 1$

15. $(2 - i)(2 + i) = 5$

16. If $x^2 = -9$, then $x = \pm 3i$.

7.6 Exercises

⟨ **Study Tips** ⟩

- When studying for a midterm or final, review the material in the order it was originally presented. This strategy will help you to see connections between the ideas.
- Studying the oldest material first will give top priority to material that you might have forgotten.

⟨2⟩ **Addition, Subtraction, and Multiplication**

Find the indicated sums and differences of complex numbers. See Example 1.

1. $(2 + 3i) + (-4 + 5i)$

2. $(-1 + 6i) + (5 - 4i)$

3. $(2 - 3i) - (6 - 7i)$

4. $(2 - 3i) - (6 - 2i)$

5. $(-1 + i) + (-1 - i)$

6. $(-5 + i) + (-5 - i)$

7. $(-2 - 3i) - (6 - i)$

8. $(-6 + 4i) - (2 - i)$

Find each product. Express each answer in the form $a + bi$. See Example 2.

9. $3(2 + 5i)$

10. $4(1 - 3i)$

11. $2i(i - 5)$

12. $3i(2 - 6i)$

13. $-4i(3 - i)$

14. $-5i(2 + 3i)$

15. $(2 + 3i)(4 + 6i)$

16. $(2 + i)(3 + 4i)$

17. $(-1 + i)(2 - i)$

18. $(3 - 2i)(2 - 5i)$

19. $(-1 - 2i)(2 + i)$

20. $(1 - 3i)(1 + 3i)$

21. $(5 - 2i)(5 + 2i)$

22. $(4 + 3i)(4 + 3i)$

23. $(1 - i)(1 + i)$

24. $(2 + 6i)(2 - 6i)$

25. $(4 + 2i)(4 - 2i)$

26. $(4 - i)(4 + i)$

Find the indicated powers of complex numbers.
See Example 3.

27. $(3i)^2$ **28.** $(5i)^2$

29. $(-5i)^2$ **30.** $(-9i)^2$

31. $(2i)^4$ **32.** $(-2i)^3$

33. i^9 **34.** i^{12}

35. i^{18} **36.** i^{33}

37. i^{25} **38.** i^{31}

⟨3⟩ Division of Complex Numbers

Find the product of the given complex number and its
conjugate. See Example 4.

39. $3 + 5i$ **40.** $3 + i$

41. $1 - 2i$ **42.** $4 - 6i$

43. $-2 + i$ **44.** $-3 - 2i$

45. $2 - i\sqrt{3}$ **46.** $\sqrt{5} - 4i$

Find each quotient. Express each answer in the form a + bi.
See Example 5.

47. $\dfrac{3}{4 + i}$ **48.** $\dfrac{6}{7 - 2i}$

 49. $\dfrac{2 + i}{3 - 2i}$ **50.** $\dfrac{3 + 5i}{2 - i}$

51. $\dfrac{4 + 3i}{i}$ **52.** $\dfrac{5 - 6i}{3i}$

53. $\dfrac{2 + 6i}{2}$ **54.** $\dfrac{9 - 3i}{-6}$

55. $\dfrac{1 + i}{3i - 2}$ **56.** $\dfrac{2 + i}{i + 5}$

57. $\dfrac{6}{3i}$ **58.** $\dfrac{8}{-2i}$

⟨4⟩ Square Roots of Negative Numbers

Write each expression in the form a + bi, where a and b are
real numbers. See Example 6.

59. $\sqrt{-25}$ **60.** $\sqrt{-81}$

61. $2 + \sqrt{-4}$ **62.** $3 + \sqrt{-9}$

63. $2\sqrt{-9} + 5$ **64.** $3\sqrt{-16} + 2$

65. $7 - \sqrt{-6}$ **66.** $\sqrt{-5} + 3$

67. $\sqrt{-8} + \sqrt{-18}$ **68.** $2\sqrt{-20} - \sqrt{-45}$

69. $\dfrac{2 + \sqrt{-12}}{2}$ **70.** $\dfrac{-6 - \sqrt{-18}}{3}$

71. $\dfrac{-4 - \sqrt{-24}}{4}$ **72.** $\dfrac{8 + \sqrt{-20}}{-4}$

73. $\sqrt{-2} \cdot \sqrt{-6}$ **74.** $\sqrt{-3} \cdot \sqrt{-15}$

75. $\sqrt{-3} \cdot \sqrt{-27}$ **76.** $\sqrt{-3} \cdot \sqrt{-7}$

77. $\dfrac{\sqrt{8}}{\sqrt{-4}}$ **78.** $\dfrac{\sqrt{6}}{\sqrt{-2}}$

⟨5⟩ Imaginary Solutions to Equations

Find the imaginary solutions to each equation. See Example 7.

79. $x^2 = -36$ **80.** $x^2 + 4 = 0$

81. $x^2 = -12$ **82.** $x^2 = -25$

83. $2x^2 + 5 = 0$ **84.** $3x^2 + 4 = 0$

85. $3x^2 + 6 = 0$ **86.** $x^2 + 1 = 0$

Miscellaneous

Write each expression in the form a + bi, where a and b are
real numbers.

87. $(2 - 3i)(3 + 4i)$ **88.** $(2 - 3i)(2 + 3i)$

89. $(2 - 3i) + (3 + 4i)$ **90.** $(3 - 5i) - (2 - 7i)$

91. $\dfrac{2 - 3i}{3 + 4i}$ **92.** $\dfrac{-3i}{3 - 6i}$

93. $i(2 - 3i)$ **94.** $-3i(4i - 1)$

95. $(-3i)^2$ **96.** $(-2i)^6$

97. $\sqrt{-12} + \sqrt{-3}$ **98.** $\sqrt{-49} - \sqrt{-25}$

99. $(2 - 3i)^2$

100. $(5 + 3i)^2$

101. $\dfrac{-4 + \sqrt{-32}}{2}$

102. $\dfrac{-2 - \sqrt{-27}}{-6}$

Getting More Involved

103. *Writing*

Explain why $2 - i$ is a solution to

$$x^2 - 4x + 5 = 0.$$

104. *Cooperative learning*

Work with a group to verify that $-1 + i\sqrt{3}$ and $-1 - i\sqrt{3}$ satisfy the equation

$$x^3 - 8 = 0.$$

In the complex number system there are three cube roots of 8. What are they?

105. *Discussion*

What is wrong with using the product rule for radicals to get

$$\sqrt{-4} \cdot \sqrt{-4} = \sqrt{(-4)(-4)} = \sqrt{16} = 4?$$

What is the correct product?

7 Wrap-Up

Summary

Powers and Roots		Examples
nth roots	If $a = b^n$ for a positive integer n, then b is an nth root of a.	2 and -2 are fourth roots of 16.
Principal root	The positive even root of a positive number	The principal fourth root of 16 is 2.
Radical notation	If n is a positive even integer and a is positive, then the symbol $\sqrt[n]{a}$ denotes the principal nth root of a. If n is a positive odd integer, then the symbol $\sqrt[n]{a}$ denotes the nth root of a. If n is any positive integer, then $\sqrt[n]{0} = 0$. If x is any real number, then use is absolute value when an even root of an exponential expression has an odd exponent.	$\sqrt[4]{16} = 2$ $\sqrt[4]{16} \neq -2$ $\sqrt[3]{-8} = -2, \sqrt[3]{8} = 2$ $\sqrt[5]{0} = 0, \sqrt[6]{0} = 0$ $\sqrt{x^2} = |x|, \sqrt{x^6} = |x^3|$ $\sqrt{x^4} = x^2, \sqrt[4]{x^4} = |x|$
Domain of a radical function	The set of all real numbers that can be used in place of the variable in the radical expression defining the function	$f(x) = \sqrt{x}$, domain $[0, \infty)$ $f(x) = \sqrt[3]{x-1}$, domain $(-\infty, \infty)$ $f(x) = \sqrt[4]{x-5}$, domain $[5, \infty)$
Definition of $a^{1/n}$	If n is any positive integer, then $a^{1/n} = \sqrt[n]{a}$, provided that $\sqrt[n]{a}$ is a real number.	$8^{1/3} = \sqrt[3]{8} = 2$ $(-4)^{1/2}$ is not real.
Definition of $a^{m/n}$	If m and n are positive integers, then $a^{m/n} = (a^{1/n})^m$, provided that $a^{1/n}$ is a real number.	$8^{2/3} = (8^{1/3})^2 = 2^2 = 4$ $(-16)^{3/4}$ is not real.
Definition of $a^{-m/n}$	If m and n are positive integers and $a \neq 0$, then $a^{-m/n} = \frac{1}{a^{m/n}}$, provided that $a^{1/n}$ is a real number.	$8^{-2/3} = \frac{1}{8^{2/3}} = \frac{1}{4}$

Rules for Radicals		Examples
Product rule for radicals	Provided that all roots are real, $\sqrt[n]{ab} = \sqrt[n]{a} \cdot \sqrt[n]{b}$.	$\sqrt{2} \cdot \sqrt{3} = \sqrt{6}$ $\sqrt{4x} = 2\sqrt{x}$

Quotient rule for radicals	Provided that all roots are real and $b \neq 0$, $$\sqrt[n]{\frac{a}{b}} = \frac{\sqrt[n]{a}}{\sqrt[n]{b}}.$$	$$\sqrt{\frac{5}{9}} = \frac{\sqrt{5}}{3}$$ $$\sqrt{10} \div \sqrt{5} = \sqrt{2}$$
Simplified radical form for radicals of index n	A simplified radical of index n has 1. *no* perfect nth powers as factors of the radicand, 2. *no* fractions inside the radical, and 3. *no* radicals in the denominator.	$$\sqrt{20} = \sqrt{4 \cdot 5} = 2\sqrt{5}$$ $$\sqrt{\frac{3}{2}} = \frac{\sqrt{3}}{\sqrt{2}}$$ $$\frac{\sqrt{3}}{\sqrt{2}} = \frac{\sqrt{3}}{\sqrt{2}} \cdot \frac{\sqrt{2}}{\sqrt{2}} = \frac{\sqrt{6}}{2}$$

Rules for Rational Exponents		**Examples**
If a and b are nonzero real numbers and r and s are rational numbers, then the following rules hold, provided all expressions represent real numbers.		
Product rule	$a^r \cdot a^s = a^{r+s}$	$3^{1/4} \cdot 3^{1/2} = 3^{3/4}$
Quotient rule	$\dfrac{a^r}{a^s} = a^{r-s}$	$\dfrac{x^{3/4}}{x^{1/4}} = x^{1/2}$
Power of a power rule	$(a^r)^s = a^{rs}$	$(2^{1/2})^{-1/2} = 2^{-1/4}$ $(x^{3/4})^4 = x^3$
Power of a product rule	$(ab)^r = a^r b^r$	$(a^2 b^6)^{1/2} = ab^3$
Power of a quotient rule	$\left(\dfrac{a}{b}\right)^r = \dfrac{a^r}{b^r}$	$\left(\dfrac{8}{x^6}\right)^{2/3} = \dfrac{4}{x^4}$

Equations		**Examples**
Equations with radicals and exponents	1. In raising each side of an equation to an even power, we can create an equation that gives extraneous solutions. We must check.	$\sqrt{x} = -3$ $x = 9$
	2. When applying the even-root property, remember that there is a positive and a negative root.	$x^2 = 36$ $x = \pm 6$
	3. For equations with rational exponents, raise each side to a positive or a negative power first, and then apply the even- or odd-root property.	$x^{-2/3} = 4$ $(x^{-2/3})^{-3} = 4^{-3}$ $x^2 = \dfrac{1}{64}$ $x = \pm\dfrac{1}{8}$

Complex Numbers		Examples
Complex numbers	Numbers of the form $a + bi$, where a and b are real numbers: $i = \sqrt{-1}$, $i^2 = -1$	$2 + 3i$ $-6i$ $\sqrt{2} + i$
Complex conjugates	Complex numbers of the form $a + bi$ and $a - bi$: Their product is the real number $a^2 + b^2$.	$(2 + 3i)(2 - 3i) = 2^2 + 3^2$ $= 13$
Complex number operations	Add, subtract, and multiply as algebraic expressions with i being the variable. Simplify using $i^2 = -1$.	$(2 + 5i) + (4 - 2i) = 6 + 3i$ $(2 + 5i) - (4 - 2i) = -2 + 7i$ $(2 + 5i)(4 - 2i) = 18 + 16i$
	Divide complex numbers by multiplying the numerator and denominator by the conjugate of the denominator.	$(2 + 5i) \div (4 - 2i)$ $= \dfrac{(2 + 5i)(4 + 2i)}{(4 - 2i)(4 + 2i)}$ $= \dfrac{-2 + 24i}{20} = -\dfrac{1}{10} + \dfrac{6}{5}i$
Square root of a negative number	For any positive real number b, $\sqrt{-b} = i\sqrt{b}$.	$\sqrt{-9} = i\sqrt{9} = 3i$
Imaginary solutions to equations	In the complex number system, $x^2 = k$ for any real k is equivalent to $x = \pm\sqrt{k}$.	$x^2 = -25$ $x = \pm\sqrt{-25} = \pm 5i$

Enriching Your Mathematical Word Power

Fill in the blank.

1. A number b such that $b^n = a$ is a(n) ___ root of a.

2. The _____ of a is a^2.

3. A number b such that $b^3 = a$ is a(n) ____ root of a.

4. The positive even root of a positive number is the _____ root of the number.

5. If n is an odd number and $b^n = a$, then b is a(n) ___ root of a.

6. In the notation $\sqrt[n]{a}$, n is the _____ of the radical.

7. In the notation $\sqrt[n]{a}$, a is the _____.

8. The set of all real numbers that can be used in place of the variable in a radical function is the _____.

9. A(n) _____ exponent is an exponent that is an integer.

10. A(n) _____ exponent is an exponent that is a rational number.

11. Radicals with the same index and the same radicand are ____ radicals.

12. A number of the form $a + bi$ is a(n) _____ number.

13. In the complex number system, i is the _____ unit.

14. A number of the form $a + bi$ where $b \neq 0$ is a(n) _____ number.

15. The numbers $a + bi$ and $a - bi$ are complex _____.

Review Exercises

7.1 Radicals

Find each root. Variables represent any real numbers. Use absolute value when necessary.

1. $\sqrt{81}$

2. $\sqrt[4]{16}$

3. $\sqrt[3]{27}$

4. $\sqrt[5]{32}$

5. $\sqrt[3]{-27}$

6. $\sqrt[5]{-32}$

7. $\sqrt{100}$

8. $\sqrt[3]{1000}$

9. $\sqrt{y^2}$

10. $\sqrt[4]{y^{12}}$

11. $\sqrt[3]{a^3}$

12. $\sqrt[5]{b^{10}}$

13. $\sqrt[4]{n^{24}}$

14. $\sqrt{m^{32}}$

15. $\sqrt[4]{n^{20}}$

16. $\sqrt{m^{30}}$

Simplify each radical expression. Assume all variables represent positive real numbers.

17. $\sqrt{72}$

18. $\sqrt{48}$

19. $\sqrt{x^{12}}$

20. $\sqrt{a^{10}}$

21. $\sqrt[3]{x^6}$

22. $\sqrt[3]{a^9}$

23. $\sqrt{2x^9}$

24. $\sqrt{3a^7}$

25. $\sqrt{8w^5}$

26. $\sqrt{20n^{25}}$

27. $\sqrt[3]{16x^4}$

28. $\sqrt[3]{54b^5}$

29. $\sqrt[4]{a^9b^5}$

30. $\sqrt[4]{32m^{11}}$

31. $\sqrt{\dfrac{x^3}{16}}$

32. $\sqrt{\dfrac{12a^3}{25}}$

Find the domain of each radical function. Use interval notation.

33. $f(x) = \sqrt{2x - 5}$

34. $f(x) = \sqrt{3x + 12}$

35. $f(x) = \sqrt[3]{7x - 1}$

36. $f(x) = \sqrt[3]{9 - 2x}$

37. $f(x) = \sqrt[4]{-3x + 1}$

38. $f(x) = \sqrt[4]{-5x - 1}$

39. $f(x) = \sqrt{\dfrac{1}{2}x + 1}$

40. $f(x) = \sqrt{\dfrac{2}{3}x - 2}$

7.2 Rational Exponents

Simplify the expressions involving rational exponents. Assume all variables represent positive real numbers. Write your answers with positive exponents.

41. $(-27)^{-2/3}$

42. $-25^{3/2}$

43. $(2^6)^{1/3}$

44. $(5^2)^{1/2}$

45. $100^{-3/2}$

46. $1000^{-2/3}$

47. $\dfrac{3x^{-1/2}}{3^{-2}x^{-1}}$

48. $\dfrac{(x^2y^{-3}z)^{1/2}}{x^{1/2}yz^{-1/2}}$

49. $(a^{1/2}b)^3(ab^{1/4})^2$

50. $(t^{-1/2})^{-2}(t^{-2}v^2)$

51. $(x^{1/2}y^{1/4})(x^{1/4}y)$

52. $(a^{1/3}b^{1/6})^2(a^{1/3}b^{2/3})$

7.3 Adding, Subtracting, and Multiplying Radicals

Perform the operations and simplify. Assume the variables represent positive real numbers.

53. $\sqrt{13} \cdot \sqrt{13}$

54. $\sqrt[3]{14} \cdot \sqrt[3]{14} \cdot \sqrt[3]{14}$

55. $\sqrt{27} + \sqrt{45} - \sqrt{75}$

56. $\sqrt{12} - \sqrt{50} + \sqrt{72}$

57. $3\sqrt{2}\,(5\sqrt{2} - 7\sqrt{3})$

58. $-2\sqrt{a}\,(\sqrt{a} - \sqrt{ab^6})$

59. $(2 - \sqrt{3})(3 + \sqrt{2})$

60. $(2\sqrt{x} - \sqrt{y})(\sqrt{x} + \sqrt{y})$

7.4 Quotients, Powers, and Rationalizing Denominators

Perform the operations and simplify.

61. $5 \div \sqrt{2}$

62. $(10\sqrt{6}) \div (2\sqrt{2})$

63. $\sqrt{\dfrac{2}{5}}$

64. $\sqrt{\dfrac{1}{6}}$

65. $\sqrt[3]{\dfrac{2}{3}}$

66. $\sqrt[3]{\dfrac{1}{9}}$

67. $\dfrac{2}{\sqrt{3x}}$

68. $\dfrac{3}{\sqrt{2y}}$

69. $\dfrac{\sqrt{10y^3}}{\sqrt{6}}$

70. $\dfrac{\sqrt{5x^5}}{\sqrt{8}}$

71. $\dfrac{3}{\sqrt[3]{2a}}$

72. $\dfrac{a}{\sqrt[3]{a^2}}$

73. $\dfrac{5}{\sqrt[4]{3x^2}}$

74. $\dfrac{b}{\sqrt[4]{a^2b^3}}$

75. $(\sqrt{3})^4$

76. $(-2\sqrt{x})^9$

77. $\dfrac{2 - \sqrt{8}}{2}$

78. $\dfrac{-3 - \sqrt{18}}{-6}$

79. $\dfrac{\sqrt{6}}{1 - \sqrt{3}}$

80. $\dfrac{\sqrt{15}}{2 + \sqrt{5}}$

81. $\dfrac{2\sqrt{3}}{3\sqrt{6} - \sqrt{12}}$

82. $\dfrac{-\sqrt{xy}}{3\sqrt{x} + \sqrt{xy}}$

83. $(2w\sqrt[3]{2w^2})^6$

84. $(m\sqrt[4]{m^3})^8$

7.5 Solving Equations with Radicals and Exponents

Find all real solutions to each equation.

85. $x^2 = 16$

86. $w^2 = 100$

87. $(a - 5)^2 = 4$

88. $(m - 7)^2 = 25$

89. $(a + 1)^2 = 5$

90. $(x + 5)^2 = 3$

91. $(m + 1)^2 = -8$

92. $(w + 4)^2 = 16$

93. $\sqrt{m - 1} = 3$

94. $3\sqrt{x + 5} = 12$

95. $\sqrt[3]{2x + 9} = 3$

96. $\sqrt[4]{2x - 1} = 2$

97. $w^{2/3} = 4$

98. $m^{-4/3} = 16$

99. $(m + 1)^{1/3} = 5$

100. $(w - 3)^{-2/3} = 4$

101. $\sqrt{x - 3} = \sqrt{x + 2} - 1$

102. $\sqrt{x^2 + 3x + 6} = 4$

103. $\sqrt{5x - x^2} = \sqrt{6}$

104. $\sqrt{x + 4} - 2\sqrt{x - 1} = -1$

105. $\sqrt{x + 7} - 2\sqrt{x} = -2$

106. $\sqrt{x} - \sqrt{x - 1} = 1$

107. $2\sqrt{x} - \sqrt{x - 3} = 3$

108. $1 + \sqrt{x + 7} = \sqrt{2x + 7}$

7.6 Complex Numbers

Perform the indicated operations. Write answers in the form a + bi.

109. $(2 - 3i)(-5 + 5i)$

110. $(2 + i)(5 - 2i)$

111. $(2 + i) + (5 - 4i)$

112. $(2 + i) + (3 - 6i)$

113. $(1 - i) - (2 - 3i)$

114. $(3 - 2i) - (1 - i)$

115. $\dfrac{6 + 3i}{3}$

116. $\dfrac{8 + 12i}{4}$

117. $\dfrac{4 - \sqrt{-12}}{2}$

118. $\dfrac{6 + \sqrt{-18}}{3}$

119. $\dfrac{2 - 3i}{4 + i}$

120. $\dfrac{3 + i}{2 - 3i}$

121. $(-2i)^4$

122. $(-2i)^5$

123. i^{14}

124. i^{21}

Find the imaginary solutions to each equation.

125. $x^2 + 100 = 0$

126. $25a^2 + 3 = 0$

127. $2b^2 + 9 = 0$

128. $3y^2 + 8 = 0$

Miscellaneous

Determine whether each equation is true or false, and explain your answer. An equation involving variables should be marked true only if it is an identity. Do not use a calculator.

129. $2^3 \cdot 3^2 = 6^5$

130. $16^{1/4} = 4^{1/2}$

131. $(\sqrt{2})^3 = 2\sqrt{2}$

132. $\sqrt[3]{9} = 3$

133. $8^{200} \cdot 8^{200} = 64^{200}$

134. $\sqrt{295} \cdot \sqrt{295} = 295$

135. $4^{1/2} = \sqrt{2}$

136. $\sqrt{a^2} = |a|$

137. $5^2 \cdot 5^2 = 25^4$

138. $\sqrt{6} \div \sqrt{2} = \sqrt{3}$

139. $\sqrt{w^{10}} = w^5$

140. $\sqrt{a^{16}} = a^4$

141. $\sqrt{x^6} = x^3$

142. $\sqrt[6]{16} = \sqrt[3]{4}$

143. $\sqrt{x^8} = x^4$

144. $\sqrt[9]{2^6} = 2^{2/3}$

145. $\sqrt{16} = 2$

146. $2^{1/2} \cdot 2^{1/4} = 2^{3/4}$

147. $2^{600} = 4^{300}$

148. $\sqrt{2} \cdot \sqrt[4]{2} = \sqrt[6]{2}$

149. $\dfrac{2 + \sqrt{6}}{2} = 1 + \sqrt{6}$

150. $\dfrac{4 + 2\sqrt{3}}{2} = 2 + \sqrt{3}$

151. $\sqrt{\dfrac{4}{6}} = \dfrac{2}{3}$

152. $8^{200} \cdot 8^{200} = 8^{400}$

153. $81^{2/4} = 81^{1/2}$

154. $(-64)^{2/6} = (-64)^{1/3}$

155. $(a^4 b^2)^{1/2} = |a^2 b|$

156. $\left(\dfrac{a^2}{b^6}\right)^{1/2} = \dfrac{|a|}{b^3}$

Solve each problem.

157. *Falling objects.* Neglecting air resistance, the number of feet s that an object falls from rest during t seconds is given by the function $s = 16t^2$. How long would it take an object to reach the earth if it is dropped from 12,000 feet?

158. *Timber.* Anne is pulling on a 60-foot rope attached to the top of a 48-foot tree while Walter is cutting the tree at its base. How far from the base of the tree is Anne standing?

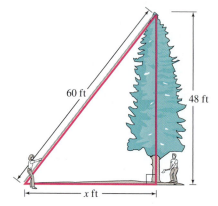

60 ft   48 ft

x ft

Figure for Exercise 158

159. *Guy wire.* If a guy wire of length 40 feet is attached to an antenna at a height of 30 feet, then how far from the base of the antenna is the wire attached to the ground?

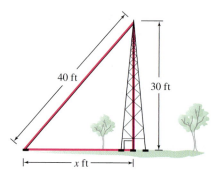

40 ft   30 ft

x ft

Figure for Exercise 159

160. *Touchdown.* Suppose at the kickoff of a football game, the receiver catches the football at the left side of the goal line and runs for a touchdown diagonally across the field. How many yards would he run? (A football field is 100 yards long and 160 feet wide.)

161. *Long guy wires.* The manufacturer of an antenna recommends that guy wires from the top of the antenna to the ground be attached to the ground at a distance from the base equal to the height of the antenna. How long would the guy wires be for a 200-foot antenna?

162. *Height of a post.* Betty observed that the lamp post in front of her house casts a shadow of length 8 feet when the angle of inclination of the sun is 60 degrees. How tall is the lamp post? (In a 30-60-90 right triangle, the side opposite 30 is one-half the length of the hypotenuse.)

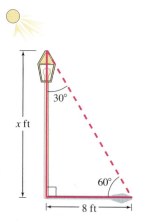

30°

x ft

60°

8 ft

Figure for Exercise 162

163. *Manufacturing a box.* A cubic box has a volume of 40 cubic feet. The amount of recycled cardboard that it takes to make the six-sided box is 10% larger than the surface area of the box. Find the exact amount of recycled cardboard used in manufacturing the box.

164. *Shipping parts.* A cubic box with a volume of 32 cubic feet is to be used to ship some machine parts. All of the parts are small except for a long, straight steel connecting rod. What is the maximum length of a connecting rod that will fit into this box?

165. *Health care costs.* The total annual cost of health care in the United States grew from \$993.3 billion in 1995 to \$2151.1 billion in 2009 (*Statistical Abstract of the United States*, www.census.gov).

Figure for Exercise 165

a) Find the average annual rate of growth r for that period by solving $2151.1 = 993.3(1 + r)^{14}$.

b) Estimate the total annual cost of health care in 2015 by reading the accompanying graph.

166. *Population growth.* The function $P = P_0(1 + r)^n$ gives the population P at the end of an n-year time period, where P_0 is the initial population and r is the average annual growth rate. The U.S. population grew from

248.7 million in 1990 to 307.8 million in 2009 (U.S. Census Bureau). Find r for that period.

167. *Landing speed.* Aircraft engineers determine the proper landing speed V (in feet per second) for an airplane from the function

$$V = \sqrt{\frac{841L}{CS}},$$

where L is the gross weight of the aircraft in pounds, C is the coefficient of lift, and S is the wing surface area in square feet. Rewrite the formula so that the expression on the right-hand side is in simplified radical form.

168. *Spillway capacity.* Civil engineers use the formula

$$Q = 3.32LH^{3/2}$$

to find the maximum discharge that the dam (a broad-crested weir) shown in the figure can pass before the water breaches its abutments (*Standard Handbook for Civil Engineers*, 1968). In the formula, Q is the discharge in cubic feet per second, L is the length of the spillway in feet, and H is the depth of the spillway. Find Q given that $L = 60$ feet and $H = 5$ feet. Find H given that $Q = 3000$ cubic feet per second and $L = 70$ feet.

Figure for Exercise 168

Chapter 7 Test

Find each root. Variables represent any real numbers. Use absolute value when necessary.

1. $\sqrt{36}$

2. $\sqrt[3]{-125}$

3. $\sqrt{t^2}$

4. $\sqrt[3]{p^3}$

5. $\sqrt[4]{w^8}$

6. $\sqrt[4]{w^{12}}$

Simplify each expression. Assume all variables represent positive numbers.

7. $8^{2/3}$

8. $4^{-3/2}$

9. $\sqrt{21} \div \sqrt{7}$

10. $2\sqrt{5} \cdot 3\sqrt{5}$

11. $\sqrt{20} + \sqrt{5}$

12. $\sqrt{5} + \dfrac{1}{\sqrt{5}}$

13. $2^{1/2} \cdot 2^{1/2}$

14. $\sqrt{72}$

15. $\sqrt{\dfrac{5}{12}}$

16. $\dfrac{6 + \sqrt{18}}{6}$

17. $(2\sqrt{3} + 1)(\sqrt{3} - 2)$

18. $\sqrt[4]{32a^5 y^8}$

19. $\dfrac{1}{\sqrt[3]{2x^2}}$

20. $\sqrt{\dfrac{8a^9}{b^3}}$

21. $\sqrt[3]{-27x^9}$

22. $\sqrt{20m^3}$

23. $x^{1/2} \cdot x^{1/4}$

24. $\left(9y^4 x^{1/2}\right)^{1/2}$

25. $\sqrt[3]{40x^7}$

26. $(4 + \sqrt{3})^2$

Find the domain of each radical function. Use interval notation.

27. $f(x) = \sqrt{4 - x}$

28. $f(x) = \sqrt[3]{5x - 3}$

Rationalize the denominator and simplify.

29. $\dfrac{2}{5 - \sqrt{3}}$

30. $\dfrac{\sqrt{6}}{4\sqrt{3} + \sqrt{2}}$

Write each expression in the form a + bi.

31. $(3 - 2i)(4 + 5i)$

32. $i^4 - i^5$

33. $\dfrac{3 - i}{1 + 2i}$

34. $\dfrac{-6 + \sqrt{-12}}{8}$

Find all real or imaginary solutions to each equation.

35. $(x - 2)^2 = 49$

36. $2\sqrt{x + 4} = 3$

37. $w^{2/3} = 4$

38. $9y^2 + 16 = 0$

39. $\sqrt{2x^2 + x - 12} = x$

40. $\sqrt{x - 1} + \sqrt{x + 4} = 5$

Show a complete solution to each problem.

41. Find the exact length of the side of a square whose diagonal is 3 feet.

42. Two positive numbers differ by 11, and their square roots differ by 1. Find the numbers.

43. If the perimeter of a rectangle is 20 feet and the diagonal is $2\sqrt{13}$ feet, then what are the length and width?

44. The average radius R of the orbit of a planet around the sun is determined by $R = T^{2/3}$, where T is the number of years for one orbit and R is measured in astronomical units (AU). If it takes Pluto 248.530 years to make one orbit of the sun, then what is the average radius of the orbit of Pluto? If the average radius of the orbit of Neptune is 30.08 AU, then how many years does it take Neptune to complete one orbit of the sun?

*Making*Connections | **A Review of Chapters 1–7**

Evaluate each expression

1. $3 + 2\sqrt{14 - 2 \cdot 5}$

2. $4 - 3|5 - 2 \cdot 4|$

3. $5 - 2(6 - 2 \cdot 4^2)$

4. $\sqrt{13^2 - 12^2} + 6$

5. $\sqrt[3]{6^2 - 3^2} - 2^5$

6. $\sqrt[3]{4(7 + 3^2)} - 2^3$

7. $(4 + 3^2) \div |5 - 2 \cdot 9|$

8. $\sqrt{9 + 16} - |9 - 16|$

9. $\sqrt{(-30)^2 - 4 \cdot 9 \cdot 25}$

10. $\sqrt{(-23)^2 - 4 \cdot 12 \cdot 5}$

Which elements of
$$\left\{-5, -\sqrt[3]{2}, -\frac{1}{9}, 0, 1, \sqrt{4}, 2.99, \pi, \frac{33}{4}, 2 + 3i\right\}$$
are members of these sets?

11. Whole numbers

12. Natural numbers

13. Integers

14. Rational numbers

15. Irrational numbers

16. Real numbers

17. Imaginary numbers

18. Complex numbers

Fill in the blank.

19. Zero is the _____ identity.

20. One is the _____ identity.

21. According to the _____ property of addition, $a + b = b + a$ for all real numbers a and b.

22. According to the _____ property of multiplication, $ab = ba$ for all real numbers a and b.

23. According to the _____ property of addition, $a + (b + c) = (a + b) + c$ for all real numbers a, b, and c.

24. According to the _____ property, $a(b + c) = ab + ac$ for all real numbers a, b, and c.

25. Every real number a has a(n) _____ inverse $-a$ such that $a + (-a) = 0$.

26. Every nonzero real number a has a(n) _____ inverse $\frac{1}{a}$ such that $a \cdot \frac{1}{a} = 1$.

Find all real solutions to each equation or inequality. For the inequalities, also sketch the graph of the solution set.

27. $3(x - 2) + 5 = 7 - 4(x + 3)$

28. $\sqrt{6x + 7} = 4$

29. $|2x + 5| > 1$

30. $8x^3 - 27 = 0$

31. $2x - 3 > 3x - 4$

32. $\sqrt{2x - 3} - \sqrt{3x + 4} = 0$

33. $\dfrac{w}{3} + \dfrac{w - 4}{2} = \dfrac{11}{2}$

34. $2(x + 7) - 4 = x - (10 - x)$

35. $(x + 7)^2 = 25$

36. $a^{-1/2} = 4$

37. $x - 3 > 2$ or $x < 2x + 6$

38. $a^{-2/3} = 16$ **39.** $3x^2 - 1 = 0$

40. $5 - 2(x - 2) = 3x - 5(x - 2) - 1$

41. $|3x - 4| < 5$

42. $3x - 1 = 0$ **43.** $\sqrt{y - 1} = 9$

44. $|5(x - 2) + 1| = 3$

45. $0.06x - 0.04(x - 20) = 2.8$

46. $|3x - 1| > -2$ **47.** $\dfrac{3\sqrt{2}}{x} = \dfrac{\sqrt{3}}{4\sqrt{5}}$

48. $\dfrac{\sqrt{x} - 4}{x} = \dfrac{1}{\sqrt{x} + 5}$

49. $\dfrac{3\sqrt{2} + 4}{\sqrt{2}} = \dfrac{x\sqrt{18}}{3\sqrt{2} + 2}$

50. $\dfrac{x}{2\sqrt{5} - \sqrt{2}} = \dfrac{2\sqrt{5} + \sqrt{2}}{x}$

51. $\dfrac{\sqrt{2x} - 5}{x} = \dfrac{-3}{\sqrt{2x} + 5}$

52. $\dfrac{\sqrt{6} + 2}{x} = \dfrac{2}{\sqrt{6} + 4}$

53. $\dfrac{x - 1}{\sqrt{6}} = \dfrac{\sqrt{6}}{x}$

54. $\dfrac{x + 3}{\sqrt{10}} = \dfrac{\sqrt{10}}{x}$

55. $\dfrac{1}{x} - \dfrac{1}{x - 1} = -\dfrac{1}{6}$

56. $\dfrac{1}{x^2 - 2x} + \dfrac{1}{x} = \dfrac{2}{3}$

The expression $\dfrac{-b + \sqrt{b^2 - 4ac}}{2a}$ will be used in Chapter 8 to solve quadratic equations. Evaluate it for the given values of a, b, and c.

57. $a = 1, b = 2, c = -15$

58. $a = 1, b = 8, c = 12$

59. $a = 2, b = 5, c = -3$

60. $a = 6, b = 7, c = -3$

Solve each problem.

61. *Popping corn.* If 1 gram of popcorn with moisture content $x\%$ is popped in a hot-air popper, then the volume of popped corn v (in cubic centimeters) that results is modeled by the formula

$$v = -94.8 + 21.4x - 0.761x^2.$$

a) Use the formula to find the volume that results when 1 gram of popcorn with moisture content 11% is popped.

b) Use the accompanying graph to estimate the moisture content that will produce the maximum volume of popped corn.

c) Use the graph to estimate the maximum possible volume for popping 1 gram of popcorn in a hot-air popper.

Figure for Exercise 61

Critical Thinking | For Individual or Group Work | Chapter 7

These exercises can be solved by a variety of techniques, which may or may not require algebra. So be creative and think critically. Explain all answers. Answers are in the Instructor's Edition of this text.

1. ***Tricky square.*** Start with a square and write any integer at each vertex. (a) At the midpoint of each side write the absolute value of the difference between the numbers at the endpoints of that side. (b) Connect the midpoints to obtain another square. Repeat (a) and (b) to obtain a sequence of nested squares as shown in the accompanying figure. What numbers will you always end up with?

Figure for Exercise 1

2. ***Planning ahead.*** Thaddeus takes one month to build a kayak (K) and two months to build a canoe (C).

Photo for Exercise 2

While planning ahead for 1 month, he notes that there is only one thing to do that will not result in any partially finished boats; that is, build one kayak. For planning 2 months ahead there are two possibilities, KK or C. For a 3-month plan there are three possibilities, KKK or KC or CK.

   a) Find the number of possibilities for a 4-month plan, a 5-month plan, and a 6-month plan by listing the possibilities. Look for a pattern.

   b) Find the number of possibilities for a 7-month plan and an 8-month plan without making a list.

3. ***Five coins.*** Place five coins on a table with heads facing downward. On each move you must turn over exactly three coins. What is the minimum number of moves necessary to get all five heads facing upward?

4. ***Rotating tires.*** Helen bought a new car with four tires and a full-size spare. If she rotated the tires so that each tire would have the same amount of wear, then how many miles were on each tire when her odometer showed 40,000 miles?

5. ***Cutting pizza.*** What is the largest number of pieces of pizza you can get if you cut a circular pizza with five straight cuts? What is the largest number of pieces of pizza you can get if you cut a circular pizza with seven straight cuts?

6. ***Mysterious rectangle.*** The length of a rectangle is a two-digit number with identical digits (*aa*). The width of the rectangle is one-tenth of the length (*a.a*). The perimeter is numerically twice as large as the area. Find the length, width, perimeter, and area.

7. ***Finding squares.*** Evaluate the expression

   $$100^2 - 99^2 + 98^2 - 97^2 + 96^2 - \cdots - 3^2 + 2^2 - 1^2$$

   without using a calculator.

8. ***Five-digit sum.*** Find the sum of all five-digit numbers that are formed by using the digits 1, 2, 3, 4, and 5 once and only once.

8

Quadratic Equations, Functions, and Inequalities

Is it possible to measure beauty? For thousands of years artists and philosophers have been challenged to answer this question. The seventeenth-century philosopher John Locke said, "Beauty consists of a certain composition of color and figure causing delight in the beholder." Over the centuries many architects, sculptors, and painters have searched for beauty in their work by exploring numerical patterns in various art forms.

Today many artists and architects still use the concepts of beauty given to us by the ancient Greeks. One principle, called the Golden Rectangle, concerns the most pleasing proportions of a rectangle. The Golden Rectangle appears in nature as well as in many cultures. Examples of it can be seen in Leonardo da Vinci's *Proportions of the Human Figure* as well as in Indonesian temples and Chinese pagodas. Perhaps one of the best-known examples of the Golden Rectangle is in the façade and floor plan of the Parthenon, built in Athens in the fifth century B.C.

In Exercise 89 of Section 8.3 we will see that the principle of the Golden Rectangle is based on a proportion that we can solve using the quadratic formula.

8.1   Factoring and Completing the Square

Factoring and the even-root property were used to solve quadratic equations in Chapters 5, 6, and 7. In this section we first review those methods. Then you will learn the method of completing the square, which can be used to solve any quadratic equation.

⟨1⟩ Review of Factoring

A quadratic equation has the form $ax^2 + bx + c = 0$, where a, b, and c are real numbers with $a \neq 0$. In Section 5.8 we solved quadratic equations by factoring and then applying the zero factor property.

Zero Factor Property

The equation $ab = 0$ is equivalent to the compound equation

$$a = 0 \quad \text{or} \quad b = 0.$$

Of course we can only use the factoring method when we can factor the quadratic polynomial. To solve a quadratic equation by factoring we use the following strategy.

Strategy for Solving Quadratic Equations by Factoring

1. Write the equation with 0 on one side.
2. Factor the other side.
3. Use the zero factor property to set each factor equal to zero.
4. Solve the simpler equations.
5. Check the answers in the original equation.

E X A M P L E   **1**

Solving a quadratic equation by factoring
Solve $3x^2 - 4x = 15$ by factoring.

Solution

Subtract 15 from each side to get 0 on the right-hand side:

$$3x^2 - 4x - 15 = 0$$
$$(3x + 5)(x - 3) = 0 \quad \text{Factor the left-hand side.}$$
$$3x + 5 = 0 \quad \text{or} \quad x - 3 = 0 \quad \text{Zero factor property}$$
$$3x = -5 \quad \text{or} \quad x = 3$$
$$x = -\frac{5}{3}$$

The solution set is $\left\{-\frac{5}{3}, 3\right\}$. Check the solutions in the original equation.

⟨ **Helpful Hint** ⟩

After you have factored the quadratic polynomial, use FOIL to check that you have factored correctly before proceeding to the next step.

> Now do Exercises 1–10

⟨2⟩ Review of the Even-Root Property

In Chapter 7 we solved some simple quadratic equations by using the even-root property, which we restate as follows:

Even-Root Property

Suppose n is a positive even integer.

      If $k > 0$, then $x^n = k$ is equivalent to $x = \pm\sqrt[n]{k}$.

      If $k = 0$, then $x^n = k$ is equivalent to $x = 0$.

      If $k < 0$, then $x^n = k$ has no real solution.

By the even-root property $x^2 = 4$ is equivalent to $x = \pm 2$, $x^2 = 0$ is equivalent to $x = 0$, and $x^2 = -4$ has no real solutions.

E X A M P L E   2

Solving a quadratic equation by the even-root property

Solve $(a - 1)^2 = 9$.

Solution

By the even-root property $x^2 = k$ is equivalent to $x = \pm\sqrt{k}$.

$$(a - 1)^2 = 9$$

$$a - 1 = \pm\sqrt{9} \quad \text{Even-root property}$$

$$a - 1 = 3 \quad \text{or} \quad a - 1 = -3$$

$$a = 4 \quad \text{or} \quad a = -2$$

Check these solutions in the original equation. The solution set is $\{-2, 4\}$.

> Now do Exercises 11–20

⟨ **Helpful Hint** ⟩

The area of an x by x square and two x by 3 rectangles is $x^2 + 6x$. The area needed to "complete the square" in this figure is 9:

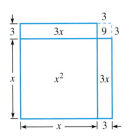

⟨3⟩ Completing the Square

We cannot solve every quadratic by factoring because not all quadratic polynomials can be factored. However, we can write any quadratic equation in the form of Example 2 and then apply the even-root property to solve it. This method is called **completing the square.**

    The essential part of completing the square is to recognize a perfect square trinomial when given its first two terms. For example, if we are given $x^2 + 6x$, how do we recognize that these are the first two terms of the perfect square trinomial $x^2 + 6x + 9$? To answer this question, recall that $x^2 + 6x + 9$ is a perfect square trinomial because it is the square of the binomial $x + 3$:

$$(x + 3)^2 = x^2 + 2 \cdot 3x + 3^2 = x^2 + 6x + 9$$

Notice that the 6 comes from multiplying 3 by 2 and the 9 comes from squaring the 3. So to find the missing 9 in $x^2 + 6x$, divide 6 by 2 to get 3, and then square 3 to get 9. This procedure can be used to find the last term in any perfect square trinomial in which the coefficient of x^2 is 1.

Rule for Finding the Last Term

The last term of a perfect square trinomial is the square of one-half of the coefficient of the middle term. In symbols, the perfect square trinomial whose first two terms are $x^2 + bx$ is $x^2 + bx + \left(\dfrac{b}{2}\right)^2$.

E X A M P L E   **3**

Finding the last term

Find the perfect square trinomial whose first two terms are given.

a) $x^2 + 8x$ **b)** $x^2 - 5x$ **c)** $x^2 + \dfrac{4}{7}x$ **d)** $x^2 - \dfrac{3}{2}x$

Solution

a) One-half of 8 is 4, and 4 squared is 16. So the perfect square trinomial is

$$x^2 + 8x + 16.$$

b) One-half of -5 is $-\dfrac{5}{2}$, and $-\dfrac{5}{2}$ squared is $\dfrac{25}{4}$. So the perfect square trinomial is

$$x^2 - 5x + \dfrac{25}{4}.$$

c) Since $\dfrac{1}{2} \cdot \dfrac{4}{7} = \dfrac{2}{7}$ and $\dfrac{2}{7}$ squared is $\dfrac{4}{49}$, the perfect square trinomial is

$$x^2 + \dfrac{4}{7}x + \dfrac{4}{49}.$$

d) Since $\dfrac{1}{2}\left(-\dfrac{3}{2}\right) = -\dfrac{3}{4}$ and $\left(-\dfrac{3}{4}\right)^2 = \dfrac{9}{16}$, the perfect square trinomial is

$$x^2 - \dfrac{3}{2}x + \dfrac{9}{16}.$$

> Now do Exercises 21–28

CAUTION The rule for finding the last term applies only to perfect square trinomials with $a = 1$. A trinomial such as $9x^2 + 6x + 1$ is a perfect square trinomial because it is $(3x + 1)^2$, but the last term is certainly not the square of one-half the coefficient of the middle term.

Another essential step in completing the square is to write the perfect square trinomial as the square of a binomial. Recall that

$$a^2 + 2ab + b^2 = (a + b)^2$$

and

$$a^2 - 2ab + b^2 = (a - b)^2.$$

E X A M P L E   **4**

Factoring perfect square trinomials

Factor each trinomial.

a) $x^2 + 12x + 36$ **b)** $y^2 - 7y + \dfrac{49}{4}$

c) $z^2 - \dfrac{4}{3}z + \dfrac{4}{9}$

Solution

a) The trinomial $x^2 + 12x + 36$ is of the form $a^2 + 2ab + b^2$ with $a = x$ and $b = 6$. So,

$$x^2 + 12x + 36 = (x + 6)^2.$$

Check by squaring $x + 6$.

b) The trinomial $y^2 - 7y + \frac{49}{4}$ is of the form $a^2 - 2ab + b^2$ with $a = y$ and $b = \frac{7}{2}$. So,

$$y^2 - 7y + \frac{49}{4} = \left(y - \frac{7}{2}\right)^2.$$

Check by squaring $y - \frac{7}{2}$.

c) The trinomial $z^2 - \frac{4}{3}z + \frac{4}{9}$ is of the form $a^2 - 2ab + b^2$ with $a = z$ and $b = -\frac{2}{3}$. So,

$$z^2 - \frac{4}{3}z + \frac{4}{9} = \left(z - \frac{2}{3}\right)^2.$$

Now do Exercises 29–36

In Example 5, we use the skills that we learned in Examples 2, 3, and 4 to solve the quadratic equation $ax^2 + bx + c = 0$ with $a = 1$ by the method of completing the square. This method works only if $a = 1$ because the method for completing the square developed in Examples 2, 3, and 4 works only for $a = 1$.

E X A M P L E   5

Completing the square with $a = 1$

Solve $x^2 + 6x + 5 = 0$ by completing the square.

Solution

The perfect square trinomial whose first two terms are $x^2 + 6x$ is

$$x^2 + 6x + 9.$$

So we move 5 to the right-hand side of the equation, and then add 9 to each side to create a perfect square on the left side:

$$
\begin{aligned}
x^2 + 6x &= -5 && \text{Subtract 5 from each side.} \\
x^2 + 6x + 9 &= -5 + 9 && \text{Add 9 to each side to get} \\
&&& \text{a perfect square trinomial.} \\
(x + 3)^2 &= 4 && \text{Factor the left-hand side.} \\
x + 3 &= \pm\sqrt{4} && \text{Even-root property}
\end{aligned}
$$

$$
\begin{array}{ccc}
x + 3 = 2 & \text{or} & x + 3 = -2 \\
x = -1 & \text{or} & x = -5
\end{array}
$$

Check in the original equation:

$$(-1)^2 + 6(-1) + 5 = 0$$

and

$$(-5)^2 + 6(-5) + 5 = 0$$

The solution set is $\{-1, -5\}$.

Now do Exercises 37–44

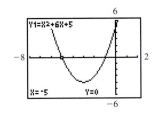

CAUTION All of the perfect square trinomials that we have used so far had a leading coefficient of 1. If $a \neq 1$, then we must divide each side of the equation by a to get an equation with a leading coefficient of 1.

The strategy for solving a quadratic equation by completing the square is stated in the following box.

> ### Strategy for Solving Quadratic Equations by Completing the Square
>
> 1. If $a \neq 1$, then divide each side of the equation by a.
> 2. Get only the x^2 and the x terms on the left-hand side.
> 3. Add to each side the square of $\frac{1}{2}$ the coefficient of x.
> 4. Factor the left-hand side as the square of a binomial.
> 5. Apply the even-root property.
> 6. Solve for x.
> 7. Simplify.

E X A M P L E 6

Completing the square with $a \neq 1$

Solve $2x^2 + 3x - 2 = 0$ by completing the square.

Solution

For completing the square, the coefficient of x^2 must be 1. So we first divide each side of the equation by 2:

$$\frac{2x^2 + 3x - 2}{2} = \frac{0}{2} \qquad \text{Divide each side by 2.}$$

$$x^2 + \frac{3}{2}x - 1 = 0 \qquad \text{Simplify.}$$

$$x^2 + \frac{3}{2}x = 1 \qquad \begin{array}{l}\text{Get only } x^2 \text{ and } x \text{ terms on the} \\ \text{left-hand side.}\end{array}$$

$$x^2 + \frac{3}{2}x + \frac{9}{16} = 1 + \frac{9}{16} \qquad \text{One-half of } \tfrac{3}{2} \text{ is } \tfrac{3}{4}, \text{ and } \left(\tfrac{3}{4}\right)^2 = \tfrac{9}{16}.$$

$$\left(x + \frac{3}{4}\right)^2 = \frac{25}{16} \qquad \text{Factor the left-hand side.}$$

$$x + \frac{3}{4} = \pm\sqrt{\frac{25}{16}} \qquad \text{Even-root property}$$

$$x + \frac{3}{4} = \frac{5}{4} \qquad \text{or} \qquad x + \frac{3}{4} = -\frac{5}{4}$$

$$x = \frac{2}{4} = \frac{1}{2} \qquad \text{or} \qquad x = -\frac{8}{4} = -2$$

Check these values in the original equation. The solution set is $\left\{-2, \frac{1}{2}\right\}$.

> **Now do Exercises 45–46**

‹ **Calculator Close-Up** ›

Note that the x-intercepts for the graph of the function

$$y = 2x^2 + 3x - 2$$

are $(-2, 0)$ and $\left(\frac{1}{2}, 0\right)$:

In Examples 5 and 6, the solutions were rational numbers, and the equations could have been solved by factoring. In Example 7, the solutions are irrational numbers, and factoring will not work.

EXAMPLE 7

A quadratic equation with irrational solutions
Solve $x^2 - 3x - 6 = 0$ by completing the square.

Solution

Because $a = 1$, we first get the x^2 and x terms on the left-hand side:

$$x^2 - 3x - 6 = 0$$

$$x^2 - 3x \qquad = 6 \qquad \text{Add 6 to each side.}$$

$$x^2 - 3x + \frac{9}{4} = 6 + \frac{9}{4} \qquad \text{One-half of } -3 \text{ is } -\frac{3}{2}, \text{ and } \left(-\frac{3}{2}\right)^2 = \frac{9}{4}.$$

$$\left(x - \frac{3}{2}\right)^2 = \frac{33}{4} \qquad 6 + \frac{9}{4} = \frac{24}{4} + \frac{9}{4} = \frac{33}{4}$$

$$x - \frac{3}{2} = \pm\sqrt{\frac{33}{4}} \qquad \text{Even-root property}$$

$$x = \frac{3}{2} \pm \frac{\sqrt{33}}{2} \qquad \text{Add } \frac{3}{2} \text{ to each side.}$$

$$x = \frac{3 \pm \sqrt{33}}{2}$$

The solution set is $\left\{ \frac{3 + \sqrt{33}}{2}, \frac{3 - \sqrt{33}}{2} \right\}$.

> **Now do Exercises 47–56**

⟨4⟩ Radicals and Rational Expressions

Examples 8 and 9 show equations that are not originally in the form of quadratic equations. However, after simplifying these equations, we get quadratic equations. Even though completing the square can be used on any quadratic equation, factoring and the square root property are usually easier and we can use them when applicable. In Examples 8 and 9, we will use the most appropriate method.

EXAMPLE 8

An equation containing a radical
Solve $x + 3 = \sqrt{153 - x}$.

Solution

Square both sides of the equation to eliminate the radical:

$$x + 3 = \sqrt{153 - x} \qquad \text{The original equation}$$

$$(x + 3)^2 = \left(\sqrt{153 - x}\right)^2 \qquad \text{Square each side.}$$

$$x^2 + 6x + 9 = 153 - x \qquad \text{Simplify.}$$

$$x^2 + 7x - 144 = 0$$

$$(x - 9)(x + 16) = 0 \qquad \text{Factor.}$$

$$x - 9 = 0 \quad \text{or} \quad x + 16 = 0 \qquad \text{Zero factor property}$$

$$x = 9 \quad \text{or} \qquad x = -16$$

<< **Calculator Close-Up** >>

You can provide graphical support for the solution to Example 8 by graphing

$$y_1 = x + 3$$

and

$$y_2 = \sqrt{153 - x}.$$

It appears that the only point of intersection occurs when $x = 9$.

Because we squared each side of the original equation, we must check for extraneous roots. Let $x = 9$ in the original equation:

$$9 + 3 = \sqrt{153 - 9}$$

$$12 = \sqrt{144} \quad \text{Correct}$$

Let $x = -16$ in the original equation:

$$-16 + 3 = \sqrt{153 - (-16)}$$

$$-13 = \sqrt{169} \quad \text{Incorrect because } \sqrt{169} = 13$$

Because -16 is an extraneous root, the solution set is $\{9\}$.

> **Now do Exercises 57–60**

E X A M P L E   **9**

An equation containing rational expressions

Solve $\frac{1}{x} + \frac{3}{x-2} = \frac{5}{8}$.

Solution

The least common denominator (LCD) for x, $x - 2$, and 8 is $8x(x - 2)$.

$$\frac{1}{x} + \frac{3}{x-2} = \frac{5}{8}$$

$$8x(x-2)\frac{1}{x} + 8x(x-2)\frac{3}{x-2} = 8x(x-2)\frac{5}{8} \quad \text{Multiply each side by the LCD.}$$

$$8x - 16 + 24x = 5x^2 - 10x$$

$$32x - 16 = 5x^2 - 10x$$

$$-5x^2 + 42x - 16 = 0$$

$$5x^2 - 42x + 16 = 0 \qquad \text{Multiply each side by } -1 \text{ for easier factoring.}$$

$$(5x - 2)(x - 8) = 0 \qquad \text{Factor.}$$

$$5x - 2 = 0 \quad \text{or} \quad x - 8 = 0$$

$$x = \frac{2}{5} \quad \text{or} \quad x = 8$$

Check these values in the original equation. The solution set is $\left\{\frac{2}{5}, 8\right\}$.

> **Now do Exercises 61–64**

<5> Imaginary Solutions

In Chapter 7, we found imaginary solutions to quadratic equations using the even-root property. We can get imaginary solutions also by completing the square.

E X A M P L E   **10**

An equation with imaginary solutions

Find the complex solutions to $x^2 - 4x + 12 = 0$.

Solution

Because the quadratic polynomial cannot be factored, we solve the equation by completing the square.

<**Calculator Close-Up**>

The answer key (ANS) can be used to check imaginary answers as shown here.

```
2+2i√(2)
        2+2.828427125i
Ans²-4Ans+12
                     0
```

$$x^2 - 4x + 12 = 0 \qquad \text{The original equation}$$
$$x^2 - 4x = -12 \qquad \text{Subtract 12 from each side.}$$
$$x^2 - 4x + 4 = -12 + 4 \qquad \text{One-half of } -4 \text{ is } -2, \text{ and } (-2)^2 = 4.$$
$$(x - 2)^2 = -8$$
$$x - 2 = \pm\sqrt{-8} \qquad \text{Even-root property}$$
$$x = 2 \pm i\sqrt{8}$$
$$= 2 \pm 2i\sqrt{2}$$

Check these values in the original equation. The solution set is $\{2 \pm 2i\sqrt{2}\}$.

> **Now do Exercises 65–74**

Warm-Ups ▼

Fill in the blank.

1. In this section quadratic equations are solved by _____, the _____ property, and _____ the square.
2. If $b = 0$ in $ax^2 + bx + c = 0$, then the equation can be solved by the _____ property.
3. The last term of a perfect square trinomial is the square of one-half the coefficient of the _____ term.
4. If the leading coefficient is not 1, then the first step in completing the square is to divide both sides of the equation by the _____ .

True or false?

5. Every quadratic equation can be solved by factoring.

6. All quadratic equations have two distinct complex solutions.
7. The trinomial $x^2 + \frac{3}{2}x + \frac{9}{16}$ is a perfect square trinomial.
8. Every quadratic equation can be solved by completing the square.
9. $(x - 3)^2 = 12$ is equivalent to $x - 3 = 2\sqrt{3}$.
10. $(2x - 3)(3x + 5) = 0$ is equivalent to $x = \frac{3}{2}$ or $x = \frac{5}{3}$.
11. $x^2 = 8$ is equivalent to $x = \pm 2\sqrt{2}$.
12. To complete the square for $x^2 - 3x = 4$, add $\frac{9}{4}$ to each side.

Exercises 8.1

<**Study Tips**>

- Stay calm and confident. Take breaks when you study. Get 6 to 8 hours of sleep every night.
- Keep reminding yourself that working hard throughout the semester will really pay off in the end.

<**1**> **Review of Factoring**

Solve by factoring. See Example 1. See the Strategy for Solving Quadratic Equations by Factoring box on page 530.

1. $x^2 - x - 6 = 0$
2. $x^2 + 6x + 8 = 0$
3. $a^2 + 2a = 15$
4. $w^2 - 2w = 15$
5. $2x^2 - x - 3 = 0$
6. $6x^2 - x - 15 = 0$

7. $y^2 + 14y + 49 = 0$ **8.** $a^2 - 6a + 9 = 0$

9. $a^2 - 16 = 0$ **10.** $4w^2 - 25 = 0$

33. $z^2 - \dfrac{4}{7}z + \dfrac{4}{49}$ **34.** $m^2 - \dfrac{6}{5}m + \dfrac{9}{25}$

35. $t^2 + \dfrac{3}{5}t + \dfrac{9}{100}$ **36.** $h^2 + \dfrac{3}{2}h + \dfrac{9}{16}$

⟨2⟩ Review of the Even-Root Property

Use the even-root property to solve each equation. See Example 2.

11. $x^2 = 81$ **12.** $x^2 = \dfrac{9}{4}$

13. $x^2 = \dfrac{16}{9}$ **14.** $a^2 = 32$

15. $(x - 3)^2 = 16$ **16.** $(x + 5)^2 = 4$

17. $(z + 1)^2 = 5$ **18.** $(a - 2)^2 = 8$

 19. $\left(w - \dfrac{3}{2}\right)^2 = \dfrac{7}{4}$ **20.** $\left(w + \dfrac{2}{3}\right)^2 = \dfrac{5}{9}$

Solve by completing the square. See Examples 5–7. See the Strategy for Solving Quadratic Equations by Completing the Square box on page 534. Use your calculator to check.

37. $x^2 - 2x - 15 = 0$

38. $x^2 - 6x - 7 = 0$

39. $2x^2 - 4x = 70$

40. $3x^2 - 6x = 24$

41. $w^2 - w - 20 = 0$

42. $y^2 - 3y - 10 = 0$

43. $q^2 + 5q = 14$

44. $z^2 + z = 2$

45. $2h^2 - h - 3 = 0$

46. $2m^2 - m - 15 = 0$

⟨3⟩ Completing the Square

Find the perfect square trinomial whose first two terms are given. See Example 3.

21. $x^2 + 2x$ **22.** $m^2 + 14m$

23. $x^2 - 3x$ **24.** $w^2 - 5w$

25. $y^2 + \dfrac{1}{4}y$ **26.** $z^2 + \dfrac{3}{2}z$

27. $x^2 + \dfrac{2}{3}x$ **28.** $p^2 + \dfrac{6}{5}p$

47. $x^2 + 4x = 6$

48. $x^2 + 6x - 8 = 0$

49. $x^2 + 8x - 4 = 0$

50. $x^2 + 10x - 3 = 0$

51. $x^2 + 5x + 5 = 0$

52. $x^2 - 7x + 4 = 0$

53. $4x^2 - 4x - 1 = 0$

54. $4x^2 + 4x - 2 = 0$

Factor each perfect square trinomial. See Example 4.

29. $x^2 + 8x + 16$ **30.** $x^2 - 10x + 25$

31. $y^2 - 5y + \dfrac{25}{4}$ **32.** $w^2 + w + \dfrac{1}{4}$

55. $2x^2 + 3x - 4 = 0$

56. $2x^2 + 5x - 1 = 0$

⟨4⟩ Radicals and Rational Expressions

Solve each equation by an appropriate method. See Examples 8 and 9.

57. $\sqrt{2x + 1} = x - 1$ **58.** $\sqrt{2x - 4} = x - 14$

59. $w = \dfrac{\sqrt{w + 1}}{2}$ **60.** $y - 1 = \dfrac{\sqrt{y + 1}}{2}$

61. $\dfrac{t}{t - 2} = \dfrac{2t - 3}{t}$ **62.** $\dfrac{z}{z + 3} = \dfrac{3z}{5z - 1}$

63. $\dfrac{2}{x^2} + \dfrac{4}{x} + 1 = 0$

64. $\dfrac{1}{x^2} + \dfrac{3}{x} + 1 = 0$

⟨5⟩ Imaginary Solutions

Use completing the square to find the imaginary solutions to each equation. See Example 10.

65. $x^2 + 2x + 5 = 0$ **66.** $x^2 + 4x + 5 = 0$

67. $x^2 - 6x + 11 = 0$ **68.** $x^2 - 8x + 19 = 0$

69. $x^2 = -\dfrac{1}{2}$ **70.** $x^2 = -\dfrac{1}{8}$

71. $x^2 + 12 = 0$ **72.** $-3x^2 - 21 = 0$

73. $5z^2 - 4z + 1 = 0$ **74.** $2w^2 - 3w + 2 = 0$

Miscellaneous

Find all real or imaginary solutions to each equation. Use the method of your choice.

75. $x^2 = -121$

76. $w^2 = -225$

77. $4x^2 + 25 = 0$

78. $5w^2 - 3 = 0$

79. $\left(p + \dfrac{1}{2}\right)^2 = \dfrac{9}{4}$

80. $\left(y - \dfrac{2}{3}\right)^2 = \dfrac{4}{9}$

81. $5t^2 + 4t - 3 = 0$

82. $3v^2 + 4v - 1 = 0$

83. $m^2 + 2m - 24 = 0$

84. $q^2 + 6q - 7 = 0$

85. $(x - 2)^2 = -9$

86. $(2x - 1)^2 = -4$

87. $-x^2 + x + 6 = 0$

88. $-x^2 + x + 12 = 0$

89. $x^2 - 6x + 10 = 0$

90. $x^2 - 8x + 17 = 0$

91. $2x - 5 = \sqrt{7x + 7}$

92. $\sqrt{7x + 29} = x + 3$

93. $\dfrac{1}{x} + \dfrac{1}{x - 1} = \dfrac{1}{4}$

94. $\dfrac{1}{x} - \dfrac{2}{1 - x} = \dfrac{1}{2}$

Find the real solutions to each equation by examining the following graphs.

95. $x^2 + 2x - 15 = 0$

96. $100x^2 + 20x - 3 = 0$

97. $x^2 + 4x + 15 = 0$

98. $100x^2 - 60x + 9 = 0$

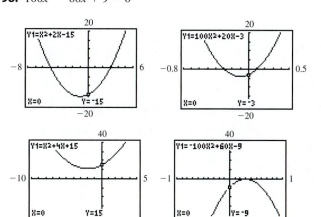

Applications

Solve each problem.

99. *Approach speed.* The formula $1211.1L = CA^2S$ is used to determine the approach speed for landing an aircraft, where L is the gross weight of the aircraft in pounds, C is the coefficient of lift, S is the surface area of the wings in square feet (ft^2), and A is approach speed in feet per second. Find A for the Piper Cheyenne, which has a gross weight of 8700 lb, a coefficient of lift of 2.81, and wing surface area of 200 ft^2.

100. *Time to swing.* The period T (time in seconds for one complete cycle) of a simple pendulum is related to the length L (in feet) of the pendulum by the formula $8T^2 = \pi^2L$. If a child is on a swing with a 10-foot chain, then how long does it take to complete one cycle of the swing?

101. *Time for a swim.* Tropical Pools figures that its monthly revenue in dollars on the sale of x aboveground pools is given by $R = 1500x - 3x^2$, where x is less than 25. What number of pools sold would provide a revenue of $17,568?

102. *Pole vaulting.* In 1981 Vladimir Poliakov (USSR) set a world record of 19 ft $\frac{3}{4}$ in. for the pole vault (www.polevault.com). To reach that height, Poliakov obtained a speed of approximately 36 feet per second on the runway. The function $h = -16t^2 + 36t$ gives his height t seconds after leaving the ground.

a) Use the formula to find the exact values of t for which his height was 18 feet.

b) Use the accompanying graph to estimate the value of t for which he was at his maximum height.

Figure for Exercise 102

c) Approximately how long was he in the air?

Getting More Involved

103. *Discussion*

Which of the following equations is not a quadratic equation? Explain your answer.

a) $\pi x^2 - \sqrt{5}x - 1 = 0$
b) $3x^2 - 1 = 0$
c) $4x + 5 = 0$
d) $0.009x^2 = 0$

104. *Exploration*

Solve $x^2 - 4x + k = 0$ for $k = 0, 4, 5,$ and 10.

a) When does the equation have only one solution?
b) For what values of k are the solutions real?
c) For what values of k are the solutions imaginary?

105. *Cooperative learning*

Write a quadratic equation of each of the following types, and then trade your equations with those of a classmate. Solve the equations, and verify that they are of the required types.

a) a single rational solution
b) two rational solutions
c) two irrational solutions
d) two imaginary solutions

106. *Exploration*

In Section 8.2 we will solve $ax^2 + bx + c = 0$ for x by completing the square. Try it now without looking ahead.

Graphing Calculator Exercises

For each equation, find approximate solutions rounded to two decimal places.

107. $x^2 - 7.3x + 12.5 = 0$

108. $1.2x^2 - \pi x + \sqrt{2} = 0$

109. $2x - 3 = \sqrt{20 - x}$

110. $x^2 - 1.3x = 22.3 - x^2$

Math *at Work*   **Financial Matters**

In the United States, over 1 million new homes are sold annually, with a median price of about $200,000. Over 17 million new cars are sold each year with a median price over $20,000. Americans are constantly saving and borrowing. Nearly everyone will need to know a monthly payment or what their savings will total over time. The answers to these questions are in the following table.

What $P Left at Compound Interest Will Grow to	What $R Deposited Periodically Will Grow to	Periodic Payment That Will Pay off a Loan of $P
$P(1 + i)^{nt}$	$R\dfrac{(1 + i)^{nt} - 1}{i}$	$P\dfrac{i}{1 - (1 + i)^{-nt}}$

In each case, n is the number of periods per year, r is the annual percentage rate (APR), t is the number of years, and i is the interest rate per period $\left(i = \dfrac{r}{n}\right)$. For periodic payments or deposits these expressions apply only if the compounding period equals the payment period. So let's see what these expressions do.

A person inherits $10,000 and lets it grow at 4% APR compounded daily for 20 years. Use the first expression with $n = 365$, $i = \dfrac{0.04}{365}$, and $t = 20$ to get $10{,}000\left(1 + \dfrac{0.04}{365}\right)^{365 \cdot 20}$ or $22,254.43, which is the amount after 20 years.

20-year $200,000 mortgage

Monthly payment ($) vs. APR (percent)

More often, people save money with periodic deposits. Suppose you deposit $100 per month at 4% compounded monthly for 20 years. Use the second expression with $R = 100$, $i = \dfrac{0.04}{12}$, $n = 12$, and $t = 20$ to get $100\dfrac{(1 + 0.04/12)^{12 \cdot 20} - 1}{0.04/12}$ or $36,677.46, which is the amount after 20 years.

Suppose that you get a 20-year $200,000 mortgage at 7% APR compounded monthly to buy an average house. Try using the third expression to calculate the monthly payment of $1550.60. See the accompanying figure.

8.2   The Quadratic Formula

In This Section

⟨1⟩ Developing the Formula
⟨2⟩ Using the Formula
⟨3⟩ Number of Solutions
⟨4⟩ Applications

Completing the square from Section 8.1 can be used to solve any quadratic equation. Here we apply this method to the general quadratic equation to get a formula for the solutions to any quadratic equation.

⟨1⟩ Developing the Formula

Start with the general form of the quadratic equation,

$$ax^2 + bx + c = 0.$$

Assume a is positive for now, and divide each side by a:

$$\frac{ax^2 + bx + c}{a} = \frac{0}{a}$$

$$x^2 + \frac{b}{a}x + \frac{c}{a} = 0$$

$$x^2 + \frac{b}{a}x = -\frac{c}{a} \qquad \text{Subtract } \frac{c}{a} \text{ from each side.}$$

One-half of $\frac{b}{a}$ is $\frac{b}{2a}$, and $\frac{b}{2a}$ squared is $\frac{b^2}{4a^2}$:

$$x^2 + \frac{b}{a}x + \frac{b^2}{4a^2} = -\frac{c}{a} + \frac{b^2}{4a^2}$$

Factor the left-hand side and get a common denominator for the right-hand side:

$$\left(x + \frac{b}{2a}\right)^2 = \frac{b^2}{4a^2} - \frac{4ac}{4a^2} \qquad\qquad \frac{c(4a)}{a(4a)} = \frac{4ac}{4a^2}$$

$$\left(x + \frac{b}{2a}\right)^2 = \frac{b^2 - 4ac}{4a^2}$$

$$x + \frac{b}{2a} = \pm\sqrt{\frac{b^2 - 4ac}{4a^2}} \qquad \text{Even-root property}$$

$$x = \frac{-b}{2a} \pm \frac{\sqrt{b^2 - 4ac}}{2a} \qquad \text{Because } a > 0,\ \sqrt{4a^2} = 2a.$$

$$x = \frac{-b \pm \sqrt{b^2 - 4ac}}{2a}$$

We assumed a was positive so that $\sqrt{4a^2} = 2a$ would be correct. If a is negative, then $\sqrt{4a^2} = -2a$, and we get

$$x = \frac{-b}{2a} \pm \frac{\sqrt{b^2 - 4ac}}{-2a}.$$

However, the negative sign can be omitted in $-2a$ because of the $\pm$ symbol preceding it. For example, the results of $5 \pm (-3)$ and 5 ± 3 are the same. So when a is negative, we get the same formula as when a is positive. It is called the **quadratic formula.**

The Quadratic Formula

The solution to $ax^2 + bx + c = 0$, with $a \neq 0$, is given by the formula

$$x = \frac{-b \pm \sqrt{b^2 - 4ac}}{2a}.$$

⟨2⟩ Using the Formula

The quadratic formula solves any quadratic equation. Simply identify a, b, and c and insert those numbers into the formula. Note that if b is positive then $-b$ (the opposite of b) is a negative number. If b is negative, then $-b$ is a *positive* number.

E X A M P L E 1

Two rational solutions

Solve $x^2 + 2x - 15 = 0$ using the quadratic formula.

Solution

To use the formula, we first identify the values of a, b, and c:

$$1x^2 + 2x - 15 = 0$$
$$\quad\ \uparrow \quad\ \uparrow \quad\ \uparrow$$
$$\quad\ a \quad\ b \quad\ c$$

The coefficient of x^2 is 1, so $a = 1$. The coefficient of $2x$ is 2, so $b = 2$. The constant term is -15, so $c = -15$. Substitute these values into the quadratic formula:

$$x = \frac{-2 \pm \sqrt{2^2 - 4(1)(-15)}}{2(1)}$$

$$= \frac{-2 \pm \sqrt{4 + 60}}{2}$$

$$= \frac{-2 \pm \sqrt{64}}{2}$$

$$= \frac{-2 \pm 8}{2}$$

$$x = \frac{-2 + 8}{2} = 3 \quad \text{or} \quad x = \frac{-2 - 8}{2} = -5$$

Check 3 and -5 in the original equation. The solution set is $\{-5, 3\}$.

> **Now do Exercises 1–8**

⟨ Calculator Close-Up ⟩

Note that the two solutions to

$$x^2 + 2x - 15 = 0$$

correspond to the two x-intercepts for the graph of the function

$$y = x^2 + 2x - 15.$$

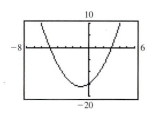

CAUTION   To identify a, b, and c for the quadratic formula, the equation must be in the standard form $ax^2 + bx + c = 0$. If it is not in that form, then you must first rewrite the equation.

E X A M P L E 2

One rational solution

Solve $4x^2 = 12x - 9$ by using the quadratic formula.

Solution

Rewrite the equation in the form $ax^2 + bx + c = 0$ before identifying a, b, and c:

$$4x^2 - 12x + 9 = 0$$

Note that the single solution to

$$4x^2 - 12x + 9 = 0$$

corresponds to the single x-intercept for the graph of the function

$$y = 4x^2 - 12x + 9.$$

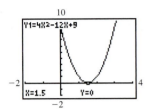

In this form we get $a = 4$, $b = -12$, and $c = 9$.

$$x = \frac{12 \pm \sqrt{(-12)^2 - 4(4)(9)}}{2(4)} \qquad \text{Because } b = -12, -b = 12.$$

$$= \frac{12 \pm \sqrt{144 - 144}}{8}$$

$$= \frac{12 \pm 0}{8}$$

$$= \frac{12}{8}$$

$$= \frac{3}{2}$$

Check $\frac{3}{2}$ in the original equation. The solution set is $\left\{\frac{3}{2}\right\}$.

Now do Exercises 9–14

Because the solutions to the equations in Examples 1 and 2 were rational numbers, these equations could have been solved by factoring. In Example 3, the solutions are irrational.

E X A M P L E   3

Two irrational solutions

Solve $\frac{1}{3}x^2 + x + \frac{1}{2} = 0$.

Solution

We could use $a = \frac{1}{3}$, $b = 1$, and $c = \frac{1}{2}$ in the quadratic formula, but it is easier to use the formula with integers. So we first multiply each side of the equation by 6, the least common denominator. Multiplying by 6 yields

$$2x^2 + 6x + 3 = 0.$$

Now let $a = 2$, $b = 6$, and $c = 3$ in the quadratic formula:

$$x = \frac{-6 \pm \sqrt{(6)^2 - 4(2)(3)}}{2(2)}$$

$$= \frac{-6 \pm \sqrt{36 - 24}}{4}$$

$$= \frac{-6 \pm \sqrt{12}}{4}$$

$$= \frac{-6 \pm 2\sqrt{3}}{4}$$

$$= \frac{2(-3 \pm \sqrt{3})}{2 \cdot 2}$$

$$= \frac{-3 \pm \sqrt{3}}{2}$$

The two irrational solutions to

$$2x^2 + 6x + 3 = 0$$

correspond to the two x-intercepts for the graph of

$$y = 2x^2 + 6x + 3.$$

Check these values in the original equation. The solution set is $\left\{\dfrac{-3 \pm \sqrt{3}}{2}\right\}$.

Now do Exercises 15–20

EXAMPLE 4

Two imaginary solutions, no real solutions

Find the complex solutions to $x^2 + x + 5 = 0$.

Solution

Let $a = 1$, $b = 1$, and $c = 5$ in the quadratic formula:

$$x = \frac{-1 \pm \sqrt{(1)^2 - 4(1)(5)}}{2(1)}$$

$$= \frac{-1 \pm \sqrt{-19}}{2}$$

$$= \frac{-1 \pm i\sqrt{19}}{2}$$

Check these values in the original equation. The solution set is $\left\{\frac{-1 \pm i\sqrt{19}}{2}\right\}$. There are no real solutions to the equation.

> Now do Exercises 21–26

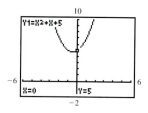
You have learned to solve quadratic equations by four different methods: the even-root property, factoring, completing the square, and the quadratic formula. The even-root property and factoring are limited to certain special equations, but you should use those methods when possible. Any quadratic equation can be solved by completing the square or using the quadratic formula. Because the quadratic formula is usually faster, it is used more often than completing the square. However, completing the square is an important skill to learn. It will be used in the study of conic sections later in this text.

Summary of Methods for Solving $ax^2 + bx + c = 0$

Method	Comments	Examples
Even-root property	Use when $b = 0$.	$(x - 2)^2 = 8$ $x - 2 = \pm\sqrt{8}$
Factoring	Use when the polynomial can be factored.	$x^2 + 5x + 6 = 0$ $(x + 2)(x + 3) = 0$
Quadratic formula	Solves any quadratic equation	$x^2 + 5x + 3 = 0$ $x = \dfrac{-5 \pm \sqrt{25 - 4(3)}}{2}$
Completing the square	Solves any quadratic equation, but quadratic formula is faster	$x^2 - 6x + 7 = 0$ $x^2 - 6x + 9 = -7 + 9$ $(x - 3)^2 = 2$

‹3› Number of Solutions

The quadratic equations in Examples 1 and 3 had two real solutions each. In each of those examples, the value of $b^2 - 4ac$ was positive. In Example 2, the quadratic equation had only one solution because the value of $b^2 - 4ac$ was zero. In Example 4, the quadratic equation had no real solutions because $b^2 - 4ac$ was negative. Because $b^2 - 4ac$ determines the kind and number of solutions to a quadratic equation, it is called the **discriminant.**

> ### Number of Solutions to a Quadratic Equation
> The quadratic equation $ax^2 + bx + c = 0$ with $a \neq 0$ has
> *two* real solutions if $b^2 - 4ac > 0$,
> *one* real solution if $b^2 - 4ac = 0$, and
> *no* real solutions (two imaginary solutions) if $b^2 - 4ac < 0$.

E X A M P L E   5

Using the discriminant
Use the discriminant to determine the number of real solutions to each quadratic equation.

a) $x^2 - 3x - 5 = 0$

b) $x^2 = 3x - 9$

c) $4x^2 - 12x + 9 = 0$

Solution

a) For $x^2 - 3x - 5 = 0$, use $a = 1$, $b = -3$, and $c = -5$ in $b^2 - 4ac$:

$$b^2 - 4ac = (-3)^2 - 4(1)(-5) = 9 + 20 = 29$$

Because the discriminant is positive, there are two real solutions to this quadratic equation.

b) Rewrite $x^2 = 3x - 9$ as $x^2 - 3x + 9 = 0$. Then use $a = 1$, $b = -3$, and $c = 9$ in $b^2 - 4ac$:

$$b^2 - 4ac = (-3)^2 - 4(1)(9) = 9 - 36 = -27$$

Because the discriminant is negative, the equation has no real solutions. It has two imaginary solutions.

c) For $4x^2 - 12x + 9 = 0$, use $a = 4$, $b = -12$, and $c = 9$ in $b^2 - 4ac$:

$$b^2 - 4ac = (-12)^2 - 4(4)(9) = 144 - 144 = 0$$

Because the discriminant is zero, there is only one real solution to this quadratic equation.

Now do Exercises 27–42

⟨4⟩ Applications
With the quadratic formula we can easily solve problems whose solutions are irrational numbers. When the solutions are irrational numbers, we usually use a calculator to find rational approximations and to check.

E X A M P L E   6

Area of a tabletop
The area of a rectangular tabletop is 6 square feet. If the width is 2 feet shorter than the length, then what are the dimensions?

Figure 8.1

Solution

Let x be the length and $x - 2$ be the width, as shown in Fig. 8.1. Because the area is 6 square feet and $A = LW$, we can write the equation

$$x(x - 2) = 6$$

or

$$x^2 - 2x - 6 = 0.$$

Because this equation cannot be factored, we use the quadratic formula with $a = 1$, $b = -2$, and $c = -6$:

$$x = \frac{2 \pm \sqrt{(-2)^2 - 4(1)(-6)}}{2(1)}$$

$$= \frac{2 \pm \sqrt{28}}{2} = \frac{2 \pm 2\sqrt{7}}{2} = 1 \pm \sqrt{7}$$

Because $1 - \sqrt{7}$ is a negative number, it cannot be the length of a tabletop. If $x = 1 + \sqrt{7}$, then $x - 2 = 1 + \sqrt{7} - 2 = \sqrt{7} - 1$. Checking the product of $\sqrt{7} + 1$ and $\sqrt{7} - 1$, we get

$$(\sqrt{7} + 1)(\sqrt{7} - 1) = 7 - 1 = 6.$$

The exact length is $\sqrt{7} + 1$ feet, and the width is $\sqrt{7} - 1$ feet. Using a calculator, we find that the approximate length is 3.65 feet and the approximate width is 1.65 feet.

> **Now do Exercises 71–90**

 ## Warm-Ups ▼

Fill in the blank.

1. The _____ formula can be used to solve any quadratic equation.

2. The _____ is $b^2 - 4ac$.

3. In the _____ number system every quadratic equation has at least one solution.

4. If $b^2 - 4ac = 0$, then the quadratic equation has ____ real solution.

5. If $b^2 - 4ac > 0$, then the quadratic equation has ____ real solutions.

6. If $b^2 - 4ac < 0$, then the quadratic equation has ____ imaginary solutions.

True or false?

7. Completing the square is used to develop the quadratic formula.

8. The quadratic formula will not work on $x^2 - 3 = 0$.

9. If $a = 2$, $b = -3$, and $c = -4$, then $b^2 - 4ac = 41$.

10. If $x^2 + 4x - 5 = 0$, then $x = \dfrac{-4 \pm \sqrt{16 - 4(-5)}}{2}$.

11. If $3x^2 - 5x + 9 = 0$, then $x = \dfrac{5 \pm \sqrt{25 - 4(3)(9)}}{2}$.

12. If $mx^2 + nx + p = 0$ and $m \neq 0$, then
$$x = \frac{-n \pm \sqrt{n^2 - 4mp}}{2m}.$$

‹2› Using the Formula

Solve each equation by using the quadratic formula. See Example 1.

1. $x^2 - 3x + 2 = 0$

2. $x^2 - 7x + 12 = 0$

3. $x^2 + 5x + 6 = 0$

4. $x^2 + 4x + 3 = 0$

5. $y^2 + y = 6$

6. $m^2 + 2m = 8$

7. $-6z^2 + 7z + 3 = 0$

8. $-8q^2 - 2q + 1 = 0$

Solve each equation by using the quadratic formula. See Example 2.

9. $4x^2 - 4x + 1 = 0$

10. $4x^2 - 12x + 9 = 0$

11. $-9x^2 + 6x - 1 = 0$

12. $-9x^2 + 24x - 16 = 0$

13. $9 + 24x + 16x^2 = 0$

14. $4 + 20x = -25x^2$

Solve each equation by using the quadratic formula. See Example 3.

15. $v^2 + 8v + 6 = 0$

16. $p^2 + 6p + 4 = 0$

17. $-x^2 - 5x + 1 = 0$

18. $-x^2 - 3x + 5 = 0$

19. $\frac{1}{3}t^2 - t + \frac{1}{6} = 0$

20. $\frac{3}{4}x^2 - 2x + \frac{1}{2} = 0$

Solve each equation by using the quadratic formula. See Example 4.

21. $2t^2 - 6t + 5 = 0$

22. $2y^2 + 1 = 2y$

23. $-2x^2 + 3x = 6$

24. $-3x^2 - 2x - 5 = 0$

25. $\frac{1}{2}x^2 + 13 = 5x$

26. $\frac{1}{4}x^2 + \frac{17}{4} = 2x$

‹3› Number of Solutions

Find $b^2 - 4ac$ and the number of real solutions to each equation. See Example 5.

27. $x^2 - 6x + 2 = 0$

28. $x^2 + 6x + 9 = 0$

29. $-2x^2 + 5x - 6 = 0$

30. $-x^2 + 3x - 4 = 0$

31. $4m^2 + 25 = 20m$

32. $v^2 = 3v + 5$

33. $y^2 - \frac{1}{2}y + \frac{1}{4} = 0$

34. $\frac{1}{2}w^2 - \frac{1}{3}w + \frac{1}{4} = 0$

35. $-3t^2 + 5t + 6 = 0$

36. $9m^2 + 16 = 24m$

37. $9 - 24z + 16z^2 = 0$

38. $12 - 7x + x^2 = 0$

39. $5x^2 - 7 = 0$

40. $-6x^2 - 5 = 0$

41. $x^2 = x$

42. $-3x^2 + 7x = 0$

Miscellaneous

Solve by the method of your choice. See the Summary of Methods for Solving $ax^2 + bx + c = 0$ on page 545.

43. $\frac{1}{4}y^2 + y = 1$

44. $\frac{1}{2}x^2 + x = 1$

45. $\frac{1}{3}x^2 + \frac{1}{2}x = \frac{1}{3}$

46. $\frac{4}{9}w^2 + 1 = \frac{5}{3}w$

47. $3y^2 + 2y - 4 = 0$

48. $2y^2 - 3y - 6 = 0$

49. $\frac{w}{w-2} = \frac{w}{w-3}$

50. $\frac{y}{3y-4} = \frac{2}{y+4}$

51. $\dfrac{9(3x-5)^2}{4}=1$ **52.** $\dfrac{25(2x+1)^2}{9}=0$

53. $25-\dfrac{1}{3}x^2=0$ **54.** $\dfrac{49}{2}-\dfrac{1}{4}x^2=0$

55. $1+\dfrac{20}{x^2}=\dfrac{8}{x}$ **56.** $\dfrac{34}{x^2}=\dfrac{6}{x}-1$

57. $(x-8)(x+4)=-42$ **58.** $(x-10)(x-2)=-20$

59. $y=\dfrac{3(2y+5)}{8(y-1)}$ **60.** $z=\dfrac{7z-4}{12(z-1)}$

 Use the quadratic formula and a calculator to solve each equation. Round answers to three decimal places and check your answers.

61. $x^2+3.2x-5.7=0$

62. $x^2+7.15x+3.24=0$

63. $x^2-7.4x+13.69=0$

64. $1.44x^2+5.52x+5.29=0$

65. $1.85x^2+6.72x+3.6=0$

66. $3.67x^2+4.35x-2.13=0$

67. $3x^2+14{,}379x+243=0$

68. $x^2+12{,}347x+6741=0$

69. $x^2+0.00075x-0.0062=0$

70. $4.3x^2-9.86x-3.75=0$

⟨4⟩ Applications

Find the exact solution(s) to each problem. If the solution(s) are irrational, then also find approximate solution(s) to the nearest tenth. See Example 6.

71. *Missing numbers.* Find two positive real numbers that differ by 1 and have a product of 16.

72. *Missing numbers.* Find two positive real numbers that differ by 2 and have a product of 10.

73. *More missing numbers.* Find two real numbers that have a sum of 6 and a product of 4.

74. *More missing numbers.* Find two real numbers that have a sum of 8 and a product of 2.

75. *Bulletin board.* The length of a bulletin board is 1 foot more than the width. The diagonal has a length of $\sqrt{3}$ feet (ft). Find the length and width of the bulletin board.

76. *Diagonal brace.* The width of a rectangular gate is 2 meters (m) larger than its height. The diagonal brace measures $\sqrt{6}$ m. Find the width and height.

Figure for Exercise 76

77. *Area of a rectangle.* The length of a rectangle is 4 ft longer than the width, and its area is 10 square feet (ft²). Find the length and width.

78. *Diagonal of a square.* The diagonal of a square is 2 m longer than a side. Find the length of a side.

If an object is given an initial velocity of v_0 feet per second from a height of s_0 feet, then its height S after t seconds is given by the formula $S=-16t^2+v_0 t+s_0$.

79. *Projected pine cone.* If a pine cone is projected upward at a velocity of 16 ft/sec from the top of a 96-foot pine tree, then how long does it take to reach the earth?

80. *Falling pine cone.* If a pine cone falls from the top of a 96-foot pine tree, then how long does it take to reach the earth?

81. *Tossing a ball.* A ball is tossed into the air at 10 ft/sec from a height of 5 feet. How long does it take to reach the earth?

82. *Time in the air.* A ball is tossed into the air from a height of 12 feet at 16 ft/sec. How long does it take to reach the earth?

83. *Penny tossing.* If a penny is thrown downward at 30 ft/sec from the bridge at Royal Gorge, Colorado, how long does it take to reach the Arkansas River 1000 ft below?

84. *Foul ball.* Suppose Charlie O'Brien of the Braves hits a baseball straight upward at 150 ft/sec from a height of 5 ft.

     a) Use the formula to determine how long it takes the ball to return to the earth.

b) Use the accompanying graph to estimate the maximum height reached by the ball.

Figure for Exercise 84

Solve each problem.

85. *Kitchen countertop.* A 30 in. by 40 in. countertop for a work island is to be covered with green ceramic tiles, except for a border of uniform width as shown in the figure. If the area covered by the green tiles is 704 square inches (in.^{2}), then how wide is the border?

Figure for Exercise 85

86. *Recovering an investment.* The manager at Cream of the Crop bought a load of watermelons for $200. She priced the melons so that she would make $1.50 profit on each melon. When all but 30 had been sold, the manager had recovered her initial investment. How many did she buy originally?

87. *Baby shower.* A group of office workers plans to share equally the $100 cost of giving a baby shower for a coworker. If they can get six more people to share the cost, then the cost per person will decrease by $15. How many people are in the original group?

88. *Sharing cost.* The members of a flying club plan to share equally the cost of a $200,000 airplane. The members want to find five more people to join the club so that the cost per

person will decrease by $2000. How many members are currently in the club?

89. *Farmer's delight.* The manager of Farmer's Delight bought a load of watermelons for $750 and priced the watermelons so that he would make a profit of $2 on each melon. When all but 100 of the melons had been sold, he broke even. How many did he buy originally?

90. *Traveling club.* The members of a traveling club plan to share equally the cost of a $150,000 motor home. If they can find 10 more people to join and share the cost, then the cost per person will decrease by $1250. How many members are there originally in the club?

Getting More Involved

91. *Discussion*

Find the solutions to $6x^2 + 5x - 4 = 0$. Is the sum of your solutions equal to $-\frac{b}{a}$? Explain why the sum of the solutions to any quadratic equation is $-\frac{b}{a}$. (*Hint:* Use the quadratic formula.)

92. *Discussion*

Use the result of Exercise 91 to check whether $\left\{\frac{2}{3}, \frac{1}{3}\right\}$ is the solution set to $9x^2 - 3x - 2 = 0$. If this solution set is not correct, then what is the correct solution set?

93. *Discussion*

What is the product of the two solutions to $6x^2 + 5x - 4 = 0$? Explain why the product of the solutions to any quadratic equation is $\frac{c}{a}$.

94. *Discussion*

Use the result of Exercise 93 to check whether $\left\{\frac{9}{2}, -2\right\}$ is the solution set to $2x^2 - 13x + 18 = 0$. If this solution set is not correct, then what is the correct solution set?

Graphing Calculator Exercises

Determine the number of real solutions to each equation by examining the calculator graph of $y = ax^2 + bx + c$. Use the discriminant to check your conclusions.

95. $x^2 - 6.33x + 3.7 = 0$

96. $1.8x^2 + 2.4x - 895 = 0$

97. $4x^2 - 67.1x + 344 = 0$

98. $-2x^2 - 403 = 0$

99. $-x^2 + 30x - 226 = 0$

100. $16x^2 - 648x + 6562 = 0$

| 8.3 | **More on Quadratic Equations** |

In This Section

⟨1⟩ **Writing a Quadratic Equation with Given Solutions**

⟨2⟩ **Using the Discriminant in Factoring**

⟨3⟩ **Equations Quadratic in Form**

⟨4⟩ **Applications**

In this section, we use the ideas and methods of the previous sections to explore additional topics involving quadratic equations.

⟨1⟩ Writing a Quadratic Equation with Given Solutions

Not every quadratic equation can be solved by factoring, but the factoring method can be used (in reverse) to write a quadratic equation with given solutions.

E X A M P L E   1

Writing a quadratic given the solutions

Write a quadratic equation that has each given pair of solutions.

a) $4, -6$ b) $-\sqrt{2}, \sqrt{2}$ c) $-3i, 3i$

Solution

a) Reverse the factoring method using solutions 4 and -6:

$$x = 4 \quad \text{or} \quad x = -6$$
$$x - 4 = 0 \quad \text{or} \quad x + 6 = 0$$
$$(x - 4)(x + 6) = 0 \quad \text{Zero factor property}$$
$$x^2 + 2x - 24 = 0 \quad \text{Multiply the factors.}$$

b) Reverse the factoring method using solutions $-\sqrt{2}$ and $\sqrt{2}$:

$$x = -\sqrt{2} \quad \text{or} \quad x = \sqrt{2}$$
$$x + \sqrt{2} = 0 \quad \text{or} \quad x - \sqrt{2} = 0$$
$$(x + \sqrt{2})(x - \sqrt{2}) = 0 \quad \text{Zero factor property}$$
$$x^2 - 2 = 0 \quad \text{Multiply the factors.}$$

c) Reverse the factoring method using solutions $-3i$ and $3i$:

$$x = -3i \quad \text{or} \quad x = 3i$$
$$x + 3i = 0 \quad \text{or} \quad x - 3i = 0$$
$$(x + 3i)(x - 3i) = 0 \quad \text{Zero factor property}$$
$$x^2 - 9i^2 = 0 \quad \text{Multiply the factors.}$$
$$x^2 + 9 = 0 \quad \text{Note: } i^2 = -1$$

Now do Exercises 1–12

⟨ **Calculator Close-Up** ⟩

The graph of $y = x^2 + 2x - 24$ supports the conclusion in Example 1(a) because the graph crosses the x-axis at $(4, 0)$ and $(-6, 0)$.

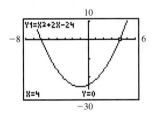

The process in Example 1 can be shortened somewhat if we observe the correspondence between the solutions to the equation and the factors.

Correspondence Between Solutions and Factors

If a and b are solutions to a quadratic equation, then the equation is equivalent to

$$(x - a)(x - b) = 0.$$

So if 2 and -3 are solutions to a quadratic equation, then the quadratic equation is $(x - 2)(x + 3) = 0$ or $x^2 + x - 6 = 0$. If the solutions are fractions, it is not necessary to use fractions in the factors. For example, if $\frac{2}{3}$ is a solution, then $3x - 2$ is a factor because $3x - 2 = 0$ is equivalent to $x = \frac{2}{3}$. If $-\frac{1}{5}$ is a solution, then $5x + 1$ is a factor because $5x + 1 = 0$ is equivalent to $x = -\frac{1}{5}$. So if $\frac{2}{3}$ and $-\frac{1}{5}$ are solutions to a quadratic equation, then the equation is $(3x - 2)(5x + 1) = 0$ or $15x^2 - 7x - 2 = 0$.

⟨2⟩ Using the Discriminant in Factoring

The quadratic formula $x = \dfrac{-b \pm \sqrt{b^2 - 4ac}}{2a}$ gives the solutions to the quadratic equation $ax^2 + bx + c = 0$. If a, b, and c are integers and $b^2 - 4ac$ is a perfect square, then $\sqrt{b^2 - 4ac}$ is a whole number and the quadratic formula produces solutions that are rational. The quadratic equations with rational solutions are precisely the ones that we solve by factoring. So we can use the discriminant $b^2 - 4ac$ to determine whether a quadratic polynomial is prime.

Identifying Prime Quadratic Polynomials Using $b^2 - 4ac$

Let $ax^2 + bx + c$ be a quadratic polynomial with integral coefficients having a greatest common factor of 1. The quadratic polynomial is prime if and only if the discriminant $b^2 - 4ac$ is *not* a perfect square.

E X A M P L E   2

Using the discriminant

Use the discriminant to determine whether each polynomial can be factored.

a) $6x^2 + x - 15$ 　　　　　　 b) $5x^2 - 3x + 2$

Solution

a) Use $a = 6$, $b = 1$, and $c = -15$ to find $b^2 - 4ac$:

$$b^2 - 4ac = 1^2 - 4(6)(-15) = 361$$

Because $\sqrt{361} = 19$, $6x^2 + x - 15$ can be factored. Using the *ac* method, we get

$$6x^2 + x - 15 = (2x - 3)(3x + 5).$$

b) Use $a = 5$, $b = -3$, and $c = 2$ to find $b^2 - 4ac$:

$$b^2 - 4ac = (-3)^2 - 4(5)(2) = -31$$

Because the discriminant is not a perfect square, $5x^2 - 3x + 2$ is prime.

> Now do Exercises 13–24

⟨3⟩ Equations Quadratic in Form

An equation in which an expression appears in place of x in $ax^2 + bx + c = 0$ is called **quadratic in form**. So,

$$3(x - 7)^2 + (x - 7) + 8 = 0,$$
$$-2(x^2 + 3)^2 - (x^2 + 3) + 1 = 0, \quad \text{and}$$
$$-7x^4 + 5x^2 - 6 = 0$$

are quadratic in form. Note that last equation is quadratic in form because it could be written as $-7(x^2)^2 + 5(x^2) - 6$, where x^2 is used in place of x. To solve an equation that is quadratic in form we replace the expression with a single variable and then solve the resulting quadratic equation, as shown in Example 3.

EXAMPLE   **3**

An equation quadratic in form

Solve $(x + 15)^2 - 3(x + 15) - 18 = 0$.

Solution

Note that $x + 15$ and $(x + 15)^2$ both appear in the equation. Let $a = x + 15$ and substitute a for $x + 15$ in the equation:

$$(x + 15)^2 - 3(x + 15) - 18 = 0$$

$$a^2 - 3a - 18 = 0$$

$$(a - 6)(a + 3) = 0 \quad \text{Factor.}$$

$$a - 6 = 0 \quad \text{or} \quad a + 3 = 0$$

$$a = 6 \quad \text{or} \quad a = -3$$

$$x + 15 = 6 \quad \text{or} \quad x + 15 = -3 \quad \text{Replace } a \text{ by } x + 15.$$

$$x = -9 \quad \text{or} \quad x = -18$$

Check in the original equation. The solution set is $\{-18, -9\}$.

> **Now do Exercises 25–30**

In Example 4, we have a fourth-degree equation that is quadratic in form. Note that the fourth-degree equation has four solutions.

EXAMPLE   **4**

A fourth-degree equation

Solve $x^4 - 6x^2 + 8 = 0$.

Solution

Note that x^4 is the square of x^2. If we let $w = x^2$, then $w^2 = x^4$. Substitute these expressions into the original equation.

$$x^4 - 6x^2 + 8 = 0$$

$$w^2 - 6w + 8 = 0 \quad \text{Replace } x^4 \text{ by } w^2 \text{ and } x^2 \text{ by } w.$$

$$(w - 2)(w - 4) = 0 \quad \text{Factor.}$$

$$w - 2 = 0 \quad \text{or} \quad w - 4 = 0$$

$$w = 2 \quad \text{or} \quad w = 4$$

$$x^2 = 2 \quad \text{or} \quad x^2 = 4 \quad \text{Substitute } x^2 \text{ for } w.$$

$$x = \pm\sqrt{2} \quad \text{or} \quad x = \pm 2 \quad \text{Even-root property}$$

Check. The solution set is $\{-2, -\sqrt{2}, \sqrt{2}, 2\}$.

> **Now do Exercises 31–38**

‹ **Helpful Hint** ›

The fundamental theorem of algebra says that the number of solutions to a polynomial equation is less than or equal to the degree of the polynomial. This famous theorem was proved by Carl Friedrich Gauss when he was a young man.

CAUTION   If you replace x^2 by w, do not quit when you find the values of w. If the variable in the original equation is x, then you must solve for x.

E X A M P L E   **5**

A quadratic within a quadratic

Solve $(x^2 + 2x)^2 - 11(x^2 + 2x) + 24 = 0$.

Solution

Note that $x^2 + 2x$ and $(x^2 + 2x)^2$ appear in the equation. Let $a = x^2 + 2x$ and substitute.

$$a^2 - 11a + 24 = 0$$
$$(a - 8)(a - 3) = 0 \quad \text{Factor.}$$

$a - 8 = 0$	or	$a - 3 = 0$	

$a - 8 = 0$ or $a - 3 = 0$

$a = 8$ or $a = 3$

$x^2 + 2x = 8$ or $x^2 + 2x = 3$ Replace a by $x^2 + 2x$.

$x^2 + 2x - 8 = 0$ or $x^2 + 2x - 3 = 0$

$(x - 2)(x + 4) = 0$ or $(x + 3)(x - 1) = 0$

$x - 2 = 0$ or $x + 4 = 0$ or $x + 3 = 0$ or $x - 1 = 0$

$x = 2$ or $x = -4$ or $x = -3$ or $x = 1$

Check. The solution set is $\{-4, -3, 1, 2\}$.

> Now do Exercises 39–44

‹ **Calculator Close-Up** ›

The four x-intercepts on the graph of $y = (x^2 + 2x)^2 - 11(x^2 + 2x) + 24$ support the conclusion in Example 5.

Example 6 involves a fractional exponent. To identify this type of equation as quadratic in form, recall how to square an expression with a fractional exponent. For example, $(x^{1/2})^2 = x$, $(x^{1/4})^2 = x^{1/2}$, and $(x^{1/3})^2 = x^{2/3}$.

E X A M P L E   **6**

A fractional exponent

Solve $x - 9x^{1/2} + 14 = 0$.

Solution

Note that the square of $x^{1/2}$ is x. Let $w = x^{1/2}$; then $w^2 = (x^{1/2})^2 = x$. Now substitute w and w^2 into the original equation:

$$w^2 - 9w + 14 = 0$$
$$(w - 7)(w - 2) = 0$$

$w - 7 = 0$ or $w - 2 = 0$

$w = 7$ or $w = 2$

$x^{1/2} = 7$ or $x^{1/2} = 2$ Replace w by $x^{1/2}$.

$x = 49$ or $x = 4$ Square each side.

Because we squared each side, we must check for extraneous roots. First evaluate $x - 9x^{1/2} + 14$ for $x = 49$:

$$49 - 9 \cdot 49^{1/2} + 14 = 49 - 9 \cdot 7 + 14 = 0$$

Now evaluate $x - 9x^{1/2} + 14$ for $x = 4$:

$$4 - 9 \cdot 4^{1/2} + 14 = 4 - 9 \cdot 2 + 14 = 0$$

Because each solution checks, the solution set is $\{4, 49\}$.

> Now do Exercises 45–52

CAUTION   An equation of quadratic form with variable x must contain a power of x and its square. Equations such as $x^4 - 5x^3 + 6 = 0$ or $x^{1/2} - 3x^{1/3} - 18 = 0$ are not quadratic in form and cannot be solved by substitution.

‹4› **Applications**

Applied problems often result in quadratic equations that cannot be factored. For such equations we use the quadratic formula to find exact solutions and a calculator to find decimal approximations for the exact solutions.

E X A M P L E   **7**

Figure 8.2

Changing area

Marvin's flower bed is rectangular in shape with a length of 10 feet and a width of 5 feet (ft). He wants to increase the length and width by the same amount to obtain a flower bed with an area of 75 square feet (ft²). What should the amount of increase be?

Solution

Let x be the amount of increase. The length and width of the new flower bed are $x + 10$ ft and $x + 5$ ft, as shown in Fig. 8.2. Because the area is to be 75 ft², we have

$$(x + 10)(x + 5) = 75.$$

Write this equation in the form $ax^2 + bx + c = 0$:

$$x^2 + 15x + 50 = 75$$

$$x^2 + 15x - 25 = 0 \quad \text{Get 0 on the right.}$$

$$x = \frac{-15 \pm \sqrt{225 - 4(1)(-25)}}{2(1)}$$

$$= \frac{-15 \pm \sqrt{325}}{2} = \frac{-15 \pm 5\sqrt{13}}{2}$$

Because the value of x must be positive, the exact increase is

$$\frac{-15 + 5\sqrt{13}}{2} \text{ feet.}$$

Using a calculator, we can find that x is approximately 1.51 ft. If $x = 1.51$ ft, then the new length is 11.51 ft, and the new width is 6.51 ft. The area of a rectangle with these dimensions is 74.93 ft². Of course, the approximate dimensions do not give an area of exactly 75 ft².

Now do Exercises 79–86

E X A M P L E   **8**

Mowing the lawn

It takes Carla 1 hour longer to mow the lawn than it takes Sharon to mow the lawn. If they can mow the lawn in 5 hours working together, then how long would it take each girl by herself?

Solution

If Sharon can mow the lawn by herself in x hours, then she works at the rate of $\frac{1}{x}$ of the lawn per hour. If Carla can mow the lawn by herself in $x + 1$ hours, then she works at the rate of $\frac{1}{x+1}$ of the lawn per hour. We can use a table to list all of the important quantities.

	Rate	Time	Work
Sharon	$\dfrac{1}{x} \dfrac{\text{lawn}}{\text{hr}}$	5 hr	$\dfrac{5}{x}$ lawn
Carla	$\dfrac{1}{x+1} \dfrac{\text{lawn}}{\text{hr}}$	5 hr	$\dfrac{5}{x+1}$ lawn

‹ **Helpful Hint** ›

Note that the equation concerns the portion of the job done by each girl. We could have written an equation about the rates at which the two girls work. Because they can finish the lawn together in 5 hours, they are mowing together at the rate of $\frac{1}{5}$ lawn per hour. So,

$$\frac{1}{x} + \frac{1}{x+1} = \frac{1}{5}.$$

Because they complete the lawn in 5 hours, the portion of the lawn done by Sharon and the portion done by Carla have a sum of 1:

$$\frac{5}{x} + \frac{5}{x+1} = 1$$

$$x(x+1)\,\frac{5}{x} + x(x+1)\,\frac{5}{x+1} = x(x+1)1 \quad \text{Multiply by the LCD.}$$

$$5x + 5 + 5x = x^2 + x$$

$$10x + 5 = x^2 + x$$

$$-x^2 + 9x + 5 = 0$$

$$x^2 - 9x - 5 = 0$$

$$x = \frac{9 \pm \sqrt{(-9)^2 - 4(1)(-5)}}{2(1)}$$

$$= \frac{9 \pm \sqrt{101}}{2}$$

Using a calculator, we find that $\frac{9 - \sqrt{101}}{2}$ is negative. So Sharon's time alone is

$$\frac{9 + \sqrt{101}}{2} \text{ hours.}$$

To find Carla's time alone, we add 1 hour to Sharon's time:

$$\frac{9 + \sqrt{101}}{2} + 1 = \frac{9 + \sqrt{101}}{2} + \frac{2}{2} = \frac{11 + \sqrt{101}}{2} \text{ hours}$$

Sharon's time alone is approximately 9.525 hours, and Carla's time alone is approximately 10.525 hours.

> **Now do Exercises 87–90**

Warm-Ups

Fill in the blank.

1. If d is a solution to a quadratic equation, then $x - d$ is a _____ of the quadratic polynomial.

2. If $b^2 - 4ac$ is not a perfect square, then $ax^2 + bx + c$ is a _____ polynomial.

3. An equation that is quadratic after a substitution is called _____ in form.

4. If m and n are _____ to a quadratic equation, then $(x - m)(x - n) = 0$ is a quadratic with those solutions.

True or false?

5. The equation $(x - 4)(x + 5) = 0$ is a quadratic equation with solutions 4 and -5.

6. If $w = x^{1/6}$, then $w^2 = x^{1/3}$.

7. If $y = 2^{1/2}$, then $y^2 = 2^{1/4}$.

8. To solve $x^4 - 5x^2 + 6 = 0$ by substitution, let $w = x^2$.

9. To solve $x^5 - 3x^3 - 10 = 0$ by substitution, let $w = x^3$.

10. If Ann's boat goes 10 mph in still water, then against a 5-mph current it will go 2 mph.

11. If Elvia drives 300 miles in x hours, then her rate is $\dfrac{300}{x}$ mph.

12. If John paints a 100-foot fence in x hours, then his rate is $\dfrac{100}{x}$ of the fence per hour.

Exercises 8.3

For each given pair of numbers find a quadratic equation with integral coefficients that has the numbers as its solutions. See Example 1.

1. $3, -7$ **2.** $-8, 2$

3. $4, 1$ **4.** $3, 2$

5. $\sqrt{5}, -\sqrt{5}$ **6.** $-\sqrt{7}, \sqrt{7}$

7. $4i, -4i$ **8.** $-3i, 3i$

9. $i\sqrt{2}, -i\sqrt{2}$ **10.** $3i\sqrt{2}, -3i\sqrt{2}$

11. $\dfrac{1}{2}, \dfrac{1}{3}$ **12.** $-\dfrac{1}{5}, -\dfrac{1}{2}$

‹2› **Using the Discriminant in Factoring**

Use the discriminant to determine whether each quadratic polynomial can be factored, then factor the ones that are not prime. See Example 2.

13. $x^2 + 9$ **14.** $x^2 - 9$

15. $2x^2 - x + 4$ **16.** $2x^2 + 3x - 5$

17. $2x^2 + 6x - 5$ **18.** $3x^2 + 5x - 1$

19. $6x^2 + 19x - 36$ **20.** $8x^2 + 6x - 27$

21. $4x^2 - 5x - 12$ **22.** $4x^2 - 27x + 45$

23. $8x^2 - 18x - 45$ **24.** $6x^2 + 9x - 16$

‹3› **Equations Quadratic in Form**

Find all real solutions to each equation. See Example 3.

25. $(x - 1)^2 - 2(x - 1) - 8 = 0$

26. $(m + 3)^2 + 5(m + 3) - 14 = 0$

27. $(2a - 1)^2 + 2(2a - 1) - 8 = 0$

28. $(3a + 2)^2 - 3(3a + 2) = 10$

29. $(w - 1)^2 + 5(w - 1) + 5 = 0$

30. $(2x - 1)^2 - 4(2x - 1) + 2 = 0$

Find all real solutions to each equation. See Example 4.

31. $x^4 - 13x^2 + 36 = 0$

32. $x^4 - 20x^2 + 64 = 0$

33. $x^6 - 28x^3 + 27 = 0$

34. $x^6 - 3x^3 - 4 = 0$

35. $x^4 - 14x^2 + 45 = 0$

36. $x^4 + 2x^2 = 15$

37. $x^6 + 7x^3 = 8$ **38.** $a^6 + 6a^3 = 16$

Find all real solutions to each equation. See Example 5.

39. $(x^2 + 1)^2 - 11(x^2 + 1) = -10$

40. $(x^2 + 2)^2 - 11(x^2 + 2) = -30$

41. $(x^2 + 2x)^2 - 7(x^2 + 2x) + 12 = 0$

42. $(x^2 + 3x)^2 + (x^2 + 3x) - 20 = 0$

43. $(y^2 + y)^2 - 8(y^2 + y) + 12 = 0$

44. $(w^2 - 2w)^2 + 24 = 11(w^2 - 2w)$

Find all real solutions to each equation. See Example 6.

45. $x - 3x^{1/2} + 2 = 0$

46. $x^{1/2} - 3x^{1/4} + 2 = 0$

47. $x^{2/3} + 4x^{1/3} + 3 = 0$

48. $x^{2/3} - 3x^{1/3} - 10 = 0$

49. $x^{1/2} - 5x^{1/4} + 6 = 0$

50. $2x - 5\sqrt{x} + 2 = 0$

51. $2x - 5x^{1/2} - 3 = 0$ **52.** $x^{1/4} + 2 = x^{1/2}$

Find all real solutions to each equation.

53. $x^{-2} + x^{-1} - 6 = 0$ **54.** $x^{-2} - 2x^{-1} = 8$

55. $x^{1/6} - x^{1/3} + 2 = 0$ **56.** $x^{2/3} - x^{1/3} - 20 = 0$

57. $\left(\dfrac{1}{y - 1}\right)^2 + \left(\dfrac{1}{y - 1}\right) = 6$

58. $\left(\dfrac{1}{w + 1}\right)^2 - 2\left(\dfrac{1}{w + 1}\right) - 24 = 0$

59. $2x^2 - 3 - 6\sqrt{2x^2 - 3} + 8 = 0$

60. $x^2 + x + \sqrt{x^2 + x} - 2 = 0$

61. $x^{-2} - 2x^{-1} - 1 = 0$ **62.** $x^{-2} - 6x^{-1} + 6 = 0$

Miscellaneous

Find all real and imaginary solutions to each equation.

63. $w^2 + 4 = 0$ **64.** $w^2 + 9 = 0$

65. $a^4 + 6a^2 + 8 = 0$

66. $b^4 + 13b^2 + 36 = 0$

67. $m^4 - 16 = 0$

68. $t^4 - 4 = 0$

69. $16b^4 - 1 = 0$

70. $b^4 - 81 = 0$

71. $x^3 + 1 = 0$

72. $x^3 - 1 = 0$

73. $x^3 + 8 = 0$

74. $x^3 - 27 = 0$

75. $a^{-2} - 2a^{-1} + 5 = 0$

76. $b^{-2} - 4b^{-1} + 6 = 0$

77. $(2x - 1)^2 - 2(2x - 1) + 5 = 0$

78. $(4x - 1)^2 - 6(4x - 1) + 25 = 0$

⟨4⟩   Applications

Find the exact solution to each problem. If the exact solution is an irrational number, then also find an approximate decimal solution. See Examples 7 and 8.

79. *Country singers.* Harry and Gary are traveling to Nashville to make their fortunes. Harry leaves on the train at 8:00 A.M. and Gary travels by car, starting at 9:00 A.M. To complete the 300-mile trip and arrive at the same time as Harry, Gary travels 10 miles per hour (mph) faster than the train. At what time will they both arrive in Nashville?

80. *Gone fishing.* Debbie traveled by boat 5 miles upstream to fish in her favorite spot. Because of the 4-mph current, it took her 20 minutes longer to get there than to return. How fast will her boat go in still water?

81. *Cross-country cycling.* Erin was traveling across the desert on her bicycle. Before lunch she traveled 60 miles (mi); after lunch she traveled 46 mi. She put in 1 hour more after lunch than before lunch, but her speed was 4 mph slower than before. What was her speed before lunch and after lunch?

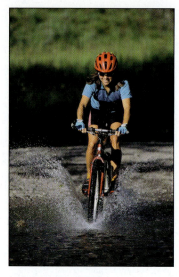

Photo for Exercise 81

82. *Extreme hardship.* Kim starts to walk 3 mi to school at 7:30 A.M. with a temperature of 0°F. Her brother Bryan starts at 7:45 A.M. on his bicycle, traveling 10 mph faster than Kim. If they get to school at the same time, then how fast is each one traveling?

83. *American pie.* John takes 3 hours longer than Andrew to peel 500 pounds (lb) of apples. If together they can peel 500 lb of apples in 8 hours, then how long would it take each one working alone?

84. *On the half shell.* It takes Brent 1 hour longer than Calvin to shuck a sack of oysters. If together they shuck a sack of oysters in 45 minutes, then how long would it take each one working alone?

85. *The growing garden.* Eric's garden is 20 ft by 30 ft. He wants to increase the length and width by the same amount to have a 1000-ft² garden. What should be the new dimensions of the garden?

86. *Open-top box.* Thomas is going to make an open-top box by cutting equal squares from the four corners of an 11 inch by 14 inch sheet of cardboard and folding up the sides. If the area of the base is to be 80 square inches, then what size square should be cut from each corner?

Figure for Exercise 86

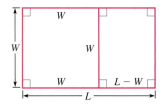

Figure for Exercise 89

87. *Pumping the pool.* It takes pump A 2 hours less time than pump B to empty a certain swimming pool. Pump A is started at 8:00 A.M., and pump B is started at 11:00 A.M. If the pool is still half full at 5:00 P.M., then how long would it take pump A working alone?

88. *Time off for lunch.* It usually takes Eva 3 hours longer to do the monthly payroll than it takes Cicely. They start working on it together at 9:00 A.M., and at 5:00 P.M. they have 90% of it done. If Eva took a 2-hour lunch break while Cicely had none, then how much longer will it take for them to finish the payroll working together?

89. *Golden Rectangle.* One principle used by the ancient Greeks to get shapes that are pleasing to the eye in art and architecture was the Golden Rectangle. If a square is removed from one end of a Golden Rectangle, as shown in the figure, the sides of the remaining rectangle are proportional to the original rectangle. So the length and width of the original rectangle satisfy

$$\frac{L}{W} = \frac{W}{L - W}.$$

If the length of a Golden Rectangle is 10 meters, then what is its width?

90. *Golden painting.* An artist wants her painting to be in the shape of a golden rectangle. If the length of the painting is 36 inches, then what should be the width? See Exercise 89.

Getting More Involved

91. *Exploration*

  a) Given that $P(x) = x^4 + 6x^2 - 27$, find $P(3i)$, $P(-3i)$, $P(\sqrt{3})$, and $P(-\sqrt{3})$.

  b) What can you conclude about the values $3i$, $-3i$, $\sqrt{3}$, and $-\sqrt{3}$ and their relationship to each other?

92. *Cooperative learning*

Work with a group to write a quadratic equation that has each given pair of solutions.

  a) $3 + \sqrt{5}, 3 - \sqrt{5}$ **b)** $4 - 2i, 4 + 2i$

  c) $\dfrac{1 + i\sqrt{3}}{2}, \dfrac{1 - i\sqrt{3}}{2}$

Graphing Calculator Exercises

Solve each equation by locating the x-intercepts on a calculator graph. Round approximate answers to two decimal places.

93. $(5x - 7)^2 - (5x - 7) - 6 = 0$

94. $x^4 - 116x^2 + 1600 = 0$

95. $(x^2 + 3x)^2 - 7(x^2 + 3x) + 9 = 0$

96. $x^2 - 3x^{1/2} - 12 = 0$

Mid-Chapter **Quiz** | Sections 8.1 through 8.3 | **Chapter 8**

Solve each equation by factoring.

1. $x^2 - 4x - 32 = 0$ **2.** $6x^2 - 5x + 1 = 0$

Solve using the even-root property.

3. $x^2 = \dfrac{16}{25}$ **4.** $(w + 3)^2 = 6$

Solve by completing the square.

5. $x^2 - 4x = 1$ **6.** $2z^2 + z = 1$

Solve by using the quadratic formula.

7. $2x^2 - 5x + 2 = 0$

8. $2h^2 + 4h + 1 = 0$

Solve by any method.

9. $(x + 7)^2 - 5(x + 7) + 6 = 0$

10. $x^4 - 6x^2 + 5 = 0$

11. $x - 2x^{1/2} - 8 = 0$

12. $\sqrt{2x - 3} = \dfrac{x - 4}{2}$

Miscellaneous.

13. Find the imaginary solutions to $x^2 - 10x + 26 = 0$.

14. Find the discriminant for the equation $3x^2 - x + 5 = 0$.

15. Find a quadratic equation that has -3 and 8 as its solutions.

8.4   Quadratic Functions and Their Graphs

In This Section

⟨1⟩ **Quadratic Functions**

⟨2⟩ **Graphing Quadratic Functions**

⟨3⟩ **The Vertex and Intercepts**

⟨4⟩ **Applications**

We have seen *quadratic functions* on several occasions in this text, but we have not yet defined the term. In this section, we study quadratic functions and their graphs.

⟨1⟩ Quadratic Functions

If y is determined from x by a formula involving a quadratic polynomial, then we say that y is a *quadratic function of* x. Recall that we can use $f(x)$ or y for the dependent variable.

> **Quadratic Function**
>
> A **quadratic function** is a function of the form
>
> $$f(x) = ax^2 + bx + c,$$
>
> where a, b, and c are real numbers and $a \neq 0$.

Without the term ax^2, this function would be a linear function. That is why we specify that $a \neq 0$.

E X A M P L E   1   **Finding ordered pairs of a quadratic function**

Complete each ordered pair so that it satisfies the given equation.

a) $f(x) = x^2 - x - 6$; $(2, \quad)$, $(\quad, 0)$

b) $s = -16t^2 + 48t + 84$; $(0, \quad)$, $(\quad, 20)$

Solution

a) If $x = 2$, then $f(2) = 2^2 - 2 - 6 = -4$. So the ordered pair is $(2, -4)$. To find x when $y = 0$, replace $f(x)$ by 0 and solve the resulting quadratic equation:

$$x^2 - x - 6 = 0$$
$$(x - 3)(x + 2) = 0$$
$$x - 3 = 0 \quad \text{or} \quad x + 2 = 0$$
$$x = 3 \quad \text{or} \quad x = -2$$

The ordered pairs are $(-2, 0)$ and $(3, 0)$.

b) If $t = 0$, then $s = -16 \cdot 0^2 + 48 \cdot 0 + 84 = 84$. The ordered pair is $(0, 84)$. To find t when $s = 20$, replace s by 20 and solve the equation for t:

$$-16t^2 + 48t + 84 = 20$$
$$-16t^2 + 48t + 64 = 0 \qquad \text{Subtract 20 from each side.}$$
$$t^2 - 3t - 4 = 0 \qquad \text{Divide each side by } -16.$$
$$(t - 4)(t + 1) = 0 \qquad \text{Factor.}$$
$$t - 4 = 0 \quad \text{or} \quad t + 1 = 0 \quad \text{Zero factor property}$$
$$t = 4 \quad \text{or} \quad t = -1$$

The ordered pairs are $(-1, 20)$ and $(4, 20)$.

> **Now do Exercises 1–6**

CAUTION When variables other than x and y are used, the independent variable is the first coordinate of an ordered pair, and the dependent variable is the second coordinate. In Example 1(b), t is the independent variable and first coordinate because s depends on t by the formula $s = -16t^2 + 48t + 84$.

⟨2⟩ Graphing Quadratic Functions

Any real number may be used for x in $f(x) = ax^2 + bx + c$. So the domain (the set of x-coordinates) for any quadratic function is the set of all real numbers, $(-\infty, \infty)$. The range (the set of y-coordinates) can be determined from the graph. All quadratic functions have graphs that are similar in shape. The graph of any quadratic function is called a **parabola.**

The parabola in Example 2 is said to **open upward.** In Example 3 we see a parabola that **opens downward.** If $a > 0$ in the equation $y = ax^2 + bx + c$, then the parabola opens upward. If $a < 0$, then the parabola opens downward.

E X A M P L E   2

Graphing the simplest quadratic function

Graph the function $f(x) = x^2$, and state the domain and range.

Solution

Make a table of values for x and y:

x	-2	-1	0	1	2
$y = x^2$	4	1	0	1	4

‹ **Calculator Close-Up** ›

This close-up view of $y = x^2$ shows how rounded the curve is at the bottom. When drawing a parabola by hand, be sure to draw it smoothly.

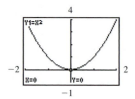

See Fig. 8.3 for the graph. The domain is the set of all real numbers, $(-\infty, \infty)$, because we can use any real number for x. From the graph we see that the smallest y-coordinate of the function is 0. So the range is the set of real numbers that are greater than or equal to 0, $[0, \infty)$.

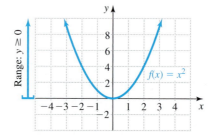

Figure 8.3

Now do Exercises 7–12

Note the symmetry of the parabola in Fig. 8.3. If the paper was folded along the y-axis, the two sides of the parabola would come together. The point $(-1, 1)$ would match up with $(1, 1)$, the point $(-2, 4)$ would match up with $(2, 4)$, and so on.

E X A M P L E   **3**

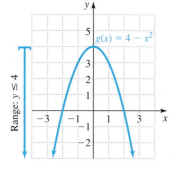

Figure 8.4

A quadratic function

Graph the function $g(x) = 4 - x^2$, and state the domain and range.

Solution

We plot enough points to get the correct shape of the graph:

x	-2	-1	0	1	2
$y = 4 - x^2$	0	3	4	3	0

See Fig. 8.4 for the graph. The domain is the set of all real numbers, $(-\infty, \infty)$. From the graph we see that the largest y-coordinate is 4. So the range is $(-\infty, 4]$.

Now do Exercises 13–22

‹3› The Vertex and Intercepts

The lowest point on a parabola that opens upward or the highest point on a parabola that opens downward is called the **vertex.** The y-coordinate of the vertex is the *minimum* y-coordinate or **minimum value** of the function if the parabola opens upward, and it is the *maximum* y-coordinate or **maximum value** of the function if the parabola opens downward. For $f(x) = x^2$ the vertex is $(0, 0)$, and 0 is the minimum value of the function. For $g(x) = 4 - x^2$ the vertex is $(0, 4)$, and 4 is the maximum value of the function.

If $y = ax^2 + bx + c$ has x-intercepts, they can be found by solving $ax^2 + bx + c = 0$ by the quadratic formula. The vertex is midway between the x-intercepts as shown in Fig. 8.5. Note that in the quadratic formula

$$x = \frac{-b \pm \sqrt{b^2 - 4ac}}{2a},$$

$\sqrt{b^2 - 4ac}$ is added and subtracted from the numerator of $\frac{-b}{2a}$. So $\left(\frac{-b}{2a}, 0\right)$ is the point midway between the x-intercepts and the vertex has the same x-coordinate. Even if the parabola has no x-intercepts, the x-coordinate of the vertex is still $\frac{-b}{2a}$.

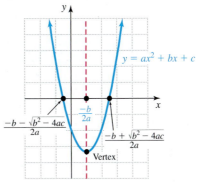

Figure 8.5

Vertex of a Parabola

The x-coordinate of the vertex of $f(x) = ax^2 + bx + c$ is $\frac{-b}{2a}$, provided $a \neq 0$.
Find the y-coordinate of the vertex by evaluating $f\left(\frac{-b}{2a}\right)$.

When you graph a parabola, you should always locate the vertex because it is the point at which the graph "turns around." With the vertex and several nearby points you can see the correct shape of the parabola.

EXAMPLE 4

Using the vertex in graphing a quadratic function

Graph $f(x) = -x^2 - x + 2$, and state the domain and range.

Solution

First find the x-coordinate of the vertex:

$$x = \frac{-b}{2a} = \frac{-(-1)}{2(-1)} = \frac{1}{-2} = -\frac{1}{2}$$

Now find $f\left(-\frac{1}{2}\right)$:

$$f\left(-\frac{1}{2}\right) = -\left(-\frac{1}{2}\right)^2 - \left(-\frac{1}{2}\right) + 2 = -\frac{1}{4} + \frac{1}{2} + 2 = \frac{9}{4}$$

The vertex is $\left(-\frac{1}{2}, \frac{9}{4}\right)$. Now find a few points on either side of the vertex:

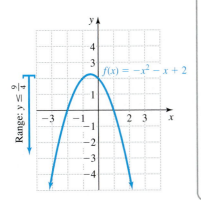

x	-2	-1	$-\frac{1}{2}$	0	1
$f(x) = -x^2 - x + 2$	0	2	$\frac{9}{4}$	2	0

Sketch a parabola through these points as in Fig. 8.6. The domain is $(-\infty, \infty)$. Because the graph goes no higher than $\frac{9}{4}$, the range is $\left(-\infty, \frac{9}{4}\right]$.

Now do Exercises 23–30

Figure 8.6

The y-intercept for the parabola $y = ax^2 + bx + c$ is the point that has 0 as its x-coordinate. If we let $x = 0$, we get $y = a(0)^2 + b(0) + c = c$. So the y-intercept is $(0, c)$. To find the x-intercepts let $y = 0$ and solve $ax^2 + bx + c = 0$. A parabola may have 0, 1, or 2 x-intercepts, depending on the number of solutions to this equation.

> **Finding Intercepts**
>
> The y-intercept for $y = ax^2 + bx + c$ is $(0, c)$.
>
> To find the x-intercepts solve $ax^2 + bx + c = 0$.

E X A M P L E **5**

Using the intercepts in graphing a quadratic function

Find the vertex and intercepts, and sketch the graph of each function.

a) $f(x) = x^2 - 2x - 8$ **b)** $s = -16t^2 + 64t$

Solution

a) Use $x = \frac{-b}{2a}$ to get $x = 1$ as the x-coordinate of the vertex. If $x = 1$, then

$$f(1) = 1^2 - 2 \cdot 1 - 8$$
$$= -9.$$

So the vertex is $(1, -9)$. If $x = 0$, then

$$f(0) = 0^2 - 2 \cdot 0 - 8$$
$$= -8.$$

The y-intercept is $(0, -8)$. To find the x-intercepts, replace $f(x)$ by 0:

$$x^2 - 2x - 8 = 0$$
$$(x - 4)(x + 2) = 0$$

$$x - 4 = 0 \quad \text{or} \quad x + 2 = 0$$
$$x = 4 \quad \text{or} \quad x = -2$$

The x-intercepts are $(-2, 0)$ and $(4, 0)$. The graph is shown in Fig. 8.7.

b) Because s is expressed as a function of t, the first coordinate is t. Use $t = \frac{-b}{2a}$ to get

$$t = \frac{-64}{2(-16)} = 2.$$

If $t = 2$, then

$$s = -16 \cdot 2^2 + 64 \cdot 2$$
$$= 64.$$

So the vertex is $(2, 64)$. If $t = 0$, then

$$s = -16 \cdot 0^2 + 64 \cdot 0$$
$$= 0.$$

So the s-intercept is $(0, 0)$. To find the t-intercepts, replace s by 0:

$$-16t^2 + 64t = 0$$
$$-16t(t - 4) = 0$$

$$-16t = 0 \quad \text{or} \quad t - 4 = 0$$
$$t = 0 \quad \text{or} \quad t = 4$$

The t-intercepts are $(0, 0)$ and $(4, 0)$. The graph is shown in Fig. 8.8.

Figure 8.7

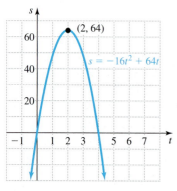

Figure 8.8

> **Now do Exercises 31–46**

⟨ **Calculator Close-Up** ⟩

You can find the vertex of a parabola with a calculator by using either the maximum or minimum feature. First graph the parabola as shown.

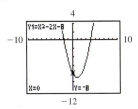

Because this parabola opens upward, the y-coordinate of the vertex is the minimum

y-coordinate on the graph. Press CALC and choose minimum.

The calculator will ask for a left bound, a right bound, and a guess. For the left bound choose a point to the left of the vertex by

moving the cursor to the point and pressing ENTER. For the right bound choose a point to the right of the vertex. For the guess choose a point close to the vertex.

⟨4⟩ Applications

In applications, we are often interested in finding the maximum or minimum value of a variable. If the graph of a quadratic function opens downward, then the maximum value of the dependent variable is the second coordinate of the vertex. If the parabola opens upward, then the minimum value of the dependent variable is the second coordinate of the vertex.

E X A M P L E   **6**

Finding the maximum height

If a projectile is launched with an initial velocity of v_0 feet per second from an initial height of s_0 feet, then its height $s(t)$ in feet is determined by the quadratic function $s(t) = -16t^2 + v_0t + s_0$, where t is the time in seconds. If a ball is tossed upward with velocity 64 feet per second from a height of 5 feet, then what is the maximum height reached by the ball?

Solution

The height $s(t)$ of the ball for any time t is given by $s(t) = -16t^2 + 64t + 5$. Because the maximum height occurs at the vertex of the parabola, we use $t = \frac{-b}{2a}$ to find the vertex:

$$t = \frac{-64}{2(-16)} = 2$$

Now use $t = 2$ to find the second coordinate of the vertex:

$$s(2) = -16(2)^2 + 64(2) + 5 = 69$$

The maximum height reached by the ball is 69 feet. See Fig. 8.9.

Figure 8.9

$s(t) = -16t^2 + 64t + 5$

(2, 69)

> Now do Exercises 55–63

Warm-Ups ▼

Fill in the blank.

1. A _____ function has the form $f(x) = ax^2 + bx + c$.
2. The graph of $y = ax^2 + bx + c$ with $a \neq 0$ is a _____.
3. The graph of a quadratic function with $a > 0$ opens _____.
4. The graph of a quadratic function with $a < 0$ opens _____.
5. The _____ of any quadratic function is $(-\infty, \infty)$.
6. The _____ is the highest point on a parabola that opens downward or the lowest point on a parabola that opens upward.
7. The x-coordinate of the vertex for $y = ax^2 + bx + c$ is _____.

True or false?

8. The ordered pair $(-2, -1)$ satisfies $f(x) = x^2 - 5$.

9. The y-intercept for $g(x) = x^2 - 3x + 9$ is $(9, 0)$.

10. The x-intercepts for $y = x^2 - 5$ are $\left(\sqrt{5}, 0\right)$ and $\left(-\sqrt{5}, 0\right)$.

11. The graph of $f(x) = x^2 - 12$ opens upward.

12. The graph of $y = 4 + x^2$ opens downward.

13. The parabola $y = x^2 + 1$ has no x-intercepts.

14. The y-intercept for $g(x) = ax^2 + bx + c$ is $(0, c)$.

15. If $w = -2t^2 + 9$, then the maximum value of w is 9.

8.4   Exercises

‹ Study Tips ›

- Get an early start studying for your final exams.
- If you have several final exams, it can be difficult to find the time to prepare for all of them in the last couple of days.

‹1› Quadratic Functions

Complete each ordered pair so that it satisfies the given equation. See Example 1.

1. $f(x) = x^2$ $(4, \quad), (\quad, 9)$

2. $f(x) = -x^2$ $(-9, \quad), (\quad, -4)$

3. $f(x) = x^2 - x - 12$ $(3, \quad), (\quad, 0)$

4. $f(x) = -\dfrac{1}{2}x^2 - x + 1$ $(0, \quad), (\quad, -3)$

5. $s = -16t^2 + 32t$ $(4, \quad), (\quad, 0)$

6. $a = b^2 + 4b + 5$ $(-2, \quad), (\quad, 2)$

‹2› Graphing Quadratic Functions

Determine whether the graph of each quadratic function opens upward or downward. See Examples 2 and 3.

7. $f(x) = x^2 + 5$ **8.** $f(x) = 2x^2 + x - 1$

9. $y = -3x^2 + 4x + 2$ **10.** $y = -x^2 + 3$

11. $f(x) = (-2x + 3)^2$ **12.** $f(x) = (5 - x)^2$

Graph each quadratic function, and state its domain and range. See Examples 2 and 3.

13. $f(x) = x^2 + 2$

14. $g(x) = x^2 - 4$

15. $y = \dfrac{1}{2}x^2 - 4$

16. $y = \frac{1}{3}x^2 - 6$

22. $h(x) = (x + 3)^2$

17. $f(x) = -2x^2 + 5$

18. $g(x) = -x^2 - 1$

19. $y = -\frac{1}{3}x^2 + 5$

20. $y = -\frac{1}{2}x^2 + 3$

21. $h(x) = (x - 2)^2$

⟨3⟩ **The Vertex and Intercepts**

Find the vertex for the graph of each quadratic function.
See Example 4.

23. $f(x) = x^2 - 9$ **24.** $f(x) = x^2 + 12$

25. $y = x^2 - 4x + 1$ **26.** $y = x^2 + 8x - 3$

27. $f(x) = -2x^2 + 20x + 1$ **28.** $f(x) = -3x^2 + 18x - 7$

29. $y = x^2 - x + 1$ **30.** $y = 3x^2 - 2x + 1$

Find all intercepts for the graph of each quadratic function.
See Example 5.

31. $f(x) = 16 - x^2$ **32.** $f(x) = x^2 - 9$

33. $y = x^2 - 2x - 15$ **34.** $y = x^2 - x - 6$

35. $f(x) = -4x^2 + 12x - 9$ **36.** $f(x) = -2x^2 - x + 3$

Find the vertex and intercepts for each quadratic function.
Sketch the graph, and state the domain and range.
See Examples 4 and 5.

37. $f(x) = x^2 - x - 2$

38. $f(x) = x^2 + 2x - 3$

39. $g(x) = x^2 + 2x - 8$

40. $g(x) = x^2 + x - 6$

41. $y = -x^2 - 4x - 3$

42. $y = -x^2 - 5x - 4$

43. $h(x) = -x^2 + 3x + 4$

44. $h(x) = -x^2 - 2x + 8$

45. $a = b^2 - 6b - 16$

46. $v = -u^2 - 8u + 9$

Find the maximum or minimum value of y for each function.

47. $y = x^2 - 8$

48. $y = 33 - x^2$

49. $y = -3x^2 + 14$

50. $y = 6 + 5x^2$

51. $y = x^2 + 2x + 3$

52. $y = x^2 - 2x + 5$

53. $y = -2x^2 - 4x$

54. $y = -3x^2 + 24x$

⟨4⟩ **Applications**

Solve each problem. See Example 6.

55. *Maximum height.* If a baseball is projected upward from ground level with an initial velocity of 64 feet per second, then its height is a function of time, given by $s(t) = -16t^2 + 64t$. Graph this function for $0 \leq t \leq 4$. What is the maximum height reached by the ball?

56. *Maximum height.* If a soccer ball is kicked straight up with an initial velocity of 32 feet per second, then its height above the earth is a function of time given by $s(t) = -16t^2 + 32t$. Graph this function for $0 \leq t \leq 2$. What is the maximum height reached by this ball?

57. *Minimum cost.* It costs Acme Manufacturing C dollars per hour to operate its golf ball division. An analyst has determined that C is related to the number of golf balls produced per hour, x, by the equation $C = 0.009x^2 - 1.8x + 100$. What number of balls per hour should Acme produce to minimize the cost per hour of manufacturing these golf balls?

58. *Maximum profit.* A chain store manager has been told by the main office that daily profit, P, is related to the number of clerks working that day, x, according to the equation $P = -25x^2 + 300x$. What number of clerks will maximize the profit, and what is the maximum possible profit?

59. *Maximum area.* Jason plans to fence a rectangular area with 100 meters of fencing. He has written the formula $A = w(50 - w)$ to express the area in terms of the width w. What is the maximum possible area that he can enclose with his fencing?

Photo for Exercise 59

60. *Minimizing cost.* A company uses the function $C(x) = 0.02x^2 - 3.4x + 150$ to model the unit cost in dollars for producing x stabilizer bars. For what number of bars is the unit cost at its minimum? What is the unit cost at that level of production?

61. *Air pollution.* The amount of nitrogen dioxide A in parts per million (ppm) that was present in the air in the city of Homer on a certain day in June is modeled by the function

$$A(t) = -2t^2 + 32t + 12,$$

where t is the number of hours after 6:00 A.M. Use this function to find the time at which the nitrogen dioxide level was at its maximum.

62. *Stabilization ratio.* The stabilization ratio (births/deaths) for South and Central America can be modeled by the function

$$y = -0.0012x^2 + 0.074x + 2.69$$

where y is the number of births divided by the number of deaths in the year $1950 + x$ (World Resources Institute, www.wri.org).

a) Use the graph to estimate the year in which the stabilization ratio was at its maximum.
b) Use the function to find the year in which the stabilization ratio was at its maximum.
c) What was the maximum stabilization ratio from part (b)?
d) What is the significance of a stabilization ratio of 1?

Figure for Exercise 62

63. *Suspension bridge.* The cable of the suspension bridge shown in the figure hangs in the shape of a parabola with equation $y = 0.0375x^2$, where x and y are in meters. What is the height of each tower above the roadway? What is the length z for the cable bracing the tower?

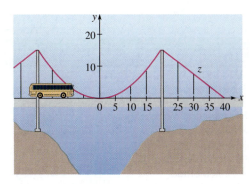

Figure for Exercise 63

Getting More Involved

64. *Exploration*

   a) Write the function $y = 3(x - 2)^2 + 6$ in the form $y = ax^2 + bx + c$, and find the vertex of the parabola using the formula $x = \dfrac{-b}{2a}$.

   b) Repeat part (a) with the functions $y = -4(x - 5)^2 - 9$ and $y = 3(x + 2)^2 - 6$.

   c) What is the vertex for a parabola that is written in the form $y = a(x - h)^2 + k$? Explain your answer.

a) $y = 100x^2 - 30x + 2$

b) $y = x^2 - 110x + 3000$

c) $y = 999x - 10 - 10x^2$

Graphing Calculator Exercises

65. Graph $y = x^2$, $y = \frac{1}{2}x^2$, and $y = 2x^2$ on the same coordinate system. What can you say about the graph of $y = ax^2$?

66. Graph $y = x^2$, $y = (x - 3)^2$, and $y = (x + 3)^2$ on the same coordinate system. How does the graph of $y = (x - h)^2$ compare to the graph of $y = x^2$?

67. The equation $x = y^2$ is equivalent to $y = \pm\sqrt{x}$. Graph both $y = \sqrt{x}$ and $y = -\sqrt{x}$ on a graphing calculator. How does the graph of $x = y^2$ compare to the graph of $y = x^2$?

68. Graph each of the following equations by solving for y.

   a) $x = y^2 - 1$

   b) $x = -y^2$

   c) $x^2 + y^2 = 4$

69. Graph each quadratic function using a viewing window that contains the vertex and all intercepts. Answers may vary.

70. Determine the approximate vertex, domain, range, and x-intercepts for each quadratic function.

   a) $y = 3.2x^2 - 5.4x + 1.6$

   b) $y = -1.09x^2 + 13x + 7.5$

8.5   Quadratic Inequalities

In This Section

⟨1⟩ Solving Quadratic Inequalities Graphically

⟨2⟩ Solving Quadratic Inequalities with the Test-Point Method

⟨3⟩ Applications

In this section, we solve inequalities involving quadratic polynomials. We use two methods, which are based on the graphs of the corresponding quadratic functions.

⟨1⟩ Solving Quadratic Inequalities Graphically

An inequality involving a quadratic polynomial is called a *quadratic* inequality.

Quadratic Inequality

A **quadratic inequality** has one of the forms

$$ax^2 + bx + c > 0, \qquad\qquad ax^2 + bx + c \geq 0,$$
$$ax^2 + bx + c < 0, \qquad\qquad ax^2 + bx + c \leq 0,$$

where a, b, and c are real numbers with $a \neq 0$.

To solve $ax^2 + bx + c > 0$ we can examine the graph of the corresponding quadratic function $y = ax^2 + bx + c$. The values of x that satisfy the inequality are the same as the values of x for which $y > 0$ on the graph of $y = ax^2 + bx + c$. We use the following strategy for the **graphical method.**

Strategy for the Graphical Method

1. Rewrite the inequality (if necessary) so that 0 is on the right side and a quadratic polynomial is on the left side.

2. Find the roots to the quadratic polynomial.

3. Plot the x-intercepts using the roots from (2), and graph the parabola passing through the x-intercepts.

4. Read the solution set to the inequality from the graph.

E X A M P L E **1**

Solving quadratic inequalities graphically

Solve each quadratic inequality. Write the solution set in interval notation and graph it.

a) $x^2 + 3x > 10$ **b)** $x^2 - 2x - 1 \leq 0$ **c)** $-x^2 + 3 \geq 0$

Solution

a) Rewrite the inequality as $x^2 + 3x - 10 > 0$. Then find the roots to the quadratic polynomial:

$$x^2 + 3x - 10 = 0$$
$$(x + 5)(x - 2) = 0$$
$$x + 5 = 0 \quad \text{or} \quad x - 2 = 0$$
$$x = -5 \quad \text{or} \quad x = 2$$

The graph of $y = x^2 + 3x - 10$ is a parabola that opens upward with x-intercepts at $(-5, 0)$ and $(2, 0)$ as shown in Fig. 8.10. The y-coordinates on the parabola are negative between the intercepts and positive to the left and right of the intercepts. Since $y = x^2 + 3x - 10$, whenever y is positive $x^2 + 3x - 10$ is positive and the inequality is satisfied. So the solution set to the inequality is $(-\infty, -5) \cup (2, \infty)$. The graph of the solution set is shown in Fig. 8.11.

b) Find the roots to the quadratic polynomial using the quadratic formula:

$$x^2 - 2x - 1 = 0$$
$$x = \frac{-(-2) \pm \sqrt{(-2)^2 - 4(1)(-1)}}{2(1)}$$
$$= \frac{2 \pm \sqrt{8}}{2} = \frac{2 \pm 2\sqrt{2}}{2} = 1 \pm \sqrt{2}$$

Figure 8.10

Figure 8.11

Figure 8.12

Figure 8.13

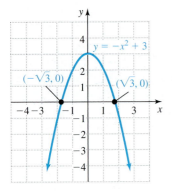

Figure 8.14

The graph of $y = x^2 - 2x - 1$ is a parabola that opens upward with x-intercepts at $\left(1 - \sqrt{2}, 0\right)$ and $\left(1 + \sqrt{2}, 0\right)$ as shown in Fig. 8.12. The y-coordinates on the parabola are negative between the intercepts, positive to the left and right of the intercepts, and zero at the intercepts. Because the inequality symbol is $\leq$, the solution set includes the roots to the polynomial. So the solution set to the inequality is $\left[1 - \sqrt{2}, 1 + \sqrt{2}\right]$. The graph of the solution set is shown in Fig. 8.13.

c) Find the roots to the quadratic polynomial:

$$-x^2 + 3 = 0$$
$$-x^2 = -3$$
$$x^2 = 3$$
$$x = \pm\sqrt{3}$$

The graph of $y = -x^2 + 3$ is a parabola that opens downward with x-intercepts at $\left(-\sqrt{3}, 0\right)$ and $\left(\sqrt{3}, 0\right)$ as shown in Fig. 8.14. The y-coordinates on the parabola are greater than or equal to zero whenever x is in the interval $\left[-\sqrt{3}, \sqrt{3}\right]$, and that is the solution set to $-x^2 + 3 \geq 0$. The graph of the solution set is shown in Fig. 8.15.

Figure 8.15

Now do Exercises 1–18

The graphs of the corresponding quadratic polynomials in Example 1 all had two x-intercepts. In Example 2 the graphs have fewer than two x-intercepts.

E X A M P L E   2

Solving quadratic inequalities graphically

Solve each quadratic inequality. Write the solution set in interval notation and graph it.

a) $x^2 - 4x + 4 \leq 0$ **b)** $x^2 + 2x + 3 > 0$ **c)** $-x^2 - 4 \geq 0$

Solution

a) Find the roots to the quadratic polynomial:

$$x^2 - 4x + 4 = 0$$
$$(x - 2)^2 = 0$$
$$x - 2 = 0$$
$$x = 2$$

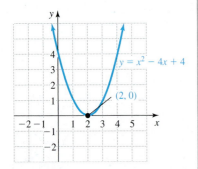

Figure 8.16

The graph of $y = x^2 - 4x + 4$ is a parabola that opens upward with x-intercept at $(2, 0)$ as shown in Fig. 8.16. The y-coordinates on the parabola are positive except when $x = 2$. At $x = 2$ the y-coordinate is zero. So there is only one value for x that satisfies $x^2 - 4x + 4 \leq 0$, and that is $x = 2$. So the solution set to the inequality is $\{2\}$. The graph of the solution set is shown in Fig. 8.17.

Figure 8.17

Figure 8.18

Figure 8.19

b) Find the roots to the quadratic polynomial using the quadratic formula:

$$x^2 + 2x + 3 = 0$$

$$x = \frac{-2 \pm \sqrt{2^2 - 4(1)(3)}}{2(1)} = \frac{-2 \pm \sqrt{-8}}{2}$$

Since the radical contains a negative number, there are no real solutions to the equation and no x-intercepts. The graph of $y = x^2 + 2x + 3$ is a parabola that opens upward from its vertex $(-1, 2)$ as shown in Fig. 8.18. Since all y-coordinates on this graph are positive for any value of x, the solution set to $x^2 + 2x + 3 > 0$ is the set of all real numbers, $(-\infty, \infty)$. The graph of the solution set is shown in Fig. 8.19.

c) Find the roots to the quadratic polynomial:

$$-x^2 - 4 = 0$$
$$-x^2 = 4$$
$$x^2 = -4$$
$$x = \pm\sqrt{-4} = \pm 2i$$

Since there are no real solutions to this equation, there are no x-intercepts for the graph of $y = -x^2 - 4$. The graph of $y = -x^2 - 4$ opens downward from its vertex $(0, -4)$ as shown in Fig. 8.20. The y-coordinates on the parabola are negative for every value of x. So there are no values of x that would make $-x^2 - 4 \geq 0$ and the solution set for the inequality is the empty set, $\varnothing$.

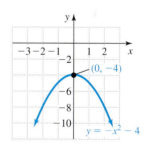

Figure 8.20

> **Now do Exercises 19–30**

⟨2⟩ Solving Quadratic Inequalities with the Test-Point Method

The **test-point method** is a variation of the graphical method, but we don't graph the parabola. We have seen that the y-coordinates on a parabola can change sign only at an x-intercept. So we find the x-intercepts (if there are any) and then test points in the intervals determined by the intercepts to see if they satisfy the inequality. Here is the strategy.

Strategy for the Test-Point Method

1. Rewrite the inequality (if necessary) with 0 on the right.
2. Solve the quadratic equation that results from replacing the inequality symbol with the equals symbol.
3. Locate the solutions to the quadratic equation on a number line.
4. Select one test point in each interval determined by the solutions.
5. Check to see whether each test point satisfies the original inequality.
6. Write the solution set using interval notation.

EXAMPLE   **3**

Solving quadratic inequalities with the test-point method

Solve each quadratic inequality. Write the solution set in interval notation and graph it.

a) $x^2 - x - 6 \leq 0$ **b)** $x^2 - 4x > 6$ **c)** $x^2 + 6x + 10 \geq 0$

Solution

a) We can solve the quadratic equation $x^2 - x - 6 = 0$ by factoring:

$$(x - 3)(x + 2) = 0$$
$$x - 3 = 0 \quad \text{or} \quad x + 2 = 0$$
$$x = 3 \quad \text{or} \quad x = -2$$

Figure 8.21

Locate -2 and 3 on the number line, and select three test points as shown in Fig. 8.21. We have chosen the points $-5, 0,$ and 5. Now test $-5, 0,$ and 5 in the original inequality $x^2 - x - 6 \leq 0$:

$$(-5)^2 - (-5) - 6 \leq 0 \qquad \text{Incorrect}$$
$$0^2 - 0 - 6 \leq 0 \qquad \text{Correct}$$
$$5^2 - 5 - 6 \leq 0 \qquad \text{Incorrect}$$

Figure 8.22

Of the three test points, only 0 satisfies the inequality. So the solution set is the interval containing 0. Since the inequality symbol includes equality, -2 and 3 are included in the solution set $[-2, 3]$. The graph of the solution set is shown in Fig. 8.22.

b) First rewrite the inequality as $x^2 - 4x - 6 > 0$. Then solve $x^2 - 4x - 6 = 0$ using the quadratic formula:

$$x = \frac{-(-4) \pm \sqrt{(-4)^2 - 4(1)(-6)}}{2(1)} = \frac{4 \pm \sqrt{40}}{2} = 2 \pm \sqrt{10}$$

Now $2 - \sqrt{10} \approx -1.2$ and $2 + \sqrt{10} \approx 5.2$. Plot these points on a number line, and select three test points as shown in Fig. 8.23. We have chosen the points $-2, 0,$ and 7. Now test $-2, 0,$ and 7 in the original inequality $x^2 - 4x > 6$:

$$(-2)^2 - 4(-2) > 6 \qquad \text{Correct}$$
$$0^2 - 4(0) > 6 \qquad \text{Incorrect}$$
$$7^2 - 4(7) > 6 \qquad \text{Correct}$$

Figure 8.23

So the inequality is satisfied on the intervals containing -2 and 7. The solution set is $\left(-\infty, 2 - \sqrt{10}\right) \cup \left(2 + \sqrt{10}, \infty\right)$, and its graph is shown in Fig. 8.24.

Figure 8.24

c) Solve $x^2 + 6x + 10 = 0$ using the quadratic formula:

$$x = \frac{-6 \pm \sqrt{6^2 - 4(1)(10)}}{2(1)} = \frac{-6 \pm \sqrt{-4}}{2}$$

Since there is a negative number inside the radical, there are no real solutions to the equation. So there are no points to plot on the number line. There is just one interval to consider, and that is $(-\infty, \infty)$. We can select any point in $(-\infty, \infty)$ as a test point. Let's try 2:

$$2^2 + 6(2) + 10 \geq 0 \qquad \text{Correct}$$

Since the inequality is satisfied for $x = 2$, it is satisfied for every real number and the solution set is $(-\infty, \infty)$. The graph is shown in Fig. 8.25.

Figure 8.25

| Now do Exercises 31–44 |

If there are no solutions to the quadratic equation, then the quadratic polynomial does not change sign. The solution set is either all real numbers or no real numbers. If the inequality in Example 3(c) was not satisfied when $x = 2$, then the solution set would have been the empty set. In fact the solution set to $x^2 + 6x + 10 \leq 0$ is the empty set, $\varnothing$.

⟨3⟩ Applications

Example 4 shows how a quadratic inequality can be used to solve a problem.

E X A M P L E   **4**

Making a profit

Charlene's daily profit P (in dollars) for selling x magazine subscriptions is determined by the formula

$$P = -x^2 + 80x - 1500.$$

For what values of x is her profit positive?

Solution

The profit is positive whenever $-x^2 + 80x - 1500 > 0$. Find the solutions to the corresponding quadratic equation:

$$-x^2 + 80x - 1500 = 0$$
$$x^2 - 80x + 1500 = 0$$
$$(x - 30)(x - 50) = 0$$
$$x - 30 = 0 \qquad \text{or} \qquad x - 50 = 0$$
$$x = 30 \qquad \text{or} \qquad x = 50$$

Figure 8.26

Locate 30 and 50 on a number line as shown in Fig. 8.26, and select 0, 40, and 60 as test points. Check the test points in the original inequality $-x^2 + 80x - 1500 > 0$:

$$-0^2 + 80(0) - 1500 > 0 \qquad \text{Incorrect}$$
$$-40^2 + 80(40) - 1500 > 0 \qquad \text{Correct}$$
$$-60^2 + 80(60) - 1500 > 0 \qquad \text{Incorrect}$$

Since 40 satisfies the inequality, every point between 30 and 50 also satisfies the inequality. So for a positive profit, she must sell between 30 and 50 magazine subscriptions.

| Now do Exercises 65–70 |

Warm-Ups ▼

Fill in the blank.

1. A(n) _____ inequality has the form $ax^2 + bx + c > 0$.

2. A quadratic inequality can be solved by the _____ method or the _____ method.

True or false?

3. The solution set to $x^2 > 4$ is $(2, \infty)$.

4. To solve $x^2 + x - 2 < 0$ by graphing, we graph $y = x^2 + x - 2$.

5. In solving quadratic inequalities, we must get 0 on one side.

6. To solve $(x - 3)(x + 5) \geq 0$ using test points, the test points could be -5, 0, and 3.

7. We can't solve quadratic inequalities that do not factor.

8. The parabola $y = x^2 + 4$ has no x-intercepts.

8.5 Exercises

⟨ Study Tips ⟩

- Keep track of your time for one entire week. Account for every half hour.
- You should be sleeping 50 to 60 hours per week and studying 1 to 2 hours for every credit hour you are taking. For a 3-credit-hour class, you should be studying 3 to 6 hours per week.

⟨1⟩ Solving Quadratic Inequalities Graphically

Use the graphical method to solve each inequality. State the solution set using interval notation, and graph it. See Example 1 and the Strategy for the Graphical Method on page 571.

1. $x^2 + x - 6 < 0$

2. $x^2 - 3x - 4 \geq 0$

3. $z^2 - 16 < 0$

4. $y^2 - 4 > 0$

5. $x^2 - 2x - 8 \leq 0$

6. $x^2 + x - 12 \leq 0$

7. $2u^2 + 5u \geq 12$

8. $2v^2 + 7v < 4$

9. $4x^2 - 8x \geq 0$

10. $x^2 + x > 0$

11. $5x - 10x^2 < 0$

12. $3x - x^2 > 0$

13. $x^2 - 5 > 0$

14. $x^2 - 3 < 0$

15. $x^2 - 2x - 5 \leq 0$

16. $x^2 - 2x - 4 > 0$

17. $2x^2 - 6x + 3 \geq 0$

18. $2x^2 - 8x + 3 < 0$

Use the graphical method to solve each inequality. State the solution set using interval notation, and graph it. See Example 2.

19. $x^2 + 6x + 9 \geq 0$

20. $x^2 + 10x + 25 \geq 0$

21. $x^2 + 4 < 4x$

22. $x^2 < 8x - 16$

23. $4x^2 - 20x + 25 \leq 0$

24. $9x^2 + 12x + 4 \leq 0$

25. $25x^2 + 10x + 1 > 0$

26. $16x^2 - 16x + 4 > 0$

27. $x^2 + 5x + 12 \geq 0$

28. $x^2 + 3x + 9 > 0$

29. $2x^2 + 5x + 5 < 0$

30. $-3x^2 + x - 6 \geq 0$

⟨2⟩ **Solving Quadratic Inequalities with the Test-Point Method**

Use the test-point method to solve each inequality. State the solution set using interval notation, and graph it. See Example 3 and the Strategy for the Test-Point Method on page 573.

31. $x^2 - 4x - 12 > 0$

32. $x^2 + 7x - 18 \leq 0$

33. $x^2 + 3x < 40$

34. $x^2 - 15x > 16$

35. $x^2 + 8x + 17 \geq 0$

36. $x^2 + 10x + 27 \leq 0$

37. $9x - 4x^2 > x$

38. $5x - x^2 \leq x$

39. $x^2 + 4 > 6x$

40. $x^2 - 2 \leq 4x$

41. $-5x^2 + 2x \leq 4$

42. $3x - 5 \le 3x^2$

43. $y^2 - 3y - 9 \le 0$

44. $z^2 - 5z - 7 < 0$

Miscellaneous

Solve each inequality. State the solution set using interval notation when possible.

45. $x^2 > 0$

46. $x^2 \ge 0$

47. $x^2 + 4 \ge 0$

48. $x^2 + 1 \le 0$

49. $x^2 \le 9$

50. $x^2 \ge 36$

51. $16 - x^2 > 0$

52. $9 - x^2 < 0$

53. $x^2 - 4x \ge 0$

54. $4x^2 - 9 > 0$

55. $3(2w^2 - 5) < w$

56. $6(y^2 - 2) + y < 0$

57. $z^2 \ge 4(z + 3)$

58. $t^2 < 3(2t - 3)$

59. $(q + 4)^2 > 10q + 31$

60. $(2p + 4)(p - 1) < (p + 2)^2$

61. $\frac{1}{2}x^2 \ge 4 - x$

62. $\frac{1}{2}x^2 \le x + 12$

63. $0.23x^2 + 6.5x + 4.3 < 0$

64. $0.65x^2 + 3.2x + 5.1 > 0$

⟨3⟩ Applications

Solve each problem by using a quadratic inequality. See Example 4.

65. *Positive profit.* The monthly profit P (in dollars) that Big Jim makes on the sale of x mobile homes is determined by the formula $P = x^2 + 5x - 50$. For what values of x is his profit positive?

66. *Profitable fruitcakes.* Sharon's revenue R (in dollars) on the sale of x fruitcakes is determined by the formula $R = 50x - x^2$. Her cost C (in dollars) for producing x fruitcakes is given by the formula $C = 2x + 40$. For what values of x is Sharon's profit positive? (Profit = revenue − cost.)

If an object is given an initial velocity straight upward of v_0 feet per second from a height of s_0 feet, then its altitude S after t seconds is given by the formula

$$S = -16t^2 + v_0 t + s_0.$$

67. *Flying high.* An arrow is shot straight upward with a velocity of 96 feet per second (ft/sec) from an altitude of 6 feet. For how many seconds is this arrow more than 86 feet high?

68. *Putting the shot.* In 1978 Udo Beyer (East Germany) set a world record in the shot put of 72 ft 8 in. If Beyer had projected the shot straight upward with a velocity of 30 ft/sec from a height of 5 ft, then for what values of t would the shot be under 15 ft high?

If a projectile is fired at a 45° angle from a height of s_0 feet with initial velocity v_0 ft/sec, then its altitude S in feet after t seconds is given by

$$S = -16t^2 + \frac{v_0}{\sqrt{2}}t + s_0.$$

69. *Siege and garrison artillery.* An 8-inch mortar used in the Civil War fired a 44.5-lb projectile from ground level a distance of 3600 ft when aimed at a 45° angle (Harold R. Peterson, *Notes on Ordinance of the American Civil War*). The accompanying graph shows the altitude of the projectile when it is fired with a velocity of $240\sqrt{2}$ ft/sec.

a) Use the graph to estimate the maximum altitude reached by the projectile.

b) Use the graph to estimate approximately how long the altitude of the projectile was greater than 864 ft.

c) Use the formula to determine the length of time for which the projectile had an altitude of more than 864 ft.

Figure for Exercise 69

70. **Seacoast artillery.** The 13-inch mortar used in the Civil War fired a 220-lb projectile a distance of 12,975 ft when aimed at a 45° angle. If the 13-inch mortar was fired from a hill 100 ft above sea level with an initial velocity of 644 ft/sec, then for how long was the projectile more than 800 ft above sea level?

Figure for Exercise 70

Getting More Involved

71. *Cooperative learning*

Work in a small group to solve each inequality for x, given that h and k are real numbers with $h < k$.

a) $(x - h)(x - k) < 0$

b) $(x - h)(x - k) > 0$

c) $(x + h)(x + k) < 0$

d) $(x + h)(x + k) \geq 0$

72. *Cooperative learning*

Work in a small group to solve $ax^2 + bx + c > 0$ for x in each case.

a) $b^2 - 4ac = 0$ and $a > 0$

b) $b^2 - 4ac = 0$ and $a < 0$

c) $b^2 - 4ac < 0$ and $a > 0$

d) $b^2 - 4ac < 0$ and $a < 0$

e) $b^2 - 4ac > 0$ and $a > 0$

f) $b^2 - 4ac > 0$ and $a < 0$

Chapter

8 Wrap-Up

Summary

Quadratic Equations

		Examples
Quadratic equation	An equation of the form $ax^2 + bx + c = 0$, where a, b, and c are real numbers, with $a \neq 0$	$x^2 = 11$ $(x - 5)^2 = 99$ $x^2 + 3x - 20 = 0$
Methods for solving quadratic equations	**Factoring:** Factor the quadratic polynomial, and then set each factor equal to 0.	$x^2 + x - 6 = 0$ $(x + 3)(x - 2) = 0$ $x + 3 = 0$ or $x - 2 = 0$
	The even-root property: If $x^2 = k$ ($k > 0$), then $x = \pm\sqrt{k}$. If $x^2 = 0$, then $x = 0$. There are no real solutions to $x^2 = k$ for $k < 0$.	$(x - 5)^2 = 10$ $x - 5 = \pm\sqrt{10}$
	Completing the square: Take one-half of middle term, square it, and then add it to each side.	$x^2 + 6x = -4$ $x^2 + 6x + 9 = -4 + 9$ $(x + 3)^2 = 5$
	Quadratic formula: If $ax^2 + bx + c = 0$ with $a \neq 0$, then $x = \dfrac{-b \pm \sqrt{b^2 - 4ac}}{2a}$.	$2x^2 + 3x - 5 = 0$ $x = \dfrac{-3 \pm \sqrt{3^2 - 4(2)(-5)}}{2(2)}$
Number of solutions	Determined by the discriminant $b^2 - 4ac$: $b^2 - 4ac > 0$ 2 real solutions	$x^2 + 6x - 12 = 0$ $6^2 - 4(1)(-12) > 0$
	$b^2 - 4ac = 0$ 1 real solution	$x^2 + 10x + 25 = 0$ $10^2 - 4(1)(25) = 0$
	$b^2 - 4ac < 0$ no real solutions, 2 imaginary solutions	$x^2 + 2x + 20 = 0$ $2^2 - 4(1)(20) < 0$
Writing equations	To write an equation with given solutions, reverse the steps in solving an equation by factoring.	$x = 2$ or $x = -3$ $(x - 2)(x + 3) = 0$ $x^2 + x - 6 = 0$
Factoring	The quadratic polynomial $ax^2 + bx + c$ (with integral coefficients) can be factored if and only if $b^2 - 4ac$ is a perfect square.	$2x^2 - 11x + 12$ $b^2 - 4ac = 25$ $(2x - 3)(x - 4)$
Equations quadratic in form	Use substitution to convert to a quadratic.	$x^4 + 3x^2 - 10 = 0$ Let $a = x^2$ $a^2 + 3a - 10 = 0$

Graphing Quadratic Functions

Examples

Parabola

The graph of the quadratic function
$f(x) = ax^2 + bx + c$ with $a \neq 0$ is a parabola.

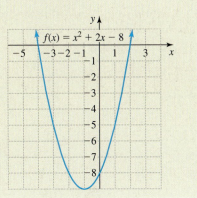

Properties of parabolas

If $a > 0$, then the parabola opens upward.
If $a < 0$, then the parabola opens downward.
The first coordinate of the vertex is $\frac{-b}{2a}$.

The second coordinate of the vertex is the minimum y-value if $a > 0$ or the maximum y-value if $a < 0$.

The x-intercepts are found by solving
$ax^2 + bx + c = 0$.
Let $x = 0$ to find the y-intercept.

$y = x^2 + 2x - 8$
Opens upward

$x = \frac{-b}{2a} = \frac{-2}{2(1)} = -1$

Vertex: $(-1, -9)$
Minimum y-value: -9

x-intercepts: $(-4, 0), (2, 0)$

y-intercept: $(0, -8)$

Quadratic Inequalities

Examples

Quadratic inequality

An inequality involving a quadratic polynomial

$2x^2 - 7x + 6 \geq 0$
$x^2 - 4x - 5 < 0$

The graphical method

Graph the corresponding quadratic function, and determine the solution to the inequality from the graph.

To solve $x^2 - 5x + 6 < 0$
graph $y = x^2 - 5x + 6$.
Solution set: $(2, 3)$

The test-point method

Test points in the intervals on the number line that are determined by the roots of the quadratic polynomial.

To solve $x^2 - 4 > 0$ plot -2 and 2 on a number line. Test $-5, 0$, and 5 in the inequality.
Solution set: $(-\infty, -2) \cup (2, \infty)$

Enriching Your Mathematical Word Power

Fill in the blank.

1. A _____ equation has the form $ax^2 + bx + c = 0$ where $a \neq 0$.

2. The polynomial $a^2 + 2ab + b^2$ is a _____ trinomial.

3. Finding the third term of a perfect square trinomial when given the first two is _____ the square.

4. The _____ formula is $x = \dfrac{-b \pm \sqrt{b^2 - 4ac}}{2a}$.

5. The _____ is $b^2 - 4ac$.

6. A _____ function has the form $y = ax^2 + bx + c$ with $a \neq 0$.

7. An equation that becomes quadratic after a substitution is quadratic in _____.

8. The inequality $ax^2 + bx + c > 0$ with $a \neq 0$ is a _____ inequality.

9. A number used to check whether an inequality is satisfied is a ____ point.

Review Exercises

8.1 Factoring and Completing the Square

Solve by factoring.

1. $x^2 - 2x - 15 = 0$

2. $x^2 - 2x - 24 = 0$

3. $2x^2 + x = 15$

4. $2x^2 + 7x = 4$

5. $w^2 - 25 = 0$

6. $a^2 - 121 = 0$

7. $4x^2 - 12x + 9 = 0$

8. $x^2 - 12x + 36 = 0$

Solve by using the even-root property.

9. $x^2 = 12$ **10.** $x^2 = 20$

11. $(x - 1)^2 = 9$ **12.** $(x + 4)^2 = 4$

13. $(x - 2)^2 = \dfrac{3}{4}$ **14.** $(x - 3)^2 = \dfrac{1}{4}$

15. $4x^2 = 9$ **16.** $2x^2 = 3$

Solve by completing the square.

17. $x^2 - 6x + 8 = 0$

18. $x^2 + 4x + 3 = 0$

19. $x^2 - 5x + 6 = 0$

20. $x^2 - x - 6 = 0$

21. $2x^2 - 7x + 3 = 0$

22. $2x^2 - x = 6$

23. $x^2 + 4x + 1 = 0$

24. $x^2 + 2x - 2 = 0$

8.2 The Quadratic Formula

Solve by the quadratic formula.

25. $x^2 - 3x - 10 = 0$

26. $x^2 - 5x - 6 = 0$

27. $6x^2 - 7x = 3$

28. $6x^2 = x + 2$

29. $x^2 + 4x + 2 = 0$

30. $x^2 + 6x = 2$

31. $3x^2 + 1 = 5x$

32. $2x^2 + 3x - 1 = 0$

Find the value of the discriminant and the number of real solutions to each equation.

33. $25x^2 - 20x + 4 = 0$ **34.** $16x^2 + 1 = 8x$

35. $x^2 - 3x + 7 = 0$ **36.** $3x^2 - x + 8 = 0$

37. $2x^2 + 1 = 5x$ **38.** $-3x^2 + 6x - 2 = 0$

Find the complex solutions to the quadratic equations.

39. $2x^2 - 4x + 3 = 0$

40. $2x^2 - 6x + 5 = 0$

41. $2x^2 + 3 = 3x$

42. $x^2 + x + 1 = 0$

43. $3x^2 + 2x + 2 = 0$

44. $x^2 + 2 = 2x$

45. $\dfrac{1}{2}x^2 + 3x + 8 = 0$

46. $\dfrac{1}{2}x^2 - 5x + 13 = 0$

8.3 More on Quadratic Equations

Use the discriminant to determine whether each quadratic polynomial can be factored, and then factor the ones that are not prime.

47. $8x^2 - 10x - 3$

48. $18x^2 + 9x - 2$

49. $4x^2 - 5x + 2$

50. $6x^2 - 7x - 4$

51. $8y^2 + 10y - 25$

52. $25z^2 - 15z - 18$

Write a quadratic equation that has each given pair of solutions.

53. $-3, -6$

54. $4, -9$

55. $-5\sqrt{2}, 5\sqrt{2}$

56. $-2i\sqrt{3}, 2i\sqrt{3}$

Find all real solutions to each equation.

57. $x^6 + 7x^3 - 8 = 0$

58. $8x^6 + 63x^3 - 8 = 0$

59. $x^4 - 13x^2 + 36 = 0$

60. $x^4 + 7x^2 + 12 = 0$

61. $(x^2 + 3x)^2 - 28(x^2 + 3x) + 180 = 0$

62. $(x^2 + 1)^2 - 8(x^2 + 1) + 15 = 0$

63. $x^2 - 6x + 6\sqrt{x^2 - 6x} - 40 = 0$

64. $x^2 - 3x - 3\sqrt{x^2 - 3x} + 2 = 0$

65. $t^{-2} + 5t^{-1} - 36 = 0$

66. $a^{-2} + a^{-1} - 6 = 0$

67. $w - 13\sqrt{w} + 36 = 0$

68. $4a - 5\sqrt{a} + 1 = 0$

8.4 Quadratic Functions and Their Graphs

Find the vertex and intercepts for each quadratic function, and sketch its graph.

69. $f(x) = x^2 - 6x$

70. $f(x) = x^2 + 4x$

71. $g(x) = x^2 - 4x - 12$

72. $g(x) = x^2 + 2x - 24$

73. $h(x) = -2x^2 + 8x$

74. $h(x) = -3x^2 + 6x$

75. $y = -x^2 + 2x + 3$

76. $y = -x^2 - 3x - 2$

Find the domain and range of each quadratic function.

77. $f(x) = x^2 + 4x + 1$

78. $f(x) = x^2 - 6x + 2$

79. $y = -2x^2 - x + 4$

80. $y = -3x^2 + 2x + 7$

8.5 Quadratic Inequalities

Solve each inequality. State the solution set using interval notation and graph it.

81. $a^2 + a > 6$

82. $x^2 - 5x + 6 > 0$

83. $x^2 - x - 20 \le 0$

84. $a^2 + 2a \le 15$

85. $w^2 - w < 0$

86. $x - x^2 \le 0$

87. $2x^2 + 5x \ge 3$

88. $3x^2 - 4 < x$

89. $x^2 + 2x + 4 \ge 0$

90. $10x - x^2 \le 28$

91. $x^2 - 10x + 25 \le 0$

92. $4x \ge 4x^2 + 1$

93. $x^2 - 2x + 10 < 0$
94. $-x^2 + 4x - 5 > 0$

Miscellaneous

In Exercises 95–106, find all real or imaginary solutions to each equation.

95. $144x^2 - 120x + 25 = 0$

96. $49x^2 + 9 = 42x$

97. $(2x + 3)^2 + 7 = 12$

98. $6x = -\dfrac{19x + 25}{x + 1}$

99. $1 + \dfrac{20}{9x^2} = \dfrac{8}{3x}$

100. $\dfrac{x - 1}{x + 2} = \dfrac{2x - 3}{x + 4}$

101. $\sqrt{3x^2 + 7x - 30} = x$

102. $\dfrac{x^4}{3} = x^2 + 6$

103. $2(2x + 1)^2 + 5(2x + 1) = 3$

104. $(w^2 - 1)^2 + 2(w^2 - 1) = 15$

105. $x^{1/2} - 15x^{1/4} + 50 = 0$

106. $x^{-2} - 9x^{-1} + 18 = 0$

Find exact and approximate solutions to each problem.

107. *Missing numbers.* Find two positive real numbers that differ by 4 and have a product of 4.

108. *One on one.* Find two positive real numbers that differ by 1 and have a product of 1.

109. *Big screen TV.* On a 19-inch diagonal measure television picture screen, the height is 4 inches less than the width. Find the height and width.

19 in.

$x - 4$ in.

x in.

Figure for Exercise 109

110. *Boxing match.* A boxing ring is in the shape of a square, 20 ft on each side. How far apart are the fighters when they are in opposite corners of the ring?

111. *Students for a Clean Environment.* A group of environmentalists plans to print a message on an 8 inch by 10 inch paper. If the typed message requires 24 square inches of paper and the group wants an equal border on all sides, then how wide should the border be?

STOP THE POLLUTION!

10 in.

8 in.

Figure for Exercise 111

112. *Winston works faster.* Winston can mow his dad's lawn in 1 hour less than it takes his brother Willie. If they take 2 hours to mow it when working together, then how long would it take Winston working alone?

113. *Ping Pong.* The table used for table tennis is 4 ft longer than it is wide and has an area of 45 ft². What are the dimensions of the table?

Figure for Exercise 113

114. *Swimming pool design.* An architect has designed a motel pool within a rectangular area that is fenced on three sides as shown in the figure. If she uses 60 yards of fencing to enclose an area of 352 square yards, then what are the dimensions marked L and W in the figure? Assume L is greater than W.

Figure for Exercise 114

115. *Minimizing cost.* The unit cost in dollars for manufacturing n starters is given by $C(n) = 0.004n^2 - 3.2n + 660$. What is the unit cost when 390 starters are manufactured? For what number of starters is the unit cost at a minimum?

116. *Maximizing profit.* The total profit (in dollars) for sales of x rowing machines is given by $P(x) = -0.2x^2 + 300x - 200$. What is the profit if 500 are sold? For what value of x will the profit be at a maximum?

117. *Decathlon champion.* For 1989 and 1990 Dave Johnson had the highest decathlon score in the world. When Johnson reached a speed of 32 ft/sec on the pole vault runway, his height above the ground t seconds after leaving the ground was given by $h = -16t^2 + 32t$. (The elasticity of the pole converts the horizontal speed into vertical speed.) Find the value of t for which his height was 12 ft.

118. *Time of flight.* Use the information from Exercise 117 to determine how long Johnson was in the air. For how long was he more than 14 ft in the air?

119. *Golden ratio.* The ancient Greeks believed that a rectangle had the most pleasing shape when the ratio of its length to width was the *golden ratio*. To find the golden ratio remove a 1 by 1 square from a 1 by x rectangle as shown in the diagram. The ratio of the length to width of the small rectangle that remains should be equal to the ratio of the length to width of the original rectangle. So,

$$\frac{x}{1} = \frac{1}{x - 1}.$$

Find x (the golden ratio) to three decimal places.

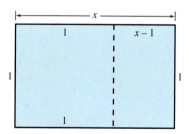

Figure for Exercise 119

Chapter 8 Test

Calculate the value of $b^2 - 4ac$, and state how many real solutions each equation has.

1. $2x^2 - 3x + 2 = 0$ **2.** $-3x^2 + 5x - 1 = 0$

3. $4x^2 - 4x + 1 = 0$

Solve by using the quadratic formula.

4. $2x^2 + 5x - 3 = 0$

5. $x^2 + 6x + 6 = 0$

Solve by completing the square.

6. $x^2 + 10x + 25 = 0$

7. $2x^2 + x - 6 = 0$

Solve by any method.

8. $x(x + 1) = 12$

9. $a^4 - 5a^2 + 4 = 0$

10. $x - 2 - 8\sqrt{x - 2} + 15 = 0$

Find the complex solutions to the quadratic equations.

11. $x^2 + 36 = 0$

12. $x^2 + 6x + 10 = 0$

13. $3x^2 - x + 1 = 0$

Graph each quadratic function. Identify the domain and range, vertex, intercepts, and the maximum or minimum y-value.

14. $f(x) = 16 - x^2$

15. $g(x) = x^2 - 3x$

Write a quadratic equation that has each given pair of solutions.

16. $-4, 6$

17. $-5i, 5i$

Solve each inequality. State and graph the solution set.

18. $w^2 + 3w < 18$

19. $x^2 - 2x \geq 1$

20. $x^2 - 6x + 13 > 0$

21. $x - x^2 \geq 4$

Find the exact solution to each problem.

22. The length of a rectangle is 2 ft longer than the width. If the area is 16 ft^2, then what are the length and width?

23. A new computer can process a company's monthly payroll in 1 hour less time than the old computer. To really save time, the manager used both computers and finished the payroll in 3 hours. How long would it take the new computer to do the payroll by itself?

24. The height in feet for a ball thrown upward at 48 feet per second is given by $s(t) = -16t^2 + 48t$, where t is the time in seconds after the ball is tossed. What is the maximum height that the ball will reach?

*Making*Connections | **A Review of Chapters 1–8**

Evaluate each expression.

1. $\sqrt{\dfrac{16}{81}}$

2. $\sqrt[3]{\dfrac{8}{27}}$

3. $(100 + 21)^{1/2}$

4. $(4 + 4)^{2/3}$

5. $8^{1/6} \cdot 8^{1/2}$

6. $\dfrac{32^{1/10}}{32^{-3/10}}$

7. $(4^{1/2} + 36^{1/2})^{-2/3}$

8. $(3^{-2} + 5^{-2} - 9^{-1})^{3/2}$

Factor Completely.

9. $y^2 + 97y - 300$

10. $20y^2 - 7y - 3$

11. $6a^3 - 60a^2 + 150a$

12. $b^3 + 2b^2 + 4b + 8$

13. $ab - ay^2 - by^2 + y^4$

14. $-2m^3 - 16$

Solve each equation.

15. $2x - 15 = 0$

16. $2x^2 - 15 = 0$

17. $2x^2 + x - 15 = 0$

18. $2x^2 + 4x - 15 = 0$

19. $|4x + 11| = 3$

20. $|4x^2 + 11x| = 3$

21. $\sqrt{x} = x - 6$

22. $(2x - 5)^{2/3} = 4$

Solve each inequality. State the solution set using interval notation.

23. $1 - 2x < 5 - x$

24. $(1 - 2x)(5 - x) \le 0$

25. $x^2 \ge x$

26. $5x^2 + 3 \le 0$

27. $3x - 1 < 5$ and $-3 \le x$

28. $x - 3 < 1$ or $2x \ge 8$

Solve each equation for y.

29. $2x - 3y = 9$

30. $\dfrac{y - 3}{x + 2} = -\dfrac{1}{2}$

31. $3y^2 + cy + d = 0$

32. $my^2 - ny = w$

33. $\dfrac{1}{3}x - \dfrac{2}{5}y = \dfrac{5}{6}$

34. $y - 3 = -\dfrac{2}{3}(x - 4)$

Let $m = \dfrac{y_2 - y_1}{x_2 - x_1}$. *Find the value of m for each of the following choices of* $x_1, x_2, y_1,$ *and* y_2.

35. $x_1 = 2, x_2 = 5, y_1 = 3, y_2 = 7$

36. $x_1 = -3, x_2 = 4, y_1 = 5, y_2 = -6$

37. $x_1 = 0.3, x_2 = 0.5, y_1 = 0.8, y_2 = 0.4$

38. $x_1 = \dfrac{1}{2}, x_2 = \dfrac{1}{3}, y_1 = \dfrac{3}{5}, y_2 = -\dfrac{4}{3}$

Solve each problem.

39. *Ticket prices.* If the price of a concert ticket goes up, then the number sold will go down, as shown in the figure. If you use the formula $n = 48{,}000 - 400p$ to predict the number sold depending on the price p, then how many will be sold at \$20 per ticket? How many will be sold at \$25 per ticket? Use the bar graph to estimate the price if 35,000 tickets were sold.

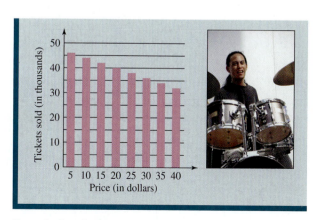

Figure for Exercise 39

40. *Increasing revenue.* Even though the number of tickets sold for a concert decreases with increasing price, the revenue generated does not necessarily decrease. Use the formula $R = p(48{,}000 - 400p)$ to determine the revenue when the price is \$20 and when the price is \$25. What price would produce a revenue of \$1.28 million? Use the graph to find the price that determines the maximum revenue.

Figure for Exercise 40

Critical **Thinking** | **For Individual or Group Work** | **Chapter 8**

These exercises can be solved by a variety of techniques, which may or may not require algebra. So be creative and think critically. Explain all answers. Answers are in the Instructor's Edition of this text.

1. *Wagon wheel.* A wagon wheel is placed against a wall as shown in the accompanying figure. One point on the edge of the wheel is 5 inches from the ground and 10 inches from the wall. What is the radius of the wheel?

Figure for Exercise 1

2. *Comparing jobs.* Bob has two job offers with a starting salary of $100,000 per year and monthly paychecks. The Atlanta employer will raise his annual salary by $2000 at the end of every year, while the Chicago employer will raise his annual salary by $1000 at the end of every six months.

a) Which job is the better deal?
b) How much more will Bob have made at the end of 10 years with the better deal?

3. *Floor tiles.* A square floor is tiled using 121 square floor tiles. Only whole tiles are used. How many tiles are neither diagonal tiles nor edge tiles?

4. *Counting days.* If the first day of this century was January 1, 2000, then how many days are there in this century? (A year is a leap year if it is divisible by 4, unless it's divisible by 100, in which case it isn't, unless it's divisible by 400, in which case it is.)

5. *Planting trees.* How can you plant 10 trees in five rows with four trees in each row?

6. *Counting rectangles.* How many rectangles of any size are there on an 8 by 8 checker board?

7. *Chime time.* The clock in the bell tower at Webster College chimes every hour on the hour: once at 1 o'clock, twice at 2 o'clock, and so on. The clock takes 5 seconds to chime at 4 o'clock and 15 seconds to chime at 10 o'clock. The time needed to chime 1 o'clock is negligible. What is the total number of seconds needed for the clock to do all of its chiming in a 24-hour period starting at 1 P.M.?

Photo for Exercise 7

8. *Arranging digits.* In how many ways can you arrange the digits 8, 7, 6, and 3 to form a four-digit number divisible by 9, using each digit once and only once?

9

Additional Function Topics

Working in a world of numbers, designers of racing boats blend art with science to design attractive boats that are also fast and safe. If the sail area is increased, the boat will go faster but will be less stable in open seas. If the displacement is increased, the boat will be more stable but slower. Increasing length increases speed but reduces stability. To make yacht racing both competitive and safe, racing boats must satisfy complex systems of rules, many of which involve mathematical formulas.

After the 1988 mismatch between Dennis Conner's catamaran and New Zealander Michael Fay's 133-foot monohull, an international group of yacht designers rewrote the America's Cup rules to ensure the fairness of the race. In addition to hundreds of pages of other rules, every yacht must satisfy the basic inequality

$$\frac{L + 1.25\sqrt{S} - 9.8\sqrt[3]{D}}{0.679} \le 24.000,$$

which balances the length L, the sail area S, and the displacement D.

In the 1979 Fastnet Race 15 sailors lost their lives. After *Exide Challenger*'s carbon-fiber keel snapped off, Tony Bullimore spent 4 days inside the overturned hull before being rescued by the Australian navy. Yacht racing is a dangerous sport. To determine the general performance and safety of a yacht, designers calculate the displacement-length ratio, the sail area-displacement ratio, the ballast-displacement ratio, and the capsize screening value.

In Exercises 83 and 84 of Section 9.3 we will see how composition of functions is used to define the displacement-length ratio and the sail area-displacement ratio.

9.1   Graphs of Functions and Relations

Functions were introduced in Section 3.5. In this section, we will study the graphs of several types of functions. We graphed linear functions in Chapter 3 and quadratic functions in Chapter 8, but for completeness we will review them here.

⟨1⟩   Linear and Constant Functions

Linear functions get their name from the fact that their graphs are straight lines.

> **Linear Function**
>
> A **linear function** is a function of the form
> $$f(x) = mx + b,$$
> where m and b are real numbers with $m \neq 0$.

The graph of the linear function $f(x) = mx + b$ is exactly the same as the graph of the linear equation $y = mx + b$. If $m = 0$, then we get $f(x) = b$, which is called a **constant function.** If $m = 1$ and $b = 0$, then we get the function $f(x) = x$, which is called the **identity function.** When we graph a function given in function notation, we can label the vertical axis as $f(x)$ rather than y.

E X A M P L E   **1**

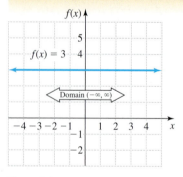

Figure 9.1

Graphing a constant function

Graph $f(x) = 3$, and state the domain and range.

Solution

The graph of $f(x) = 3$ is the same as the graph of $y = 3$, which is the horizontal line in Fig. 9.1. Since any real number can be used for x in $f(x) = 3$ and since the line in Fig. 9.1 extends without bounds to the left and right, the domain is the set of all real numbers, $(-\infty, \infty)$. Since the only y-coordinate for $f(x) = 3$ is 3, the range is $\{3\}$.

> Now do Exercises 1–2

The domain and range of a function can be determined from the formula or the graph. However, the graph is usually very helpful for understanding domain and range.

E X A M P L E   **2**

Graphing a linear function

Graph the function $f(x) = 3x - 4$, and state the domain and range.

Solution

The y-intercept is $(0, -4)$ and the slope of the line is 3. We can use the y-intercept and the slope to draw the graph in Fig. 9.2. Since any real number can be used for x in $f(x) = 3x - 4$, and since the line in Fig. 9.2 extends without bounds to the left and right, the domain is the set of all real numbers, $(-\infty, \infty)$. Since the graph extends without bounds upward and downward, the range is the set of all real numbers, $(-\infty, \infty)$.

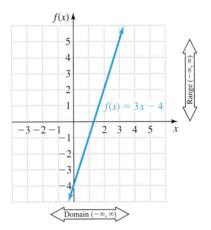

Figure 9.2

Now do Exercises 3–10

‹2› Absolute Value Functions

The equation $y = |x|$ defines a function because every value of x determines a unique value of y. We call this function the absolute value function.

> **Absolute Value Function**
> The **absolute value function** is the function defined by
> $$f(x) = |x|.$$

To graph the absolute value function, we simply plot enough ordered pairs of the function to see what the graph looks like.

E X A M P L E   3

The absolute value function
Graph $f(x) = |x|$, and state the domain and range.

Solution

To graph this function, we find points that satisfy the equation $f(x) = |x|$.

x	-2	-1	0	1	2
$f(x) = |x|$	2	1	0	1	2

Plotting these points, we see that they lie along the V-shaped graph shown in Fig. 9.3 on page 594. Since any real number can be used for x in $f(x) = |x|$ and since the graph extends without bounds to the left and right, the domain is $(-\infty, \infty)$. Because $|x|$ is never

‹ **Helpful Hint** ›

The most important feature of an absolute value function is its V-shape. If we had plotted only points in the first quadrant, we would not have seen the V-shape. So for an absolute value function we always plot enough points to see the V-shape.

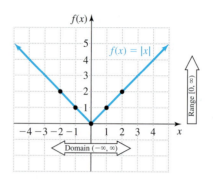

Figure 9.3

negative, the graph does not go below the *x*-axis. So the range is the set of nonnegative real numbers, $[0, \infty)$.

Now do Exercises 11–12

Many functions involving absolute value have graphs that are V-shaped, as in Fig. 9.3. To graph functions involving absolute value, we must choose points that determine the correct shape and location of the V-shaped graph.

E X A M P L E   **4**

Other functions involving absolute value

Graph each function, and state the domain and range.

a) $f(x) = |x| - 2$ **b)** $g(x) = |2x - 6|$

Solution

a) Choose values for *x* and find $f(x)$.

x	-2	-1	0	1	2
$f(x) = |x| - 2$	0	-1	-2	-1	0

Plot these points and draw a V-shaped graph through them as shown in Fig. 9.4. The domain is $(-\infty, \infty)$, and the range is $[-2, \infty)$.

Figure 9.4

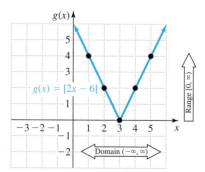

Figure 9.5

b) Make a table of values for x and $g(x)$.

x	1	2	3	4	5
$g(x) = \lvert 2x - 6 \rvert$	4	2	0	2	4

Draw the graph as shown in Fig. 9.5. The domain is $(-\infty, \infty)$, and the range is $[0, \infty)$.

> Now do Exercises 13–20

⟨3⟩ Quadratic Functions

A function defined by a second-degree polynomial is a *quadratic function*.

> **Quadratic Function**
>
> A **quadratic function** is a function of the form
> $$f(x) = ax^2 + bx + c,$$
> where a, b, and c are real numbers, with $a \neq 0$.

In Chapter 8 we learned that the graph of any quadratic function is a parabola, which opens upward or downward. The vertex of a parabola is the lowest point on a parabola that opens upward or the highest point on a parabola that opens downward. Parabolas will be discussed again when we study conic sections later in this text.

E X A M P L E   5

A quadratic function

Graph the function $g(x) = 4 - x^2$, and state the domain and range.

Solution

We plot enough points to get the correct shape of the graph.

x	-2	-1	0	1	2
$g(x) = 4 - x^2$	0	3	4	3	0

See Fig. 9.6 for the graph. The domain is $(-\infty, \infty)$. From the graph we see that the largest y-coordinate is 4. So the range is $(-\infty, 4]$.

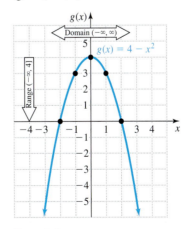

Figure 9.6

> Now do Exercises 21–28

‹ **Calculator Close-Up** ›

You can find the vertex of a parabola with a calculator. For example, graph
$$y = -x^2 - x + 2.$$
Then use the maximum feature, which is found in the CALC menu. For the left bound pick a point to the left of the vertex; for the right bound pick a point to the right of the vertex; and for the guess pick a point near the vertex.

‹4› **Square-Root Functions**

We define the square-root function as follows.

Square-Root Function

The **square-root function** is the function defined by

$$f(x) = \sqrt{x}.$$

Because squaring and square root are inverse operations, the graph of $f(x) = \sqrt{x}$ is related to the graph of $f(x) = x^2$. Recall that there are two square roots of every positive real number, but the radical symbol represents only the positive root. That is why we get only half of a parabola, as shown in Example 6.

EXAMPLE 6

Square-root functions

Graph each equation, and state the domain and range.

a) $y = \sqrt{x}$ b) $y = \sqrt{x + 3}$

Solution

a) The graph of the equation $y = \sqrt{x}$ and the graph of the function $f(x) = \sqrt{x}$ are the same. Because $\sqrt{x}$ is a real number only if $x \geq 0$, the domain of this function is the set of nonnegative real numbers. The following ordered pairs are on the graph:

x	0	1	4	9
$y = \sqrt{x}$	0	1	2	3

The graph goes through these ordered pairs as shown in Fig. 9.7. Note that x is chosen from the nonnegative numbers. The domain is $[0, \infty)$ and the range is $[0, \infty)$.

b) Note that $\sqrt{x + 3}$ is a real number only if $x + 3 \geq 0$, or $x \geq -3$. So we make a table of ordered pairs in which $x \geq -3$:

x	-3	-2	1	6
$y = \sqrt{x + 3}$	0	1	2	3

The graph goes through these ordered pairs as shown in Fig. 9.8. The domain is $[-3, \infty)$ and the range is $[0, \infty)$.

Figure 9.7

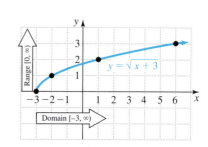

Figure 9.8

Now do Exercises 29–36

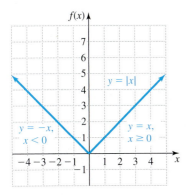

Figure 9.9

⟨5⟩ Piecewise Functions

Most of our functions are defined by a single formula, but functions can be defined by different formulas for different regions of the domain. Such functions are called **piecewise functions.** The simplest example of a piecewise function is the absolute value function. The graph of $f(x) = |x|$ is the straight line $y = x$ to the right of the y-axis, and it is the straight line $y = -x$ to the left of the y-axis, as shown in Fig. 9.9. So $f(x) = |x|$ could be written as

$$f(x) = \begin{cases} x & \text{for} \quad x \geq 0 \\ -x & \text{for} \quad x < 0. \end{cases}$$

In Example 7, we graph some piecewise functions.

E X A M P L E   **7**

Graphing piecewise functions

Graph each function.

a) $f(x) = \begin{cases} \dfrac{1}{2}x & \text{for} \quad x \geq 0 \\ -3x & \text{for} \quad x < 0 \end{cases}$

b) $f(x) = \begin{cases} \sqrt{x} & \text{for} \quad x \geq 0 \\ -x + 2 & \text{for} \quad x < 0 \end{cases}$

Solution

a) For $x \geq 0$, we graph the line $y = \frac{1}{2}x$. For $x < 0$, we graph the line $y = -3x$. Make a table of ordered pairs for each.

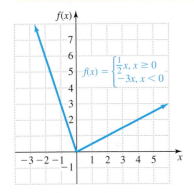

Figure 9.10

$x \,(x \geq 0)$	0	2	4	6
$y = \frac{1}{2}x$	0	1	2	3

$x \,(x < 0)$	-0.1	-1	-2	-3
$y = -3x$	0.3	3	6	9

Plot these ordered pairs, and draw the lines as shown in Fig. 9.10. Note that both lines "start" at the origin and neither line extends below the x-axis.

b) For $x \geq 0$, we graph the curve $y = \sqrt{x}$. For $x < 0$, we graph the line $y = -x + 2$. Make a table of ordered pairs for each.

$x \,(x \geq 0)$	0	1	4	9
$y = \sqrt{x}$	0	1	2	3

$x \,(x < 0)$	-0.1	-1	-2	-3
$y = -x + 2$	2.1	3	4	5

Plot these ordered pairs and sketch the graph, as shown in Fig. 9.11. Note that the line comes right up to the point $(0, 2)$ but does not include it. So the point is shown with a hollow circle. The point $(0, 0)$ is included on the curve. So it is shown with a solid circle.

> Now do Exercises 37–44

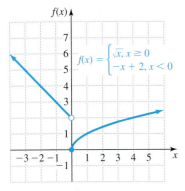

Figure 9.11

⟨6⟩ Graphing Relations

A function is a set of ordered pairs in which no two have the same first coordinate and different second coordinates. A relation is any set of ordered pairs. The domain of a relation is the set of x-coordinates of the ordered pairs, and the range of a relation is the set of y-coordinates of the ordered pairs. In Example 8, we graph the relation $x = y^2$. Note that this relation is not a function because ordered pairs such as $(4, 2)$ and $(4, -2)$ satisfy $x = y^2$.

E X A M P L E   8

Figure 9.12

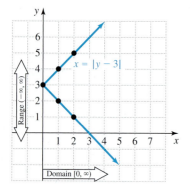

Figure 9.13

Graphing relations that are not functions
Graph each relation and state the domain and range.

a) $x = y^2$

b) $x = |y - 3|$

Solution

a) Because the equation $x = y^2$ expresses x in terms of y, it is easier to choose the y-coordinate first and then find the x-coordinate:

$x = y^2$	4	1	0	1	4
y	−2	−1	0	1	2

Figure 9.12 shows the graph. The domain is $[0, \infty)$ and the range is $(-\infty, \infty)$.

b) Again we select values for y first and find the corresponding x-coordinates:

$x = \|y - 3\|$	2	1	0	1	2
y	1	2	3	4	5

Plot these points as shown in Fig. 9.13. The domain is $[0, \infty)$ and the range is $(-\infty, \infty)$.

> **Now do Exercises 45–56**

Note that $y = x^2$ is a function and $x = y^2$ is not a function because we have agreed that x is *always* the independent variable and y the dependent variable. You can determine the y-coordinate from x in $y = x^2$, but y can't be determined from x in $x = y^2$. If we use variables other than x and y, then we must know which is the independent variable to decide if the equation is a function. For example, if $W = a^2$ and a is the independent variable, then W is a function of a.

Warm-Ups ▼

Fill in the blank.

1. A _____ function has the form $f(x) = mx + b$ where $m \neq 0$.

2. A _____ function has the form $f(x) = k$ where k is a real number.

3. The _____ function is $f(x) = x$.

4. The graph of a quadratic function is a _____.

5. An _____ function has a V-shaped graph.

True or false?

6. The graph of a function is a picture of all of the ordered pairs of the function.

7. The domain of $f(x) = 3$ is $(-\infty, \infty)$.

8. The range of a quadratic function is $(-\infty, \infty)$.

9. The y-axis and the $f(x)$-axis are the same.

10. The domain of $x = y^2$ is $[0, \infty)$.

11. The domain of $f(x) = \sqrt{x - 1}$ is $(1, \infty)$.

12. The domain of a quadratic function is $(-\infty, \infty)$.

⟨1⟩ Linear and Constant Functions

Graph each function, and state its domain and range. See Examples 1 and 2.

1. $h(x) = -2$

2. $f(x) = 4$

3. $f(x) = 2x - 1$

4. $g(x) = x + 2$

5. $g(x) = \dfrac{1}{2}x + 2$

6. $h(x) = \dfrac{2}{3}x - 4$

7. $y = -\dfrac{2}{3}x + 3$

8. $y = -\dfrac{3}{4}x + 4$

9. $y = -0.3x + 6.5$

10. $y = 0.25x - 0.5$

⟨2⟩ Absolute Value Functions

Graph each absolute value function, and state its domain and range. See Examples 3 and 4.

11. $f(x) = |x| + 1$

12. $g(x) = |x| - 3$

13. $h(x) = |x + 1|$

14. $f(x) = |x - 2|$

15. $g(x) = |3x|$

16. $h(x) = |-2x|$

17. $f(x) = |2x - 1|$ **18.** $y = |2x - 3|$ **27.** $y = 6 - x^2$ **28.** $y = -2x^2 + 3$

19. $f(x) = |x - 2| + 1$ **20.** $y = |x - 1| + 2$

⟨4⟩ Square-Root Functions

Graph each square-root function, and state its domain and range. See Example 6.

29. $g(x) = 2\sqrt{x}$ **30.** $g(x) = \sqrt{x} - 1$

⟨3⟩ Quadratic Functions

Graph each quadratic function, and state its domain and range. See Example 5.

21. $y = x^2$ **22.** $y = -x^2$

31. $f(x) = \sqrt{x - 1}$ **32.** $f(x) = \sqrt{x + 1}$

23. $g(x) = x^2 + 2$ **24.** $f(x) = x^2 - 4$

33. $h(x) = -\sqrt{x}$ **34.** $h(x) = -\sqrt{x - 1}$

25. $f(x) = 2x^2$ **26.** $h(x) = -3x^2$ **35.** $y = \sqrt{x} + 2$ **36.** $y = 2\sqrt{x} + 1$

⟨5⟩ Piecewise Functions

Graph each piecewise function. See Example 7.

37. $f(x) = \begin{cases} x & \text{for } x \geq 0 \\ -4x & \text{for } x < 0 \end{cases}$ **38.** $f(x) = \begin{cases} 3x + 1 & \text{for } x \geq 0 \\ -x + 1 & \text{for } x < 0 \end{cases}$

47. $x = -y^2$

48. $x = 1 - y^2$

39. $f(x) = \begin{cases} 2 & \text{for } x > 1 \\ -2 & \text{for } x \leq 1 \end{cases}$ **40.** $f(x) = \begin{cases} 3 & \text{for } x > -2 \\ -4 & \text{for } x \leq -2 \end{cases}$

49. $x = 5$

50. $x = -3$

41. $f(x) = \begin{cases} \sqrt{x} & \text{for } x > 1 \\ x + 3 & \text{for } x \leq 1 \end{cases}$ **42.** $f(x) = \begin{cases} \sqrt{x - 2} & \text{for } x \geq 3 \\ 6 - x & \text{for } x < 3 \end{cases}$

51. $x + 9 = y^2$

52. $x + 3 = |y|$

43. $f(x) = \begin{cases} \sqrt{x} & \text{for } 0 \leq x \leq 4 \\ x - 4 & \text{for } x > 4 \end{cases}$

53. $x = \sqrt{y}$

54. $x = -\sqrt{y}$

44. $f(x) = \begin{cases} \sqrt{x + 1} & \text{for } -1 \leq x \leq 3 \\ x - 5 & \text{for } x > 3 \end{cases}$

55. $x = (y - 1)^2$

56. $x = (y + 2)^2$

⟨6⟩ Graphing Relations

Graph each relation, and state its domain and range. See Example 8.

45. $x = |y|$

46. $x = -|y|$

Miscellaneous

Graph each function, and state the domain and range.

57. $f(x) = 1 - |x|$

58. $h(x) = \sqrt{x - 3}$

59. $y = (x - 3)^2 - 1$

60. $y = x^2 - 2x - 3$

61. $y = |x + 3| + 1$

62. $f(x) = -2x + 4$

63. $y = \sqrt{x} - 3$

64. $y = 2|x|$

65. $y = 3x - 5$

66. $g(x) = (x + 2)^2$

67. $y = -x^2 + 4x - 4$

68. $y = -2|x - 1| + 4$

Classify each function as either a linear, constant, quadratic, square-root, or absolute value function.

69. $f(x) = \sqrt{x - 3}$

70. $f(x) = |x| + 5$

71. $f(x) = 4$

72. $f(x) = 4x - 7$

73. $f(x) = 4x^2 - 7$

74. $f(x) = -3$

75. $f(x) = 5 + \sqrt{x}$

76. $f(x) = |x - 99|$

77. $f(x) = 99x - 100$

78. $f(x) = -5x^2 + 8x + 2$

Graphing Calculator Exercises

79. Graph the function $f(x) = \sqrt{x^2}$, and explain what this graph illustrates.

80. Graph the function $f(x) = \frac{1}{x}$, and state the domain and range.

81. Graph $y = x^2$, $y = \frac{1}{2}x^2$, and $y = 2x^2$ on the same coordinate system. What can you say about the graph of $y = ax^2$ for $a > 0$?

82. Graph $y = x^2$, $y = x^2 + 2$, and $y = x^2 - 3$ on the same screen. What can you say about the position of $y = x^2 + k$ relative to $y = x^2$?

83. Graph $y = x^2$, $y = (x + 5)^2$, and $y = (x - 2)^2$ on the same screen. What can you say about the position of $y = (x - h)^2$ relative to $y = x^2$?

84. You can graph the relation $x = y^2$ by graphing the two functions $y = \sqrt{x}$ and $y = -\sqrt{x}$. Try it and explain why this works.

85. Graph $y = (x - 3)^2$, $y = |x - 3|$, and $y = \sqrt{x - 3}$ on the same coordinate system. How does the graph of $y = f(x - h)$ compare to the graph of $y = f(x)$?

9.2 Transformations of Graphs

In This Section

⟨1⟩ Horizontal Translation
⟨2⟩ Stretching and Shrinking
⟨3⟩ Reflecting
⟨4⟩ Vertical Translation
⟨5⟩ Multiple Transformations

If a, h, and k are real numbers with $a \neq 0$, then the graph of $y = af(x - h) + k$ is a **transformation** of the graph of $y = f(x)$. All of the transformations of a function form a **family of functions.** For example, all functions of the form $y = a(x - h)^2 + k$ form the **square or quadratic family** because they are transformations of $y = x^2$. The **absolute-value family** consists of functions of the form $y = a|x - h| + k$, and the **square-root family** consists of functions of the form $y = a\sqrt{x - h} + k$. Understanding families of functions makes graphing easier because all of the functions in a family have similar graphs. The graph of any function in the square family is a **parabola.** We will now see what effect each of the numbers a, h, and k has on the graph of the original function $y = f(x)$.

⟨1⟩ Horizontal Translation

According to the order of operations, to find y in $y = af(x - h) + k$, we first subtract h from x, evaluate $f(x - h)$, multiply by a, and then add k. The order is important here, and we look at the effects of these numbers in the order h, a, and k.

Consider the graphs of $f(x) = \sqrt{x}$, $g(x) = \sqrt{x - 2}$, and $h(x) = \sqrt{x + 6}$ shown in Fig. 9.14. In the expression $\sqrt{x - 2}$, subtracting 2 is the first operation to perform. So every point on the graph of g is exactly two units to the right of a corresponding point on the graph of f. (We must start with a larger value of x to get the same y-coordinate because we first subtract 2.) Every point on the graph of h is exactly six units to the left of a corresponding point on the graph of f.

⟨ **Calculator Close-Up** ⟩

Note that for a translation of six units to the left, $x + 6$ must be written in parentheses on a graphing calculator.

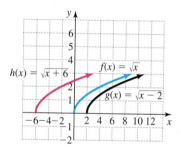

Figure 9.14

Translating to the Right or Left

If $h > 0$, then the graph of $y = f(x - h)$ is a **translation to the right** of the graph of $y = f(x)$.

If $h < 0$, then the graph of $y = f(x - h)$ is a **translation to the left** of the graph of $y = f(x)$.

E X A M P L E   **1**

Horizontal Translation
Sketch the graph of each function, and state the domain and range.

a) $f(x) = (x - 2)^2$ **b)** $f(x) = |x + 3|$

Solution

a) The graph of $f(x) = (x - 2)^2$ is a translation two units to the right of the familiar graph of $f(x) = x^2$. Calculate a few ordered pairs for accuracy. The points $(2, 0)$, $(0, 4)$, and $(4, 4)$ are on the graph in Fig. 9.15. Since any real number can be used in place of x in $(x - 2)^2$, the domain is $(-\infty, \infty)$. Since the graph extends upward from $(2, 0)$, the range is $[0, \infty)$.

b) The graph of $f(x) = |x + 3|$ is a translation three units to the left of the familiar graph of $f(x) = |x|$. The points $(0, 3)$, $(-3, 0)$, and $(-6, 3)$ are on the graph in Fig. 9.16. Since any real number can be used in place of x in $|x + 3|$, the domain is $(-\infty, \infty)$. Since the graph extends upward from $(-3, 0)$, the range is $[0, \infty)$.

Figure 9.15

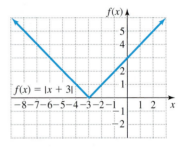

Figure 9.16

Now do Exercises 1–8

⟨2⟩ Stretching and Shrinking

Consider the graphs of $f(x) = x^2$, $g(x) = 2x^2$, and $h(x) = \frac{1}{2}x^2$ shown in Fig. 9.17. Every point on $g(x) = 2x^2$ corresponds to a point directly below on the graph of $f(x) = x^2$. The y-coordinate on g is exactly twice as large as the corresponding y-coordinate on f. This situation occurs because, in the expression $2x^2$, multiplying by 2 is the last operation performed. Every point on h corresponds to a point directly above on f, where the y-coordinate on h is half as large as the y-coordinate on f. The factor 2 has stretched the graph of f to form the graph of g, and the factor $\frac{1}{2}$ has shrunk the graph of f to form the graph of h.

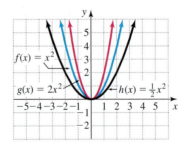

Figure 9.17

Stretching and Shrinking

If $a > 1$, then the graph of $y = af(x)$ is obtained by **stretching** the graph of $y = f(x)$. If $0 < a < 1$, then the graph of $y = af(x)$ is obtained by **shrinking** the graph of $y = f(x)$.

Note that the last operation to be performed in stretching or shrinking is multiplication by a. Whereas the function $g(x) = 2\sqrt{x}$ is obtained by stretching $f(x) = \sqrt{x}$ by a factor of 2, $h(x) = \sqrt{2x}$ is not.

E X A M P L E   2

Stretching and shrinking

Graph the functions $f(x) = \sqrt{x}$, $g(x) = 2\sqrt{x}$, and $h(x) = \frac{1}{2}\sqrt{x}$ on the same coordinate system.

Solution

The graph of g is obtained by stretching the graph of f, and the graph of h is obtained by shrinking the graph of f. The graph of f includes the points $(0, 0)$, $(1, 1)$, and $(4, 2)$. The graph of g includes the points $(0, 0)$, $(1, 2)$, and $(4, 4)$. The graph of h includes the points $(0, 0)$, $(1, 0.5)$, and $(4, 1)$. The graphs are shown in Fig. 9.18.

‹ **Calculator Close-Up** ›

The following calculator screen shows the curves $y = \sqrt{x}, y = 2\sqrt{x}$, $y = 3\sqrt{x}$, and so on, through $y = 10\sqrt{x}$.

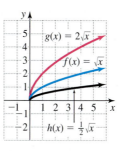

Figure 9.18

Now do Exercises 9–16

‹ **Calculator Close-Up** ›

With a graphing calculator, you can quickly see the result of modifying the formula for a function. If you have a graphing calculator, use it to graph the functions in the examples. Experimenting with it will help you to understand the ideas in this section.

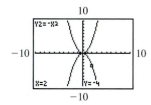

‹3› Reflecting

Consider the graphs of $f(x) = x^2$ and $g(x) = -x^2$ shown in Fig. 9.19. Notice that the graph of g is a mirror image of the graph of f. For any value of x we compute the y-coordinate of an ordered pair of f by squaring x. For an ordered pair of g we square first and then find the opposite because of the order of operations. This gives a correspondence between the ordered pairs of f and the ordered pairs of g. For every ordered pair on the graph of f there is a corresponding ordered pair directly below it on the graph of g, and these ordered pairs are the same distance from the x-axis. We say that the graph of g is obtained by reflecting the graph of f in the x-axis or that g is a reflection of the graph of f.

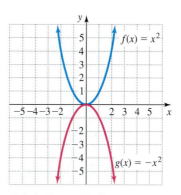

Figure 9.19

> **Reflection**
>
> The graph of $y = -f(x)$ is a **reflection** in the x-axis of the graph of $y = f(x)$.

E X A M P L E   **3**

Reflection

Sketch the graphs of each pair of functions on the same coordinate system.

a) $f(x) = \sqrt{x}$, $g(x) = -\sqrt{x}$ **b)** $f(x) = |x|$, $g(x) = -|x|$

Solution

In each case the graph of g is a reflection in the x-axis of the graph of f. Recall that we graphed the square-root function and the absolute value function in Section 9.1. Figures 9.20 and 9.21 show the graphs for these functions.

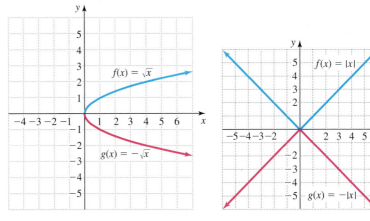

Figure 9.20 **Figure 9.21**

Now do Exercises 17–24

⟨4⟩ Vertical Translation

Consider the graphs of the functions $f(x) = \sqrt{x}$, $g(x) = \sqrt{x} + 2$, and $h(x) = \sqrt{x} - 6$ shown in Fig. 9.22. In the expression $\sqrt{x} + 2$, adding 2 is the last operation to perform. So every point on the graph of g is exactly two units above a corresponding point on the graph of f, and g has the same shape as the graph of f. Every point on the graph of h is exactly six units below a corresponding point on the graph of f. The graph of g is an upward translation of the graph of f, and the graph of h is a downward translation of the graph of f.

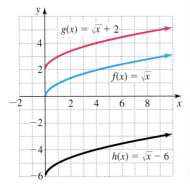

Figure 9.22

> **Translating Upward or Downward**
>
> If $k > 0$, then the graph of $y = f(x) + k$ is an **upward translation** of the graph of $y = f(x)$.
>
> If $k < 0$, then the graph of $y = f(x) + k$ is a **downward translation** of the graph of $y = f(x)$.

EXAMPLE **4**

Vertical translation

Graph the function $f(x) = |x| - 6$ and state the domain and range.

Solution

The graph of $f(x) = |x| - 6$ is a translation six units downward of the familiar graph of $f(x) = |x|$. Calculate a few ordered pairs for accuracy. The ordered pairs $(0, -6)$, $(1, -5)$, and $(-1, -5)$ are on the graph in Fig. 9.23. Since any real number can be used in place of x in $|x| - 6$, the domain is $(-\infty, \infty)$. Since the graph extends upward from $(0, -6)$, the range is $[-6, \infty)$.

Now do Exercises 25–30

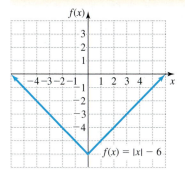

Figure 9.23

⟨5⟩ Multiple Transformations

When graphing a function containing more than one transformation, perform the transformations in the following order:

Strategy for Graphing $y = af(x - h) + k$

To graph $y = af(x - h) + k$, start with the graph of $y = f(x)$ and perform

1. horizontal translation   (right for $h > 0$ and left for $h < 0$)
2. stretching/shrinking   (stretch for $a > 1$ and shrink for $0 < a < 1$)
3. reflection   [reflect in x-axis for $a < 0$ or $y = -f(x)$]
4. vertical translation   (up for $k > 0$ and down for $k < 0$).

Note that the order in which you reflect, stretch, or shrink does not matter. It does matter that you do vertical translation last. For example, if $y = x^2$ is reflected in the x-axis and then moved up two units, the equation is $y = -x^2 + 2$. If it is moved up two units and then reflected in the x-axis, the equation is $y = -(x^2 + 2)$ or $y = -x^2 - 2$. A change in the order produces different functions.

EXAMPLE **5**

A multiple transformation of $y = \sqrt{x}$

Graph the function $y = -2\sqrt{x - 3}$, and state the domain and range.

Solution

You can check Example 5 by graphing $y = -2\sqrt{x - 3}$ with a graphing calculator.

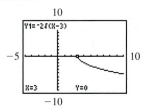

Start with the graph of $y = \sqrt{x}$ through $(0, 0)$, $(1, 1)$, and $(4, 2)$, as shown in Fig. 9.24. Translate it three units to the right to get the graph of $y = \sqrt{x - 3}$. Stretch this graph by a factor of two to get the graph of $y = 2\sqrt{x - 3}$ shown in Fig. 9.24. Now reflect in the x-axis to get the graph of $y = -2\sqrt{x - 3}$. To get an accurate graph calculate a few points on the final graph as follows:

x	3	4	7
$y = -2\sqrt{x - 3}$	0	-2	-4

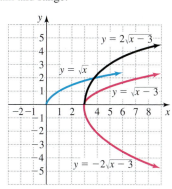

Figure 9.24

Since $x - 3$ must be nonnegative in the expression $-2\sqrt{x - 3}$, we must have $x - 3 \geq 0$ and $x \geq 3$. So the domain is $[3, \infty)$. Since the graph extends downward from the point $(3, 0)$, the range is $(-\infty, 0]$.

Now do Exercises 31–32

The graph of $y = x^2$ is a parabola opening upward with vertex $(0, 0)$. The graph of a function of the form $y = a(x - h)^2 + k$ is a transformation of $y = x^2$ and is also a parabola. It opens upward if $a > 0$ and downward if $a < 0$. Its vertex is (h, k). In Example 6, we graph a multiple transformation of $y = x^2$.

EXAMPLE   **6**

A multiple transformation of the parabola $y = x^2$

Graph the function $y = -2(x + 3)^2 + 4$, and state the domain and range.

Solution

Think of the parabola $y = x^2$ through $(-1, 1)$, $(0, 0)$, and $(1, 1)$. To get the graph of $y = -2(x + 3)^2 + 4$, translate it three units to the left, stretch by a factor of two, reflect in the x-axis, and finally translate upward four units. The graph is a stretched parabola opening downward from the vertex $(-3, 4)$ as shown in Fig. 9.25. To get an accurate graph calculate a few points around the vertex as follows:

x	-5	-4	-3	-2	-1
$y = -2(x + 3)^2 + 4$	-4	2	4	2	-4

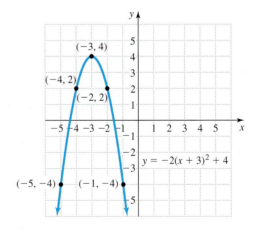

Figure 9.25

Since any real number can be used for x in $-2(x + 3)^2 + 4$, the domain is $(-\infty, \infty)$. Since the graph extends downward from $(-3, 4)$, the range is $(-\infty, 4]$.

Now do Exercises 33–34

Understanding transformations helps us to see the location of the graph of a function. To get an accurate graph we must still calculate ordered pairs that satisfy the equation. However, if we know where to expect the graph, it is easier to choose appropriate ordered pairs.

EXAMPLE **7**

A multiple transformation of the absolute value function $y = |x|$

Graph the function $y = \frac{1}{2}|x - 4| - 1$, and state the domain and range.

Solution

Think of the V-shaped graph of $y = |x|$ through $(-1, 1)$, $(0, 0)$, and $(1, 1)$. To get the graph of $y = \frac{1}{2}|x - 4| - 1$, translate $y = |x|$ to the right four units, shrink by a factor of $\frac{1}{2}$, and finally translate downward one unit.

The graph is shown in Fig. 9.26. To get an accurate graph, calculate a few points around the lowest point on the V-shaped graph as follows:

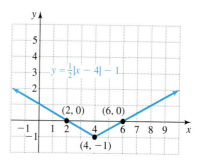

Figure 9.26

x	2	4	6
$y = \frac{1}{2}|x - 4| - 1$	0	-1	0

Since any real number can be used for x in $\frac{1}{2}|x - 4| - 1$, the domain is $(-\infty, \infty)$. Since the graph extends upward from $(4, -1)$, the range is $[-1, \infty)$.

> **Now do Exercises 35–48**

Warm-Ups ▼

Fill in the blank.

1. The graph of $y = -f(x)$ is a _____ of the graph of $y = f(x)$.

2. The graph of $y = f(x) + k$ for $k > 0$ is a(n) _____ of the graph of $y = f(x)$.

3. The graph of $y = f(x) + k$ for $k < 0$ is a(n) _____ of the graph of $y = f(x)$.

4. The graph of $y = f(x - h)$ for $h > 0$ is a translation to the _____ of the graph of $y = f(x)$.

5. The graph of $y = f(x - h)$ for $h < 0$ is a translation to the _____ of the graph of $y = f(x)$.

6. The graph of $y = af(x)$ is a _____ of the graph of $y = f(x)$ if $a > 1$.

7. The graph of $y = af(x)$ is a _____ of the graph of $y = f(x)$ if $0 < a < 1$.

True or false?

8. The graph of $f(x) = -x^2$ is a reflection in the x-axis of the graph of $f(x) = x^2$.

9. The graph of $y = |x - 3|$ lies 3 units to the left of the graph of $y = |x|$.

10. The graph of $y = |x| - 3$ lies 3 units below the graph of $y = |x|$.

11. The graph of $f(x) = 2$ is a reflection in the x-axis of the graph of $f(x) = -2$.

12. The graph of $y = -2x^2$ can be obtained by stretching and reflecting the graph of $y = x^2$.

13. The graph of $y = \sqrt{x - 3} + 5$ has the same shape as the graph of $y = \sqrt{x}$.

‹ **1** › **Horizontal Translation**

Graph each function, and state the domain and range. See Example 1.

1. $y = x + 3$

2. $y = x - 1$

11. $y = \dfrac{1}{5}x$

12. $y = 5x$

3. $f(x) = (x - 3)^2$

4. $f(x) = (x + 1)^2$

13. $f(x) = 3\sqrt{x}$

14. $f(x) = \dfrac{1}{3}\sqrt{x}$

5. $f(x) = \sqrt{x - 1}$

6. $f(x) = \sqrt{x + 6}$

15. $y = \dfrac{1}{4}|x|$

16. $y = 4|x|$

7. $f(x) = |x + 2|$

8. $f(x) = |x - 4|$

‹ **3** › **Reflecting**

Sketch the graphs of each pair of functions on the same coordinate system. See Example 3.

17. $f(x) = \sqrt{2x},$
$g(x) = -\sqrt{2x}$

18. $y = x, y = -x$

‹ **2** › **Stretching and Shrinking**

Use stretching and shrinking to graph each function, and state the domain and range. See Example 2.

9. $f(x) = 3x^2$

10. $f(x) = \dfrac{1}{3}x^2$

19. $f(x) = x^2 + 1,$
$g(x) = -(x^2 + 1)$

20. $f(x) = |x| + 1,$
$g(x) = -|x| - 1$

29. $y = |x| + 2$

30. $y = |x| - 4$

21. $y = \sqrt{x} - 2,$
$y = -\sqrt{x} - 2$

22. $y = |x - 1|,$
$y = -|x - 1|$

⟨5⟩ **Multiple Transformations**

Sketch the graph of each function, and state the domain and range. See Examples 5–7. See the Strategy for Graphing $y = af(x - h) + k$ on page 607.

31. $y = \sqrt{x - 2} + 1$ **32.** $y = -\sqrt{x + 3}$

23. $f(x) = x - 3,$
$g(x) = 3 - x$

24. $f(x) = x^2 - 2,$
$g(x) = 2 - x^2$

33. $f(x) = (x + 3)^2 - 5$ **34.** $f(x) = -2x^2$

⟨4⟩ **Vertical Translation**

Graph each function, and state the domain and range. See Example 4.

25. $y = \sqrt{x} + 1$ **26.** $y = \sqrt{x} - 3$

35. $y = -|x + 3|$ **36.** $y = |x - 2| + 1$

27. $f(x) = x^2 - 4$ **28.** $f(x) = x^2 + 2$

37. $y = -\sqrt{x + 1} - 2$ **38.** $y = -3\sqrt{x + 4} + 6$

39. $y = -2|x - 3| + 4$ **40.** $y = 3|x - 1| + 2$

Match each function with its graph a–h.

49. $y = 2 + \sqrt{x}$ **50.** $y = \sqrt{2 + x}$

51. $y = 2\sqrt{x}$ **52.** $y = \sqrt{\dfrac{x}{2}}$

53. $y = \dfrac{1}{2}\sqrt{x}$ **54.** $y = 2 + \sqrt{x - 2}$

55. $y = -2\sqrt{x}$ **56.** $y = \sqrt{-x}$

41. $y = -2x + 3$ **42.** $y = 3x - 1$

a)

b)

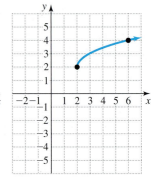

43. $y = 2(x + 3)^2 + 1$ **44.** $y = 2(x + 1)^2 - 2$

c)

d)

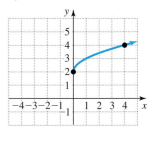

45. $y = -2(x - 4)^2 + 2$ **46.** $y = -2(x - 1)^2 + 3$

e)

f)

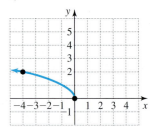

47. $y = -3(x - 1)^2 + 6$ **48.** $y = 3(x + 2)^2 - 6$

g)

h)

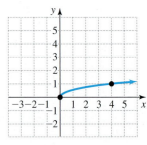

Getting More Involved

57. If the graph of $y = x^2$ is translated eight units upward, then what is the equation of the curve at that location?

58. If the graph of $y = x^2$ is translated six units to the right, then what is the equation of the curve at that location?

59. If the graph of $y = \sqrt{x}$ is translated five units to the left, then what is the equation of the curve at that location?

60. If the graph of $y = \sqrt{x}$ is translated four units downward, then what is the equation of the curve at that location?

61. If the graph of $y = |x|$ is translated three units to the left and then five units upward, then what is the equation of the curve at that location?

62. If the graph of $y = |x|$ is translated four units downward and then nine units to the right, then what is the equation of the curve at that location?

🖩 Graphing Calculator Exercises

63. Graph $f(x) = |x|$ and $g(x) = |x - 20| + 30$ on the same screen of your calculator. What transformations will transform the graph of f into the graph of g?

64. Graph $f(x) = (x + 3)^2$, $g(x) = x^2 + 3^2$, and $h(x) = x^2 + 6x + 9$ on the same screen of your calculator.

a) Which two of these functions has the same graph? Why are they the same?

b) Is it true that $(x + 3)^2 = x^2 + 9$ for all real numbers x?

c) Describe each graph in terms of a transformation of the graph of $y = x^2$.

Math *at Work* | Sailboat Design

Mention sailing and your mind drifts to exotic locations, azure seas with soothing tropical breezes, crystal-clear waters, and dazzling white sand. But sailboat designers live in a world of computers, numbers, and formulas. Some of the measurements and formulas used to describe the sailing characteristics and stability of sailboats are the maximum hull speed formula, the sail area-displacement ratio, and the motion-comfort ratio.

To estimate the theoretical maximum hull speed (M) in knots, designers use the formula $M = 1.34\sqrt{LWL}$, where LWL is the loaded waterline length (the length of the hull at the waterline). See the accompanying figure.

The sail area-displacement ratio r indicates how fast the boat is in light wind. It is given by $r = \frac{A}{D^{2/3}}$, where A is the sail area in square feet and D is the displacement in cubic feet. Values of r range from 10 to 15 for cruisers and above 24 for high-performance racers.

The motion-comfort ratio MCR, created by boat designer Ted Brewer, predicts the speed of the upward and downward motion of the boat as it encounters waves. The faster the motion, the more uncomfortable the passengers. If D is the displacement in pounds, LWL the loaded waterline length in feet, LOA the length overall, and B the beam (width) in feet, then

$$MCR = \frac{D}{\frac{2}{3} B^{3/4} \left(\frac{7}{10} LWL + \frac{1}{3} LOA \right)}.$$

As the displacement increases, MCR increases. As the length and beam increases, MCR decreases. MCR should be in the low 30's for a boat with an LOA of 42 feet.

Maximum hull speed (knots) vs. Loaded waterline (feet); $M = 1.34\sqrt{LWL}$

9.3   Combining Functions

In this section you will learn how to combine functions to obtain new functions.

⟨1⟩ Basic Operations with Functions

An entrepreneur plans to rent a stand at a farmers market for \$25 per day to sell strawberries. If she buys x flats of berries for \$5 per flat and sells them for \$9 per flat, then her daily cost in dollars can be written as a function of x:

$$C(x) = 5x + 25$$

Assuming she sells as many flats as she buys, her revenue in dollars is also a function of x:

$$R(x) = 9x$$

Because profit is revenue minus cost, we can find a function for the profit by subtracting the functions for cost and revenue:

$$P(x) = R(x) - C(x)$$
$$= 9x - (5x + 25)$$
$$= 4x - 25$$

The function $P(x) = 4x - 25$ expresses the daily profit as a function of x. Since $P(6) = -1$ and $P(7) = 3$, the profit is negative if 6 or fewer flats are sold and positive if 7 or more flats are sold.

   In the example of the entrepreneur we subtracted two functions to find a new function. In other cases we may use addition, multiplication, or division to combine two functions. For any two given functions we can define the sum, difference, product, and quotient functions as follows.

Sum, Difference, Product, and Quotient Functions

Given two functions f and g, the functions $f + g$, $f - g$, $f \cdot g$, and $\frac{f}{g}$ are defined as follows:

Sum function:	$(f + g)(x) = f(x) + g(x)$
Difference function:	$(f - g)(x) = f(x) - g(x)$
Product function:	$(f \cdot g)(x) = f(x) \cdot g(x)$
Quotient function:	$\left(\dfrac{f}{g}\right)(x) = \dfrac{f(x)}{g(x)}$ provided that $g(x) \neq 0$

The domain of the function $f + g$, $f - g$, $f \cdot g$, or $\frac{f}{g}$ is the intersection of the domain of f and the domain of g. For the function $\frac{f}{g}$ we also rule out any values of x for which $g(x) = 0$.

E X A M P L E   **1**

Operations with functions

Let $f(x) = 4x - 12$ and $g(x) = x - 3$. Find the following:

a) $(f + g)(x)$ **b)** $(f - g)(x)$

c) $(f \cdot g)(x)$ **d)** $\left(\dfrac{f}{g}\right)(x)$

‹ **Helpful Hint** ›

Note that we use $f + g$, $f - g$, $f \cdot g$, and f/g to name these functions only because there is no application in mind here. We generally use a single letter to name functions after they are combined as we did when using P for the profit function rather than $R - C$.

Solution

a) $(f + g)(x) = f(x) + g(x)$
$= 4x - 12 + x - 3$
$= 5x - 15$

b) $(f - g)(x) = f(x) - g(x)$
$= 4x - 12 - (x - 3)$
$= 3x - 9$

c) $(f \cdot g)(x) = f(x) \cdot g(x)$
$= (4x - 12)(x - 3)$
$= 4x^2 - 24x + 36$

d) $\left(\dfrac{f}{g}\right)(x) = \dfrac{f(x)}{g(x)} = \dfrac{4x - 12}{x - 3} = \dfrac{4(x - 3)}{x - 3} = 4 \qquad$ for $x \neq 3$.

> **Now do Exercises 1–4**

E X A M P L E   2

Evaluating a sum function

Let $f(x) = 4x - 12$ and $g(x) = x - 3$. Find $(f + g)(2)$.

Solution

In Example 1(a) we found a general formula for the function $f + g$, namely, $(f + g)(x) = 5x - 15$. If we replace x by 2, we get

$$(f + g)(2) = 5(2) - 15$$
$$= -5.$$

We can also find $(f + g)(2)$ by evaluating each function separately and then adding the results. Because $f(2) = -4$ and $g(2) = -1$, we get

$$(f + g)(2) = f(2) + g(2)$$
$$= -4 + (-1)$$
$$= -5.$$

> **Now do Exercises 5–12**

‹ **Helpful Hint** ›

The difference between the first four operations with functions and composition is like the difference between parallel and series in electrical connections. Components connected in parallel operate simultaneously and separately. If components are connected in series, then electricity must pass through the first component to get to the second component.

‹2› **Composition**

A salesperson's monthly salary is a function of the number of cars he sells: $1000 plus $50 for each car sold. If we let S be his salary and n be the number of cars sold, then S in dollars is a function of n:

$$S = 1000 + 50n$$

Each month the dealer contributes $100 plus 5% of his salary to a profit-sharing plan. If P represents the amount put into profit sharing, then P (in dollars) is a function of S:

$$P = 100 + 0.05S$$

Now P is a function of S, and S is a function of n. Is P a function of n? The value of n certainly determines the value of P. In fact, we can write a formula for P in terms of n by substituting one formula into the other:

$$P = 100 + 0.05S$$
$$= 100 + 0.05(1000 + 50n) \quad \text{Substitute } S = 1000 + 50n.$$
$$= 100 + 50 + 2.5n \qquad\qquad \text{Distributive property}$$
$$= 150 + 2.5n$$

Now P is written as a function of n, bypassing S. We call this idea **composition of functions.**

EXAMPLE **3**

The composition of two functions

Given that $y = x^2 - 2x + 3$ and $z = 2y - 5$, write z as a function of x.

Solution

Replace y in $z = 2y - 5$ by $x^2 - 2x + 3$:

$$z = 2y - 5$$
$$= 2(x^2 - 2x + 3) - 5 \quad \text{Replace } y \text{ by } x^2 - 2x + 3.$$
$$= 2x^2 - 4x + 1$$

The equation $z = 2x^2 - 4x + 1$ expresses z as a function of x.

| Now do Exercises 13–22 |

A composition of functions is simply one function followed by another. The output of the first function is the input for the second. For example, let $f(x) = x - 3$ and $g(x) = x^2$. If we start with 5, then $f(5) = 5 - 3 = 2$. Now use 2 as the input for g, $g(2) = 2^2 = 4$. So $g(f(5)) = 4$. The function that pairs 5 with 4 is called the *composition* of g and f, and we write $(g \circ f)(5) = 4$. Since we subtracted 3 first and then squared, a formula for $g \circ f$, is $(g \circ f)(x) = (x - 3)^2$. If we apply g first and then f, we get a different function, $(f \circ g)(x) = x^2 - 3$, the composition of f and g.

⟨ **Helpful Hint** ⟩

A composition of functions can be viewed as two function machines where the output of the first is the input of the second.

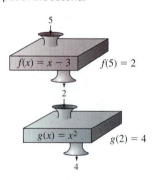

Composition of Functions

The **composition** of f and g is denoted $f \circ g$ and is defined by the equation

$$(f \circ g)(x) = f(g(x)),$$

provided that $g(x)$ is in the domain of f.

The notation $f \circ g$ is read as "the composition of f and g" or "f compose g." The diagram in Fig. 9.27 shows a function g pairing numbers in its domain with numbers in its range. If the range of g is contained in or equal to the domain of f, then f pairs the second coordinates of g with numbers in the range of f. The composition function $f \circ g$ is a rule for pairing numbers in the domain of g directly with numbers in the range of f, bypassing the middle set. The domain of the function $f \circ g$ is the domain of g (or a subset of it), and the range of $f \circ g$ is the range of f (or a subset of it).

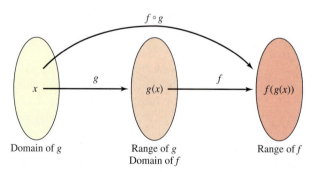

Figure 9.27

CAUTION The order in which functions are written is important in composition. For the function $f \circ g$ the function f is applied to $g(x)$. For the function $g \circ f$ the function g is applied to $f(x)$. The function closest to the variable x is applied first.

EXAMPLE 4

Evaluating compositions

Let $f(x) = 3x - 2$ and $g(x) = x^2 + 2x$. Evaluate each of the following expressions.

a) $g(f(3))$ b) $f(g(-4))$ c) $(g \circ f)(2)$ d) $(f \circ g)(2)$

Solution

a) Because $f(3) = 3(3) - 2 = 7$, we have
$$g(f(3)) = g(7) = 7^2 + 2 \cdot 7 = 63.$$
So $g(f(3)) = 63$.

b) Because $g(-4) = (-4)^2 + 2(-4) = 8$, we have
$$f(g(-4)) = f(8) = 3(8) - 2 = 22.$$
So $f(g(-4)) = 22$.

c) Because $(g \circ f)(2) = g(f(2))$ we first find $f(2)$:
$$f(2) = 3(2) - 2 = 4$$
Because $f(2) = 4$, we have
$$(g \circ f)(2) = g(f(2)) = g(4) = 4^2 + 2(4) = 24.$$
So $(g \circ f)(2) = 24$.

d) Because $(f \circ g)(2) = f(g(2))$, we first find $g(2)$:
$$g(2) = 2^2 + 2(2) = 8$$
Because $g(2) = 8$, we have
$$(f \circ g)(2) = f(g(2)) = f(8) = 3(8) - 2 = 22.$$
So $(f \circ g)(2) = 22$.

> Now do Exercises 23–36

‹ **Calculator Close-Up** ›

Set $y_1 = 3x - 2$ and $y_2 = x^2 + 2x$. You can find the composition for Examples 4(c) and 4(d) by evaluating $y_2(y_1(2))$ and $y_1(y_2(2))$. Note that the order in which you evaluate the functions is critical.

```
Y₂(Y₁(2))
              24
Y₁(Y₂(2))
              22
```

In Example 4, we found specific values of compositions of two functions. In Example 5, we find a general formula for the two functions from Example 4.

EXAMPLE 5

Finding formulas for compositions

Let $f(x) = 3x - 2$ and $g(x) = x^2 + 2x$. Find the following.

a) $(g \circ f)(x)$ b) $(f \circ g)(x)$

Solution

a) Since $f(x) = 3x - 2$, we replace $f(x)$ with $3x - 2$:

$(g \circ f)(x) = g(f(x))$

$\qquad\qquad = g(3x - 2)$ Replace $f(x)$ with $3x - 2$.

$\qquad\qquad = (3x - 2)^2 + 2(3x - 2)$ Replace x in $g(x) = x^2 + 2x$ with $3x - 2$.

$\qquad\qquad = 9x^2 - 12x + 4 + 6x - 4$ Simplify.

$\qquad\qquad = 9x^2 - 6x$

So $(g \circ f)(x) = 9x^2 - 6x$.

b) Since $g(x) = x^2 + 2x$, we replace $g(x)$ with $x^2 + 2x$:

$$(f \circ g)(x) = f(g(x)) \qquad \text{Definition of composition}$$
$$= f(x^2 + 2x) \qquad \text{Replace } g(x) \text{ with } x^2 + 2x.$$
$$= 3(x^2 + 2x) - 2 \qquad \text{Replace } x \text{ in } f(x) = 3x - 2 \text{ with } x^2 + 2x.$$
$$= 3x^2 + 6x - 2 \qquad \text{Simplify.}$$

So $(f \circ g)(x) = 3x^2 + 6x - 2$.

> **Now do Exercises 37–46**

Notice that in Example 4(c) and (d), $(g \circ f)(2) \neq (f \circ g)(2)$. In Example 5(a) and (b) we see that $(g \circ f)(x)$ and $(f \circ g)(x)$ have different formulas defining them. In general, $f \circ g \neq g \circ f$. However, in Section 9.4 we will see some functions for which the composition in either order results in the same function.

It is often useful to view a complicated function as a composition of simpler functions. For example, the function $Q(x) = (x - 3)^2$ consists of two operations, subtracting 3 and squaring. So Q can be described as a composition of the functions $f(x) = x - 3$ and $g(x) = x^2$. To check this, we find $(g \circ f)(x)$:

$$(g \circ f)(x) = g(f(x))$$
$$= g(x - 3)$$
$$= (x - 3)^2$$

We can express the fact that Q is the same as the composition function $g \circ f$ by writing $Q = g \circ f$ or $Q(x) = (g \circ f)(x)$.

E X A M P L E   6

Expressing a function as a composition of simpler functions

Let $f(x) = x - 2$, $g(x) = 3x$, and $h(x) = \sqrt{x}$. Write each of the following functions as a composition, using f, g, and h.

a) $F(x) = \sqrt{x - 2}$

b) $H(x) = x - 4$

c) $K(x) = 3x - 6$

Solution

a) The function F consists of first subtracting 2 from x and then taking the square root of that result. So $F = h \circ f$. Check this result by finding $(h \circ f)(x)$:

$$(h \circ f)(x) = h(f(x)) = h(x - 2) = \sqrt{x - 2}$$

b) Subtracting 4 from x can be accomplished by subtracting 2 from x and then subtracting 2 from that result. So $H = f \circ f$. Check by finding $(f \circ f)(x)$:

$$(f \circ f)(x) = f(f(x)) = f(x - 2) = x - 2 - 2 = x - 4$$

c) Notice that $K(x) = 3(x - 2)$. The function K consists of subtracting 2 from x and then multiplying the result by 3. So $K = g \circ f$. Check by finding $(g \circ f)(x)$:

$$(g \circ f)(x) = g(f(x)) = g(x - 2) = 3(x - 2) = 3x - 6$$

> **Now do Exercises 47–56**

CAUTION In Example 6(a) we have $F = h \circ f$ because in F we subtract 2 before taking the square root. If we had the function $G(x) = \sqrt{x} - 2$, we would take the square root before subtracting 2. So $G = f \circ h$. Notice how important the order of operations is here.

In Example 7, we see functions for which the composition is the identity function. Each function undoes what the other function does. We will study functions of this type further in Section 9.4.

EXAMPLE 7

Composition of functions

Show that $(f \circ g)(x) = x$ for each pair of functions.

a) $f(x) = 2x - 1$ and $g(x) = \dfrac{x + 1}{2}$

b) $f(x) = x^3 + 5$ and $g(x) = (x - 5)^{1/3}$

Solution

a) $(f \circ g)(x) = f(g(x)) = f\left(\dfrac{x + 1}{2}\right)$

$$= 2\left(\dfrac{x + 1}{2}\right) - 1$$
$$= x + 1 - 1$$
$$= x$$

b) $(f \circ g)(x) = f(g(x)) = f\left((x - 5)^{1/3}\right)$

$$= \left((x - 5)^{1/3}\right)^3 + 5$$
$$= x - 5 + 5$$
$$= x$$

> Now do Exercises 57–64

Warm-Ups ▼

Fill in the blank.

1. The function $(f + g)(x)$ is the _____ of the functions $f(x)$ and $g(x)$.

2. The function $(f - g)(x)$ is the _____ of the functions $f(x)$ and $g(x)$.

3. The function $(f \cdot g)(x)$ is the _____ of the functions $f(x)$ and $g(x)$.

4. The function $(f/g)(x)$ is the _____ of the functions $f(x)$ and $g(x)$.

5. The function $(f \circ g)(x)$ is the _____ of the functions $f(x)$ and $g(x)$.

True or false?

6. For the composition of f and g, the function f is evaluated after g.

7. If $f(x) = x - 2$ and $g(x) = x + 3$, then $(f - g)(x) = -5$.

8. If $f(x) = x^2 - 2x$ and $g(x) = 3x + 9$, then $(f + g)(x) = x^2 + x + 9$.

9. If $f(x) = x^2$ and $g(x) = x + 2$, then $(f \circ g)(x) = x^2 + 2$.

10. If $f(x) = 3x$ and $g(x) = \dfrac{x}{3}$, then $(f \circ g)(x) = x$.

11. The function $f \circ g$ and $g \circ f$ are always the same.

12. If $F(x) = (x - 1)^2$, $h(x) = x - 1$, and $g(x) = x^2$, then $F = g \circ h$.

‹1› **Basic Operations with Functions**

Let $f(x) = 4x - 3$ and $g(x) = x^2 - 2x$. Find the following. See Examples 1 and 2.

1. $(f + g)(x)$ **2.** $(f - g)(x)$

3. $(f \cdot g)(x)$ **4.** $\left(\dfrac{f}{g}\right)(x)$

5. $(f + g)(3)$ **6.** $(f + g)(2)$
7. $(f - g)(-3)$ **8.** $(f - g)(-2)$
9. $(f \cdot g)(-1)$ **10.** $(f \cdot g)(-2)$
11. $\left(\dfrac{f}{g}\right)(4)$ **12.** $\left(\dfrac{f}{g}\right)(-2)$

‹2› **Composition**

Use the two functions to write y as a function of x. See Example 3.

13. $y = 2a, \ a = 3x$
14. $y = w^2, \ w = 5x$
15. $y = 3a - 2, \ a = 2x - 6$
16. $y = 2c + 3, \ c = -3x + 4$
17. $y = 2d + 1, \ d = \dfrac{x + 1}{2}$
18. $y = -3d + 2, \ d = \dfrac{2 - x}{3}$
19. $y = m^2 - 1, \ m = x + 1$
20. $y = n^2 - 3n + 1, \ n = x + 2$
21. $y = \dfrac{a - 3}{a + 2}, \ a = \dfrac{2x + 3}{1 - x}$
22. $y = \dfrac{w + 2}{w - 5}, \ w = \dfrac{5x + 2}{x - 1}$

Let $f(x) = 2x - 3$, $g(x) = x^2 + 3x$, and $h(x) = \frac{x + 3}{2}$. Find the following. See Examples 4 and 5.

23. $(g \circ f)(1)$ **24.** $(f \circ g)(-2)$
25. $(f \circ g)(1)$ **26.** $(g \circ f)(-2)$
27. $(f \circ f)(4)$ **28.** $(h \circ h)(3)$
29. $(h \circ f)(5)$ **30.** $(f \circ h)(0)$

31. $(f \circ h)(5)$ **32.** $(h \circ f)(0)$
33. $(g \circ h)(-1)$ **34.** $(h \circ g)(-1)$
35. $(f \circ g)(2.36)$ **36.** $(h \circ f)(23.761)$
37. $(g \circ f)(x)$ **38.** $(g \circ h)(x)$
39. $(f \circ g)(x)$ **40.** $(h \circ g)(x)$

41. $(h \circ f)(x)$ **42.** $(f \circ h)(x)$
43. $(f \circ f)(x)$ **44.** $(g \circ g)(x)$
45. $(h \circ h)(x)$ **46.** $(f \circ f \circ f)(x)$

Let $f(x) = \sqrt{x}$, $g(x) = x^2$, and $h(x) = x - 3$. Write each of the following functions as a composition using f, g, or h. See Example 6.

47. $F(x) = \sqrt{x - 3}$ **48.** $N(x) = \sqrt{x} - 3$

49. $G(x) = x^2 - 6x + 9$ **50.** $P(x) = x$ for $x \geq 0$

51. $H(x) = x^2 - 3$ **52.** $M(x) = x^{1/4}$

53. $J(x) = x - 6$ **54.** $R(x) = \sqrt{x^2 - 3}$

55. $K(x) = x^4$ **56.** $Q(x) = \sqrt{x^2 - 6x + 9}$

Show that $(f \circ g)(x) = x$ and $(g \circ f)(x) = x$ for each given pair of functions. See Example 7.

57. $f(x) = 3x + 5, \ g(x) = \dfrac{x - 5}{3}$

58. $f(x) = 3x - 7, \ g(x) = \dfrac{x + 7}{3}$

59. $f(x) = x^3 - 9, \ g(x) = \sqrt[3]{x + 9}$
60. $f(x) = x^3 + 1, \ g(x) = \sqrt[3]{x - 1}$
61. $f(x) = \dfrac{x - 1}{x + 1}, \ g(x) = \dfrac{x + 1}{1 - x}$

62. $f(x) = \dfrac{x + 1}{x - 3}, \ g(x) = \dfrac{3x + 1}{x - 1}$

63. $f(x) = \dfrac{1}{x}, \ g(x) = \dfrac{1}{x}$ **64.** $f(x) = 2x^3, \ g(x) = \left(\dfrac{x}{2}\right)^{1/3}$

Miscellaneous

Let $f(x) = x^2$ and $g(x) = x + 5$. Determine whether each of these statements is true or false.

65. $f(3) = 9$

66. $g(3) = 8$

67. $(f + g)(4) = 21$

68. $(f - g)(0) = 5$

69. $(f \cdot g)(3) = 72$

70. $(f/g)(0) = 5$

71. $(f \circ g)(2) = 14$

72. $(g \circ f)(7) = 54$

73. $f(g(x)) = x^2 + 25$

74. $(g \circ f)(x) = x^2 + 5$

75. If $h(x) = x^2 + 10x + 25$, then $h = f \circ g$.

76. If $p(x) = x^2 + 5$, then $p = g \circ f$.

Applications

Solve each problem.

77. *Area.* A square gate in a wood fence has a diagonal brace with a length of 10 feet.

   a) Find the area of the square gate.

   b) Write a formula for the area of a square as a function of the length of its diagonal.

78. *Perimeter.* Write a formula for the perimeter of a square as a function of its area.

79. *Profit function.* A plastic bag manufacturer has determined that the company can sell as many bags as it can produce each month. If it produces x thousand bags in a month, the revenue is $R(x) = x^2 - 10x + 30$ dollars, and the cost is $C(x) = 2x^2 - 30x + 200$ dollars. Use the fact that profit is revenue minus cost to write the profit as a function of x.

80. *Area of a sign.* A sign is in the shape of a square with a semicircle of radius x adjoining one side and a semicircle of diameter x removed from the opposite side. If the sides of the square are length $2x$, then write the area of the sign as a function of x.

Figure for Exercise 80

81. *Junk food expenditures.* Suppose the average family spends 25% of its income on food, $F = 0.25I$, and 10% of each food dollar on junk food, $J = 0.10F$. Write J as a function of I.

82. *Area of an inscribed circle.* A pipe of radius r must pass through a square hole of area M as shown in the figure. Write the cross-sectional area of the pipe A as a function of M.

Figure for Exercise 82

83. *Displacement-length ratio.* To find the displacement-length ratio D for a sailboat, first find x, where $x = (L/100)^3$ and L is the length at the waterline in feet (www.sailing.com). Next find D, where $D = (d/2240)/x$ and d is the displacement in pounds.

Figure for Exercise 83

   a) For the Pacific Seacraft 40, $L = 30$ ft 3 in. and $d = 24{,}665$ pounds. Find D.

   b) For a boat with a displacement of 25,000 pounds, write D as a function of L.

   c) The graph for the function in part (b) is shown in the accompanying figure. For a fixed displacement, does the displacement-length ratio increase or decrease as the length increases?

84. *Sail area-displacement ratio.* To find the sail area-displacement ratio S, first find y, where $y = (d/64)^{2/3}$ and d

is the displacement in pounds. Next find S, where $S = A/y$ and A is the sail area in square feet.

a) For the Pacific Seacraft 40, $A = 846$ square feet (ft^2) and $d = 24{,}665$ pounds. Find S.

b) For a boat with a sail area of 900 ft^2, write S as a function of d.

c) For a fixed sail area, does S increase or decrease as the displacement increases?

Getting More Involved

85. *Discussion*

Let $f(x) = \sqrt{x} - 4$ and $g(x) = \sqrt{x}$. Find the domains of f, g, and $g \circ f$.

86. *Discussion*

Let $f(x) = \sqrt{x} - 4$ and $g(x) = \sqrt{x} - 8$. Find the domains of f, g, and $f + g$.

Graphing Calculator Exercises

87. Graph $y_1 = x$, $y_2 = \sqrt{x}$, and $y_3 = x + \sqrt{x}$ in the same screen. Find the domain and range of $y_3 = x + \sqrt{x}$ by examining its graph. (On some graphing calculators you can enter y_3 as $y_3 = y_1 + y_2$.)

88. Graph $y_1 = |x|$, $y_2 = |x - 3|$, and $y_3 = |x| + |x - 3|$. Find the domain and range of $y_3 = |x| + |x - 3|$ by examining its graph.

Mid-Chapter **Quiz** | Sections 9.1 through 9.3 | Chapter 9

Graph each relation and state its domain and range.

1. $f(x) = 2x - 4$

2. $g(x) = |2x - 4|$

3. $h(x) = -x^2 + 3$

4. $y = \sqrt{x + 6} - 5$

5. $y = \begin{cases} x + 1 & \text{for } x \geq 0 \\ 2 & \text{for } x < 0 \end{cases}$

6. $x = y^2 - 2$

7. $y = -2(x - 1)^2 + 3$

Let $f(x) = 5x - 2$ and $g(x) = x^2 - 3x$. Find the following.

8. $(f + g)(x)$

9. $(f - g)(2)$

10. $(f \cdot g)(5)$

11. $\left(\dfrac{f}{g}\right)(4)$

12. $(f \circ g)(3)$

13. $(f \circ g)(x)$

Miscellaneous.

14. Let $f(x) = \sqrt{x}$, $g(x) = x - 8$, and $F(x) = \sqrt{x - 8}$. Which is correct: $F = f \circ g$ or $F = g \circ f$?

15. Let $A = \pi r^2$ and $r = \dfrac{d}{2}$. Express A as a function of d.

16. If the graph of $y = x^2$ is translated 2 units to the left and 4 units upward, then what is the equation of the curve in its final position?

17. If the graph of $y = |x|$ is stretched by a factor of 2, translated 3 units to the left, and reflected in the x-axis, then what is the equation of the curve in its final position?

18. If the graph of $y = \sqrt{x}$ is shrunk by a factor of $\frac{1}{2}$, translated 9 units to the left and 5 units downward, then what is the equation of the curve in its final position?

9.4   Inverse Functions

In Section 9.3, we introduced the idea of a pair of functions such that $(f \circ g)(x) = x$ and $(g \circ f)(x) = x$. Each function reverses what the other function does. In this section we explore that idea further.

⟨1⟩ Inverse of a Function

You can buy a 6-, 7-, or 8-foot conference table in the K-LOG Catalog for $299, $329, or $349, respectively. The set

$$f = \{(6, 299), (7, 329), (8, 349)\}$$

gives the price as a function of the length. We use the letter f as a name for this set or function, just as we use the letter f as a name for a function in the function notation. In the function f, lengths in the domain $\{6, 7, 8\}$ are paired with prices in the range $\{299, 329, 349\}$. The **inverse** of the function f, denoted f^{-1}, is a function whose ordered pairs are obtained from f by interchanging the x- and y-coordinates:

$$f^{-1} = \{(299, 6), (329, 7), (349, 8)\}$$

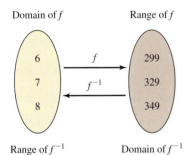

Domain of f Range of f

Range of f^{-1} Domain of f^{-1}

Figure 9.28

We read f^{-1} as "f inverse." The domain of f^{-1} is $\{299, 329, 349\}$, and the range of f^{-1} is $\{6, 7, 8\}$. The inverse function reverses what the function does: it pairs prices in the range of f with lengths in the domain of f. For example, to find the cost of a 7-foot table, we use the function f to get $f(7) = 329$. To find the length of a table costing $349, we use the function f^{-1} to get $f^{-1}(349) = 8$. Of course, we could find the length of a $349 table by looking at the function f, but f^{-1} is a function whose input is price and whose output is length. In general, *the domain of f^{-1} is the range of f, and the range of f^{-1} is the domain of f.* See Fig. 9.28.

> **CAUTION** The -1 in f^{-1} is not read as an exponent. It does not mean $\dfrac{1}{f}$.

The cost per ink cartridge is a function of the number of boxes of ink cartridges purchased:

$$g = \{(1, 4.85), (2, 4.60), (3, 4.60), (4, 4.35)\}$$

If we interchange the first and second coordinates in the ordered pairs of this function, we get

$$\{(4.85, 1), (4.60, 2), (4.60, 3), (4.35, 4)\}.$$

Consider the universal product codes (UPC) and the prices for all of the items in your favorite grocery store. The price of an item is a function of the UPC because every UPC determines a price. This function is not invertible because you cannot determine the UPC from a given price.

This set of ordered pairs is not a function because it contains ordered pairs with the same first coordinates and different second coordinates. So g does not have an inverse function. A function is **invertible** if you obtain a function when the coordinates of all ordered pairs are reversed. So f is invertible and g is not invertible.

Any function that pairs more than one number in the domain with the same number in the range is not invertible, because the set is not a function when the ordered pairs are reversed. So we turn our attention to functions where each member of the domain corresponds to one member of the range and vice versa.

One-to-One Function

If a function is such that no two ordered pairs have different x-coordinates and the same y-coordinate, then the function is called a **one-to-one** function.

In a one-to-one function each member of the domain corresponds to just one member of the range, and each member of the range corresponds to just one member of the domain. *Functions that are one-to-one are invertible functions.*

Inverse Function

The inverse of a one-to-one function f is the function f^{-1}, which is obtained from f by interchanging the coordinates in each ordered pair of f.

E X A M P L E   1

Identifying invertible functions

Determine whether each function is invertible. If it is invertible, then find the inverse function.

a) $f = \{(2, 4), (-2, 4), (3, 9)\}$

b) $g = \left\{ \left(2, \dfrac{1}{2}\right), \left(5, \dfrac{1}{5}\right), \left(7, \dfrac{1}{7}\right) \right\}$

c) $h = \{(3, 5), (7, 9)\}$

Solution

a) Since $(2, 4)$ and $(-2, 4)$ have the same y-coordinate, this function is not one-to-one, and it is not invertible.

b) This function is one-to-one, and so it is invertible.

$$g^{-1} = \left\{ \left(\dfrac{1}{2}, 2\right), \left(\dfrac{1}{5}, 5\right), \left(\dfrac{1}{7}, 7\right) \right\}$$

c) This function is invertible, and $h^{-1} = \{(5, 3), (9, 7)\}$.

> **Now do Exercises 1–10**

You learned to use the vertical-line test in Section 3.5 to determine whether a graph is the graph of a function. The **horizontal-line test** is a similar visual test for determining whether a function is invertible. If a horizontal line crosses a graph two (or more) times, as in Fig. 9.29, then there are two points on the graph, say (x_1, y) and (x_2, y), that have different x-coordinates and the same y-coordinate. So the function is not one-to-one, and the function is not invertible.

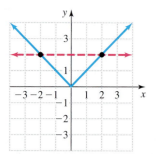

Figure 9.29

Horizontal-Line Test

A function is invertible if and only if no horizontal line crosses its graph more than once.

EXAMPLE **2**

Using the horizontal-line test

Determine whether each function is invertible by examining its graph.

a)

b)
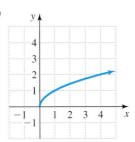

‹ **Helpful Hint** ›

Tests such as the vertical-line test and the horizontal-line test are certainly not accurate in all cases. We discuss these tests to get a visual idea of what graphs of functions and invertible functions look like.

Solution

a) This function is not invertible because a horizontal line can be drawn so that it crosses the graph at $(2, 4)$ and $(-2, 4)$.

b) This function is invertible because every horizontal line that crosses the graph crosses it only once.

Now do Exercises 11–14

‹ 2 › Identifying Inverse Functions

Consider the one-to-one function $f(x) = 3x$. The inverse function must reverse the ordered pairs of the function. Because division by 3 undoes multiplication by 3, we could guess that $g(x) = \frac{x}{3}$ is the inverse function. To verify our guess, we can use the following rule for determining whether two given functions are inverses of each other.

> ### Identifying Inverse Functions
>
> Functions f and g are inverses of each other if and only if
> $$(g \circ f)(x) = x \text{ for every number } x \text{ in the domain of } f \text{ and}$$
> $$(f \circ g)(x) = x \text{ for every number } x \text{ in the domain of } g.$$

In Example 3, we verify that $f(x) = 3x$ and $g(x) = \frac{x}{3}$ are inverses.

EXAMPLE **3**

Identifying inverse functions

Determine whether the functions f and g are inverses of each other.

a) $f(x) = 3x$ and $g(x) = \dfrac{x}{3}$ **b)** $f(x) = 2x - 1$ and $g(x) = \dfrac{1}{2}x + 1$

c) $f(x) = x^2$ and $g(x) = \sqrt{x}$

Solution

a) Find $g \circ f$ and $f \circ g$:

$$(g \circ f)(x) = g(f(x)) = g(3x) = \frac{3x}{3} = x$$

$$(f \circ g)(x) = f(g(x)) = f\left(\frac{x}{3}\right) = 3 \cdot \frac{x}{3} = x$$

Because each of these equations is true for any real number x, f and g are inverses of each other. We write $g = f^{-1}$ or $f^{-1}(x) = \frac{x}{3}$.

b) Find the composition of g and f:

$$(g \circ f)(x) = g(f(x))$$
$$= g(2x - 1) = \frac{1}{2}(2x - 1) + 1 = x + \frac{1}{2}$$

So f and g are not inverses of each other.

c) If x is any real number, we can write

$$(g \circ f)(x) = g(f(x))$$
$$= g(x^2) = \sqrt{x^2} = |x|.$$

The domain of f is $(-\infty, \infty)$, and $|x| \neq x$ if x is negative. So g and f are not inverses of each other. Note that $f(x) = x^2$ is not a one-to-one function, since both $(3, 9)$ and $(-3, 9)$ are ordered pairs of this function. Thus, $f(x) = x^2$ does not have an inverse.

| Now do Exercises 15–22 |

⟨3⟩ Switch-and-Solve Strategy

If an invertible function is defined by a list of ordered pairs, as in Example 1, then the inverse function is found by simply interchanging the coordinates in the ordered pairs. If an invertible function is defined by a formula, then the inverse function must reverse or undo what the function does. Because the inverse function interchanges the roles of x and y, we interchange x and y in the formula and then solve the new formula for y to undo what the original function did. The steps to follow in this **switch-and-solve** strategy are given in the following box and illustrated in Examples 4 and 5.

Strategy for Finding f^{-1} by Switch-and-Solve

1. Replace $f(x)$ by y.

2. Interchange x and y.

3. Solve the equation for y.

4. Replace y by $f^{-1}(x)$.

E X A M P L E 4

The switch–and–solve strategy
Find the inverse of $h(x) = 2x + 1$.

Solution

First write the function as $y = 2x + 1$, and then interchange x and y:

$$y = 2x + 1$$
$$x = 2y + 1 \qquad \text{Interchange } x \text{ and } y.$$
$$x - 1 = 2y \qquad \text{Solve for } y.$$
$$\frac{x - 1}{2} = y$$
$$h^{-1}(x) = \frac{x - 1}{2} \qquad \text{Replace } y \text{ by } h^{-1}(x).$$

We can verify that h and h^{-1} are inverses by using composition:

$$(h^{-1} \circ h)(x) = h^{-1}(h(x)) = h^{-1}(2x + 1) = \frac{2x + 1 - 1}{2} = \frac{2x}{2} = x$$

$$(h \circ h^{-1})(x) = h(h^{-1}(x)) = h\left(\frac{x - 1}{2}\right) = 2 \cdot \frac{x - 1}{2} + 1 = x - 1 + 1 = x$$

> **Now do Exercises 23–36**

E X A M P L E   5

The switch-and-solve strategy

If $f(x) = \dfrac{x + 1}{x - 3}$, find $f^{-1}(x)$.

Solution

Replace $f(x)$ by y, interchange x and y, and then solve for y:

$$y = \frac{x + 1}{x - 3} \qquad \text{Use } y \text{ in place of } f(x).$$

$$x = \frac{y + 1}{y - 3} \qquad \text{Switch } x \text{ and } y.$$

$$x(y - 3) = y + 1 \qquad \text{Multiply each side by } y - 3.$$

$$xy - 3x = y + 1 \qquad \text{Distributive property}$$

$$xy - y = 3x + 1$$

$$y(x - 1) = 3x + 1 \qquad \text{Factor out } y.$$

$$y = \frac{3x + 1}{x - 1} \qquad \text{Divide each side by } x - 1.$$

$$f^{-1}(x) = \frac{3x + 1}{x - 1} \qquad \text{Replace } y \text{ by } f^{-1}(x).$$

To check, compute $(f \circ f^{-1})(x)$:

$$(f \circ f^{-1})(x) = f\left(\frac{3x + 1}{x - 1}\right) = \frac{\dfrac{3x + 1}{x - 1} + 1}{\dfrac{3x + 1}{x - 1} - 3} = \frac{(x - 1)\left(\dfrac{3x + 1}{x - 1} + 1\right)}{(x - 1)\left(\dfrac{3x + 1}{x - 1} - 3\right)}$$

$$= \frac{3x + 1 + 1(x - 1)}{3x + 1 - 3(x - 1)} = \frac{4x}{4} = x$$

You should check that $(f^{-1} \circ f)(x) = x$.

> **Now do Exercises 37–40**

If we use the switch-and-solve strategy to find the inverse of $f(x) = x^3$, then we get $f^{-1}(x) = x^{1/3}$. For $h(x) = 6x$ we have $h^{-1}(x) = \frac{x}{6}$. The inverse of $k(x) = x - 9$ is $k^{-1}(x) = x + 9$. For each of these functions there is an appropriate operation of arithmetic that undoes what the function does.

If a function involves two operations, the inverse function undoes those operations in the opposite order from which the function does them. For example, the function $g(x) = 3x - 5$ multiplies x by 3 and then subtracts 5 from that result. To undo these operations, we add 5 and then divide the result by 3. So,

$$g^{-1}(x) = \frac{x + 5}{3}.$$

Note that $g^{-1}(x) \neq \frac{x}{3} + 5$.

⟨4⟩ Even Roots or Even Powers

We need to use special care in finding inverses for functions that involve even roots or even powers. We saw in Example 3(c) that $f(x) = x^2$ is not the inverse of $g(x) = \sqrt{x}$. However, because $g(x) = \sqrt{x}$ is a one-to-one function, it has an inverse. The domain of g is $[0, \infty)$, and the range is $[0, \infty)$. So the inverse of g must have domain $[0, \infty)$ and range $[0, \infty)$. See Fig. 9.30. The only reason that $f(x) = x^2$ is not the inverse of g is that it has the wrong domain. So to write the inverse function, we must use the appropriate domain:

$$g^{-1}(x) = x^2 \qquad \text{for} \quad x \geq 0$$

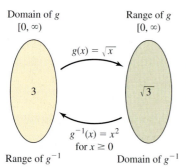

Domain of g
$[0, \infty)$

Range of g
$[0, \infty)$

$g(x) = \sqrt{x}$

3

$\sqrt{3}$

$g^{-1}(x) = x^2$
for $x \geq 0$

Range of g^{-1} Domain of g^{-1}

Figure 9.30

Note that by restricting the domain of g^{-1} to $[0, \infty)$, g^{-1} is one-to-one. With this restriction it is true that $(g \circ g^{-1})(x) = x$ and $(g^{-1} \circ g)(x) = x$ for every nonnegative number x.

E X A M P L E   6

Inverse of a function with an even exponent

Find the inverse of the function $f(x) = (x - 3)^2$ for $x \geq 3$.

Solution

Because of the restriction $x \geq 3$, f is a one-to-one function with domain $[3, \infty)$ and range $[0, \infty)$. The domain of the inverse function is $[0, \infty)$, and its range is $[3, \infty)$. Use the switch-and-solve strategy to find the formula for the inverse:

$$y = (x - 3)^2$$
$$x = (y - 3)^2$$
$$y - 3 = \pm\sqrt{x}$$
$$y = 3 \pm \sqrt{x}$$

Because the inverse function must have range $[3, \infty)$, we use the formula $f^{-1}(x) = 3 + \sqrt{x}$. Because the domain of f^{-1} is assumed to be $[0, \infty)$, no restriction is required on x.

Now do Exercises 41–48

⟨5⟩ Graphs of f and f^{-1}

Consider $f(x) = x^2$ for $x \geq 0$ and $f^{-1}(x) = \sqrt{x}$. Their graphs are shown in Fig. 9.31. Notice the symmetry. If we folded the paper along the line $y = x$, the two graphs would coincide.

If a point (a, b) is on the graph of the function f, then (b, a) must be on the graph of $f^{-1}(x)$. See Fig. 9.32. The points (a, b) and (b, a) lie on opposite sides of the

Figure 9.31

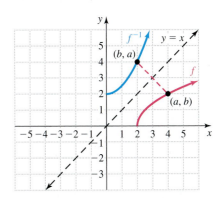

Figure 9.32

diagonal line $y = x$ and are the same distance from it. For this reason the graphs of f and f^{-1} are symmetric with respect to the line $y = x$.

E X A M P L E   7

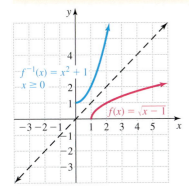

Figure 9.33

Inverses and their graphs

Find the inverse of the function $f(x) = \sqrt{x - 1}$, and graph f and f^{-1} on the same pair of axes.

Solution

To find f^{-1}, first switch x and y in the formula $y = \sqrt{x - 1}$:

$$x = \sqrt{y - 1}$$
$$x^2 = y - 1 \quad \text{Square both sides.}$$
$$x^2 + 1 = y$$

Because the range of f is the set of nonnegative real numbers $[0, \infty)$, we must restrict the domain of f^{-1} to be $[0, \infty)$. Thus, $f^{-1}(x) = x^2 + 1$ for $x \geq 0$. The two graphs are shown in Fig. 9.33.

> **Now do Exercises 49–58**

Warm-Ups ▼

Fill in the blank.

1. The _____ of a function is a function with the same ordered pairs except that the coordinates are reversed.

2. The domain of f^{-1} is the _____ of f.

3. The range of f^{-1} is the _____ of f.

4. A function is _____ if no two ordered pairs have the same second coordinates with different first coordinates.

5. The graphs of f and f^{-1} are _____ with respect to the line $y = x$.

6. If a _____ line can be drawn so that it crosses the graph of a function more than once, then the function is not one-to-one.

True or false?

7. The inverse of $\{(1, 3), (2, 5)\}$ is $\{(3, 1), (2, 5)\}$.

8. The function $f(x) = 3$ is one-to-one.

9. Only one-to-one functions are invertible.

10. The function $f(x) = x^4$ is invertible.

11. If $f(x) = -x$, then $f^{-1}(x) = -x$.

12. If h is invertible and $h(7) = -95$, then $h^{-1}(-95) = 7$.

13. If $f(x) = 4x + 5$, then $f^{-1}(x) = \dfrac{x - 5}{4}$.

14. If $g(x) = 3x - 6$, then $g^{-1}(x) = \dfrac{1}{3}x + 2$.

9.4 Exercises

‹1› Inverse of a Function

Determine whether each function is invertible. If it is invertible, then find the inverse. See Example 1.

1. $\{(1, 3), (2, 9)\}$

2. $\{(0, 5), (-2, 0)\}$

3. $\{(-3, 3), (-2, 2), (0, 0), (2, 2)\}$

4. $\{(1, 1), (2, 8), (3, 27)\}$

5. $\{(16, 4), (9, 3), (0, 0)\}$

6. $\{(-1, 1), (-3, 81), (3, 81)\}$

7. $\{(0, 5), (5, 0), (6, 0)\}$

8. $\{(3, -3), (-2, 2), (1, -1)\}$

9. $\{(0, 0), (2, 2), (9, 9)\}$

10. $\{(9, 1), (2, 1), (7, 1), (0, 1)\}$

Determine whether each function is invertible by examining the graph of the function. See Example 2.

 11.

12.

13.

14.

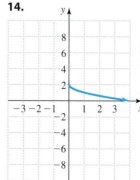

‹2› Identifying Inverse Functions

Determine whether each pair of functions f and g are inverses of each other. See Example 3.

15. $f(x) = 2x$ and $g(x) = 0.5x$

16. $f(x) = 3x$ and $g(x) = 0.33x$

17. $f(x) = 2x - 10$ and $g(x) = \dfrac{1}{2}x + 5$

18. $f(x) = 3x + 7$ and $g(x) = \dfrac{x - 7}{3}$

19. $f(x) = -x$ and $g(x) = -x$

20. $f(x) = \dfrac{1}{x}$ and $g(x) = \dfrac{1}{x}$

21. $f(x) = x^4$ and $g(x) = x^{1/4}$

22. $f(x) = |2x|$ and $g(x) = \left|\dfrac{x}{2}\right|$

‹3› Switch-and-Solve Strategy

Find f^{-1}. Check that $(f \circ f^{-1})(x) = x$ and $(f^{-1} \circ f)(x) = x$. See Examples 4 and 5. See the Strategy for Finding f^{-1} by Switch-and-Solve box on page 626.

23. $f(x) = 5x$

24. $h(x) = -3x$

25. $g(x) = x - 9$

26. $j(x) = x + 7$

27. $k(x) = 5x - 9$

28. $r(x) = 2x - 8$

29. $m(x) = \dfrac{2}{x}$

30. $s(x) = \dfrac{-1}{x}$

31. $f(x) = \sqrt[3]{x - 4}$

32. $f(x) = \sqrt[3]{x + 2}$

33. $f(x) = \dfrac{3}{x - 4}$

34. $f(x) = \dfrac{2}{x + 1}$

35. $f(x) = \sqrt[3]{3x + 7}$

36. $f(x) = \sqrt[3]{7 - 5x}$

37. $f(x) = \dfrac{x+1}{x-2}$

38. $f(x) = \dfrac{1-x}{x+3}$

52. $f(x) = x^2 + 3$ for $x \geq 0$

39. $f(x) = \dfrac{x+1}{3x-4}$

40. $g(x) = \dfrac{3x+5}{2x-3}$

⟨**4**⟩ **Even Roots or Even Powers**

53. $f(x) = 5x$

Find the inverse of each function. See Example 6.

41. $p(x) = \sqrt[4]{x}$

42. $v(x) = \sqrt[6]{x}$

43. $f(x) = (x-2)^2$ for $x \geq 2$

44. $g(x) = (x+5)^2$ for $x \geq -5$

45. $f(x) = x^2 + 3$ for $x \geq 0$

46. $f(x) = x^2 - 5$ for $x \geq 0$

54. $f(x) = \dfrac{x}{4}$

47. $f(x) = \sqrt{x+2}$

48. $f(x) = \sqrt{x-4}$

⟨**5**⟩ **Graphs of f and f^{-1}**

Find the inverse of each function, and graph f and f^{-1} on the same pair of axes. See Example 7.

55. $f(x) = x^3$

49. $f(x) = 2x + 3$

56. $f(x) = 2x^3$

50. $f(x) = -3x + 2$

57. $f(x) = \sqrt{x-2}$

51. $f(x) = x^2 - 1$ for $x \geq 0$

58. $f(x) = \sqrt{x + 3}$

Miscellaneous

Find the inverse of each function.

59. $f(x) = 2x$

60. $f(x) = x - 1$

61. $f(x) = 2x - 1$

62. $f(x) = 2(x - 1)$

63. $f(x) = \sqrt[3]{x}$

64. $f(x) = 2\sqrt[3]{x}$

65. $f(x) = \sqrt[3]{x - 1}$

66. $f(x) = \sqrt[3]{2x - 1}$

67. $f(x) = 2\sqrt[3]{x} - 1$

68. $f(x) = 2\sqrt[3]{x - 1}$

For each pair of functions, find $(f^{-1} \circ f)(x)$.

69. $f(x) = x^3 - 1$ and $f^{-1}(x) = \sqrt[3]{x + 1}$

70. $f(x) = 2x^3 + 1$ and $f^{-1}(x) = \sqrt[3]{\dfrac{x - 1}{2}}$

71. $f(x) = \dfrac{1}{2}x - 3$ and $f^{-1}(x) = 2x + 6$

72. $f(x) = 3x - 9$ and $f^{-1}(x) = \dfrac{1}{3}x + 3$

73. $f(x) = \dfrac{1}{x} + 2$ and $f^{-1}(x) = \dfrac{1}{x - 2}$

74. $f(x) = 4 - \dfrac{1}{x}$ and $f^{-1}(x) = \dfrac{1}{4 - x}$

75. $f(x) = \dfrac{x + 1}{x - 2}$ and $f^{-1}(x) = \dfrac{2x + 1}{x - 1}$

76. $f(x) = \dfrac{3x - 2}{x + 2}$ and $f^{-1}(x) = \dfrac{2x + 2}{3 - x}$

Applications

Solve each problem.

77. *Accident reconstruction.* The distance that it takes a car to stop is a function of the speed and the drag factor. The drag factor is a measure of the resistance between the tire and the road surface. The formula $S = \sqrt{30LD}$ is used to determine the minimum speed S [in miles per hour (mph)] for a car that has left skid marks of length L feet (ft) on a surface with drag factor D.

a) Find the minimum speed for a car that has left skid marks of length 50 ft where the drag factor is 0.75.

b) Does the drag factor increase or decrease for a road surface when it gets wet?

c) Write L as a function of S for a road surface with drag factor 1.

Figure for Exercise 77

78. *Area of a circle.* Let x be the radius of a circle and $h(x)$ be the area of the circle. Write a formula for $h(x)$ in terms of x. What does x represent in the notation $h^{-1}(x)$? Write a formula for $h^{-1}(x)$.

79. *Vehicle cost.* At Bill Hood Ford in Hammond a sales tax of 9% of the selling price x and a \$125 title and license fee are added to the selling price to get the total cost of a vehicle. Find the function $T(x)$ that the dealer uses to get the total cost as a function of the selling price x. Citizens National Bank will not include sales tax or fees in a loan. Find the function $T^{-1}(x)$ that the bank can use to get the selling price as a function of the total cost x.

80. *Carpeting cost.* At the Windrush Trace apartment complex all living rooms are square, but the length of x feet may vary. The cost of carpeting a living room is \$18 per square yard plus a \$50 installation fee. Find the function $C(x)$ that gives the total cost of carpeting a living room of length x. The manager has an invoice for the total cost of a living room carpeting job but does not know in which apartment it was done. Find the function $C^{-1}(x)$ that gives the length of a living room as a function of the total cost of the carpeting job x.

Getting More Involved

81. *Discussion*

Let $f(x) = x^n$ for n a positive integer. For which values of n is f an invertible function? Explain.

82. *Discussion*

Suppose f is a function with range $(-\infty, \infty)$ and g is a function with domain $(0, \infty)$. Is it possible that g and f are inverse functions? Explain.

Graphing Calculator Exercises

83. Most graphing calculators can form compositions of functions. Let $f(x) = x^2$ and $g(x) = \sqrt{x}$. To graph the composition $g \circ f$, let $y_1 = x^2$ and $y_2 = \sqrt{y_1}$. The graph of y_2 is the graph of $g \circ f$. Use the graph of y_2 to determine whether f and g are inverse functions.

84. Let $y_1 = x^3 - 4$, $y_2 = \sqrt[3]{x + 4}$, and $y_3 = \sqrt[3]{y_1 + 4}$. The function y_3 is the composition of the first two functions. Graph all three functions on the same screen. What do the graphs indicate about the relationship between y_1 and y_2?

9.5 Variation

In This Section

⟨1⟩ **Direct, Inverse, and Joint Variation**

⟨2⟩ **Finding the Variation Constant**

⟨3⟩ **Finding a New Value for the Dependent Variable**

⟨4⟩ **Applications**

If $y = 3x$, then as x varies so does y. Certain functions are customarily expressed in terms of variation. In this section, you will learn to write formulas for those functions from verbal descriptions of the functions.

⟨1⟩ Direct, Inverse, and Joint Variation

In a community with an 8% sales tax rate, the amount of tax, t (in dollars), is a function of the amount of the purchase, a (in dollars). This function is expressed by the formula

$$t = 0.08a.$$

If the amount increases, then the tax increases. If a decreases, then t decreases. In this situation we say that t *varies directly with a, or t is directly proportional to a.* The constant tax rate, 0.08, is called the **variation constant** or **proportionality constant.** Notice that t is just a simple linear function of a. We are merely introducing some new terms to express an old idea.

⟨ **Calculator Close-Up** ⟩

The graph of $t = 0.08a$ shows that the tax increases as the amount increases.

40	Y1=0.08X

X=250   Y=20

0   500

> **Direct Variation**
>
> The statement y **varies directly as** x, or y **is directly proportional to** x, means that
> $$y = kx$$
> for some constant, k. The constant, k, is a fixed nonzero real number.

In making a 500-mile trip by car, the time it takes is a function of the speed of the car. The greater the speed, the less time it will take. If you decrease the speed, the time increases. We say that the time is *inversely proportional* to the speed. Using the formula $D = RT$ or $T = \dfrac{D}{R}$, we can write

$$T = \frac{500}{R}.$$

In general, we make the following definition.

‹ **Calculator Close-Up** ›

The graph of $T = 500/R$ shows the time decreasing as the rate increases. For 50 mph the time is 10 hours, whereas for 100 mph the time is 5 hours.

Inverse Variation

The statement **y varies inversely as x**, or **y is inversely proportional to x**, means that

$$y = \frac{k}{x}$$

for some nonzero constant, k.

CAUTION   Be sure to understand the difference between direct and inverse variation. If y varies directly as x (with $k > 0$), then as x increases, y increases. If y varies inversely as x (with $k > 0$), then as x increases, y decreases.

On a deposit of \$5000 in a savings account, the interest earned, I, depends on the rate, r, and the time, t. Assuming the interest is simple interest, we can use the formula $I = Prt$ to write

$$I = 5000rt.$$

The variable I is a function of two independent variables, r and t. In this case we say that I *varies jointly* as r and t.

Joint Variation

The statement **y varies jointly as x and z**, or **y is jointly proportional to x and z**, means that

$$y = kxz$$

for some nonzero constant, k.

E X A M P L E   1

Writing the formula

Write a formula that expresses the relationship described in each statement. Use k as the variation constant.

a) w varies directly as t.

b) x is inversely proportional to p.

c) z varies jointly as x and q.

Solution

a) Since w varies directly as t, we have $w = kt$.

b) Since x is inversely proportional to p, we have $x = \dfrac{k}{p}$.

c) Since z varies jointly as x and q, we have $z = kxq$.

Now do Exercises 1–6

A combination of direct and inverse variation is called **combined variation**. We can also use powers and roots in a statement involving variation.

EXAMPLE 2

Writing a combined variation formula

Write a formula that expresses the relationship described in each statement. Use k as the variation constant.

    a) a varies directly as m and inversely as n.

    b) w is inversely proportional to the square of t.

    c) v varies directly as a and inversely as the square root of y.

Solution

    a)   Since a varies directly as m and inversely as n, we have $a = \frac{km}{n}$.

    b)   Since w is inversely proportional to the square of t, we have $w = \frac{k}{t^2}$.

    c)   Since v varies directly as a and inversely as the square root of y, we have $v = \frac{ka}{\sqrt{y}}$.

| Now do Exercises 7–22 |

CAUTION   The variation terms never signify addition or subtraction. We always use multiplication unless we see the word "inversely." In that case, we divide.

⟨2⟩ Finding the Variation Constant

If we know the values of all variables in a variation statement, we can find the value of the constant and write a formula using the value of the constant rather than an unknown constant k.

EXAMPLE 3

Finding the variation constant

Find the variation constant, and write a function that expresses the relationship described in each statement.

    a) a varies directly as x and $a = 6$ when $x = 2$.

    b) w is inversely proportional to y and $w = 10$ when $y = 2$.

    c) v varies directly as a and inversely as the square root of y and $v = 12$ when $a = 2$ and $y = 9$.

Solution

    a)   Since a varies directly as x, we have $a = kx$. Since $a = 6$ when $x = 2$, we have $6 = k(2)$ or $k = 3$. So we can write the function as $a = 3x$.

    b)   Since w is inversely proportional to y, we have $w = \frac{k}{y}$. Since $w = 10$ when $y = 2$, we have $10 = \frac{k}{2}$ or $k = 20$. So we can write the function as $w = \frac{20}{y}$.

    c)   Since v varies directly as a and inversely as the square root of y, we have $v = \frac{ka}{\sqrt{y}}$. Since $v = 12$ when $a = 2$ and $y = 9$, we have $12 = \frac{k(2)}{\sqrt{9}}$ or $k = 18$. So we can write the function as $v = \frac{18a}{\sqrt{y}}$.

| Now do Exercises 23–32 |

⟨3⟩ Finding a New Value for the Dependent Variable

If we know "benchmark" values for all of the variables in a variation statement, then we can find the value of the constant, as we did in Example 3, and new values for the dependent variable.

E X A M P L E   **4**

Finding a new value for the dependent variable

Find the requested values.

a) If y varies directly as x and $y = 12$ when $x = 9$, find y when $x = 21$.

b) If s is inversely proportional to t and $s = 15$ when $t = 7$, find s when $t = 3$.

Solution

a) Since y varies directly as x, we have $y = kx$. Since $y = 12$ when $x = 9$, we have $12 = k(9)$ or $k = \frac{4}{3}$. So we can write the function as $y = \frac{4}{3}x$. Now when $x = 21$, we have $y = \frac{4}{3}(21) = 28$.

b) Since s is inversely proportional to t, we have $s = \frac{k}{t}$. Since $s = 15$ when $t = 7$, we have $15 = \frac{k}{7}$ or $k = 105$. So we can write the function as $s = \frac{105}{t}$. Now if $t = 3$, we have $s = \frac{105}{3} = 35$.

> **Now do Exercises 33–40**

⟨4⟩ Applications

Examples 5, 6, and 7 illustrate applications of the language of variation.

E X A M P L E   **5**

Direct variation application

In a downtown office building the monthly rent for an office is directly proportional to the size of the office. If a 420-square-foot office rents for $1260 per month, then what is the rent for a 900-square-foot office?

Solution

Because the rent, R, varies directly with the area of the office, A, we have

$$R = kA.$$

Because a 420-square-foot office rents for $1260, we can substitute to find k:

$$1260 = k \cdot 420$$
$$3 = k$$

Now that we know the value of k, we can write

$$R = 3A.$$

To get the rent for a 900-square-foot office, insert 900 into this formula:

$$R = 3 \cdot 900$$
$$= 2700$$

So a 900-square-foot office rents for $2700 per month.

> **Now do Exercises 41–44**

EXAMPLE **6**

Joint variation application

The labor cost for installing ceramic floor tile varies jointly with the length and width of the room in which it is installed. If the labor cost for a 9 foot by 12 foot room is $324, then what is the labor cost for a 6 foot by 8 foot room?

Solution

Since the labor cost C varies jointly with the length L and width W, we have $C = kLW$ for some constant k. Use $C = 324$, $L = 12$, and $W = 9$ in this formula to find k:

$$324 = k \cdot 12 \cdot 9$$
$$324 = 108k$$
$$k = \frac{324}{108} = 3$$

So the labor cost is $3 per square foot and the formula is $C = 3LW$. Now find C when $L = 8$ and $W = 6$:

$$C = 3LW$$
$$= 3 \cdot 8 \cdot 6 = 144$$

So the labor cost for a 6 foot by 8 foot room is $144.

> **Now do Exercises 45–48**

EXAMPLE **7**

Combined variation application

The time t that it takes to frame a house varies directly with the size of the house s in square feet and inversely with the number of framers n working on the job. If three framers can complete a 2500-square-foot house in 6 days, then how long will it take six framers to complete a 4500-square-foot house?

Solution

Because t varies directly with s and inversely with n, we have

$$t = \frac{ks}{n}.$$

Substitute $t = 6$, $s = 2500$, and $n = 3$ into this equation to find k:

$$6 = \frac{k \cdot 2500}{3}$$
$$18 = 2500k$$
$$0.0072 = k$$

Now use $k = 0.0072$, $s = 4500$, and $n = 6$ to find t:

$$t = \frac{0.0072 \cdot 4500}{6}$$
$$= 5.4$$

So six framers can frame a 4500-square-foot house in 5.4 days.

> **Now do Exercises 49–55**

Warm-Ups ▼

Fill in the blank.

1. If y varies _____ as x, then $y = kx$ for some nonzero constant k.

2. If y varies _____ as x, then $y = \dfrac{k}{x}$ for some nonzero constant k.

3. The number k is the constant of _____.

4. If y varies _____ as x and z, then $y = kxz$ for some nonzero constant k.

5. "Varies directly" means the same as directly _____.

6. "Varies inversely" means the same as inversely _____.

True or false?

7. If $a = 5b$, then a varies directly as b.

8. If $n = \dfrac{70}{m}$, then n varies directly as m.

9. If a is jointly proportional to b and c, then $a = bc$.

10. If $a = \dfrac{8b}{c}$, then a varies directly as b and inversely as c.

11. If $m = \dfrac{19z}{\sqrt{n}}$, then a varies directly as z and inversely as n.

12. If $b = \dfrac{16}{z^2}$, then b varies inversely as the square of z.

9.5 Exercises

‹ Study Tips ›

- Many schools have study skills centers that offer courses, workshops, and individual help on how to study.
- A search for "study skill" on the World Wide Web will turn up an endless amount of useful information.

‹1› Direct, Inverse, and Joint Variation

Write a formula that expresses the relationship described by each statement. Use k for the constant of variation. See Examples 1 and 2.

1. a varies directly as m.

2. w varies directly with P.

3. d varies inversely with e.

4. y varies inversely as x.

5. I varies jointly as r and t.

6. q varies jointly as w and v.

7. m is directly proportional to the square of p.

8. g is directly proportional to the cube of r.

9. B is directly proportional to the cube root of w.

10. F is directly proportional to the square of m.

11. t is inversely proportional to the square of x.

12. y is inversely proportional to the square root of z.

13. v varies directly as m and inversely as n.

14. b varies directly as the square of n and inversely as the square root of v.

Determine whether each equation represents direct, inverse, joint, or combined variation.

15. $y = \dfrac{78}{x}$

16. $y = \dfrac{\pi}{x}$

17. $y = \dfrac{1}{2}x$

18. $y = \dfrac{x}{4}$

19. $y = \dfrac{3x}{w}$

20. $y = \dfrac{4t^2}{\sqrt{x}}$

21. $y = \dfrac{1}{3}xz$

22. $y = 99qv$

‹2› Finding the Variation Constant

Find the proportionality constant, and write a formula that expresses the indicated variation. See Example 3.

23. y varies directly as x, and $y = 6$ when $x = 4$.

24. m varies directly as w, and $m = \frac{1}{3}$ when $w = \frac{1}{4}$.

25. A varies inversely as B, and $A = 10$ when $B = 3$.

26. c varies inversely as d, and $c = 0.31$ when $d = 2$.

27. m varies inversely as the square root of p, and $m = 12$ when $p = 9$.

28. s varies inversely as the square root of v, and $s = 6$ when $v = \frac{3}{2}$.

29. A varies jointly as t and u, and $A = 6$ when $t = 5$ and $u = 3$.

30. N varies jointly as the square of p and the cube of q, and $N = 72$ when $p = 3$ and $q = 2$.

31. y varies directly as x and inversely as z, and $y = 2.37$ when $x = \pi$ and $z = \sqrt{2}$.

32. a varies directly as the square root of m and inversely as the square of n, and $a = 5.47$ when $m = 3$ and $n = 1.625$.

‹3› Finding a New Value for the Dependent Variable

Find the requested values. See Example 4.

33. If y varies directly as x, and $y = 7$ when $x = 5$, find y when $x = -3$.

34. If n varies directly as p, and $n = 0.6$ when $p = 0.2$, find n when $p = \sqrt{2}$.

35. If w varies inversely as z, and $w = 6$ when $z = 2$, find w when $z = -8$.

36. If p varies inversely as q, and $p = 5$ when $q = \sqrt{3}$, find p when $q = 5$.

37. If A varies jointly as F and T, and $A = 6$ when $F = 3\sqrt{2}$ and $T = 4$, find A when $F = 2\sqrt{2}$ and $T = \frac{1}{2}$.

38. If j varies jointly as the square of r and the cube of v, and $j = -3$ when $r = 2\sqrt{3}$ and $v = \frac{1}{2}$, find j when $r = 3\sqrt{5}$ and $v = 2$.

39. If D varies directly with t and inversely with the square of s, and $D = 12.35$ when $t = 2.8$ and $s = 2.48$, find D when $t = 5.63$ and $s = 6.81$.

40. If M varies jointly with x and the square of v, and $M = 39.5$ when $x = \sqrt{10}$ and $v = 3.87$, find M when $x = \sqrt{30}$ and $v = 7.21$.

‹4› Applications

Solve each problem. See Examples 5–7.

41. Lawn maintenance. At Larry's Lawn Service the cost of lawn maintenance varies directly with the size of the lawn. If the monthly maintenance on a 4000-square-foot lawn is \$280, then what is the maintenance fee for a 6000-square-foot lawn?

42. Weight of the iguana. The weight of an iguana is directly proportional to its length. If a 4-foot iguana weighs 30 pounds, then how much should a 5-foot iguana weigh?

43. Gas laws. The volume of a gas in a cylinder at a fixed temperature is inversely proportional to the weight on the piston. If the gas has a volume of 6 cubic centimeters (cm^3) for a weight of 30 kilograms (kg), then what would the volume be for a weight of 20 kg?

44. Selling software. A software vendor sells a software package at a price that is inversely proportional to the number of packages sold per month. When they are selling 900 packages per month, the price is \$80 each. If they sell 1000 packages per month, then what should the new price be?

45. Costly culvert. The price of an aluminum culvert is jointly proportional to its radius and length. If a 12-foot culvert with a 6-inch radius costs \$324, then what is the price of a 10-foot culvert with an 8-inch radius?

46. Pricing plastic. The cost of a piece of PVC water pipe varies jointly as its diameter and length. If a 20-foot pipe with a diameter of 1 inch costs \$6.80, then what will be the cost of a 10-foot pipe with a $\frac{3}{4}$-inch diameter?

47. Reinforcing rods. The price of a steel rod varies jointly as the length and the square of the diameter. If an 18-foot rod with a 2-inch diameter costs \$12.60, then what is the cost of a 12-foot rod with a 3-inch diameter?

48. Pea soup. The weight of a cylindrical can of pea soup varies jointly with the height and the square of the radius. If a 4-inch-high can with a 1.5-inch radius weighs 16 ounces, then what is the weight of a 5-inch-high can with a radius of 3 inches?

49. Falling objects. The distance an object falls in a vacuum varies directly with the square of the time it is falling. In the first 0.1 second after an object is dropped, it falls 0.16 feet.

a) Find the formula that expresses the distance d an object falls as a function of the time it is falling t.

b) How far does an object fall in the first 0.5 second after it is dropped?

c) How long does it take for a watermelon to reach the ground when dropped from a height of 100 feet?

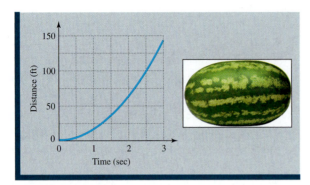

Figure for Exercise 49

50. *Making Frisbees.* The cost of material used in making a Frisbee varies directly with the square of the diameter. If it costs the manufacturer $0.45 for the material in a Frisbee with a 9-inch diameter, then what is the cost for the material in a 12-inch-diameter Frisbee?

51. *Using leverage.* The basic law of leverage is that the force required to lift an object is inversely proportional to the length of the lever. If a force of 2000 pounds applied 2 feet from the pivot point would lift a car, then what force would be required at 10 feet to lift the car?

52. *Resistance.* The resistance of a wire varies directly with the length and inversely as the square of the diameter. If a wire of length 20 feet and diameter 0.1 inch has a resistance of 2 ohms, then what is the resistance of a 30-foot wire with a diameter of 0.2 inch?

53. *Computer programming.* The time t required to complete a programming job varies directly with the complexity of the job and inversely with the number n of programmers working on the job. The complexity c is an arbitrarily assigned number between 1 and 10, with 10 being the most complex. It takes 8 days for a team of three programmers to complete a job with complexity 6. How long will it take five programmers to complete a job with complexity 9?

54. *Breakfast cereal.* On average a family of three eats a 12-ounce box of breakfast cereal in 8 days. The number of days required varies directly with the size of the box and inversely with the number of family members. How long does it take for a family of four to eat an 18-ounce box of cereal?

55. *Bicycle gear ratio.* A bicycle's gear ratio G varies jointly with the number of teeth on the chain ring N (by the pedals) and the diameter of the wheel d, and inversely with the number of teeth on the cog c (on the rear wheel). A bicycle with 27-inch-diameter wheels, 26 teeth on the cog, and 52 teeth on the chain ring has a gear ratio of 54.

a) Find a formula that expresses the gear ratio as a function of N, d, and c.

b) What is the gear ratio for a bicycle with 26-inch-diameter wheels, 42 teeth on the chain ring, and 13 teeth on the cog?

c) A five-speed bicycle with 27-inch-diameter wheels and 44 teeth on the chain ring has gear ratios of 52, 59, 70, 79, and 91. Find the number of teeth on the cog (a whole number) for each gear ratio.

d) For a fixed wheel size and chain ring, does the gear ratio increase or decrease as the number of teeth on the cog increases?

Figure for Exercise 55

Graphing Calculator Exercises

56. To see the difference between direct and inverse variation, graph $y_1 = 2x$ and $y_2 = \frac{2}{x}$ using $0 \le x \le 5$ and $0 \le y \le 10$. Which of these functions is increasing and which is decreasing?

57. Graph $y_1 = 2\sqrt{x}$ and $y_2 = \frac{2}{\sqrt{x}}$ by using $0 \le x \le 5$ and $0 \le y \le 10$. At what point in the first quadrant do the curves cross? Which function is increasing and which is decreasing? Which represents direct variation and which represents inverse variation?

Wrap-Up

Summary

Types of Functions		Examples
Linear function	$y = mx + b$ or $f(x) = mx + b$ for $m \neq 0$ Domain $(-\infty, \infty)$, range $(-\infty, \infty)$ If $m = 0$, $y = b$ is a constant function. Domain $(-\infty, \infty)$, range $\{b\}$	$f(x) = 2x - 3$
Absolute value function	$y = |x|$ or $f(x) = |x|$ Domain $(-\infty, \infty)$, range $[0, \infty)$	$f(x) = |x + 5|$
Quadratic function	$f(x) = ax^2 + bx + c$ for $a \neq 0$	$f(x) = x^2 - 4x + 3$
Square-root function	$f(x) = \sqrt{x}$ Domain $[0, \infty)$, range $[0, \infty)$	$f(x) = \sqrt{x - 4}$

Transformations of Graphs		Examples
Reflecting	The graph of $y = -f(x)$ is a reflection in the x-axis of the graph of $y = f(x)$.	The graph of $y = -x^2$ is a reflection of the graph of $y = x^2$.

Translating	The graph of $y = f(x) + k$ is
	  k units above $y = f(x)$ if $k > 0$ or
	  $|k|$ units below $y = f(x)$ if $k < 0$.
	
	The graph of $y = f(x - h)$ is
	  h units to the right of $y = f(x)$ if $h > 0$ or
	  $|h|$ units to the left of $y = f(x)$ if $h < 0$.
		The graph of $y = x^2 + 3$ is three units above $y = x^2$, and $y = x^2 - 3$ is three units below $y = x^2$.
		The graph of $y = (x - 3)^2$ is three units to the right of $y = x^2$, and $y = (x + 3)^2$ is three units to the left.

Stretching and shrinking	The graph of $y = af(x)$ is obtained by stretching (if $a > 1$) or shrinking (if $0 < a < 1$) the graph of $y = f(x)$.	The graph of $y = 5x^2$ is obtained by stretching $y = x^2$, and $y = 0.1x^2$ is obtained by shrinking $y = x^2$.

Combining Functions		Examples
Sum	$(f + g)(x) = f(x) + g(x)$	For $f(x) = x^2$ and $g(x) = x + 1$ $(f + g)(x) = x^2 + x + 1$
Difference	$(f - g)(x) = f(x) - g(x)$	$(f - g)(x) = x^2 - x - 1$
Product	$(f \cdot g)(x) = f(x) \cdot g(x)$	$(f \cdot g)(x) = x^3 + x^2$

Quotient	$\left(\dfrac{f}{g}\right)(x) = \dfrac{f(x)}{g(x)}, g(x) \neq 0$	$\left(\dfrac{f}{g}\right)(x) = \dfrac{x^2}{x+1}$
Composition of functions	$(g \circ f)(x) = g(f(x))$ $(f \circ g)(x) = f(g(x))$	$(g \circ f)(x) = g(x^2) = x^2 + 1$ $(f \circ g)(x) = f(x+1)$ $\qquad\qquad = x^2 + 2x + 1$

Inverse Functions		**Examples**
One-to-one function	A function in which no two ordered pairs have different x-coordinates and the same y-coordinate	$f = \{(2, 20), (3, 30)\}$
Inverse function	The inverse of a one-to-one function f is the function f^{-1}, which is obtained from f by interchanging the coordinates in each ordered pair of f. The domain of f^{-1} is the range of f, and the range of f^{-1} is the domain of f.	$f^{-1} = \{(20, 2), (30, 3)\}$
Horizontal-line test	If there is a horizontal line that crosses the graph of a function more than once, then the function is not invertible.	
Function notation for inverse	Two functions f and g are inverses of each other if and only if both of the following conditions are met. 1. $(g \circ f)(x) = x$ for every number x in the domain of f. 2. $(f \circ g)(x) = x$ for every number x in the domain of g.	$f(x) = x^3 + 1$ $f^{-1}(x) = \sqrt[3]{x-1}$
Switch-and-solve strategy for finding f^{-1}	1. Replace $f(x)$ by y. 2. Interchange x and y. 3. Solve for y. 4. Replace y by $f^{-1}(x)$.	$y = x^3 + 1$ $x = y^3 + 1$ $x - 1 = y^3$ $y = \sqrt[3]{x-1}$ $f^{-1}(x) = \sqrt[3]{x-1}$
Graphs of f and f^{-1}	Graphs of inverse functions are symmetric with respect to the line $y = x$.	

The Language of Variation		**Examples**
Direct	y varies directly as x, $y = kx$	$z = 5m$
Inverse	y varies inversely as x, $y = \dfrac{k}{x}$	$a = \dfrac{1}{c}$
Joint	y varies jointly as x and z, $y = kxz$	$V = 6LW$
Combined	y varies directly as x and inversely as z, $y = \dfrac{kx}{z}$	$S = \dfrac{3A}{B}$

Enriching Your Mathematical Word Power

Fill in the blank.

1. A _____ function has the form $f(x) = k$.

2. A _____ function has the form $f(x) = mx + b$ with $m \neq 0$.

3. A _____ function has the form $f(x) = ax^2 + bx + c$ with $a \neq 0$.

4. A _____ value function has the form $f(x) = |x|$.

5. The _____ of f and g is $f \circ g$ where $(f \circ g)(x) = f(g(x))$.

6. The ____ of f and g is $f + g$ where $(f + g)(x) = f(x) + g(x)$.

7. The _____ of f is a function in which the ordered pairs of f are reversed.

8. A function in which no two ordered pairs have the same second coordinates and different first coordinates is a(n) _____ function.

9. The _____-line test is a visual method for determining whether a graph is the graph of a function.

10. The _____-line test is a visual method for determining whether a function is one-to-one.

11. If $y = kx$ for some constant k, then y varies _____ as x.

12. If $y = k/x$ for some constant k, then y varies _____ as x.

13. If $y = kxz$ for some constant k, then y varies _____ as x and z.

14. The graph of $y = -f(x)$ is a(n) _____ in the x-axis of the graph of $y = f(x)$.

15. The graph of $y = f(x) + c$ for $c > 0$ is an upward _____ of the graph of $y = f(x)$.

16. The graph of $y = f(x - c)$ for $c > 0$ is a translation to the ____ of the graph of $y = f(x)$.

Review Exercises

9.1 Graphs of Functions and Relations
Graph each function, and state the domain and range.

1. $f(x) = 3x - 4$

2. $y = 0.3x$

3. $h(x) = |x| - 2$

4. $y = |x - 2|$

5. $y = x^2 - 2x + 1$

6. $g(x) = x^2 - 2x - 15$

7. $k(x) = \sqrt{x} + 2$

8. $y = \sqrt{x - 2}$

9. $y = 30 - x^2$

10. $y = 4 - x^2$

11. $f(x) = \begin{cases} \sqrt{x+4} & \text{for } -4 \le x \le 0 \\ x+2 & \text{for } x > 0 \end{cases}$

12. $f(x) = \begin{cases} \sqrt{x+1} & \text{for } -1 \le x \le 3 \\ x-1 & \text{for } x > 3 \end{cases}$

Graph each relation, and state its domain and range.

13. $x = 2$

14. $x = y^2 - 1$

15. $x = |y| + 1$

16. $x = \sqrt{y - 1}$

9.2 Transformations of Graphs

Sketch the graph of each function, and state the domain and range.

17. $y = \sqrt{x}$

18. $y = -\sqrt{x}$

19. $y = -2\sqrt{x}$

20. $y = 2\sqrt{x}$

21. $y = \sqrt{x - 2}$

22. $y = \sqrt{x + 2}$

23. $y = \frac{1}{2}\sqrt{x}$

24. $y = \sqrt{x - 1} + 2$

25. $y = -\sqrt{x + 1} + 3$

26. $y = 3\sqrt{x + 4} - 5$

9.3 Combining Functions

Let $f(x) = 3x + 5$, $g(x) = x^2 - 2x$, and $h(x) = \frac{x - 5}{3}$. Find the following.

27. $f(-3)$

28. $h(-4)$

29. $(h \circ f)(\sqrt{2})$

30. $(f \circ h)(\pi)$

31. $(g \circ f)(2)$

32. $(g \circ f)(x)$

33. $(f + g)(3)$

34. $(f - g)(x)$

35. $(f \cdot g)(x)$

36. $\left(\dfrac{f}{g}\right)(1)$

37. $(f \circ f)(0)$

38. $(f \circ f)(x)$

Let $f(x) = |x|$, $g(x) = x + 2$, and $h(x) = x^2$. Write each of the following functions as a composition of functions, using f, g, or h.

39. $F(x) = |x + 2|$

40. $G(x) = |x| + 2$

41. $H(x) = x^2 + 2$

42. $K(x) = x^2 + 4x + 4$

43. $I(x) = x + 4$

44. $J(x) = x^4 + 2$

9.4 Inverse Functions

Determine whether each function is invertible. If it is invertible, find the inverse.

45. $\{(-2, 4), (2, 4)\}$

46. $\{(1, 1), (3, 3)\}$

47. $f(x) = 8x$

48. $i(x) = -\dfrac{x}{3}$

49. $g(x) = 13x - 6$

50. $h(x) = \sqrt[3]{x - 6}$

51. $j(x) = \dfrac{x + 1}{x - 1}$

52. $k(x) = |x| + 7$

53. $m(x) = (x - 1)^2$

54. $n(x) = \dfrac{3}{x}$

Find the inverse of each function, and graph f and f^{-1} on the same pair of axes.

55. $f(x) = 3x - 1$

56. $f(x) = 2 - x^2$ for $x \geq 0$

57. $f(x) = \dfrac{x^3}{2}$

58. $f(x) = -\dfrac{1}{4}x$

9.5 Variation

Solve each variation problem.

59. If y varies directly as m and $y = -3$ when $m = \dfrac{1}{4}$, find y when $m = -2$.

60. If a varies inversely as b and $a = 6$ when $b = -3$, find a when $b = 4$.

61. If c varies directly as m and inversely as n, and $c = 20$ when $m = 10$ and $n = 4$, find c when $m = 6$ and $n = -3$.

62. If V varies jointly as h and the square of r, and $V = 32$ when $h = 6$ and $r = 3$, find V when $h = 3$ and $r = 4$.

Miscellaneous

Solve each problem.

63. *Falling object.* If a ball is dropped from a tall building, then the distance traveled by the ball in t seconds varies directly as the square of the time t. If the ball travels 144 feet (ft) in 3 seconds, then how far does it travel in 4 seconds?

64. *Studying or partying.* Evelyn's grade on a math test varies directly with the number of hours spent studying and inversely with the number of hours spent partying during the 24 hours preceding the test. If she scored a 90 on a test after she studied 10 hours and partied 2 hours, then what should she score after studying 4 hours and partying 6 hours?

65. *Inscribed square.* Given that B is the area of a square inscribed in a circle of radius r and area A, write B as a function of A.

66. *Area of a window.* A window is in the shape of a square of side s, with a semicircle of diameter s above it. Write a function that expresses the total area of the window as a function of s.

Figure for Exercise 66

67. *Composition of functions.* Given that $a = 3k + 2$ and $k = 5w - 6$, write a as a function of w.

68. *Volume of a cylinder.* The volume of a cylinder with a fixed height of 10 centimeters (cm) is given by $V = 10\pi r^2$, where r is the radius of the circular base. Write the volume as a function of the area of the base, A.

69. *Square formulas.* Write the area of a square A as a function of the length of a side of the square s. Write the length of a side of a square as a function of the area.

70. *Circle formulas.* Write the area of a circle A as a function of the radius of the circle r. Write the radius of a circle as a function of the area of the circle. Write the area as a function of the diameter d.

Chapter 9 Test

Sketch the graph of each function or relation and state the domain and range.

1. $f(x) = -\dfrac{2}{3}x + 1$

2. $y = |x| - 4$

3. $g(x) = x^2 + 2x - 8$

4. $x = y^2$

5. $f(x) = \begin{cases} \sqrt{x} & \text{for } x \geq 0 \\ -x - 3 & \text{for } x < 0 \end{cases}$

6. $y = -|x - 2|$

7. $y = \sqrt{x + 5} - 2$

Let $f(x) = -2x + 5$ and $g(x) = x^2 + 4$. Find the following.

8. $f(-3)$

9. $(g \circ f)(-3)$

10. $f^{-1}(11)$

11. $f^{-1}(x)$

12. $(g + f)(x)$

13. $(f \cdot g)(1)$

14. $(f^{-1} \circ f)(1776)$

15. $(f/g)(2)$

16. $(f \circ g)(x)$

17. $(g \circ f)(x)$

Let $f(x) = x - 7$ and $g(x) = x^2$. Write each of the following functions as a composition of functions using f and g.

18. $H(x) = x^2 - 7$

19. $W(x) = x^2 - 14x + 49$

Determine whether each function is invertible. If it is invertible, find the inverse.

20. $\{(2, 3), (4, 3), (1, 5)\}$

21. $\{(2, 3), (3, 4), (4, 5)\}$

Find the inverse of each function.

22. $f(x) = x - 5$

23. $f(x) = 3x - 5$

24. $f(x) = \sqrt[3]{x} + 9$

25. $f(x) = \dfrac{2x + 1}{x - 1}$

Solve each problem.

26. The volume of a sphere varies directly as the cube of the radius. If a sphere with radius 3 feet (ft) has a volume of 36π cubic feet (ft^3), then what is the volume of a sphere with a radius of 2 ft?

27. Suppose y varies directly as x and inversely as the square root of z. If $y = 12$ when $x = 7$ and $z = 9$, then what is the proportionality constant?

28. The cost of a Persian rug varies jointly as the length and width of the rug. If the cost is $2256 for a 6 foot by 8 foot rug, then what is the cost of a 9 foot by 12 foot rug?

*Making*Connections | **A Review of Chapters 1–9**

Simplify each expression.

1. $125^{-2/3}$

2. $\left(\dfrac{8}{27}\right)^{-1/3}$

3. $\sqrt{18} - \sqrt{8}$

4. $x^5 \cdot x^3$

5. $16^{1/4}$

6. $\dfrac{x^{12}}{x^3}$

Find the real solution set to each equation.

7. $x^2 = 9$

8. $x^2 = 8$

9. $x^2 = x$

10. $x^2 - 4x - 6 = 0$

11. $x^{1/4} = 3$

12. $x^{1/6} = -2$

13. $|x| = 8$

14. $|5x - 4| = 21$

15. $x^3 = 8$

16. $(3x - 2)^3 = 27$

17. $\sqrt{2x - 3} = 9$

18. $\sqrt{x - 2} = x - 8$

Sketch the graph of each set.

19. $\{(x, y) \mid y = 5\}$

20. $\{(x, y) \mid y = 2x - 5\}$

21. $\{(x, y) \mid x = 5\}$

22. $\{(x, y) \mid 3y = x\}$

23. $\{(x, y) \mid y = 5x^2\}$

24. $\{(x, y) \mid y = -2x^2\}$

Find the missing coordinates in each ordered pair so that the ordered pair satisfies the given equation.

25. $(2, \ ), (3, \ ), (\ , 2), (\ , 16), \quad 2^x = y$

26. $\left(\dfrac{1}{2}, \ \right), (-1, \ ), (\ , 16), (\ , 1), \quad 4^x = y$

Find the domain of each expression.

27. $\sqrt{x}$

28. $\sqrt{6 - 2x}$

29. $\dfrac{5x - 3}{x^2 + 1}$

30. $\dfrac{x - 3}{x^2 - 10x + 9}$

Solve each system of equations, and state whether the system is independent, dependent, or inconsistent.

31. $4x - 9y = -1$
$2x + 12y = 5$

32. $10x + 20y = 143$
   $y = x + 5.2$

33. $3x - 9y = 6$
   $y = \dfrac{1}{3}x - \dfrac{2}{3}$

34. $x = 5y + 12$
   $y = \dfrac{1}{5}x - 7$

35. $x + y - z = -5$
   $x - 2y + z = 11$
   $3x - y - z = 3$

36. $2x + y + 3z = 2$
   $x + y + z = 5$
   $3x + 2y + 4z = 6$

Perform the indicated operations.

37. $\dfrac{11}{15} - \dfrac{5}{12}$

38. $\dfrac{12}{35} \cdot \dfrac{10}{21}$

39. $\dfrac{4}{9} \div \dfrac{14}{15}$

40. $\dfrac{2}{3x^2} + \dfrac{1}{6x}$

41. $\dfrac{5xy^3}{6a^3b^3} \cdot \dfrac{2a^3b^7}{5xy^5}$

42. $\dfrac{12a^2 - 48}{a^2 + a - 6} \div \dfrac{9a + 18}{a^2 - 9}$

Solve each problem.

43. *Capital cost and operating cost.* To decide when to replace company cars, an accountant looks at two cost components: capital cost and operating cost. The capital cost C (the difference between the original cost and the salvage value) for a certain car is $3000 plus $0.12 for each mile that the car is driven.

   a) Write the capital cost C as a linear function of x, the number of miles that the car is driven.

   b) The operating cost P is $0.15 per mile initially and increases linearly to $0.25 per mile when the car reaches

Figure for Exercise 43(a)

Figure for Exercise 43(b)

100,000 miles. Write P as a function of x, the number of miles that the car is driven.

44. *Total cost.* The accountant in Exercise 43 uses the function $T = \dfrac{C}{x} + P$ to find the total cost per mile.

   a) Find T for $x = 20,000$, $30,000$, and $90,000$.

   b) Sketch a graph of the total cost function.

   c) The accountant has decided to replace the car when T reaches $0.38 for the second time. At what mileage will the car be replaced?

   d) For what values of x is T less than or equal to $0.38?

*Critical*Thinking | For Individual or Group Work | Chapter 9

These exercises can be solved by a variety of techniques, which may or may not require algebra. So be creative and think critically. Explain all answers. Answers are in the Instructor's Edition of this text.

1. Ant parade. An ant marches from point *A* to point *B* on the cylindrical garbage can shown in the accompanying figure. The can is 1 foot in diameter and 2 feet high. If the ant makes two complete revolutions of the can in a perfect spiral, then exactly how far did he travel?

Figure for Exercise 1

2. Connecting points. Draw a circle, and pick any three points on the circle.

   a) How many line segments can be drawn connecting these points?
   b) How many line segments can be drawn connecting four points on a circle? Five points? Six points?
   c) How many line segments can be drawn connecting *n* points on a circle?

3. Summing the digits. Find the sum of the digits in the standard form of the number $2^{2005} \cdot 5^{2007}$.

4. Consecutive odd numbers. Find three consecutive odd whole numbers such that the sum of their squares is a four-digit whole number whose digits are all the same.

5. Reversible prime numbers. The prime number 13 has an interesting property. When its digits are reversed, the new number 31 is also prime. Find the sum of all prime

numbers greater than 10 yet less than 125 that have this property.

6. Circles and squares. Start with a square piece of paper. Draw the largest possible circle inside the square. Cut out the circle and keep it. Now draw the largest possible square inside the circle. Cut out the square and keep it. What is the ratio of the area of the original square to the area of the final square? If you repeat this process six more times, then what is the ratio of the area of the original square to the area of the final square?

7. Perpendicular hands. What are the first two times (to the nearest second) after 12 noon for which the minute hand and hour hand of a clock are perpendicular to each other?

Photo for Exercise 7

8. Going broke. Albert and Zelda agreed to play a game. If heads appeared on the toss of an ordinary coin, Zelda had to double the amount of money that Albert had. If the result was tails, then Albert had to pay Zelda $24. As it turned out, the coin came up heads, tails, heads, tails, heads, tails. Then Albert was broke. How much money did Albert start with?

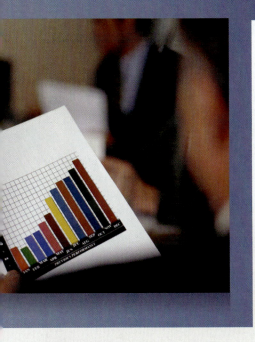

Polynomial and Rational Functions

New products are coming to market daily. Television ads are constantly touting the features and benefits of new exercise devices, diets, can openers, bag sealers, vacuum cleaners, cars, trucks, drugs, makeup, toothpaste, and so on. By some estimates 90% of new products are failures. But how can products fail to make money? The answer to that question lies in the tremendous cost of research, development, manufacturing, distribution, and advertising.

It now takes an average of 14.8 years and $350 million to get a new drug to market. However, the payoff can be quick and tremendous. At around $9 per pill, sales of a new miracle drug can top $1 billion in the first year.

Mercedes-Benz introduced a new sport utility vehicle in 1998 that took 5 years to design. The company spent $700 million on design, $300 million to build the Alabama plant for manufacturing the SUV, $400 million to design the engine, and $450 million to construct a plant for making the engine—not to mention the cost of manufacturing each car. Mercedes-Benz expects to sell 40,000 of these vehicles the first year at around $35,000 each. Making a profit will take a while.

For any product, profits are made when the average cost of the product is below the price at which the company sells the product.

In Exercises 77 and 78 of Section 10.5 we will see how rational functions can be used to model average cost.

10.1   The Factor Theorem

We have worked with polynomials on many occasions in this text. In this section, we study functions defined by polynomials and learn to solve some higher-degree polynomial equations.

⟨1⟩ The Factor Theorem

If a polynomial is used to define a function, then the function is called a **polynomial function.** For example, the functions

$$m(x) = 2, \quad n(x) = 3x - 6, \quad h(x) = 5x^2 - 2x + 6, \quad \text{and} \quad j(x) = x^3$$

are polynomial functions. The degree of a polynomial function is the degree of the polynomial used to define it. The polynomials m, n, h, and j have degrees 0, 1, 2, and 3, respectively. If $f(c) = 0$, then c is called a **root** or **zero** of the function f.

Consider the polynomial function $P(x) = x^2 + 2x - 15$. We can find the zeros of the function by solving the equation $P(x) = 0$:

$$x^2 + 2x - 15 = 0$$
$$(x + 5)(x - 3) = 0$$

$$x + 5 = 0 \quad \text{or} \quad x - 3 = 0$$
$$x = -5 \quad \text{or} \quad x = 3$$

⟨ **Helpful Hint** ⟩

Note that the zeros of the polynomial function are factors of the constant term -15.

Because $x + 5$ is a factor of $x^2 + 2x - 15$, -5 is a solution to the equation $x^2 + 2x - 15 = 0$ and a zero of the function $P(x) = x^2 + 2x - 15$. We can check that -5 is a zero of $P(x) = x^2 + 2x - 15$ as follows:

$$P(-5) = (-5)^2 + 2(-5) - 15$$
$$= 25 - 10 - 15$$
$$= 0$$

Because $x - 3$ is a factor of the polynomial, 3 is also a solution to the equation $x^2 + 2x - 15 = 0$ and a zero of the polynomial function. Check that $P(3) = 0$:

$$P(3) = 3^2 + 2 \cdot 3 - 15$$
$$= 9 + 6 - 15$$
$$= 0$$

Every linear factor of the polynomial corresponds to a zero of the polynomial function, and every zero of the polynomial function corresponds to a linear factor.

Now suppose $P(x)$ represents an arbitrary polynomial. If $x - c$ is a factor of the polynomial $P(x)$, then c is a solution to the equation $P(x) = 0$, and so $P(c) = 0$. If we divide $P(x)$ by $x - c$ and the remainder is 0, we must have

$$P(x) = (x - c)(\text{quotient}). \quad \text{Dividend equals the divisor times the quotient.}$$

If the remainder is 0, then $x - c$ is a factor of $P(x)$.

The **factor theorem** summarizes these ideas.

> **The Factor Theorem**
> The following statements are equivalent for any polynomial $P(x)$.
>
> **1.** The remainder is zero when $P(x)$ is divided by $x - c$.
> **2.** $x - c$ is a factor of $P(x)$.
> **3.** c is a solution to $P(x) = 0$.
> **4.** c is a zero of the function $P(x)$, or $P(c) = 0$.

To say that statements are equivalent means that the truth of any one of them implies that the others are true.

According to the factor theorem, if we want to determine whether a given number c is a zero of a polynomial function, we can divide the polynomial by $x - c$. The remainder is zero if and only if c is a zero of the polynomial function. The quickest way to divide by $x - c$ is to use synthetic division from Section 6.5.

E X A M P L E   **1**

Using the factor theorem

Use synthetic division to determine whether 2 is a zero of
$P(x) = x^3 - 3x^2 + 5x - 2$.

‹ Calculator Close-Up ›

You can perform the multiply-and-add steps for synthetic division with a graphing calculator as shown here.

```
Ans*2+ -3
                    1
                   -1
Ans*2+5
                    3
Ans*2+ -2
                    4
```

Solution

By the factor theorem, 2 is a zero of the function if and only if the remainder is zero when $P(x)$ is divided by $x - 2$. We can use synthetic division to determine the remainder. If we divide by $x - 2$, we use 2 on the left in synthetic division along with the coefficients $1, -3, 5, -2$ from the polynomial:

$$
\begin{array}{r|rrrr}
2 & 1 & -3 & 5 & -2 \\
  &   & 2 & -2 & 6 \\
\hline
  & 1 & -1 & 3 & 4
\end{array}
$$

Because the remainder is 4, 2 is not a zero of the function.

> Now do Exercises 1–12

E X A M P L E   **2**

Using the factor theorem

Use synthetic division to determine whether -4 is a solution to the equation $2x^4 - 28x^2 + 14x - 8 = 0$.

Solution

By the factor theorem, -4 is a solution to the equation if and only if the remainder is zero when $P(x)$ is divided by $x + 4$. When dividing by $x + 4$, we use -4 in the synthetic division:

$$
\begin{array}{r|rrrrr}
-4 & 2 & 0 & -28 & 14 & -8 \\
   &   & -8 & 32 & -16 & 8 \\
\hline
   & 2 & -8 & 4 & -2 & 0
\end{array}
$$

Because the remainder is zero, -4 is a solution to $2x^4 - 28x^2 + 14x - 8 = 0$.

> Now do Exercises 13–22

In Example 3, we use the factor theorem to determine whether a given binomial is a factor of a polynomial.

E X A M P L E   3

Using the factor theorem

Use synthetic division to determine whether $x + 4$ is a factor of $x^3 + 3x^2 + 16$.

Solution

According to the factor theorem, $x + 4$ is a factor of $x^3 + 3x^2 + 16$ if and only if the remainder is zero when the polynomial is divided by $x + 4$. Use synthetic division to determine the remainder:

$$
\begin{array}{r|rrrr}
-4 & 1 & 3 & 0 & 16 \\
   &   & -4 & 4 & -16 \\
\hline
   & 1 & -1 & 4 & 0
\end{array}
$$

Because the remainder is zero, $x + 4$ is a factor, and the polynomial can be written as

$$x^3 + 3x^2 + 16 = (x + 4)(x^2 - x + 4).$$

Because $x^2 - x + 4$ is a prime polynomial, the factoring is complete.

> **Now do Exercises 23–36**

⟨2⟩ Solving Polynomial Equations

The techniques used to solve polynomial equations of degree 3 or higher are not as straightforward as those used to solve linear equations and quadratic equations. Example 4 shows how the factor theorem can be used to solve a third-degree polynomial equation.

E X A M P L E   4

Solving a third-degree equation

Suppose the equation $x^3 - 4x^2 - 17x + 60 = 0$ is known to have a solution that is an integer between -3 and 3 inclusive. Find the solution set.

⟨ **Helpful Hint** ⟩

How did we know where to find a solution to the equation in Example 4? One way to get a good idea of where the solutions are is to graph

$$y = x^3 - 4x^2 - 17x + 60.$$

Every x-intercept on this graph corresponds to a solution to the equation.

Solution

Because one of the numbers $-3, -2, -1, 0, 1, 2,$ and 3 is a solution to the equation, we can use synthetic division with these numbers until we discover which one is a solution. We arbitrarily select 1 to try first:

$$
\begin{array}{r|rrrr}
1 & 1 & -4 & -17 & 60 \\
  &   & 1 & -3 & -20 \\
\hline
  & 1 & -3 & -20 & 40
\end{array}
$$

Because the remainder is 40, 1 is not a solution to the equation. Next try 2:

$$
\begin{array}{r|rrrr}
2 & 1 & -4 & -17 & 60 \\
  &   & 2 & -4 & -42 \\
\hline
  & 1 & -2 & -21 & 18
\end{array}
$$

Because the remainder is not zero, 2 is not a solution to the equation. Next try 3:

$$
\begin{array}{r|rrrr}
3 & 1 & -4 & -17 & 60 \\
  &   & 3 & -3 & -60 \\
\hline
  & 1 & -1 & -20 & 0
\end{array}
$$

The remainder is zero, so 3 is a solution to the equation, and $x - 3$ is a factor of the polynomial. (If 3 had not produced a remainder of zero, then we would have tried $-3, -2, -1,$ and 0.) The other factor is the quotient, $x^2 - x - 20$.

$$x^3 - 4x^2 - 17x + 60 = 0$$
$$(x - 3)(x^2 - x - 20) = 0 \quad \text{Use the results of synthetic division to factor.}$$
$$(x - 3)(x - 5)(x + 4) = 0 \quad \text{Factor completely.}$$
$$x - 3 = 0 \quad \text{or} \quad x - 5 = 0 \quad \text{or} \quad x + 4 = 0$$
$$x = 3 \quad \text{or} \quad x = 5 \quad \text{or} \quad x = -4$$

Check each of these solutions in the original equation. The solution set is $\{3, 5, -4\}$.

Now do Exercises 37–46

Warm-Ups ▼

Fill in the blank.

1. If $f(a) = 0$, then a is a(n) _____ or _____ of f.
2. If two statements are either both true or both false, then they are _____.
3. To divide a polynomial by $x - c$ quickly, we use _____ division.
4. If the _____ is zero when $P(x)$ is divided by $x - c$, then $P(c) = 0$.
5. If $x - c$ is a(n) _____ of the polynomial $P(x)$, then $P(c) = 0$.

True or false?

6. To divide $x^3 - 4x^2 - 3$ by $x - 5$, use 5 in the synthetic division.

7. To divide $x^4 - x - 1$ by $x + 7$, use -7 in the synthetic division.
8. If $P(x) = 3x^3 - 5x^2 - 2x + 2$, then $P(2) = 0$.
9. If $x^3 - 8$ is divided by $x - 2$, then the remainder is 0.
10. The binomial $x - 1$ is a factor of $x^{35} - 3x^{24} + 2x^{18}$.
11. The binomial $x + 1$ is a factor of $x^3 - 3x^2 + x + 5$.
12. If $x^3 - 5x + 4$ is divided by $x - 1$, then the remainder is 0.

Exercises 10.1

‹ **Study Tips** ›

• If you are having the kind of success in school that you desire, congratulations.
• If you are not having the success you want, do something about it. What you do now will affect you the rest of your life.

‹1› **The Factor Theorem**

Determine whether each given value of x is a zero of the given function. See Example 1.

1. $x = 2$, $P(x) = x - 2$
2. $x = 3$, $P(x) = x - 3$

3. $x = 7$, $P(x) = x + 7$
4. $x = -5$, $P(x) = x - 5$
5. $x = 4$, $P(x) = x^2 - x - 10$
6. $x = 5$, $P(x) = x^2 - 5x$
7. $x = 1$, $P(x) = x^3 - x^2 + x - 1$

8. $x = -2$, $P(x) = -2x^3 - 5x^2 + 3x + 10$

9. $x = -3$, $P(x) = -x^4 - 3x^3 - 2x^2 + 18$

10. $x = 4$, $P(x) = x^4 - x^2 - 8x - 16$

11. $x = 2$, $P(x) = 2x^3 - 4x^2 - 5x + 9$

12. $x = -3$, $P(x) = x^3 + 5x^2 + 2x + 1$

Use synthetic division to determine whether each given value of x is a solution to the given equation. See Example 2.

13. $x = 2$, $x^2 - 6x + 8 = 0$

14. $x = 3$, $3x^2 + x - 30 = 0$

15. $x = -3$, $x^3 + 5x^2 + 2x - 12 = 0$

16. $x = -5$, $x^2 - 3x - 40 = 0$

17. $x = -2$, $x^4 + 3x^3 - 5x^2 - 10x + 5 = 0$

18. $x = -3$, $-x^3 - 4x^2 + x + 12 = 0$

19. $x = 4$, $-2x^4 + 30x^2 + 5x + 12 = 0$

20. $x = 6$, $x^4 + x^3 - 40x^2 - 72 = 0$

21. $x = 3$, $0.8x^2 - 0.3x - 6.3 = 0$

22. $x = 5$, $6.2x^2 - 28.2x - 41.7 = 0$

Use synthetic division to determine whether the first polynomial is a factor of the second. If it is, then factor the polynomial completely. See Example 3.

23. $x - 1$, $x^2 - 5x + 4$

24. $x - 2$, $x^2 + 2x - 8$

25. $x + 3$, $-3x^2 - 5x + 18$

26. $x + 2$, $-2x^2 - x + 4$

27. $x - 3$, $x^3 - 6x - 9$

28. $x + 2$, $x^3 - 6x - 4$

29. $x + 5$, $x^3 + 9x^2 + 23x + 15$

30. $x - 3$, $x^4 - 9x^2 + x - 7$

31. $x - 2$, $x^3 - 8x^2 + 4x - 6$

32. $x + 5$, $x^3 + 125$

33. $x + 1$, $x^4 + x^3 - 8x - 8$

34. $x - 2$, $x^3 - 6x^2 + 12x - 8$

35. $x - 0.5$, $2x^3 - 3x^2 - 11x + 6$

36. $x - \dfrac{1}{3}$, $3x^3 - 10x^2 - 27x + 10$

⟨2⟩ Solving Polynomial Equations

Solve each equation, given that at least one of the solutions to each equation is an integer between −5 and 5. See Example 4.

37. $x^3 - 13x + 12 = 0$

38. $x^3 + 2x^2 - 5x - 6 = 0$

39. $2x^3 - 9x^2 + 7x + 6 = 0$

40. $6x^3 + 13x^2 - 4 = 0$

41. $2x^3 - 3x^2 - 50x - 24 = 0$

42. $x^3 - 7x^2 + 2x + 40 = 0$

43. $x^3 + 5x^2 + 3x - 9 = 0$

44. $x^3 + 6x^2 + 12x + 8 = 0$

45. $x^4 - 4x^3 + 3x^2 + 4x - 4 = 0$

46. $x^4 + x^3 - 7x^2 - x + 6 = 0$

Getting More Involved

47. *Exploration*

We can find the zeros of a polynomial function by solving a polynomial equation. We can also work backward to find a polynomial function that has given zeros.

a) Write a first-degree polynomial function whose zero is −2.

b) Write a second-degree polynomial function whose zeros are 5 and −5.

c) Write a third-degree polynomial function whose zeros are 1, −3, and 4.

d) Is there a polynomial function with any given number of zeros? What is its degree?

48. *Exploration*

For each equation find the value of k given that 3 satisfies the equation.

a) $x^4 - 3x^3 + 5x^2 - 7x + k = 0$

b) $x^4 - x^3 - 2x^2 + kx - k = 0$

c) $5x^3 - kx^2 - kx - 3k = 0$

Graphing Calculator Exercises

49. The x-coordinate of each x-intercept on the graph of a polynomial function is a zero of the polynomial function. Find the zeros of each function from its graph. Use synthetic division to check that the zeros found on your calculator really are zeros of the function.

a) $P(x) = x^3 - 2x^2 - 5x + 6$

b) $P(x) = 12x^3 - 20x^2 + x + 3$

50. With a graphing calculator an equation can be solved without the kind of hint that was given for Exercises 37–46. Solve each of the following equations by examining the graph of a corresponding function. Use synthetic division to check.

a) $x^3 - 4x^2 - 7x + 10 = 0$

b) $8x^3 - 20x^2 - 18x + 45 = 0$

10.2   Zeros of a Polynomial Function

In This Section

⟨1⟩ **The Remainder Theorem**

⟨2⟩ **The Fundamental Theorem of Algebra**

⟨3⟩ **The Rational Root Theorem**

In this section, we continue our study of zeros of polynomial functions, which we began in Section 10.1. An essential theorem in this regard is the remainder theorem, which was introduced in Section 6.5. For completeness we review and restate the remainder theorem here.

⟨1⟩   The Remainder Theorem

The factor theorem in Section 10.1 indicates what happens when a polynomial is divided by $x - c$ and the remainder is zero. The remainder theorem gives us more information about the remainder.

> **The Remainder Theorem**
>
> If R is the remainder when a polynomial $P(x)$ is divided by $x - c$, then $R = P(c)$.

E X A M P L E   1

Evaluating a polynomial function

Let $f(x) = x^3$ and $g(x) = x^3 - 3x^2 + 5x - 12$. Use synthetic division to find the following function values.

   a) $f(-2)$

   b) $g(4)$

⟨ Calculator Close-Up ⟩

With a calculator you can evaluate a function by using Y= to define

$$y_1 = x^3 - 3x^2 + 5x - 12.$$

```
Plot1 Plot2 Plot3
\Y1■X^3-3X²+5X-1
2
\Y2=
\Y3=
\Y4=
\Y5=
\Y6=
```

Then you can use the function variables to evaluate the function.

```
Y1(4)
              24
```

Solution

   a) Use synthetic division to divide x^3 by $x + 2$ as follows:

$$
\begin{array}{r|rrrr}
-2 & 1 & 0 & 0 & 0 \\
   &   & -2 & 4 & -8 \\
\hline
   & 1 & -2 & 4 & -8
\end{array}
$$

The synthetic division shows that the quotient is $x^2 - 2x + 4$ and the remainder is -8. So $f(-2) = -8$. The answer is correct because $f(-2) = (-2)^3 = -8$.

   b) Use synthetic division to divide $x^3 - 3x^2 + 5x - 12$ by $x - 4$ as follows:

$$
\begin{array}{r|rrrr}
4 & 1 & -3 & 5 & -12 \\
  &   & 4 & 4 & 36 \\
\hline
  & 1 & 1 & 9 & 24
\end{array}
$$

The synthetic division shows that the quotient is $x^2 + x + 9$ and the remainder is 24. So $g(4) = 24$. We can check this answer by evaluating the polynomial $x^3 - 3x^2 + 5x - 12$ for $x = 4$. We get $4^3 - 3(4^2) + 5(4) - 12 = 24$.

> **Now do Exercises 1–12**

‹ **Helpful Hint** ›

If we only wanted the value of the function, we probably would not use synthetic division. We use synthetic division because it gives us the value of the polynomial, $f(c)$, and the quotient of division by $x - c$.

Note that we now have two different ways to find the value of a polynomial $P(x)$ when $x = c$. We can find the remainder when the polynomial is divided by $x - c$, or we can evaluate the polynomial by using c in place of x. In Example 1(a) it is easier to find $(-2)^3$ than to do synthetic division to get $f(-2) = -8$. In Example 1(b) synthetic division is easier because there are fewer arithmetic operations in the synthetic division than in evaluating the polynomial. For polynomials of higher degree there may be an even bigger difference between the number of operations in evaluating the polynomial and finding the remainder of synthetic division.

In Example 2, we use synthetic division and the remainder theorem to determine whether a given number is a zero to a polynomial function.

E X A M P L E   2

Zeros of polynomial functions

Determine whether each given number is a zero of the polynomial function following the number.

   a) $2, f(x) = -6x^3 + 15x + 3$

   b) $-2, g(x) = x^4 - 3x^3 + 5x^2 - 6x - 72$

‹ **Calculator Close-Up** ›

You can use a calculator to perform the arithmetic in synthetic division as shown here.

```
              -6
Ans*2
              -12
Ans*2+15
              -9
Ans*2+3
              -15
```

Solution

a) Use synthetic division to find $f(2)$ as follows:

$$
\begin{array}{r|rrrr}
2 & -6 & 0 & 15 & 3 \\
  &   & -12 & -24 & -18 \\
\hline
  & -6 & -12 & -9 & -15
\end{array}
$$

So $f(2) = -15$, and 2 is not a zero of the polynomial function.

b) Use synthetic division to find $g(-2)$ as follows:

$$
\begin{array}{r|rrrrr}
-2 & 1 & -3 & 5 & -6 & -72 \\
   &   & -2 & 10 & -30 & 72 \\
\hline
   & 1 & -5 & 15 & -36 & 0
\end{array}
$$

So $g(-2) = 0$, and -2 is a zero of the polynomial function.

> Now do Exercises 13–24

CAUTION To evaluate a polynomial for $x = -2$, use -2 in synthetic division just as we use -2 in place of x in the polynomial. If -2 is a zero of the polynomial, then $x - (-2)$ or $x + 2$ is a factor of the polynomial. Note that if $x + 2 = 0$, then $x = -2$.

‹2› The Fundamental Theorem of Algebra

It is easy to determine whether a number is a zero of a polynomial function. But does every polynomial function have a zero? The answer comes from a theorem that is extremely important in algebra, the fundamental theorem of algebra.

> **The Fundamental Theorem of Algebra**
>
> If $P(x)$ is a polynomial function of positive degree, then $P(x)$ has at least one zero in the set of complex numbers.

Note that a polynomial function of zero degree might not have any zeros. For example, $P(x) = 7$ has no zeros. We already know how to find the zeros of many polynomial functions.

EXAMPLE   **3**

Finding all zeros of a polynomial function

Find all real and imaginary zeros of each polynomial function.

a) $f(x) = 3x - 5$

b) $g(x) = x^3 - 3x^2 + 4x - 12$

‹ **Calculator Close-Up** ›

The zeros of a function correspond to the x-intercepts on the graph of the function. With a graphing calculator you can check numerically and graphically that $\frac{5}{3}$ is a zero of $f(x) = 3x - 5$.

Solution

a) To find the zeros of the function, we solve the equation $3x - 5 = 0$:

$$3x - 5 = 0$$
$$3x = 5$$
$$x = \frac{5}{3}$$

The only zero of the function $f(x) = 3x - 5$ is $\frac{5}{3}$.

b) To find the zeros of the function, we solve $x^3 - 3x^2 + 4x - 12 = 0$:

$$x^3 - 3x^2 + 4x - 12 = 0$$
$$x^2(x - 3) + 4(x - 3) = 0$$
$$(x^2 + 4)(x - 3) = 0$$

$$x^2 + 4 = 0 \quad \text{or} \quad x - 3 = 0$$
$$x^2 = -4 \quad \text{or} \quad x = 3$$
$$x = \pm 2i \quad \text{or} \quad x = 3$$

The zeros of g are $-2i$, $2i$, and 3.

Now do Exercises 25–38

‹ **Calculator Close-Up** ›

The real zeros of a function correspond to the x-intercepts on the graph of the function. Because this function has only one real zero, its graph has only one x-intercept.

‹3› **The Rational Root Theorem**

We can find the zeros of any first-degree polynomial function because we can solve any linear equation. We can find the zeros of any second-degree polynomial function because we can solve any quadratic equation. Keep in mind that the zeros of a function $P(x)$ are exactly the same as the solutions to the equation $P(x) = 0$. In Example 2(b), we found the zeros of a third-degree polynomial function, but it is generally not so easy to factor a third- or higher-degree polynomial. The rational root theorem describes the rational zeros of a polynomial function.

The Rational Root Theorem

If $P(x) = a_n x^n + a_{n-1} x^{n-1} + a_{n-2} x^{n-2} + \cdots + a_1 x + a_0$ is a polynomial function with integral coefficients ($a_n \neq 0$ and $a_0 \neq 0$) and p/q (in its lowest terms) is a rational zero of $P(x)$, then p is a factor of the constant term a_0, and q is a factor of the leading coefficient a_n.

To understand the rational root theorem, consider the quadratic function $P(x) = 6x^2 - x - 12$. We find the zeros of the function as follows:

$$6x^2 - x - 12 = 0$$
$$(2x - 3)(3x + 4) = 0$$
$$2x - 3 = 0 \quad \text{or} \quad 3x + 4 = 0$$
$$2x = 3 \quad \text{or} \quad 3x = -4$$
$$x = \frac{3}{2} \quad \text{or} \quad x = -\frac{4}{3}$$

The solutions to the equation are $\frac{3}{2}$ and $-\frac{4}{3}$. For the zero $\frac{3}{2}$, notice that 3 is a factor of 12 and 2 is a factor of 6. For the zero $-\frac{4}{3}$, notice that 4 is a factor of 12 and 3 is a factor of 6. Of course, these observations are not surprising since we used these facts to factor the quadratic polynomial in the first place. Note that there are a lot of other factors of 12 and 6 for which the ratio is not a zero of this function because a quadratic function has at most two distinct zeros. The rational root theorem does not identify exactly which numbers are zeros of a function, it only gives the possibilities for the rational zeros.

E X A M P L E   **4**

Finding possible rational zeros
Find all possible rational zeros for each polynomial function.

a) $f(x) = 2x^3 - 3x^2 - 11x + 6$ **b)** $g(x) = 3x^3 - 8x^2 - 8x + 8$

⟨ **Helpful Hint** ⟩

Remember that the rational numbers listed here are only possibilities for the rational roots. There might not be any rational roots.

Solution

a) If the rational number p/q is a zero of $f(x)$, then p must be a factor of 6 and q must be a factor of 2. The factors of 6 are 1, 2, 3, and 6. The factors of 2 are 1 and 2. For simplicity we have listed only the positive factors. If we take each factor of 6 and divide by 1, we get $\frac{1}{1}, \frac{2}{1}, \frac{3}{1}$, and $\frac{6}{1}$. If we take each factor of 6 and divide by 2, we get $\frac{1}{2}, \frac{2}{2}, \frac{3}{2}$, and $\frac{6}{2}$. Simplify the ratios, eliminate duplications, and put in the negative factors to get

$$\pm 1, \quad \pm 2, \quad \pm 3, \quad \pm 6, \quad \pm\frac{1}{2}, \quad \text{and} \quad \pm\frac{3}{2}$$

as the possible rational zeros to the function $f(x)$.

b) If the rational number p/q is a zero of $g(x)$, then p must be a factor of 8 and q must be a factor of 3. The factors of 8 are 1, 2, 4, and 8. The factors of 3 are 1 and 3. If we take all possible ratios of a factor of 8 over a factor of 3, we get

$$\pm 1, \quad \pm 2, \quad \pm 4, \quad \pm 8, \quad \pm \frac{1}{3}, \quad \pm \frac{2}{3}, \quad \pm \frac{4}{3}, \quad \text{and} \quad \pm \frac{8}{3}$$

as the possible rational zeros of the function $g(x)$.

Now do Exercises 39–46

A polynomial function might or might not have any rational zeros, but with the rational root theorem we have a place to start in the problem of finding all of the zeros of a polynomial function. Synthetic division is used to determine whether a possible rational zero is actually a zero of the function.

EXAMPLE 5

Finding all zeros of a polynomial function

Find all of the real and imaginary zeros for each polynomial function of Example 4.

a) $f(x) = 2x^3 - 3x^2 - 11x + 6$

b) $g(x) = 3x^3 - 8x^2 - 8x + 8$

Solution

a) The possible rational zeros are listed in Example 4(a) for the function $f(x)$. Now use synthetic division to check each possible zero to see whether it is actually a zero. Try $\frac{1}{2}$ first. (If $\frac{1}{2}$ does not produce a remainder of 0, then we would try another number from the list of possible rational zeros.)

$$
\begin{array}{r|rrrr}
\frac{1}{2} & 2 & -3 & -11 & 6 \\
  &   & 1 & -1 & -6 \\
\hline
  & 2 & -2 & -12 & 0
\end{array}
$$

Since the remainder is 0, the remainder theorem indicates that $\frac{1}{2}$ is a zero of the function and a root of the equation $2x^3 - 3x^2 - 11x + 6 = 0$. By the factor theorem, $x - \frac{1}{2}$ is a factor of the polynomial, and the other factor is the quotient:

$$2x^3 - 3x^2 - 11x + 6 = 0$$

$$\left(x - \frac{1}{2}\right)\left(2x^2 - 2x - 12\right) = 0$$

$$(2x - 1)\left(x^2 - x - 6\right) = 0$$

$$(2x - 1)(x - 3)(x + 2) = 0$$

$$2x - 1 = 0 \quad \text{or} \quad x - 3 = 0 \quad \text{or} \quad x + 2 = 0$$

$$x = \frac{1}{2} \quad \text{or} \quad x = 3 \quad \text{or} \quad x = -2$$

The zeros of the function f are $\frac{1}{2}$, 3, and -2.

b) We listed the possible rational roots in Example 4(b) for the function $g(x)$. We will first check $\frac{2}{3}$ to see whether it produces a remainder of 0:

$$
\begin{array}{r|rrrr}
\frac{2}{3} & 3 & -8 & -8 & 8 \\
  &   & 2 & -4 & -8 \\
\hline
  & 3 & -6 & -12 & 0
\end{array}
$$

‹ Calculator Close-Up ›

The graph of a polynomial function can give us a good idea of which possible rational zeros are actually zeros of the function. The graph can also be used to support the answers that are obtained algebraically. In this case the graph has x-intercepts at $(-2, 0)$, $(0.5, 0)$, and $(3, 0)$, which supports the algebraic answers.

‹ **Calculator Close-Up** ›

Note how the graph of the function supports the solutions obtained algebraically. When the function is evaluated at $1 + \sqrt{5}$, or approximately 3.236068, the calculator does not get exactly 0 because it cannot perform this computation exactly. However, the answer 2×10^{-12} is very close to 0.

Since the remainder is 0, by the remainder theorem $\frac{2}{3}$ is a zero of the function and a root of the equation $3x^3 - 8x^2 - 8x + 8 = 0$. By the factor theorem we know that $x - \frac{2}{3}$ is a factor of the polynomial and the quotient is the other factor:

$$3x^3 - 8x^2 - 8x + 8 = 0$$

$$\left(x - \frac{2}{3}\right)(3x^2 - 6x - 12) = 0$$

$$(3x - 2)(x^2 - 2x - 4) = 0$$

$$3x - 2 = 0 \quad \text{or} \quad x^2 - 2x - 4 = 0$$

$$x = \frac{2}{3} \quad \text{or} \quad x = \frac{2 \pm \sqrt{20}}{2}$$

$$x = \frac{2}{3} \quad \text{or} \quad x = 1 \pm \sqrt{5}$$

There is one rational root and two irrational roots to the equation. So the zeros of the function $g(x)$ are $\frac{2}{3}$, $1 + \sqrt{5}$, and $1 - \sqrt{5}$.

> **Now do Exercises 47–56**

Note that in Example 5(a), all of the zeros were rational. We could have found all three by continuing to check the possible rational roots using synthetic division. In Example 5(b), we would be wasting time if we continued to check the possible rational roots because there is only one. When we get to the point in the equation at which one of the factors of the polynomial is a quadratic polynomial, it is best either to factor the quadratic polynomial or to use the quadratic formula to find the remaining solutions to the equation.

Warm-Ups ▼

Fill in the blank.

1. If $g(b) = 0$, then b is a(n) _____ or _____ of g.

2. A function defined by a polynomial is a(n) _____ function.

3. If a polynomial $P(x)$ is divided by $x - c$, then the _____ is $P(c)$.

4. If a rational number p/q (in lowest terms) is a zero of a polynomial, then p is a factor of the _____ term and q is a factor of the _____ coefficient.

5. If $P(x)$ is a polynomial function of positive degree, then $P(x)$ has at least one zero in the set of _____ numbers.

True or false?

6. If $y = x^3 - 4x^2 + 3x + 7$, then to find y when $x = 3$, use -3 in synthetic division.

7. If $P(x) = x^3 - 6x^2 + 3x - 7$ is divided by $x - 5$, then $P(5)$ is equal to the remainder.

8. If $x^4 - 1$ is divided by $x - 2$, then the remainder is 15.

9. The polynomial function $P(x) = 5$ has at least one zero.

10. If $P(x) = x^3 - 3x^2 + 5x - 14$, then $P(3) = 0$.

11. If $P(x) = 5x^3 - 5x^2 + 7x - 9$, then $\frac{1}{3}$ is not a zero of the polynomial.

‹1› **The Remainder Theorem**

Let $f(x) = x^4 - 1$, $g(x) = x^3 - 3x^2 + 5$, and $h(x) = 4x^4 - 3x^2 + 3x - 1$. Find the following function values by using two different methods. See Example 1.

1. $f(1)$ **2.** $f(-3)$

3. $f(-2)$ **4.** $f(5)$

5. $g(1)$ **6.** $g(-1)$

7. $g(-2)$ **8.** $g(2)$

9. $h\left(\dfrac{1}{2}\right)$ **10.** $h\left(-\dfrac{1}{2}\right)$

 11. $h\left(\dfrac{3}{2}\right)$ **12.** $h\left(-\dfrac{3}{2}\right)$

Determine whether each given number is a zero of the polynomial function following the number. See Example 2.

13. 1, $P(x) = x^2 - 2$

14. 5, $P(x) = x^2 - 3x$

15. -2, $P(x) = x^2 - x - 6$

16. -3, $P(x) = x^2 - 3x - 18$

17. 3, $f(x) = 3x^3 - 6x^2 - 4x - 15$

18. -2, $f(x) = 3x^3 - 6x^2 - 4x - 15$

19. -2, $P(x) = x^3 + 2x^2 + 3x + 2$

20. -1, $P(x) = x^3 + 2x^2 + 3x + 2$

21. -1, $g(x) = x^4 + 3x^3 + 5x^2 + 5x + 2$

22. 3, $g(x) = x^4 + 3x^3 + 5x^2 + 5x + 2$

23. $\dfrac{1}{2}$, $h(x) = x^3 + 3x^2 - 4x + 1$

24. $-\dfrac{1}{2}$, $h(x) = x^3 + 3x^2 - 4x + 1$

‹2› **The Fundamental Theorem of Algebra**

Find all real and imaginary zeros of each polynomial function. See Example 3.

25. $f(x) = 5x - 9$

26. $h(x) = 3x + 12$

27. $k(x) = x^2 - 5x + 6$

28. $m(x) = x^2 - x - 12$

29. $P(t) = t^2 + 9$

30. $W(t) = t^2 - 2t + 6$

31. $H(x) = x^3 - 3x^2 + 4x - 12$

32. $A(x) = x^3 + x^2 + x + 1$

33. $K(s) = 2s^3 - 5s^2 - 3s$

34. $J(s) = 2s^3 + s^2 - 3s$

35. $L(x) = -x^3 - x^2 + 7x + 7$

36. $Q(x) = -3x^3 + 3x^2 + 2x - 2$

37. $M(x) = x^3 - 1$

38. $N(x) = x^3 - 8$

‹3› **The Rational Root Theorem**

Find all possible rational zeros for each polynomial function. See Example 4.

39. $P(x) = x^2 - 5x + 2$

40. $P(x) = x^2 - 4x + 3$

41. $P(x) = 2x^3 - 5x^2 + x + 2$

42. $P(x) = x^3 - 2x^2 - 5x + 6$

43. $P(x) = x^3 + 9x^2 + 26x + 24$

44. $P(x) = 3x^3 + 4x^2 - 5x - 2$

45. $P(x) = x^3 + 5x^2 - 6$

46. $P(x) = 3x^3 + 16x^2 - 8$

Find all of the real and imaginary zeros for each polynomial function. See Example 5.

47. $P(x) = 2x^3 - 5x^2 + x + 2$

48. $P(x) = x^3 - 2x^2 - 5x + 6$

49. $P(x) = x^3 + 9x^2 + 26x + 24$

50. $P(x) = 3x^3 + 4x^2 - 5x - 2$

51. $P(x) = x^3 + 5x^2 - 6$

52. $P(x) = 3x^3 + 16x^2 - 8$

53. $P(x) = x^3 + 2x^2 - 16$

54. $P(x) = x^3 + 6x^2 + 13x + 10$

55. $P(x) = 24x^3 - 26x^2 + 9x - 1$

56. $P(x) = 12x^3 - 20x^2 + x + 3$

Applications

Solve each problem.

57. The total profit in dollars on the sale of x Electronic Tummy Trimmers is given by the polynomial function $P(x) = x^3 - 40x^2 + 400x$. Find the profit when 10 are sold. How many must be sold to get a profit of 0 dollars?

58. The velocity in feet per second (ft/sec) of a rocket t seconds (sec) after launching is given by the polynomial function $v(t) = t^3 - 20t^2 + 110t$. What is the velocity of the rocket 10 sec after launching? For what value of t does the rocket have 0 velocity?

Figure for Exercise 58

Graphing Calculator Exercises

Find all real zeros to each polynomial function by graphing the function and locating the x-intercepts.

59. $f(x) = x^3 - 0.2x^2 - 0.05x + 0.006$

60. $f(x) = x^3 - 13x^2 + 32x - 20$

61. $f(x) = x^3 - 60x^2 + 1100x - 6000$

62. $f(x) = x^3 - 3x^2 - 2200x + 50,000$

Math *at Work* | Cube Root Trick

Anyone can determine that $28^3 = 21{,}952$ or that $\sqrt[3]{21{,}952} = 28$ with a calculator. But wouldn't your friends be impressed if you could find the cube root of 21,952 *without* paper, pencil, or a calculator? With the cube root trick you can do it. This trick takes just a bit of work to master, but it is worth it if you like to show off.

First, memorize the first nine perfect cubes. You probably already know some of these:

$$1^3 = 1, 2^3 = 8, 3^3 = 27, 4^3 = 64, 5^3 = 125, 6^3 = 216, 7^3 = 343, 8^3 = 512, 9^3 = 729$$

Next memorize how the ones digits in the perfect cubes are paired with the original numbers. The ones digits for 1^3, 4^3, 5^3, 6^3, and 9^3 are 1, 4, 5, 6, and 9, respectively, which are the same as the number cubed. The ones digits for 2^3, 3^3, 7^3, and 8^3 are 8, 7, 3, and 2. They can be obtained by subtracting from 10. For example, the ones digit in 8^3 is $10 - 8$ or 2.

Now here is the trick. Ask a friend to secretly select a two digit number, use a calculator to find its cube, and report the result. Suppose the friend picks 73 and then tells you the cube, 389,017. You can instantly determine the original number (the cube root) as follows.

- To the left of the comma you see 389. Now from memory you know that the largest perfect cube less than 389 is 343 and the cube root of 343 is 7. So the tens digit is 7.
- The ones digit in 389,017 is 7, which is paired with $10 - 7$ or 3.

So the cube root of 389,017 is 73.

As another example, the friend tells you 636,056. The largest perfect cube less than 636 is 512 and the cube root of 512 is 8. Since 636,056 has ones digit 6, the cube root of 636,056 is 86.

10.3   The Theory of Equations

The zeros of a polynomial function $P(x)$ are the same as the roots of the polynomial equation $P(x) = 0$. Remember that one of our main goals in algebra is to keep expanding our knowledge of solving equations. In this section we will learn several facts that are useful in solving polynomial equations.

In This Section

⟨1⟩ **The Number of Roots to a Polynomial Equation**

⟨2⟩ **The Conjugate Pairs Theorem**

⟨3⟩ **Descartes' Rule of Signs**

⟨4⟩ **Bounds on the Roots**

⟨1⟩ The Number of Roots to a Polynomial Equation

In solving a polynomial equation by factoring, we find that a factor may occur more than once.

Multiplicity

If the factor $x - c$ occurs n times in the complete factorization of the polynomial $P(x)$, then we say that c is a root of the equation $P(x) = 0$ with **multiplicity** n.

For example, the equation $x^2 - 10x + 25 = 0$ is equivalent to $(x - 5)(x - 5) = 0$. The only root to this equation is 5. Since the factor $x - 5$ occurs twice, we say that 5 is a root with multiplicity 2. Counting multiplicity, every quadratic equation has two roots.

Consider a polynomial equation $P(x) = 0$ of positive degree n. By the fundamental theorem of algebra there is at least one complex root c_1 to this equation. By the factor theorem $P(x) = 0$ is equivalent to $(x - c_1)Q_1(x) = 0$, where $Q_1(x)$ is a polynomial with degree $n - 1$ [the quotient when $P(x)$ is divided by $x - c_1$]. By the fundamental theorem of algebra there is at least one complex root c_2 to $Q_1(x) = 0$. By the factor theorem $P(x) = 0$ can be written as $(x - c_1)(x - c_2)Q_2(x) = 0$, where $Q_2(x)$ is a polynomial with degree $n - 2$. Continuing this reasoning n times, we get a quotient polynomial that has 0 degree, n factors for $P(x)$, and n complex roots, not necessarily all different. We have just proved the following theorem.

n-Root Theorem

If $P(x) = 0$ is a polynomial equation with real or complex coefficients and positive degree n, then counting multiplicities, $P(x) = 0$ has n complex roots.

Note that the n-root theorem also means that a polynomial function of positive degree n has n zeros, counting multiplicities.

E X A M P L E **1**

Finding all roots to a polynomial equation

State the degree of each polynomial equation. Find all of the real and imaginary roots to each equation, stating multiplicity when it is greater than 1.

a) $6x^5 + 24x^3 = 0$ **b)** $(x - 3)^2(x + 4)^5 = 0$

Solution

a) The equation is a fifth-degree equation. We can solve it by factoring:

$$6x^3(x^2 + 4) = 0$$

$$6x^3 = 0 \quad \text{or} \quad x^2 + 4 = 0$$
$$x^3 = 0 \quad \text{or} \quad x^2 = -4$$
$$x = 0 \quad \text{or} \quad x = \pm 2i$$

The roots are $\pm 2i$ and 0. Since 0 is a root with multiplicity 3, counting multiplicities there are five roots.

b) If we would multiply the factors in this polynomial equation, then the highest power of x would be 7. So the degree of the equation is 7. There are only two distinct roots to the equation, 3 and -4. We say that 3 is a root with multiplicity 2 and -4 is a root with multiplicity 5. So counting multiplicities, there are seven roots to the equation.

> Now do Exercises 1–10

‹ **Calculator Close-Up** ›

You can check Example 1 by examining the graphs shown here.

The real roots to the equations correspond to the *x*-intercepts.

‹2› The Conjugate Pairs Theorem

The solutions to the quadratic equation $x^2 - 2x + 5 = 0$ are the complex numbers $1 - 2i$ and $1 + 2i$. These numbers are conjugates of one another. The quadratic formula guarantees that complex solutions of quadratic equations with real coefficients occur in conjugate pairs. This situation also occurs for polynomial equations of higher degree.

Conjugate Pairs Theorem

If $P(x) = 0$ is a polynomial equation with real coefficients and the complex number $a + bi$ $(b \neq 0)$ is a root, then the complex number $a - bi$ is also a root.

EXAMPLE **2**

Finding an equation with given roots

Find a polynomial equation with real coefficients that has 2 and $1 - i$ as roots.

Solution

Since the polynomial is to have real coefficients, the imaginary roots occur in conjugate pairs. So a polynomial with these two roots actually must have at least three roots: 2, $1 - i$, and $1 + i$. Since each root of the equation comes from a factor of the polynomial, we can write the following equation:

$$(x - 2)(x - [1 - i])(x - [1 + i]) = 0$$

$$(x - 2)(x^2 - 2x + 2) = 0 \quad \color{blue}{(1 - i)(1 + i) = 1 - i^2 = 2}$$

$$x^3 - 4x^2 + 6x - 4 = 0$$

This equation has the required solutions and the smallest degree. Any multiple of this equation would also have the required solutions but would not be as simple.

> Now do Exercises 11–28

‹3› **Descartes' Rule of Signs**

None of the theorems in this chapter tells us how to find all of the n roots to a polynomial equation of degree n. The theorems and rules presented here add to our knowledge of polynomial equations and help us to solve more equations. Descartes' rule of signs is a method for looking at a polynomial equation and estimating the number of positive, negative, and imaginary solutions.

When a polynomial is written in descending order, a variation of sign occurs when the signs of consecutive terms change. For example, if $P(x) = 3x^5 - 7x^4 - 8x^3 - x^2 + 3x - 9$, there are sign changes in going from the first to the second terms, from the fourth to the fifth terms, and from the fifth to the sixth terms. So there are three variations in sign for $P(x)$. Descartes' rule requires that we also count the variations in sign for $P(-x)$ after it is simplified:

$$P(-x) = 3(-x)^5 - 7(-x)^4 - 8(-x)^3 - (-x)^2 + 3(-x) - 9$$
$$= -3x^5 - 7x^4 + 8x^3 - x^2 - 3x - 9$$

In the polynomial $P(-x)$, the signs of the terms change from the second to the third terms, and then the signs change again from the third to the fourth terms. So there are two variations in sign for $P(-x)$.

Descartes' Rule of Signs

If $P(x) = 0$ is a polynomial equation with real coefficients, then the number of positive roots of the equation is either equal to the number of variations in sign of $P(x)$ or less than that by an even number. The number of negative roots of the equation is either equal to the number of variations in sign of $P(-x)$ or less than that by an even number.

EXAMPLE 3

Discussing the possibilities for the roots

Discuss the possibilities for the roots to $2x^3 - 5x^2 - 6x + 4 = 0$.

Solution

The number of variations of sign in $P(x) = 2x^3 - 5x^2 - 6x + 4$ is 2. According to Descartes' rule, the number of positive roots is either 2 or 0. Since $P(-x) = 2(-x)^3 - 5(-x)^2 - 6(-x) + 4 = -2x^3 - 5x^2 + 6x + 4$, there is only one variation of sign in $P(-x)$. So there is exactly one negative root. If only one negative root exists, then the other two roots must be positive or imaginary. The number of imaginary roots is determined by the number of positive and negative roots because the total number of roots must be three. The following table summarizes these two possibilities.

⟨ **Calculator Close-Up** ⟩

The following graph shows us that there are actually two positive roots, one negative root, and no imaginary roots to the equation in Example 3.

Number of Positive Roots	Number of Negative Roots	Number of Imaginary Roots
2	1	0
0	1	2

Now do Exercises 29–34

EXAMPLE 4

Discussing the possibilities for the roots

Discuss the possibilities for the roots to $3x^4 - 5x^3 - x^2 - 8x + 4 = 0$.

Solution

The number of variations of sign in $P(x) = 3x^4 - 5x^3 - x^2 - 8x + 4$ is 2. According to Descartes' rule, there are either two or no positive roots to the equation. Since $P(-x) = 3(-x)^4 - 5(-x)^3 - (-x)^2 - 8(-x) + 4 = 3x^4 + 5x^3 - x^2 + 8x + 4$, there are two variations of sign in $P(-x)$. So the number of negative roots is either 2 or 0. Each line of the following table gives a possible distribution of the type of roots to the equation.

⟨ **Calculator Close-Up** ⟩

From the graph it appears that there are actually two positive roots, no negative roots, and two imaginary roots to the equation in Example 4.

Number of Positive Roots	Number of Negative Roots	Number of Imaginary Roots
2	2	0
2	0	2
0	2	2
0	0	4

Now do Exercises 35–40

Descartes' rule of signs adds to our knowledge of the roots of an equation. It is especially helpful when the number of variations of sign is zero or one. If there are no variations in sign for $P(x)$, then there are no positive roots. If there is one variation of sign in $P(x)$, then we know that one positive root exists.

‹4› **Bounds on the Roots**

The next theorem on roots has to do with determining the size of the roots.

> **Theorem on Bounds**
>
> Suppose $P(x)$ is a polynomial with real coefficients and a positive leading coefficient, and synthetic division with c is performed.
>
> - If $c > 0$ and all terms in the bottom row are nonnegative, then no number greater than c can be a root of $P(x) = 0$.
> - If $c < 0$ and the terms in the bottom row alternate in sign, then no number less than c can be a root of $P(x) = 0$.

If there are no roots greater than c, then c is called an **upper bound** for the roots. If there are no roots less than c, then c is called a **lower bound** for the roots. If 0 appears in the bottom row of the synthetic division, we may consider it as a positive or negative term in determining whether the signs alternate.

EXAMPLE 5

Integral bounds for the roots

Use the theorem on bounds to establish the best integral bounds for the roots of $2x^3 - 5x^2 - 6x + 4 = 0$.

Solution

Try synthetic division with the integers 1, 2, 3, and so on. The first integer for which all terms on the bottom row are nonnegative is the best upper bound for the roots:

$$
\begin{array}{r|rrrr}
1 & 2 & -5 & -6 & 4 \\
  &   & 2 & -3 & -9 \\
\hline
  & 2 & -3 & -9 & -5
\end{array}
\qquad
\begin{array}{r|rrrr}
2 & 2 & -5 & -6 & 4 \\
  &   & 4 & -2 & -16 \\
\hline
  & 2 & -1 & -8 & -12
\end{array}
$$

$$
\begin{array}{r|rrrr}
3 & 2 & -5 & -6 & 4 \\
  &   & 6 & 3 & -9 \\
\hline
  & 2 & 1 & -3 & -5
\end{array}
\qquad
\begin{array}{r|rrrr}
4 & 2 & -5 & -6 & 4 \\
  &   & 8 & 12 & 24 \\
\hline
  & 2 & 3 & 6 & 28
\end{array}
$$

By the theorem on bounds no number greater than 4 can be a root to the equation. Now try synthetic division with the integers -1, -2, -3, and so on. The first negative integer for which the terms on the bottom row alternate in sign is the best lower bound for the roots:

$$
\begin{array}{r|rrrr}
-1 & 2 & -5 & -6 & 4 \\
   &   & -2 & 7 & -1 \\
\hline
   & 2 & -7 & 1 & 3
\end{array}
\qquad
\begin{array}{r|rrrr}
-2 & 2 & -5 & -6 & 4 \\
   &   & -4 & 18 & -24 \\
\hline
   & 2 & -9 & 12 & -20
\end{array}
$$

By the theorem on bounds no number less than -2 can be a root to the equation. So all of the real roots to this equation are between -2 and 4.

Now do Exercises 41–48

In Example 6, we will use all of the information available to find all of the solutions to a polynomial equation.

E X A M P L E   **6**

Finding all solutions to a polynomial equation
Find all of the solutions to $2x^3 - 5x^2 - 6x + 4 = 0$.

Solution

In Example 3, we saw that this equation has either two positive roots and one negative root or one negative root and two imaginary roots. In Example 5, we saw that all of the real roots to this equation are between -2 and 4. From the rational root theorem we have ± 1, ± 2, ± 4, and $\pm \frac{1}{2}$ as the possible rational roots. Since there must be one negative root and it must be greater than -2, the only possible numbers from the list are -1 and $-\frac{1}{2}$. So start by checking $-\frac{1}{2}$ and -1 with synthetic division:

‹ **Calculator Close-Up** ›

Because all x-intercepts are between -2 and 4, the graph supports the conclusion that all of the roots to the equation are between -2 and 4.

$$
\begin{array}{r|rrrr}
-\frac{1}{2} & 2 & -5 & -6 & 4 \\
 & & -1 & 3 & \frac{3}{2} \\
\hline
 & 2 & -6 & -3 & \frac{11}{2}
\end{array}
\qquad
\begin{array}{r|rrrr}
-1 & 2 & -5 & -6 & 4 \\
 & & -2 & 7 & -1 \\
\hline
 & 2 & -7 & 1 & 3
\end{array}
$$

Since neither -1 nor $-\frac{1}{2}$ is a root, the negative root must be irrational. Since there might be two positive roots smaller than 4, we check $\frac{1}{2}$, 1, and 2:

$$
\begin{array}{r|rrrr}
\frac{1}{2} & 2 & -5 & -6 & 4 \\
 & & 1 & -2 & -4 \\
\hline
 & 2 & -4 & -8 & 0
\end{array}
$$

Since $\frac{1}{2}$ is a root of the equation, $x - \frac{1}{2}$ is a factor of the polynomial:

$$\left(x - \frac{1}{2}\right)\left(2x^2 - 4x - 8\right) = 0$$

$$(2x - 1)\left(x^2 - 2x - 4\right) = 0$$

$$2x - 1 = 0 \quad \text{or} \quad x^2 - 2x - 4 = 0$$

$$x = \frac{1}{2} \quad \text{or} \quad x = \frac{2 \pm \sqrt{4 - 4(1)(-4)}}{2} = 1 \pm \sqrt{5}$$

There are two positive roots, $\frac{1}{2}$ and $1 + \sqrt{5}$. The negative root is $1 - \sqrt{5}$. Note that the roots guaranteed by Descartes' rule of signs are real numbers but not necessarily rational numbers.

> **Now do Exercises 49–60**

Warm-Ups ▼

Fill in the blank.

1. The _____ of a root c is the number of times that $x - c$ occurs as a factor.

2. A polynomial of degree n has __ roots counting multiplicities.

3. For a polynomial equation with real coefficients, the imaginary roots occur in _____ pairs.

4. For a polynomial equation $P(x) = 0$ the number of _____ roots is either equal to the number of variations in sign of $P(x)$ or less than that by an even number.

5. For a polynomial equation $P(x) = 0$ the number of _____ roots is either equal to the number of variations in sign of $P(-x)$ or less than that by an even number.

6. If there are no roots to a polynomial equation greater than c, then c is an _____ for the roots.

7. If there are no roots to a polynomial equation less than c, then c is a(n) _____ for the roots.

True or false?

8. The number 3 is a root to $x^2 - 9 = 0$ with multiplicity 2.

9. Counting multiplicities, $x^8 = 1$ has eight solutions in the set of complex numbers.

10. The number $\frac{2}{3}$ is a root with multiplicity 4 for $(3x - 2)^4(x^2 + 2x + 1) = 0$.

11. The number 2 is a root of multiplicity 3 for $(x - 2)^3(x^2 + x - 6) = 0$.

12. The complex numbers $2 - 3i$ and $-2 + 3i$ are conjugates.

13. If $1 + i$ is a root to $x^3 + x^2 - 4x + 6 = 0$, then $1 - i$ is also a root.

14. Both $-1 - i\sqrt{2}$ and $1 - i\sqrt{2}$ are solutions to $7x^3 - 5x^2 + 6x - 8 = 0$.

15. There are no positive solutions to $x^3 + 5x^2 + 6x + 1 = 0$.

16. There are two imaginary solutions to $x^3 - 5 = 0$.

Exercises 10.3

⟨ **Study Tips** ⟩

- Don't worry about the final exam. If you have been studying all semester, there is nothing to worry about.
- Most students perform about the same on the final as they have all semester. A few get all fired up and outperform, and a few get lazy and underperform.

⟨ 1 ⟩ **The Number of Roots to a Polynomial Equation**

State the degree of each polynomial equation. Find all of the real and imaginary roots to each equation. State the multiplicity of a root when it is greater than 1. See Example 1.

1. $x^5 - 4x^3 = 0$
2. $x^6 + 9x^4 = 0$
3. $x^4 + 2x^3 + x^2 = 0$
4. $x^5 - 4x^4 + 4x^3 = 0$

5. $x^4 - 6x^2 + 9 = 0$

6. $x^4 - 8x^2 + 16 = 0$

7. $(x - 1)^2(x + 2)^2 = 0$

8. $(2x + 1)^2(3x - 5)^4 = 0$

VIDEO 9. $x^4 - 2x^2 + 1 = 0$

10. $4x^4 - 4x^2 + 1 = 0$

⟨2⟩ The Conjugate Pairs Theorem

Find a polynomial equation with real coefficients that has the given roots. See Example 2.

11. $-3, 3$

12. $-4, 2$

13. $-2i, 2i$

14. $-4i, 4i$

15. $1, 2i$

16. $4, -3i$

17. $3, 2 - i$

18. $-4, 3 + i$

19. $-2, i$

20. $4, -i$

21. $0, i\sqrt{2}$

22. $-3, i\sqrt{3}$

23. $i, 1 - i$

24. $2i, -i$

25. $1, 2$

26. $\frac{1}{2}, -1$

27. $-1, 2, 3$

28. $-2, 3, 2$

⟨3⟩ Descartes' Rule of Signs

Discuss the possibilities for the roots to each equation. Do not solve the equation. See Examples 3 and 4.

29. $x^3 + 3x^2 + 5x + 7 = 0$

30. $2x^3 - 3x^2 + 5x - 6 = 0$

31. $-2x^3 - x^2 + 3x + 2 = 0$

32. $x^3 + x^2 - 5x - 1 = 0$

33. $x^4 - x^3 + x^2 - x + 1 = 0$

34. $x^4 - 1 = 0$

35. $x^4 + x^2 + 1 = 0$

36. $x^6 + 3x^4 + 2x^2 + 6 = 0$

37. $x^3 + x - 1 = 0$

38. $-x^4 + 3x^3 + 5x + 5 = 0$

39. $x^5 + x^3 + 3x = 0$

40. $x^3 - 5x^2 + 6x = 0$

⟨4⟩ Bounds on the Roots

Establish the best integral bounds for the roots of each equation according to the theorem on bounds. See Example 5.

41. $x^4 - 5x^2 + 7 = 0$

42. $2x^3 - x^2 - 7x + 7 = 0$

43. $2x^3 - 5x^2 + 9x - 18 = 0$

44. $x^2 - 7x - 16 = 0$

45. $x^2 + x - 13 = 0$

46. $x^3 - 15x + 25 = 0$

47. $2x^3 - 13x^2 + 25x - 14 = 0$

48. $x^3 - 6x^2 + 11x - 6 = 0$

Use the rational root theorem, Descartes' rule of signs, and the theorem on bounds as aids in finding all solutions to each equation. See Example 6.

49. $x^3 + x + 10 = 0$

50. $x^3 - 7x^2 + 17x - 15 = 0$

51. $2x^3 - 5x^2 - 6x + 4 = 0$

52. $3x^3 - 17x^2 + 12x + 6 = 0$

53. $4x^3 - 6x^2 - 2x + 1 = 0$

54. $x^3 + 5x^2 - 20x - 42 = 0$

55. $x^4 - 5x^3 + 5x^2 + 5x - 6 = 0$

56. $x^4 - 2x^3 + 5x^2 - 8x + 4 = 0$

57. $x^4 - 7x^3 + 17x^2 - 17x + 6 = 0$

58. $x^4 + 7x^3 + 17x^2 + 17x + 6 = 0$

59. $x^6 - x^5 + 2x^4 - 2x^3 - 15x^2 + 15x = 0$

60. $2x^6 + 4x^5 + x^4 + 2x^3 - x^2 - 2x = 0$

Applications

Solve each problem.

61. *Cylindrical tank.* Willard is designing a cylindrical tank with cone-shaped ends. The length of the cylinder is to be 20 feet (ft) larger than the radius of the cylinder, and the height of the cone is 2 ft. If the volume of the tank is 984π cubic feet (ft³), then what is the radius of the cylinder?

Figure for Exercise 61

62. *Storage tank.* Dr. Hu is designing a chemical storage tank in the shape of a cylinder with hemispherical ends. If the length of the cylinder is to be 20 ft larger than its radius and the volume is to be 3321π ft³, then what is the radius?

Figure for Exercise 62

63. *Frozen specimens.* A box of frozen specimens measures 4 inches by 5 inches by 3 inches. It is wrapped in an insulating material of uniform thickness for shipment. The volume of the box including the insulating material is 120 cubic inches (in.³). How thick is the insulation?

64. *Cereal box.* An independent marketing research agency has determined that the best box for breakfast cereal has a height that is 6 inches (in.) larger than its thickness and a width that is 5 in. larger than its thickness. If such a box is to have a volume of 112 in.³, then what should the thickness be?

Graphing Calculator Exercises

Find all real roots to each polynomial equation by graphing the corresponding function and locating the x-intercepts.

65. $x^4 - 12x^2 + 10 = 0$

66. $x^5 + x^4 - 7x^3 - 7x^2 + 12x + 12 = 0$

67. $x^6 - 9x^4 + 20x^2 - 12 = 0$

68. $4x^5 + 16x^4 - 5x^3 - 20x^2 + x + 8 = 0$

Mid-Chapter **Quiz** | Sections 10.1 through 10.3 | **Chapter 10**

Use synthetic division to determine whether the first polynomial is a factor of the second.

1. $x - 2$, $x^3 - 6x^2 + 7x + 2$

2. $x + 3$, $x^4 - 5x^2 + 14x + 4$

Solve each equation, given that one of the solutions is 3.

3. $x^3 - 11x^2 + 36x - 36 = 0$

4. $x^3 - 5x^2 + 3x + 9 = 0$

Find all real and imaginary zeros for each function.

5. $P(x) = x^3 - 27$

6. $P(x) = x^3 - 6x^2 + 11x - 6$

Find a polynomial equation with real coefficients that has the given roots.

7. $2, -3, 1$

8. $4, 2 + i$

Discuss the possibilities for the roots to each equation according to Descartes' rule of signs.

9. $x^3 + x^2 - 3x + 5 = 0$

10. $x^4 + x^3 + 9 = 0$

Miscellaneous.

11. Let $f(x) = x^4 + 5x^2 - 8x + 1$. Find $f(-2)$ using two different methods.

12. List the possible rational zeros for the function $f(x) = 6x^3 - 5x^2 + 7x - 1$ according to the rational root theorem.

10.4 Graphs of Polynomial Functions

In This Section

⟨1⟩ **Symmetry**

⟨2⟩ **Behavior at the x-Intercepts**

⟨3⟩ **Sketching Graphs of Polynomial Functions**

⟨4⟩ **Solving Polynomial Inequalities**

The graph of a polynomial function of degree 0 or 1 is a straight line, and the graph of a second-degree polynomial function is a parabola. In this section, we will concentrate on graphs of polynomial functions of degree larger than 2.

⟨1⟩ Symmetry

Consider the graph of the quadratic function $f(x) = x^2$ shown in Fig. 10.1. Notice that both $(2, 4)$ and $(-2, 4)$ are on the graph. In fact, $f(x) = f(-x)$ for any value of x. We get the same y-coordinate whether we evaluate the function at a number or its opposite. This fact causes the graph to be symmetric about the y-axis. If we folded the paper along the y-axis, the two halves of the graph would coincide.

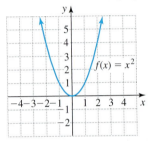

Figure 10.1

Symmetric about the y-Axis

If $f(x)$ is a function such that $f(x) = f(-x)$ for any value of x in its domain, then the graph of the function is said to be **symmetric about the y-axis.**

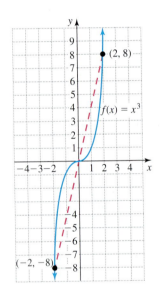

Figure 10.2

Consider the graph of $f(x) = x^3$ shown in Fig. 10.2. It is not symmetric about the y-axis like the graph of $f(x) = x^2$, but it has a different kind of symmetry. On the graph of $f(x) = x^3$ we find the points $(2, 8)$ and $(-2, -8)$. In this case $f(x)$ and $f(-x)$ are not equal, but $f(-x) = -f(x)$. Notice that the points $(2, 8)$ and $(-2, -8)$ are the same distance from the origin and lie on a line through the origin.

Symmetric about the Origin

If $f(x)$ is a function such that $f(-x) = -f(x)$ for any value of x in its domain, then the graph of the function is said to be **symmetric about the origin.**

E X A M P L E   **1**

Determining the symmetry of a graph

Discuss the symmetry of the graph of each polynomial function.

a) $f(x) = 5x^3 - x$ **b)** $f(x) = 2x^4 - 3x^2$ **c)** $f(x) = x^2 - 3x + 6$

Solution

a) Since $f(-x) = 5(-x)^3 - (-x) = -5x^3 + x$, we have $f(-x) = -f(x)$. So the graph is symmetric about the origin.

b) Since $f(-x) = 2(-x)^4 - 3(-x)^2 = 2x^4 - 3x^2$, we have $f(x) = f(-x)$. So the graph is symmetric about the y-axis.

c) In this case $f(-x) = (-x)^2 - 3(-x) + 6 = x^2 + 3x + 6$. So $f(-x) \neq f(x)$ and $f(-x) \neq -f(x)$. This graph has neither type of symmetry.

> Now do Exercises 1–18

‹ **Calculator Close-Up** ›

We can use graphs to check the conclusions about symmetry that were arrived at algebraically in Example 1. The graph of $f(x) = 5x^3 - x$ appears to be symmetric about the origin.

The graph of $f(x) = 2x^4 - 3x^2$ appears to be symmetric about the y-axis.

The graph of $f(x) = x^2 - 3x + 6$ does not appear to have either type of symmetry.

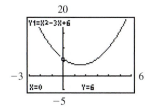

‹2› **Behavior at the x-Intercepts**

Each of the graphs of $f(x) = x^2$ and $f(x) = x^3$ has one x-intercept $(0, 0)$. If we use a positive or negative number very close to 0 for x in $f(x) = x^2$, the y-coordinate is positive because the power on x is even. That is why the graph just touches the x-axis at $(0, 0)$ but does not cross the x-axis there. If we use a positive number very close to 0 for x in $f(x) = x^3$, we get a positive y-coordinate. If we use a negative number very close to 0 for x in $f(x) = x^3$, we get a negative y-coordinate because the power on x is odd. That is why the graph crosses the x-axis at $(0, 0)$. In general, we have the following theorem.

Behavior at the x-Intercepts

Suppose $x - c$ is a factor of a polynomial function. The graph of the function crosses the x-axis at $(c, 0)$ if $x - c$ occurs an odd number of times and does not cross the x-axis at $(c, 0)$ if $x - c$ occurs an even number of times.

E X A M P L E   2

Behavior at the x-intercepts

Find the x-intercepts and discuss the behavior of the graph of each polynomial function at its x-intercepts.

a) $f(x) = (x - 1)^2(x - 3)$ b) $f(x) = x^3 + 2x^2 - 3x$

Solution

a) The x-intercepts are found by solving $(x - 1)^2(x - 3) = 0$. They are $(1, 0)$ and $(3, 0)$. The graph does not cross the x-axis at $(1, 0)$ but does cross the x-axis at $(3, 0)$.

b) The x-intercepts are found by solving $x^3 + 2x^2 - 3x = 0$. By factoring, we get $x(x + 3)(x - 1) = 0$. The x-intercepts are $(0, 0)$, $(-3, 0)$, and $(1, 0)$. Since each factor occurs an odd number of times, the graph crosses the x-axis at each of the x-intercepts.

> Now do Exercises 19–30

‹ **Calculator Close-Up** ›

The graphs of the functions in Example 2 support the conclusions about the behavior at the x-intercepts.

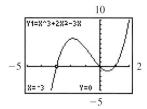

‹ 3 › Sketching Graphs of Polynomial Functions

We can use the ideas of symmetry and behavior at the x-intercepts to sketch a graph of a polynomial function. You can use the following strategy to assist you in graphing polynomial functions.

Strategy for Graphing Polynomial Functions

1. Check for symmetry. If $f(-x) = f(x)$, then there is y-axis symmetry. If $f(-x) = -f(x)$, the graph is symmetric about the origin.
2. Find the y-intercept by evaluating $f(0)$.
3. Find the x-intercepts by solving $f(x) = 0$.
4. Factor the polynomial completely.
5. If the x-intercept corresponds to a factor with an odd power, the graph crosses the x-axis. If the power is even, the curve touches but does not cross the x-axis.
6. Plot the intercepts and a few more points, and then sketch the curve through them.

EXAMPLE 3

Graphing a polynomial function

Sketch the graph of each polynomial function.

a) $f(x) = x^3 - 5x^2 + 7x - 3$ **b)** $f(x) = x^4 - 2x^2 + 1$

‹ **Calculator Close-Up** ›

This calculator graph supports the conclusions made in Example 3(a).

‹ **Calculator Close-Up** ›

This calculator graph supports the conclusions made in Example 3(b).

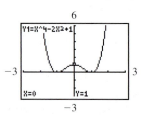

Solution

a) Since $f(-x) = -x^3 - 5x^2 - 7x - 3$, the graph has neither type of symmetry. By Descartes' rule of signs there are no negative roots to the equation $x^3 - 5x^2 + 7x - 3 = 0$, and the number of positive roots is either 3 or 1. So the only possible rational roots are 1 and 3:

$$
1 \, \underline{\left|\begin{array}{rrrr} 1 & -5 & 7 & -3 \\ & 1 & -4 & 3 \\ \hline 1 & -4 & 3 & 0 \end{array}\right.}
$$

From the synthetic division we get $f(x) = (x - 1)(x^2 - 4x + 3)$, and if we factor again, we get $f(x) = (x - 1)^2(x - 3)$. The x-intercepts are $(1, 0)$ and $(3, 0)$. The discussion of Example 2(a) applies to this function. The points $(0, -3)$, $(2, -1)$, and $(4, 9)$ are also on the graph. The graph is shown in Fig. 10.3.

b) Since $f(-x) = (-x)^4 - 2(-x)^2 + 1 = x^4 - 2x^2 + 1$, we have $f(x) = f(-x)$. So the graph is symmetric about the y-axis. We can factor the polynomial as follows:

$$
\begin{aligned}
f(x) &= x^4 - 2x^2 + 1 \\
&= (x^2 - 1)(x^2 - 1) \\
&= (x - 1)(x + 1)(x - 1)(x + 1) \\
&= (x - 1)^2(x + 1)^2
\end{aligned}
$$

The x-intercepts are $(1, 0)$ and $(-1, 0)$. Since each factor for these intercepts has an even power, the graph does not cross the x-axis at the intercepts. The points $(0, 1)$, $(2, 9)$, and $(-2, 9)$ are also on the graph shown in Fig. 10.4.

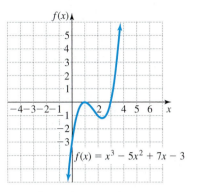

Figure 10.3   **Figure 10.4**

> Now do Exercises 39–50

⟨4⟩ Solving Polynomial Inequalities

An inequality such as $x^3 - 3x > 0$ is a **polynomial inequality.** The graphical method that we used for quadratic inequalities in Section 8.5 can be used also with polynomial inequalities. We can read the solution to the inequality from the graph of $y = x^3 - 3x$ provided we know all of the x-intercepts. Any value of x for which $y > 0$ on the graph is a solution to the inequality.

E X A M P L E   4

Solving a polynomial inequality with the graphical method

Solve each polynomial inequality. Write the solution set in interval notation, and graph it.

a) $x^3 - 3x > 0$ **b)** $(x - 1)(x + 2)(x - 3) \le 0$

Solution

a) To solve $x^3 - 3x > 0$ graph $y = x^3 - 3x$. We can determine the solution set to the inequality from the graph if we know the x-intercepts. So first find them by solving $x^3 - 3x = 0$:

$$x^3 - 3x = 0$$
$$x(x^2 - 3) = 0$$

$$x = 0 \quad \text{or} \quad x^2 - 3 = 0$$
$$x = 0 \quad \text{or} \quad x^2 = 3$$
$$x = 0 \quad \text{or} \quad x = \pm\sqrt{3}$$

Figure 10.5

Figure 10.6

The x-intercepts are $\left(-\sqrt{3}, 0\right)$, $(0, 0)$, and $\left(\sqrt{3}, 0\right)$. The graph in Fig. 10.5 crosses the x-axis at each intercept. The inequality is satisfied for any x that corresponds to a positive y-coordinate on this graph. So the solution set to the inequality is $\left(-\sqrt{3}, 0\right) \cup \left(\sqrt{3}, \infty\right)$ and its graph is shown in Fig. 10.6.

b) To solve $(x - 1)(x + 2)(x - 3) \le 0$, graph $y = (x - 1)(x + 2)(x - 3)$. To find the x-intercepts we solve $(x - 1)(x + 2)(x - 3) = 0$. The x-intercepts are $(1, 0)$, $(-2, 0)$, and $(3, 0)$. The graph in Fig. 10.7 on the next page crosses the x-axis at each intercept.

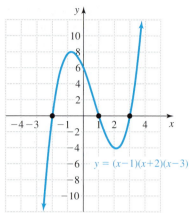

Figure 10.7

The inequality is satisfied for any x that corresponds to a negative or zero y-coordinate on this graph. So the solution set is $(-\infty, -2] \cup [1, 3]$ and its graph is shown in Fig. 10.8.

Figure 10.8

Now do Exercises 57–66

The **test point method** that we used for quadratic inequalities in Section 8.5 can be used also with polynomial inequalities. For this method we first find all of the roots to the polynomial, as in the graphical method. Instead of graphing the polynomial function, we plot the roots on a number line and then test a point in each interval determined by the roots.

E X A M P L E   5

Solving a polynomial inequality with the test point method
Solve each polynomial inequality. Write the solution set in interval notation, and graph it.

a) $x^4 - 16 \leq 0$ **b)** $x^4 - 3x^3 - 18x^2 \geq 0$

Solution

a) Find the roots to $x^4 - 16 = 0$:

$$x^4 - 16 = 0$$
$$(x^2 - 4)(x^2 + 4) = 0$$
$$x^2 - 4 = 0 \quad \text{or} \quad x^2 + 4 = 0$$
$$x^2 = 4 \qquad\qquad x^2 = -4$$
$$x = \pm 2$$

The only real solutions to the equation are -2 and 2. Locate these numbers on a number line as in Fig. 10.9.

Select the test points -3, 0, and 3. Test them in the original inequality $x^4 - 16 \leq 0$.

$$(-3)^4 - 16 \leq 0 \qquad \text{Incorrect}$$
$$(0)^4 - 16 \leq 0 \qquad \text{Correct}$$
$$(3)^4 - 16 \leq 0 \qquad \text{Incorrect}$$

Figure 10.9

Figure 10.10

Since 0 is the only test point that satisfies the inequality, the interval containing 0 is the solution set to the equation. Because of the $\leq$ symbol we include the endpoints. The solution set is $[-2, 2]$, and its graph is shown in Fig. 10.10.

b) Find the roots to $x^4 - 3x^3 - 18x^2 = 0$:

$$x^2(x^2 - 3x - 18) = 0$$

$$x^2(x - 6)(x + 3) = 0$$

$$x^2 = 0 \quad \text{or} \quad x - 6 = 0 \quad \text{or} \quad x + 3 = 0$$

$$x = 0 \quad \text{or} \quad x = 6 \quad \text{or} \quad x = -3$$

Now locate -3, 0, and 6 on a number line as in Fig. 10.11.

Figure 10.11

Select the test points -5, -1, 2, and 7. Use a calculator to test them in the original inequality $x^4 - 3x^3 - 18x^2 \geq 0$:

$$(-5)^4 - 3\,(-5)^3 - 18(-5)^2 \leq 0 \qquad \textcolor{blue}{\text{Incorrect}}$$

$$(-1)^4 - 3(-1)^3 - 18(-1)^2 \leq 0 \qquad \textcolor{blue}{\text{Correct}}$$

$$(2)^4 - 3(2)^3 - 18(2)^2 \leq 0 \qquad \textcolor{blue}{\text{Correct}}$$

$$(7)^4 - 3(7)^3 - 18(7)^2 \leq 0 \qquad \textcolor{blue}{\text{Incorrect}}$$

The inequality is satisfied on the intervals containing -1 and 2, which are $[-3, 0]$ and $[0, 6]$. Since the symbol is $\leq$, the endpoints of the intervals are included. The solution set is $[-3, 0] \cup [0, 6]$, which is simplified to $[-3, 6]$. The graph of the solution set is shown in Fig. 10.12.

Figure 10.12

> Now do Exercises 67–74

Warm-Ups ▼

Fill in the blank.

1. The graph of $y = f(x)$ is symmetric about the _____ if $f(x) = f(-x)$ for all x in the domain of f.

2. The graph of $y = f(x)$ is symmetric about the _____ if $f(-x) = -f(x)$ for all x in the domain of f.

3. The graph of a polynomial function $P(x)$ crosses the x-axis at c if $x - c$ occurs a(n) ____ number of times in the prime factorization of P.

4. The graph of a polynomial function $P(x)$ touches but does not cross the x-axis at c if $x - c$ occurs a(n) _____ number of times in the prime factorization of P.

True or false?

5. The graph of $f(x) = x^3 - x$ is symmetric about the y-axis.

6. The graph of $y = 2x - 1$ is symmetric about the origin.

7. If $f(x) = 3x$, then $f(x) = f(-x)$ for any real number x.

8. If $f(x) = 3x^4 - 5x^3 + 2x^2 - 6x + 7$, then $f(-x) = 3x^4 + 5x^3 + 2x^2 + 6x + 7$.

9. There is only one x-intercept for the graph of $f(x) = x^2 - 4x + 4$.

10. The graph of $y = (x - 1)^2(x + 4)^4$ does not cross the x-axis at either of its intercepts.

Exercises

‹ **Study Tips** ›

• If you haven't been working with a group all semester, form a study group for the final exam.
• You might even ask your instructor to meet with your study group.

‹1› Symmetry

Discuss the symmetry of the graph of each polynomial function. See Example 1.

1. $f(x) = 2x$

2. $f(x) = -x$

3. $f(x) = -2x^2$

4. $f(x) = x^2 + 1$

5. $f(x) = 2x^3$

6. $f(x) = -x^3$

7. $f(x) = x^4$

8. $f(x) = x^4 - x$

9. $f(x) = x^3 - 5x + 1$

10. $f(x) = 5x^3 + 7x$

11. $f(x) = 6x^6 - 3x^2 - x$

12. $f(x) = x^6 - x^4 + x^2 - 8$

13. $f(x) = (x - 3)^2$

14. $f(x) = 3(x + 2)^2$

15. $f(x) = (x^2 - 5)^3$

16. $f(x) = (x^2 + 1)^2$

17. $f(x) = x$

18. $f(x) = -3x$

‹2› Behavior at the *x*-Intercepts

Find the x-intercepts, and discuss the behavior of the graph of each polynomial function at its x-intercepts. See Example 2.

19. $f(x) = (x - 5)^2$

20. $f(x) = 3x - 2$

21. $f(x) = x^2 - 3x - 4$

22. $f(x) = 9x^2 - 12x + 4$

23. $f(x) = x^4$

24. $f(x) = x^4 - 1$

25. $f(x) = (x - 3)^2(x + 5)$

26. $f(x) = (2x + 1)^3$

27. $f(x) = x^3 - x^2$

28. $f(x) = x^3 - x^2 - x + 1$

29. $f(x) = 2x^3 + 3x^2 - 1$

30. $f(x) = x^3 + 3x^2 - 4$

Match each polynomial function with its graph a–h.

31. $f(x) = -2x + 3$

32. $f(x) = -2x^2 + 3$

33. $f(x) = -2x^3 + 3$

34. $f(x) = -2x^2 + 4x + 3$

35. $f(x) = -x^4 + 3$

36. $f(x) = x^3 - 3x^2$

37. $f(x) = x^3 + 3x^2 - x - 3$

38. $f(x) = \frac{1}{2}x^4 - 3$

a)

b)

c)

d)

e)

f)

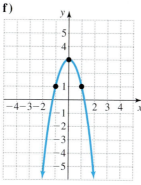

43. $f(x) = x^3 - 2x^2$

44. $f(x) = x^3 - 4x$

45. $f(x) = (x - 1)^2(x + 1)^2$

46. $f(x) = (x + 2)^2(x - 1)$

g)

h)

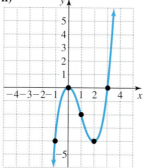

47. $f(x) = x^3 - 5x^2 + 7x - 3$ **48.** $f(x) = x^3 + 2x^2 - 3x$

⟨3⟩ **Sketching Graphs of Polynomial Functions**

Sketch the graph of each polynomial function.
See Example 3.

See the Strategy for Graphing Polynomial Functions box on page 676.

39. $f(x) = 2x - 6$

40. $f(x) = -3x + 3$

49. $f(x) = x^4 - 4x^3 + 4x^2$ **50.** $f(x) = -x^4 + 6x^3 - 9x^2$

Graphing Calculator Exercises

Sketch the graph of each polynomial function. First graph the function on a calculator and use the calculator graph as a guide.

51. $f(x) = x - 20$

52. $f(x) = (x - 20)^2$

41. $f(x) = -x^2$

42. $f(x) = x^2 - 3$

53. $f(x) = (x - 20)^2(x + 30)$ **54.** $f(x) = (x - 20)^2(x + 30)^2$

64. $-x^4 + 6x^3 - 9x^2 \geq 0$

65. $(x - 1)^2(x + 1)^2 \leq 0$

66. $(x + 2)^2(x - 1) \leq 0$

55. $f(x) = (x - 20)^2(x + 30)^2x$

Solve each polynomial inequality using the test point method. State the solution set using interval notation, and graph it. See Example 5.

67. $x^4 - 81 > 0$

68. $x^4 - 1 < 0$

69. $x^4 - x^3 - 6x^2 \leq 0$

56. $f(x) = (x - 20)^2(x + 30)^2x^2$

70. $x^4 + 2x^3 - 8x^2 \geq 0$

71. $x^3 + 6x^2 - 4x - 24 > 0$

72. $x^3 + 5x^2 - 9x - 45 < 0$

⟨4⟩ Solving Polynomial Inequalities

73. $x^4 - 10x^2 + 9 \leq 0$

Solve each polynomial inequality using the graphical method. State the solution set using interval notation and graph it. See Example 4.

74. $x^4 - 18x^2 + 32 \geq 0$

57. $x^3 - 4x > 0$

58. $x^3 - 16x < 0$

Getting More Involved

59. $(x - 3)(x - 5)(x + 2) \leq 0$

In each case, find a polynomial function whose graph behaves in the required manner. Answers may vary.

75. The graph has only one x-intercept at $(3, 0)$ and crosses the x-axis there.

60. $(x + 4)(x + 1)(x - 6) \geq 0$

76. The graph has only one x-intercept at $(3, 0)$ but does not cross the x-axis there.

77. The graph has only two x-intercepts at $(-2, 0)$ and $(1, 0)$. It crosses the x-axis at $(-2, 0)$ but does not cross at $(1, 0)$.

61. $x^3 - 2x^2 \geq 0$

62. $-x^3 + 5x^2 \leq 0$

78. The graph has only two x-intercepts at $(5, 0)$ and $(-6, 0)$. It does not cross the x-axis at either x-intercept.

63. $x^4 - 4x^3 + 4x^2 \geq 0$

10.5 Graphs of Rational Functions

We first studied rational expressions in Chapter 6. In this section, we will study functions that are defined by rational expressions.

In This Section

⟨1⟩ **Rational Functions**

⟨2⟩ **Asymptotes**

⟨3⟩ **Sketching Graphs of Rational Functions**

⟨4⟩ **Rational Inequalities**

⟨1⟩ Rational Functions

A rational expression was defined in Chapter 6 as a ratio of two polynomials. If a ratio of two polynomials is used to define a function, then the function is called a rational function.

> **Rational Function**
>
> If $P(x)$ and $Q(x)$ are polynomials with no common factor and $f(x) = \frac{P(x)}{Q(x)}$ for $Q(x) \neq 0$, then $f(x)$ is called a **rational function.**

The domain of a rational function is the set of all real numbers except those that cause the denominator to have a value of 0.

E X A M P L E 1

Domain of a rational function

Find the domain of each rational function.

a) $f(x) = \dfrac{x - 3}{x - 1}$

b) $g(x) = \dfrac{2x - 3}{x^2 - 4}$

Solution

a) Since $x - 1 = 0$ only for $x = 1$, the domain of f is the set of all real numbers except 1, $(-\infty, 1) \cup (1, \infty)$.

b) Since $x^2 - 4 = 0$ for $x = \pm 2$, the domain of g is the set of all real numbers excluding 2 and -2, $(-\infty, -2) \cup (-2, 2) \cup (2, \infty)$.

> Now do Exercises 1–6

⟨2⟩ Asymptotes

⟨ **Calculator Close-Up** ⟩

If the viewing window is too large, a rational function will appear to touch its asymptotes.

Because the asymptotes are an important feature of a rational function, we should draw it so that it approaches but does not touch its asymptotes.

Consider the simplest rational function $f(x) = 1/x$. Its domain does not include 0, but 0 is an important number for the graph of this function. The behavior of the graph of f when x is very close to 0 is what interests us. For this function the y-coordinate is the reciprocal of the x-coordinate. When the x-coordinate is close to 0, the y-coordinate is far from 0. Consider the following tables of ordered pairs that satisfy $f(x) = 1/x$:

\hspace{2em} $x > 0$			\hspace{2em} $x < 0$	
x	y		x	y
0.1	10		-0.1	-10
0.01	100		-0.01	-100
0.001	1000		-0.001	-1000
0.0001	10,000		-0.0001	$-10,000$

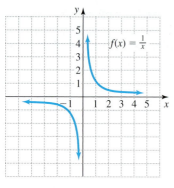

Figure 10.13

As x gets closer and closer to 0 from above 0, the value of y gets larger and larger. We say that y goes to positive infinity. As x gets closer and closer to 0 from below 0, the values of y are negative but $|y|$ gets larger and larger. We say that y goes to negative infinity. The graph of f gets closer and closer to the vertical line $x = 0$, and so $x = 0$ is called a **vertical asymptote.** On the other hand, as $|x|$ gets larger and larger, y gets closer and closer to 0. The graph approaches the x-axis as x goes to infinity, and so the x-axis is a **horizontal asymptote** for the graph of f. See Fig. 10.13 for the graph of $f(x) = 1/x$.

In general, a rational function has a vertical asymptote for every number excluded from the domain of the function. The horizontal asymptotes are determined by the behavior of the function when $|x|$ is large.

E X A M P L E   **2**

Horizontal and vertical asymptotes

Find the horizontal and vertical asymptotes for each rational function.

a) $f(x) = \dfrac{3}{x^2 - 1}$ **b)** $g(x) = \dfrac{x}{x^2 - 4}$ **c)** $h(x) = \dfrac{2x + 1}{x + 3}$

Solution

a) The denominator $x^2 - 1$ has a value of 0 if $x = \pm 1$. So the lines $x = 1$ and $x = -1$ are vertical asymptotes. If $|x|$ is very large, the value of $\dfrac{3}{x^2 - 1}$ is approximately 0. So the x-axis is a horizontal asymptote.

b) The denominator $x^2 - 4$ has a value of 0 if $x = \pm 2$. So the lines $x = 2$ and $x = -2$ are vertical asymptotes. If $|x|$ is very large, the value of $\dfrac{x}{x^2 - 4}$ is approximately 0. So the x-axis is a horizontal asymptote.

c) The denominator $x + 3$ has a value of 0 if $x = -3$. So the line $x = -3$ is a vertical asymptote. If $|x|$ is very large, the value of $h(x)$ is not approximately 0. To understand the value of $h(x)$, we change the form of the rational expression by using long division:

$$\begin{array}{r} 2 \\ x + 3 \overline{)2x + 1} \\ \underline{2x + 6} \\ -5 \end{array}$$

Writing the rational expression as quotient $+ \dfrac{\text{remainder}}{\text{divisor}}$, we get $h(x) = \dfrac{2x + 1}{x + 3} = 2 + \dfrac{-5}{x + 3}$. If $|x|$ is very large, $\dfrac{-5}{x + 3}$ is approximately 0, and so the y-coordinate is approximately 2. The line $y = 2$ is a horizontal asymptote.

> **Now do Exercises 7–12**

Example 2 illustrates two important facts about horizontal asymptotes. If the degree of the numerator is less than the degree of the denominator, then the x-axis is the horizontal asymptote. For example, $y = \dfrac{x - 4}{x^2 - 7}$ has the x-axis as a horizontal asymptote. If the degree of the numerator is equal to the degree of the denominator, then the ratio of the leading coefficients determines the horizontal asymptote. For example, $y = \dfrac{2x - 7}{3x - 5}$ has $y = \dfrac{2}{3}$ as its horizontal asymptote. The remaining case is when the degree of the numerator is greater than the degree of the denominator. This case is discussed next.

Each rational function of Example 2 had one horizontal asymptote and a vertical asymptote for each number that caused the denominator to be 0. The horizontal asymptote $y = 0$ occurs because as $|x|$ gets larger and larger, the y-coordinate gets closer and closer to 0. Some rational functions have a nonhorizontal line for an asymptote. An asymptote that is neither horizontal nor vertical is called an **oblique asymptote** or **slant asymptote.**

E X A M P L E   3

Finding an oblique asymptote

Determine all of the asymptotes for

$$g(x) = \frac{2x^2 + 3x - 5}{x + 2}.$$

Solution

If $x + 2 = 0$, then $x = -2$. So the line $x = -2$ is a vertical asymptote. Use long division to rewrite the function as quotient $+\frac{\text{remainder}}{\text{divisor}}$:

$$g(x) = \frac{2x^2 + 3x - 5}{x + 2} = 2x - 1 + \frac{-3}{x + 2}$$

If $|x|$ is large, the value of $\frac{-3}{x+2}$ is approximately 0. So when $|x|$ is large, the value of $g(x)$ is approximately $2x - 1$. The line $y = 2x - 1$ is an oblique asymptote for the graph of g.

> **Now do Exercises 13–14**

We can summarize this discussion of asymptotes with the following strategy for finding asymptotes for a rational function.

Strategy for Finding Asymptotes for a Rational Function

Suppose $f(x) = \frac{P(x)}{Q(x)}$ is a rational function with the degree of $Q(x)$ at least 1.

1. Solve the equation $Q(x) = 0$. The graph of f has a vertical asymptote corresponding to each solution to the equation.

2. If the degree of $P(x)$ is less than the degree of $Q(x)$, then the x-axis is a horizontal asymptote.

3. If the degree of $P(x)$ is equal to the degree of $Q(x)$, then find the ratio of the leading coefficients. The horizontal line through that ratio is the horizontal asymptote.

4. If the degree of $P(x)$ is one larger than the degree of $Q(x)$, then use division to rewrite the function as

$$\text{quotient} + \frac{\text{remainder}}{\text{divisor}}.$$

The equation formed by setting y equal to the quotient gives us an oblique asymptote.

⟨3⟩ Sketching Graphs of Rational Functions

The most important feature of a rational function is its asymptotes. The asymptotes are guidelines for the graph of a rational function. Use the following strategy to assist you in graphing rational functions.

Strategy for Graphing Rational Functions

1. Check for symmetry. If $f(-x) = f(x)$, then there is y-axis symmetry. If $f(-x) = -f(x)$, the graph is symmetric about the origin.

2. Find the y-intercept if it has one by evaluating $f(0)$.

3. Find the x-intercepts if it has any by solving $f(x) = 0$.

4. Find all asymptotes and graph them.

5. Plot the intercepts and enough points to see how the graph approaches its asymptotes.

6. Sketch the curve using the asymptotes as guidelines.

E X A M P L E   4

Graphing a rational function

Sketch the graph of each rational function.

a) $f(x) = \dfrac{3}{x^2 - 1}$ **b)** $g(x) = \dfrac{x}{x^2 - 4}$

Solution

a) From Example 2(a) we know that the lines $x = 1$ and $x = -1$ are vertical asymptotes and the x-axis is a horizontal asymptote. Draw the vertical asymptotes on the graph with dashed lines. Since all of the powers of x are even, $f(x) = f(-x)$, and the graph is symmetric about the y-axis. The ordered pairs $(0, -3)$, $(0.9, -15.789)$, $(1.1, 14.286)$, $(2, 1)$, and $\left(3, \frac{3}{8}\right)$ are also on the graph. Use the symmetry to sketch the graph shown in Fig. 10.14.

b) Draw the vertical asymptotes $x = 2$ and $x = -2$ from Example 2(b) as dashed lines. The x-axis is a horizontal asymptote. Because $f(-x) = -f(x)$, the graph is symmetric about the origin. The ordered pairs $(0, 0)$, $\left(1, -\frac{1}{3}\right)$, $(1.9, -4.872)$, $(2.1, 5.122)$, $\left(3, \frac{3}{5}\right)$, and $\left(4, \frac{1}{3}\right)$ are on the graph. Use the symmetry to get the graph shown in Fig. 10.15.

‹ **Calculator Close-Up** ›

This calculator graph supports the graph drawn in Fig. 10.15. Remember that the calculator graph can be misleading. The vertical lines drawn by the calculator are not part of the graph of the function.

Figure 10.14

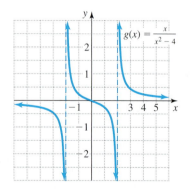

Figure 10.15

Now do Exercises 15–26

EXAMPLE   **5**

Graphing a rational function

Sketch the graph of each rational function.

a) $h(x) = \dfrac{2x + 1}{x + 3}$ b) $g(x) = \dfrac{2x^2 + 3x - 5}{x + 2}$

Solution

a) Draw the vertical asymptote $x = -3$ and the horizontal asymptote $y = 2$ from Example 2(c) as dashed lines. The points $(-2, -3)$, $\left(0, \frac{1}{3}\right)$, $\left(-\frac{1}{2}, 0\right)$, $(7, 1.5)$, $(-4, 7)$, and $(-13, 2.5)$ are on the graph shown in Fig. 10.16.

‹ **Calculator Close-Up** ›

This calculator graph supports the graph drawn in Fig. 10.16. Note that if x is -3, there is no y-coordinate because $x = -3$ is the vertical asymptote.

Figure 10.16

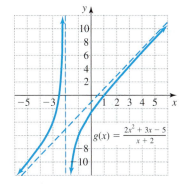

Figure 10.17

b) Draw the vertical asymptote $x = -2$ and the oblique asymptote $y = 2x - 1$ from Example 3 as dashed lines. The points $(-1, -6)$, $\left(0, -\frac{5}{2}\right)$, $(1, 0)$, $(4, 6.5)$, and $(-2.5, 0)$ are on the graph shown in Fig. 10.17.

> Now do Exercises 27–32

‹**4**› **Rational Inequalities**

Inequalities involving rational expressions are **rational inequalities.** We can solve rational inequalities using the graphical method or the test point method as we did for quadratic inequalities in Section 8.5 and polynomial inequalities in Section 10.4.

EXAMPLE   **6**

Solving a rational inequality graphically

Solve each rational inequality. Write the solution set in interval notation, and graph it.

a) $\dfrac{x - 1}{x + 2} > 0$ b) $\dfrac{x + 4}{x - 2} \leq 3$

Solution

a) To solve the rational inequality, graph $y = \frac{x-1}{x+2}$. The vertical asymptote is $x = -2$. The x-intercept is $(1, 0)$ and the y-intercept is $\left(0, -\frac{1}{2}\right)$. The graph is shown in Fig. 10.18 on the next page. The values of x that satisfy the inequality are the same values of x for which $y > 0$ on the graph in the figure. The y-coordinates in the figure are positive for x in the interval $(-\infty, -2)$ and also in the interval $(1, \infty)$. So the solution set is $(-\infty, -2) \cup (1, \infty)$ and the graph of the solution set is shown in Fig. 10.19.

Figure 10.18

Figure 10.19

Figure 10.20

Figure 10.21

b) First rewrite the inequality so that 0 is on the right:

$$\frac{x + 4}{x - 2} \leq 3$$

$$\frac{x + 4}{x - 2} - 3 \leq 0 \qquad \text{Subtract 3 from each side.}$$

$$\frac{x + 4}{x - 2} - \frac{3(x - 2)}{x - 2} \leq 0 \qquad \text{Get a common denominator.}$$

$$\frac{-2x + 10}{x - 2} \leq 0 \qquad \text{Get a single rational expression.}$$

Now graph $y = \frac{-2x + 10}{x - 2}$. The vertical asymptote is $x = 2$. Find the x-intercept by solving $-2x + 10 = 0$. The x-intercept is $(5, 0)$. The y-intercept is $(0, -5)$. The graph is shown in Fig. 10.20. From the graph we see that the y-coordinates are less than 0 for x in the interval $(-\infty, 2)$ and less than or equal to zero for x in the interval $[5, \infty)$. So the solution set is $(-\infty, 2) \cup [5, \infty)$. The graph of the solution set is shown in Fig. 10.21.

> **Now do Exercises 47–58**

When solving an equation involving rational expressions, we multiply each side by the least common denominator. When solving a rational inequality, we do not use that technique. If we multiply each side of an inequality by a negative number, then the inequality symbol is reversed. But it is not reversed if we multiply by a positive number. If the LCD involves a variable, then the value of the LCD could be positive or negative. We won't know what to do with the inequality symbol if we multiply by an LCD containing a variable.

Example 7 illustrates the **test point method.** The key fact here is that the y-coordinates on the graph of a rational function can change sign only at an x-intercept or a vertical asymptote. The x-intercepts are found by setting the numerator equal to zero, and the vertical asymptotes are found by setting the denominator equal to zero. We locate these x-values on a number line and then test a point from each interval that is determined by them.

E X A M P L E   7

Solving a rational inequality with test points
Solve each rational inequality. Write the solution set in interval notation, and graph it.

a) $\dfrac{4 - x}{2x - 1} \leq 0$ **b)** $\dfrac{x - 1}{x + 2} > \dfrac{x}{x - 4}$ **c)** $\dfrac{x^2 + 2}{x^2 + 2x + 4} \geq 0$

Figure 10.22

Solution
a) First solve $4 - x = 0$ to get $x = 4$. Then solve $2x - 1 = 0$ to get $x = \frac{1}{2}$. Plot $\frac{1}{2}$ and 4 on a number line as shown in Fig. 10.22. We put a 0 above 4 and a U above $\frac{1}{2}$ because the value of the rational expression is 0 when $x = 4$ and undefined if $x = \frac{1}{2}$. Select three test points, say 0, 2, and 6. Now try each point in the original inequality $\frac{4 - x}{2x - 1} \leq 0$:

$$\frac{4 - 0}{2(0) - 1} \leq 0 \qquad \text{Correct}$$

$$\frac{4 - 2}{2(2) - 1} \leq 0 \qquad \text{Incorrect}$$

$$\frac{4 - 6}{2(6) - 1} \leq 0 \qquad \text{Correct}$$

Figure 10.23

Since 0 and 6 satisfy the inequality, the solution set consists of the intervals containing 0 and 6. Since the inequality symbol is $\leq$, we include 4 in the solution set. Note that $\frac{1}{2}$ does not satisfy the inequality. The solution set is the interval $\left(-\infty, \frac{1}{2}\right) \cup [4, \infty)$. The graph of the solution set is shown in Fig. 10.23.

b) First rewrite the inequality with 0 on the right.

$$\frac{x-1}{x+2} > \frac{x}{x-4}$$

$$\frac{x-1}{x+2} - \frac{x}{x-4} > 0$$

$$\frac{(x-1)(x-4)}{(x+2)(x-4)} - \frac{x(x+2)}{(x-4)(x+2)} > 0$$

$$\frac{x^2 - 5x + 4 - (x^2 + 2x)}{(x+2)(x-4)} > 0$$

$$\frac{-7x + 4}{(x+2)(x-4)} > 0$$

Figure 10.24

Now solve $-7x + 4 = 0$ to get $x = \frac{4}{7}$. The denominator is zero if $x = -2$ or $x = 4$. Plot -2, $\frac{4}{7}$, and 4 on a number line as in Fig. 10.24. Select a test point in each interval determined by these three numbers. We have chosen -4, 0, 3, and 5. Evaluate the original inequality at the test points:

$$\frac{-4-1}{-4+2} > \frac{-4}{-4-4} \qquad \text{Correct}$$

$$\frac{0-1}{0+2} > \frac{0}{0-4} \qquad \text{Incorrect}$$

$$\frac{3-1}{3+2} > \frac{3}{3-4} \qquad \text{Correct}$$

$$\frac{5-1}{5+2} > \frac{5}{5-4} \qquad \text{Incorrect}$$

Figure 10.25

Since -4 and 3 satisfy the inequality, the solution set consists of the intervals containing -4 and 3. Since the inequality symbol is $>$, no endpoints are included in the intervals. The solution set is the interval $(-\infty, -2) \cup \left(\frac{4}{7}, 4\right)$. The graph of the solution set is shown in Fig. 10.25.

c) If $x^2 + 2 = 0$, then $x^2 = -2$ and there is no real solution to this equation. If $x^2 + 2x + 4 = 0$, then

$$x = \frac{-2 \pm \sqrt{2^2 - 4(1)(4)}}{2(1)} = \frac{-2 \pm \sqrt{-12}}{2}$$

and again there is no real solution. So the solution set is either all real numbers or the empty set. To decide, test 0 in $\frac{x^2 + 2}{x^2 + 2x + 4} \geq 0$:

$$\frac{0^2 + 2}{0^2 + 2(0) + 4} \geq 0 \qquad \text{Correct}$$

Figure 10.26

Since the inequality is satisfied at the test point, it is satisfied for all real numbers. The solution set is $(-\infty, \infty)$, and the graph of the solution set is shown in Fig. 10.26.

Now do Exercises 59–74

Warm-Ups ▼

Fill in the blank.

1. A(n) _____ function has the form $f(x) = P(x)/Q(x)$ where $P(x)$ and $Q(x)$ are polynomials with no common factor and $Q(x) \neq 0$.

2. The _____ of a rational function is all real numbers except those that cause the denominator to be 0.

3. A(n) _____ asymptote is a vertical line that is approached by the graph of a rational function.

4. A(n) _____ asymptote is a horizontal line approached by the graph of a rational function.

5. A(n) _____ or _____ asymptote is a nonvertical nonhorizontal line approached by the graph of a rational function.

True or false?

6. The domain of $f(x) = \dfrac{1}{x - 9}$ is $x = 9$.

7. The domain of $f(x) = \dfrac{x - 1}{x - 2}$ is $(-\infty, -2) \cup (-2, 1) \cup (1, \infty)$.

8. The line $x = 2$ is the only vertical asymptote for the graph of $f(x) = \dfrac{1}{x^2 - 4}$.

9. The x-axis is a horizontal asymptote for the graph of $f(x) = \dfrac{x^2 - 3x + 5}{x^3 - 9x}$.

10. The line $y = 2x - 5$ is an asymptote for the graph of $f(x) = 2x - 5 + \dfrac{1}{x}$.

11. The graph of $f(x) = \dfrac{x^2}{x^2 - 9}$ is symmetric about the y-axis.

12. The solution set to $\dfrac{4}{x - 3} \geq 0$ is $(3, \infty)$.

10.5   Exercises

‹1› Rational Functions

Find the domain of each rational function. See Example 1.

1. $f(x) = \dfrac{2}{x - 1}$

2. $f(x) = \dfrac{-2}{x + 3}$

3. $f(x) = \dfrac{x^2 - 1}{x}$

4. $f(x) = \dfrac{-2x + 3}{x^2}$

5. $f(x) = \dfrac{5}{x^2 - 16}$

6. $f(x) = \dfrac{x + 12}{x^2 - x - 6}$

‹2› Asymptotes

Determine all asymptotes for the graph of each rational function. See Examples 2 and 3. See the Strategy for Finding Asymptotes for a Rational Function box on page 685.

7. $f(x) = \dfrac{7}{x + 4}$

8. $f(x) = \dfrac{-8}{x - 9}$

9. $f(x) = \dfrac{1}{x^2 - 16}$

10. $f(x) = \dfrac{-2}{x^2 - 5x + 6}$

11. $f(x) = \dfrac{5x}{x - 7}$

12. $f(x) = \dfrac{3x + 8}{x - 2}$

13. $f(x) = \dfrac{2x^2}{x - 3}$

14. $f(x) = \dfrac{3x^2 + 2}{x + 1}$

⟨3⟩ Sketching Graphs of Rational Functions

Match each rational function with its graph a–h.

15. $f(x) = -\dfrac{2}{x}$ **16.** $f(x) = -\dfrac{1}{x - 2}$

17. $f(x) = \dfrac{x}{x - 2}$ **18.** $f(x) = \dfrac{x - 2}{x}$

19. $f(x) = \dfrac{1}{x^2 - 2x}$ **20.** $f(x) = \dfrac{x^2}{x^2 - 4}$

21. $f(x) = -\dfrac{x + 4}{2}$ **22.** $f(x) = \dfrac{x^2 + 2x + 1}{x}$

e)

f)

g)

h)

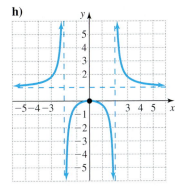

Determine all asymptotes, and sketch the graph of each function. See Examples 4 and 5. See the Strategy for Graphing Rational Functions box on page 686.

23. $f(x) = \dfrac{2}{x + 4}$ **24.** $f(x) = \dfrac{-3}{x - 1}$

a)

b)

c)

d)

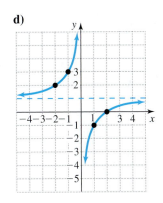

25. $f(x) = \dfrac{x}{x^2 - 9}$

26. $f(x) = \dfrac{-2}{x^2 + x - 2}$

35. $f(x) = \dfrac{2x - 3}{x^2 + x - 6}$

36. $f(x) = \dfrac{x}{x^2 + 4x + 4}$

27. $f(x) = \dfrac{2x - 1}{x + 3}$

28. $f(x) = \dfrac{5 - 2x}{x - 2}$

37. $f(x) = \dfrac{x + 1}{x^2}$

38. $f(x) = \dfrac{x - 1}{x^2}$

29. $f(x) = \dfrac{x^2 - 3x + 1}{x}$

30. $f(x) = \dfrac{x^3 + 1}{x^2}$

39. $f(x) = \dfrac{2x - 1}{x^3 - 9x}$

40. $f(x) = \dfrac{2x^2 + 1}{x^3 - x}$

31. $f(x) = \dfrac{3x^2 - 2x}{x - 1}$

32. $f(x) = \dfrac{-x^2 + 5x - 5}{x - 3}$

41. $f(x) = \dfrac{x}{x^2 - 1}$

42. $f(x) = \dfrac{x}{x^2 + x - 2}$

Find all asymptotes, x-intercepts, and y-intercepts for the graph of each rational function, and sketch the graph of the function.

33. $f(x) = \dfrac{1}{x^2}$

34. $f(x) = \dfrac{2}{x^2 - 4x + 4}$

43. $f(x) = \dfrac{2}{x^2 + 1}$

44. $f(x) = \dfrac{x}{x^2 + 1}$

45. $f(x) = \dfrac{x^2}{x + 1}$ **46.** $f(x) = \dfrac{x^2}{x - 1}$

57. $\dfrac{x - 3}{x + 5} \geq 2$

58. $\dfrac{x + 2}{x - 6} \leq 3$

⟨4⟩ Rational Inequalities

Solve each rational inequality using the graphical method. State the solution set using interval notation, and graph it. See Example 6.

Solve each rational inequality using the test point method. State the solution set using interval notation, and graph it. See Example 7.

59. $\dfrac{x - 4}{x + 2} \geq 0$

47. $\dfrac{1}{x} > 0$

60. $\dfrac{x - 3}{x + 5} < 0$

48. $\dfrac{1}{x} \leq 0$

61. $\dfrac{x - 2}{x + 3} < 1$

49. $\dfrac{x}{x - 3} > 0$

62. $\dfrac{x - 3}{x + 4} > 2$

50. $\dfrac{a}{a + 2} > 0$

63. $\dfrac{3}{x + 2} > \dfrac{1}{x + 1}$

51. $\dfrac{x + 2}{x} \leq 0$

64. $\dfrac{1}{x + 1} < \dfrac{1}{x - 1}$

52. $\dfrac{w - 6}{w} \leq 0$

53. $\dfrac{t - 3}{t + 6} > 0$

65. $\dfrac{2}{x - 5} > \dfrac{1}{x + 4}$

54. $\dfrac{x - 2}{2x + 5} < 0$

66. $\dfrac{3}{x + 2} > \dfrac{2}{x - 1}$

55. $\dfrac{x}{x + 2} > -1$

VIDEO **67.** $\dfrac{m}{m - 5} + \dfrac{3}{m - 1} > 0$

56. $\dfrac{x + 3}{x} \leq -2$

68. $\dfrac{p}{p - 16} + \dfrac{2}{p - 6} \leq 0$

69. $\dfrac{x}{x-3} \le \dfrac{-8}{x-6}$

70. $\dfrac{x}{x+20} > \dfrac{2}{x+8}$

71. $\dfrac{x^2}{x^2+4} \ge 0$

72. $\dfrac{x^2+2x+3}{x^2} > 0$

73. $\dfrac{x^2}{x^2+9} < 0$

74. $\dfrac{x^2+4x+5}{x^2+1} \le 0$

Applications

Solve each problem.

75. *Oscillating modulators.* The number of oscillating modulators produced by a factory in t hours is given by the polynomial function $n(t) = t^2 + 6t$ for $t \ge 1$. The cost in dollars of operating the factory for t hours is given by the function $c(t) = 36t + 500$ for $t \ge 1$. The average cost per modulator is given by the rational function $f(t) = \dfrac{36t + 500}{t^2 + 6t}$ for $t \ge 1$. Graph the function f. What is the average cost per modulator at time $t = 20$ and time $t = 30$? What can you conclude about the average cost per modulator after a long period of time?

76. *Nonoscillating modulators.* The number of nonoscillating modulators produced by a factory in t hours is given by the polynomial function $n(t) = 16t$ for $t \ge 1$. The cost in dollars of operating the factory for t hours is given by the function $c(t) = 64t + 500$ for $t \ge 1$. The average cost per modulator is given by the rational function $f(t) = \dfrac{64t + 500}{16t}$ for $t \ge 1$. Graph the function f. What is the

average cost per modulator at time $t = 10$ and $t = 20$? What can you conclude about the average cost per modulator after a long period of time?

77. *Average cost of an SUV.* Mercedes-Benz spent $700 million to design its new SUV (Motor Trend, www.motortrend.com). If it costs $25,000 to manufacture each SUV, then the average cost per vehicle in dollars when x vehicles are manufactured is given by the rational function

$$A(x) = \frac{25{,}000x + 700{,}000{,}000}{x}.$$

a) What is the horizontal asymptote for the graph of this function?
b) What is the average cost per vehicle when 50,000 vehicles are made?
c) For what number of vehicles is the average cost $30,000?
d) Graph this function for x ranging from 0 to 100,000.

78. *Average cost of a pill.* Assuming Pfizer spent a typical $350 million to develop its latest miracle drug and $0.10 each to make the pills, then the average cost per pill in dollars when x pills are made is given by the rational function

$$A(x) = \frac{0.10x + 350{,}000{,}000}{x}.$$

a) What is the horizontal asymptote for the graph of this function?
b) What is the average cost per pill when 100 million pills are made?
c) For what number of pills is the average cost per pill $2?

d) Graph this function for x ranging from 0 to 150 million.

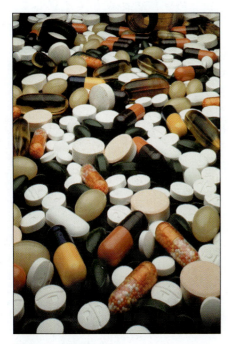

Photo for Exercise 78

Graphing Calculator Exercises

Sketch the graph of each pair of functions in the same coordinate system. What do you observe in each case?

79. $f(x) = x^2$, $g(x) = x^2 + 1/x$

80. $f(x) = x^2$, $g(x) = x^2 + 1/x^2$

81. $f(x) = |x|$, $g(x) = |x| + 1/x$

82. $f(x) = |x|$, $g(x) = |x| + 1/x^2$

83. $f(x) = \sqrt{x}$, $g(x) = \sqrt{x} + 1/x$

84. $f(x) = x^3$, $g(x) = x^3 + 1/x^2$

Getting More Involved

In each case find a rational function whose graph has the required asymptotes. Answers may vary.

85. The graph has the x-axis as a horizontal asymptote and the y-axis as a vertical asymptote.

86. The graph has the x-axis as a horizontal asymptote and the line $x = 2$ as a vertical asymptote.

87. The graph has the x-axis as a horizontal asymptote and lines $x = 3$ and $x = -1$ as vertical asymptotes.

88. The graph has the line $y = 2$ as a horizontal asymptote and the line $x = 1$ as a vertical asymptote.

Chapter 10 Wrap-Up

Summary

Polynomial Functions		Examples
Polynomial function	A function defined by a polynomial	$P(x) = x^2 - x - 12$
Root or zero of a function	A number c such that $f(c) = 0$	Since $P(4) = 4^2 - 4 - 12 = 0$, 4 is a root or zero of the function P.
Factor theorem	The following are equivalent for $P(x)$, a polynomial in x. 1. The remainder is zero when $P(x)$ is divided by $x - c$. 2. $x - c$ is a factor of $P(x)$. 3. c is a solution to $P(x) = 0$. 4. c is a zero of the function $P(x)$, or $P(c) = 0$.	$P(x) = x^2 - x - 2$ $P(x) = (x - 2)(x + 1)$ $P(-1) = 0, P(2) = 0$ -1 and 2 both satisfy $x^2 - x - 2 = 0$
Remainder theorem	If R is the remainder when a polynomial $P(x)$ is divided by $x - c$, then $R = P(c)$.	$P(x) = x^2 - x - 12, P(2) = -10$ $\dfrac{x^2 - x - 12}{x - 2} = x + 1 + \dfrac{-10}{x - 2}$
Fundamental theorem of algebra	If $P(x)$ is a polynomial function of positive degree, then $P(x)$ has at least one complex zero.	$H(x) = x^7 - \pi x^6 + 99x^5 + x - \sqrt{2}$ has at least one complex zero.
Rational root theorem	If $P(x)$ is a polynomial with integral coefficients, then any rational zero is a factor of the constant term divided by a factor of the leading coefficient.	The possible rational zeros for $P(x) = 2x^2 - 3x + 1$ are ± 1 and $\pm\dfrac{1}{2}$.
Multiplicity	If $x - c$ occurs n times in the complete factorization of $P(x)$, then c is a root of $P(x) = 0$ with multiplicity n.	If $P(x) = (x - 2)^2$, then 2 is a root of $(x - 2)^2 = 0$ with multiplicity 2.
n-root theorem	If $P(x) = 0$ is a polynomial equation with real or complex coefficients and positive degree n, then counting multiplicity $P(x) = 0$ has n complex roots.	$x^3 - 5x = 0$ has three roots. $x^4 - 1 = 0$ has four roots.
Conjugate pairs theorem	If $P(x) = 0$ is a polynomial equation with real coefficients and $a + bi$ $(b \neq 0)$ is a root, then $a - bi$ is also a root.	The roots to $x^2 - 6x + 13 = 0$ are $3 + 2i$ and $3 - 2i$.

Descartes' rule of signs	If $P(x) = 0$ is a polynomial equation with real coefficients, then the number of positive roots of the equation is either equal to the number of variations of sign of $P(x)$ or less than that by an even number. The number of negative roots of the equation is either equal to the number of variations in sign of $P(-x)$ or less than that by an even number.	The number of positive roots to $x^3 - x^2 + x - 5 = 0$ is either 3 or 1. Since $P(-x) = -x^3 - x^2 - x - 5$, there are no negative roots.
Theorem on bounds	Suppose $P(x)$ is a polynomial with real coefficients and a positive leading coefficient, and synthetic division with c is performed. If $c > 0$ and all terms in the bottom row are nonnegative, then no number larger than c can be a root of $P(x) = 0$. If $c < 0$ and the terms in the bottom row alternate in sign, then no number less than c can be a root of $P(x) = 0$.	$P(x) = 2x^2 + 7x - 4$ $\begin{array}{r\|rrr} 1 & 2 & 7 & -4 \\ & & 2 & 9 \\ \hline & 2 & 9 & 5 \end{array}$ $\begin{array}{r\|rrr} -5 & 2 & 7 & -4 \\ & & -10 & 15 \\ \hline & 2 & -3 & 11 \end{array}$ Any real root of $P(x) = 0$ is between -5 and 1.

Symmetry

		Examples
Symmetric about the y-axis	If $f(x)$ is a function such that $f(x) = f(-x)$ for any value of x in its domain, then the graph of f is symmetric about the y-axis.	If $f(x) = x^4 - x^2$, then $f(-x) = (-x)^4 - (-x)^2 = x^4 - x^2$, and $f(x) = f(-x)$.
Symmetric about the origin	If $f(x)$ is a function such that $f(-x) = -f(x)$ for any value of x in its domain, then the graph of f is symmetric about the origin.	If $f(x) = x^3 + x$, then $f(-x) = (-x)^3 + (-x) = -x^3 - x$, and $f(-x) = -f(x)$.

Rational Functions

		Examples
Rational function	If $P(x)$ and $Q(x)$ are polynomials with no common factor and $f(x) = \dfrac{P(x)}{Q(x)}$ for $Q(x) \neq 0$, then $f(x)$ is a rational function.	$f(x) = \dfrac{x^2 - 1}{3x - 2},\ f(x) = \dfrac{1}{x - 3}$
Finding asymptotes for a rational function $f(x) = \dfrac{P(x)}{Q(x)}$	1. The graph of f has a vertical asymptote for each solution to the equation $Q(x) = 0$. 2. If the degree of $P(x)$ is less than the degree of $Q(x)$, then the x-axis is a horizontal asymptote. 3. If the degree of $P(x)$ is equal to the degree of $Q(x)$, then the horizontal asymptote is determined by the ratio of the leading coefficients.	$f(x) = \dfrac{1}{x - 2}$ Vertical: $x = 2$ Horizontal: x-axis $f(x) = \dfrac{x}{x - 2}$ Vertical: $x = 2$ Horizontal: $y = 1$

4. If the degree of $P(x)$ is one larger than the degree of $Q(x)$, then use long division to find the quotient of $P(x)$ and $Q(x)$.	$f(x) = \dfrac{2x^2 + 3x - 5}{x + 2} = 2x - 1 + \dfrac{-3}{x + 2}$ Vertical: $x = -2$ Oblique: $y = 2x - 1$

Inequalities		Examples
Polynomial inequality	An inequality involving a polynomial	$x^3 - x > 0$
Rational inequality	An inequality involving a rational expression	$\dfrac{1}{x - 2} \le 0$
Methods for solving	Use either the graphical method or the test point method.	$x^3 - x > 0$ $(-1, 0) \cup (1, \infty)$

Enriching Your Mathematical Word Power

Fill in the blank.

1. A function defined by a polynomial is a(n) _____ function.
2. A(n) ____ or ____ of a function is a number c such that $f(c) = 0$.
3. The number of times a linear factor occurs in a polynomial is its _____.
4. A number M such that no root is greater than M is an _____ bound for the roots.

5. A number N such that no root is less than N is a(n) _____ bound for the roots.
6. If $f(-x) = f(x)$, then the graph of f is _____ about the y-axis.
7. If $f(-x) = -f(x)$, then the graph of f is symmetric about the _____.
8. A ratio of two polynomial functions is a(n) _____ function.
9. A line that is approached by a curve is a(n) _____.
10. A(n) _____ asymptote is neither horizontal nor vertical.

Review Exercises

10.1 The Factor Theorem

Given that either 2 or -2 is a solution to each of the following equations, solve each equation.

1. $x^3 - 4x^2 - 11x + 30 = 0$
2. $x^3 - 2x^2 - 6x + 12 = 0$
3. $x^3 - 5x^2 - 2x + 24 = 0$
4. $x^3 + 7x^2 + 4x - 12 = 0$

Determine whether the first polynomial is a factor of the second. If it is, then factor the polynomial completely.

5. $x - 3,\quad x^3 + 4x^2 - 11x - 30$
6. $x + 4,\quad x^3 + x^2 - 10x + 8$
7. $x + 2,\quad x^3 - 5x - 6$
8. $x - 1,\quad 2x^3 - 5x^2 + 2x - 9$

10.2 Zeros of a Polynomial Function

For each polynomial find $P(-2)$ and $P\left(\frac{1}{2}\right)$ in two different ways.

9. $P(x) = 4x^3 - 2x^2 + 3x + 1$
10. $P(x) = 2x^3 - 3x^2 + 5x - 2$

Use the rational root theorem to list all possible rational roots for each polynomial function.

11. $f(x) = 3x^3 - 5x^2 + 7x - 4$
12. $f(x) = 2x^4 - 5x^2 + 3x - 5$
13. $f(x) = 4x^4 - 5x^2 - 3x + 3$
14. $f(x) = 6x^3 + x^2 - 25x - 9$

Find all real and imaginary zeros for each polynomial function.

15. $f(x) = 2x - 3$

16. $f(x) = 5$

17. $f(x) = x^2 - 100$

18. $f(x) = 2x^2 - 1$

19. $f(x) = x^3 - 64$

20. $f(x) = x^3 - 27$

21. $f(x) = x^4 - 64$

22. $f(x) = 4x^4 - 1$

23. $f(x) = 12x^3 - 4x^2 - 3x + 1$

24. $f(x) = 18x^3 + 9x^2 - 2x - 1$

10.3 The Theory of Equations

Find a polynomial equation with real coefficients that has the given roots.

25. $\dfrac{1}{2}, 2$

26. $\dfrac{1}{3}, -4$

27. $5 - 2i$

28. $3 + 2i$

29. $3, 1 + 3i$

30. $-5, 2 - 4i$

Use Descartes' rule of signs to discuss the possibilities for the roots to each equation. Do not solve the equation.

31. $x^6 + x^4 + x^2 = 0$

32. $2x^3 - 3x^2 - 5x - 9 = 0$

33. $x^3 - 5x^2 + 3x - 9 = 0$

34. $x^5 + x^3 + 5x = 0$

Establish the best integral bounds for the roots of each equation according to the theorem on bounds.

35. $x^2 - 2x - 4 = 0$

36. $x^2 - 2x - 6 = 0$

37. $x^3 - 3x^2 - 5x - 6 = 0$

38. $2x^3 - 4x^2 - 12x - 7 = 0$

Find all real and imaginary solutions to each equation, stating multiplicity when it is greater than 1.

39. $2x^3 - 19x^2 + 49x - 20 = 0$

40. $3x^3 + 2x^2 - 37x + 12 = 0$

41. $8x^3 - 12x^2 - 26x + 15 = 0$

42. $12x^3 - 16x^2 - 5x + 3 = 0$

43. $4x^3 - 24x^2 + 25x - 25 = 0$

44. $2x^3 + 4x^2 - 5x + 3 = 0$

45. $4x^4 - 28x^3 + 59x^2 - 30x - 9 = 0$

46. $2x^4 - 6x^3 + 5x^2 - 1 = 0$

47. $x^3 - 3x^2 + 3x - 1 = 0$

48. $x^4 - 4x^3 + 6x^2 - 4x + 1 = 0$

10.4 Graphs of Polynomial Functions

Discuss the symmetry of the graph of each function.

49. $f(x) = 3x^4 - 2x^2$

50. $f(x) = x^7 - 3x^5 + x^3$

51. $f(x) = -x^3 + 3x$

52. $f(x) = -2x^4 + 6x^2 + 3$

Find the x-intercepts and y-intercept for the graph of each function, and sketch the graph.

53. $f(x) = x^3 - 3x + 2$

54. $f(x) = -x^3 - 3x^2 + 4$

55. $f(x) = -x^4 + 2x^2 - 1$

67. $x^3 - x^2 - 9x + 9 \geq 0$

68. $x^3 - 4x^2 - 16x + 64 \leq 0$

10.5 Graphs of Rational Functions
Find the domain of each rational function.

56. $f(x) = x^4 - 2x^2$

69. $f(x) = \dfrac{x^2 - 1}{2x + 3}$

70. $f(x) = \dfrac{3x + 2}{x^2 - x - 12}$

71. $f(x) = \dfrac{1}{x^2 + 9}$

57. $f(x) = \dfrac{1}{2}x^3 - 2x$

72. $f(x) = \dfrac{x - 4}{x^2 - 9}$

Find all asymptotes for each rational function, and sketch the graph of the function.

73. $f(x) = \dfrac{2}{x - 3}$

74. $f(x) = \dfrac{-1}{x + 1}$

58. $f(x) = x^3 - 6x^2 + 8x$

Solve each polynomial inequality. State the solution set using interval notation, and graph it.

75. $f(x) = \dfrac{x}{x^2 - 4}$

76. $f(x) = \dfrac{x^2}{x^2 - 4}$

59. $\dfrac{1}{2}x^3 - 2x \leq 0$

60. $x^3 - 6x^2 + 8x \geq 0$

61. $x^4 - 2x^2 \geq 0$

62. $-x^4 + 2x^2 - 1 \geq 0$

77. $f(x) = \dfrac{2x - 1}{x - 1}$

78. $f(x) = \dfrac{-x - 1}{x}$

63. $x^4 - 2x^2 \leq 0$

64. $-x^4 + 2x^2 - 1 \leq 0$

65. $x^3 - x^2 - 9x + 9 < 0$

66. $x^3 - 4x^2 - 16x + 64 > 0$

79. $f(x) = \dfrac{x^2 - 2x + 1}{x - 2}$ **80.** $f(x) = \dfrac{-x^2 + x + 2}{x - 1}$ **90.** $\dfrac{x}{x - 6} \geq \dfrac{x}{x + 3}$

Miscellaneous

Sketch the graph of each function.

91. $f(x) = 3$ **92.** $f(x) = 2x - 3$

Solve each rational inequality. State the solution set using interval notation, and graph it.

81. $\dfrac{x + 3}{x - 4} \geq 0$

93. $f(x) = x^2 - 3$ **94.** $f(x) = 3 - x^2$

82. $\dfrac{x + 5}{x - 7} \leq 0$

83. $\dfrac{x^2 - 4}{x} > 0$

95. $f(x) = \dfrac{1}{x^2 - 3}$ **96.** $f(x) = \dfrac{2x}{(x - 1)(x + 2)}$

84. $\dfrac{x^2}{x^2 - 16} < 0$

85. $\dfrac{x}{x + 5} < 2$

97. $f(x) = x(x - 1)(x + 2)$ **98.** $f(x) = x^3 - 4x^2 + 4x$

86. $\dfrac{x + 4}{x - 1} \leq 4$

87. $\dfrac{x^2 + 4}{x} \leq 0$

Solve each inequality. State the solution set using interval notation.

88. $\dfrac{x^2}{x^2 + 8} \geq 0$

99. $4 - 2x > 0$

100. $2x - 3 > 0$

101. $x^2 - 3 \leq 0$

89. $\dfrac{2x}{x + 5} \leq \dfrac{x}{x - 2}$

102. $3 - x^2 \leq 0$

103. $\dfrac{1}{x^2 - 3} < 0$

104. $\dfrac{x}{(x-1)(x+2)} \geq 0$

105. $x(x-1)(x+2) < 0$

106. $x^3 - 4x^2 + 4x \geq 0$

107. $\dfrac{x-4}{x+3} \leq 0$

108. $\dfrac{2x-1}{x+5} \geq 0$

109. $(x-2)(x+1)(x-5) \geq 0$

110. $(x-1)(x+2)(2x-5) < 0$

111. $x^3 + 3x^2 - x - 3 < 0$

112. $x^3 + 5x^2 - 4x - 20 \geq 0$

Graphing Calculator Exercises

Match the given inequalities with their solution sets (a through d) by examining a table or a graph.

113. $x^2 - 2x - 8 < 0$

114. $x^2 - 3x > 54$

115. $\dfrac{x}{x-2} > 2$

116. $\dfrac{3}{x-2} < \dfrac{5}{x+2}$

a. $(-2, 2) \cup (8, \infty)$

b. $(2, 4)$

c. $(-2, 4)$

d. $(-\infty, -6) \cup (9, \infty)$

Solve each problem.

117. *Golf trophy.* Tanya is designing a trophy for a golf tournament. The trophy consists of a solid bronze sphere mounted on a solid bronze square base. The diameter of the sphere will equal the length of the side of the square, and the thickness of the base will be π inches (in.). If she wants to use 72π cubic inches (in.^{3}) of bronze for the trophy, then what is the radius of the sphere?

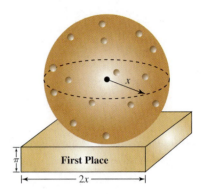

Figure for Exercise 117

118. Joshua is designing a metal silo for storing grain. The silo consists of a cylindrical base with a cone-shaped roof. He wants the radius of the base to be equal to its height and the height of the cone to be 5 feet (ft). If the silo is to have a volume of 864π ft^3, then what is the radius of the base?

Figure for Exercise 118

Chapter 10 Test

Find the following values for $f(x) = 3x^4 + 5x^3 - 10x^2 + 3x - 2$ using two different methods.

1. $f(2)$

2. $f(-3)$

Find all possible rational zeros for each polynomial function.

3. $f(x) = x^4 + x^3 - 7x^2 + 5x - 6$

4. $f(x) = 2x^3 - 5x^2 + 9x - 3$

Find all of the real and imaginary zeros for each polynomial function.

5. $f(x) = x^3 - 13x + 12$

6. $f(x) = 2x^3 - 9x^2 + 14x - 5$

Write a polynomial equation with real coefficients that has the given roots.

7. $-1, 1, 2$

8. $2, 3 - i$

Solve each problem.

9. Is $x + 1$ a factor of $2x^4 - 5x^3 + 3x^2 + 6x - 4$? Explain.

10. Factor $x^3 - 12x^2 + 47x - 60$ completely.

Use Descartes' rule of signs to discuss the possibilities for the roots to each equation. Do not solve the equation.

11. $2x^3 - 6x^2 + 5x - 9 = 0$

12. $x^5 + 5x^3 + x = 0$

18. $f(x) = \dfrac{1}{x^2 - 4x + 4}$

Establish the best integral bounds for the roots of each equation according to the theorem on bounds.

13. $4x^2 - 4x - 15 = 0$

14. $2x^3 - 3x^2 - 7x + 4 = 0$

19. $f(x) = \dfrac{2x - 3}{x - 2}$

Determine whether the graph of each function is symmetric about the y-axis or the origin.

15. $f(x) = x^3 - 16x$

20. $f(x) = x^3 - x^2 - 4x + 4$

16. $f(x) = x^4 - 25x^2 + 1$

Graph each function.

17. $f(x) = (x + 2)(x - 2)^2$

Solve each inequality. State the solution set using interval notation.

21. $x^3 + 4x^2 - 32x > 0$

22. $\dfrac{x}{x^2 - 25} \le 0$

Making **Connections** | **A Review of Chapters 1–10**

Find all real and imaginary solutions to each equation.

1. $2x - 3 = x - 5$

2. $(2x - 3)(x - 5) = 0$

3. $\dfrac{2x - 3}{x - 5} = 0$

4. $2x^2 - 3 = 0$

5. $\dfrac{3}{2x - 3} = \dfrac{2}{x - 5}$

6. $2x^4 + 7x^2 - 15 = 0$

7. $\dfrac{1}{x} + \dfrac{3}{x - 5} = -\dfrac{1}{2}$

8. $30x^3 - 29x^2 - 22x - 3 = 0$

Sketch the graph of each function.

9. $y = 2x$

10. $y = 2 - x$

11. $y = x - 2$

12. $y = \dfrac{1}{x - 2}$

13. $y = x^2 - 2$

14. $y = (x - 2)^2$

15. $y = \dfrac{1}{(x - 2)^2}$

16. $y = x^3 - 2$

17. $y = x^3 - x$

18. $y = x^4 - x^2$

19. $y = \sqrt{x + 5} - 2$

20. $y = -|x - 2|$

Let $f(x) = x^2 - 3$, and $g(x) = 4x + 1$. Find the following.

21. $f(-5)$

22. $g(-5)$

23. $g^{-1}(-19)$

24. $g^{-1}(5)$

25. $f(g(2))$

26. $g(f(2))$

27. $(f \circ g)(-2)$

28. $(g \circ f)(4)$

29. $(f + g)(1)$

30. $(g - f)(6)$

Solve each inequality. State the solution set using interval notation.

31. $4x - 9 > 11$

32. $5 - 2x \leq 15$

33. $|7 - 2x| \leq 3$

34. $|2x - 1| > 5$

35. $x^2 - 3 \geq 0$

36. $4x^2 + 4x - 3 < 0$

37. $x^3 - 4x > 0$

38. $x^3 - 4x^2 > 0$

39. $x^4 - 4x^2 < 0$

40. $x^3 + 3x^2 - 4x - 12 \leq 0$

Critical **Thinking** | For Individual or Group Work | **Chapter 10**

These exercises can be solved by a variety of techniques, which may or may not require algebra. So be creative and think critically. Explain all answers. Answers are in the Instructor's Edition of this text.

1. *Knight moves.* Draw a 3 by 3 chess board on paper and place two pennies (P) and two nickels (N) in the corners as shown in part (a) of the figure. Move the Ns to the positions of the Ps and the Ps to the position of the Ns using the moves that a knight can make in chess (one space vertically followed by two spaces horizontally or one space horizontally followed by two spaces vertically). If you allow a P or an N to make more than one move on a given turn, then it takes six turns. Try it. Find the minimum number of turns required to interchange the coins starting with the arrangement in (b).

         (a)   (b)

Figure for Exercise 1

2. *Friedman numbers.* A Friedman number is a positive integer that can be written in some nontrivial way using its own digits together with the elementary operations $(+, -, \cdot, \div$, exponents, and grouping symbols). For example, $25 = 5^2$, and $126 = 21 \cdot 6$. The only two-digit Friedman number is 25. Show that 121 and 125 are Friedman numbers. There are 13 three-digit Friedman numbers. Find the other 10 three-digit Friedman numbers.

3. *Large Friedman numbers.* Show that 123,456,789 and 987,654,321 are Friedman numbers.

4. *Year numbers.* Using all four of the digits in the current year and only those digits, write expressions for the integers from 1 through 100. You may use grouping symbols, and the operations of addition, subtraction, multiplication, division, powers, roots, and factorial, but no two-digit numbers or decimal points. For example if the year is 2005, then $5^0 + 0 \cdot 2 = 1$, $5^0 + 2^0 = 2$, and so on. See how far you can go. Vary the problem by trying another year (say 1776), or allowing decimal points, or two-digit numbers.

5. *Real numbers.* Two real numbers have a sum of 200 and a product of 50. What is the sum of their reciprocals?

6. *Telling time.* Find the first time after 11 A.M. for which the minute hand and hour hand of a clock form a perfect right angle. Find the time to the nearest tenth of a second. The answer is not 11:10.

7. *Identity crisis.* Determine the value of a that will make this equation an identity.

$$\frac{1}{x-1} + \frac{2}{1-x} + \frac{3}{x-1} + \frac{4}{1-x} + \frac{5}{x-1} + \frac{6}{1-x}$$

$$+ \frac{7}{x-1} + \frac{8}{1-x} + \frac{9}{x-1} + \frac{10}{1-x} = \frac{a}{x-1}$$

8. *Difference of two squares.* Let $a = 7^{6006} + 7^{-6006}$ and $b = 7^{6006} - 7^{-6006}$. Find $a^2 - b^2$.

Exponential and Logarithmic Functions

Water is one of the essentials of life, yet it is something that most of us take for granted. Among other things, the U.S. Geological Survey (U.S.G.S.) studies freshwater. For over 50 years the Water Resources Division of the U.S.G.S. has been gathering basic data about the flow of both freshwater and saltwater from streams and groundwater surfaces. This division collects, compiles, analyzes, verifies, organizes, and publishes data gathered from groundwater data collection networks in each of the 50 states, Puerto Rico, and the Trust Territories. Records of stream flow, groundwater levels, and water quality provide hydrological information needed by local, state, and federal agencies as well as the private sector.

There are many instances of the importance of the data collected by the U.S.G.S. For example, before 1987 the Tangipahoa River in Louisiana was used extensively for swimming and boating. In 1987 data gathered by the U.S.G.S. showed that fecal coliform levels in the river exceeded safe levels. Consequently, Louisiana banned recreational use of the river. Other studies by the Water Resources Division include the results of pollutants on salt marsh environments and the effect that salting highways in winter has on our drinking water supply.

May 3, 1953
Record Flood 50,500 ft³/sec

In Exercises 87 and 88 of Section 11.2 you will see how data from the U.S.G.S. is used in a logarithmic function to measure water quality.

11.1   Exponential Functions and Their Applications

We have studied functions such as

$$f(x) = x^2, \qquad g(x) = x^3, \qquad \text{and} \qquad h(x) = x^{1/2}.$$

For these functions the variable is the base. In this section, we discuss functions that have a variable as an exponent. These functions are called *exponential functions*.

⟨1⟩ Exponential Functions

Some examples of exponential functions are

$$f(x) = 2^x, \qquad f(x) = \left(\frac{1}{2}\right)^x, \qquad \text{and} \qquad f(x) = 3^x.$$

Exponential Function

An **exponential function** is a function of the form

$$f(x) = a^x,$$

where $a > 0$, $a \neq 1$, and x is a real number.

We rule out the base 1 in the definition because $f(x) = 1^x$ is the same as the constant function $f(x) = 1$. Zero is not used as a base because $0^x = 0$ for any positive x and nonpositive powers of 0 are undefined. Negative numbers are not used as bases because an expression such as $(-4)^x$ is not a real number if $x = \frac{1}{2}$.

E X A M P L E   **1**

Evaluating exponential functions

Let $f(x) = 2^x$, $g(x) = \left(\frac{1}{4}\right)^{1-x}$, and $h(x) = -3^x$. Find the following:

a) $f\left(\frac{3}{2}\right)$ **b)** $f(-3)$ **c)** $g(3)$ **d)** $h(2)$

Solution

a) $f\left(\frac{3}{2}\right) = 2^{3/2} = \sqrt{2^3} = \sqrt{8} = 2\sqrt{2}$

b) $f(-3) = 2^{-3} = \dfrac{1}{2^3} = \dfrac{1}{8}$

c) $g(3) = \left(\frac{1}{4}\right)^{1-3} = \left(\frac{1}{4}\right)^{-2} = 4^2 = 16$

d) $h(2) = -3^2 = -9$ Note that $-3^2 \neq (-3)^2$.

Now do Exercises 1–12

For many applications of exponential functions we use base 10 or another base called e. The number e is an irrational number that is approximately 2.718. We will see how e is used in compound interest in Example 9 of this section. Base 10 will be used in Section 11.2. Base 10 is called the **common base,** and base e is called the **natural base.**

E X A M P L E   2

Most graphing calculators have keys for the functions 10^x and e^x.

Base 10 and base e

Let $f(x) = 10^x$ and $g(x) = e^x$. Find the following, and round approximate answers to four decimal places:

a) $f(3)$ **b)** $f(1.51)$ **c)** $g(0)$ **d)** $g(2)$

Solution

a) $f(3) = 10^3 = 1000$

b) $f(1.51) = 10^{1.51} \approx 32.3594$ Use the 10^x key on a calculator.

c) $g(0) = e^0 = 1$

d) $g(2) = e^2 \approx 7.3891$ Use the e^x key on a calculator.

> Now do Exercises 13–20

In the definition of an exponential function no restrictions were placed on the exponent x because the domain of an exponential function is the set of all real numbers. So both rational and irrational numbers can be used as the exponent. We have been using rational numbers for exponents since Chapter 7, but we have not yet seen an irrational number as an exponent. Even though we do not formally define irrational exponents in this text, an irrational number such as π can be used as an exponent, and you can evaluate an expression such as 2^π by using a calculator. Try it:

$$2^\pi \approx 8.824977827$$

> **Domain**
>
> The domain of an exponential function is the set of all real numbers.

‹2› Graphing Exponential Functions

Even though the domain of an exponential function is the set of all real numbers, we can graph an exponential function by evaluating it for just a few integers.

E X A M P L E   3

Exponential functions with base greater than 1

Sketch the graph of each function.

a) $f(x) = 2^x$ **b)** $g(x) = 3^x$

Solution

a) We first make a table of ordered pairs that satisfy $f(x) = 2^x$:

x	-2	-1	0	1	2	3
$f(x) = 2^x$	$\frac{1}{4}$	$\frac{1}{2}$	1	2	4	8

As x increases, 2^x increases: $2^4 = 16$, $2^5 = 32$, $2^6 = 64$, and so on. As x decreases, the powers of 2 are getting closer and closer to 0, but always remain positive: $2^{-3} = \frac{1}{8}$, $2^{-4} = \frac{1}{16}$, $2^{-5} = \frac{1}{32}$, and so on. So as x decreases, the graph approaches but does not touch the x-axis. Because the domain of the function is $(-\infty, \infty)$

The graph of $f(x) = 2^x$ on a calculator appears to touch the x-axis. When drawing this graph by hand, make sure that it does not touch the x-axis. Use zoom to see that the curve is always above the x-axis.

Figure 11.1

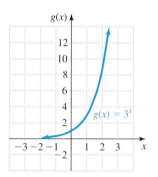

Figure 11.2

we draw the graph in Fig. 11.1 as a smooth curve through the points in the table. Since the powers of 2 are always positive, the range is $(0, \infty)$.

b) Make a table of ordered pairs that satisfy $g(x) = 3^x$:

x	-2	-1	0	1	2	3
$g(x) = 3^x$	$\frac{1}{9}$	$\frac{1}{3}$	1	3	9	27

Draw a smooth curve through the points indicated in the table. As x increases, 3^x increases. As x decreases, 3^x gets closer and closer to 0, but does not reach 0. So the graph shown in Fig. 11.2 approaches but does not touch the x-axis. The domain is $(-\infty, \infty)$ and the range is $(0, \infty)$.

> **Now do Exercises 25–26**

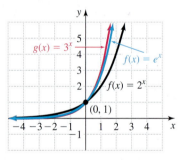

Figure 11.3

The curves in Figs. 11.1 and 11.2 are said to approach the x-axis **asymptotically,** and the x-axis is called an **asymptote** for the curves. Every exponential function has a horizontal asymptote.

Because $e \approx 2.718$, the graph of $f(x) = e^x$ lies between the graphs of $f(x) = 2^x$ and $g(x) = 3^x$, as shown in Fig. 11.3. Note that all three functions have the same domain and range and the same y-intercept. We summarize these ideas as follows:

$f(x) = a^x$ with $a > 1$

1. The y-intercept of the curve is $(0, 1)$.
2. The domain is $(-\infty, \infty)$ and the range is $(0, \infty)$.
3. The x-axis is an asymptote for the curve.
4. The y-values increase as we go from left to right on the curve.

EXAMPLE   **4**

Exponential functions with base between 0 and 1
Graph each function.

a) $f(x) = \left(\dfrac{1}{2}\right)^x$

b) $f(x) = 4^{-x}$

‹ Calculator Close-Up ›

The graph of $y = (1/2)^x$ is a reflection of the graph of $y = 2^x$.

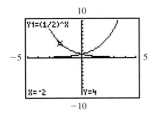

Solution

a) First make a table of ordered pairs that satisfy $f(x) = \left(\frac{1}{2}\right)^x$:

x	-2	-1	0	1	2	3
$f(x) = \left(\frac{1}{2}\right)^x$	4	2	1	$\frac{1}{2}$	$\frac{1}{4}$	$\frac{1}{8}$

As x increases, $\left(\frac{1}{2}\right)^x$ decreases, getting closer and closer to 0. Draw a smooth curve through these points as shown in Fig. 11.4.

Figure 11.4

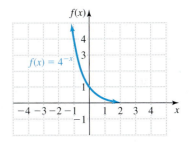

Figure 11.5

b) Because $4^{-x} = \left(\frac{1}{4}\right)^x$, we make a table for $f(x) = \left(\frac{1}{4}\right)^x$:

x	-2	-1	0	1	2	3
$f(x) = \left(\frac{1}{4}\right)^x$	16	4	1	$\frac{1}{4}$	$\frac{1}{16}$	$\frac{1}{64}$

As x increases, $\left(\frac{1}{4}\right)^x$, or 4^{-x}, decreases, getting closer and closer to 0. Draw a smooth curve through these points as shown in Fig. 11.5.

> **Now do Exercises 27−30**

Notice the similarities and differences between the exponential functions with $a > 1$ and those with base between 0 and 1. The main points are summarized as follows:

> **$f(x) = a^x$ with $0 < a < 1$**
>
> 1. The y-intercept of the curve is $(0, 1)$.
> 2. The domain is $(-\infty, \infty)$ and the range is $(0, \infty)$.
> 3. The x-axis is an asymptote for the curve.
> 4. The y-values decrease as we go from left to right on the curve.

CAUTION An exponential function can be written in more than one form. For example, $f(x) = \left(\frac{1}{2}\right)^x$ is the same as $f(x) = \frac{1}{2^x}$, or $f(x) = 2^{-x}$.

‹3› Transformations of Exponential Functions

We discussed transformation of functions in Section 9.2. In Example 5, we will graph some transformations of $f(x) = a^x$. Any transformation of an exponential function can be called an exponential function also.

Transformations of $f(x) = a^x$

Use transformations to graph each exponential function.

a) $f(x) = -2^x$ **b)** $f(x) = \frac{1}{3} \cdot 2^x + 1$ **c)** $f(x) = 2^{x-3} - 4$

Solution

a) The graph of $f(x) = -2^x$ is a reflection in the x-axis of the graph of $f(x) = 2^x$. Calculate a few ordered pairs for accuracy:

x	-1	0	1	2
$y = -2^x$	$-\frac{1}{2}$	-1	-2	-4

Plot these ordered pairs, and draw a curve through them as shown in Fig. 11.6.

b) To graph $f(x) = \frac{1}{3} \cdot 2^x + 1$, shrink the graph of $y = 2^x$ by a factor of $\frac{1}{3}$ and translate it upward one unit. Calculate a few ordered pairs for accuracy:

x	-1	0	1	2
$y = \frac{1}{3} \cdot 2^x + 1$	$\frac{7}{6}$	$\frac{4}{3}$	$\frac{5}{3}$	$\frac{7}{3}$

Plot these ordered pairs, and draw a curve through them as shown in Fig. 11.7.

Figure 11.6

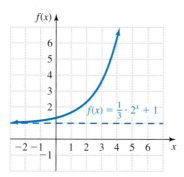

Figure 11.7   **Figure 11.8**

c) To graph $f(x) = 2^{x-3} - 4$ move $f(x) = 2^x$ to the right 3 units and down 4 units. Calculate a few ordered pairs for accuracy:

x	2	3	4	5
$y = 2^{x-3} - 4$	-3.5	-3	-2	0

Plot these ordered pairs, and draw a curve through them as shown in Fig. 11.8.

> **Now do Exercises 35–46**

⟨4⟩ Exponential Equations

In Chapter 9, we used the horizontal-line test to determine whether a function is one-to-one. Because no horizontal line can cross the graph of an exponential function more than once, exponential functions are one-to-one functions. For an exponential function one-to-one means that *if two exponential expressions with the same base are equal, then the exponents are equal.* If $2^x = 2^y$, then $x = y$.

> ### One-to-One Property of Exponential Functions
> For $a > 0$ and $a \neq 1$,
> $$\text{if} \quad a^m = a^n, \quad \text{then} \quad m = n.$$

In Example 6, we use the one-to-one property to solve equations involving exponential functions.

EXAMPLE 6

‹ Calculator Close-Up ›

You can see the solution to $2^{2x-1} = 8$ by graphing $y_1 = 2^{2x-1}$ and $y_2 = 8$. The x-coordinate of the point of intersection is the solution to the equation.

‹ Calculator Close-Up ›

The equation $9^{|x|} = 3$ has two solutions because the graphs of $y_1 = 9^{|x|}$ and $y_2 = 3$ intersect twice.

Using the one-to-one property
Solve each equation.

a) $2^{2x-1} = 8$ **b)** $9^{|x|} = 3$ **c)** $\dfrac{1}{8} = 4^x$

Solution

a) Because 8 is 2^3, we can write each side as a power of the same base, 2:

$$2^{2x-1} = 8 \qquad \text{Original equation}$$
$$2^{2x-1} = 2^3 \qquad \text{Write each side as a power of the same base.}$$
$$2x - 1 = 3 \qquad \text{One-to-one property}$$
$$2x = 4$$
$$x = 2$$

Check: $2^{2 \cdot 2 - 1} = 2^3 = 8$. The solution set is $\{2\}$.

b) Because $9 = 3^2$, we can write each side as a power of 3:

$$9^{|x|} = 3 \qquad \text{Original equation}$$
$$(3^2)^{|x|} = 3^1$$
$$3^{2|x|} = 3^1 \qquad \text{Power of a power rule}$$
$$2|x| = 1 \qquad \text{One-to-one property}$$
$$|x| = \frac{1}{2}$$
$$x = \pm\frac{1}{2} \qquad \text{Since } \left|-\frac{1}{2}\right| = \left|\frac{1}{2}\right| = \frac{1}{2}, \text{ there are two solutions to } |x| = \frac{1}{2}.$$

Check $x = \pm\frac{1}{2}$ in the original equation. The solution set is $\left\{-\frac{1}{2}, \frac{1}{2}\right\}$.

c) Because $\frac{1}{8} = 2^{-3}$ and $4 = 2^2$, we can write each side as a power of 2:

$$\frac{1}{8} = 4^x \qquad \text{Original equation}$$
$$2^{-3} = (2^2)^x \qquad \text{Write each side as a power of 2.}$$
$$2^{-3} = 2^{2x} \qquad \text{Power of a power rule}$$
$$2x = -3 \qquad \text{One-to-one property}$$
$$x = -\frac{3}{2}$$

Check $x = -\frac{3}{2}$ in the original equation. The solution set is $\left\{-\frac{3}{2}\right\}$.

> **Now do Exercises 47–60**

The one-to-one property is also used to find the first coordinate when given the second coordinate of an exponential function.

E X A M P L E   **7**

Finding the x-coordinate in an exponential function

Let $f(x) = 2^x$ and $g(x) = \left(\frac{1}{2}\right)^{1-x}$. Find x if

a) $f(x) = 32$

b) $g(x) = 8$

Solution

a) Because $f(x) = 2^x$ and $f(x) = 32$, we can find x by solving $2^x = 32$:

$$2^x = 32$$

$$2^x = 2^5 \quad \text{Write both sides as a power of the same base.}$$

$$x = 5 \quad \text{One-to-one property}$$

b) Because $g(x) = \left(\frac{1}{2}\right)^{1-x}$ and $g(x) = 8$, we can find x by solving $\left(\frac{1}{2}\right)^{1-x} = 8$:

$$\left(\frac{1}{2}\right)^{1-x} = 8$$

$$\left(2^{-1}\right)^{1-x} = 2^3 \quad \text{Because } \tfrac{1}{2} = 2^{-1} \text{ and } 8 = 2^3$$

$$2^{x-1} = 2^3 \quad \text{Power of a power rule}$$

$$x - 1 = 3 \quad \text{One-to-one property}$$

$$x = 4$$

> Now do Exercises 61–72

⟨5⟩ Applications

The simple interest formula $A = P + Prt$ gives the amount A after t years for a principal P invested at simple interest rate r. If an investment is earning **compound interest,** then interest is periodically paid into the account and the interest that is paid also earns interest. To compute the amount of an account earning compound interest, the simple interest formula is used repeatedly. For example, if an account earns 6% compounded quarterly and the amount at the beginning of the first quarter is $5000, we apply the simple interest formula with $P = \$5000$, $r = 0.06$, and $t = \frac{1}{4}$ to find the amount in the account at the end of the first quarter:

$$A = P + Prt$$

$$= P(1 + rt) \qquad \text{Factor.}$$

$$= 5000\left(1 + 0.06 \cdot \frac{1}{4}\right) \qquad \text{Substitute.}$$

$$= 5000(1.015)$$

$$= \$5075$$

To repeat this computation for another quarter, we multiply $5075 by 1.015. If A represents the amount in the account at the end of n quarters, we can write A as an exponential function of n:

$$A = \$5000(1.015)^n$$

In general, the amount A is given by the following formula.

Compound Interest Formula

If P represents the principal, i the interest rate per period, n the number of periods, and A the amount at the end of n periods, then

$$A = P(1 + i)^n.$$

EXAMPLE **8**

Compound interest formula

If $350 is deposited in an account paying 12% compounded monthly, then how much is in the account at the end of 6 years and 6 months?

Solution

Interest is paid 12 times per year, so the account earns $\frac{1}{12}$ of 12%, or 1% each month, for 78 months. So $i = 0.01$, $n = 78$, and $P = \$350$:

$$A = P(1 + i)^n$$
$$A = \$350(1.01)^{78}$$
$$\approx \$760.56$$

Now do Exercises 77–82

⟨ **Calculator Close-Up** ⟩

Graph $y = 350(1.01)^x$ to see the growth of the $350 deposit in Example 8 over time. After 360 months, it is worth $12,582.37.

If we shorten the length of the time period (yearly, quarterly, monthly, daily, hourly, etc.), the number of periods n increases while the interest rate for the period decreases. As n increases, the amount A also increases but will not exceed a certain amount. That certain amount is the amount obtained from *continuous compounding* of the interest. It is shown in more advanced courses that the following formula gives the amount when interest is compounded continuously.

⟨ **Helpful Hint** ⟩

Compare Examples 8 and 9 to see the difference between compounded monthly and compounded continuously. Although there is not much difference to an individual investor, there could be a large difference to the bank. Rework Examples 8 and 9 using $50 million as the deposit.

Continuous-Compounding Formula

If P is the principal or beginning balance, r is the annual percentage rate compounded continuously, t is the time in years, and A is the amount or ending balance, then

$$A = Pe^{rt}.$$

CAUTION The value of t in the continuous-compounding formula must be in years. For example, if the time is 1 year and 3 months, then $t = 1.25$ years. If the time is 3 years and 145 days, then

$$t = 3 + \frac{145}{365}$$

$$\approx 3.3973 \text{ years.}$$

E X A M P L E   9

Continuous-compounding formula

If $350 is deposited in an account paying 12% compounded continuously, then how much is in the account after 6 years and 6 months?

‹ **Calculator Close-Up** ›

Graph $y = 350e^{0.12x}$ to see the growth of the $350 deposit in Example 9 over time. After 30 years, it is worth $12,809.38.

Solution

Use $r = 12\%$, $t = 6.5$ years, and $P = \$350$ in the formula for compounding interest continuously:

$$A = Pe^{rt}$$

$$= 350e^{(0.12)(6.5)}$$

$$= 350e^{0.78}$$

$$\approx \$763.52 \quad \text{Use the } e^x \text{ key on a scientific calculator.}$$

Note that compounding continuously amounts to a few dollars more than compounding monthly did in Example 8.

Now do Exercises 83–90

Warm-Ups ▼

Fill in the blank.

1. An _____ function has the form $f(x) = a^x$, where $a > 0$ and $a \neq 1$.

2. The _____ of an exponential function is all real numbers.

3. Base e is the _____ base.

4. Base 10 is the _____ base.

5. The _____ property states that if $a^m = a^n$, then $m = n$.

6. The formula $A = P(1 + i)^n$ is for _____ interest.

7. The formula $A = Pe^{rt}$ is used when interest is compounded _____.

True or false?

8. If $f(x) = 4^x$, then $f\left(-\dfrac{1}{2}\right) = -2$.

9. If $f(x) = \left(\dfrac{1}{3}\right)^x$, then $f(-1) = 3$.

10. The functions $f(x) = \left(\dfrac{1}{2}\right)^x$ and $f(x) = 2^{-x}$ have the same graph.

11. The function $f(x) = 2^x$ is invertible.

12. The graph of $y = 2^x$ has an x-intercept.

13. The y-intercept for $f(x) = e^x$ is $(0, 1)$.

14. The expression $2^{\sqrt{2}}$ is undefined.

‹ Study Tips ›

- Study for the final exam by reworking all of your old test questions.
- It might have been a couple of months since you last worked a certain type of problem. Don't assume that you can do it correctly now just because you did it correctly a long time ago.

‹1› Exponential Functions

Let $f(x) = 4^x$, $g(x) = \left(\frac{1}{3}\right)^{x+1}$, and $h(x) = -2^x$. Find the following. See Example 1.

1. $f(2)$

2. $f(-1)$

3. $f\left(\frac{1}{2}\right)$

4. $f\left(-\frac{3}{2}\right)$

5. $g(-2)$

6. $g(1)$

7. $g(0)$

8. $g(-3)$

9. $h(0)$

10. $h(3)$

11. $h(-2)$

12. $h(-4)$

 Let $h(x) = 10^x$ and $j(x) = e^x$. Find the following. Use a calculator as necessary, and round approximate answers to three decimal places. See Example 2.

13. $h(0)$

14. $h(-1)$

15. $h(2)$

16. $h(3.4)$

17. $j(1)$

18. $j(3.5)$

19. $j(-2)$

20. $j(0)$

Fill in the missing entries in each table.

21.

x	-2	-1	0	1	2
4^x					

22.

x	-2	-1	0	1	2
5^x					

23.

x	-2	-1	0	1	2
$\left(\frac{1}{3}\right)^x$					

24.

x	-2	-1	0	1	2
$\left(\frac{1}{5}\right)^x$					

‹2› Graphing Exponential Functions

Sketch the graph of each function. See Examples 3 and 4.

25. $f(x) = 4^x$

26. $g(x) = 5^x$

27. $h(x) = \left(\frac{1}{3}\right)^x$

28. $i(x) = \left(\frac{1}{5}\right)^x$

29. $y = 10^x$

30. $y = (0.1)^x$

Fill in the missing entries in each table.

31.

x	-4	-3	-2	-1	0
10^{x+2}					

32.

x	-2	-1	$-\frac{1}{2}$	0	1
3^{2x+1}					

33.

x	-2	-1	0	1	2
-2^x					

34.

x	0	1	2	3	4
-2^{x-2}					

45. $f(x) = e^{-x} + 2$

46. $f(x) = e^{-x} - 1$

⟨3⟩ Transformations of Exponential Functions

Use transformations to help you sketch the graph of each function. See Example 5.

35. $f(x) = -3^x$

36. $f(x) = -10^x$

37. $f(x) = \dfrac{1}{2} \cdot 3^x$

38. $f(x) = -2 \cdot 3^x$

39. $f(x) = -3^x + 2$

40. $f(x) = -3^x - 4$

41. $f(x) = 3^{x-2} + 1$

42. $f(x) = 3^{x+1} - 2$

43. $f(x) = 10^x + 2$

44. $f(x) = -10^x + 3$

⟨4⟩ Exponential Equations

Solve each equation. See Example 6.

47. $2^x = 64$

48. $3^x = 9$

49. $10^x = 0.001$

50. $10^{2x} = 0.1$

51. $2^x = \dfrac{1}{4}$

52. $3^x = \dfrac{1}{9}$

53. $\left(\dfrac{2}{3}\right)^{x-1} = \dfrac{9}{4}$

54. $\left(\dfrac{1}{4}\right)^{3x} = 16$

55. $5^{-x} = 25$

56. $10^{-x} = 0.01$

57. $-2^{1-x} = -8$

58. $-3^{2-x} = -81$

59. $10^{|x|} = 1000$

60. $3^{|2x-5|} = 81$

Let $f(x) = 2^x$, $g(x) = \left(\dfrac{1}{3}\right)^x$, and $h(x) = 4^{2x-1}$. Find x in each case. See Example 7.

61. $f(x) = 4$

62. $f(x) = \dfrac{1}{4}$

63. $f(x) = 4^{2/3}$

64. $f(x) = 1$

65. $g(x) = 9$

66. $g(x) = \dfrac{1}{9}$

67. $g(x) = 1$

68. $g(x) = \sqrt{3}$

69. $h(x) = 16$

70. $h(x) = \dfrac{1}{2}$

71. $h(x) = 1$

72. $h(x) = \sqrt{2}$

Fill in the missing entries in each table.

73.

x	−5		0		4
2^x		$\dfrac{1}{8}$		2	

74.

x	−4		0		3
3^x		$\dfrac{1}{9}$		3	

75.

x		−2		1	
$\left(\dfrac{1}{2}\right)^x$	8		1		$\dfrac{1}{32}$

76.

x		-1		2	
$\left(\dfrac{1}{10}\right)^x$	100		1		$\dfrac{1}{1000}$

⟨5⟩ Applications

 Solve each problem. See Example 8.

77. Compounding quarterly. If $6000 is deposited in an account paying 5% compounded quarterly, then what amount will be in the account after 10 years?

78. Compounding quarterly. If $400 is deposited in an account paying 10% compounded quarterly, then what amount will be in the account after 7 years?

79. Bond fund. Fidelity's Municipal Income Fund (www.fidelity.com) returned an average of 4.72% annually from 1999 to 2009.

  a) How much was an investment of $10,000 in this fund in 1999 worth in 2009 at 4.72% compounded annually?

  b) Use the accompanying graph to estimate the year in which the $10,000 investment in 1999 would be worth $20,000 if it continued to return 4.72% annually.

Figure for Exercise 79

80. Slow growth. Fidelity's Contrafund returned an average of 1.98% annually from 1999 to 2009. How much was an investment of $10,000 in this fund in 1999 worth in 2009?

81. Depreciating knowledge. The value of a certain textbook seems to decrease according to the formula $V = 45 \cdot 2^{-0.9t}$, where V is the value in dollars and t is the age of the book in years. What is the book worth when it is new? What is it worth when it is 2 years old?

82. Mosquito abatement. In a Minnesota swamp in the springtime the number of mosquitoes per acre appears to grow according to the formula $N = 10^{0.1t\,+2}$, where t is the number of days since the last frost. What is the size of the mosquito population at times $t = 10$, $t = 20$, and $t = 30$?

Solve each problem. See Example 9.

83. Compounding continuously. If $500 is deposited in an account paying 7% compounded continuously, then how much will be in the account after 3 years?

84. Compounding continuously. If $7000 is deposited in an account paying 8% compounded continuously, then what will it amount to after 4 years?

85. One year's interest. How much interest will be earned the first year on $80,000 on deposit in an account paying 7.5% compounded continuously?

86. Partial year. If $7500 is deposited in an account paying 6.75% compounded continuously, then how much will be in the account after 5 years and 215 days?

87. Radioactive decay. The number of grams of a certain radioactive substance present at time t is given by the formula $A = 300 \cdot e^{-0.06t}$, where t is the number of years. Find the amount present at time $t = 0$. Find the amount present after 20 years. Use the graph to estimate the number of years that it takes for one-half of the substance to decay. Will the substance ever decay completely?

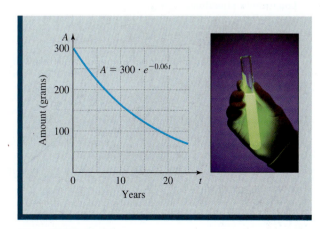

Figure for Exercise 87

88. *Population growth.* The population of a certain country appears to be growing according to the formula $P = 20 \cdot e^{0.1t}$, where P is the population in millions and t is the number of years since 1990. What was the population in 1990? What will the population be in the year 2010?

89. *Man overboard.* The difference in temperature between a warm human body (98.6°F) and a cold ocean (48.6°F) is given by the function $D = 50e^{-0.03t}$, where D is in degrees Fahrenheit and t is time in minutes. What is the difference between the body and the ocean for $t = 0$? What is the difference for $t = 15$? What is the ocean temperature at $t = 15$? What is the temperature of the human body at $t = 15$?

90. *Cooking a turkey.* The difference in temperature between a hot oven (350°F) and a cold turkey (38°F) is given by the function $D = 312e^{-0.12t}$, where D is in degrees Fahrenheit and t is time in hours. What is the difference between the turkey and the oven for $t = 0$? What is the difference for $t = 4$? What is the oven temperature at $t = 4$? What is the temperature of the turkey at $t = 4$?

Getting More Involved

91. *Exploration*

An approximate value for e can be found by adding the terms in the following infinite sum:

$$1 + \frac{1}{1} + \frac{1}{2 \cdot 1} + \frac{1}{3 \cdot 2 \cdot 1} + \frac{1}{4 \cdot 3 \cdot 2 \cdot 1} + \cdots$$

Use a calculator to find the sum of the first four terms. Find the difference between the sum of the first four terms and e. (For e, use all of the digits that your calculator gives for e^1.) What is the difference between e and the sum of the first eight terms?

Graphing Calculator Exercises

92. Graph $y_1 = 2^x$, $y_2 = e^x$, and $y_3 = 3^x$ on the same coordinate system. Which point do all three graphs have in common?

93. Graph $y_1 = 3^x$, $y_2 = 3^{x-1}$, and $y_3 = 3^{x-2}$ on the same coordinate system. What can you say about the graph of $y = 3^{x-h}$ for any real number h?

11.2   Logarithmic Functions and Their Applications

In This Section

⟨1⟩ Logarithmic Functions

⟨2⟩ Graphing Logarithmic Functions

⟨3⟩ Logarithmic Equations

⟨4⟩ Applications

In Section 11.1, you learned that exponential functions are one-to-one functions. Because they are one-to-one functions, they have inverse functions. In this section we study the inverses of the exponential functions.

⟨1⟩ Logarithmic Functions

We define $\log_a(x)$ as *the exponent that is used on the base a to obtain the result x.* Read the expression $\log_a(x)$ as "the base a logarithm of x." The expression $\log_a(x)$ is called a **logarithm.** If the *exponent* 3 is used on the *base* 2, then the *result* is 8 ($2^3 = 8$). So,

$$\log_2(8) = 3.$$

Base   Result   Exponent

Because $5^2 = 25$, the exponent used to obtain 25 with base 5 is 2 and $\log_5(25) = 2$. Because $2^{-5} = \frac{1}{32}$, the exponent used to obtain $\frac{1}{32}$ with base 2 is -5 and $\log_2\left(\frac{1}{32}\right) = -5$. From these examples, we see that the definition of $\log_a(x)$ can also be stated as follows:

> **Definition of $\log_a(x)$**
>
> For any $a > 0$ and $a \neq 1$,
>
> $$y = \log_a(x) \qquad \text{if and only if} \qquad a^y = x.$$

Note that the base of a logarithm must be a positive number and it cannot be 1.

E X A M P L E   1

Using the definition of logarithm

Write each logarithmic equation as an exponential equation and each exponential equation as a logarithmic equation.

    a) $\log_5(125) = 3$

    b) $6 = \log_{1/4}(x)$

    c) $\left(\dfrac{1}{2}\right)^m = 8$

    d) $7 = 3^z$

Solution

    a) "The base-5 logarithm of 125 equals 3" means that 3 is the exponent on 5 that produces 125. So $5^3 = 125$.

    b) The equation $6 = \log_{1/4}(x)$ is equivalent to $\left(\dfrac{1}{4}\right)^6 = x$ by the definition of logarithm.

    c) The equation $\left(\dfrac{1}{2}\right)^m = 8$ is equivalent to $\log_{1/2}(8) = m$.

    d) The equation $7 = 3^z$ is equivalent to $\log_3(7) = z$.

> **Now do Exercises 1–12**

The inverse of the base-a exponential function $f(x) = a^x$ is the **base-a logarithmic function** $f^{-1}(x) = \log_a(x)$. For example, $f(x) = 2^x$ and $f^{-1}(x) = \log_2(x)$ are inverse functions, as shown in Fig. 11.9. Each function undoes the other.

$$f(5) = 2^5 = 32 \quad \text{and} \quad f^{-1}(32) = \log_2(32) = 5.$$

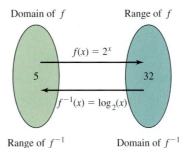

Figure 11.9

To evaluate logarithmic functions remember that a logarithm is an exponent: $\log_a(x)$ is the exponent that is used on the base a to obtain x.

EXAMPLE   **2**

Finding logarithms
Evaluate each logarithm.

a) $\log_5(25)$

b) $\log_2\left(\dfrac{1}{8}\right)$

c) $\log_{1/2}(4)$

d) $\log_{10}(0.001)$

e) $\log_9(3)$

‹ **Helpful Hint** ›

When we write $C(x) = 12x$, we may think of C as a variable and write $C = 12x$, or we may think of C as the name of a function, the cost function. In $y = \log_a(x)$ we are thinking of $\log_a$ only as the name of the function that pairs an x-value with a y-value.

Solution

a) The number $\log_5(25)$ is the exponent that is used on the base 5 to obtain 25. Because $25 = 5^2$, we have $\log_5(25) = 2$.

b) The number $\log_2\left(\dfrac{1}{8}\right)$ is the power of 2 that gives us $\frac{1}{8}$. Because $\frac{1}{8} = 2^{-3}$, we have $\log_2\left(\dfrac{1}{8}\right) = -3$.

c) The number $\log_{1/2}(4)$ is the power of $\frac{1}{2}$ that produces 4. Because $4 = \left(\frac{1}{2}\right)^{-2}$, we have $\log_{1/2}(4) = -2$.

d) Because $0.001 = 10^{-3}$, we have $\log_{10}(0.001) = -3$.

e) Because $9^{1/2} = 3$, we have $\log_9(3) = \frac{1}{2}$.

> **Now do Exercises 13–22**

There are two bases for logarithms that are used more frequently than the others: They are 10 and e. The base-10 logarithm is called the **common logarithm** and is usually written as $\log(x)$. The base-e logarithm is called the **natural logarithm** and is usually written as $\ln(x)$. Most scientific calculators have function keys for $\log(x)$ and $\ln(x)$. The simplest way to obtain a common or natural logarithm is to use a scientific calculator.

In Example 3, we find natural and common logarithms of certain numbers without a calculator.

EXAMPLE   **3**

Finding common and natural logarithms
Evaluate each logarithm.

a) $\log(1000)$

b) $\ln(e)$

c) $\log\left(\dfrac{1}{10}\right)$

‹ **Calculator Close-Up** ›

A graphing calculator has keys for the common logarithm (LOG) and the natural logarithm (LN).

Solution

a) Because $10^3 = 1000$, we have $\log(1000) = 3$.

b) Because $e^1 = e$, we have $\ln(e) = 1$.

c) Because $10^{-1} = \frac{1}{10}$, we have $\log\left(\dfrac{1}{10}\right) = -1$.

> **Now do Exercises 23–34**

The domain of the exponential function $y = 2^x$ is $(-\infty, \infty)$, and its range is $(0, \infty)$. Because the logarithmic function $y = \log_2(x)$ is the inverse of $y = 2^x$, the domain of $y = \log_2(x)$ is $(0, \infty)$, and its range is $(-\infty, \infty)$.

CAUTION The domain of $y = \log_a(x)$ for $a > 0$ and $a \neq 1$ is $(0, \infty)$. So expressions such as $\log_2(-4)$, $\log_{1/3}(0)$, and $\ln(-1)$ are undefined, because -4, 0, and -1 are not in the domain $(0, \infty)$.

⟨2⟩ Graphing Logarithmic Functions

In Chapter 9, we saw that the graphs of a function and its inverse function are symmetric about the line $y = x$. Because the logarithm functions are inverses of exponential functions, their graphs are also symmetric about $y = x$.

E X A M P L E   **4**

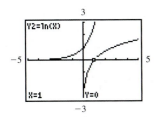
A logarithmic function with base greater than 1

Sketch the graph of $g(x) = \log_2(x)$ and compare it to the graph of $y = 2^x$.

Solution

Make a table of ordered pairs for $g(x) = \log_2(x)$ using positive numbers for x:

x	$\frac{1}{4}$	$\frac{1}{2}$	1	2	4	8
$g(x) = \log_2(x)$	-2	-1	0	1	2	3

Draw a curve through these points as shown in Fig. 11.10. The graph of the inverse function $y = 2^x$ is also shown in Fig. 11.10 for comparison. Note the symmetry of the two curves about the line $y = x$.

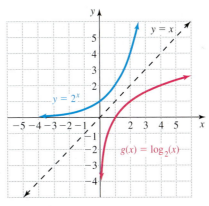

Figure 11.10

Now do Exercises 43–46

All logarithm functions with bases greater than 1 have graphs that are similar to the one in Fig. 11.10. In general, these functions have the following characteristics, as shown in Fig. 11.11.

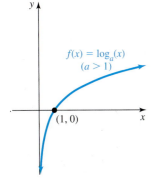

Figure 11.11

$f(x) = \log_a(x)$ with $a > 1$

1. The x-intercept of the curve is $(1, 0)$.

2. The domain is $(0, \infty)$ and the range is $(-\infty, \infty)$.

3. The y-axis is an asymptote for the curve.

4. The y-values increase as we go from left to right on the curve.

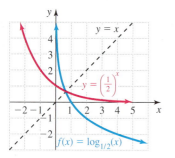

Figure 11.12

E X A M P L E   **5**

A logarithmic function with base less than 1

Sketch the graph of $f(x) = \log_{1/2}(x)$, and compare it to the graph of $y = \left(\frac{1}{2}\right)^x$.

Solution

Make a table of ordered pairs for $f(x) = \log_{1/2}(x)$ using positive numbers for x:

x	$\frac{1}{4}$	$\frac{1}{2}$	1	2	4	8
$f(x) = \log_{1/2}(x)$	2	1	0	-1	-2	-3

The curve through these points is shown in Fig. 11.12. The graph of the inverse function $y = \left(\frac{1}{2}\right)^x$ is also shown in Fig. 11.12 for comparison. Note the symmetry with respect to the line $y = x$.

Now do Exercises 47–50

All logarithm functions with bases between 0 and 1 have graphs that are similar to the one in Fig. 11.12. In general, these functions have the following characteristics, as shown in Fig. 11.13.

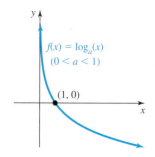

Figure 11.13

$f(x) = \log_a(x)$ with $0 < a < 1$

1. The x-intercept of the curve is $(1, 0)$.
2. The domain is $(0, \infty)$ and the range is $(-\infty, \infty)$.
3. The y-axis is an asymptote for the curve.
4. The y-values decrease as we go from left to right on the curve.

Figures 11.11 and 11.13 illustrate the fact that $y = \log_a(x)$ and $y = a^x$ are inverse functions for any base a. For any given exponential or logarithmic function the inverse function can be easily obtained from the definition of logarithm.

E X A M P L E   **6**

Inverses of logarithmic and exponential functions

Find the inverse of each function.

a) $f(x) = 10^x$ **b)** $g(x) = \log_3(x)$

Solution

a) To find any inverse function we switch the roles of x and y. So $y = 10^x$ becomes $x = 10^y$. Now $x = 10^y$ is equivalent to $y = \log_{10}(x)$. So the inverse of $f(x) = 10^x$ is $y = \log(x)$ or $f^{-1}(x) = \log(x)$.

b) In $g(x) = \log_3(x)$ or $y = \log_3(x)$ we switch x and y to get $x = \log_3(y)$. Now $x = \log_3(y)$ is equivalent to $y = 3^x$. So the inverse of $g(x) = \log_3(x)$ is $y = 3^x$ or $g^{-1}(x) = 3^x$.

Now do Exercises 51–56

⟨3⟩ **Logarithmic Equations**

In Section 11.1, we learned that the exponential functions are one-to-one functions. Because logarithmic functions are inverses of exponential functions, they are one-to-one functions also. For a base-a logarithmic function *one-to-one means that if the base-a logarithms of two numbers are equal, then the numbers are equal.*

> **One-to-One Property of Logarithms**
>
> For $a > 0$ and $a \neq 1$,
>
> $$\text{if} \quad \log_a(m) = \log_a(n), \quad \text{then} \quad m = n.$$

The one-to-one property of logarithms and the definition of logarithms are the two basic tools that we use to solve equations involving logarithms. We use these tools in Example 7.

EXAMPLE 7

Logarithmic equations

Solve each equation.

a) $\log_3(x) = -2$ 　　　 b) $\log_x(8) = -3$ 　　　 c) $\log(x^2) = \log(4)$

Solution

a) Use the definition of logarithms to rewrite the logarithmic equation as an equivalent exponential equation:

$$\log_3(x) = -2$$
$$3^{-2} = x \quad \text{Definition of logarithm}$$
$$\frac{1}{9} = x$$

Because $3^{-2} = \frac{1}{9}$ or $\log_3\left(\frac{1}{9}\right) = -2$, the solution set is $\left\{\frac{1}{9}\right\}$.

b) Use the definition of logarithms to rewrite the logarithmic equation as an equivalent exponential equation:

$$\log_x(8) = -3$$
$$x^{-3} = 8 \qquad\qquad \text{Definition of logarithm}$$
$$(x^{-3})^{-1} = 8^{-1} \qquad \text{Raise each side to the } -1 \text{ power.}$$
$$x^3 = \frac{1}{8}$$
$$x = \sqrt[3]{\frac{1}{8}} = \frac{1}{2} \qquad \text{Odd-root property}$$

Because $\left(\frac{1}{2}\right)^{-3} = 2^3 = 8$ or $\log_{1/2}(8) = -3$, the solution set is $\left\{\frac{1}{2}\right\}$.

c) To write an equation equivalent to $\log(x^2) = \log(4)$, we use the one-to-one property of logarithms:

$$\log(x^2) = \log(4)$$
$$x^2 = 4 \qquad \text{One-to-one property of logarithms}$$
$$x = \pm 2 \qquad \text{Even-root property}$$

If $x = \pm 2$, then $x^2 = 4$ and $\log(4) = \log(4)$. The solution set is $\{-2, 2\}$.

> **Now do Exercises 57–68**

CAUTION If we have equality of two logarithms with the same base, we use the one-to-one property to eliminate the logarithms. If we have an equation with only one logarithm, such as $\log_a(x) = y$, we use the definition of logarithm to write $a^y = x$ and to eliminate the logarithm.

⟨4⟩ Applications

The definition of logarithm indicates that $y = \log_a(x)$ if and only if $a^y = x$. If the base is e, then the definition indicates that

$$y = \ln(x) \quad \text{if and only if} \quad e^y = x.$$

In Example 8, we use the definition of logarithm to solve a problem involving the continuous-compounding formula

$$A = Pe^{rt},$$

where A is the amount after t years of an investment of P dollars at annual percentage rate r compounded continuously.

E X A M P L E   8

Finding the time with continuous compounding

How long does it take for $80 to grow to $240 at 12% annual percentage rate compounded continuously?

Solution

Use $r = 0.12$, $P = \$80$, and $A = \$240$ in the formula $A = Pe^{rt}$ to get $240 = 80e^{0.12t}$. Now use the definition of logarithm to solve for t:

$$240 = 80e^{0.12t}$$

$$3 = e^{0.12t} \qquad \text{Divide each side by 80.}$$

$$0.12t = \ln(3) \qquad \begin{array}{l}\text{Definition of logarithm:}\\ y = e^x \text{ means } x = \ln(y)\end{array}$$

$$t = \frac{\ln(3)}{0.12} \qquad \text{Divide each side by 0.12.}$$

$$t \approx 9.155$$

The time is approximately 9.155 years. Multiply 365 by 0.155 to get approximately 57 days. So the time is 9 years and 57 days to the nearest day.

> **Now do Exercises 79–90**

Note that we can also use the technique of Example 8 to solve a continuous-compounding problem in which the rate is the only unknown quantity.

Warm-Ups ▼

Fill in the blank.

1. The inverse for an exponential function is a _____ function.

2. A _____ logarithm is a base-10 logarithm.

3. A _____ logarithm is a base-e logarithm.

4. The _____ of $f(x) = \log_a(x)$ is $(0, \infty)$.

5. The _____ property states that if $\log_a(m) = \log_a(n)$, then $m = n$.

6. The graphs of $f(x) = 2^x$ and $g(x) = \log_2(x)$ are _____ about the line $y = x$.

7. The expression $\log_a(x)$ is the _____ that is used on base a to obtain x.

True or false?

8. The equation $a^3 = 2$ is equivalent to $\log_a(2) = 3$.

9. If (a, b) satisfies $y = 8^x$, then (a, b) satisfies $y = \log_8(x)$.

10. The inverse of $y = 5^x$ is $y = \log_5(x)$.

11. If $f(x) = \ln(x)$, then $f^{-1}(x) = e^x$.

12. $\log_{25}(5) = 2$

13. $\log(-10) = 1$

14. $\log_{1/2}(32) = -5$

15. $10^{\log(19)} = 19$

‹1› Logarithmic Functions

Write each exponential equation as a logarithmic equation and each logarithmic equation as an exponential equation. See Example 1.

1. $\log_2(8) = 3$

2. $\log_{10}(10) = 1$

3. $10^2 = 100$

4. $5^3 = 125$

5. $y = \log_5(x)$

6. $m = \log_b(N)$

7. $2^a = b$

8. $a^3 = c$

9. $\log_3(x) = 10$

10. $\log_c(t) = 4$

11. $e^3 = x$

12. $m = e^x$

Evaluate each logarithm. See Examples 2 and 3.

13. $\log_2(4)$

14. $\log_2(1)$

15. $\log_2(16)$

16. $\log_4(16)$

17. $\log_2(64)$

18. $\log_8(64)$

19. $\log_4(64)$

20. $\log_{64}(64)$

21. $\log_2\left(\dfrac{1}{4}\right)$

22. $\log_2\left(\dfrac{1}{8}\right)$

23. $\log(100)$

24. $\log(1)$

25. $\log(0.01)$

26. $\log(10{,}000)$

27. $\log_{1/3}\left(\dfrac{1}{3}\right)$

28. $\log_{1/3}\left(\dfrac{1}{9}\right)$

29. $\log_{1/3}(27)$

30. $\log_{1/3}(1)$

31. $\log_{25}(5)$

32. $\log_{16}(4)$

33. $\ln(e^2)$

34. $\ln\left(\dfrac{1}{e}\right)$

Use a calculator to evaluate each logarithm. Round answers to four decimal places.

35. $\log(5)$

36. $\log(0.03)$

37. $\ln(6.238)$

38. $\ln(0.23)$

Fill in the missing entries in each table.

39.

x	$\frac{1}{9}$	$\frac{1}{3}$	1	3	9
$\log_3(x)$					

40.

x	$\frac{1}{100}$	$\frac{1}{10}$	1	10	100
$\log_{10}(x)$					

41.

x	16	4	1	$\frac{1}{4}$	$\frac{1}{16}$
$\log_{1/4}(x)$					

42.

x	9	3	1	$\frac{1}{3}$	$\frac{1}{9}$
$\log_{1/3}(x)$					

‹2› Graphing Logarithmic Functions

Sketch the graph of each function. See Examples 4 and 5.

43. $f(x) = \log_3(x)$

44. $g(x) = \log_{10}(x)$

45. $y = \log_4(x)$

46. $y = \log_5(x)$

47. $h(x) = \log_{1/4}(x)$

48. $y = \log_{1/3}(x)$

49. $y = \log_{1/5}(x)$

50. $y = \log_{1/6}(x)$

Find the inverse of each function. See Example 6.

51. $f(x) = 6^x$ **52.** $f(x) = 4^x$

53. $f(x) = \ln(x)$ **54.** $f(x) = \log(x)$

55. $f(x) = \log_{1/2}(x)$ **56.** $f(x) = \log_{1/4}(x)$

⟨3⟩ Logarithmic Equations

Solve each equation. See Example 7.

57. $x = \left(\dfrac{1}{2}\right)^{-2}$ **58.** $x = 16^{-1/2}$

59. $5 = 25^x$ **60.** $0.1 = 10^x$

61. $\log(x) = -3$ **62.** $\log(x) = 5$

63. $\log_x(36) = 2$ **64.** $\log_x(100) = 2$

65. $\log_x(5) = -1$ **66.** $\log_x(16) = -2$

 67. $\log(x^2) = \log(9)$ **68.** $\ln(2x - 3) = \ln(x + 1)$

 Use a calculator to solve each equation. Round answers to four decimal places.

69. $3 = 10^x$ **70.** $10^x = 0.03$

71. $10^x = \dfrac{1}{2}$ **72.** $75 = 10^x$

73. $e^x = 7.2$ **74.** $e^{3x} = 0.4$

Fill in the missing entries in each table.

75.

x	$\dfrac{1}{4}$		1		16
$\log_2(x)$		-1		2	

76.

x	$\dfrac{1}{125}$		1		625
$\log_5(x)$		-2		1	

77.

x		4		$\dfrac{1}{2}$	
$\log_{1/2}(x)$	-4		0		2

78.

x		6		$\dfrac{1}{36}$	
$\log_{1/6}(x)$	-2		0		3

⟨4⟩ Applications

Solve each problem. See Example 8. Use a calculator as necessary.

79. *Double your money.* How long does it take $5000 to grow to $10,000 at 12% compounded continuously?

80. *Half the rate.* How long does it take $5000 to grow to $10,000 at 6% compounded continuously?

81. *Earning interest.* How long does it take to earn $1000 in interest on a deposit of $6000 at 8% compounded continuously?

82. *Lottery winnings.* How long does it take to earn $1000 interest on a deposit of one million dollars at 9% compounded continuously?

83. *Investing.* An investment of $10,000 in Bonavista Energy in 2000 grew to $15,431 in 2009.

    a) Assuming that the investment grew continuously, what was the annual growth rate?

    b) If Bonavista Energy continued to grow continuously at the rate from part (a), then what would the investment be worth in 2020?

84. *Investing.* An investment of $10,000 in Baytex Energy in 2002 was worth $18,125 in 2009.

    a) Assuming that the investment grew continuously, what was the annual rate?

    b) If Baytex Energy continued to grow continuously at the rate from part (a), then what would the investment be worth in 2015?

In chemistry the pH of a solution is defined by

$$\text{pH} = -\log_{10}[H+],$$

where H+ is the hydrogen ion concentration of the solution in moles per liter. Distilled water has a pH of approximately 7. A solution with a pH under 7 is called an acid, and one with a pH over 7 is called a base.

85. *Tomato juice.* Tomato juice has a hydrogen ion concentration of $10^{-4.1}$ mole per liter (mol/L). Find the pH of tomato juice.

86. *Stomach acid.* The gastric juices in your stomach have a hydrogen ion concentration of 10^{-1} mol/L. Find the pH of your gastric juices.

87. *Neuse River* pH. The hydrogen ion concentration of a water sample from the Neuse River at New Bern, North Carolina, was 1.58×10^{-7} mol/L (wwwnc.usgs.gov). What was the pH of this water sample?

88. *Roanoke River* pH. The hydrogen ion concentration of a water sample from the Roanoke River at Janesville, North Carolina, was 1.995×10^{-7} mol/L (wwwnc.usgs.gov). What was the pH of this water sample?

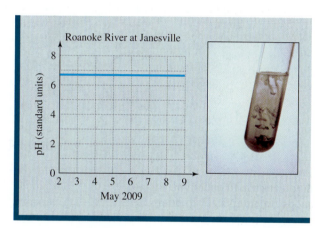

Figure for Exercise 88

Solve each problem.

89. *Sound level.* The level of sound in decibels (dB) is given by the formula

$$L = 10 \cdot \log(I \times 10^{12}),$$

where I is the intensity of the sound in watts per square meter. If the intensity of the sound at a rock concert is 0.001 watt per square meter at a distance of 75 meters from the stage, then what is the level of the sound at this point in the audience?

90. *Logistic growth.* If a rancher has one cow with a contagious disease in a herd of 1000, then the time in days t for n of the cows to become infected is modeled by

$$t = -5 \cdot \ln\left(\frac{1000 - n}{999n}\right).$$

Find the number of days that it takes for the disease to spread to 100, 200, 998, and 999 cows. This model, called a *logistic growth model,* describes how a disease can spread very rapidly at first and then very slowly as nearly all of the population has become infected. See the accompanying figure.

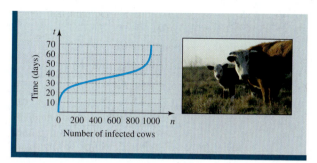

Figure for Exercise 90

Getting More Involved

91. *Discussion*

Use the switch-and-solve method from Chapter 9 to find the inverse of the function $f(x) = 5 + \log_2(x - 3)$. State the domain and range of the inverse function.

92. *Discussion*

Find the inverse of the function $f(x) = 2 + e^{x+4}$. State the domain and range of the inverse function.

Graphing Calculator Exercises

93. *Composition of inverses.* Graph the functions $y = \ln(e^x)$ and $y = e^{\ln(x)}$. Explain the similarities and differences between the graphs.

94. *The population bomb.* The population of the earth is growing continuously with an annual rate of about 1.6%. If the present population is 6 billion, then the function $y = 6e^{0.016x}$ gives the population in billions x years from now. Graph this function for $0 \leq x \leq 200$. What will the population be in 100 years and in 200 years?

| **Math** *at* **Work** | **Drug Administration** |

When a drug is taken continuously or intermittently, plasma concentrations of the drug increase. Over time, the rate of increase slows and eventually reaches a plateau. As concentration increases, the rate of elimination increases until a point is reached at which the amount of drug being eliminated from the body equals the amount being administered (steady state).

The time to reach steady state depends on the half-life of the drug. The half-life of a drug is the time it takes for the plasma concentration to be reduced by one-half. See the accompanying figure. The basic rule is that after administering a drug for a period equal to the half-life of the drug, the plasma concentration will be halfway between the starting concentration and steady state. This rule holds for any starting concentration. Mathematically, steady state is a limit and it is never reached. It is usually assumed that when a drug reaches 90% or more of steady state it is at steady state. It takes 3.3 half-lives of drug administration to reach 90% of steady state.

The half-life $t_{1/2}$ of a drug depends on the patient and is calculated from two plasma levels separated by a time interval. The first plasma level or peak (P) is measured after the drug has been fully distributed. The second plasma level or trough (T) is measured at some interval later (t). From P, T, and t, the elimination constant k is found by $k = \frac{\ln(P) - \ln(T)}{t}$. The half-life is then found using $t_{1/2} = \frac{\ln(2)}{k}$. When the dosing interval is much longer than the half-life, there is more time for elimination between doses and accumulation is small. When the dosing interval is much shorter than the half-life, there is little time for elimination and more accumulation of the drug.

Mid-Chapter **Quiz** | Sections 11.1 through 11.2 | **Chapter 11**

Fill in the missing entries in each table.

1.

x	-2		0		3
2^x		$\frac{1}{2}$		4	

2.

x	-4		1		4
$\left(\frac{1}{2}\right)^x$		4		$\frac{1}{4}$	

3.

x	16	1	$\frac{1}{4}$	$\frac{1}{16}$
$\log_4(x)$				

4.

x	32	16	1	$\frac{1}{8}$
$\log_{1/2}(x)$				

Solve each equation.

5. $3^x = 81$

6. $\left(\frac{1}{2}\right)^{3x} = 8$

7. $10^x = 0.01$

8. $\log_x(64) = 2$

9. $\log(x) = 4$

Find the inverse of each function.

10. $f(x) = 10^x$ **11.** $g(x) = \log_8(x)$ **16.** $y = \log_2(x)$

Rewrite the exponential equation as a logarithmic equation and the logarithmic equation as an exponential equation.

12. $M = \log_5(W)$ **13.** $a^3 = y$

Graph each function, and state its domain and range.

14. $f(x) = 2^x + 1$

Miscellaneous.

17. Use a calculator to find $g(-2.3)$ to three decimal places if $g(x) = e^x$.

18. If \$4000 is invested at 6% compounded quarterly, then what amount will be in the account after 5 years?

19. If \$8000 is invested at 4.3% compounded continuously, then what will it amount to after $12\frac{1}{2}$ years?

15. $g(x) = \left(\dfrac{1}{2}\right)^x - 3$

20. How long to the nearest day does it take for \$5000 to grow to \$8000 at 5% compounded continuously?

    11.3 **Properties of Logarithms**

In This Section

⟨1⟩ **The Inverse Properties**

⟨2⟩ **The Product Rule for Logarithms**

⟨3⟩ **The Quotient Rule for Logarithms**

⟨4⟩ **The Power Rule for Logarithms**

⟨5⟩ **Using the Properties**

The properties of logarithms are very similar to the properties of exponents because *logarithms are exponents*. In this section, we use the properties of exponents to write some properties of logarithms. The properties will be used in solving logarithmic equations in Section 11.4.

⟨1⟩ **The Inverse Properties**

An exponential function and logarithmic function with the same base are inverses of each other. For example, the logarithm of 32 base 2 is 5 and the fifth power of 2 is 32. In symbols, we have

$$2^{\log_2(32)} = 2^5 = 32.$$

If we raise 3 to the fourth power, we get 81; and if we find the base-3 logarithm of 81, we get 4. In symbols, we have

$$\log_3(3^4) = \log_3(81) = 4.$$

We can state the inverse relationship between exponential and logarithm functions in general with the following inverse properties:

> **The Inverse Properties**
> 1. $\log_a(a^x) = x$ for any real number x.
> 2. $a^{\log_a(x)} = x$ for any positive real number x.

E X A M P L E   1

Using the inverse properties
Simplify each expression.

a) $\ln(e^5)$ **b)** $2^{\log_2(8)}$

Solution
a) Using the first inverse property, we get $\ln(e^5) = 5$.

b) Using the second inverse property, we get $2^{\log_2(8)} = 8$.

> Now do Exercises 1–8

‹ **Calculator Close-Up** ›

You can illustrate the product rule for logarithms with a graphing calculator.

```
log(7)+log(8)
         1.748188027
log(56)
         1.748188027
```

⟨2⟩ **The Product Rule for Logarithms**

Using the product rule for exponents and the inverse property $a^{\log_a(x)} = x$ we have

$$a^{\log_a M + \log_a N} = a^{\log_a M} a^{\log_a N} = M \cdot N.$$

By the definition of logarithm, that power of a that produces $M \cdot N$ is the base a logarithm of $M \cdot N$. So,

$$\log_a(M \cdot N) = \log_a M + \log_a N.$$

This last equation is called the **product rule for logarithms.** It says that *the logarithm of a product of two numbers is equal to the sum of their logarithms,* provided all logarithms are defined and all have the same base.

> **The Product Rule for Logarithms**
> For $M > 0$ and $N > 0$,
> $$\log_a(M \cdot N) = \log_a M + \log_a N.$$

E X A M P L E   2

Using the product rule for logarithms
Write each expression as a single logarithm.

a) $\log_2(7) + \log_2(5)$ **b)** $\ln(\sqrt{2}) + \ln(\sqrt{3})$

Solution
a) $\log_2(7) + \log_2(5) = \log_2(35)$ Product rule for logarithms

b) $\ln(\sqrt{2}) + \ln(\sqrt{3}) = \ln(\sqrt{6})$ Product rule for logarithms

> Now do Exercises 9–20

⟨3⟩ The Quotient Rule for Logarithms

Using the quotient rule for exponents and the inverse property $a^{\log_a(x)} = x$ we have

$$a^{\log_a M - \log_a N} = \frac{a^{\log_a M}}{a^{\log_a N}} = \frac{M}{N}.$$

By the definition of logarithm, the power of a that produces $\frac{M}{N}$ is the base a logarithm of $\frac{M}{N}$. So,

$$\log_a\left(\frac{M}{N}\right) = \log_a M - \log_a N.$$

This last equation is called the **quotient rule for logarithms.** It says that *the logarithm of a quotient of two numbers is equal to the difference of their logarithms,* provided all logarithms are defined and all have the same base.

> **The Quotient Rule for Logarithms**
>
> For $M > 0$ and $N > 0$,
>
> $$\log_a\left(\frac{M}{N}\right) = \log_a M - \log_a N.$$

‹ Calculator Close-Up ›

You can illustrate the quotient rule for logarithms with a graphing calculator.

```
ln(99/2)
            3.90197267
ln(99)-ln(2)
            3.90197267
```

E X A M P L E   3

Using the quotient rule for logarithms

Write each expression as a single logarithm.

    a) $\log_2(3) - \log_2(7)$

    b) $\ln(w^8) - \ln(w^2)$

Solution

    a) $\log_2(3) - \log_2(7) = \log_2\left(\frac{3}{7}\right)$ Quotient rule for logarithms

    b) $\ln(w^8) - \ln(w^2) = \ln\left(\frac{w^8}{w^2}\right)$ Quotient rule for logarithms

                           $= \ln(w^6)$ Quotient rule for exponents

> **Now do Exercises 21–32**

‹ Calculator Close-Up ›

You can illustrate the power rule for logarithms with a graphing calculator.

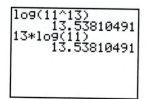

```
log(11^13)
           13.53810491
13*log(11)
           13.53810491
```

⟨4⟩ The Power Rule for Logarithms

Using the power rule for exponents and the inverse property $a^{\log_a(x)} = x$ we have

$$a^{N \cdot \log_a M} = \left(a^{\log_a M}\right)^N = M^N.$$

By the definition of logarithm, the power of a that produces M^N is the base a logarithm of M^N. So,

$$\log_a(M^N) = N \cdot \log_a M.$$

This last equation is called the **power rule for logarithms.** It says that *the logarithm of a power of a number is equal to the power times the logarithm of the number,* provided all logarithms are defined.

The Power Rule for Logarithms

For $M > 0$,

$$\log_a(M^N) = N \cdot \log_a M.$$

E X A M P L E   4

Using the power rule for logarithms

Rewrite each logarithm in terms of log(2).

    a) $\log(2^{10})$ **b)** $\log(\sqrt{2})$ **c)** $\log\left(\dfrac{1}{2}\right)$

Solution

    a) $\log(2^{10}) = 10 \cdot \log(2)$ Power rule for logarithms

    b) $\log(\sqrt{2}) = \log(2^{1/2})$ Write $\sqrt{2}$ as a power of 2.

               $= \dfrac{1}{2}\log(2)$ Power rule for logarithms

    c) $\log\left(\dfrac{1}{2}\right) = \log(2^{-1})$ Write $\frac{1}{2}$ as a power of 2.

               $= -1 \cdot \log(2)$ Power rule for logarithms

               $= -\log(2)$

Now do Exercises 33–38

⟨5⟩ Using the Properties

We have already seen many properties of logarithms. There are three properties that we have not yet formally stated. Because $a^1 = a$ and $a^0 = 1$, we have $\log_a(a) = 1$ and $\log_a(1) = 0$ for any positive number a. If we apply the quotient rule to $\log_a(1/N)$, we get

$$\log_a\left(\frac{1}{N}\right) = \log_a(1) - \log_a(N) = 0 - \log_a(N) = -\log_a(N).$$

So $\log_a\left(\frac{1}{N}\right) = -\log_a(N)$. These three new properties along with all of the other properties of logarithms are summarized as follows.

Properties of Logarithms

If M, N, and a are positive numbers, $a \neq 1$, then

  1. $\log_a(a) = 1$

  2. $\log_a(1) = 0$

  3. $\log_a(a^x) = x$ for any real number x.

                                             Inverse properties

  4. $a^{\log_a(x)} = x$ for any positive real number x.

  5. $\log_a(MN) = \log_a(M) + \log_a(N)$ Product rule

  6. $\log_a\left(\dfrac{M}{N}\right) = \log_a(M) - \log_a(N)$ Quotient rule

  7. $\log_a\left(\dfrac{1}{N}\right) = -\log_a(N)$

  8. $\log_a(M^N) = N \cdot \log_a(M)$.   Power rule

We have already seen several ways in which to use the properties of logarithms. In Examples 5, 6, and 7 we see more uses of the properties. First we use the rules of logarithms to write the logarithm of a complicated expression in terms of logarithms of simpler expressions.

E X A M P L E   **5**

Using the properties of logarithms
Rewrite each expression in terms of $\log(2)$ and/or $\log(3)$.

a) $\log(6)$ **b)** $\log(16)$ **c)** $\log\left(\dfrac{9}{2}\right)$ **d)** $\log\left(\dfrac{1}{3}\right)$

Solution

a) $\log(6) = \log(2 \cdot 3)$

$= \log(2) + \log(3)$ Product rule

b) $\log(16) = \log(2^4)$

$= 4 \cdot \log(2)$ Power rule

c) $\log\left(\dfrac{9}{2}\right) = \log(9) - \log(2)$ Quotient rule

$= \log(3^2) - \log(2)$

$= 2 \cdot \log(3) - \log(2)$ Power rule

d) $\log\left(\dfrac{1}{3}\right) = -\log(3)$ Property 7

Now do Exercises 39–50

‹ **Calculator Close-Up** ›

Examine the values of $\log(9/2)$, $\log(9) - \log(2)$, and $\log(9)/\log(2)$.

```
log(9/2)
       .6532125138
log(9)-log(2)
       .6532125138
log(9)/log(2)
       3.169925001
```

CAUTION   Do not confuse $\dfrac{\log(9)}{\log(2)}$ with $\log\left(\dfrac{9}{2}\right)$. We can use the quotient rule to write $\log\left(\dfrac{9}{2}\right) = \log(9) - \log(2)$, but $\dfrac{\log(9)}{\log(2)} \neq \log(9) - \log(2)$. The expression $\dfrac{\log(9)}{\log(2)}$ means $\log(9) \div \log(2)$. Use your calculator to verify these two statements.

The properties of logarithms can be used to combine several logarithms into a single logarithm (as in Examples 2 and 3) or to write a logarithm of a complicated expression in terms of logarithms of simpler expressions.

E X A M P L E   **6**

Using the properties of logarithms
Rewrite each expression as a sum or difference of multiples of logarithms.

a) $\log\left(\dfrac{xz}{y}\right)$ **b)** $\log_3\left(\dfrac{(x-3)^{2/3}}{\sqrt{x}}\right)$

Solution

a) $\log\left(\dfrac{xz}{y}\right) = \log(xz) - \log(y)$ Quotient rule

$= \log(x) + \log(z) - \log(y)$ Product rule

b) $\log_3\left(\dfrac{(x-3)^{2/3}}{\sqrt{x}}\right) = \log_3((x-3)^{2/3}) - \log_3(x^{1/2})$ Quotient rule

$\qquad\qquad\qquad\qquad = \dfrac{2}{3}\log_3(x-3) - \dfrac{1}{2}\log_3(x)$ Power rule

Now do Exercises 51–62

In Example 7, we use the properties of logarithms to convert expressions involving several logarithms into a single logarithm. The skills we are learning here will be used to solve logarithmic equations in Section 11.4.

EXAMPLE 7

Combining logarithms

Rewrite each expression as a single logarithm.

a) $\dfrac{1}{2}\log(x) - 2 \cdot \log(x+1)$ **b)** $3 \cdot \log(y) + \dfrac{1}{2}\log(z) - \log(x)$

Solution

a) $\dfrac{1}{2}\log(x) - 2 \cdot \log(x+1) = \log(x^{1/2}) - \log((x+1)^2)$ Power rule

$\qquad\qquad\qquad\qquad\qquad = \log\left(\dfrac{\sqrt{x}}{(x+1)^2}\right)$ Quotient rule

b) $3 \cdot \log(y) + \dfrac{1}{2}\log(z) - \log(x) = \log(y^3) + \log(\sqrt{z}) - \log(x)$ Power rule

$\qquad\qquad\qquad\qquad\qquad = \log(y^3 \cdot \sqrt{z}) - \log(x)$ Product rule

$\qquad\qquad\qquad\qquad\qquad = \log\left(\dfrac{y^3\sqrt{z}}{x}\right)$ Quotient rule

Now do Exercises 63–74

Warm-Ups ▼

Fill in the blank.

1. The _____ rule for logarithms states that $\log_a(MN) = \log_a(M) + \log_a(N)$.

2. The _____ rule for logarithms states that $\log_a(M/N) = \log_a(M) - \log_a(N)$.

3. The _____ rule for logarithms states that $\log_a(M^N) = N \cdot \log_a(M)$.

4. The _____ properties of logarithms state that $\log_a(a^x) = x$ and $a^{\log_a(x)} = x$.

True or false?

5. $\log_2\left(\dfrac{x^2}{8}\right) = \log_2(x^2) - 3$

6. $\dfrac{\log(100)}{\log(10)} = \log(100) - \log(10)$

7. $\dfrac{\log(100)}{10} = \log(10)$

8. $\ln(\sqrt{2}) = \dfrac{\ln(2)}{2}$

9. $\ln(1) = e$

10. $\ln(2) + \ln(3) - \ln(7) = \ln\left(\dfrac{6}{7}\right)$

11. $\ln(8) = 3 \cdot \ln(2)$

12. $e^{\ln(7)} = 7$

‹1› The Inverse Properties

Simplify each expression. See Example 1.

1. $\log_2(2^{10})$

2. $\ln(e^9)$

3. $5^{\log_5(19)}$

4. $10^{\log(2.3)}$

5. $\log(10^8)$

6. $\log_4(4^5)$

7. $e^{\ln(4.3)}$

8. $3^{\log_3(5.5)}$

‹2› The Product Rule for Logarithms

Assume all variables involved in logarithms represent numbers for which the logarithms are defined. Write each expression as a single logarithm and simplify. See Example 2.

9. $\log(3) + \log(7)$

10. $\ln(5) + \ln(4)$

11. $\log_3(\sqrt{5}) + \log_3(\sqrt{x})$

12. $\ln(\sqrt{x}) + \ln(\sqrt{y})$

13. $\log(x^2) + \log(x^3)$

14. $\ln(a^3) + \ln(a^5)$

15. $\ln(2) + \ln(3) + \ln(5)$

16. $\log_2(x) + \log_2(y) + \log_2(z)$

17. $\log(x) + \log(x + 3)$

18. $\ln(x - 1) + \ln(x + 1)$

19. $\log_2(x - 3) + \log_2(x + 2)$

20. $\log_3(x - 5) + \log_3(x - 4)$

‹3› The Quotient Rule for Logarithms

Write each expression as a single logarithm. See Example 3.

21. $\log(8) - \log(2)$

22. $\ln(3) - \ln(6)$

23. $\log_2(x^6) - \log_2(x^2)$

24. $\ln(w^9) - \ln(w^3)$

25. $\log(\sqrt{10}) - \log(\sqrt{2})$

26. $\log_3(\sqrt{6}) - \log_3(\sqrt{3})$

27. $\ln(4h - 8) - \ln(4)$

28. $\log(3x - 6) - \log(3)$

29. $\log_2(w^2 - 4) - \log_2(w + 2)$

30. $\log_3(k^2 - 9) - \log_3(k - 3)$

31. $\ln(x^2 + x - 6) - \ln(x + 3)$

32. $\ln(t^2 - t - 12) - \ln(t - 4)$

‹4› The Power Rule for Logarithms

Write each expression in terms of $\log(3)$. See Example 4.

33. $\log(27)$

34. $\log\left(\dfrac{1}{9}\right)$

35. $\log(\sqrt{3})$

36. $\log(\sqrt[4]{3})$

37. $\log(3^x)$

38. $\log(3^{-99})$

‹5› Using the Properties

Rewrite each expression in terms of $\log(3)$ and/or $\log(5)$. See Example 5.

39. $\log(15)$

40. $\log(9)$

41. $\log\left(\dfrac{5}{3}\right)$

42. $\log\left(\dfrac{3}{5}\right)$

43. $\log(25)$

44. $\log\left(\dfrac{1}{27}\right)$

45. $\log(75)$

46. $\log(0.6)$

47. $\log\left(\dfrac{1}{3}\right)$

48. $\log(45)$

49. $\log(0.2)$

50. $\log\left(\dfrac{9}{25}\right)$

Rewrite each expression as a sum or a difference of multiples of logarithms. See Example 6.

51. $\log(xyz)$

52. $\log(3y)$

53. $\log_2(8x)$

54. $\log_2(16y)$

55. $\ln\!\left(\dfrac{x}{y}\right)$

56. $\ln\!\left(\dfrac{z}{3}\right)$

57. $\log(10x^2)$

58. $\log(100\sqrt{x})$

59. $\log_5\!\left(\dfrac{(x-3)^2}{\sqrt{w}}\right)$

60. $\log_3\!\left(\dfrac{(y+6)^3}{y-5}\right)$

61. $\ln\!\left(\dfrac{yz\sqrt{x}}{w}\right)$

62. $\ln\!\left(\dfrac{(x-1)\sqrt{w}}{x^3}\right)$

Rewrite each expression as a single logarithm. See Example 7.

63. $\log(x) + \log(x-1)$

64. $\log_2(x-2) + \log_2(5)$

65. $\ln(3x-6) - \ln(x-2)$

66. $\log_3(x^2-1) - \log_3(x-1)$

67. $\ln(x) - \ln(w) + \ln(z)$

68. $\ln(x) - \ln(3) - \ln(7)$

69. $3 \cdot \ln(y) + 2 \cdot \ln(x) - \ln(w)$

70. $5 \cdot \ln(r) + 3 \cdot \ln(t) - 4 \cdot \ln(s)$

71. $\dfrac{1}{2}\log(x-3) - \dfrac{2}{3}\log(x+1)$

72. $\dfrac{1}{2}\log(y-4) + \dfrac{1}{2}\log(y+4)$

73. $\dfrac{2}{3}\log_2(x-1) - \dfrac{1}{4}\log_2(x+2)$

74. $\dfrac{1}{2}\log_3(y+3) + 6 \cdot \log_3(y)$

Determine whether each equation is true or false.

75. $\log(56) = \log(7) \cdot \log(8)$

76. $\log\!\left(\dfrac{5}{9}\right) = \dfrac{\log(5)}{\log(9)}$

77. $\log_2(4^2) = (\log_2(4))^2$

78. $\ln(4^2) = (\ln(4))^2$

79. $\ln(25) = 2 \cdot \ln(5)$

80. $\ln(3e) = 1 + \ln(3)$

81. $\dfrac{\log_2(64)}{\log_2(8)} = \log_2(8)$

82. $\dfrac{\log_2(16)}{\log_2(4)} = \log_2(4)$

83. $\log\!\left(\dfrac{1}{3}\right) = -\log(3)$

84. $\log_2(8 \cdot 2^{59}) = 62$

85. $\log_2(16^5) = 20$

86. $\log_2\!\left(\dfrac{5}{2}\right) = \log_2(5) - 1$

87. $\log(10^3) = 3$

88. $\log_3(3^7) = 7$

89. $\log(100 + 3) = 2 + \log(3)$

90. $\dfrac{\log_7(32)}{\log_7(8)} = \dfrac{5}{3}$

Applications

Solve each problem.

91. *Richter scale.* The Richter scale rating of an earthquake is given by the formula $r = \log(I) - \log(I_0)$, where I is the *intensity* of the earthquake and I_0 is the intensity of a small "benchmark" earthquake. Use the appropriate property of logarithms to rewrite this formula using a single logarithm. Find r if $I = 100 \cdot I_0$.

92. *Diversity index.* The U.S.G.S. measures the quality of a water sample by using the diversity index d, given by

$$d = -[p_1 \cdot \log_2(p_1) + p_2 \cdot \log_2(p_2) + \cdots + p_n \cdot \log_2(p_n)],$$

where n is the number of different taxons (biological classifications) represented in the sample and p_1 through p_n are the percentages of organisms in each of the n taxons. The value of d ranges from 0 when all organisms in the water sample are the same to some positive number when all organisms in the sample are different. If two-thirds of the organisms in a water sample are in one taxon and one-third of the organisms are in a second taxon, then $n = 2$ and

$$d = -\left[\dfrac{2}{3}\log_2\!\left(\dfrac{2}{3}\right) + \dfrac{1}{3}\log_2\!\left(\dfrac{1}{3}\right)\right].$$

Use the properties of logarithms to write the expression on the right-hand side as $\log_2\!\left(\dfrac{3\sqrt[3]{2}}{2}\right)$. (In Section 11.4 you will learn how to evaluate a base-2 logarithm using a calculator.)

Getting More Involved

93. *Discussion*

Which of the following equations is an identity? Explain.

a) $\ln(3x) = \ln(3) \cdot \ln(x)$
b) $\ln(3x) = \ln(3) + \ln(x)$
c) $\ln(3x) = 3 \cdot \ln(x)$
d) $\ln(3x) = \ln(x^3)$

94. *Discussion*

Which of the following expressions is not equal to $\log(5^{2/3})$? Explain.

a) $\dfrac{2}{3}\log(5)$ **b)** $\dfrac{\log(5) + \log(5)}{3}$

c) $\left(\log(5)\right)^{2/3}$ **d)** $\dfrac{1}{3}\log(25)$

Graphing Calculator Exercises

95. Graph the functions $y_1 = \ln\!\left(\sqrt{x}\right)$ and $y_2 = 0.5 \cdot \ln(x)$ on the same screen. Explain your results.

96. Graph the functions $y_1 = \log(x)$, $y_2 = \log(10x)$, $y_3 = \log(100x)$, and $y_4 = \log(1000x)$ using the viewing window $-2 \le x \le 5$ and $-2 \le y \le 5$. Why do these curves appear as they do?

97. Graph the function $y = \log(e^x)$. Explain why the graph is a straight line. What is its slope?

11.4   Solving Equations and Applications

In This Section

⟨1⟩ **Logarithmic Equations**

⟨2⟩ **Exponential Equations**

⟨3⟩ **Changing the Base**

⟨4⟩ **Strategy for Solving Equations**

⟨5⟩ **Applications**

We solved some equations involving exponents and logarithms in Sections 11.1 and 11.2. In this section, we use the properties of exponents and logarithms to solve more complex equations.

⟨1⟩ Logarithmic Equations

The main tool that we have for solving logarithmic equations is the definition of logarithms: $y = \log_a(x)$ if and only if $a^y = x$. We can use the definition to rewrite any equation that has only one logarithm as an equivalent exponential equation.

E X A M P L E 1

A logarithmic equation with only one logarithm

Solve $\log(x + 3) = 2$.

Solution

Write the equivalent exponential equation:

$$\log(x + 3) = 2 \qquad \text{Original equation}$$
$$10^2 = x + 3 \qquad \text{Definition of logarithm}$$
$$100 = x + 3$$
$$97 = x$$

Check: $\log(97 + 3) = \log(100) = 2$. The solution set is $\{97\}$.

> **Now do Exercises 1–8**

In Example 2, we use the product rule for logarithms to write a sum of two logarithms as a single logarithm.

Using the product rule to solve an equation

Solve $\log_2(x + 3) + \log_2(x - 3) = 4$.

Solution

Rewrite the sum of the logarithms as the logarithm of a product:

$$\log_2(x + 3) + \log_2(x - 3) = 4 \qquad \text{Original equation}$$
$$\log_2[(x + 3)(x - 3)] = 4 \qquad \text{Product rule}$$
$$\log_2[x^2 - 9] = 4 \qquad \text{Multiply the binomials.}$$
$$x^2 - 9 = 2^4 \qquad \text{Definition of logarithm}$$
$$x^2 - 9 = 16$$
$$x^2 = 25$$
$$x = \pm 5 \qquad \text{Even-root property}$$

To check, first let $x = -5$ in the original equation:

$$\log_2(-5 + 3) + \log_2(-5 - 3) = 4$$
$$\log_2(-2) + \log_2(-8) = 4 \qquad \text{Incorrect}$$

Because the domain of any logarithm function is the set of positive real numbers, these logarithms are undefined. Now check $x = 5$ in the original equation:

$$\log_2(5 + 3) + \log_2(5 - 3) = 4$$
$$\log_2(8) + \log_2(2) = 4$$
$$3 + 1 = 4 \qquad \text{Correct}$$

The solution set is $\{5\}$.

> **Now do Exercises 9–16**

> **CAUTION**   Always check that your solutions to a logarithmic equation do not produce undefined logarithms in the original equation.

Using the one-to-one property of logarithms

Solve $\log(x) + \log(x - 1) = \log(8x - 12) - \log(2)$.

‹ Calculator Close-Up ›

Graph

$$y_1 = \log(x) + \log(x - 1)$$

and

$$y_2 = \log(8x - 12) - \log(2)$$

to see the two solutions to the equation in Example 3.

Solution

Apply the product rule to the left-hand side and the quotient rule to the right-hand side to get a single logarithm on each side:

$$\log(x) + \log(x - 1) = \log(8x - 12) - \log(2)$$
$$\log[x(x - 1)] = \log\left(\frac{8x - 12}{2}\right) \qquad \text{Product rule; quotient rule}$$
$$\log(x^2 - x) = \log(4x - 6) \qquad \text{Simplify.}$$
$$x^2 - x = 4x - 6 \qquad \text{One-to-one property of logarithms}$$
$$x^2 - 5x + 6 = 0$$
$$(x - 2)(x - 3) = 0$$
$$x - 2 = 0 \qquad \text{or} \qquad x - 3 = 0$$
$$x = 2 \qquad \text{or} \qquad x = 3$$

Neither $x = 2$ nor $x = 3$ produces undefined terms in the original equation. Use a calculator to check that they both satisfy the original equation. The solution set is $\{2, 3\}$.

Now do Exercises 17–22

CAUTION   The product rule, quotient rule, and power rule do not eliminate logarithms from equations. To do so, we use the definition to change $y = \log_a(x)$ into $a^y = x$ or the one-to-one property to change $\log_a(m) = \log_a(n)$ into $m = n$.

‹2› Exponential Equations

If an equation has a single exponential expression, we can write the equivalent logarithmic equation.

E X A M P L E   **4**

A single exponential expression
Find the exact solution to $2^x = 10$.

Solution

The equivalent logarithmic equation is

$$x = \log_2(10).$$

The solution set is $\{\log_2(10)\}$. The number $\log_2(10)$ is the exact solution to the equation. Later in this section you will learn how to use the base-change formula to find an approximate value for an expression of this type.

Now do Exercises 23–26

In Section 11.1 we solved some exponential equations by writing each side as a power of the same base and then applying the one-to-one property of exponential functions. We review that method in Example 5.

E X A M P L E   **5**

Powers of the same base
Solve $2^{(x^2)} = 4^{3x-4}$.

Solution

We can write each side as a power of the same base:

$$2^{(x^2)} = \left(2^2\right)^{3x-4} \qquad \text{Because } 4 = 2^2$$
$$2^{(x^2)} = 2^{6x-8} \qquad \text{Power of a power rule}$$
$$x^2 = 6x - 8 \qquad \text{One-to-one property of}$$
$$x^2 - 6x + 8 = 0 \qquad \text{exponential functions}$$
$$(x - 4)(x - 2) = 0$$
$$x - 4 = 0 \quad \text{or} \quad x - 2 = 0$$
$$x = 4 \quad \text{or} \quad x = 2$$

Check $x = 2$ and $x = 4$ in the original equation. The solution set is $\{2, 4\}$.

Now do Exercises 27–30

For some exponential equations we cannot write each side as a power of the same base as we did in Example 5. In this case, we take a logarithm of each side and simplify, using the rules for logarithms.

E X A M P L E   6

Exponential equation with two different bases

Find the exact and approximate solution to $2^{x-1} = 3^x$.

Solution

Since we want an approximate solution, we must use base 10 or base e, which are both available on a calculator. Either one will work here. We will use base 10:

$2^{x-1} = 3^x$	Original equation
$\log(2^{x-1}) = \log(3^x)$	Take log of each side.
$(x-1)\log(2) = x \cdot \log(3)$	Power rule
$x \cdot \log(2) - \log(2) = x \cdot \log(3)$	Distributive property
$x \cdot \log(2) - x \cdot \log(3) = \log(2)$	Get all x-terms on one side.
$x[\log(2) - \log(3)] = \log(2)$	Factor out x.
$x = \dfrac{\log(2)}{\log(2) - \log(3)}$	Exact solution
$x \approx -1.7095$	Approximate solution

You can use a calculator to check -1.7095 in the original equation.

> **Now do Exercises 31–36**

⟨3⟩ Changing the Base

Scientific calculators have an x^y key for computing any power of any base, in addition to the function keys for computing 10^x and e^x. For logarithms we have the keys ln and log, but there are no function keys for logarithms using other bases. To solve this problem, we develop a formula for expressing a base-a logarithm in terms of base-b logarithms.

If $y = \log_a(M)$, then $a^y = M$. Now we solve $a^y = M$ for y, using base-b logarithms:

$a^y = M$	
$\log_b(a^y) = \log_b(M)$	Take the base-b logarithm of each side.
$y \cdot \log_b(a) = \log_b(M)$	Power rule
$y = \dfrac{\log_b(M)}{\log_b(a)}$	Divide each side by $\log_b(a)$.

Because $y = \log_a(M)$, we can write $\log_a(M)$ in terms of base-b logarithms.

Base-Change Formula

If a and b are positive numbers not equal to 1 and M is positive, then

$$\log_a(M) = \frac{\log_b(M)}{\log_b(a)}.$$

In words, we take the logarithm with the new base and divide by the logarithm of the old base. The most important use of the base-change formula is to find base-a logarithms using a calculator. If the new base is 10 or e, then

$$\log_a(M) = \frac{\log(M)}{\log(a)} = \frac{\ln(M)}{\ln(a)}.$$

E X A M P L E   **7**

Using the base-change formula

Find $\log_7(99)$ to four decimal places.

Solution

Use the base-change formula with $a = 7$ and $b = 10$:

$$\log_7(99) = \frac{\log(99)}{\log(7)} \approx 2.3614$$

Check by finding $7^{2.3614}$ with your calculator. Note that we also have

$$\log_7(99) = \frac{\ln(99)}{\ln(7)} \approx 2.3614.$$

> **Now do Exercises 37–44**

⟨4⟩ Strategy for Solving Equations

There is no formula that will solve every equation in this section. However, we have a strategy for solving exponential and logarithmic equations. The following list summarizes the ideas that we need for solving these equations.

Strategy for Solving Exponential and Logarithmic Equations

1. If the equation has a single logarithm or a single exponential expression, rewrite the equation using the definition $y = \log_a(x)$ if and only if $a^y = x$.

2. Use the properties of logarithms to combine logarithms as much as possible.

3. Use the one-to-one properties:
   a) If $\log_a(m) = \log_a(n)$, then $m = n$.
   b) If $a^m = a^n$, then $m = n$.

4. To get an approximate solution of an exponential equation, take the common or natural logarithm of each side of the equation.

‹5› **Applications**

In compound interest problems, logarithms are used to find the time it takes for money to grow to a specified amount.

E X A M P L E   **8**

Finding the time

If $500 is deposited into an account paying 8% compounded quarterly, then in how many quarters will the account have $1000 in it?

‹ Helpful Hint ›

When we get $2 = (1.02)^n$, we can use the definition of log as in Example 8 or take the natural log of each side:

$$\ln(2) = \ln(1.02^n)$$
$$\ln(2) = n \cdot \ln(1.02)$$
$$n = \frac{\ln(2)}{\ln(1.02)}$$

In either way we arrive at the same solution.

Solution

We use the compound interest formula $A = P(1 + i)^n$ with a principal of $500, an amount of $1000, and an interest rate of 2% each quarter:

$$A = P(1 + i)^n$$

$1000 = 500(1.02)^n$ Substitute.

$2 = (1.02)^n$ Divide each side by 500.

$n = \log_{1.02}(2)$ Definition of logarithm

$\quad = \dfrac{\ln(2)}{\ln(1.02)}$ Base-change formula

$\quad \approx 35.0028$ Use a calculator.

It takes approximately 35 quarters, or 8 years and 9 months, for the initial investment to be worth $1000. Note that we could also solve $2 = (1.02)^n$ by taking the common or natural logarithm of each side. Try it.

> Now do Exercises 79–82

Radioactive substances decay continuously over time in the same manner as money grows continuously with the continuous-compounding formula from Section 11.1. The model for radioactive decay is

$$A = A_0 e^{rt},$$

where A is the amount of the substance present at time t, r is the decay rate, and A_0 is the amount present at time $t = 0$. Note that this formula is actually the same as the continuous-compounding formula, but since the amount is decreasing, the rate r is a negative number.

E X A M P L E   **9**

Finding the rate in radioactive decay

The number of grams of a radioactive substance that is present in an old bone after t years is given by

$$A = 8e^{rt},$$

where r is the decay rate. How many grams of the radioactive substance were present when the bone was in a living organism at time $t = 0$? If it took 6300 years for the radioactive substance to decay from 8 grams to 4 grams, then what is the decay rate?

Solution

If $t = 0$, then $A = 8e^{r \cdot 0} = 8e^0 = 8 \cdot 1 = 8$. So the bone contained 8 grams of the substance when it was in a living organism. Now use $A = 4$ and $t = 6300$ in the formula $A = 8e^{rt}$ and solve for r:

$$4 = 8e^{6300r}$$

$$0.5 = e^{6300r} \qquad \text{Divide each side by 8.}$$

$$6300r = \ln(0.5) \qquad \text{Definition of logarithm}$$

$$r = \frac{\ln(0.5)}{6300} \qquad \text{Divide each side by 6300.}$$

$$r \approx -1.1 \times 10^{-4} \text{ or } -0.00011$$

Note that the rate is negative because the substance is decaying.

> **Now do Exercises 83–94**

 ## Warm-Ups ▼

Fill in the blank.

1. The equations $a^y = x$ and $\log_a(x) = y$ are _____.
2. According to the _____ formula $\log_a(x) = \ln(x)/\ln(a)$.

True or false?

3. If $\log(x - 2) + \log(x + 2) = 7$, then $\log(x^2 - 4) = 7$.
4. If $\log(3x - 2) = \log(x + 2)$, then $3x - 2 = x + 2$.

5. If $e^{x-6} = e^{x^2-5x}$, then $x - 6 = x^2 - 5x$.
6. If $2^{3x-1} = 3^{5x-4}$, then $3x - 1 = 5x - 4$.
7. If $\log_2(x^2 - 2x + 5) = 3$, then $x^2 - 2x + 5 = 8$.

8. If $5^x = 23$, then $x \cdot \ln(5) = \ln(23)$.
9. $\dfrac{\ln(2)}{\ln(6)} = \dfrac{\log(2)}{\log(6)}$
10. $\log_3(5) = \dfrac{\ln(3)}{\ln(5)}$

Exercises

11.4

‹ Study Tips ›

- Always study math with a pencil and paper. Just sitting back and reading the text rarely works.
- A good way to study the examples in the text is to cover the solution with a piece of paper and see how much of the solution you can write on your own.

‹1› Logarithmic Equations

Solve each equation. See Examples 1 and 2.

1. $\log(x + 100) = 3$
2. $\log(x - 5) = 2$

3. $\log_2(x + 1) = 3$
4. $\log_3(x^2) = 4$
5. $3 \log_2(x + 1) - 2 = 13$
6. $4 \log_3(2x) - 1 = 7$

7. $12 + 2\ln(x) = 14$

8. $23 = 3\ln(x - 1) + 14$

9. $\log(x) + \log(5) = 1$

10. $\ln(x) + \ln(3) = 0$

11. $\log_2(x - 1) + \log_2(x + 1) = 3$

12. $\log_3(x - 4) + \log_3(x + 4) = 2$

13. $\log_2(x - 1) - \log_2(x + 2) = 2$

14. $\log_4(8x) - \log_4(x - 1) = 2$

15. $\log_2(x - 4) + \log_2(x + 2) = 4$

16. $\log_6(x + 6) + \log_6(x - 3) = 2$

Solve each equation. See Example 3.

17. $\ln(x) + \ln(x + 5) = \ln(x + 1) + \ln(x + 3)$

18. $\log(x) + \log(x + 5) = 2 \cdot \log(x + 2)$

19. $\log(x + 3) + \log(x + 4) = \log(x^3 + 13x^2) - \log(x)$

20. $\log(x^2 - 1) - \log(x - 1) = \log(6)$

21. $2 \cdot \log(x) = \log(20 - x)$

22. $2 \cdot \log(x) + \log(3) = \log(2 - 5x)$

⟨2⟩ Exponential Equations

Solve each equation. See Examples 4 and 5.

23. $3^x = 7$ **24.** $2^{x-1} = 5$

25. $e^{2x} = 7$ **26.** $e^{x+3} = 2$

27. $2^{3x+4} = 4^{x-1}$ **28.** $9^{2x-1} = 27^{1/2}$

29. $\left(\dfrac{1}{3}\right)^x = 3^{1+x}$ **30.** $4^{3x} = \left(\dfrac{1}{2}\right)^{1-x}$

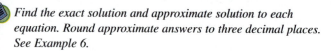 *Find the exact solution and approximate solution to each equation. Round approximate answers to three decimal places. See Example 6.*

31. $2^x = 3^{x+5}$

32. $e^x = 10^x$

33. $5^{x+2} = 10^{x-4}$

34. $3^{2x} = 6^{x+1}$

35. $8^x = 9^{x-1}$

36. $5^{x+1} = 8^{x-1}$

⟨3⟩ Changing the Base

Use the base-change formula to find each logarithm to four decimal places. See Example 7.

37. $\log_2(3)$ **38.** $\log_3(5)$

39. $\log_3\left(\dfrac{1}{2}\right)$ **40.** $\log_5(2.56)$

 41. $\log_{1/2}(4.6)$ **42.** $\log_{1/3}(3.5)$

43. $\log_{0.1}(0.03)$ **44.** $\log_{0.2}(1.06)$

⟨4⟩ Strategy for Solving Equations

For each equation, find the exact solution and an approximate solution when appropriate. Round approximate answers to three decimal places. See the Strategy for Solving Exponential and Logarithmic Equations box on page 687.

45. $x \cdot \ln(2) = \ln(7)$

46. $x \cdot \log(3) = \log(5)$

47. $3x - x \cdot \ln(2) = 1$

48. $2x + x \cdot \log(5) = \log(7)$

49. $3^x = 5$

50. $2^x = \dfrac{1}{3}$

51. $2^{x-1} = 9$

52. $10^{x-2} = 6$

53. $3^x = 20$

54. $2^x = 128$

55. $\log_3(x) + \log_3(5) = 1$

56. $\log(x) - \log(3) = \log(6)$

57. $8^x = 2^{x+1}$

58. $2^x = 5^{x+1}$

59. $\log_2(1 - x) = 2$

60. $\log_5(-x) = 3$

61. $\log_3(1 - x) + \log_3(2x + 13) = 3$

62. $\log_2(3 - x) + \log_2(x + 9) = 5$

63. $\ln(2x - 1) - \ln(x + 1) = \ln(5)$

64. $\log(x - 4) - \log(x + 5) = 1$

65. $\log_3(x - 14) - \log_3(x - 6) = 2$

66. $\log_3(7 - x^2) - \log_3(1 - x) = 1$

67. $\log(x + 1) + \log(x - 2) = 1$

68. $\log_2(x^2 - 8) - \log_2(x^2 - 5) = 2$

69. $2 \cdot \ln(x) = \ln(2) + \ln(5x - 12)$

70. $\ln(8 - x^3) - \ln(2 - x) = \ln(2x + 5)$

71. $\log_3(x^3 + 16x^2) - \log_3(x) = \log_3(36)$

72. $\ln(x) + \ln(x - 2) = \ln(x + 2) + \ln(x - 3)$

73. $\log(x) + \log(x + 5) = 2 \cdot \log(x + 2)$

74. $\log_2(x^2 - 9) - \log_2(x + 3) = \log_2(12)$

75. $\log_7(x^2 + 6x + 8) - \log_7(x + 2) = \log_7(3)$

76. $3 \cdot \log_5(x) = 2 \cdot \log_5(x)$

77. $\ln(6) + 2 \cdot \ln(x) = \ln(38x - 30) - \ln(2)$

78. $3 \cdot \ln(x + 1) = \ln(x + 1) + \ln(x^2 - x + 1)$

⟨5⟩ Applications

Solve each problem. See Examples 8 and 9.

79. *Finding the time.* How many months does it take for $1000 to grow to $1500 in an account paying 12% compounded monthly?

80. *Finding the time.* How many years does it take for $25 to grow to $100 in an account paying 8% compounded annually?

81. *Finding days.* How many days does it take for a deposit of $100 to grow to $105 at 3% annual percentage rate compounded daily? Round to the nearest day.

82. *Finding quarters.* How many quarters does it take for a deposit of $500 to grow to $600 at 2% annual percentage rate compounded quarterly? Round to the nearest quarter.

83. *Radioactive decay.* The number of grams of a radioactive substance that is present in an old piece of cloth after t years is given by

$$A = 10e^{-0.0001t}.$$

How many grams of the radioactive substance did the cloth contain when it was made at time $t = 0$? If the cloth now contains only 4 grams of the substance, then when was the cloth made?

84. *Finding the decay rate.* The number of grams of a radioactive substance that is present in an old log after t years is given by

$$A = 5e^{rt},$$

where r is the decay rate. How many grams of the radioactive substance were present when the log was alive at time $t = 0$? If it took 5000 years for the substance to decay from 5 grams to 2 grams, then what is the decay rate?

85. *Going with the flow.* The flow y [in cubic feet per second (ft³/sec)] of the Tangipahoa River at Robert, Louisiana, is modeled by the exponential function $y = 114.308e^{0.265x}$, where x is the depth in feet. Find the flow when the depth is 15.8 feet.

Figure for Exercises 85 and 86

86. *Record flood.* Use the formula of Exercise 85 to find the depth of the Tangipahoa River at Robert, Louisiana, on May 3, 1953, when the flow reached an all-time record of 50,500 ft³/sec (U.S.G.S., waterdata.usgs.gov).

87. *Above the poverty level.* In a certain country the number of people above the poverty level is currently 28 million and growing 5% annually. Assuming the population is growing continuously, the population P (in millions), t years from now, is determined by the formula $P = 28e^{0.05t}$. In how many years will there be 40 million people above the poverty level?

88. *Below the poverty level.* In the same country as in Exercise 87, the number of people below the poverty level is currently 20 million and growing 7% annually. This population (in millions), t years from now, is determined by the formula $P = 20e^{0.07t}$. In how many years will there be 40 million people below the poverty level?

89. *Fifty-fifty.* For this exercise, use the information given in Exercises 87 and 88. In how many years will the number of people above the poverty level equal the number of people below the poverty level?

90. *Golden years.* In a certain country there are currently 100 million workers and 40 million retired people. The population of workers is decreasing according to the formula $W = 100e^{-0.01t}$, where t is in years and W is in millions. The population of retired people is increasing according to the formula $R = 40e^{0.09t}$, where t is in years and R is in millions. In how many years will the number of workers equal the number of retired people?

91. *Ions for breakfast.* Orange juice has a pH of 3.7. What is the hydrogen ion concentration of orange juice? (See Exercises 85–88 of Section 11.2.)

Figure for Exercises 91 and 92

92. *Ions in your veins.* Normal human blood has a pH of 7.4. What is the hydrogen ion concentration of normal human blood?

93. *Diversity index.* In Exercise 92 of Section 11.3 we expressed the diversity index d for a certain water sample as

$$d = \log_2\left(\frac{3\sqrt[3]{2}}{2}\right).$$

Use the base-change formula and a calculator to calculate the value of d. Round the answer to four decimal places.

94. *Quality water.* In a certain water sample, 5% of the organisms are in one taxon, 10% are in a second taxon, 20% are in a third taxon, 15% are in a fourth taxon, 23% are in a fifth taxon, and the rest are in a sixth taxon. Use the formula given in Exercise 92 of Section 11.3 with $n = 6$ to find the diversity index of the water sample.

Getting More Involved

95. *Exploration*

Logarithms were designed to solve equations that have variables in the exponents, but logarithms can be used to solve certain polynomial equations. Consider the following example:

$$x^5 = 88$$

$$5 \cdot \ln(x) = \ln(88)$$

$$\ln(x) = \frac{\ln(88)}{5} \approx 0.895467$$

$$x = e^{0.895467} \approx 2.4485$$

Solve $x^3 = 12$ by taking the natural logarithm of each side. Round the approximate solution to four decimal places. Solve $x^3 = 12$ without using logarithms, and compare with your previous answer.

96. *Discussion*

Determine whether each logarithm is positive or negative without using a calculator. Explain your answers.

a) $\log_2(0.45)$
b) $\ln(1.01)$
c) $\log_{1/2}(4.3)$
d) $\log_{1/3}(0.44)$

Graphing Calculator Exercises

97. Graph $y_1 = 2^x$ and $y_2 = 3^{x-1}$ on the same coordinate system. Use the intersect feature of your calculator to find the point of intersection of the two curves. Round to two decimal places.

98. Bob invested $1000 at 6% compounded continuously. At the same time Paula invested $1200 at 5% compounded monthly. Write two functions that give the amounts of Bob's and Paula's investments after x years. Graph these functions on a graphing calculator. Use the intersect feature of your graphing calculator to find the approximate value of x for which the investments are equal in value.

99. Graph the functions $y_1 = \log_2(x)$ and $y_2 = 3^{x-4}$ on the same coordinate system, and use the intersect feature to find the points of intersection of the curves. Round to two decimal places. [*Hint:* To graph $y = \log_2(x)$, use the base-change formula to write the function as $y = \ln(x)/\ln(2)$.]

Wrap-Up

Summary

Exponential and Logarithmic Functions		**Examples**
Exponential function	A function of the form $f(x) = a^x$ for $a > 0$ and $a \neq 1$	$f(x) = 3^x$
Logarithmic function	A function of the form $f(x) = \log_a(x)$ for $a > 0$ and $a \neq 1$ $\quad\quad y = \log_a(x)$ if and only if $a^y = x$.	$f(x) = \log_2(x)$ $\log_3(8) = x \leftrightarrow 3^x = 8$
Common logarithm	Base-10: $f(x) = \log(x)$	$\log(100) = 2$ because $100 = 10^2$.
Natural logarithm	Base-e: $f(x) = \ln(x)$ $e \approx 2.718$	$\ln(e) = 1$ because $e^1 = e$.
Inverse functions	$f(x) = a^x$ and $g(x) = \log_a(x)$ are inverse functions.	If $f(x) = e^x$, then $f^{-1}(x) = \ln(x)$.

Properties		**Examples**
M, N, and a are positive numbers with $a \neq 1$. $\quad\quad \log_a(a) = 1 \quad\quad \log_a(1) = 0$		$\log_5(5) = 1, \log_5(1) = 0$
Inverse properties	$\log_a(a^x) = x$ for any real number x. $a^{\log_a(x)} = x$ for any positive real number x.	$\log(10^7) = 7, e^{\ln(3.4)} = 3.4$
Product rule	$\log_a(MN) = \log_a(M) + \log_a(N)$	$\ln(3x) = \ln(3) + \ln(x)$
Quotient rule	$\log_a\left(\dfrac{M}{N}\right) = \log_a(M) - \log_a(N)$ $\log_a\left(\dfrac{1}{N}\right) = -\log_a(N)$	$\ln\left(\dfrac{2}{3}\right) = \ln(2) - \ln(3)$ $\ln\left(\dfrac{1}{3}\right) = -\ln(3)$
Power rule	$\log_a(M^N) = N \cdot \log_a(M)$	$\log(x^3) = 3 \cdot \log(x)$
Base-change formula	$\log_a(M) = \dfrac{\log_b(M)}{\log_b(a)}$	$\log_3(5) = \dfrac{\ln(5)}{\ln(3)}$

Equations Involving Logarithms and Exponents		**Examples**
Strategy	1. If there is a single logarithm or a single exponential expression, rewrite the equation using the definition of logarithms: $y = \log_a(x)$ if and only if $a^y = x$.	$2^x = 3$ and $x = \log_2(3)$ are equivalent.

2. Use the properties of logarithms to combine logarithms as much as possible.

$$\log(x) + \log(x - 3) = 1$$
$$\log(x^2 - 3x) = 1$$

3. Use the one-to-one properties:
   a) If $\log_a(m) = \log_a(n)$, then $m = n$.
   b) If $a^m = a^n$, then $m = n$.

$$\ln(x) = \ln(5 - x),$$
$$x = 5 - x$$
$$2^{3x} = 2^{5x-7}, 3x = 5x - 7$$

4. To get an approximate solution, take the common or natural logarithm of each side of an exponential equation.

$$2^x = 3, \ln(2^x) = \ln(3)$$
$$x \cdot \ln(2) = \ln(3)$$
$$x = \frac{\ln(3)}{\ln(2)}$$

Enriching Your Mathematical Word Power

Fill in the blank.

1. A(n) _____ function has the form $f(x) = a^x$ where $a > 0$ and $a \neq 1$.

2. A base-10 logarithm is a(n) _____ logarithm.

3. A base-e logarithm is a(n) _____ logarithm.

4. The set of first coordinates of a relation is its _____.

5. With _____ interest, interest is periodically paid into the account so that the interest earns interest.

6. With _____ compounding, the formula $A = Pe^{rt}$ is used to compute the amount.

7. The _____ logarithm of x is the power of a that produces x.

8. If $a^x = y$, then x is the base-a _____ of y.

Review Exercises

11.1 Exponential Functions and Their Applications

Use $f(x) = 5^x$, $g(x) = 10^{x-1}$, and $h(x) = \left(\frac{1}{4}\right)^x$ for Exercises 1–28. Find the following.

1. $f(-2)$

2. $f(0)$

3. $f(3)$

4. $f(4)$

5. $g(1)$

6. $g(-1)$

7. $g(0)$

8. $g(3)$

9. $h(-1)$

10. $h(2)$

11. $h\left(\frac{1}{2}\right)$

12. $h\left(-\frac{1}{2}\right)$

Find x in each case.

13. $f(x) = 25$

14. $f(x) = -\frac{1}{125}$

15. $g(x) = 1000$

16. $g(x) = 0.001$

17. $h(x) = 32$

18. $h(x) = 8$

19. $h(x) = \frac{1}{16}$

20. $h(x) = 1$

Find the following.

21. $f(1.34)$

22. $f(-3.6)$

23. $g(3.25)$

24. $g(4.87)$

25. $h(2.82)$

26. $h(\pi)$

27. $h(\sqrt{2})$

28. $h\left(\frac{1}{3}\right)$

Sketch the graph of each function.

29. $f(x) = 5^x$

30. $g(x) = e^x$

31. $y = \left(\dfrac{1}{5}\right)^x$

32. $y = e^{-x}$

33. $f(x) = 3^{-x}$

34. $f(x) = -3^{x-1}$

35. $y = 1 + 2^x$

36. $y = 1 - 2^x$

11.2 Logarithmic Functions and Their Applications

Write each exponential equation as a logarithmic equation and each logarithmic equation as an exponential equation.

37. $10^m = n$ **38.** $b = a^5$

39. $h = \log_k(t)$ **40.** $\log_v(5) = u$

Let $f(x) = \log_2(x)$, $g(x) = \log(x)$, and $h(x) = \log_{1/2}(x)$. Find the following.

41. $f\left(\dfrac{1}{8}\right)$ **42.** $f(64)$

43. $g(0.1)$ **44.** $g(1)$

45. $g(100)$ **46.** $h\left(\dfrac{1}{8}\right)$

47. $h(1)$ **48.** $h(4)$

49. x, if $f(x) = 8$ **50.** x, if $g(x) = 3$

51. $f(77)$ **52.** $g(88.4)$

53. $h(33.9)$ **54.** $h(0.05)$

55. x, if $f(x) = 2.475$ **56.** x, if $g(x) = 1.426$

For each function f, find f^{-1}, and sketch the graphs of f and f^{-1} on the same set of axes.

57. $f(x) = 10^x$

58. $f(x) = \log_8(x)$

59. $f(x) = e^x$

60. $f(x) = \log_3(x)$

11.3 Properties of Logarithms

Rewrite each expression as a sum or a difference of multiples of logarithms.

61. $\log(x^2y)$

62. $\log_3(x^2 + 2x)$

63. $\ln(16)$

64. $\log\left(\dfrac{y}{\sqrt{x}}\right)$

65. $\log_5\left(\dfrac{1}{x}\right)$

66. $\ln\left(\dfrac{xy}{z}\right)$

Rewrite each expression as a single logarithm.

67. $\dfrac{1}{2}\log(x + 2) - 2 \cdot \log(x - 1)$

68. $3 \cdot \ln(x) + 2 \cdot \ln(y) - \dfrac{1}{3}\ln(z)$

11.4 Solving Equations and Applications

Find the exact solution to each equation.

69. $\log_2(x) = 8$

70. $\log_3(x) = 0.5$

71. $\log_2(8) = x$

72. $3^x = 8$

73. $x^3 = 8$

74. $3^2 = x$

75. $\log_x(27) = 3$

76. $\log_x(9) = -\dfrac{1}{3}$

77. $x \cdot \ln(3) - x = \ln(7)$

78. $x \cdot \log(8) = x \cdot \log(4) + \log(9)$

79. $3^x = 5^{x-1}$

80. $5^{(2x^2)} = 5^{3-5x}$

81. $4^{2x} = 2^{x+1}$

82. $\log(12) = \log(x) + \log(7 - x)$

83. $\ln(x + 2) - \ln(x - 10) = \ln(2)$

84. $2 \cdot \ln(x + 3) = 3 \cdot \ln(4)$

85. $\log(x) - \log(x - 2) = 2$

86. $\log_2(x) = \log_2(x + 16) - 1$

 Use a calculator to find an approximate solution to each of the following. Round your answers to four decimal places.

87. $6^x = 12$

88. $5^x = 8^{3x+2}$

89. $3^{x+1} = 5$

90. $\log_3(x) = 2.634$

Miscellaneous

Solve each problem.

91. *Compounding annually.* What does $10,000 invested at 11.5% compounded annually amount to after 15 years?

92. *Doubling time.* How many years does it take for an investment to double at 6.5% compounded annually?

93. *Decaying substance.* The amount, A, of a certain radioactive substance remaining after t years is given by the formula $A = A_0 e^{-0.0003t}$, where A_0 is the initial amount. If we have 218 grams of this substance today, then how much of it will be left 1000 years from now?

94. *Wildlife management.* The number of white-tailed deer in the Hiawatha National Forest is believed to be growing according to the function

$$P = 517 + 10 \cdot \ln(8t + 1),$$

where t is the time in years from the year 2000.

a) What is the size of the population in 2000?

b) In what year will the population reach 600?

c) Does the population as shown on the accompanying graph appear to be growing faster during the period 2000 to 2005 or during the period 2005 to 2010?

d) What is the average rate of change of the population for each period in part (c)?

where I and E are in millions of dollars and t is the number of years after 2000.

a) What are the values of imports and exports in 2000?

b) Use the accompanying graph to estimate the year in which imports will equal exports.

c) Algebraically find the year in which imports will equal exports.

Figure for Exercise 94

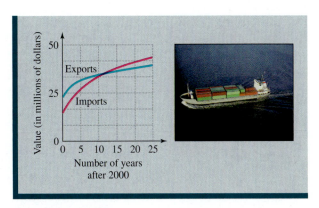

Figure for Exercise 96

95. *Comparing investments.* Melissa deposited $1000 into an account paying 5% annually; on the same day Frank deposited $900 into an account paying 7% compounded continuously. Find the number of years that it will take for the amounts in the accounts to be equal.

96. *Imports and exports.* The value of imports for a small Central American country is believed to be growing according to the function

$$I = 15 \cdot \log(16t + 33),$$

and the value of exports appears to be growing according to the function

$$E = 30 \cdot \log(t + 3),$$

97. *Finding river flow.* The U.S.G.S. measures the water height h (in feet above sea level) for the Tangipahoa River at Robert, Louisiana, and then finds the flow y [in cubic feet per second (ft^3/sec)], using the formula

$$y = 114.308e^{0.265(h-6.87)}.$$

Find the flow when the river at Robert is 20.6 ft above sea level.

98. *Finding the height.* Rewrite the formula in Exercise 97 to express h as a function of y. Use the new formula to find the water height above sea level when the flow is 10,000 ft^3/sec.

Chapter 11 Test

Let $f(x) = 5^x$ and $g(x) = \log_5(x)$. Find the following.

1. $f(2)$

2. $f(-1)$

3. $f(0)$

4. $g(125)$

5. $g(1)$

6. $g\left(\dfrac{1}{5}\right)$

Sketch the graph of each function.

7. $y = 2^x$

8. $f(x) = \log_2(x)$

Suppose $\log_a(M) = 6$ and $\log_a(N) = 4$. Find the following.

13. $\log_a(MN)$

14. $\log_a\left(\dfrac{M^2}{N}\right)$

15. $\dfrac{\log_a(M)}{\log_a(N)}$

9. $y = \left(\dfrac{1}{3}\right)^x$

16. $\log_a(a^3M^2)$

17. $\log_a\left(\dfrac{1}{N}\right)$

Find the exact solution to each equation.

18. $3^x = 12$

10. $g(x) = \log_{1/3}(x)$

19. $\log_3(x) = \dfrac{1}{2}$

20. $5^x = 8^{x-1}$

21. $\log(x) + \log(x + 15) = 2$

22. $2 \cdot \ln(x) = \ln(3) + \ln(6 - x)$

11. $f(x) = -2^x + 3$

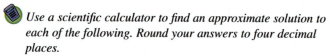 *Use a scientific calculator to find an approximate solution to each of the following. Round your answers to four decimal places.*

23. Solve $20^x = 5$.

24. Solve $\log_3(x) = 2.75$.

12. $f(x) = 2^{x-3} - 1$

25. The number of bacteria present in a culture at time t is given by the formula $N = 10e^{0.4t}$, where t is in hours. How many bacteria are present initially? How many are present after 24 hours?

26. How many hours does it take for the bacteria population of Problem 25 to double?

*Making*Connections | A Review of Chapters 1–11

Find the exact solution to each equation.

1. $(x - 3)^2 = 8$

2. $\log_2(x - 3) = 8$

3. $2^{x-3} = 8$

4. $2x - 3 = 8$

5. $|x - 3| = 8$

6. $\sqrt{x - 3} = 8$

7. $\log_2(x - 3) + \log_2(x) = \log_2(18)$

8. $2 \cdot \log_2(x - 3) = \log_2(5 - x)$

9. $\dfrac{1}{2}x - \dfrac{2}{3} = \dfrac{3}{4}x + \dfrac{1}{5}$

10. $3x^2 - 6x + 2 = 0$

Find the inverse of each function.

11. $f(x) = \dfrac{1}{3}x$

12. $g(x) = \log_3(x)$

13. $f(x) = 2x - 4$

14. $h(x) = \sqrt{x}$

15. $j(x) = \dfrac{1}{x}$

16. $k(x) = 5^x$

17. $m(x) = e^{x-1}$

18. $n(x) = \ln(x)$

Sketch the graph of each equation.

19. $y = 2x$

20. $y = 2^x$

21. $y = x^2$

22. $y = \log_2(x)$

23. $y = \dfrac{1}{2}x - 4$

24. $y = |2 - x|$

25. $y = 2 - x^2$

26. $y = e^2$

Simplify each expression.

27. $\dfrac{1}{25} + \dfrac{1}{5} + \dfrac{1}{4} + \dfrac{1}{2}$

28. $\dfrac{1}{100} + \dfrac{1}{50} + \dfrac{1}{5} + \dfrac{1}{4} + \dfrac{1}{2}$

29. $-6^{-1} + 5^{-1}$

30. $27^{4/3}$

31. $\sqrt[4]{625}$

32. $\sqrt[5]{-32}$

33. $\sqrt{80}$

34. $\sqrt[3]{80}$

35. $\dfrac{6\sqrt{2}}{\sqrt{3}}$

36. $\dfrac{20\sqrt{2}}{\sqrt[3]{5}}$

37. $\dfrac{\sqrt{3}\sqrt{14}}{\sqrt{21}}$

38. $\dfrac{\sqrt{8} + \sqrt{50}}{\sqrt{18} + \sqrt{32}}$

Solve each problem.

39. *Civilian labor force.* The number of workers in the civilian labor force can be modeled by the linear function

$$n(t) = 1.51t + 125.5$$

or by the exponential function

$$n(t) = 125.6e^{0.011t},$$

where *t* is the number of years since 1990 and $n(t)$ is in millions of workers (Bureau of Labor Statistics, www.bls.gov).

a) Graph both functions on the same coordinate system for $0 \le t \le 30$.

b) What does each model predict for the value of *n* in 2010?

c) What does each model predict for the value of *n* in the present year? Which model's prediction is closest to the actual size of the present civilian labor force?

40. *Measuring ocean depths.* In this exercise you will see how a geophysicist uses sound reflection to measure the depth of the ocean. Let *v* be the speed of sound through the water and d_1 be the depth of the ocean below the ship, as shown in the accompanying figure.

a) The time it takes for sound to travel from the ship at point *S* straight down to the ocean floor at point B_1 and back to point *S* is 0.270 second. Write d_1 as a function of *v*.

b) It takes 0.432 second for sound to travel from point *S* to point B_2 and then to a receiver at *R*, which is towed 500 meters behind the ship. Assuming $d_2 = d_3$, write d_2 as a function of *v*.

c) Use the Pythagorean theorem to find *v*. Then find the ocean depth d_1.

Figure for Exercise 40

Critical Thinking | For Individual or Group Work | Chapter 11

These exercises can be solved by a variety of techniques, which may or may not require algebra. So be creative and think critically. Explain all answers. Answers are in the Instructor's Edition of this text.

1. Tennis time. Tennis balls are sold in a cylindrical container that contains three balls. Assume that the balls just fit into the container as shown in the accompanying figure. What is the ratio of the amount of space in the container that is occupied by the balls to the amount of space that is not occupied by the balls?

Figure for Exercise 1

2. Planting trees. A landscaper planted 7 trees so that they were arranged in 6 rows with 3 trees in each row. How did she do this?

3. Division problem. Start with any three-digit number and write the number twice to form a six-digit number. Divide the six-digit number by 7. Divide the answer by 11. Finally, divide the last answer by 13. What do you notice? Explain why this works.

4. Totaling 25. How many ways are there to add three different positive integers and get a sum of 25? Do not count rearrangements of the integers. For example, count 1, 2, and 22 as one possibility, but do not count 2, 22, and 1 as another.

5. Temple of Gloom. The famous explorer Indiana Smith wants to cross a desert on foot. He plans to hire some men to help him carry supplies on the journey. However, the journey takes 6 days, but Smith and his helpers can each

Photo for Exercise 5

carry only a 4-day supply of food and water. Of course, every day, each man must consume a 1-day supply of food and water or he will die. Devise a plan for getting Smith across the desert without anyone dying and using the minimum number of helpers.

6. Counting zeros. How many zeros are at the end of the number $(5^5)!$?

7. Perfect computers. Of 6000 computers coming off a manufacturer's assembly line, every third computer had a hardware problem, every fourth computer had a software problem, and every tenth computer had a cosmetic defect. The remaining computers were perfect and were shipped to Wal-Mart. How many were shipped to Wal-Mart?

8. Leap frog. In Martha's garden is a circular pond with a diameter of 100 feet. A frog with an average leap of $2\frac{1}{4}$ feet is sitting on a lily pad in the exact center of the pond. If the lily pads are all in the right places, then what is the minimum number of leaps required for the frog to jump out of the pond.

Nonlinear Systems and the Conic Sections

With a cruising speed of 1540 miles per hour, the Concorde was the fastest commercial aircraft ever built. First flown in 1969, the Concorde could fly from London to New York in about 3 hours. However, the Concordes never made a profit and were all taken out of service in 2003, which ended the age of supersonic commercial air travel.

Perhaps the biggest problem for the Concorde was that it was generally prohibited from flying over land areas because of the noise. Any jet flying faster than the speed of sound creates a cone-shaped wave in the air on which there is a momentary change in air pressure. This change in air pressure causes a thunderlike sonic boom. When the jet is traveling parallel to the ground, the cone-shaped wave intersects the ground along one branch of a hyperbola. People on the ground hear the boom as the hyperbola passes them.

In this chapter, we will discuss curves, including the hyperbola, that occur when a geometric plane intersects a cone.

In Exercise 60 of Section 12.4 you will see how the altitude of the aircraft is related to the width of the area where the sonic boom is heard.

12.1   Nonlinear Systems of Equations

In This Section

⟨1⟩ Solving by Elimination

⟨2⟩ Applications

We studied systems of linear equations in Chapter 4. In this section, we turn our attention to nonlinear systems of equations.

⟨1⟩ Solving by Elimination

An equation whose graph is not a straight line is a **nonlinear equation.** For example,

$$y = x^2, \qquad y = \sqrt{x}, \qquad y = |x|, \qquad y = 2^x, \qquad \text{and} \qquad y = \log_2(x)$$

are nonlinear equations. A **nonlinear system** is a system of equations in which there is at least one nonlinear equation. We use the same techniques for solving nonlinear systems that we use for linear systems. Graphing the equations is used to explain the number of solutions to the system, but is generally not an accurate method for solving systems of equations. Eliminating a variable by either substitution or addition is used for solving linear or nonlinear systems.

E X A M P L E   1

A parabola and a line

Solve the system of equations, and draw the graph of each equation on the same coordinate system:

$$y = x^2 - 1$$
$$x + y = 1$$

Solution

We can eliminate y by substituting $y = x^2 - 1$ into $x + y = 1$:

$$x + y = 1$$
$$x + \left(x^2 - 1\right) = 1 \qquad \text{Substitute } x^2 - 1 \text{ for } y.$$
$$x^2 + x - 2 = 0$$
$$(x - 1)(x + 2) = 0$$
$$x - 1 = 0 \qquad \text{or} \qquad x + 2 = 0$$
$$x = 1 \qquad \text{or} \qquad x = -2$$

Replace x by 1 and -2 in $y = x^2 - 1$ to find the corresponding values of y:

$$y = (1)^2 - 1 \qquad y = (-2)^2 - 1$$
$$y = 0 \qquad \qquad y = 3$$

Check that each of the points $(1, 0)$ and $(-2, 3)$ satisfies both of the original equations. The solution set is $\{(1, 0), (-2, 3)\}$. If we solve $x + y = 1$ for y, we get $y = -x + 1$. The line $y = -x + 1$ has y-intercept $(0, 1)$ and slope -1. The graph of $y = x^2 - 1$ is a parabola with vertex $(0, -1)$. Of course, $(1, 0)$ and $(-2, 3)$ are on both graphs. The two graphs are shown in Fig. 12.1.

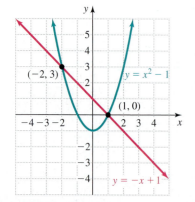

Figure 12.1

Now do Exercises 1–10

Nonlinear systems often have more than one solution, and drawing the graphs helps us to understand why. However, it is not necessary to draw the graphs to solve the system, as shown in Example 2.

E X A M P L E   **2**

Solving a system algebraically with substitution

Solve the system:

$$x^2 + y^2 + 2y = 3$$
$$x^2 - y = 5$$

Solution

If we substitute $y = x^2 - 5$ into the first equation to eliminate y, we will get a fourth-degree equation to solve. Instead, we can eliminate the variable x by writing $x^2 - y = 5$ as $x^2 = y + 5$. Now replace x^2 by $y + 5$ in the first equation:

$$x^2 + y^2 + 2y = 3$$
$$(y + 5) + y^2 + 2y = 3$$
$$y^2 + 3y + 5 = 3$$
$$y^2 + 3y + 2 = 0$$
$$(y + 2)(y + 1) = 0 \quad \text{Solve by factoring.}$$

$$y + 2 = 0 \quad \text{or} \quad y + 1 = 0$$
$$y = -2 \quad \text{or} \quad y = -1$$

Let $y = -2$ in the equation $x^2 = y + 5$ to find the corresponding x:

$$x^2 = -2 + 5$$
$$x^2 = 3$$
$$x = \pm\sqrt{3}$$

Now let $y = -1$ in the equation $x^2 = y + 5$ to find the corresponding x:

$$x^2 = -1 + 5$$
$$x^2 = 4$$
$$x = \pm 2$$

Check these values in the original equations. The solution set is

$$\{(\sqrt{3}, -2), (-\sqrt{3}, -2), (2, -1), (-2, -1)\}.$$

The graphs of these two equations intersect at four points.

> **Now do Exercises 11–20**

E X A M P L E   **3**

Solving a system with the addition method

Solve each system:

a) $x^2 - y^2 = 5$
   $x^2 + y^2 = 7$

b) $\dfrac{2}{x} + \dfrac{1}{y} = \dfrac{1}{5}$

   $\dfrac{1}{x} - \dfrac{3}{y} = \dfrac{1}{3}$

Solution

a) We can eliminate y by adding the equations:

$$x^2 - y^2 = 5$$
$$\underline{x^2 + y^2 = 7}$$
$$2x^2 \quad\quad = 12$$
$$x^2 = 6$$
$$x = \pm\sqrt{6}$$

Since $x^2 = 6$, the second equation yields $6 + y^2 = 7$, $y^2 = 1$, and $y = \pm 1$. If $x^2 = 6$ and $y^2 = 1$, then both of the original equations are satisfied. The solution set is

$$\{(\sqrt{6}, 1), (\sqrt{6}, -1), (-\sqrt{6}, 1), (-\sqrt{6}, -1)\}.$$

b) Usually with equations involving rational expressions we first multiply by the least common denominator (LCD), but this would make the given system more complicated. So we will just use the addition method to eliminate y:

$$\frac{6}{x} + \frac{3}{y} = \frac{3}{5} \quad\quad \text{Eq. (1) multiplied by 3}$$

$$\frac{1}{x} - \frac{3}{y} = \frac{1}{3} \quad\quad \text{Eq. (2)}$$

$$\overline{}$$

$$\frac{7}{x} \quad\quad = \frac{14}{15} \quad\quad \frac{3}{5} + \frac{1}{3} = \frac{14}{15}$$

$$14x = 7 \cdot 15$$

$$x = \frac{7 \cdot 15}{14} = \frac{15}{2}$$

To find y, substitute $x = \frac{15}{2}$ into Eq. (1):

$$\frac{2}{\dfrac{15}{2}} + \frac{1}{y} = \frac{1}{5}$$

$$\frac{4}{15} + \frac{1}{y} = \frac{1}{5} \quad\quad \frac{2}{\dfrac{15}{2}} = 2 \cdot \frac{2}{15} = \frac{4}{15}$$

$$15y \cdot \frac{4}{15} + 15y \cdot \frac{1}{y} = 15y \cdot \frac{1}{5} \quad\quad \text{Multiply each side by the LCD, 15y.}$$

$$4y + 15 = 3y$$

$$y = -15$$

Check that $x = \frac{15}{2}$ and $y = -15$ satisfy both original equations. The solution set is $\left\{\left(\frac{15}{2}, -15\right)\right\}.$

Now do Exercises 21–36

A system of nonlinear equations might involve exponential or logarithmic functions. To solve such systems, you will need to recall some facts about exponents and logarithms.

EXAMPLE **4**

A system involving logarithms

Solve the system

$$y = \log_2(x + 28)$$
$$y = 3 + \log_2(x)$$

Solution

Eliminate y by substituting $\log_2(x + 28)$ for y in the second equation:

$$\log_2(x + 28) = 3 + \log_2(x) \quad \text{Eliminate } y.$$
$$\log_2(x + 28) - \log_2(x) = 3 \qquad\qquad \text{Subtract } \log_2(x) \text{ from each side.}$$
$$\log_2\left(\frac{x + 28}{x}\right) = 3 \qquad\qquad \text{Quotient rule for logarithms}$$
$$\frac{x + 28}{x} = 8 \qquad\qquad \text{Definition of logarithm}$$
$$x + 28 = 8x \qquad\qquad \text{Multiply each side by } x.$$
$$28 = 7x \qquad\qquad \text{Subtract } x \text{ from each side.}$$
$$4 = x \qquad\qquad \text{Divide each side by 7.}$$

If $x = 4$, then $y = \log_2(4 + 28) = \log_2(32) = 5$. Check $(4, 5)$ in both equations. The solution to the system is $\{(4, 5)\}$.

> **Now do Exercises 37–42**

‹2› Applications

Example 5 shows a geometric problem that can be solved with a system of nonlinear equations.

EXAMPLE **5**

Nonlinear equations in applications

A 15-foot ladder is leaning against a wall so that the distance from the bottom of the ladder to the wall is one-half of the distance from the top of the ladder to the ground. Find the distance from the top of the ladder to the ground.

Solution

Let x be the number of feet from the bottom of the ladder to the wall and y be the number of feet from the top of the ladder to the ground (see Fig. 12.2 on page 764). We can write two equations involving x and y:

$$x^2 + y^2 = 15^2 \quad \text{Pythagorean theorem}$$
$$y = 2x$$

Solve by substitution:

$$x^2 + (2x)^2 = 225 \quad \text{Replace } y \text{ by } 2x.$$
$$x^2 + 4x^2 = 225$$
$$5x^2 = 225$$
$$x^2 = 45$$
$$x = \pm\sqrt{45} = \pm3\sqrt{5}$$

‹ Calculator Close-Up ›

To see the solutions, graph

$$y_1 = \sqrt{15^2 - x^2},$$
$$y_2 = -\sqrt{15^2 - x^2}, \text{and}$$
$$y_3 = 2x.$$

The line intersects the circle twice.

Figure 12.2

Because x represents distance, x must be positive. So $x = 3\sqrt{5}$. Because $y = 2x$, we get $y = 6\sqrt{5}$. The distance from the top of the ladder to the ground is $6\sqrt{5}$ feet.

> Now do Exercises 43–46

Example 6 shows how a nonlinear system can be used to solve a problem involving work.

EXAMPLE   **6**

Nonlinear equations in applications

A large fish tank at the Gulf Aquarium can usually be filled in 10 minutes using pumps A and B. However, pump B can pump water in or out at the same rate. If pump B is inadvertently run in reverse, then the tank will be filled in 30 minutes. How long would it take each pump to fill the tank by itself?

< **Helpful Hint** >

Note that we could write equations about the rates. Pump A's rate is $\frac{1}{a}$ tank per minute, B's rate is $\frac{1}{b}$ tank per minute, and together their rate is $\frac{1}{10}$ tank per minute or $\frac{1}{30}$ tank per minute.

$$\frac{1}{a} + \frac{1}{b} = \frac{1}{10}$$

$$\frac{1}{a} - \frac{1}{b} = \frac{1}{30}$$

Solution

Let a represent the number of minutes that it takes pump A to fill the tank alone and b represent the number of minutes it takes pump B to fill the tank alone. The rate at which pump A fills the tank is $\frac{1}{a}$ of the tank per minute, and the rate at which pump B fills the tank is $\frac{1}{b}$ of the tank per minute. Because the work completed is the product of the rate and time, we can make the following table when the pumps work together to fill the tank:

	Rate	Time	Work
Pump A	$\frac{1}{a}$ $\frac{\text{tank}}{\text{min}}$	10 min	$\frac{10}{a}$ tank
Pump B	$\frac{1}{b}$ $\frac{\text{tank}}{\text{min}}$	10 min	$\frac{10}{b}$ tank

Note that each pump fills a fraction of the tank and those fractions have a sum of 1:

$$(1) \qquad \frac{10}{a} + \frac{10}{b} = 1$$

In the 30 minutes in which pump B is working in reverse, A puts in $\frac{30}{a}$ of the tank whereas B takes out $\frac{30}{b}$ of the tank. Since the tank still gets filled, we can write the following equation:

$$(2) \qquad \frac{30}{a} - \frac{30}{b} = 1$$

Multiply Eq. (1) by 3 and add the result to Eq. (2) to eliminate b:

$$\frac{30}{a} + \frac{30}{b} = 3 \qquad \text{Eq. (1) multiplied by 3}$$

$$\frac{30}{a} - \frac{30}{b} = 1 \qquad \text{Eq. (2)}$$

$$\overline{\phantom{\frac{30}{a}}}$$

$$\frac{60}{a} = 4$$

$$4a = 60$$

$$a = 15$$

Use $a = 15$ in Eq. (1) to find b:

$$\frac{10}{15} + \frac{10}{b} = 1$$

$$\frac{10}{b} = \frac{1}{3} \qquad \text{Subtract } \tfrac{10}{15} \text{ from each side.}$$

$$b = 30$$

So pump A fills the tank in 15 minutes working alone, and pump B fills the tank in 30 minutes working alone.

> Now do Exercises 47–56

Warm-Ups ▼

Fill in the blank.

1. The graph of a(n) _____ equation is not a straight line.
2. The _____ of a nonlinear system can show us the number of solutions and the approximate value of the solutions.
3. We solve nonlinear systems using _____ and _____.

True or false?

4. The graph of $y = x^2$ is a parabola.
5. The graph of $y = |x|$ is a straight line.

6. The point $(3, -4)$ satisfies both $x^2 + y^2 = 25$ and $y = \sqrt{5x + 1}$.
7. There is no solution to the system $y = x^2 + 2$ and $y = x$.
8. The graphs of $y = \sqrt{x}$ and $y = -x - 2$ intersect at a single point.
9. If Bob paints a fence in x hours, then he paints $\frac{1}{x}$ of the fence per hour.
10. The area of a right triangle is one-half the product of the lengths of its legs.

‹ 1 › **Solving by Elimination**

Solve each system, and graph both equations on the same set of axes. See Example 1.

1. $y = x^2$
$x + y = 6$

2. $y = x^2 - 1$
$x + y = 11$

9. $y = -x^2 + 1$
$y = x^2$

10. $y = x^2$
$y = \sqrt{x}$

3. $y = |x|$
$2y - x = 6$

4. $y = |x|$
$3y = x + 6$

Solve each system. See Examples 2 and 3.

11. $xy = 6$
$x = 2$

12. $xy = 1$
$y = 3$

13. $xy = 1$
$y = x$

14. $y = x^2$
$y = x$

5. $y = \sqrt{2x}$
$x - y = 4$

6. $y = \sqrt{x}$
$x - y = 6$

15. $y = x^2$
$y = 2$

16. $xy = 3$
$y = x$

17. $x^2 + y^2 = 25$
$y = x^2 - 5$

18. $x^2 + y^2 = 25$
$y = x + 1$

19. $xy - 3x = 8$
$y = x + 1$

20. $xy + 2x = 9$
$x - y = 2$

7. $4x - 9y = 9$
$xy = 1$

8. $2x + 2y = 3$
$xy = -1$

21. $xy - x = 8$
$xy + 3x = -4$

22. $2xy - 3x = -1$
$xy + 5x = -7$

23. $x^2 + y^2 = 8$
$x^2 - y^2 = 2$

24. $y^2 - 2x^2 = 1$
$y^2 + 2x^2 = 5$

25. $x^2 + 2y^2 = 8$
$2x^2 - y^2 = 1$

26. $2x^2 + 3y^2 = 8$
$3x^2 + 2y^2 = 7$

27. $\dfrac{1}{x} - \dfrac{1}{y} = 5$
$\dfrac{2}{x} + \dfrac{1}{y} = -3$

28. $\dfrac{2}{x} - \dfrac{3}{y} = \dfrac{1}{2}$
$\dfrac{3}{x} + \dfrac{1}{y} = \dfrac{1}{2}$

29. $\dfrac{2}{x} - \dfrac{1}{y} = \dfrac{5}{12}$
$\dfrac{1}{x} - \dfrac{3}{y} = -\dfrac{5}{12}$

30. $\dfrac{3}{x} - \dfrac{2}{y} = 5$
$\dfrac{4}{x} + \dfrac{3}{y} = 18$

31. $x^2y = 20$
$xy + 2 = 6x$

32. $y^2x = 3$
$xy + 1 = 6x$

33. $x^2 + xy - y^2 = -11$
$x + y = 7$

34. $x^2 + xy + y^2 = 3$
$y = 2x - 5$

35. $3y - 2 = x^4$
$y = x^2$

36. $y - 3 = 2x^4$
$y = 7x^2$

Solve the following systems involving logarithmic and exponential functions. See Example 4.

37. $y = \log_2(x - 1)$
$y = 3 - \log_2(x + 1)$

38. $y = \log_3(x - 4)$
$y = 2 - \log_3(x + 4)$

39. $y = \log_2(x - 1)$
$y = 2 + \log_2(x + 2)$

40. $y = \log_4(8x)$
$y = 2 + \log_4(x - 1)$

41. $y = 2^{3x+4}$
$y = 4^{x-1}$

42. $y = 4^{3x}$
$y = \left(\dfrac{1}{2}\right)^{1-x}$

⟨**2**⟩ **Applications**

Solve each problem by using a system of two equations in two unknowns. See Examples 5 and 6.

43. *Known hypotenuse.* Find the lengths of the legs of a right triangle whose hypotenuse is $\sqrt{15}$ feet and whose area is 3 square feet.

44. *Known diagonal.* A small television is advertised to have a picture with a diagonal measure of 5 inches and a viewing area of 12 square inches (in.^{2}). What are the length and width of the screen?

Figure for Exercise 44

45. *House of seven gables.* Vincent has plans to build a house with seven gables. The plans call for an attic vent in the shape of an isosceles triangle in each gable. Because of the slope of the roof, the ratio of the height to the base of each triangle must be 1 to 4. If the vents are to provide a total ventilating area of 3500 in.^{2}, then what should be the height and base of each triangle?

Figure for Exercise 45

46. *Known perimeter.* Find the lengths of the sides of a triangle whose perimeter is 6 feet (ft) and whose angles are 30°, 60°, and 90° (see inside the front cover of the book).

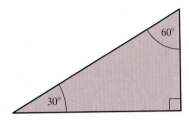

Figure for Exercise 46

47. *Filling a tank.* Pump A can either fill a tank or empty it in the same amount of time. If pump A and pump B are working together, the tank can be filled in 6 hours. When pump A was inadvertently left in the drain position while pump B was trying to fill the tank, it took 12 hours to fill the tank. How long would it take either pump working alone to fill the tank?

48. *Cleaning a house.* Roxanne either cleans the house or messes it up at the same rate. When Roxanne is cleaning with her mother, they can clean up a completely messed up house in 6 hours. If Roxanne is not cooperating, it takes her mother 9 hours to clean the house, with Roxanne continually messing it up. How long would it take her mother to clean the entire house if Roxanne were sent to her grandmother's house?

49. *Cleaning fish.* Jan and Beth work in a seafood market that processes 200 pounds of catfish every morning.

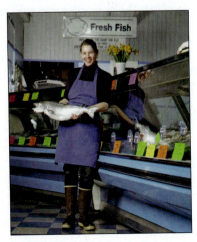

Photo for Exercise 49

On Monday, Jan started cleaning catfish at 8:00 A.M. and finished cleaning 100 pounds just as Beth arrived. Beth then took over and finished the job at 8:50 A.M. On Tuesday they both started at 8 A.M. and worked together to finish the job at 8:24 A.M. On Wednesday, Beth was sick. If Jan is the faster worker, then how long did it take Jan to complete all of the catfish by herself?

50. *Building a patio.* Richard has already formed a rectangular area for a flagstone patio, but his wife Susan is unsure of the size of the patio they want. If the width is increased by 2 ft, then the area is increased by 30 square feet (ft²). If the width is increased by 1 ft and the length by 3 ft, then the area is increased by 54 ft². What are the dimensions of the rectangle that Richard has already formed?

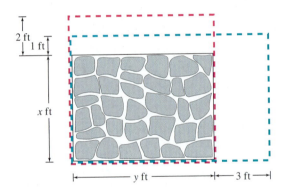

Figure for Exercise 50

51. *Fencing a rectangle.* If 34 ft of fencing is used to enclose a rectangular area of 72 ft², then what are the dimensions of the area?

52. *Real numbers.* Find two numbers that have a sum of 8 and a product of 10.

53. *Imaginary numbers.* Find two complex numbers whose sum is 8 and whose product is 20.

54. *Imaginary numbers.* Find two complex numbers whose sum is −6 and whose product is 10.

55. *Making a sign.* Rico's Sign Shop has a contract to make a sign in the shape of a square with an isosceles triangle on top of it, as shown in the figure. The contract calls for a total height of 10 ft with an area of 72 ft². How long should Rico make the side of the square, and what should be the height of the triangle?

Figure for Exercise 55

56. *Designing a box.* Angelina is designing a rectangular box of 120 cubic inches that is to contain new Eaties breakfast cereal. The box must be 2 inches thick so that it is easy to hold. It must have 184 square inches of surface area to provide enough space for all of the special offers and coupons. What should be the dimensions of the box?

 Graphing Calculator Exercises

57. Solve each system by graphing each pair of equations on a graphing calculator and using the intersect feature to estimate the point of intersection. Find the coordinates of each intersection to the nearest hundredth.

a) $y = e^x - 4$
$y = \ln(x + 3)$

b) $3^{y-1} = x$
$y = x^2$

c) $x^2 + y^2 = 4$
$y = x^3$

12.2   The Parabola

In This Section

⟨1⟩ **The Distance and Midpoint Formulas**

⟨2⟩ **The Parabola**

⟨3⟩ **Changing Forms**

⟨4⟩ **Parabolas Opening to the Right or Left**

The **conic sections** are the four curves that are obtained by intersecting a cone and a plane as in Fig. 12.3. The figure explains why the parabola, ellipse, circle, and hyperbola are called conic sections, but it does not help us find equations for the curves. To develop equations for these curves we will redefine them more precisely using distance between points. So we will first discuss the distance formula.

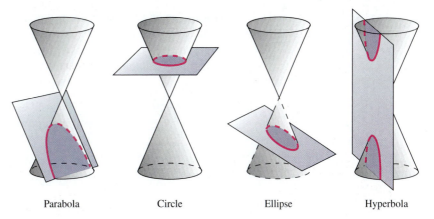

Parabola   Circle   Ellipse   Hyperbola

Figure 12.3

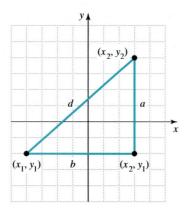

Figure 12.4

⟨1⟩ **The Distance and Midpoint Formulas**

Consider the points (x_1, y_1) and (x_2, y_2), as shown in Fig. 12.4. The distance between these points is the length of the hypotenuse of a right triangle as shown in the figure. The length of side a is $y_2 - y_1$, and the length of side b is $x_2 - x_1$. Using the Pythagorean theorem, we can write

$$d^2 = (x_2 - x_1)^2 + (y_2 - y_1)^2.$$

If we apply the even-root property and omit the negative square root (because the distance is positive), we can express this formula as follows.

> ### Distance Formula
> The distance d between (x_1, y_1) and (x_2, y_2) is given by the formula
> $$d = \sqrt{(x_2 - x_1)^2 + (y_2 - y_1)^2}.$$

E X A M P L E   1

Using the distance formula
Find the length of the line segment with endpoints $(-8, -10)$ and $(6, -4)$.

Solution
Let $(x_1, y_1) = (-8, -10)$ and $(x_2, y_2) = (6, -4)$. Now substitute the appropriate values into the distance formula:

$$
\begin{aligned}
d &= \sqrt{[6 - (-8)]^2 + [-4 - (-10)]^2} \\
&= \sqrt{(14)^2 + (6)^2} \\
&= \sqrt{196 + 36} \\
&= \sqrt{232} \\
&= \sqrt{4 \cdot 58} \\
&= 2\sqrt{58} \quad \text{Simplified form}
\end{aligned}
$$

The exact length of the segment is $2\sqrt{58}$.

> Now do Exercises 1–10

The **midpoint** of a line segment is a point that is on the line segment and equidistant from the endpoints. We use the notation $(\bar{x}, \bar{y})$ (read "x bar, y bar") for the midpoint of a line segment. The midpoint is found by "averaging" the x-coordinates and y-coordinates of the endpoints, in the same manner that you would average two test scores:

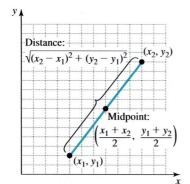

Figure 12.5

> ### Midpoint Formula
> The midpoint of the line segment with endpoints (x_1, y_1) and (x_2, y_2) is given by
> $$(\bar{x}, \bar{y}) = \left(\frac{x_1 + x_2}{2}, \frac{y_1 + y_2}{2} \right).$$

The length of a line segment is the distance between its endpoints, and it is given by the distance formula. See Fig. 12.5.

E X A M P L E   **2**

Finding the midpoint and length of a line segment

Find the midpoint and length of the line segment with endpoints (1, 7) and (5, 4).

Solution

Use the midpoint formula with $(x_1, y_1) = (1, 7)$ and $(x_2, y_2) = (5, 4)$:

$$(\bar{x}, \bar{y}) = \left(\frac{1 + 5}{2}, \frac{7 + 4}{2}\right) = \left(3, \frac{11}{2}\right)$$

Use the distance formula to find the length of the line segment:

$$\sqrt{(x_2 - x_1)^2 + (y_2 - y_1)^2} = \sqrt{(5 - 1)^2 + (4 - 7)^2}$$
$$= \sqrt{16 + 9}$$
$$= \sqrt{25} = 5$$

Note that $(x_1, y_1) = (5, 4)$ and $(x_2, y_2) = (1, 7)$ gives the same midpoint and length. Try it.

Now do Exercises 11–18

‹2› The Parabola

In Section 8.4 we called the graph of $y = ax^2 + bx + c$ a parabola. In this section, you will see that the following geometric definition describes the same curve as the equation.

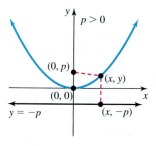

Parabola

Focus   Vertex

Directrix

Figure 12.6

> **Parabola**
>
> Given a line (the **directrix**) and a point not on the line (the **focus**), the set of all points in the plane that are equidistant from the point and the line is called a **parabola.**

In Section 8.4 we defined the vertex as the highest point on a parabola that opens downward or the lowest point on a parabola that opens upward. We learned that $x = -b/(2a)$ gives the x-coordinate of the vertex. We can also describe the vertex of a parabola as the midpoint of the line segment that joins the focus and directrix, perpendicular to the directrix. See Fig. 12.6.

The focus of a parabola is important in applications. When parallel rays of light travel into a parabolic reflector, they are reflected toward the focus, as in Fig. 12.7. This property is used in telescopes to see the light from distant stars. If the light source is at the focus, as in a searchlight, the light is reflected off the parabola and projected outward in a narrow beam. This reflecting property is also used in camera lenses, satellite dishes, and eavesdropping devices.

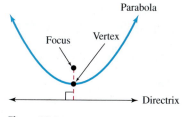

Focus

Figure 12.7

To develop an equation for a parabola, given the focus and directrix, choose the point $(0, p)$, where $p > 0$, as the focus and the line $y = -p$ as the directrix, as shown in Fig. 12.8. The vertex of this parabola is $(0, 0)$. For an arbitrary point (x, y) on the parabola the distance to the directrix is the distance from (x, y) to $(x, -p)$. The distance to the focus is the distance between (x, y) and $(0, p)$. We use the fact that these distances are equal to write the equation of the parabola:

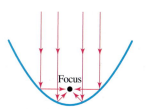

y

$p > 0$

$(0, p)$

(x, y)

$(0, 0)$

x

$y = -p$

$(x, -p)$

Figure 12.8

$$\sqrt{(x - 0)^2 + (y - p)^2} = \sqrt{(x - x)^2 + (y - (-p))^2}$$

To simplify the equation, first remove the parentheses inside the radicals:

$$\sqrt{x^2 + y^2 - 2py + p^2} = \sqrt{y^2 + 2py + p^2}$$

$$x^2 + y^2 - 2py + p^2 = y^2 + 2py + p^2 \quad \text{Square each side.}$$

$$x^2 = 4py \quad \text{Subtract } y^2 \text{ and } p^2 \text{ from each side.}$$

$$y = \frac{1}{4p}x^2$$

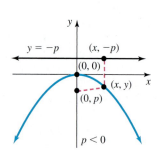

$y = -p$ $(x, -p)$

$(0, 0)$

(x, y) x

$(0, p)$

$p < 0$

Figure 12.9

So the parabola with focus $(0, p)$ and directrix $y = -p$ for $p > 0$ has equation $y = \frac{1}{4p}x^2$. This equation has the form $y = ax^2 + bx + c$, where $a = \frac{1}{4p}$, $b = 0$, and $c = 0$.

If the focus is $(0, p)$ with $p < 0$ and the directrix is $y = -p$, then the parabola opens downward, as shown in Fig. 12.9. Deriving the equation using the distance formula again yields $y = \frac{1}{4p}x^2$.

The simplest parabola, $y = x^2$, has vertex $(0, 0)$. The transformation $y = a(x - h)^2 + k$ is also a parabola and its vertex is (h, k). The focus and directrix of the transformation are found as follows:

Parabolas in the Form $y = a(x - h)^2 + k$

The graph of the equation $y = a(x - h)^2 + k$ $(a \neq 0)$ is a parabola with vertex (h, k), focus $(h, k + p)$, and directrix $y = k - p$, where $a = \frac{1}{4p}$. If $a > 0$, the parabola opens upward; if $a < 0$, the parabola opens downward.

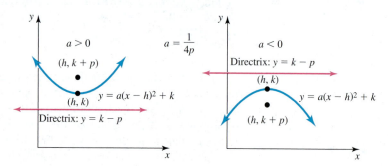

Figure 12.10

Figure 12.10 shows the location of the focus and directrix for parabolas with vertex (h, k) and opening either upward or downward. Note that the location of the focus and directrix determine the value of a and the shape and opening of the parabola.

CAUTION   For a parabola that opens upward, $p > 0$, and the focus $(h, k + p)$ is above the vertex (h, k). For a parabola that opens downward, $p < 0$, and the focus $(h, k + p)$ is below the vertex (h, k). In either case, the distance from the vertex to the focus and the vertex to the directrix is $|p|$.

In Example 3 we find the vertex, focus, and directrix from an equation of a parabola. In Example 4 we find the equation given the focus and directrix.

E X A M P L E **3**

Finding the vertex, focus, and directrix, given an equation

Find the vertex, focus, and directrix for the parabola $y = x^2$.

Solution

Compare $y = x^2$ to the general formula $y = a(x - h)^2 + k$. We see that $h = 0$, $k = 0$, and $a = 1$. So the vertex is $(0, 0)$. Because $a = 1$, we can use $a = \frac{1}{4p}$ to get

$$1 = \frac{1}{4p},$$

or $p = \frac{1}{4}$. Use $(h, k + p)$ to get the focus $\left(0, \frac{1}{4}\right)$. Use the equation $y = k - p$ to get $y = -\frac{1}{4}$ as the equation of the directrix. See Fig. 12.11.

> Now do Exercises 19–26

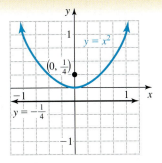

Figure 12.11

E X A M P L E **4**

Finding an equation, given a focus and directrix

Find the equation of the parabola with focus $(-1, 4)$ and directrix $y = 3$.

Solution

Because the vertex is halfway between the focus and directrix, the vertex is $\left(-1, \frac{7}{2}\right)$. See Fig. 12.12. The distance from the vertex to the focus is $\frac{1}{2}$. Because the focus is above the vertex, p is positive. So $p = \frac{1}{2}$, and $a = \frac{1}{4p} = \frac{1}{2}$.

The equation is

$$y = \frac{1}{2}(x - (-1))^2 + \frac{7}{2}.$$

Convert to $y = ax^2 + bx + c$ form as follows:

$$y = \frac{1}{2}(x + 1)^2 + \frac{7}{2}$$

$$y = \frac{1}{2}(x^2 + 2x + 1) + \frac{7}{2}$$

$$y = \frac{1}{2}x^2 + x + 4$$

> Now do Exercises 27–36

Figure 12.12

The graph of $y = x^2$ shown in Fig. 12.11 is **symmetric about the y-axis** because the two halves of the parabola would coincide if the paper were folded on the y-axis. In general, the vertical line through the vertex is the **axis of symmetry** for the

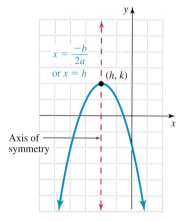

Figure 12.13

parabola. See Fig. 12.13. In the form $y = ax^2 + bx + c$, the x-coordinate of the vertex is $-b/(2a)$ and the equation of the axis of symmetry is $x = -b/(2a)$. In the form $y = a(x - h)^2 + k$, the vertex is (h, k) and the equation for the axis of symmetry is $x = h$.

⟨3⟩ Changing Forms

Since there are two forms for the equation of a parabola, it is sometimes useful to change from one form to the other. To change from $y = a(x - h)^2 + k$ to the form $y = ax^2 + bx + c$, we square the binomial and combine like terms, as in Example 4. To change from $y = ax^2 + bx + c$ to the form $y = a(x - h)^2 + k$, we complete the square, as in Example 5.

EXAMPLE   **5**

Converting $y = ax^2 + bx + c$ to $y = a(x - h)^2 + k$

Write $y = 2x^2 - 4x + 5$ in the form $y = a(x - h)^2 + k$, and identify the vertex, focus, directrix, and axis of symmetry of the parabola.

Solution

Use completing the square to rewrite the equation:

$$y = 2(x^2 - 2x) + 5$$

$$y = 2(x^2 - 2x + 1 - 1) + 5 \quad \text{Complete the square.}$$

$$y = 2(x^2 - 2x + 1) - 2 + 5 \quad \text{Move } 2(-1) \text{ outside the parentheses.}$$

$$y = 2(x - 1)^2 + 3$$

The vertex is $(1, 3)$. Because $a = \dfrac{1}{4p}$, we have

$$\frac{1}{4p} = 2,$$

and $p = \frac{1}{8}$. Because the parabola opens upward, the focus is $\frac{1}{8}$ unit above the vertex at $\left(1, 3\frac{1}{8}\right)$, or $\left(1, \frac{25}{8}\right)$, and the directrix is the horizontal line $\frac{1}{8}$ unit below the vertex, $y = 2\frac{7}{8}$ or $y = \frac{23}{8}$. The axis of symmetry is $x = 1$.

> Now do Exercises 37–44

⟨ **Calculator Close-Up** ⟩

The graphs of

$$y_1 = 2x^2 - 4x + 5$$

and

$$y_2 = 2(x - 1)^2 + 3$$

appear to be identical. This supports the conclusion that the equations are equivalent.

CAUTION   Be careful when you complete a square within parentheses as in Example 5. For another example, consider the equivalent equations

$$y = -3(x^2 + 4x),$$

$$y = -3(x^2 + 4x + 4 - 4),$$

and

$$y = -3(x + 2)^2 + 12.$$

EXAMPLE **6**

Finding the features of a parabola in the form $y = ax^2 + bx + c$

Find the vertex, focus, directrix, and axis of symmetry of the parabola $y = -3x^2 + 9x - 5$, and determine whether the parabola opens upward or downward.

Solution

The x-coordinate of the vertex is

$$x = \frac{-b}{2a} = \frac{-9}{2(-3)} = \frac{-9}{-6} = \frac{3}{2}.$$

To find the y-coordinate of the vertex, let $x = \frac{3}{2}$ in $y = -3x^2 + 9x - 5$:

$$y = -3\left(\frac{3}{2}\right)^2 + 9\left(\frac{3}{2}\right) - 5 = -\frac{27}{4} + \frac{27}{2} - 5 = \frac{7}{4}$$

The vertex is $\left(\frac{3}{2}, \frac{7}{4}\right)$. Because $a = -3$, the parabola opens downward. To find the focus, use $-3 = \frac{1}{4p}$ to get $p = -\frac{1}{12}$. The focus is $\frac{1}{12}$ of a unit below the vertex at $\left(\frac{3}{2}, \frac{7}{4} - \frac{1}{12}\right)$ or $\left(\frac{3}{2}, \frac{5}{3}\right)$. The directrix is the horizontal line $\frac{1}{12}$ of a unit above the vertex, $y = \frac{7}{4} + \frac{1}{12}$ or $y = \frac{11}{6}$. The equation of the axis of symmetry is $x = \frac{3}{2}$.

> Now do Exercises 45–54

‹4› Parabolas Opening to the Right or Left

If we interchange x and y in the equation $y = a(x - h)^2 + k$, we get the equation $x = a(y - k)^2 + h$, which is a parabola opening to the right or left.

> **Parabolas in the Form $x = a(y - k)^2 + h$**
>
> The graph of $x = a(y - k)^2 + h$ $(a \neq 0)$ is a parabola with vertex (h, k), focus $(h + p, k)$, and directrix $x = h - p$, where $a = \frac{1}{4p}$. If $a > 0$, the parabola opens to the right; if $a < 0$, the parabola opens to the left.

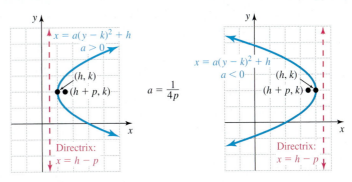

Figure 12.14

Figure 12.14 shows the location of the focus and directrix for parabolas with vertex (h, k) and opening either right or left. The location of the focus and directrix determine the value of a and the shape and opening of the parabola. Note that a and p have the same sign because $a = \frac{1}{4p}$.

The equation $x = ay^2 + by + c$ could be converted to the form $x = a(y - k)^2 + h$ from which the vertex, focus, and directrix could be determined. Without converting we can determine that the graph of $x = ay^2 + by + c$ opens to the right for $a > 0$ and to the left for $a < 0$. The y-coordinate of the vertex is $\frac{-b}{2a}$. The x-coordinate of the vertex can be determined by substituting $\frac{-b}{2a}$ for y in $x = ay^2 + by + c$.

E X A M P L E   7

Graphing a parabola opening to the right

Find the vertex, focus, and directrix for the parabola $x = \frac{1}{2}(y - 2)^2 + 1$, and sketch the graph.

Solution

In the form $x = a(y - k)^2 + h$, the vertex is (h, k). So the vertex for $x = \frac{1}{2}(y - 2)^2 + 1$ is $(1, 2)$. Since $a = \frac{1}{4p}$ and $a = \frac{1}{2}$, we have $p = \frac{1}{2}$ and the focus is $\left(\frac{3}{2}, 2\right)$. The directrix is the vertical line $x = \frac{1}{2}$. Find a few points that satisfy $x = \frac{1}{2}(y - 2)^2 + 1$ as follows:

$x = \frac{1}{2}(y - 2)^2 + 1$	3	$\frac{3}{2}$	1	$\frac{3}{2}$	3
y	0	1	2	3	4

Sketch the graph through these points, as shown in Fig. 12.15.

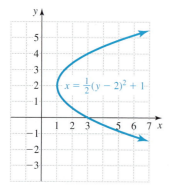

Figure 12.15

Now do Exercises 55–60

Warm-Ups ▼

Fill in the blank.

1. A(n) _____ is the set of all points in a plane that are equidistant from a given line and a fixed point not on the line.

2. The _____ of a parabola is the fixed point in the definition.

3. The _____ is the given line in the definition.

4. The _____ of a parabola is the midpoint of the line segment joining the focus and directrix, perpendicular to the directrix.

5. The distance from the _____ to the _____ is $|p|$, where $a = \frac{1}{4p}$.

6. We can convert $y = ax^2 + bx + c$ to the form $y = a(x - h)^2 + k$ by _____.

True or false?

7. If the focus of a parabola is (0, 4) and the directrix is $y = 1$, then the vertex is (0, 3).

8. The focus for $y = \frac{1}{4}x^2 + 1$ is (0, 2).

9. The vertex for $y - 3 = 5(x - 4)^2$ is (4, 3).

10. The parabola $y = 2x - x^2 + 9$ opens upward.

11. The vertex for $y = x^2$ is the y-intercept.

12. A parabola with vertex (2, 3) and focus (2, 4) has no x-intercepts.

‹1› The Distance and Midpoint Formulas

Find the distance between each given pair of points. See Example 1.

1. $(2, 1)$, $(5, 5)$

2. $(3, 2)$, $(8, 14)$

3. $(4, -3)$, $(5, -2)$

4. $(-1, 5)$, $(-2, 6)$

5. $(6, 5)$, $(4, 2)$

6. $(7, 3)$, $(5, 1)$

7. $(3, 5)$, $(1, -3)$

8. $(6, 2)$, $(3, -5)$

 9. $(4, -2)$, $(-3, -6)$

10. $(-2, 3)$, $(1, -4)$

Find the midpoint and length of the line segment with the given endpoints. See Example 2.

11. $(0, 0)$ and $(6, 8)$

12. $(0, 0)$ and $(-6, 8)$

13. $(2, 5)$ and $(5, 1)$

14. $(1, 7)$ and $(5, 10)$

15. $(-2, 4)$ and $(6, -2)$

16. $(-3, 5)$ and $(3, -3)$

17. $(-1, 4)$ and $(1, 1)$

18. $(-3, -4)$ and $(-6, 1)$

‹2› The Parabola

Find the vertex, focus, and directrix for each parabola. See Example 3.

19. $y = 2x^2$

20. $y = \dfrac{1}{2}x^2$

21. $y = -\dfrac{1}{4}x^2$

22. $y = -\dfrac{1}{12}x^2$

23. $y = \dfrac{1}{2}(x - 3)^2 + 2$

24. $y = \dfrac{1}{4}(x + 2)^2 - 5$

25. $y = -(x + 1)^2 + 6$

26. $y = -3(x - 4)^2 + 1$

Find the equation of the parabola with the given focus and directrix. See Example 4.

27. Focus $(0, 2)$, directrix $y = -2$

28. Focus $(0, -3)$, directrix $y = 3$

29. Focus $\left(0, -\dfrac{1}{2}\right)$, directrix $y = \dfrac{1}{2}$

30. Focus $\left(0, \dfrac{1}{8}\right)$, directrix $y = -\dfrac{1}{8}$

31. Focus $(3, 2)$, directrix $y = 1$

32. Focus $(-4, 5)$, directrix $y = 4$

33. Focus $(1, -2)$, directrix $y = 2$

34. Focus $(2, -3)$, directrix $y = 1$

35. Focus $(-3, 1.25)$, directrix $y = 0.75$

36. Focus $\left(5, \dfrac{17}{8}\right)$, directrix $y = \dfrac{15}{8}$

‹3› Changing Forms

Write each equation in the form $y = a(x - h)^2 + k$. Identify the vertex, focus, directrix, and axis of symmetry of each parabola. See Example 5.

37. $y = x^2 - 6x + 1$

38. $y = x^2 + 4x - 7$

39. $y = 2x^2 + 12x + 5$

40. $y = 3x^2 + 6x - 7$

41. $y = -2x^2 + 16x + 1$

42. $y = -3x^2 - 6x + 7$

43. $y = 5x^2 + 40x$

44. $y = -2x^2 + 10x$

Find the vertex, focus, directrix, and axis of symmetry of each parabola (without completing the square), and determine whether the parabola opens upward or downward. See Example 6.

45. $y = x^2 - 4x + 1$

46. $y = x^2 - 6x - 7$

47. $y = -x^2 + 2x - 3$

48. $y = -x^2 + 4x + 9$

49. $y = 3x^2 - 6x + 1$

50. $y = 2x^2 + 4x - 3$

51. $y = -x^2 - 3x + 2$

52. $y = -x^2 + 3x - 1$

53. $y = 3x^2 + 5$

54. $y = -2x^2 - 6$

‹4› Parabolas Opening to the Right or Left

Find the vertex, focus, and directrix for each parabola. See Example 7.

55. $x = (y - 2)^2 + 3$

56. $x = (y + 3)^2 - 1$

57. $x = \frac{1}{4}(y - 1)^2 - 2$

58. $x = \frac{1}{4}(y + 1)^2 + 2$

59. $x = -\frac{1}{2}(y - 2)^2 + 4$

60. $x = -\frac{1}{2}(y + 1)^2 - 1$

Miscellaneous

Sketch the graph of each parabola.

61. $y = (x - 2)^2 + 3$ **62.** $y = (x + 3)^2 - 1$

63. $y = -2(x - 1)^2 + 3$

64. $y = -\dfrac{1}{2}(x + 1)^2 + 5$

71. $y = x^2 - 2$
$y = 2x - 3$

72. $y = x^2 + x - 6$
$y = 7x - 15$

65. $x = (y - 2)^2 + 3$

66. $x = (y + 3)^2 - 1$

73. $y = x^2 + 3x - 4$
$y = -x^2 - 2x + 8$

67. $x = -2(y - 1)^2 + 3$

68. $x = -\dfrac{1}{2}(y + 1)^2 + 5$

74. $y = x^2 + 2x - 8$
$y = -x^2 - x + 12$

Graph both equations of each system on the same coordinate axes. Use elimination of variables to find all points of intersection.

69. $y = -x^2 + 3$
$y = x^2 + 1$

75. $y = x^2 + 3x - 4$
$y = 2x + 2$

70. $y = x^2 - 3$
$y = -x^2 + 5$

76. $y = x^2 + 5x + 6$
$y = x + 11$

Solve each problem.

77. Find all points of intersection of the parabola $y = x^2 - 2x - 3$ and the x-axis.

78. Find all points of intersection of the parabola $y = 80x^2 - 33x + 255$ and the y-axis.

79. Find all points of intersection of the parabola $y = 0.01x^2$ and the line $y = 4$.

80. Find all points of intersection of the parabola $y = 0.02x^2$ and the line $y = x$.

81. Find all points of intersection of the parabolas $y = x^2$ and $x = y^2$.

82. Find all points of intersection of the parabolas $y = x^2$ and $y = (x - 3)^2$.

Applications

Solve each problem.

83. *Pipeline charges.* Ewing Oil paid a subcontractor $84 per yard for laying a pipe in a west Texas oil field. The pipe connects wells located at $(185, 234)$ and $(-215, -352)$ in the oil field coordinate system shown in the figure. The units in the figure are yards.

a) What was the cost to the nearest dollar for this project?

b) What is the location of the valve installed at the midpoint?

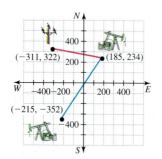

Figure for Exercises 83 and 84

84. *Electricity charges.* Texas Power installed a power line from a transformer at $(-311, 322)$ to the well at $(185, 234)$ as shown in the figure for $116 per yard.

a) What was the cost to the nearest dollar for the power line?

b) What is the location of the pole used at the midpoint?

85. *World's largest telescope.* The largest reflecting telescope in the world is the 6-meter (m) reflector on Mount Pastukhov in Russia. The accompanying figure shows a cross section of a parabolic mirror 6 m in diameter with the vertex at the origin and the focus at $(0, 15)$. Find the equation of the parabola.

Figure for Exercise 85

86. *Arecibo observatory.* The largest radio telescope in the world uses a 1000-ft parabolic dish, suspended in a valley in Arecibo, Puerto Rico. The antenna hangs above the vertex of the dish on cables stretching from two towers. The accompanying figure shows a cross section of the parabolic dish and the towers. Assuming the vertex is at $(0, 0)$, find the equation for the parabola. Find the

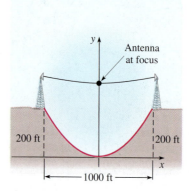

Figure for Exercise 86

distance from the vertex to the antenna located at the focus.

b) What is the equation of its axis of symmetry?

c) Sketch the graphs $x = 2(y - 3)^2 + 1$ and $x = -(y + 1)^2 + 2$.

Getting More Involved

87. *Exploration*

Consider the parabola with focus $(p, 0)$ and directrix $x = -p$ for $p > 0$. Let (x, y) be an arbitrary point on the parabola. Write an equation expressing the fact that the distance from (x, y) to the focus is equal to the distance from (x, y) to the directrix. Rewrite the equation in the form $x = ay^2$, where $a = \dfrac{1}{4p}$.

88. *Exploration*

In general, the graph of $x = a(y - k)^2 + h$ for $a \neq 0$ is a parabola opening left or right with vertex at (h, k).

a) For which values of a does the parabola open to the right, and for which values of a does it open to the left?

Graphing Calculator Exercises

89. Graph $y = x^2$ using the viewing window with $-1 \leq x \leq 1$ and $0 \leq y \leq 1$. Next graph $y = 2x^2 - 1$ using the viewing window $-2 \leq x \leq 2$ and $-1 \leq y \leq 7$. Explain what you see.

90. Graph $y = x^2$ and $y = 6x - 9$ in the viewing window $-5 \leq x \leq 5$ and $-5 \leq y \leq 20$. Does the line appear to be tangent to the parabola? Solve the system $y = x^2$ and $y = 6x - 9$ to find all points of intersection for the parabola and the line.

12.3 The Circle

In This Section

⟨1⟩ The Equation of a Circle

⟨2⟩ Equations Not in Standard Form

⟨3⟩ Systems of Equations

In this section, we continue the study of the conic sections with a discussion of the circle.

⟨1⟩ The Equation of a Circle

A circle is obtained by cutting a cone, as was shown in Fig. 12.3. We can also define a circle using points and distance, as we did for the parabola.

> **Circle**
>
> A **circle** is the set of all points in a plane that lie a fixed distance from a given point in the plane. The fixed distance is called the **radius,** and the given point is called the **center.**

We can use the distance formula of Section 12.2 to write an equation for the circle with center (h, k) and radius r, shown in Fig. 12.16. If (x, y) is a point on the circle, its distance from the center is r. So,

$$\sqrt{(x - h)^2 + (y - k)^2} = r.$$

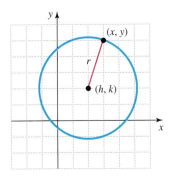

Figure 12.16

We square both sides of this equation to get the **standard form** for the equation of a circle.

> **Standard Equation for a Circle**
> The graph of the equation
> $$(x - h)^2 + (y - k)^2 = r^2$$
> with $r > 0$, is a circle with center (h, k) and radius r.

Note that a circle centered at the origin with radius r ($r > 0$) has the standard equation

$$x^2 + y^2 = r^2.$$

E X A M P L E   1

Finding the equation, given the center and radius
Write the equation for the circle with the given center and radius.

    a) Center $(0, 0)$, radius 2

    b) Center $(-1, 2)$, radius 4

Solution

a) The center at $(0, 0)$ means that $h = 0$ and $k = 0$ in the standard equation. So the equation is $(x - 0)^2 + (y - 0)^2 = 2^2$, or $x^2 + y^2 = 4$. The circle with radius 2 centered at the origin is shown in Fig. 12.17.

b) The center at $(-1, 2)$ means that $h = -1$ and $k = 2$. So,

$$[x - (-1)]^2 + [y - 2]^2 = 4^2.$$

Simplify this equation to get

$$(x + 1)^2 + (y - 2)^2 = 16.$$

The circle with center $(-1, 2)$ and radius 4 is shown in Fig. 12.18.

Figure 12.17

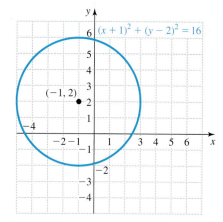

Figure 12.18

> Now do Exercises 1–12

CAUTION The equations $(x - 1)^2 + (y + 3)^2 = -9$ and $(x - 1)^2 + (y + 3)^2 = 0$ might look like equations of circles, but they are not. The first equation is not satisfied by any ordered pair of real numbers because the left-hand side is nonnegative for any x and y. The second equation is satisfied only by the point $(1, -3)$.

EXAMPLE   **2**

Finding the center and radius, given the equation

Determine the center and radius of the circle $x^2 + (y + 5)^2 = 2$.

Solution

We can write this equation as

$$(x - 0)^2 + [y - (-5)]^2 = (\sqrt{2})^2.$$

In this form we see that the center is $(0, -5)$ and the radius is $\sqrt{2}$.

> Now do Exercises 13–22

EXAMPLE   **3**

Graphing a circle

Find the center and radius of $(x - 1)^2 + (y + 2)^2 = 9$, and sketch the graph.

Solution

The graph of this equation is a circle with center $(1, -2)$ and radius 3. See Fig. 12.19 for the graph.

> Now do Exercises 23–32

Figure 12.19

‹ Calculator Close-Up ›

To graph the circle in Example 3, graph

$$y_1 = -2 + \sqrt{9 - (x - 1)^2}$$

and

$$y_2 = -2 - \sqrt{9 - (x - 1)^2}.$$

To get the circle to look round, you must use the same unit length on each axis. Most calculators have a *square* feature that automatically adjusts the window to use the same unit length on each axis.

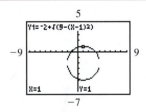

‹2› Equations Not in Standard Form

It is not easy to recognize that $x^2 - 6x + y^2 + 10y = -30$ is the equation of a circle, but it is. In Example 4, we convert this equation into the standard form for a circle by completing the squares for the variables x and y.

EXAMPLE   **4**

Converting to standard form

Find the center and radius of the circle given by the equation

$$x^2 - 6x + y^2 + 10y = -30.$$

‹ Helpful Hint ›

What do circles and lines have in common? They are the two simplest graphs to draw. We have compasses to make our circles look good and rulers to make our lines look good.

Solution

To complete the square for $x^2 - 6x$, we add 9, and for $y^2 + 10y$, we add 25. To get an equivalent equation, we must add on both sides:

$$x^2 - 6x + \qquad y^2 + 10y \qquad = -30$$
$$x^2 - 6x + 9 + y^2 + 10y + 25 = -30 + 9 + 25 \qquad \text{Add 9 and 25 to both sides.}$$
$$(x - 3)^2 + (y + 5)^2 = 4 \qquad\qquad \text{Factor the trinomials on the left-hand side.}$$

From the standard form we see that the center is $(3, -5)$ and the radius is 2.

> Now do Exercises 33–44

⟨3⟩ **Systems of Equations**

We first solved systems of nonlinear equations in two variables in Section 12.1. We found the points of intersection of two graphs without drawing the graphs. Here we will solve systems involving circles, parabolas, and lines. In Example 5, we find the points of intersection of a line and a circle.

E X A M P L E **5**

Intersection of a line and a circle

Graph both equations of the system

$$(x - 3)^2 + (y + 1)^2 = 9$$

$$y = x - 1$$

on the same coordinate axes, and solve the system by elimination of variables.

Solution

The graph of the first equation is a circle with center $(3, -1)$ and radius 3. The graph of the second equation is a straight line with slope 1 and y-intercept $(0, -1)$. Both graphs are shown in Fig. 12.20. To solve the system by elimination, we substitute $y = x - 1$ into the equation of the circle:

$$(x - 3)^2 + (x - 1 + 1)^2 = 9$$

$$(x - 3)^2 + x^2 = 9$$

$$x^2 - 6x + 9 + x^2 = 9$$

$$2x^2 - 6x = 0$$

$$x^2 - 3x = 0$$

$$x(x - 3) = 0$$

$$x = 0 \quad \text{or} \quad x = 3$$

$$y = -1 \qquad y = 2 \quad \text{Because } y = x - 1$$

Check $(0, -1)$ and $(3, 2)$ in the original system and with the graphs in Fig. 12.20. The solution set is $\{(0, -1), (3, 2)\}$.

Now do Exercises 45–50

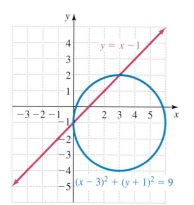

Figure 12.20

Warm-Ups ▼

Fill in the blank.

1. A(n) _____ is the set of all points in a plane that lie at a fixed distance from a fixed point.

2. The _____ of a circle is the fixed point in the definition.

3. The _____ is the fixed distance in the definition of a circle.

4. The equation $(x - h)^2 + (y - k)^2 = r^2$ (for $r > 0$) is the standard equation for a(n) _____ with _____ (h, k) and _____ r.

True or false?

5. The radius of a circle can be any nonzero real number.

6. The coordinates for the center satisfy the equation for the circle.

7. The center for $x^2 + y^2 = 4$ is the origin.

8. The radius for $x^2 + y^2 = 9$ is 9.

9. The center for $(x - 3)^2 + (y + 4)^2 = 25$ is $(3, -4)$.

10. The center for $x^2 + y^2 + 6y - 4 = 0$ is on the y-axis.

‹ 1 › **The Equation of a Circle**

Write the standard equation for each circle with the given center and radius. See Example 1.

1. Center $(0, 0)$, radius 4

2. Center $(0, 0)$, radius 3

3. Center $(0, 3)$, radius 5

4. Center $(2, 0)$, radius 3

5. Center $(1, -2)$, radius 9

6. Center $(-3, 5)$, radius 4

7. Center $(0, 0)$, radius $\sqrt{3}$

8. Center $(0, 0)$, radius $\sqrt{2}$

9. Center $(-6, -3)$, radius $\frac{1}{2}$

10. Center $(-3, -5)$, radius $\frac{1}{4}$

11. Center $\left(\frac{1}{2}, \frac{1}{3}\right)$, radius 0.1

12. Center $\left(-\frac{1}{2}, 3\right)$, radius 0.2

Find the center and radius for each circle. See Example 2.

13. $x^2 + y^2 = 1$

14. $x^2 + (y - 1)^2 = 9$

15. $(x - 3)^2 + (y - 5)^2 = 2$

16. $(x + 3)^2 + (y - 7)^2 = 6$

17. $x^2 + \left(y - \frac{1}{2}\right)^2 = \frac{1}{2}$

18. $5x^2 + 5y^2 = 5$

19. $4x^2 + 4y^2 = 9$

20. $9x^2 + 9y^2 = 49$

21. $3 - y^2 = (x - 2)^2$

22. $9 - x^2 = (y + 1)^2$

Sketch the graph of each equation. See Example 3.

23. $x^2 + y^2 = 9$

24. $x^2 + y^2 = 16$

25. $x^2 + (y - 3)^2 = 9$

26. $(x - 4)^2 + y^2 = 16$

27. $(x + 1)^2 + (y - 1)^2 = 2$

28. $(x - 2)^2 + (y + 2)^2 = 8$

29. $(x - 4)^2 + (y + 3)^2 = 16$

30. $(x - 3)^2 + (y - 7)^2 = 25$

31. $\left(x - \frac{1}{2}\right)^2 + \left(y + \frac{1}{2}\right)^2 = \frac{1}{4}$

32. $\left(x + \frac{1}{3}\right)^2 + y^2 = \frac{1}{9}$

⟨2⟩ Equations Not in Standard Form

Rewrite each equation in the standard form for the equation of a circle, and identify its center and radius. See Example 4.

33. $x^2 + 4x + y^2 + 6y = 0$

34. $x^2 - 10x + y^2 + 8y = 0$

35. $x^2 - 2x + y^2 - 4y - 3 = 0$

36. $x^2 - 6x + y^2 - 2y + 9 = 0$

37. $x^2 + y^2 = 8y + 10x - 32$

38. $x^2 + y^2 = 8x - 10y$

39. $x^2 - x + y^2 + y = 0$

40. $x^2 - 3x + y^2 = 0$

41. $x^2 - 3x + y^2 - y = 1$

42. $x^2 - 5x + y^2 + 3y = 2$

43. $x^2 - \dfrac{2}{3}x + y^2 + \dfrac{3}{2}y = 0$

44. $x^2 + \dfrac{1}{3}x + y^2 - \dfrac{2}{3}y = \dfrac{1}{9}$

⟨3⟩ Systems of Equations

Graph both equations of each system on the same coordinate axes. Solve the system by elimination of variables to find all points of intersection of the graphs. See Example 5.

45. $x^2 + y^2 = 10$
$y = 3x$

46. $x^2 + y^2 = 4$
$y = x - 2$

47. $x^2 + y^2 = 9$
$y = x^2 - 3$

48. $x^2 + y^2 = 4$
$y = x^2 - 2$

49. $(x - 2)^2 + (y + 3)^2 = 4$
$y = x - 3$

50. $(x + 1)^2 + (y - 4)^2 = 17$
$y = x + 2$

Miscellaneous

Solve each problem.

51. Determine all points of intersection of the circle $(x - 1)^2 + (y - 2)^2 = 4$ with the y-axis.

52. Determine the points of intersection of the circle $x^2 + (y - 3)^2 = 25$ with the x-axis.

53. Find the radius of the circle that has center $(2, -5)$ and passes through the origin.

54. Find the radius of the circle that has center $(-2, 3)$ and passes through $(3, -1)$.

55. Determine the equation of the circle that is centered at $(2, 3)$ and passes through $(-2, -1)$.

56. Determine the equation of the circle that is centered at $(3, 4)$ and passes through the origin.

57. Find all points of intersection of the circles $x^2 + y^2 = 9$ and $(x - 5)^2 + y^2 = 9$.

63. $y = \sqrt{1 - x^2}$

64. $y = -\sqrt{1 - x^2}$

58. A donkey is tied at the point $(2, -3)$ on a rope of length 12. Turnips are growing at the point $(6, 7)$. Can the donkey reach them?

59. *Volume of a flute.* The volume of air in a flute is a critical factor in determining its pitch. A cross section of a Renaissance flute in C is shown in the accompanying figure. If the length of the flute is 2874 millimeters, then what is the volume of air in the flute [to the nearest cubic millimeter (mm^3)]? (*Hint:* Use the formula for the volume of a cylinder.)

Getting More Involved

65. *Cooperative learning*

The equation of a circle is a special case of the general equation $Ax^2 + Bx + Cy^2 + Dy = E$, where A, B, C, D, and E are real numbers. Working in small groups, find restrictions that must be placed on A, B, C, D, and E so that the graph of this equation is a circle. What does the graph of $x^2 + y^2 = -9$ look like?

66. *Discussion*

Suppose lighthouse A is located at the origin and lighthouse B is located at coordinates $(0, 6)$. The captain of a ship has determined that the ship's distance from lighthouse A is 2 and its distance from lighthouse B is 5. What are the possible coordinates for the location of the ship?

The units for x and y are millimeters.

$x^2 + y^2 = 193.21$

$x^2 + y^2 = 83.72$
(Bore hole)

Figure for Exercises 59 and 60

60. *Flute reproduction.* To make the smaller C# flute, Friedrich von Huene multiplies the length and cross-sectional area of the flute of Exercise 59 by 0.943. Find the equation for the bore hole (centered at the origin) and the volume of air in the C# flute.

Graph each equation.

61. $x^2 + y^2 = 0$

62. $x^2 - y^2 = 0$

Graphing Calculator Exercises

Graph each relation on a graphing calculator by solving for y and graphing two functions.

67. $x^2 + y^2 = 4$

68. $(x - 1)^2 + (y + 2)^2 = 1$

69. $x = y^2$

70. $x = (y + 2)^2 - 1$

71. $x = y^2 + 2y + 1$

72. $x = 4y^2 + 4y + 1$

Mid-Chapter **Quiz** | Sections 12.1 through 12.3 | **Chapter 12**

Solve each system.

1. $x - y = -6$
$y = x^2$

2. $y = |x - 2|$
$y = \dfrac{1}{2}x + 2$

3. $x^2 + y^2 = 7$
$x^2 - y^2 = 1$

4. $x^2 + y^2 = 13$
$y = x^2 - 1$

Find the vertex, focus, directrix, axis of symmetry, and opening for each parabola.

5. $y = \dfrac{1}{8}x^2 - 2x + 1$

6. $y = -(x - 3)^2 + 4$

7. $x = (y - 2)^2$

Miscellaneous.

8. Find the distance between the points $(-2, 6)$ and $(-4, 5)$.

9. Find the midpoint of the line segment with endpoints $(-5, 7)$ and $(-9, -3)$.

10. Write the equation $y = 4x^2 - 8x + 2$ in the form $y = a(x - h)^2 + k$.

11. Write the standard equation for a circle with center $(-4, 5)$ and radius 10.

12. Find the center and radius of the circle $y^2 + (x - 2)^2 = 7$.

13. Write $x^2 + 4x + y^2 - 10y = 1$ in the standard form for the equation of a circle.

14. Find two numbers that have a sum of 10 and a product of 12.

12.4 The Ellipse and Hyperbola

In this section, we study the remaining two conic sections: the ellipse and the hyperbola.

In This Section

‹1› The Ellipse
‹2› The Hyperbola

‹1› The Ellipse

An ellipse can be obtained by intersecting a plane and a cone, as was shown in Fig. 12.3. We can also give a definition of an ellipse in terms of points and distance.

> **Ellipse**
>
> An **ellipse** is the set of all points in a plane such that the sum of their distances from two fixed points is a constant. Each fixed point is called a **focus** (plural: foci).

Figure 12.21

An easy way to draw an ellipse is illustrated in Fig. 12.21. A string is attached at two fixed points, and a pencil is used to take up the slack. As the pencil is moved around the paper, the sum of the distances of the pencil point from the two fixed points remains constant. Of course, the length of the string is that constant. You may wish to try this.

Like the parabola, the ellipse also has interesting reflecting properties. All light or sound waves emitted from one focus are reflected off the ellipse to concentrate at the other focus (see Fig. 12.22). This property is used in light fixtures where a concentration of light at a point is desired or in a whispering gallery such as Statuary Hall in the U.S. Capitol Building.

The orbits of the planets around the sun and satellites around the earth are elliptical. For the orbit of the earth around the sun, the sun is at one focus. For the elliptical path of an earth satellite, the earth is at one focus and a point in space is the other focus.

Figure 12.23 shows an ellipse with foci $(c, 0)$ and $(-c, 0)$. The origin is the center of this ellipse. In general, the **center** of an ellipse is a point midway between the foci. The ellipse in Fig. 12.23 has x-intercepts at $(a, 0)$ and $(-a, 0)$ and y-intercepts at

Figure 12.22

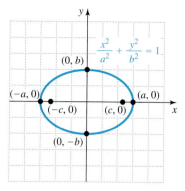

Figure 12.23

(0, b) and (0, $-b$). The distance formula can be used to write the following equation for this ellipse. (See Exercise 61.)

Equation of an Ellipse Centered at the Origin

An ellipse centered at (0, 0) with foci at ($\pm c$, 0) and constant sum $2a$ has equation

$$\frac{x^2}{a^2} + \frac{y^2}{b^2} = 1,$$

where a, b, and c are positive real numbers with $c^2 = a^2 - b^2$.

To draw a "nice-looking" ellipse, we would locate the foci and use string as shown in Fig. 12.21. We can get a rough sketch of an ellipse centered at the origin by using the x- and y-intercepts only.

E X A M P L E   1

Graphing an ellipse

Find the x- and y-intercepts for the ellipse, and sketch its graph.

$$\frac{x^2}{9} + \frac{y^2}{4} = 1$$

Solution

To find the y-intercepts, let $x = 0$ in the equation:

$$\frac{0}{9} + \frac{y^2}{4} = 1$$

$$\frac{y^2}{4} = 1$$

$$y^2 = 4$$

$$y = \pm 2$$

To find the x-intercepts, let $y = 0$. We get $x = \pm 3$. The four intercepts are (0, 2), (0, -2), (3, 0), and (-3, 0). Plot the intercepts, and draw an ellipse through them as in Fig. 12.24.

‹ **Calculator Close-Up** ›

To graph the ellipse in Example 1, graph

$$y_1 = \sqrt{4 - 4x^2/9}$$

and

$$y_2 = -y_1.$$

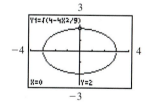

‹ **Helpful Hint** ›

When sketching ellipses or circles by hand, use your hand like a compass and rotate your paper as you draw the curve.

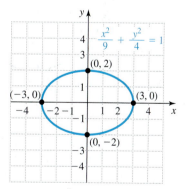

Figure 12.24

Now do Exercises 1–14

Ellipses, like circles, may be centered at any point in the plane. To get the equation of an ellipse centered at (h, k), we replace x by $x - h$ and y by $y - k$ in the equation of the ellipse centered at the origin.

Equation of an Ellipse Centered at (h, k)

An ellipse centered at (h, k) has equation

$$\frac{(x - h)^2}{a^2} + \frac{(y - k)^2}{b^2} = 1,$$

where a and b are positive real numbers.

EXAMPLE   2

An ellipse with center (h, k)

Sketch the graph of the ellipse:

$$\frac{(x - 1)^2}{9} + \frac{(y + 2)^2}{4} = 1$$

Solution

The graph of this ellipse is exactly the same size and shape as the ellipse

$$\frac{x^2}{9} + \frac{y^2}{4} = 1,$$

which was graphed in Example 1. However, the center for

$$\frac{(x - 1)^2}{9} + \frac{(y + 2)^2}{4} = 1$$

is $(1, -2)$. The denominator 9 is used to determine that the ellipse passes through points that are three units to the right and three units to the left of the center: $(4, -2)$ and $(-2, -2)$. See Fig. 12.25. The denominator 4 is used to determine that the ellipse passes through points that are two units above and two units below the center: $(1, 0)$ and $(1, -4)$. We draw an ellipse using these four points, just as we did for an ellipse centered at the origin.

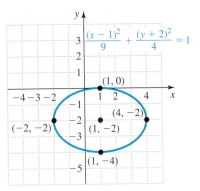

Figure 12.25

Now do Exercises 15–20

⟨2⟩ The Hyperbola

A hyperbola is the curve that occurs at the intersection of a cone and a plane, as was shown in Fig. 12.3 in Section 12.2. A hyperbola can also be defined in terms of points and distance.

Hyperbola

A **hyperbola** is the set of all points in the plane such that the difference of their distances from two fixed points (foci) is constant.

Like the parabola and the ellipse, the hyperbola also has reflecting properties. If a light ray is aimed at one focus, it is reflected off the hyperbola and goes to the other focus, as shown in Fig. 12.26. Hyperbolic mirrors are used in conjunction with parabolic mirrors in telescopes.

Figure 12.26

Figure 12.27

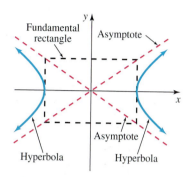

Figure 12.28

The definitions of a hyperbola and an ellipse are similar, and so are their equations. However, their graphs are very different. Figure 12.27 shows a hyperbola in which the distance from a point on the hyperbola to the closer focus is N and the distance to the farther focus is M. The value $M - N$ is the same for every point on the hyperbola.

A hyperbola has two parts called **branches.** These branches look like parabolas, but they are not parabolas. The branches of the hyperbola shown in Fig. 12.28 get closer and closer to the dashed lines, called **asymptotes,** but they never intersect them. The asymptotes are used as guidelines in sketching a hyperbola. The asymptotes are found by extending the diagonals of the **fundamental rectangle,** shown in Fig. 12.28. The key to drawing a hyperbola is getting the fundamental rectangle and extending its diagonals to get the asymptotes. You will learn how to find the fundamental rectangle from the equation of a hyperbola. The hyperbola in Fig. 12.28 opens to the left and right.

If we start with foci at $(\pm c, 0)$ and a positive number a, then we can use the definition of a hyperbola to derive the following equation of a hyperbola in which the constant difference between the distances to the foci is $2a$.

Equation of a Hyperbola Centered at (0, 0) Opening Left and Right

A hyperbola centered at $(0, 0)$ with foci $(c, 0)$ and $(-c, 0)$ and constant difference $2a$ has equation

$$\frac{x^2}{a^2} - \frac{y^2}{b^2} = 1,$$

where a, b, and c are positive real numbers such that $c^2 = a^2 + b^2$.

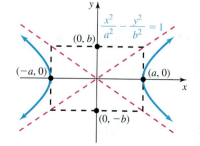

Figure 12.29

The graph of a general equation for a hyperbola is shown in Fig. 12.29. Notice that the fundamental rectangle extends to the x-intercepts along the x-axis and extends b units above and below the origin along the y-axis. Use the following procedure for graphing a hyperbola centered at the origin and opening to the left and to the right.

> ### Strategy for Graphing a Hyperbola Centered at the Origin, Opening Left and Right
>
> To graph the hyperbola $\frac{x^2}{a^2} - \frac{y^2}{b^2} = 1$:
>
> **1.** Locate the x-intercepts at $(a, 0)$ and $(-a, 0)$.
> **2.** Draw the fundamental rectangle through $(\pm a, 0)$ and $(0, \pm b)$.
> **3.** Draw the extended diagonals of the rectangle to use as asymptotes.
> **4.** Draw the hyperbola to the left and right approaching the asymptotes.

EXAMPLE 3

A hyperbola opening left and right

Sketch the graph of $\frac{x^2}{36} - \frac{y^2}{9} = 1$, and find the equations of its asymptotes.

Solution

The x-intercepts are $(6, 0)$ and $(-6, 0)$. Draw the fundamental rectangle through these x-intercepts and the points $(0, 3)$ and $(0, -3)$. Extend the diagonals of the fundamental rectangle to get the asymptotes. Now draw a hyperbola passing through the x-intercepts and approaching the asymptotes as shown in Fig. 12.30. From the graph in Fig. 12.30 we see that the slopes of the asymptotes are $\frac{1}{2}$ and $-\frac{1}{2}$. Because the y-intercept for both asymptotes is the origin, their equations are $y = \frac{1}{2}x$ and $y = -\frac{1}{2}x$.

‹ **Calculator Close-Up** ›

To graph the hyperbola and its asymptotes from Example 3, graph

$$y_1 = \sqrt{x^2/4 - 9}, \quad y_2 = -y_1,$$
$$y_3 = 0.5x, \quad \text{and} \quad y_4 = -y_3.$$

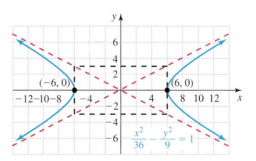

Figure 12.30

> Now do Exercises 21–22

If the variables x and y are interchanged in the equation of the hyperbola, then the hyperbola opens up and down.

> ### Equation of a Hyperbola Centered at (0, 0) Opening Up and Down
>
> A hyperbola centered at $(0, 0)$ with foci $(0, c)$ and $(0, -c)$ and constant difference $2b$ has equation
>
> $$\frac{y^2}{b^2} - \frac{x^2}{a^2} = 1,$$
>
> where a, b, and c are positive real numbers such that $c^2 = a^2 + b^2$.

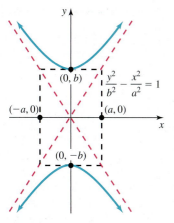

Figure 12.31

The graph of the general equation for a hyperbola opening up and down is shown in Fig. 12.31. Notice that the fundamental rectangle extends to the y-intercepts along the y-axis and extends a units to the left and right of the origin along the x-axis. The procedure for graphing a hyperbola opening up and down follows.

Strategy for Graphing a Hyperbola Centered at the Origin, Opening Up and Down

To graph the hyperbola $\frac{y^2}{b^2} - \frac{x^2}{a^2} = 1$:

1. Locate the y-intercepts at $(0, b)$ and $(0, -b)$.
2. Draw the fundamental rectangle through $(0, \pm b)$ and $(\pm a, 0)$.
3. Draw the extended diagonals of the rectangle to use as asymptotes.
4. Draw the hyperbola opening up and down approaching the asymptotes.

E X A M P L E   **4**

‹ **Helpful Hint** ›

We could include here general formulas for the equations of the asymptotes, but that is not necessary. It is easier first to draw the asymptotes as suggested and then to figure out their equations by looking at the graph.

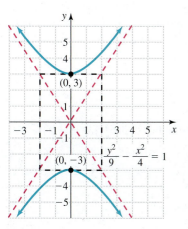

Figure 12.32

A hyperbola opening up and down

Graph the hyperbola $\frac{y^2}{9} - \frac{x^2}{4} = 1$, and find the equations of its asymptotes.

Solution

If $y = 0$, we get

$$-\frac{x^2}{4} = 1$$

$$x^2 = -4.$$

Because this equation has no real solution, the graph has no x-intercepts. Let $x = 0$ to find the y-intercepts:

$$\frac{y^2}{9} = 1$$

$$y^2 = 9$$

$$y = \pm 3$$

The y-intercepts are $(0, 3)$ and $(0, -3)$, and the hyperbola opens up and down. From $a^2 = 4$ we get $a = 2$. So the fundamental rectangle extends to the intercepts $(0, 3)$ and $(0, -3)$ on the y-axis and to the points $(2, 0)$ and $(-2, 0)$ along the x-axis. We extend the diagonals of the rectangle and draw the graph of the hyperbola as shown in Fig. 12.32. From the graph in Fig. 12.32 we see that the asymptotes have slopes $\frac{3}{2}$ and $-\frac{3}{2}$. Because the y-intercept for both asymptotes is the origin, their equations are $y = \frac{3}{2}x$ and $y = -\frac{3}{2}x$.

Now do Exercises 23–28

E X A M P L E   **5**

A hyperbola not in standard form

Sketch the graph of the hyperbola $4x^2 - y^2 = 4$.

Solution

First write the equation in standard form. Divide each side by 4 to get

$$x^2 - \frac{y^2}{4} = 1.$$

There are no y-intercepts. If $y = 0$, then $x = \pm 1$. The hyperbola opens left and right with x-intercepts at $(1, 0)$ and $(-1, 0)$. The fundamental rectangle extends to the intercepts along the x-axis and to the points $(0, 2)$ and $(0, -2)$ along the y-axis. We extend the diagonals of the rectangle for the asymptotes and draw the graph as shown in Fig. 12.33.

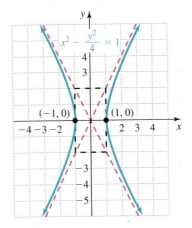

Figure 12.33

Now do Exercises 29–32

Like circles and ellipses, hyperbolas may be centered at any point in the plane. To get the equation of a hyperbola centered at (h, k), we replace x by $x - h$ and y by $y - k$ in the equation of the hyperbola centered at the origin.

Equation of a Hyperbola Centered at (h, k)

A hyperbola centered at (h, k) has one of the following equations depending on which way it opens.

Opening left and right:

$$\frac{(x - h)^2}{a^2} - \frac{(y - k)^2}{b^2} = 1$$

Opening up and down:

$$\frac{(y - k)^2}{b^2} - \frac{(x - h)^2}{a^2} = 1$$

EXAMPLE 6

Graphing a hyperbola centered at (h, k)

Graph the hyperbola $\frac{(x-3)^2}{16} - \frac{(y+1)^2}{4} = 1$.

Solution

This hyperbola is centered at $(3, -1)$ and opens left and right. It is a transformation of the graph of $\frac{x^2}{16} - \frac{y^2}{4} = 1$. The fundamental rectangle for $\frac{x^2}{16} - \frac{y^2}{4} = 1$ is centered at the origin and goes through $(\pm 4, 0)$ and $(0, \pm 2)$. So draw a fundamental rectangle centered at $(3, -1)$ that extends four units to the right and left and two units up and down as shown in Fig. 12.34. Draw the asymptotes through the vertices of the fundamental rectangle and the hyperbola opening to the left and right.

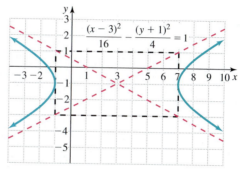

Figure 12.34

Now do Exercises 33–38

Warm-Ups ▼

Fill in the blank.

1. A(n) _____ is the set of all points in a plane such that the sum of their distances from two fixed points is constant.

2. The ____ of an ellipse are the two fixed points in the definition.

3. The _____ of an ellipse is the point that is midway between the foci.

4. A(n) _____ is the set of all points in a plane such that the difference of their distances from two fixed points is constant.

5. A hyperbola is made up of two separate curves called _____.

6. A hyperbola approaches two lines called _____.

True or false?

7. The graph of $\frac{x^2}{36} + \frac{y^2}{25} = 1$ is an ellipse.

8. The x-intercepts for $\frac{x^2}{36} + \frac{y^2}{25} = 1$ are $(5, 0)$ and $(-5, 0)$.

9. If the foci of an ellipse coincide, then the ellipse is a circle.

10. The graph of $\frac{x^2}{36} - \frac{y^2}{25} = 1$ is a hyperbola.

11. The x-intercepts for $\frac{x^2}{36} - \frac{y^2}{25} = 1$ are $(6, 0)$ and $(-6, 0)$.

12. The graph of $4x^2 - y^2 = 4$ is a hyperbola.

‹ 1 › **The Ellipse**

Sketch the graph of each ellipse. See Example 1.

1. $\dfrac{x^2}{9} + \dfrac{y^2}{4} = 1$

2. $\dfrac{x^2}{9} + \dfrac{y^2}{16} = 1$

3. $\dfrac{x^2}{9} + y^2 = 1$

4. $x^2 + \dfrac{y^2}{4} = 1$

5. $\dfrac{x^2}{36} + \dfrac{y^2}{25} = 1$

6. $\dfrac{x^2}{25} + \dfrac{y^2}{49} = 1$

7. $\dfrac{x^2}{24} + \dfrac{y^2}{5} = 1$

8. $\dfrac{x^2}{6} + \dfrac{y^2}{17} = 1$

9. $9x^2 + 16y^2 = 144$

10. $9x^2 + 25y^2 = 225$

11. $25x^2 + y^2 = 25$

12. $x^2 + 16y^2 = 16$

13. $4x^2 + 9y^2 = 1$

14. $25x^2 + 16y^2 = 1$

Sketch the graph of each ellipse. See Example 2.

15. $\dfrac{(x - 3)^2}{4} + \dfrac{(y - 1)^2}{9} = 1$

16. $\dfrac{(x + 5)^2}{49} + \dfrac{(y - 2)^2}{25} = 1$

17. $\dfrac{(x + 1)^2}{16} + \dfrac{(y - 2)^2}{25} = 1$

18. $\dfrac{(x - 3)^2}{36} + \dfrac{(y + 4)^2}{64} = 1$

19. $(x - 2)^2 + \dfrac{(y + 1)^2}{36} = 1$ **20.** $\dfrac{(x + 3)^2}{9} + (y + 1)^2 = 1$ **29.** $9x^2 - 16y^2 = 144$ **30.** $9x^2 - 25y^2 = 225$

⟨2⟩   The Hyperbola

31. $x^2 - y^2 = 1$ **32.** $y^2 - x^2 = 1$

Graph each hyperbola, and write the equations of its asymptotes. See Examples 3–5. See the Strategies for Graphing a Hyperbola boxes on pages 736 and 737.

21. $\dfrac{x^2}{4} - \dfrac{y^2}{9} = 1$ **22.** $\dfrac{x^2}{16} - \dfrac{y^2}{9} = 1$

Sketch the graph of each hyperbola. See Example 6.

33. $\dfrac{(x - 2)^2}{4} - (y + 1)^2 = 1$ **34.** $(x + 3)^2 - \dfrac{(y - 1)^2}{4} = 1$

23. $\dfrac{y^2}{4} - \dfrac{x^2}{25} = 1$ **24.** $\dfrac{y^2}{9} - \dfrac{x^2}{16} = 1$

25. $\dfrac{x^2}{25} - y^2 = 1$ **26.** $x^2 - \dfrac{y^2}{9} = 1$

35. $\dfrac{(x + 1)^2}{16} - \dfrac{(y - 1)^2}{9} = 1$ **36.** $\dfrac{(x - 2)^2}{9} - \dfrac{(y + 2)^2}{16} = 1$

27. $x^2 - \dfrac{y^2}{25} = 1$ **28.** $\dfrac{x^2}{9} - y^2 = 1$

37. $\dfrac{(y - 2)^2}{9} - \dfrac{(x - 4)^2}{4} = 1$ **38.** $\dfrac{(y + 3)^2}{16} - \dfrac{(x + 1)^2}{9} = 1$

Miscellaneous

Determine whether the graph of each equation is a circle, parabola, ellipse, or hyperbola.

39. $y = x^2 + 1$

40. $x^2 + y^2 = 1$

41. $x^2 - y^2 = 1$

42. $4x^2 + y^2 = 1$

43. $\dfrac{x^2}{2} + y^2 = 1$

44. $x^2 - \dfrac{y^2}{9} = 1$

45. $(x - 2)^2 + (y - 4)^2 = 9$

46. $(x - 2)^2 + y = 9$

Graph both equations of each system on the same coordinate axes. Use elimination of variables to find all points of intersection.

47. $\dfrac{x^2}{4} + \dfrac{y^2}{9} = 1$

$x^2 - \dfrac{y^2}{9} = 1$

48. $x^2 - \dfrac{y^2}{4} = 1$

$\dfrac{x^2}{9} + \dfrac{y^2}{4} = 1$

49. $\dfrac{x^2}{4} + \dfrac{y^2}{16} = 1$

$x^2 + y^2 = 1$

50. $x^2 + \dfrac{y^2}{9} = 1$

$x^2 + y^2 = 4$

51. $x^2 + y^2 = 4$

$x^2 - y^2 = 1$

52. $x^2 + y^2 = 16$

$x^2 - y^2 = 4$

53. $x^2 + 9y^2 = 9$

$x^2 + y^2 = 4$

54. $x^2 + y^2 = 25$

$x^2 + 25y^2 = 25$

55. $x^2 + 9y^2 = 9$
  $y = x^2 - 1$

56. $4x^2 + y^2 = 4$
  $y = 2x^2 - 2$

57. $9x^2 - 4y^2 = 36$
  $2y = x - 2$

58. $25y^2 - 9x^2 = 225$
  $y = 3x + 3$

Applications

Solve each problem.

59. *Marine navigation.* The loran (long-range navigation) system is used by boaters to determine their location at sea. The loran unit on a boat measures the difference in time that it takes for radio signals from pairs of fixed points to reach the boat. The unit then finds the equations of two hyperbolas that pass through the location of the boat. Suppose a boat is located in the first quadrant at the intersection of $x^2 - 3y^2 = 1$ and $4y^2 - x^2 = 1$.

  a) Use the accompanying graph to approximate the location of the boat.
  b) Algebraically find the exact location of the boat.

Figure for Exercise 59

60. *Sonic boom.* An aircraft traveling at supersonic speed creates a cone-shaped wave that intersects the ground along a hyperbola, as shown in the accompanying figure. A thunderlike sound is heard at any point on the hyperbola. This sonic boom travels along the ground, following the aircraft. The area where the sonic boom is most noticeable is called the *boom carpet*. The width of the boom carpet is roughly five times the altitude of the aircraft. Suppose the equation of the hyperbola in the figure is

$$\frac{x^2}{400} - \frac{y^2}{100} = 1,$$

where the units are miles and the width of the boom carpet is measured 40 miles behind the aircraft. Find the altitude of the aircraft.

Figure for Exercise 60

Getting More Involved

61. *Cooperative learning*

Let (x, y) be an arbitrary point on an ellipse with foci $(c, 0)$ and $(-c, 0)$ for $c > 0$. The following equation expresses the fact that the distance from (x, y) to $(c, 0)$ plus the distance from (x, y) to $(-c, 0)$ is the constant value $2a$ (for $a > 0$):

$$\sqrt{(x - c)^2 + (y - 0)^2} + \sqrt{(x - (-c))^2 + (y - 0)^2} = 2a$$

Working in groups, simplify this equation. First get the radicals on opposite sides of the equation, and then square both sides twice to eliminate the square roots. Finally, let $b^2 = a^2 - c^2$ to get the equation

$$\frac{x^2}{a^2} + \frac{y^2}{b^2} = 1.$$

62. Cooperative learning

Let (x, y) be an arbitrary point on a hyperbola with foci $(c, 0)$ and $(-c, 0)$ for $c > 0$. The following equation expresses the fact that the distance from (x, y) to $(c, 0)$ minus the distance from (x, y) to $(-c, 0)$ is the constant value $2a$ (for $a > 0$):

$$\sqrt{(x - c)^2 + (y - 0)^2} - \sqrt{(x - (-c))^2 + (y - 0)^2} = 2a$$

Working in groups, simplify the equation. You will need to square both sides twice to eliminate the square roots. Finally, let $b^2 = c^2 - a^2$ to get the equation

$$\frac{x^2}{a^2} - \frac{y^2}{b^2} = 1.$$

Graphing Calculator Exercises

63. Graph $y_1 = \sqrt{x^2 - 1}$, $y_2 = -\sqrt{x^2 - 1}$, $y_3 = x$, and $y_4 = -x$ to get the graph of the hyperbola $x^2 - y^2 = 1$ along with its asymptotes. Use the viewing window $-3 \leq x \leq 3$ and $-3 \leq y \leq 3$. Notice how the branches of the hyperbola approach the asymptotes.

64. Graph the same four functions in Exercise 63, but use $-30 \leq x \leq 30$ and $-30 \leq y \leq 30$ as the viewing window. What happened to the hyperbola?

Math *at Work*   Kepler's Laws

With great patience, Danish astronomer Tycho Brahe (1546–1601) made very careful observations of the motion of the planets in the sky. Brahe tried to explain the orbits of the planets using circles. His assistant, Johannes Kepler (1571–1630), studied Tycho's tables and came up with three laws that better explained the motion of the planets. Kepler's first law went contrary to Brahe's theory and states that each planet moves around the sun in an elliptical orbit with the sun at one focus of the ellipse.

The second law states that the line joining a planet with the sun sweeps out equal areas in equal times. A planet moves faster when it is closer to the sun and slower when it is far from the sun. So the planet illustrated in the accompanying figure moves from A to B in the same time that it moves from C to D, even though the distance from A to B is greater. According to Kepler's law, the shaded areas in the figure are equal.

The third law states that the square of the period of a planet orbiting the sun is equal to the cube of the mean distance from the planet to the sun. In symbols, $P^2 = a^3$, where P is the number of earth years that it takes for the planet to orbit the sun, and a is the mean distance from the planet to the sun in astronomical units (AU). (One AU is the mean distance from the earth to the sun.) $P^2 = a^3$ can be written as $P = a^{3/2}$ or $a = P^{2/3}$ and used to find the period or the distance. For example, the period of Mars is observed to be 1.88 years. So the mean distance from Mars to the sun is $1.88^{2/3}$ or 1.53 AU. The mean distance from Pluto to the sun is observed to be 39.44 AU, so Pluto takes $39.44^{3/2}$ or 247.69 years to complete one orbit of the sun.

Equal areas in
equal times

12.5 Second-Degree Inequalities

In this section we graph second-degree inequalities and systems of inequalities involving second-degree inequalities.

⟨1⟩ Graphing a Second-Degree Inequality

A second-degree inequality is an inequality involving squares of at least one of the variables. Changing the equal sign to an inequality symbol for any of the equations of the conic sections gives us a second-degree inequality. Second-degree inequalities are graphed in the same manner as linear inequalities.

EXAMPLE 1

A second-degree inequality

Graph the inequality $y < x^2 + 2x - 3$.

Solution

We first graph $y = x^2 + 2x - 3$. This parabola has x-intercepts at $(1, 0)$ and $(-3, 0)$, y-intercept at $(0, -3)$, and vertex at $(-1, -4)$. The graph of the parabola is drawn with a dashed line, as shown in Fig. 12.35. The graph of the parabola divides the plane into two regions. Every point on one side of the parabola satisfies the inequality $y < x^2 + 2x - 3$, and every point on the other side satisfies the inequality $y > x^2 + 2x - 3$. To determine which side is which, we test a point that is not on the parabola, say $(0, 0)$. Because

$$0 < 0^2 + 2 \cdot 0 - 3$$

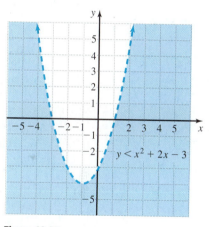

Figure 12.35

is false, the region not containing the origin is shaded, as in Fig. 12.35.

Now do Exercises 1–6

EXAMPLE 2

A second-degree inequality

Graph the inequality $x^2 + y^2 \leq 9$.

Solution

The graph of $x^2 + y^2 = 9$ is a circle of radius 3 centered at the origin. The circle divides the plane into two regions. Every point in one region satisfies $x^2 + y^2 < 9$, and every point in the other region satisfies $x^2 + y^2 > 9$. To identify the regions, we pick a point and test it. Select $(0, 0)$. The inequality

$$0^2 + 0^2 < 9$$

is true. Because $(0, 0)$ is inside the circle, all points inside the circle satisfy $x^2 + y^2 < 9$. Points outside the circle satisfy $x^2 + y^2 > 9$. Because the inequality symbol is $\leq$ the circle is included in the solution set. So the circle is drawn as a solid curve as shown in Fig. 12.36, and the area inside the circle is shaded.

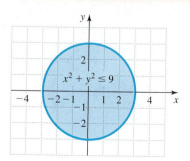

Figure 12.36

Now do Exercises 7–10

E X A M P L E   **3**

A second-degree inequality

Graph the inequality $\frac{x^2}{4} - \frac{y^2}{9} > 1$.

Solution

First graph the hyperbola $\frac{x^2}{4} - \frac{y^2}{9} = 1$. Because the hyperbola shown in Fig. 12.37 divides the plane into three regions, we select a test point in each region and check to see whether it satisfies the inequality. Testing the points $(-3, 0)$, $(0, 0)$, and $(3, 0)$ gives us the inequalities

$$\frac{(-3)^2}{4} - \frac{0^2}{9} > 1, \qquad \frac{0^2}{4} - \frac{0^2}{9} > 1, \qquad \text{and} \qquad \frac{3^2}{4} - \frac{0^2}{9} > 1.$$

Because only the first and third inequalities are correct, we shade only the regions containing $(3, 0)$ and $(-3, 0)$, as shown in Fig. 12.37.

> Now do Exercises 11–22

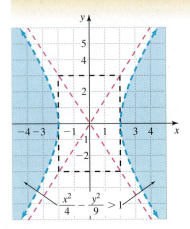

Figure 12.37

⟨2⟩ Systems of Inequalities

A point is in the solution set to a system of inequalities if it satisfies all inequalities of the system. We graph a system of inequalities by first determining the graph of each inequality and then finding the intersection of the graphs.

E X A M P L E   **4**

Systems of second-degree inequalities

Graph the system of inequalities:

$$\frac{y^2}{4} - \frac{x^2}{9} > 1 \qquad \frac{x^2}{9} + \frac{y^2}{16} < 1$$

Solution

Figure 12.38(a) shows the graph of the first inequality. Figure 12.38(b) shows the graphs of both inequalities on the same coordinate system. Points that are shaded for both inequalities in Fig. 12.38(b) satisfy the system. Figure 12.38(c) shows the graph of the system.

> Now do Exercises 27–46

(a)

(b)

(c)

Figure 12.38

Warm-Ups ▼

Fill in the blank.

1. The graph of $x^2 + 9y^2 = 1$ is a(n) _____.
2. The graph of $x^2 - y^2 = 9$ is a(n) _____.
3. The graph of $x^2 + y^2 = 9$ is a(n) _____.
4. The graph of $x^2 + y = 9$ is a(n) _____.
5. The graph of $x + y = 9$ is a(n) ____.

True or false?

6. The point $(0, 0)$ satisfies $2x^2 - y < 3$.
7. The graph of $y > x^2 - 3x + 2$ is a parabola.
8. We can use the origin as a test point when graphing $x^2 > y$.
9. The graph of $x^2 + y^2 < 4$ is the region inside a circle of radius 2.
10. The point $(0, 4)$ satisfies $x^2 - y^2 < 1$ and $y > x^2 - 2x + 3$.

Exercises   12.5

‹ **Study Tips** ›

• Don't be discouraged by the amount of material in this text that you did not cover in this course.
• Textbooks are written for a wide audience. Most instructors skip some topics.

‹1› **Graphing a Second-Degree Inequality**

Graph each inequality. See Examples 1–3.

1. $y > x^2$
2. $y \leq x^2 + 1$
3. $y < x^2 - x$
4. $y > x^2 + x$
5. $y > x^2 - x - 2$
6. $y < x^2 + x - 6$
7. $x^2 + y^2 < 9$
8. $x^2 + y^2 > 16$

9. $x^2 + 4y^2 > 4$

10. $4x^2 + y^2 \leq 4$

19. $y^2 - x^2 \leq 1$

20. $x^2 - y^2 > 1$

11. $4x^2 - 9y^2 < 36$

12. $25x^2 - 4y^2 > 100$

21. $x > y$

22. $x < 2y - 1$

13. $(x - 2)^2 + (y - 3)^2 < 4$

14. $(x + 1)^2 + (y - 2)^2 > 1$

⟨2⟩ Systems of Inequalities

Determine whether the ordered pair $(3, -4)$ *satisfies each system of inequalities.*

23. $x^2 + y^2 \leq 25$
   $y \leq x^2$

24. $x^2 - y^2 < 1$
   $y < x - 5$

25. $x - y > 1$
   $y > (x - 2)^2 + 3$

15. $x^2 + y^2 > 1$

16. $x^2 + y^2 < 25$

26. $4x^2 + y^2 \leq 36$
   $x^2 + y^2 \geq 25$

Graph the solution set to each system of inequalities. See Example 4.

27. $x^2 + y^2 < 9$
   $y > x$

28. $x^2 + y^2 > 1$
   $x > y$

17. $4x^2 - y^2 > 4$

18. $x^2 - 9y^2 \leq 9$

29. $x^2 - y^2 > 1$
$\quad\ x^2 + y^2 < 4$

30. $y^2 - x^2 < 1$
$\quad\ x^2 + y^2 > 9$

37. $x - y < 0$
$\quad\ y + x^2 < 1$

38. $y + 1 > x^2$
$\quad\ x + y < 2$

31. $y > x^2 + x$
$\quad\ y < 5$

32. $y > x^2 + x - 6$
$\quad\ y < x + 3$

39. $y < 5x - x^2$
$\quad\ x^2 + y^2 < 9$

40. $y < x^2 + 5x$
$\quad\ x^2 + y^2 < 16$

33. $y \geq x + 2$
$\quad\ y \leq 2 - x$

34. $y \geq 2x - 3$
$\quad\ y \leq 3 - 2x$

41. $y \geq 3$
$\quad\ x \leq 1$

42. $x > -3$
$\quad\ y < 2$

35. $4x^2 - y^2 < 4$
$\quad\ x^2 + 4y^2 > 4$

36. $x^2 - 4y^2 < 4$
$\quad\ x^2 + 4y^2 > 4$

43. $4y^2 - 9x^2 < 36$
$\quad\ x^2 + y^2 < 16$

44. $25y^2 - 16x^2 < 400$
$\quad\ x^2 + y^2 > 4$

45. $y < x^2$
$x^2 + y^2 < 1$

46. $y > x^2$
$4x^2 + y^2 < 4$

Photo for Exercise 47

Solve the problem.

47. *Buried treasure.* An old pirate on his deathbed gave the following description of where he had buried some treasure on a deserted island: "Starting at the large palm tree, I walked to the north and then to the east, and there I buried the treasure. I walked at least 50 paces to get to that spot, but I was not more than 50 paces, as the crow flies, from the large palm tree. I am sure that I walked farther in the northerly direction than in the easterly direction." With the large palm tree at the origin and the positive *y*-axis pointing to the north, graph the possible locations of the treasure.

Graphing Calculator Exercises

48. Use graphs to find an ordered pair that is in the solution set to the system of inequalities:

$$y > x^2 - 2x + 1$$
$$y < -1.1(x - 4)^2 + 5$$

Verify that your answer satisfies both inequalities.

49. Use graphs to find the solution set to the system of inequalities:

$$y > 2x^2 - 3x + 1$$
$$y < -2x^2 - 8x - 1$$

Wrap-Up

Summary

Nonlinear Systems		**Examples**
Nonlinear systems in two variables	Use substitution or addition to eliminate variables. Nonlinear systems may have several points in the solution set.	$y = x^2$ $x^2 + y^2 = 4$ Substitution: $y + y^2 = 4$

The Distance and Midpoint Formulas		**Examples**
Distance formula	The distance between (x_1, y_1) and (x_2, y_2) is $\sqrt{(x_2 - x_1)^2 + (y_2 - y_1)^2}$.	Distance between $(1, -2)$ and $(3, -4)$ is $\sqrt{2^2 + (-2)^2}$ or $2\sqrt{2}$.
Midpoint formula	The midpoint of the line segment with endpoints (x_1, y_1) and (x_2, y_2) is $(\bar{x}, \bar{y}) = \left(\dfrac{x_1 + x_2}{2}, \dfrac{y_1 + y_2}{2} \right)$.	If $(x_1, y_1) = (1, -2)$ and $(x_2, y_2) = (7, 8)$, then $(\bar{x}, \bar{y}) = (4, 3)$.

Parabola		**Examples**
$y = a(x - h)^2 + k$	Opens upward for $a > 0$, downward for $a < 0$ Vertex at (h, k) To find focus and directrix, use $a = \dfrac{1}{4p}$. Distance from vertex to focus or directrix is $|p|$.	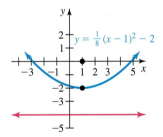
$x = a(y - k)^2 + h$	Opens right for $a > 0$, left for $a < 0$ Vertex at (h, k) To find focus and directrix use $a = \dfrac{1}{4p}$. Distance from vertex to focus or directrix is $|p|$.	

$y = ax^2 + bx + c$

Opens upward for $a > 0$, downward for $a < 0$
The x-coordinate of the vertex is $\frac{-b}{2a}$.
Find the y-coordinate of the vertex by evaluating
$y = ax^2 + bx + c$ for $x = \frac{-b}{2a}$.

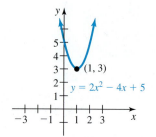

$x = ay^2 + by + c$

Opens right for $a > 0$, left for $a < 0$
The y-coordinate of the vertex is $\frac{-b}{2a}$.
Find the x-coordinate of the vertex by evaluating
$x = ay^2 + by + c$ for $y = \frac{-b}{2a}$.

Circle

Examples

Centered at origin
$x^2 + y^2 = r^2$

Center $(0, 0)$
Radius r (for $r > 0$)

Arbitrary center
$(x - h)^2 + (y - k)^2 = r^2$

Center (h, k)
Radius r (for $r > 0$)

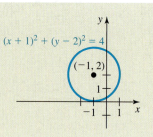

Ellipse

Examples

Centered at origin

$\dfrac{x^2}{a^2} + \dfrac{y^2}{b^2} = 1$

Center: $(0, 0)$
x-intercepts: $(a, 0)$ and $(-a, 0)$
y-intercepts: $(0, b)$ and $(0, -b)$
Foci: $(\pm c, 0)$ if $a^2 > b^2$ and $c^2 = a^2 - b^2$
$\quad\quad (0, \pm c)$ if $b^2 > a^2$ and $c^2 = b^2 - a^2$

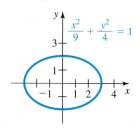

Arbitrary center   Center: (h, k)

$$\frac{(x-h)^2}{a^2} + \frac{(y-k)^2}{b^2} = 1$$

Hyperbola

Opening left and right

Centered at origin: $\dfrac{x^2}{a^2} - \dfrac{y^2}{b^2} = 1$

Center: $(0, 0)$

x-intercepts: $(a, 0)$ and $(-a, 0)$

y-intercepts: none

Centered at (h, k): $\dfrac{(x-h)^2}{a^2} - \dfrac{(y-k)^2}{b^2} = 1$

Examples

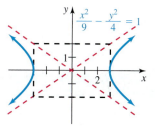

Opening up and down

Centered at origin: $\dfrac{y^2}{b^2} - \dfrac{x^2}{a^2} = 1$

Center: $(0, 0)$

x-intercepts: none

y-intercepts: $(0, b)$ and $(0, -b)$

Centered at (h, k): $\dfrac{(y-k)^2}{b^2} - \dfrac{(x-h)^2}{a^2} = 1$

Second-Degree Inequalities

Solution set for
a single inequality

Graph the boundary curve obtained by replacing
the inequality symbol by the equal sign.
Use test points to determine which regions satisfy
the inequality.

Examples

$x^2 + y^2 < 16$

| Solution set for a system of inequalities | Graph the boundary curves. Then select a test point in each region. Shade only the regions for which the test point satisfies all inequalities of the system. | $x^2 + y^2 < 16$
$y > x^2 - 1$

$x^2 + y^2 < 16$ and
$y > x^2 - 1$ |

Enriching Your Mathematical Word Power

Fill in the blank.

1. The graph of a _____ equation is not a straight line.

2. A(n) _____ is the set of all points in a plane that are equidistant from a fixed point and a fixed line.

3. The _____ is the fixed line in the definition of a parabola.

4. The _____ is the fixed point in the definition of a parabola.

5. The _____ of a parabola is the midpoint of the line segment joining the focus and directrix, perpendicular to the directrix.

6. The curves obtained by intersecting a cone and a plane are the _____ sections.

7. The _____ of symmetry of a parabola goes through the focus and directrix.

8. A(n) _____ is the set of all points in a plane that are a fixed distance from a fixed point.

9. The fixed point in the definition of a circle is its _____.

10. A(n) _____ is the set of all points in a plane such that the sum of their distances from two fixed points is constant.

11. The fixed points in the definition of an ellipse are its _____.

12. A(n) _____ is the set of all points in a plane such that the difference of their distances from two fixed points is constant.

13. The two lines that are approached by a hyperbola are its _____.

14. The points at which a hyperbola intersects the x- or y-axis are its _____.

Review Exercises

12.1 Nonlinear Systems of Equations
Graph both equations on the same set of axes, and then determine the points of intersection of the graphs by solving the system.

1. $y = x^2$
   $y = -2x + 15$

2. $y = \sqrt{x}$
   $y = \dfrac{1}{3}x$

3. $y = 3x$
   $y = \dfrac{1}{x}$

4. $y = |x|$
   $y = -3x + 5$

Solve each system.

5. $xy = 9$
  $y = x$

6. $y = x^2$
  $y = 2x$

7. $x^2 + y^2 = 4$
  $y = \dfrac{1}{3}x^2$

8. $12y^2 - 4x^2 = 9$
  $x = y^2$

9. $x^2 + y^2 = 34$
  $y = x + 2$

10. $y = 2x + 1$
  $xy - y = 5$

11. $y = \log(x - 3)$
  $y = 1 - \log(x)$

12. $y = \left(\dfrac{1}{2}\right)^x$
  $y = 2^{x-1}$

13. $x^4 = 2(12 - y)$
  $y = x^2$

14. $x^2 + 2y^2 = 7$
  $x^2 - 2y^2 = -5$

12.2 The Parabola

Find the distance between each pair of points.

15. $(1, 1), (3, 3)$

16. $(1, 2), (4, 5)$

17. $(-4, 6), (2, -8)$

18. $(-3, -5), (5, -7)$

Find the midpoint and length of the line segment with the given endpoints.

19. $(8, -2)$ and $(2, 6)$

20. $(-9, 4)$ and $(-3, -4)$

21. $(2, -2)$ and $(3, 1)$

22. $(0, 3)$ and $(-1, -1)$

Determine the vertex, axis of symmetry, focus, and directrix for each parabola.

23. $y = x^2 + 3x - 18$

24. $y = x - x^2$

25. $y = x^2 + 3x + 2$

26. $y = -x^2 - 3x + 4$

27. $y = -\dfrac{1}{2}(x - 2)^2 + 3$

28. $y = \dfrac{1}{4}(x + 1)^2 - 2$

Write each equation in the form $y = a(x - h)^2 + k$, and identify the vertex of the parabola.

29. $y = 2x^2 - 8x + 1$

30. $y = -2x^2 - 6x - 1$

31. $y = -\dfrac{1}{2}x^2 - x + \dfrac{1}{2}$

32. $y = \dfrac{1}{4}x^2 + x - 9$

12.3 The Circle

Determine the center and radius of each circle, and sketch its graph.

33. $x^2 + y^2 = 100$

34. $x^2 + y^2 = 20$

35. $(x - 2)^2 + (y + 3)^2 = 81$ **36.** $x^2 + 2x + y^2 = 8$

Sketch the graph of each hyperbola.

47. $\dfrac{x^2}{49} - \dfrac{y^2}{36} = 1$

48. $\dfrac{y^2}{25} - \dfrac{x^2}{49} = 1$

37. $9y^2 + 9x^2 = 4$ **38.** $x^2 + 4x + y^2 - 6y - 3 = 0$

49. $4x^2 - 25y^2 = 100$

Write the standard equation for each circle with the given center and radius.

39. Center $(0, 3)$, radius 6

40. Center $(0, 0)$, radius $\sqrt{6}$

50. $6y^2 - 16x^2 = 96$

41. Center $(2, -7)$, radius 5

42. Center $\left(\dfrac{1}{2}, -3\right)$, radius $\dfrac{1}{2}$

12.4 The Ellipse and Hyperbola

Sketch the graph of each ellipse.

43. $\dfrac{x^2}{36} + \dfrac{y^2}{49} = 1$ **44.** $\dfrac{x^2}{25} + y^2 = 1$

12.5 Second-Degree Inequalities

Graph each inequality.

51. $4x - 2y > 3$

45. $25x^2 + 4y^2 = 100$ **46.** $6x^2 + 4y^2 = 24$

52. $y < x^2 - 3x$

53. $y^2 < x^2 - 1$

54. $y^2 < 1 - x^2$

55. $4x^2 + 9y^2 > 36$

56. $x^2 + y > 2x - 1$

Graph the solution set to each system of inequalities.

57. $y < 4x - x^2$
    $x^2 + y^2 < 9$

58. $x^2 - y^2 < 1$
    $y < 1$

59. $4x^2 + 9y^2 > 36$
    $x^2 + y^2 < 9$

60. $y^2 - x^2 > 4$
    $y^2 + 16x^2 < 16$

Miscellaneous

Identify each equation as the equation of a straight line, parabola, circle, hyperbola, or ellipse. Try to do these without rewriting the equations.

61. $x^2 = y^2 + 1$ **62.** $x = y + 1$

63. $x^2 = 1 - y^2$ **64.** $x^2 = y + 1$

65. $x^2 + x = 1 - y^2$ **66.** $(x - 3)^2 + (y + 2)^2 = 7$

67. $x^2 + 4x = 6y - y^2$ **68.** $4x + 6y = 1$

69. $\dfrac{x^2}{3} - \dfrac{y^2}{5} = 1$ **70.** $x^2 + \dfrac{y^2}{3} = 1$

71. $4y^2 - x^2 = 8$ **72.** $9x^2 + y = 9$

Sketch the graph of each equation.

73. $x^2 = 4 - y^2$ **74.** $x^2 = 4y^2 + 4$

75. $x^2 = 4y + 4$ **76.** $x = 4y + 4$

77. $x^2 = 4 - 4y^2$ **78.** $x^2 = 4y - y^2$

79. $x^2 = 4 - (y - 4)^2$ **80.** $(x - 2)^2 + (y - 4)^2 = 4$

88. Vertex $(1, 2)$ and focus $\left(1, \frac{3}{2}\right)$

89. Vertex $(0, 0)$, passing through $(3, 2)$, and opening upward

90. Vertex $(1, 3)$, passing through $(0, 0)$, and opening downward

Write the equation of the circle with the given features.

81. Centered at the origin and passing through $(3, 4)$

82. Centered at $(2, -3)$ and passing through $(-1, 4)$

83. Centered at $(-1, 5)$ with radius 6

84. Centered at $(0, -3)$ and passing through the origin

Write the equation of the parabola with the given features.

85. Focus $(1, 4)$ and directrix $y = 2$

86. Focus $(-2, 1)$ and directrix $y = 5$

87. Vertex $(0, 0)$ and focus $\left(0, \frac{1}{4}\right)$

Solve each system of equations.

91. $x^2 + y^2 = 25$
$y = -x + 1$

92. $x^2 - y^2 = 1$
$x^2 + y^2 = 7$

93. $4x^2 + y^2 = 4$
$x^2 - y^2 = 21$

94. $y = x^2 + x$
$y = -x^2 + 3x + 12$

Solve each problem.

95. *Perimeter of a rectangle.* A rectangle has a perimeter of 16 feet and an area of 12 square feet. Find its length and width.

96. *Tale of two circles.* Find the radii of two circles such that the difference in areas of the two is 10π square inches and the difference in radii of the two is 2 inches.

Chapter 12 Test

Sketch the graph of each equation.

1. $x^2 + y^2 = 25$

2. $\dfrac{x^2}{16} - \dfrac{y^2}{25} = 1$

3. $y^2 + 4x^2 = 4$

4. $y = x^2 + 4x + 4$

5. $y^2 - 4x^2 = 4$

Graph the solution set to each system of inequalities.

10. $x^2 + y^2 < 9$
$x^2 - y^2 > 1$

6. $y = -x^2 - 2x + 3$

11. $y < -x^2 + x$
$y < x - 4$

Sketch the graph of each inequality.

7. $x^2 - y^2 < 9$

Solve each system of equations.

12. $y = x^2 - 2x - 8$
$y = 7 - 4x$

13. $x^2 + y^2 = 12$
$y = x^2$

Solve each problem.

14. Find the distance between $(-1, 4)$ and $(1, 6)$.

15. Find the midpoint and length of the line segment with endpoints $(2, 0)$ and $(-3, -1)$.

16. Find the center and radius of the circle $x^2 + 2x + y^2 + 10y = 10$.

17. Find the vertex, focus, and directrix of the parabola $y = x^2 + x + 3$. State the axis of symmetry and whether the parabola opens up or down.

8. $x^2 + y^2 > 9$

18. Write the equation $y = \frac{1}{2}x^2 - 3x - \frac{1}{2}$ in the form $y = a(x - h)^2 + k$.

19. Write the equation of a circle with center $(-1, 3)$ that passes through $(2, 5)$.

9. $y > x^2 - 9$

20. Find the length and width of a rectangular room that has an area of 108 square feet and a perimeter of 42 ft.

*Making*Connections | A Review of Chapters 1–12

Sketch the graph of each equation.

1. $y = 9x - x^2$

2. $y = 9x$

3. $y = (x - 9)^2$

4. $y^2 = 9 - x^2$

5. $y = 9x^2$

6. $y = |9x|$

7. $4x^2 + 9y^2 = 36$

8. $4x^2 - 9y^2 = 36$

9. $y = 9 - x$

10. $y = 9^x$

Find the following products.

11. $(x + 2y)^2$

12. $(x + y)(x^2 + 2xy + y^2)$

13. $(a + b)^3$

14. $(a - 3b)^2$

15. $(2a + 1)(3a - 5)$

16. $(x - y)(x^2 + xy + y^2)$

Factor completely.

17. $a^3 + ab^2$

18. $a^3 - ab^2$

19. $2x^2 - 6x - 36$

20. $32x^2 + 8x - 60$

21. $mx^2 + 2x^2 - 9m - 18$

22. $2x^4 + 54x$

Solve each system of equations.

23. $2x - 3y = -4$
     $x + 2y = 5$

24. $x^2 + y^2 = 25$
     $x + y = 7$

25. $2x - y + z = 7$
     $x - 2y - z = 2$
     $x + y + z = 2$

26. $y = x^2$
     $y - 2x = 3$

Solve each formula for the specified variable.

27. $ax + b = 0$, for x

28. $wx^2 + dx + m = 0$, for x

29. $A = \dfrac{1}{2}h(B + b)$, for B

30. $\dfrac{1}{x} + \dfrac{1}{y} = \dfrac{1}{2}$, for x

31. $L = m + mxt$, for m

32. $y = 3a\sqrt{t}$, for t

Solve each problem.

33. Write the equation of the line in slope-intercept form that goes through the points $(2, -3)$ and $(-4, 1)$.

34. Write the equation of the line in slope-intercept form that contains the origin and is perpendicular to the line $2x - 4y = 5$.

35. Write the equation of the circle that has center $(2, 5)$ and passes through the point $(-1, -1)$.

36. Find the center and radius of the circle $x^2 + 3x + y^2 - 6y = 0$.

Perform the computations with complex numbers.

37. $2i(3 + 5i)$

38. i^6

39. $(2i - 3) + (6 - 7i)$

40. $(3 + i\sqrt{2})^2$

41. $(2 - 3i)(5 - 6i)$

42. $(3 - i) + (-6 + 4i)$

43. $(5 - 2i)(5 + 2i)$

44. $(2 - 3i) \div (2i)$

45. $(4 + 5i) \div (1 - i)$

46. $\dfrac{4 - \sqrt{-8}}{2}$

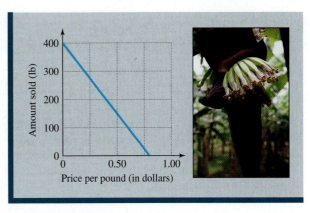

Figure for Exercise 47

Solve.

47. *Going bananas.* Salvadore has observed that when bananas are \$0.30 per pound (lb), he sells 250 lb per day, and when bananas are \$0.40 per lb, he sells only 200 lb per day.

a) Assume the number of pounds sold, q, is a linear function of the price per pound, x, and find that function.

b) Salvadore's daily revenue in dollars is the product of the number of pounds sold and the price per pound. Write the revenue as a function of x.

c) Graph the revenue function.

d) What price per pound maximizes his revenue?

e) What is his maximum possible revenue?

Critical **Thinking** | For Individual or Group Work | Chapter 12

These exercises can be solved by a variety of techniques, which may or may not require algebra. So be creative and think critically. Explain all answers. Answers are in the Instructor's Edition of this text.

1. *Semicircles.* Start with any right triangle, and draw three semicircles so that each has one side of the triangle as its diameter as shown in the accompanying figure. Show that the total area of the two smaller semicircles is equal to the area of the largest semicircle.

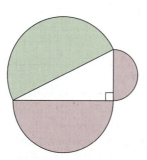

Figure for Exercise 1

2. *Huge integer.* The value of the expression $16^9 \cdot 5^{25}$ is an integer. How many digits does it have?

3. *Sevens galore.* How many seven-digit whole numbers contain the number seven at least once?

4. *Ten-digit surprise.* Use the digits 0 through 9 once each to construct a 10-digit number such that the first n digits (counting from the left) form a number divisible by n, for each n from 1 through 10. For example, for 3428, 3 is divisible by 1, 34 is divisible by 2, 342 is divisible by 3, and 3428 is divisible by 4, but 3428 is not a 10-digit number.

5. *Cattle drive.* A group of cowboys is driving a herd of cattle across the plains at a constant rate. The cowboys always keep the herd in the shape of a square that is 1 kilometer on each side. One of the cowboys starts at the left rear of the square/herd and rides his four-wheeler around the perimeter of the square at a constant rate in the same time that the herd advances 1 kilometer. How far does this cowboy travel?

Photo for Exercise 5

6. *Counting game.* A teacher plays a counting game with his students. The first student says 1. The second student says 2 and 3. The third student says 4, 5, and 6. The fourth says 7, 8, 9, and 10. This pattern continues with the fifth student saying the next five counting numbers, and so on. Find a formula for the sum of the numbers said by the kth student. $\left[\text{Hint: The sum of the first } n \text{ counting numbers is } \frac{n(n+1)}{2}.\right]$

7. *Numerical palindrome.* A numerical palindrome is a positive integer with at least two digits that reads the same forward or backward. For example, 55 and 343 are numerical palindromes. How many numerical palindromes are there less than 1000?

8. *Fractional chickens.* If 1.5 chickens lay 1.5 eggs in 1.5 days, then how many eggs do 3.5 chickens lay in 3 days?

Sequences and Series

Everyone realizes the importance of investing for the future. Some people go to great pains to study the markets and to make wise investment decisions. Some stay away from investing because they do not want to take chances. However, the most important factor in investing is making regular investments (*Money,* www.money.com). According to *Money,* if you had invested $5000 in the stock market every year at the market high for that year (the worst time to invest) for the last 40 years, your investment would be worth $2.8 million today.

A sequence of periodic investments earning a fixed rate of interest can be thought of as a geometric sequence. In this chapter you will learn how to find the sum of a geometric sequence and to calculate the future value of a sequence of periodic investments.

Growth of $5000
investment per year

Amount (in thousands of dollars)

150

100

50

0 1 2 3 4 5 6 7 8 9 10
Time (years)

In Exercise 54 of Section 13.4 you will calculate the value of $5000 invested each year for 10 years in Fidelity's Magellan Fund.

13.1   Sequences

The word "sequence" is a familiar word. We may speak of a sequence of events or say that something is out of sequence. In this section, we give the mathematical definition of a sequence.

⟨1⟩ Sequences

In mathematics we think of a sequence as a list of numbers. Each number in the sequence is called a **term** of the sequence. There is a first term, a second term, a third term, and so on. For example, the daily high temperature readings in Minot, North Dakota, for the first 10 days in January can be thought of as a finite sequence with 10 terms:

$$-9, -2, 8, -11, 0, 6, 14, 1, -5, -11$$

The set of all positive even integers,

$$2, 4, 6, 8, 10, 12, 14, \ldots,$$

can be thought of as an infinite sequence.

To give a precise definition of sequence, we use the terminology of functions. The list of numbers is the range of the function.

> **Sequence**
>
> A **finite sequence** is a function whose domain is the set of positive integers less than or equal to some fixed positive integer. An **infinite sequence** is a function whose domain is the set of all positive integers.

When the domain is apparent, we will refer to either a finite sequence or an infinite sequence simply as a sequence. For the independent variable of the function we will usually use n (for natural number) rather than x. For the dependent variable we write a_n (read "a sub n") rather than y. We call a_n the ***n*th term,** or the **general term,** of the sequence. Rather than use the $f(x)$ notation for functions, we will define sequences with formulas. When n is used as a variable, we will assume it represents natural numbers only.

E X A M P L E   1

Listing terms of a finite sequence

List all of the terms of each finite sequence.

a) $a_n = n^2$ for $1 \leq n \leq 5$ **b)** $a_n = \dfrac{1}{n + 2}$ for $1 \leq n \leq 4$

⟨ **Calculator Close-Up** ⟩

We can define the sequence with the Y= key and make a list of the terms.

X	Y1	
1	1	
2	4	
3	9	
4	16	
5	25	
6	36	
7	49	

Y1☐X²

Solution

a) Using the natural numbers from 1 through 5 in $a_n = n^2$, we get

$$a_1 = 1^2 = 1,$$
$$a_2 = 2^2 = 4,$$
$$a_3 = 3^2 = 9,$$
$$a_4 = 4^2 = 16,$$

and

$$a_5 = 5^2 = 25.$$

The five terms of this sequence are 1, 4, 9, 16, and 25. We often refer to the listing of the terms of the sequence as the sequence.

b) Using the natural numbers from 1 through 4 in $a_n = \frac{1}{n+2}$, we get the terms

$$a_1 = \frac{1}{1+2} = \frac{1}{3},$$

$$a_2 = \frac{1}{2+2} = \frac{1}{4},$$

$$a_3 = \frac{1}{5},$$

and

$$a_4 = \frac{1}{6}.$$

The four terms of the sequence are $\frac{1}{3}, \frac{1}{4}, \frac{1}{5},$ and $\frac{1}{6}$.

> Now do Exercises 1–14

E X A M P L E   2

Listing terms of an infinite sequence

List the first three terms of the infinite sequence whose nth term is

$$a_n = \frac{(-1)^n}{2^{n+1}}.$$

‹ Calculator Close-Up ›

Some calculators have a sequence feature that allows you to specify the formula and which terms to evaluate. We can even get the terms as fractions.

```
seq((-1)^n/2^(n+
1),n,1,3)▶Frac
{-1/4 1/8 -1/16}
```

Solution

Using the natural numbers 1, 2, and 3 in the formula for the nth term yields

$$a_1 = \frac{(-1)^1}{2^{1+1}} = -\frac{1}{4}, \qquad a_2 = \frac{(-1)^2}{2^{2+1}} = \frac{1}{8}, \qquad \text{and} \qquad a_3 = \frac{(-1)^3}{2^{3+1}} = -\frac{1}{16}.$$

We write the sequence as follows:

$$-\frac{1}{4}, \frac{1}{8}, -\frac{1}{16}, \dots$$

> Now do Exercises 15–22

‹2›   Finding a Formula for the nth Term

We often know the terms of a sequence and want to write a formula that will produce those terms. To write a formula for the nth term of a sequence, examine the terms and look for a pattern. Each term is a function of the term number. The first term corresponds to $n = 1$, the second term corresponds to $n = 2$, and so on.

E X A M P L E   3

A familiar sequence

Write the general term for the infinite sequence

$$3, 5, 7, 9, 11, \dots.$$

‹ **Helpful Hint** ›

Finding a formula for a sequence could be extremely difficult. For example, there is no known formula that will produce the sequence of prime numbers:

$$2, 3, 5, 7, 11, 13, 17, 19, \ldots$$

Solution

The even numbers are all multiples of 2 and can be represented as $2n$. Because each odd number is 1 more than an even number, a formula for the nth term might be

$$a_n = 2n + 1.$$

To be sure, we write out a few terms using the formula:

$$a_1 = 2(1) + 1 = 3$$
$$a_2 = 2(2) + 1 = 5$$
$$a_3 = 2(3) + 1 = 7$$

So the general term is $a_n = 2n + 1$.

Now do Exercises 23–24

CAUTION   There can be more than one formula that produces the given terms of a sequence. For example, the sequence

$$1, 2, 4, \ldots$$

could have nth term $a_n = 2^{n-1}$ or $a_n = \frac{1}{2}n^2 - \frac{1}{2}n + 1$. The first three terms for both of these sequences are identical, but their fourth terms are different.

E X A M P L E   4

A sequence with alternating signs

Write the general term for the infinite sequence

$$1, -\frac{1}{4}, \frac{1}{9}, -\frac{1}{16}, \ldots .$$

Solution

To obtain the alternating signs, we use powers of -1. Because any even power of -1 is positive and any odd power of -1 is negative, we use $(-1)^{n+1}$. The denominators are the squares of the positive integers. So the nth term of this infinite sequence is given by the formula

$$a_n = \frac{(-1)^{n+1}}{n^2}.$$

Check this sequence by using this formula to find the first four terms.

Now do Exercises 25–36

In Example 5 we use a sequence to model a physical situation.

E X A M P L E   5

The bouncing ball

Suppose a ball always rebounds $\frac{2}{3}$ of the height from which it falls and the ball is dropped from a height of 6 feet. Write a sequence whose terms are the heights from which the ball falls. What is a formula for the nth term of this sequence?

Solution

On the first fall the ball travels 6 feet (ft), as shown in Fig. 13.1. On the second fall it travels $\frac{2}{3}$ of 6, or 4 ft. On the third fall it travels $\frac{2}{3}$ of 4, or $\frac{8}{3}$ ft, and so on. We write the sequence as follows:

$$6, 4, \frac{8}{3}, \frac{16}{9}, \frac{32}{27}, \ldots$$

The nth term can be written by using powers of $\frac{2}{3}$:

$$a_n = 6\left(\frac{2}{3}\right)^{n-1}$$

6 ft

4 ft

$\frac{8}{3}$ ft

Now do Exercises 37–44

Figure 13.1

Warm-Ups ▼

Fill in the blank.

1. A list of numbers is a(n) _____.
2. Each number in a sequence is a(n) _____ .
3. A function whose domain is the set of positive integers less than or equal to some fixed positive integer is a(n) _____ sequence.
4. A function whose domain is the set of all positive integers is a(n) _____ sequence.

True or false?

5. The nth term of $2, 4, 6, 8, \ldots$ is $a_n = 2n$.
6. The nth term of $1, 3, 5, 7, \ldots$ is $a_n = 2n - 1$.
7. The nth term of $-1, 4, -9, 16, \ldots$ is $a_n = (-1)^{n+1}n^2$.
8. A sequence is a function.
9. The sixth term of $a_n = (-1)^{n+1}2^n$ is 64.
10. The tenth term of $2, 4, 8, 16, 32, \ldots$ is 1024.

Exercises

13.1

‹ 1 › Sequences

List all terms of each finite sequence. See Example 1.

1. $a_n = 2n$ for $1 \leq n \leq 5$
2. $a_n = 2n - 1$ for $1 \leq n \leq 4$
3. $a_n = n^2$ for $1 \leq n \leq 8$
4. $a_n = -n^2$ for $1 \leq n \leq 4$

5. $b_n = \dfrac{(-1)^n}{n}$ for $1 \leq n \leq 10$

6. $b_n = \dfrac{(-1)^{n+1}}{n}$ for $1 \leq n \leq 6$

7. $c_n = (-2)^{n-1}$ for $1 \leq n \leq 5$

8. $c_n = (-3)^{n-2}$ for $1 \le n \le 5$

9. $a_n = 2^{-n}$ for $1 \le n \le 6$

10. $a_n = 2^{-n+2}$ for $1 \le n \le 5$

11. $b_n = 2n - 3$ for $1 \le n \le 7$

12. $b_n = 2n + 6$ for $1 \le n \le 7$

13. $c_n = n^{-1/2}$ for $1 \le n \le 5$

14. $c_n = n^{1/2}2^{-n}$ for $1 \le n \le 4$

Write the first four terms of the infinite sequence whose nth term is given. See Example 2.

15. $a_n = \dfrac{1}{n^2 + n}$

16. $b_n = \dfrac{1}{(n + 1)(n + 2)}$

17. $b_n = \dfrac{1}{2n - 5}$

18. $a_n = \dfrac{4}{2n + 5}$

19. $c_n = (-1)^n(n - 2)^2$

20. $c_n = (-1)^n(2n - 1)^2$

21. $a_n = \dfrac{(-1)^{2n}}{n^2}$

22. $a_n = (-1)^{2n+1}2^{n-1}$

‹2› Finding a Formula for the *n*th Term

Write a formula for the general term of each infinite sequence. See Examples 3 and 4.

23. $1, 3, 5, 7, 9, \ldots$

24. $5, 7, 9, 11, 13, \ldots$

25. $1, -1, 1, -1, \ldots$

26. $-1, 1, -1, 1, \ldots$

27. $0, 2, 4, 6, 8, \ldots$

28. $4, 6, 8, 10, 12, \ldots$

29. $3, 6, 9, 12, \ldots$

30. $4, 8, 12, 16, \ldots$

31. $4, 7, 10, 13, \ldots$

32. $3, 7, 11, 15, \ldots$

33. $-1, 2, -4, 8, -16, \ldots$

34. $1, -3, 9, -27, \ldots$

35. $0, 1, 4, 9, 16, \ldots$

36. $0, 1, 8, 27, 64, \ldots$

Solve each problem. See Example 5.

37. *Football penalties.* A football is on the 8-yard line, and five penalties in a row are given that move the ball half the distance to the (closest) goal. Write a sequence of five terms that specify the location of the ball after each penalty.

38. *Infestation.* Leona planted 9 acres of soybeans, but by the end of each week, insects had destroyed one-third of the acreage that was healthy at the beginning of the week. How many acres does she have left after 6 weeks?

39. *Constant rate of increase.* The MSRP for a well-equipped 2009 Ford F-250 Lariat 4WD Super Duty Super Cab was $43,568 (www.edmunds.com). Suppose that the price of this model increases by 5% each year. Find the price to the nearest dollar for the 2010 through 2014 models.

Figure for Exercise 39

40. *Constant increase.* The MSRP for a well-equipped 2009 BMW was $89,488 (www.edmunds.com). Suppose that the price of this model increases by $1000 each year. Find the prices for the 2010 through 2014 models.

41. *Economic impact.* To assess the economic impact of a factory on a community, economists consider the annual amount the factory spends in the community, then the portion of the money that is respent in the community, then the portion of the respent money that is respent in the community, and so on. Suppose a garment manufacturer spends $1 million annually in its community and 80% of all money received in the community is respent in the community. Find the first four terms of the economic impact sequence.

Figure for Exercise 41

42. *Less impact.* The rate at which money is respent in a community varies from community to community. Find the first four terms of the economic impact sequence for the manufacturer in Exercise 41, assuming only 50% of money received in the community is respent in the community.

43. *Fabric design.* A fabric designer must take into account the capability of textile machines to produce material with vertical repeats. A textile machine can be set up for a vertical repeat every $\frac{27}{n}$ inches (in.), where n is a natural number. Write the first five terms of the sequence $a_n = \frac{27}{n}$, which gives the possible vertical repeats for a textile machine.

44. *Musical tones.* The note middle C on a piano is tuned so that the string vibrates at 262 cycles per second, or 262 hertz (Hz). The C note one octave higher is tuned to 524 Hz. The tuning for the 11 notes in between using the method called *equal temperament* is determined by the sequence $a_n = 262 \cdot 2^{n/12}$. Find the tuning for the 11 notes in between. Round to the nearest whole Hz.

Getting More Involved

45. *Discussion*

Everyone has two (biological) parents, four grandparents, eight great-grandparents, 16 great-great-grandparents, and so on. If we put the word "great" in front of the word

"grandparents" 35 times, then how many of this type of relative do you have? Is this more or less than the present population of the earth? Give reasons for your answers.

46. *Discussion*

If you deposit 1 cent into your piggy bank on September 1 and each day thereafter deposit twice as much as on the previous day, then how much will you be depositing on September 30? The total amount deposited for the month can be found without adding up all 30 deposits. Look at how the amount on deposit is increasing each day and see whether you can find the total for the month. Give reasons for your answers.

47. *Cooperative learning*

Working in groups, have someone in each group make up a formula for a_n, the nth term of a sequence, but do not show it to the other group members. Write the terms of the sequence on a piece of paper one at a time. After each term is given, ask whether anyone knows the next term. When the group can correctly give the next term, ask for a formula for the nth term.

48. *Exploration*

Find a real-life sequence in which all of the terms are the same. Find one in which each term after the first is one larger than the previous term. Find out what the sequence of fines is on your campus for your first, second, third, and fourth parking ticket.

49. *Exploration*

Consider the sequence whose nth term is $a_n = (0.999)^n$.

a) Calculate a_{100}, a_{1000}, and $a_{10,000}$.

b) What happens to a_n as n gets larger and larger?

50. *Exploration*

The first two terms of the Fibonacci sequence are 0 and 1. Every term thereafter is the sum of the two previous terms. So the third term is 1, the fourth term is 2, the fifth term is 3, and the sixth term is 5. So the first 6 terms of the Fibonacci sequence are 0, 1, 1, 2, 3, 5.

a) Write the first 10 terms of the Fibonacci sequence.

b) Find an application of the Fibonacci sequence by doing a search on the Internet.

Math *at Work*   Piano Tuning

If middle C on a piano has a frequency of 261 cycles per second or 261 hertz (Hz), then the C note one octave higher is 522 Hz. But what should be the frequencies of the 11 notes in between? On a violin, the frequencies of the notes are selected by the musician as the instrument is played, but with a piano the frequency is selected by the piano tuner.

One method, the Just scale, uses the naturally occurring overtone series for systems such as vibrating strings or air columns. All the notes are related by rational numbers. Because the ratio of the frequencies of successive notes is not constant, the tuning depends on the scale you are using. For example, the tunings for C major and for D major are different.

The equal-tempered scale was developed as a compromise scale for keyboard instruments played in many keys. The equal-tempered system uses a constant ratio of $2^{1/12}$. So playing in any key sounds equally good or equally bad, depending on your point of view. The accompanying table shows the ratio of the frequency of each note in the C major scale to middle C and the frequencies of the notes for Just and equal temperament. For this chart middle C was chosen as 261.6256 so that A would be 440 Hz in the equal-tempered scale. For example, D is $\frac{9}{8} \cdot 261.6256$ for the Just scale or $2^{2/12} \cdot 261.6256$ for equal temperament. Note that the frequencies of the notes differ by as much as 4 Hz in the two scales. Since a human ear can hear a difference of less than 1 Hz, it is easy to hear the difference between these two scales.

Note	Just Scale Ratio	Equal Temperament Ratio	Just Scale (Hz)	Equal Temperament (Hz)
C	1	1	261.63	261.63
C#	25/24	$2^{1/12}$	272.53	277.18
D	9/8	$2^{2/12}$	294.33	293.66
E♭	6/5	$2^{3/12}$	313.95	311.13
E	5/4	$2^{4/12}$	327.03	329.63
F	4/3	$2^{5/12}$	348.83	349.23
F#	45/32	$2^{6/12}$	367.91	369.99
G	3/2	$2^{7/12}$	392.44	392.00
A♭	8/5	$2^{8/12}$	418.60	415.30
A	5/3	$2^{9/12}$	436.04	440.00
B♭	9/5	$2^{10/12}$	470.93	466.16
B	15/8	$2^{11/12}$	490.55	493.88
C	2	2	523.25	523.25

13.2   Series

If you make a sequence of bank deposits, then you might be interested in the total value of the terms of the sequence. Of course, if the sequence has only a few terms, you can simply add them. In Sections 13.3 and 13.4, we will develop formulas that give the sum of the terms for certain finite and infinite sequences. In this section you will first learn a notation for expressing the sum of the terms of a sequence.

⟨1⟩ Summation Notation

To describe the sum of the terms of a sequence, we use **summation notation.** The Greek letter Σ (sigma) is used to indicate sums. For example, the sum of the first five terms of the sequence $a_n = n^2$ is written as

$$\sum_{n=1}^{5} n^2.$$

You can read this notation as "the sum of n^2 for n between 1 and 5, inclusive." To find the sum, we let n take the values 1 through 5 in the expression n^2:

$$\sum_{n=1}^{5} n^2 = 1^2 + 2^2 + 3^2 + 4^2 + 5^2$$
$$= 1 + 4 + 9 + 16 + 25$$
$$= 55$$

In this context the letter n is the **index of summation.** Other letters may also be used. For example, the expressions

$$\sum_{n=1}^{5} n^2, \quad \sum_{j=1}^{5} j^2, \quad \text{and} \quad \sum_{i=1}^{5} i^2$$

all have the same value. Note that i is used as a variable here and not as an imaginary number.

E X A M P L E   **1**

Evaluating a sum in summation notation
Find the value of the expression

$$\sum_{i=1}^{3} (-1)^i(2i + 1).$$

Solution
Replace i by 1, 2, and 3, and then add the results:

$$\sum_{i=1}^{3} (-1)^i(2i + 1) = (-1)^1[2(1) + 1] + (-1)^2[2(2) + 1] + (-1)^3[2(3) + 1]$$
$$= -3 + 5 - 7$$
$$= -5$$

> Now do Exercises 1–14

⟨2⟩ Series

The sum of the terms of the sequence 1, 4, 9, 16, 25 is written as

$$1 + 4 + 9 + 16 + 25.$$

This expression is called a *series*. It indicates that we are to add the terms of the given sequence. The sum, 55, is the sum of the series.

> **Series**
>
> The indicated sum of the terms of a sequence is called a **series.**

Just as a sequence may be finite or infinite, a series may be finite or infinite. In this section we discuss finite series only. In Section 13.4 we will discuss one type of infinite series.

Summation notation is a convenient notation for writing a series.

EXAMPLE 2

Converting to summation notation

Write the series in summation notation:

$$2 + 4 + 6 + 8 + 10 + 12 + 14$$

Solution

The general term for the sequence of positive even integers is $2n$. If we let n take the values from 1 through 7, then $2n$ ranges from 2 through 14. So,

$$2 + 4 + 6 + 8 + 10 + 12 + 14 = \sum_{n=1}^{7} 2n.$$

> Now do Exercises 15–16

EXAMPLE 3

Converting to summation notation

Write the series

$$\frac{1}{2} - \frac{1}{3} + \frac{1}{4} - \frac{1}{5} + \frac{1}{6} - \frac{1}{7} + \cdots + \frac{1}{50}$$

in summation notation.

‹ Helpful Hint ›

A series is called an *indicated sum* because the addition is indicated but not actually being performed. The sum of a series is the real number obtained by actually performing the indicated addition.

Solution

For this series we let n be 2 through 50. The expression $(-1)^n$ produces alternating signs. The series is written as

$$\sum_{n=2}^{50} \frac{(-1)^n}{n}.$$

> Now do Exercises 17–30

‹3› Changing the Index

In Example 3 we saw the index go from 2 through 50, but this is arbitrary. A series can be written with the index starting at any given number.

EXAMPLE 4

Changing the index

Rewrite the series

$$\sum_{i=1}^{6} \frac{(-1)^i}{i^2}$$

with an index j, where j starts at 0.

Solution

Because i starts at 1 and j starts at 0, we have $i = j + 1$. Because i ranges from 1 through 6 and $i = j + 1$, j must range from 0 through 5. Now replace i by $j + 1$ in the summation notation:

$$\sum_{j=0}^{5} \frac{(-1)^{j+1}}{(j+1)^2}$$

Check that these two series have exactly the same six terms.

Now do Exercises 37–46

Warm-Ups ▼

Fill in the blank.

1. _____ notation provides a way of writing a sum without writing out all of the terms.

2. A(n) _____ is the indicated sum of the terms of a sequence.

3. A(n) _____ series is the sum of a finite sequence.

4. A(n) _____ series is the sum of an infinite sequence.

True or false?

5. There are eight terms in $\sum_{i=2}^{10} i^3$.

6. $\sum_{i=2}^{10} i^3 = \sum_{j=3}^{11} (j-1)^3$

7. $\sum_{i=1}^{2} (-1)^i 2^i = 2$

8. $\sum_{i=1}^{5} 4 = 20$

9. $\sum_{i=1}^{5} 3i = 3\left(\sum_{i=1}^{5} i\right)$

10. $\sum_{i=1}^{3} (2i + 1) = \left(\sum_{i=1}^{3} 2i\right) + 1$

Exercises 13.2

‹ **Study Tips** ›

- The language of algebra is important. If you don't understand the question, it is difficult to answer it.
- To reinforce your algebra vocabulary use *Enriching Your Mathematical Word Power,* which appears at the end of every chapter.

‹1› **Summation Notation**

Find the sum of each series. See Example 1.

1. $\sum_{i=1}^{5} i$

2. $\sum_{i=1}^{6} 2i$

3. $\sum_{i=1}^{4} i^2$

4. $\sum_{j=0}^{3} (j+1)^2$

5. $\sum_{j=0}^{5} (2j - 1)$

6. $\sum_{i=1}^{6} (2i - 3)$

7. $\sum_{i=1}^{5} 2^{-i}$

8. $\sum_{i=1}^{5} (-2)^{-i}$

9. $\sum_{i=1}^{10} 5i^0$

10. $\sum_{j=1}^{20} 3$

11. $\displaystyle\sum_{i=1}^{3} (i-3)(i+1)$ **12.** $\displaystyle\sum_{i=0}^{5} i(i-1)(i-2)(i-3)$

13. $\displaystyle\sum_{j=1}^{10} (-1)^j$ **14.** $\displaystyle\sum_{j=1}^{11} (-1)^j$

⟨2⟩ Series

Write each series in summation notation. Use the index i, and let i begin at 1 in each summation. See Examples 2 and 3.

15. $1 + 2 + 3 + 4 + 5 + 6$

16. $2 + 4 + 6 + 8 + 10$

17. $-1 + 3 - 5 + 7 - 9 + 11$

18. $1 - 3 + 5 - 7 + 9$

19. $1 + 4 + 9 + 16 + 25 + 36$

20. $1 + 8 + 27 + 64 + 125$

21. $\dfrac{1}{3} + \dfrac{1}{4} + \dfrac{1}{5} + \dfrac{1}{6}$

22. $1 - \dfrac{1}{2} + \dfrac{1}{3} - \dfrac{1}{4} + \dfrac{1}{5} - \dfrac{1}{6}$

23. $\ln(2) + \ln(3) + \ln(4)$

24. $e^1 + e^2 + e^3 + e^4$

25. $a_1 + a_2 + a_3 + a_4$

26. $a^2 + a^3 + a^4 + a^5$

27. $x_3 + x_4 + x_5 + \cdots + x_{50}$

28. $y_1 + y_2 + y_3 + \cdots + y_{30}$

29. $w_1 + w_2 + w_3 + \cdots + w_n$

30. $m_1 + m_2 + m_3 + \cdots + m_k$

Write out the terms of each series.

31. $\displaystyle\sum_{i=1}^{6} x^i$

32. $\displaystyle\sum_{i=1}^{5} (-1)^i x^{i-1}$

33. $\displaystyle\sum_{j=0}^{3} (-1)^j x_j$

34. $\displaystyle\sum_{j=1}^{5} \dfrac{1}{x_j}$

35. $\displaystyle\sum_{i=1}^{3} i x^i$

36. $\displaystyle\sum_{i=1}^{5} \dfrac{x}{i}$

⟨3⟩ Changing the Index

Complete the rewriting of each series using the new index as indicated. See Example 4.

37. $\displaystyle\sum_{i=1}^{5} i^2 = \sum_{j=0}$ **38.** $\displaystyle\sum_{i=1}^{6} i^3 = \sum_{j=0}$

39. $\displaystyle\sum_{i=0}^{12} (2i-1) = \sum_{j=1}$ **40.** $\displaystyle\sum_{i=1}^{3} (3i+2) = \sum_{j=0}$

41. $\displaystyle\sum_{i=4}^{8} \dfrac{1}{i} = \sum_{j=1}$ **42.** $\displaystyle\sum_{i=5}^{10} 2^{-i} = \sum_{j=1}$

43. $\displaystyle\sum_{i=1}^{4} x^{2i+3} = \sum_{j=0}$ **44.** $\displaystyle\sum_{i=0}^{2} x^{3-2i} = \sum_{j=1}$

45. $\displaystyle\sum_{i=1}^{n} x^i = \sum_{j=0}$ **46.** $\displaystyle\sum_{i=0}^{n} x^{-i} = \sum_{j=1}$

Applications

Use a series to model the situation in each of the following problems.

47. *Leap frog.* A frog with a vision problem is 1 yard away from a dead cricket. He spots the cricket and jumps halfway to the cricket. After the frog realizes that he has not reached the cricket, he again jumps halfway to the cricket. Write a series in summation notation to describe how far the frog has moved after nine such jumps.

48. *Compound interest.* Cleo deposited $1000 at the beginning of each year for 5 years into an account paying 10% interest compounded annually. Write a series using summation notation to describe how much she has in the account at the end of the fifth year. Note that the first $1000 will receive interest for 5 years, the second $1000 will receive interest for 4 years, and so on.

49. *Total economic impact.* In Exercise 41 of Section 13.1 we described a factory that spends $1 million annually in a community in which 80% of all money received in the community is respent in the community. Use summation notation to write the sum of the first four terms of the economic impact sequence for the factory.

50. *Total earnings.* Suppose you earn $1 on January 1, $2 on January 2, $3 on January 3, and so on. Use summation

notation to write the sum of your earnings for the entire month of January.

Getting More Involved

51. *Discussion*

What is the difference between a sequence and a series?

52. *Discussion*

For what values of n is $\displaystyle\sum_{i=1}^{n} \frac{1}{i} > 4$?

13.3   **Arithmetic Sequences and Series**

In This Section

⟨1⟩ **Arithmetic Sequences**
⟨2⟩ **Arithmetic Series**

We defined sequences and series in Sections 13.1 and 13.2. In this section you will study a special type of sequence known as an arithmetic sequence. You will also study the series corresponding to this sequence.

⟨1⟩ **Arithmetic Sequences**

Consider the following sequence:

$$5, 9, 13, 17, 21, \ldots$$

This sequence is called an arithmetic sequence because of the pattern for the terms. Each term is 4 larger than the previous term.

⟨ **Helpful Hint** ⟩

Arithmetic used as an adjective (ar-ith-met′-ic) is pronounced differently from arithmetic used as a noun (a-rith′-me-tic). Arithmetic (the adjective) is accented similarly to geometric.

> **Arithmetic Sequence**
>
> A sequence in which each term after the first is obtained by adding a fixed amount to the previous term is called an **arithmetic sequence.**

The fixed amount is called the **common difference** and is denoted by the letter d. If a_1 is the first term, then the second term is $a_1 + d$. The third term is $a_1 + 2d$, the fourth term is $a_1 + 3d$, and so on.

> **Formula for the *n*th Term of an Arithmetic Sequence**
>
> The nth term, a_n, of an arithmetic sequence with first term a_1 and common difference d is
>
> $$a_n = a_1 + (n - 1)d.$$

E X A M P L E   **1**

The nth term of an arithmetic sequence

Write a formula for the nth term of the arithmetic sequence

$$5, 9, 13, 17, 21, \ldots .$$

Solution

Each term of the sequence after the first is 4 more than the previous term. Because the common difference is 4 and the first term is 5, the nth term is given by

$$a_n = 5 + (n - 1)4.$$

We can simplify this expression to get

$$a_n = 4n + 1.$$

Check a few terms: $a_1 = 4(1) + 1 = 5$, $a_2 = 4(2) + 1 = 9$, and $a_3 = 4(3) + 1 = 13$.

> Now do Exercises 1–8

In Example 2, the common difference is negative.

E X A M P L E   **2**

An arithmetic sequence of decreasing terms

Write a formula for the nth term of the arithmetic sequence

$$4, 1, -2, -5, -8, \ldots .$$

Solution

Each term is 3 less than the previous term, so $d = -3$. Because $a_1 = 4$, we can write the nth term as

$$a_n = 4 + (n - 1)(-3),$$

or

$$a_n = -3n + 7.$$

Check a few terms: $a_1 = -3(1) + 7 = 4$, $a_2 = -3(2) + 7 = 1$, and $a_3 = -3(3) + 7 = -2$.

> Now do Exercises 9–16

In Example 3, we find some terms of an arithmetic sequence using a given formula for the nth term.

E X A M P L E   **3**

Writing terms of an arithmetic sequence

Write the first five terms of the sequence in which $a_n = 3 + (n - 1)6$.

Solution

Let n take the values from 1 through 5, and find a_n:

$$a_1 = 3 + (1 - 1)6 = 3$$
$$a_2 = 3 + (2 - 1)6 = 9$$
$$a_3 = 3 + (3 - 1)6 = 15$$

$$a_4 = 3 + (4 - 1)6 = 21$$
$$a_5 = 3 + (5 - 1)6 = 27$$

Notice that $a_n = 3 + (n - 1)6$ gives the general term for an arithmetic sequence with first term 3 and common difference 6. Because each term after the first is 6 more than the previous term, the first five terms that we found are correct.

> Now do Exercises 17–30

The formula $a_n = a_1 + (n - 1)d$ involves four variables: a_1, a_n, n, and d. If we know the values of any three of these variables, we can find the fourth.

EXAMPLE **4**

Finding a missing term of an arithmetic sequence

Find the twelfth term of the arithmetic sequence whose first term is 2 and whose fifth term is 14.

Solution

Before finding the twelfth term, we use the given information to find the missing common difference. Let $n = 5$, $a_1 = 2$, and $a_5 = 14$ in the formula $a_n = a_1 + (n - 1)d$ to find d:

$$14 = 2 + (5 - 1)d$$
$$14 = 2 + 4d$$
$$12 = 4d$$
$$3 = d$$

Now use $a_1 = 2$, $d = 3$, and $n = 12$ in $a_n = a_1 + (n - 1)d$ to find a_{12}:

$$a_{12} = 2 + (12 - 1)3$$
$$a_{12} = 35$$

> Now do Exercises 31–38

⟨2⟩ Arithmetic Series

The indicated sum of an arithmetic sequence is called an **arithmetic series.** For example, the series

$$2 + 4 + 6 + 8 + 10 + \cdots + 54$$

is an arithmetic series because there is a common difference of 2 between the terms.

We can find the actual sum of this arithmetic series without adding all of the terms. Write the series in increasing order, and below that write the series in decreasing order. We then add the corresponding terms:

$$S = 2 + 4 + 6 + 8 + \cdots + 52 + 54$$
$$S = 54 + 52 + 50 + 48 + \cdots + 4 + 2$$
$$\overline{2S = 56 + 56 + 56 + 56 + \cdots + 56 + 56}$$

Now, how many times does 56 appear in the sum on the right? Because

$$2 + 4 + 6 + \cdots + 54 = 2 \cdot 1 + 2 \cdot 2 + 2 \cdot 3 + \cdots + 2 \cdot 27,$$

there are 27 terms in this sum. Because 56 appears 27 times on the right, we have $2S = 27 \cdot 56$, or

$$S = \frac{27 \cdot 56}{2} = 27 \cdot 28 = 756.$$

If $S_n = a_1 + a_2 + a_3 + \cdots + a_n$ is any arithmetic series, then we can find its sum using the same technique. Rewrite S_n as follows:

$$S_n = a_1 \qquad\quad + (a_1 + d) + (a_1 + 2d) + \cdots + a_n$$
$$\underline{S_n = a_n \qquad\quad + (a_n - d) + (a_n - 2d) + \cdots + a_1}$$
$$2S_n = (a_1 + a_n) + (a_1 + a_n) + (a_1 + a_n) + \cdots + (a_1 + a_n) \quad \text{Add.}$$

Because $(a_1 + a_n)$ appears n times on the right, we have $2S_n = n(a_1 + a_n)$. Divide each side by 2 to get the following formula.

Sum of an Arithmetic Series

The sum, S_n, of the first n terms of an arithmetic series with first term a_1 and nth term a_n, is given by

$$S_n = \frac{n}{2}(a_1 + a_n).$$

E X A M P L E   5

The sum of an arithmetic series

Find the sum of the positive integers from 1 to 100 inclusive.

Solution

The described series, $1 + 2 + 3 + \cdots + 100$, has 100 terms. So we can use $n = 100$, $a_1 = 1$, and $a_n = 100$ in the formula for the sum of an arithmetic series:

$$S_n = \frac{n}{2}(a_1 + a_n)$$

$$S_{100} = \frac{100}{2}(1 + 100)$$

$$= 50(101) = 5050$$

> **Now do Exercises 39–40**

‹ **Helpful Hint** ›

Legend has it that Carl F. Gauss knew this formula when he was in grade school. Gauss's teacher told him to add up the numbers from 1 through 100 for busy work. He immediately answered 5050.

E X A M P L E   6

The sum of an arithmetic series

Find the sum of the series

$$12 + 16 + 20 + \cdots + 84.$$

Solution

This series is an arithmetic series with $a_n = 84$, $a_1 = 12$, and $d = 4$. To get the number of terms, n, we use $a_n = a_1 + (n - 1)d$:

$$84 = 12 + (n - 1)4$$

$$84 = 8 + 4n$$

$$76 = 4n$$

$$19 = n$$

Now find the sum of these 19 terms:

$$S_{19} = \frac{19}{2}(12 + 84) = 912$$

Now do Exercises 41–52

Warm-Ups ▼

Fill in the blank.

1. A sequence in which each term after the first is obtained by adding a fixed amount to the previous term is a(n) _____ sequence.

2. For the sequence $a_n = a_1 + (n - 1)d$, d is the _____.

3. The indicated sum of an arithmetic sequence is an arithmetic _____.

4. The sum of the first n terms of a(n) _____ series with first term a_1 and nth term a_n is given by $S_n = \frac{n}{2}(a_1 + a_n)$.

True or false?

5. The common difference for 3, 1, −1, −3, . . . is 2.

6. The sequence 2, 4, 2, 4, 2, 4, . . . is an arithmetic sequence.

7. If $a_1 = 5$ and $a_3 = 10$ in an arithmetic sequence, then $a_4 = 15$.

8. If $a_1 = 6$ and $a_3 = 2$ in an arithmetic sequence, then $a_2 = 4$.

9. The series $\sum_{i=1}^{5}(3 + 2i)$ is an arithmetic series.

10. $\sum_{i=1}^{n} i = \frac{n(n + 1)}{2}$

Exercises 13.3

‹ Study Tips ›

- Write a summary of the topics that were covered in each chapter.
- Use the *Summary* that appears at the end of each chapter in this text as a guide.

‹ 1 › Arithmetic Sequences

Write a formula for the nth term of each arithmetic sequence. See Examples 1 and 2.

1. 2, 4, 6, 8, 10, . . .
2. 1, 3, 5, 7, 9, . . .
3. 0, 6, 12, 18, 24, . . .
4. 0, 5, 10, 15, 20, . . .
5. 7, 12, 17, 22, 27, . . .

6. 4, 15, 26, 37, 48, . . .
7. −4, −2, 0, 2, 4, . . .
8. −3, 0, 3, 6, 9, . . .
9. 5, 1, −3, −7, −11, . . .
10. 8, 5, 2, −1, −4, . . .
11. −2, −9, −16, −23, . . .
12. −5, −7, −9, −11, −13, . . .
13. −3, −2.5, −2, −1.5, −1, . . .
14. −2, −1.25, −0.5, 0.25, . . .

15. $-6, -6.5, -7, -7.5, -8, \ldots$

16. $1, 0.5, 0, -0.5, -1, \ldots$

Write the first five terms of the arithmetic sequence whose nth term is given. See Example 3.

17. $a_n = 9 + (n - 1)4$

18. $a_n = 13 + (n - 1)6$

19. $a_n = 7 + (n - 1)(-2)$

20. $a_n = 6 + (n - 1)(-3)$

21. $a_n = -4 + (n - 1)3$

22. $a_n = -19 + (n - 1)12$

23. $a_n = -2 + (n - 1)(-3)$

24. $a_n = -1 + (n - 1)(-2)$

25. $a_n = -4n - 3$

26. $a_n = -3n + 1$

27. $a_n = 0.5n + 4$

28. $a_n = 0.3n + 1$

29. $a_n = 20n + 1000$

30. $a_n = -600n + 4000$

Find the indicated part of each arithmetic sequence. See Example 4.

31. Find the eighth term of the sequence that has a first term of 9 and a common difference of 6.

32. Find the twelfth term of the sequence that has a first term of -2 and a common difference of -3.

33. Find the common difference if the first term is 6 and the twentieth term is 82.

34. Find the common difference if the first term is -8 and the ninth term is -64.

35. If the common difference is -2 and the seventh term is 14, then what is the first term?

36. If the common difference is 5 and the twelfth term is -7, then what is the first term?

37. Find the sixth term of the sequence that has a fifth term of 13 and a first term of -3.

38. Find the eighth term of the sequence that has a sixth term of -42 and a first term of 3.

⟨2⟩ Arithmetic Series

Find the sum of each given series. See Examples 5 and 6.

39. $1 + 2 + 3 + \cdots + 48$

40. $1 + 2 + 3 + \cdots + 12$

41. $8 + 10 + 12 + \cdots + 36$

42. $9 + 12 + 15 + \cdots + 72$

43. $-1 + (-7) + (-13) + \cdots + (-73)$

44. $-7 + (-12) + (-17) + \cdots + (-72)$

45. $-6 + (-1) + 4 + 9 + \cdots + 64$

46. $-9 + (-1) + 7 + \cdots + 103$

47. $20 + 12 + 4 + (-4) + \cdots + (-92)$

48. $19 + 1 + (-17) + \cdots + (-125)$

49. $\displaystyle\sum_{i=1}^{12} (3i - 7)$

50. $\displaystyle\sum_{i=1}^{7} (-4i + 6)$

51. $\displaystyle\sum_{i=1}^{11} (-5i + 2)$

52. $\displaystyle\sum_{i=1}^{19} (3i - 5)$

Applications

Solve each problem using the ideas of arithmetic sequences and series.

53. *Increasing salary.* If a lab technician has a salary of \$22,000 her first year and is due to get a \$500 raise each year, then what will her salary be in her seventh year?

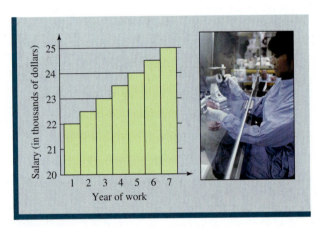

Figure for Exercise 53

54. *Seven years of salary.* What is the total salary for 7 years of work for the lab technician of Exercise 53?

55. *Light reading.* On the first day of October an English teacher suggests to his students that they read five pages of a novel and every day thereafter increase their daily reading by two pages. If his students follow this suggestion, then how many pages will they read during October?

56. *Heavy penalties.* If an air-conditioning system is not completed by the agreed upon date, the contractor pays a penalty of \$500 for the first day that it is overdue, \$600 for the second day, \$700 for the third day, and so on. If the system is completed 10 days late, then what is the total amount of the penalties that the contractor must pay?

Getting More Involved

57. *Discussion*

Which of the following sequences is not an arithmetic sequence? Explain your answer.

a) $\frac{1}{2}, 1, \frac{3}{2}, \ldots$

b) $\frac{1}{2}, \frac{1}{3}, \frac{1}{4}, \ldots$

c) $5, 0, -5, \ldots$

d) $2, 3, 4, \ldots$

58. *Discussion*

What is the smallest value of n for which

$$\sum_{i=1}^{n} \frac{i}{2} > 50?$$

Mid-Chapter Quiz | Sections 13.1 through 13.3 | Chapter 13

List all terms of each finite sequence.

1. $a_n = 3n - 4$ for $1 \leq n \leq 6$

2. $a_n = \dfrac{(-1)^n}{n^2}$ for $1 \leq n \leq 4$

List the first four terms of each infinite sequence.

3. $a_n = \dfrac{1}{n(n+1)}$

4. $a_n = (-1)^{2n-1}(n-1)^2$

Find a formula for the general term of each infinite sequence.

5. $3, 5, 7, 9, \ldots$

6. $4, -8, 12, -16, \ldots$

7. $1, 4, 9, 16, 25, \ldots$

Find the sum of each series.

8. $\displaystyle\sum_{i=1}^{3} (-i)^3$

9. $\displaystyle\sum_{i=1}^{10} (-1)^{i-1}$

10. $\displaystyle\sum_{i=1}^{4} i(i-2)(i-4)$

11. $\displaystyle\sum_{i=1}^{50} (2i+1)$

12. $0 + 2 + 4 + 6 + \cdots + 76$

13. $-2 + (-5) + (-8) + \cdots + (-32)$

Write each series in summation notation.

14. $2 + 4 + 6 + 8 + 10$

15. $1 + (-4) + 9 + (-16) + 25$

Write a formula for the nth term of each arithmetic sequence.

16. $0, 4, 8, 12, 16, 20, \ldots$

17. $6, 1, -4, -9, -14, \ldots$

Miscellaneous.

18. Rewrite the series $\displaystyle\sum_{j=5}^{20} (2j-3)$ in summation notation with i starting at 1.

19. Find the ninth term of the arithmetic sequence that has a first term of 3 and a common difference of 5.

20. Find the fifth term of the arithmetic sequence that has a first term of 9 and an eleventh term of 49.

13.4   Geometric Sequences and Series

In This Section

⟨1⟩ **Geometric Sequences**

⟨2⟩ **Finite Geometric Series**

⟨3⟩ **Infinite Geometric Series**

⟨4⟩ **Applications**

In Section 13.3, you studied the arithmetic sequences and series. In this section, you will study sequences in which each term is a *multiple* of the term preceding it. You will also learn how to find the sum of the corresponding series.

⟨1⟩ **Geometric Sequences**

In an arithmetic sequence such as $2, 4, 6, 8, 10, \ldots$ there is a common difference between consecutive terms. In a geometric sequence there is a common ratio between consecutive terms. The following table contains several geometric sequences and the common ratios between consecutive terms.

Geometric Sequence	Common Ratio
$3, 6, 12, 24, 48, \ldots$	2
$27, 9, 3, 1, \dfrac{1}{3}, \ldots$	$\dfrac{1}{3}$
$1, -10, 100, -1000, \ldots$	-10

Note that every term after the first term of each geometric sequence can be obtained by multiplying the previous term by the common ratio.

> **Geometric Sequence**
>
> A sequence in which each term after the first is obtained by multiplying the preceding term by a constant is called a **geometric sequence.**

The constant is denoted by the letter r and is called the **common ratio.** If a_1 is the first term, then the second term is $a_1 r$. The third term is $a_1 r^2$, the fourth term is $a_1 r^3$, and so on. We can write a formula for the nth term of a geometric sequence by following this pattern.

> **Formula for the nth Term of a Geometric Sequence**
>
> The nth term, a_n, of a geometric sequence with first term a_1 and common ratio r is
> $$a_n = a_1 r^{n-1}.$$

The first term and the common ratio determine all of the terms of a geometric sequence.

EXAMPLE 1

Finding the nth term

Write a formula for the nth term of the geometric sequence

$$6, 2, \frac{2}{3}, \frac{2}{9}, \ldots.$$

Solution

We can obtain the common ratio by dividing any term after the first by the term preceding it. So,

$$r = 2 \div 6 = \frac{1}{3}.$$

Because each term after the first is $\frac{1}{3}$ of the term preceding it, the nth term is given by

$$a_n = 6\left(\frac{1}{3}\right)^{n-1}.$$

Check a few terms: $a_1 = 6\left(\frac{1}{3}\right)^{1-1} = 6$, $a_2 = 6\left(\frac{1}{3}\right)^{2-1} = 2$, and $a_3 = 6\left(\frac{1}{3}\right)^{3-1} = \frac{2}{3}$.

> **Now do Exercises 1–6**

EXAMPLE **2**

Finding the *n*th term

Find a formula for the *n*th term of the geometric sequence

$$2, -1, \frac{1}{2}, -\frac{1}{4}, \ldots .$$

Solution

We obtain the ratio by dividing a term by the term preceding it:

$$r = -1 \div 2 = -\frac{1}{2}$$

Each term after the first is obtained by multiplying the preceding term by $-\frac{1}{2}$. The formula for the *n*th term is

$$a_n = 2\left(-\frac{1}{2}\right)^{n-1}.$$

Check a few terms: $a_1 = 2\left(-\frac{1}{2}\right)^{1-1} = 2$, $a_2 = 2\left(-\frac{1}{2}\right)^{2-1} = -1$, and $a_3 = 2\left(-\frac{1}{2}\right)^{3-1} = \frac{1}{2}$.

Now do Exercises 7–12

In Example 3, we use the formula for the *n*th term to write some terms of a geometric sequence.

EXAMPLE **3**

Writing the terms

Write the first five terms of the geometric sequence whose *n*th term is

$$a_n = 3(-2)^{n-1}.$$

Solution

Let *n* take the values 1 through 5 in the formula for the *n*th term:

$$a_1 = 3(-2)^{1-1} = 3$$
$$a_2 = 3(-2)^{2-1} = -6$$
$$a_3 = 3(-2)^{3-1} = 12$$
$$a_4 = 3(-2)^{4-1} = -24$$
$$a_5 = 3(-2)^{5-1} = 48$$

Notice that $a_n = 3(-2)^{n-1}$ gives the general term for a geometric sequence with first term 3 and common ratio -2. Because every term after the first can be obtained by multiplying the previous term by -2, the terms 3, -6, 12, -24, and 48 are correct.

Now do Exercises 13–20

The formula for the *n*th term involves four variables: a_n, a_1, r, and n. If we know the value of any three of them, we can find the value of the fourth.

EXAMPLE 4

Finding a missing term

Find the first term of a geometric sequence whose fourth term is 8 and whose common ratio is $\frac{1}{2}$.

Solution

Let $a_4 = 8$, $r = \frac{1}{2}$, and $n = 4$ in the formula $a_n = a_1 r^{n-1}$:

$$8 = a_1 \left(\frac{1}{2}\right)^{4-1}$$

$$8 = a_1 \cdot \frac{1}{8}$$

$$64 = a_1$$

So the first term is 64.

Now do Exercises 21–26

⟨2⟩ Finite Geometric Series

Consider the following series:

$$1 + 2 + 4 + 8 + 16 + \cdots + 512$$

The terms of this series are the terms of a finite geometric sequence. The indicated sum of a geometric sequence is called a **geometric series.**

We can find the actual sum of this finite geometric series by using a technique similar to the one used for the sum of an arithmetic series. Let

$$S = 1 + 2 + 4 + 8 + \cdots + 256 + 512.$$

Because the common ratio is 2, multiply each side by -2:

$$-2S = -2 - 4 - 8 - \cdots - 512 - 1024$$

Adding the last two equations eliminates all but two of the terms on the right:

$$
\begin{array}{ll}
S = 1 + 2 + 4 + 8 + \cdots + 256 + 512 & \\
\underline{-2S = \quad -2 - 4 - 8 - \cdots \qquad\qquad - 512 - 1024} & \\
-S = 1 \qquad\qquad\qquad\qquad\qquad\qquad - 1024 & \text{Add.} \\
-S = -1023 & \\
S = 1023 &
\end{array}
$$

If $S_n = a_1 + a_1 r + a_1 r^2 + \cdots + a_1 r^{n-1}$ is any geometric series, we can find the sum in the same manner. Multiplying each side of this equation by $-r$ yields

$$-rS_n = -a_1 r - a_1 r^2 - a_1 r^3 - \cdots - a_1 r^n.$$

If we add S_n and $-rS_n$, all but two of the terms on the right are eliminated:

$$
\begin{array}{ll}
S_n = a_1 + a_1 r + a_1 r^2 + \cdots \qquad\qquad + a_1 r^{n-1} & \\
\underline{-rS_n = \quad -a_1 r - a_1 r^2 - a_1 r^3 - \cdots \qquad\qquad - a_1 r^n} & \\
S_n - rS_n = a_1 \qquad\qquad\qquad\qquad\qquad\qquad - a_1 r^n & \text{Add.} \\
(1 - r)S_n = a_1(1 - r^n) & \text{Factor out common factors.}
\end{array}
$$

Now divide each side of this equation by $1 - r$ to get the formula for S_n.

> ### Sum of n Terms of a Geometric Series
> If S_n represents the sum of the first n terms of a geometric series with first term a_1 and common ratio r ($r \neq 1$), then
> $$S_n = \frac{a_1(1 - r^n)}{1 - r}.$$

EXAMPLE 5

The sum of a finite geometric series

Find the sum of the series

$$\frac{1}{3} + \frac{1}{9} + \frac{1}{27} + \cdots + \frac{1}{729}.$$

Solution

The first term is $\frac{1}{3}$, and the common ratio is $\frac{1}{3}$. So the nth term can be written as

$$a_n = \frac{1}{3}\left(\frac{1}{3}\right)^{n-1}.$$

We can use this formula to find the number of terms in the series:

$$\frac{1}{729} = \frac{1}{3}\left(\frac{1}{3}\right)^{n-1}$$

$$\frac{1}{729} = \left(\frac{1}{3}\right)^{n}$$

Because $3^6 = 729$, we have $n = 6$. (Of course, you could use logarithms to solve for n.) Now use the formula for the sum of six terms of this geometric series:

$$S_6 = \frac{\frac{1}{3}\left[1 - \left(\frac{1}{3}\right)^6\right]}{1 - \frac{1}{3}} = \frac{\frac{1}{3}\left[1 - \frac{1}{729}\right]}{\frac{2}{3}}$$

$$= \frac{1}{3} \cdot \frac{728}{729} \cdot \frac{3}{2}$$

$$= \frac{364}{729}$$

> **Now do Exercises 27–32**

EXAMPLE 6

The sum of a finite geometric series

Find the sum of the series

$$\sum_{i=1}^{12} 3(-2)^{i-1}.$$

Solution

This series is geometric with first term 3, ratio -2, and $n = 12$. We use the formula for the sum of the first 12 terms of a geometric series:

$$S_{12} = \frac{3[1 - (-2)^{12}]}{1 - (-2)} = \frac{3[-4095]}{3} = -4095.$$

> **Now do Exercises 33–38**

‹ Calculator Close-Up ›

Experiment with your calculator to see what happens to r^n as n gets larger and larger.

```
.99^100
            .3660323413
.99^1000
      4.317124741E-5
.99^10000
      2.24877485E-44
```

⟨3⟩   Infinite Geometric Series

Consider how a very large value of n affects the formula for the sum of a finite geometric series,

$$S_n = \frac{a_1(1 - r^n)}{1 - r}.$$

If $|r| < 1$, then the value of r^n gets closer and closer to 0 as n gets larger and larger. For example, if $r = \frac{2}{3}$ and $n = 10, 20,$ and 100, then

$$\left(\frac{2}{3}\right)^{10} \approx 0.0173415, \quad \left(\frac{2}{3}\right)^{20} \approx 0.0003007, \quad \text{and} \quad \left(\frac{2}{3}\right)^{100} \approx 2.460 \times 10^{-18}.$$

Because r^n is approximately 0 for large values of n, $1 - r^n$ is approximately 1. If we replace $1 - r^n$ by 1 in the expression for S_n, we get

$$S_n \approx \frac{a_1}{1 - r}.$$

So as n gets larger and larger, the sum of the first n terms of the infinite geometric series

$$a_1 + a_1 r + a_1 r^2 + \cdots$$

gets closer and closer to $\frac{a_1}{1 - r}$, provided that $|r| < 1$. Therefore, we say that $\frac{a_1}{1 - r}$ is the sum of *all* of the terms of the infinite geometric series.

Sum of an Infinite Geometric Series

If $a_1 + a_1 r + a_1 r^2 + \cdots$ is an infinite geometric series, with $|r| < 1$, then the sum S of all of the terms of this series is given by

$$S = \frac{a_1}{1 - r}.$$

E X A M P L E   7

Sum of an infinite geometric series

Find the sum

$$\frac{1}{2} + \frac{1}{4} + \frac{1}{8} + \frac{1}{16} + \cdots.$$

‹ Helpful Hint ›

You can imagine this series in a football game. The Bears have the ball on the Lions' 1-yard line. The Lions continually get penalties that move the ball one-half of the distance to the goal. Theoretically, the ball will never reach the goal, but the total distance it moves will get closer and closer to 1 yard.

Solution

This series is an infinite geometric series with $a_1 = \frac{1}{2}$ and $r = \frac{1}{2}$. Because $r < 1$, we have

$$S = \frac{\dfrac{1}{2}}{1 - \dfrac{1}{2}} = 1.$$

> **Now do Exercises 39–44**

For an infinite series the index of summation i takes the values 1, 2, 3, and so on, without end. To indicate that the values for i keep increasing without bound, we say that i *takes the values from 1 through* ∞ (infinity). Note that the symbol "∞" does not represent a number. Using the ∞ symbol, we can write the indicated sum of an infinite geometric series (with $|r| < 1$) by using summation notation as follows:

$$a_1 + a_1 r + a_1 r^2 + \cdots = \sum_{i=1}^{\infty} a_1 r^{i-1}$$

E X A M P L E   8

Sum of an infinite geometric series

Find the value of the sum

$$\sum_{i=1}^{\infty} 8 \left(\frac{3}{4} \right)^{i-1}.$$

Solution

This series is an infinite geometric series with first term 8 and common ratio $\frac{3}{4}$. So,

$$S = \frac{8}{1 - \dfrac{3}{4}} = 8 \cdot \frac{4}{1} = 32.$$

> **Now do Exercises 45–52**

⟨4⟩ Applications

E X A M P L E   9

Follow the bouncing ball

Suppose a ball always rebounds $\frac{2}{3}$ of the height from which it falls and the ball is dropped from a height of 6 feet. Find the total distance that the ball travels.

Solution

The ball falls 6 feet (ft) and rebounds 4 ft; it then falls 4 ft and rebounds $\frac{8}{3}$ ft. The following series gives the total distance that the ball falls:

$$F = 6 + 4 + \frac{8}{3} + \frac{16}{9} + \cdots$$

The distance that the ball rebounds is given by the following series:

$$R = 4 + \frac{8}{3} + \frac{16}{9} + \cdots$$

Each of these series is an infinite geometric series with ratio $\frac{2}{3}$. Use the formula for an infinite geometric series to find each sum:

$$F = \frac{6}{1 - \dfrac{2}{3}} = 6 \cdot \frac{3}{1} = 18 \text{ ft}, \qquad R = \frac{4}{1 - \dfrac{2}{3}} = 4 \cdot \frac{3}{1} = 12 \text{ ft}$$

The total distance traveled by the ball is the sum of F and R, 30 ft.

> **Now do Exercises 53–54**

One of the most important applications of geometric series is in calculating the value of an annuity. An **annuity** is a sequence of periodic payments. The payments might be loan payments or investments.

E X A M P L E **10**

Value of an annuity

A deposit of $1000 is made at the beginning of each year for 30 years and earns 6% interest compounded annually. What is the value of this annuity at the end of the thirtieth year?

Solution

The last deposit earns interest for only 1 year. So at the end of the thirtieth year it amounts to $1000(1.06). The next to last deposit earns interest for 2 years and amounts to $1000(1.06)^2$. The first deposit earns interest for 30 years and amounts to $1000(1.06)^{30}$. So the value of the annuity at the end of the thirtieth year is the sum of the finite geometric series

$$1000(1.06) + 1000(1.06)^2 + 1000(1.06)^3 + \cdots + 1000(1.06)^{30}.$$

Use the formula for the sum of 30 terms of a finite geometric series with $a_1 = 1000(1.06)$ and $r = 1.06$:

$$S_{30} = \frac{1000(1.06)\left(1 - (1.06)^{30}\right)}{1 - 1.06} \approx \$83{,}801.68$$

So 30 annual deposits of $1000 each amount to $83,801.68.

> Now do Exercises 55–58

 ## Warm-Ups ▼

Fill in the blank.

1. A(n) _____ sequence is one in which each term after the first is obtained by multiplying the preceding term by a constant.

2. The nth term of a geometric sequence is $a_1 r^{n-1}$, where a_1 is the first term and r is the _____.

3. A geometric _____ is the indicated sum of a geometric sequence.

4. The sum of the first n terms of a _____ series with first term a_1 and common ratio r is given by $S_n = \dfrac{a_1(1 - r^n)}{1 - r}$.

5. The sum of an _____ geometric series with first term a_1 and common ratio r is given by $S = \dfrac{a_1}{1 - r}$ provided $|r| < 1$.

True or false?

6. The sequence 2, 6, 24, 120, . . . is a geometric sequence.

7. The sequence $a_n = 2^n$ is geometric.

8. The common ratio for $a_n = 3(0.5)^{n-1}$ is 0.5.

9. The first term for $a_n = 3(2)^{-n+3}$ is 12.

10. The common ratio for $a_n = 3(2)^{-n+3}$ is $\dfrac{1}{2}$.

11. $\displaystyle\sum_{i=1}^{5} 6(0.3)^{i-1} = \dfrac{6(1 - (0.3)^5)}{1 - 0.3}$

12. $10 + 5 + \dfrac{5}{2} + \cdots = \dfrac{10}{1 - 0.5}$

‹1› Geometric Sequences

Write a formula for the nth term of each geometric sequence. See Examples 1 and 2.

1. $1, 2, 4, 8, \ldots$

2. $1, 3, 9, 27, \ldots$

3. $\frac{1}{3}, 1, 3, 9, \ldots$

4. $\frac{1}{4}, 2, 16, \ldots$

5. $64, 8, 1, \ldots$

6. $100, 10, 1, \ldots$

7. $8, -4, 2, -1, \ldots$

8. $-9, 3, -1, \ldots$

9. $2, -4, 8, -16, \ldots$

10. $-\frac{1}{2}, 2, -8, 32, \ldots$

11. $-\frac{1}{3}, -\frac{1}{4}, -\frac{3}{16}, \ldots$

12. $-\frac{1}{4}, -\frac{1}{5}, -\frac{4}{25}, \ldots$

Write the first five terms of the geometric sequence with the given nth term. See Example 3.

13. $a_n = 2\left(\frac{1}{3}\right)^{n-1}$

14. $a_n = -5\left(\frac{1}{2}\right)^{n-1}$

15. $a_n = (-2)^{n-1}$

16. $a_n = \left(-\frac{1}{3}\right)^{n-1}$

17. $a_n = 2^{-n}$

18. $a_n = 3^{-n}$

19. $a_n = (0.78)^n$

20. $a_n = (-0.23)^n$

Find the required part of each geometric sequence. See Example 4.

21. Find the first term of the geometric sequence that has fourth term 40 and common ratio 2.

22. Find the first term of the geometric sequence that has fifth term 4 and common ratio $\frac{1}{2}$.

23. Find r for the geometric sequence that has $a_1 = 6$ and $a_4 = \frac{2}{9}$.

24. Find r for the geometric sequence that has $a_1 = 1$ and $a_4 = -27$.

25. Find a_4 for the geometric sequence that has $a_1 = -3$ and $r = \frac{1}{3}$.

26. Find a_5 for the geometric sequence that has $a_1 = -\frac{2}{3}$ and $r = -\frac{2}{3}$.

‹2› Finite Geometric Series

Find the sum of each geometric series. See Examples 5 and 6.

27. $\frac{1}{2} + \frac{1}{4} + \frac{1}{8} + \cdots + \frac{1}{512}$

28. $1 + \frac{1}{3} + \frac{1}{9} + \cdots + \frac{1}{81}$

29. $\frac{1}{2} - \frac{1}{4} + \frac{1}{8} - \frac{1}{16} + \frac{1}{32}$

30. $3 - 1 + \frac{1}{3} - \frac{1}{9} + \frac{1}{27} - \frac{1}{81}$

31. $30 + 20 + \frac{40}{3} + \cdots + \frac{1280}{729}$

32. $9 - 6 + 4 - \cdots - \frac{128}{243}$

33. $\sum_{i=1}^{10} 5(2)^{i-1}$

34. $\sum_{i=1}^{7} (10{,}000)(0.1)^{i-1}$

35. $\sum_{i=1}^{6} (0.1)^i$

36. $\sum_{i=1}^{5} (0.2)^i$

37. $\sum_{i=1}^{6} 100(0.3)^i$

38. $\sum_{i=1}^{7} 36(0.5)^i$

⟨3⟩ Infinite Geometric Series

Find the sum of each infinite geometric series. See Examples 7 and 8.

39. $\dfrac{1}{8} + \dfrac{1}{16} + \dfrac{1}{32} + \cdots$

40. $\dfrac{1}{9} + \dfrac{1}{27} + \dfrac{1}{81} + \cdots$

41. $3 + 2 + \dfrac{4}{3} + \cdots$

42. $2 + 1 + \dfrac{1}{2} + \cdots$

43. $4 - 2 + 1 - \dfrac{1}{2} + \cdots$

44. $16 - 12 + 9 - \dfrac{27}{4} + \cdots$

45. $\displaystyle\sum_{i=1}^{\infty} (0.3)^i$

46. $\displaystyle\sum_{i=1}^{\infty} (0.2)^i$

47. $\displaystyle\sum_{i=1}^{\infty} 3(0.5)^{i-1}$

48. $\displaystyle\sum_{i=1}^{\infty} 7(0.4)^{i-1}$

49. $\displaystyle\sum_{i=1}^{\infty} 3(0.1)^i$

50. $\displaystyle\sum_{i=1}^{\infty} 6(0.1)^i$

51. $\displaystyle\sum_{i=1}^{\infty} 12(0.01)^i$

52. $\displaystyle\sum_{i=1}^{\infty} 72(0.01)^i$

⟨4⟩ Applications

Use the ideas of geometric series to solve each problem. See Examples 9 and 10.

53. *Retirement fund.* Suppose a deposit of $2000 is made at the beginning of each year for 45 years into an account paying 12% compounded annually. What is the amount of this annuity at the end of the forty-fifth year?

54. *World's largest mutual fund.* If you had invested $5000 at the beginning of each year for the past 10 years in the Fidelity's Magellan Fund, you would have averaged 12.46% compounded annually (www.fidelity.com). Find the amount of this annuity at the end of the tenth year.

Growth of $5000/year

Figure for Exercise 54

55. *Big saver.* Suppose you deposit one cent into your piggy bank on the first day of December and, on each day of December after that, you deposit twice as much as on the previous day. How much will you have in the bank after the last deposit?

56. *Big family.* Consider yourself, your parents, your grandparents, your great-grandparents, your great-great-grandparents, and so on, back to your grandparents with the word "great" used in front 40 times. What is the total number of people you are considering?

57. *Total economic impact.* In Exercise 41 of Section 13.1 we described a factory that spends $1 million annually in a community in which 80% of the money received is respent in the community. Economists assume the money is respent again and again at the 80% rate. The total economic impact of the factory is the total of all of this spending. Find an approximation for the total by using the formula for the sum of an infinite geometric series with a rate of 80%.

58. *Less impact.* Repeat Exercise 57, assuming money is respent again and again at the 50% rate.

Getting More Involved

59. *Discussion*

Which of the following sequences is not a geometric sequence? Explain your answer.

a) $1, 2, 4, \ldots$ **b)** $0.1, 0.01, 0.001, \ldots$

c) $-1, 2, -4, \ldots$ **d)** $2, 4, 6, \ldots$

60. *Discussion*

The repeating decimal number 0.44444 . . . can be written as

$$\frac{4}{10} + \frac{4}{100} + \frac{4}{1000} + \cdots,$$

an infinite geometric series. Find the sum of this geometric series.

61. *Discussion*

Write the repeating decimal number 0.24242424 . . . as an infinite geometric series. Find the sum of the geometric series.

13.5   Binomial Expansions

In Chapter 5, you learned how to square a binomial. In this section, you will study higher powers of binomials.

In This Section

⟨1⟩ **Some Examples**

⟨2⟩ **Obtaining the Coefficients**

⟨3⟩ **The Binomial Theorem**

⟨1⟩ Some Examples

We know that $(x + y)^2 = x^2 + 2xy + y^2$. To find $(x + y)^3$, we multiply $(x + y)^2$ by $x + y$:

$$(x + y)^3 = \left(x^2 + 2xy + y^2\right)(x + y)$$
$$= \left(x^2 + 2xy + y^2\right)x + \left(x^2 + 2xy + y^2\right)y$$
$$= x^3 + 2x^2y + xy^2 + x^2y + 2xy^2 + y^3$$
$$= x^3 + 3x^2y + 3xy^2 + y^3$$

The sum $x^3 + 3x^2y + 3xy^2 + y^3$ is called the **binomial expansion** of $(x + y)^3$. If we again multiply by $x + y$, we will get the binomial expansion of $(x + y)^4$. This method is rather tedious. However, if we examine these expansions, we can find a pattern and learn how to find binomial expansions without multiplying.

Consider the following binomial expansions:

$$(x + y)^0 = 1$$
$$(x + y)^1 = x + y$$
$$(x + y)^2 = x^2 + 2xy + y^2$$
$$(x + y)^3 = x^3 + 3x^2y + 3xy^2 + y^3$$
$$(x + y)^4 = x^4 + 4x^3y + 6x^2y^2 + 4xy^3 + y^4$$
$$(x + y)^5 = x^5 + 5x^4y + 10x^3y^2 + 10x^2y^3 + 5xy^4 + y^5$$

Observe that the exponents on the variable x are decreasing, whereas the exponents on the variable y are increasing, as we read from left to right. Also notice that the sum of the exponents in each term is the same for that entire line. For instance, in the fourth expansion the terms x^4, x^3y, x^2y^2, xy^3, and y^4 all have exponents with a sum of 4. If we continue the pattern, the expansion of $(x + y)^6$ will have seven terms containing x^6, x^5y, x^4y^2, x^3y^3, x^2y^4, xy^5, and y^6. Now we must find the pattern for the coefficients of these terms.

⟨2⟩ Obtaining the Coefficients

If we write out only the coefficients of the expansions that we already have, we can easily see a pattern. This triangular array of coefficients for the binomial expansions is called **Pascal's triangle.**

$$
\begin{array}{ccccccccccc}
 & & & & & 1 & & & & & \\
 & & & & 1 & & 1 & & & & \\
 & & & 1 & & 2 & & 1 & & & \\
 & & 1 & & 3 & & 3 & & 1 & & \\
 & 1 & & 4 & & 6 & & 4 & & 1 & \\
1 & & 5 & & 10 & & 10 & & 5 & & 1
\end{array}
$$

$(x + y)^0 = 1$
$(x + y)^1 = 1x + 1y$
$(x + y)^2 = 1x^2 + 2xy + 1y^2$
$(x + y)^3 = 1x^3 + 3x^2y + 3xy^2 + 1y^3$
$(x + y)^4 = 1x^4 + 4x^3y + 6x^2y^2 + 4xy^3 + 1y^4$
Coefficients in $(x + y)^5$

Notice that each line starts and ends with a 1 and that each entry of a line is the sum of the two entries above it in the previous line. For instance, $4 = 3 + 1$, and $10 = 6 + 4$. Following this pattern, the sixth and seventh lines of coefficients are

$$
\begin{array}{ccccccccc}
1 & & 6 & & 15 & & 20 & & 15 & & 6 & & 1 \\
1 & & 7 & & 21 & & 35 & & 35 & & 21 & & 7 & & 1.
\end{array}
$$

Pascal's triangle gives us an easy way to get the coefficients for the binomial expansion with small powers, but it is impractical for larger powers. For larger powers we use a formula involving **factorial notation.**

> ### *n*! (*n* factorial)
> If n is a positive integer, $n!$ (read "n factorial") is defined to be the product of all of the positive integers from 1 through n. We also define $0!$ to be 1.

‹ **Calculator Close-Up** ›

You can evaluate the coefficients using either the factorial notation or $_nC_r$. The factorial symbol and $_nC_r$ are found in the MATH menu under PRB.

```
4!/(1!3!)
              4
4 nCr 3
              4
```

For example, $3! = 3 \cdot 2 \cdot 1 = 6$, and $5! = 5 \cdot 4 \cdot 3 \cdot 2 \cdot 1 = 120$.

Before we state a general formula, consider how the coefficients for $(x + y)^4$ are found by using factorials:

$$\frac{4!}{4!\,0!} = \frac{4 \cdot 3 \cdot 2 \cdot 1}{4 \cdot 3 \cdot 2 \cdot 1 \cdot 1} = 1 \quad \text{Coefficient of } x^4 \text{ (or } x^4y^0\text{)}$$

$$\frac{4!}{3!\,1!} = \frac{4 \cdot 3 \cdot 2 \cdot 1}{3 \cdot 2 \cdot 1 \cdot 1} = 4 \quad \text{Coefficient of } x^3y$$

$$\frac{4!}{2!\,2!} = \frac{4 \cdot 3 \cdot 2 \cdot 1}{2 \cdot 1 \cdot 2 \cdot 1} = 6 \quad \text{Coefficient of } x^2y^2$$

$$\frac{4!}{1!\,3!} = \frac{4 \cdot 3 \cdot 2 \cdot 1}{1 \cdot 3 \cdot 2 \cdot 1} = 4 \quad \text{Coefficient of } xy^3$$

$$\frac{4!}{0!\,4!} = \frac{4 \cdot 3 \cdot 2 \cdot 1}{1 \cdot 4 \cdot 3 \cdot 2 \cdot 1} = 1 \quad \text{Coefficient of } y^4 \text{ (or } x^0y^4\text{)}$$

Note that each expression has $4!$ in the numerator, with factorials in the denominator corresponding to the exponents on x and y.

‹3› The Binomial Theorem

We now summarize these ideas in the **binomial theorem.**

> ### The Binomial Theorem
> In the expansion of $(x + y)^n$ for a positive integer n, there are $n + 1$ terms, given by the following formula:
>
> $$(x + y)^n = \frac{n!}{n!\,0!}x^n + \frac{n!}{(n-1)!\,1!}x^{n-1}y + \frac{n!}{(n-2)!\,2!}x^{n-2}y^2 + \cdots + \frac{n!}{0!\,n!}y^n$$

The notation $\binom{n}{r}$ is often used in place of $\frac{n!}{(n-r)!\,r!}$ in the binomial expansion. Using this notation, we write the expansion as

$$(x + y)^n = \binom{n}{0}x^n + \binom{n}{1}x^{n-1}y + \binom{n}{2}x^{n-2}y^2 + \cdots + \binom{n}{n}y^n.$$

Another notation for $\frac{n!}{(n-r)!r!}$ is $_nC_r$. Using this notation, we have

$$(x + y)^n = {}_nC_0 x^n + {}_nC_1 x^{n-1}y + {}_nC_2 x^{n-2}y^2 + \cdots + {}_nC_n y^n.$$

E X A M P L E   **1**

Calculating the binomial coefficients

Evaluate each expression.

a) $\dfrac{7!}{4!\,3!}$

b) $\dfrac{10!}{8!\,2!}$

Solution

a) $\dfrac{7!}{4!\,3!} = \dfrac{7 \cdot 6 \cdot 5 \cdot 4 \cdot 3 \cdot 2 \cdot 1}{4 \cdot 3 \cdot 2 \cdot 1 \cdot 3 \cdot 2 \cdot 1} = \dfrac{7 \cdot 6 \cdot 5}{3 \cdot 2 \cdot 1} = 35$

b) $\dfrac{10!}{8!\,2!} = \dfrac{10 \cdot 9 \cdot 8 \cdot 7 \cdot 6 \cdot 5 \cdot 4 \cdot 3 \cdot 2 \cdot 1}{8 \cdot 7 \cdot 6 \cdot 5 \cdot 4 \cdot 3 \cdot 2 \cdot 1 \cdot 2 \cdot 1} = \dfrac{10 \cdot 9}{2 \cdot 1} = 45$

> **Now do Exercises 1–6**

E X A M P L E   **2**

Using the binomial theorem

Write out the first three terms of $(x + y)^9$.

Solution

$$(x + y)^9 = \frac{9!}{9!\,0!}x^9 + \frac{9!}{8!\,1!}x^8 y + \frac{9!}{7!\,2!}x^7 y^2 + \cdots = x^9 + 9x^8 y + 36x^7 y^2 + \cdots$$

> **Now do Exercises 7–12**

E X A M P L E   **3**

Using the binomial theorem

Write the binomial expansion for $(x^2 - 2a)^5$.

Solution

We expand a difference by writing it as a sum and using the binomial theorem:

$$(x^2 - 2a)^5 = \left(x^2 + (-2a)\right)^5$$

$$= \frac{5!}{5!\,0!}(x^2)^5 + \frac{5!}{4!\,1!}(x^2)^4(-2a)^1 + \frac{5!}{3!\,2!}(x^2)^3(-2a)^2 + \frac{5!}{2!\,3!}(x^2)^2(-2a)^3$$

$$+ \frac{5!}{1!\,4!}(x^2)^1(-2a)^4 + \frac{5!}{0!\,5!}(-2a)^5$$

$$= x^{10} - 10x^8 a + 40x^6 a^2 - 80x^4 a^3 + 80x^2 a^4 - 32a^5$$

> **Now do Exercises 13–28**

EXAMPLE **4**

Finding a specific term

Find the fourth term of the expansion of $(a + b)^{12}$.

Solution

The variables in the first term are $a^{12}b^0$, those in the second term are $a^{11}b^1$, those in the third term are $a^{10}b^2$, and those in the fourth term are a^9b^3. So,

$$\frac{12!}{9!\,3!}a^9b^3 = 220a^9b^3.$$

The fourth term is $220a^9b^3$.

Now do Exercises 29–32

‹ **Calculator Close-Up** ›

Because $_nC_r = \frac{n!}{(n-r)!\,r!}$, we have

$$_{12}C_9 = \frac{12!}{3!\,9!} \quad \text{and} \quad _{12}C_3 = \frac{12!}{9!\,3!}.$$

So there is more than one way to compute $12!/(9!\,3!)$:

```
12!/(9!3!)
                   220
12 nCr 9
                   220
12 nCr 3
                   220
```

Using the ideas of Example 4, we can write a formula for any term of a binomial expansion.

Formula for the *k*th Term of $(x + y)^n$

For k ranging from 1 to $n + 1$, the kth term of the expansion of $(x + y)^n$ is given by the formula

$$\frac{n!}{(n - k + 1)!(k - 1)!}x^{n-k+1}y^{k-1}.$$

EXAMPLE **5**

Finding a specific term

Find the sixth term of the expansion of $\left(a^2 - 2b\right)^7$.

Solution

If $k = 6$ and $n = 7$, then $n - k + 1 = 2$ and $k - 1 = 5$. Use these values in the formula for the kth term:

$$\frac{7!}{2!\,5!}(a^2)^2(-2b)^5 = 21a^4(-32b^5) = -672a^4b^5$$

Now do Exercises 33–36

We can think of the binomial expansion as a finite series. Using summation notation, we can write the binomial theorem as follows.

The Binomial Theorem (Using Summation Notation)

For any positive integer n,

$$(x + y)^n = \sum_{i=0}^{n} \frac{n!}{(n - i)!\,i!}x^{n-i}y^i \quad \text{or} \quad (x + y)^n = \sum_{i=0}^{n} \binom{n}{i}x^{n-i}y^i.$$

EXAMPLE   **6**

Using summation notation

Write $(a + b)^5$ using summation notation.

Solution

Use $n = 5$ in the binomial theorem:

$$(a + b)^5 = \sum_{i=0}^{5} \frac{5!}{(5 - i)!\,i!} a^{5-i} b^i$$

Now do Exercises 37–40

Warm-Ups ▼

Fill in the blank.

1. The sum obtained for a power of a binomial is a(n)
   _____.

2. _____ triangle gives the coefficients for $(a + b)^n$ for $n = 1, 2, 3$, and so on.

3. The product of the positive integers from 1 through n is called n _____.

4. The _____ theorem indicates how to expand $(a + b)^n$.

True or false?

5. There are 12 terms in the expansion of $(a + b)^{12}$.

6. The sum of the coefficients in the expansion of $(a + b)^4$ is 16.

7. The sum of the coefficients in the expansion of $(a + b)^n$ is n^2.

8. $(a + b)^3 = \sum_{i=0}^{3} \frac{3!}{(3 - i)!\,i!} a^{3-i} b^i$

9. $0! = 1!$

10. $\dfrac{7!}{5!\,2!} = 21$

Exercises 13.5

‹ **Study Tips** ›

- You cannot read a math book like you read a novel. Every sentence says something about the subject.
- You can relax and read a novel passively, but a math text requires more concentration and retention.

‹ 3 › **The Binomial Theorem**

Evaluate each expression. See Example 1.

1. $\dfrac{4!}{4!\,0!}$

2. $\dfrac{5!}{5!\,0!}$

3. $\dfrac{5!}{2!\,3!}$

4. $\dfrac{6!}{5!\,1!}$

5. $\dfrac{8!}{5!\,3!}$

6. $\dfrac{9!}{2!\,7!}$

Use the binomial theorem to expand each binomial. See Examples 2 and 3.

7. $(x + 1)^3$

8. $(y + 1)^4$

9. $(a + 2)^3$

10. $(b + 3)^3$

11. $(r + t)^5$

12. $(r + t)^6$

13. $(m - n)^3$

14. $(m - n)^4$

15. $(x + 2a)^3$

16. $(a + 3b)^4$

17. $(x^2 - 2)^4$

18. $(x^2 - a^2)^5$

 19. $(x - 1)^7$

20. $(x + 1)^6$

Write out the first four terms in the expansion of each binomial. See Examples 2 and 3.

21. $(a - 3b)^{12}$

22. $(x - 2y)^{10}$

23. $(x^2 + 5)^9$

24. $(x^2 + 1)^{20}$

25. $(x - 1)^{22}$

26. $(2x - 1)^8$

27. $\left(\dfrac{x}{2} + \dfrac{y}{3}\right)^{10}$

28. $\left(\dfrac{a}{2} + \dfrac{b}{5}\right)^8$

Find the indicated term of the binomial expansion. See Examples 4 and 5.

29. $(a + w)^{13}$, 6th term

30. $(m + n)^{12}$, 7th term

31. $(m - n)^{16}$, 8th term

32. $(a - b)^{14}$, 6th term

33. $(x + 2y)^8$, 4th term

34. $(3a + b)^7$, 4th term

35. $(2a^2 - b)^{20}$, 7th term

36. $(a^2 - w^2)^{12}$, 5th term

Write each expansion using summation notation. See Example 6.

37. $(a + m)^8$

38. $(z + w)^{13}$

39. $(a - 2x)^5$

40. $(w - 3m)^7$

Getting More Involved

41. *Discussion*

Find the trinomial expansion for $(a + b + c)^3$ by using $x = a$ and $y = b + c$ in the binomial theorem.

42. *Discussion*

Find the fourth term in the binomial expansion for $(x + y)^{120}$. Find the fifth term in the binomial expansion for $(x - 2y)^{100}$. Did you have any trouble computing the coefficients?

Chapter

13   Wrap-Up

Summary

Sequences and Series		**Examples**
Sequence	Finite—A function whose domain is the set of positive integers less than or equal to a fixed positive integer	3, 5, 7, 9, 11 $a_n = 2n + 1$ for $1 \leq n \leq 5$
	Infinite—A function whose domain is the set of positive integers	2, 4, 6, 8, . . . $a_n = 2n$ for $n = 1, 2, 3, \ldots$
Series	The indicated sum of a sequence	$2 + 4 + 6 + \cdots + 50$
Summation notation	$\displaystyle\sum_{i=1}^{n} a_i = a_1 + a_2 + a_3 + \cdots + a_n$	$\displaystyle\sum_{i=1}^{25} 2i = 2 + 4 + \cdots + 50$

Arithmetic Sequences and Series		**Examples**
Arithmetic sequence	Each term after the first is obtained by adding a fixed amount to the previous term.	6, 11, 16, 21, . . . Fixed amount, d, is 5.
nth term	The nth term of an arithmetic sequence is $a_n = a_1 + (n - 1)d$.	If $a_1 = 6$ and $d = 5$, then $a_n = 6 + (n - 1)5$.
Arithmetic series	The sum of an arithmetic sequence	$6 + 11 + 16 + 21$
Sum of first n terms	$S_n = \dfrac{n}{2}(a_1 + a_n)$	$S_4 = \dfrac{4}{2}(6 + 21) = 54$

Geometric Sequences and Series		**Examples**
Geometric sequence	Each term after the first is obtained by multiplying the preceding term by a constant.	2, 6, 18, 54, . . . Constant, r, is 3.
nth term	The nth term of a geometric sequence is $a_n = a_1 r^{n-1}$.	$a_1 = 2, r = 3$ $a_n = 2 \cdot 3^{n-1}$
Geometric series (finite)	The indicated sum of a finite geometric sequence. $a_1 + a_1 r + a_1 r^2 + \cdots + a_1 r^{n-1}$	$2 + 6 + 18 + 54 + 162$
Sum of first n terms	$S_n = \dfrac{a_1(1 - r^n)}{1 - r}$	$a_1 = 2, r = 3, n = 5$ $S_5 = \dfrac{2(1 - 3^5)}{1 - 3} = 242$

Geometric series (infinite)	$a_1 + a_1 r + a_1 r^2 + a_1 r^3 + \cdots$	$8 + 4 + 2 + 1 + \dfrac{1}{2} + \cdots$
Sum of an infinite geometric series	$S = \dfrac{a_1}{1 - r}$, provided that $|r| < 1$	$a_1 = 8,\ r = \dfrac{1}{2}$ $S = \dfrac{8}{1 - \dfrac{1}{2}} = 16$
Factorial notation	The notation $n!$ represents the product of the positive integers from 1 through n.	$5! = 5 \cdot 4 \cdot 3 \cdot 2 \cdot 1 = 120$ $0! = 1$
Binomial theorem	$(x + y)^n = \dfrac{n!}{n!\,0!}x^n + \dfrac{n!}{(n-1)!\,1!}x^{n-1}y$ $+ \dfrac{n!}{(n-2)!\,2!}x^{n-2}y^2 + \cdots + \dfrac{n!}{0!\,n!}y^n$ Using summation notation: $(x + y)^n = \displaystyle\sum_{i=0}^{n} \dfrac{n!}{(n-i)!\,i!}x^{n-i}y^i = \sum_{i=0}^{n} \binom{n}{i}x^{n-i}y^i$	$(x + y)^3 = x^3 + 3x^2y + 3xy^2 + y^3$

kth term of $(x + y)^n$	$\dfrac{n!}{(n-k+1)!\,(k-1)!}x^{n-k+1}y^{k-1}$	Third term of $(a + b)^{10}$ is $\dfrac{10!}{8!\,2!}a^8 b^2 = 45 a^8 b^2$.

Enriching Your Mathematical Word Power

Fill in the blank.

1. A(n) _____ sequence is a function whose domain is the set of positive integers less than or equal to some fixed positive integer.

2. A(n) _____ sequence is a function whose domain is the set of all positive integers.

3. A(n) _____ is the indicated sum of a sequence.

4. A(n) _____ sequence is one in which each term after the first is obtained by adding a fixed amount to the previous term.

5. A(n) _____ sequence is one in which each term after the first is obtained by multiplying the preceding term by a constant.

6. A(n) _____ series is the indicated sum of an arithmetic sequence.

7. A(n) _____ series is the indicated sum of a geometric sequence.

8. The expression obtained by raising a binomial to a whole number power is a binomial _____.

9. _____ triangle is a triangular array of coefficients for binomial expansions.

10. The product of the positive integers from 1 through n is called n _____.

Review Exercises

13.1 Sequences

List all terms of each finite sequence.

1. $a_n = n^3$ for $1 \le n \le 5$

2. $b_n = (n - 1)^4$ for $1 \le n \le 4$

3. $c_n = (-1)^n (2n - 3)$ for $1 \le n \le 6$

4. $d_n = (-1)^{n-1}(3 - n)$ for $1 \le n \le 7$

Write the first three terms of the infinite sequence whose nth term is given.

5. $a_n = -\dfrac{1}{n}$

6. $b_n = \dfrac{(-1)^n}{n^2}$

7. $b_n = \dfrac{(-1)^{2n}}{2n + 1}$

8. $a_n = \dfrac{-1}{2n - 3}$

9. $c_n = \log_2(2^{n+3})$

10. $c_n = \ln(e^{2n})$

13.2 Series

Find the sum of each series.

11. $\displaystyle\sum_{i=1}^{3} i^3$ **12.** $\displaystyle\sum_{i=0}^{4} 6$

13. $\displaystyle\sum_{n=1}^{5} n(n - 1)$ **14.** $\displaystyle\sum_{j=0}^{3} (-2)^j$

Write each series in summation notation. Use the index i, and let i begin at 1.

15. $\dfrac{1}{4} + \dfrac{1}{6} + \dfrac{1}{8} + \cdots$

16. $\dfrac{1}{3} + \dfrac{1}{4} + \dfrac{1}{5} + \cdots$

17. $0 + 1 + 4 + 9 + 16 + \cdots$

18. $-1 + 2 - 3 + 4 - 5 + 6 - \cdots$

19. $x_1 - x_2 + x_3 - x_4 + \cdots$

20. $-x^2 + x^3 - x^4 + x^5 - \cdots$

13.3 Arithmetic Sequences and Series

Write the first four terms of the arithmetic sequence with the given nth term.

21. $a_n = 6 + (n - 1)5$

22. $a_n = -7 + (n - 1)4$

23. $a_n = -20 + (n - 1)(-2)$

24. $a_n = 10 + (n - 1)(-2.5)$

25. $a_n = 1000n + 2000$

26. $a_n = -500n + 5000$

Write a formula for the nth term of each arithmetic sequence.

27. $\dfrac{1}{3}, \dfrac{2}{3}, 1, \dfrac{4}{3}, \ldots$

28. $10, 6, 2, -2, \ldots$

29. $2, 4, 6, 8, \ldots$

30. $20, 10, 0, -10, \ldots$

Find the sum of each arithmetic series.

31. $1 + 2 + 3 + \cdots + 24$

32. $-5 + (-2) + 1 + 4 + \cdots + 34$

33. $\dfrac{1}{6} + \dfrac{1}{2} + \dfrac{5}{6} + \dfrac{7}{6} + \cdots + \dfrac{11}{2}$

34. $-3 - 6 - 9 - 12 - \cdots - 36$

35. $\displaystyle\sum_{i=1}^{7} (2i - 3)$

36. $\displaystyle\sum_{i=1}^{6} [12 + (i - 1)5]$

13.4 Geometric Sequences and Series

Write the first four terms of the geometric sequence with the given nth term.

37. $a_n = 3\left(\dfrac{1}{2}\right)^{n-1}$

38. $a_n = 6\left(-\dfrac{1}{3}\right)^{n}$

39. $a_n = 2^{1-n}$

40. $a_n = 5(10)^{n-1}$

41. $a_n = 23(10)^{-2n}$

42. $a_n = 4(10)^{-n}$

Write a formula for the nth term of each geometric sequence.

43. $\dfrac{1}{2}, 3, 18, \ldots$

44. $-6, 2, -\dfrac{2}{3}, \dfrac{2}{9}, \ldots$

45. $\dfrac{7}{10}, \dfrac{7}{100}, \dfrac{7}{1000}, \ldots$

46. $2, 2x, 2x^2, 2x^3, \ldots$

Find the sum of each geometric series.

47. $\dfrac{1}{3} + \dfrac{1}{9} + \dfrac{1}{27} + \dfrac{1}{81}$

48. $2 + 4 + 8 + 16 + \cdots + 512$

49. $\displaystyle\sum_{i=1}^{10} 3(10)^{-i}$

50. $\displaystyle\sum_{i=1}^{5} (0.1)^i$

51. $\dfrac{1}{4} + \dfrac{1}{12} + \dfrac{1}{36} + \dfrac{1}{108} + \cdots$

52. $12 + (-6) + 3 + \left(-\dfrac{3}{2}\right) + \cdots$

53. $\displaystyle\sum_{i=1}^{\infty} 18\left(\dfrac{2}{3}\right)^{i-1}$

54. $\displaystyle\sum_{i=1}^{\infty} 9(0.1)^i$

13.5 Binomial Expansions

Use the binomial theorem to expand each binomial.

55. $(m + n)^5$

56. $(2m - y)^4$

57. $(a^2 - 3b)^3$

58. $\left(\dfrac{x}{2} + 2a\right)^5$

Find the indicated term of the binomial expansion.

59. $(x + y)^{12}$, 5th term

60. $(x - 2y)^9$, 5th term

61. $(2a - b)^{14}$, 3rd term

62. $(a + b)^{10}$, 4th term

Write each expression in summation notation.

63. $(a + w)^7$

64. $(m - 3y)^9$

Miscellaneous

Identify each sequence as an arithmetic sequence, a geometric sequence, or neither.

65. $1, 3, 6, 10, 15, \ldots$

66. $9, 12, 16, \dfrac{64}{3}, \ldots$

67. $9, 12, 15, 18, \ldots$

68. $2, 4, 8, 16, \ldots$

69. $0, 2, 4, 6, 8, \ldots$

70. $0, 3, 9, 27, 81, \ldots$

Solve each problem.

71. Find the common ratio for the geometric sequence with first term 6 and fourth term $\dfrac{1}{30}$.

72. Find the common difference for an arithmetic sequence with first term 6 and fourth term 36.

73. Write out all of the terms of the series
$$\sum_{i=1}^{5} \frac{(-1)^i}{i!}.$$

74. Write out the first eight rows of Pascal's triangle.

75. Write out all of the terms of the series
$$\sum_{i=0}^{5} \frac{5!}{(5 - i)!\,i!}\, a^{5-i}b^i.$$

76. Write out all of the terms of the series
$$\sum_{i=0}^{8} \frac{8!}{(8 - i)!\,i!}\, x^{8-i}y^i.$$

77. How many terms are there in the expansion of $(a + b)^{25}$?

78. Calculate $\dfrac{12!}{8!4!}$.

79. If \$3000 is deposited at the beginning of each year for 16 years into an account paying 10% compounded annually, then what is the value of the annuity at the end of the 16th year?

80. If \$3000 is deposited at the beginning of each year for 8 years into an account paying 10% compounded annually, then what is the value of the annuity at the end of the eighth year? How does the value of the annuity in this exercise compare to that of Exercise 79?

81. If one deposit of \$3000 is made into an account paying 10% compounded annually, then how much will be in the account at the end of 16 years? Note that a single deposit is not an annuity.

Chapter 13 Test

List the first four terms of the sequence whose nth term is given.

1. $a_n = -10 + (n-1)6$

2. $a_n = 5(0.1)^{n-1}$

3. $a_n = \dfrac{(-1)^n}{n!}$

4. $a_n = \dfrac{2n-1}{n^2}$

Write a formula for the nth term of each sequence.

5. $7, 4, 1, -2, \ldots$

6. $-25, 5, -1, \dfrac{1}{5}, \ldots$

7. $2, -4, 6, -8, 10, -12, \ldots$

8. $1, 4, 9, 16, 25, \ldots$

Write out all of the terms of each series.

9. $\displaystyle\sum_{i=1}^{5} (2i+3)$

10. $\displaystyle\sum_{i=1}^{6} 5(2)^{i-1}$

11. $\displaystyle\sum_{i=0}^{4} \dfrac{4!}{(4-i)!\,i!} m^{4-i} q^i$

Find the sum of each series.

12. $\displaystyle\sum_{i=1}^{20} (6+3i)$

13. $\displaystyle\sum_{i=1}^{5} 10 \left(\dfrac{1}{2}\right)^{i-1}$

14. $\displaystyle\sum_{i=1}^{\infty} 0.35(0.93)^{i-1}$

15. $2 + 4 + 6 + \cdots + 200$

16. $\dfrac{1}{4} + \dfrac{1}{8} + \dfrac{1}{16} + \cdots$

17. $2 + 1 + \dfrac{1}{2} + \dfrac{1}{4} + \cdots + \dfrac{1}{128}$

Solve each problem.

18. Find the common ratio for the geometric sequence that has first term 3 and fifth term 48.

19. Find the common difference for the arithmetic sequence that has first term 1 and twelfth term 122.

20. Find the fifth term in the expansion of $(r-t)^{15}$.

21. Find the fourth term in the expansion of $(a^2 - 2b)^8$.

22. If \$800 is deposited at the beginning of each year for 25 years into an account earning 10% compounded annually, then what is the value of this annuity at the end of the 25th year?

*Making*Connections | A Review of Chapters 1–13

Let $f(x) = x^2 - 3$, $g(x) = 2x - 1$, $h(x) = 2^x$, *and* $m(x) = \log_2(x)$. *Find the following.*

1. $f(3)$

2. $f(n)$

3. $f(x + h)$

4. $f(x) - g(x)$

5. $g(f(3))$

6. $(f \circ g)(2)$

7. $m(16)$

8. $(h \circ m)(32)$

9. $h(-1)$

10. $h^{-1}(8)$

11. $m^{-1}(0)$

12. $(m \circ h)(x)$

Solve each variation problem.

13. If y varies directly as x, and $y = -6$ when $x = 4$, find y when $x = 9$.

14. If a varies inversely as b, and $a = 2$ when $b = -4$, find a when $b = 3$.

15. If y varies directly as w and inversely as t, and $y = 16$ when $w = 3$ and $t = -4$, find y when $w = 2$ and $t = 3$.

16. If y varies jointly as h and the square of r, and $y = 12$ when $h = 2$ and $r = 3$, find y when $h = 6$ and $r = 2$.

Sketch the graph of each inequality or system of inequalities.

17. $2x - 3y > 6$

18. $x > 3$ and $x + y < 0$

19. $|x - y| \geq 2$

20. $y < -2x + 3$ and $y > 2^x$

21. $|y + 2x| < 1$

22. $x^2 + y^2 < 4$

23. $x^2 - y^2 < 1$

24. $y < \log_2(x)$

25. $x^2 + 2y < 4$

42. *Consecutive integers.* Find three consecutive integers whose sum is 231.

43. *Loan shark.* Vinnie lent Julius some money at 22% simple interest. Frank wants to borrow twice as much as Julius and is willing to pay a higher rate. If Vinnie wants a 25% return after one year on the total lent to these two friends, then at what rate should he lend the money to Frank?

26. $\dfrac{x^2}{4} + \dfrac{y^2}{9} < 1$ and $y > x^2$

44. *Whiteout.* Justin drove for 3 hours in a snowstorm. After it stopped snowing, he drove 4 more hours, averaging 10 mph more than he did in the snowstorm. If he drove 460 miles altogether, then what was his average speed in the snowstorm?

Perform the indicated operation and simplify. Write answers with positive exponents.

45. *Mixed nuts.* Sue has 50 pounds of mixed nuts without peanuts that sell for $8 per pound. Peanuts sell for $1 per pound. How many pounds of peanuts should she put into her mixed nuts to get the price down to $5 per pound?

27. $\dfrac{a}{b} + \dfrac{b}{a}$

28. $1 - \dfrac{3}{y}$

29. $\dfrac{x-2}{x^2-9} - \dfrac{x-4}{x^2-2x-3}$

46. *Predicting heights of preschoolers.* A popular model in pediatrics for predicting the height of preschoolers is the JENNS model. According to this model, if $h(x)$ is the height [in centimeters (cm)] at age x (in years) for $0.25 \le x \le 6$, then

$$h(x) = 79.041 + 6.39x - e^{(3.261 - 0.993x)}.$$

30. $\dfrac{x^2-16}{2x+8} \cdot \dfrac{4x^2+16x+64}{x^3-16}$

a) Find the predicted height in inches for a child of age 4 years, 3 months.

31. $\dfrac{(a^2b)^3}{(ab^2)^4} \cdot \dfrac{ab^3}{a^{-4}b^2}$

b) Graph the function as shown in the accompanying figure, using a graphing calculator.

32. $\dfrac{x^2y}{(xy)^3} \div \dfrac{xy^2}{x^2y^4}$

c) Use your graphing calculator to find the age to the nearest tenth of a year for a child who has a height of 80 cm.

Simplify.

33. $8^{2/3}$

34. $16^{-5/4}$

35. $-4^{1/2}$

36. $27^{-2/3}$

37. -2^{-3}

38. $2^{-3/5} \cdot 2^{-7/5}$

39. $5^{-2/3} \div 5^{1/3}$

40. $\left(9^{1/2} + 4^{1/2}\right)^2$

Solve each problem.

41. *Trapezoid.* The area of a trapezoid is 30 square inches. If the height is 3 inches and one of the parallel sides is 12 inches, then what is the length of the other parallel side?

Figure for Exercise 46

Critical **Thinking** | For Individual or Group Work | **Chapter 13**

These exercises can be solved by a variety of techniques, which may or may not require algebra. So be creative and think critically. Explain all answers. Answers are in the Instructor's Edition of this text.

1. ***Tiling a floor.*** Red and white floor tiles are used to make the arrangements shown in the accompanying figure. How many red tiles would appear in the 20th figure in this sequence?

Figure for Exercise 1

2. ***Rolling dice.*** A pair of dice is rolled. What is the most likely difference between the number of dots showing on the top faces?

Photo for Exercise 2

3. ***Jocks, nerds, and turkeys.*** At Ridgemont High there are 30 jocks, 20 nerds, and some turkeys. Every nerd is a turkey. One-half of the jocks are turkeys. One-half of the turkeys are nerds. No jock is a nerd. How many turkeys are there? How many turkeys are neither nerds nor jocks?

4. ***Mind reading.*** A man and a woman are on an airplane chatting about their families. The woman says that she has three children, the age of each child is a counting number, the product of their ages is 72, and the sum of their ages is the same as the flight number. The man checks his ticket for the flight number, does a bit of figuring, and says that he needs more information to determine the ages. The woman then points to the peanuts that they are munching on and says that the oldest is allergic to peanuts. The man then tells the woman the correct ages of her children. What are the ages? Explain your answer.

5. ***Five-letter takeout.*** Take out five letters from the list

AFLIVGEELEBTRTEARS.

The remaining letters will form a common English word. What is it?

6. ***Heads and tails.*** A bag contains three coins. One coin has heads on both sides, one has tails on both sides, and one has heads on one side and tails on the other. A single coin accidentally falls onto the floor and you observe heads on that coin, but you cannot see the other side or the other two coins in the bag. What is the probability that the other side of the coin on the floor is heads?

7. ***Adjoining ones.*** Find a positive integer such that adjoining a 1 at both ends of it increases its value by 14,789. (Adjoining a 1 at both ends of 5 would produce 151 and increase its value by 146.)

8. ***Ending digits.*** What are the last two digits (tens and ones) of 3^{1234}?

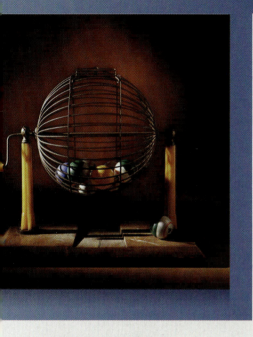

Chapter

14

Counting and Probability

Lotteries and gaming are as old as history records. Homer described how Agamemnon had his soldiers cast lots to see who would face Hector. In the Bible, Moses divided the lands among the tribes by casting lots.

Today 36 of the 50 states have lotteries and they are prosperous. Nationwide, state-run lotteries generate $26.6 billion in income annually—sales minus commissions. Available proceeds—money left over for the states after paying out prizes and paying for administration of the games—totals $10.1 billion.

But what are your chances of winning a lottery? In this chapter we will study the basic ideas of counting and probability.

14.1 Counting and Permutations

14.2 Combinations

14.3 Probability

We will find the probability of winning a lottery when we do Exercises 37 and 38 in Section 14.3.

14.1   Counting and Permutations

Although the main topic of this chapter is probability, we first study counting. We all know how to count, but here we will learn methods for counting the number of ways in which a sequence of events can occur.

⟨1⟩ The Fundamental Counting Principle

A new car is available in 3 different colors with 2 optional engines, V6 or V8. For example, you can get a red car with a V6 engine or a red car with a V8 engine. If the car comes in red, blue, or gray, then how many different cars are available? We can make a diagram showing all of the different possibilities. See Fig. 14.1. This diagram is called a **tree diagram.** Considering only color and engine type, we count 6 different cars that are available from the tree diagram. Of course, we can obtain 6 by multiplying 3 and 2. This example illustrates the fundamental counting principle.

Red — V6
      V8
Blue — V6
      V8
Gray — V6
      V8

Figure 14.1

Fundamental Counting Principle

If event A has m different outcomes and event B has n different outcomes, then there are mn different ways for events A and B to occur.

The fundamental counting principle can also be used for more than two events.

EXAMPLE 1

The fundamental counting principle

At Windy's Hamburger Palace you can get a single, double, or triple burger. You also have a choice of whether to include pickles, mustard, ketchup, onions, tomatoes, lettuce, or cheese. How many different hamburgers are available at Windy's?

Solution

There are 3 outcomes to the event of choosing the amount of meat. For each of the condiments there are 2 outcomes: whether or not to include it. So the number of different hamburgers is $3 \cdot 2 \cdot 2 \cdot 2 \cdot 2 \cdot 2 \cdot 2 \cdot 2 = 384$.

> Now do Exercises 1–4

⟨ **Helpful Hint** ⟩

When counting the number of different possibilities, it must be clear what "different" means. In one problem we might count a hamburger and fries differently from fries and a hamburger because we are interested in the order in which they occur. In another problem we will consider them as the same because they are the same meal.

EXAMPLE 2

The fundamental counting principle

How many different license plates are possible if each plate consists of 2 letters followed by a 4-digit number? Assume repetitions in the letters or numbers are allowed and any of the 10 digits may be used in each place of the 4-digit number.

Solution

Since there are 26 choices for each of the 2 letters and 10 choices for each of the numbers, by the fundamental counting principle the number of license plates is $26 \cdot 26 \cdot 10 \cdot 10 \cdot 10 \cdot 10 = 6{,}760{,}000$.

> Now do Exercises 5–6

⟨2⟩ Permutations

The number of ways in which an event can occur often depends on what has already occurred. In Examples 1 and 2, each choice was independent of the previous choices. In Example 3, we will count permutations in which the number of choices depends on what has already been chosen. A **permutation** is any ordering or arrangement of distinct objects in a linear manner.

E X A M P L E   3

Permutations of n objects

How many different ways are there for 10 people to stand in line to buy basketball tickets?

Solution

For the event of choosing the first person in line we have 10 outcomes. For the event of choosing the second person to put in line only 9 people are left, so we have only 9 outcomes. For the third person we have 8 outcomes, and so on. So the number of permutations of 10 people is $10 \cdot 9 \cdot 8 \cdot 7 \cdot 6 \cdot 5 \cdot 4 \cdot 3 \cdot 2 \cdot 1 = 10! = 3{,}628{,}800$.

> Now do Exercises 7–8

The number of arrangements of the 10 people in Example 3 is referred to as the number of permutations of 10 people taken 10 at a time, and we use the notation $P(10, 10)$ to represent this number. We have the following theorem.

> **Permutations of n Things n at a Time**
>
> The notation $P(n, n)$ represents the number of permutations of n things taken n at a time, and $P(n, n) = n!$.

⟨ **Calculator Close-Up** ⟩

You can use factorial notation on a calculator or the calculator's built-in formula for permutations to find $P(10, 3)$.

```
10!/7!
              720
10 nPr 3
              720
```

Sometimes we are interested in permutations in which all of the objects are not used. For example, how many ways are there to fill the offices of president, vice president, and treasurer in a club of 10 people, assuming no one holds more than one office? For the event of choosing the president there are 10 possible outcomes. Since no one can hold 2 offices, there are only 9 possibilities for vice president, and only 8 possibilities for treasurer. So the number of ways to fill the offices is $10 \cdot 9 \cdot 8 = 720$. We say that the number of permutations of 10 people taken 3 at a time is 720. We use the notation $P(10, 3)$ to represent the number of permutations of 10 things taken 3 at a time. Notice that

$$P(10, 3) = \frac{10!}{7!} = \frac{\mathbf{10 \cdot 9 \cdot 8} \cdot 7 \cdot 6 \cdot 5 \cdot 4 \cdot 3 \cdot 2 \cdot 1}{7 \cdot 6 \cdot 5 \cdot 4 \cdot 3 \cdot 2 \cdot 1} = 10 \cdot 9 \cdot 8 = 720.$$

In general, we have the following theorem.

> **Permutations of *n* Things *r* at a Time**
>
> The notation $P(n, r)$ represents the number of permutations of *n* things taken *r* at a time, and $P(n, r) = \dfrac{n!}{(n-r)!}$ for $0 \le r \le n$.

Even though we do not usually take *n* things 0 at a time, we allow 0 in the formula. For example,

$$P(8, 0) = \frac{8!}{8!} = 1 \qquad \text{and} \qquad P(0, 0) = \frac{0!}{0!} = \frac{1}{1} = 1.$$

E X A M P L E   **4**

Permutations of *n* things *r* at a time

Eight prize winners will be randomly selected from 25 people attending a sales meeting. There will be a first, second, third, fourth, fifth, sixth, seventh, and eighth prize, each prize being of lesser value than the one before it. In how many different ways can the prizes be awarded, assuming no one gets more than one prize?

Solution

The number of ways in which these prizes can be awarded is precisely the number of permutations of 25 things taken 8 at a time,

$$P(25, 8) = \frac{25!}{(25-8)!} = \frac{25!}{17!} = 25 \cdot 24 \cdot 23 \cdot 22 \cdot 21 \cdot 20 \cdot 19 \cdot 18$$

$$\approx 4.3609 \times 10^{10}.$$

> **Now do Exercises 9–12**

Warm-Ups ▼

Fill in the blank.

1. A(n) ____ diagram shows the number of different ways that a sequence of events can occur.

2. If _____ A has *m* different outcomes and event *B* has *n* different outcomes, then there are ___ ways for events *A* and *B* to occur.

3. A(n) _____ is an arrangement of distinct objects in a linear manner.

4. The number of _____ of *n* things taken *r* at a time is $\dfrac{n!}{(n-r)!}$.

True or false?

5. The number of permutations of 5 things taken 2 at a time is 20.

6. The number of ways for 4 people to line up for a group picture is 24.

7. The number of different ways to mark the answers to a 10-question multiple-choice test in which each question has 5 choices is 10^5.

8. The number of different ways to mark the answers to a 10-question true-false test is 2^{10}.

9. $P(100, 98) = 9900$

10. $\dfrac{100!}{98} = 9900$

‹ Study Tips ›

- If there is more than one way presented to solve a problem, learn them all.
- If you know more than one way to solve a problem, you will increase your chances of a correct solution and you will have a better understanding of the problem.

‹1› The Fundamental Counting Principle

Solve each problem. See Examples 1 and 2.

1. A parcel delivery truck can take any one of 3 different roads from Clarksville to Leesville and any one of 4 different roads from Leesville to Jonesboro. How many different routes are available from Clarksville to Jonesboro?

2. A car can be ordered in any one of 8 different colors, with 3 different engine sizes, 2 different transmissions, 3 different body styles, 4 different interior designs, and 5 different stereos. How many different cars are available?

3. A poker hand consists of 5 cards drawn from a deck of 52. How many different poker hands are there consisting of an ace, king, queen, jack, and ten?

Photo for Exercise 3

4. A certain card game consists of 3 cards drawn from a poker deck of 52. How many different 3-card hands are there containing one heart, one diamond, and one club?

5. Wendy's Hamburgers once advertised that there were 256 different hamburgers available at Wendy's. This number was obtained by using the fundamental counting principle and considering whether to include each one of several different options on the burger. How many different optional items were used to get this number?

6. A pizza can be ordered with your choice of one of 4 different meats or no meat. You also have a choice of whether to include green peppers, onions, mushrooms, anchovies, or black olives. How many different pizzas can be ordered?

Photo for Exercise 6

‹2› Permutations

Solve each problem. See Examples 3 and 4.

7. Randall has homework in mathematics, history, art, literature, and chemistry but cannot decide in which order to attack these subjects. How many different orders are possible?

8. Zita has packages to pick up at 8 different locations. How many different ways are there for her to pick up the packages?

9. Yesha has 12 schools to visit this week. In how many different ways can she pick a first, second, and third school to visit on Monday?

10. In how many ways can a history professor randomly assign exactly one A, one B, one C, one D, and six F's to a class of 10 students?

11. The program director for an independent television station has 34 one-hour shows available for Monday night prime time. How many different schedules are possible for the 7:00 to 10:00 P.M. time period?

12. A disc jockey must choose 8 songs from the top 40 to play in the next 30-minute segment of his show. How many different arrangements are possible for this segment?

Miscellaneous

Solve each counting problem.

13. How many different ways are there to mark the answers to a 20-question multiple-choice test in which each question has 4 possible answers?

14. A bookstore manager wants to make a window display that consists of a mathematics book, a history book, and an economics book in that order. He has 13 different mathematics books, 10 different history books, and 5 different economics books from which to choose. How many different displays are possible?

15. In how many ways can the 37 seats on a commuter flight be filled from the 39 people holding tickets?

16. If a couple has decided on 6 possible first names for their baby and 5 possible middle names, then how many ways are there for them to name their baby?

17. A developer builds houses with 3 different exterior styles. You have your choice of 3, 4, or 5 bedrooms, a fireplace, 2 or 3 baths, and 5 different kitchen designs. How many different houses are available?

18. How many different seven-digit phone numbers are available in Wentworth if the first three digits must be 286?

19. How many different seven-digit phone numbers are available in Creekside if the first digit cannot be a 0?

20. In how many different ways can a Mercedes, a Cadillac, and a Ford be awarded to 3 people chosen from the 9 finalists in a contest?

21. How many different ways are there to seat 7 students in a row?

Photo for Exercise 21

22. A supply boat must stop at 9 oil rigs in the Gulf of Mexico. How many different routes are possible?

23. How many different license plates can be formed by using 3 digits followed by a single letter followed by 3 more digits? How many if the single letter can occur anywhere except last?

24. How many different license plates can be formed by using any 3 letters followed by any 3 digits? How many if we allow either the 3 digits or the 3 letters to come first?

25. Make a list of all of the subsets of the set $\{a, b, c\}$. How many are there?

26. Use the fundamental counting principle to find the number of subsets of the set $\{a, b, c, d, e, f\}$.

27. How many ways are there to mark the answers to a test that consists of 10 true-false questions followed by 10 multiple-choice questions with 5 options each?

28. A fraternity votes on whether to accept each of 5 pledges. How many different outcomes are possible for the vote?

29. Make a list of all of the ways to arrange the letters in the word MILK. How many arrangements should be in your list?

30. Make a list of all of the permutations of the letters A, B, C, D, and E taken 3 at a time. How many permutations should be in your list?

Evaluate each expression.

31. $4!$

32. $0!$

33. $(4 - 4)!$

34. $5!$

35. $P(8, 0)$

36. $P(8, 8)$

37. $P(8, 3)$

38. $P(17, 4)$

39. $P(52, 0)$

40. $P(34, 1)$

41. $\dfrac{P(10, 4)}{4!}$

42. $\dfrac{P(8, 3)}{3!}$

43. $\dfrac{P(12, 3)}{3!}$

44. $\dfrac{P(15, 6)}{6!}$

45. $\dfrac{14!}{3!\,11!}$

46. $\dfrac{18!}{17!\,1!}$

47. $\dfrac{98!}{95!\,3!}$

48. $\dfrac{87!}{83!\,4!}$

14.2 Combinations

In Section 14.1, we learned the fundamental counting principle and applied it to finding the number of permutations of n objects taken r at a time. We will now learn how to count the number of combinations of n objects taken r at a time.

⟨1⟩ **Combinations**

Consider the problem of awarding 2 identical scholarships to 2 students among 4 finalists: Ahmadi, Butler, Chen, and Davis. Since the scholarships are identical, we do not count Ahmadi and Butler as different from Butler and Ahmadi. We can easily list all possible choices of 2 students from the 4 possibilities A, B, C, and D:

$$\{A, B\}\quad \{A, C\}\quad \{A, D\}\quad \{B, C\}\quad \{B, D\}\quad \{C, D\}$$

It is convenient to use set notation to list these choices because in set notation we have $\{A, B\} = \{B, A\}$. Actually, we want the number of subsets or **combinations** of size 2 from a set of 4 elements. This number is referred to as the number of combinations of 4 things taken 2 at a time, and we use the notation $C(4, 2)$ to represent it. We have $C(4, 2) = 6$.

If we had a first and second prize to give to 2 of 4 finalists, then $P(4, 2) = 4 \cdot 3 = 12$ is the number of ways to award the prizes. If the prizes are identical, then we do not count AB as different from BA, or AC as different from CA, and so on. So we divide $P(4, 2)$ by 2!, the number of permutations of the 2 prize winners, to get $C(4, 2)$.

In general, the number of subsets of size r taken from a set of size n can be found by dividing $P(n, r)$ by $r!$:

$$C(n, r) = \frac{P(n, r)}{r!}$$

Since $P(n, r) = \frac{n!}{(n-r)!}$, we can write this expression as follows:

$$C(n, r) = \frac{P(n, r)}{r!} = \frac{\dfrac{n!}{(n-r)!}}{r!} = \frac{n!}{(n-r)!r!}$$

The notation $\binom{n}{r}$ is also used for $C(n, r)$. We summarize these results in the following theorem.

> **Combinations of n Things r at a Time**
>
> The number of combinations of n things taken r at a time (or the number of subsets of size r from a set of n elements) is given by the formula
>
> $$C(n, r) = \binom{n}{r} = \frac{n!}{(n-r)!r!} \qquad \text{for} \quad 0 \le r \le n.$$

E X A M P L E 1

Combinations of *n* things *r* at a time

In how many ways can a committee of 4 people be chosen from a group of 12?

Solution

Choosing 4 people from a group of 12 is the same as choosing a subset of size 4 from a set of 12 elements. So the number of ways to choose the committee is the number of combinations of 12 things taken 4 at a time:

$$C(12, 4) = \frac{12!}{8! \, 4!} = \frac{12 \cdot 11 \cdot 10 \cdot 9 \cdot 8 \cdot 7 \cdot 6 \cdot 5 \cdot 4 \cdot 3 \cdot 2 \cdot 1}{8 \cdot 7 \cdot 6 \cdot 5 \cdot 4 \cdot 3 \cdot 2 \cdot 1 \cdot 4 \cdot 3 \cdot 2 \cdot 1} = 495$$

> Now do Exercises 1–8

E X A M P L E 2

Finding a binomial coefficient

What is the coefficient of a^4b^2 in the binomial expansion of $(a + b)^6$?

Solution

Write $(a + b)^6 = (a + b)(a + b)(a + b)(a + b)(a + b)(a + b)$. The terms of the product come from all of the different ways there are to select either a or b from each of the 6 factors and to multiply the selections. The number of ways to pick 4 factors for the selection of a is $C(6, 4) = \frac{6!}{2! \, 4!} = 15$. From the remaining 2 we select b. So in the binomial expansion of $(a + b)^6$ we find the term $15a^4b^2$.

> Now do Exercises 9–10

‹ **Calculator Close-Up** ›

You can use factorial notation to calculate $C(12, 4)$ or your calculator's built-in formula for combinations. With factorial notation you must use parentheses around the denominator.

```
12!/(8!*4!)
              495
12 nCr 4
              495
```

Note that as in Example 2, the coefficients of the terms in any binomial expansion are combinations. The coefficient of $a^r b^{n-r}$ in $(a + b)^n$ is $C(n, r)$.

‹2› Permutations, Combinations, or Neither

How do you know when to use the permutation formula and when to use the combination formula? The key to answering this question is understanding what each formula counts. In either case we are choosing from a group of distinct objects. Compare the following examples.

E X A M P L E 3

Permutations of *n* things *r* at a time

How many ways are there to choose a president, a vice president, and a treasurer from a group of 9 people, assuming no one holds more than one office?

Solution

There are 9 choices for the president, 8 choices for the vice president, and 7 choices for the treasurer. The number of ways to make these choices is $P(9, 3) = 9 \cdot 8 \cdot 7 = 504$.

> Now do Exercises 11–12

Note that in Example 3 we are not just counting the number of ways to select 3 people from 9. In the number $P(9, 3)$, every permutation of the 3 people selected is counted.

EXAMPLE   4

Combinations of n things r at a time

How many ways are there to choose 3 people to receive a $100 prize from a group of 9, assuming no one receives more than one prize?

Solution

Since each person chosen from the 9 is given the same treatment, the number of choices is the same as the number of subsets of size 3 from a set of 9 elements.

$$C(9, 3) = \frac{9!}{6!\,3!} = \frac{9 \cdot 8 \cdot 7}{3 \cdot 2 \cdot 1} = 84.$$

> Now do Exercises 13–14

Note that in Example 4 we do not count different arrangements of the 3 people selected. Example 5 is neither a permutation nor a combination.

EXAMPLE   5

The fundamental counting principle

How many ways are there to answer a 9-question multiple-choice test in which each question has 3 possible answers?

Solution

There are 3 outcomes to the event of choosing the first answer, 3 outcomes to the event of choosing the second answer, and so on. By the fundamental counting principle there are $3^9 = 19,683$ ways to answer the 9 questions.

> Now do Exercises 15–26

⟨ **Helpful Hint** ⟩

You cannot solve counting problems by picking out the numbers in the problem and performing a computation to get the answer. This approach to Example 5 will lead you to wrong answers, such as $\frac{9!}{3!6!}$, 9^3, or $3 \cdot 9$. You must think about the problem.

Note that in choosing the answer to each question in Example 5, we may repeat answers, so we are not choosing from a set of distinct objects as in permutations and combinations. In Example 6, we use both the fundamental counting principle and the permutation formula.

EXAMPLE   6

The fundamental counting principle and permutations

Josephine has 4 different mathematics books, 3 different art books, and 5 different music books that she plans to display on a shelf. If she plans to keep each type of book together and put the types in the order mathematics–art–music, then how many different arrangements are possible?

Solution

First observe that the mathematics books can be arranged in $P(4, 4) = 4! = 24$ ways, the art books in $P(3, 3) = 3! = 6$ ways, and the music books in $P(5, 5) = 5! = 120$ ways. Now we use the fundamental counting principle to get $24 \cdot 6 \cdot 120 = 17,280$ as the number of ways to arrange the books on the shelf.

> Now do Exercises 27–32

〈3〉 Labeling

In a labeling problem, we count the number of ways to put labels on distinct objects.

E X A M P L E **7**

The fundamental counting principle and combinations

Twelve students have volunteered to help clean up a small oil spill. The project director needs 3 bird washers, 4 rock wipers, and 5 sand cleaners. In how many ways can these jobs (labels) be assigned to these 12 students?

Solution

Since the 3 bird washers all get the same label, the number of ways to select the 3 students is $C(12, 3)$. The number of ways to select 4 rock wipers from the remaining 9 students is $C(9, 4)$. The number of ways to select the 5 sand cleaners from the remaining 5 students is $C(5, 5)$. By the fundamental counting principle the number of ways to make all 3 selections is

$$C(12, 3) \cdot C(9, 4) \cdot C(5, 5) = \frac{12!}{9! \, 3!} \cdot \frac{9!}{5! \, 4!} \cdot \frac{5!}{0! \, 5!} = \frac{12!}{3! \, 4! \, 5!} = 27,720.$$

> Now do Exercises 33–36

Note that in Example 7, there were 12 distinct objects to be labeled with 3 labels of one type, 4 labels of another type, and 5 labels of a third type, and the number of ways to assign those labels was found to be $\frac{12!}{3! \, 4! \, 5!}$. Instead of using combinations and the fundamental counting principle as in Example 7, we can use the following theorem.

Labeling

In a **labeling problem** n distinct objects are each to be given a label with each object getting exactly one label. If there are r_1 labels of the first type, r_2 labels of the second type, . . . , and r_k labels of the kth type, where $r_1 + r_2 + \cdots + r_k = n$, then the number of ways to assign the labels to the objects is $\dfrac{n!}{r_1! \, r_2! \cdots r_k!}$.

EXAMPLE **8**

A labeling problem

How many different arrangements are there for the 11 letters in the word MISSISSIPPI?

Solution

This problem is a labeling problem if we think of the 11 positions for the letters as 11 distinct objects to be labeled. There is 1 M-label, 4 S-labels, 4 I-labels, and 2 P-labels. So the number of ways to arrange the letters in MISSISSIPPI is $\frac{11!}{1!4!4!2!} = 34,650$.

> **Now do Exercises 37–42**

EXAMPLE **9**

Coefficients in a trinomial expansion

What is the coefficient of a^3b^2c in the expansion of $(a + b + c)^6$?

Solution

The terms of the product $(a + b + c)^6$ come from all of the different ways there are to select a, b, or c from each of the 6 distinct factors and to multiply the selections. The number of ways to pick 3 factors from which we select a, 2 factors from which we select b, and 1 factor from which we select c is $\frac{6!}{3!2!1!} = \frac{6 \cdot 5 \cdot 4 \cdot 3 \cdot 2 \cdot 1}{3 \cdot 2 \cdot 1 \cdot 2 \cdot 1 \cdot 1} = 60$. So in the expansion we have the term $60a^3b^2c$.

> **Now do Exercises 43–46**

The coefficients of the terms in an expansion such as that of Example 9 are called **multinomial coefficients.** Determining the coefficients is a labeling problem.

We may think of permutation and combination problems as being very different, but they are both labeling problems. For example, to find the number of subsets of size 3 from a set of size 5, we are assigning 3 I-labels and 2 O-labels (I for in and O for out) to the 5 distinct objects of the set. Note that $\frac{5!}{3!2!} = C(5, 3)$. To find the number of ways to give a **F**irst, **S**econd, and **T**hird prize to 3 of 10 people, we are assigning 1 **F**-label, 1 **S**-label, 1 **T**-label, and 7 **N**-labels (N for no prize) to the 10 distinct people. Note that $\frac{10!}{1!1!1!7!} = P(10, 3)$.

 Warm-Ups ▼

Fill in the blank.

1. A(n) _____ of n things taken r at a time is a subset of size r taken from a set of size n.

2. The number of _____ of n things taken r at a time is $\dfrac{n!}{r!(n-r)!}$.

3. In a(n) _____ problem, n objects are each given a label, some of which may be identical.

4. The _____ coefficients are the coefficients in the expansion of a power of three or more letters.

True or false?

5. The number of ways to choose 2 questions to answer out of 3 questions on a test is $C(3, 2)$.

6. There are $P(10, 3)$ ways to randomly choose 3 students out of 10 to be crossing guards.

7. $P(9, 4) = (4!) \cdot C(9, 4)$

8. $P(8, 3) = \dbinom{8}{5}$

9. $C(100, 1) = 100$

10. $C(12, 4) = C(4, 12)$

11. $\dfrac{8!}{1!1!1!5!} = P(8, 3)$

‹1› Combinations

Solve each problem. See Examples 1 and 2.

1. How many 5-card poker hands are possible when you are dealt 5 cards from a deck of 52?

2. How many 13-card bridge hands are possible when you are dealt 13 cards from a deck of 52?

3. How many ways are there to select 3 candidates from the 5 finalists for an in-depth interview?

4. How many ways are there to select 12 welders to be laid off from 30 welders employed at the Ingalls Shipyard?

5. How many ways are there for a health inspector to select 5 restaurants to visit from a list of 20 restaurants?

6. The water inspector in drought-stricken Marin County randomly selects 10 homes for inspection from a list of 25 suspected violators of the rationing laws. How many ways are there to pick the 10 homes?

7. In the Florida Lottery you can win a lot of money for merely selecting 6 different numbers from the numbers 1 through 49. How many different ways are there to select the 6 numbers?

8. In the game Fantasy Five you can win by selecting 5 different numbers from the numbers 1 through 39. How many ways are there to select the 5 numbers?

9. What is the coefficient of w^3y^4 in the expansion of $(w + y)^7$?

10. What is the coefficient of a^5z^9 in the expansion of $(a + z)^{14}$?

‹2› Permutations, Combinations, or Neither

Solve each problem. See Examples 3–5.

11. The Dean's Search Committee must choose 3 candidates from a list of 6 and rank them as first, second, and third. How many different outcomes are possible?

12. The Provost's Search Committee must choose 3 candidates from a list of 8 and submit the names to the president unranked. How many different outcomes are possible?

13. Charlotte must write on any 3 of the 8 essay questions on the final exam in History 201. How many ways are there for her to pick the questions?

14. How many ways are there for Murphy to mark the answers to 8 multiple-choice questions, each of which has 5 possible answers?

15. In how many different ways can Professor Reyes return his examination papers to a class of 12 students?

16. In how many different ways can Professor Lee return her examination papers to a class of 12 students if she always returns the best paper first and the worst paper last?

17. How many ways are there to select 5 seats from a 150-seat auditorium to be occupied by 5 plain-clothes officers?

18. For her final exam in Literature 302 Charlotte is allowed to omit any 5 of the 8 essay questions. How many ways are there for her to omit the 5 questions?

19. How many distinct chords (line segments with endpoints on the circle) are determined by 3 points lying on a circle? By 4 points? By 5 points? By n points?

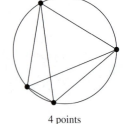

3 points   4 points

Figure for Exercise 19

20. How many distinct triangles are determined by 5 points lying on a circle, where the vertices of each triangle are chosen from the 5 points?

21. If the outcome of tossing a pair of dice is thought of as an ordered pair of numbers, then how many ordered pairs are there?

22. Francene is eligible to take 4 different mathematics classes, 5 different history classes, 6 different psychology classes, and 3 different literature classes. If her schedule is to consist of one class from each category, then how many different schedules are possible for her?

23. The outcome "heads" or "tails" is recorded on each toss of a coin. If we think of the outcome for 3 tosses as an ordered triple, then how many outcomes are there for 3 tosses of a coin?

24. A coin and a die are tossed. How many different outcomes are possible?

25. Explain why $C(n, r) = C(n, n - r)$.

26. Is $P(n, r) = P(n, n - r)$? Explain your answer.

Solve each problem. See Example 6.

27. Juanita is eligible to take 4 different mathematics classes and 5 different history classes. If her schedule is to consist of two classes from each category, then how many different schedules are possible for her?

28. A committee consisting of 2 men and 1 woman is to be formed from a department consisting of 8 men and 3 women. How many different committees are possible?

29. In a class of 20 students the teacher randomly assigns 4 A's, 5 B's, and 11 C's. In how many ways can these grades be assigned?

30. How many ways are there to seat 3 boys and 3 girls in a row with no 2 boys and no 2 girls sitting next to each other?

31. How many different orders are there for 4 marching bands and 4 floats to parade if a marching band must lead the parade and we cannot have 2 bands in a row or 2 floats in a row?

Photo for Exercise 31

32. In how many ways can 8 students be seated in a row if 2 of them must be seated next to one another?

VIDEO

‹3› **Labeling**

Solve each problem. See Examples 7–9.

33. A teacher randomly assigns 6 A's, 3 B's, and 7 C's to a batch of 16 term papers. In how many different ways can she do this?

34. The 15 volunteers for the annual Bluegrass Festival are to be divided into 3 groups of 5 each and assigned to tickets, parking, and cleanup. In how many ways can these assignments be made?

35. Bob, Carol, Ted, and Alice are playing bridge. How many different ways are there to give each of them 13 cards from the deck of 52? Is the answer larger or smaller than the number of seconds in a trillion years?

36. Bret, Bart, and Mad Dog are playing poker. How many different ways are there to give each of them 5 cards from a deck of 52?

37. How many different arrangements are there for the letters in the word TOYOTA? ALGEBRA? STATISTICS?

38. How many different orders are possible for 3 identical contemporary, 4 identical traditional, and 6 identical colonial homes to march in a parade of homes?

39. Twenty salespeople are getting new cars. How many ways are there to assign 6 identical Chevrolets, 5 identical Fords, 8 identical Buicks, and 1 Lincoln to them?

40. Twenty different Fords are being sent to 4 Ford dealers. In how many ways can Bill Poole Ford get 3 cars, Heritage Ford get 5 cars, Mid-City Ford get 5 cars, and Northside Ford get 7 cars?

41. Nine sofas are to be discounted at Sofa City with discounts as large as 75% off the manufacturers suggested retail price. (Sorry, none sold to dealers.) How many ways are there to mark 4 of the sofas 10% off, 4 of them 15% off, and 1 of them 75% off?

42. A combination for a safe is a sequence of 4 numbers with no repetitions selected from the integers 1 through 99.

Photo for Exercise 42

How many combinations are possible? Is combination the proper word to use here?

43. What is the coefficient of $a^3b^2c^5$ in the expansion of $(a + b + c)^{10}$?

44. What is the coefficient of $wx^2y^3z^9$ in the expansion of $(w + x + y + z)^{15}$?

45. What is the coefficient of a^3b^5 in the expansion of $(a + b + c)^8$?

46. What is the coefficient of x^2z^5 in the expansion of $(x + y + z)^7$?

Miscellaneous

Evaluate each expression without using a calculator.

47. $\dfrac{8!}{3! \, 5!}$

48. $\dfrac{8!}{2! \, 6!}$

49. $\dfrac{8!}{5!}$

50. $\dfrac{9!}{7!}$

51. $C(4, 0)$

52. $C(4, 1)$

53. $C(5, 2)$

54. $C(5, 3)$

55. $P(5, 2)$

56. $P(5, 3)$

Mid-Chapter **Quiz** | Sections 14.1 through 14.2 | **Chapter 14**

Solve each problem.

1. For \$7 you can get a salad, entree, and dessert at the cafeteria. If there are 4 salads, 8 entrees, and 3 desserts available, then how many different meals are available?

2. In a game of poker you are first given two cards from a deck of 52. How many possibilities are there for those two cards?

3. From the five finalists in a beauty contest, two are randomly chosen to be the queen and the alternate. In how many ways can this be done?

4. How many possible ways are there to mark the answers to a true-false quiz consisting of five questions?

5. A group of 12 friends are going to a restaurant to eat. Four of them are to go in a cab and the rest must take the subway. In how many ways can the four that go in the cab be chosen?

6. What is the coefficient of a^3b^5 in the expansion of $(a + b)^8$?

Evaluate each expression.

7. $C(10, 3)$

8. $P(8, 3)$

9. $8!$

10. $\dfrac{34!}{30! \, 4!}$

11. $\dfrac{103!}{101!}$

14.3 Probability

In This Section

⟨1⟩ **The Probability of an Event**

⟨2⟩ **The Addition Rule**

⟨3⟩ **Complementary Events**

⟨4⟩ **Odds**

In Sections 14.1 and 14.2, we were concerned with counting the number of different outcomes to an experiment. We now use those counting techniques to find probabilities.

⟨1⟩ The Probability of an Event

In probability, an **experiment** is a process such as tossing a coin, rolling a die, drawing a poker hand from a deck, or arranging people in a line. A **sample space** is the set of all possible outcomes to an experiment. An **event** is a subset of a sample space. For example, if we toss a coin, then the sample space consists of two equally likely outcomes, heads and tails. We write $S = \{H, T\}$. The subset $E = \{H\}$ is the event of getting heads when the coin is tossed. We use $n(S)$ to represent the number of equally likely outcomes in the sample space S and $n(E)$ to represent the number of outcomes in the event E. For the example of tossing a coin, $n(S) = 2$ and $n(E) = 1$.

The Probability of an Event

If S is a sample space of equally likely outcomes to an experiment and the event E is a subset of S, then the probability of E, $P(E)$, is defined to be

$$P(E) = \frac{n(E)}{n(S)}.$$

When $S = \{H, T\}$ and $E = \{H\}$,

$$P(E) = \frac{n(E)}{n(S)} = \frac{1}{2}.$$

So the probability of getting heads on a single toss of a coin is $\frac{1}{2}$. If E is the event of getting 2 heads on a single toss of a coin, then $n(E) = 0$ and $P(E) = \frac{0}{2} = 0$. If E is the event of getting fewer than 2 heads on a single toss of a coin, then for either outcome H or T we have fewer than 2 heads. So $E = \{H, T\}$, $n(E) = 2$, and $P(E) = \frac{2}{2} = 1$. Note that the probability of an event is a number between 0 and 1 inclusive, 1 being the probability of an event that is certain to occur and 0 being the probability of an event that is impossible to occur.

EXAMPLE **1**

Rolling a die

What is the probability of getting a number larger than 4 when a single die is rolled?

Solution

When we roll a die, we count the number of dots showing on the upper face of the die. So the sample space of equally likely outcomes is $S = \{1, 2, 3, 4, 5, 6\}$. Since only 5 and 6 are larger than 4, $E = \{5, 6\}$. According to the definition of probability,

$$P(E) = \frac{n(E)}{n(S)} = \frac{2}{6} = \frac{1}{3}.$$

> Now do Exercises 1–2

EXAMPLE **2**

Tossing coins

What is the probability of getting at least one head when a pair of coins is tossed?

Solution

Since there are 2 equally likely outcomes for the first coin and 2 equally likely outcomes for the second coin, by the fundamental counting principle there are 4 equally likely outcomes to the experiment of tossing a pair of coins. We can list the outcomes as ordered pairs: $S = \{(H, H), (H, T), (T, H), (T, T)\}$. Since 3 of these outcomes result in at least one head, $E = \{(H, H), (H, T), (T, H)\}$, and $n(E) = 3$. So,

$$P(E) = \frac{n(E)}{n(S)} = \frac{3}{4}.$$

> Now do Exercises 3–4

EXAMPLE 3

Rolling a pair of dice

What is the probability of getting a sum of 6 when a pair of dice is rolled?

Solution

Since there are 6 equally likely outcomes for each die, there are 36 equally likely outcomes to the experiment of rolling the pair. We can list the 36 outcomes as ordered pairs:

$$S = \{(1, 1), (1, 2), (1, 3), (1, 4), \mathbf{(1, 5)}, (1, 6),$$
$$(2, 1), (2, 2), (2, 3), \mathbf{(2, 4)}, (2, 5), (2, 6),$$
$$(3, 1), (3, 2), \mathbf{(3, 3)}, (3, 4), (3, 5), (3, 6),$$
$$(4, 1), \mathbf{(4, 2)}, (4, 3), (4, 4), (4, 5), (4, 6),$$
$$\mathbf{(5, 1)}, (5, 2), (5, 3), (5, 4), (5, 5), (5, 6),$$
$$(6, 1), (6, 2), (6, 3), (6, 4), (6, 5), (6, 6)\}$$

"The sum of the numbers is 6" describes the event $E = \{(5, 1), (4, 2), (3, 3), (2, 4), (1, 5)\}$. So,

$$P(E) = \frac{n(E)}{n(S)} = \frac{5}{36}.$$

> Now do Exercises 5–12

⟨2⟩ The Addition Rule

In tossing a pair of dice, let A be the event that doubles occurs and B be the event that the sum is 4. We can list the following events and their probabilities:

$$A = \{(1, 1), (2, 2), (3, 3), (4, 4), (5, 5), (6, 6)\} \text{ and } P(A) = \frac{6}{36}$$

$$B = \{(3, 1), (2, 2), (1, 3)\} \text{ and } P(B) = \frac{3}{36}$$

$$A \cup B = \{(1, 1), (2, 2), (3, 3), (4, 4), (5, 5), (6, 6), (3, 1), (1, 3)\}$$
$$\text{and } P(A \cup B) = \frac{8}{36}$$

$$A \cap B = \{(2, 2)\} \text{ and } P(A \cap B) = \frac{1}{36}$$

Note that the probability of doubles or a sum of 4, $P(A \cup B)$, is $\frac{8}{36}$ and

$$\frac{8}{36} = \frac{6}{36} + \frac{3}{36} - \frac{1}{36}.$$

This equation makes sense because there is one outcome, (2, 2), that is in both the events A and B. This example illustrates the addition rule.

The Addition Rule

If A and B are any events in a sample space, then

$$P(A \cup B) = P(A) + P(B) - P(A \cap B).$$

If $P(A \cap B) = 0$, then A and B are called **mutually exclusive** events and

$$P(A \cup B) = P(A) + P(B).$$

Note that for mutually exclusive events it is impossible for both events to occur. The addition rule for mutually exclusive events is a special case of the general addition rule.

E X A M P L E   4

The addition rule
At Downtown College 60% of the students are commuters (C), 50% are female (F), and 30% are female commuters. If a student is selected at random, what is the probability that the student is either a female or a commuter?

Solution
By the addition rule the probability of selecting either a female or a commuter is $P(F \cup C) = P(F) + P(C) - P(F \cap C) = 0.50 + 0.60 - 0.30 = 0.80$.

| Now do Exercises 13–14 |

E X A M P L E   5

The addition rule with dice
In rolling a pair of dice, what is the probability that the sum is 12 or at least one die shows a 2?

Solution
Let A be the event that the sum is 12 and B be the event that at least one die shows a 2. Since A occurs on only one of the 36 equally likely outcomes (see Example 3), $P(A) = \frac{1}{36}$. Since B occurs on 11 of the equally likely outcomes, $P(B) = \frac{11}{36}$. Since A and B are mutually exclusive, we have

$$P(A \cup B) = P(A) + P(B)$$
$$= \frac{1}{36} + \frac{11}{36} = \frac{12}{36} = \frac{1}{3}.$$

| Now do Exercises 15–18 |

⟨3⟩ **Complementary Events**

If the probability of rain today is 60%, then the probability that it does not rain is 40%. Rain and not rain are called complementary events. There is no possibility that both occur, and one of them must occur. If A is an event, then $\overline{A}$ (read "A bar" or "A complement") represents the complement of the event A. Note that complementary events are mutually exclusive, but mutually exclusive events are not necessarily complementary.

Complementary Events
Two events A and $\overline{A}$ are called **complementary events** if $A \cap \overline{A} = \varnothing$ and $P(A) + P(\overline{A}) = 1$.

E X A M P L E   6

Complementary events
What is the probability of getting a number less than or equal to 4 when rolling a single die?

Solution
We saw in Example 1 that getting a number larger than 4 when rolling a single die has probability $\frac{1}{3}$. The complement to getting a number larger than 4 is getting a number less than or equal to 4. So the probability of getting a number less than or equal to 4 is $\frac{2}{3}$.

| Now do Exercises 19–20 |

EXAMPLE **7**

Complementary events

If the probability that White Lightning will win the Kentucky Derby is 0.15, then what is the probability that White Lightning does not win the Kentucky Derby?

Solution

Let W be winning the Kentucky Derby and N be not winning the Kentucky Derby. Since W and N are complementary events, we have $P(W) + P(N) = 1$. So $P(N) = 1 - P(W) = 1 - 0.15 = 0.85$.

> Now do Exercises 21–22

⟨4⟩ Odds

If the probability is $\frac{2}{3}$ that the Giants win the Super Bowl and $\frac{1}{3}$ that they lose, then they are twice as likely to win as they are to lose. We say that the odds in favor of the Giants winning the Super Bowl are 2 to 1. Notice that odds are not probabilities. Odds are ratios of probabilities. We usually write odds as ratios of whole numbers.

> **Odds**
>
> If A is any event, then the **odds in favor of A** is the ratio $P(A)$ to $P(\overline{A})$ and the **odds against A** is the ratio of $P(\overline{A})$ to $P(A)$.

EXAMPLE **8**

Determining odds

What are the odds in favor of getting a sum of 6 when rolling a pair of dice? What are the odds against a sum of 6?

Solution

In Example 3 we found the probability of a sum of 6 to be $\frac{5}{36}$. So the probability of the complement (the sum is not 6) is $\frac{31}{36}$. The odds in favor of getting a sum of 6 are $\frac{5}{36}$ to $\frac{31}{36}$. Multiply each fraction by 36 to get the odds 5 to 31. The odds against a sum of 6 are 31 to 5.

> Now do Exercises 23–26

EXAMPLE **9**

Determining probability given the odds

If the odds in favor of Daddy's Darling winning the third race at Delta Downs are 4 to 1, then what is the probability that Daddy's Darling wins the third race?

⟨ **Helpful Hint** ⟩

Odds and probability are often confused, even by people who write lottery tickets. If the probability of winning a lottery is $\frac{1}{100}$, then the probability of losing is $\frac{99}{100}$, and the odds in favor of winning are 1 to 99. Many lottery tickets will state (incorrectly) that the odds in favor of winning are 1 to 100.

Solution

Since 4 to 1 is the ratio of the probability of winning to not winning, the probability of winning is four times as large as the probability of not winning. Let $P(\overline{W}) = x$ and $P(W) = 4x$. Since $P(W) + P(\overline{W}) = 1$, we have $4x + x = 1$, or $5x = 1$, or $x = \frac{1}{5}$. So the probability of winning is $\frac{4}{5}$.

> Now do Exercises 27–36

We can write the idea found in Example 9 as a strategy for converting from odds to probabilities.

Strategy for Converting from Odds to Probability

If the odds in favor of event E are a to b, then

$$P(E) = \frac{a}{a+b} \quad \text{and} \quad P(\overline{E}) = \frac{b}{a+b}.$$

Warm-Ups ▼

Fill in the blank.

1. A(n) _____ is a process for which the outcomes are uncertain.

2. A(n) _____ is the set of all possible outcomes to an experiment.

3. A(n) _____ is a subset of a sample space.

4. According to the _____ rule $P(A \cup B) = P(A) + P(B) - P(A \cap B)$ for any events A and B.

5. If $A \cap B = \varnothing$ and $P(A) + P(B) = 1$, then A and B are _____ events.

6. The _____ in favor of an event is the ratio of the probability of the event to the probability of the complement of the event.

True or false?

7. If two coins are tossed, then the sample space consists of 4 equally likely outcomes.

8. If three coins are tossed, then the sample space consists of 6 equally likely outcomes.

9. The probability of getting at least one tail when a single coin is tossed twice is 0.75.

10. The probability of getting a number greater than 1 when a single die is tossed is $\frac{5}{6}$.

11. If the odds in favor of the Giants winning the game are 2 to 3, then the probability that the Giants win is $\frac{2}{3}$.

12. If the probability of rain today is 80%, then the odds in favor of rain today are 4 to 1.

Exercises 14.3

‹ Study Tips ›

- Everyone learns and studies differently. So all of these *Study Tips* will not apply to everyone.
- If you find something in these tips that applies to you or helps you, then that is good.

‹1› The Probability of an Event

Solve each probability problem. See Examples 1–3.

1. If a single die is tossed, then what is the probability of getting

   a) a number larger than 3?

   b) a number less than or equal to 5?

   c) a number other than 6?

   d) a number larger than 7?

   e) a number smaller than 9?

2. If a single coin is tossed once, then what is the probability of getting

   a) tails?

   b) fewer than two heads?

   c) exactly three heads?

3. If a pair of coins is tossed, then what is the probability of getting

   a) exactly two heads?
   b) at least one tail?
   c) exactly two tails?
   d) at most one tail?

4. If a single coin is tossed twice, then what is the probability of getting

   a) heads followed by tails?
   b) two heads in a row?
   c) a tail on the second toss?
   d) exactly one tail?

5. If a pair of dice is tossed, then what is the probability of getting

   a) a pair of 2's?
   b) at least one 2?
   c) a sum of 7?
   d) a sum greater than 1?
   e) a sum less than 2?

6. If a single die is tossed twice, then what is the probability of getting

   a) a 1 followed by a 2?
   b) a sum of 3?
   c) a 6 on the second toss?
   d) no more than two 5's?
   e) an even number followed by an odd number?

7. A ball is selected at random from a jar containing 3 red balls, 4 yellow balls, and 5 green balls. What is the probability that

   a) the ball is red?
   b) the ball is not yellow?
   c) the ball is either red or green?
   d) the ball is neither red nor green?

8. A committee consists of 1 Democrat, 5 Republicans, and 6 independents. If one person is randomly selected from the committee to be the chairperson, then what is the probability that

   a) the person is a Democrat?
   b) the person is either a Democrat or a Republican?
   c) the person is not a Republican?

Figure for Exercise 7

9. A jar contains 10 balls numbered 1 through 10. Two balls are randomly selected one at a time without replacement. What is the probability that

   a) 1 is selected first and 2 is selected second?
   b) the sum of the numbers selected is 3?
   c) the sum of the numbers selected is 6?

10. A small company consists of a president, a vice president, and 14 salespeople. If 2 of the 16 people are randomly selected to win a Hawaiian vacation, then what is the probability that none of the salespeople is a winner?

11. If a 5-card poker hand is drawn from a deck of 52, then what is the probability that

   a) the hand contains the ace, king, queen, jack, and ten of spades?
   b) the hand contains one 2, one 3, one 4, one 5, and one 6?

12. If 5 people with different names and different weights randomly line up to buy concert tickets, then what is the probability that

   a) they line up in alphabetical order?
   b) they line up in order of increasing weight?

⟨2⟩ The Addition Rule

Use the addition rule to solve each problem. See Examples 4 and 5.

13. Among the drivers insured by American Insurance, 65% are women, 38% of the drivers are in a high-risk category, and 24% of the drivers are high-risk women. If a driver is randomly selected from that company, what is the probability that the driver is either high-risk or a woman?

14. What is the probability of getting either a sum of 7 or at least one 4 in the toss of a pair of dice?

15. A couple plans to have 3 children. Assuming males and females are equally likely, what is the probability that they have either 3 boys or 3 girls?

16. What is the probability of getting a sum of 10 or a sum of 5 in the toss of a pair of dice?

17. What is the probability of getting either a heart or an ace when drawing a single card from a deck of 52 cards?

18. What is the probability of getting either a heart or a spade when drawing a single card from a deck of 52 cards?

⟨3⟩ Complementary Events

Solve each problem. See Examples 6 and 7.

19. If the probability of surviving a head-on car accident at 55 mph is 0.005, then what is the probability of not surviving?

20. If the probability of a tax return not being audited by the IRS is 0.97, then what is the probability of a tax return being audited?

21. A pair of dice is tossed. What is the probability of

   a) getting a pair of 4's?

   b) not getting a pair of 4's?

   c) getting at least one number that is not a 4?

22. Three coins are tossed. What is the probability of

   a) getting three heads?

   b) not getting three heads?

   c) getting at least one tail?

⟨4⟩ Odds

Solve each problem. See Examples 8 and 9.

23. If the probability is 60% that the eye of Hurricane Edna comes ashore within 30 miles of Charleston, then what are the odds in favor of the eye of Edna coming ashore within 30 miles of Charleston?

24. If the probability that a Sidewinder missile hits its target is $\frac{8}{9}$, then what are the odds

   a) in favor of the Sidewinder hitting its target?

   b) against the Sidewinder hitting its target?

25. If the probability that the stock market goes up tomorrow is $\frac{3}{5}$, then what are the odds

   a) in favor of the stock market going up tomorrow?

   b) against the stock market going up tomorrow?

26. If the probability of a coal miners' strike this year is $\frac{9}{10}$, then what are the odds

   a) in favor of a strike?

   b) against a strike?

Photo for Exercise 23

27. If the odds are 3 to 1 in favor of the Black Hawks winning their next game, then

   a) what are the odds against the Black Hawks winning their next game?

   b) what is the probability that the Black Hawks win their next game?

28. If the odds are 5 to 1 against the Democratic presidential nominee winning the election, then

   a) what are the odds in favor of the Democrat winning the election?

   b) what is the probability that the Democrat wins the election?

29. What are the odds in favor of getting exactly 2 heads in 3 tosses of a coin?

30. What are the odds in favor of getting a 6 in a single toss of a die?

31. What are the odds in favor of getting a sum of 8 when tossing a pair of dice?

32. What are the odds in favor of getting at least one 6 when tossing a pair of dice?

Photo for Exercise 32

33. If one million lottery tickets are sold and only one of them is the winning ticket, then what are the odds in favor of winning if you hold a single ticket?

34. What are the odds in favor of winning a lottery where you must choose 6 numbers from the numbers 1 through 49?

35. If the odds in favor of getting 5 heads in 5 tosses of a coin are 1 to 31, then what is the probability of getting 5 heads in 5 tosses of a coin?

36. If the odds against Smith winning the election are 2 to 5, then what is the probability that Smith wins the election?

Getting More Involved

37. In the Louisiana Lottery a player chooses 6 numbers from the numbers 1 through 44. You win the big prize if the 6 chosen numbers match the 6 winning numbers chosen on Saturday night.

    a) What is the probability that you choose all 6 winning numbers?

    b) What is the probability that you do not get all 6 winning numbers?

    c) What are the odds in favor of winning the big prize with a single entry?

38. In the Louisiana Power Ball a player chooses 5 numbers from the numbers 1 through 49 and one number (the power ball) from 1 through 42.

    a) How many ways are there to choose the 5 numbers and choose the power ball?

    b) What is the probability of winning the big prize in the Louisiana Power Ball?

    c) What are the odds in favor of winning the big prize?

Math *at Work*   Roulette

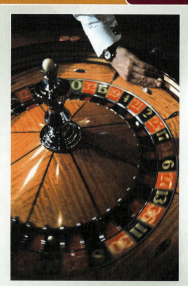

Thinking about going to a casino to play roulette and get rich? There's something you should know before you put your money on the table.

The roulette wheel is divided into 38 numbered slots. Two of these slots are green, 18 are red, and 18 are black. The wheel is spun and a ball is dropped onto its outside edge. When the wheel stops, the ball randomly drops into 1 of the 38 slots. Players bet on which slot they believe the ball will land in. If you put your money on the number 10 slot, you have probability 1/38 of winning and 37/38 of losing. If you bet that the ball will land in any of the 18 red slots, the safest bet in roulette, your chances are 18/38 that you win and 20/38 that you lose.

Both you and the casino owners are gambling, but what gives the casino the advantage? It is not the probabilities of winning or losing, it is the payoff. Betting $10 on red you have 18/38 probability of winning $10 and 20/38 probability that you lose your $10. No one knows where the ball will land on any individual spin of the wheel, but the probabilities tell you that in the long run, say 38 bets on red, you will win 18 times and lose 20 times, which means the result is losing $20. So losing $20 in 38 bets is an average of $20/38 or about 53 cents per bet. Another way to get the 53 cents is to multiply each result by the probability of that result and add: $10(18/38) + (−$10)(20/38) ≈ −53 cents. This means that you lose 53 cents and the casino makes 53 cents for every $10 bet on red. The results for all other possible bets can be calculated as well, and they are all in favor of the casino.

If you spend an hour at the roulette wheel, you might end up a winner or a loser. But the casino is not taking any chances. Since the casino is playing the game for the long run, the casino cannot lose.

Wrap-Up

Summary

Counting		Examples
Fundamental counting principle	If event A has m different outcomes and event B has n different outcomes, then there are mn different ways for events A and B to occur.	If a pie can be ordered with 2 different crusts and 5 fillings, then 10 pies are available.
Permutations of n things r at a time	The number of permutations of n things taken r at a time is $P(n, r) = \dfrac{n!}{(n - r)!}$ for $0 \le r \le n$. $P(n, n) = n!$ and $P(n, 0) = 1$.	The number of ways to choose Miss America and the First Runner Up from 5 finalists is $P(5, 2) = 5 \cdot 4$. The number of ways for 5 people to stand in line is $P(5, 5) = 5!$.
Combinations of n things r at a time	The number of combinations of n things taken r at a time is $C(n, r) = \dfrac{n!}{(n - r)!\, r!}$ for $0 \le r \le n$. $C(n, n) = 1$ and $C(n, 0) = 1$.	The number of ways to rate 2 of 5 people as excellent is $C(5, 2) = \dfrac{5!}{3!\,2!}$. $C(5, 5) = \dfrac{5!}{0!\,5!} = 1,\ C(5, 0) = \dfrac{5!}{5!\,0!} = 1$
Labeling	One label is given to each of n distinct objects. If there are r_1 labels of type 1, r_2 labels of type 2, $\ldots$, and r_k labels of type k, then the number of ways to label is $\dfrac{n!}{r_1!\,r_2!\,\ldots\,r_k!}$.	If 5 students are randomly given 2 A's, 2 B's, and 1 C, then the number of ways to assign these grades is $\dfrac{5!}{2!\,2!\,1!}$.

Coefficients		Examples
Binomial coefficients	The coefficient of the term $a^r b^{n-r}$ in the expansion of $(a + b)^n$ is $C(n, r)$ for $0 \le r \le n$.	The coefficient of $a^3 b^2$ in $(a + b)^5$ is $C(5, 3) = \dfrac{5!}{3!\,2!} = 10$.
Multinomial coefficients	Determining the coefficients in a multinominal expansion is a labeling problem.	The coefficient of $a^2 b^2 c^3$ in $(a + b + c)^7$ is $\dfrac{7!}{2!\,2!\,3!} = 210$.

Probability		Examples
Probability of an event	If S is a sample space of equally likely outcomes to an experiment and E is a subset of S, then $P(E) = \dfrac{n(E)}{n(S)}$.	If a single die is tossed, then $S = \{1, 2, 3, 4, 5, 6\}$. If $A = \{4, 5, 6\}$, then $P(A) = \dfrac{3}{6}$.

Mutually exclusive events	Two events are called mutually exclusive when it is impossible for both to occur.	If a die is tossed once, then {1} and {2} are mutually exclusive.
Addition rule	If A and B are any events in a sample space, then $P(A \cup B) = P(A) + P(B) - P(A \cap B)$. If A and B are mutually exclusive events in a sample space, then $P(A \cup B) = P(A) + P(B)$.	If $A = \{4, 5, 6\}$ and $B = \{3, 4\}$, then $P(A \cup B) = \dfrac{3}{6} + \dfrac{2}{6} - \dfrac{1}{6} = \dfrac{4}{6}$. If $C = \{1\}$ and $D = \{2\}$, then $P(C \cup D) = \dfrac{1}{6} + \dfrac{1}{6} = \dfrac{2}{6}$.
Complementary events	Two events A and $\overline{A}$ are complementary events if $A \cap \overline{A} = \varnothing$ and $P(A) + P(\overline{A}) = 1$.	If $C = \{1\}$, then $\overline{C} = \{2, 3, 4, 5, 6\}$ and $P(C) + P(\overline{C}) = \dfrac{1}{6} + \dfrac{5}{6} = 1$.

Odds		**Examples**
Odds	For an event A the odds in favor of A is the ratio of $P(A)$ to $P(\overline{A})$, and the odds against A is the ratio $P(\overline{A})$ to $P(A)$.	The odds in favor of getting 1 on a single toss of a die are 1 to 5, and the odds against getting 1 are 5 to 1.
Converting from odds to probability	If the odds in favor of E are a to b, then $P(E) = \dfrac{a}{a + b}$ and $P(\overline{E}) = \dfrac{b}{a + b}$.	Odds in favor of $E = \{1, 2, 3, 4\}$ in tossing a die are 2 to 1. So, $P(E) = \dfrac{2}{2 + 1} = \dfrac{2}{3}$ and $P(\overline{E}) = \dfrac{1}{3}$.

Enriching Your Mathematical Word Power

Fill in the blank.

1. A(n) ____ diagram is used to show all of the possibilities.

2. A(n) _____ is a linear arrangement.

3. A(n) _____ is a group or subset.

4. The coefficients in the expansion of a polynomial of more than two terms are _____ coefficients.

5. A(n) _____ is a process with an uncertain outcome.

6. The _____ space is the set of all possible outcomes to an experiment.

7. A(n) _____ is a subset of a sample space.

8. Two events that cannot both occur are mutually _____ events.

9. Two events that cannot both occur but one of them must occur are _____ events.

10. The ratio of the probability of an event occurring to the probability of the event not occurring are the ____ in favor of the event.

Review Exercises

14.1 Counting and Permutations
Solve each problem.

1. How many different ways are there to mark the answers to a 15-question multiple-choice test in which each question has 3 possible answers?

2. Eight airplanes are scheduled to depart at 1 o'clock. In how many ways can they line up for departure?

3. John is trying to find all 3-letter words that can be formed without repetition using the letters in the word UNITED. He plans to write down all "possible" 3-letter words and then check each one with a dictionary to see whether it is actually a word in the English language. How many "possible" 3-letter words are there?

4. The sales manager for an insurance company must select a team of 2 agents to give a presentation. If the manager has 9 male agents and 7 female agents available, and the team must consist of one man and one woman, then how many different teams are possible?

5. A candidate for city council is going to place one newspaper advertisement, one radio advertisement, and one television advertisement. If there are 3 newspapers, 12 radio stations, and 4 television stations available, then in how many ways can the advertisements be placed?

6. A ship has 11 different flags available. A signal consists of 3 flags displayed on a vertical pole. How many different signals are possible?

Photo for Exercise 6

14.2 Combinations
Solve each problem.

7. A travel agent offers a vacation in which you can visit any 4 cities, chosen from Paris, Rome, London, Istanbul, Monte Carlo, Vienna, Madrid, and Berlin. How many different vacations are possible, if the order in which the cities are visited is not important?

8. How many 5-element subsets are there from the set {a, b, c, d, e, f, g}?

9. A city council consists of 5 Democrats and 4 Republicans. In how many ways can 4 council members be selected by the mayor to go to a convention in San Diego if the mayor
  a) may choose any 4?
  b) must choose 4 Democrats?
  c) must choose 2 Democrats and 2 Republicans?

10. In a 5-card poker hand a full house is 3 cards of one kind and 2 cards of another. How many full houses are there consisting of three kings and two jacks?

11. What is the coefficient of x^7y^3 in the expansion of $(x + y)^{10}$?

12. What is the coefficient of a^4b^3 in the expansion of $(a + b)^7$?

13. What is the coefficient of $w^2x^3y^6$ in the expansion of $(w + x + y)^{11}$?

14. What is the coefficient of x^5y^7 in the expansion of $(w + x + y)^{12}$?

15. How many different arrangements are there for the letters in the word ARKANSAS? MARCH?

16. Eight Toyota pickups of different colors are on sale through Saturday only. How many ways are there for the dealer to mark 2 of them $7000, 3 of them $8000, and 3 of them $9000?

17. A couple plans to have 6 children. How many different families are possible, considering the sex of each child and the order of birth?

18. The Galaxy Theater has 3 horror films to show on Saturday night. How many different triple features are possible?

14.3 Probability
Solve each problem.

19. If Miriam randomly marks the answers to a 12-question true-false test, then what is the probability that she gets all 12 correct? What is the probability that she gets all 12 wrong?

20. If a couple plans to have 6 children, then what is the probability that they will have 6 girls?

21. There are 6 red, 5 green, and 3 yellow jelly beans in a jar. If a jelly bean is selected at random, then what is the probability that
  a) the selected bean is green?
  b) the selected bean is either yellow or red?
  c) the selected bean is blue?
  d) the selected bean is not purple?

22. A pair of dice is tossed. What is the probability that
  a) at least one die shows an even number?
  b) the sum is an even number?
  c) the sum is 6?
  d) the sum is 6 and at least one die shows an even number?
  e) the sum is 6 or at least one die shows an even number?

23. Suppose a couple plans to have 3 children. What are the odds in favor of having 3 boys?

24. Suppose a pair of dice is tossed. What are the odds in favor of the sum being 6?

25. If 60% of the voters in Massachusetts are Democrats and a single voter is drawn at random, then what are the odds in favor of that voter being a Democrat?

26. Seventy percent of the students at Wellknown University have money as their primary goal in life. If a student is selected at random, then what are the odds against selecting one who has money as his or her primary goal in life?

27. At Wellknown University 60% of the students are enrolled in an English class, 70% are enrolled in a mathematics class, and 40% are enrolled in both. If a student is selected at random, then what is the probability that the student is enrolled in either mathematics or English?

28. If a pair of dice is tossed, then what is the probability that the sum is either 5 or 7?

Miscellaneous

Evaluate each expression.

29. $8!$

30. $1!$

31. $\dfrac{5!}{3!}$

32. $\dfrac{7!}{(7-3)!}$

33. $\dfrac{9!}{3!\,3!\,3!}$

34. $\dfrac{10!}{0!\,2!\,3!\,5!}$

35. $C(8, 6)$

36. $\dbinom{12}{3}$

37. $P(8, 4)$

38. $P(4, 4)$

39. $C(8, 1)$

40. $C(12, 0)$

Chapter 14 Test

Solve each problem.

1. The Turbo ZXS computer comes with your choice of any one of 4 color monitors, 6 sizes of hard disks, and 3 different keyboards. How many different configurations are possible for this model?

2. One gold, one silver, and one bronze medal will be awarded to 3 students who are randomly chosen from the 20 who are entered in the sack race at the Oak Ridge Elementary School picnic. In how many ways can the 3 prizes be awarded?

3. A tour of New York City includes visits to any 3 attractions chosen from the Empire State Building, the United Nations, the Statue of Liberty, the Stock Exchange, and Central Park. How many different tours are possible, if the order in which the attractions are visited is not important?

Photo for Exercise 2

Photo for Exercise 3

4. From a class of 12 boys and 8 girls, the teacher must choose 2 boys and 2 girls for crossing guards. How many teams of 4 crossing guards are possible?

5. How many ways are there for Mother Nature to select 4 workers randomly out of the 12 available workers in the office to get the flu?

6. What is the coefficient of a^3b^8 in the expansion of $(a + b)^{11}$?

7. What is the coefficient of $w^4x^3y^2z^5$ in the expansion of $(w + x + y + z)^{14}$?

8. How many different arrangements are there for the 6 letters in the word ALASKA?

9. What is the probability that at least one head is showing when a pair of coins is tossed?

10. What is the probability that at least one head is showing or at least one tail is showing when a pair of coins is tossed?

11. One card is to be drawn from a deck of 52. What are the odds in favor of getting a spade?

12. The Fujimotos plan to have 3 children. What are the odds against having 3 boys?

13. At the start of the season professional odds makers in Las Vegas put the odds in favor of the Saints going to the Super Bowl at 1 to 20.

   a) What are the odds against the Saints going to the Super Bowl?

   b) What is the probability that the Saints go to the Super Bowl?

   c) What is the probability that the Saints do not go to the Super Bowl?

14. A pair of dice is tossed. Let A be the event that at least one die shows a 3 and B be the event that the sum is 7. Find the following probabilities.

   a) $P(A)$ **b)** $P(B)$

   c) $P(A \cap B)$ **d)** $P(A \cup B)$

Evaluate each expression.

15. $C(100, 2)$

16. $P(200, 3)$

17. $\dfrac{10!}{3!\,4!\,3!}$

MakingConnections | A Review of Chapters 1–14

Evaluate each expression if $x = -5$ and $y = 4$.

1. $-3x - 6y$

2. $2y^2 - x^2$

3. $\dfrac{x - y}{y - x}$

4. $\dfrac{x^2 - y^2}{x - y}$

Write each union or intersection of intervals as a single interval.

5. $(-3, 5) \cup (4, 9)$

6. $(-1, 5) \cup (2, 3)$

7. $(-2, 4) \cap (3, 9)$

8. $(-\infty, 6] \cap (-\infty, 9)$

In each case, express a as a function of b.

9. $a + b = 5$

10. $3b - 6a = 12$

11. $ab - 4 = a$

12. $\dfrac{1}{a} + \dfrac{1}{b} = 6$

Solve each equation.

13. $-\dfrac{3}{4}x = \dfrac{1}{2}$

14. $\dfrac{3}{4}x + \dfrac{3}{8} = \dfrac{9}{2}$

15. $0.05x = 420 - x$

16. $\dfrac{0.5x - 8}{0.02} = 10(x + 20)$

17. $1 - 3(4x + 12) = 99 - 2(6x + 1)$

18. $4 - 6(2x + 2) = -5 - 3(4x + 1)$

19. $|x| - 6 = 4$

20. $|x - 6| = 4$

21. $|x - 6| = 0$

22. $|x| + 6 = 4$

Solve each inequality. State the solution set using interval notation.

23. $3x - 6 \geq 0$

24. $6 - 3x < 9$

25. $4x > 12$ and $2x < 12$

26. $x - 3 > -6$ or $x + 5 < 10$

27. $-6 < 3x < 12$

28. $-4 \leq \dfrac{2x}{9} < 2$

29. $|8x| > 16$

30. $|x - 1| \leq 11$

31. $x^2 - 4 \geq 0$

32. $x^2 - x - 6 < 0$

33. $x^3 - x > 0$

34. $x^3 - x^2 \geq 0$

35. $x^4 - x^2 \leq 0$

36. $x^3 + x^2 - x - 1 < 0$

Perform the indicated operations.

37. $3 - 6(x - 4)$

38. $(6 - 5x)(3x - 4)$

39. $(x - 6)(x^2 - 2x + 1)$

40. $\dfrac{5}{2x} + \dfrac{6}{3x}$

41. $\dfrac{9}{x - 2} - \dfrac{7}{x^2 - 4}$

42. $\dfrac{6}{x - 3} \cdot \dfrac{x^2 - 9}{4x}$

Factor each expression completely.

43. $9w^4 + 18w^3 - 72w^2$

44. $18n^3 + 12n^2 + 3n$

45. $-6a^2 - 24ab - 24b^2$

46. $2a^2 + am + 2m + 4a$

47. $16y^3 - 250$

48. $a^3 - 5a^2 - 9a + 45$

Which elements of
$$\left\{ -6, -\sqrt[3]{7}, -\dfrac{5}{8}, 0, 3, \sqrt{3}, 2.9, \pi, \dfrac{32}{4}, 2 - i \right\}$$

are members of these sets?

49. Whole numbers

50. Natural numbers

51. Integers

52. Rational numbers

53. Irrational numbers

54. Real numbers

55. Imaginary numbers

56. Complex numbers

Evaluate each expression.

57. $\sqrt{\dfrac{144}{49}}$

58. $\sqrt[4]{\dfrac{1}{81}}$

59. $(16 + 9)^{-1/2}$

60. $8^{2/3} + 4^{-1/2}$

61. $9^{3/8} \cdot 9^{1/8}$

62. $\dfrac{16^{3/16}}{16^{-1/16}}$

Solve each system of equations, and state whether the system is independent, dependent, or inconsistent.

63. $6x - 9y = -6$
$2x + y = 2$

64. $8x + 6y = 27$
$y = x + 1$

65. $x - 9y = 18$
$y = \dfrac{1}{9}x - 2$

66. $x = 12y + 8$
$y = \dfrac{1}{12}x - \dfrac{1}{3}$

67. $2x + y - z = -4$
$x - 2y + z = 9$
$3x - y - z = 1$

68. $x + y + z = 2$
$x + 2y + 3z = 8$
$2x + 3y + 4z = 6$

Let $f(x) = 3x^2 - 1$, and $g(x) = x + 5$. Find the following.

69. $f(-4)$

70. $g(-2)$

71. $g^{-1}(3)$

72. $g^{-1}(-9)$

73. $(f \circ g)(-3)$

74. $(g \circ f)(-1)$

75. $(f + g)(2)$

76. $(g - f)(-1)$

Solve each problem.

77. The perimeter of a rectangular field is 860 yards. If the width is 10 yards less than the length, then what is the length?

78. The area of a rectangular mirror is 768 square inches. If the length is 32 inches, then what is the width?

79. A television is advertised at 20% off the regular price. If the advertised price is $1360, then what is the regular price?

80. A university has 12,000 students of which 4% are minorities. How many more minority students must the university enroll so that the university has 10% minority students?

81. The area of a rectangular coffee table is 1620 square inches and the length is 45 inches. What is the width?

82. The diagonal measure of a new television screen is 1 inch greater than the length and 2 inches greater than the width. Find the length and width.

83. A ball is thrown straight upward. Its height above the ground in feet at time t seconds is given by $h(t) = -16t^2 + 64t$.

    a) What is the height of the ball 1 second after it is thrown?

    b) For how long is the ball in the air?

84. How long to the nearest day does it takes for $1000 to grow to $2000 at 5% compounded continuously?

Critical Thinking | For Individual or Group Work | Chapter 14

These exercises can be solved by a variety of techniques, which may or may not require algebra. So be creative and think critically. Explain all answers. Answers are in the Instructor's Edition of this text.

1. *Table game.* Place one of the integers from 1 through 9 in each cell of table (a) of the accompanying figure. Do not use an integer more than once. Integers in cells that touch must differ by more than 1. Repeat this process with table (b).

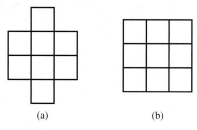

(a)   (b)

Figure for Exercise 1

2. *Let's Make a Deal.* Monty Hall shows you three boxes. He tells you that one of the boxes contains a diamond ring and the other two are empty. Monty knows which one contains the ring. He lets you choose a box but not open it. He then opens one of the unchosen boxes and shows you that it is empty. He then gives you the opportunity to trade the originally chosen box for the other unopened box. What should you do?

3. *Going to class.* Al, Bob, and Coddy must all get to an 8 A.M. class that is 6 miles from their house. They average 3 mph walking or 30 mph riding Al's motorcycle. Of course the motorcycle holds only 2 people.

a) What is the minimum amount of time needed to get all of them to class?
b) What if there is a fourth person and only one motorcycle?
c) If we keep increasing the number of people, what happens to the minimum time?

4. *Factorial fever.* The number 53! is the product of the positive integers from 1 through 53. This number is huge. With what digit does 53! end? How many times does that digit appear consecutively at the end of the number?

5. *Angle bisectors.* The angle bisectors of any triangle meet at a single point. If the hypotenuse of a 30-60-90 triangle is 4 units, then what is the exact distance from the vertex of the right angle to the point where the angle bisectors meet?

6. *Powers of i.* Evaluate
$$i^{0!} + i^{1!} + i^{2!} + \cdots + i^{100!}.$$

7. *Summing integers.* Find the exact sum of all positive 10-digit integers.

8. *Open and shut case.* At Brentwood High the lockers are numbered 1 through 500 in order down a long hallway. All lockers are closed, and the students are standing by their lockers. A student "changes the status of a locker" by opening a closed locker or closing an open locker. The first student changes the status of every locker starting with his. Then the second student changes the status of every other locker starting with hers. Then the third student changes the status of every third locker starting with his. Then the fourth student changes the status of every fourth locker starting with hers. This exercise continues through the five-hundredth student. Which lockers will be open when this exercise is finished?

Photo for Exercise 8

Appendix A

Geometry Review Exercises

(Answers are at the end of the answer section in this text.)

1. Find the perimeter of a triangle whose sides are 3 in., 4 in., and 5 in.

2. Find the area of a triangle whose base is 4 ft and height is 12 ft.

3. If two angles of a triangle are 30° and 90°, then what is the third angle?

4. If the area of a triangle is 36 ft^2 and the base is 12 ft, then what is the height?

5. If the side opposite 30° in a 30-60-90 right triangle is 10 cm, then what is the length of the hypotenuse?

6. Find the area of a trapezoid whose height is 12 cm and whose parallel sides are 4 cm and 20 cm.

7. Find the area of the right triangle that has sides of 6 ft, 8 ft, and 10 ft.

8. If a right triangle has sides of 5 ft, 12 ft, and 13 ft, then what is the length of the hypotenuse?

9. If the hypotenuse of a right triangle is 50 cm and the length of one leg is 40 cm, then what is the length of the other leg?

10. Is a triangle with sides of 5 ft, 10 ft, and 11 ft a right triangle?

11. What is the area of a triangle with sides of 7 yd, 24 yd, and 25 yd?

12. Find the perimeter of a parallelogram in which one side is 9 in. and another side is 6 in.

13. Find the area of a parallelogram which has a base of 8 ft and a height of 4 ft.

14. If one side of a rhombus is 5 km, then what is its perimeter?

15. Find the perimeter and area of a rectangle whose width is 18 in. and length is 2 ft.

16. If the width of a rectangle is 8 yd and its perimeter is 60 yd, then what is its length?

17. The radius of a circle is 4 ft. Find its area to the nearest tenth of a square foot.

18. The diameter of a circle is 12 ft. Find its circumference to the nearest tenth of a foot.

19. A right circular cone has radius 4 cm and height 9 cm. Find its volume to the nearest hundredth of a cubic centimeter.

20. A right circular cone has a radius 12 ft and a height of 20 ft. Find its lateral surface area to the nearest hundredth of a square foot.

21. A shoe box has a length of 12 in., a width of 6 in., and a height of 4 in. Find its volume and surface area.

22. The volume of a rectangular solid is 120 cm^3. If the area of its bottom is 30 cm^2, then what is its height?

23. What are the area and perimeter of a square in which one of the sides is 10 mi long?

24. Find the perimeter of a square whose area is 25 km^2.

25. Find the area of a square whose perimeter is 26 cm.

26. A sphere has a radius of 2 ft. Find its volume to the nearest thousandth of a cubic foot and its surface area to the nearest thousandth of a square foot.

27. A can of soup (right circular cylinder) has a radius of 2 in. and a height of 6 in. Find its volume to the nearest tenth of a cubic inch and total surface area to the nearest tenth of a square inch.

28. If one of two complementary angles is 34°, then what is the other angle?

29. If the perimeter of an isosceles triangle is 29 cm and one of the equal sides is 12 cm, then what is the length of the shortest side of the triangle?

30. A right triangle with sides of 6 in., 8 in., and 10 in., is similar to another right triangle that has a hypotenuse of 25 in. What are the lengths of the other two sides in the second triangle?

31. If one of two supplementary angles is 31°, then what is the other angle?

32. Find the perimeter of an equilateral triangle in which one of the sides is 4 km.

33. Find the length of a side of an equilateral triangle that has a perimeter of 30 yd.

Answers to Selected Exercises

Chapter 1

Section 1.1 Warm-Ups
1. Set   **2.** Finite   **3.** Venn   **4.** Subset   **5.** Empty
6. Intersection   **7.** Union   **8.** False   **9.** False   **10.** False
11. True   **12.** True   **13.** True   **14.** True   **15.** False
16. True   **17.** True

Section 1.1 Exercises
1. True   **3.** True   **5.** False   **7.** True   **9.** True
11. $\{1, 2, 3, 4, 5, 7, 9\}$ **13.** $\{1, 3, 5\}$ **15.** $\{1, 2, 3, 4, 5, 6, 8\}$
17. A **19.** $\varnothing$ **21.** A **23.** $=$ **25.** $\cup$ **27.** $\cap$ **29.** $\notin$ **31.** $\in$
33. True   **35.** True   **37.** True   **39.** True   **41.** False   **43.** True
45. $\{2, 3, 4, 5, 6, 7, 8\}$ **47.** $\{3, 5\}$ **49.** $\{1, 2, 3, 4, 5, 6, 8\}$
51. $\{2, 3, 4, 5\}$ **53.** $\{2, 3, 4, 5, 7\}$ **55.** $\{2, 3, 4, 5\}$ **57.** $\{2, 3, 4, 5, 7\}$
59. $\{2, 4, 6, \ldots, 18\}$ **61.** $\{13, 15, 17, \ldots\}$ **63.** $\{6, 8, 10, \ldots, 78\}$
65. $\{1, 2, 3, 4, 5, 6, 7, 8, 9\}$ **67.** $\{4, 5\}$ **69.** $\{2, 4, 5, 6, 8\}$
71. $\{1, 2, 3, 4, 5, 6, 7, 8, 9\}$ **73.** $\{5\}$ **75.** $\{4, 6, 8\}$
77. $\{x \mid x$ is a natural number between 2 and 7$\}$
79. $\{x \mid x$ is an odd natural number greater than 4$\}$
81. $\{x \mid x$ is an even natural number between 5 and 83$\}$
83. 13   **85.** No
87. a) $3 \in \{1, 2, 3\}$ **b)** $\{3\} \subseteq \{1, 2, 3\}$ **c)** $\varnothing \neq \{\varnothing\}$

Section 1.2 Warm-Ups
1. Integers   **2.** Natural   **3.** Rational   **4.** Terminating, repeating
5. Irrational   **6.** Real   **7.** Circumference   **8.** Union   **9.** False
10. True   **11.** False   **12.** False   **13.** True   **14.** True   **15.** True
16. False   **17.** True   **18.** True

Section 1.2 Exercises
1. False   **3.** True   **5.** True   **7.** False   **9.** False
11. $\{0, 1, 2, 3, 4, 5\}$
13. $\{-4, -3, -2, -1, 0, 1, \ldots\}$
15. $\{1, 2, 3, 4\}$
17. $\{-2, -1, 0, 1, 2, 3, 4\}$
19. Rational   **21.** Rational   **23.** Rational   **25.** Irrational
27. Rational   **29.** All
31. $\left\{0, \dfrac{8}{2}\right\}$ **33.** $\left\{-3, -\dfrac{5}{2}, -0.025, 0, 3\dfrac{1}{2}, \dfrac{8}{2}\right\}$ **35.** True   **37.** False
39. True   **41.** False   **43.** False   **45.** False   **47.** True   **49.** $\subseteq$
51. $\not\subseteq$ **53.** $\subseteq$ **55.** $\subseteq$ **57.** $\subseteq$ **59.** $\in$ **61.** $\in$ **63.** $\in$

65. $\notin$ **67.** $\subseteq$ **69.** $\subseteq$
71. $(1, \infty)$
73. $(-\infty, -1)$
75. $(3, 4)$
77. $[0, 2]$
79. $[1, 3)$
81. $[5, 7]$ **83.** $(-3, 0]$ **85.** $[60, \infty)$ **87.** $(-\infty, -5)$ **89.** $(1, 9)$
91. $(2, 3)$ **93.** $(-2, \infty)$ **95.** $(0, 2)$ **97.** $[2, 9]$ **99.** $[2, 6)$
101. False   **103.** True   **105.** False   **107.** True

Section 1.3 Warm-Ups
1. Additive   **2.** Absolute   **3.** Opposites   **4.** Negative   **5.** Sum
6. Product   **7.** Factors   **8.** Reciprocal   **9.** Quotient
10. Undefined   **11.** True   **12.** True   **13.** False   **14.** True
15. True   **16.** False   **17.** False   **18.** True   **19.** False   **20.** True

Section 1.3 Exercises
1. 9   **3.** 7   **5.** -4 **7.** 17   **9.** -5 **11.** 4
13. -7 **15.** -2 **17.** -10 **19.** -26 **21.** -2 **23.** -5
25. 0   **27.** $\dfrac{1}{10}$ **29.** $-\dfrac{1}{6}$ **31.** -14.98 **33.** -2.71 **35.** 2.803
37. -0.2649 **39.** -3 **41.** -11 **43.** 13   **45.** -6 **47.** -9
49. 23   **51.** 1   **53.** $-\dfrac{1}{2}$ **55.** 1.97   **57.** 7.3   **59.** -50.73
61. 1.27   **63.** -75 **65.** $\dfrac{1}{6}$ **67.** -0.09 **69.** 0.2   **71.** $\dfrac{1}{20}$ or 0.05
73. $-\dfrac{5}{6}$ **75.** $-\dfrac{10}{3}$ **77.** -2 **79.** -37.5 **81.** -0.08 **83.** 0.25
85. 0   **87.** Undefined   **89.** -12 **91.** $\dfrac{3}{5}$ **93.** -91.25
95. 17,000   **97.** 0   **99.** Undefined   **101.** -49 **103.** -7
105. 15   **107.** -342 **109.** $\dfrac{3}{4}$ **111.** -3 **113.** 20
115. 0   **117.** -55 **119.** -1 **121.** $-\dfrac{39}{2}$ **123.** 27.99
125. -29.3 **127.** -0.7 **129.** $44,400, addition
131. 20°F, subtraction   **133.** 1014 feet, subtraction

Mid-Chapter Quiz 1.1–1.3
1. True   **2.** False   **3.** False   **4.** True   **5.** True   **6.** False
7. True   **8.** False   **9.** True   **10.** True

11.
4 5 6 7 8

12.
4 5 6 7 8

13.
−1 0 1 2 3 4

14.
1 2 3 4

15. $(2, 8]$ **16.** $[4, 6)$ **17.** $-\dfrac{3}{8}$ **18.** $-\dfrac{1}{12}$ **19.** $\dfrac{1}{10}$

20. $-\dfrac{3}{4}$ **21.** 14   **22.** -1 **23.** -36 **24.** 5   **25.** -17

26. -100 **27.** 66   **28.** -7

Section 1.4 Warm-Ups

1. Arithmetic   **2.** Algebraic   **3.** Sum   **4.** Difference
5. Quotient   **6.** Exponential   **7.** Base   **8.** Square   **9.** Principal
10. Subscript   **11.** False   **12.** True   **13.** False   **14.** True
15. False   **16.** True   **17.** False   **18.** False   **19.** True   **20.** False

Section 1.4 Exercises

1. -22 **3.** -8 **5.** -14 **7.** 32   **9.** 1   **11.** $\dfrac{1}{9}$ **13.** 7

15. 10   **17.** 8   **19.** 13   **21.** -26 **23.** -8 **25.** 9   **27.** 17

29. $\dfrac{1}{24}$ **31.** 58   **33.** -25 **35.** -200 **37.** 40   **39.** -25

41. 16   **43.** -7.5 **45.** -22.4841 **47.** -1.9602 **49.** -276.48

51. -2 **53.** -1 **55.** -6 **57.** 0   **59.** Undefined   **61.** -7

63. $-\dfrac{4}{3}$ **65.** -8 **67.** 5   **69.** $-\dfrac{5}{2}$ **71.** 4   **73.** $\dfrac{10}{9}$ **75.** $\dfrac{3}{4}$

77. -2.67 **79.** 41   **81.** 27   **83.** 1   **85.** 9   **87.** -1 **89.** 26

91. $-\dfrac{3}{2}$ **93.** 17   **95.** -46 **97.** 3   **99.** 41

101. 26 beats per minute, age 43   **103.** 104 feet
105. a) \$60,000   b) \$60,776.47   **107.** \$5500, \$5441.96
109. a) 1275   b) 22,140   c) 216,225   d) 44,100   e) 166,375

Section 1.5 Warm-Ups

1. Commutative   **2.** Associative   **3.** Distributive   **4.** Additive
5. Multiplicative   **6.** Additive   **7.** Multiplicative   **8.** Product
9. True   **10.** True   **11.** True   **12.** False   **13.** False   **14.** False
15. True   **16.** False   **17.** True   **18.** False

Section 1.5 Exercises

1. 1   **3.** -14 **5.** -24 **7.** -1.7 **9.** -19.8 **11.** $4x - 24$
13. $3a + at$ **15.** $-2w + 10$ **17.** $2x + y$ **19.** $2x + 4$
21. $2(m + 5)$ **23.** $5(x - 1)$ **25.** $3(y - 5)$ **27.** $3(a + 3)$
29. $w(b + 1)$ **31.** 2   **33.** 1   **35.** $\dfrac{1}{6}$ **37.** 4   **39.** $-\dfrac{10}{7}$ **41.** $-\dfrac{5}{9}$
43. 0.6200   **45.** 0.7326   **47.** Commutative property of addition
49. Distributive property   **51.** Associative property of multiplication
53. Multiplicative inverse property
55. Commutative property of multiplication
57. Multiplicative identity property   **59.** Distributive property
61. Additive inverse property   **63.** Multiplication property of zero
65. Distributive property   **67.** $w + 5$
69. $(5x)y$ **71.** $\dfrac{1}{2}(x - 1)$ **73.** $3(2x + 3)$ **75.** 1   **77.** 0   **79.** 4

Section 1.6 Warm-Ups

1. Term   **2.** Like   **3.** Distributive   **4.** Simplify   **5.** Coefficient
6. True   **7.** True   **8.** False   **9.** False   **10.** False   **11.** True
12. False   **13.** True   **14.** True   **15.** False

Section 1.6 Exercises

1. 9000   **3.** 1   **5.** 527   **7.** 470   **9.** 38   **11.** 48,000
13. 0   **15.** 398   **17.** 1   **19.** 1700   **21.** 374   **23.** 0   **25.** $2n$
27. $7w$ **29.** $-11mw^2$ **31.** $-3x$ **33.** $9ay$ **35.** $8mn$ **37.** $-2kz^6$
39. $28t$ **41.** $10x^2$ **43.** h^2 **45.** $-28w$ **47.** $-x + x^2$ **49.** $25k^2$
51. y **53.** $2y$ **55.** $3x^3$ **57.** $x^2y + 5x$ **59.** $-x + 2$

61. $\dfrac{1}{2}xt - 5$ **63.** $-3a + 1$ **65.** $10 - x$ **67.** $3m + 1$

69. $-12b + at$ **71.** $2t^2 - 3w$ **73.** $y^2 + z$ **75.** $9x + 8$
77. $-9x + 10$ **79.** $12x - 2$ **81.** $-9x^2 + 5$ **83.** $9x^2 - 18x + 16$
85. $-7k^3 - 17$ **87.** $0.96x - 2$ **89.** $0.06x - 1.5$ **91.** $-4k + 16$

93. $-4xy - 22$ **95.** $-29w^2$ **97.** $-6a^2w^2$ **99.** $2x^2y + \dfrac{1}{3}$

101. $\dfrac{1}{4}m^2 - m$ **103.** $4t^3 + 3t^2 - 1$ **105.** $2xyz + xy - 3z$

107. $3s + 6$ ft   **109.** $\dfrac{13}{3}x$ m, $\dfrac{7}{6}x^2$ m^2

Enriching Your Mathematical Word Power

1. Term   **2.** Like   **3.** Variable   **4.** Additive   **5.** Order
6. Absolute   **7.** Integers   **8.** Natural   **9.** Whole   **10.** Rational
11. Irrational   **12.** Additive   **13.** Multiplicative
14. Dividend, divisor, quotient

Review Exercises

1. True   **3.** False   **5.** True   **7.** False   **9.** True   **11.** True
13. False   **15.** True   **17.** False   **19.** True   **21.** $\{0, 1, 31\}$
23. $\{-1, 0, 1, 31\}$ **25.** $\{-\sqrt{2}, \sqrt{3}, \pi\}$

27. $(0, \infty)$
−2 −1 0 1 2

29. $(5, 6)$
3 4 5 6 7 8

31. $[-1, 2)$
−2 −1 0 1 2 3 4

33. $(0, 5)$ **35.** $(3, 4)$ **37.** $[2, 8)$ **39.** 5   **41.** -12 **43.** -24
45. 2   **47.** $-\dfrac{1}{6}$ **49.** 10   **51.** 9.96   **53.** -4 **55.** 0   **57.** -4
59. 0   **61.** 39   **63.** 50   **65.** 121   **67.** 7   **69.** -19 **71.** 16
73. 23   **75.** 5   **77.** -1 **79.** 0   **81.** Undefined   **83.** 0.76
85. 1   **87.** 1   **89.** -8 **91.** 1   **93.** -35 **95.** 2
97. 5   **99.** -1 **101.** Commutative property of addition
103. Distributive property
105. Associative property of multiplication
107. Multiplicative identity property
109. Multiplicative inverse property
111. Multiplication property of zero
113. Additive identity property
115. Additive inverse property   **117.** $3w + 3$ **119.** $-x - 5$
121. $6x + 15$ **123.** $3(x - 2a)$ **125.** $7(x + 1)$ **127.** $p(1 - t)$
129. $a(b + 1)$ **131.** $7a + 2$ **133.** $-t - 2$ **135.** $-2a + 4$
137. $4x + 33$ **139.** $-0.8x - 0.48$ **141.** $-0.05x - 1.85$

143. $\dfrac{1}{4}x + 4$ **145.** $-3x^2 - 2x + 1$

147. 0, additive inverse, multiplication property of zero
149. 7680, distributive
151. 48, associative property of addition, additive inverse, additive
identity

153. 0, distributive, additive inverse
155. 47, associative property of multiplication, multiplicative inverse
157. -24, commutative property of multiplication, associative property of multiplication
159. 0, additive inverse, multiplication property of zero
161. $4x + 6$ feet, $x^2 + 3x$ square feet   **163. a)** \$84,547   **b)** 2019

Chapter 1 Test

1. $\{2, 3, 4, 5, 6, 7, 8, 10\}$ **2.** $\{6, 7\}$ **3.** $\{4, 6, 8, 10\}$ **4.** $\{0, 8\}$
5. $\{-4, 0, 8\}$ **6.** $\left\{-4, -\dfrac{1}{2}, 0, 1.65, 8\right\}$ **7.** $\{-\sqrt{3}, \sqrt{5}, \pi\}$
8.

$$-3 \;\; -2 \;\; -1 \;\;\; 0 \;\;\; 1 \;\;\; 2 \;\;\; 3 \;\;\; 4 \;\;\; 5$$

9.

$$-3 \;\; -2 \;\; -1 \;\;\; 0 \;\;\; 1 \;\;\; 2 \;\;\; 3 \;\;\; 4 \;\;\; 5$$

10. $(-\infty, 4)$ **11.** $[4, 8)$ **12.** -9 **13.** 8   **14.** -11 **15.** -1.98
16. -2 **17.** -4 **18.** $\dfrac{1}{18}$ **19.** -12 **20.** 7   **21.** -3
22. 0   **23.** 4780   **24.** 240   **25.** 40   **26.** 7   **27.** 0
28. Distributive property   **29.** Commutative property of multiplication
30. Associative property of addition   **31.** Additive inverse property
32. Commutative property of multiplication   **33.** $11m - 3$
34. $0.95x + 2.9$ **35.** $\dfrac{3}{4}x - \dfrac{5}{4}$ **36.** $5x^2$ **37.** $3x^2 + 2x - 1$
38. $5(x - 8)$ **39.** $7(t - 1)$
40. $4x - 8$, $x^2 - 4x$, 28 feet, 45 square feet   **41.** 12.6 billion

Chapter 2

Section 2.1 Warm-Ups
1. Equation   **2.** Satisfies   **3.** Equivalent   **4.** Linear   **5.** LCD
6. Identity   **7.** Conditional   **8.** Inconsistent   **9.** False   **10.** True
11. False   **12.** True   **13.** True   **14.** True   **15.** True   **16.** True
17. True   **18.** True

Section 2.1 Exercises
1. Yes   **3.** Yes   **5.** No   **7.** $\{21\}$ **9.** $\{4\}$ **11.** $\{14\}$
13. $\{-87\}$ **15.** $\{-24\}$ **17.** $\left\{\dfrac{3}{2}\right\}$ **19.** $\{-1\}$ **21.** $\{-3\}$ **23.** $\left\{\dfrac{5}{2}\right\}$
25. $\{18\}$ **27.** $\{18\}$ **29.** $\{0\}$ **31.** $\left\{-\dfrac{28}{3}\right\}$ **33.** $\left\{-\dfrac{28}{5}\right\}$ **35.** $\{2\}$
37. $\{12\}$ **39.** $\{-7\}$ **41.** $\{12\}$ **43.** $\{6\}$ **45.** $\{90\}$ **47.** $\{1000\}$
49. $\{800\}$ **51.** $\varnothing$, inconsistent   **53.** R, identity   **55.** $\{1\}$, conditional
57. R, identity   **59.** R, identity   **61.** $\varnothing$, inconsistent   **63.** R, identity
65. $\{-4\}$, conditional   **67.** $\{1\}$ **69.** $\varnothing$ **71.** R **73.** $\left\{\dfrac{5}{18}\right\}$
75. $\{0\}$ **77.** $\left\{\dfrac{10}{29}\right\}$ **79.** $\{3\}$ **81.** $\{16\}$ **83.** $\left\{\dfrac{3}{4}\right\}$ **85.** $\{-15\}$
87. $\{6\}$ **89.** $\{6\}$ **91.** $\{2.94\}$ **93.** $\{53,191.49\}$ **95.** $\{1.77\}$
97. a) 42.2 million   **b)** 2010   **c)** Increasing

Section 2.2 Warm-Ups
1. Formula   **2.** Solving   **3.** Function   **4.** Literal   **5.** Volume
6. Terms   **7.** False   **8.** True   **9.** True   **10.** True   **11.** True
12. True   **13.** False   **14.** False   **15.** True   **16.** False

Section 2.2 Exercises
1. $t = \dfrac{I}{Pr}$ **3.** $C = \dfrac{5}{9}(F - 32)$ **5.** $W = \dfrac{A}{L}$ **7.** $b_1 = 2A - b_2$

9. $L = \dfrac{P - 2W}{2}$ or $L = \dfrac{1}{2}P - W$ **11.** $h = \dfrac{V}{\pi r^2}$ **13.** $y = -\dfrac{2}{3}x + 3$
15. $y = x - 4$ **17.** $y = \dfrac{3}{2}x - 6$ **19.** $y = \dfrac{1}{2}x + \dfrac{1}{2}$ **21.** $t = \dfrac{A - P}{Pr}$
23. $a = \dfrac{1}{b + 1}$ **25.** $y = \dfrac{12}{1 - x}$ **27.** $x = \dfrac{6}{w^2 - y^2 - z^2}$
29. 12.472   **31.** 34.932   **33.** 0.539   **35.** $\dfrac{1}{3}$ **37.** $\dfrac{13}{2}$
39. -1 **41.** 4   **43.** -4.4507 **45.** $\dfrac{5}{8}$ **47.** $-\dfrac{7}{2}$ **49.** -10
51. $-\dfrac{8}{3}$ **53.** 4   **55.** 15%   **57.** 104%   **59.** One-half year
61. $A = \pi r^2$ **63.** $r = \dfrac{C}{2\pi}$ **65.** $W = \dfrac{P - 2L}{2}$ or $W = \dfrac{1}{2}P - L$
67. 5.75 yards   **69.** 7.2 feet   **71.** 5 feet   **73.** 15 feet
75. 14 inches   **77.** 168 feet   **79.** 1.5 meters   **81.** 3979 miles
83. 4.24 inches   **85.** 95,232 pounds   **87.** 200 feet   **89.** \$1200
91. 496 minutes   **93. a)** 2457   **b)** 420

Section 2.3 Warm-Ups
1. Consecutive   **2.** Odd, even   **3.** Perimeter   **4.** Addition
5. Uniform   **6.** Subtraction   **7.** Multiplication   **8.** False
9. True   **10.** False   **11.** False   **12.** True   **13.** True   **14.** False

Section 2.3 Exercises
1. $x, x + 2$ **3.** $x, 10 - x$ **5.** $x, x + 2$ or $x, x - 2$
7. $0.85x$ **9.** $3x$ miles   **11.** $4x + 10$
13. 27, 28, 29   **15.** 82, 84, 86   **17.** 63, 65   **19.** -13 **21.** 23
23. 42   **25.** Length 5 ft, width 3 ft   **27.** Length 10 in., width 3 in.
29. Length 16 cm, width 6 cm   **31.** Length 6 feet, width 4 feet
33. Width 11.25 feet, length 27.5 feet   **35.** 161 feet, 312 feet, 211 feet
37. \$4000 at 5%, \$8000 at 9%   **39.** \$1500 at 6%, \$2500 at 10%
41. \$80,000   **43.** \$200,000   **45.** $\dfrac{40}{3}$ gallons   **47.** 24 pounds
49. 3.75 gal of 5% alcohol, 2.25 gal of 13% alcohol
51. 1.36 ounces   **53.** 40 mph   **55.** 15 mph   **57.** 50 mph
59. 20 mph   **61.** \$86,957   **63.** \$8450   **65.** 21 feet
67. Length 13 m, width 6 m   **69.** 8 hours   **71.** \$7.20
73. Length 18 cm, width 6 cm   **75.** Packers 35, Chiefs 10
77. \$8.67 per pound   **79.** 30 pounds   **81.** $\dfrac{40}{7}$ quarts
83. 38 cents   **85.** Brian \$14,400, Daniel \$7,200, Raymond \$3,800
87. 22, 23   **89.** 7.5 hours   **91.** 20 meters by 20 meters
93. \$500 at 8%, \$2500 at 10%
95. 2 gallons of 5% solution, 3 gallons of 10% solution
97. 510 gallons   **99.** Todd 46, Darla 32

Mid-Chapter Quiz 2.1–2.3
1. $\{-10\}$ **2.** $\{-20\}$ **3.** $\{1\}$ **4.** $\{4\}$ **5.** $\{950\}$ **6.** $\{-200\}$
7. Identity   **8.** Conditional equation   **9.** Conditional equation
10. Inconsistent equation
11. $x = \dfrac{c - b}{a}$ **12.** $x = \dfrac{ma - 3b}{m - 3}$
13. $A = S^2$ **14.** \$15,800   **15.** 12 yd   **16.** 3   **17.** 42, 43, 44
18. \$10,000 at 15%, \$40,000 at 4%

Section 2.4 Warm-Ups
1. Less   **2.** Less than, equal   **3.** Linear   **4.** Reversed   **5.** $\geq$
6. Solution set   **7.** $[2, \infty)$ **8.** Equivalent   **9.** Trichotomy

10. False   **11.** False   **12.** True   **13.** True   **14.** True   **15.** True
16. True   **17.** False   **18.** True   **19.** False

Section 2.4 Exercises
1. False   **3.** True   **5.** True   **7.** True
9. Yes   **11.** No   **13.** No
15. $(-\infty, -1]$

17. $(20, \infty)$

19. $[3, \infty)$

21. $(-\infty, 2.3)$

23. $>$ **25.** $>$ **27.** $\leq$ **29.** $>$ **31.** $\leq$
33. $(-\infty, 2)$

35. $(-2, \infty)$

37. $[-4, \infty)$

39. $(-\infty, -2)$

41. $(5, \infty)$

43. $[1, \infty)$

45. $[-3, \infty)$

47. $(13, \infty)$

49. $[-1, \infty)$

51. $(-\infty, 4]$

53. $\left(\dfrac{2}{3}, \infty\right)$

55. $\left(\dfrac{13}{3}, \infty\right)$

57. $\varnothing$

59. $(-\infty, \infty)$

61. $(-\infty, \infty)$

63. $\varnothing$

65. $(-\infty, \infty)$

67. $\varnothing$

69. $x = $ Tony's height, $x > 6$ feet
71. $s = $ Wilma's salary, $s < \$80{,}000$
73. $v = $ speed of the Concorde, $v \leq 1450$ mph
75. $a = $ amount Julie can afford, $a \leq \$400$
77. $b = $ Burt's height, $b \leq 5$ feet
79. $t = $ Tina's hourly wage, $t \leq \$8.20$
81. $x = $ price of car, $x < \$9100$
83. $x = $ price of truck, $x \geq \$9100.92$
85. a) Decreasing   **b)** 2011
87. $x = $ final exam score, $x \geq 77$
89. $x = $ the price of A-Mart jeans, $x < \$16.67$
91. a) $[8, \infty)$ **b)** $(-\infty, -6)$ **c)** $(2, \infty)$ **d)** $(-\infty, -12)$ **e)** $(2, \infty)$

Section 2.5 Warm-Ups
1. Compound   **2.** And   **3.** Or   **4.** And   **5.** Intersection
6. Union   **7.** True   **8.** True   **9.** False   **10.** True   **11.** True
12. True   **13.** False   **14.** False   **15.** True   **16.** True

Section 2.5 Exercises
1. No   **3.** Yes   **5.** No   **7.** No   **9.** Yes   **11.** Yes   **13.** Yes
15.

17.

19.

21.

23. $\varnothing$

25.

27. $\varnothing$

29. $(-\infty, 1) \cup (10, \infty)$

31. $(9, \infty)$

33. $(-6, \infty)$

35. $(1, 4]$

37. $(-\infty, \infty)$

39. $\varnothing$

41. $(-4, 2)$

43. $(4, 7)$

45. $[-3, 2)$

47. $\left(-\dfrac{7}{3}, 3\right]$

49. $(-1, 5)$

51. $[2, 3]$

53. $(2, \infty)$ **55.** $(-\infty, 5)$ **57.** $[2, 4]$ **59.** $(-\infty, \infty)$ **61.** $\varnothing$
63. $[4, 5)$ **65.** $[1, 6]$ **67.** $x > 2$ **69.** $x < 3$
71. $x > 2$ or $x \leq -1$ **73.** $-2 \leq x < 3$ **75.** $x \geq -3$
77. $x = $ final exam score, $73 \leq x \leq 86.5$
79. $(50{,}000, \infty)$ **81.** $(-\infty, 20) \cup (30, \infty)$
83. $x = $ price of truck, $\$11{,}033 \leq x \leq \$13{,}811$
85. $x = $ number of cigarettes on the run, $4 \leq x \leq 18$
87. a) 1,226,950   **b)** 2011   **c)** 2019   **d)** 2011
89. $-b < x < -a$ provided $a < b$
91. a) $(12, 32)$ **b)** $(-20, 10]$ **c)** $(0, 9)$ **d)** $[-3, -1]$

Section 2.6 Warm-Ups
1. Absolute value   **2.** Two   **3.** No   **4.** One   **5.** All   **6.** No
7. $(-3, 3)$ **8.** Equivalent   **9.** True   **10.** False   **11.** True

12. False   **13.** True   **14.** True   **15.** True   **16.** True
17. False   **18.** True

Section 2.6 Exercises

1. $\{-5, 5\}$ **3.** $\{2, 4\}$ **5.** $\{-3, 9\}$ **7.** $\left\{-\dfrac{8}{3}, \dfrac{16}{3}\right\}$ **9.** $\{12\}$

11. $\varnothing$ **13.** $\{-20, 80\}$ **15.** $\{0, 5\}$ **17.** $\{0.143, 1.298\}$

19. $\{-2, 2\}$ **21.** $\{-9, 9\}$ **23.** $\{-4, 2\}$ **25.** $\{-11, 5\}$ **27.** $\{0, 3\}$

29. $\left\{-6, \dfrac{4}{3}\right\}$ **31.** $\{1, 3\}$ **33.** $(-\infty, \infty)$ **35.** $|x| < 2$ **37.** $|x| > 3$

39. $|x| \le 1$ **41.** $|x| \ge 2$ **43.** No   **45.** Yes   **47.** No

49. $(-\infty, -6) \cup (6, \infty)$

51. $[-2, 2]$

53. $(-3, 3)$

55. $(-\infty, -1] \cup [5, \infty)$

57. $(-\infty, -8) \cup (8, \infty)$

59. $[-5, 3]$

61. $\left(-\dfrac{1}{2}, \dfrac{9}{2}\right)$

63. $[-2, 12]$

65. $\left(-\infty, -\dfrac{9}{2}\right] \cup \left[\dfrac{15}{2}, \infty\right)$

67. $(-\infty, 0) \cup (0, \infty)$

69. $\{0\}$

71. $(-\infty, \infty)$

73. $\varnothing$

75. $(-\infty, \infty)$

77. $(-\infty, -3) \cup (-1, \infty)$ **79.** $(-4, 4)$ **81.** $(-1, 1)$
83. $(0.255, 0.847)$ **85.** 1401 or 1429
87. Between 121 and 133 pounds
89. a) Between 34% and 44%   **b)** $|x - 0.39| < 0.05$
91. a) 1 second   **b)** 1 second   **c)** $0.5 < t < 1.5$
93. a) $(-\infty, \infty)$ **b)** $(-\infty, \infty)$ **c)** All real numbers except $n = 0$

Enriching Your Mathematical Word Power

1. Equation   **2.** Linear   **3.** Identity   **4.** Conditional
5. Inconsistent   **6.** Equivalent   **7.** Literal, formula
8. Function   **9.** Uniform   **10.** Least   **11.** Inequality
12. Equivalent   **13.** Compound

Review Exercises

1. $\{8\}$ **3.** $\left\{-\dfrac{3}{2}\right\}$ **5.** R **7.** $\varnothing$ **9.** $\{0\}$ **11.** $\{20\}$ **13.** $\{5\}$

15. $\{5\}$ **17.** $x = \dfrac{-b}{a}$ **19.** $x = \dfrac{2}{c - a}$ **21.** $x = \dfrac{P}{mw}$ **23.** $x = -a$

25. $y = \dfrac{3}{2}x + 3$ **27.** $y = -\dfrac{1}{3}x + 4$ **29.** $y = 2x - 20$

31. Length 14 inches, width 8.5 inches
33. Wife $27,000, Roy $35,000   **35.** $9500   **37.** 11 nickels, 4 dimes
39. 15 miles

41. $(-3, \infty)$

43. $(-\infty, -5)$

45. $(0, \infty)$

47. $(-\infty, -8]$

49. $\left(-\infty, \dfrac{11}{2}\right)$

51. $[48, \infty)$

53. $(-\infty, -4) \cup (1, \infty)$

55. $(0, 9)$

57. $(0, \infty)$

59. $(-\infty, 4)$

61. $\varnothing$

63. $(-\infty, \infty)$

65. $\left[-\dfrac{17}{2}, \dfrac{13}{2}\right]$

67. $[1, \infty)$ **69.** $(3, 6)$ **71.** $(-\infty, \infty)$ **73.** $[-2, -1]$
75. $\{-14, 14\}$

77. $\{3\}$

79. $\varnothing$

81. $\{-1, 2\}$

83. $(-\infty, -4] \cup [4, \infty)$

85. $(-\infty, -4) \cup (14, \infty)$

87. $\varnothing$

89. $(-\infty, \infty)$

91. $(-\infty, 1) \cup (3, \infty)$

93. $x = $ rental price, $3 \le x \le $5   **95.** $(40.2, 53.6)$ **97.** 81 or 91
99. $50,000 accountant, $60,000 employees

101. Washington County 1200, Cade County 2400   **103.** $x > 1$

105. $|x - 2| = 0$ **107.** $|x| = 3$ **109.** $x \leq -1$

111. $|x| \leq 2$ **113.** $x \leq 2$ or $x \geq 7$ **115.** $|x| > 3$

117. $5 < x < 7$ or $|x - 6| < 1$ **119.** $|x| > 0$

Chapter 2 Test

1. $\{-4\}$ **2.** R **3.** $\{-6, 6\}$ **4.** $\{2, 5\}$ **5.** $y = \dfrac{2}{5}x - 4$

6. $y = \dfrac{5}{1 - 3x}$

7. $[4, 8]$

8. $(-\infty, -7) \cup (13, \infty)$

9. $(5, \infty)$

10. $\left(-8, -\dfrac{1}{2}\right)$

11. $[-5, 3)$

12. $(-\infty, 15)$

13. $\varnothing$ **14.** $(-\infty, \infty)$ **15.** $\varnothing$ **16.** $\{2.5\}$ **17.** $\varnothing$ **18.** $(-\infty, \infty)$

19. $\varnothing$ **20.** $(-\infty, \infty)$ **21.** $\{100\}$ **22.** 13 meters

23. 14 inches   **24.** $300   **25.** 30 liters

26. $|x - 28,000| > 3,000$ where x is Brenda's salary, Brenda makes more than $31,000 or less than $25,000.

Making Connections Chapters 1–2

1. $11x$ **2.** $30x^2$ **3.** $3x + 1$ **4.** $4x - 3$ **5.** 899   **6.** 961

7. 841   **8.** 25   **9.** 13   **10.** -25

11. 5   **12.** -4 **13.** $-2x + 13$ **14.** 60   **15.** 72   **16.** -9

17. $-3x^3$ **18.** 1   **19.** $\{0\}$ **20.** R **21.** $\{0\}$ **22.** $\{1\}$

23. $\left\{-\dfrac{1}{3}\right\}$ **24.** $\{1\}$ **25.** $(-\infty, \infty)$ **26.** $\{1000\}$ **27.** $\left\{-\dfrac{17}{5}, 1\right\}$

28. $\varnothing$ **29.** -22 **30.** -22 **31.** 21   **32.** 21   **33.** 49   **34.** 49

35. 1   **36.** 1   **37.** $(1, 8)$ **38.** $(-1, 7)$ **39.** $(30, 40)$ **40.** $(55, 88]$

41. $(7, \infty)$ **42.** $(5, 9)$ **43.** $(-\infty, 10)$ **44.** $(-\infty, \infty)$

45. a) 87,500   **b)** $C_r = 4500 + 0.06x$, $C_b = 8000 + 0.02x$
    c) 87,500   **d)** Buying is $1300 cheaper   **e)** $(75,000, 100,000)$

Chapter 3

Section 3.1 Warm-Ups

1. Origin   **2.** Ordered pair   **3.** x-intercept   **4.** y-intercept

5. Horizontal   **6.** Vertical   **7.** Dependent   **8.** Linear   **9.** False

10. True   **11.** False   **12.** False   **13.** True   **14.** True   **15.** True

16. True   **17.** False   **18.** True

Section 3.1 Exercises

1. I   **3.** III   **5.** y-axis   **7.** IV   **9.** II   **11.** x-axis   **13.** y-axis

Graph for **1–13** odd

15.

17.

19.

21.

23.

25.

27.

29.

31.

33.

35.

37.

39.

41.

43.

45.

47.

49.

51.

53. $(0, 1), (1/3, 0)$ **55.** $(0, -2), (0.005, 0)$

57. $(0, 300), (600, 0)$ **59.** $(0, 23.54), (-5.53, 0)$

61. $(0, 50), (50, 0)$ **63.** $(5, 0), (0, -3)$ **65.** $(0, 0)$

67. $\left(-\dfrac{1}{2}, 0\right)$ **69.** $\left(0, -\dfrac{2}{3}\right)$ **71.** $\left(\dfrac{1}{4}, 0\right), \left(0, \dfrac{1}{2}\right)$

73. $(2, 0), (3, -3)$ **75.** $(-4, -33), (22, 6)$

77. a) \$25,310   **b)** \$624

c)

79. a) \$146   **b)**

81. a) \$11.45   **b)** The number of toppings on a \$14.45 pizza is 11.

83. $n + 2b = 100$, 35 binders

85. a) Her weekly cost, revenue, and profit are \$517.50, \$1275, and \$757.50.   **b)** 1100. She had a profit of \$995 on selling 1100 roses.   **c)** 995. The difference between revenue and cost is \$995, which is her profit.

Section 3.2 Warm-Ups

1. Slope   **2.** Rise   **3.** Run   **4.** Horizontal   **5.** Vertical
6. Rise, run   **7.** Perpendicular   **8.** Parallel   **9.** True   **10.** False
11. True   **12.** False   **13.** True   **14.** True   **15.** False   **16.** True

Section 3.2 Exercises

1. $\dfrac{2}{3}$ **3.** Undefined   **5.** 0   **7.** -1 **9.** $\dfrac{3}{2}$ **11.** -1

13. $-\dfrac{5}{3}$ **15.** $\dfrac{4}{7}$ **17.** 5   **19.** $-\dfrac{5}{3}$ **21.** $-\dfrac{3}{5}$ **23.** $-\dfrac{2}{5}$

25. -3 **27.** 0   **29.** Undefined   **31.** 0.169

33. 3

35. $\frac{8}{3}$

37. $\frac{3}{7}$

39. $-\frac{1}{2}$

41. $-\frac{5}{4}$ T

43. 2

45. Perpendicular   **47.** Neither   **49.** Parallel   **51.** Neither
53. Perpendicular   **55.** Yes   **57.** No   **59.** No
61. a) 0.705 or $705 per year   **b)** Approximately $30,000   **c)** $30,695
63. (3, 5), (0, −7)   **65.** $-\frac{5}{2}$ **67.** −4.049
69. A horizontal line has a zero slope and a vertical line has undefined slope.
71. $-2, \frac{1}{2}$, perpendicular
73. Increasing m makes the graph increase faster. The slopes of these lines
are 1, 2, 3, and 4.

Section 3.3 Warm-Ups
1. Slope-intercept   **2.** Standard form   **3.** Point-slope   **4.** Slope
5. y-intercept   **6.** Standard   **7.** Slope-intercept   **8.** True
9. False   **10.** True   **11.** True   **12.** True   **13.** False   **14.** True
15. True

Section 3.3 Exercises

1. $y = \frac{1}{2}x + 2$ **3.** $y = -2$ **5.** $x = 1$ **7.** $y = -x$

9. $y = \frac{3}{2}x - 3$ **11.** $y = -x + 2$

13. $y = -\frac{2}{5}x + \frac{1}{5}, -\frac{2}{5}, \left(0, \frac{1}{5}\right)$

15. $y = 3x - 2, 3, (0, -2)$ **17.** $y = 2, 0, (0, 2)$

19. $y = 3x - 1, 3, (0, -1)$

21. $y = \frac{1}{3}x + \frac{7}{12}, \frac{1}{3}, \left(0, \frac{7}{12}\right)$

23. $y = 0.01x + 6057, 0.01, (0, 6057)$

25.

27.

29.

31.

33.

35.

37.

39.

41. $x - 3y = 6$ **43.** $x - 2y = -13$ **45.** $2x - 6y = 11$
47. $5x + 6y = 890$ **49.** $y = 2x - 7$ **51.** $y = -\frac{1}{2}x + 2$
53. $y = 12$ **55.** $y = 20x$ **57.** $y = 3x - 8$
59. $y = \frac{1}{3}x + \frac{4}{3}$ **61.** $y = -\frac{1}{2}x + 4$ **63.** $y = \frac{13}{2}x$
65. $y = 6$ **67.** $y = 4x - 8$ **69.** $y = -\frac{3}{4}x$ **71.** $4x + y = 14$
73. $2x + y = 6$ **75.** $x - 2y = -10$ **77.** $2x + y = 4$
79. $2x - y = -5$ **81.** $x + 2y = 7$ **83.** $2x + y = 3$
85. $3x - y = -9$ **87.** $y = 5$ **89.** Perpendicular

91. Parallel   **93.** Neither   **95.** Perpendicular

97. a) $t = \frac{7}{6}s + 60$ **b)** 95°F

c)

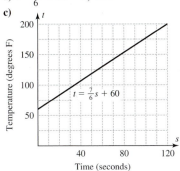

99. a) $y = 0.4x - 774$ **b)** 32 billion tons

101. a) $w = 310.77d - 1610.43$ **b)** 953 ft³/sec   **c)** Increasing

103. a) (4, 0), (0, 6)   **b)** (a, 0), (0, b)   **c)** $\dfrac{x}{-5} + \dfrac{y}{3} = 1$

   d) Horizontal lines, vertical lines, and lines through (0, 0)

107. The lines intersect at (50, 97)

Mid-Chapter Quiz 3.1–3.3

1. −6, 12, −2, 6   **2.** 0, 12, 9, −12

3.

4.

5.

6.

7.

8.
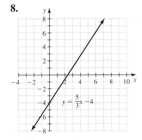

9. −5, (0, 2)   **10.** 0, (0, 6)   **11.** $\dfrac{3}{8}$, (0, −2)   **12.** $x + 50y = 250$

13. $2x - 3y = -54$ **14.** $y = \dfrac{1}{3}x + 1$ **15.** $y = -5x + 6$

16. $y = -\dfrac{5}{3}x - 4$ **17.** $y = 5$ **18.** $y = \dfrac{4}{5}x + \dfrac{13}{5}$

19.

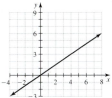

20. Perpendicular

Section 3.4 Warm-Ups

1. Linear   **2.** Dashed   **3.** Solid   **4.** Above   **5.** Below   **6.** Left
7. Right   **8.** True   **9.** False   **10.** True   **11.** True   **12.** True
13. True   **14.** False   **15.** True   **16.** True   **17.** False

Section 3.4 Exercises

1.

3.

5.

7.

9.

11.

13.

15.

17.

$2x - 3y < 5$

19.

$x + y + 3 \geq 0$

45.

$x - 2y \leq 4$ or $2x - 3y \leq 6$

47.

$y > 2$ and $x < 3$

21.
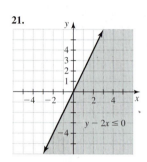
$y - 2x \leq 0$

23.

$3x - 2y > 0$

49.

$y \geq x$ and $x \leq 2$

51.

$2x < y + 3$ or $y > 2 - x$

25.

$\frac{1}{2}x + \frac{1}{3}y < 1$

27. $(-2, 5), (-6, -4)$
29. $(-2, 5)$
31. $(-6, -4)$
33. $(1, 3), (-2, 5), (-6, -4)$
35. $(7, -8)$

53.
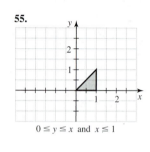
$x - 1 < y < x + 3$

55.

$0 \leq y \leq x$ and $x \leq 1$

37.
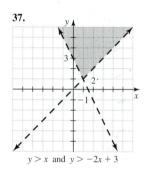
$y > x$ and $y > -2x + 3$

39.

$y < x + 3$ or $y > -x + 2$

57.

$1 \leq x \leq 3$ and $2 \leq y \leq 5$

59.
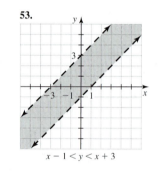
$|x + y| < 2$

41.

$x - 4y < 0$ and $3x + 2y \geq 6$

43.
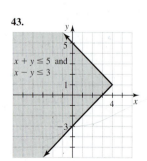
$x + y \leq 5$ and $x - y \leq 3$

61.
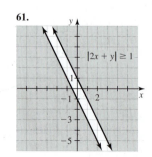
$|2x + y| \geq 1$

63.
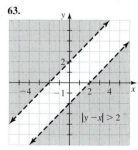
$|y - x| > 2$

65.

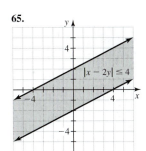

$|x - 2y| \le 4$

67.

$|x| > 2$

69.

$|y| < 1$

71.

$|x| < 2$ and $|y| < 3$

73.

$|x - 3| < 1$ and $|y - 2| < 1$

75. Not the empty set
77. ∅
79. Not the empty set
81. Not the empty set
83. ∅
85. ∅
87. ∅
89. Not the empty set

91.

Full-size

Compact

$x \ge 0,\ y \ge 0,\ 3x + 4y \le 24$

93.

Full-size

Compact

$x \ge 0,\ y \ge x,\ 3x + 4y \le 24$

95.

$h \le 187 - 0.85a,$
$h \ge 154 - 0.70a,$
$a \ge 20,\ a \le 75$

97.

$d \ge 0,\ t \ge 0,$
$3d + 10t \le 90$

Section 3.5 Warm-Ups

1. Relation   **2.** Function   **3.** Function   **4.** Domain   **5.** Range
6. Dependent   **7.** True   **8.** False   **9.** True   **10.** True
11. False   **12.** True   **13.** False   **14.** True   **15.** True

Section 3.5 Exercises

1. Yes   **3.** No   **5.** Yes   **7.** No   **9.** $C = 0.50t + 5$
11. $T = 1.09S$ **13.** $C = 2\pi r$ **15.** $P = 4s$ **17.** $A = 5h$
19. Yes   **21.** Yes   **23.** No   **25.** Yes   **27.** Yes   **29.** No
31. No   **33.** Yes   **35.** $(2, 1), (2, -1)$ **37.** $(8, 4), (8, -4)$
39. $(0, 1), (0, -1)$ **41.** $(16, 2), (16, -2)$ **43.** $(3, 1), (3, -1)$
45. Yes   **47.** No   **49.** Yes   **51.** No   **53.** Yes   **55.** No
57. No   **59.** Yes   **61.** No   **63.** No   **65.** Yes   **67.** No
69. $\{4, 7\}, \{1\}$ **71.** $\{2\}, \{3, 5, 7\}$ **73.** $(-\infty, \infty), (-\infty, \infty)$
75. $(-\infty, \infty), (-\infty, \infty)$ **77.** $[2, \infty), [0, \infty)$ **79.** $[0, \infty), [0, \infty)$
81. -2 **83.** 10   **85.** -12 **87.** 1   **89.** 2.236
91. 2   **93.** 0   **95.** -10 **97. a)** 192 ft   **b)** 0 ft
99. $A = s^2$ or $A(s) = s^2$ **101.** $C(x) = 3.98x$, \$11.94
103. $C(n) = 14.95 + 0.50n$, \$17.95

Enriching Your Mathematical Word Power

1. Graph   **2.** Origin   **3.** x-coordinate   **4.** y-intercept
5. Independent   **6.** Dependent   **7.** Slope   **8.** Standard
9. Slope   **10.** Point   **11.** Linear   **12.** Function   **13.** Relation
14. Domain   **15.** Range   **16.** Function

Review Exercises

1. III   **3.** x-axis   **5.** y-axis   **7.** IV
9. $(0, 2), \left(\frac{2}{3}, 0\right), (4, -10), \left(\frac{5}{3}, -3\right)$
11. 1   **13.** $\frac{3}{7}$ **15.** $\frac{3}{8}$ **17.** $\frac{7}{11}$

19. -1

21. 2

23. $-\frac{3}{2}$

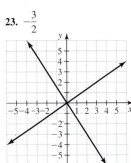

25. $-3, (0, 4)$
27. $\frac{2}{3}, \left(0, \frac{7}{3}\right)$
29. $2x - 3y = 12$
31. $x - 2y = -5$
33. $x - 2y = 7$
35. $3x + 4y = 18$
37. $y = 5$

39.

$y = 2x - 3$

41.

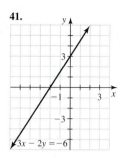

$3x - 2y = -6$

59.

$4x - 2y \geq 6$

61.

$5x - 2y < 9$

43.

$y = 13$

45.

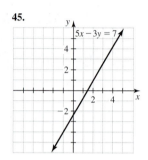

$5x - 3y = 7$

63.

$y > 3$ and $y - x < 5$

65.

$3x + 2y \geq 8$ or $3x - 2y \leq 6$

47.

$5x + 4y = 100$

49.

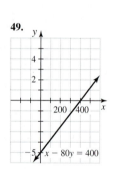

$5x - 80y = 400$

67.

$|x + 2y| < 10$

69.

$|x| \leq 5$

51.

$y > 3x - 2$

53.

$x - y \leq 5$

71.

$|y - x| > 2$

73. No   **75.** Yes   **77.** Yes
79. No   **81.** {3, 4, 5}, {1, 5, 9}
83. $(-\infty, \infty), (-\infty, \infty)$
85. $[-5, \infty), [0, \infty)$ **87.** -5
89. -6 **91.** $-\dfrac{21}{4}$
93. $3x - y = 6$ **95.** $x + 2y = 7$
97. $x = 2$ **99.** $y = 0$
101. $2x - y = -6$
103. $3x - y = -6$ **105.** $y = 5$

55.

$3x > 2$

57.

$4y \leq 0$

111. a) $h = 220 - a$ **b)** 180 beats per minute   **c)** Decreases
113. 62 days

$C = 26 + 17d$

(30, 536)

(1, 43)

Chapter 3 Test

1. $(0, 5), \left(\frac{5}{2}, 0\right), \left(\frac{13}{2}, -8\right)$ **2.** $-\frac{6}{5}$

3. $\frac{8}{5}, (0, 2)$ **5.** $V = -2000a + 22,000$ **6.** $x + 2y = 6$

7. $4x + y = -7$ **8.** $5x + 3y = 19$ **9.** $x - 2y = -4$

10.   **11.**

12.   **13.**

14.   **15.**

16.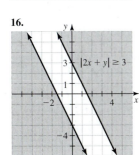

17. Yes **18.** 11
19. $[7, \infty), [0, \infty)$
20. $S = 0.50n + 3$
21. 6 ft

Making Connections   A Review of Chapters 1–3

1. 128 **2.** 64 **3.** 49 **4.** −29 **5.** −5 **6.** −5 **7.** $12t^2$

8. $7t$ **9.** $x + 2$ **10.** $7y$ **11.** $7x - 32$ **12.** $-21x^2 + 8x$

13. $\{27\}$ **14.** $\{200\}$ **15.** $\left\{\frac{7}{3}\right\}$ **16.** $\left\{\frac{4}{3}, \frac{10}{3}\right\}$ **17.** $\varnothing$ **18.** $\left\{0, \frac{14}{3}\right\}$

19. **20.**

21.   **22.**

23. **24.**

25.   **26.**

27. **28.**

29. $a = \dfrac{10}{b}$ **30.** $a = b + 10$ **31.** $a = \dfrac{10 - 5b}{3}$ **32.** $a = \dfrac{3}{5 - b}$

33. 175 yards **34.** 29 inches **35.** \$188,000 **36.** 1000 beads
37. a) $b = 500a - 24,000$, $b = 667a - 34,689$, $b = 800a - 43,600$
  b) \$8000 **c)** 69 **d)** The slopes 500, 667, and 800 indicate the additional amount per year received beyond the basic amount in each category.

Chapter 4

Section 4.1 Warm-Ups

1. Independent **2.** Parallel **3.** Graphing **4.** Substitution
5. Conditional **6.** Dependent **7.** Inconsistent **8.** True
9. False **10.** False **11.** True **12.** True **13.** True **14.** False

Section 4.1 Exercises

1. Yes **3.** No **5.** Yes **7.** No **9.** $\{(1, 2)\}$ **11.** $\{(0, -1)\}$
13. $\{(2, -1)\}$ **15.** $\{(-1, 2)\}$ **17.** $\{(x, y) \mid x + 2y = 8\}$ **19.** $\varnothing$
21. $\varnothing$ **23.** c **25.** b **27.** $\{(7.25, 6.65)\}$ **29.** $\{(12, 15)\}$
31. $\{(52, -116)\}$ **33.** $\{(110, 244)\}$ **35.** $\{(84, 712.4)\}$
37. $\{(3, 11)\}$, independent **39.** $\{(-2, 4)\}$, independent
41. $\{(6, 4)\}$, independent **43.** $\{(8, 3)\}$, independent
45. $\{(-3, 2)\}$, independent **47.** $\{(20, 10)\}$, independent
49. $\{(5, -1)\}$, independent **51.** $\{(7, 7)\}$, independent
53. $\{(15, 25)\}$, independent **55.** $\{(x, y) \mid y = 2x - 5\}$, dependent
57. $\varnothing$, inconsistent **59.** $\{(0, 0)\}$, independent
61. $\{(x, y) \mid 3y - 2x = -3\}$, dependent **63.** $\varnothing$, inconsistent
65. $\left\{\left(\dfrac{6}{17}, \dfrac{15}{17}\right)\right\}$ **67.** $\left\{\left(\dfrac{9}{2}, -\dfrac{1}{2}\right)\right\}$ **69.** $\left\{\left(\dfrac{1}{2}, \dfrac{1}{4}\right)\right\}$ **71.** $\left\{\left(\dfrac{3}{2}, -\dfrac{5}{2}\right)\right\}$
73. $\left\{\left(-\dfrac{2}{9}, \dfrac{1}{6}\right)\right\}$ **75.** $\left\{\left(-\dfrac{1}{7}, \dfrac{2}{7}\right)\right\}$ **77.** $\left\{\left(-\dfrac{1}{14}, \dfrac{5}{28}\right)\right\}$

79. $\{(0.8, 0.7)\}$ **81.** Length 27 ft, width 15 ft   **83.** Length 10 ft, width 4 ft
85. 3.5 and 6.5   **87.** -9.5 and 10.5
89. 120 tickets for $200, 80 tickets for $250
91. 55 tickets for $6, 110 tickets for $11
93. $30,000 at 5%, $10,000 at 8%
95. $31,818 in gold fund, $68,182 in stock fund
97. 12.5 L of 5% solution, 37.5 L of 25% solution
99. -12 and 14   **101.** 94 toasters, 6 vacation coupons
103. State tax $3553, federal tax $28,934   **105.** $20,000
107. a) $500,000   **b)** $300,000   **c)** 20,000   **d)** $400,000   **109.** a

Section 4.2 Warm-Ups
1. Addition   **2.** Conditional   **3.** Dependent   **4.** Inconsistent
5. Conditional   **6.** Identity   **7.** Inconsistent   **8.** True   **9.** False
10. False   **11.** False   **12.** True   **13.** True   **14.** True

Section 4.2 Exercises
1. $\{(4, 3)\}$ **3.** $\{(1, -2)\}$ **5.** $\{(5, -7)\}$
7. $\left\{\left(\frac{3}{8}, -\frac{31}{8}\right)\right\}$ **9.** $\{(-1, 3)\}$ **11.** $\left\{\left(\frac{7}{9}, \frac{2}{3}\right)\right\}$ **13.** $\{(-1, -3)\}$
15. $\{(-2, -5)\}$ **17.** $\{(22, 26)\}$ **19.** Yes   **21.** No   **23.** Yes
25. $\varnothing$, inconsistent   **27.** $\{(x, y) \mid 5x - y = 1\}$, dependent
29. $\left\{\left(\frac{5}{2}, 0\right)\right\}$, independent   **31.** $\{(12, 6)\}$ **33.** $\{(-8, 6)\}$
35. $\{(16, 12)\}$ **37.** $\left\{\left(\frac{1}{2}, \frac{1}{3}\right)\right\}$ **39.** $\{(12, 7)\}$ **41.** $\{(400, 800)\}$
43. $\{(1.5, 1.25)\}$ **45.** $\left\{\left(\frac{3}{4}, \frac{2}{3}\right)\right\}$ **47.** $\{(5, 6)\}$ **49.** $\{(2, -17)\}$
51. $\{(0, 1)\}$ **53.** $\{(3, 4)\}$ **55.** $\left\{\left(\frac{1}{2}, \frac{1}{3}\right)\right\}$ **57.** $\varnothing$ **59.** $\{(x, y) \mid y = x\}$
61. $a = -1$ **63.** $a = 2, b = -1$ **65.** $1.40   **67.** 1380 students
69. 31 dimes, 4 nickels
71. a) 20 pounds chocolate, 30 pounds peanut butter
    b) 20 pounds chocolate, 30 pounds peanut butter
73. 4 hours   **75.** 80%   **77.** Width 150 meters, length 200 meters

Section 4.3 Warm-Ups
1. Linear   **2.** Ordered   **3.** Solution   **4.** Addition, substitution
5. Plane   **6.** Independent   **7.** False   **8.** False   **9.** True
10. False   **11.** True   **12.** False   **13.** False   **14.** False

Section 4.3 Exercises
1. $\{(2, 3, 4)\}$ **3.** $\{(2, 3, 5)\}$ **5.** $\{(1, 2, 3)\}$ **7.** $\{(1, 2, -1)\}$
9. $\{(1, 3, 2)\}$ **11.** $\{(1, -5, 3)\}$ **13.** $\{(-1, 2, -1)\}$
15. $\{(-1, -2, 4)\}$ **17.** $\{(1, 3, 5)\}$ **19.** $\{(3, 4, 5)\}$
21. $\{(x, y, z) \mid x + y - z = 2\}$ **23.** $\varnothing$ **25.** $\varnothing$ **27.** $\varnothing$
29. $\{(x, y, z) \mid -x + 2y - 3z = -6\}$ **31.** $\varnothing$
33. $\{(x, y, z) \mid 5x + 4y - 2z = 150\}$ **35.** $\{(0.1, 0.3, 2)\}$
37. Yes   **39.** No   **41.** $\{-4, 4\}$, independent
43. $\{(x, y) \mid x + y = 4\}$, dependent   **45.** $\varnothing$, inconsistent
47. $\{(3, 2, 1)\}$, independent   **49.** $\{(x, y, z) \mid x - y + z = 1\}$, dependent
51. Chevrolet $20,000, Ford $22,000, Toyota $24,000
53. First 10 hr, second 12 hr, third 14 hr
55. $1500 stocks, $4500 bonds, $6000 mutual fund
57. Anna 108 pounds, Bob 118 pounds, Chris 92 pounds
59. 3 nickels, 6 dimes, 4 quarters
61. $24,000 teaching, $18,000 painting, $6000 royalties
63. Edwin 24, father 51, grandfather 84

Mid-Chapter Quiz 4.1–4.3
1. Yes   **2.** No   **3.** No   **4.** $\{(3, 2)\}$ **5.** $\{(4, -4)\}$ **6.** $\{(2, 1)\}$
7. $\{(7, -1)\}$ **8.** $\varnothing$ **9.** $\{(5, 8)\}$ **10.** $\{(1, -1)\}$

11. $\{(x, y) \mid x + 2y = 4\}$ **12.** Inconsistent   **13.** Independent
14. Dependent   **15.** $\{(12, 8)\}$ **16.** $\{(50, 40)\}$ **17.** $\{(2, 2, 4)\}$
18. $\varnothing$ **19.** $\{(x, y, z) \mid x - y + 3z = 4\}$

Section 4.4 Warm-Ups
1. Matrix   **2.** Row   **3.** Column   **4.** Size   **5.** Element
6. Augmented   **7.** Gauss-Jordan   **8.** True   **9.** False   **10.** True
11. False   **12.** True   **13.** False

Section 4.4 Exercises
1. 2×2 **3.** 3×2 **5.** 1×3 **7.** 2×1
9. $\left[\begin{array}{cc|c} 2 & -3 & 9 \\ -3 & 1 & -1 \end{array}\right]$ **11.** $\left[\begin{array}{ccc|c} 1 & -1 & 1 & 1 \\ 1 & 1 & -2 & 3 \\ 0 & 1 & -3 & 4 \end{array}\right]$
13. $5x + y = -1$
     $2x - 3y = 0$
15. $x = 6$
     $-x + z = -3$
     $x + y = 1$
17. $\left[\begin{array}{ccc|c} 1 & 1 & 7 & 2 \\ 0 & 2 & 4 & 1 \\ 0 & 0 & 5 & 10 \end{array}\right]$ **19.** $\left[\begin{array}{cc|c} 1 & 3 & 4 \\ 3 & 5 & 9 \end{array}\right]$ **21.** $\left[\begin{array}{cc|c} 1 & 0 & -2 \\ 0 & 1 & 3 \end{array}\right]$
23. $\left[\begin{array}{cc|c} 1 & 0 & 6 \\ 0 & 2 & 4 \end{array}\right]$ **25.** $\left[\begin{array}{cc|c} 1 & 3 & 4 \\ 2 & -4 & 3 \end{array}\right]$ **27.** $\left[\begin{array}{cc|c} 1 & 0 & -3 \\ 0 & 2 & 1 \end{array}\right]$
29. $\left[\begin{array}{cc|c} 1 & 2 & 3 \\ 0 & 7 & 11 \end{array}\right]$ **31.** $R_1 \leftrightarrow R_2$ **33.** $\frac{1}{5}R_2 \to R_2$ **35.** $\{(1, 4)\}$
37. $\{(-8, 2)\}$ **39.** $\{(5, -2)\}$ **41.** $\{(1, 2)\}$ **43.** $\{(4, 5)\}$
45. $\{(1, -1)\}$ **47.** $\{(7, 6)\}$ **49.** $\{(2, 4, 2)\}$ **51.** $\{(1, 2, 3)\}$
53. $\{(1, 1, 1)\}$ **55.** $\{(1, 2, 0)\}$ **57.** $\{(1, 0, 1)\}$
59. $\{(-44, 12)\}$ **61.** $\{(15, 32, 10)\}$ **63.** $\{(1, 2, 5, 8)\}$
65. $\{(x, y) \mid x - 5y = 11\}$ **67.** $\varnothing$ **69.** $\{(x, y) \mid x + 2y = 1\}$
71. $\{(x, y, z) \mid x - y + z = 1\}$ **73.** $\varnothing$ **75.** $\{(-4, 3)\}$, independent
77. $\{(x, y) \mid -x + 3y = 4\}$, dependent   **79.** $\varnothing$, inconsistent
81. $\{(-3, -2, -1)\}$, independent
83. $\{(x, y, z) \mid x - y - z = 1\}$, dependent   **85.** 5 and 7
87. Length 11 in., width 8.5 in.   **89.** Buys for $14, sells for $16
91. 45 four-wheel cars, 2 three-wheel cars, and 3 two-wheel motorcycles
93. $x = $4, y = $5, z = $2

Section 4.5 Warm-Ups
1. Determinant   **2.** Cramer's   **3.** Independent   **4.** Minor   **5.** Sign
6. True   **7.** False   **8.** False   **9.** True   **10.** True   **11.** False

Section 4.5 Exercises
1. -1 **3.** -3 **5.** -14 **7.** 0.4   **9.** $\{(2, 6)\}$ **11.** $\{(-8, 8)\}$
13. $\{(1, -3)\}$ **15.** $\{(1, 1)\}$ **17.** $\left\{\left(\frac{23}{13}, \frac{9}{13}\right)\right\}$ **19.** $\{(10, 15)\}$
21. $\left\{\left(\frac{27}{4}, \frac{13}{2}\right)\right\}$ **23.** 19.41   **25.** $\frac{7}{288}$ **27.** $\{(4.8, 1.6)\}$
29. $\{(2.83, 8.66)\}$ **31.** 11   **33.** 4   **35.** 3   **37.** 1   **39.** -7
41. -1 **43.** 9   **45.** 5   **47.** 22   **49.** 6   **51.** 70
53. 25   **55.** $\{(1, 2, 3)\}$ **57.** $\{(-1, 1, 2)\}$ **59.** $\{(-3, 2, 1)\}$
61. $\left\{\left(\frac{3}{2}, \frac{1}{2}, 2\right)\right\}$ **63.** $\{(0, 1, -1)\}$ **65.** $\{(1.1, 1.2, 1.3)\}$
67. a) 9 servings peas, 11 servings beets
    b) 9 servings peas, 11 servings beets
69. Milk $2.40, magazine $2.25   **71.** 12 singles, 10 doubles

73. Gary 39, Harry 34   **75.** Square 10 feet, triangle $\frac{40}{3}$ feet

77. 10 gallons of 10% solution, 20 gallons of 25% solution

79. Mimi 36 pounds, Mitzi 32 pounds, Cassandra 107 pounds

81. 39°, 51°, 90°   **83.** $x = \$1.29$, $y = \$1.19$, $z = \$1.39$

85. Use another method.   **87.** No. These are nonlinear equations.

Section 4.6 Warm-Ups

1. Constraint   **2.** Linear programming   **3.** Linear   **4.** Vertex

5. False   **6.** False   **7.** True   **8.** True   **9.** True   **10.** False

Section 4.6 Exercises

1.

3.

5.

7.

9.

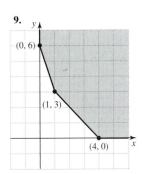

11. $x \geq 0$, $y \geq 0$, $x + 2y \leq 30$, $4x + 3y \leq 60$

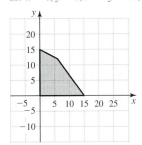

13. 46   **15.** 88   **17.** 128   **19.** 9   **21.** 59   **23.** 21   **25.** 18

27. a) 0, 320,000, 510,000, 450,000   **b)** 30 TV ads and 60 radio ads

29. 6 doubles, 4 triples   **31.** 0 doubles, 8 triples

33. 1.75 cups Doggie Dinner, 5.5 cups Puppy Power

35. 10 cups Doggie Dinner, 0 cups Puppy Power

37. Laundromat $8000, car wash $16,000

Enriching Your Mathematical Word Power

1. System   **2.** Independent   **3.** Inconsistent   **4.** Dependent

5. Consistent   **6.** Addition   **7.** Linear   **8.** Matrix

9. Augmented   **10.** Size   **11.** Determinant   **12.** Sign

13. Constraints   **14.** Linear

Review Exercises

1. $\{(1, 1)\}$, independent   **3.** $\{(x, y) \mid x + 2y = 4\}$, dependent

5. $\varnothing$, inconsistent   **7.** $\{(-3, 2)\}$, independent   **9.** $\varnothing$, inconsistent

11. $\{(x, y) \mid 2x - y = 3\}$, dependent   **13.** $\{(30, 12)\}$, independent

15. $\left\{\left(\frac{1}{5}, \frac{2}{5}\right)\right\}$, independent   **17.** $\{(-1, 5)\}$, independent

19. $\{(x, y) \mid 3x - 2y = 12\}$, dependent   **21.** $\varnothing$, inconsistent

23. $\left\{\left(2, -\frac{1}{3}\right)\right\}$, independent   **25.** $\{(20, 60)\}$, independent

27. $\{(2, 4, 6)\}$ **29.** $\{(1, -3, 2)\}$ **31.** $\varnothing$

33. $\{(x, y, z) \mid x - 2y + z = 8\}$ **35.** $\{(3, 4)\}$ **37.** $\{(2, -4)\}$

39. $\{(1, 1, 2)\}$ **41.** 2   **43.** -0.2 **45.** $\{(-1, -2)\}$ **47.** $\{(2, 1)\}$

49. 58   **51.** -30 **53.** $\{(1, 2, -3)\}$

55.

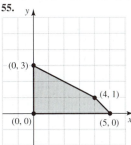

57. 30   **59.** Width 13 feet, length 28 feet   **61.** 78

63. 36 minutes   **65.** 4 liters of A, 8 liters of B, 8 liters of C

67. Three servings of each

Chapter 4 Test

1. $\{(1, 3)\}$ **2.** $\left\{\left(\frac{5}{2}, -3\right)\right\}$ **3.** $\{(x, y) \mid y = x - 5\}$

4. $\{(-1, 3)\}$ **5.** $\varnothing$ **6.** Inconsistent   **7.** Dependent

8. Independent   **9.** $\{(1, -2, -3)\}$ **10.** $\{(2, 5)\}$

11. $\{(3, 1, 1)\}$ **12.** -18 **13.** -2 **14.** $\{(-1, 2)\}$

15. $\{(2, -2, 1)\}$ **16.** Singles $79, doubles $99

17. Jill 17 hours, Karen 14 hours, Betsy 62 hours   **18.** 44

Making Connections   A Review of Chapters 1–4

1. -81 **2.** 7   **3.** 73   **4.** 5.94   **5.** $-t - 3$ **6.** $-0.9x + 0.9$

7. $3x^2 + 2x - 1$ **8.** y **9.** $y = \frac{3}{5}x - \frac{7}{5}$ **10.** $y = \frac{C}{D}x - \frac{W}{D}$

11. $y = \frac{K}{W - C}$ **12.** $y = \frac{bw - 2A}{b}$ **13.** $\left\{\frac{2}{3}\right\}$ **14.** $\left\{\frac{3}{40}\right\}$

15. $\{-32\}$ **16.** $\{18\}$ **17.** $\varnothing$ **18.** All real numbers

19. $\{-1000, 1000\}$ **20.** $\left\{\frac{1}{3}, 5\right\}$ **21.** $\left\{\frac{9}{2}\right\}$ **22.** $\varnothing$ **23.** $[6, \infty)$

24. $(2, \infty)$ **25.** $(3, 21)$ **26.** $(-\infty, \infty)$ **27.** $(-2, 4)$ **28.** $(-29, 16]$

29. $(-\infty, 1) \cup (15, \infty)$ **30.** $\left[-\dfrac{5}{6}, 1\right]$ **31.** $(-\infty, \infty)$ **32.** $\varnothing$

33. $\{(4, -1)\}$ **34.** $\{(500, 700)\}$ **35.** $\{(x, y) \mid x + 17 = 5y\}$ **36.** $\varnothing$

37. $y = \dfrac{5}{9}x + 55$ **38.** $y = -\dfrac{11}{6}x + \dfrac{2}{3}$ **39.** $y = 5x + 26$

40. $y = \dfrac{1}{2}x + 5$ **41.** $y = 5$ **42.** $x = -7$

43. a) Machine A
 b) Machine B \$0.04 per copy, machine A \$0.03 per copy
 c) The slopes 0.04 and 0.03 are the per copy cost for each machine.
 d) B: $y = 0.04x + 2000$, A: $y = 0.03x + 4000$
 e) 200,000

Chapter 5

Section 5.1 Warm-Ups
 1. Exponential   **2.** Product   **3.** Quotient   **4.** Scientific   **5.** Zero
 6. True   **7.** False   **8.** False   **9.** False   **10.** True   **11.** False
 12. True   **13.** True   **14.** True   **15.** False

Section 5.1 Exercises
 1. $4, -4, 4, \dfrac{1}{4}, -\dfrac{1}{4}, \dfrac{1}{4}$ **3.** $8, -8, -8, \dfrac{1}{8}, -\dfrac{1}{8}, -\dfrac{1}{8}$

 5. $\dfrac{1}{25}, -\dfrac{1}{25}, \dfrac{1}{25}, 25, -25, 25$ **7.** $7, -7, -7, \dfrac{1}{7}, -\dfrac{1}{7}, -\dfrac{1}{7}$

 9. $\dfrac{1}{4}, \dfrac{1}{4}, -\dfrac{1}{4}, 4, 4, -4$ **11.** $\dfrac{8}{27}, -\dfrac{8}{27}, -\dfrac{8}{27}, \dfrac{27}{8}, -\dfrac{27}{8}, -\dfrac{27}{8}$

 13. 2^{17} **15.** $-\dfrac{6}{x^6}$ **17.** $\dfrac{21}{b^{10}}$ **19.** $1, -1, 1, -1$ **21.** $1, 2, 1$

 23. $3s, 3, 1$ **25.** 2 **27.** 32 **29.** $\dfrac{100}{3}$ **31.** $\dfrac{8y^2}{5x^2}$ **33.** $\dfrac{1}{48x^2}$

 35. x^2 **37.** 3^8 **39.** $\dfrac{1}{3a^3}$ **41.** $\dfrac{-3}{w^8}$ **43.** $3^8 w^6$ **45.** $\dfrac{xy}{2}$ **47.** 9

 49. -14 **51.** $\dfrac{1}{16}$ **53.** $\dfrac{3}{8}$ **55.** $\dfrac{4}{9}$ **57.** $\dfrac{10}{x^3}$ **59.** a **61.** $\dfrac{-2x^2y^4}{3}$

 63. 3 **65.** -2 **67.** -4 **69.** -3 **71.** 486,000,000
 73. 0.00000237   **75.** 4,000,000   **77.** 0.000005   **79.** 3.2×10^5
 81. 7.1×10^{-7} **83.** 7.03×10^{-5} **85.** 2.05×10^7 **87.** 6×10^3
 89. 7.5×10^{-1} **91.** 3×10^{22} **93.** -1.2×10^{13} **95.** 1.578×10^5
 97. 9.187×10^{-5} **99.** 3.828×10^{30} **101.** 4.910×10^{11} feet
 103. 3.833×10^7 seconds   **105.** 2.6 lb/person/day

Section 5.2 Warm-Ups
 1. Power   **2.** Product   **3.** Quotient   **4.** Principal   **5.** Annually
 6. Present   **7.** False   **8.** True   **9.** True   **10.** True   **11.** False
 12. False   **13.** True   **14.** True   **15.** True   **16.** True

Section 5.2 Exercises
 1. 64 **3.** y^{10} **5.** $\dfrac{1}{x^8}$ **7.** m^{18} **9.** 1 **11.** $\dfrac{1}{x^2}$ **13.** $81y^2$

 15. $25w^6$ **17.** $\dfrac{x^9}{y^6}$ **19.** $\dfrac{b^2}{9a^2}$ **21.** $\dfrac{6x^3}{y}$ **23.** $\dfrac{1}{8a^3b^4}$ **25.** $\dfrac{w^3}{8}$

 27. $-\dfrac{27a^3}{64}$ **29.** $\dfrac{x^2y^2}{4}$ **31.** $\dfrac{y^2}{9x^6}$ **33.** $\dfrac{25}{4}$ **35.** 4 **37.** $-\dfrac{27}{8x^3}$

 39. $\dfrac{27y^3}{8x^6}$ **41.** 5^{6t} **43.** 2^{6w^2} **45.** 7^{m+3} **47.** 8^{5a+11} **49.** $6x^9$

 51. $-8x^6$ **53.** $\dfrac{3z}{x^2y}$ **55.** $-\dfrac{3}{2}$ **57.** $\dfrac{4x^6}{9}$ **59.** $-\dfrac{x^2}{2}$ **61.** $\dfrac{y^3}{8x^3}$

63. $\dfrac{b^{14}}{5a^7}$ **65.** $\dfrac{x^6}{16y^8}$ **67.** $3ac^8$ **69.** 2^{11} **71.** 2^{-4} **73.** 2^{6n}

75. 25 **77.** $\dfrac{3}{4}$ **79.** 850.559   **81.** 1.533   **83.** \$56,197.12

85. \$2958.64   **87. a)** 75.1 years   **b)** 81.4 years   **91.** d

93. a)

 b) (6.116, 20,000)   **c)** 6.116 years

Section 5.3 Warm-Ups
 1. Term   **2.** Coefficient   **3.** Constant   **4.** Polynomial   **5.** Degree
 6. Distributive   **7.** False   **8.** False   **9.** False   **10.** True
 11. False   **12.** True   **13.** True   **14.** True   **15.** False

Section 5.3 Exercises
 1. Yes   **3.** No   **5.** Yes   **7.** No   **9.** $4, -8$, binomial
 11. 0, 0, monomial   **13.** 7, 0, monomial   **15.** 6, 1, trinomial
 17. 80 **19.** -29 **21.** 0 **23.** $3a + 2$ **25.** $5xy + 25$
 27. $2x - 9$ **29.** $2x^3 - x^2 - 2x - 8$ **31.** $11x^2 - 2x - 9$
 33. $x + 5$ **35.** $-6x^2 + 5x + 2$ **37.** $2x$ **39.** $-15x^6$
 41. $x^3 - 2x^2$ **43.** $-3x + 2$ **45.** $15x^4y^4 - 20x^3y^3$ **47.** $x^2 - 4$
 49. $2x^3 - x^2 + x - 6$ **51.** $-10x^2 + 15x$ **53.** $x^2 + 10x + 25$
 55. $2x^2 + 9x - 18$ **57.** $x^3 - y^3$ **59.** $2x - 5$ **61.** $4a^2 - 11a - 4$
 63. $4w^2 - 8w - 7$ **65.** $x^3 - 8$ **67.** $xz - wz + 2xw - 2w^2$
 69. $x^4 - x^2 + 4x - 4$ **71.** $14.4375x^2 - 29.592x - 9.72$
 73. $15.369x^2 + 1.88x - 0.866$ **75.** $\dfrac{3}{4}x + \dfrac{3}{2}$ **77.** $-\dfrac{1}{2}x^2 + x$
 79. $3x^2 - 21x + 1$ **81.** $-4x^2 - 62x - 168$
 83. $x^4 - m^2 - 4m - 4$ **85.** $-4a^{2m} - 4a^m + 5$ **87.** $x^{2n} + 2x^n - 3$
 89. $5z^{3w} - z^{1+w}$ **91.** $x^{6r} + y^3$ **93.** \$75   **95.** \$49.35, \$15.61
 97. a) 6.2 years   **b)** no   **c)** 2037

Section 5.4 Warm-Ups
 1. Distributive   **2.** FOIL   **3.** FOIL   **4.** Special   **5.** Difference
 6. Expanding   **7.** True   **8.** True   **9.** True   **10.** False
 11. False   **12.** True   **13.** False   **14.** True

Section 5.4 Exercises
 1. $x^2 + 8x + 15$ **3.** $x^2 + 2x - 8$ **5.** $2x^2 + 7x + 3$
 7. $2a^2 - 7a - 15$ **9.** $4x^4 - 49$ **11.** $2x^6 + 7x^3 - 4$
 13. $w^2 + 5wz - 6z^2$ **15.** $9k^2 + 6kt - 8t^2$ **17.** $xy - 3y + xw - 3w$
 19. $m^2 + 6m + 9$ **21.** $a^2 - 8a + 16$ **23.** $4w^2 + 4w + 1$
 25. $9t^2 - 30tu + 25u^2$ **27.** $x^2 + 2x + 1$ **29.** $a^2 - 6ay^3 + 9y^6$
 31. $w^2 - 81$ **33.** $w^6 - y^2$ **35.** $49 - 4x^2$ **37.** $9x^4 - 4$
 39. $25a^6 - 4b^2$ **41.** $m^2 + 2mt + t^2 - 25$ **43.** $y^2 - r^2 - 10r - 25$
 45. $4y^2 - 4yt + t^2 + 12y - 6t + 9$ **47.** $9h^2 + 6hk - 6h + k^2 - 2k + 1$
 49. $x^3 + 3x^2 + 3x + 1$ **51.** $w^3 - 6w^2 + 12w - 8$
 53. $8x^3 + 12x^2 + 6x + 1$ **55.** $27x^3 - 27x^2 + 9x - 1$
 57. $x^4 + 4x^3 + 6x^2 + 4x + 1$ **59.** $h^4 - 12h^3 + 54h^2 - 108h + 81$
 61. $x^2 + 3x - 54$ **63.** $25 - x^2$ **65.** $6x^2 + 7ax - 20a^2$
 67. $2t^2 + 2tw - 3t - 3w$ **69.** $9x^4 + 12x^3y^3 + 4y^6$
 71. $6y^2 - 4y - 10$ **73.** $4m^2 - 28m + 49$ **75.** $49x^2 + 42x + 9$
 77. $36y^3 + 12y^2 + y$ **79.** $2ah + h^2$ **81.** $x^3 + 6x^2 + 12x + 8$
 83. $y^3 + 9y^2 + 27y + 27$ **85.** $-3x^2 + 34x - 75$
 87. $16.32x^2 - 10.47x - 17.55$ **89.** $12.96y^2 + 31.68y + 19.36$

91. $x^{3m} + 2x^{2m} + 3x^m + 6$ **93.** $a^{3n+1} + a^{2n+1} - 3a^{n+1}$
95. $a^{2m} + 2a^{m+n} + a^{2n}$ **97.** $15y^{3m} + 24y^{2m}z^k + 20y^mz^{3-k} + 32z^3$
99. $A(x) = x^2 + 4x + 3$ **101. a)** $A(x) = 4x^2 - 36x + 80$ **b)** 66.24 km^2
103. a) $V(x) = 4x^3 - 20x^2 + 24x$ **b)** 5.9 ft^3
105. a) $a^3 + 3a^2b + 3ab^2 + b^3$ **b)** $a^4 + 4a^3b + 6a^2b^2 + 4ab^3 + b^4$
$a^5 + 5a^4b + 10a^3b^2 + 10a^2b^3 + 5ab^4 + b^5$ **c)** Five terms are in $(a+b)^4$ and six in $(a+b)^5$.   **d)** There are $n+1$ terms in $(a+b)^n$.

Mid-Chapter Quiz 5.1–5.4

1. -16 **2.** $\dfrac{1}{16}$ **3.** $-\dfrac{1}{5}$ **4.** x^6 **5.** a^7b^5 **6.** $\dfrac{32b^{15}}{a^{25}}$
7. $\dfrac{1}{a^{18}b^{30}}$ **8.** $\dfrac{4y^6}{27x^7}$ **9.** 5.6×10^{-4} **10.** 5×10^5
11. $-3x^2 - 5x + 6$ **12.** $13x^2 - x - 6$ **13.** $-10x^4y^4$
14. $-12x^3 + 18x^2$ **15.** $w^3 + 8$ **16.** $-12x^2 + 21x + 2$
17. $-12x^2 + 29x - 14$ **18.** $16a^2 - 24ab + 9b^2$ **19.** $81a^2 - 25$
20. $a^3 + 15a^2 + 75a + 125$ **21.** -65 **22.** -19

Section 5.5 Warm-Ups

1. Prime   **2.** Completely   **3.** Greatest common factor   **4.** Linear
5. Quadratic   **6.** Prime   **7.** True   **8.** True   **9.** True   **10.** True
11. False   **12.** True   **13.** False   **14.** False

Section 5.5 Exercises

1. 12   **3.** 3   **5.** $6xy$ **7.** $x(x^2 - 5)$ **9.** $12w(4x + 3y)$
11. $2x(x^2 - 2x + 3)$ **13.** $12a^3b^2(3b^4 - 2a + 5a^2b)$
15. $2(x - y), -2(-x + y)$ **17.** $3x(2x - 1), -3x(-2x + 1)$
19. $w^2(-w + 3), -w^2(w - 3)$ **21.** $a(-a^2 + a - 7), -a(a^2 - a + 7)$
23. $(x - 6)(a + b)$ **25.** $(y - 4)(x - 3)$ **27.** $(y - 1)^2(y + z)$
29. $(a - b)^3$ **31.** $(x + y)(a + 3)$ **33.** $(x - 3)(y + 1)$
35. $(a - b)(4 - c)$ **37.** $(x - 1)(y - 6)$ **39.** $(x - 2)(x + 8)$
41. $(x - 10)(x + 10)$ **43.** $(2y - 7)(2y + 7)$
45. $(3x - 5a)(3x + 5a)$ **47.** $(12wz - h)(12wz + h)$ **49.** $(x - 10)^2$
51. $(2m - 1)^2$ **53.** $(w - t)^2$ **55.** $2(x + 2)(x - 2)$ **57.** $x(x + 5)^2$
59. $(2x + 1)^2$ **61.** $(x + 3)(x + 7)$ **63.** $3y(2y + 1)$ **65.** $(2x - 5)^2$
67. $(2x - 3)(x - 2)$ **69.** $a(3a + w)(3a - w)$ **71.** $-5(a - 3)^2$
73. $-3y(y + 3)^2$ **75.** $-7(ab + 1)(ab - 1)$ **77.** $(x - h)(7 - h)$
79. $(x + 3)(a - 2)(a + 2)$ **81.** $(a + 3)(b + 3)(b - 3)$
83. $4(b + 2)(a + 1)(a - 1)$ **85.** 9   **87.** 20   **89.** 16
91. a) $b - 3$ inches   **b)** 4050 cubic inches (in.^{3})   **c)** 30 inches

Section 5.6 Warm-Ups

1. Sum, product   **2.** Sum, product   **3.** Greatest common factor
4. Substitution   **5.** True   **6.** False   **7.** False   **8.** False   **9.** True
10. False   **11.** False   **12.** False   **13.** True

Section 5.6 Exercises

1. $(x + 1)(x + 3)$ **3.** $(a + 10)(a + 5)$ **5.** $(y - 7)(y + 2)$
7. $(x - 2)(x - 4)$ **9.** $(a - 9)(a - 3)$ **11.** $(a + 10)(a - 3)$
13. $(3w + 1)(2w + 1)$ **15.** $(2x + 1)(x - 3)$ **17.** $(2x + 5)(2x + 3)$
19. $(2x - 1)(3x - 1)$ **21.** $(3y + 1)(4y - 1)$ **23.** $(6a - 5)(a + 1)$
25. $(2x - 1)(x + 8)$ **27.** $(3b + 5)(b - 7)$ **29.** $(3w - 4)(2w + 3)$
31. $(4x - 1)(x - 1)$ **33.** $(5m - 2)(m + 3)$ **35.** $(3y + 4)(2y - 5)$
37. $(x^5 - 3)(x^5 + 3)$ **39.** $(z^6 - 3)^2$ **41.** $2x(x^3 + 2)^2$
43. $x(2x^2 + 1)^2$ **45.** $a(x^2 + 2)(x^2 + 6)$ **47.** $(a^n - 1)(a^n + 1)$
49. $(a^r + 3)^2$ **51.** $3(a^w - 5)^2$ **53.** $k(k^w - 5)^2$
55. $(x^3 + 5)(x^3 - 7)$ **57.** $(a^{10} - 10)^2$ **59.** $-2a(3a^2 + 1)(2a^2 + 1)$
61. $(x^a + 5)(x^a - 3)$ **63.** $(x^a - y^b)(x^a + y^b)$ **65.** $(x^4 - 3)(x^4 + 2)$
67. $x^a(x - 1)(x + 1)$ **69.** $(x^a + 3)^2$ **71.** $2(x + 5)^2$
73. $a(a - 6)(a + 6)$ **75.** $5(2a - 1)(a + 6)$ **77.** $2(x + 8y)(x - 8y)$

79. $-3(3x + 1)(x - 4)$ **81.** $m^3(m + 10)^2$ **83.** $(3x + 4)(2x + 5)$
85. $y(3y - 4)^2$ **87.** $(r - 4s)(r - 2s)$ **89.** $m(m + 1)(m + 2)$
91. $m(2m + n)(3m - 2n)$ **93.** $(3m - 5n)(3m + 5n)$
95. $5(a + 6)(a - 2)$ **97.** $-2(w - 10)(w + 1)$
99. $x^2(w + 10)(w - 10)$ **101.** $9(3x + 1)(3x - 1)$
103. $(4x + 5)(2x - 3)$ **105.** a and b
107. a) $(x + 5)^2$ **b)** $(x - 5)^2$ **c)** $(x + 25)(x + 1)$ **d)** $(x + 5)(x - 5)$
e) Not factorable
109. $(a + 3)(x + 4)$ **111.** $(x - 2)(a + 4)$ **113.** $(m - 4)(b - 5)$
115. $(n - a)(x + c)$ **117.** $(x - y)(r - w)$ **119.** $(x - t)(t - a)$
121. $(2q - 1)(h + 4)$ **123.** $(aw - 3t)(y - 2)$ **125.** $(x^2 + 7)(x - a)$
127. $(m^2 - 5)(m^2 + p)$ **129.** $(y + 1)(y + 2)$ **131.** $(x + 3)(x + 7)$
133. $(a + 9)(a + 6)$ **135.** $(y - 2)(y + 5)$ **137.** $(w - 5)(w + 3)$
139. $(b - 2)(b + 8)$ **141.** $(a - 11)(a + 3)$ **143.** $(a - 3)(a - 6)$
145. $(x - 3)(x - 8)$ **147.** $(y - 10)(y - 13)$ **149.** $(2w + 1)(w + 3)$
151. $(2x + 1)(x - 5)$ **153.** $(3x + 1)(x + 8)$ **155.** $(3x - 1)(x + 9)$
157. $(5y + 1)(y + 3)$ **159.** $(5y - 1)(y - 4)$ **161.** $(7a - 1)(a + 1)$
163. $(7a - 1)(a - 1)$ **165.** $(2w + 1)(w + 11)$
167. $(2w - 1)(w + 11)$

Section 5.7 Warm-Ups

1. Difference of two squares   **2.** Difference of two cubes
3. Sum of two cubes   **4.** Grouping   **5.** Perfect square
6. Sum of two squares   **7.** False   **8.** True   **9.** True   **10.** False
11. False   **12.** True   **13.** False   **14.** True   **15.** True

Section 5.7 Exercises

1. Prime   **3.** Not prime   **5.** Not prime   **7.** Prime   **9.** Prime
11. Prime   **13.** $(a - 1)(a^2 + a + 1)$ **15.** $(w + 3)(w^2 - 3w + 9)$
17. $(2x - 1)(4x^2 + 2x + 1)$ **19.** $(4x + 5)(16x^2 - 20x + 25)$
21. $(2a - 3b)(4a^2 + 6ab + 9b^2)$ **23.** $(x^2 - 2)(x^4 + 2x^2 + 4)$
25. $(a^2 - 5)^2$ **27.** $(x^2 - 2)(x - 2)(x + 2)$
29. $3m(m - 2)(m^2 + 2m + 4)$
31. $2(y - 2)(y^2 + 2y + 4)(y + 2)(y^2 - 2y + 4)$ **33.** $2(4a^2 + 3)(4a^2 - 3)$
35. $3(x - 2)(3x - 4)$ **37.** $(x + 3)(x - 2)(x^2 - x + 6)$
39. $(m + 5)(m + 1)$ **41.** $(3y - 7)(y + 4)$
43. $(y - 3)(y + 3)(y - 1)(y + 1)$ **45.** $(x - 2)(x + b)$
47. $(x - y)(x + a)$ **49.** $(x + 1 - a)(x + 1 + a)$
51. $(x - 2 - w)(x - 2 + w)$ **53.** $(x + 2 - z)(x + 2 + z)$
55. $(3x - 4)^2$ **57.** $(3x - 1)(4x - 3)$ **59.** $3a(a + 3)(a^2 - 3a + 9)$
61. $2(x^2 + 16)$ **63.** Prime   **65.** $(x + y - 1)(x + y + 1)$
67. $ab(a - b)(a + b)$ **69.** $(x - 2)(x + 2)(x^2 + 4)$
71. $(x + 2)(x - 2)(x^2 + 2x + 4)$ **73.** $n(m + n)^2$
75. $(m + n)(2 + w)$ **77.** $(2w + 3)(2w - 1)$ **79.** $(t^2 + 7)(t^2 - 3)$
81. $-a(a + 10)(a - 3)$ **83.** $(a - w)(a + w)(a^2 + w^2)$
85. $(y + 2)(y + 6)$ **87.** $-2(w - 5)(w + 5)(w^2 + 25)$
89. $4(a^2 + 4)$ **91.** $8a(a^2 + 1)$ **93.** $(w + 2)(w + 8)$
95. $a(2w - 3)^2$ **97.** $(x - 3y)^2$ **99.** $3x^2(x - 5)(x + 5)$
101. $n(m - 1)(m^2 + m + 1)$ **103.** $2(3x + 5)(2x - 3)$
105. $2(a^3 - 16)$ **107.** $(x + y)(x^2 - xy + y^2)(x - y)(x^2 + xy + y^2)$
109. $(a^m - 1)(a^{2m} + a^m + 1)$ **111.** $(a^w - b^{2n})(a^{2w} + a^wb^{2n} + b^{4n})$
113. $(x^m - 2)(x^{2m} + 2x^m + 4)$ **115.** $(a^m - b)(a^{2m} + a^mb + b^2)$
117. $(t^n - 2)(t^n + 2)(t^{2n} + 4)$ **119.** $a(a^n - 5)(a^n + 3)$
121. $(a^n - 3)(a^n + b)$ **125.** $(a + 4)^2$ **127.** $(6 - y)(6 + y)$
129. $(a - 4)(a + 4)$ **131.** $(w + 9)^2$ **133.** $a(a - 2)$
135. $(2w + 9)^2$ **137.** $(x - 1)(x + 1)$ **139.** $(w + 3)(w^2 - 3w + 9)$
141. $(a - 4)(w - b)$ **143.** $(z - 3)(w - 5)$ **145.** $(2b - a)^2$
147. $(a - 3)(a^2 + 3a + 9)$ **149.** $3(z - 5)^2$ **151.** $2(b - 2)(b^2 + 2b + 4)$
153. $-2a(a - 9)^2$ **155.** $(z - 2)(z + 2)(z^2 + 4)$ **157.** $a(a^2 + b^2)$
159. $(w - 2)(w + 3a)$ **161.** $(a + 5)^2$ **163.** $(5b + 3)^2$

165. $(12 - y)(12 + y)$ **167.** $(3a - z)(3a + z)$ **169.** $(w - 13)(3w + 1)$
171. $(m - 3)(m + 7)$ **173.** $(b^2 - y)(b^2 + y)$ **175.** $(z^3 - 7)(z^3 + 7)$
177. $a^3(a^2 + 4)$ **179.** $3(5w + 4)^2$ **181.** $(ax - b)(ax + b)$
183. $(b - y)(x + z)$ **185.** $(z - 5)(z^2 + 5z + 25)$ **187.** $(3x - 1)(9x + 1)$
189. $(bn - y^2)(bn + y^2)$ **191.** $-3(2a - 1)(a - 5)$ **193.** $(2ab - w)^2$
195. $(t - 3)(t^2 + 3t + 9)$ **197.** $3(z^3 - 5)^2$ **199.** $5(b - 2)(b^2 + 2b + 4)$
201. $-2a^2(a - 5)^2$ **203.** $(a - 2)(a + 2)(a^2 + 4)$ **205.** $b^2(a^2 + b^2)$
207. $(x^2 + 5)(x^2 - 3)$ **209.** $(w^2 - 2)(w^2 + 3a)$

Section 5.8 Warm-Ups

1. Zero factor property   **2.** Quadratic   **3.** Hypotenuse
4. Pythagorean   **5.** Factoring   **6.** False   **7.** True   **8.** True
9. True   **10.** False   **11.** False   **12.** True

Section 5.8 Exercises

1. $\{-4, 5\}$ **3.** $\left\{\dfrac{5}{2}, -\dfrac{4}{3}\right\}$ **5.** $\{-5, 2\}$ **7.** $\{-5, 0, 5\}$ **9.** $\{-7, 2\}$

11. $\{0, 7\}$ **13.** $\{-4, 5\}$ **15.** $\{3\}$ **17.** $\left\{-5, \dfrac{3}{2}\right\}$ **19.** $\{-4, 5\}$

21. $\{3, 7\}$ **23.** $\{4\}$ **25.** $\left\{-4, \dfrac{1}{5}\right\}$ **27.** $\{-3, 4\}$ **29.** $\left\{-4, \dfrac{5}{2}\right\}$

31. $\{-2, 0, 2\}$ **33.** $\left\{-1, 0, \dfrac{1}{4}\right\}$ **35.** $\{-5, -4, 5\}$ **37.** $\{-1, 1, 2\}$

39. $\{-3, -1, 1, 3\}$ **41.** $\{-8, -6, 4, 6\}$ **43.** $\{-4, -2, 0\}$

45. $\{-7, -3, 1\}$ **47.** $\left\{-\dfrac{3}{2}, 2\right\}$ **49.** $\{-6, -3, -2, 1\}$ **51.** $\{-6, 1\}$

53. $\{-5, 2\}$ **55.** $\{-5, -4, 0\}$ **57.** $\{-3, 0, 3\}$ **59.** $\left\{-3, \dfrac{1}{2}, 3\right\}$

61. $\left\{-\dfrac{1}{3}, \dfrac{1}{4}, \dfrac{1}{2}\right\}$ **63.** $\left\{\dfrac{3}{2}\right\}$ **65.** $\{-a, -b\}$ **67.** $\left\{-\dfrac{a}{3}\right\}$ **69.** $\{0, -b\}$

71. $\{-b, b\}$ **73.** $\left\{-\dfrac{3}{a}, 1\right\}$ **75.** Width 4 inches, length 6 inches

77. 4 and 9   **79.** $L = 43$ in., $W = 22$ in.
81. Width 5 feet, length 12 feet
83. a) 4 seconds   **b)** 4 seconds   **c)** 64 feet   **d)** 2 seconds
85. a) 1225 ft   **b)** 20.3125 sec   **87.** 3 sec
89. Width 5 feet, length 12 feet   **91.** 12 feet   **93.** -3 or 4
95. Length 8 inches, width 6 inches   **97.** Length 20 feet, width 12 feet
99. 91/20 chi

Enriching Your Mathematical Word Power

1. Polynomial   **2.** Degree   **3.** Leading   **4.** Monomial
5. Binomial   **6.** Trinomial   **7.** FOIL   **8.** Prime   **9.** Prime
10. Factoring   **11.** Completely   **12.** Greatest   **13.** Scientific
14. Sum   **15.** Difference   **16.** Quadratic   **17.** Zero

Review Exercises

1. 2   **3.** 36   **5.** $-\dfrac{1}{27}$ **7.** 1   **9.** $\dfrac{8}{x^3}$ **11.** $\dfrac{1}{y^2}$ **13.** a^7 **15.** $\dfrac{2}{x^4}$

17. 8,360,000   **19.** 0.00057   **21.** 8.07×10^6 **23.** 7.09×10^{-4}

25. 1×10^{15} **27.** 2×10^1 **29.** 1×10^2 **31.** $\dfrac{1}{a}$ **33.** $\dfrac{n^2}{m^{16}}$

35. $\dfrac{81}{16}$ **37.** $\dfrac{25}{36}$ **39.** $-\dfrac{4}{3ab}$ **41.** $\dfrac{b^{14}}{a^7}$ **43.** 5^{6w-1} **45.** 7^{15a-40}

47. $8w + 2$ **49.** $-6x + 3$ **51.** $x^3 - 4x^2 + 8x - 8$
53. $-4xy + 22z$ **55.** $5m^5 - m^3 + 2m^2$ **57.** $x^2 + 4x - 21$
59. $z^2 - 25y^2$ **61.** $m^2 + 16m + 64$ **63.** $w^2 - 10xw + 24x^2$
65. $k^2 - 6k + 9$ **67.** $m^4 - 25$ **69.** $3(x - 2)$ **71.** $-4(-a + 5)$
73. $-w(w - 3)$ **75.** $(y - 9)(y + 9)$ **77.** $(2x + 7)^2$ **79.** $(t - 9)^2$
81. $(x + 4)(x + 10)$ **83.** $(x - 10)(x + 3)$ **85.** $(w - 7)(w + 4)$

87. $(2m + 7)(m - 1)$ **89.** $m(m^3 - 5)(m^3 + 2)$
91. $5(x + 2)(x^2 - 2x + 4)$ **93.** $(3x + 2)(3x + 1)$
95. $(x + 1)^2(x - 1)$ **97.** $-y(x - 4)(x + 4)$ **99.** $-ab^2(a - 1)^2$
101. $(x - 1)(x^2 + 9)$ **103.** $(x - 2)(x + 2)(x^2 + 3)$
105. $a^3(a - 1)(a^2 + a + 1)$ **107.** $(t - 5)(t^2 + 5t + 25)$
109. $2(2 - 3x)(4 + 6x + 9x^2)$ **111.** $y^2(5x^3 - y^3)$ **113.** $-2(2m + 3)^2$
115. $(2x - 7)(2x + 1)$ **117.** $(x + 2)(x^2 - 2x + 4)(x - 1)(x^2 + x + 1)$
119. $(a^2 - 11)(a^2 - 12)$ **121.** $(x^k - 7)(x^k + 7)$
123. $(m^a - 3)(m^a + 1)$ **125.** $(3z^k - 2)^2$ **127.** $(y^a - b)(y^a + c)$

129. $\{0, 5\}$ **131.** $\{0, 5\}$ **133.** $\left\{-\dfrac{1}{2}, 5\right\}$ **135.** $\{-5, -1, 1\}$

137. $\{-3, -1, 1, 3\}$ **139.** $\{a, -b\}$ **141.** $\{ab\}$ **143.** 6 feet
145. 7 meters by 9 meters   **147.** 7 in. by 24 in.
149. a) 68.7 years   **b)** 6.4 years   **151. a)** 15%   **b)** $12,196.46

Chapter 5 Test

1. $\dfrac{1}{9}$ **2.** 36   **3.** 8   **4.** $12x^7$ **5.** $4y^{12}$ **6.** $64a^6b^3$ **7.** $\dfrac{27}{x^6}$

8. $\dfrac{2a^3}{b^3}$ **9.** 3,240,000,000   **10.** 0.0008673   **11.** 2.4×10^{-5}

12. 2×10^{-13} **13.** $3x^3 + 3x^2 - 2x + 3$ **14.** $-2x^2 - 8x - 3$
15. $x^3 - 5x^2 + 13x - 14$ **16.** $x^3 - 6x^2 + 12x - 8$
17. $2x^2 - 11x - 21$ **18.** $x^2 - 12x + 36$ **19.** $4x^2 + 20x + 25$
20. $9y^4 - 25$ **21.** $(a - 6)(a + 4)$ **22.** $(2x + 7)^2$
23. $3(m - 2)(m^2 + 2m + 4)$ **24.** $2y(x - 4)(x + 4)$
25. $(2m + 3)(6m + 5)$ **26.** $(2x^5 - 3)(x^5 + 4)$ **27.** $(2x + 3)(a - 5)$
28. $(x - 1)(x + 1)(x^2 + 4)$ **29.** $(a - 1)(a + 1)(a^2 + 1)$

30. $\left\{-5, \dfrac{3}{2}\right\}$ **31.** $\{-5\}$ **32.** $\{-2, 0, 2\}$ **33.** $\{-4, -3, 2, 3\}$

34. $\{a, -2b\}$ **35.** Width 8 inches, height 6 inches
36. 38.0, 10.8, 7.9   **37. a)** 96 ft, 96 ft   **b)** 5 sec   **38.** 20 ft by 24 ft

Making Connections   A Review of Chapters 1–5

1. 16   **2.** -8 **3.** $\dfrac{1}{16}$ **4.** 2   **5.** 1   **6.** $\dfrac{1}{6}$ **7.** $\dfrac{9}{4}$ **8.** $-\dfrac{9}{4}$

9. $-\dfrac{4}{3}$ **10.** 1   **11.** $\dfrac{25}{64}$ **12.** $\dfrac{1}{64}$ **13.** $\dfrac{16}{9}$ **14.** 8   **15.** $\dfrac{9}{8}$

16. $\dfrac{9}{4}$ **17.** $\dfrac{1}{12}$ **18.** 49   **19.** 64   **20.** 8   **21.** -29 **22.** $\dfrac{11}{30}$

23. $\{200\}$ **24.** $\left\{\dfrac{9}{5}\right\}$ **25.** $\left\{-9, \dfrac{3}{2}\right\}$ **26.** $\left\{-\dfrac{15}{2}, 0\right\}$ **27.** $\left\{2, \dfrac{8}{5}\right\}$

28. $\left\{\dfrac{9}{5}\right\}$ **29.** $\varnothing$ **30.** $\{-3, -2, 1, 2\}$ **31.** $\left\{-5, \dfrac{1}{2}\right\}$ **32.** $\left\{-\dfrac{2}{3}, \dfrac{4}{3}\right\}$

33. 5×10^{10} **34.** 2.05×10^5 **35.** $-18x + 24$ **36.** $-13x + 24$
37. $15x^2 - 38x + 24$ **38.** $15x^3 - 38x^2 + 29x - 6$ **39.** $3x - 2$
40. $2x - 2$ **41.** 65 meters   **42.** 12 inches   **43.** 12 feet
44. 8 inches   **45.** 12 inches, 5 inches
46. a) $20,000   **b)** $2500   **c)** $(20,000, 20,000)$ **d)** $0

Chapter 6

Section 6.1 Warm-Ups

1. Rational expression   **2.** Domain   **3.** Basic principle   **4.** Reduce
5. Build up   **6.** True   **7.** False   **8.** True   **9.** False   **10.** False
11. True   **12.** True   **13.** True

Section 6.1 Exercises

1. $(-\infty, 1) \cup (1, \infty)$ **3.** $(-\infty, 0) \cup (0, \infty)$
5. $(-\infty, -2) \cup (-2, 2) \cup (2, \infty)$ **7.** $(-\infty, \infty)$

9. $-3, -2$ **11.** $-3, 0, 2$ **13.** $\dfrac{2}{19}$ **15.** $\dfrac{1}{5}$ **17.** $\dfrac{x+1}{2}$

19. $-\dfrac{3}{5}$ **21.** $\dfrac{b}{a^2}$ **23.** $\dfrac{-w}{3x^2 y}$ **25.** $\dfrac{b^2}{1+a}$ **27.** $-\dfrac{1}{2}$

29. $\dfrac{x+2}{x}$ **31.** $a^2 + ab + b^2$ **33.** $\dfrac{x^2-1}{x^2+1}$ **35.** $\dfrac{6x+2}{2x+5}$

37. $\dfrac{x^2+7x-4}{(x-4)(x+4)}$ **39.** $\dfrac{a+y}{b-5}$ **41.** $\dfrac{2x-6}{x+5}$ **43.** $\dfrac{10}{50}$

45. $\dfrac{3xy^2}{3x^2 y^3}$ **47.** $\dfrac{5x-5}{x^2-2x+1}$ **49.** $\dfrac{6x+15}{4x^2-25}$ **51.** $\dfrac{-3}{-6x-6}$

53. $\dfrac{5a}{a}$ **55.** $\dfrac{x^2+x-2}{x^2+2x-3}$ **57.** $\dfrac{-7}{1-x}$

59. $\dfrac{3x^2-6x+12}{x^3+8}$ **61.** $\dfrac{2x^2+9x+10}{6x^2+13x-5}$

63. $(-\infty, -4) \cup (-4, \infty)$, $R(3) = \dfrac{4}{7}$, $R(-4)$ undefined

65. $\left(-\infty, \dfrac{4}{3}\right) \cup \left(\dfrac{4}{3}, \infty\right)$, $H(-2) = \dfrac{1}{10}$, $H\!\left(\dfrac{4}{3}\right)$ undefined

67. $(-\infty, -2) \cup (-2, 3) \cup (3, \infty)$, $W(-2)$ undefiend,

$\qquad W(3)$ undefined, $W(1) = -\dfrac{1}{2}$

69. $\dfrac{7}{21}$ **71.** $\dfrac{10}{2}$ **73.** $\dfrac{3a}{a^2}$ **75.** $\dfrac{-2}{b-a}$ **77.** $\dfrac{2x+2}{x^2-1}$

79. $\dfrac{-2}{3-w}$ **81.** $\dfrac{x+2}{3}$ **83.** $\dfrac{a+1}{a}$ **85.** $\dfrac{x^2+x+1}{x^3-1}$

87. $S(x) = \dfrac{500}{x}$ **89. a)** $C(x) = \dfrac{150}{x}$ **b)** \$30, \$15, \$5

91. a) $A(n) = \dfrac{0.50n+45}{n}$ dollars   **b)** 7.5 cents   **c)** Decreases
d) Increases

93. a) $p(n) = \dfrac{0.053n^2 - 0.64n + 6.71}{3.43n + 87.24}$ **b)** 7.7%, 18.5%, 41.4%
95. The value of $R(x)$ gets closer and closer to $\dfrac{1}{2}$.

Section 6.2 Warm-Ups
1. Multiply   **2.** Opposites   **3.** Divisor   **4.** False   **5.** False
6. True   **7.** True   **8.** True   **9.** True   **10.** False   **11.** True
12. False

Section 6.2 Exercises
1. $\dfrac{5}{11}$ **3.** $\dfrac{4}{5}$ **5.** $\dfrac{ab}{4}$ **7.** $\dfrac{1}{2}$ **9.** $\dfrac{x+1}{x^2+1}$ **11.** $\dfrac{1}{a-b}$

13. $\dfrac{a+2}{2}$ **15.** $-\dfrac{2}{3}$ **17.** $-7a-14$ **19.** $6x^2-x-1$ **21.** $\dfrac{3}{2}$

23. $\dfrac{63}{5}$ **25.** $\dfrac{6b}{5c^8}$ **27.** $\dfrac{w}{w-1}$ **29.** $\dfrac{2}{x-y}$ **31.** $\dfrac{2}{x}$

33. $2x-2y$ **35.** $2x+10$ **37.** $\dfrac{a-b}{6}$ **39.** $3a-3b$ **41.** $\dfrac{5x}{6}$

43. 3 **45.** $\dfrac{1}{12}$ **47.** $\dfrac{2x}{3}$ **49.** -1 **51.** $b-a$ **53.** -2

55. $\dfrac{a+b}{a}$ **57.** $\dfrac{x-9}{2}$ **59.** -2 **61.** $2a+2b$ **63.** $\dfrac{3x}{5y}$

65. $\dfrac{3a}{10b}$ **67.** $\dfrac{7x^2}{3x+2}$ **69.** $-\dfrac{a^6 b^2}{8c^2}$ **71.** $\dfrac{2m^8 n^2}{3}$ **73.** $\dfrac{2x-3}{4(2x-1)}$

75. $\dfrac{h-3}{5h+1}$ **77.** $\dfrac{-3a-1}{2}$ **79.** $\dfrac{k+m}{m-k}$ **81.** $\dfrac{y^b}{x^a}$ **83.** $\dfrac{x^a-1}{x^a+2}$

85. $\dfrac{m^k+1}{m^k+2}$ **87.** 7.1%   **89.** $\dfrac{75}{x}$ miles   **91.** e

Section 6.3 Warm-Ups
1. Least common multiple   **2.** Least common denominator
3. Denominators   **4.** Multiplication, division   **5.** False   **6.** False
7. True   **8.** True   **9.** False   **10.** True   **11.** False   **12.** True

Section 6.3 Exercises
1. $4x$ **3.** $-x$ **5.** $\dfrac{-x+1}{x}$ **7.** $\dfrac{5}{2}$ **9.** $\dfrac{2}{x-3}$ **11.** $\dfrac{3x-5}{x+3}$

13. 120 **15.** 396 **17.** $30x^3 y$ **19.** $a^3 b^5 c^2$ **21.** $x(x+2)(x-2)$

23. $12(a+2)$ **25.** $(x-1)(x+1)^2$ **27.** $x(x-4)(x+4)(x+2)$

29. $(2x+3)(3x+4)(3x-4)$ **31.** $\dfrac{17}{140}$ **33.** $\dfrac{1}{144}$ **35.** $\dfrac{3w+5z}{w^2 z^2}$

37. $\dfrac{2x-1}{24}$ **39.** $\dfrac{11x}{10a}$ **41.** $\dfrac{9-4xy}{4y}$ **43.** $\dfrac{-2a-14}{a(a+2)}$

45. $\dfrac{3a-b}{(a-b)(a+b)}$ **47.** 0 **49.** $\dfrac{3}{x+1}$ **51.** $\dfrac{4x+9}{(x+3)(x-3)}$

53. $\dfrac{1}{x-3}$ **55.** $\dfrac{11}{2(x-2)}$ **57.** $\dfrac{-x+3}{(x-1)(x+2)(x+3)}$

59. $\dfrac{3x^2+5x-3}{(x+1)(2x-1)(3x-1)}$ **61.** $\dfrac{8x-2}{x(x-1)(x+2)}$ **63.** $\dfrac{7}{12}$

65. $-\dfrac{19}{40}$ **67.** $\dfrac{5x}{6}$ **69.** $\dfrac{3a-2b}{3b}$ **71.** $\dfrac{3a+2}{3}$ **73.** $\dfrac{a+3}{a}$

75. $\dfrac{3}{x}$ **77.** $\dfrac{8x+3}{12x}$ **79.** -3 **81.** $\dfrac{-13}{15(x-2)}$

83. $\dfrac{-4x}{(x+1)(x-1)(2x+1)}$ **85.** b **87.** $\dfrac{x^2+10x}{(5x-2)^2}$ **89.** $\dfrac{4x^2+9}{x(2x-3)}$

91. $\dfrac{-w^2+2w+8}{(w+3)(w^2-3w+9)}$ **93.** $\dfrac{a-3}{a^2+2a+4}$

95. $\dfrac{-w^3-2w^2-5w+6}{(w-2)(w+2)(w^2+2w+4)}$

97. $\dfrac{x^3+x^2+2x+1}{(x-1)(x+1)(x^2+x+1)}$

99. a) $C(x) = \dfrac{16x+8}{x(x+1)}$ **b)** $6\dfrac{2}{3}$ claims

101. a) $M(x) = \dfrac{3x+60}{x}$ **b)** 15 subscriptions

103. a) $T(x) = \dfrac{300x+500}{x(x+5)}$ **b)** 4 hours 24 minutes

Section 6.4 Warm-Ups
1. Complex fraction   **2.** LCD   **3.** False   **4.** True   **5.** True
6. True   **7.** True   **8.** True   **9.** True

Section 6.4 Exercises
1. $\dfrac{6}{5}$ **3.** $\dfrac{10}{3}$ **5.** -8 **7.** $\dfrac{6x-4}{6x+9}$ **9.** $\dfrac{a^2 b+3a}{a+b^2}$ **11.** $\dfrac{a^2+ab}{a-b}$

13. $\dfrac{x-3y}{x+y}$ **15.** $\dfrac{60m-3m^2}{4(2m+9)}$ **17.** $\dfrac{a^2-ab}{b^2}$ **19.** $\dfrac{xy+x^2}{y^2(1-x)(1+x)}$

21. $\dfrac{x+2}{x-2}$ **23.** $\dfrac{y^2-y-2}{(y-1)(3y+4)}$ **25.** $\dfrac{4x-10}{x-4}$ **27.** $\dfrac{6w-3}{2w^2+w-4}$

29. $\dfrac{2b-a}{a-3b}$ **31.** $\dfrac{-y^2-12}{y^2-3}$ **33.** $\dfrac{3a-7}{5a-2}$ **35.** $\dfrac{3m^2-12m+12}{(m-3)(2m-1)}$

37. $\dfrac{2x^2+4x+5}{2(2x-3)(x+1)}$ **39.** $\dfrac{yz+wz}{w(y+z)}$ **41.** $\dfrac{x}{x+1}$ **43.** $\dfrac{a^2+b^2}{ab^3}$

45. $\dfrac{a-1}{a}$ **47.** $\dfrac{1}{x^2 - x + 1}$ **49.** $2m - 3$ **51.** $-\dfrac{1}{ab}$ **53.** $-x^3 y^3$

55. $x - 2$ **57.** $\dfrac{xy}{x+y}$ **59.** $-1.7333, -\dfrac{26}{15}$ **61.** $0.1667, \dfrac{1}{6}$

63. 47.4%   **65.** 49.5 mph   **67.** 49.5 mph

71. $\dfrac{2x+1}{3x+2}, x = 0, -1, -\dfrac{1}{2}, -\dfrac{2}{3}$

Mid-Chapter Quiz 6.1–6.4

1. $\dfrac{3}{7}$ **2.** $\dfrac{4x-1}{4}$ **3.** $\dfrac{w-1}{2}$ **4.** $\dfrac{-2a+6}{3}$ **5.** $\dfrac{9}{5}$ **6.** $\dfrac{3x^3 z^2}{5y^2}$

7. $\dfrac{5a+15}{4}$ **8.** $\dfrac{7}{b^4}$ **9.** $\dfrac{9}{2(x-3)}$ **10.** $\dfrac{8s^2}{21}$ **11.** $\dfrac{m-1}{2m}$ **12.** $\dfrac{25}{42}$

13. $\dfrac{4a+5b^2}{a^2 b^3}$ **14.** $\dfrac{3x^2 + 4x}{(x+1)^2}$ **15.** $\dfrac{-3y}{(y+5)(y+2)}$ **16.** $\dfrac{bc+ac+ab}{abc}$

17. 33   **18.** $\dfrac{3x^2 - x - 10}{x^2 - 3}$ **19.** $-\dfrac{1}{2}$ **20.** 4   **21.** $-\dfrac{3}{2}$

22. $(-\infty, 9) \cup (9, \infty)$

Section 6.5 Warm-Ups

1. Dividend, divisor, quotient   **2.** Descending   **3.** Dividend
4. Less than   **5.** Synthetic   **6.** Divisor, dividend
7. True   **8.** True   **9.** True   **10.** False   **11.** False
12. True   **13.** True

Section 6.5 Exercises

1. $12x^4$ **3.** -2 **5.** $2b - 3$ **7.** $x + 1$ **9.** $-5x^2 + 4x - 3$
11. $\dfrac{7}{2}x^2 - 2x$ **13.** $4x - 2, 0$ **15.** $2, -3$ **17.** $\dfrac{2}{3}x + \dfrac{1}{3}, 4x - 3$
19. $2x^2 + x, -6x - 7$ **21.** $x + 5, -2$ **23.** $x - 4, 8$
25. $x^2 - 2x + 4, 0$ **27.** $a^2 + 2a + 8, 11$ **29.** $x^2 - 2x + 3, -6$
31. $x^3 + 3x^2 + 6x + 11, 21$ **33.** $-3x^2 - 1, x - 4$ **35.** $3x + 5, -1$
37. $2b - 5, -2$ **39.** $x^2 + x - 2, 0$ **41.** $x - 5, 0$ **43.** $2 + \dfrac{10}{x-5}$
45. $x - 1 + \dfrac{1}{2x+1}$ **47.** $x^2 - 2x + 4 + \dfrac{-8}{x+2}$ **49.** $x + \dfrac{2}{x}$
51. $x - 6 + \dfrac{2}{2x+1}$ **53.** $3x^2 - x - 1 + \dfrac{6}{x-1}$
55. $6x^2 + 12x + 20 + \dfrac{45}{x-2}$ **57.** $x^2 - x + 2 + \dfrac{-2}{x+1}$
59. $x + 2, 9$ **61.** $2x - 6, 11$ **63.** $x^2 - 3x, -3$
65. $3x^3 + 9x^2 + 12x + 43, 120$ **67.** $x^4 + x^3 + x^2 + x + 1, 0$
69. $x^2 - 2x - 1, 8$ **71.** $x^2 + 2x + 10, 55$ **73.** No
75. Yes, $(x-4)(x-2)$ **77.** Yes, $(w-3)(w^2 + 3w + 9)$
79. No   **81.** Yes, $(y-2)(y^2 - 2y + 2)$ **83.** -15 **85.** 9   **87.** 28
89. a) $AC(x) = 0.03x + 300$ **b)** No
   c) Because $AC(x)$ is very close to 300 for x less than 15, the graph
   looks horizontal.
91. $x - 1$ feet   **93.** 6333.3 cubic meters

Section 6.6 Warm-Ups

1. LCD   **2.** Extraneous   **3.** Proportion   **4.** Means   **5.** Extremes
6. Extremes-means   **7.** False   **8.** True   **9.** False   **10.** True
11. False   **12.** False

Section 6.6 Exercises

1. $\{-24\}$ **3.** $\left\{\dfrac{22}{15}\right\}$ **5.** $\{5\}$ **7.** $\{1, 6\}$ **9.** $\{20, 25\}$ **11.** $\varnothing$

13. $\{-2\}$ **15.** $\varnothing$ **17.** $\{-5, 1\}$ **19.** $\left\{\dfrac{8}{3}\right\}$ **21.** $\left\{-\dfrac{3}{4}\right\}$
23. $\left\{-\dfrac{14}{5}\right\}$ **25.** $\{20\}$ **27.** $\{-3, 3\}$ **29.** $\{-5, 6\}$ **31.** $\varnothing$
33. $\left\{\dfrac{11}{2}\right\}$ **35.** $\{0\}$ **37.** $\{-6, 6\}$ **39.** $\{4, 5\}$ **41.** $\{8\}$ **43.** $\{2, 4\}$
45. $\{-1\}$ **47.** $\{5\}$ **49.** $\{-3, 3\}$ **51.** $\{8\}$ **53.** $\left\{\dfrac{25}{2}\right\}$ **55.** $\varnothing$
57. $\{4\}$ **59.** $\{-5, 2\}$ **61.** $\left\{\dfrac{16}{3}\right\}$ **63.** $\dfrac{3x+16}{4x}$ **65.** $\left\{\dfrac{8}{5}\right\}$
67. $\dfrac{-5x+8}{4x}$ **69.** $(-\infty, 0) \cup (0, \infty)$, identity
71. $\varnothing$, inconsistent equation   **73.** $\{1\}$, conditional equation
75. $(-\infty, 0) \cup (0, 1) \cup (1, \infty)$, identity   **77.** $\varnothing$, inconsistent equation
79. 27 inches   **81.** \$166,666.67   **83.** 138
85. Width 132 cm, length 154 cm   **87.** 20%, 96%
89. a) \$17,142.86   **b)** \$57,142.86
91. To solve the equation, multiply each side by the LCD. To find the
   sum, build up each rational expression so that its denominator is
   the LCD.

Section 6.7 Warm-Ups

1. Formula   **2.** Solved   **3.** Uniform   **4.** Work   **5.** False
6. True   **7.** False   **8.** True   **9.** True   **10.** False

Section 6.7 Exercises

1. $y = -\dfrac{4}{3}x + 1$ **3.** $y = \dfrac{1}{M}$ **5.** $y = \dfrac{aw}{a^2 + w^2}$ **7.** $y = \dfrac{b}{h-3}$
9. $f = \dfrac{F}{M}$ **11.** $a = \dfrac{2b}{b-2}$ **13.** $x = \dfrac{y}{4-y}$ **15.** $R_1 = \dfrac{RR_2}{R_2 - R}$
17. $T_1 = \dfrac{P_1 V_1 T_2}{P_2 V_2}$ **19.** $h = \dfrac{3V}{4\pi r^2}$ **21.** $\dfrac{1}{2}$ **23.** $\dfrac{4}{3}$ **25.** $\dfrac{144}{5}$
27. -0.047 **29.** 4 mph   **31.** 60 mph
33. Patrick 24 minutes; Guy 36 minutes, 100 mph, no
35. 10 mph   **37.** 6 hours   **39.** 60 minutes   **41.** 30 minutes
43. 10 pounds apples, 12 pounds oranges   **45.** 6 ohms   **47.** 125 mm
49. 5 workers   **51. a)** 80   **b)** $C(n) = \dfrac{24,000}{n}$ **53.** 25.255 days

Enriching Your Mathematical Word Power

1. Rational   **2.** Domain   **3.** Lowest   **4.** Reduced   **5.** Equivalent
6. Complex   **7.** Building up   **8.** Least   **9.** Extraneous   **10.** Ratio
11. Synthetic   **12.** Proportion   **13.** Extremes   **14.** Means
15. Extremes-means

Review Exercises

1. $(-\infty, 1) \cup (1, \infty)$ **3.** $(-\infty, -1) \cup (-1, 2) \cup (2, \infty)$
5. $R(5) = 0, R(1)$ undefined   **7.** $H(-2) = -\dfrac{1}{2}, H(-1)$ undefined
9. $\dfrac{c^2}{a^2 b}$ **11.** $\dfrac{4x^2}{3y}$ **13.** $-a$ **15.** $\dfrac{3}{2}$ **17.** $6x(x-2)$ **19.** $12a^5 b^3$
21. $\dfrac{3x+11}{2(x-3)(x+3)}$ **23.** $\dfrac{aw-5b}{a^2 b^2}$ **25.** $\dfrac{3}{x+5}$ **27.** $\dfrac{21}{10(x-6)}$
29. $\dfrac{7-3y}{4y-3}$ **31.** $\dfrac{b^3 - a^2}{ab}$ **33.** $x^2 + 3x - 5, 0$
35. $m^3 - m^2 + m - 1, 0$ **37.** $a^6 + 2a^3 + 4, 0$ **39.** $m^2 + 2m - 6, 0$
41. $x + 1 + \dfrac{-4}{x-1}$ **43.** $3 + \dfrac{6}{x-2}$ **45.** Yes   **47.** Yes   **49.** No
51. Yes   **53.** $\left\{-\dfrac{16}{3}\right\}$ **55.** $\varnothing$ **57.** $\{10\}$ **59.** $y = mx + b$

61. $x = \dfrac{2}{2w - 1}$ **63.** $m = \dfrac{Fr}{v^2}$ **65.** $r = \dfrac{3A}{2\pi h}$ **67.** $y = 2x - 17$

69. $q = \dfrac{pf}{p - f}$ **71.** 30,000   **73.** 10 hours   **75.** 400 hours

77. a) $B(x) = \dfrac{240x - 720}{x(x - 6)}$ **b)** 30 bushels   **79.** $\dfrac{10 + 7y}{6xy}$

81. $\dfrac{8a + 10}{(a - 5)(a + 5)}$ **83.** $\{-8, 8\}$ **85.** $-\dfrac{3}{2}$ **87.** $\{-9, 9\}$

89. $\dfrac{1}{x - m}$ **91.** $\dfrac{8a + 20}{(a - 5)(a + 5)(a + 1)}$ **93.** $\dfrac{-15a + 10}{2(a + 2)(a + 3)(a - 3)}$

95. $\{2\}$ **97.** $\left\{-\dfrac{5}{2}\right\}$ **99.** $-\dfrac{1}{3}$ **101.** $\dfrac{x - 2}{3(x + 1)}$

103. $\dfrac{3a^2 + 7a + 16}{(a - 2)(a^2 + 2a + 4)}$ **105.** $\dfrac{3 - x}{(x + 1)(x + 3)}$ **107.** $\dfrac{a + w}{a + 4}$

109. $\{-6, 8\}$ **111.** 18   **113.** -3 **115.** $4x$ **117.** $10x$

119. $\dfrac{1}{3}$ **121.** $a + 3$ **123.** $\dfrac{3}{10}$ **125.** $\dfrac{5a}{6}$

Chapter 6 Test

1. $(-\infty, 4/3) \cup (4/3, \infty)$ **2.** $(-\infty, -3) \cup (-3, 3) \cup (3, \infty)$

3. $(-\infty, \infty)$ **4.** $R\left(\dfrac{1}{2}\right) = 0$, $R(3)$ undefined   **5.** $\dfrac{3a^3}{2b}$ **6.** $\dfrac{-x - y}{2(x - y)}$

7. $-\dfrac{1}{36}$ **8.** $\dfrac{7y^2 + 3}{y}$ **9.** $\dfrac{5}{a - 9}$ **10.** $\dfrac{4a + 3b}{24a^2b^2}$ **11.** $\dfrac{a^3}{30b^2}$

12. $-\dfrac{3}{a + b}$ **13.** $\dfrac{1}{(x - 1)(x + 1)}$ **14.** $\dfrac{-4x + 2}{(x + 2)(x - 2)(x - 5)}$

15. $\dfrac{m + 1}{3}$ **16.** $\left\{\dfrac{12}{7}\right\}$ **17.** $\{4, 10\}$ **18.** $\{-2, 2\}$ **19.** $t = \dfrac{a^2}{W}$

20. $b = \dfrac{2a}{a - 2}$ **21.** $\dfrac{16}{3(3 - 2x)}$ **22.** $w - m$ **23.** $\dfrac{3a^2}{2}$

24. $3x + 2, -8$ **25.** $-1, 0$ **26.** $5 + \dfrac{-15}{x + 3}$ **27.** $x + 5 + \dfrac{4}{x - 2}$

28. 24 minutes   **29.** 30 miles   **30.** 9

31. a) $T(x) = \dfrac{50x - 60}{x(x - 3)}$ **b)** 5 hours

Making Connections   A Review of Chapters 1–6

1. $\left\{\dfrac{15}{4}\right\}$ **2.** $\{-4, 4\}$ **3.** $\left\{\dfrac{12}{5}\right\}$ **4.** $\{-6, 4\}$ **5.** $\left\{\dfrac{1}{4}\right\}$ **6.** $\{6\}$

7. $\left\{\dfrac{1}{2}\right\}$ **8.** $\left\{\dfrac{3}{2}\right\}$ **9.** $\{-3, 3\}$ **10.** $\{-8, 3\}$ **11.** $\left\{\dfrac{1}{3}, \dfrac{2}{3}\right\}$

12. $\{-3, 9\}$ **13.** $\left\{-\dfrac{9}{2}, \dfrac{1}{2}\right\}$ **14.** $\varnothing$ **15.** $y = \dfrac{C - Ax}{B}$

16. $y = -\dfrac{1}{3}x + \dfrac{4}{3}$ **17.** $y = \dfrac{C}{A - B}$ **18.** $y = A$ or $y = -A$

19. $y = 2A - 2B$ **20.** $y = \dfrac{2AC}{2B + C}$ **21.** $y = \dfrac{3}{4}x - \dfrac{3}{2}$

22. $y = A$ or $y = 2$ **23.** $y = \dfrac{2A - BC}{B}$ **24.** $y = B$ or $y = -C$

25. $12x^{13}$ **26.** $3x^5 + 15x^8$ **27.** $25x^{12}$ **28.** $27a^9b^6$ **29.** $-4a^6b^6$

30. $\dfrac{1}{32x^{10}}$ **31.** $\dfrac{27x^{12}y^{15}}{8}$ **32.** $\dfrac{a^2}{4b^6c^2}$ **33.** $\dfrac{ab + a^2b^4}{b + a^2}$ **34.** $a + b$

35. $4p^2(p^2 + 3p + 8)$ **36.** $3m(5m - 3)(m - 1)$ **37.** $-3(2a - 7)^2$

38. $-2(b + 12)(b - 12)$ **39.** $(a + w)(q + 3)$ **40.** $(a + 2b)(2m - 3n)$

41. $-7(x + 1)(x^2 - x + 1)$ **42.** $2(a - 3)(a^2 + 3a + 9)$ **43.** 8 and 10

44. 13 and 15   **45. a)** 576 feet   **b)** 20 seconds

46. a) 2188 calories   **b)** Increases   **c)** $B = 9.56W + 918.714$
    d) $B = 2328 - 4.68A$

Chapter 7

Section 7.1 Warm-Ups

1. nth root   **2.** Principal   **3.** Product   **4.** Quotient   **5.** True
6. False   **7.** True   **8.** True   **9.** False   **10.** True   **11.** True
12. False

Section 7.1 Exercises

1. 6   **3.** 10   **5.** -3 **7.** 2   **9.** -2 **11.** 2   **13.** 10

15. Not a real number   **17.** $|m|$ **19.** x^8 **21.** y^3 **23.** y^5

25. m **27.** $|w^3|$ **29.** $|b^9|$ **31.** y^6 **33.** $3\sqrt{y}$ **35.** $2a$

37. x^2y **39.** $m^6\sqrt{5}$ **41.** $2\sqrt[3]{y}$ **43.** $a^2\sqrt[3]{3}$ **45.** $2\sqrt{5}$

47. $5\sqrt{2}$ **49.** $6\sqrt{2}$ **51.** $2\sqrt[3]{5}$ **53.** $3\sqrt[3]{3}$ **55.** $2\sqrt[4]{3}$ **57.** $2\sqrt[3]{3}$

59. $a\sqrt{a}$ **61.** $3a^3\sqrt{2}$ **63.** $2x^2\sqrt{5xy}$ **65.** $2m\sqrt[3]{3m}$ **67.** $2a\sqrt[4]{2a}$

69. $2x\sqrt[5]{2x}$ **71.** $4xy^4z^3\sqrt{3xz}$ **73.** $\dfrac{\sqrt{t}}{2}$ **75.** $\dfrac{25}{4}$ **77.** $\sqrt{10}$

79. $\dfrac{\sqrt[3]{t}}{2}$ **81.** $\dfrac{-2x^2}{y}$ **83.** $\dfrac{2a^3}{3}$ **85.** $\dfrac{2\sqrt{3}}{5}$ **87.** $\dfrac{3\sqrt{3}}{4}$

89. $\dfrac{a\sqrt[3]{a}}{5}$ **91.** $\dfrac{3\sqrt[3]{3}}{2b}$ **93.** $\dfrac{x\sqrt[4]{x^3}}{y^2}$ **95.** $\dfrac{a\sqrt[4]{a}}{2b^3}$ **97.** $[2, \infty)$

99. $(-\infty, \infty)$ **101.** $(-\infty, 3]$ **103.** $\left[-\dfrac{1}{2}, \infty\right)$

105. a. $-4°F$ **b.** $-10°F$ **107. a)** $t = \dfrac{\sqrt{h}}{4}$ **b)** $\dfrac{\sqrt{10}}{2}$ sec   **c)** 100 ft

109. 5.8 knots   **111. a.** 114.1 ft/sec   **b.** 77.8 mph

113. a) Yes   **b)** No   **c)** Yes   **d)** Yes   **115.** Arithmetic mean

Section 7.2 Warm-Ups

1. nth root   **2.** mth power   **3.** Reciprocal   **4.** Even, negative
5. True   **6.** False   **7.** False   **8.** True   **9.** True   **10.** True
11. True   **12.** True

Section 7.2 Exercises

1. $7^{1/4}$ **3.** $(5x)^{1/2}$ **5.** $\sqrt[5]{9}$ **7.** $\sqrt{a}$ **9.** 5   **11.** -5

13. 2   **15.** Not a real number   **17.** $w^{7/3}$ **19.** $2^{-10/3}$ **21.** $\sqrt[4]{\dfrac{1}{w^3}}$

23. $\sqrt{(ab)^3}$ **25.** 25   **27.** 125   **29.** $\dfrac{1}{81}$ **31.** $\dfrac{1}{64}$ **33.** $-\dfrac{1}{3}$

35. Not a real number   **37.** $3^{7/12}$ **39.** 1   **41.** $\dfrac{1}{2}$ **43.** 2   **45.** 6

47. 4   **49.** 81   **51.** $\dfrac{1}{4}$ **53.** $\dfrac{9}{8}$ **55.** $|x|$ **57.** a^4 **59.** y

61. $|3x^3y|$ **63.** $\left|\dfrac{3x^3}{y^5}\right|$ **65.** $x^{3/4}$ **67.** $\dfrac{y^{3/2}}{x^{1/4}}$ **69.** $\dfrac{1}{w^{8/3}}$

71. $12x^8$ **73.** $\dfrac{a^2}{b}$ **75.** $8w^{13/4}$ **77.** 9   **79.** $-\dfrac{1}{8}$ **81.** $\dfrac{1}{625}$

83. $2^{1/4}$ **85.** 3   **87.** 3   **89.** $\dfrac{4}{9}$ **91.** Not a real number

93. $\dfrac{4}{3}$ **95.** $-\dfrac{216}{125}$ **97.** $3x^{9/2}$ **99.** $\dfrac{a^2}{27}$ **101.** $a^{5/4}b$

103. $k^{9/2}m^4$ **105.** 1.2599   **107.** -1.4142 **109.** 2   **111.** 2.5

113. $a^{3m/4}$ **115.** $a^{2m/15}$ **117.** a^nb^m **119.** $a^{4m}b^{2n}$

121. a) 13 in.   **b)** $\sqrt{3}$ or 1.73 in.   **123.** 274.96 m²   **125.** 15.7%

127. 6.12%   **129. a)** Identity   **b)** Identity   **c)** Not an identity

Section 7.3 Warm-Ups

1. Like   **2.** Distributive   **3.** Index   **4.** Conjugate   **5.** False
6. True   **7.** False   **8.** True   **9.** False   **10.** True
11. False   **12.** True

Section 7.3 Exercises

1. $-\sqrt{3}$ **3.** $9\sqrt{7x}$ **5.** $5\sqrt[3]{2}$ **7.** $4\sqrt{3} - 2\sqrt{5}$ **9.** $5\sqrt[3]{x}$

11. $2\sqrt[3]{x} - \sqrt{2x}$ **13.** $2\sqrt{2} + 2\sqrt{7}$ **15.** $5\sqrt{2}$ **17.** 0

19. $-\sqrt{2}$ **21.** $x\sqrt{5x} + 2x\sqrt{2}$ **23.** $7\sqrt[3]{3}$ **25.** $-4\sqrt[4]{3}$

27. $ty\sqrt[3]{2t}$ **29.** $\sqrt{15}$ **31.** $30\sqrt{2}$ **33.** $6a\sqrt{14}$ **35.** $3\sqrt[3]{3}$

37. 12 **39.** $2x^3\sqrt{10x}$ **41.** $\dfrac{x\sqrt[4]{x^3}}{3}$ **43.** $6\sqrt{2} + 18$

45. $5\sqrt{2} - 2\sqrt{5}$ **47.** $3\sqrt[3]{t^2} - t\sqrt[3]{3}$ **49.** $-7 - 3\sqrt{3}$ **51.** 2

53. $-8 - 6\sqrt{5}$ **55.** $-6 + 9\sqrt{2}$ **57.** -1 **59.** 3 **61.** 19

63. 13 **65.** $25 - 9x$ **67.** $\sqrt[6]{3^5}$ **69.** $\sqrt[6]{5^7}$ **71.** $\sqrt[6]{500}$

73. $\sqrt[12]{432}$ **75.** $11\sqrt{3}$ **77.** $10\sqrt{30}$ **79.** $8 - \sqrt{7}$ **81.** $16w$

83. $3x^2\sqrt{2x}$ **85.** $28 + \sqrt{10}$ **87.** $\dfrac{8\sqrt{2}}{15}$ **89.** 17

91. $9 + 6\sqrt{x} + x$ **93.** $25x - 30\sqrt{x} + 9$ **95.** $x + 3 + 2\sqrt{x + 2}$

97. $-\sqrt{w}$ **99.** $a\sqrt{a}$ **101.** $3x^2\sqrt{x}$ **103.** $13x\sqrt[3]{2x}$

105. $\sqrt[8]{32x^5}$ **107.** $3\sqrt{2}$ square feet (ft²)   **109.** $\dfrac{9\sqrt{2}}{2}$ ft²   **111.** No

113. a) $(y - \sqrt{3})(y + \sqrt{3}), (\sqrt{2}a - \sqrt{7})(\sqrt{2}a + \sqrt{7})$
b) $\{\pm 2\sqrt{2}\}$ **c)** $\{\pm\sqrt{a}\}$

Mid-Chapter Quiz 7.1–7.3

1. 8 **2.** -3 **3.** $2\sqrt{30}$ **4.** $2\sqrt[3]{7}$ **5.** $2x^3\sqrt{3x}$ **6.** $2ab^4\sqrt[3]{3b}$

7. $\dfrac{\sqrt{w}}{4}$ **8.** $\dfrac{2x\sqrt{2x}}{3}$ **9.** 9 **10.** 1000 **11.** -64 **12.** $\dfrac{1}{125}$

13. $-2\sqrt{3} - 4\sqrt{6}$ **14.** $9\sqrt{5}$ **15.** $12\sqrt{35}$ **16.** 54 **17.** $5x\sqrt[3]{x^2}$

18. $(-\infty, 2]$ **19.** $[0, \infty)$ **20.** $(-\infty, \infty)$ **21.** $\sqrt[6]{32}$

Section 7.4 Warm-Ups

1. Irrational   **2.** Rationalizing   **3.** Squares   **4.** Radical
5. Denominator   **6.** True   **7.** True   **8.** False   **9.** False
10. True   **11.** True   **12.** True

Section 7.4 Exercises

1. $\dfrac{2\sqrt{5}}{5}$ **3.** $\dfrac{\sqrt{21}}{7}$ **5.** $\dfrac{\sqrt{2}}{2}$ **7.** $\dfrac{\sqrt[3]{150}}{5}$ **9.** $\dfrac{\sqrt{15}}{6}$ **11.** $\dfrac{1}{2}$

13. $\dfrac{\sqrt{2}}{2}$ **15.** $\dfrac{\sqrt[3]{18}}{3}$ **17.** $\dfrac{\sqrt[3]{14}}{2}$ **19.** $\dfrac{\sqrt{xy}}{y}$ **21.** $\dfrac{a\sqrt{ab}}{b^4}$

23. $\dfrac{\sqrt{3ab}}{3b}$ **25.** $\dfrac{\sqrt[3]{ab^2}}{b}$ **27.** $\dfrac{\sqrt[3]{20b}}{2b}$ **29.** $\sqrt{3}$ **31.** $\dfrac{\sqrt{15}}{5}$

33. $\dfrac{3\sqrt{2}}{10}$ **35.** $\dfrac{\sqrt{2}}{3}$ **37.** $\dfrac{\sqrt[3]{3a}}{3}$ **39.** $\sqrt[3]{10}$ **41.** 2 **43.** $\dfrac{2}{w}$

45. $2 + \sqrt{5}$ **47.** $1 - \sqrt{3}$ **49.** $2\sqrt{2} - 2$ **51.** $\dfrac{\sqrt{11} + \sqrt{5}}{2}$

53. $\dfrac{1 + \sqrt{6} + \sqrt{2} + \sqrt{3}}{2}$ **55.** $\dfrac{2\sqrt{3} - \sqrt{6}}{3}$ **57.** $\dfrac{6\sqrt{6} + 2\sqrt{15}}{13}$

59. $128\sqrt{2}$ **61.** $x^2\sqrt{x}$ **63.** $-27x^4\sqrt{x}$ **65.** $8x^5$ **67.** $4\sqrt[3]{25}$

69. x^4 **71.** $\dfrac{\sqrt{6} + 2\sqrt{2}}{2}$ **73.** $2\sqrt{6}$ **75.** $\dfrac{\sqrt{2}}{2}$ **77.** $\dfrac{2 - \sqrt{2}}{5}$

79. $\dfrac{1 + \sqrt{3}}{2}$ **81.** $a - 3\sqrt{a}$ **83.** $4a\sqrt{a} + 4a$ **85.** $12m$

87. $4xy^2z$ **89.** $m - m^2$ **91.** $5x\sqrt{x}$ **93.** $8m^4\sqrt[4]{8m^2}$

95. $\sqrt{x} + 3$ **97.** $\dfrac{3k - 3\sqrt{7k}}{k - 7}$ **99.** $2 + 8\sqrt{2}$ **101.** $\dfrac{3\sqrt{2} + 2\sqrt{3}}{6}$

103. $7\sqrt{2} - 1$ **105.** $\dfrac{4x + 4\sqrt{x}}{x - 4}$ **107.** $\dfrac{x + \sqrt{x}}{x(1 - x)}$

109. a) $x^3 - 2$ **b)** $(x + \sqrt[3]{5})(x^2 - \sqrt[3]{5}x + \sqrt[3]{25})$ **c)** 3
d) $(\sqrt[3]{a} + \sqrt[3]{b})(\sqrt[3]{a^2} - \sqrt[3]{ab} + \sqrt[3]{b^2})$,
$(\sqrt[3]{a} - \sqrt[3]{b})(\sqrt[3]{a^2} + \sqrt[3]{ab} + \sqrt[3]{b^2})$

Section 7.5 Warm-Ups

1. Odd   **2.** Even   **3.** Extraneous   **4.** Even   **5.** False   **6.** True
7. False   **8.** True   **9.** True   **10.** True

Section 7.5 Exercises

1. $\{-10\}$ **3.** $\left\{\dfrac{1}{2}\right\}$ **5.** $\{1\}$ **7.** $\{-2\}$ **9.** $\{-5, 5\}$

11. $\{-2\sqrt{5}, 2\sqrt{5}\}$ **13.** No real solution   **15.** $\{-1, 7\}$

17. $\{-1 - 2\sqrt{2}, -1 + 2\sqrt{2}\}$ **19.** $\{-\sqrt{10}, \sqrt{10}\}$ **21.** $\{3\}$

23. $\{-2, 2\}$ **25.** $\{52\}$ **27.** $\left\{\dfrac{9}{4}\right\}$ **29.** $\{9\}$ **31.** $\{3\}$ **33.** $\{3\}$

35. $\{-5, 3\}$ **37.** $\{1\}$ **39.** $\varnothing$ **41.** $\{1, 2\}$ **43.** $\{9\}$ **45.** $\{4\}$

47. $\{2\}$ **49.** $\{6\}$ **51.** $\{1, 5\}$ **53.** $\{7\}$ **55.** $\{-5\}$ **57.** $\varnothing$

59. $\{0\}$ **61.** $\{-3\sqrt{3}, 3\sqrt{3}\}$ **63.** $\left\{-\dfrac{1}{27}, \dfrac{1}{27}\right\}$ **65.** $\{512\}$

67. $\left\{\dfrac{1}{81}\right\}$ **69.** $\left\{0, \dfrac{2}{3}\right\}$ **71.** $\left\{\dfrac{4 - \sqrt{2}}{4}, \dfrac{4 + \sqrt{2}}{4}\right\}$

73. No real solution   **75.** $\{-\sqrt{2}, \sqrt{2}\}$ **77.** $\{-5\}$

79. No real solution   **81.** $\{-9\}$ **83.** $\left\{\dfrac{5}{4}\right\}$ **85.** $\varnothing$

87. $\left\{-\dfrac{2}{3}, 2\right\}$ **89.** $\{-2 - 2\sqrt[4]{2}, -2 + 2\sqrt[4]{2}\}$ **91.** $\{0\}$ **93.** $\left\{\dfrac{1}{2}\right\}$

95. $4\sqrt{2}$ feet   **97.** $5\sqrt{2}$ feet   **99.** 50 feet

101. a) 2 **b)** $\sqrt{3}$ **c)** $\dfrac{\sqrt{3}}{2}$ **103. a)** 1.89 **b)** $d = \dfrac{64b^3}{C^3}$
c) $d > 19,683$ pounds

105. $\sqrt[5]{32}$ meters   **107.** $\sqrt{73}$ kilometers (km)

109. a. $S = P(1 + r)^n$ **b.** $P = S(1 + r)^{-n}$ **111.** 9.5 AU

113. $\{-1.8, 1.8\}$ **115.** $\{4.993\}$ **117.** $\{-26.372, 26.372\}$

Section 7.6 Warm-Ups

1. Complex   **2.** Imaginary   **3.** Complex   **4.** Conjugate
5. Product   **6.** Real   **7.** Subset   **8.** True   **9.** False   **10.** False
11. True   **12.** True   **13.** False   **14.** True   **15.** True   **16.** True

Section 7.6 Exercises

1. $-2 + 8i$ **3.** $-4 + 4i$ **5.** -2 **7.** $-8 - 2i$ **9.** $6 + 15i$

11. $-2 - 10i$ **13.** $-4 - 12i$ **15.** $-10 + 24i$ **17.** $-1 + 3i$

19. $-5i$ **21.** 29 **23.** 2 **25.** 20 **27.** -9 **29.** -25 **31.** 16

33. i **35.** -1 **37.** i **39.** 34 **41.** 5 **43.** 5 **45.** 7

47. $\dfrac{12}{17} - \dfrac{3}{17}i$ **49.** $\dfrac{4}{13} + \dfrac{7}{13}i$ **51.** $3 - 4i$ **53.** $1 + 3i$

55. $\dfrac{1}{13} - \dfrac{5}{13}i$ **57.** $-2i$ **59.** $5i$ **61.** $2 + 2i$ **63.** $5 + 6i$

65. $7 - i\sqrt{6}$ **67.** $5i\sqrt{2}$ **69.** $1 + i\sqrt{3}$ **71.** $-1 - \dfrac{1}{2}i\sqrt{6}$

73. $-2\sqrt{3}$ **75.** -9 **77.** $-i\sqrt{2}$ **79.** $\{\pm 6i\}$ **81.** $\{\pm 2i\sqrt{3}\}$

83. $\left\{\pm\dfrac{i\sqrt{10}}{2}\right\}$ **85.** $\{\pm i\sqrt{2}\}$ **87.** $18 - i$ **89.** $5 + i$

91. $-\dfrac{6}{25} - \dfrac{17}{25}i$ **93.** $3 + 2i$ **95.** -9 **97.** $3i\sqrt{3}$

99. $-5 - 12i$ **101.** $-2 + 2i\sqrt{2}$

Enriching Your Mathematical Word Power
1. nth **2.** Square **3.** Cube **4.** Principal **5.** Odd **6.** Index
7. Radicand **8.** Domain **9.** Integral **10.** Rational **11.** Like
12. Complex **13.** Imaginary **14.** Imaginary **15.** Conjugates

Review Exercises
1. 9 **3.** 3 **5.** -3 **7.** 10 **9.** $|y|$ **11.** a **13.** n^6

15. $|n^5|$ **17.** $6\sqrt{2}$ **19.** x^6 **21.** x^2 **23.** $x^4\sqrt{2x}$

25. $2w^2\sqrt{2w}$ **27.** $2x\sqrt[3]{2x}$ **29.** $a^2b\sqrt[4]{ab}$ **31.** $\dfrac{x\sqrt{x}}{4}$ **33.** $[2.5, \infty)$

35. $(-\infty, \infty)$ **37.** $\left(-\infty, \dfrac{1}{3}\right]$ **39.** $[-2, \infty)$ **41.** $\dfrac{1}{9}$ **43.** 4

45. $\dfrac{1}{1000}$ **47.** $27x^{1/2}$ **49.** $a^{7/2}b^{7/2}$ **51.** $x^{3/4}y^{5/4}$ **53.** 13

55. $3\sqrt{5} - 2\sqrt{3}$ **57.** $30 - 21\sqrt{6}$ **59.** $6 - 3\sqrt{3} + 2\sqrt{2} - \sqrt{6}$

61. $\dfrac{5\sqrt{2}}{2}$ **63.** $\dfrac{\sqrt{10}}{5}$ **65.** $\dfrac{\sqrt[3]{18}}{3}$ **67.** $\dfrac{2\sqrt{3x}}{3x}$ **69.** $\dfrac{y\sqrt{15y}}{3}$

71. $\dfrac{3\sqrt[3]{4a^2}}{2a}$ **73.** $\dfrac{5\sqrt[4]{27x^2}}{3x}$ **75.** 9 **77.** $1 - \sqrt{2}$ **79.** $\dfrac{-\sqrt{6} - 3\sqrt{2}}{2}$

81. $\dfrac{3\sqrt{2} + 2}{7}$ **83.** $256w^{10}$ **85.** $\{-4, 4\}$ **87.** $\{3, 7\}$

89. $\{-1 - \sqrt{5}, -1 + \sqrt{5}\}$ **91.** No real solution **93.** $\{10\}$
95. $\{9\}$ **97.** $\{-8, 8\}$ **99.** $\{124\}$ **101.** $\{7\}$ **103.** $\{2, 3\}$
105. $\{9\}$ **107.** $\{4\}$ **109.** $5 + 25i$ **111.** $7 - 3i$ **113.** $-1 + 2i$

115. $2 + i$ **117.** $2 - i\sqrt{3}$ **119.** $\dfrac{5}{17} - \dfrac{14}{17}i$ **121.** 16 **123.** -1

125. $\{\pm 10i\}$ **127.** $\left\{\pm\dfrac{3i\sqrt{2}}{2}\right\}$ **129.** False **131.** True

133. True **135.** False **137.** False **139.** False **141.** False
143. True **145.** False **147.** True **149.** False **151.** False
153. True **155.** True **157.** $5\sqrt{30}$ or approximately 27.4 seconds
159. $10\sqrt{7}$ feet **161.** $200\sqrt{2}$ feet **163.** $26.4\sqrt[3]{25}$ ft^2
165. a) 5.7% **b)** $3000 billion or $3 trillion **167.** $V = \dfrac{29\sqrt{LCS}}{CS}$

Chapter 7 Test
1. 6 **2.** -5 **3.** $|t|$ **4.** p **5.** w^2 **6.** $|w^3|$ **7.** 4 **8.** $\dfrac{1}{8}$

9. $\sqrt{3}$ **10.** 30 **11.** $3\sqrt{5}$ **12.** $\dfrac{6\sqrt{5}}{5}$ **13.** 2 **14.** $6\sqrt{2}$

15. $\dfrac{\sqrt{15}}{6}$ **16.** $\dfrac{2 + \sqrt{2}}{2}$ **17.** $4 - 3\sqrt{3}$ **18.** $2ay^2\sqrt[4]{2a}$

19. $\dfrac{\sqrt[3]{4x}}{2x}$ **20.** $\dfrac{2a^4\sqrt{2ab}}{b^2}$ **21.** $-3x^3$ **22.** $2m\sqrt{5m}$ **23.** $x^{3/4}$

24. $3y^2x^{1/4}$ **25.** $2x^2\sqrt[3]{5x}$ **26.** $19 + 8\sqrt{3}$ **27.** $(-\infty, 4]$

28. $(-\infty, \infty)$ **29.** $\dfrac{5 + \sqrt{3}}{11}$ **30.** $\dfrac{6\sqrt{2} - \sqrt{3}}{23}$ **31.** $22 + 7i$

32. $1 - i$ **33.** $\dfrac{1}{5} - \dfrac{7}{5}i$ **34.** $-\dfrac{3}{4} + \dfrac{1}{4}i\sqrt{3}$ **35.** $\{-5, 9\}$ **36.** $\left\{-\dfrac{7}{4}\right\}$

37. $\{-8, 8\}$ **38.** $\left\{\pm\dfrac{4}{3}i\right\}$ **39.** $\{3\}$ **40.** $\{5\}$ **41.** $\dfrac{3\sqrt{2}}{2}$ feet

42. 25 and 36 **43.** Length 6 ft, width 4 ft **44.** 39.53 AU, 164.97 years

Making Connections   A Review of Chapters 1–7
1. 7 **2.** -5 **3.** 57 **4.** 11 **5.** -29 **6.** -4 **7.** 1
8. -2 **9.** 0 **10.** 17 **11.** $\{0, 1, \sqrt{4}\}$ **12.** $\{1, \sqrt{4}\}$

13. $\{-5, 0, 1, \sqrt{4}\}$ **14.** $\left\{-5, -\dfrac{1}{9}, 0, 1, \sqrt{4}, 2.99, \dfrac{33}{4}\right\}$

15. $\{-\sqrt[3]{2}, \pi\}$ **16.** All except $2 + 3i$ **17.** $\{2 + 3i\}$

18. All **19.** Additive **20.** Multiplicative **21.** Commutative
22. Commutative **23.** Associative **24.** Distributive

25. Additive **26.** Multiplicative **27.** $\left\{-\dfrac{4}{7}\right\}$ **28.** $\left\{\dfrac{3}{2}\right\}$

29. $(-\infty, -3) \cup (-2, \infty)$
$-5\ -4\ -3\ -2\ -1\ \ 0$

30. $\left\{\dfrac{3}{2}\right\}$ **31.** $(-\infty, 1)$ **32.** $\varnothing$
$-3\ -2\ -1\ \ 0\ \ 1\ \ 2\ \ 3$

33. $\{9\}$ **34.** $\varnothing$ **35.** $\{-12, -2\}$ **36.** $\left\{\dfrac{1}{16}\right\}$

37. $(-6, \infty)$ **38.** $\left\{-\dfrac{1}{64}, \dfrac{1}{64}\right\}$
$-8\ -7\ -6\ -5\ -4\ -3\ -2$

39. $\left\{-\dfrac{\sqrt{3}}{3}, \dfrac{\sqrt{3}}{3}\right\}$ **40.** R **41.** $\left(-\dfrac{1}{3}, 3\right)$
$-2\ -1\ \ 0\ \ 1\ \ 2\ \ 3\ \ 4$

42. $\left\{\dfrac{1}{3}\right\}$ **43.** $\{82\}$ **44.** $\left\{\dfrac{6}{5}, \dfrac{12}{5}\right\}$ **45.** $\{100\}$ **46.** R **47.** $\{4\sqrt{30}\}$

48. $\{400\}$ **49.** $\left\{\dfrac{13 + 9\sqrt{2}}{3}\right\}$ **50.** $\{-3\sqrt{2}, 3\sqrt{2}\}$ **51.** $\{5\}$

52. $\{7 + 3\sqrt{6}\}$ **53.** $\{-2, 3\}$ **54.** $\{-5, 2\}$ **55.** $\{-2, 3\}$ **56.** $\left\{\dfrac{1}{2}, 3\right\}$

57. 3 **58.** -2 **59.** $\dfrac{1}{2}$ **60.** $\dfrac{1}{3}$

61. a) 48.5 cm^3 **b)** 14% **c)** 56 cm^3

Chapter 8

Section 8.1 Warm-Ups
1. Factoring, even-root, completing **2.** Even-root **3.** Middle
4. Leading coefficient **5.** False **6.** False **7.** True **8.** True
9. False **10.** False **11.** True **12.** True

Section 8.1 Exercises
1. $\{-2, 3\}$ **3.** $\{-5, 3\}$ **5.** $\left\{-1, \dfrac{3}{2}\right\}$ **7.** $\{-7\}$ **9.** $\{-4, 4\}$

11. $\{-9, 9\}$ **13.** $\left\{-\dfrac{4}{3}, \dfrac{4}{3}\right\}$ **15.** $\{-1, 7\}$

17. $\{-1 - \sqrt{5}, -1 + \sqrt{5}\}$ **19.** $\left\{\dfrac{3 - \sqrt{7}}{2}, \dfrac{3 + \sqrt{7}}{2}\right\}$

21. $x^2 + 2x + 1$ **23.** $x^2 - 3x + \dfrac{9}{4}$ **25.** $y^2 + \dfrac{1}{4}y + \dfrac{1}{64}$

27. $x^2 + \dfrac{2}{3}x + \dfrac{1}{9}$ **29.** $(x + 4)^2$ **31.** $\left(y - \dfrac{5}{2}\right)^2$ **33.** $\left(z - \dfrac{2}{7}\right)^2$

35. $\left(t + \dfrac{3}{10}\right)^2$ **37.** $\{-3, 5\}$ **39.** $\{-5, 7\}$ **41.** $\{-4, 5\}$

43. $\{-7, 2\}$ **45.** $\left\{-1, \dfrac{3}{2}\right\}$ **47.** $\{-2 - \sqrt{10}, -2 + \sqrt{10}\}$

49. $\{-4 - 2\sqrt{5}, -4 + 2\sqrt{5}\}$ **51.** $\left\{\dfrac{-5 + \sqrt{5}}{2}, \dfrac{-5 - \sqrt{5}}{2}\right\}$

53. $\left\{\dfrac{1 - \sqrt{2}}{2}, \dfrac{1 + \sqrt{2}}{2}\right\}$ **55.** $\left\{\dfrac{-3 - \sqrt{41}}{4}, \dfrac{-3 + \sqrt{41}}{4}\right\}$

57. $\{4\}$ **59.** $\left\{\dfrac{1 + \sqrt{17}}{8}\right\}$ **61.** $\{1, 6\}$ **63.** $\{-2 - \sqrt{2}, -2 + \sqrt{2}\}$

65. $\{-1 - 2i, -1 + 2i\}$ **67.** $\{3 + i\sqrt{2}, 3 - i\sqrt{2}\}$ **69.** $\left\{\pm\dfrac{i\sqrt{2}}{2}\right\}$

71. $\{-2i\sqrt{3}, 2i\sqrt{3}\}$ **73.** $\left\{\dfrac{2 \pm i}{5}\right\}$ **75.** $\{\pm 11i\}$ **77.** $\left\{-\dfrac{5}{2}i, \dfrac{5}{2}i\right\}$

79. $\{-2, 1\}$ **81.** $\left\{\dfrac{-2 - \sqrt{19}}{5}, \dfrac{-2 + \sqrt{19}}{5}\right\}$ **83.** $\{-6, 4\}$

85. $\{2 \pm 3i\}$ **87.** $\{-2, 3\}$ **89.** $\{3 - i, 3 + i\}$ **91.** $\{6\}$

93. $\left\{\dfrac{9 - \sqrt{65}}{2}, \dfrac{9 + \sqrt{65}}{2}\right\}$ **95.** $\{-5, 3\}$ **97.** $\varnothing$

99. 136.9 ft/sec   **101.** 12   **103.** c   **107.** $\{4.56, 2.74\}$

109. $\{3.53\}$

Section 8.2 Warm-Ups

1. Quadratic   **2.** Discriminant   **3.** Complex   **4.** One
5. Two   **6.** Two   **7.** True   **8.** False   **9.** True   **10.** True
11. False   **12.** True

Section 8.2 Exercises

1. $\{1, 2\}$ **3.** $\{-3, -2\}$ **5.** $\{-3, 2\}$ **7.** $\left\{-\dfrac{1}{3}, \dfrac{3}{2}\right\}$ **9.** $\left\{\dfrac{1}{2}\right\}$

11. $\left\{\dfrac{1}{3}\right\}$ **13.** $\left\{-\dfrac{3}{4}\right\}$ **15.** $\{-4 \pm \sqrt{10}\}$ **17.** $\left\{\dfrac{-5 \pm \sqrt{29}}{2}\right\}$

19. $\left\{\dfrac{3 \pm \sqrt{7}}{2}\right\}$ **21.** $\left\{\dfrac{3 \pm i}{2}\right\}$ **23.** $\left\{\dfrac{3 \pm i\sqrt{39}}{4}\right\}$ **25.** $\{5 \pm i\}$

27. 28, 2   **29.** $-23, 0$ **31.** 0, 1   **33.** $-\dfrac{3}{4}, 0$ **35.** 97, 2

37. 0, 1   **39.** 140, 2   **41.** 1, 2   **43.** $\{-2 \pm 2\sqrt{2}\}$ **45.** $\left\{-2, \dfrac{1}{2}\right\}$

47. $\left\{\dfrac{-1 \pm \sqrt{13}}{3}\right\}$ **49.** $\{0\}$ **51.** $\left\{\dfrac{13}{9}, \dfrac{17}{9}\right\}$ **53.** $\{\pm 5\sqrt{3}\}$

55. $\{4 \pm 2i\}$ **57.** $\{2 \pm i\sqrt{6}\}$ **59.** $\left\{-\dfrac{3}{4}, \dfrac{5}{2}\right\}$ **61.** $\{-4.474, 1.274\}$

63. $\{3.7\}$ **65.** $\{-2.979, -0.653\}$ **67.** $\{-4792.983, -0.017\}$

69. $\{-0.079, 0.078\}$ **71.** $\dfrac{1 + \sqrt{65}}{2}$ and $\dfrac{-1 + \sqrt{65}}{2}$, or 4.5 and 3.5

73. $3 + \sqrt{5}$ and $3 - \sqrt{5}$, or 5.2 and 0.8

75. $W = \dfrac{-1 + \sqrt{5}}{2} \approx 0.6$ ft, $L = \dfrac{1 + \sqrt{5}}{2} \approx 1.6$ ft

77. $W = -2 + \sqrt{14} \approx 1.7$ ft, $L = 2 + \sqrt{14} \approx 5.7$ ft

79. 3 sec   **81.** $\dfrac{5 + \sqrt{105}}{16}$ or 1.0 sec   **83.** 7.0 sec   **85.** 4 in.

87. 4   **89.** 250 melons   **95.** 2   **97.** 0   **99.** 0

Section 8.3 Warm-Ups

1. Factor   **2.** Prime   **3.** Quadratic   **4.** Solutions   **5.** True
6. True   **7.** False   **8.** True   **9.** False   **10.** False
11. True   **12.** False

Section 8.3 Exercises

1. $x^2 + 4x - 21 = 0$ **3.** $x^2 - 5x + 4 = 0$ **5.** $x^2 - 5 = 0$
7. $x^2 + 16 = 0$ **9.** $x^2 + 2 = 0$ **11.** $6x^2 - 5x + 1 = 0$
13. Prime   **15.** Prime   **17.** Prime   **19.** $(3x - 4)(2x + 9)$ **21.** Prime
23. $(4x - 15)(2x + 3)$ **25.** $\{-1, 5\}$ **27.** $\left\{-\dfrac{3}{2}, \dfrac{3}{2}\right\}$
29. $\left\{\dfrac{-3 \pm \sqrt{5}}{2}\right\}$ **31.** $\{\pm 2, \pm 3\}$ **33.** $\{1, 3\}$ **35.** $\{\pm\sqrt{5}, \pm 3\}$
37. $\{-2, 1\}$ **39.** $\{0, \pm 3\}$ **41.** $\{-1 \pm \sqrt{5}, -3, 1\}$
43. $\{-3, -2, 1, 2\}$ **45.** $\{1, 4\}$ **47.** $\{-27, -1\}$ **49.** $\{16, 81\}$
51. $\{9\}$ **53.** $\left\{-\dfrac{1}{3}, \dfrac{1}{2}\right\}$ **55.** $\{64\}$ **57.** $\left\{\dfrac{2}{3}, \dfrac{3}{2}\right\}$

59. $\left\{\pm\dfrac{\sqrt{14}}{2}, \pm\dfrac{\sqrt{38}}{2}\right\}$ **61.** $\{-1 + \sqrt{2}, -1 - \sqrt{2}\}$ **63.** $\{\pm 2i\}$

65. $\{\pm i\sqrt{2}, \pm 2i\}$ **67.** $\{\pm 2, \pm 2i\}$ **69.** $\left\{\pm\dfrac{1}{2}, \pm\dfrac{i}{2}\right\}$

71. $\left\{\dfrac{1 \pm i\sqrt{3}}{2}, -1\right\}$ **73.** $\{1 \pm i\sqrt{3}, -2\}$ **75.** $\left\{\dfrac{1 \pm 2i}{5}\right\}$

77. $\{1 \pm i\}$ **79.** 2:00 P.M.

81. Before $-5 + \sqrt{265}$ or 11.3 mph, after $-9 + \sqrt{265}$ or 7.3 mph

83. Andrew $\dfrac{13 + \sqrt{265}}{2}$ or 14.6 hours, John $\dfrac{19 + \sqrt{265}}{2}$ or 17.6 hours

85. Length $5 + 5\sqrt{41}$ or 37.02 ft, width $-5 + 5\sqrt{41}$ or 27.02 ft

87. $14 + 2\sqrt{58}$ or 29.2 hours   **89.** $-5 + 5\sqrt{5}$ or 6.2 meters

93. $\{1, 2\}$ **95.** $\{-4.25, -3.49, 0.49, 1.25\}$

Mid-Chapter Quiz 8.1–8.3

1. $\{-4, 8\}$ **2.** $\left\{\dfrac{1}{3}, \dfrac{1}{2}\right\}$ **3.** $\left\{\pm\dfrac{4}{5}\right\}$ **4.** $\{-3 \pm \sqrt{6}\}$ **5.** $\{2 \pm \sqrt{5}\}$

6. $\left\{-1, \dfrac{1}{2}\right\}$ **7.** $\left\{\dfrac{1}{2}, 2\right\}$ **8.** $\left\{\dfrac{-2 \pm \sqrt{2}}{2}\right\}$ **9.** $\{-5, -4\}$

10. $\{\pm 1, \pm\sqrt{5}\}$ **11.** $\{16\}$ **12.** $\{14\}$ **13.** $\{5 \pm i\}$ **14.** -59

15. $x^2 - 5x - 24 = 0$

Section 8.4 Warm-Ups

1. Quadratic   **2.** Parabola   **3.** Upward   **4.** Downward
5. Domain   **6.** Vertex   **7.** $-b/(2a)$ **8.** True   **9.** False
10. True   **11.** True   **12.** False   **13.** True   **14.** True   **15.** True

Section 8.4 Exercises

1. $(4, 16), (3, 9), (-3, 9)$ **3.** $(3, -6), (4, 0), (-3, 0)$
5. $(4, -128), (0, 0), (2, 0)$ **7.** Upward   **9.** Downward
11. Upward

13. Domain $(-\infty, \infty)$, range $[2, \infty)$

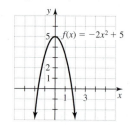

15. Domain $(-\infty, \infty)$, range $[-4, \infty)$

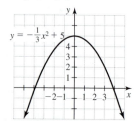

17. Domain $(-\infty, \infty)$, range $(-\infty, 5]$

19. Domain $(-\infty, \infty)$, range $(-\infty, 5]$

21. Domain $(-\infty, \infty)$, range $[0, \infty)$

$h(x) = (x - 2)^2$

23. $(0, -9)$ **25.** $(2, -3)$

27. $(5, 51)$ **29.** $\left(\dfrac{1}{2}, \dfrac{3}{4}\right)$

31. $(0, 16)$, $(-4, 0)$, $(4, 0)$

33. $(0, -15)$, $(-3, 0)$, $(5, 0)$

35. $(0, -9)$, $\left(\dfrac{3}{2}, 0\right)$

65. The graph of $y = ax^2$ gets narrower as a gets larger.

67. The graph of $y = x^2$ has the same shape as $x = y^2$.

69. (a)

(b)

37. Vertex $\left(\dfrac{1}{2}, -\dfrac{9}{4}\right)$, intercepts $(0, -2)$, $(-1, 0)$, $(2, 0)$, domain $(-\infty, \infty)$, range $\left[-\dfrac{9}{4}, \infty\right)$

$f(x) = x^2 - x - 2$

39. Vertex $(-1, -9)$, intercepts $(0, -8)$, $(-4, 0)$, $(2, 0)$, domain $(-\infty, \infty)$, range $[-9, \infty)$

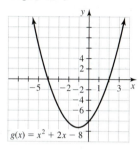

$g(x) = x^2 + 2x - 8$

(c)

Section 8.5 Warm-Ups

1. Quadratic   **2.** Graphical, test-point   **3.** False   **4.** True   **5.** True   **6.** False   **7.** False   **8.** True

Section 8.5 Exercises

1. $(-3, 2)$

3. $(-4, 4)$

41. Vertex $(-2, 1)$, intercepts $(0, -3)$, $(-1, 0)$, $(-3, 0)$, domain $(-\infty, \infty)$, range $(-\infty, 1]$

$y = -x^2 - 4x - 3$

43. Vertex $\left(\dfrac{3}{2}, \dfrac{25}{4}\right)$, intercepts $(0, 4)$, $(4, 0)$, $(-1, 0)$, domain $(-\infty, \infty)$, range $\left(-\infty, \dfrac{25}{4}\right]$

$h(x) = -x^2 + 3x + 4$

5. $[-2, 4]$

7. $\left(-\infty, -4\right] \cup \left[\dfrac{3}{2}, \infty\right)$

9. $(-\infty, 0] \cup [2, \infty)$

11. $(-\infty, 0) \cup \left(\dfrac{1}{2}, \infty\right)$

45. Vertex $(3, -25)$, intercepts $(0, -16)$, $(8, 0)$, $(-2, 0)$, domain $(-\infty, \infty)$, range $[-25, \infty)$

$a = b^2 - 6b - 16$

47. Minimum -8

49. Maximum 14

51. Minimum 2

53. Maximum 2

55. Maximum 64 feet

$s(t) = -16t^2 + 64t$

13. $(-\infty, -\sqrt{5}) \cup (\sqrt{5}, \infty)$

15. $[1 - \sqrt{6}, 1 + \sqrt{6}]$

17. $\left(-\infty, \dfrac{3 - \sqrt{3}}{2}\right] \cup \left[\dfrac{3 + \sqrt{3}}{2}, \infty\right)$

19. $(-\infty, \infty)$

21. $\varnothing$

57. 100   **59.** 625 square meters   **61.** 2 P.M.   **63.** 15 meters, 25 meters

23. $\left\{\dfrac{5}{2}\right\}$

25. $\left(-\infty, -\dfrac{1}{5}\right) \cup \left(-\dfrac{1}{5}, \infty\right)$

27. $(-\infty, \infty)$

29. $\varnothing$

31. $(-\infty, -2) \cup (6, \infty)$

33. $(-8, 5)$

35. $(-\infty, \infty)$

37. $(0, 2)$

39. $\left(-\infty, 3 - \sqrt{5}\right) \cup \left(3 + \sqrt{5}, \infty\right)$

41. $(-\infty, \infty)$

43. $\left[\dfrac{3 - 3\sqrt{5}}{2}, \dfrac{3 + 3\sqrt{5}}{2}\right]$

45. $(-\infty, 0) \cup (0, \infty)$ **47.** $(-\infty, \infty)$ **49.** $[-3, 3]$ **51.** $(-4, 4)$

53. $(-\infty, 0] \cup [4, \infty)$ **55.** $\left(-\dfrac{3}{2}, \dfrac{5}{3}\right)$ **57.** $(-\infty, -2] \cup [6, \infty)$

59. $(-\infty, -3) \cup (5, \infty)$ **61.** $(-\infty, -4] \cup [2, \infty)$ **63.** $(-27.58, -0.68)$

65. Greater than 5, or 6, 7, 8, . . .   **67.** 4 seconds

69. a) 900 ft   **b)** 3 seconds   **c)** 3 seconds

71. a) (h, k) **b)** $(-\infty, h) \cup (k, \infty)$ **c)** $(-k, -h)$
d) $(-\infty, -k] \cup [-h, \infty)$

Enriching Your Mathematical Word Power

1. Quadratic   **2.** Perfect square   **3.** Completing   **4.** Quadratic
5. Discriminant   **6.** Quadratic   **7.** Form   **8.** Quadratic   **9.** Test

Review Exercises

1. $\{-3, 5\}$ **3.** $\left\{-3, \dfrac{5}{2}\right\}$ **5.** $\{-5, 5\}$ **7.** $\left\{\dfrac{3}{2}\right\}$ **9.** $\{\pm 2\sqrt{3}\}$

11. $\{-2, 4\}$ **13.** $\left\{\dfrac{4 \pm \sqrt{3}}{2}\right\}$ **15.** $\left\{\pm\dfrac{3}{2}\right\}$ **17.** $\{2, 4\}$ **19.** $\{2, 3\}$

21. $\left\{\dfrac{1}{2}, 3\right\}$ **23.** $\{-2 \pm \sqrt{3}\}$ **25.** $\{-2, 5\}$ **27.** $\left\{-\dfrac{1}{3}, \dfrac{3}{2}\right\}$

29. $\{-2 \pm \sqrt{2}\}$ **31.** $\left\{\dfrac{5 \pm \sqrt{13}}{6}\right\}$ **33.** 0, 1   **35.** $-19, 0$

37. 17, 2   **39.** $\left\{\dfrac{2 \pm i\sqrt{2}}{2}\right\}$ **41.** $\left\{\dfrac{3 \pm i\sqrt{15}}{4}\right\}$ **43.** $\left\{\dfrac{-1 \pm i\sqrt{5}}{3}\right\}$

45. $\{-3 \pm i\sqrt{7}\}$ **47.** $(4x + 1)(2x - 3)$ **49.** Prime

51. $(4y - 5)(2y + 5)$ **53.** $x^2 + 9x + 18 = 0$ **55.** $x^2 - 50 = 0$

57. $\{-2, 1\}$ **59.** $\{\pm 2, \pm 3\}$ **61.** $\{-6, -5, 2, 3\}$ **63.** $\{-2, 8\}$

65. $\left\{-\dfrac{1}{9}, \dfrac{1}{4}\right\}$ **67.** $\{16, 81\}$

69. Vertex $(3, -9)$, intercepts $(0, 0)$, $(6, 0)$

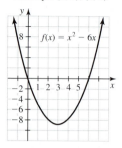

71. Vertex $(2, -16)$, intercepts $(0, -12)$, $(-2, 0)$, and $(6, 0)$

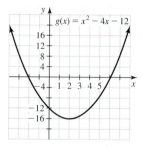

73. Vertex $(2, 8)$, intercepts $(0, 0)$, $(4, 0)$

75. Vertex $(1, 4)$, intercepts $(0, 3)$, $(-1, 0)$, $(3, 0)$

77. Domain $(-\infty, \infty)$, range $[-3, \infty)$
79. Domain $(-\infty, \infty)$, range $(-\infty, 4.125]$
81. $(-\infty, -3) \cup (2, \infty)$ **83.** $[-4, 5]$

85. $(0, 1)$ **87.** $(-\infty, -3] \cup \left[\dfrac{1}{2}, \infty\right)$

89. $(-\infty, \infty)$

91. $\{5\}$

93. $\varnothing$ **95.** $\left\{\dfrac{5}{12}\right\}$ **97.** $\left\{\dfrac{-3 \pm \sqrt{5}}{2}\right\}$ **99.** $\left\{\dfrac{4 \pm 2i}{3}\right\}$ **101.** $\left\{\dfrac{5}{2}\right\}$

103. $\left\{-2, -\dfrac{1}{4}\right\}$ **105.** $\{625, 10{,}000\}$

107. $-2 + 2\sqrt{2}$ and $2 + 2\sqrt{2}$, or 0.83 and 4.83

109. Width $\dfrac{4 + \sqrt{706}}{2}$ or 15.3 inches, height $\dfrac{-4 + \sqrt{706}}{2}$ or 11.3 inches

111. 2 inches   **113.** Width 5 ft, length 9 ft   **115.** \$20.40, 400

117. 0.5 second and 1.5 seconds   **119.** 1.618

Chapter 8 Test

1. $-7, 0$ **2.** 13, 2   **3.** 0, 1   **4.** $\left\{-3, \dfrac{1}{2}\right\}$ **5.** $\{-3 \pm \sqrt{3}\}$

6. $\{-5\}$ **7.** $\left\{-2, \dfrac{3}{2}\right\}$ **8.** $\{-4, 3\}$ **9.** $\{\pm 1, \pm 2\}$ **10.** $\{11, 27\}$

11. $\{\pm 6i\}$ **12.** $\{-3 \pm i\}$ **13.** $\left\{\dfrac{1 \pm i\sqrt{11}}{6}\right\}$

14. Domain $(-\infty, \infty)$, range $(-\infty, 16]$, vertex $(0, 16)$, intercepts $(0, 16)$, $(-4, 0)$, $(4, 0)$, maximum y-value 16

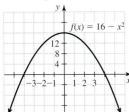

15. Domain $(-\infty, \infty)$, range $[-2.25, \infty)$, vertex $(1.5, -2.25)$, intercepts $(0, 0)$, $(3, 0)$, minimum y-value -2.25

16. $x^2 - 2x - 24 = 0$ **17.** $x^2 + 25 = 0$
18. $(-6, 3)$

19. $\left(-\infty, 1 - \sqrt{2}\right] \cup \left[1 + \sqrt{2}, \infty\right)$

20. $(-\infty, \infty)$

21. $\varnothing$ **22.** Width $-1 + \sqrt{17}$ ft, length $1 + \sqrt{17}$ ft

23. $\dfrac{5 + \sqrt{37}}{2}$ or 5.5 hours   **24.** 36 feet

Making Connections   A Review of Chapters 1–8

1. $\dfrac{4}{9}$ **2.** $\dfrac{2}{3}$ **3.** 11   **4.** 4   **5.** 4   **6.** 4   **7.** $\dfrac{1}{4}$ **8.** $\dfrac{1}{125}$

9. $(y - 3)(y + 100)$ **10.** $(5y - 3)(4y + 1)$ **11.** $6a(a - 5)^2$

12. $(b^2 + 4)(b + 2)$ **13.** $(a - y^2)(b - y^2)$

14. $-2(m + 2)(m^2 - 2m + 4)$ **15.** $\left\{\dfrac{15}{2}\right\}$ **16.** $\left\{\pm\dfrac{\sqrt{30}}{2}\right\}$

17. $\left\{-3, \dfrac{5}{2}\right\}$ **18.** $\left\{\dfrac{-2 \pm \sqrt{34}}{2}\right\}$ **19.** $\left\{-\dfrac{7}{2}, -2\right\}$

20. $\left\{-3, \dfrac{1}{4}, \dfrac{-11 \pm \sqrt{73}}{8}\right\}$ **21.** $\{9\}$ **22.** $\left\{-\dfrac{3}{2}, \dfrac{13}{2}\right\}$

23. $(-4, \infty)$ **24.** $\left[\dfrac{1}{2}, 5\right]$ **25.** $(-\infty, 0] \cup [1, \infty)$ **26.** $\varnothing$

27. $[-3, 2)$ **28.** $(-\infty, \infty)$ **29.** $y = \dfrac{2}{3}x - 3$ **30.** $y = -\dfrac{1}{2}x + 2$

31. $y = \dfrac{-c \pm \sqrt{c^2 - 12d}}{6}$ **32.** $y = \dfrac{n \pm \sqrt{n^2 + 4mw}}{2m}$

33. $y = \dfrac{5}{6}x - \dfrac{25}{12}$ **34.** $y = -\dfrac{2}{3}x + \dfrac{17}{3}$ **35.** $\dfrac{4}{3}$ **36.** $-\dfrac{11}{7}$

37. -2 **38.** $\dfrac{58}{5}$ **39.** 40,000, 38,000, $32.50

40. $800,000, $950,000, $40 or $80, $60

Chapter 9

Section 9.1 Warm-Ups
1. Linear   **2.** Constant   **3.** Identity   **4.** Parabola
5. Absolute value   **6.** True   **7.** True   **8.** False   **9.** True
10. True   **11.** False   **12.** True

Section 9.1 Exercises
1. $(-\infty, \infty)$, $\{-2\}$

3. $(-\infty, \infty)$, $(-\infty, \infty)$

5. $(-\infty, \infty)$, $(-\infty, \infty)$

7. $(-\infty, \infty)$, $(-\infty, \infty)$
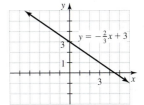

9. $(-\infty, \infty)$, $(-\infty, \infty)$

11. $(-\infty, \infty)$, $[1, \infty)$
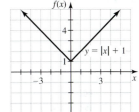

13. $(-\infty, \infty)$, $[0, \infty)$
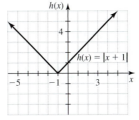

15. $(-\infty, \infty)$, $[0, \infty)$
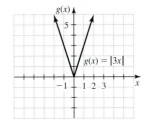

17. $(-\infty, \infty)$, $[0, \infty)$

19. $(-\infty, \infty)$, $[1, \infty)$

21. $(-\infty, \infty)$, $[0, \infty)$

23. $(-\infty, \infty)$, $[2, \infty)$

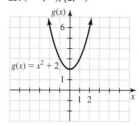

$g(x) = x^2 + 2$

41.

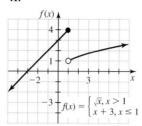

$f(x) = \begin{cases} \sqrt{x}, x > 1 \\ x + 3, x \le 1 \end{cases}$

43.

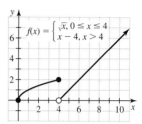

$f(x) = \begin{cases} \sqrt{x}, 0 \le x \le 4 \\ x - 4, x > 4 \end{cases}$

25. $(-\infty, \infty)$, $[0, \infty)$

$f(x) = 2x^2$

27. $(-\infty, \infty)$, $(-\infty, 6]$

$y = 6 - x^2$

45. $[0, \infty)$, $(-\infty, \infty)$

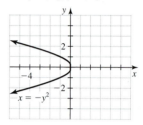

$x = |y|$

47. $(-\infty, 0]$, $(-\infty, \infty)$

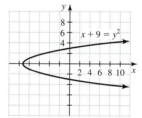

$x = -y^2$

29. $[0, \infty)$, $[0, \infty)$

$g(x) = 2\sqrt{x}$

31. $[1, \infty)$, $[0, \infty)$

$f(x) = \sqrt{x - 1}$

49. $\{5\}$, $(-\infty, \infty)$

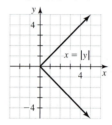

$x = 5$

51. $[-9, \infty)$, $(-\infty, \infty)$

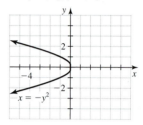

$x + 9 = y^2$

33. $[0, \infty)$, $(-\infty, 0]$

$h(x) = -\sqrt{x}$

35. $[0, \infty)$, $[2, \infty)$

$y = \sqrt{x} + 2$

53. $[0, \infty)$, $[0, \infty)$

$x = \sqrt{y}$

55. $[0, \infty)$, $(-\infty, \infty)$

$x = (y - 1)^2$

37.

$f(x) = \begin{cases} x, x \ge 0 \\ -4x, x < 0 \end{cases}$

39.

$f(x) = \begin{cases} 2, x > 1 \\ -2, x \le 1 \end{cases}$

57. $(-\infty, \infty)$, $(-\infty, 1]$

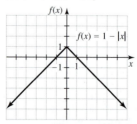

$f(x) = 1 - |x|$

59. $(-\infty, \infty)$, $[-1, \infty)$

$y = (x - 3)^2 - 1$

61. $(-\infty, \infty)$, $[1, \infty)$

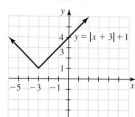

63. $[0, \infty)$, $[-3, 0)$

9. $(-\infty, \infty)$, $[0, \infty)$

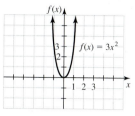

11. $(-\infty, \infty)$, $(-\infty, \infty)$

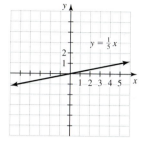

65. $(-\infty, \infty)$, $(-\infty, \infty)$

67. $(-\infty, \infty)$, $(-\infty, 0]$

13. $[0, \infty)$, $[0, \infty)$

15. $(-\infty, \infty)$, $[0, \infty)$

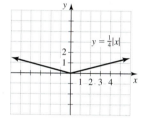

69. Square-root   **71.** Constant   **73.** Quadratic
75. Square-root   **77.** Linear
79. The graph of $f(x) = \sqrt{x^2}$ is the same as the graph of $f(x) = |x|$.
81. For large values of a the graph gets narrower, and for smaller values of a the graph gets broader.
83. The graph of $y = (x - h)^2$ moves to the right for $h > 0$ and to the left for $h < 0$.
85. The graph of $y = f(x - h)$ lies to the right of the graph of $y = f(x)$ when $h > 0$.

17.

19.

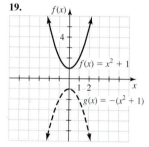

Section 9.2 Warm-Ups

1. Reflection   **2.** Upward translation   **3.** Downward translation
4. Right   **5.** Left   **6.** Stretching   **7.** Shrinking   **8.** True
9. False   **10.** True   **11.** True   **12.** True   **13.** True

Section 9.2 Exercises

1. $(-\infty, \infty)$, $(-\infty, \infty)$

3. $(-\infty, \infty)$, $[0, \infty)$

21.

23.

25. $[0, \infty)$, $[1, \infty)$

27. $(-\infty, \infty)$, $[-4, \infty)$

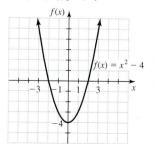

5. $[1, \infty)$, $[0, \infty)$

7. $(-\infty, \infty)$, $[2, \infty)$

29. $(-\infty, \infty)$, $[2, \infty)$

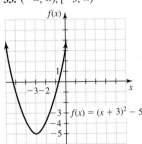

31. $[2, \infty)$, $[1, \infty)$

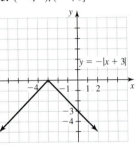

33. $(-\infty, \infty)$, $[-5, \infty)$

35. $(-\infty, \infty)$, $(-\infty, 0]$

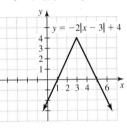

37. $[-1, \infty)$, $(-\infty, -2]$

39. $(-\infty, \infty)$, $(-\infty, 4]$

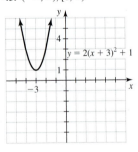

41. $(-\infty, \infty)$, $(-\infty, \infty)$

43. $(-\infty, \infty)$, $[1, \infty)$

45. $(-\infty, \infty)$, $(-\infty, 2]$

47. $(-\infty, \infty)$, $(-\infty, 6]$

49. d   **51.** e   **53.** h   **55.** c   **57.** $y = x^2 + 8$
59. $y = \sqrt{x + 5}$ **61.** $y = |x + 3| + 5$
63. Move f to the right 20 units and upward 30 units.

Section 9.3 Warm-Ups
1. Sum   **2.** Difference   **3.** Product   **4.** Quotient   **5.** Composition
6. True   **7.** True   **8.** True   **9.** False   **10.** True   **11.** False
12. True

Section 9.3 Exercises
1. $x^2 + 2x - 3$ **3.** $4x^3 - 11x^2 + 6x$ **5.** 12   **7.** -30 **9.** -21
11. $\dfrac{13}{8}$ **13.** $y = 6x$ **15.** $y = 6x - 20$ **17.** $y = x + 2$
19. $y = x^2 + 2x$ **21.** $y = x$ **23.** -2 **25.** 5   **27.** 7   **29.** 5
31. 5   **33.** 4   **35.** 22.2992   **37.** $4x^2 - 6x$ **39.** $2x^2 + 6x - 3$
41. x **43.** $4x - 9$ **45.** $\dfrac{x + 9}{4}$ **47.** $F = f \circ h$ **49.** $G = g \circ h$
51. $H = h \circ g$ **53.** $J = h \circ h$ **55.** $K = g \circ g$ **65.** True   **67.** False
69. True   **71.** False   **73.** False   **75.** True   **77. a)** 50 ft^2 **b)** $A = \dfrac{d^2}{2}$
79. $P(x) = -x^2 + 20x - 170$ **81.** $J = 0.025I$
83. a) 397.8   **b)** $D = \dfrac{1.116 \times 10^7}{L^3}$ **c)** Decreases
85. $[0, \infty)$, $[0, \infty)$, $[16, \infty)$ **87.** $[0, \infty)$, $[0, \infty)$

Mid-Chapter Quiz 9.1–9.3
1. $(-\infty, \infty)$, $(-\infty, \infty)$

2. $(-\infty, \infty)$, $[0, \infty)$

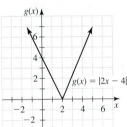

3. $(-\infty, \infty)$, $(-\infty, 3]$

4. $[-6, \infty)$, $[-5, \infty)$

5. $(-\infty, \infty)$, $[1, \infty)$

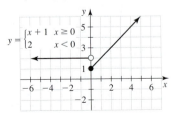

6. $[-2, \infty), (-\infty, \infty)$

7. $(-\infty, \infty), (-\infty, 3]$

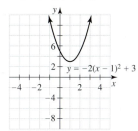

57. $f^{-1}(x) = x^2 + 2$ for $x \geq 0$

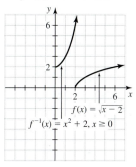

8. $x^2 + 2x - 2$ **9.** 10   **10.** 230   **11.** $\dfrac{9}{2}$ **12.** -2

13. $5x^2 - 15x - 2$ **14.** $F = f \circ g$ **15.** $A = \dfrac{\pi d^2}{4}$

16. $y = (x + 2)^2 + 4$ **17.** $y = -2|x + 3|$ **18.** $y = \dfrac{1}{2}\sqrt{x + 9} - 5$

Section 9.4 Warm-Ups
1. Inverse   **2.** Range   **3.** Domain   **4.** One-to-one   **5.** Symmetric
6. Horizontal   **7.** False   **8.** False   **9.** True   **10.** False   **11.** True
12. True   **13.** True   **14.** True

Section 9.4 Exercises
1. Yes, $\{(3, 1), (9, 2)\}$ **3.** No   **5.** Yes, $\{(4, 16), (3, 9), (0, 0)\}$
7. No   **9.** Yes, $\{(0, 0), (2, 2), (9, 9)\}$ **11.** No   **13.** Yes   **15.** Yes
17. Yes   **19.** Yes   **21.** No   **23.** $f^{-1}(x) = \dfrac{x}{5}$ **25.** $g^{-1}(x) = x + 9$
27. $k^{-1}(x) = \dfrac{x + 9}{5}$ **29.** $m^{-1}(x) = \dfrac{2}{x}$ **31.** $f^{-1}(x) = x^3 + 4$
33. $f^{-1}(x) = \dfrac{3}{x} + 4$ **35.** $f^{-1}(x) = \dfrac{x^3 - 7}{3}$ **37.** $f^{-1}(x) = \dfrac{2x + 1}{x - 1}$
39. $f^{-1}(x) = \dfrac{1 + 4x}{3x - 1}$ **41.** $p^{-1}(x) = x^4$ for $x \geq 0$ **43.** $f^{-1}(x) = 2 + \sqrt{x}$
45. $f^{-1}(x) = \sqrt{x - 3}$ **47.** $f^{-1}(x) = x^2 - 2$ for $x \geq 0$
49. $f^{-1}(x) = \dfrac{1}{2}x - \dfrac{3}{2}$

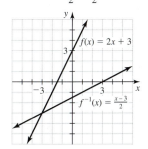

51. $f^{-1}(x) = \sqrt{x + 1}$

53. $f^{-1}(x) = \dfrac{x}{5}$

55. $f^{-1}(x) = \sqrt[3]{x}$

59. $f^{-1}(x) = \dfrac{x}{2}$ **61.** $f^{-1}(x) = \dfrac{x + 1}{2}$ **63.** $f^{-1}(x) = x^3$
65. $f^{-1}(x) = x^3 + 1$ **67.** $f^{-1}(x) = \left(\dfrac{x + 1}{2}\right)^3$ **69.** $(f^{-1} \circ f)(x) = x$
71. $(f^{-1} \circ f)(x) = x$ **73.** $(f^{-1} \circ f)(x) = x$ **75.** $(f^{-1} \circ f)(x) = x$
77. a) 33.5 mph   **b)** Decreases   **c)** $L = \dfrac{S^2}{30}$
79. $T(x) = 1.09x + 125$, $T^{-1}(x) = \dfrac{x - 125}{1.09}$
81. An odd positive integer   **83.** Not inverses

Section 9.5 Warm-Ups
1. Directly   **2.** Inversely   **3.** Proportionality   **4.** Jointly
5. Proportional   **6.** Proportional   **7.** True   **8.** False   **9.** False
10. True   **11.** False   **12.** True

Section 9.5 Exercises
1. $a = km$ **3.** $d = k/e$ **5.** $I = krt$ **7.** $m = kp^2$ **9.** $B = k\sqrt[3]{w}$
11. $t = \dfrac{k}{x^2}$ **13.** $v = \dfrac{km}{n}$ **15.** Inverse   **17.** Direct   **19.** Combined
21. Joint   **23.** $y = \dfrac{3}{2}x$ **25.** $A = \dfrac{30}{B}$ **27.** $m = \dfrac{36}{\sqrt{p}}$ **29.** $A = \dfrac{2}{5}tu$
31. $y = \dfrac{1.067x}{z}$ **33.** $-\dfrac{21}{5}$ **35.** $-\dfrac{3}{2}$ **37.** $\dfrac{1}{2}$ **39.** 3.293   **41.** \$420
43. 9 cm³   **45.** \$360   **47.** \$18.90
49. a) $d = 16t^2$ **b)** 4 feet   **c)** 2.5 seconds
51. 400 pounds   **53.** 7.2 days
55. a) $G = \dfrac{Nd}{c}$ **b)** 84   **c)** 23, 20, 17, 15, 13   **d)** Decreases
57. $(1, 1)$, y_1 increasing, y_2 decreasing, y_1 direct variation, y_2 inverse variation

Enriching Your Mathematical Word Power
1. Constant   **2.** Linear   **3.** Quadratic   **4.** Absolute
5. Composition   **6.** Sum   **7.** Inverse   **8.** One-to-one
9. Vertical   **10.** Horizontal   **11.** Directly   **12.** Inversely
13. Jointly   **14.** Reflection   **15.** Translation   **16.** Left

Review Exercises
1. $(-\infty, \infty), (-\infty, \infty)$

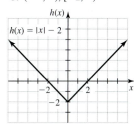

3. $(-\infty, \infty), [-2, \infty)$

5. $(-\infty, \infty), [0, \infty)$

7. $[0, \infty), [2, \infty)$

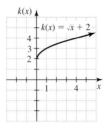

23. $[0, \infty), [0, \infty)$

25. $[-1, \infty), (-\infty, 3]$

9. $(-\infty, \infty), (-\infty, 30]$

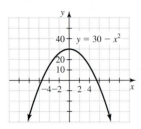

11. $[-4, \infty), [0, \infty)$

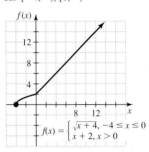

27. -4 **29.** $\sqrt{2}$ **31.** 99 **33.** 17 **35.** $3x^3 - x^2 - 10x$

37. 20 **39.** $F = f \circ g$ **41.** $H = g \circ h$ **43.** $I = g \circ g$

45. No   **47.** Yes, $f^{-1}(x) = x/8$ **49.** Yes, $g^{-1}(x) = \dfrac{x + 6}{13}$

51. Yes, $j^{-1}(x) = \dfrac{x + 1}{x - 1}$ **53.** No

55. $f^{-1}(x) = \dfrac{1}{3}x + \dfrac{1}{3}$ **57.** $f^{-1}(x) = \sqrt[3]{2x}$

13. $\{2\}, (-\infty, \infty)$

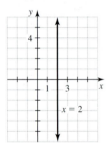

15. $[1, \infty), (-\infty, \infty)$

59. 24 **61.** -16 **63.** 256 ft   **65.** $B = \dfrac{2A}{\pi}$ **67.** $a = 15w - 16$

69. $A = s^2, s = \sqrt{A}$

Chapter 9 Test

1. $(-\infty, \infty), (-\infty, \infty)$

2. $(-\infty, \infty), [-4, \infty)$

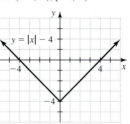

17. $[0, \infty), [0, \infty)$

19. $[0, \infty), (-\infty, 0]$

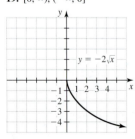

21. $[2, \infty), [0, \infty)$

3. $(-\infty, \infty), [-9, \infty)$

4. $[0, \infty), (-\infty, \infty)$

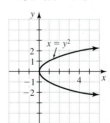

5. $(-\infty, \infty), (-3, \infty)$

$f(x) = \begin{cases} \sqrt{x}, x \geq 0 \\ -x - 3, x < 0 \end{cases}$

6. $(-\infty, \infty), (-\infty, 0]$

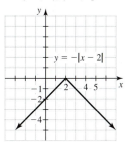

$y = -|x - 2|$

7. $[-5, \infty), [-2, \infty)$

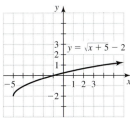

$y = \sqrt{x + 5} - 2$

8. 11   **9.** 125   **10.** -3 **11.** $-\dfrac{1}{2}x + \dfrac{5}{2}$ **12.** $x^2 - 2x + 9$

13. 15   **14.** 1776   **15.** $\dfrac{1}{8}$ **16.** $-2x^2 - 3$ **17.** $4x^2 - 20x + 29$

18. $H = f \circ g$ **19.** $W = g \circ f$ **20.** Not invertible

21. $\{(3, 2), (4, 3), (5, 4)\}$ **22.** $f^{-1}(x) = x + 5$ **23.** $f^{-1}(x) = \dfrac{x + 5}{3}$

24. $f^{-1}(x) = (x - 9)^3$ **25.** $f^{-1}(x) = \dfrac{x + 1}{x - 2}$ **26.** $\dfrac{32\pi}{3}$ ft³

27. $\dfrac{36}{7}$ **28.** \$5076

Making Connections   A Review of Chapters 1–9

1. $\dfrac{1}{25}$ **2.** $\dfrac{3}{2}$ **3.** $\sqrt{2}$ **4.** x^8 **5.** 2   **6.** x^9 **7.** $\{\pm 3\}$

8. $\{\pm 2\sqrt{2}\}$ **9.** $\{0, 1\}$ **10.** $\{2 \pm \sqrt{10}\}$ **11.** $\{81\}$ **12.** $\varnothing$

13. $\{\pm 8\}$ **14.** $\left\{-\dfrac{17}{5}, 5\right\}$ **15.** $\{2\}$ **16.** $\left\{\dfrac{5}{3}\right\}$ **17.** $\{42\}$ **18.** $\{11\}$

19.

$y = 5$

20.

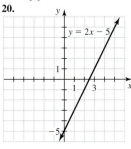

$y = 2x - 5$

21.

$x = 5$

22.

$3y = x$

23.

$y = 5x^2$

24.

$y = -2x^2$

25. $(2, 4), (3, 8), (1, 2), (4, 16)$ **26.** $\left(\dfrac{1}{2}, 2\right), \left(-1, \dfrac{1}{4}\right), (2, 16), (0, 1)$

27. $[0, \infty)$ **28.** $(-\infty, 3]$ **29.** $(-\infty, \infty)$ **30.** $(-\infty, 1) \cup (1, 9) \cup (9, \infty)$

31. $\left\{\left(\dfrac{1}{2}, \dfrac{1}{3}\right)\right\}$, independent   **32.** $\{(1.3, 6.5)\}$, independent

33. $\left\{(x, y) \mid y = \dfrac{1}{3}x - \dfrac{2}{3}\right\}$, dependent   **34.** $\varnothing$, inconsistent

35. $\{(2, -2, 5)\}$, independent   **36.** $\varnothing$, inconsistent   **37.** $\dfrac{19}{60}$

38. $\dfrac{8}{49}$ **39.** $\dfrac{10}{21}$ **40.** $\dfrac{4 + x}{6x^2}$ **41.** $\dfrac{b^4}{3y^2}$ **42.** $\dfrac{4a - 12}{3}$

43. a) $C = 0.12x + 3000$

    b) $P = 1 \times 10^{-6}x + 0.15$

44. a) \$0.44, \$0.40, \$0.39   **b)**

    c) 60,000 miles

    d) $[50,000, 60,000]$

$T = 0.27 + \dfrac{3000}{x} + \dfrac{x}{10^6}$

Chapter 10

Section 10.1 Warm-Ups

1. Zero, root   **2.** Equivalent   **3.** Synthetic   **4.** Remainder

5. Factor   **6.** True   **7.** True   **8.** False   **9.** True   **10.** True

11. True   **12.** True

Section 10.1 Exercises

1. Yes   **3.** No   **5.** No   **7.** Yes   **9.** Yes   **11.** No   **13.** Yes

15. Yes   **17.** No   **19.** Yes   **21.** Yes   **23.** $(x - 1)(x - 4)$ **25.** No

27. $(x - 3)(x^2 + 3x + 3)$ **29.** $(x + 5)(x + 3)(x + 1)$ **31.** No

33. $(x + 1)(x - 2)(x^2 + 2x + 4)$ **35.** $(2x - 1)(x - 3)(x + 2)$

37. $\{-4, 1, 3\}$ **39.** $\left\{-\dfrac{1}{2}, 2, 3\right\}$ **41.** $\left\{-4, -\dfrac{1}{2}, 6\right\}$ **43.** $\{-3, 1\}$

45. $\{-1, 1, 2\}$ **47. a)** $f(x) = x + 2$ **b)** $f(x) = x^2 - 25$

    c) $f(x) = (x - 1)(x + 3)(x - 4)$

    d) Yes. The degree is greater than or equal to the number of zeros.

49. a) $\{-2, 1, 3\}$ **b)** $\left\{-\dfrac{1}{3}, \dfrac{1}{2}, \dfrac{3}{2}\right\}$

Section 10.2 Warm-Ups

1. Zero, root   **2.** Polynomial   **3.** Remainder   **4.** Constant, leading

5. Complex   **6.** False   **7.** True   **8.** True   **9.** False   **10.** False

11. True

Section 10.2 Exercises

1. 0   **3.** 15   **5.** 3   **7.** -15 **9.** 0   **11.** 17   **13.** No

15. Yes   **17.** Yes   **19.** No   **21.** Yes   **23.** No   **25.** $\dfrac{9}{5}$

27. 2, 3   **29.** $3i, -3i$ **31.** $-2i, 2i, 3$ **33.** $-\dfrac{1}{2}, 0, 3$ **35.** $-1, \pm\sqrt{7}$

37. $\dfrac{-1 \pm i\sqrt{3}}{2}, 1$ **39.** $\pm 1, \pm 2$ **41.** $\pm 1, \pm 2, \pm\dfrac{1}{2}$

43. $\pm 1, \pm 2, \pm 3, \pm 4, \pm 6, \pm 8, \pm 12, \pm 24$ **45.** $\pm 1, \pm 2, \pm 3, \pm 6$

47. $-\dfrac{1}{2}, 1, 2$ **49.** $-4, -3, -2$ **51.** $-3 - \sqrt{3}, -3 + \sqrt{3}, 1$

53. $-2 \pm 2i, 2$ **55.** $\dfrac{1}{2}, \dfrac{1}{3}, \dfrac{1}{4}$ **57.** \$1000, 0 or 20   **59.** $-0.2, 0.1, 0.3$

61. 10, 20, 30

Section 10.3 Warm-Ups

1. Multiplicity   **2.** n **3.** Conjugate   **4.** Positive   **5.** Negative
6. Upper bound   **7.** Lower bound   **8.** False   **9.** True   **10.** True
11. False   **12.** False   **13.** True   **14.** False   **15.** True   **16.** True

Section 10.3 Exercises

1. Degree 5, -2, 2, 0 multiplicity 3
3. Degree 4, 0 multiplicity 2, -1 multiplicity 2
5. Degree 4, $-\sqrt{3}$ multiplicity 2, $\sqrt{3}$ multiplicity 2
7. Degree 4, 1 multiplicity 2, -2 multiplicity 2
9. Degree 4, -1 multiplicity 2, 1 multiplicity 2
11. $x^2 - 9 = 0$ **13.** $x^2 + 4 = 0$ **15.** $x^3 - x^2 + 4x - 4 = 0$
17. $x^3 - 7x^2 + 17x - 15 = 0$ **19.** $x^3 + 2x^2 + x + 2 = 0$
21. $x^3 + 2x = 0$ **23.** $x^4 - 2x^3 + 3x^2 - 2x + 2 = 0$
25. $x^2 - 3x + 2 = 0$ **27.** $x^3 - 4x^2 + x + 6 = 0$
29. 3 negative, or 1 negative and 2 imaginary
31. 1 positive and 2 negative, or 1 positive and 2 imaginary
33. 4 positive, or 2 positive and 2 imaginary, or 4 imaginary
35. No positive, no negative, 4 imaginary   **37.** 1 positive and 2 imaginary
39. 4 imaginary and 0   **41.** Between -3 and 3   **43.** Between -1 and 3
45. Between -5 and 4   **47.** Between -1 and 7   **49.** $-2, 1 \pm 2i$

51. $\dfrac{1}{2}, 1 \pm \sqrt{5}$ **53.** $-\dfrac{1}{2}, \dfrac{2 \pm \sqrt{2}}{2}$ **55.** $-1, 1, 2, 3$ **57.** 1, 2, 3

59. $0, 1, \pm i\sqrt{5}, \pm\sqrt{3}$ **61.** 6 ft   **63.** $\dfrac{1}{2}$ in.

65. 3.3315, 0.9492, -0.9492, -3.3315
67. $-2.4495, -1.4142, -1, 1, 1.4142, 2.4495$

Mid-Chapter Quiz 10.1–10.3

1. Yes   **2.** No   **3.** $\{2, 3, 6\}$ **4.** $\{-1, 3\}$ **5.** $\left\{3, \dfrac{-3 \pm 3i\sqrt{3}}{2}\right\}$

6. $\{1, 2, 3\}$ **7.** $x^3 - 7x + 6 = 0$ **8.** $x^3 - 8x^2 + 21x - 20 = 0$
9. 1 neg, 2 pos; 1 neg, 2 imag   **10.** 0 neg, 4 imag; 2 neg, 2 imag
11. 53   **12.** $\pm\left(\dfrac{1}{6}, \dfrac{1}{3}, \dfrac{1}{2}, 1\right)$

Section 10.4 Warm-Ups

1. y-axis   **2.** Origin   **3.** Odd   **4.** Even   **5.** False   **6.** False
7. False   **8.** True   **9.** True   **10.** True

Section 10.4 Exercises

1. Symmetric about origin   **3.** Symmetric about y-axis
5. Symmetric about origin   **7.** Symmetric about y-axis
9. Neither symmetry   **11.** Neither symmetry   **13.** Neither symmetry

15. Symmetric about y-axis   **17.** Symmetric about origin
19. Does not cross at $(5, 0)$ **21.** Crosses at $(4, 0)$ and $(-1, 0)$
23. Does not cross at $(0, 0)$
25. Crosses at $(-5, 0)$, does not cross at $(3, 0)$
27. Crosses at $(1, 0)$, does not cross at $(0, 0)$
29. Crosses at $\left(\dfrac{1}{2}, 0\right)$, does not cross at $(-1, 0)$

31. d   **33.** a   **35.** c   **37.** e

39.

41.

43.

45.

47.

49.

51.

53.

55.

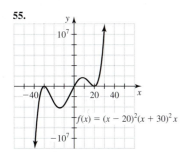

57. $(-2, 0) \cup (2, \infty)$

59. $(-\infty, -2] \cup [3, 5]$

61. $\{0\} \cup [2, \infty)$ **63.** $(-\infty, \infty)$

65. $\{-1, 1\}$ **67.** $(-\infty, -3) \cup (3, \infty)$

69. $[-2, 3]$ **71.** $(-6, -2) \cup (2, \infty)$

73. $[-3, -1] \cup [1, 3]$

75. $f(x) = x - 3$
77. $f(x) = (x + 2)(x - 1)^2$

Section 10.5 Warm-Ups
 1. Rational   **2.** Domain   **3.** Vertical   **4.** Horizontal
 5. Oblique, slant   **6.** False   **7.** False   **8.** False   **9.** True
 10. True   **11.** True   **12.** True

Section 10.5 Exercises
 1. $(-\infty, 1) \cup (1, \infty)$ **3.** $(-\infty, 0) \cup (0, \infty)$
 5. $(-\infty, -4) \cup (-4, 4) \cup (4, \infty)$
 7. Vertical: $x = -4$; horizontal: x-axis
 9. Vertical: $x = 4, x = -4$; horizontal: x-axis
 11. Vertical: $x = 7$; horizontal: $y = 5$
 13. Vertical: $x = 3$; oblique: $y = 2x + 6$ **15.** c   **17.** b   **19.** g   **21.** f
 23. $x = -4$, x-axis   **25.** $x = 3, x = -3$, x-axis

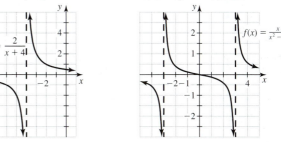

 27. $x = -3, y = 2$ **29.** y-axis, $y = x - 3$

 31. $x = 1, y = 3x + 1$ **33.** $x = 0, y = 0$

 35. $x = -3, x = 2, y = 0, \left(0, \frac{1}{2}\right), \left(\frac{3}{2}, 0\right)$

37. $x = 0$, $y = 0$, $(-1, 0)$

39. $x = 0$, $x = \pm 3$, $y = 0$, $\left(\frac{1}{2}, 0\right)$

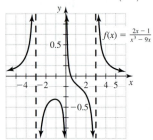

41. $x = \pm 1$, $y = 0$, $(0, 0)$ **43.** $y = 0$, $(0, 2)$

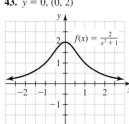

45. $x = -1$, $y = x - 1$, $(0, 0)$

47. $(0, \infty)$

49. $(-\infty, 0) \cup (3, \infty)$

51. $[-2, 0)$

53. $(-\infty, -6) \cup (3, \infty)$

55. $(-\infty, -2) \cup (-1, \infty)$

57. $[-13, -5)$

59. $(-\infty, -2) \cup [4, \infty)$

61. $(-3, \infty)$ **63.** $(-2, -1) \cup \left(-\frac{1}{2}, \infty\right)$

65. $(-13, -4) \cup (5, \infty)$

67. $(-\infty, -5) \cup (1, 3) \cup (5, \infty)$ **69.** $[-6, 3) \cup [4, 6)$

71. $(-\infty, \infty)$ **73.** $\varnothing$

75. $f(20) = \$2.35$, $f(30) = \$1.46$, average approaches 0

77. a) $y = 25{,}000$ **b)** $\$39{,}000$ **c)** $140{,}000$
d)

79, 81, and 83. The graph of $f(x)$ is an asymptote for the graph of $g(x)$.

85. $f(x) = 1/x$ **87.** $f(x) = \dfrac{1}{(x - 3)(x + 1)}$

Enriching Your Mathematical Word Power

1. Polynomial   **2.** Root, zero   **3.** Multiplicity   **4.** Upper
5. Lower   **6.** Symmetric   **7.** Origin   **8.** Rational
9. Asymptote   **10.** Oblique

Review Exercises

1. $\{-3, 2, 5\}$ **3.** $\{-2, 3, 4\}$ **5.** $(x - 3)(x + 2)(x + 5)$
7. No   **9.** $-45, \dfrac{5}{2}$ **11.** $\pm 1, \pm 2, \pm 4, \pm \dfrac{1}{3}, \pm \dfrac{2}{3}, \pm \dfrac{4}{3}$
13. $\pm 1, \pm 3, \pm \dfrac{1}{2}, \pm \dfrac{3}{2}, \pm \dfrac{1}{4}, \pm \dfrac{3}{4}$ **15.** $\dfrac{3}{2}$ **17.** ± 10

19. $4, -2 \pm 2i\sqrt{3}$ **21.** $\pm 2\sqrt{2}, \pm 2i\sqrt{2}$ **23.** $\dfrac{1}{3}, \pm \dfrac{1}{2}$

25. $2x^2 - 5x + 2 = 0$ **27.** $x^2 - 10x + 29 = 0$

29. $x^3 - 5x^2 + 16x - 30 = 0$ **31.** 4 imaginary and 0 with multiplicity 2

33. 3 positive, or 1 positive and 2 imaginary   **35.** Between -2 and 4

37. Between -2 and 5   **39.** $\dfrac{1}{2}, 4, 5$ **41.** $-\dfrac{3}{2}, \dfrac{1}{2}, \dfrac{5}{2}$ **43.** $5, \dfrac{1}{2} \pm i$

45. 3 with multiplicity 2, $\dfrac{1 \pm \sqrt{2}}{2}$ **47.** 1 with multiplicity 3

49. Symmetric about y-axis   **51.** Symmetric about origin

53. $(1, 0), (-2, 0), (0, 2)$

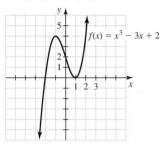

55. $(1, 0), (-1, 0), (0, -1)$

57. $(0, 0), (2, 0), (-2, 0)$

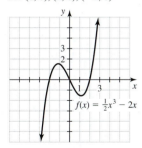

59. $(-\infty, -2] \cup [0, 2]$

61. $\left(-\infty, -\sqrt{2}\right] \cup \{0\} \cup \left[\sqrt{2}, \infty\right)$

63. $\left[-\sqrt{2}, \sqrt{2}\right]$

65. $(-\infty, -3) \cup (1, 3)$

67. $[-3, 1] \cup [3, \infty)$

69. $\left(-\infty, -\dfrac{3}{2}\right) \cup \left(-\dfrac{3}{2}, \infty\right)$ **71.** $(-\infty, \infty)$

73. $x = 3$, x-axis

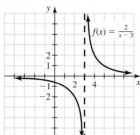

75. $x = 2, x = -2$, x-axis

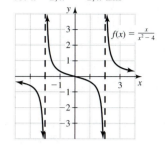

77. $x = 1, y = 2$

79. $x = 2, y = x$

81. $(-\infty, -3] \cup (4, \infty)$

83. $(-2, 0) \cup (2, \infty)$

85. $(-\infty, -10) \cup (-5, \infty)$

87. $(-\infty, 0)$

89. $(-5, 0] \cup (2, 9]$

91.

93.

95.

97.

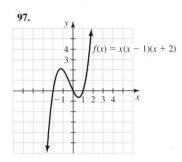

$f(x) = x(x - 1)(x + 2)$

99. $(-\infty, 2)$ **101.** $\left[-\sqrt{3}, \sqrt{3}\right]$ **103.** $\left(-\sqrt{3}, \sqrt{3}\right)$
105. $(-\infty, -2) \cup (0, 1)$ **107.** $(-3, 4]$ **109.** $[-1, 2] \cup [5, \infty)$
111. $[-\infty, -3] \cup (-1, 1)$ **113.** c   **115.** b   **117.** 3 in.

Chapter 10 Test

1. 52   **2.** 7   **3.** $\pm 1, \pm 2, \pm 3, \pm 6$ **4.** $\pm 1, \pm 3, \pm\dfrac{1}{2}, \pm\dfrac{3}{2}$

5. $-4, 1, 3$ **6.** $\dfrac{1}{2}, 2 \pm i$ **7.** $x^3 - 2x^2 - x + 2 = 0$

8. $x^3 - 8x^2 + 22x - 20 = 0$ **9.** Yes   **10.** $(x - 3)(x - 4)(x - 5)$
11. 3 positive, or 1 positive and 2 imaginary
12. 4 imaginary roots and 0   **13.** Between -2 and 3
14. Between -2 and 3   **15.** Symmetry about origin
16. Symmetry about y-axis
17.

$f(x) = (x + 2)(x - 2)^2$

18.

$f(x) = \dfrac{1}{x^2 - 4x + 4}$

19.

$f(x) = \dfrac{2x - 3}{x - 2}$

20.

$f(x) = x^3 - x^2 - 4x + 4$

21. $(-8, 0) \cup (4, \infty)$
22. $(-\infty, -5) \cup [0, 5)$

Making Connections   A Review of Chapters 1–10

1. -2 **2.** $\dfrac{3}{2}, 5$ **3.** $\dfrac{3}{2}$ **4.** $\pm\dfrac{\sqrt{6}}{2}$ **5.** -9 **6.** $\pm\dfrac{\sqrt{6}}{2}, \pm i\sqrt{5}$

7. $-5, 2$ **8.** $-\dfrac{1}{3}, -\dfrac{1}{5}, \dfrac{3}{2}$

9.

$y = 2x$

10.

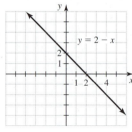

$y = 2 - x$

11.

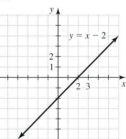

$y = x - 2$

12.

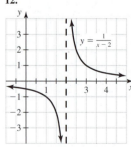

$y = \dfrac{1}{x - 2}$

13.

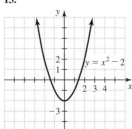

$y = x^2 - 2$

14.

$y = (x - 2)^2$

15.

$y = \frac{1}{(x-2)^2}$

16.

$y = x^3 - 2$

29.

$y = 10^x$

31. $\frac{1}{100}, \frac{1}{10}, 1, 10, 100$

33. $-\frac{1}{4}, -\frac{1}{2}, -1, -2, -4$

17.

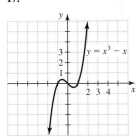

$y = x^3 - x$

18.

$y = x^4 - x^2$

35.

$f(x) = -3^x$

37.

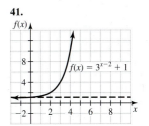

$f(x) = \frac{1}{2} \cdot 3^x$

19.

$y = \sqrt{x+5} - 2$

20.

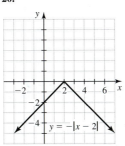

$y = -|x - 2|$

39.

$f(x) = -3^x + 2$

41.

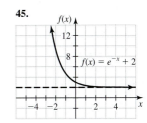

$f(x) = 3^{x-2} + 1$

43.

$f(x) = 10^x + 2$

45.

$f(x) = e^{-x} + 2$

21. 22   **22.** -19 **23.** -5 **24.** 1   **25.** 78   **26.** 5   **27.** 46
28. 53   **29.** 3   **30.** -8 **31.** $(5, \infty)$ **32.** $[-5, \infty)$ **33.** $[2, 5]$
34. $(-\infty, -2) \cup (3, \infty)$ **35.** $(-\infty, -\sqrt{3}] \cup [\sqrt{3}, \infty)$ **36.** $(-3/2, 1/2)$
37. $(-2, 0) \cup (2, \infty)$ **38.** $(4, \infty)$ **39.** $(-2, 0) \cup (0, 2)$
40. $(-\infty, -3] \cup [-2, 2]$

Chapter 11

Section 11.1 Warm-Ups

1. Exponential   **2.** Domain   **3.** Natural   **4.** Common
5. One-to-one   **6.** Compound   **7.** Continuously   **8.** False
9. True   **10.** True   **11.** True   **12.** False   **13.** True   **14.** False

Section 11.1 Exercises

1. 16   **3.** 2   **5.** 3   **7.** $\frac{1}{3}$ **9.** -1 **11.** $-\frac{1}{4}$ **13.** 1   **15.** 100

17. 2.718   **19.** 0.135   **21.** $\frac{1}{16}, \frac{1}{4}, 1, 4, 16$ **23.** $9, 3, 1, \frac{1}{3}, \frac{1}{9}$

25.

$f(x) = 4^x$

27.

$h(x) = \left(\frac{1}{3}\right)^x$

47. $\{6\}$ **49.** $\{-3\}$ **51.** $\{-2\}$ **53.** $\{-1\}$ **55.** $\{-2\}$ **57.** $\{-2\}$

59. $\{-3, 3\}$ **61.** 2   **63.** $\frac{4}{3}$ **65.** -2 **67.** 0   **69.** $\frac{3}{2}$ **71.** $\frac{1}{2}$

73. $\frac{1}{32}, -3, 1, 1, 16$ **75.** $-3, 4, 0, \frac{1}{2}, 5$ **77.** \$9861.72

79. a) \$15,859.75   **b)** 2014   **81.** \$45, \$12.92   **83.** \$616.84

85. \$6230.73   **87.** 300 grams, 90.4 grams, 12 years, no

89. 50°F, 31.9°F, 48.6°F, 80.5°F   **91.** 2.66666667, 0.0516, 2.8×10^{-5}

93. The graph of $y = 3^{x-h}$ lies h units to the right of $y = 3^x$ when $h > 0$ and $|h|$ units to the left of $y = 3^x$ when $h < 0$.

Section 11.2 Warm-Ups

1. Logarithmic   **2.** Common   **3.** Natural   **4.** Domain
5. One-to-one   **6.** Symmetric   **7.** Exponent   **8.** True   **9.** False
10. True   **11.** True   **12.** False   **13.** False   **14.** True   **15.** True

Section 11.2 Exercises

1. $2^3 = 8$ **3.** $\log(100) = 2$ **5.** $5^y = x$ **7.** $\log_2(b) = a$
9. $3^{10} = x$ **11.** $\ln(x) = 3$ **13.** 2   **15.** 4   **17.** 6   **19.** 3

21. -2 **23.** 2   **25.** -2 **27.** 1   **29.** -3 **31.** $\dfrac{1}{2}$ **33.** 2

35. 0.6990   **37.** 1.8307   **39.** $-2, -1, 0, 1, 2$ **41.** $-2, -1, 0, 1, 2$

43.

45.

47.

49.

51. $f^{-1}(x) = \log_6(x)$ **53.** $f^{-1}(x) = e^x$ **55.** $f^{-1}(x) = \left(\dfrac{1}{2}\right)^x$

57. $\{4\}$ **59.** $\left\{\dfrac{1}{2}\right\}$ **61.** $\{0.001\}$ **63.** $\{6\}$ **65.** $\left\{\dfrac{1}{5}\right\}$ **67.** $\{\pm 3\}$

69. $\{0.4771\}$ **71.** $\{-0.3010\}$ **73.** $\{1.9741\}$ **75.** $-2, \dfrac{1}{2}, 0, 4, 4$

77. $16, -2, 1, 1, \dfrac{1}{4}$ **79.** 5.776 years   **81.** 1.927 years

83. a) 4.82%   **b)** \$26,222   **85.** 4.1   **87.** 6.8   **89.** 90 dB

91. $f^{-1}(x) = 2^{x-5} + 3, (-\infty, \infty), (3, \infty)$

93. $y = \ln(e^x) = x$ for $-\infty < x < \infty$, $y = e^{\ln(x)} = x$ for $0 < x < \infty$

Mid-Chapter Quiz 11.1–11.2

1. $\dfrac{1}{4}, -1, 1, 2, 8$ **2.** $16, -2, \dfrac{1}{2}, 2, \dfrac{1}{16}$ **3.** $2, 0, -1, -2$

4. $-5, -4, 0, 3$ **5.** $\{4\}$,   **6.** $\{-1\}$ **7.** $\{-2\}$ **8.** $\{8\}$

9. $\{10,000\}$ **10.** $f^{-1}(x) = \log(x)$ **11.** $g^{-1}(x) = 8^x$

12. $W = 5^M$ **13.** $\log_a(y) = 3$

14. D: $(-\infty, \infty)$, R: $(1, \infty)$

15. D: $(-\infty, \infty)$, R: $(-3, \infty)$

16. D: $(0, \infty)$, R: $(-\infty, \infty)$

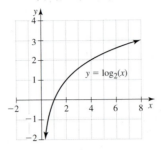

Section 11.3 Warm-Ups

1. Product   **2.** Quotient   **3.** Power   **4.** Inverse   **5.** True
6. False   **7.** False   **8.** True   **9.** False   **10.** True   **11.** True
12. True

Section 11.3 Exercises

1. 10   **3.** 19   **5.** 8   **7.** 4.3   **9.** $\log(21)$ **11.** $\log_3(\sqrt{5x})$
13. $\log(x^5)$ **15.** $\ln(30)$ **17.** $\log(x^2 + 3x)$ **19.** $\log_2(x^2 - x - 6)$
21. $\log(4)$ **23.** $\log_2(x^4)$ **25.** $\log(\sqrt{5})$ **27.** $\ln(h - 2)$

29. $\log_2(w - 2)$ **31.** $\ln(x - 2)$ **33.** $3 \log(3)$ **35.** $\dfrac{1}{2}\log(3)$

37. $x \log(3)$ **39.** $\log(3) + \log(5)$ **41.** $\log(5) - \log(3)$
43. $2 \log(5)$ **45.** $2 \log(5) + \log(3)$ **47.** $-\log(3)$ **49.** $-\log(5)$
51. $\log(x) + \log(y) + \log(z)$ **53.** $3 + \log_2(x)$ **55.** $\ln(x) - \ln(y)$

57. $1 + 2 \log(x)$ **59.** $2 \log_5(x - 3) - \dfrac{1}{2}\log_5(w)$

61. $\ln(y) + \ln(z) + \dfrac{1}{2}\ln(x) - \ln(w)$ **63.** $\log(x^2 - x)$ **65.** $\ln(3)$

67. $\ln\left(\dfrac{xz}{w}\right)$ **69.** $\ln\left(\dfrac{x^2 y^3}{w}\right)$ **71.** $\log\left(\dfrac{(x - 3)^{1/2}}{(x + 1)^{2/3}}\right)$ **73.** $\log_2\left(\dfrac{(x - 1)^{2/3}}{(x + 2)^{1/4}}\right)$

75. False   **77.** True   **79.** True   **81.** False   **83.** True   **85.** True

87. True   **89.** False   **91.** $r = \log(I/I_0), r = 2$ **93.** b

95. The graphs are the same because $\ln(\sqrt{x}) = \ln(x^{1/2}) = \dfrac{1}{2}\ln(x)$.

97. The graph is a straight line because $\log(e^x) = x \log(e) \approx 0.434x$.
The slope is $\log(e)$ or approximately 0.434.

Section 11.4 Warm-Ups

1. Equivalent   **2.** Base-change   **3.** True   **4.** True   **5.** True
6. False   **7.** True   **8.** True   **9.** True   **10.** False

Section 11.4 Exercises

1. $\{900\}$ **3.** $\{7\}$ **5.** $\{31\}$ **7.** $\{e\}$ **9.** $\{2\}$ **11.** $\{3\}$ **13.** $\varnothing$

15. $\{6\}$ **17.** $\{3\}$ **19.** $\{2\}$ **21.** $\{4\}$ **23.** $\{\log_3(7)\}$ **25.** $\left\{\dfrac{\ln(7)}{2}\right\}$

27. $\{-6\}$ **29.** $\left\{-\dfrac{1}{2}\right\}$ **31.** $\dfrac{5 \ln(3)}{\ln(2) - \ln(3)}, -13.548$

33. $\dfrac{4 + 2 \log(5)}{1 - \log(5)}, 17.932$ **35.** $\dfrac{\ln(9)}{\ln(9) - \ln(8)}, 18.655$ **37.** 1.5850

39. -0.6309 **41.** -2.2016 **43.** 1.5229   **45.** $\dfrac{\ln(7)}{\ln(2)}, 2.807$

47. $\dfrac{1}{3 - \ln(2)}, 0.433$ **49.** $\dfrac{\ln(5)}{\ln(3)}, 1.465$ **51.** $1 + \dfrac{\ln(9)}{\ln(2)}, 4.170$

53. $\dfrac{\ln(20)}{\ln(3)}, 2.727$ **55.** $\dfrac{3}{5}$ **57.** $\dfrac{1}{2}$ **59.** $\{-3\}$ **61.** $\left\{-\dfrac{7}{2}, -2\right\}$

63. ∅   **65.** ∅   **67.** {4}   **69.** {4, 6}   **71.** {2}   **73.** {4}   **75.** {−1}

77. $\left\{\dfrac{3}{2}, \dfrac{5}{3}\right\}$ **79.** 41 months   **81.** 594 days   **83.** 10 g, 9163 years ago

85. 7524 ft^3/sec   **87.** 7.1 years   **89.** 16.8 years   **91.** 2.0×10^{-4}

93. 0.9183   **95.** $\sqrt[3]{12}$ or 2.2894   **97.** (2.71, 6.54)

99. (1.03, 0.04), (4.74, 2.24)

Enriching Your Mathematical Word Power

1. Exponential   **2.** Common   **3.** Natural   **4.** Domain
5. Compound   **6.** Continuous   **7.** Base-a **8.** Logarithm

Review Exercises

1. $\dfrac{1}{25}$ **3.** 125   **5.** 1   **7.** $\dfrac{1}{10}$ **9.** 4   **11.** $\dfrac{1}{2}$ **13.** 2   **15.** 4

17. $-\dfrac{5}{2}$ **19.** 2   **21.** 8.6421   **23.** 177.828   **25.** 0.02005

27. 0.1408

29.   **31.**

33.   **35.**

37. $\log(n) = m$ **39.** $k^h = t$ **41.** −3   **43.** −1   **45.** 2

47. 0   **49.** 256   **51.** 6.267   **53.** −5.083   **55.** 5.560

57. $f^{-1}(x) = \log(x)$ **59.** $f^{-1}(x) = \ln(x)$

61. $2\log(x) + \log(y)$ **63.** $4\ln(2)$ **65.** $-\log_5(x)$ **67.** $\log\left(\dfrac{\sqrt{x+2}}{(x-1)^2}\right)$

69. {256}   **71.** {3}   **73.** {2}   **75.** {3}   **77.** $\left\{\dfrac{\ln(7)}{\ln(3)-1}\right\}$

79. $\left\{\dfrac{\ln(5)}{\ln(5)-\ln(3)}\right\}$ **81.** $\left\{\dfrac{1}{3}\right\}$ **83.** {22}   **85.** $\left\{\dfrac{200}{99}\right\}$ **87.** {1.3869}

89. {0.4650}   **91.** \$51,182.68   **93.** 161.5 grams   **95.** 5 years

97. 4347.5 ft^3/sec

Chapter 11 Test

1. 25   **2.** $\dfrac{1}{5}$ **3.** 1   **4.** 3   **5.** 0   **6.** −1

7.   **8.**

9.   **10.**

11.   **12.**

13. 10   **14.** 8   **15.** $\dfrac{3}{2}$ **16.** 15   **17.** −4

18. {$\log_3(12)$} or {$\ln(12)/\ln(3)$}   **19.** $\{\sqrt{3}\}$ **20.** $\left\{\dfrac{\ln(8)}{\ln(8)-\ln(5)}\right\}$

21. {5}   **22.** {3}   **23.** {0.5372}   **24.** {20.5156}

25. 10; 147,648   **26.** 1.733 hours

Making Connections   A Review of Chapters 1–11

1. $\left\{3 \pm 2\sqrt{2}\right\}$ **2.** {259}   **3.** {6}   **4.** $\left\{\dfrac{11}{2}\right\}$ **5.** {−5, 11}

6. {67}   **7.** {6}   **8.** {4}   **9.** $\left\{-\dfrac{52}{15}\right\}$ **10.** $\left\{\dfrac{3 \pm \sqrt{3}}{3}\right\}$

11. $f^{-1}(x) = 3x$ **12.** $g^{-1}(x) = 3^x$ **13.** $f^{-1}(x) = \dfrac{x+4}{2}$

14. $h^{-1}(x) = x^2$ for $x \geq 0$ **15.** $j^{-1}(x) = \dfrac{1}{x}$ **16.** $k^{-1}(x) = \log_5(x)$

17. $m^{-1}(x) = 1 + \ln(x)$ **18.** $n^{-1}(x) = e^x$

19.   **20.**

21.

22.

23.

24.

25.

26.

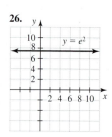

27. $\dfrac{99}{100}$ **28.** $\dfrac{49}{50}$ **29.** $\dfrac{1}{30}$ **30.** 81   **31.** 5   **32.** -2 **33.** $4\sqrt{5}$

34. $2\sqrt[3]{10}$ **35.** $2\sqrt{6}$ **36.** $4\sqrt[3]{50}$ **37.** $\sqrt{2}$ **38.** 1

39. a)

b) Linear 155.7 million, exponential 156.5 million

40. a) $d_1 = 0.135v$ **b)** $d_2 = 0.216v$

   c) $v = 1482.67$ m/sec, $d_1 = 200.2$ meters

Chapter 12

Section 12.1 Warm-Ups

 1. Nonlinear   **2.** Graph   **3.** Substitution, addition   **4.** True

 5. False   **6.** False   **7.** True   **8.** False   **9.** True   **10.** True

Section 12.1 Exercises

 1. $\{(2, 4), (-3, 9)\}$

 3. $\{(-2, 2), (6, 6)\}$

 5. $\{(8, 4)\}$

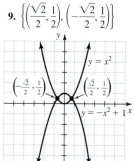

 7. $\left\{\left(-\dfrac{3}{4}, -\dfrac{4}{3}\right), \left(3, \dfrac{1}{3}\right)\right\}$

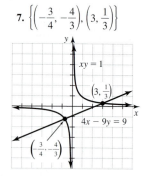

 9. $\left\{\left(\dfrac{\sqrt{2}}{2}, \dfrac{1}{2}\right), \left(-\dfrac{\sqrt{2}}{2}, \dfrac{1}{2}\right)\right\}$

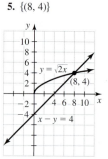

 11. $\{(2, 3)\}$
 13. $\{(-1, -1), (1, 1)\}$
 15. $\{(-\sqrt{2}, 2), (\sqrt{2}, 2)\}$
 17. $\{(0, -5), (3, 4), (-3, 4)\}$
 19. $\{(4, 5), (-2, -1)\}$
 21. $\left\{\left(-3, -\dfrac{5}{3}\right)\right\}$

 23. $\{(\sqrt{5}, \sqrt{3}), (\sqrt{5}, -\sqrt{3}), (-\sqrt{5}, \sqrt{3}), (-\sqrt{5}, -\sqrt{3})\}$
 25. $\{(\sqrt{2}, \sqrt{3}), (\sqrt{2}, -\sqrt{3}), (-\sqrt{2}, \sqrt{3}), (-\sqrt{2}, -\sqrt{3})\}$
 27. $\left\{\left(\dfrac{3}{2}, -\dfrac{3}{13}\right)\right\}$ **29.** $\{(3, 4)\}$ **31.** $\left\{\left(-\dfrac{5}{3}, \dfrac{36}{5}\right), (2, 5)\right\}$
 33. $\{(2, 5), (19, -12)\}$ **35.** $\{(\sqrt{2}, 2), (-\sqrt{2}, 2), (1, 1), (-1, 1)\}$
 37. $\{(3, 1)\}$ **39.** $\varnothing$ **41.** $\{(-6, 4^{-7})\}$ **43.** $\sqrt{3}$ ft and $2\sqrt{3}$ ft
 45. Height $5\sqrt{10}$ in., base $20\sqrt{10}$ in.
 47. Pump A 24 hours, pump B 8 hours   **49.** 40 minutes
 51. 8 ft by 9 ft   **53.** $4 - 2i$ and $4 + 2i$
 55. Side 8 ft, height of triangle 2 ft
 57. a) $(1.71, 1.55), (-2.98, -3.95)$ **b)** $(1, 1), (0.40, 0.16)$
   c) $(1.17, 1.62), (-1.17, -1.62)$

Section 12.2 Warm-Ups

 1. Parabola   **2.** Focus   **3.** Directrix   **4.** Vertex
 5. Focus, directrix   **6.** Completing the square   **7.** False   **8.** True
 9. True   **10.** False   **11.** True   **12.** True

Section 12.2 Exercises

1. 5   **3.** $\sqrt{2}$ **5.** $\sqrt{13}$ **7.** $2\sqrt{17}$ **9.** $\sqrt{65}$

11. (3, 4), 10   **13.** $\left(\frac{7}{2}, 3\right)$, 5   **15.** (2, 1), 10   **17.** $\left(0, \frac{5}{2}\right)$, $\sqrt{13}$

19. Vertex (0, 0), focus $\left(0, \frac{1}{8}\right)$, directrix $y = -\frac{1}{8}$

21. Vertex (0, 0), focus (0, −1), directrix $y = 1$

23. Vertex (3, 2), focus (3, 2.5), directrix $y = 1.5$

25. Vertex (−1, 6), focus (−1, 5.75), directrix $y = 6.25$

27. $y = \frac{1}{8}x^2$ **29.** $y = -\frac{1}{2}x^2$ **31.** $y = \frac{1}{2}x^2 - 3x + 6$

33. $y = -\frac{1}{8}x^2 + \frac{1}{4}x - \frac{1}{8}$ **35.** $y = x^2 + 6x + 10$

37. $y = (x - 3)^2 - 8$, vertex (3, −8), focus (3, −7.75),
directrix $y = -8.25$, axis $x = 3$

39. $y = 2(x + 3)^2 - 13$, vertex (−3, −13), focus (−3, −12.875),
directrix $y = -13.125$, axis $x = -3$

41. $y = -2(x - 4)^2 + 33$, vertex (4, 33), focus $\left(4, 32\frac{7}{8}\right)$, directrix $y = 33\frac{1}{8}$,
axis $x = 4$

43. $y = 5(x + 4)^2 - 80$, vertex (−4, −80), focus $\left(-4, -79\frac{19}{20}\right)$,
directrix $y = -80\frac{1}{20}$, axis $x = -4$

45. Vertex (2, −3), focus $\left(2, -2\frac{3}{4}\right)$, directrix $y = -3\frac{1}{4}$, $x = 2$, upward

47. Vertex (1, −2), focus $\left(1, -2\frac{1}{4}\right)$, directrix $y = -1\frac{3}{4}$, $x = 1$, downward

49. Vertex (1, −2), focus $\left(1, -1\frac{11}{12}\right)$, directrix $y = -2\frac{1}{12}$, $x = 1$, upward

51. Vertex $\left(-\frac{3}{2}, \frac{17}{4}\right)$, focus $\left(-\frac{3}{2}, 4\right)$, directrix $y = \frac{9}{2}$, $x = -\frac{3}{2}$, downward

53. Vertex (0, 5), focus $\left(0, 5\frac{1}{12}\right)$, directrix $y = 4\frac{11}{12}$, $x = 0$, upward

55. (3, 2), $\left(\frac{13}{4}, 2\right)$, $x = \frac{11}{4}$ **57.** (−2, 1), (−1, 1), $x = -3$

59. (4, 2), $\left(\frac{7}{2}, 2\right)$, $x = \frac{9}{2}$

61.

63.

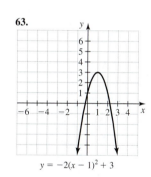

$y = -2(x - 1)^2 + 3$

65.

67.

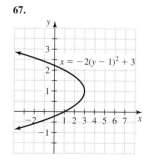

69. {(−1, 2), (1, 2)}

71. {(1, −1)}

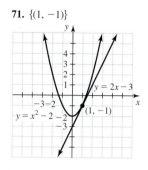

73. $\left\{\left(\frac{3}{2}, \frac{11}{4}\right), (-4, 0)\right\}$

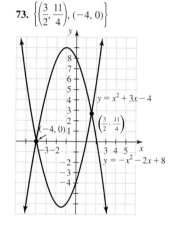

75. {(−3, −4), (2, 6)}

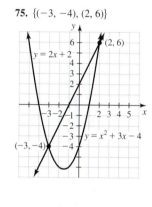

77. (3, 0), (−1, 0)   **79.** (20, 4), (−20, 4)   **81.** (0, 0), (1, 1)

83. a) $59,598   **b)** (−15, −59)   **85.** $y = \frac{1}{60}x^2$

89. The graphs have identical shapes.

Section 12.3 Warm-Ups

1. Circle   **2.** Center   **3.** Radius   **4.** Circle, center, radius
5. False   **6.** False   **7.** True   **8.** False   **9.** True   **10.** True

Section 12.3 Exercises

1. $x^2 + y^2 = 16$ **3.** $x^2 + (y - 3)^2 = 25$ **5.** $(x - 1)^2 + (y + 2)^2 = 81$

7. $x^2 + y^2 = 3$ **9.** $(x + 6)^2 + (y + 3)^2 = \frac{1}{4}$

11. $\left(x - \frac{1}{2}\right)^2 + \left(y - \frac{1}{3}\right)^2 = 0.01$ **13.** (0, 0), 1   **15.** (3, 5), $\sqrt{2}$

17. $\left(0, \frac{1}{2}\right)$, $\frac{\sqrt{2}}{2}$ **19.** (0, 0), $\frac{3}{2}$ **21.** (2, 0), $\sqrt{3}$

23.

25.

27.

$(x + 1)^2 + (y - 1)^2 = 2$

29.

$(x - 4)^2 + (y + 3)^2 = 16$

65. B and D can be any real numbers, but A must equal C, and $4AE + B^2 + D^2 > 0$. No ordered pairs satisfy $x^2 + y^2 = -9$.
67. $y = \pm\sqrt{4 - x^2}$ **69.** $y = \pm\sqrt{x}$ **71.** $y = -1 \pm \sqrt{x}$

Mid-Chapter Quiz 12.1–12.3
1. $\{(3, 9), (-2, 4)\}$ **2.** $\{(0, 2), (8, 6)\}$ **3.** $\{(2, \pm\sqrt{3}), (-2, \pm\sqrt{3})\}$
4. $\{(2, 3), (-2, 3)\}$ **5.** $(8, -7), (8, -5), y = -9, x = 8$, up
6. $(3, 4), (3, 15/4), y = 17/4, x = 3$, down
7. $(0, 2), (1/4, 2), x = -1/4, y = 2$, right   **8.** $\sqrt{5}$ **9.** $(-7, 2)$
10. $y = 4(x - 1)^2 - 2$ **11.** $(x + 4)^2 + (y - 5)^2 = 100$ **12.** $(2, 0), \sqrt{7}$
13. $(x + 2)^2 + (y - 5)^2 = 30$ **14.** $5 + \sqrt{13}, 5 - \sqrt{13}$

31.

$\left(x - \frac{1}{2}\right)^2 + \left(y + \frac{1}{2}\right)^2 = \frac{1}{4}$

33. $(x + 2)^2 + (y + 3)^2 = 13, (-2, -3), \sqrt{13}$
35. $(x - 1)^2 + (y - 2)^2 = 8, (1, 2), 2\sqrt{2}$
37. $(x - 5)^2 + (y - 4)^2 = 9, (5, 4), 3$
39. $\left(x - \frac{1}{2}\right)^2 + \left(y + \frac{1}{2}\right)^2 = \frac{1}{2}, \left(\frac{1}{2}, -\frac{1}{2}\right), \frac{\sqrt{2}}{2}$
41. $\left(x - \frac{3}{2}\right)^2 + \left(y - \frac{1}{2}\right)^2 = \frac{7}{2}, \left(\frac{3}{2}, \frac{1}{2}\right), \frac{\sqrt{14}}{2}$
43. $\left(x - \frac{1}{3}\right)^2 + \left(y + \frac{3}{4}\right)^2 = \frac{97}{144}, \left(\frac{1}{3}, -\frac{3}{4}\right), \frac{\sqrt{97}}{12}$

Section 12.4 Warm-Ups
1. Ellipse   **2.** Foci   **3.** Center   **4.** Hyperbola   **5.** Branches
6. Asymptotes   **7.** True   **8.** False   **9.** True   **10.** True
11. True   **12.** True

Section 12.4 Exercises

45. $\{(1, 3), (-1, -3)\}$

$x^2 + y^2 = 10$ $y = 3x$ $(1, 3)$ $(-1, -3)$

47. $\{(0, -3), (\sqrt{5}, 2), (-\sqrt{5}, 2)\}$

$(-\sqrt{5}, 2)$ $(\sqrt{5}, 2)$ $y = x^2 - 3$ $x^2 + y^2 = 9$

1.

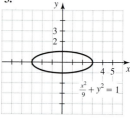

$\frac{x^2}{9} + \frac{y^2}{4} = 1$

3.

$\frac{x^2}{9} + y^2 = 1$

49. $\{(0, -3), (2, -1)\}$

$(x - 2)^2 + (y + 3)^2 = 4$ $y = x - 3$ $(2, -1)$ $(0, -3)$

51. $(0, 2 + \sqrt{3})$ and $(0, 2 - \sqrt{3})$
53. $\sqrt{29}$
55. $(x - 2)^2 + (y - 3)^2 = 32$
57. $\left(\frac{5}{2}, -\frac{\sqrt{11}}{2}\right)$ and $\left(\frac{5}{2}, \frac{\sqrt{11}}{2}\right)$
59. 755,903 mm^3

5.

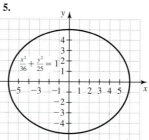

$\frac{x^2}{36} + \frac{y^2}{25} = 1$

7.

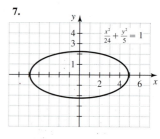

$\frac{x^2}{24} + \frac{y^2}{5} = 1$

9.

$9x^2 + 16y^2 = 144$

11.

$25x^2 + y^2 = 25$

61. $(0, 0)$ only

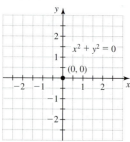

$x^2 + y^2 = 0$ $(0, 0)$

63.

$y = \sqrt{1 - x^2}$

13.

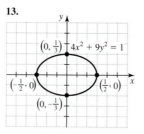

$\left(0, \frac{1}{3}\right)$ $4x^2 + 9y^2 = 1$ $\left(-\frac{1}{2}, 0\right)$ $\left(\frac{1}{2}, 0\right)$ $\left(0, -\frac{1}{3}\right)$

15.

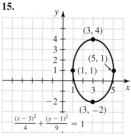

$(3, 4)$ $(5, 1)$ $(1, 1)$ $(3, -2)$ $\frac{(x - 3)^2}{4} + \frac{(y - 1)^2}{9} = 1$

17.

$$\frac{(x+1)^2}{16} + \frac{(y-2)^2}{25} = 1$$

$(-1, 7)$, $(-5, 2)$, $(3, 2)$, $(-1, -3)$

19.

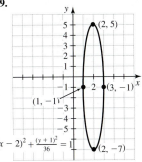

$(2, 5)$, $(1, -1)$, $(3, -1)$, $(2, -7)$

$$(x-2)^2 + \frac{(y+1)^2}{36} = 1$$

33.

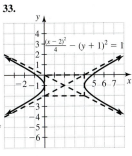

$$\frac{(x-2)^2}{4} - (y+1)^2 = 1$$

35.

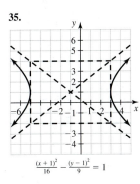

$$\frac{(x+1)^2}{16} - \frac{(y-1)^2}{9} = 1$$

37.

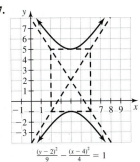

$$\frac{(y-2)^2}{9} - \frac{(x-4)^2}{4} = 1$$

39. Parabola
41. Hyperbola
43. Ellipse
45. Circle

21. $y = \pm\frac{3}{2}x$

$$\frac{x^2}{4} - \frac{y^2}{9} = 1$$

23. $y = \pm\frac{2}{5}x$

$$\frac{y^2}{4} - \frac{x^2}{25} = 1$$

25. $y = \pm\frac{1}{5}x$

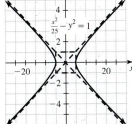

$$\frac{x^2}{25} - y^2 = 1$$

27. $y = \pm5x$

$$x^2 - \frac{y^2}{25} = 1$$

47. $\left(\frac{2\sqrt{10}}{5}, \frac{3\sqrt{15}}{5}\right)$,

$\left(\frac{2\sqrt{10}}{5}, -\frac{3\sqrt{15}}{5}\right)$,

$\left(-\frac{2\sqrt{10}}{5}, \frac{3\sqrt{15}}{5}\right)$,

$\left(-\frac{2\sqrt{10}}{5}, -\frac{3\sqrt{15}}{5}\right)$

$$x^2 - \frac{y^2}{9} = 1$$
$$\frac{x^2}{4} + \frac{y^2}{9} = 1$$

49. No points of intersection

$$x^2 + y^2 = 1$$
$$\frac{x^2}{4} + \frac{y^2}{16} = 1$$

29. $y = \pm\frac{3}{4}x$

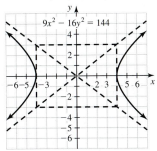

$$9x^2 - 16y^2 = 144$$

31. $y = \pm x$

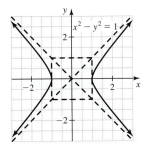

$$x^2 - y^2 = 1$$

51. $\left(\frac{\sqrt{10}}{2}, \frac{\sqrt{6}}{2}\right)$, $\left(\frac{\sqrt{10}}{2}, -\frac{\sqrt{6}}{2}\right)$,

$\left(-\frac{\sqrt{10}}{2}, \frac{\sqrt{6}}{2}\right)$,

$\left(-\frac{\sqrt{10}}{2}, -\frac{\sqrt{6}}{2}\right)$

$$x^2 + y^2 = 4$$
$$x^2 - y^2 = 1$$

53. $\left(\dfrac{3\sqrt{6}}{4}, \dfrac{\sqrt{10}}{4}\right),$

$\left(\dfrac{3\sqrt{6}}{4}, -\dfrac{\sqrt{10}}{4}\right),$

$\left(-\dfrac{3\sqrt{6}}{4}, \dfrac{\sqrt{10}}{4}\right),$

$\left(-\dfrac{3\sqrt{6}}{4}, -\dfrac{\sqrt{10}}{4}\right)$

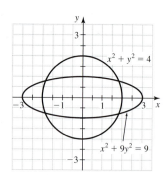

55. $\left(\dfrac{\sqrt{17}}{3}, \dfrac{8}{9}\right), \left(-\dfrac{\sqrt{17}}{3}, \dfrac{8}{9}\right),$
$(0, -1)$

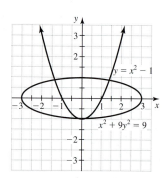

57. $(2, 0), \left(-\dfrac{5}{2}, -\dfrac{9}{4}\right)$

59. a) $(2.5, 1.5)$ **b)** $(\sqrt{7}, \sqrt{2})$

Section 12.5 Warm-Ups
 1. Ellipse   **2.** Hyperbola   **3.** Circle   **4.** Parabola   **5.** Line
 6. True   **7.** False   **8.** False   **9.** True   **10.** True

Section 12.5 Exercises
 1.

 3.

5.

7.

9.

11.

13.

15.

17.

19.

21.

23. Yes
25. No

27.

$x^2 + y^2 < 9$ and $y > x$

29.

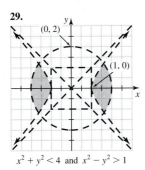

$x^2 + y^2 < 4$ and $x^2 - y^2 > 1$

47.

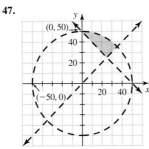

$x^2 + y^2 < 50^2$, $y > x$, $x + y > 50$

49. No solution

31.

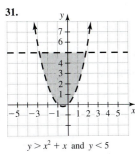

$y > x^2 + x$ and $y < 5$

33.

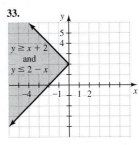

$y \geq x + 2$ and $y \leq 2 - x$

Enriching Your Mathematical Word Power

1. Nonlinear   **2.** Parabola   **3.** Directrix   **4.** Focus   **5.** Vertex
6. Conic   **7.** Axis   **8.** Circle   **9.** Center   **10.** Ellipse   **11.** Foci
12. Hyperbola   **13.** Asymptotes   **14.** Vertices

Review Exercises

1. $\{(3, 9), (-5, 25)\}$

3. $\left\{ \left(\frac{\sqrt{3}}{3}, \sqrt{3} \right), \left(-\frac{\sqrt{3}}{3}, -\sqrt{3} \right) \right\}$

35.

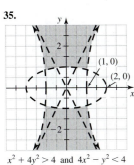

$x^2 + 4y^2 > 4$ and $4x^2 - y^2 < 4$

37.

$y + x^2 < 1$ and $y > x$

5. $\{(3, 3), (-3, -3)\}$
7. $\{(\sqrt{3}, 1), (-\sqrt{3}, 1)\}$ **9.** $\{(-5, -3), (3, 5)\}$ **11.** $\{(5, \log(2))\}$
13. $\{(2, 4), (-2, 4)\}$ **15.** $2\sqrt{2}$ **17.** $2\sqrt{58}$
19. $(5, 2), 10$ **21.** $\left(\frac{5}{2}, -\frac{1}{2} \right), \sqrt{10}$

23. Vertex $\left(-\frac{3}{2}, -\frac{81}{4} \right)$, axis of symmetry $x = -\frac{3}{2}$,
focus $\left(-\frac{3}{2}, -20 \right)$, directrix $y = -\frac{41}{2}$

25. Vertex $\left(-\frac{3}{2}, -\frac{1}{4} \right)$, axis of symmetry $x = -\frac{3}{2}$, focus $\left(-\frac{3}{2}, 0 \right)$,
directrix $y = -\frac{1}{2}$

27. Vertex $(2, 3)$, axis of symmetry $x = 2$, focus $\left(2, \frac{5}{2} \right)$,
directrix $y = \frac{7}{2}$

29. $y = 2(x - 2)^2 - 7, (2, -7)$ **31.** $y = -\frac{1}{2}(x + 1)^2 + 1, (-1, 1)$

33. $(0, 0), 10$ **35.** $(2, -3), 9$

39.

$x^2 + y^2 < 9$ and $y < 5x - x^2$

41.

$x \leq 1$ and $y \geq 3$

43.

$x^2 + y^2 < 16$ and $4y^2 - 9x^2 < 36$

45.

$x^2 + y^2 < 1$ and $y < x^2$

$x^2 + y^2 = 100$

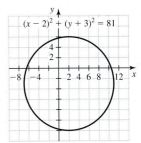

$(x - 2)^2 + (y + 3)^2 = 81$

37. $(0, 0), \dfrac{2}{3}$

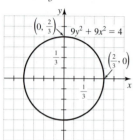

39. $x^2 + (y - 3)^2 = 36$

41. $(x - 2)^2 + (y + 7)^2 = 25$

43.

45.

47.

49.

51.

53.

55.

57.

59.

61. Hyperbola
63. Circle
65. Circle
67. Circle
69. Hyperbola
71. Hyperbola

73.

75.

77.

79.

81. $x^2 + y^2 = 25$ **83.** $(x + 1)^2 + (y - 5)^2 = 36$

85. $y = \dfrac{1}{4}(x - 1)^2 + 3$ **87.** $y = x^2$ **89.** $y = \dfrac{2}{9}x^2$

91. $\{(4, -3), (-3, 4)\}$ **93.** $\varnothing$ **95.** 6 ft, 2 ft

Chapter 12 Test

1.

2.

3.

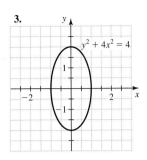
$y^2 + 4x^2 = 4$

4.

$y = x^2 + 4x + 4$

17. Vertex $\left(-\frac{1}{2}, \frac{11}{4}\right)$, focus $\left(-\frac{1}{2}, 3\right)$, directrix $y = \frac{5}{2}$, axis of symmetry $x = -\frac{1}{2}$, upward

18. $y = \frac{1}{2}(x - 3)^2 - 5$ **19.** $(x + 1)^2 + (y - 3)^2 = 13$ **20.** 12 ft, 9 ft

Making Connections   A Review of Chapters 1–12

1.

$y = 9x - x^2$

2.

$y = 9x$

5.

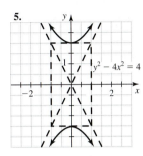
$y^2 - 4x^2 = 4$

6.

$y = -x^2 - 2x + 3$

3.

$y = (x - 9)^2$

4.

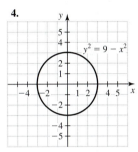
$y^2 = 9 - x^2$

7.

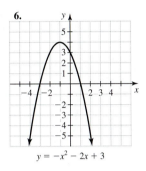
$x^2 - y^2 < 9$

8.

$x^2 + y^2 > 9$

5.

$y = 9x^2$

6.

$y = |9x|$

9.

$y > x^2 - 9$

10.

$x^2 - y^2 > 1$ and $x^2 + y^2 < 9$

7.

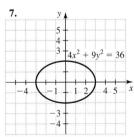
$4x^2 + 9y^2 = 36$

8.

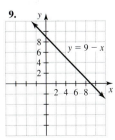
$4x^2 - 9y^2 = 36$

11.

$y < -x^2 + x$ and $y < x - 4$

12. $\{(-5, 27), (3, -5)\}$
13. $\{(\sqrt{3}, 3), (-\sqrt{3}, 3)\}$
14. $2\sqrt{2}$
15. $\left(-\frac{1}{2}, -\frac{1}{2}\right), \sqrt{26}$
16. $(-1, -5)$, 6

9.

$y = 9 - x$

10.

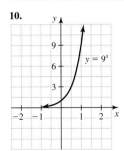
$y = 9^x$

11. $x^2 + 4xy + 4y^2$ **12.** $x^3 + 3x^2y + 3xy^2 + y^3$

13. $a^3 + 3a^2b + 3ab^2 + b^3$ **14.** $a^2 - 6ab + 9b^2$ **15.** $6a^2 - 7a - 5$

16. $x^3 - y^3$ **17.** $a(a^2 + b^2)$ **18.** $a(a + b)(a - b)$

19. $2(x + 3)(x - 6)$ **20.** $4(2x + 3)(4x - 5)$

21. $(m + 2)(x + 3)(x - 3)$ **22.** $2x(x + 3)(x^2 - 3x + 9)$

23. $\{(1, 2)\}$ **24.** $\{(3, 4), (4, 3)\}$ **25.** $\{(1, -2, 3)\}$

26. $\{(-1, 1), (3, 9)\}$ **27.** $x = -\dfrac{b}{a}$ **28.** $x = \dfrac{-d \pm \sqrt{d^2 - 4wm}}{2w}$

29. $B = \dfrac{2A - bh}{h}$ **30.** $x = \dfrac{2y}{y - 2}$ **31.** $m = \dfrac{L}{1 + xt}$

32. $t = \dfrac{y^2}{9a^2}$ **33.** $y = -\dfrac{2}{3}x - \dfrac{5}{3}$ **34.** $y = -2x$

35. $(x - 2)^2 + (y - 5)^2 = 45$ **36.** $\left(-\dfrac{3}{2}, 3\right), \dfrac{3\sqrt{5}}{2}$

37. $-10 + 6i$ **38.** -1 **39.** $3 - 5i$ **40.** $7 + 6i\sqrt{2}$ **41.** $-8 - 27i$

42. $-3 + 3i$ **43.** 29 **44.** $-\dfrac{3}{2} - i$ **45.** $-\dfrac{1}{2} + \dfrac{9}{2}i$ **46.** $2 - i\sqrt{2}$

47. a) $q = -500x + 400$ **b)** $R = -500x^2 + 400x$

c)

d) \$0.40 per pound   **e)** \$80

Chapter 13

Section 13.1 Warm-Ups

1. Sequence   **2.** Term   **3.** Finite   **4.** Infinite   **5.** True
6. True   **4.** False   **8.** True   **9.** False   **10.** True

Section 13.1 Exercises

1. 2, 4, 6, 8, 10   **3.** 1, 4, 9, 16, 25, 36, 49, 64

5. $-1, \dfrac{1}{2}, -\dfrac{1}{3}, \dfrac{1}{4}, -\dfrac{1}{5}, \dfrac{1}{6}, -\dfrac{1}{7}, \dfrac{1}{8}, -\dfrac{1}{9}, \dfrac{1}{10}$ **7.** 1, -2, 4, -8, 16

9. $\dfrac{1}{2}, \dfrac{1}{4}, \dfrac{1}{8}, \dfrac{1}{16}, \dfrac{1}{32}, \dfrac{1}{64}$ **11.** $-1, 1, 3, 5, 7, 9, 11$

13. $1, \dfrac{\sqrt{2}}{2}, \dfrac{\sqrt{3}}{3}, \dfrac{1}{2}, \dfrac{\sqrt{5}}{5}$ **15.** $\dfrac{1}{2}, \dfrac{1}{6}, \dfrac{1}{12}, \dfrac{1}{20}$ **17.** $-\dfrac{1}{3}, -1, 1, \dfrac{1}{3}$

19. $-1, 0, -1, 4$ **21.** $1, \dfrac{1}{4}, \dfrac{1}{9}, \dfrac{1}{16}$ **23.** $a_n = 2n - 1$

25. $a_n = (-1)^{n+1}$ **27.** $a_n = 2n - 2$ **29.** $a_n = 3n$ **31.** $a_n = 3n + 1$

33. $a_n = (-1)^n 2^{n-1}$ **35.** $a_n = (n - 1)^2$ **37.** $4, 2, 1, \dfrac{1}{2}, \dfrac{1}{4}$ yard line

39. \$45,746, \$48,034, \$50,435, \$52,957, \$55,605

41. \$1,000,000, \$800,000, \$640,000, \$512,000

43. 27 in., 13.5 in., 9 in., 6.75 in., 5.4 in.

45. 137,438,953,472, larger

49. a) 0.9048, 0.3677, 0.00004517   **b)** a_n goes to zero

Section 13.2 Warm-Ups

1. Summation   **2.** Series   **3.** Finite   **4.** Infinite   **5.** False
6. True   **7.** True   **8.** True   **9.** True   **10.** False

Section 13.2 Exercises

1. 15   **3.** 30   **5.** 24   **7.** $\dfrac{31}{32}$ **9.** 50   **11.** -7 **13.** 0

15. $\sum\limits_{i=1}^{6} i$ **17.** $\sum\limits_{i=1}^{6} (-1)^i(2i - 1)$ **19.** $\sum\limits_{i=1}^{6} i^2$ **21.** $\sum\limits_{i=1}^{4} \dfrac{1}{2 + i}$

23. $\sum\limits_{i=1}^{3} \ln(i + 1)$ **25.** $\sum\limits_{i=1}^{4} a_i$ **27.** $\sum\limits_{i=1}^{48} x_{i+2}$ **29.** $\sum\limits_{i=1}^{n} w_i$

31. $x + x^2 + x^3 + x^4 + x^5 + x^6$ **33.** $x_0 - x_1 + x_2 - x_3$

35. $x + 2x^2 + 3x^3$ **37.** $\sum\limits_{j=0}^{4} (j + 1)^2$ **39.** $\sum\limits_{j=1}^{13} (2j - 3)$ **41.** $\sum\limits_{j=1}^{5} \dfrac{1}{j + 3}$

43. $\sum\limits_{j=0}^{3} x^{2j+5}$ **45.** $\sum\limits_{j=0}^{n-1} x^{j+1}$ **47.** $\sum\limits_{i=1}^{9} 2^{-i}$ **49.** $\sum\limits_{i=1}^{4} 1,000,000(0.8)^{i-1}$

51. A sequence is basically a list of numbers. A series is the indicated sum of the terms of a sequence.

Section 13.3 Warm-Ups

1. Arithmetic   **2.** Common difference   **3.** Series   **4.** Arithmetic
5. False   **6.** False   **7.** False   **8.** True   **9.** True   **10.** True

Section 13.3 Exercises

1. $a_n = 2n$ **3.** $a_n = 6n - 6$ **5.** $a_n = 5n + 2$ **7.** $a_n = 2n - 6$

9. $a_n = -4n + 9$ **11.** $a_n = -7n + 5$ **13.** $a_n = 0.5n - 3.5$

15. $a_n = -0.5n - 5.5$ **17.** 9, 13, 17, 21, 25   **19.** 7, 5, 3, 1, -1

21. $-4, -1, 2, 5, 8$ **23.** $-2, -5, -8, -11, -14$

25. $-7, -11, -15, -19, -23$ **27.** 4.5, 5, 5.5, 6, 6.5

29. 1020, 1040, 1060, 1080, 1100   **31.** 51   **33.** 4   **35.** 26

37. 17   **39.** 1176   **41.** 330   **43.** -481 **45.** 435   **47.** -540

49. 150   **51.** -308 **53.** \$25,000   **55.** 1085 pages   **57.** b

Mid-Chapter Quiz 13.1–13.3

1. $-1, 2, 5, 8, 11, 14$ **2.** $-1, \dfrac{1}{4}, -\dfrac{1}{9}, \dfrac{1}{16}$ **3.** $\dfrac{1}{2}, \dfrac{1}{6}, \dfrac{1}{12}, \dfrac{1}{20}$

4. $0, -1, -4, -9$ **5.** $a_n = 2n + 1$ **6.** $a_n = (-1)^{n+1} 4n$

7. $a_n = n^2$ **8.** -36 **9.** 0   **10.** 0   **11.** 2600   **12.** 1482

13. -187 **14.** $\sum\limits_{i=1}^{5} 2i$ **15.** $\sum\limits_{i=1}^{5} (-1)^{i+1} i^2$ **16.** $a_n = 4n - 4$

17. $a_n = -5n + 11$ **18.** $\sum\limits_{i=1}^{16} (2i + 5)$ **19.** 43   **20.** 25

Section 13.4 Warm-Ups

1. Geometric   **2.** Common ratio   **3.** Series   **4.** Geometric
5. Infinite   **6.** False   **7.** True   **8.** True   **9.** True   **10.** True
11. True   **12.** True

Section 13.4 Exercises

1. $a_n = 2^{n-1}$ **3.** $a_n = \dfrac{1}{3}(3)^{n-1}$ **5.** $a_n = 64\left(\dfrac{1}{8}\right)^{n-1}$

7. $a_n = 8\left(-\dfrac{1}{2}\right)^{n-1}$ **9.** $a_n = 2(-2)^{n-1}$ **11.** $a_n = -\dfrac{1}{3}\left(\dfrac{3}{4}\right)^{n-1}$

13. $2, \dfrac{2}{3}, \dfrac{2}{9}, \dfrac{2}{27}, \dfrac{2}{81}$ **15.** $1, -2, 4, -8, 16$ **17.** $\dfrac{1}{2}, \dfrac{1}{4}, \dfrac{1}{8}, \dfrac{1}{16}, \dfrac{1}{32}$

19. 0.78, 0.6084, 0.4746, 0.3702, 0.2887   **21.** 5   **23.** $\dfrac{1}{3}$ **25.** $-\dfrac{1}{9}$

27. $\dfrac{511}{512}$ **29.** $\dfrac{11}{32}$ **31.** $\dfrac{63,050}{729}$ **33.** 5115   **35.** 0.111111

37. 42.8259   **39.** $\dfrac{1}{4}$ **41.** 9   **43.** $\dfrac{8}{3}$ **45.** $\dfrac{3}{7}$ **47.** 6   **49.** $\dfrac{1}{3}$

51. $\dfrac{4}{33}$ **53.** $3,042,435.27   **55.** $21,474,836.47   **57.** $5,000,000

59. d   **61.** $\dfrac{24}{100} + \dfrac{24}{10,000} + \dfrac{24}{1,000,000} + \cdots, \dfrac{8}{33}$

Section 13.5 Warm-Ups

1. Binomial expansion   **2.** Pascal's   **3.** Factorial   **4.** Binomial
5. False   **6.** True   **7.** False   **8.** True   **9.** True   **10.** True

Section 13.5 Exercises

1. 1   **3.** 10   **5.** 56   **7.** $x^3 + 3x^2 + 3x + 1$

9. $a^3 + 6a^2 + 12a + 8$ **11.** $r^5 + 5r^4t + 10r^3t^2 + 10r^2t^3 + 5rt^4 + t^5$

13. $m^3 - 3m^2n + 3mn^2 - n^3$ **15.** $x^3 + 6ax^2 + 12a^2x + 8a^3$

17. $x^8 - 8x^6 + 24x^4 - 32x^2 + 16$

19. $x^7 - 7x^6 + 21x^5 - 35x^4 + 35x^3 - 21x^2 + 7x - 1$

21. $a^{12} - 36a^{11}b + 594a^{10}b^2 - 5940a^9b^3$

23. $x^{18} + 45x^{16} + 900x^{14} + 10,500x^{12}$

25. $x^{22} - 22x^{21} + 231x^{20} - 1540x^{19}$ **27.** $\dfrac{x^{10}}{1024} + \dfrac{5x^9y}{768} + \dfrac{5x^8y^2}{256} + \dfrac{5x^7y^3}{144}$

29. $1287a^8w^5$ **31.** $-11,440m^9n^7$ **33.** $448x^5y^3$ **35.** $635,043,840a^{28}b^6$

37. $\displaystyle\sum_{i=0}^{8} \dfrac{8!}{(8-i)!\,i!}a^{8-i}m^i$ **39.** $\displaystyle\sum_{i=0}^{5} \dfrac{5!}{(5-i)!\,i!}a^{5-i}(-2x)^i$

41. $a^3 + b^3 + c^3 + 3a^2b + 3a^2c + 3ab^2 + 3ac^2 + 3b^2c + 3bc^2 + 6abc$

Enriching Your Mathematical Word Power

1. Finite   **2.** Infinite   **3.** Series   **4.** Arithmetic   **5.** Geometric
6. Arithmetic   **7.** Geometric   **8.** Expansion   **9.** Pascal's
10. Factorial

Review Exercises

1. 1, 8, 27, 64, 125   **3.** 1, 1, -3, 5, -7, 9   **5.** $-1, -\dfrac{1}{2}, -\dfrac{1}{3}$

7. $\dfrac{1}{3}, \dfrac{1}{5}, \dfrac{1}{7}$ **9.** 4, 5, 6   **11.** 36   **13.** 40   **15.** $\displaystyle\sum_{i=1}^{\infty} \dfrac{1}{2(i+1)}$

17. $\displaystyle\sum_{i=1}^{\infty} (i-1)^2$ **19.** $\displaystyle\sum_{i=1}^{\infty} (-1)^{i+1}x_i$ **21.** 6, 11, 16, 21

23. $-20, -22, -24, -26$ **25.** 3000, 4000, 5000, 6000

27. $a_n = \dfrac{n}{3}$ **29.** $a_n = 2n$ **31.** 300   **33.** $\dfrac{289}{6}$ **35.** 35

37. $3, \dfrac{3}{2}, \dfrac{3}{4}, \dfrac{3}{8}$ **39.** $1, \dfrac{1}{2}, \dfrac{1}{4}, \dfrac{1}{8}$ **41.** 0.23, 0.0023, 0.000023, 0.00000023

43. $a_n = \dfrac{1}{2}(6)^{n-1}$ **45.** $a_n = \dfrac{7}{10}\left(\dfrac{1}{10}\right)^{n-1}$ **47.** $\dfrac{40}{81}$ **49.** 0.3333333333

51. $\dfrac{3}{8}$ **53.** 54   **55.** $m^5 + 5m^4n + 10m^3n^2 + 10m^2n^3 + 5mn^4 + n^5$

57. $a^6 - 9a^4b + 27a^2b^2 - 27b^3$ **59.** $495x^8y^4$ **61.** $372,736a^{12}b^2$

63. $\displaystyle\sum_{i=0}^{7} \dfrac{7!}{(7-i)!\,i!}a^{7-i}w^i$ **65.** Neither   **67.** Arithmetic

69. Arithmetic   **71.** $\dfrac{1}{\sqrt[3]{180}}$ or $\dfrac{\sqrt[3]{150}}{30}$ **73.** $-1 + \dfrac{1}{2} - \dfrac{1}{6} + \dfrac{1}{24} - \dfrac{1}{120}$

75. $a^5 + 5a^4b + 10a^3b^2 + 10a^2b^3 + 5ab^4 + b^5$ **77.** 26

79. $118,634.11   **81.** $13,784.92

Chapter 13 Test

1. $-10, -4, 2, 8$ **2.** 5, 0.5, 0.05, 0.005   **3.** $-1, \dfrac{1}{2}, -\dfrac{1}{6}, \dfrac{1}{24}$

4. $1, \dfrac{3}{4}, \dfrac{5}{9}, \dfrac{7}{16}$ **5.** $a_n = 10 - 3n$ **6.** $a_n = -25\left(-\dfrac{1}{5}\right)^{n-1}$

7. $a_n = (-1)^{n-1}2n$ **8.** $a_n = n^2$ **9.** $5 + 7 + 9 + 11 + 13$

10. $5 + 10 + 20 + 40 + 80 + 160$

11. $m^4 + 4m^3q + 6m^2q^2 + 4mq^3 + q^4$ **12.** 750

13. $\dfrac{155}{8}$ **14.** 5   **15.** 10,100   **16.** $\dfrac{1}{2}$ **17.** $\dfrac{511}{128}$ **18.** ± 2

19. 11   **20.** $1365r^{11}t^4$ **21.** $-448a^{10}b^3$ **22.** $86,545.41

Making Connections   A Review of Chapters 1–13

1. 6   **2.** $n^2 - 3$ **3.** $x^2 + 2xh + h^2 - 3$ **4.** $x^2 - 2x - 2$

5. 11   **6.** 6   **7.** 4   **8.** 32   **9.** $\dfrac{1}{2}$ **10.** 3   **11.** 1   **12.** x

13. $-\dfrac{27}{2}$ **14.** $-\dfrac{8}{3}$ **15.** $-\dfrac{128}{9}$ **16.** 16

17.

18.

19.

20.

21.

22.

23.

24.

25.

$x^2 + 2y < 4$

26.

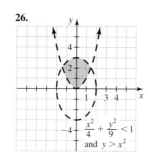

$\dfrac{x^2}{4} + \dfrac{y^2}{9} < 1$
and $y > x^2$

27. $\dfrac{a^2 + b^2}{ab}$ **28.** $\dfrac{y - 3}{y}$

29. $\dfrac{10}{(x - 3)(x + 3)(x + 1)}$

30. $\dfrac{2(x^3 - 64)}{x^3 - 16}$ **31.** $\dfrac{a^7}{b^4}$

32. 1   **33.** 4   **34.** $\dfrac{1}{32}$

35. -2 **36.** $\dfrac{1}{9}$

37. $-\dfrac{1}{8}$ **38.** $\dfrac{1}{4}$ **39.** $\dfrac{1}{5}$ **40.** 25   **41.** 8 in.   **42.** 76, 77, 78

43. 26.5%   **44.** 60 mph   **45.** 37.5 lbs

46. a) 105.8 cm or 41.7 in.   **c)** 1.3 years

Chapter 14

Section 14.1 Warm-Ups
 1. Tree   **2.** event, mn **3.** Permutation   **4.** Permutations   **5.** True
 6. True   **7.** False   **8.** True   **9.** False   **10.** True

Section 14.1 Exercises
 1. 12   **3.** 1024   **5.** 8   **7.** 120   **9.** 1320   **11.** 35,904
13. 1.0995×10^{12} **15.** 1.01989×10^{46} **17.** 180   **19.** 9,000,000
21. 5040   **23.** 26,000,000, 156,000,000   **25.** 8   **27.** 1×10^{10}
29. 24   **31.** 24   **33.** 1   **35.** 1   **37.** 336   **39.** 1   **41.** 210
43. 220   **45.** 364   **47.** 152,096

Section 14.2 Warm-Ups
 1. Combination   **2.** Combinations   **3.** Labeling   **4.** Multinomial
 5. True   **6.** False   **7.** True   **8.** False   **9.** True   **10.** False
11. True

Section 14.2 Exercises
 1. 2,598,960   **3.** 10   **5** 15,504   **7.** 13,983,816   **9.** 35
11. 120   **13.** 56   **15.** 479,001,600   **17.** 591,600,030
19. 3, 6, 10, $C(n, 2)$ **21.** 36   **23.** 8
25. $C(n, r) = \dfrac{n!}{(n - r)!r!}$ and $C(n, n - r) = \dfrac{n!}{r!(n - r)!}$ **27.** 60
29. 21,162,960   **31.** 576   **33.** 960,960   **35.** 5.36447×10^{28}, larger
37. 180, 2520, 50,400   **39.** 698,377,680   **41.** 630   **43.** 2520
45. 56   **47.** 56   **49.** 336   **51.** 1   **53.** 10   **55.** 20

Mid-Chapter Quiz 14.1–14.2
 1. 96 meals   **2.** 1326   **3.** 20   **4.** 32   **5.** 495   **6.** 56   **7.** 120
 8. 336   **9.** 40,320   **10.** 46,376   **11.** 10,506

Section 14.3 Warm-Ups
 1. Experiment   **2.** Sample space   **3.** Event   **4.** Addition
 5. Complementary   **6.** Odds   **7.** True   **8.** False   **9.** True
10. True   **11.** False   **12.** True

Section 14.3 Exercises
 1. $\dfrac{1}{2}, \dfrac{5}{6}, \dfrac{5}{6}$, 0, 1   **3.** $\dfrac{1}{4}, \dfrac{3}{4}, \dfrac{1}{4}, \dfrac{3}{4}$ **5.** $\dfrac{1}{36}, \dfrac{11}{36}, \dfrac{6}{36}$, 1, 0
 7. a) $\dfrac{1}{4}$ **b)** $\dfrac{2}{3}$ **c)** $\dfrac{2}{3}$ **d)** $\dfrac{1}{3}$ **9. a)** $\dfrac{1}{90}$ **b)** $\dfrac{1}{45}$ **c)** $\dfrac{2}{45}$
11. a) $\dfrac{1}{2,598,960}$ **b)** $\dfrac{1024}{2,598,960}$ **13.** 0.79   **15.** $\dfrac{1}{4}$ **17.** $\dfrac{4}{13}$
19. 0.995   **21. a)** $\dfrac{1}{36}$ **b)** $\dfrac{35}{36}$ **c)** $\dfrac{35}{36}$ **23.** 3 to 2
25. a) 3 to 2   **b)** 2 to 3   **27. a)** 1 to 3   **b)** $\dfrac{3}{4}$ **29.** 3 to 5
31. 5 to 31   **33.** 1 to 999,999   **35.** $\dfrac{1}{32}$
37. $\dfrac{1}{7,059,052}, \dfrac{7,059,051}{7,059,052}$, 1 to 7,059,051

Enriching Your Mathematical Word Power
 1. Tree   **2.** Permutation   **3.** Combination   **4.** Multinomial
 5. Experiment   **6.** Sample   **7.** Event   **8.** Exclusive
 9. Complementary   **10.** Odds

Review Exercises
 1. 14,348,907   **3.** 120   **5.** 144   **7.** 70   **9. a)** 126   **b)** 5   **c)** 60
11. 120   **13.** 4620   **15.** 3360, 120   **17.** 64   **19.** $\dfrac{1}{4096}, \dfrac{1}{4096}$
21. a) $\dfrac{5}{14}$ **b)** $\dfrac{9}{14}$ **c)** 0 **d)** 1   **23.** 1 to 7   **25.** 3 to 2   **27.** 0.9
29. 40,320   **31.** 20   **33.** 1680   **35.** 28   **37.** 1680   **39.** 8

Chapter 14 Test
 1. 72   **2.** 6840   **3.** 10   **4.** 1848   **5.** 495   **6.** 165
 7. 2,522,520   **8.** 120   **9.** $\dfrac{3}{4}$ **10.** 1   **11.** 1 to 3   **12.** 7 to 1
13. a) 20 to 1   **b)** $\dfrac{1}{21}$ **c)** $\dfrac{20}{21}$ **14. a)** $\dfrac{11}{36}$ **b)** $\dfrac{1}{6}$ **c)** $\dfrac{1}{18}$ **d)** $\dfrac{5}{12}$
15. 4950   **16.** 7,880,400   **17.** 4200

Making Connections   A Review of Chapters 1–14
 1. -9 **2.** 7   **3.** -1 **4.** -1 **5.** $(-3, 9)$ **6.** $(-1, 5)$
 7. $(3, 4)$ **8.** $(-\infty, 6]$ **9.** $a = 5 - b$ **10.** $a = \dfrac{1}{2}b - 2$
11. $a = \dfrac{4}{b - 1}$ **12.** $a = \dfrac{b}{6b - 1}$ **13.** $\left\{-\dfrac{2}{3}\right\}$ **14.** $\left\{\dfrac{11}{2}\right\}$ **15.** $\{400\}$
16. $\{40\}$ **17.** $\varnothing$ **18.** All real numbers   **19.** $\{-10, 10\}$
20. $\{2, 10\}$ **21.** $\{6\}$ **22.** $\varnothing$ **23.** $[2, \infty)$ **24.** $(-1, \infty)$ **25.** $(3, 6)$
26. $(-\infty, \infty)$ **27.** $(-2, 4)$ **28.** $[-18, 9)$ **29.** $(-\infty, -2) \cup (2, \infty)$
30. $[-10, 12]$ **31.** $(-\infty, -2] \cup [2, \infty)$ **32.** $(-2, 3)$
33. $(-1, 0) \cup (1, \infty)$ **34.** $\{0\} \cup [1, \infty)$ **35.** $[-1, 1]$
36. $(-\infty, -1) \cup (-1, 1)$ **37.** $-6x + 27$ **38.** $-15x^2 + 38x - 24$
39. $x^3 - 8x^2 + 13x - 6$ **40.** $\dfrac{9}{2x}$ **41.** $\dfrac{9x + 11}{x^2 - 4}$ **42.** $\dfrac{3x + 9}{2x}$
43. $9w^2(w + 4)(w + 2)$ **44.** $3n(6n^2 + 4n + 1)$ **45.** $-6(a + 2b)^2$
46. $(2a + m)(a + 2)$ **47.** $2(2y - 5)(4y^2 + 10y + 25)$
48. $(a - 5)(a + 3)(a - 3)$ **49.** $\left\{0, 3, \dfrac{32}{4}\right\}$ **50.** $\left\{3, \dfrac{32}{4}\right\}$

51. $\left\{-6, 0, 3, \dfrac{32}{4}\right\}$ **52.** $\{-6, -5/8, 0, 3, 2.9, 32/4\}$

53. $\{-\sqrt[3]{7}, \sqrt{3}, \pi\}$ **54.** All except $2 - i$ **55.** $\{2 - i\}$ **56.** All

57. $\dfrac{12}{7}$ **58.** $\dfrac{1}{3}$ **59.** $\dfrac{1}{5}$ **60.** $\dfrac{9}{2}$ **61.** 3   **62.** 2

63. $\left\{\left(\dfrac{1}{2}, 1\right)\right\}$, independent   **64.** $\{(1.5, 2.5)\}$, independent

65. $\left\{(x, y) \mid y = \dfrac{1}{9}x - 2\right\}$, dependent   **66.** $\varnothing$, inconsistent

67. $\{(1, -2, 4)\}$, independent   **68.** $\varnothing$, inconsistent   **69.** 47   **70.** 3

71. -2 **72.** -14 **73.** 11   **74.** 7   **75.** 18   **76.** 2

77. 220 yards   **78.** 24 inches   **79.** \$1700   **80.** 800   **81.** 36 inches

82. 4 inches, 3 inches   **83. a)** 48 feet   **b)** 4 seconds

84. 13 years, 315 days

Appendix A

Geometry Review Exercises

1. 12 in.   **2.** 24 ft^2 **3.** 60°   **4.** 6 ft   **5.** 20 cm   **6.** 144 cm^2

7. 24 ft^2 **8.** 13 ft   **9.** 30 cm   **10.** No   **11.** 84 yd^2 **12.** 30 in.

13. 32 ft^2 **14.** 20 km   **15.** 7 ft, 3 ft^2 **16.** 22 yd   **17.** 50.3 ft^2

18. 37.7 ft   **19.** 150.80 cm^3 **20.** 879.29 ft^2 **21.** 288 in.^{3}, 288 in.^{2}

22. 4 cm   **23.** 100 mi^2, 40 mi   **24.** 20 km   **25.** 42.25 cm^2

26. 33.510 ft^3, 50.265 ft^2 **27.** 75.4 in.^{3}, 100.5 in.^{2} **28.** 56°

29. 5 cm   **30.** 15 in. and 20 in.   **31.** 149°   **32.** 12 km   **33.** 10 yd

Index

DEFINITIONS, RULES, AND FORMULAS

Subsets of the Real Numbers

Natural Numbers = $\{1, 2, 3, \ldots\}$

Whole Numbers = $\{0, 1, 2, 3, \ldots\}$

Integers = $\{\ldots -3, -2, -1, 0, 1, 2, 3, \ldots\}$

Rational = $\left\{\dfrac{a}{b} \,\middle|\, a \text{ and } b \text{ are integers with } b \neq 0\right\}$

Irrational = $\{x \mid x \text{ is not rational}\}$

Properties of the Real Numbers

For all real numbers a, b, and c

$a + b = b + a$; $a \cdot b = b \cdot a$ Commutative

$(a + b) + c = a + (b + c)$; $(ab)c = a(bc)$ Associative

$a(b + c) = ab + ac$; $a(b - c) = ab - ac$ Distributive

$a + 0 = a$; $1 \cdot a = a$ Identity

$a + (-a) = 0$; $a \cdot \dfrac{1}{a} = 1 \ (a \neq 0)$ Inverse

$a \cdot 0 = 0$ Multiplication property of 0

Absolute Value

$$|a| = \begin{cases} a & \text{for } a \geq 0 \\ -a & \text{for } a < 0 \end{cases}$$

Order of Operations

No parentheses or absolute value present:

   1. Exponential expressions

   2. Multiplication and division

   3. Addition and subtraction

With parentheses or absolute value:

   First evaluate within each set of parentheses

   or absolute value, using the order of operations.

Exponents

$$a^0 = 1 \qquad\qquad a^{-1} = \frac{1}{a}$$

$$a^{-r} = \frac{1}{a^r} = \left(\frac{1}{a}\right)^r \qquad \frac{1}{a^{-r}} = a^r$$

$$a^r a^s = a^{r+s} \qquad\qquad \frac{a^r}{a^s} = a^{r-s}$$

$$(a^r)^s = a^{rs} \qquad\qquad (ab)^r = a^r b^r$$

$$\left(\frac{a}{b}\right)^r = \frac{a^r}{b^r} \qquad\qquad \left(\frac{a}{b}\right)^{-r} = \left(\frac{b}{a}\right)^r$$

Roots and Radicals

$$a^{1/n} = \sqrt[n]{a} \qquad\qquad a^{m/n} = \left(\sqrt[n]{a}\right)^m = \sqrt[n]{a^m}$$

$$\sqrt[n]{ab} = \sqrt[n]{a} \cdot \sqrt[n]{b} \qquad\qquad \sqrt[n]{\frac{a}{b}} = \frac{\sqrt[n]{a}}{\sqrt[n]{b}}$$

Factoring

$$a^2 + 2ab + b^2 = (a + b)^2$$

$$a^2 - 2ab + b^2 = (a - b)^2$$

$$a^2 - b^2 = (a + b)(a - b)$$

$$a^3 - b^3 = (a - b)(a^2 + ab + b^2)$$

$$a^3 + b^3 = (a + b)(a^2 - ab + b^2)$$

Rational Expressions

$$\frac{a}{b} + \frac{c}{b} = \frac{a + c}{b} \qquad\qquad \frac{a}{b} - \frac{c}{b} = \frac{a - c}{b}$$

$$\frac{ac}{bc} = \frac{a}{b} \qquad\qquad \frac{a}{b} + \frac{c}{d} = \frac{ad + bc}{bd}$$

$$\frac{a}{b} \cdot \frac{c}{d} = \frac{ac}{bd} \qquad\qquad \frac{a}{b} \div \frac{c}{d} = \frac{a}{b} \cdot \frac{d}{c}$$

$$\text{If } \frac{a}{b} = \frac{c}{d}, \text{ then } ad = bc.$$